AF780768

Seismic Assessment and Design of Structures

Seismic Assessment and Design of Structures

Editor

Maria Favvata

MDPI • Basel • Beijing • Wuhan • Barcelona • Belgrade • Manchester • Tokyo • Cluj • Tianjin

Editor
Maria Favvata
University of Patras
Greece

Editorial Office
MDPI
St. Alban-Anlage 66
4052 Basel, Switzerland

This is a reprint of articles from the Special Issue published online in the open access journal *Applied Sciences* (ISSN 2076-3417) (available at: https://www.mdpi.com/journal/applsci/special_issues/Seismic_Structural).

For citation purposes, cite each article independently as indicated on the article page online and as indicated below:

LastName, A.A.; LastName, B.B.; LastName, C.C. Article Title. *Journal Name* **Year**, *Volume Number*, Page Range.

ISBN 978-3-0365-6458-6 (Hbk)
ISBN 978-3-0365-6459-3 (PDF)

© 2023 by the authors. Articles in this book are Open Access and distributed under the Creative Commons Attribution (CC BY) license, which allows users to download, copy and build upon published articles, as long as the author and publisher are properly credited, which ensures maximum dissemination and a wider impact of our publications.
The book as a whole is distributed by MDPI under the terms and conditions of the Creative Commons license CC BY-NC-ND.

Contents

About the Editor

Maria Favvata

Maria Favvata is an Assistant Professor at the Civil Engineering Department, University of Patras, Greece. Her primary research concerns the field of the seismic assessment and design of structures: nonlinear FE modelling; nonlinear static and dynamic analysis; seismic codes; building damage; performance-based earthquake engineering; structural pounding; and the morphological effects on structures. She has authored or co-authored 41 papers published in international journals and conference proceedings (one Keynote address, and two invited address) and one book chapter, while her published work has been referred by more than 1000 papers, h-index 15.

applied sciences

MDPI

Editorial

Special Issue on Seismic Assessment and Design of Structures

Maria Favvata

Department of Civil Engineering, School of Engineering, University of Patras, 26504 Rio, Greece; mfavvata@upatras.gr

Introduction

This Special Issue gathers 29 scientific papers that capture various open and challenging issues in earthquake engineering for the assessment and design of structures. Advanced computational, analytical, numerical, and experimental studies have provided novel results and interesting discussions.

Six papers are focused on evaluating the seismic performance of structures, considering key parameters that still have not been fully understood. Thus, the first paper authored by J. Valdés-Vázquez, A. García-Soto, and M. Jaimes [1] is focused on studying the effect of the vertical seismic component in the assessment and design of a steel frame structure. The paper authored by W. An, and G. Song [2] provides an insight into the influence of near-fault vertical seismic excitation amplitude on bridge pier failure. The paper authored by M. Srbić, A. Mandić Ivanković, A. Vlašić, and G. Hrelja Kovačević [3] is focused on the seismic performance of existing bridge columns with an atypical cross-section, without seismic details and with smooth reinforcement. The vertical irregularity setback in a reinforced concrete (RC) building subjected to earthquake ground motions is the topic of interest in the paper authored by T. Maulana, B. Enkhtengis, and T. Saito [4]. A review of the methodologies regarding the effect of the seismic action's incidence angle on the performance of the structures is presented in the paper authored by I. Bugueño, J. Carvallo, and J. Vielma [5]. The sixth paper authored by A. Marchisella and G. Muciaccia [6] is focused on the evaluation of bi-axial shear demand for RC beam–column joints by comparing different numerical methods for seismic analysis.

Four papers are focused on providing new results to improve the effectiveness of dampers on the seismic mitigation performance of structures. In the first paper authored by M. Shih, and W. Sung [7] a new impulsive semi-active mass damper to mitigate the damage to facilities in high-rise buildings is presented. In the paper authored by J. You, Y. Yang, Y. Fan, and X. Zhang [8] a magnetorheological damper is proposed to analyze the multi-dimensional seismic mitigation performance and the torsional vibration characteristic of L-shaped frame structure. In the third paper authored by C. Mrad, M. Titirla, and W. Larbi [9] three types of passive energy dissipation systems tuned mass dampers, viscous dampers, and friction dampers are evaluated as strengthening solutions for RC symmetric buildings. Finally, the structural performance of a 10-story steel building with passive control systems using a single type of damper or a combination of different types of dampers is evaluated in the paper authored by A.K. Karimi, E. Moscoso Alcantara, and T. Saito [10].

The next four papers introduce new approaches for the seismic fragility assessment of structures. In the paper authored by M. Flenga, and M. Favvata [11] the seismic performance of RC structures subjected to structural pounding is assessed through displacement-based and curvature-based fragility curves. Different methodologies for developing the fragility curves of the pounding risk are evaluated, while linear and bilinear probabilistic seismic demand models are introduced. C. Contiguglia, A. Pelle, B. Briseghella, and C. Nuti [12] study the accuracy and the effectiveness of different analysis methods to assess the vulnerability of structures. For this purpose, fragility curves derived by MPA-based

Citation: Favvata, M. Special Issue on Seismic Assessment and Design of Structures. *Appl. Sci.* **2023**, *13*, 505. https://doi.org/10.3390/app13010505

Received: 23 December 2022
Accepted: 23 December 2022
Published: 30 December 2022

Copyright: © 2022 by the author. Licensee MDPI, Basel, Switzerland. This article is an open access article distributed under the terms and conditions of the Creative Commons Attribution (CC BY) license (https://creativecommons.org/licenses/by/4.0/).

cloud analysis, IMPA, and cloud analysis are compared against IDA. J. Vielma, R. Aguiar, C. Frau, and A. Zambrano review the effect of masonry infill panels on the seismic response of RC structures with characteristics similar to the typology and the mechanical and the geometric characteristics of the corner buildings of Portoviejo [13]. Fragility curves have been developed based on IDAs, and a new damage measure based on floor rotations is proposed. In the paper authored by Q. Liu, and C. Yang a copula technique for developing seismic fragility curves for an RC-isolated continuous girder bridge, is presented taking into account different damage indicators [14].

In two papers, new types of reinforcing steel for the seismic design and assessment of RC structures are presented and discussed. The first paper authored by J. Žižmond, and M. Dolšek [15] introduces an insight into the seismic performance of RC frame buildings designed by Eurocode 8 and reinforced by a recently developed dual-phase reinforcing steel (DPD2). The DPD2 is a new type of reinforcing steel that increases corrosion resistance. The seismic performance of DPD2 buildings was found to be improved compared to those designed with conventional reinforcing steel. The other paper authored by V. Kytinou, P. Kosmidou, and C. Chalioris [16] evaluates the seismic behavior of RC external beam-column joints with CFRP longitudinal bars in the beam. Experimental tests followed by FE analyses were performed. The results of the FE analyses reveal useful insights into replacing conventional reinforcement with CFRP bars.

In the next four papers, experimental results are used to evaluate and/or validate the structural performance of components such as anchors, connectors, and nuclear components. So, in the paper authored by S. Lee, and W. Jung static and dynamic experiments on the anchorages of non-structural elements are conducted using the seismic performance evaluation criteria of anchorages in accordance with the standards of the United States and Europe [17]. In the paper authored by G. Koo, T. Shin, and S. Ma the seismic isolation performance, and the dynamic characteristics of full-scale laminated rubber bearings (LRBs) are investigated through a well-established test matrix [18]. Results of shaking table tests are used by G. Koo, S. Ahn, J. Hwang, and J. Kim to validate the inelastic seismic analysis method applicable to pressure-retaining metal components in nuclear power plants [19]. An external ALC panel connector is proposed by K. Ding, D. Zong, Y. Liu, S. He, and W. Shen [20]. In this paper, experimental and numerical results between the external hooked bolt connector and the pendulous Z-panel connector are analyzed and discussed.

Also, an interesting topic that is addressed in this Special Issue is the seismic performance of retrofitted structures. In the paper authored by Y. Sonoda, H. Tamai, and H. Ikeda the effectiveness of the SRS method as a seismic retrofitting method on existing dam piers is studied using FEM analyses [21]. The practical difficulties in upgrading the structural performance of existing RC structures when retrofitting structural members by conventional RC jacketing are discussed in the paper authored by G. Manos, and K. Katakalos [22]. In this paper, an effective procedure for the shear upgrade of under-designed RC T-beams is presented. The tensile capacity of CFRP strips with or without anchors can be found by using a simple, novel laboratory test set-up devised by the authors.

Some other studies are motivated by the need for analytical modeling tools that can effectively capture the seismic behavior of substandard RC structural elements. Thus, in the paper authored by S. Pardalopoulos, S. Pantazopoulou, and G. Manolis a new assessment framework is proposed aiming to determine whether the estimated local drift demands can be tolerated without failure developing along the load resistance path of substandard RC buildings [23]. The paper authored by C. Lu, and Y. Sonoda [24] presents an analysis model that can produce crack development during the pull-out process and evaluate concrete fracture, bolt fracture, and steel–concrete bond fracture in the SPH method.

A modification of the modal-based ground motion selection method is presented in the next paper authored by Y. Liu [25]. The proposed modified method aims on improving the reliability of the nonlinear response time history analysis of RC shear wall structures. On the other hand, the paper authored by R. Zhang, D. Ye, J. Zhou, and D. Quan [26] presents a modified displacement-based method for the seismic analysis of an underground structure in the loess area. Nevertheless, in the paper authored by A. Naqi, T. Roy, and T. Saito a method for evaluating the cumulative damage of buckling-restrained braces under multi-hazard events that are expected to occur during the service life of a high-rise building, is introduced [27]. The paper authored by E. Diaz-Segura [28] provides a review on the methodologies for seismic site classification according to the Chilean regulations from their implementation in the 1930s to the most recent proposal in 2018–2021.

In the last paper authored by F. Fuentes, S. Lozano, M. Gomez, J. Vielma, and A. Lopez the dynamic characteristics of a multistory RC buildings are estimated, considering two different identification techniques [29]. Data based on earthquake acceleration time histories recorded by accelerometers placed throughout the building are used.

Funding: This research received no external funding.

Data Availability Statement: Data sharing is not applicable.

Acknowledgments: I am grateful to all contributors who made this Special Issue a success. My thanks and congratulations to all the authors for submitting their work. My sincere gratefulness to all reviewers for their effort and time spent to help authors improving their papers. Finally, I want to express my gratitude to the editorial team of Applied Sciences for their effective and friendly collaboration. I wish this collection to serve as an inspiration for future research in seismic engineering of structures.

Conflicts of Interest: The author declares no conflict of interest.

References

1. Valdés-Vázquez, J.; García-Soto, A.; Jaimes, M. Impact of the Vertical Component of Earthquake Ground Motion in the Performance Level of Steel Buildings. *Appl. Sci.* **2021**, *11*, 1925. Available online: https://www.mdpi.com/2076-3417/11/4/1925 (accessed on 22 February 2021). [CrossRef]
2. An, W.; Song, G. Transient Response of Bridge Piers to Structure Separation under Near-Fault Vertical Earthquake. *Appl. Sci.* **2021**, *11*, 4068. Available online: https://www.mdpi.com/2076-3417/11/9/4068 (accessed on 29 April 2021). [CrossRef]
3. Srbić, M.; Mandić Ivanković, A.; Vlašić, A.; Hrelja Kovačević, G. Plastic Joints in Bridge Columns of Atypical Cross-Sections with Smooth Reinforcement without Seismic Details. *Appl. Sci.* **2021**, *11*, 2658. Available online: https://www.mdpi.com/2076-3417/11/6/2658 (accessed on 16 March 2021). [CrossRef]
4. Maulana, T.; Enkhtengis, B.; Saito, T. Proposal of Damage Index Ratio for Low- to Mid-Rise Reinforced Concrete Moment-Resisting Frame with Setback Subjected to Uniaxial Seismic Loading. *Appl. Sci.* **2021**, *11*, 6754. Available online: https://www.mdpi.com/2076-3417/11/15/6754 (accessed on 22 July 2021). [CrossRef]
5. Bugueño, I.; Carvallo, J.; Vielma, J. Influence of Directionality on the Seismic Response of Typical RC Buildings. *Appl. Sci.* **2022**, *12*, 1534. Available online: https://www.mdpi.com/2076-3417/12/3/1534 (accessed on 31 January 2022). [CrossRef]
6. Marchisella, A.; Muciaccia, G. Comparative Assessment of Shear Demand for RC Beam-Column Joints under Earthquake Loading. *Appl. Sci.* **2022**, *12*, 7153. Available online: https://www.mdpi.com/2076-3417/12/14/7153 (accessed on 15 July 2022). [CrossRef]
7. Shih, M.; Sung, W. Seismic Resistance and Parametric Study of Building under Control of Impulsive Semi Active Mass Damper. *Appl. Sci.* **2021**, *11*, 2468. Available online: https://www.mdpi.com/2076-3417/11/6/2468 (accessed on 10 March 2011). [CrossRef]
8. You, J.; Yang, Y.; Fan, Y.; Zhang, X. Seismic Response Study of L-Shaped Frame Structure with Magnetorheological Dampers. *Appl. Sci.* **2022**, *12*, 5976. Available online: https://www.mdpi.com/2076-3417/12/12/5976 (accessed on 12 June 2022). [CrossRef]
9. Mrad, C.; Titirla, M.; Larbi, W. Comparison of Strengthening Solutions with Optimized Passive Energy Dissipation Systems in Symmetric Buildings. *Appl. Sci.* **2021**, *11*, 10103. Available online: https://www.mdpi.com/2076-3417/11/21/10103 (accessed on 28 October 2021). [CrossRef]
10. Karimi, A.; Moscoso Alcantara, E.; Saito, T. Reliability Analysis of Response-Controlled Buildings Using Fragility Curves. *Appl. Sci.* **2022**, *12*, 7717. Available online: https://www.mdpi.com/2076-3417/12/15/7717 (accessed on 31 July 2022). [CrossRef]
11. Flenga, M.; Favvata, M. Fragility Curves and Probabilistic Seismic Demand Models on the Seismic Assessment of RC Frames Subjected to Structural Pounding. *Appl. Sci.* **2021**, *11*, 8253. Available online: https://www.mdpi.com/2076-3417/11/17/8253 (accessed on 6 September 2021). [CrossRef]
12. Contiguglia, C.; Pelle, A.; Briseghella, B.; Nuti, C. IMPA versus Cloud Analysis and IDA: Different Methods to Evaluate Structural Seismic Fragility. *Appl. Sci.* **2022**, *12*, 3687. Available online: https://www.mdpi.com/2076-3417/12/7/3687 (accessed on 6 April 2022). [CrossRef]
13. Vielma, J.; Aguiar, R.; Frau, C.; Zambrano, A. Irregularity of the Distribution of Masonry Infill Panels and Its Effect on the Seismic Collapse of Reinforced Concrete Buildings. *Appl. Sci.* **2021**, *11*, 8691. Available online: https://www.mdpi.com/2076-3417/11/18/8691 (accessed on 17 September 2021). [CrossRef]

14. Liu, Q.; Yang, C. Seismic Damage Probability Assessment of Isolated Girder Bridges Based on Performance under Near-Field Earthquakes. *Appl. Sci.* **2021**, *11*, 9595. Available online: https://www.mdpi.com/2076-3417/11/20/9595 (accessed on 14 October 2021). [CrossRef]
15. Žižmond, J.; Dolšek, M. Seismic Design and Performance Assessment of Frame Buildings Reinforced by Dual-Phase Steel. *Appl. Sci.* **2021**, *11*, 4998. Available online: https://www.mdpi.com/2076-3417/11/11/4998 (accessed on 28 May 2021). [CrossRef]
16. Kytinou, V.; Kosmidou, P.; Chalioris, C. Numerical Analysis Exterior RC Beam-Column Joints with CFRP Bars as Beam's Tensional Reinforcement under Cyclic Reversal Deformations. *Appl. Sci.* **2022**, *12*, 7419. Available online: https://www.mdpi.com/2076-3417/12/15/7419 (accessed on 24 July 2022). [CrossRef]
17. Lee, S.; Jung, W. Evaluation of Structural Performance of Post-Installed Anchors Embedded in Cracked Concrete in Power Plant Facilities. *Appl. Sci.* **2021**, *11*, 3488. Available online: https://www.mdpi.com/2076-3417/11/8/3488 (accessed on 13 April 2021). [CrossRef]
18. Koo, G.; Shin, T.; Ma, S. Shaking Table Tests of Lead Inserted Small-Sized Laminated Rubber Bearing for Nuclear Component Seismic Isolation. *Appl. Sci.* **2021**, *11*, 4431. Available online: https://www.mdpi.com/2076-3417/11/10/4431 (accessed on 13 May 2021). [CrossRef]
19. Koo, G.; Ahn, S.; Hwang, J.; Kim, J. Shaking Table Tests to Validate Inelastic Seismic Analysis Method Applicable to Nuclear Metal Components. *Appl. Sci.* **2021**, *11*, 9264. Available online: https://www.mdpi.com/2076-3417/11/19/9264 (accessed on 6 October 2021). [CrossRef]
20. Ding, K.; Zong, D.; Liu, Y.; He, S.; Shen, W. Experimental and Finite Element Analysis of External ALC Panel Steel Frames with New Semi-Rigid Connector. *Appl. Sci.* **2021**, *11*, 10990. Available online: https://www.mdpi.com/2076-3417/11/22/10990 (accessed on 19 November 2021). [CrossRef]
21. Sonoda, Y.; Tamai, H.; Ikeda, H. Seismic Performance of Dam Piers Retrofitted with Reinforced Polymer Cement Mortar. *Appl. Sci.* **2021**, *11*, 7255. Available online: https://www.mdpi.com/2076-3417/11/16/7255 (accessed on 6 August 2021). [CrossRef]
22. Manos, G.; Katakalos, K. Reinforced Concrete Beams Retrofitted with External CFRP Strips towards Enhancing the Shear Capacity. *Appl. Sci.* **2021**, *11*, 7952. Available online: https://www.mdpi.com/2076-3417/11/17/7952 (accessed on 28 August 2021). [CrossRef]
23. Pardalopoulos, S.; Pantazopoulou, S.; Manolis, G. On the Modeling and Analysis of Brittle Failure in Existing R/C Structures Due to Seismic Loads. *Appl. Sci.* **2022**, *12*, 1602. Available online: https://www.mdpi.com/2076-3417/12/3/1602 (accessed on 2 February 2022). [CrossRef]
24. Lu, C.; Sonoda, Y. An Analytical Study on the Pull-Out Strength of Anchor Bolts Embedded in Concrete Members by SPH Method. *Appl. Sci.* **2021**, *11*, 8526. Available online: https://www.mdpi.com/2076-3417/11/18/8526 (accessed on 14 September 2021). [CrossRef]
25. Liu, Y. Modal-Based Ground Motion Selection Method for the Nonlinear Response Time History Analysis of Reinforced Concrete Shear Wall Structures. *Appl. Sci.* **2021**, *11*, 8230. Available online: https://www.mdpi.com/2076-3417/11/17/8230 (accessed on 5 September 2021). [CrossRef]
26. Zhang, R.; Ye, D.; Zhou, J.; Quan, D. Seismic Analysis Method for Underground Structure in Loess Area Based on the Modified Displacement-Based Method. *Appl. Sci.* **2021**, *11*, 11245. Available online: https://www.mdpi.com/2076-3417/11/23/11245 (accessed on 26 November 2021). [CrossRef]
27. Naqi, A.; Roy, T.; Saito, T. Time-Dependent Damage Estimation of a High-Rise Steel Building Equipped with Buckling-Restrained Brace under a Series of Earthquakes and Winds. *Appl. Sci.* **2021**, *11*, 9253. Available online: https://www.mdpi.com/2076-3417/11/19/9253 (accessed on 5 October 2021). [CrossRef]
28. Diaz-Segura, E. Evolution of Seismic Site Classification According to the Criteria in Chilean Design Codes. *Appl. Sci.* **2021**, *11*, 10754. Available online: https://www.mdpi.com/2076-3417/11/22/10754 (accessed on 15 November 2021). [CrossRef]
29. Fuentes, F.; Lozano, S.; Gomez, M.; Vielma, J.; Lopez, A. Determining the Dynamic Characteristics of a Multi-Story RC Building Located in Chile: A Comparison of the Results between the Nonparametric Spectral Analysis Method and the Parametric Stochastic Subspace Identification Method. *Appl. Sci.* **2022**, *12*, 7760. Available online: https://www.mdpi.com/2076-3417/12/15/7760 (accessed on 2 August 2022). [CrossRef]

Disclaimer/Publisher's Note: The statements, opinions and data contained in all publications are solely those of the individual author(s) and contributor(s) and not of MDPI and/or the editor(s). MDPI and/or the editor(s) disclaim responsibility for any injury to people or property resulting from any ideas, methods, instructions or products referred to in the content.

applied sciences

MDPI

Article

Impact of the Vertical Component of Earthquake Ground Motion in the Performance Level of Steel Buildings

Jesús-Gerardo Valdés-Vázquez [1,*], Adrián David García-Soto [1] and Miguel Á. Jaimes [2]

1 Department of Civil and Environmental Engineering, Universidad de Guanajuato, Juárez 77, Zona Centro, Guanajuato, Gto. 36000, Mexico; adgarcia@ugto.mx
2 Instituto de Ingeniería, UNAM, Coordinación de Ingeniería Estructural, Av. Universidad, No. 3000, Coyoacán, Mexico City C.P. 04510, Mexico; mjaimest@iingen.unam.mx
* Correspondence: valdes@ugto.mx; Tel.: +52-473-1020-100 (ext. 2216)

Featured Application: In this work, the non-linear dynamic response of buildings is studied considering both the horizontal component and the vertical component of seismic acceleration records. With an adequate discretization of the structure, it is found that the plastic rotations of the structure can change significantly, which is why the inclusion of the vertical seismic component in the assessment and design of buildings is recommended.

Abstract: This study discusses the impact of the vertical component of earthquake ground motion in the performance level of steel building subjected to earthquake excitations. Analyses are carried out for the strong column-weak beam philosophy because the structural performance is focused on these elements. A realistic steel frame is also considered to investigate the impact of including the seismic vertical component in the non-linear response of the building. The main findings of this study are: (1) When an analysis is performed by considering the horizontal and vertical components of ground motion acting simultaneously (near the causative fault), larger plastic rotations in the beams are obtained as compared to those resulting by considering only the horizontal component. (2) Due to the previous finding, if a codified criterion to inspect the steel beams performance in terms of the plastic rotation is considered, the beam performance could lie within a different acceptation criterion (i.e., from immediate occupancy to collapse prevention) if the vertical component is included in the analysis.

Keywords: non-linear dynamic analysis; seismic response; vertical component; acceptance criteria; steel buildings

Citation: Valdés-Vázquez, J.-G.; García-Soto, A.D.; Jaimes, M.Á. Impact of the Vertical Component of Earthquake Ground Motion in the Performance Level of Steel Buildings. *Appl. Sci.* **2021**, *11*, 1925. https://doi.org/10.3390/app11041925

Academic Editors: Andrea Pagliotti and Maria Favvata

Received: 30 December 2020
Accepted: 19 February 2021
Published: 22 February 2021

Publisher's Note: MDPI stays neutral with regard to jurisdictional claims in published maps and institutional affiliations.

Copyright: © 2021 by the authors. Licensee MDPI, Basel, Switzerland. This article is an open access article distributed under the terms and conditions of the Creative Commons Attribution (CC BY) license (https://creativecommons.org/licenses/by/4.0/).

1. Introduction

It has been reported in the works by Papazoglou et al. [1] and Elnashai et al. [2] that the vertical earthquake ground motion component of some earthquakes located near the causative fault can have an important detrimental impact in the structural behavior of different systems. For instance, Papazoglou et al. [1] attributed seismic failures to the effect of the vertical component of three studied earthquakes on structures. Besides possible compressive overstress or tension failure, the vertical response can lead to failures for the ultimate limit states of flexure moment and shear. The latter (i.e., shear) may explain observed failures perhaps associated with underlying vertical motion effects, because the compression can be reduced, or even mild tension could be reached, decreasing the contribution of concrete to shear resistance. The moment capacity and ductility of reinforced concrete (RC) columns can also be reduced. Unconsidered fluctuations in the axial force due to vertical motions may precipitate flexural failures. In another study, Elnashai et al. [2] based on modal analysis proposed a conservative simple approach to assess vertical seismic forces for buildings. More recently, Kim et al. [3] by means of experimental and analytical studies corroborated that the axial load level associated with the seismic vertical component has a detrimental effect on the shear capacity.

The influence of the earthquake ground motion vertical component has been found not only important in the structural response of buildings but also in that of bridges. For instance, Kim et al. [4] evaluated the effect of vertical earthquake ground motion on RC bridge piers and found that the inclusion of the vertical component importantly impacts the response of all components. They concluded that the vertical component should be taken in consideration for bridge design. Likewise, for highway overcrossings, Kunnath et al. [5] found the vertical component of ground motions as a main cause of amplification in the axial force in the columns and flexure moment in the girders. This was reported as especially critical for midspan moments in negative bending. Ancient and historical buildings have also presented damaged attributed to the vertical component of earthquakes, as observed on 7 September 2017 in historical structures located near the epicentral zone of the great Tehuantepec Mw8.2 earthquake, Mexico, Pozos-Estrada et al. [6]; it was suggested that the high frequency contents generated by this event and its vertical component led to damages to belfries, domes, towers and vaults of the historical structures.

On the other hand, it is also reported that when a modal analysis is carried out, different bending moments, axial forces and shear forces are obtained whether the lumped or consistent matrix mass approach is followed, as indicated in Valdés-Vázquez et al. [7], being the structural response (in terms of bending moments and shear forces) significantly increased for the consistent matrix mass approach if the vertical component is included. This is possibly due to the dynamic behavior of beams, which implies carrying loads (due to the vertical component of earthquakes) perpendicular to their longitudinal direction, which in turn causes bending moment and deflections, as cited in Paz et al. [8]. The two approximate methods to consider the inertial effect in the structure in general, and in the beams in particular, are the lumped mass approach (the distributed mass is concentrated in point masses) and the consistent mass approach (the point masses also include rotational effects); this latter method is consistent with the static traverse displacements of the beam, as cited in Paz et al. [8].

In addition, in the studies by Bozorgnia et al. [9] is reported that under certain conditions (i.e., short structural periods, especially for soft soils and near the epicentre) the vertical component becomes much more significant than the horizontal component of ground motion. Besides, it has been pointed out that, specifically for soft soils in the proximity of an active fault in Mexico City, the effects of the vertical component of ground motion should be incorporated in codified design, Jaimes et al. [10]. Other studies including the vertical component are given in [11–17], and some others related to bridge engineering with the vertical component are [18–21].

The effects, damage and possible collapse of structures under the action of the vertical component of ground motion due to earthquakes have been reported in the literature. For instance, the contribution of the vertical component to the response of structures is indicated by Di Sarno et al. [22], who found that the demand due to this component can be significant. The observed evidence, after large seismic events, of structural damage induced by the vertical seismic component is reported for the 1994 Northridge, California and the 1995 Kobe, Japan historical earthquakes [22]. More recently, similar observations are also reported for the 2009 L'Aquila, Italy earthquake [22]; it was indicated that the axial load in reinforced concrete columns was very large (more noticeably in compression), if both components were considered in the seismic analysis. It was stated that both components should be incorporated in the evaluation of the response of building frames, more markedly in the near-fault cases.

Compressions in columns above the balance load, as well as tensions in isolation systems and columns (over the balanced load too) have also been reported for buildings (Mazza et al. [23]) when the horizontal and vertical components of near-source earthquakes are considered. Other study related to seismically isolated systems subjected to seismic excitations from horizontal and vertical components found that including the latter is of key importance to designing pendulum isolators for two-degree of freedom systems (Landi et al. [24]). In addition, regarding base-isolated structures (irregular ones with friction pendulum), it has been found that when the vertical component is significant (near-source sites), uplift could occur if the effect of the vertical component is not accounted for (Mazza et al. [25]).

Collapse-prone frame structural types under the action of the seismic vertical component have been identified in terms of the ground motion characteristics, when the structures are subjected to multi-directional excitations (Harrington et al. [26]). The Housner intensity (Housner [27]) can be used to assess the influence of the vertical seismic component in the collapse of structures, especially structures more vulnerable to such component, as buildings with many cantilevered members and buildings which columns are subjected to large load effects.

Therefore, investigating the impact of including the earthquake ground motion vertical component in the earthquake resistant design and analysis of buildings is needed. In this study, the seismic response of steel buildings subjected to horizontal and vertical ground motions during Mexican earthquakes is presented.

The main objective of this study is to assess the impact of including the vertical component of essentially near-fault earthquake ground motions in the performance level of steel beams, by carrying out non-linear dynamic analyses on steel-frame buildings. Comparisons are given against the analysis by considering only the horizontal component, including the case of a steel frame designed by an experienced practicing engineer.

2. Hypothetical Buildings Considering the Strong-Column Weak-Beam Philosophy

To establish the framework used in this study, first the seismic response assessment of two hypothetical steel-frame buildings subjected to Mexican ground motions is considered. These structures do not necessarily correspond to minimum requirements from any code, but to structures which follows the strong-column weak-beam philosophy and are expected to exhibit a non-linear behavior under seismic loading. Nevertheless, this framework is applied later to a realistic structure designed by an experienced practicing engineer as per international standards, to validate the findings from a practical design standpoint.

For the hypothetical structures in this section, the considered steel is A-36 according to the American Institute of Steel Construction. In Figure 1 the studied buildings are shown. One is formed by a six-story steel-frame structure with columns W18 × 60 and beams W12 × 14 sections; the showed diagonals are W10 × 45 sections. A uniformly distributed load of 12 kN/m for all beams is assumed; this load plus the self-weight of the structure is used for calculating the natural vibration periods. The second is a 10-story building frame formed by W18 × 211, W16 × 31 and W10 × 45 sections for the columns, beams and diagonals, respectively. In this case, an 18 kN/m load is considered for all the beams plus the structure self-weight. The plastic hinges of steel beams to be studied are indicated by red dots in Figure 1; they are selected because the analyses indicated that they are the most critical in terms of plastic rotations (i.e., they are identified after performing all seismic dynamic analyses and rank the rotations in all the buildings). The plastic hinges shown are not the only ones (which also vary depending on the record used). For the comparisons, the study is focused on the plastic hinges with the largest rotations.

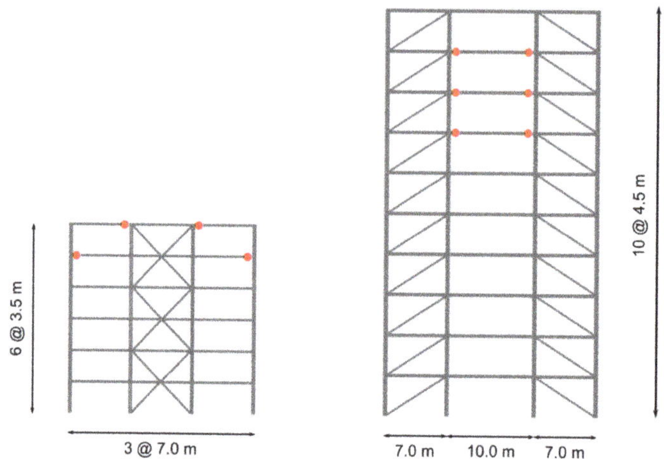

Figure 1. Steel-frame buildings and plastic hinges (red dots) considered.

Although not specific code was used for designing the frames as mentioned before, they correspond to the strong-column weak-beam philosophy contained in many relevant standards (e.g., American Society of Civil Engineers (ASCE 7–16) [28]; IBC-2018 [29]; NTCS-2017 [30]; Eurocode [31]).

For the six-story building, the first horizontal vibration period is T_{1h} = 0.398 s, the second is T_{2h} = 0.108 s and the third one is T_{3h} = 0.052 s, as shown in Figure 2. The same corresponding periods, but for the vertical response are T_{1v} = 0.173 s, T_{2v} = 0.164 s and T_{3v} = 0.060 s, respectively, as in Figure 3.

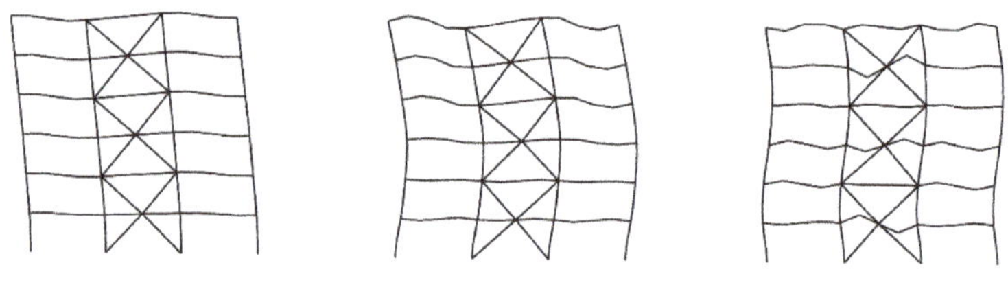

(**a**) T_{1h} = 0.398 s (**b**) T_{2h} = 0.108 s (**c**) T_{3h} = 0.052 s

Figure 2. Horizontal vibration modes for the six-story building.

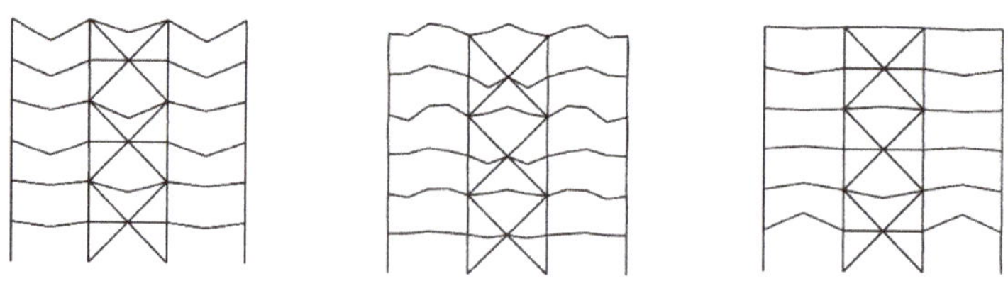

(**a**) T_{1v} = 0.173 s (**b**) T_{2v} = 0.164 s (**c**) T_{3v} = 0.060 s

Figure 3. Vertical vibration modes for the six-story building.

The computed horizontal and vertical vibration modes in the same order as previously described for the ten-story building are T_{1h} = 0.841 s, T_{2h} = 0.238 s and T_{3h} = 0.121 s, and

$T_{1v} = 0.198$ s, $T_{2v} = 0.122$ s and $T_{3v} = 0.058$ s, respectively, which are depicted in Figure 4 for the horizontal modes and Figure 5 for the vertical ones.

Based on a previous study by Valdés et al. [7], the frames and elements were selected so that they work in a range 60–80% of their capacity for gravitational load only. Therefore, when the buildings are subjected to the records reported in Table 1, the inelastic behavior of several elements can be inspected and compared for the one-versus the two-component analyses. Another possibility to guarantee that many members of the structures undergo into the inelastic range is to scale the records. This alternative is recommended in future research using reliable methods and recognizing their limitations. In the last floors, where the member sizes are normally smaller, it has been showed that when the vertical component is added to the horizontal component in the dynamic analysis using consistent masses, the shear forces and bending moments are larger in the upper levels, especially for the roof, which is not normally noticed in regular design (Valdés et al. [7]).

In general terms, two finite element-based formulations to analyze structures are available: the displacement-based formulation and the force-based formulation. For the former an adequate discretization of the structure is required, so that an adequate solution can be captured, whereas for the latter, the right solution is determined solely with one element. Therefore, the force-based formulation is preferred. However, when the vertical seismic component is included, the force-based formulation is insufficient, and an adequate discretization is warranted to accurately obtain the solution. In this study, elements by considering the force-based formulation are used, together with an adequate discretization.

Ground Motions

In order to capture the influence of the vertical component of earthquake ground motion in the performance level of steel building beams, this study uses different ground motions from Mexico listed in Table 1 (where PGAh and PGAv denote peak ground acceleration for the horizontal and vertical component, respectively) and whose response spectra are shown in Figure 6. These earthquake ground motions were selected to represent significant to non-significant vertical earthquake ground motion values (e.g., PGA_V from 0.45 to 0.094 g).

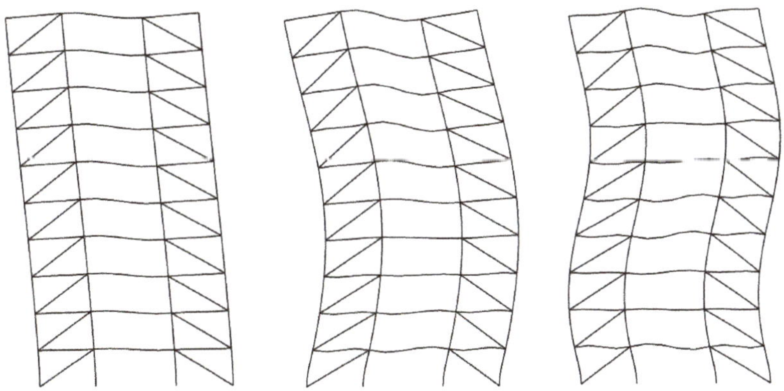

(**a**) $T_{1h} = 0.841$ s (**b**) $T_{2h} = 0.238$ s (**c**) $T_{3h} = 0.121$ s

Figure 4. Horizontal vibration modes for the ten-story building.

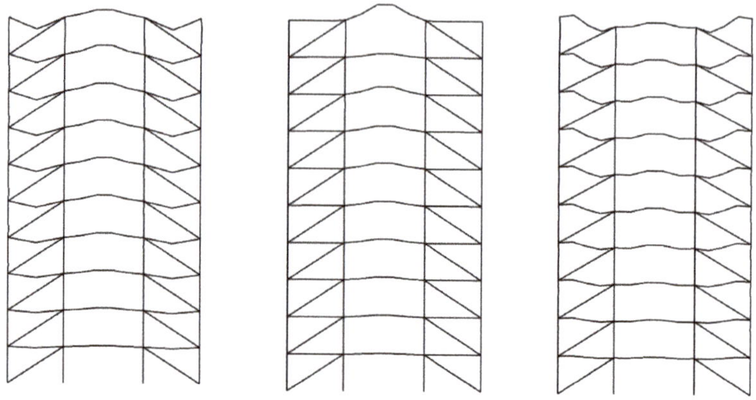

(a) $T_{1v} = 0.198$ s (b) $T_{2v} = 0.122$ s (c) $T_{3v} = 0.058$ s

Figure 5. Vertical vibration modes for the ten-story building.

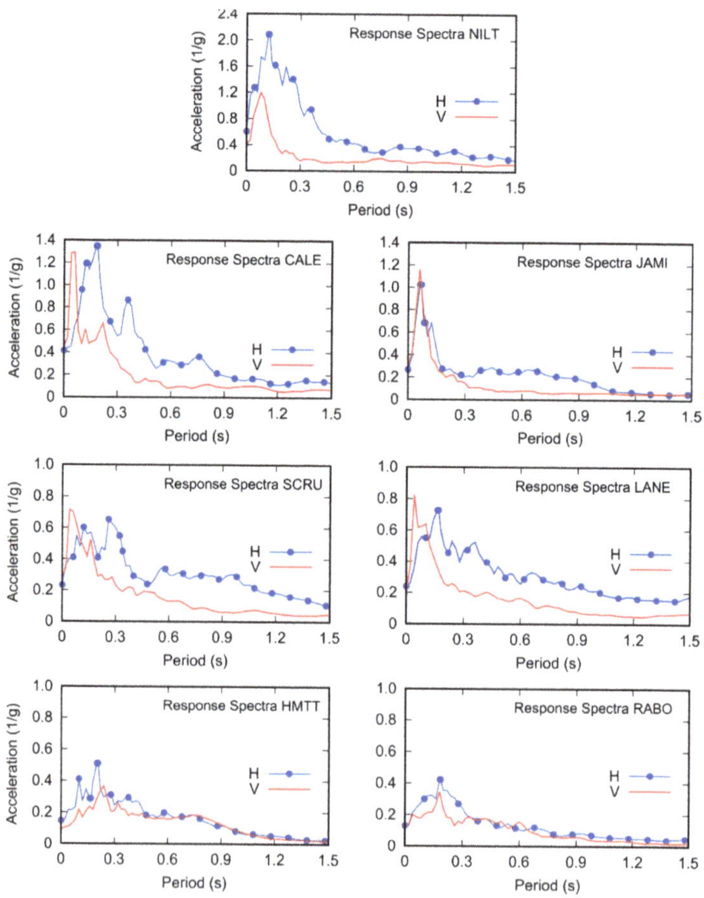

Figure 6. Response spectra used from different stations.

Table 1. Mexican ground motions used for the frames in Figure 1.

[1] Station	Direction	Date (YYYY/MM/DD)	Magnitud Mw	Rrup (km)	Lat.	Long.	H (km)	PGA_h (g)	PGA_v (g)
NILT	NS	2017/09/07	8.2	100.6	14.76	94.11	58	0.498	0.449
CALE	EW	1997/01/11	7.1	25.15	17.91	103.91	40	0.404	0.422
JAMI	EW	1999/09/30	7.4	48.74	16.03	96.96	47	0.256	0.292
SCRU	NS	2017/09/07	8.2	91.65	14.76	94.11	58	0.253	0.299
LANE	NS	1999/09/30	7.4	22.56	16.03	96.96	47	0.244	0.213
HMTT	EW	2017/09/19	7.1	48.99	18.33	98.67	57	0.147	0.094
RABO	NS	2017/09/19	7.1	14.51	18.33	98.67	57	0.132	0.120

[1] Stations designations are somewhat arbitrary acronyms of names in Spanish from different accelerometric arrays.

3. Seismic Structural Response

For the analysis of the structures, the OpenSEES program [32] was used, together with the pre- and post-processor GiD [33] and the coupling interface between both of them, GiD+OpenSees Platform [34].

They consider distributed plasticity with fibers in all the elements. In the case of beams, the P-delta effect is added. At intersections, the joint offset node with a dimension equal to half the depth of the elements is also considered. At the base, all the columns are fixed in all degrees of freedom and, in the stories, a rigid diaphragm was considered at each level. For the whole structure, the OpenSEES element type force-beam-column was used. The maximum size for the finite element mesh was 0.87 m for the columns and 1.75 m for the beams, with the idea of adding intermediate nodes in the elements to capture the vertical modes more accurately. The selected mesh size was determined by inspecting the maximum size for which no further accuracy is gained if more refined mesh sizes are used. Note also that the same mesh size was used in other study by Gremer et al. [15].

The damping ratio was determined through the Rayleigh damping method taking into account the first two horizontal vibration periods of each structure, together with the current stiffness matrix. The used element type from OpenSEES is the force-Beam-Column, considering three Gauss–Lobatto integration points. The material type is Steel02 with a yield stress Fy = 248.21 MPa, initial elastic tangent E = 200 GPa and a strain-hardening ratio b = 0.01 (as defined in OpenSEES). A 2D model was used, where first a static analysis under gravity loads was carried out, then the inertial force generated by every record was added in a second step considering a 0.01 s time step.

In this section, the results from the non-linear dynamic analyses using the records in the previous section are presented. For brevity, only flexure moments for the outermost fiber of the beam sections are depicted in moment-rotation diagrams. The employed notation is as follows: XXXX-YY-ZZ, where XXXX defines the considered event (NILT, CALE, etc.; e.g., Figure 6), YY the story (e.g., 5S for the 5th story of the building in study) where the most critical studied beam is located and ZZ denotes the side where the considered beam is situated (RS and LS for right and left side, respectively), in agreement with the plastic hinges depicted in Figure 1. In the following, results are first presented for the 6-story building and then for the 10-story building. Note that for the horizontal components in this study, only one of the orthogonal horizontal components is selected. Although every ground motion contains recordings for two horizontal and one vertical (orthogonal) components, the horizontal one leading to the most critical results when the vertical component is added was selected. This not necessarily implies that the most critical horizontal component corresponds to a larger peak ground acceleration.

3.1. Results for 6-Story Building

For the 6-story steel-frame structure the most critical plastic hinges are located at Floors 5 and 6 as shown in Figure 1. The W12 × 14 beams in this structure has a yielding flexure moment My = 60.7 kN-m and a yielding rotation θy = 0.0109 rad. For brevity, H and V are used to denote the horizontal and the vertical component, respectively.

When the 6-story building is subjected to the earthquake recorded at station NILT, the rotations and equivalent values (as function of the yielding rotation) from the non-linear dynamic analyses are 0.0293 rad (2.69θy) for H and 0.0348 rad (3.20θy) for H&V (i.e., horizontal and vertical component acting simultaneously). This represents an increase of 18.9% if the two components are used instead of only H (Figure 7, NILT-5S-RS).

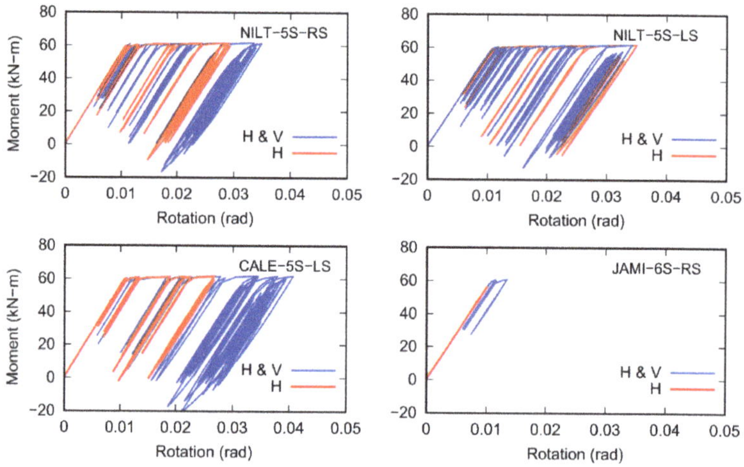

Figure 7. Non-linear response of the outer-most fiber in different beams for different earthquakes; 6-story buildings (Part 1).

As an exceptional response for NILT recording, the rotations and equivalent values considering one or the two components for the non-linear dynamic analysis are 0.0351 rad and 3.22θy (H) and 0.0342 rad and 3.15θy (H&V), respectively. In this case the change is −2.1% (i.e., the structural response actually decreased for this earthquake when both components were considered). This last result indicates that consideration of both components simultaneously can lead in some cases to less critical results (Figure 7, NILT-5S-LS). This could be explained because in some cases the vertical demand counteracts the horizontal one, for certain elements and under certain excitations, albeit this is not frequent.

For the recording of station CALE, the most critical beam is on the left side of the 5th floor. When only the horizontal seismic component is considered for the analysis, the maximum rotation value is 0.0268 rad, equivalent to 2.46θy. On the other hand, when both, the horizontal and the vertical, seismic components are considered, the maximum rotation results in 0.0407 rad, equivalent to 3.73θy. This means that the maximum rotation increased 51.6% with respect to that calculated when only the horizontal component is taken into account (Figure 7, CALE-5S-LS).

For the earthquake recording of station JAMI, this time the most critical beam is situated on the left side of the 6th floor (6th story—6S–LS). For the horizontal response, the rotations remain linear with a value of 0.0102 rad, equivalent to 0.94θy. On the other hand, when both components are considered simultaneously, the rotation is 0.0134 rad, which is equivalent to 1.23θy. This represents a 30.8% rotation increase, as compared to the horizontal component analysis alone (Figure 7, JAMI-6S-RS).

For the earthquake recorded at station SCRU (Figure 8, SCRU-6S-RS), the most critical beam is located on the right side of the 6th level. In this case for the seismic horizontal component, 0.0100 rad and 0.92θy were obtained. When both components were included, this led to 0.0303 rad and 2.79θy, representing a very significant increase of 203.2% with respect to the single-component analysis.

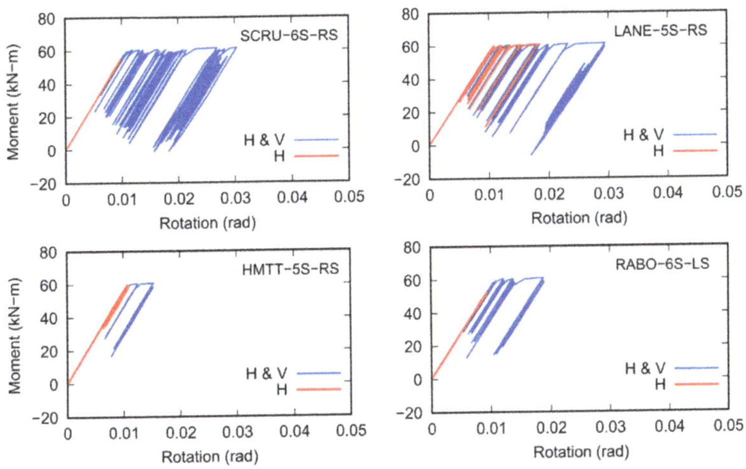

Figure 8. Non-linear response of the outer-most fiber in different beams for different earthquakes; 6-story buildings (part 2).

For the earthquake recording of station LANE, the most critical beam is on the right side in Floor 5. When only the horizontal component is considered, the maximum rotation reaches 0.0183 rad, or 1.69θy. When both components are included in the analysis, the values are 0.0295 rad and 2.71θy, respectively, which represents an increment of 60.3% in this case (Figure 8, LANE-5S-RS).

For the recording of station HMTT the rotations and equivalent values considering one or the two components for the non-linear dynamic analyses are 0.0110 rad and 1.01θy and 0.0154 rad and 1.41θy, respectively. This represents an increase of the two-component dynamic analysis in relation to the horizontal component analysis of 39.6% (Figure 8, HMTT-5S-RS).

Finally, for earthquake recorded at station RABO (Figure 8, RABO-6S-LS), the most critical beam is located on the left side of the 6th level. In this case, for the horizontal component seismic analysis, 0.0094 rad and 0.87θy were obtained. When both components were included, this led to 0.0191 rad and 1.75θy, representing a significant increase of 101.1%.

As a summary, the previous values are listed in Table 2.

Table 2. Rotations generated by horizontal earthquake components and by horizontal and vertical earthquake components acting together (6-story building).

Rotation Computed	Rotation Values for Earthquake Station Register Considered (rad)						
	NILT	**CALE**	**JAMI**	**SCRU**	**LANE**	**HMTT**	**RABO**
θ_h	0.0293 (2.69θy)	0.0268 (2.46θy)	0.0102 (0.94θy)	0.0100 (0.92θy)	0.0183 (1.69θy)	0.0110 (1.01θy)	0.0094 (0.87θy)
θ_{h+v}	0.0348 (3.20θy)	0.0407 (3.73θy)	0.0134 (1.23θy)	0.0303 (2.79θy)	0.0295 (2.71θy)	0.0154 (1.41θy)	0.0191 (1.75θy)

Figure 9 shows plastic hinges for JAMI seismic record where the red color indicates that the plastic hinge is formed. The color contour represents the plastic rotation depicted graphically with the GiD+OpenSEES interface, i.e., it shows the zones where the plastic hinges occur. The red/blue color indicates higher/lower degree of plastic rotation.

(**a**) Only horizontal component (**b**) Horizontal and vertical component

Figure 9. Plastic hinges for earthquake JAMI at the end of the seismic register. The red/blue color in the color contours indicates higher/lower degree of plastic rotation.

3.2. Results for 10-Story Building

This section is analogous to the previous one but results for the 10-story building are reported. The most critical plastic hinges were found in Levels 7, 8 and 9, as observed in Figure 1.

For the W16 × 31 beams, the yielding flexure moment is My = 192.0 kN-m, with a yielding rotation θy = 0.0114 rad. Likewise, in Figure 10 rotation diagrams for cases NILT-7S-LS and NILT-8S-RS are shown. Their respective values were (again for the one- and two-component analyses) 0.0274 rad (2.41θy) and 0.0377 rad (3.31θy), 0.0284 rad (2.50θy) and 0.0353 rad(3.10θy) corresponding to increases of 37.3% and 24.0%, respectively (i.e., increases with respect to the horizontal component analysis).

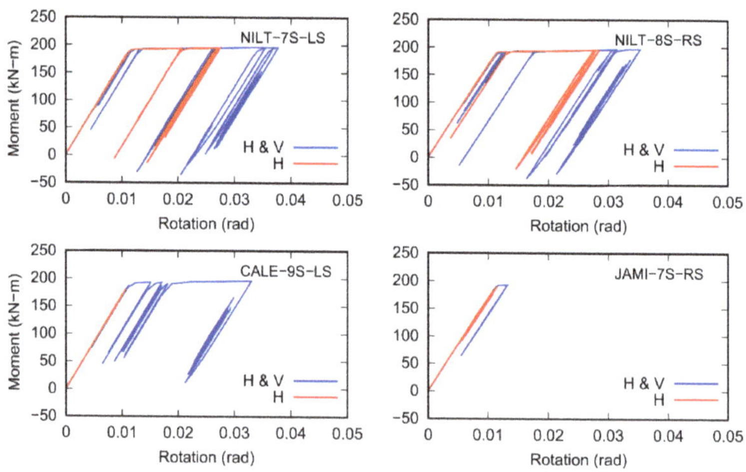

Figure 10. Non-linear response of the outer-most fiber in different beams for different earthquakes; 10-story buildings (Part 1).

For the case CALE-9S-LS, one- and two-component analysis led to rotations of 0.0107 rad (0.94θy) and 0.0330 rad (2.89θy), which is a 207.4% increase with respect to considering only the horizontal component (Figure 10).

Response to the earthquake for case JAMI-7S-RS depicted in Figure 10, led to 0.0114 rad (1.0θy) and 0.0132 rad (1.15θy) for the one- and two-component analysis, respectively. These values represent an increase of the two-component non-linear dynamic analysis in relation to its one-component counterpart of 15.0%.

Likewise, in Figure 11 rotation diagrams for SCRU-8S-LS are shown. Their respective values were (again for the one- and two-component analyses) 0.0175 rad (1.54θy) and

0.0276 rad (2.42θy), corresponding to an increase of 57.1% (i.e., increase with respect to the horizontal component analysis).

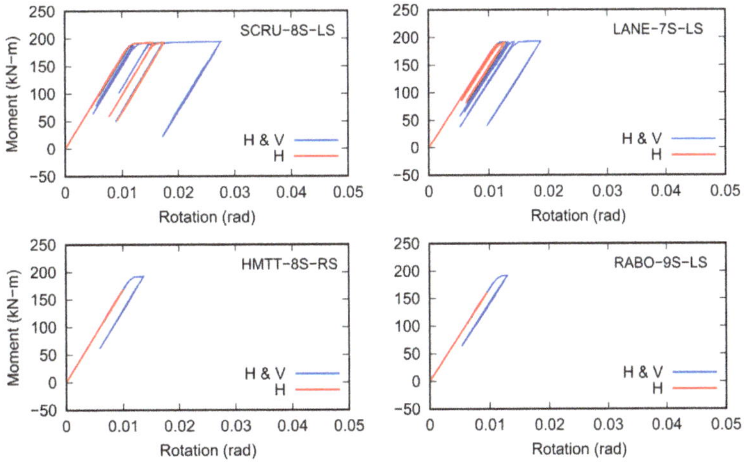

Figure 11. Non-linear response of the outer-most fiber in different beams for different earthquakes; 10-story buildings (Part 2).

The response to the earthquake of station LANE-7S-LS is depicted in Figure 11, led to 0.0130 rad (1.14θy) and 0.0178 rad (1.56θy) for the one- and two-component analysis, respectively. These values represent increases of the two-component non-linear dynamic analyses in relation to their one-component counterparts of 36.8%.

Response to the earthquake for case HMTT-8S-RS depicted in Figure 11, led to 0.0100 rad (0.88θy) and 0.0137 rad (1.20θy) for the one- and two-component analysis, respectively. These values represent increases of the two-component non-linear dynamic analyses in relation to their one-component counterparts of 36.3%.

Likewise, in Figure 11 rotation diagrams for earthquake response for the case RABO-9S-LS are shown. Their respective values were (again for the one- and two-component analyses) 0.0095 rad (0.84θy) and 0.0129 rad (1.13θy) corresponding to increases of 34.5% with respect to considering only the horizontal component.

These values are also summarized in Table 3. Figure 12 shows plastic hinges for SCRU seismic register which are colored in red.

Table 3. Rotations generated by horizontal earthquake components and by horizontal and vertical earthquake components acting together (10-story building).

Rotation	Rotation Values for Earthquake Station Register Considered (rad)						
Computed	NILT	CALE	JAMI	SCRU	LANE	HMTT	RABO
θ_h	0.0274 (2.41θy)	0.0107 (0.94θy)	0.0114 (1.0θy)	0.0175 (1.54θy)	0.0130 (1.14θy)	0.0100 (0.88θy)	0.0095 (0.84θy)
θ_{h+v}	0.0377 (3.31θy)	0.0330 (2.89θy)	0.0132 (1.15θy)	0.0276 (2.42θy)	0.0178 (1.56θy)	0.0137 (1.20θy)	0.0129 (1.13θy)

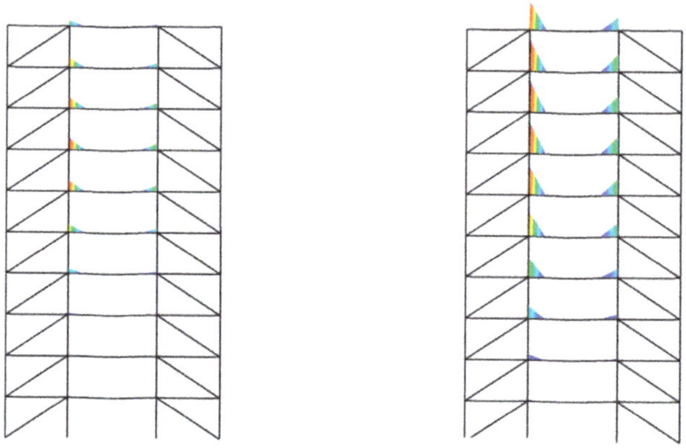

(**a**) Only horizontal component (**b**) Horizontal and vertical component

Figure 12. Plastic hinges for earthquake SCRU at the end of the seismic register. The red/blue color in the color contours indicates higher/lower degree of plastic rotation.

4. Trends of Damage as a Function of Seismic Intensity

In this section, it is investigated whether the level of a seismic intensity measure (IM) can be associated with the level of damage (in terms of rotation) of the steel beams for the considered hypothetical buildings. The pseudo-spectral acceleration, Sa, is selected to inspect possible correlations with the beams rotations, because this IM is readily available in engineering practice and easier to communicate to decision makers.

First, the horizontal accelerations for the three vibration modes leading to the highest mass participation factors Γ_h are obtained. This information is listed in Table 4 for the 6-story building. It is pointed out that not necessarily the three first modes lead to the highest participation factors; whereas this is usually the case for the horizontal component, it is not for the vertical component. Then the square root of the sum of the squares (SRSS) of the horizontal accelerations is computed (denoted as $Sa\ H_{SRSS}$) with the following expression:

$$Sa\ H_{SRSS} = \sqrt{\sum_{j=1}^{n} \frac{1}{n}\left[\Gamma_{hj} \times Sa(Th_j)\right]^2} \quad (1)$$

where n is the number of modes selected (i.e., 3 in this study), Γ_{hj} is the mass horizontal modal participation factor for the j mode and $Sa(Th_j)$ is the pseudospectral acceleration for j mode. Results are shown in Table 4.

Table 4. Accelerations due to horizontal earthquake in the 6-story building.

Period (s)	$^{1}\Gamma_{hj}$	Pseudospectral Horizontal Acceleration						
		^{2}Sa-NILT	^{2}Sa-CALE	^{2}Sa-JAMI	^{2}Sa-SCRU	^{2}Sa-LANE	^{2}Sa-HMTT	^{2}Sa-RABO
0.398	0.698	0.665	0.599	0.261	0.296	0.412	0.265	0.174
0.108	0.196	1.864	1.059	0.613	0.532	0.548	0.348	0.309
0.052	0.043	1.232	0.563	0.909	0.405	0.457	0.213	0.233
	$\Sigma\Gamma_h = 0.937$							
$Sa\ H_{SRSS}$		0.342	0.270	0.128	0.134	0.177	0.114	0.078

[1] Γ_{hj} stands for mass horizontal modal participation factor; [2] Sa denotes the pseudospectral accelerations obtained for each spectrum (i.e., NILT, CALE, etc.) in the horizontal direction.

Likewise, for the vertical accelerations with the highest modal participation (not necessarily the first three as mentioned before) $Sa\ V_{SRSS}$ is computed with:

$$Sa\ V_{SRSS} = \sqrt{\sum_{j=1}^{n} \frac{1}{n} \left[\Gamma_{vj} \times Sa\left(Tv_j\right)\right]^2} \quad (2)$$

where Γ_{vj} stands for the mass vertical modal participation factor for the vertical direction and $Sa\left(Tv_j\right)$ denotes the pseudospectral accelerations obtained for each spectrum (i.e., NILT, CALE, etc.) in the vertical direction. The values are listed in Table 5.

Table 5. Accelerations due to horizontal earthquake in the 6-story building.

Period (s)	$^2\Gamma_{vj}$	Pseudospectral Vertical Acceleration						
		^{2}Sa-NILT	^{2}Sa-CALE	^{2}Sa-JAMI	^{2}Sa-SCRU	^{2}Sa-LANE	^{2}Sa-HMTT	^{2}Sa-RABO
0.173	0.524	0.374	0.502	0.240	0.431	0.325	0.223	0.318
0.164	0.046	0.432	0.495	0.247	0.493	0.353	0.231	0.278
0.060	0.106	1.025	1.266	1.150	0.697	0.615	0.128	0.185
	$\Sigma\Gamma_v = 0.677$							
$Sa\ V_{SRSS}$		0.130	0.171	0.101	0.138	0.105	0.068	0.097

$^2\Gamma_{vj}$ stands for mass horizontal modal participation factor; ^2Sa denotes the pseudospectral accelerations obtained for each spectrum (i.e., NILT, CALE, etc.) in the vertical direction

To inspect which earthquake ground motion causes more damage when its horizontal and vertical components are simultaneously applied to the 6-story building, the information from Tables 4 and 5 is also included in Table 6. Table 6 contains a new value termed as *AF* (adjustment factor) that is calculated as the inverse of the sum of the modal participations $\Sigma\Gamma_h$ and $\Sigma\Gamma_v$, for the modes included in Tables 4 and 5, respectively; these are the inverse of 0.937 and 0.677, leading to 1.066 and 1.477, respectively. Then, $Sa_{H\&V}$ is computed with:

$$Sa_{H\&V} = \sqrt{\frac{\left((\Sigma\Gamma_h)^{-1} \times (Sa\ H_{SRSS})\right)^2 + \left((\Sigma\Gamma_v)^{-1} \times (Sa\ V_{SRSS})\right)^2}{2}} \quad (3)$$

which is used to complete the values listed in Table 6.

Table 6. Resultant accelerations of the horizontal plus vertical components (6-story building).

Component	*AF*	Sa-NILT	Sa-CALE	Sa-JAMI	Sa-SCRU	Sa-LANE	Sa-HMTT	Sa-RABO
$Sa\ H_{SRSS}$	1.066	0.342	0.270	0.128	0.134	0.177	0.114	0.078
$Sa\ V_{SRSS}$	1.477	0.130	0.171	0.101	0.138	0.105	0.068	0.097
$Sa_{H\&V}$		0.292	0.271	0.1438	0.176	0.173	0.111	0.117

The values in Table 6 and the corresponding rotations from the non-linear analyses are depicted in Figure 13 (blue points) for the beams reported in Figures 7 and 8. Additionally, in Figure 13 simple linear fits are shown to visualize possible trends. In general, Figure 13 shows that larger damage (rotation) is expected for larger horizontal Sa (Figure 13 left) and that, the same occurs, but with even higher damage when the horizontal and vertical components act simultaneously (Figure 13 right). In both cases the plastic rotation is 0.0109.

Analogous results and trends are listed in Tables 7–9 and Figure 14. The same conclusions can be drawn, except that the dispersion of the damage around the trend line is higher for the 10-story building under the action of both components (Figure 14 right). Results for other beams are not shown for brevity, but they exhibit similar trends but with a value of plastic rotation equal to 0.0114.

Table 7. Accelerations due to horizontal earthquake in the 10-story building.

Periodo (s)	Γ_{hj}	Pseudospectral Horizontal Acceleration						
		Sa-NILT	Sa-CALE	Sa-JAMI	Sa-SCRU	Sa-LANE	Sa-HMTT	Sa-RABO
0.8412	0.738	0.379	0.229	0.195	0.294	0.224	0.126	0.070
0.2383	0.157	1.383	0.733	0.245	0.457	0.521	0.301	0.320
0.1216	0.039	2.048	1.187	0.671	0.597	0.557	0.275	0.322
	$\Sigma\Gamma_h = 0.934$							
$Sa\ H_{SRSS}$		0.210	0.121	0.087	0.132	0.107	0.060	0.042

Table 8. Accelerations due to vertical earthquake in the 10-story building.

Periodo (s)	Γ_{vj}	Pseudospectral Vertical Acceleration						
		Sa-NILT	Sa-CALE	Sa-JAMI	Sa-SCRU	Sa-LANE	Sa-HMTT	Sa-RABO
0.198	0.398	0.263	0.584	0.198	0.292	0.259	0.277	0.247
0.122	0.070	0.766	0.587	0.366	0.464	0.494	0.172	0.216
0.058	0.254	1.004	1.287	1.128	0.701	0.634	0.125	0.187
	$\Sigma\Gamma_v = 0.722$							
$Sa\ V_{SRSS}$		0.162	0.233	0.172	0.124	0.112	0.066	0.063

Table 9. Resultant accelerations of the horizontal plus vertical components (10-story building).

Component	*AF*	Sa-CALE	Sa-JAMI	Sa-LANE	Sa-HMTT	Sa-NILT	Sa-RABO	Sa-SCRU
$Sa\ H_{SRSS}$	1.070	0.121	0.087	0.107	0.060	0.210	0.042	0.132
$Sa\ V_{SRSS}$	1.383	0.233	0.172	0.112	0.066	0.162	0.063	0.124
$Sa_{H\&V}$		0.245	0.181	0.136	0.080	0.224	0.070	0.157

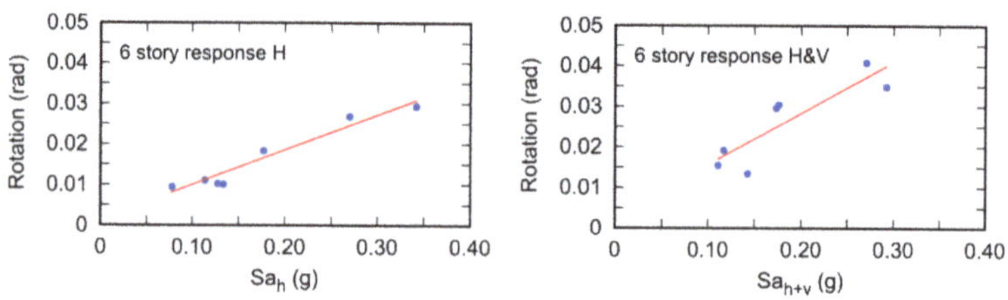

Figure 13. Rotation as a function of Sa for the 6-story building for beams studied in Figures 7 and 8; left for the horizontal component, right for the horizontal plus vertical component.

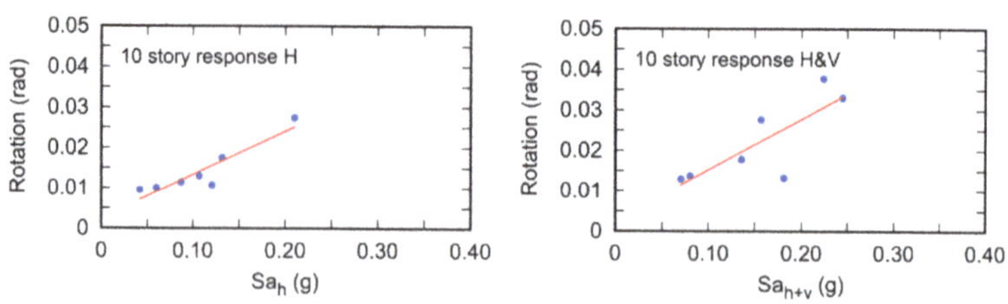

Figure 14. Rotation as a function of Sa for the 10-story building; for beams studied in Figures 10 and 11. Left for the horizontal component, right for the horizontal plus vertical component.

Before proceeding to the next section, it is emphasized that Equation (3) leads to the consistent (quantifiable) trend shown in Figure 13 because the modes with the highest modal participation (not necessarily the first three) are used and an extrapolation is carried out to consider the 100% mass participation. Other combinations of the response (different to Equation (3)) do not lead to this consistent trend. Therefore, Equation (3) is considered as a contribution of the present study.

5. Performance Level of Critical Beams

To discuss the performance level of the beams studied, guidelines from the American Society of Civil Engineering—ASCE SEI 41–17, [35] are considered. The stipulated criteria in that reference (acceptance criteria) are divided in: immediate occupancy (IO), life safety (LS) and collapse prevention (CP).

If the beams in the examples presented above, and the reference values IO = 1θy, LS = 9θy and CP = 11θy, are considered, it is noted that in no case the LS level is reached. However, it was noticed that if only the horizontal seismic component is considered, the elements remain in the elastic range in several cases, as for stations JAMI, SCRU and RABO for the 6-story building, and as for stations CALE, HMTT and RABO for the 10-story building.

As reported before, the rotations change, sometimes significantly, when both, the horizontal and the vertical, components are considered. In almost all cases there are increases in rotation ranging from 15% to up to about 207% for the two-component analysis as compared to the one-component non-linear dynamic analysis. This indicates that the response could lead to different acceptation criteria levels in the ASCE SEI 41–17, depending on whether both components are acting simultaneously.

Therefore, it is concluded that while a certain structure would reach the LS acceptation level with the horizontal component only, it could reach the CP acceptation level if both components are acting simultaneously. This is important, because it may imply that a supposedly rightly designed structure could collapse if the horizontal and vertical components act together. This is schematically illustrated for some cases in Figures 15 and 16 for the 6-story building, and Figures 17 and 18 for the 10-story building.

For instance, Figure 15 shows that for the case in the bottom-left column of the figure (case CALE-5S-LS) the rotation resulting from the horizontal component (H) is located within the LS performance level, whereas the rotation caused by both components acting together (H&V) increases the rotation, but it is also within the LS performance level. This is also observed in Figure 15 in the top-left column of the figure (case NILT-7S-LS).

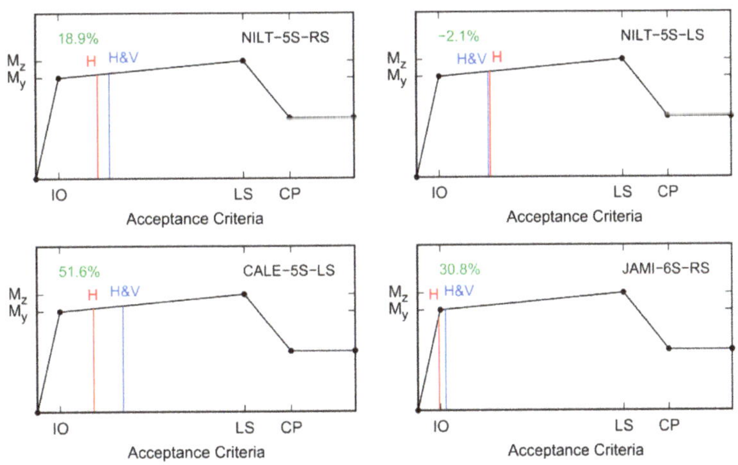

Figure 15. Examples of acceptance criteria for the 6-story building. Part 1.

Contrarily, the other cases in Figures 16 and 17 (cases SCRU-6S-RS and CALE-9S-LS, respectively) do lead to a change in the performance level from IO to LS when the vertical component is added in the analyses. These results indicate that inclusion of the effects of the vertical component of earthquake ground motion, when considered simultaneously with the horizontal component, can also have an impact in codified design. Therefore, code developers and designers should also be aware of this impact for practical purposes.

Figures 15 and 16 and Figures 17 and 18 shows the acceptance criteria of beams studied in Figures 7 and 8 and Figures 10 and 11, respectively. In all these figures, at the top-left of each individual figure the percentage difference between horizontal and horizontal + vertical acceptance criteria are shown in green.

It could be argued that the results in this section may not be useful in practice, because the structures considered in this section meet the strong-column weak-beam philosophy but are not realistically code-designed buildings. Therefore, a moment-resisting steel frame, designed by a practicing engineer to withstand seismic loads in accordance with an international code, is used in the following section to further inspect the obtained results, in terms of the seismic performance levels due to the adding effects of the vertical component.

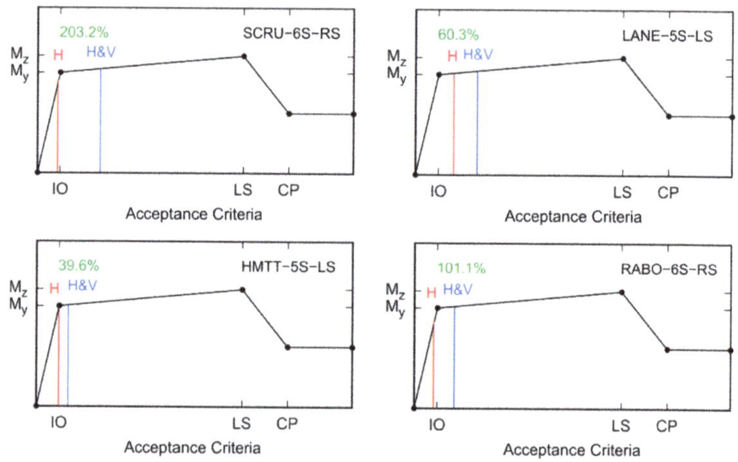

Figure 16. Examples of acceptance criteria for the 6-story building. Part 2.

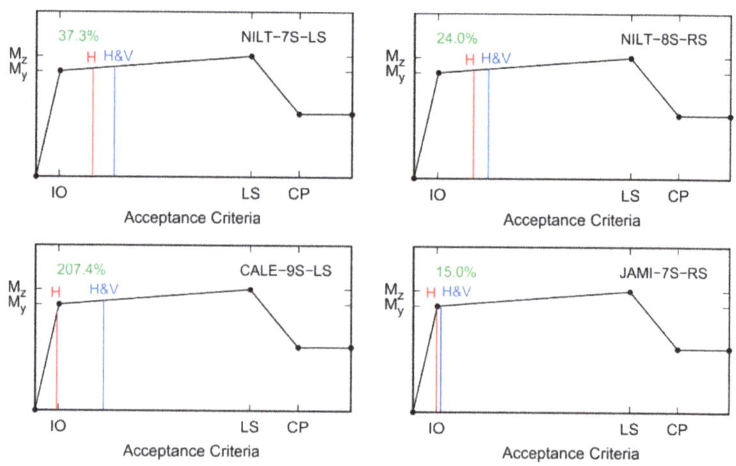

Figure 17. Examples of acceptance criteria for the 10-story building. Part 1.

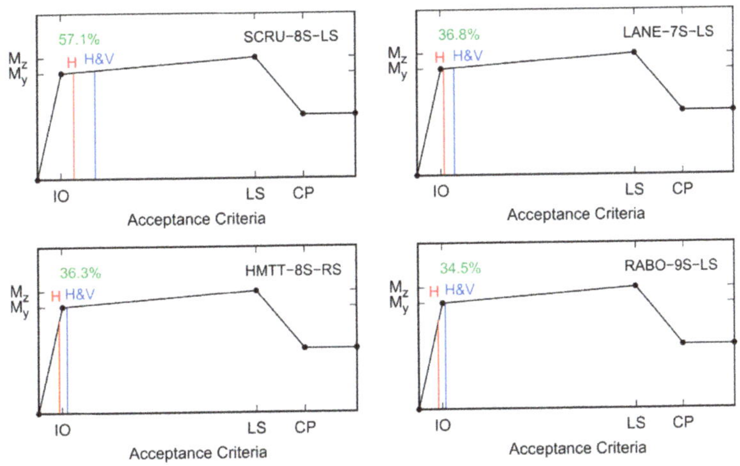

Figure 18. Examples of acceptance criteria for the 10-story building. Part 2.

6. Performance Levels of a Code-Designed Structure

In this section, the same procedure used to obtain the results shown in Figures 15–18 is applied to a structure designed by an experienced practicing engineer (Roeder et al., 1993 [36]; Schneider et al., 1993 [37]) as per the seismic provisions of the Uniform Building Code (UBC, 1988 [38]). This structure could be considered more representative of existing structures and be closer in design terms to buildings located in seismic-prone regions, in comparison to the structures considered in the previous sections.

It is the eight-story moment-resisting steel frame showed in Figure 19, where the W-Shapes of A-36 steel for beams and columns can be observed, together with the dimensions and the points with the largest rotations (red dots); depicted green symbols indicate change in the columns section (at midstory).

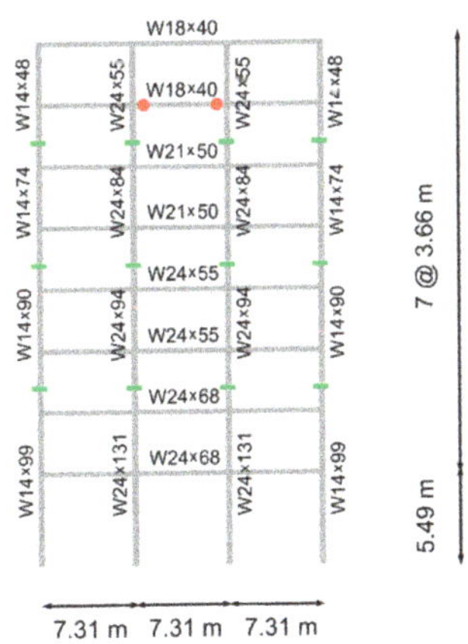

Figure 19. Realistic code-designed moment-resisting steel frame to assess seismic performance levels. Red dots denote selected plastic hinges. Green symbols indicate the section change at midstory.

This structure just meets the UBC (1988) [38] requirements for earthquake design (Roeder et al., 1993 [36]; Schneider et al., 1993 [37]). The considered gravity load (including dead load and live load) in the non-linear analyses is 52.54 kN/m at interstory levels and 43.78 kN/m at roof level. The self-weight of the elements is considered in the analyses.

These uniformly distributed loads are based on the loads per area unit reported for this structure (Roeder et al., 1993 [36]; Schneider et al., 1993 [37]) and the consideration that the frame corresponds to a central frame and that the tributary areas for each side of the beams are half of the area for a given floor.

For the eight-story building, the first horizontal vibration period is T_{1h} = 2.048 s, the second is T_{2h} = 0.738 s and the third one is T_{3h} = 0.415 s. The same corresponding periods, but for the vertical response are T_{1v} = 0.230 s, T_{2v} = 0.151 s and T_{3v} = 0.126 s, respectively.

Roeder et al., 1993 [36] reported only the fundamental period for the horizontal component. Some assumptions were made with the available information found in [36], as mentioned above. This led to some variation in the fundamental period since a period of 2.4 s is indicated in [36].

The structure in Figure 19 is to be subjected to the records in Table 1 for the non-linear dynamic analyses too, plus other four Mexican records listed in Table 10. The corresponding response spectra are depicted in Figure 20 for the additional records.

Table 10. Additional Mexican ground motions used in this section.

Station	Direction	Date	Magnitud Mw	Rrup (km)	Lat.	Long.	H (km)	PGA_h (g)	PGA_v (g)
BALC	EW	1994/12/10	6.4	59.2	17.98	101.52	50	0.185	0.202
PANG	EW	1999/09/30	7.4	50.0	16.03	96.96	47	0.114	0.096
PETA	EW	1994/12/10	6.4	74.8	17.98	101.52	50	0.208	0.086
RIOG	EW	1999/09/30	7.4	33.8	16.03	96.96	47	0.303	0.224

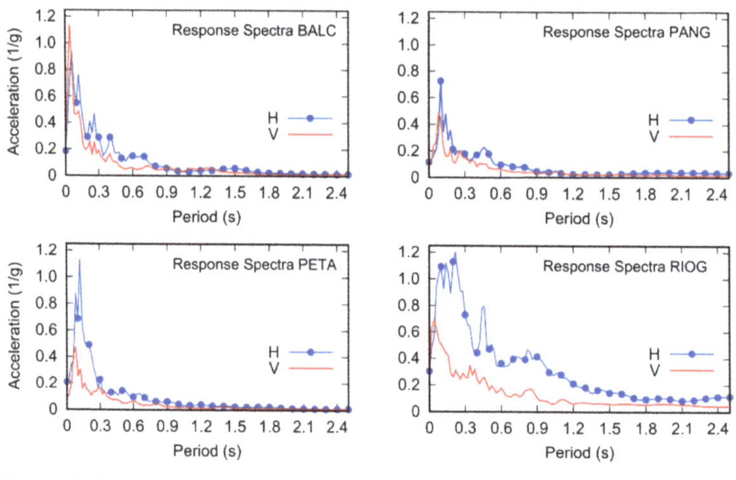

Figure 20. Response spectra for the ground motions in Table 10.

The non-linear response is computed as before, i.e., the OpenSEES program [32], the pre- and post-processor GiD [33] and the coupling interface GiD+OpenSees Platform [34] are used with the same previously stated assumptions and nomenclature, except that the steel frame in Figure 19 and its characteristics are considered. Distributed plasticity with fibers in all the elements is assumed. However, in this case, the P-delta effect is not added for simplicity and at intersections the joint offset node was not taken into account.

The results are shown in Figures 21–23 for the red dots in Figure 19. The W18 × 40 beams on the 7th floor have a yielding flexure moment My = 278.1 kN-m and a yielding rotation θy = 0.0076 rad. It is observed that the beam undergoes into the inelastic range for almost the 11 ground motions (except for the PETA-7S-LS case) and that, when both components act simultaneously, larger plastic rotations (from moderate to significant increases) are obtained.

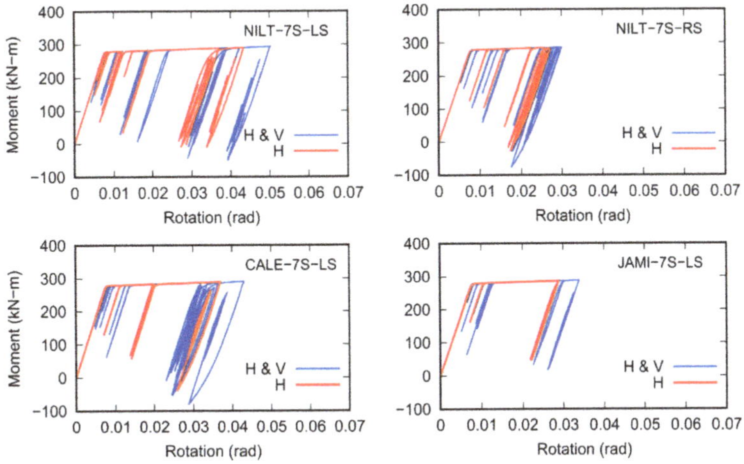

Figure 21. Non-linear response of the outer-most fiber in the considered beam for different earthquakes; 8-story building (Part 1).

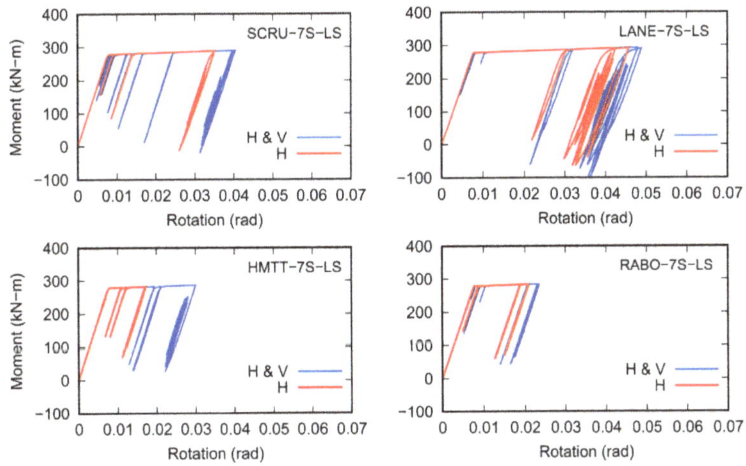

Figure 22. Non-linear response of the outer-most fiber in the considered beam for different earthquakes; 8-story building (Part 2).

These results confirmed the findings in the previous section, namely, the inclusion of the vertical component in the analyses can lead to significant larger plastic rotations. This also implies that the structure could change its performance level if the vertical component of the ground motion is considered together with the horizontal component. This is to be discussed shortly after.

The acceptance criteria for the steel frame in Figure 19 are shown in Figures 24–26. Again, the considered acceptance thresholds in this last example have values IO = 1θy, LS = 9θy and CP = 11θy. As for the previous examples, in Figures 24–26 the increasing effect due to the addition of the seismic vertical component is readily observed and, although the crossing of a threshold (in terms of the acceptance criteria) is not as clearly observed, case BALC-7S-LS can be seen as an example that including the simultaneous occurrence of the vertical component in the analysis may lead to a change from IO to LS. The increase in terms of percentage can be also significant as observed for instance for case HMTT-7S-LS in Figure 25 (as before, percentages are indicated in green in the left upper corner for each case). It is highlighted that the largest rotations of 0.0502 rad and 0.0489 rad are obtained for cases NILT-7S-Ls and LANE-7S-LS, respectively.

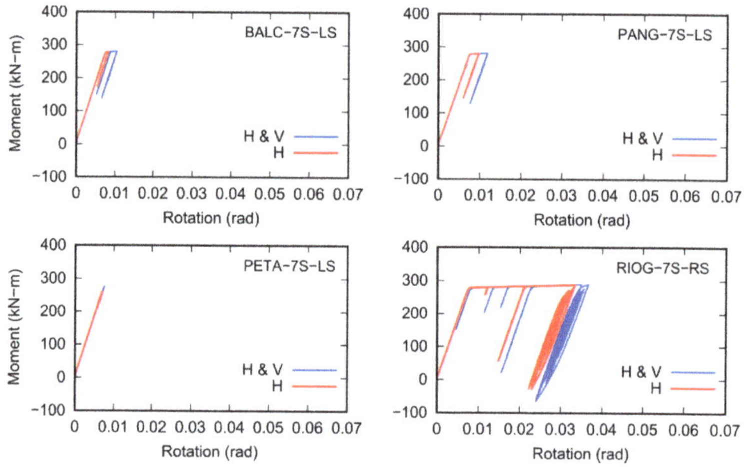

Figure 23. Non-linear response of the outer-most fiber in the considered beam for different earthquakes; 8-story building (Part 3).

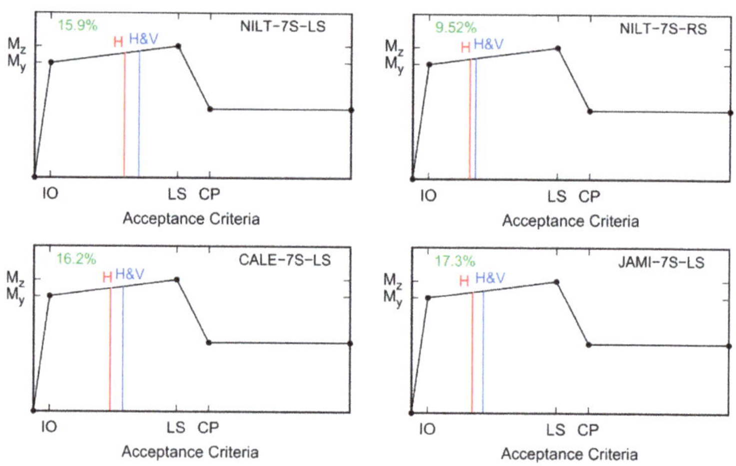

Figure 24. Examples of acceptance criteria for the 8-story building. Part 1.

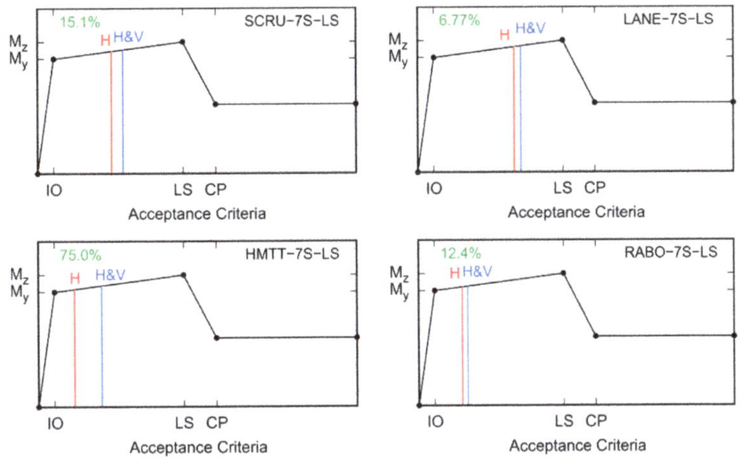

Figure 25. Examples of acceptance criteria for the 8-story building. Part 2.

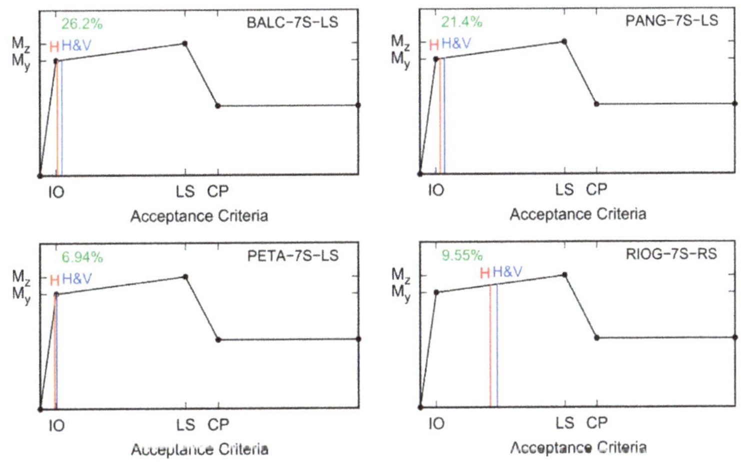

Figure 26. Examples of acceptance criteria for the 8-story building. Part 3.

In Table 11, a summary of the maximum rotations for all considered records is given.

Table 11. Rotations generated by horizontal earthquake components and by horizontal and vertical earthquake components acting together (8-story building).

Rotation	Rotation Values for Earthquake Station Register Considered (rad)										
Computed	NILT	CALE	JAMI	SCRU	LANE	HMTT	RABO	BALC	PANG	PETA	RIOG
θ_h	0.0433 (5.7θy)	0.0370 (4.9θy)	0.0289 (3.8θy)	0.0351 (4.6θy)	0.0458 (6.0θy)	0.0172 (2.3θy)	0.0209 (2.8θy)	0.0084 (1.1θy)	0.0098 (1.3θy)	0.0072 (0.95θy)	0.0335 (4.4θy)
θ_{h+v}	0.0502 (6.6θy)	0.0430 (5.7θy)	0.0339 (4.5θy)	0.0404 (5.3θy)	0.0489 (6.4θy)	0.0301 (4.0θy)	0.0235 (3.1θy)	0.0106 (1.4θy)	0.0119 (1.6θy)	0.0077 (1.01θy)	0.0367 (4.8θy)

Therefore, the structural response of the moment-resisting steel frame in Figure 19 under the seismic records in Tables 1 and 10 demonstrates that the conclusions previously stated (for the frames in Figure 1) hold also for a realistic code-designed structure subjected to real ground motions. It is, therefore, considered that the findings in this study can be of guidance for practical applications and can further highlight that the simultaneous action of the vertical component and the horizontal component (as it occurs in the real physical phenomenon) should not be disregarded in seismic design. It is acknowledged though, that the study is far from exhaustive and that further research using different steel structures configurations, standards and ground motions is required.

One last figure, Figure 27, is analogous to Figures 13 and 14, but considering the code-designed steel frame in this section. It can be observed that the increasing trends of rotations as a function of increasing Sa is also confirmed (with some more dispersion though). Notice that the pseudo-accelerations are not so large for the horizontal component alone. The reason is that the structural periods correspond to small spectral ordinates. However, when both components are considered, the spectral ordinates for the vertical component corresponds to larger spectral ordinates, shifting this case to larger Sa in Figure 27. In contrast, the difference in terms of rotations is not as noticeable for both cases depicted in Figure 27.

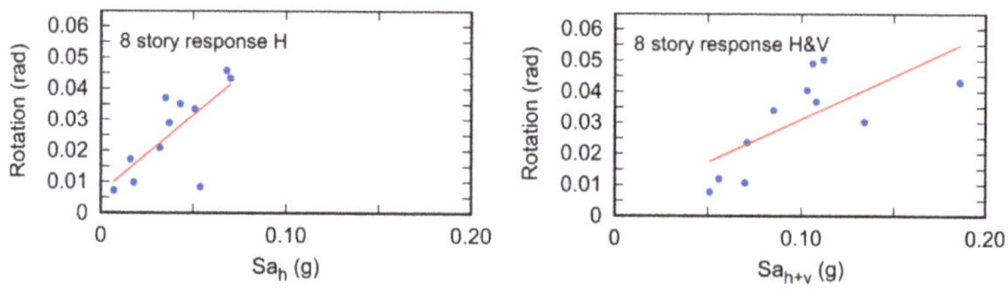

Figure 27. Rotation as a function of Sa for the 8-story building beam (see Figure 19). Left for the horizontal component, right for the horizontal plus vertical component.

7. Discussion

This study discusses the impact of the vertical earthquake ground motion component in the performance level of steel building beams. First, two hypothetical steel-frame buildings with diagonals are used as case studies to perform seismic non-linear dynamic analyses subjected to earthquake excitations from very large earthquakes that damaged many structures in Mexico. OpenSEES software is used to inspect the performance of the steel beams in terms of plastic rotations. The performance is inspected from the perspective of codified design.

It was found that the inclusion of the vertical component of earthquake ground motion, located near the causative fault, in the analysis leads to larger plastic rotations in the beams (i.e., when both, the horizontal and vertical components act simultaneously on the structure), compared to the rotations obtained by subjecting the structure to the earthquake excitations of only the horizontal component.

It had not been previously reported that most of the practical cases focus only on the horizontal component, and when the vertical component is included it does not significantly change the load effects, as described in [7]. An important change in the load effects of the structural elements is obtained when considering consistent mass matrices in the analysis, which is not common to find in commercial software [7]. Another way to achieve this change is through the discretization of the structural elements as is done with the analyses based on the finite element method.

It is concluded that designers and code developers should incorporate as criterion to perform non-linear analyses of the horizontal and vertical components acting together, because this leads in some elements to a change in the acceptation criteria in codified design to a most critical one when both components act simultaneously. This would be particularly important for places where the vertical component becomes relevant (near the source). General trends indicate that larger pseudo-spectral accelerations lead to larger plastic rotations of the steel beams. These general trends also confirm that structures subjected to the simultaneous action of horizontal and vertical components of ground motion yield larger plastic rotations (as compared to those from the one-component analysis) for the same level of pseudo-spectral acceleration.

Moreover, a more realistic structure was also considered to validate the results. It was demonstrated that when a steel frame designed by a practicing engineer is subjected to real near-fault ground motions by means of non-linear analyses, the same conclusions hold. It is acknowledged that other seismic standards, including the newest versions of worldwide used codes, more recordings representative of the seismic hazard of a given structure designed as per the applicable regulations, other structural typologies, and other materials (e.g., reinforced concrete frames) should be considered to have further insight into the findings of the present study. This is recommended for further research.

8. Conclusions

Current structural design regulations for the construction of buildings in seismic zones specify that a non-linear seismic analysis could be performed. Depending on the type of structure, it can be a non-linear static seismic analysis (pushover) or a non-linear dynamic seismic analysis. An important difference between both methods is that generally different acceptance levels result for the same structure, becoming more critical when performing a dynamic seismic analysis.

Taking this into account, in this work a non-linear dynamic seismic analysis was carried out where both the horizontal component and the vertical component of several seismic records occurred in Mexico are considered. Two hypothetical and one realistic (code-designed) steel frames are considered.

It was found that the dynamic analysis considering the horizontal and the vertical components simultaneously (as compared to using only the horizontal component), does not always lead to the largest rotation. This does not occur in most cases and could be attributed to the counteracting effect of the vertical component for certain ground motions and considered structure.

In addition, it is found that there are higher levels of acceptance (i.e., larger plastic rotations) when both seismic components (horizontal and vertical acting simultaneously) are considered than when only the horizontal component is considered, as happens in most practical cases. It is recommended to include the vertical seismic component in future earthquake assessment and design of buildings since the extra calculation effort is minimal (compared to a dynamic analysis by considering only the horizontal component).

This study contributes to the understanding of structural behavior by considering the effect of both components of ground motions from a specific tectonic environment, since Mexican earthquakes are used.

A simple quantitative expression to correlate the non-linear response (in terms of plastic rotations) and a ground motion intensity measure (pseudo-spectral acceleration) is proposed. In the proposed equation the three modes with the highest modal participation, which are not necessarily the first three, are used. Since other combinations of the response do not lead to this consistent trend, the proposed expression is considered as a contribution of the present study.

It is acknowledged that in order to formulate more robust conclusions, further studies are required by extending the analysis to other structures and by using many other recorded near-fault ground motions, while also incorporating more refined modelling assumptions for the elements and connections of realistic structures.

Other aspects which are recommended for further research are to explore a probabilistic rather than a deterministic approach (for instance to obtain median values of the response using two components, while also determining the variability in quantitative terms), as well as the use of scaled records to guarantee that many members of realistic structures undergo into the inelastic range consistent with future extraordinary seismic scenarios.

Author Contributions: Conceptualization, J.-G.V.-V., A.D.G.-S. and M.Á.J.; methodology, J.-G.V.-V., A.D.G.-S. and M.Á.J.; software, J.-G.V.-V., A.D.G.-S. and M.Á.J.; validation, J.-G.V.-V., A.D.G.-S. and M.Á.J.; formal analysis, J.-G.V.-V., A.D.G.-S. and M.Á.J.; investigation, J.-G.V.-V., A.D.G.-S. and M.Á.J.; resources, J.-G.V.-V., A.D.G.-S. and M.Á.J.; data curation, J.-G.V.-V., A.D.G.-S. and M.Á.J.; writing—original draft preparation, J.-G.V.-V., A.D.G.-S. and M.Á.J.; writing—review and editing, J.-G.V.-V., A.D.G.-S. and M.Á.J.; visualization, J.-G.V.-V., A.D.G.-S. and M.Á.J.; supervision, J.-G.V.-V., A.D.G.-S. and M.Á.J.; project administration, J.-G.V.-V. and A.D.G.-S.; funding acquisition, J.-G.V.-V. All authors have read and agreed to the published version of the manuscript.

Funding: The financial support from Universidad de Guanajuato (División de Ingenierías and Campus Guanajuato) is gratefully acknowledged.

Acknowledgments: We are grateful to the International Centre for Numerical Methods in Engineering (CIMNE) for providing us with the pre- and post-processor GiD (2020) in our CIMNE-classroom at the Civil Engineering Department. We are thankful to four anonymous reviewers for their comments, suggestions and constructive criticism, which helped to improve this study.

Conflicts of Interest: The authors declare no conflict of interest.

References

1. Papazoglou, A.J.; Elnashai, A.S. Analytical and field evidence of the damaging effect of vertical earthquake ground motion. *Earthq. Eng. Struct. D* **1996**, *25*, 1109–1137. [CrossRef]
2. Elnashai, A.S.; Papazoglou, A.J. Procedure and spectra for analysis of RC structures subjected to strong vertical earthquake loads. *J. Earthq. Eng.* **1997**, *1*, 121–155. [CrossRef]
3. Kim, S.J.; Elnashai, A.S. Seismic assessment of RC structures considering vertical ground motion. *Mid-Am. Earthq. Cent. CD* **2008**. Available online: https://www.ideals.illinois.edu/handle/2142/9454 (accessed on 20 February 2021).
4. Kim, S.J.; Holub, C.J.; Elnashai, A.S. Analytical assessment of the effect of vertical earthquake motion on RC bridge piers. *J. Struct. Eng.* **2011**, *137*, 252–260. [CrossRef]
5. Kunnath, S.K.; Erduran, E.; Chai, Y.H.; Yashinsky, M. Effect of near-fault vertical ground motions on seismic response of highway overcrossings. *J. Bridge. Eng.* **2008**, *13*, 282–290. [CrossRef]
6. Pozos-Estrada, A.; Chávez, M.M.; Jaimes, M.Á.; Arnau, O.; Guerrero, H. Damages observed in locations of Oaxaca due to the Tehuantepec Mw8. 2 earthquake, Mexico. *Nat. Hazards* **2019**, *97*, 623–641. [CrossRef]
7. Valdés-Vázquez, J.G.; García-Soto, A.D.; Luna-Rodríguez, F.J.; Valdés-Vázquez, J.F.; Hernández-Martínez, A. Modal analysis of structures using consistent mass matrices instead of concentrated mass matrices (In Spanish). In Proceedings of the XXI National Conference of Earthquake Engineering, Monterrey, NL, Mexico, 20–23 November 2019.
8. Paz, M.; Kim, Y.H. *Structural Dynamics Theory and Computation*; Springer International Publishing: Cham, Switzerland, 2019.
9. Bozorgnia, Y.; Campbell, K.W. Ground Motion Model for the Vertical-to-Horizontal (V/H) Ratios of PGA, PGV, and response spectra. *Earthq. Spectra.* **2016**, *32*, 951–978. [CrossRef]
10. Jaimes, M.A.; Ruiz-García, J. New vertical-to-horizontal ratio spectrum due to intraslab earthquakes for soft-soil sites of Mexico City. *Soil Dyn. Earthq. Eng.* **2019**, *126*, 105804. [CrossRef]
11. Di Michele, F.; Cantagallo, C.; Spacone, E. Effects of the vertical seismic component on seismic performance of an unreinforced masonry structures. *B Earthq. Eng.* **2020**, *18*, 1635–1656. [CrossRef]
12. Ruiz-García, J. Examination of the vertical earthquake ground motion component during the September 19, 2017 (Mw = 7.1) earthquake in Mexico City. *Soil Dyn. Earthq. Eng.* **2018**, *110*, 13–17. [CrossRef]
13. Badalouka, B.G.; Papadopoulos, G.A. Experimental Study of a Structure Under Stress Pulse Simulating Vertical Ground Motion. *J. Earthq. Eng.* **2008**, *12*, 341–356. [CrossRef]
14. Dolatshahi, K.M.; Rezaie, A.; Rafiee-Dehkharghani, R. Topology Optimization of Wave Barriers for Mitigation of Vertical Component of Seismic Ground Motions. *J. Earthq. Eng.* **2020**, *24*, 84–108. [CrossRef]
15. Gremer, N.; Adam, C.; Medina, R.A.; Moschen, L. Vertical peak floor accelerations of elastic moment-resisting steel frames. *B Earthq. Eng.* **2019**, *17*, 3233–3254. [CrossRef]
16. Jaimes, M.A.; García-Soto, A.D. Updated ground motion prediction model for Mexican intermediate-depth intraslab earthquakes including V/H ratios. *Earthq. Spectra.* **2020**, *36*, 1298–1330. [CrossRef]
17. Liberatore, D.; Doglioni, C.; AlShawa, O.; Atzori, S.; Sorrentino, L. Effects of coseismic ground vertical motion on masonry constructions damage during the 2016 Amatrice-Norcia (Central Italy) earthquakes. *Soil Dyn. Earthq. Eng.* **2019**, *120*, 423–435. [CrossRef]
18. Yang, H.; Yin, X. Transient responses of girder bridges with vertical poundings under near-fault vertical earthquake. *Earthq. Eng. Struct. D.* **2015**, *44*, 2637–2657. [CrossRef]
19. Dehghanpoor, A.; Thambiratnam, D.; Chan, T.; Taciroglu, E.; Kouretzis, G.; Li, Z. Coupled Horizontal and Vertical Component Analysis of Strong Ground Motions for Soil–Pile–Superstructure Systems: Application to a Bridge Pier with Soil–Structure Interaction. *J. Earthq. Eng.* **2019**. [CrossRef]

20. Chen, Y.; Kun, K.; Larkin, T.; Chouw, N. Impact of Vertical Ground Excitation on a Bridge with Footing Uplift. *J. Earthq. Eng.* **2016**, *20*, 1035–1053. [CrossRef]
21. Gülerce, Z.; Abrahamson, N.A. Vector-Valued Probabilistic Seismic Hazard Assessment for the Effects of Vertical Ground Motions on the Seismic Response of Highway Bridges. *Earthq. Spectra.* **2010**, *26*, 999–1016. [CrossRef]
22. Di Sarno, L.; Elnashai, A.S.; Manfredi, G. Assessment of RC Columns Subjected to Horizontal and Vertical Ground Motions Recorded During the 2009 L'Aquila (Italy) Earthquake. *Eng. Struct.* **2011**, *33*, 1514–1535. [CrossRef]
23. Mazza, F.; Vulcano, A. Effects of Near-Fault Ground Motions on the Nonlinear Dynamic Response of Base-Isolated RC Framed Buildings. *Earthq. Eng. Struct. D* **2012**, *41*, 211–232. [CrossRef]
24. Landi, L.; Grazi, G.; Diotallevi, P.P. Comparison of Different Models for Friction Pendulum Isolators in Structures Subjected to Horizontal and Vertical Ground Motions. *Soil Dyn. Earthq. Eng.* **2016**, *81*, 75–83. [CrossRef]
25. Mazza, F.; Mazza, M. Nonlinear Seismic Analysis of Irregular RC Framed Buildings Base-Isolated with Friction Pendulum System under Near-Fault Excitations. *Soil Dyn. Earthq. Eng.* **2016**, *90*, 299–312. [CrossRef]
26. Harrington, C.C.; Liel, A.B. Collapse assessment of moment frame buildings, considering vertical ground shaking. *Earthq. Eng. Struct. D* **2016**, *45*, 2475–2493. [CrossRef]
27. Housner, G.W. Spectrum intensities of strong motion earthquakes. In *Proceedings of the Symposium on Earthquakes and Blast Effects on Structures, Los Angeles, CA, USA, June 1952*; Earthquake Engineering Research Institute: Los Angeles, CA, USA; pp. 20–36.
28. American Society of Civil Engineers (ASCE). *ASCE Standard–ASCE/SEI 7-16: Minimum Design Loads for Buildings and Other Structures*; ASCE: New York, NY, USA, 2016.
29. *IBC-2018 International Building Code*; International Code Council: Country Club Hills, IL, USA, 2018.
30. NTCS-2017 Normas Técnicas Complementarias para Diseño por Sismo, Reglamento de Construcciones para el Distrito Federal, Gaceta Oficial del Departamento del Distrito Federal. 2017. (In Spanish)
31. Eurocode. *Eurocode 8: Design of Structures for Earthquake Resistance-Part 1: General Rules, Seismic Actions and Rules for Buildings*; The National Standards Authority of Ireland: Dublin, Ireland, 2004.
32. Mazzoni, S.; McKenna, F.; Scott, M.H.; Fenves, G.L. OpenSees Command Language Manual. University of California, Berkeley. Available online: https://opensees.berkeley.edu/ (accessed on 29 December 2020).
33. GiD. The Personal Pre- and Post-Processor. CIMNE, www.gidhome.com, 15th Eds. Available online: https://www.gidhome.com/ (accessed on 29 December 2020).
34. GiD+OpenSEES v.2.8.0. An Integrated FEA Platform. Lab of R/C and Masonry Structures, School of Civil Engineering, AUTh, Greece. Available online: http://gidopensees.rclab.civil.auth.gr/ (accessed on 29 December 2020).
35. American Society of Civil Engineers Standard. *Seismic Evaluation and Retrofit of Existing Buildings 41-17*; American Society of Civil Engineers: Reston Town Center, VA, USA, 2017.
36. Roeder, C.W.; Schneider, S.P.; Carpenter, J.D. Seismic behavior of moment-resisting steel frames: Analytical investigation. *J. Struct. Engrg.* **1993**, *119*, 1866–1884. [CrossRef]
37. Schneider, S.P.; Roeder, C.W.; Carpenter, J.E. Behavior of moment resisting steel frames: Experimental investigation. *J. Struct. Engrg.* **1993**, *119*, 1885–1902. [CrossRef]
38. ICBO. *Uniform Building Code*; International Conference of Building Officials: Whittier, CA, USA, 1988.

applied sciences

MDPI

Article

Transient Response of Bridge Piers to Structure Separation under Near-Fault Vertical Earthquake

Wenjun An [1,2,*] and Guquan Song [1]

1 School of Civil Engineering and Architecture, Nanchang University, Nanchang 330031, China; gqsong@ncu.edu.cn
2 College of Civil Engineering, Jiangxi University of Engineering, Xinyu 330046, China
* Correspondence: wjan@email.ncu.edu.cn

Abstract: Given the possible separation problem caused by the double-span continuous beam bridge under the action of the vertical earthquake, considering the wave effect, the transient wave characteristic function method and the indirect mode superposition method are used to solve the response theory of the bridge structure during the earthquake. Through the example analysis, the pier bending moment changes under different vertical excitation periods and excitation amplitudes are calculated. Calculations prove that: (1) When the seismic excitation period is close to the vertical natural vibration period of the bridge, the main girder and the bridge pier may be separated; (2) When the pier has a high height, the separation has a more significant impact on the longitudinal displacement of the bridge, but the maximum relative displacement caused by the separation is random; (3) Large-scale vertical excitation will increase the number of partitions of the structure, and at the same time increase the vertical collision force between the main girder and the pier, but the effect on the longitudinal displacement of the form is uncertain; (4) When V/H exceeds a specific value, the pier will not only be damaged by bending, but will also be damaged by axial compression.

Keywords: bridge; near-fault vertical earthquake; multiple separations; seismic excitation period; vertical excitation amplitude; bending damage; axial compression failure

Citation: An, W.; Song, G. Transient Response of Bridge Piers to Structure Separation under Near-Fault Vertical Earthquake. *Appl. Sci.* **2021**, *11*, 4068. https://doi.org/10.3390/app11094068

Academic Editor: Maria Favvata

Received: 8 April 2021
Accepted: 28 April 2021
Published: 29 April 2021

Publisher's Note: MDPI stays neutral with regard to jurisdictional claims in published maps and institutional affiliations.

Copyright: © 2021 by the authors. Licensee MDPI, Basel, Switzerland. This article is an open access article distributed under the terms and conditions of the Creative Commons Attribution (CC BY) license (https://creativecommons.org/licenses/by/4.0/).

1. Introduction

As a transportation hub, bridges cause serious damage to the road network when an earthquake occurs, which brings great difficulties to rescue work in the disaster-stricken areas. Simultaneously, this greatly affected the post disaster recovery and reconstruction work and significantly reduced the traffic function between regions [1,2]. In the past, the research on pier failure mainly focused on horizontal excitation, ignoring the influence of vertical earthquake [3–5]. However, with the progress of monitoring level in recent years, increasingly monitoring data show that the near-fault earthquake which has a sizeable vertical excitation amplitude, may cause landslides and damage to the bridge foundation [6–9]. In addition, the pier will suffer more tremendous vertical pressure, or even cause the pier axial compression damage [10–12]. Moreover, the fluctuation of axial force may also change the flexural and shear performance of the pier. In this regard, many experimental studies have been carried out [13–16]. For the main girder, the vertical seismic excitation will also increase the mid-span bending moment and even cause the bending failure of the structure [13,17]. In order to reduce the response of bridges under earthquake, a series of research have been carried out in the field of energy dissipation, such as the performance of rubber-bearing isolators under seismic events [18,19].

For the ratio of vertical and horizontal seismic acceleration, many codes simply set it as 2/3. However, these years, monitoring data show that V/H was far more than 2/3 or even 2. For example, in the Northridge earthquake, the ratio of vertical acceleration to horizontal acceleration reached 1.79, and the amplitude of vertical acceleration reached 1.18 g [20]. Under the Kobe earthquake, the peak acceleration ratio V/H is as high as 2 [21].

Analyzing nine pieces of data collected within 20 km of the Wenchuan earthquake source, the average acceleration ratio V/H is 0.89, and the maximum value is 1.2 [22].

Unlike foreign rigid frame bridges, most bridges in China use rubber bearings. The main beam is directly placed on the pier, and the structure lacks a tensile connection [23,24]. When the vertical seismic excitation is too large, the main beam and pier may be separated. In the 1989 Loms Prieta earthquake [25], a highway bridge is divided under the vertical quake, and the girder is thrown up. When the bridge collides again, the pier directly passes through the bridge deck. Besides, Tanimura [26] considered that the impact force caused by separation might damage the bearing and local pier.

In the past, the collision of bridge mainly focused on the longitudinal beam to beam collision [27], and the research on the vertical collision was less. In addition, the damping measures are set in the horizontal direction [28,29], but the effect of the vertical separation on the horizontal seismic response is ignored. Previous studies on bridge failure assumed that the structure always kept in contact and did not consider the possibility of structural separation caused by the near-fault vertical earthquake. In this study, a double-span continuous beam bridge model is established, and the forced resonance response of the bridge is calculated by using the transient wave function method [30]. The structural displacement response caused by the collision force is calculated by using the indirect modal function method [31]. The force and displacement response of the bridge are calculated by the theoretical approach, and the bending conditions of the bridge pier under different conditions are obtained.

2. Theoretical Model and Vertical Displacement Calculation

The model used in this study is a double-span continuous bridge. The calculation model is shown in Figure 1. The main beam is a prestressed box beam, and the pier is a double-column circular pier. The round lead high damping rubber bearing is used between the main girder and the piers. In the vertical direction, the hysteresis curve of the bearing is long and narrow, ignoring the damping of the vertical bearing. In the horizontal direction, the damping of the bearing is $\zeta_1 = 20\%$. In the vertical direction, the stiffness of the bearing is $K_c = 2.4 \times 10^9$ N/m; in the horizontal direction, the stiffness is $K_v = 2.4 \times 10^6$ N/m. To simplify the calculation, this study makes the following assumptions:

(1) In order to simplify the calculation, linear elastic calculation is adopted for the material of the model, ignoring the nonlinearity of the structure;
(2) Ignore the possible bearing shear failure caused by a horizontal earthquake;
(3) During seismic action, there are often stops in the lateral direction. This study only considers the coupling of vertical and longitudinal seismic activities;
(4) Ignore the difference in the arrival time of the horizontal and vertical seismic waves, assuming that the earthquakes in both directions are excited at the same time.

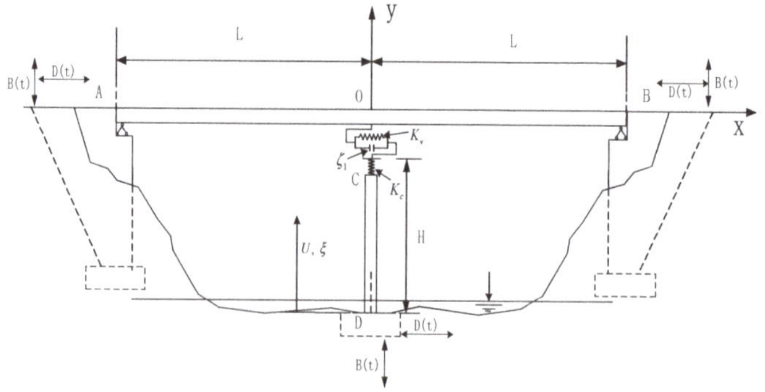

Figure 1. Model of a double-span continuous beam bridge.

2.1. Vertical Seismic Response Spectrum

The bridge is located in an area with a seismic intensity of 8^0. The reference peak value of horizontal seismic acceleration is selected as 510 gal (5.1 m/s^2). The values of horizontal seismic acceleration under different seismic excitation periods are shown in the specification [32]. The selection of the characteristic value of the vertical seismic response spectrum is given in reference [33], and the epicenter distances are 3 km, 10 km, and 20 km, respectively:

$$\lambda = V/H = \begin{cases} \alpha & T < 0.1 \\ \alpha - \beta(T - 0.1) & 0.1 \leq T < 0.3 \\ 0.5 & T \geq 0.3 \end{cases} \tag{1}$$

where T is the vertical seismic period, α is the peak value of V/H, and β is the linear attenuation coefficient. When the epicenter distance is 3 km, 10 km, and 20 km, α = 1.5, 1.4, 1.3; β = 5, 4, 3.

Complex seismic waves can be obtained by the superposition of harmonic components of each order by the Fourier expansion method. By calculating the excitation of a single harmonic component, the seismic excitation response of bridge structure in full frequency state can be obtained by superposition method. Therefore, for simplicity, this paper uses the single harmonic motion instead of the seismic excitation, and the acceleration peak value uses the seismic excitation acceleration peak value.

2.2. Theoretical Solution of Displacement Response of Bridge in Vertical Contact Stage

The wave equations of girder AB and pier CD can be described as the following:

$$\begin{array}{c} OA: \frac{\partial^2(E_bI_b\partial^2Y_1(x,t)/\partial x^2)}{\partial x^2} + \rho A_b\frac{\partial^2Y_1(x,t)}{\partial t^2} + q = 0 \\ OB: \frac{\partial^2(E_bI_b\partial^2Y_1(x,t)/\partial x^2)}{\partial x^2} + \rho A_b\frac{\partial^2Y_1(x,t)}{\partial t^2} + q = 0 \\ CD: E_rA_r\frac{\partial^2U(\xi,t)}{\partial^2\xi} - \rho A_r\frac{\partial^2U(\xi,t)}{\partial t^2} = 0 \end{array} \tag{2}$$

In these equations, $Y_1(x,t), Y_2(x,t)$ is the deflection of the beam; $U(\xi,t)$ is the axial displacement of pier.

The boundary condition of bridge displacement is:

$$\begin{array}{c} Y_1(-L,t) = 0,\ Y_2(L,t) = 0,\ U(0,t) = 0 \\ \frac{\partial^2Y_1(-L,t)}{\partial^2x} = 0,\ \frac{\partial^2Y_2(L,t)}{\partial^2x} = 0 \end{array} \tag{3}$$

and the displacement continuity condition at the middle of the main girder is:

$$Y_1(0,t) = Y_2(0,t),\ \frac{\partial Y_1(0,t)}{\partial x} = \frac{\partial Y_2(0,t)}{\partial x},\ \frac{\partial^2Y_1(0,t)}{\partial x^2} = \frac{\partial^2Y_2(0,t)}{\partial x^2} \tag{4}$$

The continuity condition of shear force and displacement between girder, pier and support are:

$$\begin{array}{c} U(H,t) - Y_1(0,t) = \delta_1 = \frac{F_c}{K_c} = -\frac{E_rA_r}{K_c}\frac{\partial U(H,t)}{\partial\xi} \\ E_rA_r\frac{\partial U(H,t)}{\partial\xi} = E_bI_b(\frac{\partial^3Y_1(0,t)}{\partial x^3} - \frac{\partial^3Y_2(0,t)}{\partial x^3}) \end{array} \tag{5}$$

Vertical displacement field $Y(x,t)$ of the girder can be divided into static displacement Y_s, rigid body displacement Y_g and dynamic deformation Y_d.

$$\begin{array}{c} Y(x,t) = Y_s(x) + Y_g(x,t) + Y_d(x,t) \\ U(\xi,t) = U_s(\xi) + U_g(\xi,t) + U_d(\xi,t) \end{array} \tag{6}$$

Dynamic displacement satisfies the wave equation, continuity condition, equilibrium differential equation, and force boundary condition.

Static displacement of the bridge is as follows:

$$\begin{gathered} Y_{1s}(x) = [q(-5L^4 + 6L^2x^2 - x^4) - 2F_c(2L^3 - x^3 - 3x^2L)]/24E_bI_b \\ Y_{2s}(x) = [q(-5L^4 + 6L^2x^2 - x^4) - 2F_c(2L^3 + x^3 - 3x^2L)]/24E_bI_b \\ U_s(\xi) = F_c\xi/E_rA_r \end{gathered} \quad (7)$$

where F_c is the axial pressure of the initial girder and the support.

The displacement of the rigid body of the bridge is as follows:

$$Y_{1g}(x,t) = Y_{2g}(x,t) = U_g(\xi,t) = B(t) \quad (8)$$

The dynamic deformation part of the structure can be expanded as an infinite series of wave mode products:

$$\begin{gathered} Y_{1d}(x,t) = \sum_{n=1}^{\infty} \varphi_{nb1}(x)q_n(t),\ Y_{2d}(x,t) = \sum_{n=1}^{\infty} \varphi_{nb2}(x)q_n(t) \\ U_d(\xi,t) = \sum_{n=1}^{\infty} \varphi_{nr}(\xi)q_n(t) \end{gathered} \quad (9)$$

The equation includes the bending wave function $\varphi_{nb1}, \varphi_{nb2}$ of the girder, the longitudinal wave function φ_{nr} of the pier, and the time function $q_n(t)$.

The wave model function is solved by the characteristic equations. The characteristic equations of the main girder and pier are, respectively:

$$\begin{aligned} OA: &\quad a^2\varphi_{nb1}{}^{(4)}(x) - \omega_n{}^2\varphi_{nb1}(x) = 0 \\ OB: &\quad a^2\varphi_{nb2}{}^{(4)}(x) - \omega_n{}^2\varphi_{nb2}(x) = 0 \\ CD: &\quad c^2\varphi_{nr}{}^{(2)}(\xi) + \omega_n{}^2\varphi_{nr}(\xi) = 0 \end{aligned} \quad (10)$$

where $\omega_n(n = 1,2,3,\cdots\cdots)$ is the natural frequency of bridge structure, $a = \sqrt{E_bI_b/\rho A_b}$ is the coefficient related to the beam flexural wave speed, $c = \sqrt{E_r/\rho}$ is the rod phase speed.

Flexural wave modes of the main girder and longitudinal wave of the pier can be described as:

$$\begin{gathered} \varphi_{nb1}(x) = A_1 \sin k_{bn}x + B_1 \cos k_{bn}x + C_1 \mathrm{sinh} k_{bn}x + D_1 \cosh k_{bn}x \\ \varphi_{nb2}(x) = A_2 \sin k_{bn}x + B_2 \cos k_{bn}x + C_2 \mathrm{sinh} k_{bn}x + D_2 \cosh k_{bn}x \\ \varphi_{nr} = E_1 \sin k_{rn}\xi + F_1 \cos k_{rn}\xi \end{gathered} \quad (11)$$

where k_{bn} and k_{rn} are the wave numbers of flexural wave and longitudinal wave respectively. $A_1, B_1, C_1, D_1, A_2, B_2, C_2, D_2,\ E_1, F_1$ are the coefficients.

The boundary conditions of the characteristic equation are as follows:

$$\begin{gathered} \varphi_{nb1}(-L) = 0,\ \varphi_{nb2}(L) = 0,\ \varphi_{nr}(0) = 0 \\ \varphi''_{nb1}(-L) = 0,\ \varphi''_{nb2}(L) = 0 \end{gathered} \quad (12)$$

The continuity conditions of feature direction are as follows:

$$\begin{gathered} \varphi_{nb1}(0) = \varphi_{nb2}(0),\ \varphi'_{nb1}(0) = \varphi'_{nb2}(0),\ \varphi''_{nb1}(0) = \varphi''_{nb2}(0) \\ \varphi_{nb1}(0) = \varphi_{nr}(H) + \frac{E_rA_r\varphi'_{nr}(H)}{K_c},\ E_bI_b(\varphi'''_{nb1}(0) - \varphi'''_{nb2}(0)) = E_rA_r\varphi'_{nr}(H) \end{gathered} \quad (13)$$

The orthogonal consistency of bending wave and longitudinal wave can be obtained by Equation (10):

$$\int_{-L}^{0} \rho A\varphi_{mb1}\varphi_{nb1}dx + \int_{0}^{L} \rho A\varphi_{mb2}\varphi_{nb2}dx + \int_{0}^{H} \rho A_r\varphi_{mr}\varphi_{nr}d\xi = \delta_{mn} \quad (14)$$

By introducing Equations (12) and (13) into Equation (11), the wave functions of the bridge structure can be obtained as follows:

$$\begin{aligned} \varphi_{nb1}(x) &= M_n A_n\left(-\frac{\sin k_{bn}(x+L)}{\cos k_{bn}L} + \frac{\mathrm{sinh}k_{bn}(x+L)}{\cosh k_{bn}L}\right) \\ \varphi_{nb2}(x) &= M_n A_n\left(\frac{\sin k_{bn}(x-L)}{\cos k_{bn}L} - \frac{\mathrm{sinh}k_{bn}(x-L)}{\cosh k_{bn}L}\right) \\ \varphi_{nr}(\xi) &= A_n \sin k_{rn}\xi \end{aligned} \tag{15}$$

By introducing Equation (15) into Equation (13) can solve for M_n and A_n.

Through the orthogonality condition, the time function $q_n(t)$ differential equation of the bridge can be obtained:

$$\omega_n^2 q_n(t) + 2\zeta\omega_n \dot{q}_n(t) + \ddot{q}_n(t) = \ddot{Q}_n(t) \tag{16}$$

By Laplace transformation, $q_n(t)$ can be obtained:

$$\begin{aligned} q_n(t) = e^{-\zeta_2\omega_n t}\left(q_n(0)\cos\omega_d t + \frac{\dot{q}_n(0)+\zeta_2\omega_n q_n(0)}{\omega_d}\sin\omega_d t\right) \\ +\frac{1}{\omega_d}\int_0^t e^{-\zeta_2\omega_n\tau}\ddot{Q}_n(\tau)\sin(\omega_d(t-\tau))d\tau \end{aligned} \tag{17}$$

In Equation (7) $\omega_d = \sqrt{1-\zeta_2^2}\omega_n$.

2.3. Corresponding Theoretical Solution of Bridge Vertical Separation Stage Displacement

In the separation process, the beam and the rod do not interact and move at their own characteristic frequencies ω_{nb} and ω_{nr}.

The displacement response of the main girder and pier can be decomposed into static displacement, rigid displacement, and dynamic deformation, and the main beam does not need to be decomposed into two segments.

$$\begin{aligned} \overline{Y}(x,t) &= \overline{Y}_s(x) + \overline{Y}_g(x,t) + \overline{Y}_d(x,t) \\ \overline{U}(\xi,t) &= \overline{U}_s(\xi,t) + \overline{U}_g(\xi,t) + \overline{U}_d(\xi,t) \end{aligned} \tag{18}$$

The vertical static displacement and rigid displacement of the bridge structure are as follows:

$$\begin{aligned} \overline{Y}_s(x) = q(-5L^4 + 6L^2x^2 - x^4)/24E_bI_b,\ \overline{U}_s(\xi) = 0 \\ \overline{Y}_g(x,t) = \overline{U}_g(\xi,t) = B(t) \end{aligned} \tag{19}$$

The wave equations of girder AB and pier CD can be described as the following:

$$\begin{aligned} AB: \frac{\partial^2(E_bI_b\partial^2 Y(x,t)/\partial x^2)}{\partial x^2} + \rho A_b\frac{\partial^2\overline{Y}(x,t)}{\partial t^2} + q = 0 \\ CD: E_rA_r\frac{\partial^2\overline{U}(\xi,t)}{\partial^2\xi} - \rho A_r\frac{\partial^2\overline{U}(\xi,t)}{\partial t^2} = 0 \end{aligned} \tag{20}$$

The wave mode functions of the main girder and pier are:

$$\begin{aligned} \overline{\varphi}_{nb}(x) = A_3\sin\overline{k}_{bn}x + B_3\cos\overline{k}_{bn}x + C_3\mathrm{sinh}\overline{k}_{bn}x + D_3\cosh\overline{k}_{bn}x \\ \overline{\varphi}_{nr} = E_2\sin\overline{k}_{rn}\xi + F_2\cos\overline{k}_{rn}\xi \end{aligned} \tag{21}$$

where $A_3, B_3, C_1, D_3, E_2, F_2$ are the coefficients.

The boundary conditions of the characteristic equation are as follows:

$$\begin{aligned} \overline{\varphi}_{nb}(-L) = 0,\ \overline{\varphi}_{nb}(L) = 0,\ \overline{\varphi}_{nr}(0) = 0 \\ \overline{\varphi''}_{nb}(-L) = 0,\ \overline{\varphi''}_{nb}(L) = 0 \end{aligned} \tag{22}$$

The wave functions of the bridge structure can be obtained as follows:

$$\begin{aligned} \overline{\varphi}_{nb}(x) &= A_{nb}\sin\overline{k}_{bn}(x+L) \\ \overline{\varphi}_{nr}(\xi) &= A_{nr}\sin\overline{k}_{rn}\xi \end{aligned} \tag{23}$$

For the main girder and pier, the wavenumber is:

$$\bar{k}_{bn} = \sqrt{\omega_{bn}/a} = n\pi/2L, \ \bar{k}_{rn} = \omega_{rn}/c = (2n-1)\pi/2H \tag{24}$$

Based on the orthogonality of the wave mode function, the coefficient can be obtained as:

$$A_{nb} = 1/\sqrt{\rho A_b L}, \ A_{nr} = 2/\sqrt{\rho A_r H} \tag{25}$$

If the separation contact phenomenon occurs many times, it can be assumed that $t^* = t - t_{2k}$ is the time variable of the k-th collision, $t^* = t - t_{2k+1}$ is the time variable of the k-th separation.

In the k-th separation process, the dynamic displacement responses of the main girder and pier are as follows:

$$\begin{aligned} q_{nb}(t^*) &= e^{-\zeta_2\omega_{b1}t^*}\left(q_{1b}(t^+_{2k+1})\cos\omega_{b1}t^* + \frac{\dot{q}_n(t^-_{2k+1})+\zeta_2\omega_{b1}q_{nb}(0)}{\omega_{b1}}\sin\omega_{b1}t^*\right) \\ &+\frac{1}{\omega_{bd}}\int_{t^+_{2k+1}}^{t^*} e^{-\zeta_2\omega_{bn}\tau}\ddot{Q}_{bn}(\tau)\sin(\omega_{bd}(t^*-\tau))d\tau \\ q_{nr}(t^*) &= e^{-\zeta_2\omega_{r1}t^*}\left(q_{1r}(t^+_{2k+1})\cos\omega_{b1}t^* + \frac{\dot{q}_n(t^-_{2k+1})+\zeta_2\omega_{r1}q_{nr}(0)}{\omega_{r1}}\sin\omega_{r1}t^*\right) \\ &+\frac{1}{\omega_{rd}}\int_{t^+_{2k+1}}^{t^*} e^{-\zeta_2\omega_{rn}\tau}\ddot{Q}_{rn}(\tau)\sin(\omega_{rd}(t^*-\tau))d\tau \end{aligned} \tag{26}$$

The initial displacement and velocity of the girder and pier structure are considered only by the first mode. By viewing the residual rate and deformation of the last process, the contact separation process is solved continuously.

2.4. Corresponding Theoretical Solution of Bridge Vertical Impact Stage Displacement

When the relative displacement between the girder and pier is less than zero, it is considered that the girder and the pier are in contact again. At the moment of collision contact, the overall frequency of the bridge cannot be calculated. The use of resonance frequency calculation will produce large dispersion, and it is difficult to ensure the convergence of the calculation results. In the process of vertical impact process, the dynamic deformation after contact collision is divided into impact force deformation X_F and dynamic wave deformation X_z. The indirect mode superposition method [24] calculates the structural displacement under the impact force. Initial contact time, $Y_s + Y_d = U_s + U_d = 0$. In the subsequent vertical collision process, the main beam and the bridge pier have no vertical contact force at the initial moment. And the static displacement of the girder and the pier is zero. It can be concluded that the dynamic displacement of the bridge at the initial moment is:

$$Y_{1d}(0, t^+_{2k}) = Y_{2d}(0, t^+_{2k}) = U_d(0, t^+_{2k}) = 0 \tag{27}$$

The collision displacements of the structure are:

$$\begin{aligned} Y_F &= \frac{\sum\limits_{i=1}^{\infty}\overline{\varphi}_{nb}(0)\dot{q}_{bn}(t^-_{2k})}{\omega_{nb}}\sin\omega_{nb}t - \sum_{n=1}^{\infty}\overline{\varphi}_{nb}(x)\int_{t_{2k}}^{t^*}Q_{nb}h_{nb}d\tau \\ U_F &= \frac{\sum\limits_{i=1}^{\infty}\overline{\varphi}_{nr}(0)\dot{q}_{rn}(t^-_{2k})}{\omega_{nr}}\sin\omega_{nr}t + \sum_{n=1}^{\infty}\overline{\varphi}_{nr}(\xi)\int_{t_{2k}}^{t^*}Q_{nr}h_{nr}d\tau \end{aligned} \tag{28}$$

where $Q_{nb} = F_p\overline{\varphi}_{nb}(x_0)$, $Q_{nr} = F_p\overline{\varphi}_{nr}(\xi_0)$ is the generalized collision force. x_0 and ξ_0 are the coordinate of the collision point of the main beam and pier, respectively. F_p is the impact force. The positive and negative signs in Equation (27) denote the relationship between force and displacement direction, respectively.

$$h_{nb} = \frac{1}{M_{nb}\omega_{nb}}\sin\omega_{nb}(t-\tau), \ h_{nr} = \frac{1}{M_{nr}\omega_{nr}}\sin\omega_{nr}(t-\tau) \tag{29}$$

where M_{nb}, M_{nr} are modal masses.

$$M_{nb} = \int_{-L}^{L} \rho A \overline{\varphi}_{nb}^{2}(x)dx,\ M_{nr} = \int_{0}^{H} \rho A_r \overline{\varphi}_{nr}^{2}(\xi)d\xi \tag{30}$$

In the stage of impact contact, the displacements of the girder and the pier are different. The impact force $F_p = 0$ when the girder and pier are separated, and the collision force $F_p > 0$. At the impact point, the displacement response is:

$$Y_F(0,t) - U_F(H,t) = F_p(t)/K_c \tag{31}$$

By introducing Equation (28) in Equation (29), the vertical load F_p and the collision contact part of bridge structure displacement Y_F, U_F can be calculated by using the step-by-step integration method.

The partial time function of dynamic deformation is as follows:

$$q_n(t) = \frac{1}{\omega_n}\int_{t_{2k}}^{t^*} \ddot{Q}_n(\tau)\sin(\omega_n(t-\tau))d\tau \tag{32}$$

The dynamic deformation of the bridge can be expressed as follows:

$$Y_{d1} = \sum_{n=1}^{\infty} \varphi_{nb1}(x)q_n(t),\ Y_{d2} = \sum_{n=1}^{\infty} \varphi_{nb2}(x)q_n(t),\ U_d = \sum_{n=1}^{\infty} \varphi_{nr}(\xi)q_n(t) \tag{33}$$

3. Calculation of Horizontal Displacement Response of Bridge

Referring to the process of solving the vertical displacement response of the bridge, the wave mode equations of the girder and pier in the longitudinal direction can be obtained as:

$$\begin{aligned} \varphi_{nb1}(x) &= A_n \sin k_{bn}x + A_n \tan k_{bn}L \cos k_{bn}x \\ \varphi_{nb2}(x) &= -A_n \sin k_{bn}x + A_n \tan k_{bn}L \cos k_{bn}x \\ \varphi_{nr}(\xi) &= M_1A_n(\sin k_{rn}\xi - \mathrm{sinh}k_{rn}\xi) + M_2A_n(\cos k_{rn}\xi - \cosh k_{rn}\xi) \end{aligned} \tag{34}$$

Refer to Equations (13) and (14) to solve for M_1, M_2, A_n.

Considering the bearing damping, the time function of the bridge displacement response is:

$$\begin{aligned} q_n(t^*) = e^{-(\zeta_1+\zeta_2)\omega_n t^*}(q_n(0)\cos\omega_d t^* + \tfrac{\dot{q}_n(0)+q_n(0)\omega_n(\zeta_1+\zeta_2)}{\omega_d}\sin\omega_d t^*) \\ +\tfrac{1}{\omega_d}\int_0^{t^*} \ddot{Q}_n(\tau)e^{-(\zeta_1+\zeta_2)\omega_n(t-\tau)}\sin\omega_d(t^*-\tau)d\tau \end{aligned} \tag{35}$$

where $\omega_d = \sqrt{[1-(\zeta_1+\zeta_2)^2}\omega_n$.

During the separation process, the wave mode functions of the main beam and bridge pier are:

$$\begin{aligned} \overline{\varphi}_{nb}(x) &= A_{nb}\sin\overline{k}_{bn}(x+L) \\ \overline{\varphi}_{nr}(\xi) &= A_{nr}(\cosh\overline{k}_{rn}\xi - \cos\overline{k}_{rn}\xi) + M_3(\mathrm{sinh}\overline{k}_{rn}\xi - \sin\overline{k}_{rn}\xi)) \end{aligned} \tag{36}$$

The time function of the girder and pier is consistent with Equation (28). The calculation process of the longitudinal displacement response of the bridge in the collision process is the same as the vertical calculation process, except that ζ_2 in Equations (27) and (30) is changed to ζ_{1+2}.

By referring to the calculation of longitudinal displacement of bridge under earthquake, the transverse wave function of main girder and pier can be calculated:

$$\begin{aligned} \varphi_{nb1}(x) &= M_{n1}A_n\left(-\frac{\sin k_{bn}(x+L)}{\cos k_{bn}L} + \frac{\text{sinh}k_{bn}(x+L)}{\cosh k_{bn}L}\right) \\ \varphi_{nb2}(x) &= M_{n1}A_n\left(\frac{\sin k_{bn}(x-L)}{\cos k_{bn}L} - \frac{\text{sinh}k_{bn}(x-L)}{\cosh k_{bn}L}\right) \\ \varphi_{nr}(\xi) &= A_n(\sin k_{rn}\xi - \text{sinh}k_{rn}\xi) + M_{n2}A_n(\cos k_{rn}\xi - \cosh k_{rn}\xi) \end{aligned} \tag{37}$$

During the separation process, the wave mode functions of the main beam and bridge pier are:

$$\begin{aligned} \overline{\varphi}_{nb}(x) &= A_{nb}\sin \overline{k}_{bn}(x+L) \\ \overline{\varphi}_{nr}(\xi) &= A_{nr}(\cosh \overline{k}_{rn}\xi - \cos \overline{k}_{rn}\xi) + M_3(\text{sinh}\overline{k}_{rn}\xi - \sin \overline{k}_{rn}\xi)) \end{aligned} \tag{38}$$

The solution of time function is referred to in Section 3.

4. Numerical Simulation and Analysis

The model used in this study is a double-span continuous bridge. The main beam is a prestressed box beam, and the pier is a double-column circular pier. Figure 2 shows the reinforcement details and cross-sectional dimensions of the bridge. The bearing adopts a high damping rubber bearing. To simplify the analysis, the equivalent parameters of the bridge can be calculated according to specifications for the design of highway reinforced concrete and prestressed concrete bridges and culverts in China [24]. According to the specification, the equivalent cross-sectional area of the bridge pier is $A_r = A_{rc} + (\alpha_y - 1)A_{ry}$, the equivalent Young's modulus of bridge pier is $E_r = (E_{rc}A_{rc} + E_{ry}A_{ry})/(A_{rc} + A_{ry})$, the equivalent moment of inertia of bridge pier is $I_r = I_{rc} + I_{ry}$. The equivalent section area of main girder is $A_b = A_{bc} + (\alpha_y - 1)A_{by} + (\alpha_p - 1)A_{bp}$, the equivalent Young's modulus of the main girder is $E_b = \left(E_{bc}A_{bc} + E_{by}A_{by} + E_{bp}A_{bp}\right)/(A_{bc}+A_{by} + A_{bp})$. The equivalent section moment of inertia of the main girder is $I_b = I_{bc} + I_{by} + I_{bp}$. The round lead high damping rubber bearing is used between the main girder and the piers.

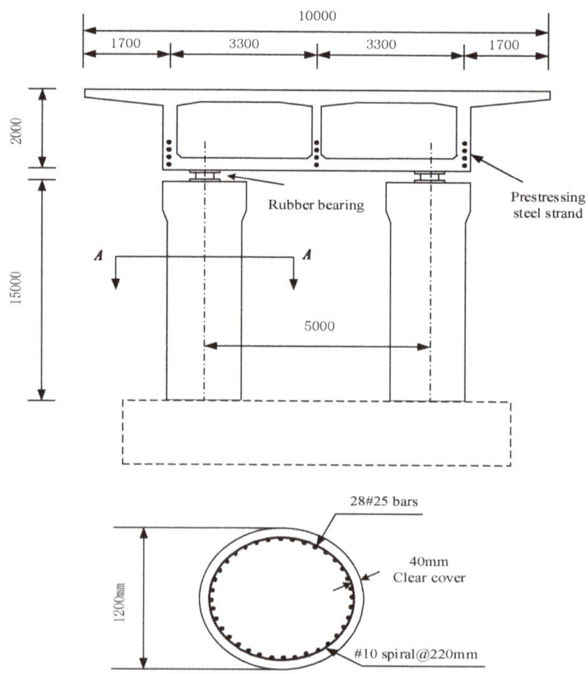

Figure 2. Dimensions and details of the bridge elevation and section drawings.

In the vertical direction, the hysteresis curve of the bearing is long and narrow, ignoring the damping of the vertical bearing. In the horizontal direction, the damping of the bearing is ζ_1 = 20%. In the vertical direction, the stiffness of the bearing is $K_c = 2.4 \times 10^9$ N/m; in the horizontal direction, the stiffness is $K_v = 2.4 \times 10^6$ N/m. To simplify the calculation, this study makes the following assumptions:

(1) When the bridge is forced to resonate, the structural force and displacement response are always calculated by elastic deformation;
(2) Ignore the possible bearing shear failure caused by a horizontal earthquake;
(3) During seismic action, there are often stops in the lateral direction. This study only considers the coupling of vertical and longitudinal seismic activities;
(4) Ignore the difference in the arrival time of the horizontal and vertical seismic waves, assuming that the earthquakes in both directions are excited at the same time.

4.1. The Influence of Near-Field Vertical Seismic Acceleration on the Structure

To use the methods in Sections 3 and 4 to calculate the response of the bridge under earthquake action, it is necessary to select an appropriate number of modes and time-step increments.

Due to the effect of structural damping, the high-order modal response has little effect on the structure. The number of modes selected in this paper is n = 5. The shortest time for the bending wave and axial wave penetration structure of the main beam is $\Delta t = L/c_b$ = 10.8×10^{-3} s. The shortest time for the pier's bending wave and axial wave to penetrate the system is $\Delta t = L/c_b = 4.38 \times 10^{-3}$ s. Therefore, the chosen time step must be less than 4.38×10^{-3} s, and the selected time step in this article is 18×10^{-3} s.

Figure 3 shows the calculation flow chart of the vertical displacement of the bridge in the case of possible vertical separation. The calculation time was 2 s. By recording the time of each separation and recontact in the vertical calculation, and substituting it into the longitudinal seismic calculation, the longitudinal displacement response of the bridge considering the separation condition was obtained.

When the seismic excitation period T = 0.2 s, two types of V/H are selected. One is 0.67 specified in the specification, and the other is given by Equation (1), where the value is 1.0. Figure 4 shows the seismic response of the bridge under two V/H values. When λ = 0.67, it can be seen that the main girder and the pier are always in contact, and the maximum vertical contact force is 20.4 MN. When λ = 1, the main girder and the bridge pier will be divided, and the main girder and the bridge pier will be separated six times in 2 s. The vertical contact force increased to 26.6 MN, an increase of 30.4%.

Compared with the vertical contact force, the separation has a more significant impact on the longitudinal displacement response of the structure. When the height of the pier is H = 15 m ($T_s > T_v$. T_s and T_v are the horizontal and vertical natural vibration periods of the bridge, respectively), the maximum longitudinal relative displacement of the pier and beam is 14.9 mm when the separation is ignored. When the structure is separated, the maximum longitudinal relative displacement increases to 29.9 mm, nearly two times.

Figure 5 shows the effect of separation on the displacement response of the structure when the pier height H = 9 m ($T_s < T_v$) and 12 m ($T_s \approx T_v$). When H = 9 m, the separation caused the maximum longitudinal displacement of the pier beam to increase from 6.06 mm to 7.09 mm, increasing by 17%. When H = 12 m, the separation caused the maximum longitudinal displacement of the pier beam to increase from 17.83 mm to 21.1 mm, increasing 21.9%. The effect of separation on the displacement response of the structure is significantly lower than when the pier height is 15 m. It can be seen that the partition has a more significant impact on the longitudinal displacement of the bridge when $T_s > T_v$.

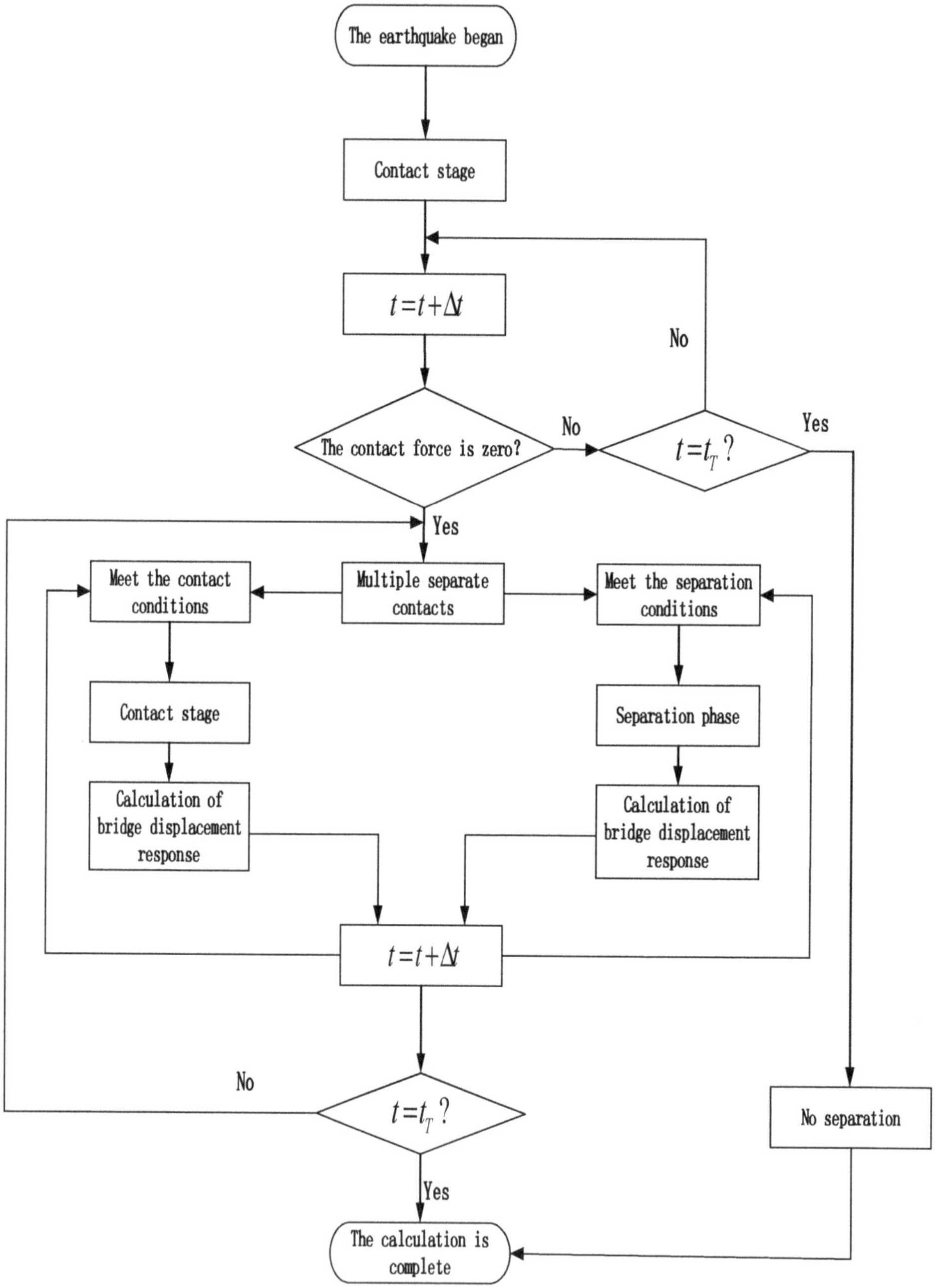

Figure 3. Flow chart showing bridge vertical displacement calculation procedure.

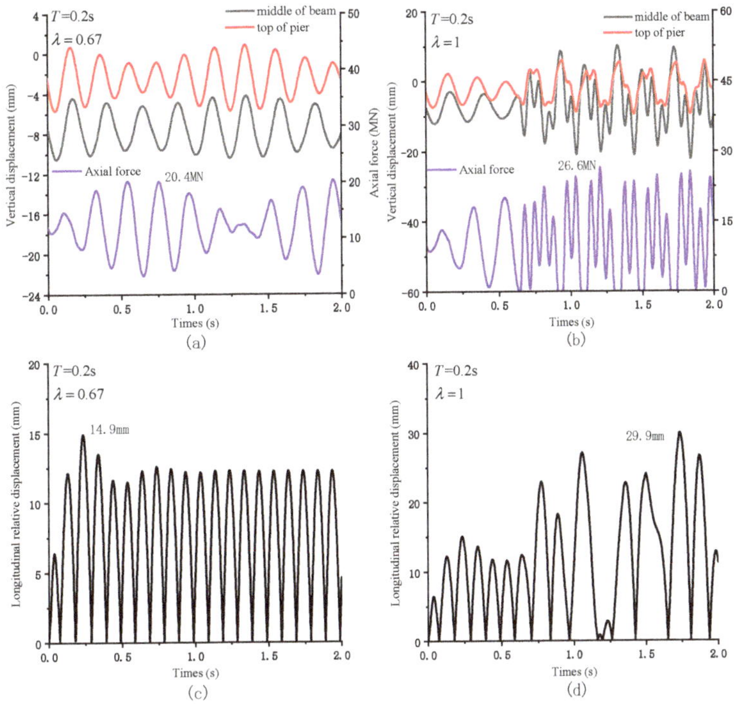

Figure 4. Seismic response of the bridge under different λ: (**a**) The vertical displacement and contact force response of the bridge when $\lambda = 0.67$; (**b**) The vertical displacement and contact force response of the bridge when $\lambda = 1$; (**c**) Longitudinal relative displacement of pier beam when $\lambda = 0.67$; (**d**) Longitudinal relative displacement of pier beam when $\lambda = 1$.

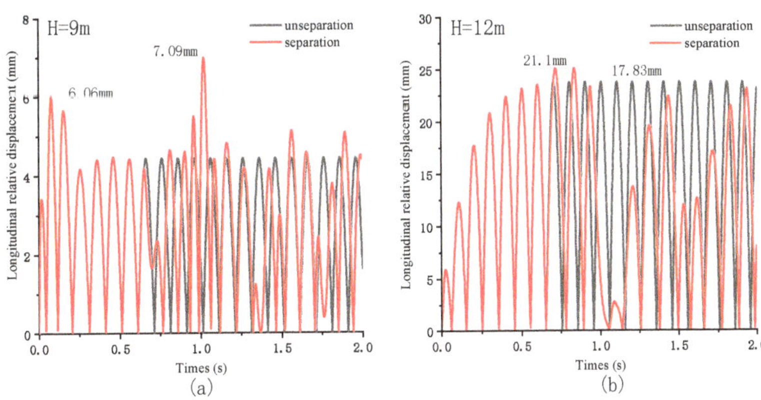

Figure 5. Longitudinal relative displacement of pier beam under different pier height: (**a**) Longitudinal relative displacement of pier beam under two conditions of separation or not when H = 9 m; (**b**) Longitudinal relative displacement of pier beam under two conditions of separation or not when H = 11 m.

Figure 6 shows the change of bending moment of bridge pier in two cases. It can be seen that when the pier is higher, the separation has a greater impact on the bending moment.

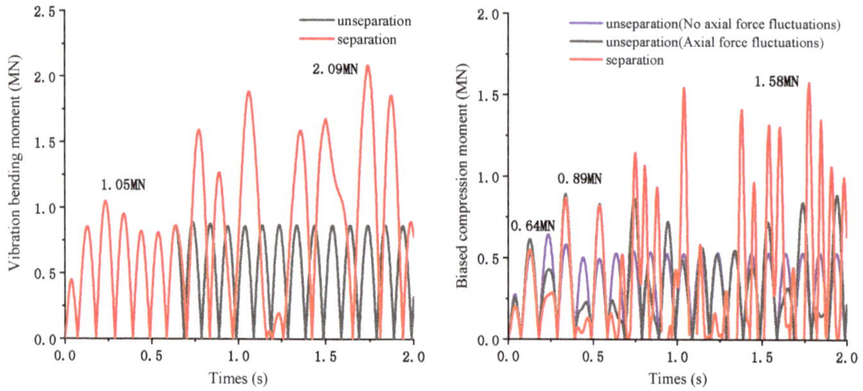

Figure 6. Bending moment change of bridge pier under two conditions: (**left**) Vibration bending moment; (**right**) Eccentric compression moment.

To calculate the bending failure of the pier, it is assumed that the pier is under eccentric compression. The collision diagram is shown in Figure 7. In order to consider the most unfavorable situation of the structure, it is assumed that the leading eccentricity and the maximum collision force occur at the same time. The shear stiffness of the bearing has no attenuation and remains unchanged within the calculation range. The bottom of the pier is composed of three bending moments: the bending moment M_z caused by forced resonance, the bending moment M_v caused by the bearing shear force at the pier bottom, and the bending moment M_c generated by eccentric collision. To simplify the calculation in this study, the influence of the plastic hinge produced at the pier bottom on the displacement of the pier top and the bending moment at the pier bottom is ignored. Ignoring the reset of the bearing during separation, the shear deformation is the longitudinal relative displacement of the pier and beam. The calculation formula of each bending moment is as follows:

$$M_c = F_{max} \times \eta \times \Delta d;\ M_v = K_v \times \Delta d \times L;\ M_z = E_r I_r \partial^2 U_d / \partial \xi^2 \quad (39)$$

where η is the magnification factor of the eccentric collision.

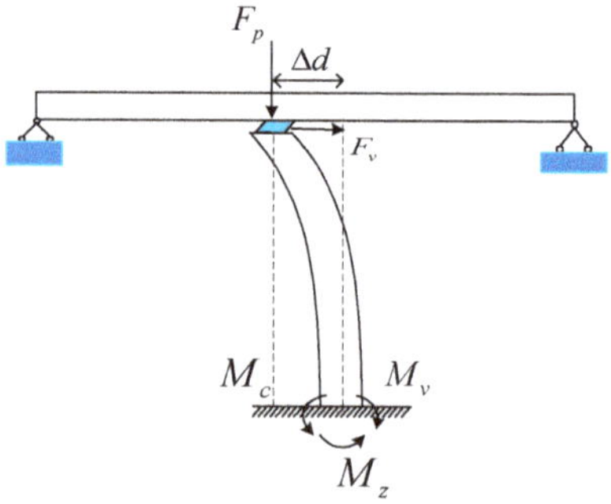

Figure 7. Force diagram at the bottom of the pier under eccentric compression.

To better study the effect of separation on the structure, three types of piers are selected here: 9 m, 12 m, and 15 m. Figure 8 shows the changes in the bending moment at the bottom of the pier at three heights. Figure 8a shows that when the pier height H = 9 m, the high amplitude vertical acceleration reduces the allowable bending moment of the pier. Whether separated or not, the piers are in a safe zone. When the pier height H = 12 m, T_s is close to the excitation period T. The forced resonance of the bridge pier produces a large bending moment, the high amplitude vertical impact force under the condition of separation reduces the allowable bending moment of the pier and causes damage to the pier. When the pier height H = 15 m, the moment of the pier is 1.66 MN and the structure is in the safe zone. When the system is separated, the bending moment increases to 2.94 MN, which increases by 77%, and the pier is damaged.

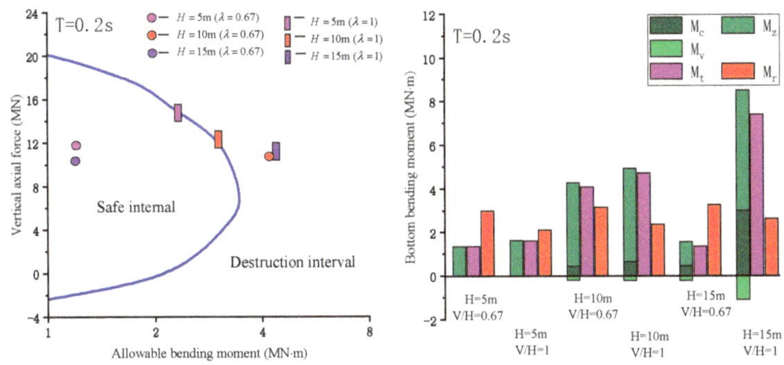

(a) Allowable and total bending moment (b) Various bending moment values

Figure 8. Various bending moments and allowable bending moments at the pier bottom: (**a**) Changes of various bending moments at the bottom of the pier at three different heights; (**b**) Longitudinal relative displacement of pier beam under different pier height.

Figure 8b shows the variation of various bending moments at the pier bottom under different conditions. It can be seen that the shear stiffness of rubber bearing is low, and the bending moment produced by bearing shear has little influence on pier failure in most cases. When H = 12 m, the pier failure is mainly due to forced resonance, and the bending moment caused by eccentric compression is significantly increased compared with H = 9 m. When H = 15 m, the seismic excitation period is far away from the longitudinal natural vibration period of the bridge, the seismic response of the bridge is low, and the structure is always in the safe range. When the structure is separated, M_c and M_z are immensely increased by the large expansion of longitudinal relative displacement. It can be seen that ignoring the bridge separation caused by high vertical acceleration not only underestimates the vertical contact force between the main girder and pier, but also may misestimate the seismic displacement response of the bridge.

4.2. Seismic Response of Bridge Pier under Different Excitation Periods

The seismic excitation period affects the bridge displacement response. The research range selected in Figure 9 is T = 0.05 s∼0.5 s. The change of seismic acceleration under a long period, the characteristic site period T_g = 0.3 s, is selected in this study. It can be seen that only when the excitation period is close to the vertical first-order natural vibration period of the bridge that the structure will separate, and the higher-order modes have little effect on the structural separation. The full longitudinal relative displacement of the pier beam will peak at $T = T_s$, but this crest may not be the maximum value. When the pier height H = 15 m, the ultimate value appears at T = 0.25 s, and the excitation period at this time is close to T_v and far away from T_s. It can be seen that the most unfavorable conditions caused by the forced vibration of the bridge piers are not entirely in the interval $T = T_s$. When the piers are higher ($T_v > T_s$), the structural separation produced in the interval $T = T_v$ will increase the seismic response of the bridge and cause the pier to be bent damaged.

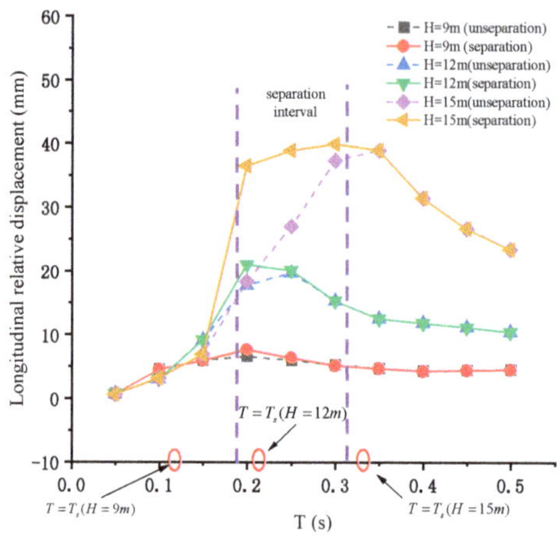

Figure 9. Longitudinal relative displacement of pier beam under different excitation periods.

The bending failure of the pier is affected by the relative displacement of the pier and beam. The change of axial force not only affects the allowable bending moment, but also changes the bending moment generated by eccentric compression. Figure 10 shows the vertical contact force between the main girder and the pier under different excitation cycles. It can be seen that the peak value of the vertical contact force is concentrated in the $T \approx T_s$ interval. Compared with the specification $\lambda = 0.67$, calculated by Equation (1), it has a more considerable peak value in a short period and a smaller peak value in an extended period. The structure separation of the bridge will occur under both conditions, but the excitation period of the structural separation calculated by Equation (1) is shorter. The maximum vertical contact force of the two is the same, in the range of 2.5 F_c ~$3F_c$.

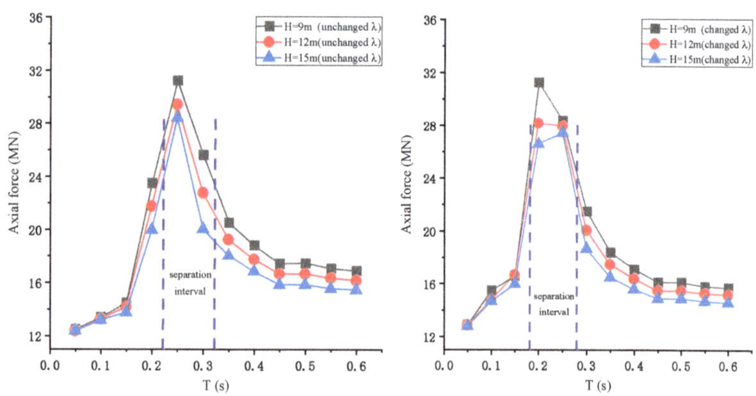

Figure 10. Various bending moments and allowable bending moments at the pier bottom: (**left**) Changes of various bending moments at the bottom of the pier at three extra heights; (**right**) Longitudinal relative displacement of pier beam under extra pier height.

Figure 11 shows the variation of the bending moment of the pier under different seismic excitation periods. Under different excitation periods, when H = 9 m, the bending moment produced by eccentric compression is shallow, which can be ignored. When H = 12 m, M_c has a peak at T close to T_s, which is due to the more enormous displacement response produced by resonance. When H = 15 m, M_c has two extremes, one is when T is close to T_s, this is because the resonance produces a larger displacement response. One is in the separation interval, which is due to the expansion of the relative displacement due to structural separation. It can be seen from Figure 11b,c that the variation trend of M_z and M_v in different seismic periods is the same as that of M_c. When the height of the pier is low, the bending moment of the pier under the earthquake action is small. When the pier height is higher, the bending moment at the bottom of the pier is greater. In addition, when considering the structure separation, there will be two wave peaks, which need to consider the pier damage in two cases.

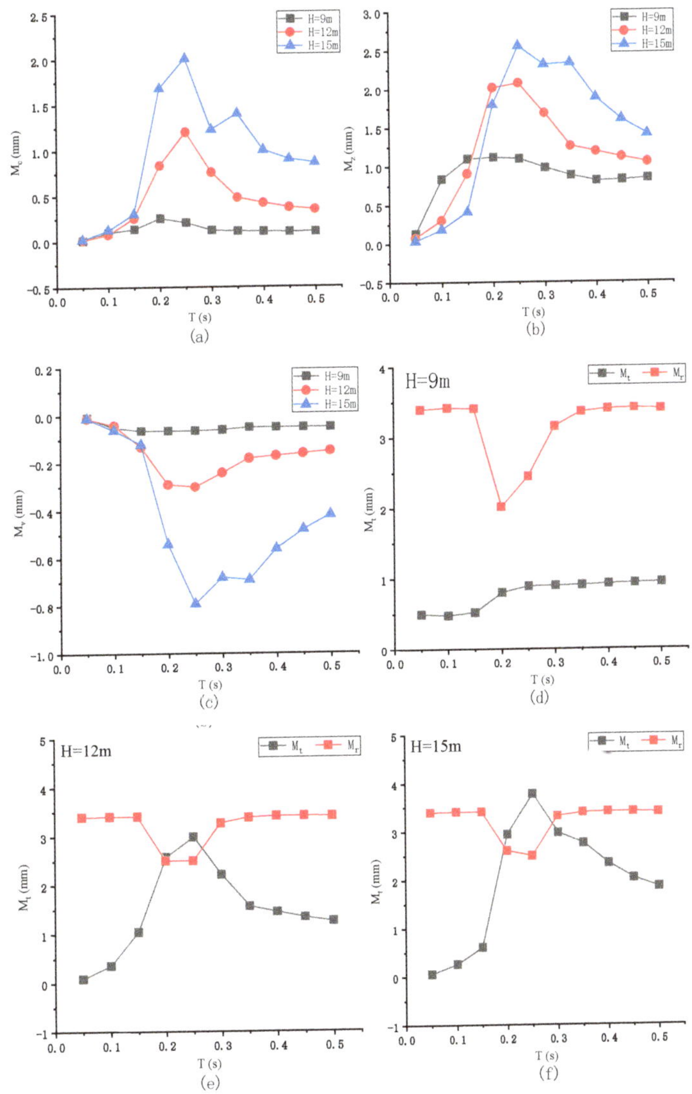

Figure 11. Various bending moments at the bottom of the pier under different periods: (**a**) Bending moment caused by eccentric collision under different excitation periods; (**b**) Bending moment generated by forced vibration under different excitation periods; (**c**) Bending moment caused by bearing shear under different excitation periods; (**d**) When H = 9 m full moment and the permissible moment pier; (**e**) When H = 12 m full moment and the permissible moment pier; (**f**) When H = 15 m full moment and the permissible moment pier.

The variation trend of the allowable bending moment of the pier with three heights under different excitation periods is the same. Only when T approaches to T_v, there is an extreme value, and the other intervals remain unchanged. With the increase of pier height, the vertical contact force decreases and the allowable bending moment increases. When H = 9 m, the total bending moment of the pier has little change. When H = 12 m and 15 m, the total bending moment of the pier increases greatly, the former is caused by resonance, and the latter is caused by structural separation. With the increase of excitation period, the total bending moment decreases gradually when H = 12 m. However, when H = 15 m, the second wave crest will appear, and the total bending moment is close to the allowable bending moment, which may cause damage to the pier.

4.3. Seismic Response of Bridge Piers under Different Vertical Excitation Amplitudes

A large number of monitoring data show that the vertical and horizontal acceleration amplitudes of near-field earthquakes are not only more than 2/3 of the code, but sometimes even more than one or even close to two. To study the influence of vertical seismic excitation amplitude on pier failure, Figure 12a shows the variation of longitudinal relative displacement and vertical contact force of pier beam with vertical seismic excitation amplitude when the length of the single-span beam is L = 38m and the pier height is H = 9 m, 12 m and 15 m, respectively. When H = 9 m and 12 m, the amplitude of V/H has little effect on the longitudinal displacement of pier beam. When H = 15 m, the maximum longitudinal relative displacement of pier beam is irregular, which is because the displacement response of the structure after separation is related to the initial position and velocity at the time of separation, and has great randomness. Compared with the longitudinal displacement, V/H has a greater impact on the vertical impact force. It can be seen from Figure 12b that with the increase of V/H, the vertical contact force of the structure increases monotonically, and the smaller the pier height, the greater the increase. When V/H exceeds a certain value, the vertical impact force exceeds the compressive strength of the pier, resulting in compression failure of the pier.

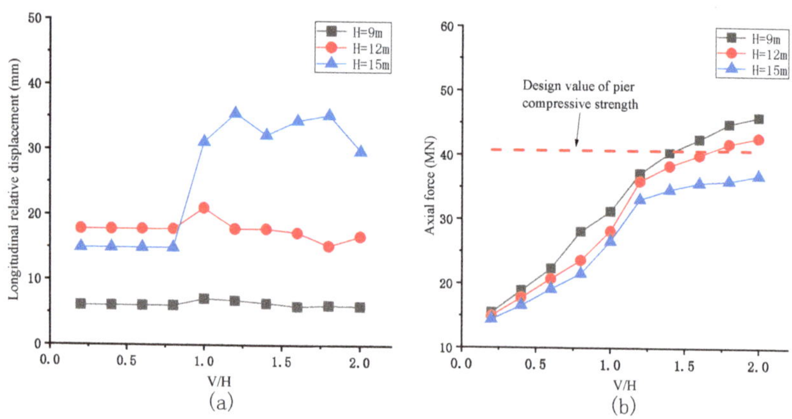

Figure 12. Seismic response of bridges under different vertical seismic excitation amplitudes: (**a**) Longitudinal relative displacement of main girder and pier under different vertical seismic excitation amplitudes; (**b**) Vertical contact force between main girder and pier under different vertical seismic excitation amplitudes.

Figure 13 shows the bending moment variation of pier under different vertical excitation amplitudes. As the amplitude of V/H increases, the total bending moment of the bridge pier shows an increasing trend. When H = 15 m, the total bending moment will increase drastically and fluctuate after the structure is separated. The allowable bending moment of the bridge pier decreases monotonously as the vertical collision force increases.

It can be seen from Figure 13 that when the height of the pier is low, with the increase of V/H, the damage of the pier is mainly due to the increase in the vertical collision force, which reduces the allowable bending moment of the pier. When the height of the pier is high, the damage is partly due to the decrease in the allowable bending moment, and partly due to the increase in the total bending moment of the pier.

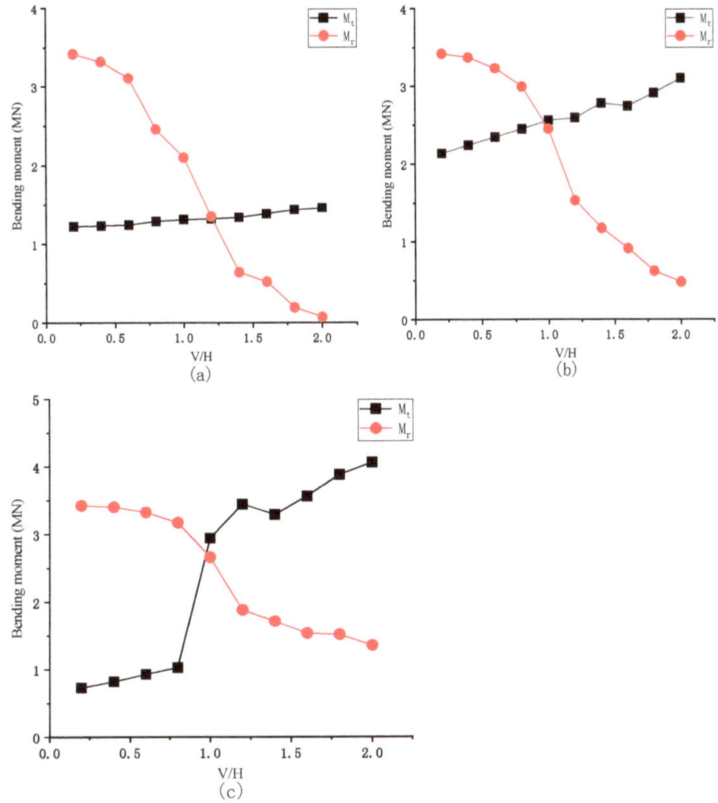

Figure 13. Bending moments at the bottom of the pier under different periods: (**a**) When H = 9 m total moment and the permissible moment pier; (**b**) When H = 12 m total moment and the permissible moment pier; (**c**) When H = 15 m total moment and the permissible moment pier.

5. Discussion

In this paper, the influence of the separation of main girder and pier on the failure of bridge pier under near-field earthquake is calculated by numerical simulation. The seismic force and displacement responses of medium- and small-span bridges under the condition of separation can be easily calculated by using this method. However, in order to simplify the calculation, the structure is assumed to be always in the elastic stage, and the plastic hinge at the bottom of the pier is ignored, resulting in a certain error. Therefore, the influence of the plastic hinge at the bottom of the pier needs to be considered in the follow-up, and further verification is obtained through finite element calculation.

6. Conclusions

In this study, the influence of near-fault vertical seismic excitation amplitude on pier failure was considered. Double-span continuous beam bridge model was established. The transient wave function expansion method and the indirect mode superposition method were used. The theory solved the longitudinal displacement and vertical contact force response of the bridge. By applying different vertical excitation amplitudes, the damage of three types of piers under different amplitudes was studied. The following conclusions were obtained by calculation:

1. For medium- and small-span rubber bearing bridges, when the designed piers were high ($T_s > T_v$), the change of longitudinal displacement caused by separation will affect the bending failure of bridge piers.
2. Only when the excitation period was close to the vertical natural vibration period can the structure separation occur, and the allowable moment of pier is the lowest. There may be two peaks in the total bending moment of pier. The excitation period was close to the longitudinal natural vibration period, and the other is when the structure is separated.
3. With the increase of V/H, the total bending moment at the bottom of pier increased slowly, the allowable bending moment decreased gradually, and the pier is damaged.
4. When V/H exceeded a specific range, the structure will be separated and a large vertical impact force will be produced. The impact force will cause the pier axial compression failure.

Author Contributions: Conceptualization, W.A. and G.S.; methodology, W.A.; software, W.A.; validation, W.A. and G.S.; formal analysis, W.A. and G.S.; investigation, W.A.; resources, W.A.; data curation, G.S.; writing—original draft preparation, W.A.; writing—review and editing, W.A.; visualization, W.A.; supervision, G.S.; project administration, G.S.; funding acquisition, G.S. Both authors have read and agreed to the published version of the manuscript.

Funding: APC was funded by Science and Technology Project of Jiangxi Provincial Education Department (GJJ202908).

Institutional Review Board Statement: Not applicable.

Informed Consent Statement: Informed consent was obtained from all subjects involved in the study.

Conflicts of Interest: The authors declare no conflict of interest.

Nomenclature

λ	the magnitude of V/H
α	the peak value of V/H
β	the linear attenuation coefficient
ρ	density of structure
E_b	modulus of elasticity of main girder
I_b	main girder section coefficient
A_b	main girder section area
E_r	modulus of elasticity of pier
I_r	pier section coefficient
A	pier section area
q	uniform load of main beam
Y_i	the deflection of the beam (i = 1,2)
Y_{is}	static displacement of the beam (i = 1,2)
Y_{id}	dynamic deformation part of the beam
U	the deflection of the pier (i = 1,2)
U_s	static displacement of the pier (i = 1,2)
U_d	dynamic deformation part of the pier
F_c	axial pressure of the initial girder and the support
φ_{nbi}	equation includes the bending wave function of the girder at contact time (i = 1,2)
φ_{nr}	the longitudinal wave functionof the pier at contact time
$\overline{\varphi_{nbi}}$	equation includes the bending wave function of the girder at separation of time (i = 1,2)

$\overline{\varphi_{nr}}$	the longitudinal wave functionof the pier at separation of time
ω_n	natural vibration frequency of the structure at contact time
ω_{nr}	main beam seismic frequency at contact time
ω_{nb}	pier seismic frequency at contact time
ω_d	natural frequency of the vibration considering the damping effect
ζ_1	bearing damping
ζ_2	structural self-damping
k_{bn}	wave numbers of flexural wave at contact time (longitudinal direction is wave numbers of longitudinal wave)
k_{rn}	wave numbers of longitudinal wave at contact time (longitudinal direction is wave numbers of flexural wave)
$\overline{k_{bn}}$	wave numbers of flexural wave at separation time (longitudinal direction is wave numbers of longitudinal wave)
$\overline{k_{rn}}$	wave numbers of longitudinal wave at separation time (longitudinal direction is wave numbers of flexural wave)
$q_n(t)$	time function of the bridge
$q_{nb}(t)$	time function of the main beam
$q_{nr}(t)$	time function of the pier
h_{nb}	impulse response function of the main beam
h_{nr}	impulse response function of the pier
M_{nb}	modal masses of the main beam
M_{nr}	modal masses of the pier

References

1. Li, H.N.; Xiao, S.Y.; Huo, L.S. Damage investigation and analysis of Engineering structures in the Wenchuan earthquake. *J. Build. Struct.* **2008**, *29*, 10–19.
2. Zhuang, W.L.; Liu, Z.; Jiang, J. Earthquake-induced damage analysis of highway bridges in Wenchuan earthquake and counter-measures. *Chin. J. Rock Mech. Eng.* **2009**, *28*, 1377–1387.
3. Xu, S.Y.; Zhang, J. Axial–shear–flexure interaction hysteretic model for RC columns under combined actions. *Eng. Struct.* **2012**, *34*, 548–563. [CrossRef]
4. Ghannoum, W.M.; Mohle, J.P. Rotation-Based Shear Failure Model for Lightly Confined RC Columns. *J. Struct. Eng.* **2012**, *138*, 1267–1278. [CrossRef]
5. Elwood, K.J. Modeling failures in existing reinforced concrete. *Can. J. Civ. Eng.* **2004**, *31*, 846–859. [CrossRef]
6. Button, M.R.; Cronin, C.J.; Mayes, R.L. Effect of Vertical Motions on Seismic Response of Highway Bridges. *J. Struct. Eng.* **2002**, *128*, 1551–1564. [CrossRef]
7. Huang, F.; Cao, Z.; Jiang, S.H.; Zhou, C.; Huang, J.; Guo, Z. Landslide susceptibility prediction based on a semi-supervised multiple-layer perceptron model. *Landslides* **2020**, *17*, 2919–2930. [CrossRef]
8. Li, W.; Fan, X.; Huang, F.; Chen, W.; Hong, H.; Huang, J.; Guo, Z. Uncertainties analysis of collapse susceptibility prediction based on remote sensing and GIS: Influences of different data-based models and connections between collapses and environmental factors. *Remote Sens.* **2020**, *12*, 4134. [CrossRef]
9. Zhu, L.; Huang, L.; Fan, L.; Huang, J.; Huang, F.; Chen, J.; Zhang, Z.; Wang, Y. Landslide Susceptibility Prediction Modeling Based on Remote Sensing and a Novel Deep Learning Algorithm of a Cascade-Parallel Recurrent Neural Network. *Sensors* **2020**, *20*, 1576. [CrossRef]
10. Wilson, T.; Chen, S.; Mahmoud, H. Analytical case study on the seismic performance of a curved and skewed reinforced concrete bridge under vertical ground motion. *Eng. Struct.* **2015**, *100*, 128–136. [CrossRef]
11. Kunnath, S.K.; Erduran, E.; Chai, Y.H.; Yashinsky, M. Effect of near-fault vertical ground motions on seismic response of high overcrossings. *J. Bridge Eng.* **2008**, *13*, 282–290. [CrossRef]
12. Yu, C.P. Effect of Vertical Earthquake Components on Bridge Response. Ph.D. Thesis, University of Texas, Austin, TX, USA, 1998.
13. Papazoglou, A.J.; Elnashai, A.S. Analytical and field evidence of the damaging effect of vertical earthquake ground motion. *Earthq. Eng. Struct. Dyn.* **1996**, *25*, 1109–1137. [CrossRef]
14. Varecac, D.; Draganic, H.; Gazic, G. Influence of the Vertical Component of Earthquake on Large Span Rc Beams. *Teh. Vjesn. Tech. Gaz.* **2010**, *17*, 357–366.
15. Kim, S.J.; Holub, C.J.; Elnashai, A.S. Experimental investigation of the behavior of RC bridge piers subjected to horizontal and vertical earthquake motion. *Eng. Struct.* **2011**, *33*, 2221–2235. [CrossRef]
16. Güllü, H.; Jaf, H.S. Full 3D nonlinear time history analysis of dynamic soil structure interaction for a historical masonry arch bridge. *Environ. Earth Sci.* **2016**, *75*, 1–17. [CrossRef]
17. Zheng, W.; Leonardo, D.O.; Jamie, E.P. Seismic response of a bridge-soil-foundation system under the combined effect of vertical and horizontal ground motions. *Earthq. Eng. Struct. Dyn.* **2012**, *4*, 545–564.

18. Rezaei Rad, A.; Banazadeh, M. Probabilistic Risk-Based Performance Evaluation of Seismically Base-Isolated Steel Structures Subjected to Far-Field Earthquakes. *Buildings* **2018**, *8*, 128. [CrossRef]
19. Sharma, A.; Jangid, R.S. Seismic Response of Base-Isolated Benchmark Building with Variable Sliding Isolators. *J. Earthq. Eng.* **2010**, *14*, 1063–1091. [CrossRef]
20. Bozorgnia, Y.; Niazi, M.; Campbell, K.W. Characteristic of free-field vertical ground motion during the Northridge earthquake. *Earthq. Spectra* **1995**, *11*, 515–526. [CrossRef]
21. Wang, G.-Q.; Zhou, X.-Y.; Zhang, P.-Z.; Igel, H. Characteristics of amplitude and duration for near fault strong ground motion from the 1999 Chi-Chi, Taiwan earthquake. *Soil Dyn. Earthq. Eng.* **2002**, *22*, 73–96. [CrossRef]
22. Wang, D.; Xie, L. Attenuation of peak ground accelerations from the great Wenchuan earthquake. *Earthq. Eng. Eng. Vib.* **2009**, *8*, 179–188. [CrossRef]
23. Wang, C.J.; Shih, M.H. Performance study of a bridge involving sliding decks and pounded abutment during a violent earthquake. *Eng. Struct.* **2007**, *29*, 802–812. [CrossRef]
24. Zuo, Y.; Sun, G.J.; Li, H.J. Comparison and Research of Unseating Prevention Measures in Seismic Codes of China and Foreign Countries. *J. Disaster Prev. Mitig. Eng.* **2016**, *36*, 617–623, 639.
25. Han, W.; Song, C.; Li, Q. Strong ground motion at meizoseisal area & safety of important engineering projects at potential earthquake region. *J. Eng. Geol.* **2004**, *12*, 346–353.
26. Tanimura, S.; Mimura, K.; Nonaka, T.; Zhu, W. Dynamic failure of structures due to the great Hanshin-Awaji earthquake. *Int. J. Impact Eng.* **2000**, *24*, 583–596. [CrossRef]
27. Kim, S.H.; Shinozuka, M. Effects of Seismically Induced Pounding at Expansion Joints of Concrete Bridges. *J. Eng. Mech.* **2003**, *129*, 1225–1234. [CrossRef]
28. Li, S.; Guo, A.; Cui, L. Pounding-induced stress wave analysis and mitigation of highway bridges under earthquake. In Proceedings of the 11th International Conference in Experimental Structural Engineering, University of Illinois Urbana-Champaign, Chmapaign, IL, USA, 1–2 August 2015.
29. DesRoches, R.; Delemont, M. Seismic retrofit of simply supported bridges using shape memory alloys. *Eng. Struct.* **2002**, *24*, 325–332. [CrossRef]
30. Yang, H.B.; Yin, X.C.; Hao, H. Theoretical Investigation of Bridge Seismic Responses with Pounding under Near-Fault Vertical Ground Motions. *Adv. Struct. Eng.* **2015**, *18*, 452–468. [CrossRef]
31. Yang, H.B.; Yin, X.C. Transient responses of girder bridges with vertical poundings under near-fault vertical earthquake. *Earthq. Eng. Struct. Dyn.* **2015**, *44*, 2637–2657. [CrossRef]
32. Xing, Y.F.; Wang, L.J. Analysis of wave propagation in the built-up structures of rod-beam and beam-beam. *J. Beijing Univ. Aeronaut. Astronaut.* **2013**, *30*, 520–523.
33. Ministry of Housing and Urban-Rural Development of the People's Republic of China. *CJJ 166-2011. Code for Seismic Design of Urban Bridges*; Ministry of Housing and Urban-Rural Development of the People's Republic of China: Beijing, China, 2011.

applied sciences

MDPI

Article

Plastic Joints in Bridge Columns of Atypical Cross-Sections with Smooth Reinforcement without Seismic Details

Mladen Srbić *, Ana Mandić Ivanković, Anđelko Vlašić and Gordana Hrelja Kovačević

Department of Structures Faculty of Civil Engineering, University of Zagreb, 10000 Zagreb, Croatia; ana.mandic.ivankovic@grad.unizg.hr (A.M.I.); andjelko.vlasic@grad.unizg.hr (A.V.); gordana.hrelja.kovacevic@grad.unizg.hr (G.H.K.)

* Correspondence: mladen.srbic@grad.unizg.hr

Abstract: In seismically active areas, knowledge of the actual behavior of bridges under seismic load is extremely important, as they are crucial elements of the transport infrastructure. To assess their seismic resistance, it is necessary to know the key indicators of their seismic response. Bridges built before the adoption of standards for seismic detailing may still contain structural reserves due to the properties of the used materials and construction approach. For example, smooth reinforcement which is found in older bridges due to the material properties, detailing principles, and lower bond strength compared to ribbed reinforcement, allows for greater deformations. In bridges, columns are vital elements employed in the dissipation of seismic energy. Their cross-sections often deviate from the regular square, rectangular, or round cross-sections, which are typically found in building. Based on the behavior of the columns in the vicinity of potential plastic joints, we can determine their deformability. This paper presents an experimental study of seismic resistance indicators around a potential plastic joint for a column with an atypical cross-section, without seismic details and with smooth reinforcement. The experimental results are compared with the numerical and analytical, but also with the experimental results on samples with ribbed reinforcement. Conclusions are made about the behavior of such column elements and their seismic resistance indicators, allowing for the application of an analytical or numerical method with realistic material and element properties and derivation of correction factors due to the effect of the smooth-reinforcement slippage from the anchorage area.

Keywords: plastic hinge region; smooth reinforcement; M/φ characteristics; atypical cross-section

Citation: Srbić, M.; Mandić Ivanković, A.; Vlašić, A.; Hrelja Kovačević, G. Plastic Joints in Bridge Columns of Atypical Cross-Sections with Smooth Reinforcement without Seismic Details. *Appl. Sci.* **2021**, *11*, 2658. https://doi.org/10.3390/app11062658

Academic Editor: Maria Favvata

Received: 16 February 2021
Accepted: 12 March 2021
Published: 16 March 2021

Publisher's Note: MDPI stays neutral with regard to jurisdictional claims in published maps and institutional affiliations.

Copyright: © 2021 by the authors. Licensee MDPI, Basel, Switzerland. This article is an open access article distributed under the terms and conditions of the Creative Commons Attribution (CC BY) license (https://creativecommons.org/licenses/by/4.0/).

1. Introduction

The assessment of the condition of existing bridges and their appropriate and optimal maintenance plan is certainly an issue that will take priority over the design of new transport infrastructure in the near future. To be more precise, the construction of much of today's European transport infrastructure began in the 1960s and more and more European bridges are approaching their planned lifetime [1,2]. This hot topic is further accentuated by exceptional events and the increasing failure of bridges in recent years, which, in addition to material damage [3,4], also has tragic consequences.

It is also worth mentioning that these structures are often located in areas of extreme seismicity. The current design approach for new structures implies that it is uneconomic to design structures that provide a fully elastic response to the design value of seismic actions. An alternative is the widely accepted structural design approach for smaller seismic design loads, while at the same time detailing them according to the rules of ductile behavior so that they can withstand significant seismic loads due to inelastic deformation. There are many existing bridges that were designed according to now outdated norms which did not contain guidelines for the design of elements for ductile behavior. Some even older bridges do not even take earthquake loads into account. The degree of ductility of these bridges is unknown. Therefore, the question of the correct seismic assessment of bridges is still open for research.

Several previous studies [5–9] indicate that structures that are not designed and detailed for seismic actions may still have some degree of seismic resistance. The loss of bond strength between concrete and smooth reinforcement, but also the better ductility of such reinforcement allows for greater deformability of the elements. In general, previous research on structural elements has focused on elements typical of high-rise buildings (beams, columns), which are produced with ribbed reinforcement [10–14] while a smaller number of studies evaluate the seismic resistance of bridges, especially girders [15–17] and arches [18–21]. By analyzing the main parameters of the seismic resistance indicators [22–24], selecting a suitable evaluation method [25], and using the currently available guidelines for the evaluation of structures [26] and recent research in this field [10–14], it is possible to determine seismic resistance reserves of bridges that have not been previously designed and detailed for such actions.

Nonlinear methods are mainly used to assess the seismic load-bearing capacity of existing bridges [25–27]. The application of such methods requires the knowledge of the actual behavior of the structure, either the actual rotational capacity of the elements under static load, if non-linear static analyses are used, or under cyclic load, if non-linear dynamic analyses are used.

The ductility of reinforced concrete elements in the cracked state is estimated based on of the effective stiffness when the yield strength is reached. A review of the literature shows two approaches in defining effective stiffness [11]. First approach is the theoretical effective stiffness obtained from the relationship between the bending moment M_y and the chord rotation θ_y when the yield point is reached. This approach is more suitable for assessing the state of existing structures. The second approach is empirical formulation which is more suitable for designing new structures. The chord rotation at the yield point depends on the sum of the rotations of all cross-sections along the height of the element. The rotation of the end section is dominant. The end section of columns fixed to the foundation, arch, or superstructure is the key component in assessing the seismic resistance of bridges.

The ratio of moment and cross-section rotation is the basis for deriving the value of the cross-section rotation parameter φ_y when the yield point is reached. Several expressions [14,28,29] can be found in the literature to calculate the approximate value of cross-sectional rotation during yielding. These expressions can be used to estimate the value of rotation when the yield point is reached for rectangular and circular sections. In bridge structures, we often use cross-sections of columns that differ from these two typical shapes, i.e., cross-sections for which a detailed M/φ analysis is required to obtain information about their rotational capacity.

Accurate data input (shape and dimensions of the cross-section, arrangement, and quantity of longitudinal and transverse reinforcement) and the properties of the materials used (stress and strain ratio) are crucial to obtain the most accurate ratio of moment to rotation of the cross-section. This refers primarily to the ratio between stress and strain, which should be considered as close as possible to the actual behavior of the material (non-linear diagrams). The accurate determination of the bending moment to rotation diagram of reinforced concrete sections is a reliable indicator of the bearing capacity of elements under seismic loads [30,31].

In simplified calculations of reinforced concrete elements, the connection between the reinforcement and the concrete is considered completely fixed. This assumption can be applied in areas where very low stresses occur between concrete and reinforcement. In areas with high shear stresses, such as in sections around cracks, relative deformations between concrete and reinforcement occur. For this reason, the slip effect must be used as a parameter when analyzing the behavior of the structural element [32].

The low bond strength between reinforcement and concrete directly affects three main deformation mechanisms: Bending, shear, and end section rotation. The results of experimental research show an average 35% higher deformation capacity of elements made with smooth reinforcement compared to elements made with ribbed reinforcement [7]. While the deformability of elements with smooth reinforcement is increased due to their better ductility properties, it simultaneously decreases due to the nonseismic detailing of such elements. Unconfined concrete has significantly lover properties in the region of plastic hinges than confined concrete. Tests have shown that the level of the rotational capacity is significantly affected by the amount of axial force applied and that the deformability of the elements decreases almost linearly under the influence of such a force [33]. An insufficient number of tests on elements with such properties suggests caution in assessing their deformability.

In addition to rectangular, square, and circular cross-sections, which are all typical for columns in buildings, bridge columns often have cross-sections whose seismic resistance indicators cannot be analyzed using the previously mentioned expressions developed by individual researchers.

The purpose of the research presented in this paper is to define the seismic resistance indicators of the columns of the existing bridge, which have an atypical cross-section with smooth reinforcement and without seismic details. The experimental part of this research extends the current base of the tested column thus providing the improvement and development of the standards for the seismic assessment of structures.

2. Experimental Testing

2.1. Specimens Information and Production

A total of 15 column specimens were prepared for experimental testing. Previous experimental researches are presented in [34–38]. Since this research focuses on indicators of seismic resistance of existing bridges, specimens of the test columns were modelled according to an over-arch column of a real bridge built in 1968. This column does not have any seismic details. Cauchy relations [39] were used to reduce the physical quantities of the real column to the size of a specimen suitable for laboratory tests. Scaling factor 4 was used. Comparative cross-sections and material properties of the actual column and specimen are shown in Figure 1.

The columns were manufactured under controlled production conditions. First the anchor base of the column including the reinforcement of the column was erected, then the column itself was concreted. This erection process resulted with a construction joint at the connection between the anchor block and the column. During the erection of the columns, samples were taken to check the material properties of the concrete and reinforcement. The static system of the test column is a console that was fixed to the laboratory floor. Several column specimens are prepared to test different types of horizontal loads (monotonic and cyclic) and different axial load levels (100, 125, and 150 kN). The columns differ according to the diameter (8, 10, and 12 mm) and the type of longitudinal reinforcement (smooth or ribbed). Details of the column samples are given in Table 1. The values of the reinforcement area coefficient (related to the cross-sectional area) and the normalized axial force (ratio of the applied axial force to the total cross-sectional resistance) are also given.

For later analysis and comparison of the results, the columns were divided into four series with common parameters. The first and largest series of nine column specimens comprised columns with the same geometric characteristics, but differing in load type and magnitude of axial force. In the second and third series, the columns differed in the type of load and the geometric characteristics of the cross-section, while the level of axial force was the same. The fourth series included columns with ribbed reinforcement for different load types and axial force levels.

By choosing the appropriate ratio of column height to section height, the dominant bending stress of the column was ensured while the shear stress was minimized. The magnitude of the applied axial force was determined by modal analysis of the global bridge model using the spectral load from the actual bridge position Figure 1. The total axial force, which is composed of the column dead weight and the seismic load, was also reduced to experimental values using Cauchy relationships. The final result of such an approach in experimental modeling is that the cross-sectional rotation in the area of plastic joint formation is the same for both the actual column and the tested specimen.

a)

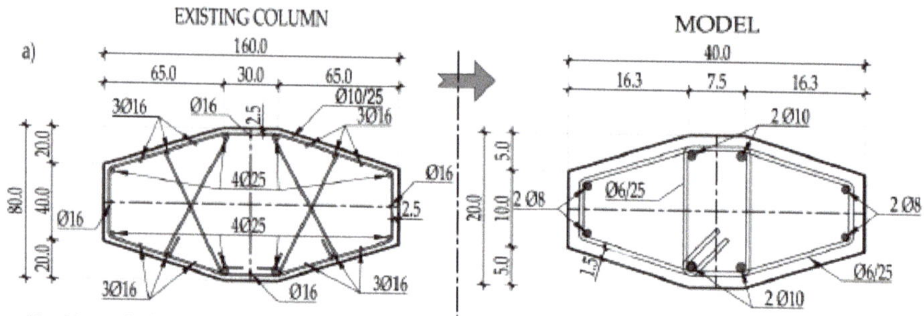

Material propertis of existing column and model

Reinforcement diameter	16×Ø16+8×Ø25	4×Ø10+4×Ø8
Reinforcement area	71,44 cm^2	5,16 cm^2
Proportion of reinforcement in concrete cross section	0.70%	0.80%
Column hight	24 m	1,85 m
Concrete	MB 450	C 35/45
Reinforcement	GA 240/360	GA 240/360

b)

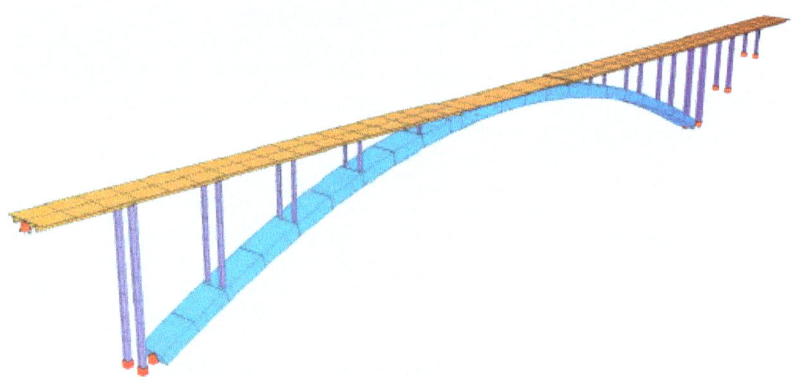

Figure 1. (**a**) The comparison of geometric and material properties of the existing column and model, (**b**) global bridge model for assessment of actions on columns.

In order to carry out the experimental investigations, it was necessary to first investigate the behavior of the column under the influence of the expected loads. The expected displacement at the top of the column and the structural behavior of the column in the area of the plastic joint (area of reinforcement plastification, concrete crack zone, and the level of compressive stresses in the concrete) are important parameters that determined the positioning of the individual measuring instruments.

Table 1. Column specimens' characteristics.

Sample Designation	Reinforcement Type	Applied Load	Reinforcement Diameter	Reinforcement Area Coefficient	Axial Load (kN)	Normalized Axial Load
AB-NS-001-1	GA 240/360	Monotone	Φ10 and Φ8	0.0081	100	0.05
AB-NS-001-2	GA 240/360	Monotone	Φ10 and Φ8	0.0081	100	0.05
AB-NS-001-3	GA 240/360	Cyclic	Φ10 and Φ8	0.0081	100	0.05
AB-NS-001-4	GA 240/360	Cyclic	Φ10 and Φ8	0.0081	100	0.05
AB-NS-001-5	GA 240/360	Monotone	Φ10 and Φ8	0.0081	150	0.07
AB-NS-001-6	GA 240/360	Monotone	Φ10 and Φ8	0.0081	150	0.07
AB-NS-001-7	GA 240/360	Cyclic	Φ10 and Φ8	0.0081	150	0.1
AB-NS-001-8	GA 240/360	Cyclic	Φ10 and Φ8	0.0081	150	0.1
AB-NS-001-9	GA 240/360	Cyclic	Φ10 and Φ8	0.0081	125	0.08
AB-NS-002-1	GA 240/360	Monotone	Φ12 and Φ10	0.0120	125	0.07
AB-NS-002-2	GA 240/360	Cyclic	Φ12 and Φ10	0.0120	125	0.07
AB-NS-003-1	GA 240/360	Monotone	Φ12 and Φ8	0.0103	125	0.07
AB-NS-003-2	GA 240/360	Cyclic	Φ12 and Φ8	0.0103	125	0.08
AB-NS-004-1	B 500B	Monotone	Φ10 and Φ8	0.0081	100	0.07
AB-NS-004-2	B 500B	Cyclic	Φ10 and Φ8	0.0081	150	0.10

Geometric features for the numerical model were taken over from the experimental model according to the dimensions. During the planning of the experiment the properties of the materials used were not yet known, so that for this numerical model the assumed values are similar to those in the documentation of the bridge project. A nonlinear numerical analysis of the column was performed, considering the material and geometrical nonlinearity, and the results obtained were important for the further planning of the experiment.

The test specimen consisted of two parts, an anchor block, and the column itself. The dimensions of the anchor block 90 × 80 × 45 cm were defined by conditions to achieve a firm fixation in the laboratory floor. All specimens had the same dimensions, with only minimal deviations that occurred during the erection. The height of the column measured from the anchor block was 185 cm. The cross-sectional shape of all columns was a polygon with maximum dimensions of 40 × 20 cm. The dimensions of the column specimens and the reinforcement details are shown in Figure 2.

The longitudinal reinforcement of the column was continuous over the entire length. The bars ended with hooks to achieve a good anchorage between the reinforcement and the concrete. The column specimens differed in the cross-section depending on the amount of the longitudinal reinforcement. A typical arrangement consists of four bars arranged around the center of the cross-section and four bars at the edges of the cross-section. Table 1 shows the diameters of the longitudinal reinforcement for each column specimen (the first diameter is for the reinforcement in the center of the cross-section and the second diameter is for the edge reinforcement). The amount of longitudinal reinforcement in the cross-section area was in the range of 0.8–1.2% and is shown in Table 1. The transverse reinforcement consisted of a four-legged stirrup of smooth bars Ø6 mm with a spacing of 25 cm. The position of the initial stirrup was shifted by half the distance of the transverse reinforcement from the cross-section with the maximum bending moment in order to reduce the effects of confinement around the concrete crushing area.

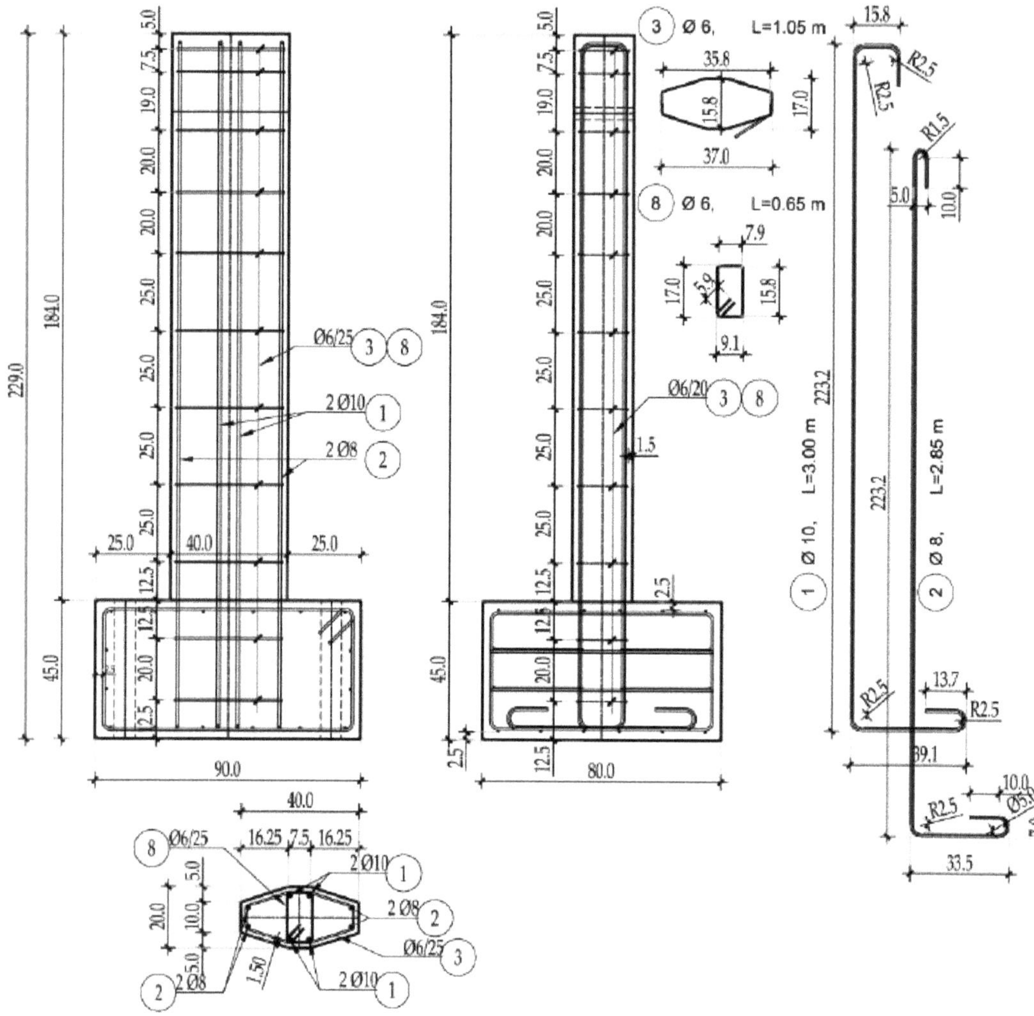

Figure 2. Column (series 001) and reinforcement details.

2.2. Test Setup and Protocol

Special equipment adapted to the dimensions of the column was used to test the column specimens in order to ensure fixed conditions in the laboratory floor and sufficient force introduction into the column. The joint axis of the fixing anchor bolts passing through the column foot are perpendicular to the axis of the horizontal press and is located in the axis of the column. Such an anchoring position did not provide sufficient fixation, so that an additional steel grid had to be erected to ensure fixation to the laboratory floor. The steel frames on which the presses are located were additionally stiffened with struts to minimize the displacement of the anchoring points of the press.

All 15 specimens were loaded with axial and horizontal force at the top of the column (end of the static cantilever system). A total of two presses were used to apply this force—a vertical press with a capacity of 1000 kN and a horizontal press with a capacity of 250 kN, both with hinged connection to the steel frame. The load transfer from both presses was achieved by a specially designed steel cap at the top of the column. The horizontal press was located at a height of 2.3 m above the laboratory floor. The horizontal press was supported by a hinged connection to the frame structure and the connection to the column cap. Due to the rotation of the column cap and thus the axial displacement of the horizontal press, it was additionally supported at its other end. The elements and dimensions of the test set-up are shown in Figure 3.

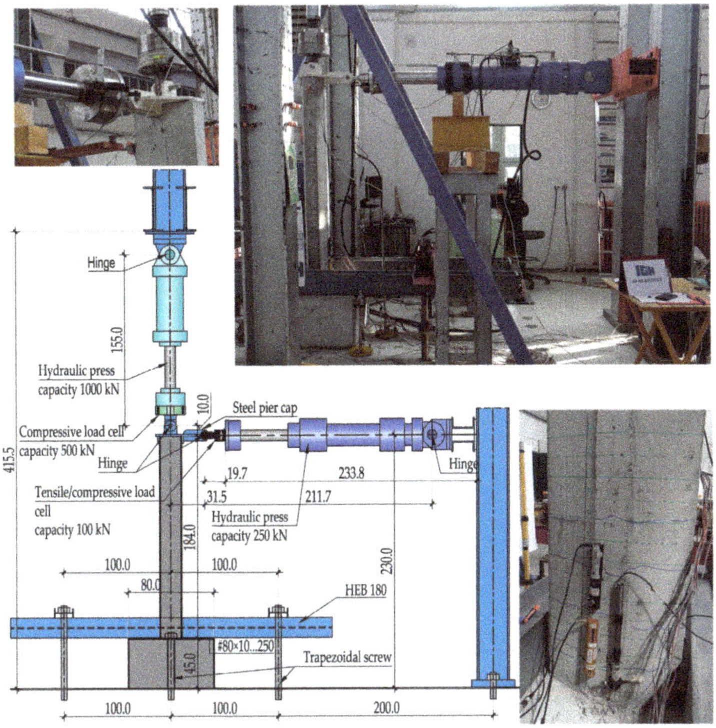

Figure 3. Test setup.

In the first phase, the desired value of the axial load was applied by a vertical press (Table 1). The vertical force was applied by means of a force control. Due to the large number of measuring points, all measuring points were checked after the force was applied and the position of a single measuring instrument was recorded. The first test phase was the same for all test specimens. In the second phase a horizontal load was then applied. The second test phase differed depending on the type of load (monotonic, cyclic). The horizontal load was applied by the displacement control method, i.e., a time interval is defined in which a certain displacement is reached. The horizontal force was measured by a compression–tension dose with a capacity of 100 kN. This type of dose had to be used because of the cyclical input of the horizontal force at the top of the column. Due to the horizontal displacement at the top of the column, the vertical press rotates and was thus stretched. This leads to a decrease in the press force, so that a constant force correction is required to keep the vertical force constant. At this stage the displacement in the horizontal press increases until the specimen fails. The effect of the rotation of the vertical press results in an additional horizontal force that acts together with the horizontal press. This effect can be seen in Figure 4.

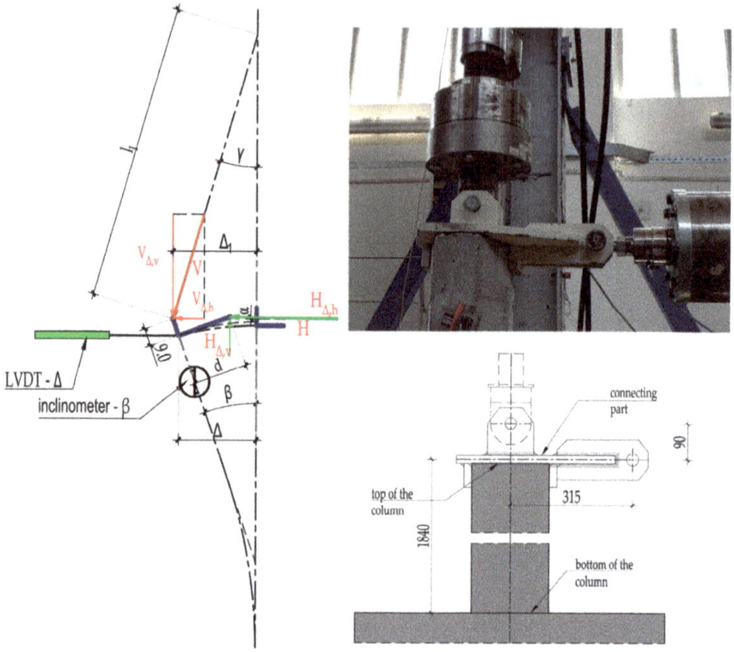

Where:
Δ – horizontal displacement at the column top
Δ_1 – horizontal displacement at the press hinge
α – rotation of the horizontal press axis
β – rotation of the column axis measured my inclinometer
γ – rotation of the vertical press axis

V – force in the vertical press
$V_{\Delta,v}$ – vertical component of the force in the vertical press
$V_{\Delta,h}$ – horizontal component of the force in the vertical press
H – horizontal press force
$H_{\Delta,v}$ – vertical component of the force in the horizontal press
$H_{\Delta,h}$ – horizontal component of the force in the horizontal press

Figure 4. Effect of the vertical press rotation to a horizontal force component.

Before starting the test, the distances of the individual press supports and the positions of the axis of the steel cap dowels in relation to the column axis were known (Figure 4). During the test the displacement of the column top Δ was measured with an LVDT measuring device, and the rotation of the axis of the column top β was measured with an inclinometer. In addition to the geometrical measurements, the components of the forces in the direction of the axis of each press were also measured. The actual vertical and horizontal force components acting on the column top were further determined according to the relationships shown in Figure 4.

2.2.1. Monotonous and Cyclic Load Application

During the testing of the specimens, the increase of the horizontal load was controlled by manually entering the displacement of the horizontal press. In the case of a monotonous load, the curve of the displacement increment was programmed so that the horizontal press rested for a certain time (1 min) after each 10 mm displacement increment in order to check the measuring points (Figure 5). These pauses during the test were also necessary to determine the development of cracks and to mark and measure their widths.

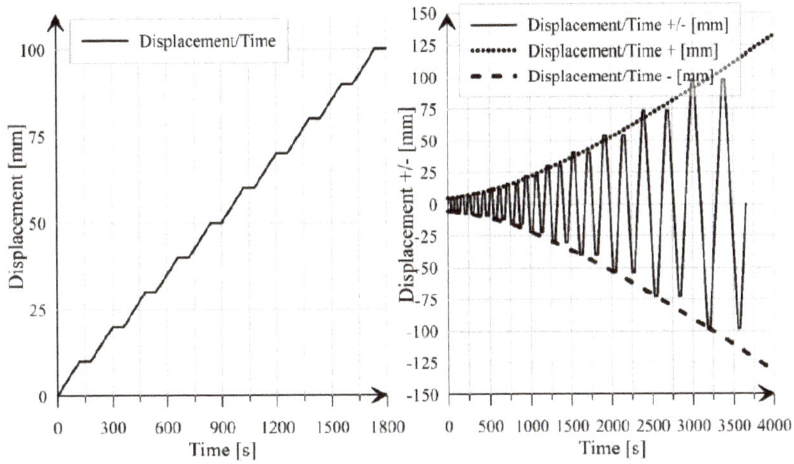

Figure 5. Load input curves by displacement control; left: monotonous load; right: cyclic load.

The quasi-static cyclic load input report was preformed according to the guidelines given in [40].

The following guidelines were adopted in the testing protocol:

- In total, two load cycles were performed for each load (displacement) amplitude;
- More than six load cycles (8 cycles) have been carried out until the lowest damage limit
- The value of the maximum amplitude was determined from the monotonous testing of the elements
- The number of load input cycles was more than 10 (22 cycles)
- Each subsequent load amplitude is approximately increased according to the following expression:

$$\alpha_{i+1} = 1.4 \times \alpha_i \quad (1)$$

where α_i is the amplitude of the previous step and α_{i+1} is the amplitude of the next step.

A cyclic curve was formed according to the above guidelines and then applied with a computer-controlled horizontal press by entering a displacement cycle from 0 to 100 mm.

The displacement was increased until the moment of failure of the concrete in the compression section at the bottom of the column. After the maximum horizontal displacement was reached and all the measuring points were recorded, the horizontal press was returned to its initial position, followed by the unloading of the vertical press.

The maximum displacement of the top of the column was 100 mm for all monotonically tested specimens. In all tested specimens, failure occurred in the compressive area of the concrete in the lower section of the column.

2.2.2. Measuring Instruments and Measured Parameters

Several different measuring instruments were used during the test (Figure 6). Strain gauges were attached to reinforcing bars prior to concreting, LVDT extensometers were placed at characteristic locations on the outer surface of the column, LVDT measuring instruments were used to control the displacements of characteristic points and to measure the relative deformations of the end section, inclinometers were used to measure column rotation and dial gauges without electronic recording were used to cheque the displacement measurements. In addition to the above mentioned measuring instruments, LVDT was also installed in the presses and measuring doses to define the magnitude of the load and displacement of the press at each test step. Most of the measuring instruments were located near an expected plastic joint.

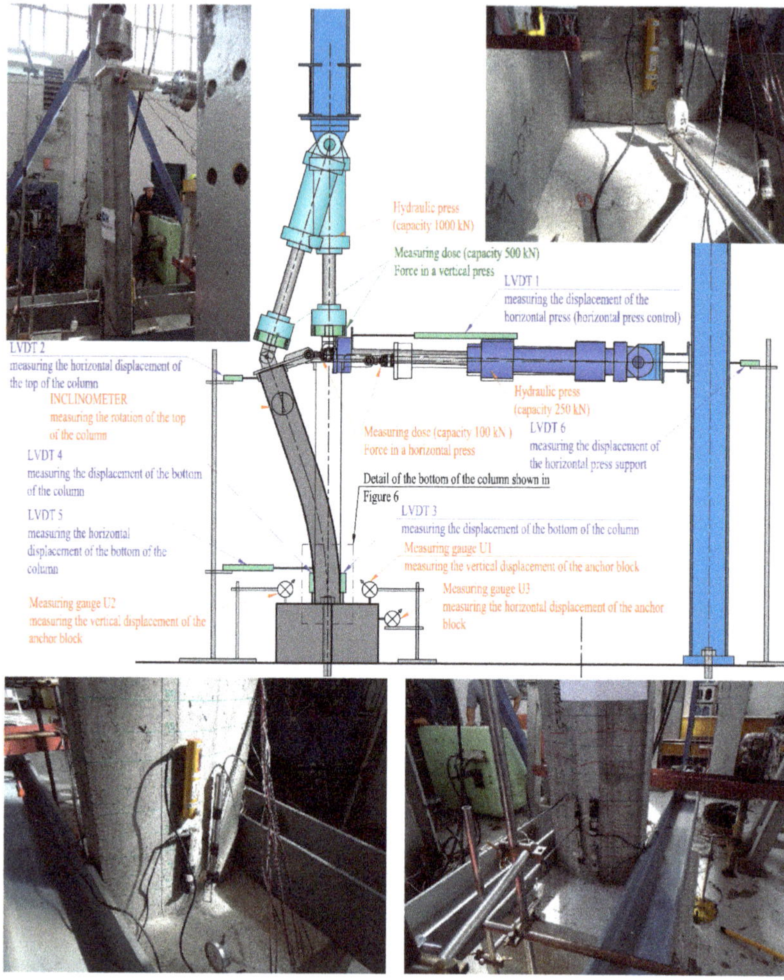

Figure 6. Positions of measuring spots.

The stress level and deformation in the longitudinal reinforcement bars was monitored by strain gages installed during the concreting of the specimens. The position and number of the strain gages were not the same for each specimen. Since there were several parameters determined by experimental analysis, the strain gages were placed in the most suitable locations for recording the results of each parameter. All columns had built-in strain gages on the rebars in the area of the end section in the compression and tension zones. In most cases one strain gage is installed in the compression zone and two strain gages in the tension zone. For monotonically tested specimens, the compression zone of the cross-section is uniquely determined, whereas for cyclically tested specimens the compression zone is considered to be the one in which compressive stresses occur in the first load cycle. For monotonically tested specimens, which are marked AB-NS-001-2, AB-NS-001-6, and AB-NS-004-1, four additional strain gages were installed at the height of the plastic joint on the tensile reinforcement at distances from the end section of 100, 200, 250, and 300 mm. Cyclically tested column specimens marked AB-NS-001-4, AB-NS-001-8, AB-NS-001-9, AB-NS-002-2, and AB-NS-003-2 had strain gage pairs (compression-tension side) at heights of 100, 200, 250, and 300 mm. An additional stress level was monitored by a strain gage installed in the anchor foot just below the end section on the columns with the designation AB-NS-001-1; AB-NS-001-3; AB-NS-001-5; AB-NS-001-7; AB-NS-001-9; AB-NS-002-1-2; and AB-NS-003-1-2. HBM K-CLY4-0030-1-120-3-020-N strain gages with a 3 mm measurement base were used. Figure 7 shows the arrangement of the measuring devices near the plastic joint. The column AB-NS-001/18/9-9 is shown in this figure, as this column had the most measurement points.

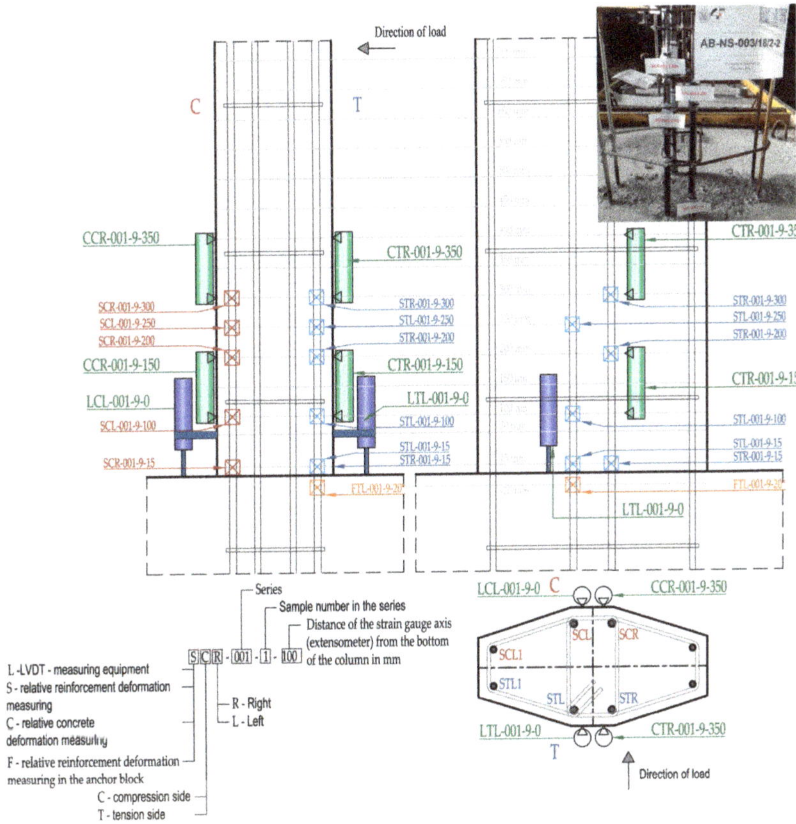

Figure 7. The position of measuring instruments in the vicinity of the plastic joint for column AB-NS-001/18/9-9.

LVDT extensometers were placed at the level of the plastic joint formation on both sides of the column to measure the relative deformations. Strain transducers with a measurement base length of 100 mm were used. A pair of strain transducers were placed at the bottom of the column to measure the relative deformations near the end section of the column. Due to its dimensions, the measuring instrument cannot be placed at the end section itself, but 150 mm from the bottom of the column. A second pair of extensometers was placed at the level of the plastic joint, the position of which was obtained from the numerical analysis. This served to obtain the relative deformations at this point and to confirm the accuracy of the numerical calculation. From the experimentally obtained values of relative deformations, the cross-sectional rotations were calculated, but also, if found, the crack width around the measured base of the extensometer was recorded.

The measurement of the displacement at the column top was performed with an LVDT instrument with a measured base of 250 mm. This displacement was also controlled by the LVDT installed in the horizontal press. Together with the inclinometer, which was mounted in the upper third of the column, this displacement provided all necessary values for the calculation and correction of the forces applied at the top of the column.

The expected rotation of the end section in relation to the base was measured with two LVDT gages mounted on the column surface with a measuring rod resting on the column base. In combination with the results of the internal strain gage and the data from this LVDT, it was possible to determine the rotation of the end section due to the loss of bond strength between the concrete and the smooth reinforcement.

2.3. Testing of Material Mechanical Properties

A total of five series of three column specimens were erected. In total, three samples were taken from each series to test the compressive and tensile strength of concrete and the static and dynamic modulus of elasticity Figure 8. The compressive and tensile strength of concrete was tested on cubes of 150 × 150 × 150 mm, the modulus of elasticity on prisms of 100 × 100 × 400 mm according to standardized methods [41–45]. The average values of the results obtained are shown in Table 2. For each series of columns, differences in properties were found. The values obtained from the tests were used in the further analysis.

Figure 8. Performance testing of concrete and steel.

Table 2. Performance of concrete.

Series	$f_{cm,cube}$ Derive from Testing (MPa)	f_{cm}= $0.8f_{cm,cube}/0.92$ (MPa)	f_{ck}= f_{cm}-8 (MPa)	$f_{ck,cube}$= fck/0.8 (MPa)	E_{cm}= Derive from Testing (GPa)
1	47.27	41.10	33.10	41.38	33.62
2	47.93	41.68	33.68	42.10	33.76
3	35.97	31.28	23.28	29.10	30.97
4	40.83	35.50	27.50	34.38	32.17
5	36.23	31.50	23.50	29.38	31.04

Smooth and ribbed reinforcement bars were both used for the reinforcement, depending on the individual column specimen. The reinforcement test is shown in Figure 8. For longitudinal bars smooth (Ø8, Ø10, and Ø12 mm profiles) and ribbed reinforcement (Ø8 and Ø10 mm profiles) were used. The mean values of the reinforcement properties derived from testing are shown in Table 3. The transverse reinforcement was made of smooth bars with a diameter of Ø6 mm.

Table 3. Performance of reinforcement.

Type of Reinforcement	Reinforcement Diameter	Es (MPa)	f_{yk} (MPa)	f_{tk} (MPa)	ε_{yk} (‰)	ε_{sh} (‰)	ε_{uk} (‰)
Smooth reinforcement GA 240/360	Ø8	2.1×10^5	375.0	492.0	1.786	8.581	119.0
	Ø10	2.1×10^5	402.0	539.0	1.914	13.95	129.0
	Ø12	2.1×10^5	346.0	470.0	1.649	13.52	141.0
Ribbed reinforcement B 500B	Ø8	2.0×10^5	597.0	676.0	2.98	12.4	42.6
	Ø10	2.0×10^5	571.0	643.0	2.855	12.0	46.43

3. Numerical and Analytical Analysis

In order to compare the experimentally obtained results, a numerical and analytical analysis of the ratio of moment and rotation of the end section was carried out. To obtain the best possible comparison, the properties of the material obtained experimentally on the samples taken during the erection of the columns were used.

The numerical analysis was performed with a commercial software professionally used for bridge structures using FEM. A non-linear numerical analysis was used, which included material and geometric non-linearity. The material properties were taken as material working law diagrams (stress/deformation ratio) for concrete (Equation (2)) and for reinforcement (Equations (3)–(5)). These expressions resulted in a good match of the constitutive laws with the experimentally obtained ones. The quantities and the position of the reinforcement in each cross-section of the column were taken in the same way as in the experimentally tested specimens. Numerical column models comprise cantilever 3D beam finite elements (mesh size 2 cm) with 6 degrees of freedom which are fixed for all translations and rotations at the point corresponding to the connection of the column and the anchor block. The beam axis was assigned the column cross-section in its center of gravity. Model of the column was then loaded horizontally in steps corresponding to those performed in the experimental tests. The load increase was carried out until the load-bearing capacity of the element was reached. The vertical axial load was entered immediately at the beginning of the calculation and remained unchanged until the end of the calculation. The amount of the axial load changes for different columns (Table 1). Due to the properties of the software package used, it was possible to perform the calculation until the maximum concrete stress value and the corresponding relative deformation was reached. Then the ratio of moment and rotation of the final cross-section was observed. Points of interest, such as the yield strength, are marked to compare the results. Only the monotonic load was applied in the numerical analysis and the results obtained were processed for comparison with experimental and analytical results.

The analytically obtained ratio of moment and rotation of the end section is presented. This ratio is described by three characteristic points in the curve, which represent the cross-sectional state due to the load increase. The first domain is up to the point where the first crack occurs, followed by the domain until the yield strength of the reinforcement is reached (or concrete compressive failure), and finally the domain until the load bearing capacity of the cross-section is reached (concrete tensile strength or compression zone failure). The domain between the characteristic points was achieved by gradually increasing the load. Non-linear material properties were used. The non-linear stress-strain ratio specified in [46] was used for concrete, while the ratio specified in [47] was used for reinforcement.

$$\frac{\sigma_c}{f_{cm}} = \frac{k \cdot \eta - \eta^2}{1 + (k-2) \cdot \eta} \tag{2}$$

where σ_c and ε_c are compressive stress and strain respectively. The coefficients η and k are given in [32].

$$0 \leq \varepsilon_s \leq \varepsilon_{yk} \sigma_s = E_s \cdot \varepsilon_s \leq f_{yk} \tag{3}$$

$$\varepsilon_{yk} \leq \varepsilon_s \leq \varepsilon_{sh} \sigma_s = f_{yk} \tag{4}$$

$$\varepsilon_{sh} \leq \varepsilon_s \leq \varepsilon_{uk} \sigma_s = f_{tk} - \left(f_{tk} - f_y\right) \left(\frac{\varepsilon_{uk} - \varepsilon_s}{\varepsilon_{uk} - \varepsilon_{sh}}\right)^2 \tag{5}$$

In Equations (3)–(5), ε_s and f_s are the reinforcement steel strain and stress, E_s is the elasticity module, f_u and ε_{su} are the stress and strain at the ultimate stress, f_y and ε_y are the stress and strain at the yielding.

As the cross-section was not seismically designed, the influence of the concrete reinforcement confinement was not considered, i.e., the stress–strain ratio of unconfined concrete was used. The method itself includes the analysis of the equilibrium of internal and external forces acting on the cross-section. The internal forces resulted from the tension and compression part of the concrete cross-section and the force in the tension and compression reinforcement, while the external forces were the axial compression force and the bending moment. The irregular shape of the cross-section was divided into a sufficient number of strips with the corresponding values of stresses and strains, which were then integrated into the resultant of the compression and tension zone of the concrete. The analytically obtained ratio agreed well with the numerically obtained results, as shown in chapter 4.4. Due to the large number of parameters that have different influences on the cross-sectional behavior, the analysis of the influence of each parameter on the ratio of moment and cross-section rotation helped in understanding the results of the experimental analysis. The method of analysis used is presented in [48].

4. Experimental Results and Discussion

The results of the monotonically and cyclically tested specimens are given below. The results are presented in the form of the ratio of moment and rotation of the cross-section at several points in the area of the plastic joint (Figure 9 and Figure 13). The width and distribution of the cracks along the height of the element is shown in Figure 11. The rotation of the cross-section was measured at three points along the height of the element, starting at the bottom of the column. The first point is located directly at the connection of column and anchor block, the second point is located at a height of 150 mm from the bottom of the column, while the third point is 350 mm from the bottom of the column.

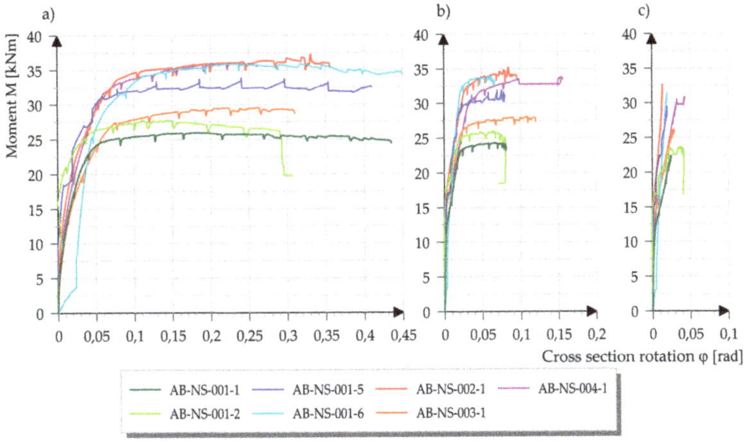

Figure 9. Moment to rotation ratio for monotonously tested column specimens: (**a**) end section; (**b**) section at 150 mm height; (**c**) section at 350 mm height.

Knowing the level of the horizontal and vertical force and the displacement and rotation of the top of the column, the bending moment in each section was determined. The bending moment of a single section includes the influence of the horizontal force component, the influence of the vertical force due to the rotation of the vertical press, and the influence of the vertical force due to the displacement of the column top (P-Δ effect).

There are many parameters that influence the result of the relationship between bending moment and section rotation. The interaction of the individual parameters makes it difficult to define the degree of influence of each parameter on the indicators of seismic resistance of the element. Therefore, in the next chapter separate results for monotonically and cyclically loaded specimens are presented and commented on in detail in order to draw final conclusions.

4.1. Monotonously Tested Specimens

In total, seven column specimens were tested with a monotonic load. The specimens AB-NS-001-1, AB-NS-001-2, AB-NS-001-5 and AB-NS-001-6 are made with smooth reinforcement and have the same cross-sectional properties (same amount of reinforcement-central reinforcement 4 × Ø10 and corner reinforcement 4 × Ø8). The first two specimens were loaded with a force of 100 kN, while the other two specimens were loaded with a force of 150 kN. The specimens AB -NS-002-1 and AB -NS-003-1 were loaded with the same force of 125 kN, but had different diameters of the longitudinal reinforcement bars. The central cross-sectional reinforcement (4 × Ø12) is the same for both specimens. The corner reinforcement is different, it is 4 × Ø10 in the first specimen, while it is 4 × Ø8 in the second specimen. The last monotonically tested specimen with the designation AB-NS-004-1 was produced with ribbed reinforcement, 4Ø10 as central reinforcement and 4Ø8 as corner reinforcement. This specimen was loaded with an axial force of 100 kN. The geometrical properties and the type of longitudinal reinforcement of each column are given in Table 1, while the mechanical properties of each type of reinforcement are shown in Table 3.

For monotonically tested specimens, the analyzed results of the moment and the rotation ratio showed a clear correlation. The main characteristic of all tested specimens is a high ductility due to the plastic deformation of the smooth reinforcement around the plastic joint resulting in a large displacement at the top of the column. The behavior of the column in three characteristic cross-sections of the monotonically tested column specimens is shown as the M/φ relationship in Figure 9.

Experimentally obtained curves show three characteristic domains of the ratio of moment and rotation of the cross-section: the domain until the occurrence of the first crack (the initial part of the diagram where the curve of the ratio of moment and rotation for all samples coincides), the linear elastic domain until the occurrence of the cross-section yield (the part of the diagram from the beginning of the separation of the curves up to the beginning of the horizontal part of each curve) and the plastic domain up to the limit state of the bearing capacity of the cross-section (the horizontal part of the diagram).

After the first crack has appeared, there is a significant separation of results for each sample. The reasons for this are differences in cross-section properties, material properties, and changes in axial force.

When the yield strength of the reinforcement is reached, we move to the third domain of the diagram, which shows a small slope of the curve tangent for all monotonically examined samples. This relationship between moment and cross-section rotation is then continued until the specimen fails.

From the ratio of moment and cross-section rotation it is clear that in the end cross-section and cross-section at a height of 150 mm there is a yield area, whereas in a cross-section at a height of 350 mm the yield area only occurred in two tested specimens (in one specimen AB-NS-004-1 with ribbed reinforcement and in one specimen AB-NS-001-2, which showed a different crack pattern). All tested specimens manufactured with smooth reinforcement have a pronounced plastic area that is five times or more larger than the elastic area. The influence of the axial force on the bearing capacity is also visible, as well as the influence on the rotation of the cross-section around the plastic joint. A higher axial force leads to a higher bending resistance of the cross-section, but also to a slightly lower rotation of the cross-section when reaching the final limit state. This can be seen from a comparison of monotonically tested columns loaded with an axial force of 100 kN (AB-NS-001-1) and columns loaded with a force of 150 kN (AB-NS-001-5). These specimens have the same cross-sectional properties but different axial force levels.

Besides the influence of the axial force, the influence of the amount of the longitudinal reinforcement is also clearly visible. The samples AB-NS-002-1 and AB-NS-003-1 were produced with different diameters in the corners of the column (AB-NS-002-1: 4 × Φ10, AB-NS-003-1: 4 × Φ8), whereby a larger amount of reinforcement is expected to increase the load-bearing capacity of the column. These two specimens were tested with the same axial force (125 kN) and were erected in the same series so that their concrete properties are the same. It should be noted that in addition to the amount of horizontal force and the amount of reinforcement (diameter), there are also differences in the mechanical properties of the installed reinforcement.

In order to compare the influence of the different types of reinforcement, the results of the experimental testing of three specimens, which were carried out with the same amount of reinforcement, are compared. Specimens AB-NS-001-1 and AB-NS-001-2 were made with smooth reinforcement, while specimen AB-NS-004-1 was made with ribbed reinforcement. The same axial force (100 kN) was applied to all three specimens by experimental testing. When examining the moment to rotation ratio of the end cross-section, we can see that the specimen made with ribbed reinforcement can reach a limit moment almost 40% higher than the specimen with smooth reinforcement with the same properties. The same can be concluded by observing the other two cross-sections in which the relative deformations have been measured, thus determining the ratio of moment and rotation of the cross-section. From this it can be concluded that the specimen made with ribbed reinforcement has a higher load bearing capacity than the specimens made with smooth reinforcement. It should be emphasized that by testing the compressive strength of the concrete on cubes of specimens made with ribbed reinforcement, the mean compressive strength was determined to be 36.2 MPa, while the same was determined for specimens with smooth reinforcement at 47.3 MPa. From this it can be concluded that the load-bearing capacity of the specimen would have been even higher if it had been made from the concrete with the same properties as those of the two specimens made with smooth reinforcement. If we compare the end cross-sectional rotations of these specimens, we can see that a specimen made with ribbed reinforcement has a significantly lower value of cross-sectional rotation on failure than specimens made with smooth reinforcement with the same geometrical properties and the same level of applied axial force. However, if we look at the rotation of other measured cross-sections along the height of the ribbed reinforced specimen, we can see that a slightly higher value of rotation was measured at failure than for smooth reinforced specimens. From this we can conclude that the cross-sectional rotation is more evenly distributed over the height of the member in the ribbed reinforcement specimen than in the specimens made with smooth reinforcement where the dominant rotation occurs in the end cross-section and in the cross-section where the first crack occurs.

Most of the monotonically examined specimens made with smooth reinforcement do not show a clear boundary between the areas in the diagram. If one considers bending moment and section rotation ratio for all tested specimens, it is not possible to determine the beginning of the reinforcement yield clearly. The reason for this is the loss of bond strength between the concrete and the reinforcement, which significantly influences the cross-sectional rotation. The gradual reduction in the stiffness of the element makes it difficult to detect the appearance of the first crack. For specimens with ribbed reinforcement, these limits are somewhat more pronounced.

The effect of gradual loss of bond strength and its influence on stiffness is visible in a constant change of the tangent slope almost from the beginning of load introduction (the ratio M-φ does not show a constant tangent slope until the first crack). This is particularly evident in the diagram of the ratio M/φ of the end section. In the two other observed cross-sections, the characteristic points in the diagram are more clearly visible (sudden jumps in the slope of the tangent of the diagram). The tested column specimen AB-NS-001-6 shows a sudden loss of stiffness immediately at the beginning of the test, followed by a strong increase in stiffness with the ratio M/φ almost vertical. This is considered to be caused by a slippage between concrete and reinforcement at the beginning of the test due to a larger unanchored part of the bar as a result of the column erection method, which then causes a decrease in stiffness. As the process progresses, the effects of slipping become less pronounced and lead to a significant increase in stiffness.

4.2. Specimens Cracking

The pattern of crack development of all monotonically tested specimens made with smooth reinforcement is characterized by the occurrence of two dominant cracks. The first crack is a crack of the previously mentioned end cross-section. The second crack occurs in most specimens near the first stirrup (125 mm from the bottom of the column).

A monotonically tested specimens showed a crack at a height of 200 mm, which differs significantly from all other specimens (the ratio of moment and cross-sectional rotation of this specimen also showed differences compared to other tested specimens). These two cracks at the bottom of the column have a constant change in width as the displacement of the top of the column increases and have a significant effect on the amount of element rotation around the plastic joint. Other cracks along the height of the column do not show significant width growth during the test (in some cases it even decreases). The crack width measurements of each specimen are shown in Figure 10. The results given apply to monotonic (designation M) and cyclic (designation C) specimens.

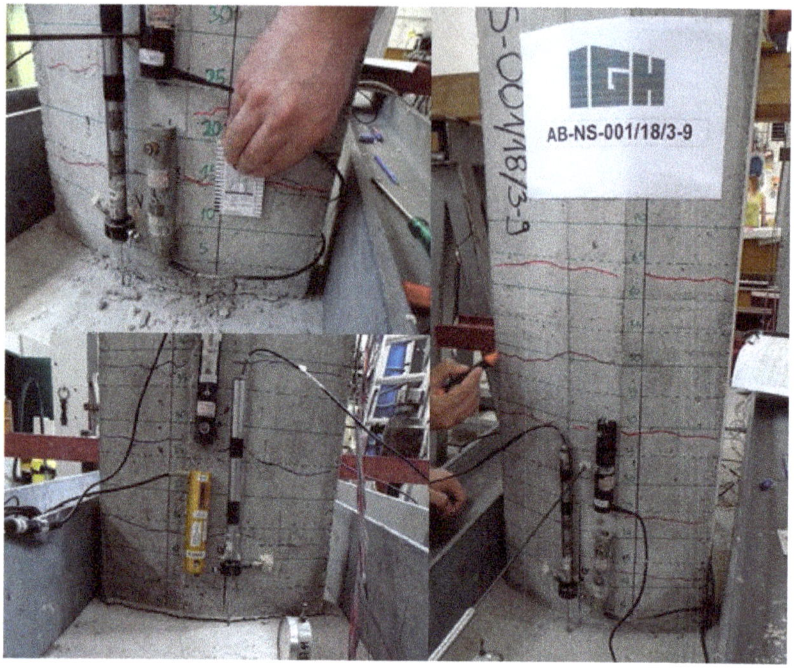

Figure 10. View of the crack in the area of plastic hinges.

The cracks shown are located on the plastic joint itself, but also in the immediate vicinity. The crack designations contain the distance of the observed crack from the bottom of the column. The figure shows the dominance of the crack at the end cross-section and the smaller but still significant influence of the first crack from the bottom of the column of all monotonically tested specimens. For monotonically tested specimens (AB-NS-004-1) made with ribbed reinforcement, the crack width at the end section is almost half of the crack width of specimens made with smooth reinforcement. The crack in the end cross-section is also significantly less pronounced compared to the first crack from the bottom of the column. The occurrence of cracks in the sample with ribbed reinforcement was observed up to a height of 1.2 m from the bottom of the column, whereas in the samples with smooth reinforcement the height of the crack reached only 0.65 m. In total, two specimens made with a larger amount of longitudinal reinforcement (AB-NS-002-1 and AB-NS-003-1) also show a smaller crack width in the end section and thus a lower dominance of the crack in relation to the adjacent crack. It should be noted that a larger amount of reinforcement was achieved by increasing the reinforcement diameter, while the position and distribution of the reinforcement over the cross-section remained almost the same (the only difference is the position of the center of gravity due to a larger bar diameter).

Figure 11 shows the width and distribution of cracks at the location of the plastic joint and in its vicinity for cyclically tested specimens. For almost all cyclically tested specimens, the dominance of the crack is visible at the end cross-section. In contrast to the monotonous tests, these tests show that the second crack from the bottom of the column is significantly less dominant and does not deviate much from the other cracks above. According to the measured crack sizes at the location of the plastic joint and in its immediate vicinity, the specimens with ribbed reinforcement behave in the same way as specimens with smooth reinforcement. The difference is only visible in the height at which cracks occur. In the case of the cyclically tested specimen with ribbed reinforcement, cracks appear up to a height of 1 m, whereas in the case of specimens with smooth reinforcement the crack height is the same as in the monotonously tested specimens and is about 0.65 m.

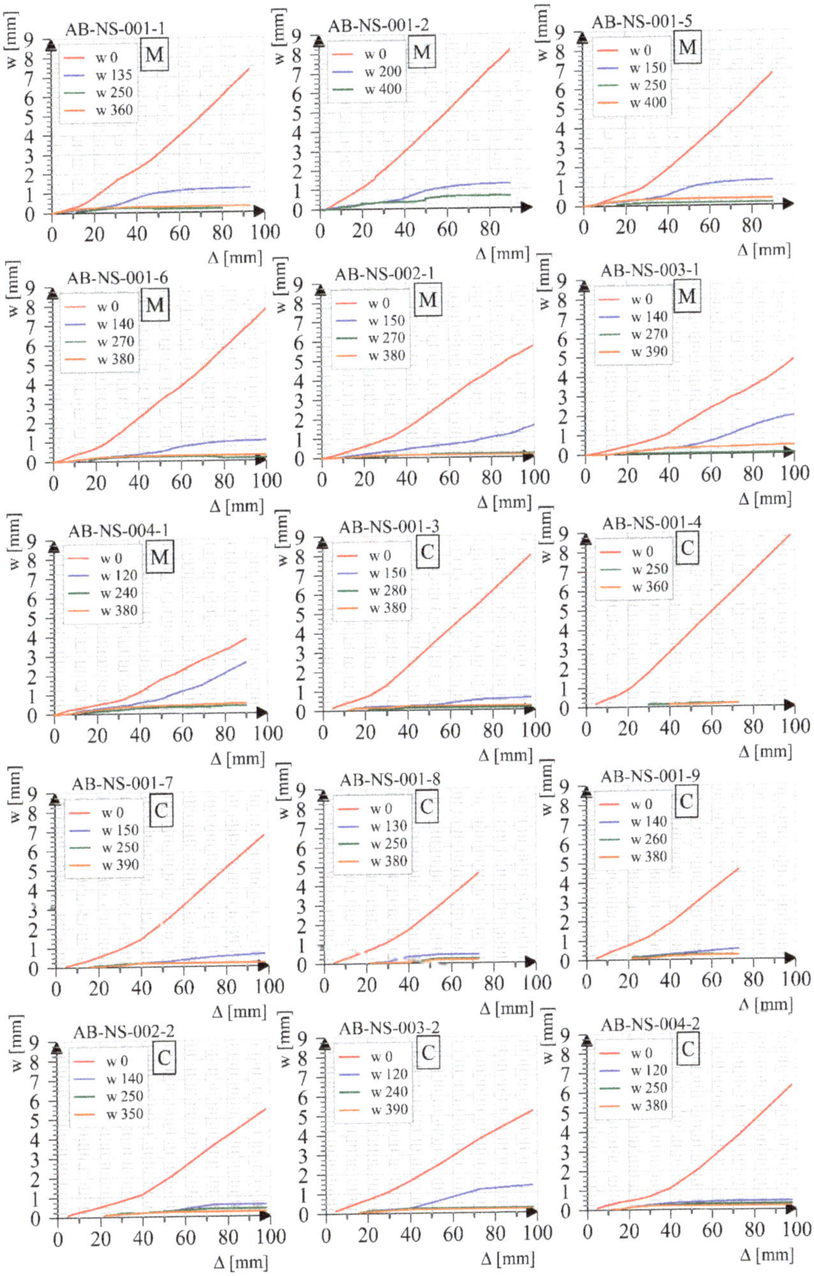

Figure 11. Crack width and position around plastic joint in relation to column top displacement.

4.3. Cyclically Tested Specimens

Cyclic load tests were carried out on eight column specimens. Figure 12 shows the hysteresis loop for one test column obtained through the ratio of the transverse load and the displacement of the column upper end. A gradual increase in the area of the hysteresis loop with an increase in the displacement cycle is visible, which results in greater energy dissipation. The low bond strength between reinforcement and concrete affects pinching effect. The gradual reduction in stiffness in each cycle is favorable in term of the seismic resistance of the columns. For cyclically tested specimens, the results were prepared in such a way that envelope curves were derived from the hysteresis curves, which connect the extreme values of the curves and thus define the relationship between moment and cross-sectional rotation, which can be analyzed further.

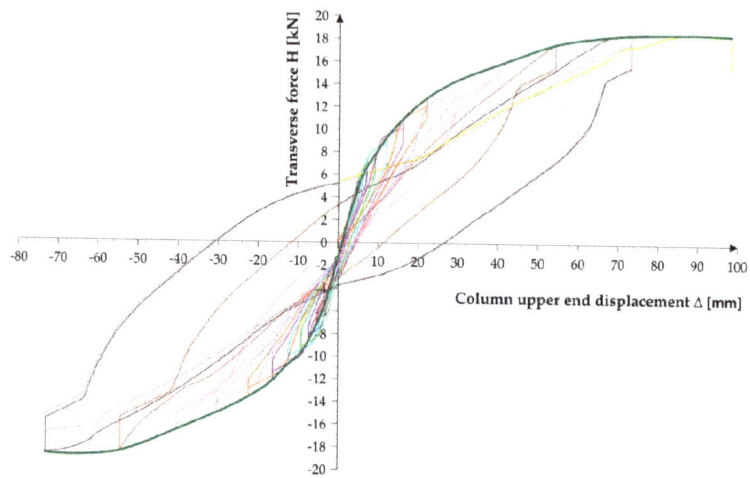

Figure 12. Hysteresis cycles in one of the specimens.

As with monotonically loaded specimens, the diagram of the ratio of moment to rotation obtained here can be divided into three characteristic areas. In the second area, the results for each specimen differ due to the geometrical and material properties of the individual specimen and the amount of axial load. Figure 13 shows the behavior of a single column specimen in three characteristic cross-sections around the plastic joint area. Their analysis shows the influence of each parameter on the load bearing capacity and deformability of the column.

The specimens AB-NS-001-3, AB-NS-001-4, AB-NS-001-7, AB-NS-001-8, and AB-NS-001-9 are made with smooth reinforcement and have the same cross-sectional properties. The longitudinal reinforcement in the central part of the cross-section is 4 × Ø10, while the reinforcement in the corners is 4 × Ø8. The differences are only in the applied axial load and the concrete properties, as they are not all from the same series. The first two specimens were loaded with an axial force of 100 kN, the other two with a force of 150 kN, while the last specimen was loaded with a force of 125 kN. For the first two specimens, the mean compressive strength of the concrete tested on cubes was 47.5 MPa, while the other three specimens were loaded with 36.0 MPa. The specimens AB-NS-002-2 and AB-NS-003-2 were loaded with the same axial force of 125 kN, but their geometrical and material properties of the cross-sections are different, as they are made with smooth reinforcement. As with the monotonically tested specimens of the same series, the differences are in the amount of the reinforcement. The longitudinal reinforcement in the middle of the cross-section is Ø12 mm. In the first specimen, the corner bars are Ø10 mm, in the second specimen these bars are Ø8 mm. The concrete compressive strength test showed a small difference in the average compressive strength for both specimens, as they were not erected in the same series. The average compressive strength of 40.8 MPa was measured in the specimen AB-NS-002-2, while the compressive strength of 36.2 MPa was measured in the specimen AB-NS-003-2. The specimen AB-NS-004-2 is made with ribbed reinforcement. The reinforcement around the center of the cross-section is Ø10 mm, while the corner reinforcement is Ø8 mm. The average compressive strength of the concrete cube is 36.2 MPa. The specimen was loaded with an axial force of 150 kN.

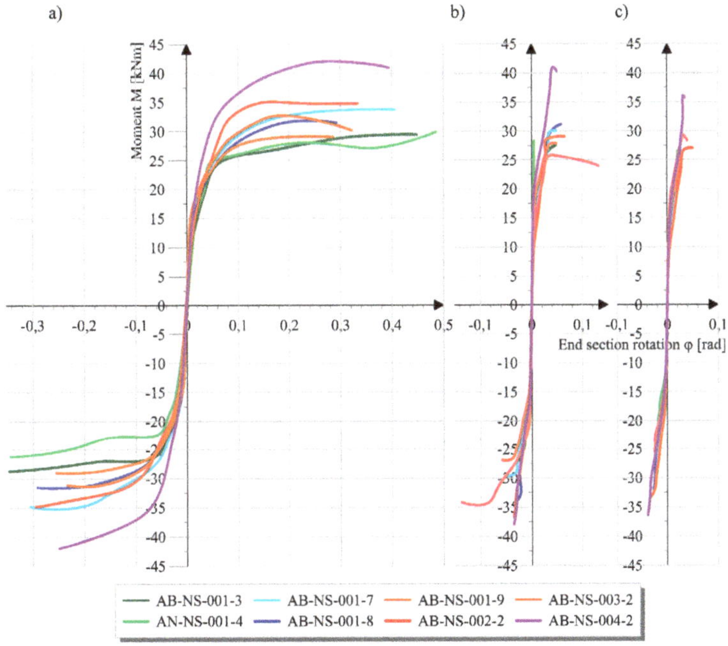

Figure 13. Moment to section rotation ratio of cyclically tested specimens: (**a**) end section; (**b**) 150 mm higher section; (**c**) 350 mm higher section.

A characteristic failure of all specimens occurs as crushing of the concrete in the compression area around the end section after large deformations of the tensile area due to the loss of bond strength between reinforcement and concrete and the yielding of the reinforcement.

As with the monotonically tested specimens, no clear boundaries are visible between the individual areas of the M/φ ratio of the end section diagram, while in the other two sections these boundaries are more pronounced. The reason for this is again the loss of bond strength between the reinforcement and the concrete, where there is no sudden change in the slope of the diagram due to the loss of stiffness after crack formation. In cyclically tested specimens, the effect of the loss of bond strength between reinforcement and concrete on the decrease in stiffness during loading with horizontal force is not visible, as was the case with a monotonically tested specimen.

The influence of the axial force on the displacement–force ratio is visible in the first four cyclically tested specimens. During the test, the specimens AB-NS-001-3 and AB-NS-001-4 were loaded with a lower axial force than the specimens AB-NS-001-7 and AB-NS-001-8, resulting in a lower cross-sectional load capacity. It should be noted that the compressive strength test showed that the first two sets of specimens had a concrete compressive strength one class higher than the other two (47.5/36.0 MPa). It can therefore be assumed that the differences would be even greater for the same concrete properties. Looking at the rotation of the cross-section when the failure is reached on the same specimens, it can be seen that for both specimens with a smaller value of the axial force, the rotation of the cross-section is higher. Specimen AB-NS-001-9 was tested with an axial force of 125 kN. The load bearing capacity of this specimen is between the specimens tested with forces of 100 and 150 kN, while its rotation is lower than that of the other specimens tested in this series, which is also due to the lower compressive strength of the concrete.

The specimens AB-NS-002-2 and AB-NS-003-2 were made with different amounts of reinforcement in the corners. The column AB -NS-002-2 column has a larger amount of reinforcement (4Φ10) than the column AB-NS-003-2 (4Φ8), so the load-bearing capacity is higher as expected.

If we compare columns AB-NS-001-7, AB-NS-001-8, and AB-NS-004-2, we can see differences in the load-bearing capacity of smooth and ribbed reinforcement of cyclically loaded columns. These columns are made with the same amount of reinforcement, have approximately the same concrete properties and are loaded with the same axial force, but the column made with ribbed reinforcement AB-NS-004-2 showed a 30% higher load bearing capacity than the column made with smooth reinforcement. If we observe the rotation of the end cross-section of specimens with the same geometric properties and the same amount of axial force applied, we see that the specimen tested with ribbed reinforcement has a slightly lower rotation on failure than the specimens produced with smooth reinforcement. The difference here is not as pronounced as in the case of monotonically tested specimens.

The results of measuring the relative deformations of the cross-section at a height of 150 and 350 mm from the bottom of the column and the corresponding values of cross-section rotation show the behavior of the column at the plastic joint. For the cross-section at a height of 150 mm from the bottom, we can see that in most of the tested specimens a third area, i.e., the area of plasticization, was present. At a height of 350 mm from the bottom, the two specimens reached the values where plastification takes place.

4.4. Comparison between Results of Analytical and Numerical Analysis

By measuring the relative deformations at the location of the extremely loaded cross-section (at the bottom of the column), the amount of cross-section rotation at which the yield strength of the reinforcement was reached was determined. The results obtained were compared with the values of the cross-section rotation when the yield strength obtained by analytical and numerical methods was reached. As the analytical and numerical results do not include the effect of slippage of the smooth reinforcement, a significant difference of the experimentally obtained cross-section rotation when reaching the yield strength is visible.

Figure 14 (above) shows the moment and rotation ratios for the end cross-section of the two monotonically tested specimens. Experimentally, analytically, and numerically obtained curves are compared in the diagram. The first specimen was made with smooth reinforcement (Figure 14 top left) and the second with ribbed reinforcement (Figure 14 top right).

The point of reaching the yield strength obtained by experimental tests is shown by a rhombus marking, while the point of reaching the analytically and numerically obtained yield strength is shown by cross and triangle markings. The cross-sectional rotation at the point of reaching the analytically and numerically obtained yield strength almost coincides, while the difference of the experimentally obtained rotation angle in comparison with the two previous analyzes is marked with the notation $\Delta\varphi$. The given diagrams clearly show the differences in the angle of rotation of the cross-section when reaching the yield strength of the specimen made with smooth reinforcement and the specimen made with ribbed reinforcement. Specimens with smooth reinforcement show a significantly greater deviation of the experiment results from the numerical or analytical analysis.

The values of the angle of rotation angle when reaching the yield strength of other monotonically tested columns are shown in Figure 15. A comparison of the results for the end section and two sections where relative deformations were measured in relation to the results obtained analytically and numerically is given. The results of the tested specimens with smooth reinforcement show on average a five-times higher rotation of the experimentally obtained end section when reaching the yield point compared to the results obtained with analytical and numerical methods which do not take into account the effect of slippage from the anchoring area (Figure 15). For specimens with ribbed reinforcement (AB-NS-004), the rotation angles of the end cross-section when reaching the yield point in the experiment have a significantly smaller difference from the results obtained by analytical and numerical analysis (two-times the rotation of the experimental result).

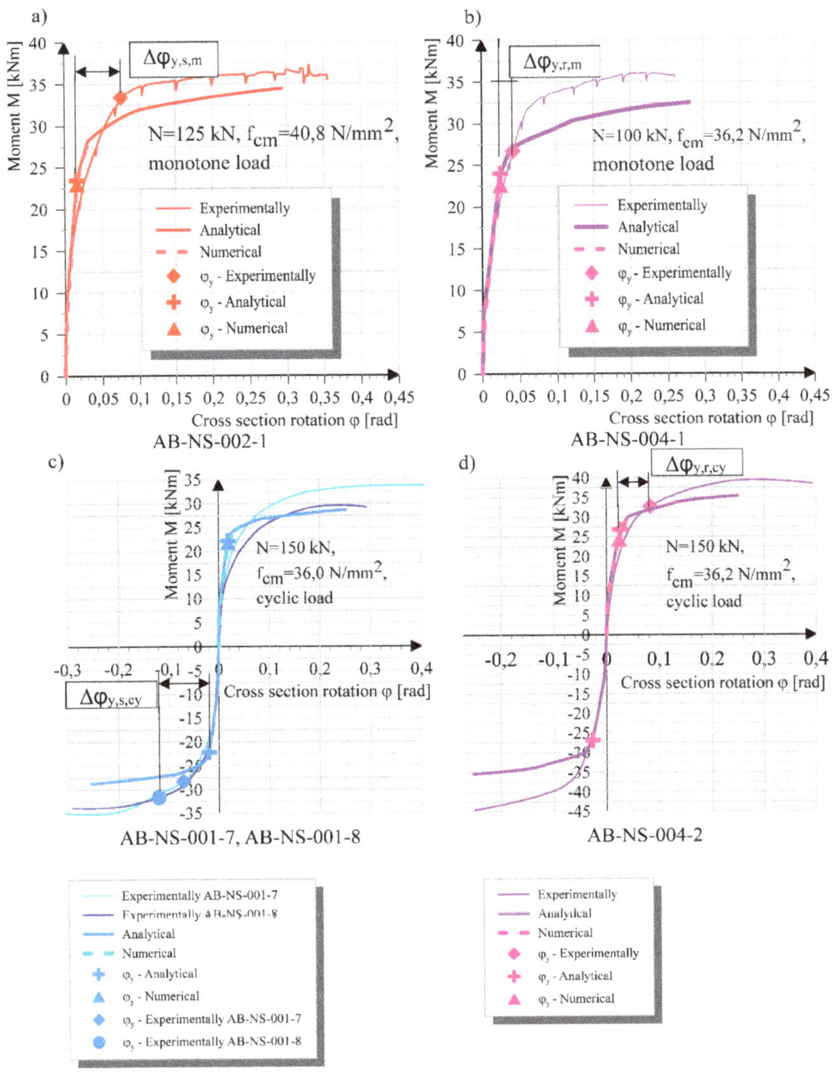

Figure 14. Moment to section rotation ratios: (**a**) monotonous specimen with smooth reinforcement; (**b**) monotonous specimen with ribbed reinforcement; (**c**) cyclical specimens with smooth reinforcement; (**d**) cyclical specimen with ribbed reinforcement.

After opening the first crack, specimens made with smooth reinforcement also show a greater deviation of the experimentally obtained moment to rotation curves compared to specimens with ribbed reinforcement. The effect of slipping of the smooth reinforcement from the anchorage area on the side of the anchor block and the column body itself influences the rotation of the cross-section as soon as the first crack occurs.

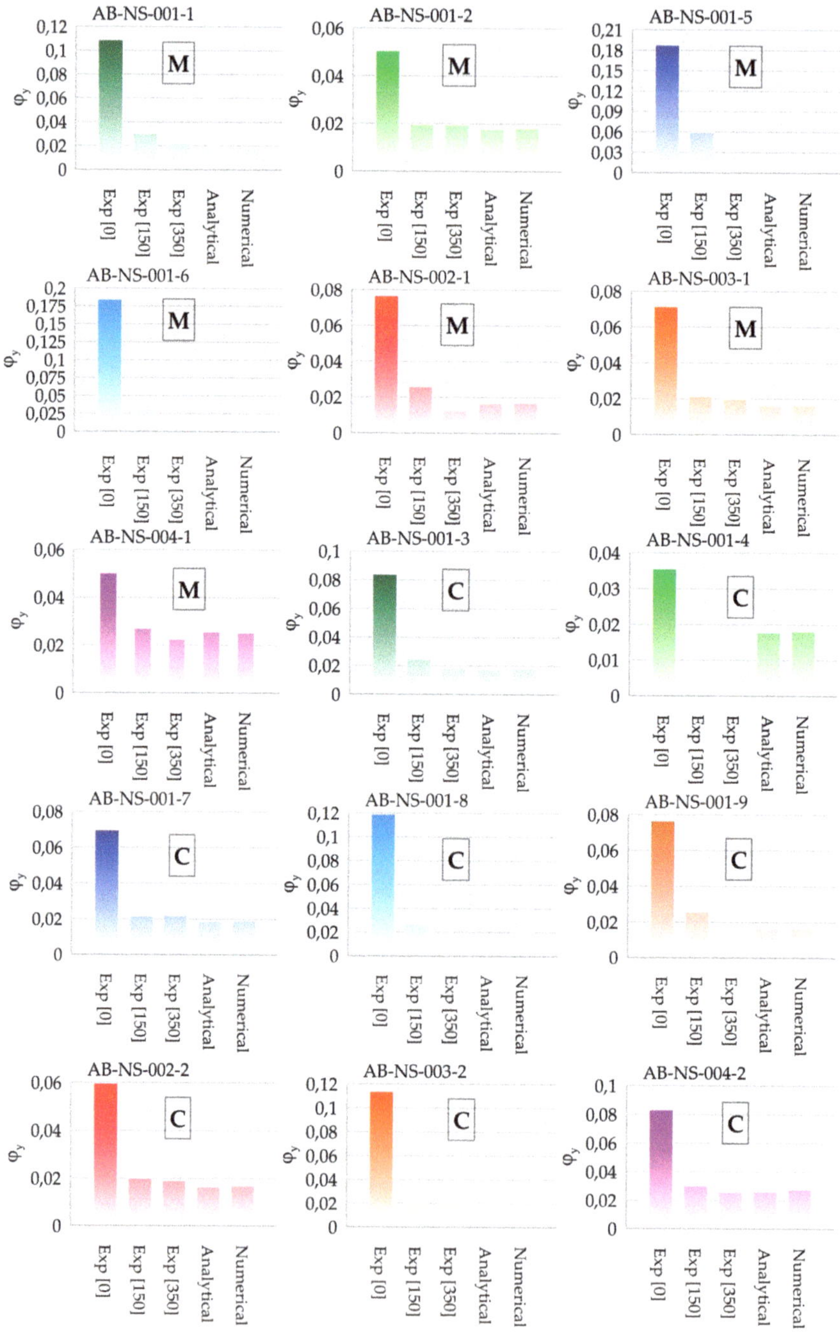

Figure 15. Section rotation when reaching yield point for monotonously (M) and cyclically (C) tested specimens.

The greater distance between the anchoring points of the smooth reinforcement leads to a greater relative deformation of the bar and thus to a greater rotation of the cross-section, which then accumulates at the end cross-section. From this it can be concluded that the ribbed reinforcement bars are better anchored in the concrete, which leads to a shorter length where relative deformation occurs, and thus to a smaller angle of rotation of the end cross-section.

To compare the ratio of moment and cross-sectional rotation of smooth and ribbed reinforcement under cyclic loading, comparative M/φ diagrams of two specimens with smooth and one specimen with ribbed reinforcement are shown in Figure 14 (below). In addition, the locations where the yield strength of the reinforcement is reached are marked according to different analysis methods. For cyclically tested specimens, the difference between the angle of rotation when the yield point is reached, which was determined by experimental analysis, and those determined numerically and analytically for smooth and ribbed reinforcement is significantly smaller than for monotonously tested specimens. This leads to the conclusion that these two types of reinforcement around the plastic joint behave similarly with regard to bond strength under cyclic loading. Alternating load cycles with a compression–tension effect causes the concrete to be crushed near the reinforcement ribs, which causes the non-anchored part of the ribbed bar to increase in size and thus behave like smooth reinforcement, so that the results for both smooth and ribbed reinforcement are closer together.

The results of measurements on cyclically tested specimens made with smooth reinforcement show similar behavior near the end cross-section as monotonically tested specimens. The rotations of the end section and the section where the first crack occurs are dominant. For cyclically tested specimens, the difference between the experimental results of the end section rotation when the yield strength is reached, and the analytical and numerical results is somewhat smaller compared to monotonically tested specimens. On average, the difference is four-and-a-half-times higher than the difference between the rotation of the experimental measurements compared to the analytical and numerical results. As with monotonically tested specimens, the reason for this discrepancy is the loss of bond strength between concrete and reinforcement in the anchorage area of the smooth reinforcement.

Differences between the cyclically tested specimens and the monotonically tested specimens can be seen in the specimens made with ribbed reinforcement. A 50% larger rotation of the end cross-section of the cyclically tested specimens can be seen in comparison to the monotonically tested specimens made with ribbed reinforcement. Since the influence of a higher axial force reduces the rotatability of the cross-section, it can be assumed that the difference would be even greater since the cyclically tested specimen was loaded with a 50% higher force. This proves that ribbed reinforcement with high ductility properties (B500 B) behaves much better under cyclic loading in terms of element ductility than under monotonic loading.

5. Final Remarks

This research defines important indicators that influence the behavior of a column with an atypical cross-section, loaded by seismic action, erected without the guidelines for anti-seismic reinforcement placement, and made with smooth reinforcement in the area of plastic joint formation. From the analytical and experimental analysis results, a great influence on the properties of the material and therefore the importance of modeling them accurately in order to define more realistically the indicators of seismic resistance was achieved. Using non-linear material properties in analytical and numerical analysis, results have been obtained that are approximately equal to those obtained in experimental analysis, with deviations due to slippage of reinforcement from the anchorage area.

For most of the specimens tested, all three methods used showed a good agreement of results in the linear range. Small deviations of the results occurred immediately after the opening of the first crack and significant deviations are visible when the yield strength is reached. The main reason for the discrepancy between the results was the slippage of the reinforcement from the anchorage area. In addition, there were deviations due to a mismatch of the stress-strain curve (material properties) after the beginning of hardening of the reinforcement steel-difference between the working law diagrams of the experimentally obtained reinforcement and the theoretical diagrams used in analytical and numerical analysis. Another reason are the deviations that occurred during the erection, such as deviations in the dimensions of the cross-section and thickness of the protective layer of the reinforcement.

Comparing monotonically and cyclically tested specimens, significant deviations of the end section rotation when reaching the yield strength are visible in relation to the analytical and numerical analysis. For all monotonically tested specimens, the rotation of the experimentally obtained end section is about five times greater than that obtained by analytical and numerical analysis. For cyclically tested samples, this deviation is somewhat smaller and is on average four-and-a-half-times greater. In the case of monotonous testing of spacemen made with ribbed reinforcement, the difference obtained is much smaller and is twice the rotation in the experimental results. When comparing the results of the specimens made with ribbed reinforcement, a 50% higher rotation of the end cross-section of the cyclically tested spacemen compared to the monotonously tested spacemen can be seen. From this it can be concluded that significant deviations of the end section rotation when reaching the yield point are due to the loss of bond strength between concrete and reinforcement.

When the final limit state is reached, the deviation of the experimental results in relation to the analytical results is much smaller than when the yield point is reached. This proves that the influence of reinforcement slippage from the anchorage area decreases as we approach the final limit state.

Depending on the arrangement and width of the cracks around the plastic joint formation area in columns with smooth reinforcement, the following can be concluded:

- First cracks appear in the early phase of loading for most of the tested specimens (up to 20 mm displacement at the top of the column).
- Discrete cracks generally developed around the area of the transverse reinforcement.
- For all specimens, cracks appear first at the location of the initiated crack (end section).
- The first crack was formed mainly at a height of 120–140 mm from the bottom of the column and, together with the crack at the bottom, represents the place of plastification of the element.
- Other cracks stop expanding after some time (when the displacement at the top of the column is 40–50 mm) and remain within the limits allowed for reinforced concrete elements.

From the above it can be concluded that the total plastic deformation is reduced to a rotation around the crack of the end section and the first crack from the bottom of the element. The same can be concluded by analyzing the rotation of three characteristic cross-sections. It can be observed that most of the plastic rotation is concentrated in the end section, while in the cross-section of the first crack the amount of plastic rotation is much smaller, and in the cross-section of the second crack (the third cross-section observed) plastification is almost non-existent in most of the specimens tested. From all the above, a conclusion is drawn about the height of the plastic joint for monotonically and cyclically tested specimens, which mainly develops to about the height of the first crack and is 0.7 h, where h is the height of the cross-section in the load direction.

For specimens with ribbed reinforcement (monotonically tested), the plastification height extends up to a height of 1.2 h, which corresponds approximately to the values given in [11]. From this it can be concluded that the given expression is calibrated to elements with ribbed reinforcement and its application to elements with smooth reinforcement is questionable. The experiment has shown that for specimens with ribbed reinforcement, the crack widths are evenly distributed over the height of the column, the larger ones at the bottom, and the smaller ones because the distance from the bottom is greater. The same applies to the distribution of rotations along the height. This explains why the plasticizing zone in specimens with ribbed reinforcement is distributed at a higher column height than in specimens with smooth reinforcement.

For specimens with ribbed reinforcement, the differences in the results of the cyclic and monotonic tests should be emphasized. The specimen subjected to the cyclic test showed a similar behavior in the area of the plastic joint as the specimen with smooth reinforcement (two cracks dominating at the bottom of the element), in contrast to the monotonically tested specimen where the results are significantly different. Alternating load cycles cause the concrete to be crushed in the vicinity of the reinforcement ribs, resulting in a larger area of the non-anchored part of the reinforcement and therefore greater relative deformation. From all the above it can be concluded that this behavior in the vicinity of the ribbed reinforcement further contributes to the ductility of elements made with highly ductile ribbed reinforcement (B 500 B).

The behavior of the elements in the vicinity of the formation of plastic joints under the influence of seismic loads has not yet been sufficiently investigated. This refers primarily to structural elements that were erected before the adoption of regulations and guidelines for the seismic calculation and design of earthquake-resistant elements, to elements that do not have adequate quantities of transverse reinforcement and reinforcement for the concrete core enclosure according to modern design guidelines. Due to the special characteristics of their structural elements, but also the way they react to seismic actions, bridge structures must be included in future guidelines for the assessment of existing structures. These guidelines will certainly become increasingly applicable due to the large number of structures that require a more detailed analysis of the load-bearing elements to determine their actual seismic resistance.

The results of this research are applicable to the analysis of the basic indicator (Diagram M/φ) of the seismic resistance of atypical cross-sections. It will allow for the application of an analytical or numerical method with realistic properties of the materials and elements and the derivation of correction factors due to the effect of slippage of the smooth reinforcement from the anchorage area.

Author Contributions: Conceptualization, M.S. and A.M.I.; methodology, M.S. and A.M.I.; software, M.S. and A.V.; validation, A.M.I., A.V., and G.H.K.; formal analysis, M.S.; investigation, M.S.; resources, M.S., A.M.I., A.V., and G.H.K.; data curation, A.M.I.; writing—original draft preparation, M.S.; writing—review and editing, A.M.I., A.V., and G.H.K.; visualization, M.S.; supervision, A.M.I. and A.V.; project administration, M.S. Please turn to the CRediT taxonomy for the term explanation. Authorship must be limited to those who have contributed substantially to the work reported. All authors have read and agreed to the published version of the manuscript.

Funding: This research received no external funding.

Institutional Review Board Statement: Not applicable.

Informed Consent Statement: Not applicable.

Data Availability Statement: The data presented in this study are available on request from the corresponding author. The data are not publicly available due to privacy restrictions.

Conflicts of Interest: The authors declare no conflict of interest. The funders had no role in the design of the study; in the collection, analyses, or interpretation of data; in the writing of the manuscript, or in the decision to publish the results.

References

1. Calvi, G.M.; Moratti, M.; O'Reilly, G.J.; Scattarreggia, N.; Monteiro, R.; Malomo, D.; Calvi, P.M.; Pinho, R. Once upon a Time in Italy: The Tale of the Morandi Bridge. *Struct. Eng. Int.* **2019**, *29*, 198–217. [CrossRef]
2. Nicodème, C.; Diamandouros, K.; Diez, J.; Durso, C.; Arampidou, K.; Nuri, A.K. Road Statistics—Yearbook 2017. *Eur. Road Fed.* **2017**, *2017*, 93.
3. Marić, M.K.; Ivanković, A.M.; Vlašić, A.; Bleiziffer, J.; Srbić, M.; Skokandić, D. Assessment of reinforcement corrosion and concrete damage on bridges using non-destructive testing. *J. Croat. Assoc. Civ. Eng.* **2019**, *71*, 843–862. [CrossRef]
4. Bernal, J.; Fenaux, M.; Moragues, A.; Reyes, E.; Gálvez, J. Study of chloride penetration in concretes exposed to high-mountain weather conditions with presence of deicing salts. *Constr. Build. Mater.* **2016**, *127*, 971–983. [CrossRef]
5. Verderame, G.M.; Fabbrocino, G.; Manfredi, G. Seismic response of r.c. columns with smooth reinforcement. Part II: Cyclic tests. *Eng. Struct.* **2008**, *30*, 2289–2300. [CrossRef]

6. Verderame, G.M.; Fabbrocino, G.; Manfredi, G. Seismic response of r.c. columns with smooth reinforcement. Part I: Monotonic tests. *Eng. Struct.* **2008**, *30*, 2277–2288. [CrossRef]
7. Verderame, G.M.; Ricci, P.; Manfredi, G.; Cosenza, E. Ultimate chord rotation of RC columns with smooth bars: Some considerations about EC8 prescriptions. *Bull. Earthq. Eng.* **2010**, *8*, 1351–1373. [CrossRef]
8. Melo, J.; Varum, H.; Costa, A. Experimental response of RC columns built with plain bars under unidirectional cyclic loading. In Proceedings of the 15th World Conference on Earthquake Engineering, Lisbon, Portugal, 24–28 September 2012.
9. Cosenza, E.; Manfredi, G.; Verderame, G.M. Capacity models of rc members with emphasis on sub-standard columns with plain bars. In Proceedings of the Eurocode 8 Perspectives from the Italian Standpoint WorkshopAt, Naple, Italy, 1 April 2009; pp. 129–144.
10. Biskinis, D.; Fardis, M.N. Flexure-controlled ultimate deformations of members with continuous or lap-spliced bars. *Struct. Concr.* **2010**, *11*, 93–108. [CrossRef]
11. Fardis, M.N. *Guidelines for Displacement-Based Design of Buildings and Bridges*; IUSS Press: Pavia, Italy, 2007.
12. Paulay, T.; Priestly, M.J.N. *Seismic Design of Reinforced Concrete and Masonry Buildings*; John Wiley & Sons: New York, NY, USA, 1992.
13. Priestley, M.J.N.; Park, R. Strength and ductility of concrete bridge columns under seismic loading. *ACI Struct. J.* **1987**, *84*, 61–76.
14. Biskinis, D.; Fardis, M.N. Deformations at flexural yielding of members with continuous or lap-spliced bars. *Struct. Concr.* **2010**, *11*, 127–138. [CrossRef]
15. Isaković, T.; Fischinger, M. Pojednostavnjene nelinearne metode proračuna betonskih mostova. *Gradjevinar* **2009**, *61*, 625–633.
16. Kowalsky, M.J. A displacement-based approach for the seismic design of continuous concrete bridges. *Earthq. Eng. Struct. Dyn.* **2002**, *31*, 719–747. [CrossRef]
17. Mancini, M. *Structural Performance Assessment of Existing R.C. Bridges in Seismic Prone Areas*; University of Naples Federico II: Portici, Italy, 2011.
18. Lu, Z.; Ge, H.; Usami, T. Applicability of pushover analysis-based seismic performance evaluation procedure for steel arch bridges. *Eng. Struct.* **2004**, *26*, 1957–1977. [CrossRef]
19. Cetinkaya, O.T.; Nakamura, S.; Takahashi, K. A static analysis-based method for estimating the maximum out-of-plane inelastic seismic response of steel arch bridges. *Eng. Struct.* **2006**, *28*, 635–647. [CrossRef]
20. Franetović, M.; Ivanković, A.M.; Radic, J. Seismic Assessment of Existing Bridges in Croatia. In Proceedings of the IABSE Conference, Rotterdam 2013: Assessment, Upgrading and Refurbishment of Infrastructures, Rotterdam, The Netherlands, 6–8 May 2013; International Association for Bridge and Structural Engineering (IABSE): Zurich, Switzerland, 2013.
21. Franetović, M.; Radić, J.; Šavor, Z. Seismic Assessment of Arch Bridge Across Slunjčica River in Slunj. In Proceedings of the 3rd Chinese-Croatian Joint Colloquium on Sustainable Arch Bridges, Zagreb, Croatia, 15–16 July 2011.
22. Mandić Ivanković, A.; Srbić, M.; Radić, J. Seismic Performance of Concrete Arch Bridges. In Proceedings of the FIB Symposium 2016: Performance—Based Approaches for Concrete Structures, Cape Town, South Africa, 21–23 November 2016; Hans, B., Ed.; Fib: Lausanne, Switzerland, 2016; pp. 237–238.
23. Ana, M.I.; Srbic, M.; Radic, J. Performance indicators in assessment of concrete arch bridges. In *Maintenance, Monitoring, Safety, Risk and Resilience of Bridges and Bridge Networks*; Bittencourt, T.N., Frangopol, D.M., Beck, A.T., Eds.; CRC Press Taylor & Francis Group: São Paulo, Brazil; Leiden, The Netherlands, 2016; pp. 301–302.
24. Mandić Ivanković, A.; Srbic, M.; Franetović, M. Performance of existing concrete arch bridges. In Proceedings of the IABSE Conference, Geneva 2015: Structural Engineering: Providing Solutions to Global Challenges, Geneva, Switzerland, 23–25 September 2015; International Association for Bridge and Structural Engineering (IABSE): Geneva, Switzerland, 2015. [CrossRef]
25. Franetović, M.; Mandić Ivanković, A.; Radić, J. Seismic assessment of existing reinforced concrete arch bridges. *J. Croat. Assoc. Civ. Eng.* **2014**, *66*, 691–703. [CrossRef]
26. Eurocode. *Eurocode 8: Design of Structures for Earthquake Resistance—Part 3—Assessment and Retrofitting of Buildings*; European Committee for Standardization CEN: Brussels, Brussels, 2004; Volume 3.
27. Kappos, A.J.; Saiidi, M.S.; Aydınoğlu, M.N.; Isaković, T. (Eds.) *Seismic Design and Assessment of Bridges: Inelastic Methods of Analysis and Case Studies*; Springer: Dordrecht, The Netherlands, 2012.
28. Eurocode. *Eurocode 8—Design of Structures for Earthquake Resistance—Part 2—Bridges*; European Committee for Standardization CEN: Brussels, Brussels, 2005; Volume 3.
29. Priestley, M.J.N. Myths and Fallacies in Earthquake Engineering, Revisited The Ninth Mallet Milne Lecture. *Bull. N. Z. Soc. Earthq. Eng.* **2003**, *26*, 329–341.
30. Ćurić, I.; Radić, J.; Franetović, M. Determination of the bending moment—curvature relationship for bridge concrete columns. *Teh. Vjesn. Tech. Gaz.* **2016**, *23*, 907–915. [CrossRef]
31. Srinivasan, C.; Nunziante, L.; Serino, G.; Carannante, F. *Seismic Design Aids for Nonlinear Analysis of Reinforced Concrete Structures*; CRC Press Taylor & Francis Group: Boca Roton, FL, USA, 2009.
32. Kwak, H.; Kim, S. Nonlinear analysis of RC beams based on moment—Curvature relation. *Comput. Struct.* **2002**, *80*, 615–628. [CrossRef]
33. Panagiotakos, T.B.; Fardis, M.N. Deformations of Reinforced Concrete Members at Yielding and Ultimate. *ACI Struct. J.* **2001**, *98*, 135–148.

34. Rasulo, A.; Pelle, A.; Lavorato, D.; Fiorentino, G.; Nuti, C.; Briseghella, B. Finite Element Analysis of Reinforced Concrete Bridge Piers Including a Flexure-Shear Interaction Model. *Appl. Sci.* **2020**, *10*, 2209. [CrossRef]
35. Kong, Q.; Robert, R.H.; Silva, P.; Mo, Y.L. Cyclic Crack Monitoring of a Reinforced Concrete Column under Simulated Pseudo-Dynamic Loading Using Piezoceramic-Based Smart Aggregates. *Appl. Sci.* **2016**, *6*, 341. [CrossRef]
36. Hwang, H.-J.; Kim, C.-S. Simplified Plastic Hinge Model for Reinforced Concrete Beam–Column Joints with Eccentric Beams. *Appl. Sci.* **2021**, *11*, 1303. [CrossRef]
37. Guo, K.; Guo, Q.; Wang, Y. Effect of Bond-Slip on Dynamic Response of FRP-Confined RC Columns with Non-Linear Damping. *Appl. Sci.* **2021**, *11*, 2124. [CrossRef]
38. Masi, A.; Santarsiero, G. Seismic Tests on RC Building Exterior Joints with Wide Beams. *Adv. Mater. Res.* **2013**, *787*, 771–777. [CrossRef]
39. Taucer, F.; Pinto, A.V. *Mock-Up Design of Reinforced Concrete Bridge Piers for PsD Testing at the ELSA Laboratory: (Vulnerability Assessment of Bridges Project)*; Institute for Systems, Informatics and Safety: Ispra, Italy, 2000.
40. *Interim Testing Protocols for Determining the Seismic Performance Characteristics of Structural and Nonstructural Components*; FEMA 461; Applied Technology Council: Redwood City, CA, USA, 2007; p. 113.
41. Eurocode. *Steel for the Reinforcement and Prestressing of Concrete—Test Methods—Part 1: Reinforcing Bars, Rods and Wire (ISO 15630-1:2019; EN ISO 15630-1:2019)*; European Committee for Standardization CEN: Brussels, Belgium, 2019.
42. Eurocode. *Testing Hardened Concrete—Part 3: Compressive Strength of Test Specimens (EN 12390-3:2009)*; European Committee for Standardization CEN: Brussels, Belgium, 2009.
43. Eurocode. *Testing Hardened Concrete—Part 6: Tensile Splitting Strength of Test Specimens (EN 12390-6:2009)*; European Committee for Standardization CEN: Brussels, Belgium, 2010.
44. Eurocode. *Testing Hardened Concrete—Part 4: Determination of Ultrasonic Pulse Velocity (EN 12504-4:2004)*; European Committee for Standardization CEN: Brussels, Belgium, 2004.
45. Eurocode. *Testing Hardened Concrete—Part 13: Determination of Secant Modulus of Elasticity in Compression (EN 12390-13:2013)*; European Committee for Standardization CEN: Brussels, Belgium, 2013.
46. Eurocode. *Eurocode 2—Design of Concrete Structures—Part. 1-1: General Rules and Rules for Buildings*; European Committee for Standardization CEN: Brussels, Brussels, 2004; Volume BS EN 1992, ISBN 978-0-580-73752-7.
47. M.J.N. Priestley, G.M.; Calvi, M.J.K. *Displacement Based Seismic Design of Structures*; IUSS Press: Pavia, Italy, 2007.
48. Srbić, M.; Mandić Ivanković, A.; Brozović, T. Bending moment curvature relationship as an indicator of seismic resistance of older bridge piers. *Gradjevinar* **2019**, *71*, 481–488. [CrossRef]

applied sciences

MDPI

Article

Proposal of Damage Index Ratio for Low- to Mid-Rise Reinforced Concrete Moment-Resisting Frame with Setback Subjected to Uniaxial Seismic Loading

Taufiq Ilham Maulana, Badamkhand Enkhtengis and Taiki Saito *

Department of Architecture and Civil Engineering, Toyohashi University of Technology, Aichi 441-8580, Japan; taufiq.ilham.maulana.mt@tut.jp (T.I.M.); badamkhand.enkhtengis.do@tut.jp (B.E.)
* Correspondence: saito.taiki.bv@tut.jp

Abstract: A vertical irregularity setback in reinforced concrete (RC) building affects its performance and response especially subjected to earthquake ground motions. It is necessary to understand how the seismic damage is established due to setbacks and avoid damage concentration on the irregularity section. The objective of this study is to propose a formula to estimate the damage distribution along the height of the setback building from a geometric measure of the degree of irregularity. First, previous experimental tests for two types of setback buildings, a towered and a stepped setback frames, were analyzed to verify the accuracy of the frame analysis. The results of the frame analysis considerably matched the experimental test results. Furthermore, to study the relationship between the degree of setback and the distribution of damage, a parametric study was conducted using 35 reinforced concrete setback frames, consisting of models with stepped setback type and towered setback type with different degrees of setback. The inelastic dynamic analyses of all the frames under three earthquake ground motions were conducted. The irregularity indices proposed in literature were adopted to express the degree of setback and the structural damage was expressed by the Park–Ang damage index. Using nonlinear regression analysis, formulas to estimate damage index ratio between two main structure parts (tower and base) from setback indices were proposed. Finally, the proposed formula was applied to the experimental test results to confirm its validity.

Keywords: setbacks; earthquake; damage index; seismic evaluation performance; damage distribution

Citation: Maulana, T.I.; Enkhtengis, B.; Saito, T. Proposal of Damage Index Ratio for Low- to Mid-Rise Reinforced Concrete Moment-Resisting Frame with Setback Subjected to Uniaxial Seismic Loading. *Appl. Sci.* **2021**, *11*, 6754. https://doi.org/10.3390/app11156754

Academic Editor: Maria Favvata

Received: 9 June 2021
Accepted: 20 July 2021
Published: 22 July 2021

Publisher's Note: MDPI stays neutral with regard to jurisdictional claims in published maps and institutional affiliations.

cc BY

Copyright: © 2021 by the authors. Licensee MDPI, Basel, Switzerland. This article is an open access article distributed under the terms and conditions of the Creative Commons Attribution (CC BY) license (https://creativecommons.org/licenses/by/4.0/).

1. Introduction

1.1. Background

In recent modern multistory building construction, the implementation of horizontal and vertical irregularities is getting popular due to different building functions and architectural needs. Vertical irregularity often regarded as a setback affects discontinuities in mass, stiffness, and structural strength distributions [1,2]. Under seismic excitations, this irregularity has a considerable effect on structure responses [3]. The existence of the setback may cause torsional response [4] and stress and damage concentration on the irregular part [5].

Studies on the seismic response of buildings with setbacks had been conducted by several researchers. Shakib and Pirizadeh [6] assessed the effect of simultaneous action of two orthogonal ground motion components on the performance of structure with different setback ratios and found that the effect of ground motion components is significant for the seismic performance of setback structures. Lin et al. [7] investigated the modal response history analysis of setback buildings and instead of using the conventional SDOF modal system, the 2DOF modal system was employed in modal response history analysis of building with specific vertical irregularities, to characterize the distinctly different modal response of tower and base structures. It was found that the 2DOF approach adequately captured the characteristic of drastic change in the peak interstory drift ratios between tower and base. Georgoussis and Mamou [8] proposed the equations to minimize the rotational response due to the mass eccentricity of setback buildings by changing the location of lateral load resisting elements. The parameters in the proposed equations involved the first mode frequency and the displacement ratio between the tower and the base structure. The proposed equations were also applied for minimizing the rotational response of medium-rise buildings with irregular plans under seismic excitations [9].

Habibi and Asadi [10] developed equations to determine the Park–Ang damage index [11,12] of setback buildings using three independent parameters, namely overall drift (which is the ratio between the maximum roof displacement to the total height of the frame), the natural period, and the irregularity indices of the building. The adopted irregularity indices were originally introduced by Karavasilis et al. [13], involving the geometry of setback buildings, such as the number of stories and bays and the height and the width of setback buildings. Varadharajan et al. [14,15] also predicted the Park–Ang damage index of setback structures using three primary parameters: the ratio of modal participation factor between irregular and regular buildings, the beam to column stiffness ratio, and the displacement ductility. Hait et al. [16] assessed the Park–Ang damage index of low to mid-rise buildings using parameters generated from dynamic analyses, such as the maximum interstory drift, the peak roof displacement, and the maximum joint rotation of the members. Hait et al. [17]. also predicted the Park–Ang damage index by multi-variable regression using the artificial neural network. In these studies, the Park–Ang damage index was computed through nonlinear dynamic analysis for both tower and base sections, and the damage distribution could be retrieved after. This procedure could take time and resources for the dynamic analysis.

The objective of this paper is to propose the formula to determine the damage distribution of setback buildings based on the geometric measure of the degree of irregularity without conducting dynamic analysis. First, to verify the accuracy of the analytical method, the experimental tests for two types of setback buildings are analyzed. The experimental tests are a towered structure of fifteenth-scale 9-story 3-bay RC frame with a setback at level 3 by Wood [18] and a quarter model of 6-story 2-bay setback RC frame by Shahrooz and Moehle [19]. Then, the relationship between damage distribution of the Park–Ang damage index and the irregularity indices are examined by the parametric study of 35 models with different setback types. Finally, the nonlinear regression equations are proposed to determine the building damage distribution from the geometrical irregularity indices. Then, these equations are validated using the previous experimental results.

1.2. Park and Ang Damage Index

1.2.1. Definition

The damage index (DI) introduced by Park and Ang [11,12] is expressed by Equation (1). This DI value is based on the structural deformation and the hysteretic energy response due to seismic excitations.

$$DI = \frac{u_m}{u_u} + \beta\frac{E_h}{F_y u_u} \tag{1}$$

where:

u_m: Maximum displacement response of structure element due to earthquake,
u_u: Ultimate displacement capacity under a monotonic loading,
E_h: Hysteretic energy dissipated by the structural element,
F_y: Yield force,
β: Non-negative parameter based on repeated loading effect.

The relationship between the DI and the physical damage appearance for RC elements suggested by Park and Ang [12] is shown in Table 1. The parameters in Equation (1) are explained in the following passages.

Table 1. Damage Index for RC elements.

Damage Degree	Damage Index	Physical Appearance
Collapse	DI > 1.0	Collapse state
Severe	0.4 < DI < 1.0	Extensive crushing of concrete, disclosure of buckled reinforcement.
Moderate	0.25 < DI < 0.4	Extensive large cracks, spalling of concrete in weaker elements.
Minor	0.1 < DI < 0.25	Minor damage, light cracking throughout.
Slight	DI < 0.1	No damage or localized minor cracking

1.2.2. Determination of u_u

The ultimate displacement u_u is formulated as follows.

$$u_u = \mu_u u_y \quad (2)$$

where:

u_u: Ultimate displacement under monotonic loading,
μ_u: Ultimate ductility factor,
u_y: Yield displacement.

The yield displacement u_y is defined for each member based on element geometry and material properties. Due to the limitation of the available references stating clearly about the ultimate ductility factor μ_u, in this study, for each member, the μ_u is taken value as 15 by considering the probable failure mode occurs at the beam and column elements to be dominated as flexural failure [11].

1.2.3. Determination of β

The parameter β by Park and Ang [12] is presented in Equations (3) and (4).

$$\beta = \left(-0.447 + 0.073\frac{l}{d} + 0.24n_0 + 0.314p_t\right) \times 0.7^{p_w} \quad (3)$$

$$n_0 = \frac{N}{f_c bd} \quad (4)$$

where:

l/d: Shear span ratio (= 1.7 if l/d less than 1.7),
n_0: Normalized axial stress (= 0.2 if n_0 less than 0.2),
p_t: Longitudinal bars ratio, in percentage (= 0.75% if p_t less than 0.75%),
p_w: Transversal bars ratio,
N: Axial load,
f_c: Concrete compressive strength,
b: Width of cross section,
d: Effective height of cross section.

1.2.4. Damage Index of Building Component

The DI value is also suggested for a part of a structure, such as the individual story and the entire structure, by considering a weighting factor that is based on the dissipated hysteretic energy (E_h) of each member or the component as shown below, that are proposed originally by Park and Ang [12] and developed by Belkacem et al. [20]. Figure 1 shows the difference between DI_{story} and $DI_{component}$. In the component weighting factor calculation, the dissipated energy (E_h) considered is only in the same story, while the calculation of story weighting factor, all energies in all stories are considered.

$$DI_{Story} = \sum_{i=1}^{n_m} \left(\lambda_{i,component} DI_{i,component}\right) \quad (5)$$

$$\lambda_{i,component} = \left[\frac{E_{h,i}}{\sum_{i=1}^{n_m} E_{h,i}} \right]_{component} \tag{6}$$

$$DI_{overall} = \sum_{j=1}^{n_k} (\lambda_{j,story} DI_{j,story}) \tag{7}$$

$$\lambda_{j,story} = \left[\frac{E_{h,j}}{\sum_{j=1}^{n_k} E_{h,j}} \right]_{story} \tag{8}$$

where:

i: number of element member, starting at 1,
j: number of story, starting at 1,
n_k: total number of building's story,
n_m: total element member (beams and columns) in the reviewed story,
$DI_{i,component}$: Damage index for each component such as beam and column,
DI_{Story}: Damage index for each story,
$DI_{overall}$: Overall damage index,
$\lambda_{i,component}$: Weighting factor of each component,
$\lambda_{j,story}$: Weighting factor of each story.

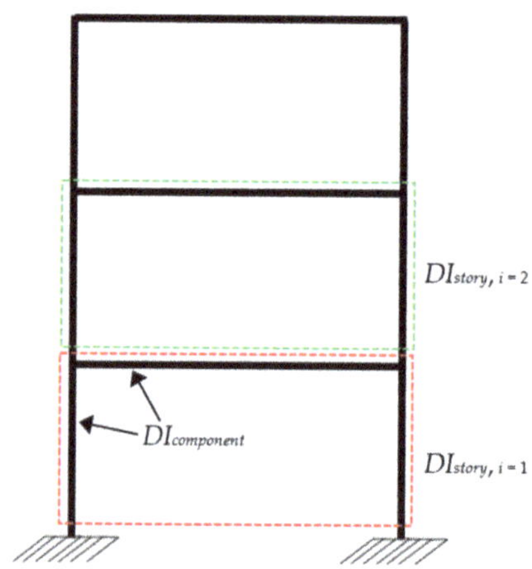

Figure 1. Illustration about damage index component and damage index story in a frame structure.

1.3. Setback Indices

Two kinds of vertical irregularities are selected to be studied: stepped and towered. The stepped type is a setback with the upper structure part at one side of the building, while the towered type is the condition when the upper structure part is in the middle region of the building. The illustration of the stepped type is shown in Figure 2a, and the towered type is shown in Figure 2b.

To consider the setback effect, two irregularity indices are adopted. These indices were initially introduced by Mazzolani and Piluso [21], which were then developed by Karavasilis et al. [13]. The indices describe the irregularity due to the presence of setbacks through simple geometrical indices φ_b and φ_s given by the following Equations (9) and (10).

$$\varphi_s = \frac{1}{n_s - 1} \times \sum_{i=1}^{i=n_s-1} \frac{L_i}{L_{i+1}} \tag{9}$$

$$\varphi_b = \frac{1}{n_b - 1} \times \sum_{i=1}^{i=n_b-1} \frac{H_i}{H_{i+1}} \tag{10}$$

where:

n_s: Number of stories of the frame,
n_b: Number of the bays of the first story of the frames,
H_i: Height of each bay from base to roof,
L_i: Total width of each story.

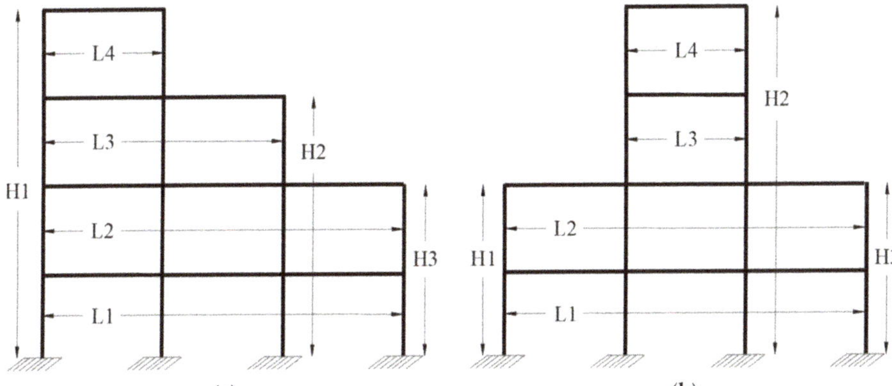

Figure 2. Frame geometry for irregularity indices definition for: (**a**) stepped type of setbacks; (**b**) towered type of setbacks.

In the case of stepped setback type, since the height H_i is counted from the left side, the value of φ_b will be different if the tower part is on the right side or on the left side. Therefore, in this study, H_i is always counted from the side that has the highest level.

2. Simulation of Setback RC Buildings Tested by Shaking Table

To verify the accuracy of the frame analysis, shaking table experimental tests of scaled setback RC buildings conducted by Wood in 1985 [18] and Shahrooz and Moehle in 1987 [19] were chosen to be numerically analyzed, and the raw data of both experimental tests were retrieved from DataCenterHub repository [22,23]. A software developed by one of the co-authors, STERA_3D, is used to perform the nonlinear dynamic structural analysis [24]. The following subchapters elaborate about the numerical tool and the detail of specimens.

2.1. Simulation Method of RC Frame Analysis

Figure 3a,b shows the interface of the STERA_3D for the test specimens of towered setback and stepped setback respectively. The RC beam element is modeled as a line element with two nonlinear flexural springs at both ends and one nonlinear shear spring at the middle as shown in Figure 4. The end displacement vector is obtained from Equation (11) as the sum of the displacement vector of each component. The dynamic earthquake response analysis is conducted by adopting Newmark-β numerical integration method [25].

$$\begin{Bmatrix} \theta_A \\ \theta_B \\ \delta_x \end{Bmatrix} = \begin{Bmatrix} \tau_A \\ \tau_B \\ \delta_x \end{Bmatrix} + \begin{Bmatrix} \varphi_A \\ \varphi_B \\ 0 \end{Bmatrix} + \begin{Bmatrix} \eta_A \\ \eta_B \\ 0 \end{Bmatrix} \tag{11}$$

where:

θ: the total rotation at the element joint,
δ_x: is the element deformation at direction x,

τ: is the elastic element rotation,
ϕ: is the nonlinear element rotation due to bending, and
η: is the nonlinear element rotation due to shear.

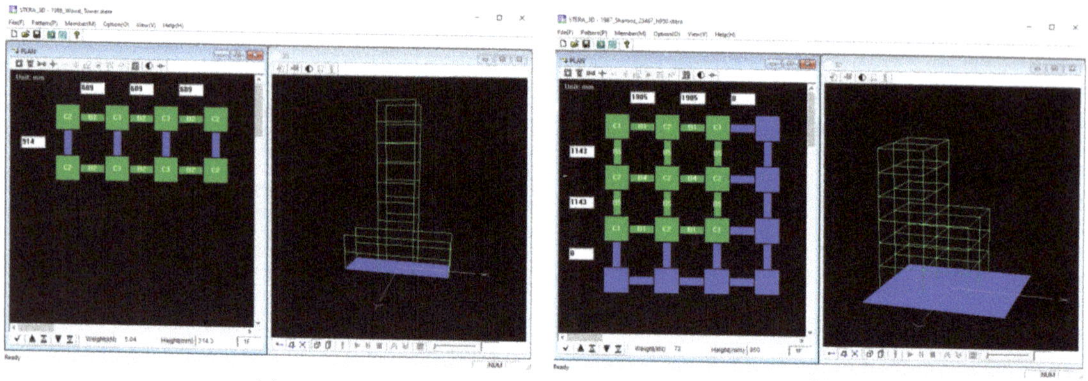

(a) (b)

Figure 3. STERA_3D interface on (**a**) a towered setback and (**b**) a stepped setback building models.

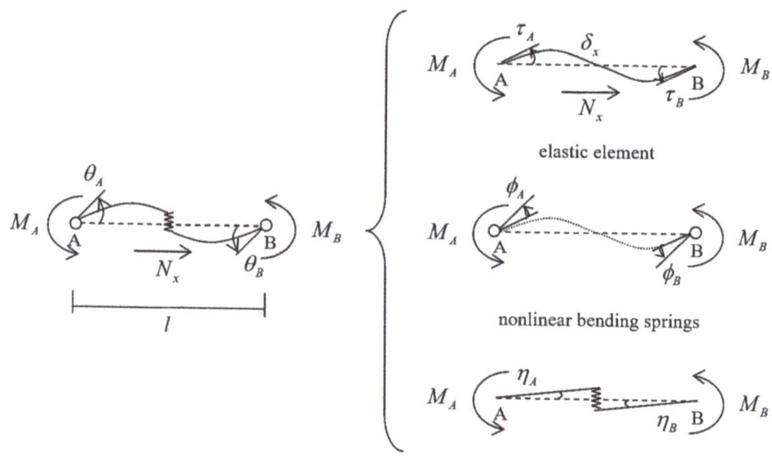

Figure 4. Elastic, nonlinear bending, and nonlinear shear springs for elements modeled by STERA_3D.

2.2. Simulation of Experimental Test of Towered Setback Structure

2.2.1. Specimen Description

The specimen tested by Wood [18] represents a scaled building with three bays and nine stories where the first story has a slightly higher elevation, and the setback location is in between level 2 and 3. The size of the experiment test frame model was 15 times smaller than the real building. The structure consists of two 2D beam-column frames, where the mass is attached to both frames using additional structures. For the first and second story, the weight is 5.04 kN, while the rest are 1.73 kN per level. The concrete compressive strength was 42.33 MPa, and the beam and column longitudinal rebar yielding strengths are 380.59 MPa and 388.17 MPa, respectively. The elevation of the specimen is presented in Figure 5.

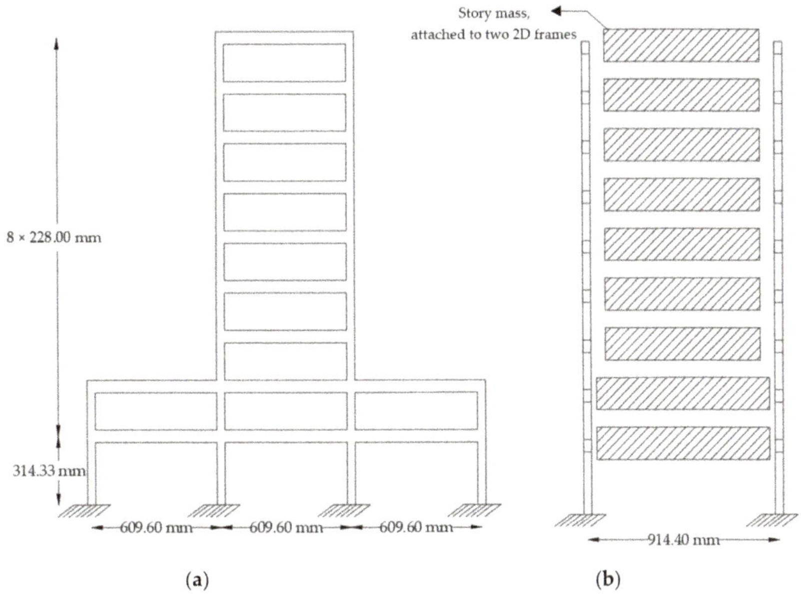

Figure 5. Experimental test specimen of towered setback structure: (**a**) front view, and (**b**) side view.

2.2.2. Input Data

For specimen of Wood [18], the input motion is only one direction parallel to the 2D frames. The input motions were a set of successive earthquake acceleration records, based on scaled 1940 El Centro NS with different magnitude. In this study, the first input of scaled 1940 El Centro NS with the peak acceleration of 382.8 cm/s^2 was used in the numerical study. The detail of input motion properties and the earthquake time history are presented in Table 2 and Figure 6.

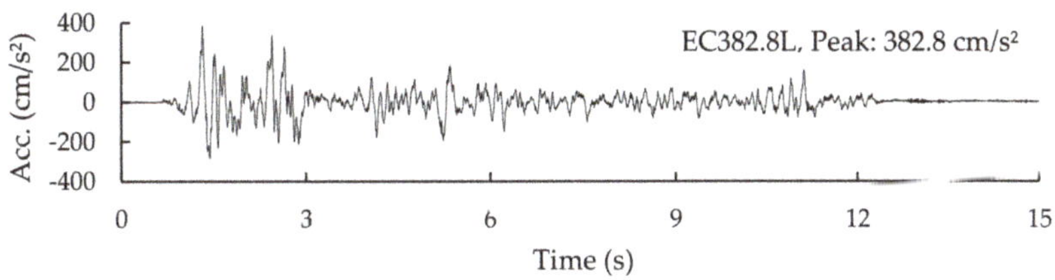

Figure 6. Time history acceleration of input motion.

Table 2. Detail of input motion properties.

No	Year	Event, Station	Component	Max. Acc. (cm/s^2)	Abbreviation	Input Direction
1	1940	scaled Imperial Valley, El Centro	NS	382.89	EC382.8L	Longitudinal

2.2.3. Results of Comparison

The maximum story drift and the maximum acceleration response of each story are shown in Figure 7a,b, and the time history of top floor displacement response is presented in Figure 8. Although there are some differences between analytical and experimental results, the trends of both displacement and acceleration responses are similar between simulation and experiment. The demonstrated results show that STERA_3D software has the sufficient accuracy to conduct numerical dynamic analyses of towered setback structures.

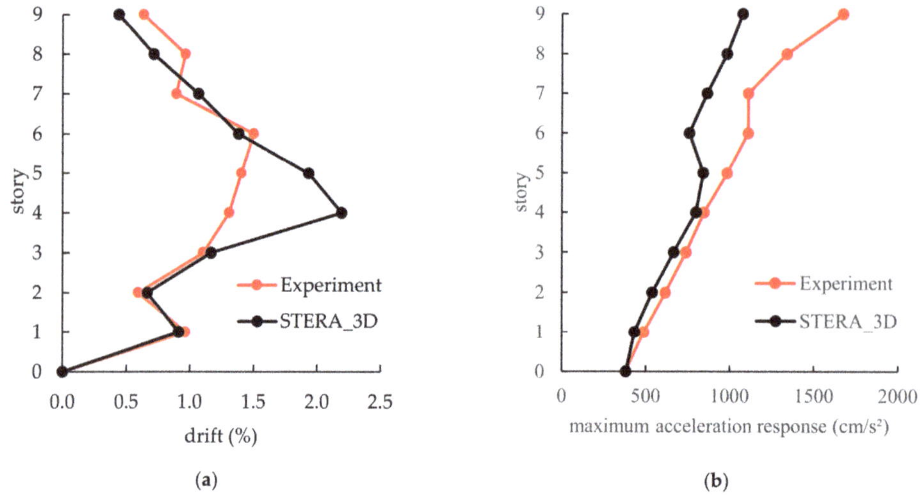

Figure 7. Comparison of experiment of towered setback structure with STERA_3D response: (**a**) interstory drift; (**b**) maximum acceleration.

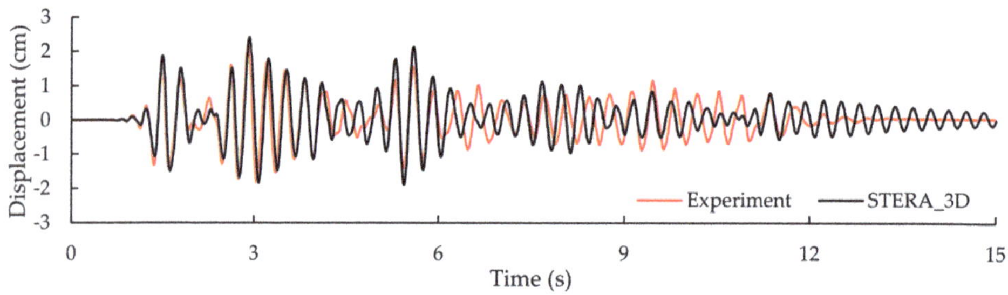

Figure 8. Roof displacement history responses of experiment of towered setback structure compared with STERA_3D.

2.3. Simulation of Experimental Test of Stepped Setback Structure

2.3.1. Specimen Description

The specimen tested by Shahrooz and Moehle [19] was designed for combined gravity and seismic effects according to the 1982 Uniform Building Code requirements to satisfy the seismic provisions of the American Concrete Institute Building Code (ACI 318-83) and the ACI-ASCE Committee 352 recommendations. The prototype is represented by the 1/4 scale model as depicted in Figure 9, in three dimensions with 2 2 bays of 1143 mm × 1905 mm floor panel size. Design concrete strength was 27.5 MPa, and all reinforcement was Grade 60 (minimum yield stress of 413 MPa). The weight of structures is 72 kN per level for first to third story and 41 kN per level for fourth to sixth story, and the inter story height is 914.4 mm.

2.3.2. Input Data

The specimen of Shahrooz and Moehle [19] was subjected to two input motions sequentially: unidirectional and bidirectional. The first stage consists of three different scaled 1940 El Centro NS ground motions, while the second phase includes three unidirectional horizontal motions inputted at an angle of 45 degrees to simulate bidirectional motions.

In this study, the numerical analysis only focuses on the longitudinal study and neglects the bidirectional simulation. The detail of input motion properties and the earthquake time histories are presented in Table 3 and Figure 10.

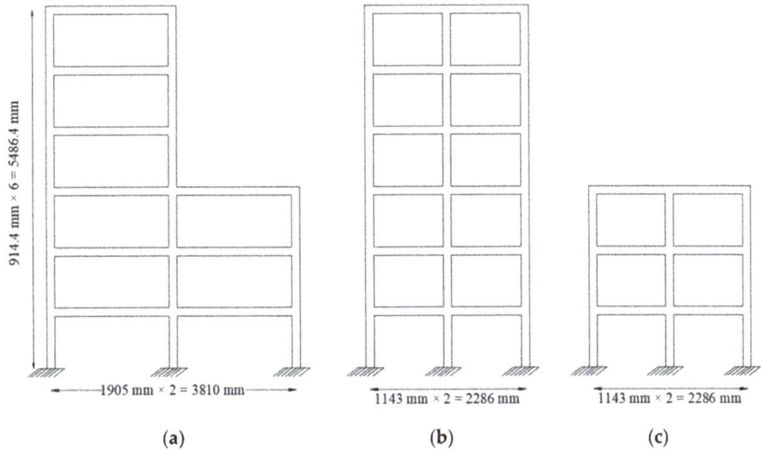

Figure 9. Elevation view of setback experimental test specimen: (**a**) front view, (**b**) side left view, and (**c**) side right view.

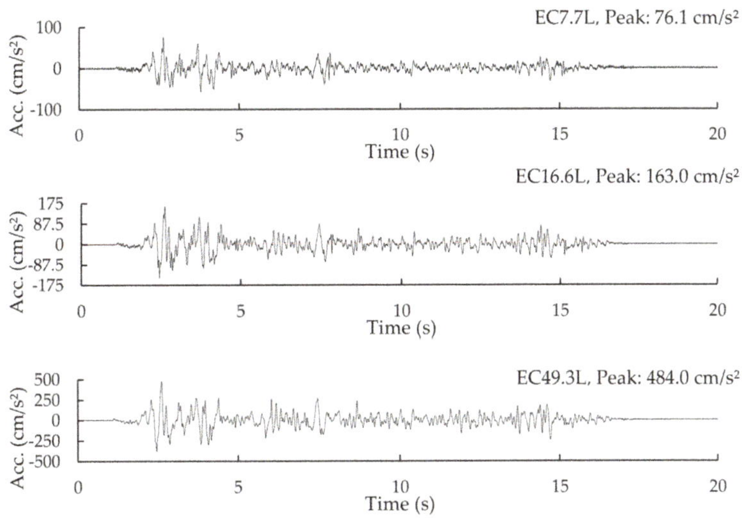

Figure 10. Time history acceleration of input motions.

Table 3. Detail of input motion properties.

No	Year	Event, Station	Component	Max. Acc. (cm/s^2)	Abbreviation	Input Direction
1	1940	scaled Imperial Valley, El Centro	NS	76.10	EC7.7L	Longitudinal
2				163.00	EC16.6L	
3				484.00	EC49.3L	

2.3.3. Results of Comparison

The relative maximum displacement response and the absolute maximum acceleration response of unidirectional experimental tests by Shahrooz and Moehle [19] and analytical tests are depicted in Figure 11a,b, and the time history of displacement response at the top floor against the last input motion is presented in Figure 12. The result shows that there are only slight differences between experiment and simulation in both acceleration and displacement response.

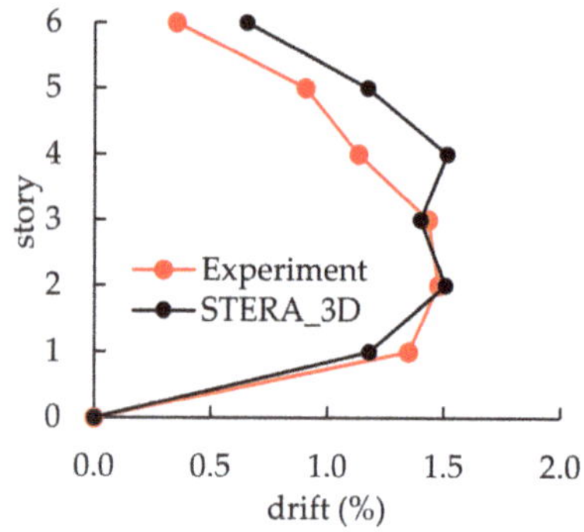

(a)

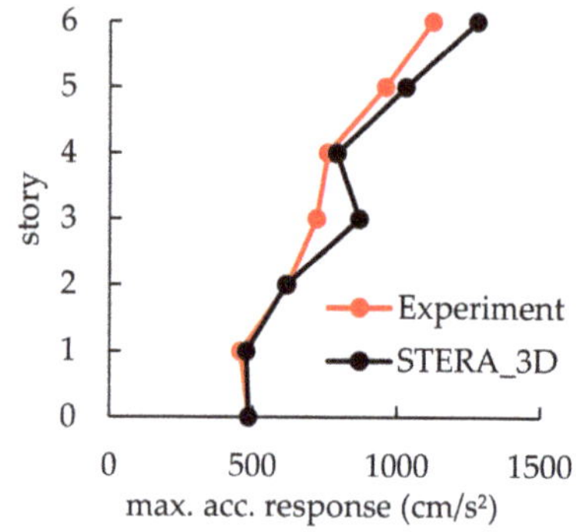

(b)

Figure 11. Comparison of experiment of stepped setback structure specimen with STERA_3D response: (**a**) interstory drift; (**b**) maximum acceleration.

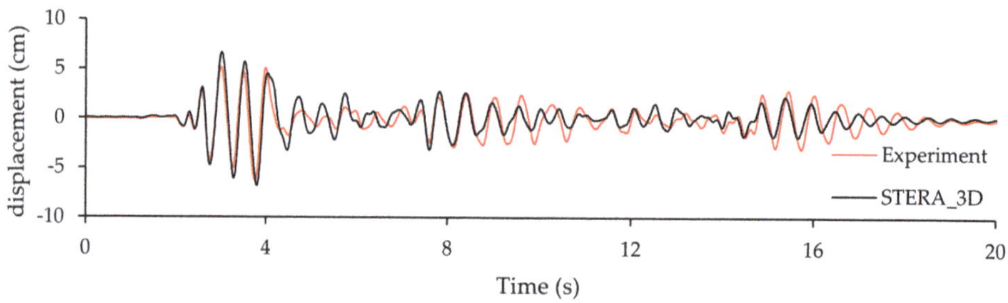

Figure 12. Roof displacement history responses of experiment of stepped setback structure compared with STERA_3D.

3. Parametric Study of Generated Setback Frames

To examine the relationship between the degree of setback and the distribution of damage, a parametric study was conducted using 35 reinforced concrete setback frames, consisting of 20 models with stepped setback type and 15 models with towered setback type with different degree of setback as depicted in Figures 13 and 14. All the frames have six stores with the number of bays ranging from two to five depending on the model. Each of generated models is represented with two numbers at the above frame illustration. The first number is the difference of the number of bays between top and base structures and the second number is the number of stories in the top structure.

The typical plan of the structure is shown in Figure 15. All the frames use one type of beam section, namely B1, and one type of column section, namely C1. The concrete compressive strength is 30 MPa, while the yield strength of both longitudinal and transversal rebars are 390 MPa. The slab thickness is 15 cm with reinforcement of D13 with the spacing of 10 cm. The section details of elements are presented in Table 4. The total gravity load implemented for every floor is 12 kN/m^2.

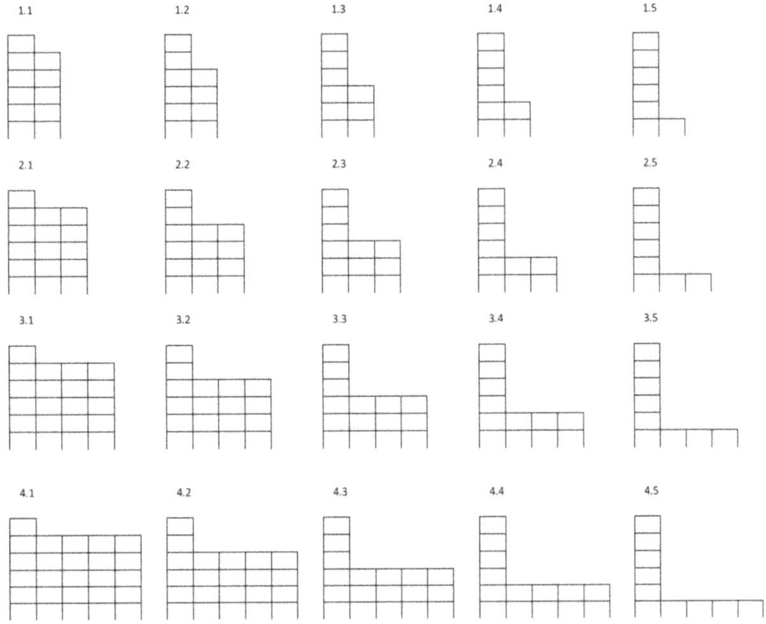

Figure 13. Generic 2D frame structures of stepped setback model.

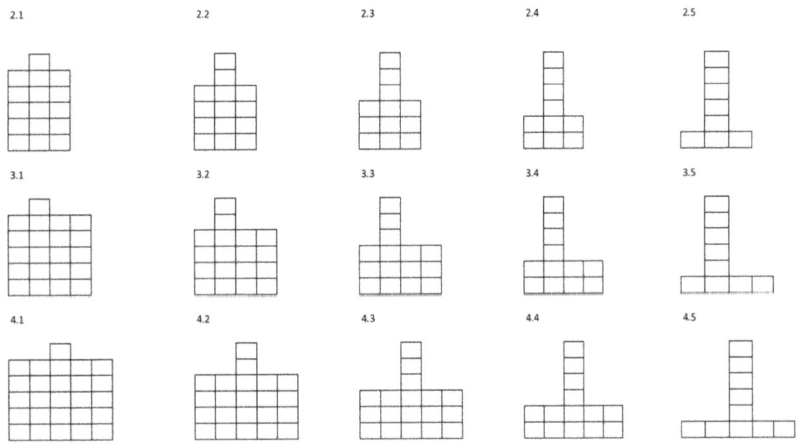

Figure 14. Generic frame structures of towered setback model.

Table 4. Section details of column and beam.

No	Element	Code	Story Level	Depth (mm)	Width (mm)	Longitudinal Rebars	Transversal Rebars
1	Column	C1	1–6	700	700	8 D-35	2 D-13 @ 10 cm
2	Beam	B1	1–6	600	300	3 D-35 on top and 3 D-35 on bottom	2 D-13 @ 15 cm

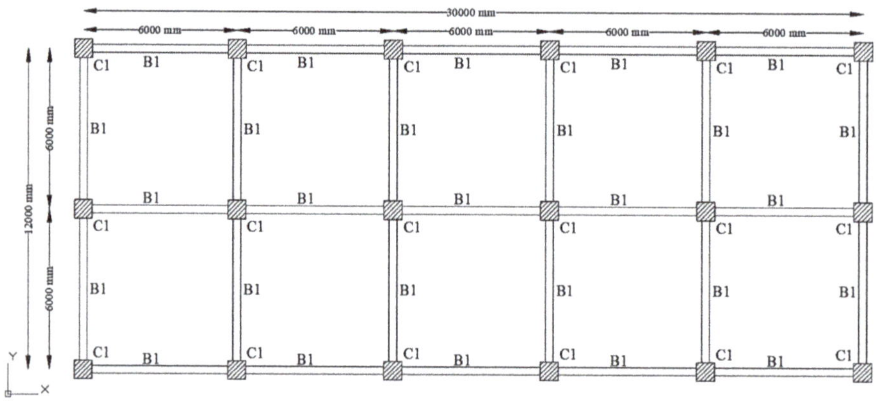

Figure 15. 2D plan of generated structure of 2 × 5 bays for 6-story building.

Nonlinear dynamic analyses were performed using STERA_3D, and the story damage indices were populated. The list of input earthquake ground motions is shown in Table 5. Ten input ground motions are recorded acceleration waves scaled to have the maximum velocity of 50 cm/s. The ground motion data are retrieved from PEER [26] and COSMOS Virtual Data Center [27]. The intensity of these input ground motions corresponds to the level of design earthquakes for the safety limit state in the Japanese seismic standard. The acceleration response spectrum is presented in Figure 16, and the wave shapes of the earthquake input motions are depicted in Figure 17.

Table 5. Selected earthquake ground motion.

No	Event	Year	Station	Component	Original Max. Acc. (cm/s^2)	Scaled Max. Acc. (cm/s^2)
1	Imperial Valley	1940	El Centro	NS	341.69	510.70
2	Kern County	1952	Taft	EW	152.69	496.50
3	Chi-chi	1999	CHY080	360 DEG	836.84	434.04
4	Northridge	1994	Arleta-Nordhoff Ave Fire Station	90 DEG	337.32	417.87
5	Loma Prieta	1989	Saratoga-Aloha Ave	0 DEG	494.45	597.96
6	Valparaiso, Chile	1985	Vina del Mar	200 DEG	355.50	578.31
7	Villita, Mexico	1985	Guerrero Array Stn VIL	N00W	125.37	318.36
8	Cape Mendocino	1992	Petrolia	0 DEG	578.14	598.43
9	Cape Mendocino	1992	Rio Dell-101/Painter St. Overpass	270 DEG	378.27	422.98
10	Kobe	1995	JMA	NS	817.80	449.80

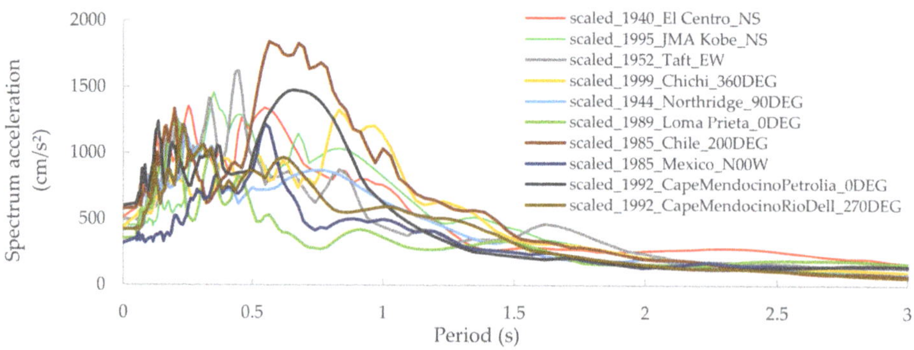

Figure 16. Acceleration response spectra (with 5% damping factor) of input earthquake motions.

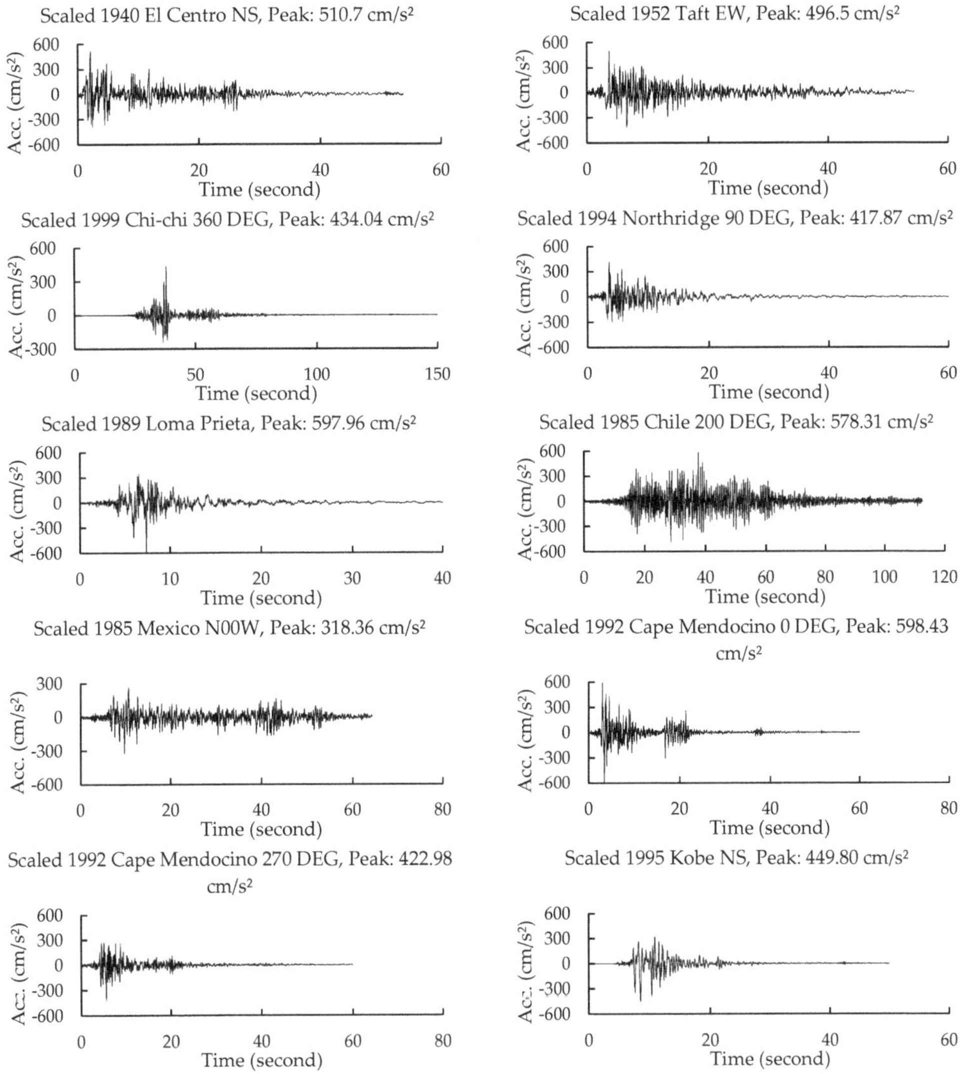

Figure 17. Time history acceleration waves of input earthquake motions.

4. Proposal of Tower-Base Damage Index Ratio

The nonlinear dynamic analysis using STERA_3D was conducted for all 35 frames with three input earthquake ground motions to calculate the damage index (*DI*) of each story. The damage index of each story is calculated using Equation (5). The following DI_{ratio} is proposed to examine the distribution of the *DI* along the height of the frame.

$$DI_{ratio} = \frac{\max(DI_{tower})}{\max(DI_{base})} \tag{12}$$

where the DI_{tower} and DI_{base} are taken as the average for the ten input ground motions. If the value of DI_{ratio} is larger than one, the damage will be concentrated at the tower structure. Conversely, if the value is less than one, the base structure will suffer more damage than the tower.

The mathematical model is developed for predicting DI_{ratio} from irregularity indices φ_s and φ_b using the nonlinear regression analysis so that the damage distribution is easily evaluated without conducting the nonlinear dynamic analysis. The relation between the calculated DI_{ratio} of all the frames and irregularity indices φ_s and φ_b are shown in Figure 18 for the stepped model and Figure 19 for the towered model type, respectively.

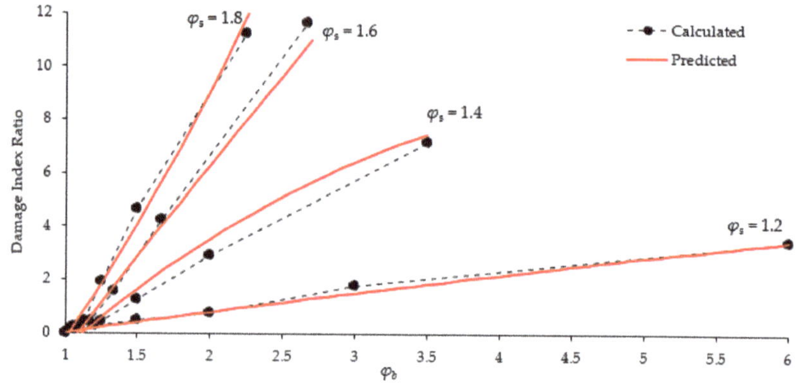

Figure 18. DI_{ratio} with different φ_s and φ_b for stepped model.

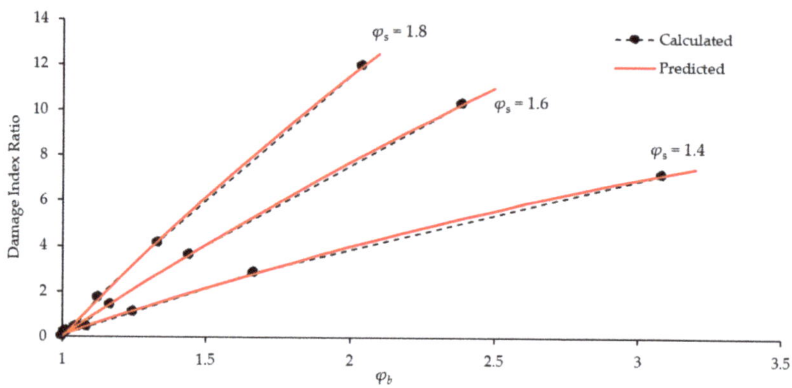

Figure 19. DI_{ratio} with different φ_s and φ_b for towered model.

Values of φ_s and φ_b equal to 1 correspond to a structure with a uniform configuration. As values of φ_s and φ_b increase, the degree of the setback in the frame structure becomes larger. A large value of DI_{ratio} indicates a relatively large damage concentration in the tower. The nonlinear regression analysis is performed to derive the quadratic polynomial equation for estimating the damage index ratios from irregularity indices, φ_s and φ_b. The obtained equations are shown in Equation (13) for stepped type and Equation (14) for towered type. The red lines in Figures 18 and 19 indicate the prediction of the damage index ratio using Equations (13) and (14). The values of R-squared of red lines, which indicate the goodness-of-fit measure for the regression model, vary between 0.865 and 0.999.

$$DI_{r,stepped} = 98.19 - 124.29\varphi_s - 105.06\varphi_b + 142.75\varphi_s\varphi_b + 38.34{\varphi_s}^2 + 25.47{\varphi_b}^2 - 37.09\varphi_s{\varphi_b}^2 - 45.44{\varphi_s}^2\varphi_b + 13.22{\varphi_s}^2{\varphi_b}^2 \quad (13)$$

$$DI_{r,towered} = -3.09 + 16.89\varphi_s + 17.22\varphi_b - 35.99\varphi_s\varphi_b - 12.88{\varphi_s}^2 - 12.53{\varphi_b}^2 + 17.34\varphi_s{\varphi_b}^2 + 19.59{\varphi_s}^2\varphi_b - 6.22{\varphi_s}^2{\varphi_b}^2 \quad (14)$$

5. Validation of Proposed Damage Index Ratio

To validate the proposed method to calculate DI_{ratio}, the towered setback frame by Wood [18] and the stepped setback frame tested by Shahrooz and Moehle [19] are examined. The input wave is the 1940 El Centro NS record scaled to have the maximum velocity of 50 cm/s; however, the time step is adapted for each specimen. The specimen of Wood [18] is subjected to the earthquake with the time step divided by 4 by rounding up the value of $\sqrt{15}$, while the specimen of Shahrooz and Moehle [19] is subjected to the earthquake with the time step divided by 2, by considering the scale factor of test specimen. Dynamic time history analysis was performed using STERA_3D to calculate the *DI* of each story. The calculated damage index ratios were compared with the predicted damage index ratios from the proposed equations and presented in Table 6.

Table 6. Numerical analysis result compared to prediction result using Regression Equation.

Frame	φ_b	φ_s	Natural Period T (s)	Calculated DI_{ratio}	Predicted DI_{ratio}	Accuracy (%)
Wood [18]	2.75	1.25	0.118	1.2691	1.3552	93.22
Shahrooz and Moehle [19]	2.00	1.20	0.248	1.1641	0.7766	66.71

The accuracy of the prediction from Equation (13) is adequate, 66% for the stepped setback and the prediction accuracy from Equation (14) is relatively high, 93% for the towered setback.

6. Conclusions

To understand better the earthquake response and damage concentration of setback buildings, numerical investigations were conducted and a practical method to evaluate the damage distribution was proposed.

First, two previous experimental tests of reinforced concrete buildings comprising a towered setback and a stepped setback were modeled and analyzed by STERA_3D and the test results were successfully simulated.

Then, to study the building damage distribution, 2D reinforced concrete frame models containing 20 stepped and 15 towered setback frames were generated to inspect the relationship between the damage index ratio and the irregularity indices under safety level earthquake input ground motions. Based on the results, two nonlinear regression equations were proposed as an alternative to the dynamic analysis procedures.

Finally, the proposed equations were applied to the previous experimental tests, and it was demonstrated that the proposed formulas could predict the damage to vertically irregular reinforced concrete moment-resisting frame with reasonable accuracy. Although the prediction formula works, it is important to be noted that the setback buildings considered in this study are limited for low- to mid-rise reinforced concrete frame buildings with uniaxial seismic loading.

Author Contributions: Conceptualization, T.S., B.E. and T.I.M.; methodology, B.E., T.I.M.; software, T.S.; writing—original draft preparation, T.I.M.; writing—review and editing, T.S.; supervision, T.S. All authors have read and agreed to the published version of the manuscript.

Funding: This research received no external funding.

Institutional Review Board Statement: Not applicable.

Informed Consent Statement: Not applicable.

Data Availability Statement: The data presented in this study are available on request from the corresponding author.

Acknowledgments: The first author appreciates AUN/SEED-Net JICA for the support through scholarship of doctoral degree.

Conflicts of Interest: The authors declare no conflict of interest.

References

1. Soni, D.P.; Mistry, B.B. Qualitative review of seismic response of vertically irregular building frames. *ISET J. Earthq. Technol.* **2006**, *43*, 121–132.
2. Mwafy, A.; Khalifa, S. Effect of vertical structural irregularity on seismic design of tall buildings. *Struct. Des. Tall. Spec. Build.* **2017**, *26*, 1–22. [CrossRef]
3. Shahrooz, B.M.; Moehle, J.P. Seismic response and design of setback buildings. *J. Struct. Eng.* **1990**, *116*, 1423–1439. [CrossRef]
4. Duan, X.N.; Chandler, A.M. Seismic torsional response and design procedures for a class of setback frame buildings. *Earthq. Eng. Struct. Dyn.* **1995**, *24*, 761–777. [CrossRef]
5. Dutta, S.C.; Das, P.K.; Sengupta, P. Seismic behaviour of irregular structures. *Struct. Eng. Int.* **2017**, *27*, 526–545. [CrossRef]
6. Shakib, H.; Pirizadeh, M. Probabilistic seismic performance assessment of setback buildings under bidirectional excitation. *J. Struct. Eng.* **2014**, *140*, 04013061. [CrossRef]
7. Lin, J.-L.; Tsaur, C.-C.; Tsai, K.-C. Two-degree-of-freedom modal response history analysis of buildings with specific vertical irregularities. *Eng. Struct.* **2019**, *184*, 505–523. [CrossRef]
8. Georgoussis, G.K.; Mamou, A. Mitigating mass eccentricity effects on the rotational response of setbacks structures: An analytical solution for linear systems. *Structures* **2020**, *28*, 1539–1556. [CrossRef]
9. Georgoussis, G.K. An approach for minimum rotational response of medium-rise asymmetric structures under seismic excitations. *Adv. Struct. Eng.* **2016**, *19*, 420–436. [CrossRef]
10. Habibi, A.; Asadi, K. Development of drift-based damage index for reinforced concrete moment resisting frames with setback. *Int. J. Civ. Eng.* **2017**, *15*, 487–498. [CrossRef]
11. Park, Y.-J.; Ang, A.H.-S. Mechanistic seismic damage model for reinforced concrete. *J. Struct. Eng.* **1985**, *111*, 722–739. [CrossRef]
12. Park, Y.-J.; Ang, A.H.-S.; Wen, Y.K. Seismic damage analysis of reinforced concrete buildings. *J. Struct. Eng.* **1985**, *111*, 740–757. [CrossRef]
13. Karavasilis, T.L.; Bazeos, N.; Beskos, D.E. Seismic response of plane steel MRF with setbacks: Estimation of inelastic deformation demands. *J. Constr. Steel Res.* **2008**, *64*, 644–654. [CrossRef]
14. Varadharajan, S.; Sehgal, V.K.; Saini, B. Determination of inelastic seismic demands of RC moment resisting setback frames. *Arch. Civ. Mech. Eng.* **2013**, *13*, 370–393. [CrossRef]
15. Varadharajan, S.; Sehgal, V.K.; Saini, B. Seismic behavior of multistory RC building frames with vertical setback irregularity. *Struct. Des. Tall. Spec. Build.* **2014**, *23*, 1345–1380. [CrossRef]
16. Hait, P.; Sil, A.; Choudhury, S. Damage assessment of low to mid rise reinforced concrete buildings considering planner irregularities. Int. J. Comput. *Methods Eng. Sci. Mech.* **2020**, *22*, 150–168.
17. Hait, P.; Sil, A.; Choudhury, S. Seismic damage assessment and prediction using artificial neural network of RC building considering irregularities. *Int. J. Struct. Integr. Maint.* **2020**, *5*, 51–69. [CrossRef]
18. Wood, S.L. Experiments to Study the Earthquake Response of Reinforced Concrete Frames with Setbacks. Ph.D. Thesis, University of Illinois at Urbana-Champaign, Champaign, IL, USA, 1985.
19. Shahrooz, B.M.; Moehle, J.P. *Experimental Study of Seismic Response of RC Setback Buildings*; National Science Foundation Report; UCB: Berkeley, CA, USA, 1987.
20. Belkacem, M.A.; Bechtoula, H.; Bourahla, N.; Belkacem, A.A. Damage index for reinforced concrete columns. *Građevinar* **2020**, *72*, 139–149.
21. Mazzolani, F.; Piluso, V. *Theory and Design of Seismic Resistant Steel Frames*; E & FN Spon: London, UK, 1996.
22. Shahrooz, B.M.; Moehle, J.P. Experimental Study of Seismic Response of Reinforced Concrete Setback Buildings. Available online: https://datacenterhub.org/deedsdv/publications/view/293 (accessed on 1 October 2020).
23. Wood, S.L.; Sozen, M. Experiments to Study the Earthquake Response of Reinforced Concrete Frames with Setbacks (NEES-2011-1068). Available online: https://datacenterhub.org/deedsdv/publications/view/285 (accessed on 1 October 2020).
24. Saito, T. Structural Earthquake Response Analysis, STERA_3D Version 10.8. Available online: http://www.rc.ace.tut.ac.jp/saito/software-e.html (accessed on 1 October 2020).
25. Newmark, N.M. A method for computation of structural dynamics. *J. Eng. Mech. Div.* **1959**, *85*, 67–94. [CrossRef]
26. PEER (Pacific Earthquake Engineering Center). Ground Motion Database. Available online: https://peer.berkeley.edu/ (accessed on 4 July 2021).
27. COSMOS (The Consortium of Organizations for Strong-Motion Observation Systems). Virtual Data Center. Available online: www.strongmotioncenter.org (accessed on 4 July 2021).

applied sciences

Article

Influence of Directionality on the Seismic Response of Typical RC Buildings

Ignacio Bugueño [1], Jorge Carvallo [2] and Juan Carlos Vielma [2,*]

[1] Carvallo Carvallo Ltda, Viña del Mar 2520000, Chile; ibugueno@carvalloi.cl
[2] Civil Engineering School, Pontificia Universidad Católica de Valparaíso, Valparaíso 2340000, Chile; jorge.carvallo@pucv.cl
* Correspondence: juan.vielma@pucv.cl; Tel.: +56-332273664

Abstract: Current seismic analysis contemplates the simultaneous use of the orthogonal components of an earthquake in order to determine the structural stresses closer to reality. This has led to these components being combined considering a fraction of them, as applying them completely would lead to excessively conservative results. However, their application is carried out considering that the direction of the components coincides with the orientation of the orthogonal axes that define the resistant structure. The assumption takes on special importance when it comes to establishing performance demands on a structure based on nonlinear time-history analysis. To establish the proportional relationship between the seismic components, the angle of incidence is used, which is one of the imponderable variables of an earthquake. In this investigation, a group of reinforced concrete structural archetypes with various typologies and regularity in plan is presented, which allow the effect of the angle of incidence in determining the maximum displacement demands to be studied. To study the response, a set of strong earthquakes recorded in Chile is used, obtaining the angle of incidence that produces the maximum displacement demands through interstory drift and roof displacement. A statistical analysis is also carried out in which the influence of the angle of incidence that produces the maximum response is studied.

Keywords: structural irregularity; incidence angle; nonlinear time-history analysis; maximum displacement; maximum interstory drift

Citation: Bugueño, I.; Carvallo, J.; Vielma, J.C. Influence of Directionality on the Seismic Response of Typical RC Buildings. *Appl. Sci.* **2022**, *12*, 1534. https://doi.org/10.3390/app12031534

Academic Editor: Maria Favvata

Received: 23 December 2021
Accepted: 28 January 2022
Published: 31 January 2022

Publisher's Note: MDPI stays neutral with regard to jurisdictional claims in published maps and institutional affiliations.

Copyright: © 2022 by the authors. Licensee MDPI, Basel, Switzerland. This article is an open access article distributed under the terms and conditions of the Creative Commons Attribution (CC BY) license (https://creativecommons.org/licenses/by/4.0/).

1. Introduction

At present, the seismic-resistant design of structures is based on the results of the seismic analysis carried out considering two components of ground acceleration, which are transformed into seismic forces that act in two directions. The problem lies in determining the proportions in which the application of the accelerations, and thus the design of seismic forces, must be carried out, as the combination of the proportions can lead to results that are either on the conservative side or on the low side. In particular, the angle of incidence considered when performing the seismic analysis becomes important when determining the demands to which the structure will be subjected (MacRae and Mattheis [1]). For example, Athanatopoulou et al. [2] concluded that the estimation of interstory drift due to bi-directional earthquakes is dependent on the chosen reference axes, which is especially important in irregular structures. On the other hand, Rigato and Medina [3] obtained variations of up to 80% between the maximum inelastic deformations calculated with angles of incidence different from the usual ones (0° and 90°). On the other hand, underestimation in engineering demand parameters (interstory drifts and displacements) can be reduced by up to 20% by performing nonlinear time-history analysis with a large number of records in normal directions and parallel to the fault (Reyes and Kalkan [4]). However, these types of solutions may not be practical in Chilean territory given the extension of the active seismic zone. Regarding the errors observed, according to Kashkooli and Banan [5], mean maximum errors have been reported for story displacement and interstory drift of up to 22% and 42%, respectively, or errors of up to 37% in roof displacements and rotations in plastic hinges (Magliulo et al. [6]). Finally, Prajwal et al. [7] reported variations in displacement demands of up to 90%.

The direction in which the earthquake interacts with the structure is arbitrary, so the structural axes do not necessarily correspond to the main axes of movement, which would present the most unfavorable situation. To study this characteristic, Rigato and Medina [3] obtained the response of asymmetric and symmetric structures by varying the degrees of inelasticity, vibration periods, and incidence angles used, observing that the response varies depending on the fundamental period, structuring, and degree of inelasticity. In the analyses, two archetypes were used, varying the angle of incidence in increments of 5° from 0° to 180°. According to the authors, the ratio between the maximum inelastic deformation for a given angle and the maximum obtained for an angle of 0° and 90° tends to increase along with the fundamental period, averaging between 1.1 and 1.6 for torsionally balanced and unbalanced models. It was concluded that the structural axes are not necessarily the most unfavorable and that the critical incidence angle varies depending on the intensity of the record. On the other hand, Lagaros [8] studied two different structures, using three records and incidence angles in increments of 5° varying from 0° to 360°. These records were used as a basis to scale an acceleration spectrum considering the first period of the structures, damping of 5%, and three levels of intensity. The response of the structures varied according to the records, intensity level, and incidence angle.

In order to study the effect of the directionality of earthquakes on structures, nonlinear response analysis has frequently been used. Cantagallo et al. [9], studied the influence of directionality on seismic response variation, including in the study the variation of the angle of incidence between 0° and 180° in increments of 22.5°, reaching the conclusion that, for the analyzed structures, the main axes underestimated the demand, especially in plan irregular structures, while in structures with double symmetry, no considerable variations were obtained. Additionally, it was observed that the effects of directionality are lower in scaled seismic records. It should be noted that only four structures were considered in this study, so these results are limited.

Fontara et al. [10] studied the influence of the orientation of the seismic action and its intensity on single-story asymmetric reinforced concrete structures. The sensitivity of the structure to the angle of incidence varied depending on the degree of incursion in a nonlinear range; as the level of damage increases, the variability in the results as a function of the angle of incidence increases as well. The results indicated that the most unfavorable case was not obtained in the structural axes; that scaling the records using the same scale factor allows for similar levels of damage regardless of whether they are recorded, correlated, or not correlated; and that, if different scale factors are used, different levels of damage are obtained at the critical angle.

The influence of the angle of incidence of an earthquake has mainly been determined using three-dimensional nonlinear time-history analysis on reinforced concrete frame structures of between one and four stories, not necessarily applying both records simultaneously. This is found in several works [3,8,10–12], among others, with increments of incidence angles ranging from 1° to 22.5°. The highest structure analyzed in order to study the angle of incidence had nine stories (Kalkan and Reyes [12]). With this, it was determined that, when performing a nonlinear time-history analysis, the records should be applied in the maximum direction, this being the direction in which there is the greatest linear response of a mass with degrees of freedom in both horizontal directions: the normal direction and that which is parallel to the fault, at sites less than 15 km away from the fault. The recommendation given is based on the low attenuation observed in propagating waves within this range. Although the previous recommendations would not represent the worst case, they would deliver higher demand parameters than usual. Recently, dynamic incremental analysis has gained special importance in evaluating the response of various structural types [13–15] or even in determining their seismic design factors [16], but this type of analysis is usually computationally expensive in all its phases.

Many authors have tried to estimate, generally, the least favorable angle of incidence or the response given by it, and to express this through formulas. An example of this has been the procedure applied to reinforced concrete structures (Wilson and Button [17]). Although this procedure is practical in design, as it is based on modal response spectrum analysis, it does not take into account the components of the earthquake when they act on the main structural axis. Meanwhile, Smerby and Der Kiureghian [18] have indicated that the previous method had certain limitations. The problem has continued to be studied by González [19], even applying linear analysis, obtaining relative errors of up to 30%. A formula for determining the critical incidence angle has been proposed by Wilson et al. [20] based on linear response spectrum analysis or modal response spectrum analysis to determine the critical angle, using five load cases and three response spectra (López and Torres [21]).

Response spectra are commonly used in the definition of seismic action, so it is advisable to study their application by explicitly using the definition of directionality (Anastassiadis and Panetsos [22]). Regarding the results, the different types of analysis used tend to significantly affect their deviation from those obtained through conventional analysis. For example, a bridge with curvature in plan studied by Gao et al. [23] using nonlinear time-history analysis, combined with seismic records with incidence angles varying between 0° and 180° in 6° increments, presented variations in force and moments of up to 45% with respect to the conventional analysis. Similar studies [24–26] that have addressed the use of different types of analysis present different results regarding the influence of the angle of incidence on the maximum responses of the structures.

Next, some of the cited works are presented in greater detail, which will allow a better understanding of the different approaches that have been applied to the problem.

1.1. Wilson and Button

They consider two response spectra applied at angles of 0° and 90°, evaluate the forces obtained using the CQC combination [17], and subsequently determine the critical angle of incidence by means of the following equation:

$$tan\ (2\theta_{cr}) = 2f_0 f_{90}\frac{1+\alpha^2}{(f_0{}^2 - f_{90}{}^2)(1-\alpha^2)} \tag{1}$$

where α is the factor that multiplies the response spectrum and f_0 and f_{90} are the responses obtained for an angle of incidence of 0° and 90°, respectively.

1.2. González

In this work, three structures were analyzed with the proposed methodology that consists of calculating the maximum response in each vibration mode using the angle of incidence of the earthquake as a parameter and subsequently combining the results with any modal combination criterion [19]. The displacement and stress results were compared using nonlinear time-history analysis and modal response spectrum analysis as a basis. It was concluded that the SRSS combination presented the best results in conjunction with the proposed method.

1.3. López and Torres

The authors propose addressing the problem of directionality through modal response spectrum analysis, solving five load cases [21].

The proposed method consists of the following:

- Solving the five spectral modal load cases to obtain the response parameters.
- Using Equation (3) to obtain the critical angle of incidence.

- Using Equation (2) to obtain the maximum responses for each critical angle of incidence.

It should be noted that each parameter can have a different critical angle, where R is the peak response of the chosen parameter, C is the participation coefficient, and θ_{crit} is the critical incidence angle, calculated using the following equations:

$$R(\theta) = \left\{ \left[\left(R^{1x}\right)^2 + \left(R^{2y}\right)^2 \right] cos^2(\theta) + \left[\left(R^{1y}\right)^2 + \left(R^{2x}\right)^2 \right] sin^2(\theta) + 2sin\ (\theta) \right.$$
$$\left. cos(\theta) \left[\sum_i \sum_j C_{ij} R_i^{1x} R_j^{1y} - \sum_i \sum_j C_{ij} R_i^{2y} R_j^{2x} \right] + \left(R^3\right)^2 \right\}^{1/2} \quad (2)$$

$$\theta_{crit} = \frac{1}{2} tan^{-1} \left(\frac{2 \sum_i \sum_j C_{ij} \left[R_i^{2y} R_j^{2x} - R_i^{1x} R_j^{1y} \right]}{\left(R^{1y}\right)^2 + \left(R^{2y}\right)^2 - \left(R^{1x}\right)^2 - \left(R^{2y}\right)^2} \right) \quad (3)$$

In Equations (2) and (3), R_j^{2x} and R_i^{2y} are the peak modal responses obtained along the structural reference axes x and y, respectively. Finally, R^3 is the peak modal response in the vertical axis z.

1.4. Menun and Der Kiureghian

The authors propose the use of the CQC3 combination as an alternative to the multi-component combination (instead of the use of SRSS, CQC, and rules of 30% or 40%) [27]. Additionally, a formula is proposed for the calculation of the critical angle, where k and l represent the axes and the seismic components considered, α is the coefficient of participation related between the modes and the indicated direction, ρ corresponds to the correlation coefficient between modes, and γ is the ratio between the design spectra used. Finally, S_{ki} correspond to the kth principal axis of the ground motion. The equations are presented below:

$$R_k^2 = \sum_i \sum_j \rho_{ij} \alpha_{kj} \alpha_{ki} S_{ki} S_{kj} \quad (4)$$

$$R_{kl}^2 = \sum_i \sum_j \rho_{ij} \alpha_{lj} \alpha_{ki} S_{ki} S_{lj} \quad (5)$$

$$\theta_{cr} = \frac{1}{2} tan^{-1} \left[\frac{\frac{2}{\gamma} R_{12}}{R_1^2 - \frac{1}{\gamma^2} R_2^2} \right] \quad (6)$$

1.5. Anastassiadis and Panetsos

This method is based on previously published considerations presented by Penzien and Watanabe [24], which are used by means of the response spectra when considering the existence of a major axis in the direction of the epicenter, intermediate transversal, and minor vertical, this being valid only when the effects of proximity to the fault are absent. The proposed seismic design is based on determining the critical orientation that provides the greatest response, on calculating maximum and minimum response values and on the application of the extreme stress or extreme force method to determine the most unfavorable combination of the resulting stresses acting on specific structural elements [22]. To determine the critical direction, the reference systems indicated in Figure 1 and the following equations should be used:

$$\theta_{cr} = \begin{cases} \theta_0 = \frac{1}{2} tan^{-1} \left(2 \frac{R_{xy}}{R_x{}^2 - R_y{}^2} \right) \\ \theta_0 + \pi/2 \end{cases} \quad (7)$$

$$\max R^2 = \frac{\left(R_x{}^2 + R_y{}^2\right)}{2} + \left(\left(\left(R_x{}^2 + R_y{}^2\right)/2 \right)^2 + R_{xy}{}^2 \right)^{0.5} \quad (8)$$

$$\min R^2 = \frac{(R_x{}^2 + R_y{}^2)}{2} - \left(\left(\left(R_x{}^2 + R_y{}^2\right)/2\right)^2 + R_{xy}{}^2\right)^{0.5} \quad (9)$$

$$R_x^2 = R_{xa}{}^2 + R_{yb}{}^2 = \sum_i \sum_j \varepsilon_{ij}\left(R_{i,xa}R_{j,xa} + R_{i,yb}R_{j,yb}\right) \quad (10)$$

$$R_y^2 = R_{xb}{}^2 + R_{ya}{}^2 = \sum_i \sum_j \varepsilon_{ij}\left(R_{i,xb}R_{j,xb} + R_{i,ya}R_{j,ya}\right) \quad (11)$$

$$R_{xy} = R_{xy,a} - R_{xy,b} = \sum_i \sum_j \varepsilon_{ij}\left(R_{i,xa}R_{j,ya} - R_{i,xb}R_{j,yb}\right) \quad (12)$$

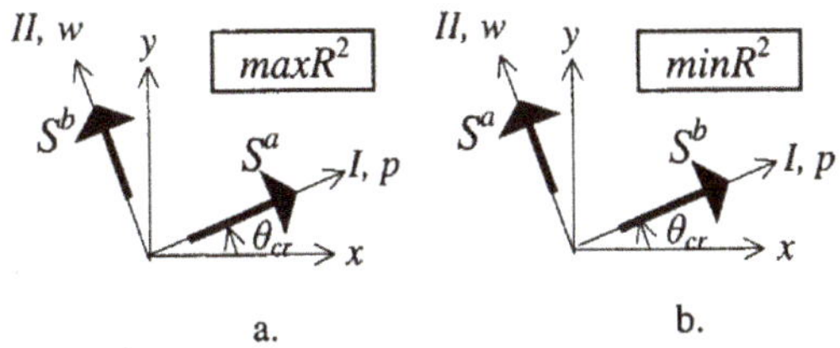

Figure 1. Reference system; (**a**) maximum reactions; (**b**) minimum reactions [22].

In Equations (10)–(12), ε_{ij} represents the correlation coefficients between the modes i and j. A graphical representation of the reactions calculated using Mohr's circle is shown in Figure 2.

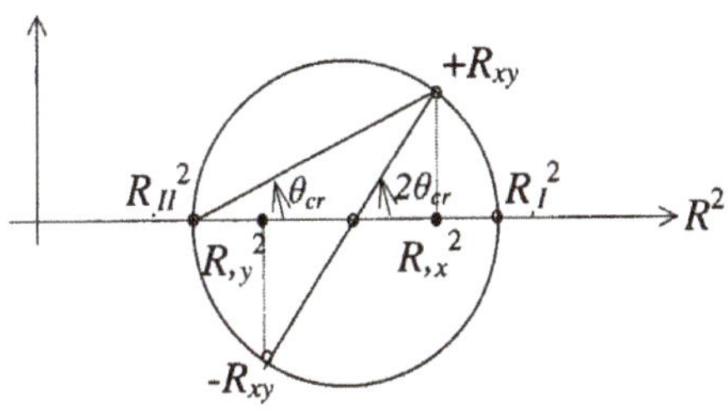

Figure 2. Representation using the circle of Mohr [22].

1.6. Dongsheng et al.

This method corresponds to a simplified version of the CQC3 method. The similar method SRSS3 was introduced for the first time by Gao et al. [23], who postulate that this method presents considerable errors if there is a high correlation between the vibration modes of the structure, otherwise it gives similar results to the CQC3 method. The equations presented below takes into account the proportionality between the design spectra in each direction of analysis [28].

$$R^{1x} = \sum_i R_i{}^{1x} R_i{}^{1x} \quad (13)$$

$$R^{1y} = \sum_i R_i{}^{1y} R_i{}^{1y} \quad (14)$$

$$\theta = \frac{1}{2} tan^{-1}\left(\frac{2\sum_i R_i^{1x} R_i^{1y}}{(R^{1x})^2 - (R^{1y})^2}\right. \quad (15)$$

$$R(\theta) = \left(\left[\left(R^{1x}\right)^2 + \left(\gamma R^{1y}\right)^2 \right] cos^2(\theta) + \left[\left(R^{1y}\right)^2 + \left(\gamma R^{1x}\right)^2 \right] sin^2(\theta) + 2sin\ (\theta) \right.$$
$$\left. cos\ (\theta)\ \left(1 - \gamma^2\right) \left[\sum_i R_i^{1x} R_i^{1y} \right] + R^{32} \right)^{1/2} \tag{16}$$

1.7. Athanapoulou et al.

The authors presented an analytical formula to determine the critical incidence angle for structures subjected to three correlated seismic components. This implies solving three time-history cases (one for each component) without the use of the model of Penzien and Watanabe [24] and assuming linear behavior, as response spectra cannot be used because correlation of the records is assumed. The object of the study was an irregular structure with five stories [26]. Figures 3 and 4 show the reference axes considered and a graphic representation of the reactions indicated.

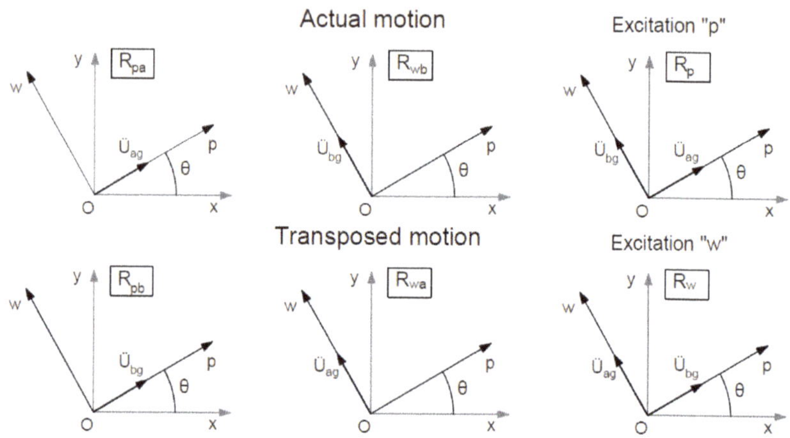

Figure 3. Displacements with respect to reference axis (p, w) [26].

The equations necessary for the calculation of the critical angle are presented below, where θ_{cr1} corresponds to the angle for which the maximum response is obtained and θ_{cr2} to the minimum response of the envelope R_O, both taking into account the signs.

$$R_O(t_{cr})^2 = R_x(t_{cr})^2 + R_y(t_{cr})^2 \tag{17}$$

$$\theta_{cr1} = tan^{-1}\left(\frac{R_y(t_{cr1})}{R_x(t_{cr1})}\right) \tag{18}$$

$$\theta_{cr2} = tan^{-1}\left(\frac{R_y(t_{cr2})}{R_x(t_{cr2})}\right) - \pi \tag{19}$$

where t_{cr1} and t_{cr2} are the critical times when the maximum positive and minimum negative displacements occur.

1.8. Lagaros

Multidirectional incremental dynamic analysis (MIDA), using pairs of records and incident angles generated by Latin hypercube sampling (LHS), has been proposed to take into account the variability of seismic excitation and the angle of incidence, focused on the implementation of incremental dynamic analysis for performance-based design [29]. One symmetric and one asymmetric structure (both of three stories) were analyzed, considering fifteen records and an incidence angle between 0° and 180°, uniformly distributed

for the generation of the sample using LHS. Subsequently, the MIDA and its respective representative MIDA curve were performed.

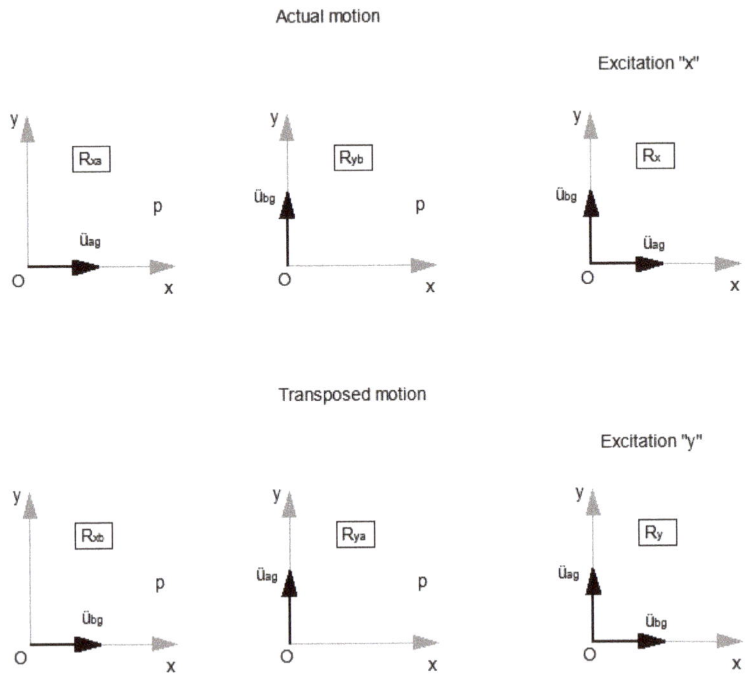

Figure 4. Displacements with respect to the reference axis (x, y) [26].

To perform the MIDA, intensity measurement must be selected, such as peak ground velocity (PGA), peak ground acceleration (PGV), a spectrum related to the damping and the first period, and so on. Then, the engineering demand parameter must be determined, which was categorized into maximum deformation, cumulative damage, and global parameters [30]. The interstory drift measurements are based on maximum displacements, and are used in this analysis as there is an established relationship between such measurements and the performance of the structures: immediate occupation, life safety, and collapse prevention (FEMA 273 [31]).

In summary, the method consists of the following:

- Selection of records;
- Selection of intensity;
- Selection of engineering demand parameter;
- Sample generation using LHS;
- Carrying out MIDA and its curves.

1.9. Summary of the Studies That Analyze the Influence of the Angle of Incidence

The studies shown so far include relevant procedures that are directly applied to determine the seismic response of structures, either at the regulatory level [27] or at the level of research carried out on structures whose importance or irregularity requires more refined procedures of analysis [29]. On the other hand, these procedures serve as a basis for questioning the regulatory application of address combinations [19,21,26], among which is the combination of 100% and 30%, or the way in which the results are combined [19,22,28] to obtain the stresses that will be applied in the structural design. These studies have served as the basis for determining the influence of the seismic action's incidence angle

on different structural types that are presented below. A numerical study will be applied to a set of representative archetypes of various structural typologies, which will make it possible to obtain the maximum demands in displacements (produced by means of a nonlinear time-history analysis combining the two components of strong earthquakes that occurred in Chile), and in which the angle of incidence has been allowed variation in order to determine under which angle the highest demands are reached.

2. Methodology

The following steps summarize the methodology applied to determine the influence of the angle of incidence on the response of structures subjected to the seismic action typified by accelerograms.

1. Definition of the archetypes to be analyzed, representative of the structural typologies conventionally used in reinforced concrete buildings.
2. Selection of demand parameters, in this case corresponding to roof displacement and interstory drift.
3. Selection of records to be used, considering strong earthquakes recorded in Chile.
4. Matching of the selected records with respect to the elastic design spectrum for a type-B soil (fractured rock, stiff or dense soil) and reduction of the duration by means of Arias intensity.
5. Application of transverse and longitudinal component of each record pair of components simultaneously to the archetypes.
6. Normalization of the results obtained based on the maximum response obtained for the archetype and record in question, considering all the angles of incidence used.
7. Combination of the engineering demand parameters by SRSS, as shown in the following equation:

$$R_{i,\theta,j} = \left[(\max|R_{i,\theta,x}|)^2 + (\max|R_{i,\theta,y}|)^2\right]^{0.5} \tag{20}$$

   where $max(R_{i,\theta,x})$ is the maximum reaction for archetype i in the angle of incidence θ in direction x, or y in the case of $max(R_{i,\theta,y})$.
8. Calculation of the demand parameters for each archetype as the average of the maximums given the records used.

3. Definition of Case Studies

To study the combined effect of the angle of incidence of the seismic action and the symmetry of the structures, a set of archetypes was defined in which different structural typologies were used that gave rise to symmetric and asymmetric configurations. It is important to clarify that the symmetry being studied refers to the orientation plane of the structural axes (x and y), as asymmetry in elevation (with respect to the z axis) is outside the scope of this study. The conventional axes found in the literature (shown in Figure 5) were used as reference axes, where x and y correspond to the orthogonal axes that define the structure and that are commonly used in its design and analysis. Meanwhile, w and p are the axes upon which the pair of records was applied to perform the nonlinear time-history analysis, with w being the longitudinal component and p the transverse component. The pair of components of each record was applied with a variation at every 22.5°, making a sweep between the 0° position and the 360° position.

A set of strong motions recorded in Chile, which have been used in a recent study [32], were used to perform the nonlinear analysis; the information of the records is summarized in Table 1. A matching process was carried out, considering a Type-B soil (fractured rock, stiff or dense soil) characterized by shear wave velocity greater than 500 m/s, modifying the records so that the average of the combined displacement spectra for all pairs of records is not less than 1.17 times the elastic spectrum of displacements of the Chilean seismic code NCh433 [33], according to the procedure defined by Achisina [34]. The above spectrum was determined for a high level of seismic hazard (0.4 g). The seismic records were matched using the SeismoMatch software [35], which allows for the ordinates of the

resulting response spectra to be adjusted to the ordinates of a spectrum specified by the user. The algorithm developed by Al Atik and Abrahamson [36], available in the software, was used. The results of the pairing process are shown in Figure 6. The significant duration of the seismic records was reduced according to the Arias intensity [37], taking into account the section equivalent to 5% to 95% of the earthquake intensity, obtaining records with different duration, with this being a variable that affects the seismic response of low-rise buildings [38].

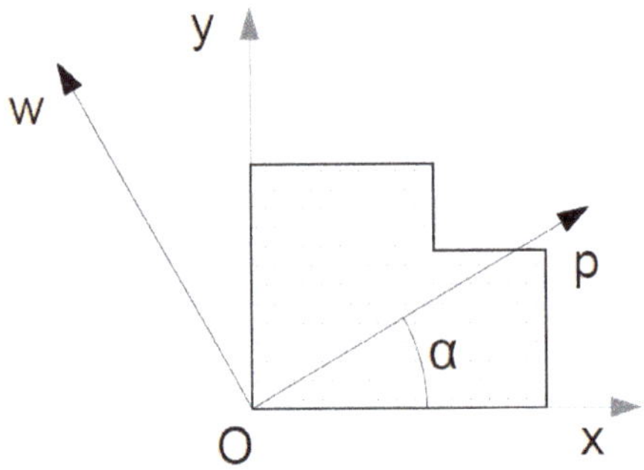

Figure 5. Definition of axes according to the angle of incidence [29].

Table 1. Set of strong motions used to perform the nonlinear response analysis [32].

Seismic Event	Mw	Epicenter Coordinates	Station	Direction	PGA (g)	PGV (cm/s)	PGD (cm)	Significant Duration (s)
Valparaíso 1985	8.0 (CSN)	33.207° S	Melipilla	Longitudinal	0.53	40.30	6.03	29.35
				Transverse	0.69	34.20	12.01	29.35
	7.4 (USGS)	71.663° W	San Isidro	Longitudinal	0.72	43.20	6.91	45.40
				Transverse	0.71	43.40	5.54	45.40
Bío-Bío 2010	8.8 (CSN)	36.122° S	Angol	Longitudinal	0.94	37.50	3.40	54.01
				Transverse	0.70	34.30	7.00	54.01
			Concepción San Pedro	Longitudinal	0.61	44.70	21.10	78.15
				Transverse	0.65	38.10	20.10	78.15
	8.8 (USGS)	72.898° W	Constitución	Longitudinal	0.54	43.30	9.80	66.49
				Transverse	0.63	68.60	13.90	66.49
Coquimbo 2015	8.4 (CSN)	31.573° S	El Pedregal	Longitudinal	0.69	30.60	19.80	52.74
				Transverse	0.57	26.20	15.60	52.74
	8.3 (USGS)	71.674° W	Tololo	Longitudinal	0.24	31.80	14.50	53.72
				Transverse	0.35	38.20	11.60	53.72

The structural archetypes proposed for the analysis suppose a representation of certain specific conditions typical of real structures [39,40]. The archetypes represent buildings with five stories, with spacing between stories of 3 m; a floor plan of 18 × 15 m with four axes distributed every 6 m in the x axis and four every 5 m in the y axis; the sections of the columns are 0.5 × 0.5 m and the walls are 0.2 m thick; the beams used have sections of 0.30 m by 0.60 m; and the slabs are 0.15 m thick (see Figure 7). A residential use of the buildings was considered. The structural elements were designed following the procedures contained in the current versions of the Chilean regulations. Finite elements with distributed plasticity were used, modeling concrete with the model proposed by Mander et al. [41] and reinforcing steel with the Menegotto and Pinto model [42]. The slabs were modeled as rigid elements. The buildings were modeled and analyzed using the SeismoStruct software [43]. The design characteristics of the element materials are summarized in Table 2.

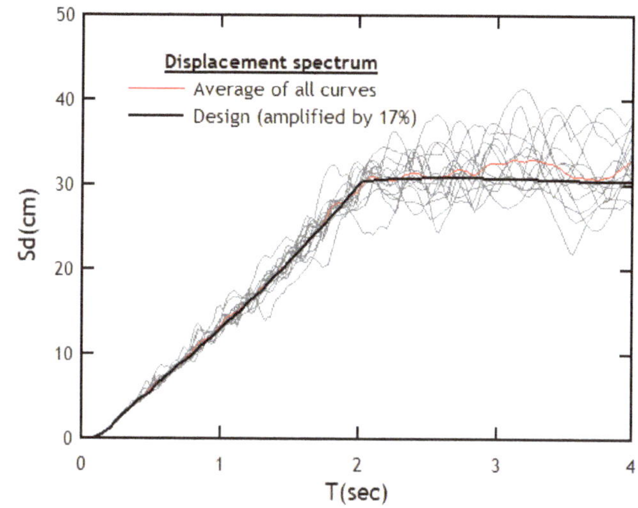

Figure 6. Design elastic spectrum in displacements, amplified by 17% with both average response spectra and those obtained for the records used in the investigation.

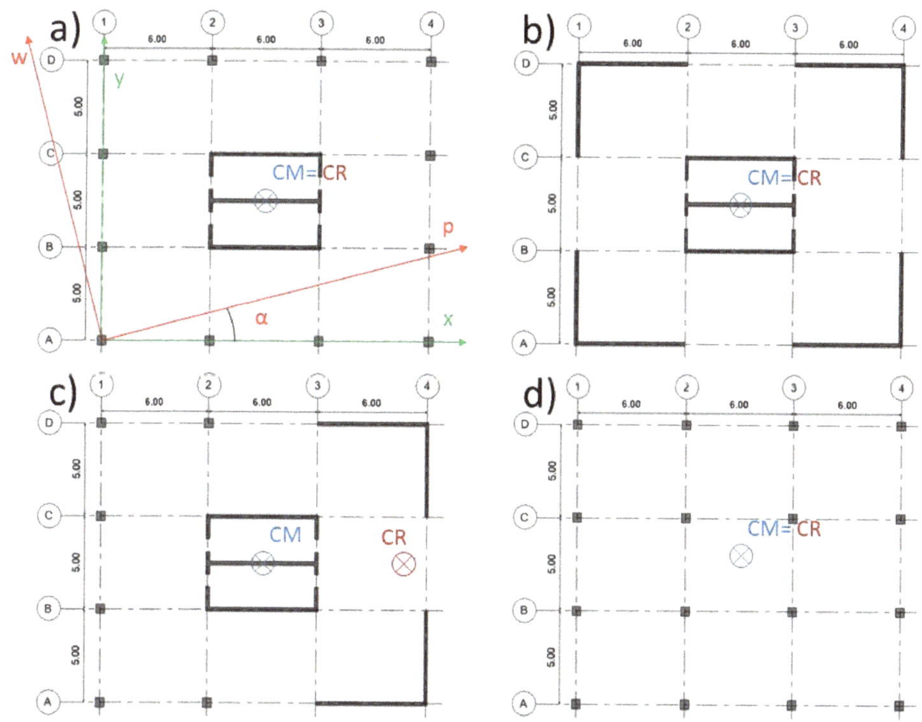

Figure 7. Plan of archetypes used for analysis of the influence of the angle of incidence: (**a**) Archetype 1 plan, torsionally flexible; (**b**) Archetype 2 plan, torsionally rigid; (**c**) Archetype 3 plan, unidirectional eccentricity; (**d**) Archetype 4 plan, flexible.

Table 2. Properties of concrete and reinforcement steel.

Material	Denomination	f_c or F_y[MPa]	E [GPa]	ν	γ[kg/m^3]
Concrete	G25	25	25.7	0.2	2500
Steel	A630-420H	420	200	0.3	7850

The four archetypes chosen correspond to reinforced concrete structures configured using different typologies, ranging from very stiff (building with shear walls, Archetype 2) to structures of intermediate stiffness (frame building with a core of shear walls, archetype 1) to flexible structures (frame building, Archetype 4). An archetype that presents a high plan irregularity was intentionally introduced, so that the effect of such irregularity could be studied along with the directionality of the applied seismic components. Figure 8 shows the isometric views of the four archetypes studied; note the irregular distribution of the shear walls of Archetype 3, which generates unidirectional eccentricity (according to the x axis). The irregularity of Archetype 3 is proposed in a way that influences the dynamic results. This strategy has been used in structures with another type of irregularity in plan [39]. All the Archetypes designed comply with the normative prescriptions that govern the maximum admissible drifts for the different structural typologies. Likewise, all the Archetypes, except naturally Archetype 3, satisfy the design requirements that limit irregularity in plan.

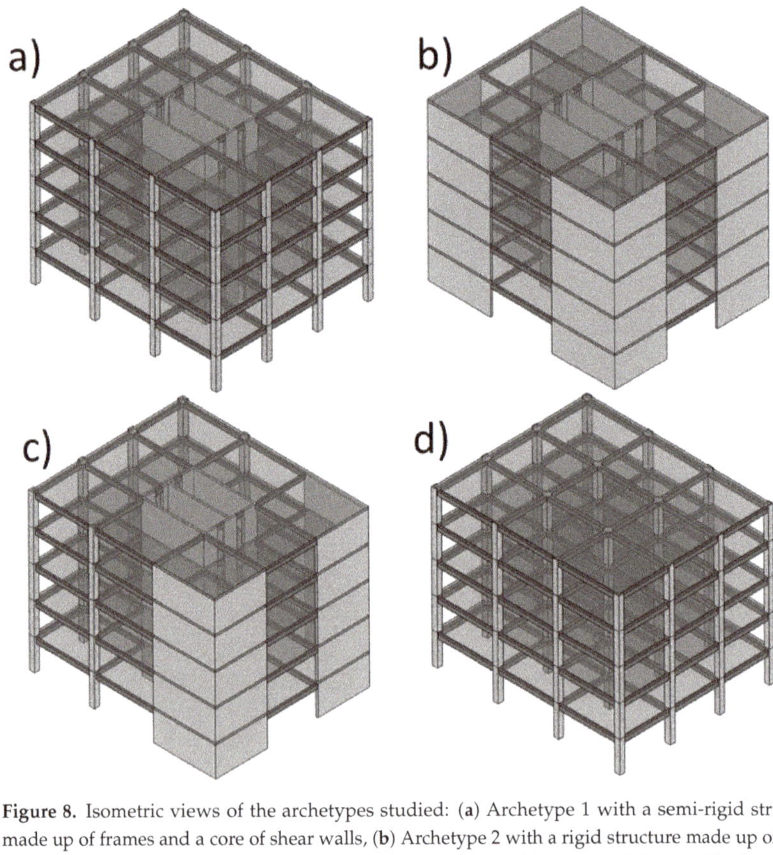

Figure 8. Isometric views of the archetypes studied: (**a**) Archetype 1 with a semi-rigid structure made up of frames and a core of shear walls, (**b**) Archetype 2 with a rigid structure made up of shear walls, (**c**) Archetype 3 with a semi-rigid structure made up of frames and shear walls distributed asymmetrically, and (**d**) Archetype 4 with a flexible structure made up of frames.

The results of the linear elastic analysis of the archetypes were obtained considering a fraction of 5% of the critical damping. Table 3 shows the periods of the first six vibration modes (obtained as a result of the dynamic analysis), specifying whether each mode is translational (along the x or y axis) or rotational (along the z axis). Figure 9 shows the design inelastic spectrum reduced by a regulatory response reduction factor of 11, and presents the vibration periods of the first modes in the x and y direction. Some of the most noteworthy findings are the special difference among the periods of Archetype 3 and the practically equal periods in both directions for Archetype 4. It should also be noted that some archetypes did not reach the minimum seismic coefficient in some directions (in which cases, the response reduction factors were adjusted to ensure compliance).

Table 3. Periods of the first six modes of vibration of the archetypes.

	Archetype							
	1		2		3		4	
Mode	**T(s)**	**Mode Type**	**T(s)**	**Mode Type**	**T(s)**	**Mode Type**	**T(s)**	**Mode Type**
1	0.239	y	0.103	y	0.17	y	0.348	y
2	0.205	z	0.077	x	0.093	x	0.346	x
3	0.116	x	0.057	z	0.074	y	0.283	z
4	0.082	y	0.037	y	0.06	y	0.12	y
5	0.07	z	0.032	z	0.042	z	0.119	x
6	0.05	y	0.03	x	0.037	z	0.097	z

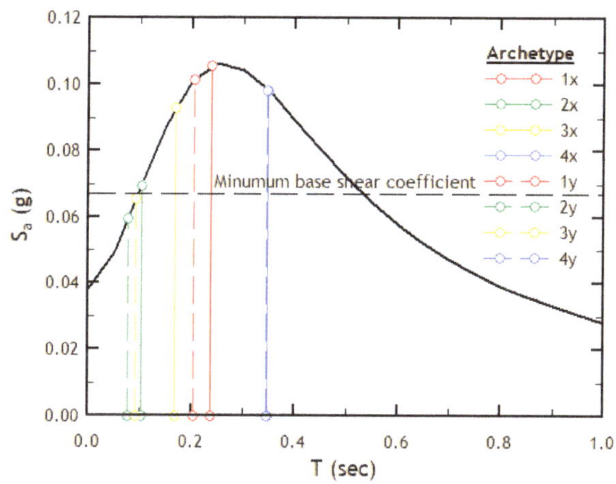

Figure 9. Inelastic design spectrum with the predominant structural periods in the directions of analysis.

The stiffness and torsional properties of the archetypes were calculated and are summarized in Table 4. The results show that there is a significant difference between the stiffness of Archetypes 1 and 3. Archetype 2 presents a less notable difference in stiffness than the previous one, while Archetype 4 presents practically the same stiffness in both directions. The imbalance in the rigidity will prove to be significant in determining the influence of the angle of incidence on the response of the archetypes. The calculated value of torsional stiffness, as expected, is significantly higher in Archetype 2 (whose structure is based on shear walls), and to a lesser extent in Archetype 3 (which has shear walls only on the right side of the plan). This irregular distribution is what produces the eccentricity of 8.7 m in this archetype. It is important to point out that Archetype 1 presents insufficient torsional rigidity, when verifying that $r_{yc} < l_s$.

Table 4. Torsional properties of the archetypes.

Parameter		Archetype 1	2	3	4
Structural eccentricity	e_x (m)	0.00	0.00	8.70	0.00
	e_y (m)	0.00	0.00	0.00	0.00
Structural stiffness	K_x (kN/m)	3,209,720.85	7,330,165.60	5,072,777.75	325,002.45
	K_y (kN/m)	737,992.79	4,069,043.81	1,610,221.60	320,719.98
Torsional stiffness	JTC (kN.m/rad)	49,469,906.05	706,687,893.34	237,687,881.87	29,340,630.20
Trosional radius	r_{xc} (m)	8.19	13.18	12.15	9.56
	r_{yc} (m)	3.93	9.82	6.85	9.50
Radius of gyration	l_s (m)	6.76	6.76	6.76	6.76

The nonlinear behavior of the archetypes was determined using fiber finite elements, with a structural damping equal to 5% of the critical damping. The structural members were incorporated into the model including the characteristics of the reinforcement, both longitudinal and transverse, previously designed according to current regulations in Chile for a high level of seismic hazard (0.4 g). In both the static and dynamic analyses, the P-Δ effect was included.

As a first approximation to the seismic response, the conventional pushover analysis was carried out in both directions for each archetype, whose capacity curves are summarized in Figure 10. The pushover analysis was carried out considering a distribution of lateral forces corresponding to the first vibration mode. This procedure, however, is not the most appropriate to determine the torsional behavior of the archetypes studied.

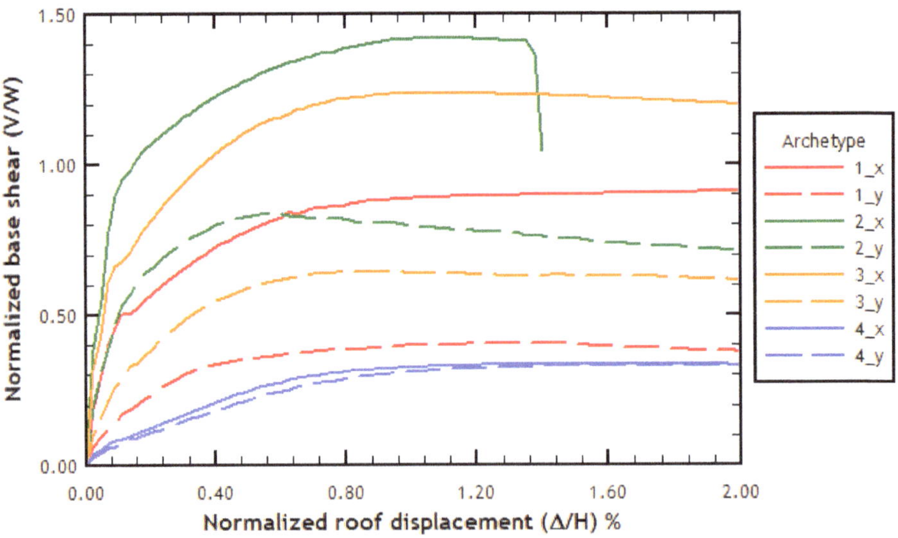

Figure 10. Capacity curves obtained from conventional pushover analysis.

The results of the pushover analysis show capacity curves in which it is possible to appreciate the difference in lateral stiffness of archetypes 1, 2, and 3 in the two directions of analysis. Note that the capacity curves were represented in a normalized way with respect to height and seismic weight, so that the response of the different archetypes can be compared.

4. Results and Discussion

Among all the results of the nonlinear time-history analysis, the two parameters of the seismic response chosen are interstory drift (usually an indicator of the damage to structural elements corresponding to a specific story) and roof displacement (which usually allows the global ductility demands on a structure to be observed).

Next, figures are presented with the average results obtained for each archetype in terms of interstory drift. All displacements and interstory drifts were calculated at the centre of mass of each story. It should be noted that these figures display both the values obtained by axes as well as the values calculated by SRSS combination. Global averages refer to the average obtained by considering all the results on the indicated axis. There is a clear trend in terms of the angle of incidence for each archetype, which does not seem to be significantly affected by the story being evaluated on the structure in question.

The results presented below involving both interstory drift and roof displacement were expressed in polar graphs. Each angle of incidence was presented considering a sweep from 0° to 360° with an increment of 22.5°. To graph the values of the interstory drift or the displacement, the maximum values were calculated for each pair of components of the corresponding earthquake considered in Table 1. Then, the maximum value, both for each earthquake and for each angle of incidence, is determined between the previous maximum values and for each direction of the response. This maximum value per direction is taken in order to serve as a value against which all previous maximum values are normalized according to the same value for all stories. It is important to mention that the maximum displacements and the maximum drifts do not all occur at the same time of application of the records used in the study. Similarly, it was not determined that these maximum values occur at the times in which the PGA of the records occurs.

4.1. Interstory Drift

The results obtained in the archetypes using the accelerograms contained in Table 1 are presented below. This part shows the interstory drift measurements, which are very important in the verification of the design as they can be correlated with the damage achieved in a specific story. Figure 11a shows that Archetype 1 presents a clear trend, with the angles of incidence of 135° and 315° producing the least response in the y axis (dotted lines). For the x axis (solid lines), the trend is not so clear, because, apart from the two angles of incidence mentioned for the y axis, it is noted that the incidence angles of 112.5° and 292.5° also present low values of interstory drift. The response obtained in the y axis is similar to that obtained by the SRSS combination (see Figure 8b). Note that, in these figures, each color corresponds to a specific story, obtaining in this case greater drift on the highest stories of the archetype.

Figure 12a,b show the results obtained in Archetype 2. Given the irregular shape of the displacements and interstory drifts, it seems that the minimums are given for angles of incidence of 135° and 315°, with a clear similarity between the SRSS combination and the results obtained by the y axis. Also note that the largest interstory drift measurements are reached for angles of incidence of 45° and 225°. For the analysis in the x axis, it is noted that the angles of incidence for which the smallest drift is obtained are the 45° and 225° angles. On the contrary, the largest interstory drifts in the y axis are reached for angles of incidence of 157.5° and 337.5°.

Regarding Archetype 3, Figure 13 shows a relatively homogeneous distribution for the results obtained along the y axis. However, on the x axis, a clear influence of the angle of incidence is observed. First, it can be seen that, on the x axis, the maximum interstory drift measurements are reached for both angles of incidence of 0° and 90°, while on the y axis, the maximum interstory drift measurements are reached for angles of incidence of 157.5° and 337.5°, with a notable reduction in interstory drift for angles of incidence of 67.5° and 247.5°. The distribution obtained by means of the SRSS combination is similar to that obtained in the results with respect to the y axis. This dissimilar behavior can be

attributed to the fact that the structure on the x axis is symmetric, while on the y axis, it is highly asymmetric.

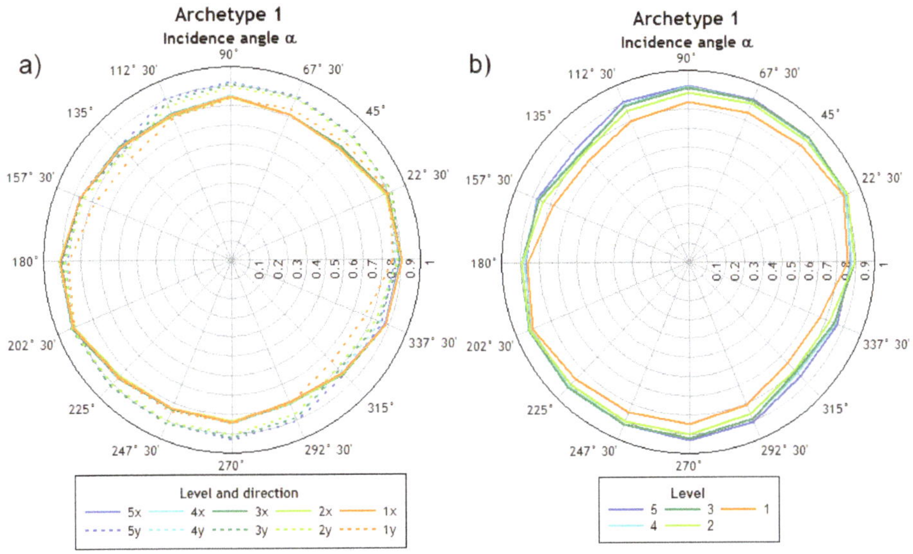

Figure 11. Mean interstory drift measurements normalized for Archetype 1 per story, (**a**) without and (**b**) with SRSS combination.

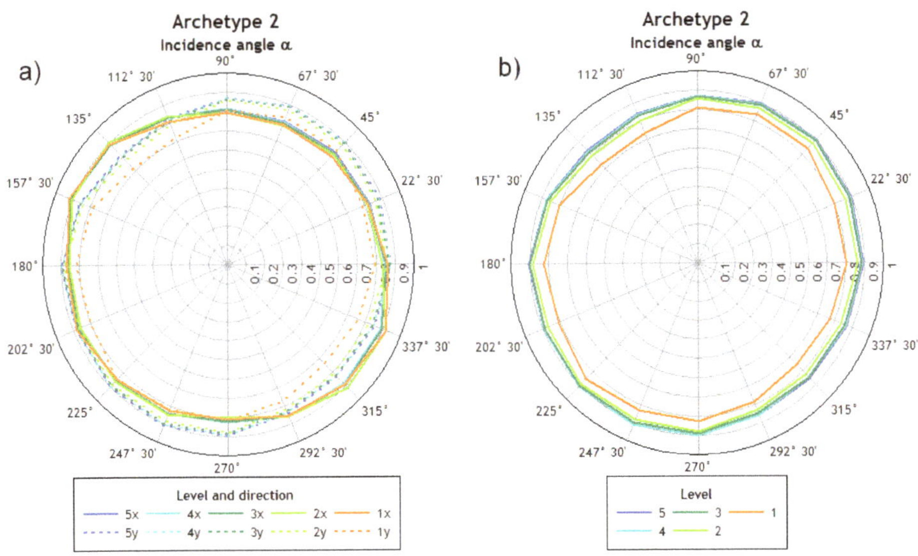

Figure 12. Mean interstory drift normalized for Archetype 2 per story, (**a**) without and (**b**) with SRSS combination.

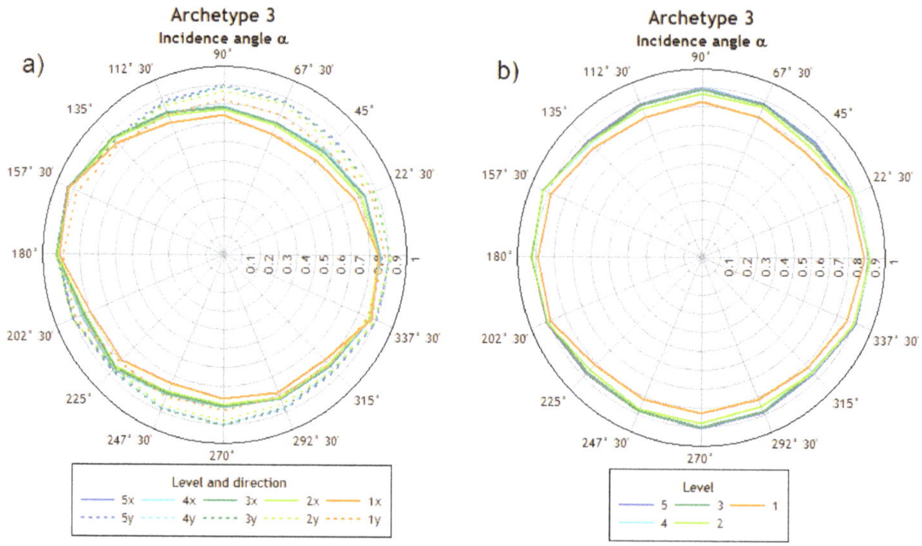

Figure 13. Mean interstory drift normalized for Archetype 3 per story, (**a**) without and (**b**) with SRSS combination.

Finally, the results obtained with Archetype 4 are presented. It is observed that this archetype presents the most homogeneous response, which coincides with the more regular structure obtained with structural frames in both directions, without the presence of eccentricities. Note especially in Figure 14b that the interstory drift values are uniform regardless of the angle of incidence, with a slight predominance of the 0° angle. Additionally, it can be mentioned that the drift distribution in this archetype maintains the tendency to increase with the height observed in the rest of the archetypes.

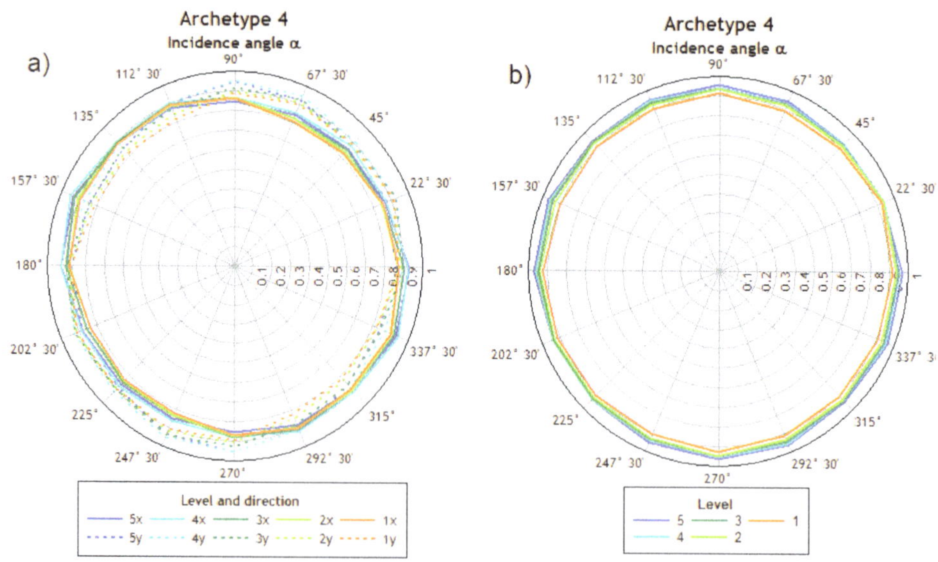

Figure 14. Mean interstory drift normalized for Archetype 4 per story, (**a**) without and (**b**) with SRSS combination.

Table 5 summarizes the results obtained from all the analyses regarding interstory drift. The results show the angles of incidence under which the maximum and minimum interstory drift occurred on both the x and y axes. Note that there is no common preferential angle of incidence for the four archetypes studied, although, for Archetypes 2, 3, and 4 on the x axis, maximum interstory drift values are maintained for an angle of incidence of 67° 30′, while the minimum drift occurs for an angle of incidence of 157° 30′, which means that there is a rotation of 90° between the angle of incidence of maximum and minimum interstory drift. The same happens for the results of Archetype 1, where interstory drift on the x axis reached maximum values for an angle of incidence of 292° 30′ and minimum values for an angle of incidence of 202° 30′. Regarding interstory drift on the y axis, it can be noted that the angle of incidence that produces the highest values is different for each of the four archetypes, while the minimum values of interstory drift occur for an angle of 135° in the case of Archetypes 1, 2, and 3, and Archetype 4 reaches minimum values under an angle of incidence of 157° 30′.

Table 5. Summary of the angles of incidence that produce the maximum and minimum drifts on the archetypes in the x and y directions.

	Min. Interstory Drift x		Max. Interstory Drift x		Min. Interstory Drift y		Max. Interstory Drift y	
Archetype	**Inc. Angle**	**Story**	**Inc. Angle**	**Story**	**Inc. Angle**	**Story**	**Inc. Angle**	**Story**
1	292° 30′	1	202° 30′	5	135°	1	270°	5
2	67° 30′	2	157° 30′	1	135°	1	247° 30′	5
3	67° 30′	1	157° 30′	5	45°	1	157° 30′	5
4	67° 30′	1	157° 30′	4	157° 30′	1	90°	4

Meanwhile, Table 6 shows the interstory drift values calculated with the two components of the analysis using the SRSS method. The values show that there is no clear trend regarding the angles of incidence that produce the maximum and minimum interstory drift for the different archetypes considered.

Table 6. Summary of the angles of incidence that produce the maximum and minimum interstory drifts on the archetypes.

Archetype	Min. Interstory Drift	Max. Interstory Drift
1	135°	270°
2	135°	247° 30′
3	315°	157° 30′
4	225°	0°

4.2. Roof Displacement

Figure 15 shows the results of the roof displacement obtained and normalized. It is observed that there is no major variation between any of the axes and the results obtained by combination. The results obtained are similar to those presented above for each story. On the other hand, the influence of the displacements reached in the y direction and on the results of the combination according to SRSS is verified.

Table 7 shows the incidence angles for which the maximum and minimum roof displacement occurs. Note the wide variety of angles of incidence that produce the maximum displacement in both directions of analysis, which shows that the different archetypes do not reach the maximum demands for a specific angle of incidence. The same can be inferred from Table 8, in which the combined roof displacement is shown using SRSS. Note the influence of the prevailing angle of incidence on the direction, which is reflected in both the minimum and maximum values.

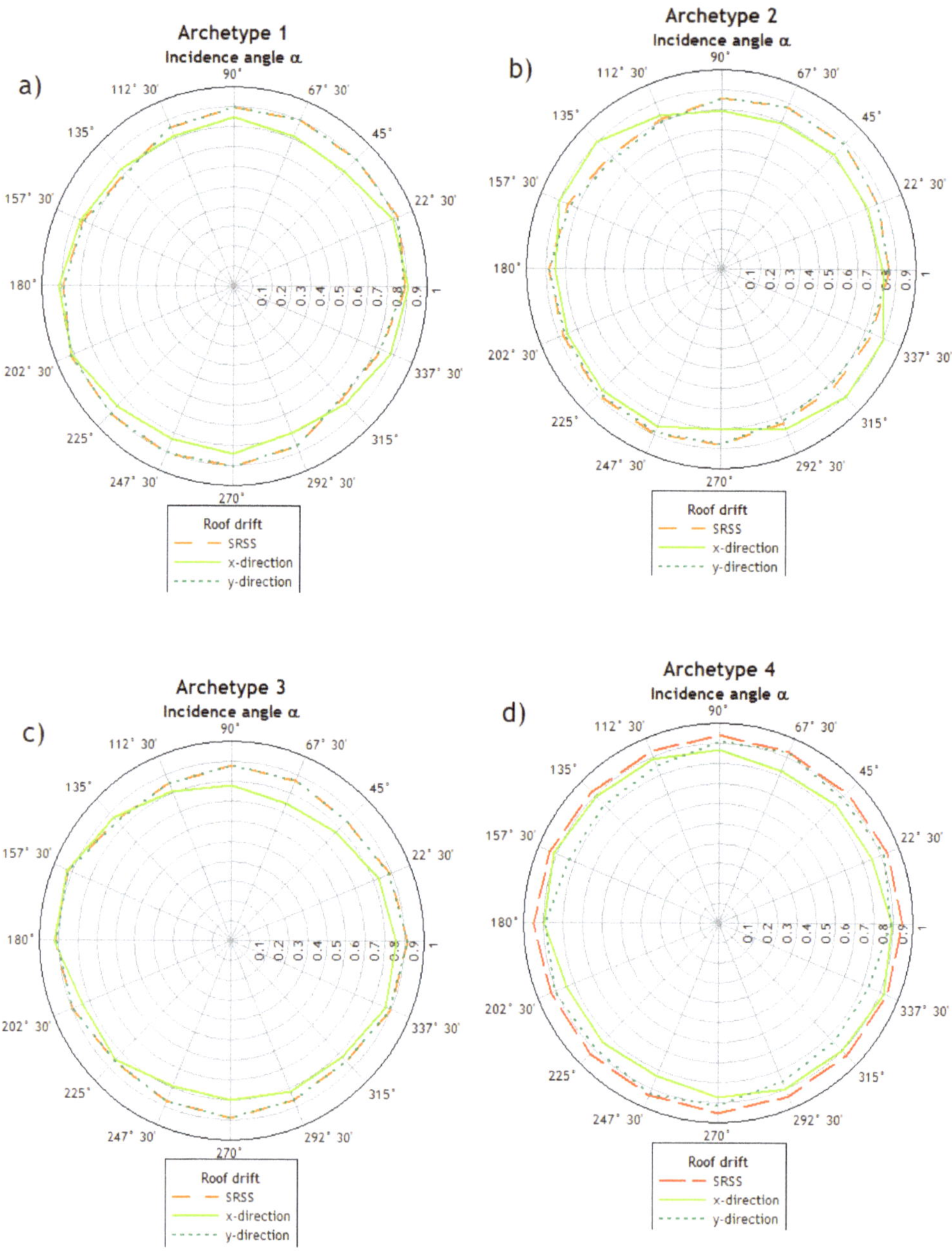

Figure 15. Mean roof displacement, SRSS combination for (**a**) Archetype 1, (**b**) Archetype 2, (**c**) Archetype 3 and (**d**) Archetype 4.

Table 7. Summary of the angles of incidence that produce the maximum and minimum roof displacement on the archetypes in each direction.

Archetype	Min. Displ. x	Max. Displ. x	Min. Displ. y	Max. Displ. y
1	292° 30′	202° 30′	135°	22° 30′
2	67° 30′	315°	135°	247° 30′
3	67° 30′	157° 30′	135°	157° 30′
4	67° 30′	337° 30′	157° 30′	247° 30′

Table 8. Summary of the angles of incidence that produce the maximum and minimum roof displacement on the archetypes, SRSS combination.

Archetype	Min. Displ.	Max. Displ.
1	135°	22° 30′
2	135°	247° 30′
3	135°	157° 30′
4	225°	270°

4.3. Comparison of Results with SRSS Combination

Figure 16a–d show the results of the interstory drifts and the global drift (Du is the roof displacement normalized to the height of the building) in the axes for different angles of incidence compared with those obtained using the SRSS combination. The results shown were obtained taking into account the interstory drifts calculated in both directions of analysis and for all angles of incidence. In these figures, the mean values and the values of 1 and 1.5 standard deviation were included. It is observed that the combination has a noteworthy impact in Archetypes 2 and 4, while in the rest, it can be considered negligible. In none of the cases presented below is there a coefficient of variation greater than 5%, which reflects some uniformity in the results obtained after their normalization. However, it is observed that the structural elements could be oversized during the design process, as the variations are noteworthy compared with not using the SRSS combination, especially in Archetypes 2 and 4. From this, the following can be inferred: in buildings with very similar stiffnesses in both directions of analysis (Archetypes 2 and 4), the results obtained applying the SRSS method show less dependence to the flexible direction of the buildings, compared with Archetypes 1 and 3, in which there is a greater difference between the stiffnesses depending on the directions of analysis (see Table 3).

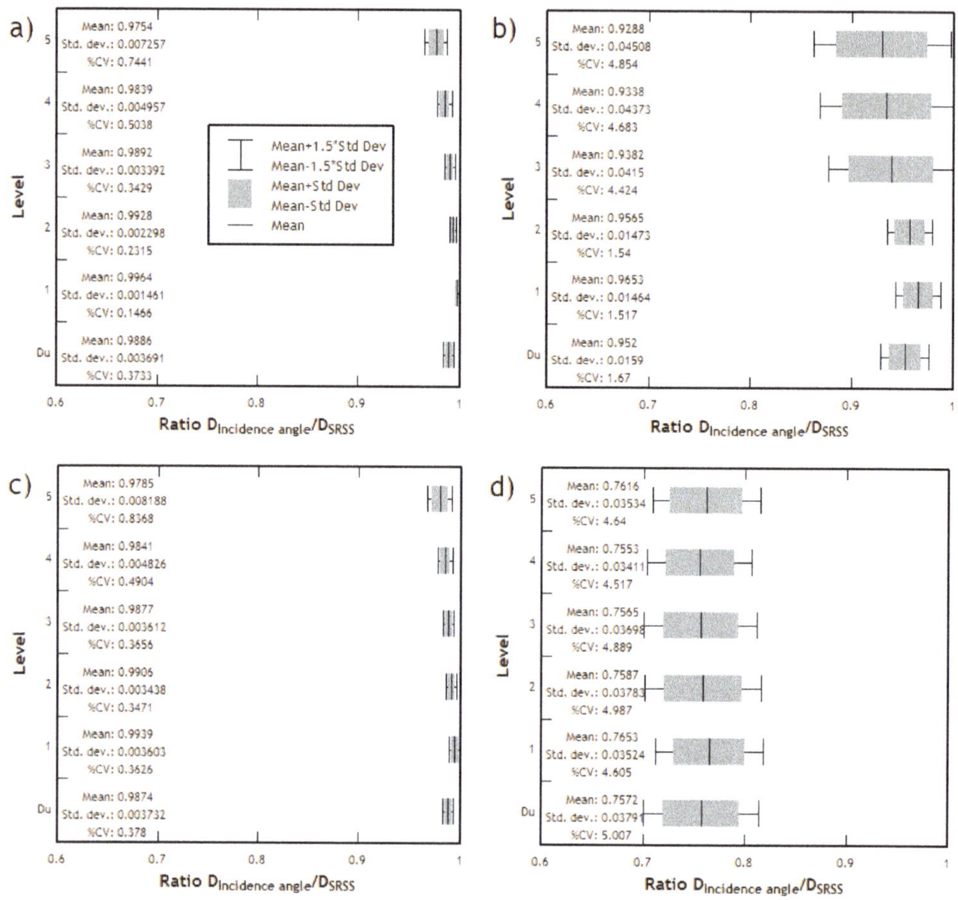

Figure 16. Ratio of the interstory drifts achieved with different angles of incidence with respect to the displacements achieved using SRSS combination for (**a**) Archetype 1, (**b**) Archetype 2, (**c**) Archetype 3, and (**d**) Archetype 4.

5. Concluding Remarks

In this study, four archetypes with different structures were used, using shear walls (Archetype 2), frames (Archetype 4), and two combinations of such considering a regular distribution in plan (Archetype 1) and an irregular one (Archetype 3). The results of the nonlinear time-history analysis indicate that there is no clear predominance in the incidence angle that produces the highest values of the chosen engineering demand parameters (roof displacement and interstory drift).

When analyzing the results obtained by calculating seismic demand parameters such as roof displacement and interstory drift, there seems to be a relationship between the proportion of the stiffnesses obtained in each direction of analysis and the influence of the angle of incidence; that is, as the proportion of the stiffnesses approaches 1, the influence of the angle of incidence on the structural response decreases.

This influence is clear when comparing the shapes of the displacement distributions of Archetype 2 (oval) and Archetype 4 (circular), as seen in Figure 9b or Figure 11b. This is also supported by the results presented in Table 4. It should be noted that this difference could also be due to the structuring of the archetypes, as Archetype 2 has a flexible axis and a rigid axis if its periods are analyzed, while Archetype 4 does not have such a clear difference between its two orthogonal directions. On the other hand, Archetype 3, having significant differences between the periods of its axes, shows more uniform distributions with its flexible y axis as a reference, thus more studies are required to reach a conclusive determination regarding the influence of the period on the angle of incidence.

In terms of roof displacement ratios obtained with various angles of incidence (calculated with SRSS combination), structures composed of shear walls and frames (Archetypes 1 and 3) present a lower dispersion than structures composed only of shear walls (Archetype 2) or frame (Archetype 4). The moment-resisting frame archetype presents the lowest mean values of roof displacement ratios, allowing us to consider that the design driven by the values obtained with the SRSS combination should lead to a conservative design of such structural typology.

Although the literature has extensively dealt with this issue, no results are found regarding the cases of shear wall structures, so it would be prudent to carry out more nonlinear time-history analysis using real structures, of different numbers of stories and varied seismic hazard levels, given that—as indicated by various authors—the seismic hazard level is essential when calculating the critical incidence angle.

In this study, the influence of a unidirectional eccentricity was also evaluated, but no substantial differences were observed with respect to other archetypes, thus the review of bidirectional eccentricities of greater magnitude is suggested in order to verify their influence in these case studies. This case corresponds to structures with large plan irregularities.

The results obtained are limited and should go further in the study of the directionality of the earthquake to verify if other parameters of seismic demand are substantially affected or not depending on the angle of incidence considered.

Author Contributions: Conceptualization, J.C.V. and J.C.; methodology, J.C.V.; software, I.B.; validation, I.B., J.C.V., and J.C.; formal analysis, I.B.; investigation, I.B. and J.C.V.; resources, I.B.; data curation, J.C.V.; writing—original draft preparation, I.B.; writing—review and editing, J.C.V.; visualization, J.C.; supervision, J.C.V. and J.C. All authors have read and agreed to the published version of the manuscript.

Funding: This research received no external funding.

Institutional Review Board Statement: Not applicable.

Informed Consent Statement: Not applicable.

Data Availability Statement: The data produced in this study are supported by the authors and may be provided to interested persons upon request to the corresponding author.

Acknowledgments: The authors of this article want to thank Pontificia Universidad Católica de Valparaíso for sponsoring the development of this study and financing its publication in Open Access format.

Conflicts of Interest: The authors declare no conflict of interest.

References

1. McRae, G.; Matteis, J. Three-Dimensional Steel Building Response to Near-Fault Motions. *J. Struct. Eng.* **2000**, *126*, 117–126. [CrossRef]
2. Athanatopoulou, A.M. Critical orientation of three correlated seismic components. *Eng. Struct.* **2005**, *27*, 301–312. [CrossRef]
3. Rigato, A.B.; Medina, R.A. Influence of angle of incidence on seismic demands for inelastic single-storey structures subjected to bi-directional ground motions. *Eng. Struct.* **2007**, *29*, 2593–2601. [CrossRef]
4. Reyes, J.C.; Kalkan, E. *Should Ground-Motion Records Be Rotated to Fault-Normal/Parallel or Maximum Direction for Response History Analysis of Buildings?* U.S. Geological Survey Open-File Report; USGS: Reston, VA, USA, 2012. Available online: http://pubs.usgs.gov/of/2012/1261/ (accessed on 2 January 2019).
5. Kashkooli, N.A.; Banan, N.R. Effect of Frame Irregularity on Accuracy of Modal Equivalent Nonlinear Static Seismic Analysis. *KSCE J. Civ. Eng.* **2013**, *17*, 1064–1072. [CrossRef]
6. Magliulo, G.; Maddaloni, G.; Petrone, C. Influence of earthquake direction on the seismic response of irregular plan RC frame buildings. *Earthq. Eng. Eng. Vib.* **2014**, *13*, 243–256. [CrossRef]

7. Prajwal, T.P.; Parvez, I.; Kamath, K. Nonlinear Analysis of Irregular Buildings Considering the Direction of Seismic Waves. *Mater. Today Proc.* **2017**, *4*, 9828–9832. [CrossRef]
8. Lagaros, N.D. The impact of the earthquake incident angle on the seismic loss estimation. *Eng. Struct.* **2010**, *32*, 1577–1589. [CrossRef]
9. Cantagallo, C.; Camata, G.; Spacone, E. The effect of the earthquake incidence angle on seismic demand of reinforced concrete structures. In Proceedings of the 15th World Conference on Earthquake Engineering, Lisboa, Portugal, 24–28 September 2012.
10. Fontara, I.K.; Kostinakis, K.G.; Manoukas, G.E.; Athanatopoulou, A.M. Parameters affecting the seismic response of buildings under bi-directional excitation. *Struct. Eng. Mech.* **2015**, *53*, 957–979. [CrossRef]
11. Kostinakis, K.G.; Athanatopoulou, A.M.; Avramidis, I.E. Evaluation of inelastic response of 3D single-story R/C frames under bi-directional excitation using different orientation schemes. *Bull. Earthq. Eng.* **2013**, *11*, 637–661. [CrossRef]
12. Kalkan, E.; Reyes, J.C. Significance of rotating ground motions on behavior of symmetric- and asymmetric-plan structures: Part I. Single-story structures. *Earthq. Spectra* **2015**, *31*, 1591–1612. [CrossRef]
13. Vielma, J.C.; Porcu, M.C.; López, N. Intensity measure based on a smooth inelastic peak period for a more effective Incremental Dynamic Analysis. *Appl. Sci.* **2020**, *10*, 8632. [CrossRef]
14. Vielma, J.C.; Porcu, M.C.; Fuentes, M. Non-linear analyses to assess the seismic performance of RC buildings retrofitted with FRP. *Int. J. Numer. Methods Calc. Des. Eng.* **2020**, *36*. Available online: https://www.scipedia.com/public/Vielma-Perez_et_al_2019a (accessed on 2 January 2019).
15. Porcu, M.C.; Vielma, J.C.; Panu, F.; Aguilar, C.; Curreli, G. Seismic Retrofit of Existing Buildings Led by Non-Linear Dynamic Analyses. *Int. J. Saf. Secur. Eng.* **2019**, *9*, 201–212. [CrossRef]
16. Vielma, J.C.; Cando, M. Calibration of the response reduction factors used in Ecuador for steel SMRF. *Bull. Int. Inst. Seismol. Earthq. Eng.* **2018**, *52*, 22–37.
17. Wilson, E.L.; Button, M.R. Three-dimensional dynamic analysis for multi-component earthquake spectra. Earthquake Engineering & Structural Dynamics. *Earthq. Eng. Struct. Dyn.* **1982**, *10*, 471–476. [CrossRef]
18. Smeby, W.; Der Kiureghian, A. Modal combinations rules for multicomponent earthquake excitation. *Earthq. Eng. Struct. Dyn.* **1985**, *13*, 1–12. [CrossRef]
19. González, P. Considering earthquake direction on seismic analysis. In Proceedings of the Tenth World Conference on Earthquake Engineering, Madrid, Spain, 19–24 July 1992.
20. Wilson, E.; Iqbal, S.; Ashraf, H.A. Clarification of the Orthogonal Effects in a Three-Dimensional Seismic Analysis. *Earthq. Spectra* **1995**, *11*, 659. [CrossRef]
21. López, O.A.; Torres, R. The critical angle of seismic incidence and the maximum structural response. *Earthq. Eng. Struct. Dyn.* **1997**, *26*, 881–894. [CrossRef]
22. Anastassiadis, K.; Avramidis, I.E.; Panetsos, P. Concurrent design forces in structures under three-component orthotropic seismic excitation. *Earthq. Spectra* **2002**, *18*, 1–17. [CrossRef]
23. Gao, X.A.; Zhou, X.Y.; Wang, L. Multi-component seismic analysis for irregular structures. In Proceedings of the 13th World Conference on Earthquake Engineering, Vancouver, BC, Canada, 1–6 August 2004.
24. Penzien, J.; Watabe, M. Characteristics of 3-dimensional earthquake ground motions. *Earthq. Eng. Struct. Dyn.* **1974**, *3*, 365–373. [CrossRef]
25. Fontara, I.K.; Kostinakis, K.G.; Athanatopoulou, A.M. Some Issues Related to the Inelastic Response of Buildings Under Bi-Directional Excitation. In Proceedings of the 15th World Conference on Earthquake Engineering, Lisboa, Portugal, 24–28 September 2012.
26. Athanapoulou, A.M.; Tsourekas, A.; Papamanolis, G. Variation of response with incident angle under two horizontal correlated seismic components. *WIT Trans. Built Environ.* **2005**, *81*, 183–192. [CrossRef]
27. Menun, C.; Der Kiureghian, A. A replacement for the 30%, 40%, and SRSS rules for multicomponent seismic analysis. *Earthq. Spectra* **1998**, *14*, 153–163. [CrossRef]
28. Dongsheng, W.; Hong-Nan, L.; Wang, G.X.; Fan, Y.F. Inelastic response spectra for bidirectional ground motions. In Proceedings of the 14th World Conference on Earthquake Engineering, Beijing, China, 12–17 October 2008.
29. Lagaros, N.D. Multicomponent incremental dynamic analysis considering variable incident angle. *Struct. Infrastruct. Eng.* **2010**, *6*, 77–94. [CrossRef]
30. Ghobarah, A.; Abou-Elfath, H.; Biddah, A. Response-based damage assessment of structures. *Earthq. Eng. Struct. Dyn.* **1999**, *28*, 79–104. [CrossRef]
31. FEMA. *FEMA 273. NEHRP Guidelines for Seismic Rehabilitation of Buildings*; FEMA: Washington, DC, USA, 1997.
32. Bravo, F.; Tapia, J. Análisis no Lineal de Comportamiento del Edificio "Mercado Puerto" Reforzado. Pontificia Universidad Católica de Valparaíso, Valparaíso, Chile. 2019. Available online: https://biblioteca.pucv.cl/ (accessed on 2 January 2019). (In Spanish).
33. INN, NCh433Of.1996 (last version released in 2009). Diseño Sísmico de Edificios, Norma Chilena Oficial, Instituto Nacional de Normalización, Santiago, Chile. 2012. Available online: http://normastecnicas.minvu.cl/ (accessed on 2 January 2019). (In Spanish).

34. Achisina. Alternative Procedure for the Seismic Analysis and Design of Tall Buildings. Santiago. 2017. Available online: https://www.achisina.cl/images/PBD/ACHISINA_Procedimiento_Alternativo_para_el_An%C3%A1lisis_y_Dise%C3%B1o_S%C3%ADsmico.pdf (accessed on 21 March 2021).
35. Seismosoft, SeismoMatch 2018—A Computer to Adjust Earthquake Accelerograms to Match a Specific Target Response Spectrum. 2018. Available online: http://www.seismosoft.com (accessed on 1 May 2018).
36. Al-Atik, L.; Abrahamson, N.A. An improved method for nonstationary spectral matching. *Earthq. Spectra* **2010**, *26*, 601–617. [CrossRef]
37. Arias, A. A measure of earthquake intensity. In *Seismic Design for Nuclear Power Plants*; Hansen, R.J., Ed.; MIT Press: Cambridge, MA, USA, 1970; pp. 438–483.
38. Martineau, M.; López, A.; Vielma, J.C. Effect of Earthquake Ground Motion Duration on the Seismic Response of a Low-Rise RC Building. *Adv. Civ. Eng.* **2020**, *2020*, 8891282. [CrossRef]
39. FEMA (Federal Emergency Management Agency). *NEHRP Guidelines for the Seismic Rehabilitation of Buildings*; FEMA: Washington, DC, USA, 1996.
40. Vielma, J.C.; Mulder, M. Improved procedure for determining the ductility of buildings under seismic loads. *Int. J. Numer. Methods Calc. Des. Eng.* **2018**, *34*, 1–27. [CrossRef]
41. Mander, J.B.; Priestley, M.J.N.; Park, R. Theoretical stress-strain model for confined concrete. *J. Struct. Eng.* **1988**, *114*, 1804–1826. [CrossRef]
42. Menegotto, M.; Pinto, P.E. Method of analysis for cyclically loaded R.C. plane frames including changes in geometry and non-elastic behaviour of elements under combined normal force and bending. In *Symposium on the Resistance and Ultimate Deformability of Structures Acted on by Well Defined Repeated Loads*; International Association for Bridge and Structural Engineering: Zurich, Switzerland, 1973; pp. 15–22.
43. Seismosoft, SeismoStruct 2018—A Computer Program for Static and Dynamic Nonlinear Analysis of Framed Structures. 2018. Available online: http://www.seismosoft.com (accessed on 1 May 2018).

applied sciences

MDPI

Article

Comparative Assessment of Shear Demand for RC Beam-Column Joints under Earthquake Loading

Angelo Marchisella * and Giovanni Muciaccia

Department of Civil and Environmental Engineering, Politecnico di Milano, 20133 Milano, Italy; giovanni.muciaccia@polimi.it
* Correspondence: angelo.marchisella@polimi.it

Abstract: This paper focuses on the evaluation of bi-axial shear demand for reinforced concrete (RC) beam–column joints assuming: (i) the SPEAR frame as a benchmark; and (ii) different structural analysis methods which share the same seismic input. A numerical model was implemented using lumped plasticity. The joints were modeled as rigid offsets of beams and columns. The shear demand at a joint is evaluated as a post-process of the beam's nodal moment. The discussion focuses on the differences between the estimated shear demand considering modal-response-spectrum analysis (MRSA), non-linear static analysis (NLSA) and non-linear time history (NLTH). Strength assessment of joints is discussed as well. Significant strength differences were recognized by using different building codes targeted to existing structures which, in general, behaved on the safe side. The elliptical shear strength domain resulted in being conservative when compared to NLTH shear demand orbits. NLSA, using modal combination, proved to estimate the larger shear demand with respect to MRSA and NLTH.

Keywords: SPEAR; RC beam-column joint; pushover

Citation: Marchisella, A.; Muciaccia, G. Comparative Assessment of Shear Demand for RC Beam-Column Joints under Earthquake Loading. *Appl. Sci.* **2022**, *12*, 7153. https://doi.org/10.3390/app12147153

Academic Editor: Maria Favvata

Received: 20 June 2022
Accepted: 13 July 2022
Published: 15 July 2022

Publisher's Note: MDPI stays neutral with regard to jurisdictional claims in published maps and institutional affiliations.

Copyright: © 2022 by the authors. Licensee MDPI, Basel, Switzerland. This article is an open access article distributed under the terms and conditions of the Creative Commons Attribution (CC BY) license (https://creativecommons.org/licenses/by/4.0/).

1. Introduction

Post-earthquakes surveys revealed that beam–column joint failure is recurrent in RC buildings showing structural deficiencies such as inadequate joint reinforcement [1]. When a beam–column joint fails prematurely, the RC frame is both not able to sustain significant displacement demand and, eventually, loses its gravity load carrying capacity because of the buckling of column reinforcement [2].

The stress state of a beam–column joint is prevalently characterized by shear developed at the boundaries of the joint and at the interfaces between the joint concrete and the beam or column reinforcement [3]. Given its shear nature, design check is based on equilibrium between shear demand and shear strength [4]. Nonetheless, compatible models, defining the shear stress vs shear distortion constitutive law, are used in more sophisticated numerical applications [5].

According to Mohele [6], a beam column joint is defined as that portion of the column within the depth of the deepest beam that frames into the column. This definition implies that more than one beam can frame into the column and, consequently, force input should be considered bi-axial.

This paper focuses on the evaluation of bi-axial shear demand for beam–column joints, comparing different numerical methods to carry out seismic structural analysis. The selected benchmark structure is the SPEAR building [7]. First, the definition of joint shear demand and joint shear strength are summarized in the section titled "Fundamentals of RC beam–column joint". Second, the section "Numerical study" presents the SPEAR frame by recalling the associated literature; details of the implemented numerical models are given as well. Third, the section "Results and Discussion" focuses on the shear demand at beam–column joints, highlighting the differences between modal-response-spectrum analysis, non-linear-static-analysis and non-linear-time-history. Strength assessment according to building codes targeted to existing structures is discussed as well. Finally, some conclusions summarize the work.

2. Fundamentals of RC Beam-Column Joints

2.1. *Joint Shear Demand*

Considering an RC seismic primary frame such as the one shown in the Figure 1, a beam–column joint panel is defined as the zone where both the beam and column converge. A remarkable definition of the joint shear demand (V_j) is due to Paulay and Priestley [8], who assumed the joint panel as a continuous part of the column (the same considerations apply if it is considered as a part of the beam). When the diagram of bending moments are characterized by a seismic prevalent condition, a sign inversion is needed, passing from the top face to the bottom face of the joint panel. In this regard, V_j can be defined as follows, by considering it as the gradient of the bending moment within the beam depth:

$$V_{jh} = \frac{V_c(H - h_b)}{h_b}, \quad (1)$$

where V_c is the column shear, H is the distance between the column's contraflexure points, h_b is the beam depth. As a result, by considering a common value of 3 m for H and 0.30 for h_b, the ratio V_{jh}/V_c amounts to 9, roughly suggesting that the shear demand in the beam–column joint has a different order of magnitude with respect to shear force either in the column or in the beam. Different results have been suggested by Paulay [8], who claimed the occurrence of a reduced gradient. In fact, V_{jh}/V_c was supposed to vary from four to six. Similarly, Mohele [6] suggested V_{jh}/V_c ranging from three to five for interior joints and half of those values for the exterior.

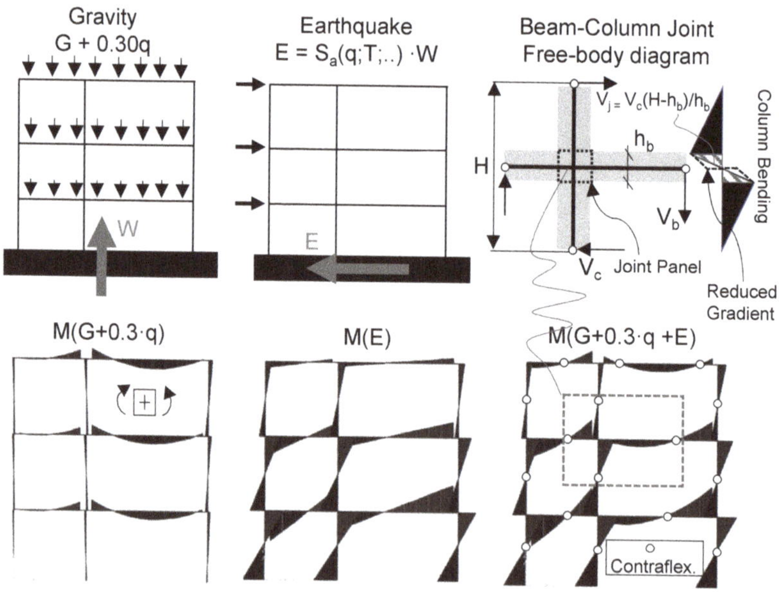

Figure 1. Beam–column joint definition for a planar moment-resisting frame under gravity and lateral forces. (Notes. The diagram of bending moment (M) are obtained from a simplified 2D analysis carried out on SPEAR frame).

The conventional definition of V_{jh}, widely adopted in design codes, is shown in Figure 2a. An interior beam–column joint is considered when the hogging moment acts on the right beam. After cutting with a horizontal plane, the horizontal free-body equilibrium requires:

$$V_{jh} = T_R + C_L - V_c, \tag{2}$$

where T_R is the tensile force of the top reinforcement on the right part, C_L is the result of compression stresses on the left part. This study assumes neglecting V_c in Equation (2). Such an assumption will always result in being conservative as long as the seismic condition is dominant for the bending moment diagram.

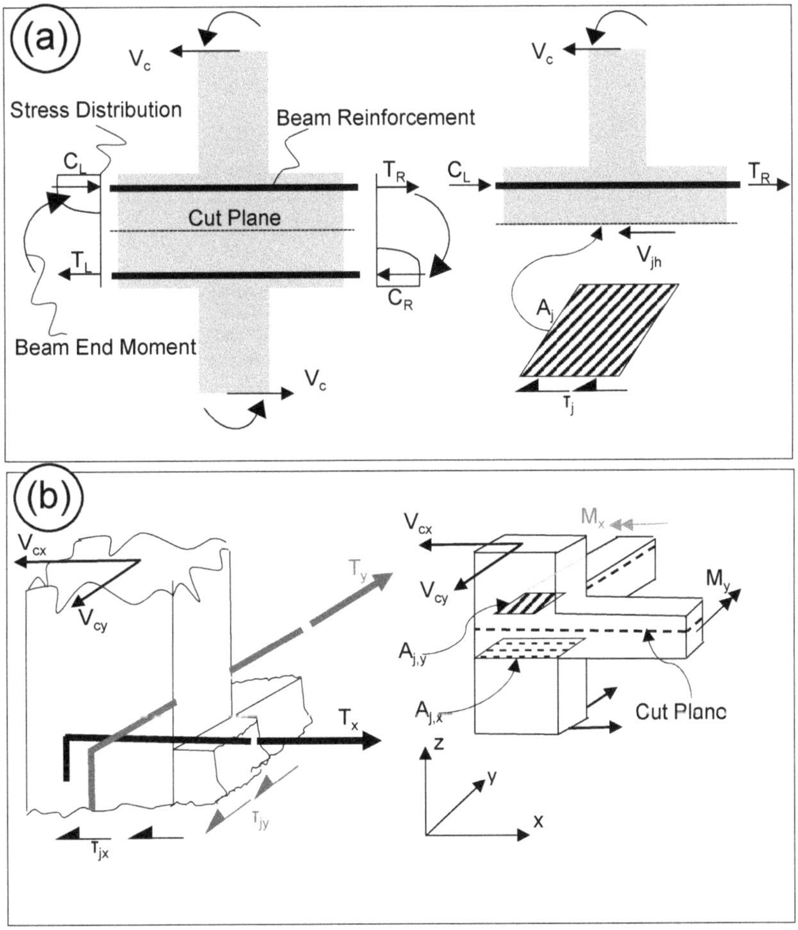

Figure 2. Horizontal joint shear (V_{jh}) definition: (**a**) Case of a 2D interior joint; (**b**) 3D corner joint.

A similar definition can apply, considering the case of a bi-dimensional joint as the one shown in Figure 2b. In this case, Equation (2) should apply both in x-z and y-z planes.

Frequently, shear stresses are used instead of shear forces. The definition is given as follows:

$$\tau_j = \frac{V_{jh}}{A_j}, \tag{3}$$

where A_j is the horizontal projection of the joint area.

According to ACI318 [9], the forces T_R and C_L, in Equation (2), should be calculated either from: (i) the maximum moments transferred between the beam and column as determined from factored-load analysis for beam–column joints with continuous beams in the direction of joint shear considered; or (ii) from the beam nominal moments strength possibly with overstrength. The latter is intended to account for: (a) the actual yield stress of a typical reinforcing bar being commonly 10 to 25% higher than the nominal value; and (b) the reinforcing bars strain hardening at member displacements only slightly larger than the yield rotation. EC8 [10] and ACI352 [11], specifying an overstrength factor (γ_{Rd}) equal to 1.20 and 1.25, respectively. As an example, by defining A_s as the area of the beam reinforcement anchored within the joint and f_y the yielding stress, the yielding force (T_y) can be written as follows:

$$T_y = A_s f_y \gamma_{Rd}. \tag{4}$$

2.2. *Joint Shear Strength*

The following section reviews the evaluation of joint shear strength according to three different building codes targeted to the assessment of existing structures, i.e., European code [12] (draft of the new generation version), New Zealand code [13] and US code [14]. The background of the analytical method is summarized in the following. A final parametrical comparison is given.

2.2.1. Ec8

The draft of the new Eurocode 8 [12] applies the principal stress criterion for the assessment of an existing RC beam–column joint. With reference to Figure 3a, the joint volume is studied in the plane stress condition. The input forces are the horizontal shear force (assumed to be equal to T) and the axial force in the column (N). The stress condition is studied via Mohr's circle as the one shown in Figure 3b. Both the cases of un-reinforced and reinforced joint are represented.

When the principal tensile stress reaches f_{ct} the joint shear stress (τ_ρ), for a reinforced joint, can be written as follows:

$$\tau_\rho = f_c \sqrt{\left(\nu\rho_{sh} + \frac{f_{ct}}{f_c}\right)\left(\nu + \frac{f_{ct}}{f_c}\right)}, \tag{5}$$

where:

- $\rho_{sh} = \frac{A_{sh}}{A_b}$ is the reinforcement ratio;
- $\nu = \frac{N}{A_c f_c}$ is the normalized axial force.

The shear stress is usually divided by the factor $\sqrt{f_c}$. Such normalization is commonly adopted by US authors and building codes, the cracking resistance of a beam–column joint being assimilated to the tensile strength of concrete which is traditionally given as a function of $\sqrt{f_c}$ [15]. Assuming ρ_{sh} is equal to zero in Equation (5) (un-reinforced joint), the tensile normalized shear strength ($v_{j,t}$) is written as follows:

$$v_{j,t} = \frac{f_{ct}}{\sqrt{f_c}}\sqrt{1 + \frac{\nu f_c}{f_{ct}}}. \tag{6}$$

The procedure to obtain the compressive normalized shear strength ($v_{j,c}$) is similar to that presented above for $v_{j,t}$. First, Mohr's circle is used to obtain the shear stress by imposing the value ηf_c for the compressive principal stress, where $\eta = 0.55[min(1;30/f_c)]^{1/3}$ is a reduction factor which takes into account the detrimental effect of transversal tensile strains. Second, $\sqrt{f_c}$ normalization is used. Finally, the expression is as follows:

$$v_{j,c} = \eta\frac{f_c}{\sqrt{f_c}}\sqrt{1 - \frac{\nu}{\eta}}. \tag{7}$$

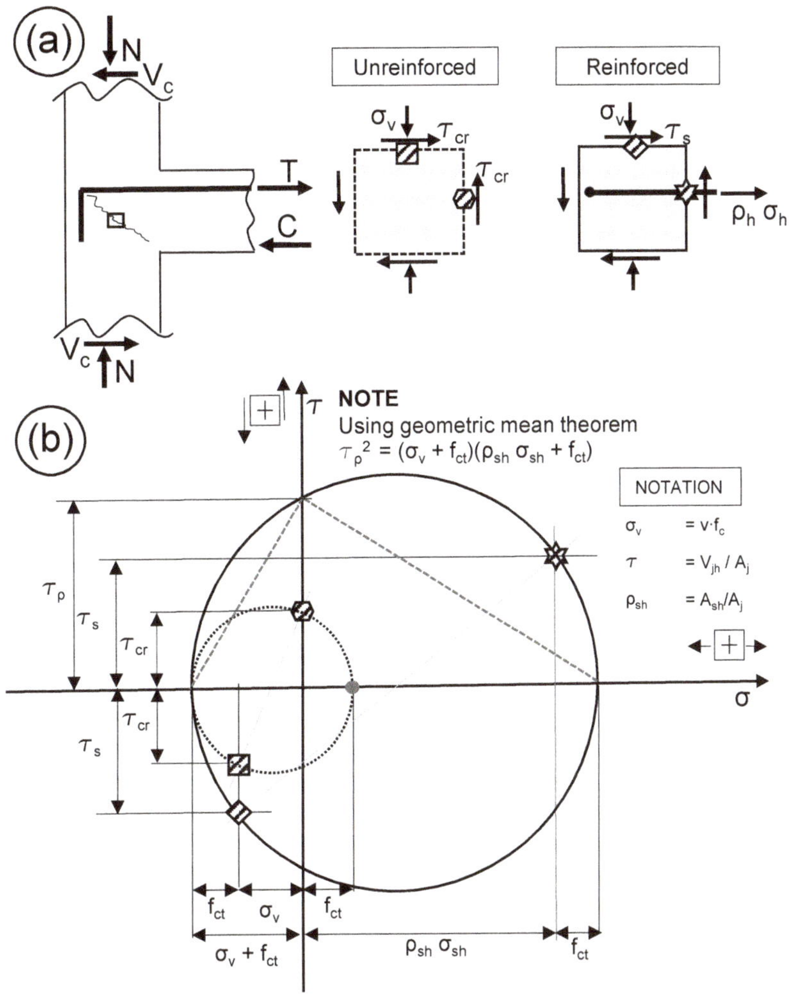

Figure 3. Principal stress method application to beam–column joint: (**a**) Free body diagram and stress state definition for un-reinforced and reinforced joint; (**b**) Mohr's Circle.

2.2.2. Nzsee-2017

Similarly to Eurocode 8, NZSEE-2017 [13] assumes the principal stress method to evaluate the joint shear strength. However, an empirical coefficient (k_j) is used: (i) to define the increased strength of interior joint with respect to exterior; and (ii) to represent the strength reduction at an imposed ductility level. The background of k_j is experimental. Further details are due to Hakuto [16]. k_j values and its effect on the Mohr's Circle are shown in Figure 4.

The normalized shear strengths, $v_{j,t}$ and $v_{j,c}$, are written as follows:

$$v_{j,t} = \frac{0.85}{\sqrt{f_c}}\sqrt{(k_j\sqrt{f_c})^2 + k_j\sqrt{f_c}\nu f_c} \tag{8}$$

$$v_{j,c} = \frac{0.85}{\sqrt{f_c}}\sqrt{(0.60f_c)^2 - 0.60f_c^2\nu}. \tag{9}$$

A reduction of 30% is prescribed for joint shear strength when the joint is subjected to bidirectional loading.

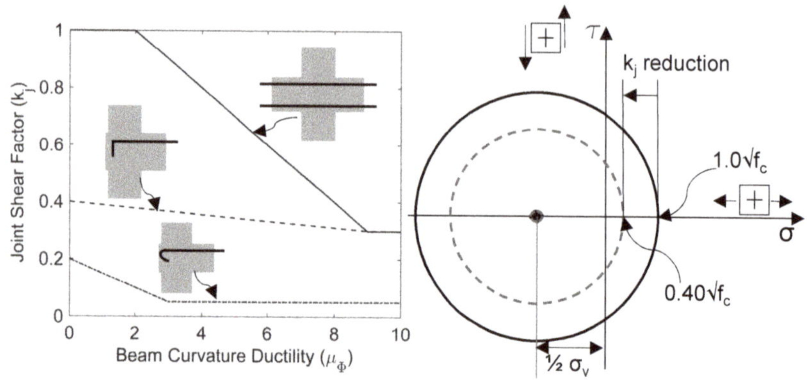

Figure 4. Definition of k_j coefficient according to New Zealand code [13].

2.2.3. Asce41-17

ASCE 41-17 [14] prescribes nominal values for normalized joint shear strength by selecting the joint type (interior or exterior) and the presence of transverse beam (defined as the beam orthogonal to the flexural plane). This definition has been adapted for three-dimensional joints, either having two-ways or three-ways converging beams, by defining the joint type in each flexural plane as shown in Figure 5.

The reduced value of the strength is given for "non-conforming" joints, i.e., joints characterized by the absence of horizontal reinforcement.

Additional provisions for bi-axial strength are given in ACI352 [11], i.e., the elliptical interaction domain is assumed.

2.2.4. Parametrical Comparison

A parametrical analysis was carried out using the predicted values of the normalized shear strength according to the reviewed building codes. The investigated parameters are: (i) the compressive strength of concrete (f_c); and (ii) the normalized axial load (ν) ranging from 10 to 60 MPa and from 0 to 0.70, respectively. The un-reinforced beam–column joints being the target of this paper, Equation (5) was not addressed. A value of k_j equal to 0.40 was assumed for NZSEE2017.

Figure 6a,b represents the EC8 results using Equations (6) and (7), respectively. The condition of ν larger than η implies the lost of meaning for Equation (7), which is then conveniently represented with zero resistance. Figure 6c,d shows the NZSEE2017 outcomes using Equations (8) and (9), respectively. Fixed values of shear strength are given by ASCE41-17 for different joint types, i.e., (INT) interior, (INT-TR) interior with transverse beam, (EXT) exterior, and (EXT-TR) exterior with transverse beam.

As expected, for un-reinforced joints, tensile strength is generally lower with respect to the compressive one. Exceptions are due to large values of the normalized axial force contemporary to low values of f_c, e.g., ν larger than 0.5 and f_c lower than 30 MPa for EC8.

The normalized shear strength calculated according to the reviewed building codes partly illustrates substantial differences to the point that a variation in the safety margin should be expected if an existing structure is assessed according to different codes as will be proved in the following numerical study.

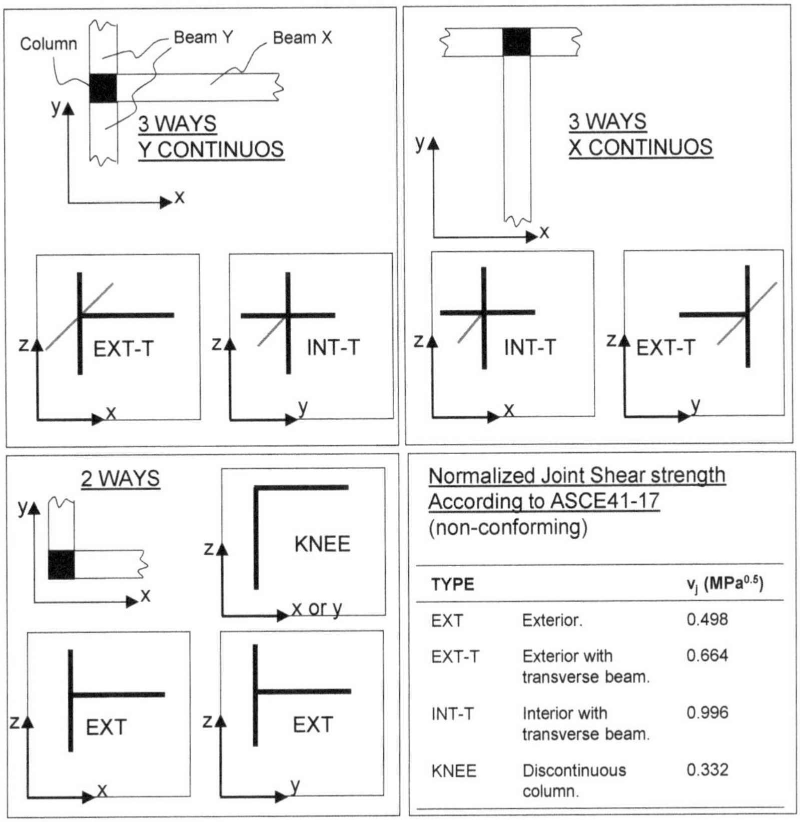

Figure 5. Beam–column classification according to ASCE41-17 [14].

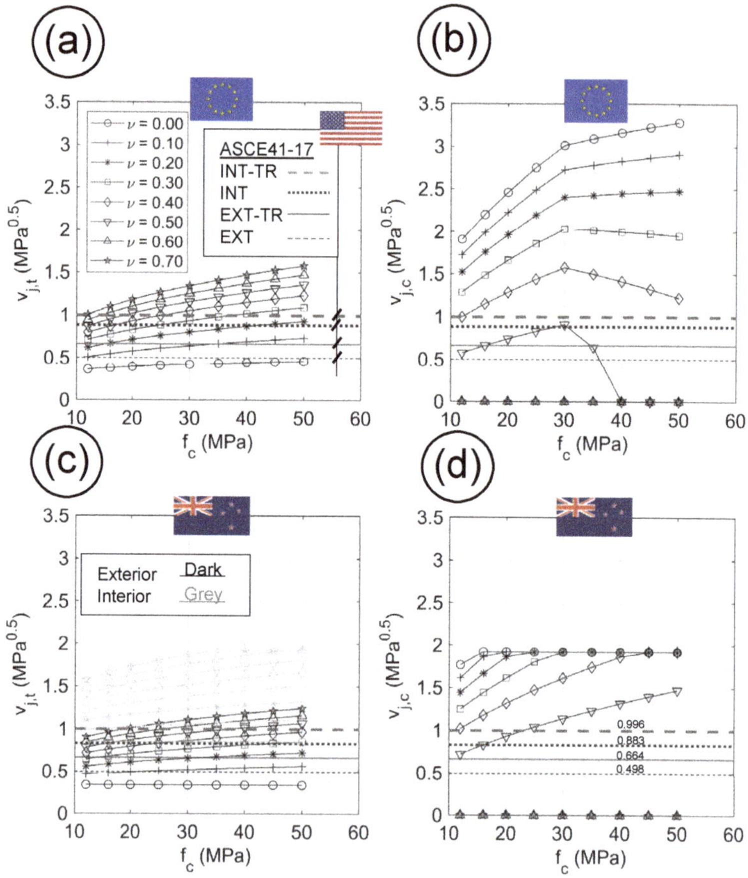

Figure 6. Results for parametrical analysis on beam–column joint strength (normalized shear strength v_j []MPa$^{0.5}$]) predicted by different building codes: (**a**) Equation (6); (**b**) Equation (7); (**c**) Equation (8); (**d**) Equation (9). All the plots includes the constant values of the prediction given by ASCE41-17 [14].

3. Numerical Study

3.1. Scope

A numerical study is presented addressing the estimated shear demand for beam–column joints of an RC 3D frame structure using different methods for the seismic structural analysis, such as modal-response-spectrum-analysis (MRSA), non-linear-static-analysis (NLSA) and non-linear-time-history (NLTH). The selected benchmark structure is the SPEAR frame which was tested in 2004 pseudo-dynamically (PsD) at the European Joint Research Centre at Ispra, Italy. Details of the structure as well as of the structural analysis methods are given in the following.

3.2. Details of SPEAR Frame

The frame was a three storey frame with two bays in both plan directions. The details of the structures are represented in Figure 7.

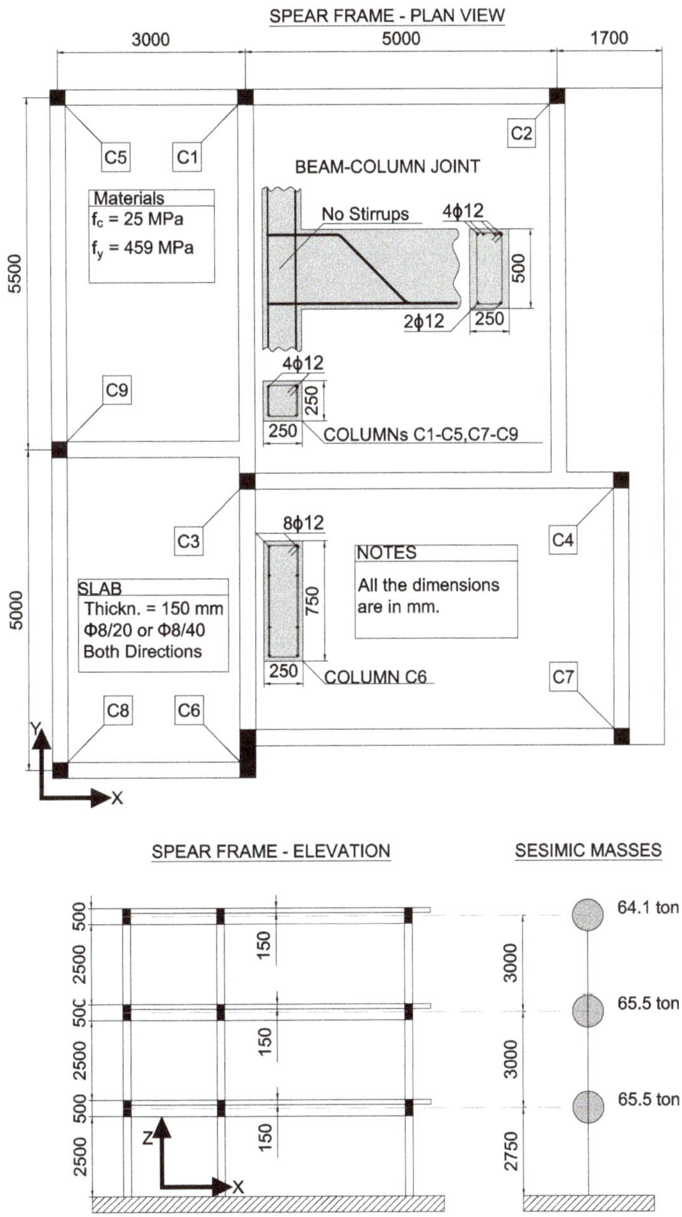

Figure 7. SPEAR frame layout.

The structure was conceived by Prof. Fardis [17] as a typical RC structure designed for gravity load only. Beams had a 250 × 500 mm cross-section, column had 250 × 250 mm cross-section, 250 × 750 mm cross-section was used for one column. Flat slab had 150 mm thickness with reinforcement mesh grid of 8 mm diameter bar spaced at 20 cm or 40 cm. Beam–column joints were characterized by the absence of stirrups and straight anchorage for the beam longitudinal reinforcement both at the bottom (two bars with diameter 12 mm) and top layers (four bars with diameter 12 mm). The estimated seismic masses were 65.5 tons for the first and second floors and 64.1 for the third floor.

The structure was tested both in the as-built condition and retrofitted using different techniques for column strengthening, i.e., fiber-reinforced polymer (FRP) or jacketing [17,18]. PsD tests were carried out with bi-axial seismic input. Montenegro 1979 Herceg–Novi ground motion record was selected. The recordings of the two orthogonal components of horizontal accelerations were modified from natural records to be compatible with the Eurocode 8 [10], Type 1 design spectrum, soil type C and 5% damping [19]. Time histories were scaled to conventional values of peak-ground-acceleration (PGA) equal to 0.15 g and 0.20 g.

During the strongest test (0.20 g of PGA), the obtained displacements at the third floor were in the order of 100 mm in both directions and 20 mrad of floor rotation. Columns were the most damaged members of the structure, especially at the second story; significant inclined cracks were observed on their compressive sides and on the tensile side at the beam–column interface [20]. The damage on the rectangular column C6 was less important, even though the crushing of concrete and cracks at the interface with beams were observed [18].

Through the years many researchers [7] addressed the SPEAR frame to validate numerical models; a survey of the most relevant studies is summarized in Table 1. NLTH analysis in conjunction with frame modeling was often adopted. In one case [21] brick finite elements were used. The validation chain usually adopted: (i) experimental results of the PsD test against NLTH; and (ii) NLTH against either NLSA or MRSA. To facilitate the validation, the modeling criteria used in this study, which are summarized in the following, were generally compliant with those assumed by the surveyed literature.

Table 1. Literature survey of numerical studies addressing SPEAR frame.

Author	Ele [(a)]	Type of Analysis			Scope of Investigation	Refs.
		MRSA	NLSA	NLTH		
Bento	F	✗	✓	✓	Validation of non-linear static procedures.	[22]
Bhatt	F	✗	✓	✓	Extension of Capacity Spectrum Method [23] to non-symmetric case.	[24]
Brun	F	✗	✗	✓	Validation of GC [25] time integration algorithms.	[26]
Di Ludovico	F	✗	✓	✓	Assessment of experimental results.	[27]
Dolsek	F	✗	✓	✗	Validation of a probabilistic seismic performance assessment.	[28]
Fajfar	F	✓	✗	✓	Definition of torsional amplification to be applied to N2 method [29].	[30,31]
Kosmopoulos	F	✓	✗	✓	Validation of chord rotation demand from MRSA.	[17]
Mola	F	✗	✓	✓	Assessment of experimental results.	[32]
Pardalopoulos	F	✗	✗	✓	Validation of lumped non-linear hinges representing brittle failures.	[33]
Reynouard	B	✗	✗	✓	Assessment of experimental results.	[21]
Rozman	F	✓	✓	✗	Comparison of seismic conforming variants of SPEAR frame.	[34]
Stratan	F	✗	✓	✓	Assessment of experimental results. Influence of modelling assumptions.	[35,36]

Notes. [(a)] Type of elements: [F] frame; [B] Bricks or shells.

3.3. Structural Modeling and Seismic Demand

A three dimensional frame structural model was developed using SAP2000 [37]. Details are given in Figure 8. Beams and columns were modeled using a lumped-plasticity [38] approach. Specifically, elastic 3D frame elements (two nodes, 12 dofs) were used in conjunction with rotational non-linear springs (plastic hinges) assigned at the same location (zero length) of the extreme nodes. The element size was kept equal to one half of the inter-storey height for the columns and half-bay for the beams. Columns were clamped at the base. Floor masses were lumped at the centers of mass. Diaphragm kinematic constraint was assigned at each floor.

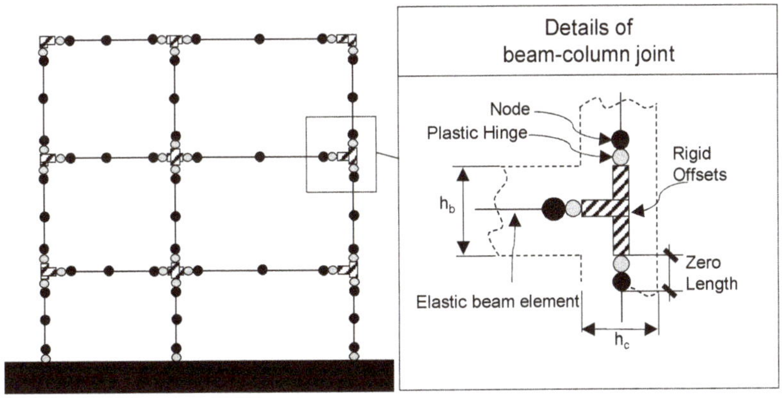

Figure 8. Modeling features.

The plastic hinges' constitutive law (moment vs rotation) was defined for each beam and columns, considering different axial load levels for the latter. In particular, a preliminary sectional analysis was carried out using Response2000 [39]. The yielding moment was evaluated from the moment–curvature diagrams (examples are given in Figure 9). Plastic rotation capacities were attributed empirically as shown in Figure 10, complying with the FEMA356 [40], similarly with what was presented by Stratan [35]. Takeda [41] model was used for the hysteretic response.

Slab contribution to the flexural capacity of the beam was neglected because it would activate only for large ductility demand [42], which was not reached at the imposed seismic level. Furthermore, a weak column condition is recognized even using the beam's nominal dimensions, as can be inferred from Figure 9. Indeed, despite the case of the strong column (cross-section 25 × 75 cm), beam bending capacity (nominal cross section 25 × 50 cm) results in being larger with respect to the columns having a 25 × 25 cm cross-section.

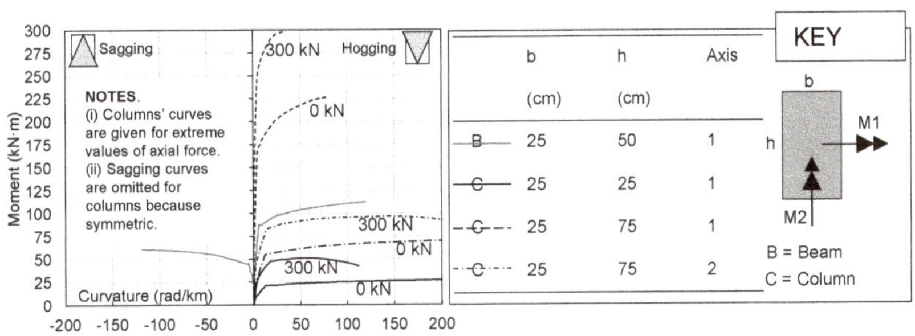

Figure 9. Moment–curvature diagrams for the cross sections of beams and columns of SPEAR frame. Results were obtained using the software Response 2000 [39].

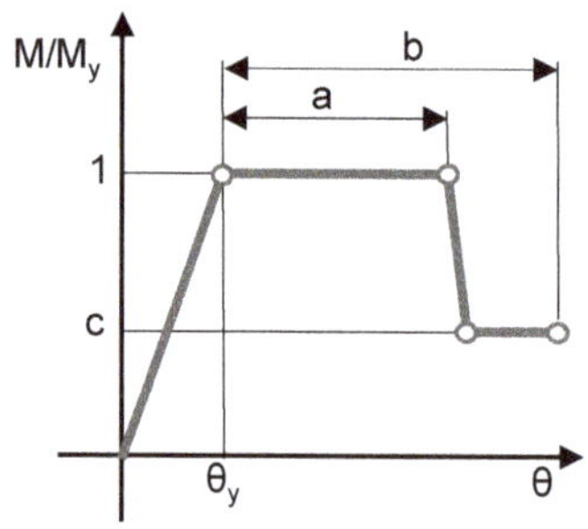

	N/Af_c	a	b	c
	(-)	(rad)		(-)
Beam	0.0	0.020	0.040	0.20
Column	0.1	0.006	0.015	0.20
	0.4	0.003	0.010	0.20

Figure 10. Definition of plastic hinges according to FEMA356 [40].

Beam–column joints were modeled as rigid. The half depth of both the column and beam was set as a rigid length. A discussion about differences in adopting center-line modeling, which assumes beams and columns that are entirely flexible, can be generally found in [43]. Moreover, Stratan [35] studied the influence for SPEAR frame's joint. In essence, higher shear demand is a consequence of rigid offsets. Center-line modeling may be used as a simple way to account for both the reduction of stiffness and strength due to additional deformation in the joint regions [44].

Different methods were used to carry out the seismic structural analysis as presented in Table 2. The seismic demand was chosen such that the comparison between different methods can be sustained as explained in the following.

NLTH assumed Herceg–Novi (1979) spectrum-compatible accelerograms as shown in Figure 11. The original time recordings, available at http://ngawest2.berkeley.edu in April 2022, were processed (using OpenSeismoMath [45]) to fit the elastic spectra (5% damping, Type I, Soil C), defined according to Eurocode 8 [10]. This procedure is consistent with what was described by Negro [19] for the execution of the PsD test.

MRSA used a design spectrum. A behavior factor (q) equal to 3.45 was assumed to be similar to that declared by Rozman [34].

NLSA simply assumed 100 mm roof displacement (center of mass), in both x and y directions, as a target displacement. This condition is comparable to the results obtained via both NLTH and MRSA. Gravity load and P-Δ effect were not taken into account.

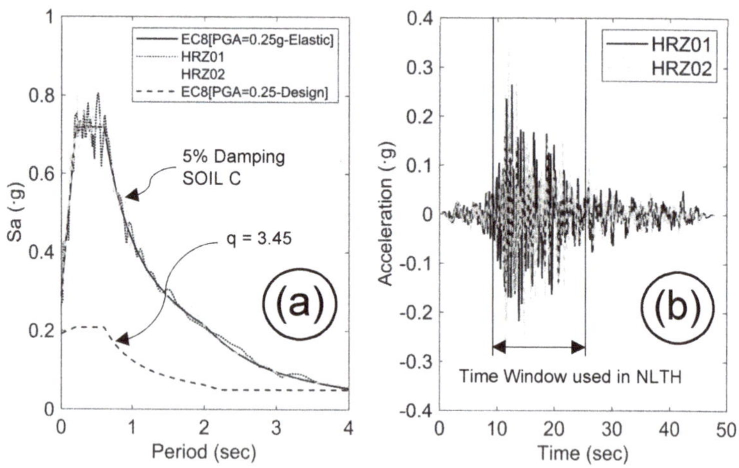

Figure 11. Seismic demand: (**a**) Response spectrum; (**b**) Herceg–Novi (1979) acceleration time histories.

Table 2. Details for structural analyses.

Method	Seismic Demand	Notes	Ref. [(a)]	Bi-Dir. [(b)]
MRSA	Design Response Spectrum with q = 3.45 according to Figure 11.	Modal Combination uses CQC rule.	[34]	30% rule.
NLSA	Capacity Spectrum with 100 mm target displacement in both directions.	Multi-Modal according to [46].	[18,36]	SRSS.
NLTH	Herceg-Novi Accelerogram (PGA = 0.25 g) according to Figure 11	Direct integration according to [47].	[33]	Simultaneous inputs.

Notes. [(a)] References used to compare the numerical results. [(b)] Method used to account for bi-directional seismic demand.

3.4. Validation

The structural analyses were validated against both experimental and numerical results derived from the literature. Results are briefly discussed, in the following, assuming the global behavior of the structure as a target, e.g., inter-story drifts, capacity curves, displacement time histories.

The influence of beam–column joints modeling was recognized by modal analysis. In particular, rigid offsets gave shorter periods for the first three modes as can be inferred from Table 3. Furthermore, the first three modes (Figure 12a) amount to almost 80% of the participating mass in both x and y directions. Clearly, the third mode involved torsional behavior, although only partly confirmed by previous numerical investigation. Nevertheless, MRSA produced comparable results with respect to Rozman [34] for drift values as shown in Figure 12b.

Table 3. Validation of Modal Analysis results.

Author/Model	Ref.	T_1 (s)	M_X (%)	M_Y (%)	T_2 (s)	M_X (%)	M_Y (%)	T_3 (s)	M_X (%)	M_Y (%)
DiLudovico	[27]	0.62	71.80	5.80	0.54	12.40	60.50	0.43	12.40	60.50
Reynouard	[21]	0.64	-	-	0.54	-	-	0.42	-	-
Stratan	[36]	0.57	-	-	0.48	-	-	0.39	-	-
Rozman	[34]	0.80	69.00	4.80	0.69	15.60	47.80	0.58	2.70	30.30
Negro (Experimental)	[19]	0.84	-	-	0.78	-	-	0.67	-	-
RIGID	This paper	0.85	55.00	0.00	0.47	2.18	56.00	0.35	31.00	27.00
FLEXI	This paper	1.24	54.00	0.00	0.64	2.10	56.00	0.50	31.00	27.00

Notes. M_X and M_Y are the participating masses in X and Y direction, respectively.

Results of NLSA are shown in Figure 13. For the sake of validation, two different load patterns were adopted, i.e., uniform distribution of forces and modal; the latter resulting in reduced lateral load bearing capacity. Plastic hinges formed in the columns. Most of the beams remained in the elastic regime. This evidence agreed with what was presented by Rozman [34].

Figure 14 shows the resulting displacement history (roof) in the y direction obtained via NLTH. The comparison with experimental outcomes shows that agreement was found for the oscillation period, the differences in peak values being in line with those obtained, for instance, by Pardalopoulos [33].

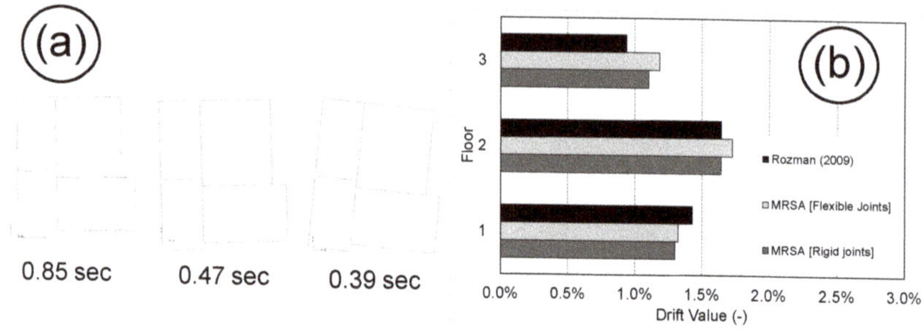

Figure 12. Validation of MRSA applied to the SPEAR frame: (**a**) Modal analysis results [Note. The first three relevant modes amount 86% and 82% of participating masses in X and Y directions, respectively]; (**b**) MRSA drifts results.

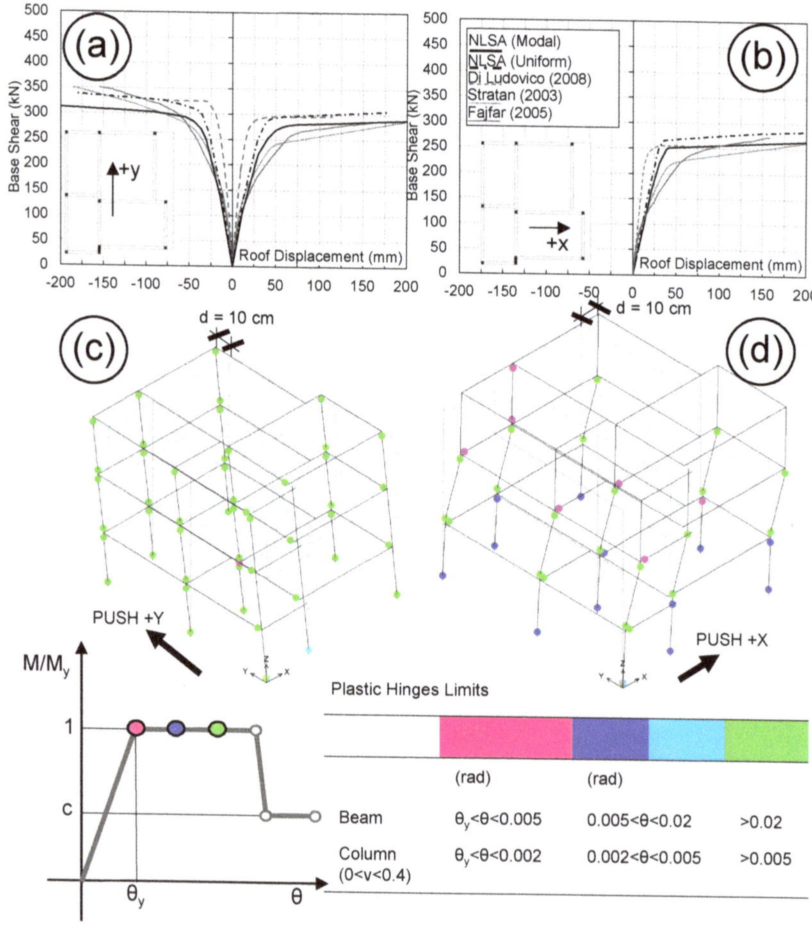

Figure 13. Validation of NLSA applied to the SPEAR frame: (**a**,**b**) capacity curves in y and x directions, respectively; (**c**,**d**) plastic hinges rotation at 10 cm roof displacement in y and x–directions, respectively. The Reader is referred to the color version of this figure.

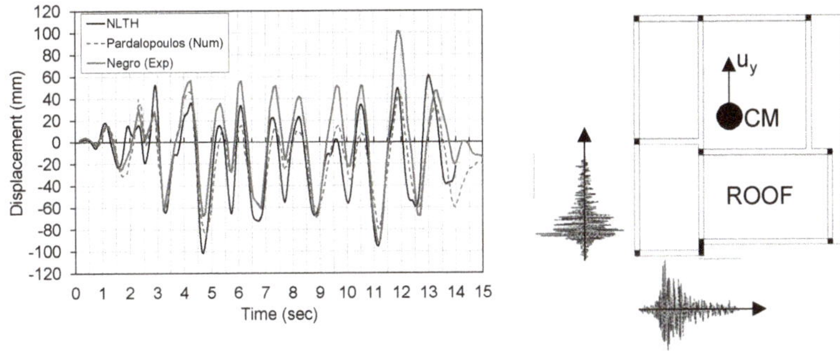

Figure 14. Validation of NLTH applied to the SPEAR frame, y–displacement at the roof.

4. Results and Discussion

4.1. Shear Demand at a Joint

The shear demand at a beam–column joint is not conventionally given as an output by commercial FEM software, thus post-processing is needed. Figure 15 shows the key aspects of the implemented post-processing. At the joint labeled "i" each converging beam delivers a bending moment, represented as arrows in the x-y plane. The horizontal shear demand (V_{jhx} and V_{jhy}) is given in each flexural plane as the summation of T-forces contribution coming from left and right side of the column neglecting the column's shear. The T forces represent the resultant of stresses at the column–face cross section under the applied bending moment. For the sake of simplicity, a fixed value of the internal lever arm is assumed equal to 0.83 times the effective depth of the beam. Different assumptions should not lead to significant changes for the final shear demand [48].

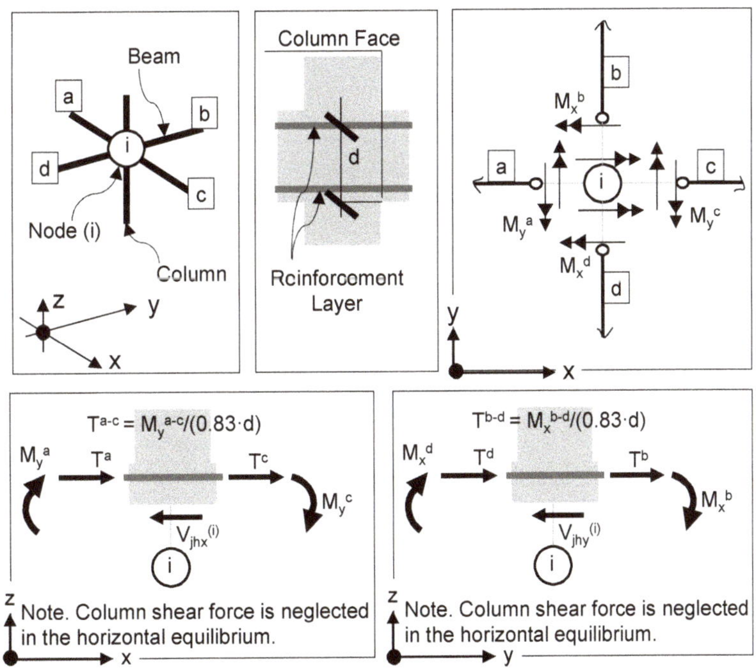

Figure 15. Post-process procedure to evaluate shear demand at beam–column joint "i". (Note "a", "b", "c", "d" labels define the converging beams at the node).

Considering a practical assessment scenario, the designer might be interested in envelope values. In this regard, the following additional assumptions either on modal combination or bi-directional seismic demand were made for the different analysis methods:

- (MRSA) Maximum values of shear demand at a joint are evaluated, considering the 30% rule for the bi-directional combination. Modal contributions were summed using the complete-quadratic combination (CQC);
- (NLSA) Pushover curves were calculated using the modal force distribution for the first three modes. Specifically, target displacements were assumed in the x-direction for the mode "1" and in the y-direction for the mode "2". Mode "3" being torsional, both the $\pm$x and $\pm$y directions of target displacements were considered. Subsequently, idealization of the pushover curve into a bi-linear curve is made. Shear demand at a joint is extracted for each idealized pushover curve, at the target displacement. The final shear demand is then computed, summing the modal contribution using CQC. This procedure is compliant with what was proposed by Chopra [46], named modal-pushover-analysis (MPA);
- (NLTH) Extreme values of shear demand at a joint are extracted from time histories. Acceleration of time histories apply simultaneously in $\pm$x and $\pm$y directions. Envelope of the results coming from different direction signs is made.

To summarize, the shear demand at each flexural plane of a beam–column joint is computed as a post-process starting from values of beams' bending moments converging at one node. Two major hypotheses were made: (i) the internal lever arm at the column face cross section of the beam is assumed to be constant; and (ii) the contribution to the horizontal equilibrium of the column's shear is neglected. The latter assumption has already been discussed in Section 2.1, defining the joint shear demand for a planar joint.

Different strategies with respect to the one proposed in this study might facilitate the output of shear demand. Usually they imply additional dofs for the numerical model. As an example, explicit joint modeling can be pursued via single or multi-springs [5]. Besides, the introduction of a stiff "shear hinge" was recently proposed by Pardalopoulos [33]. It is worth mentioning, though, that the introduction of a large stiffness, compared with the flexural stiffness of the beam, remains dubious in the light of numerical errors (the reader is refereed to paragraph 8.2.5 of [49]).

4.2. Bi-Axial Shear Demand Using NLTH

Figure 16 shows the bi-axial shear demand curves obtained via NLTH analysis for the case of +x and +y excitation. This condition is privileged in this discussion because it is equal to the pseudo-dynamic experimental input [19]. Additionally, the elliptical domain prescribed by ACI352 [11] is defined at each beam–column joint. The ellipse radii represent the joint shear strength evaluated according to ASCE41-17 [14]. Moreover, the results are compared with the yielding threshold defined according to Equation (4). For the sake of synthesis, only the first and the second floor results are discussed. The latter was characterized by the highest shear demand peaks. The third floor has been omitted because shear demand results were moderate.

As expected, the shear demand does not overcome the yielding threshold, this being inherently included in the definition backbone curves for the beams' plastic hinges. The elliptical envelope shape of the shear demand orbits is recognized for the majority of the cases. The principal axes of the envelope do not necessarily coincide with x and y directions. This result is consistent with what was proven, numerically, by Menun [50] for bi-axial bending interaction at the columns.

Cases where the shear demand exceeds the shear resistance deserve a particular explanation. In fact, the studies describing the experimental results [18,20] of the SPEAR frame did not report beam–column joint failure and diagonal cracking has been described as occasional. In this regard, and by observing that the shear resistance is exceeded especially for the cases where bi-axial interaction becomes significant (i.e., corner joint), it follows that the elliptical strength domain is conservative. Nothing can be said about a possible over-estimation of shear demand which cannot be sustained conclusively due to the impossibility to validate the obtained results against experimental references. To the authors' knowledge, only Stratan [36] attempted to numerically evaluate the shear demand at the beam–column joint of the SPEAR frame. Results were reported as mean extreme values of NLTH analysis (the reader may find the definition of mean extreme value in the Section titled "Probability Distribution for extreme values" in [51]). Comparison with those is given in the following. Besides, Pardalopoulos [33] modeled the SPEAR frame including "shear hinges" at beam–column joints but shear demands were not presented explicitly.

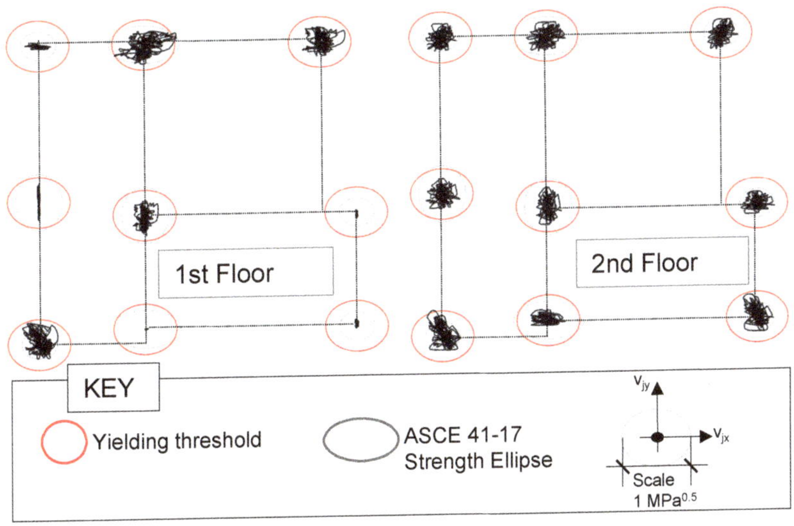

Figure 16. Bi-axial shear demand of beam–column joint obtained with NLTH (Case of +x and +y excitation). The reader is referred to the color version of this figure.

4.3. Evaluated Shear Strength

Shear strengths evaluated according to building codes reviewed in Section 2.2 are graphically presented in Figure 17. The numerical values are given in Table 4. For the sake of synthesis, both the Figure and the Table include also the shear demand which is discussed subsequently.

Shear strength has been evaluated considering the beam–column joints as being three dimensional without stirrups. The definition of joint type, at each flexural plane, according to Figure 5 applies only for ASCE41-17 and NZSEE2017. No distinction is given for EC8. A reduction of 30% has been applied to the strength values of NZSEE2017, complying with the provision for the bi-axial condition. The axial load for the column is considered at the seismic combination. Only tensile failure (e.g., Equation (6)) is reported because it has been recognized as dominant with respect to the compressive. Moreover, a constant value of 0.4 for the coefficient k_j (defined at Figure 4) in Equation (8) has been assumed according to the evidence of reduced ductility demand for the beam.

Differences between predicted values of shear strength, using different building codes, are recognized. NZSE2017 gives generally the lowest prediction with respect to EC8 and ASCE41-17. The latter prevails for knee joints. Figure 18 shows the ratio between the evaluated shear demand via NLTH (assumed as the most accurate with respect to NLSA and MRSA) with respect to predicted shear strength (demand-to-strength) as a function of the different joint type. Demand-to-strength ratio is expected to be less than one according to the experimental absence of joint shear failure. Values larger than one are recognized mostly for exterior joint type. This condition can be explained as an excess of safety margins for corner joints if the plane exterior joint type is attributed to both the flexural planes.

Probably, a larger strength should be expected as the result of the confinement offered by the beams in both directions as claimed by Kurose [52] discussing experimental evidence.

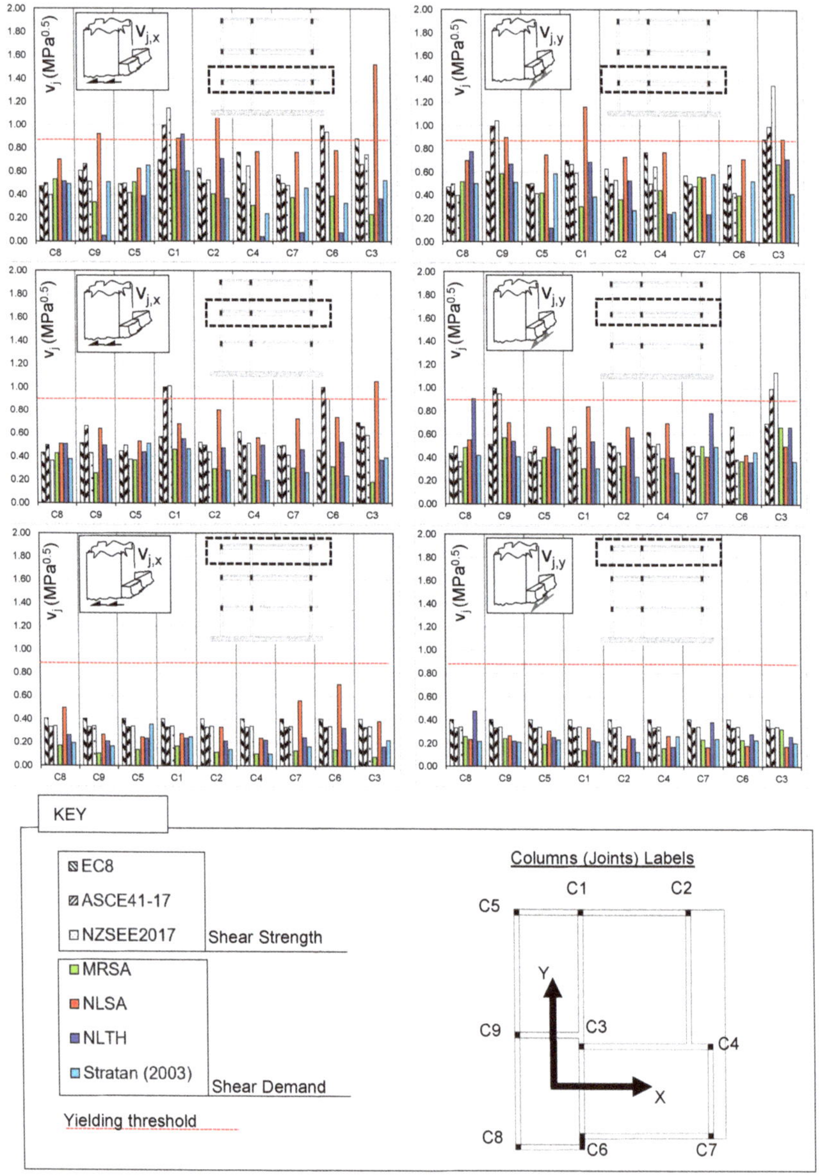

Figure 17. Evaluated joint shear demand and strength for beam–column joints of SPEAR frame. The reader is referred to the color version of this figure.

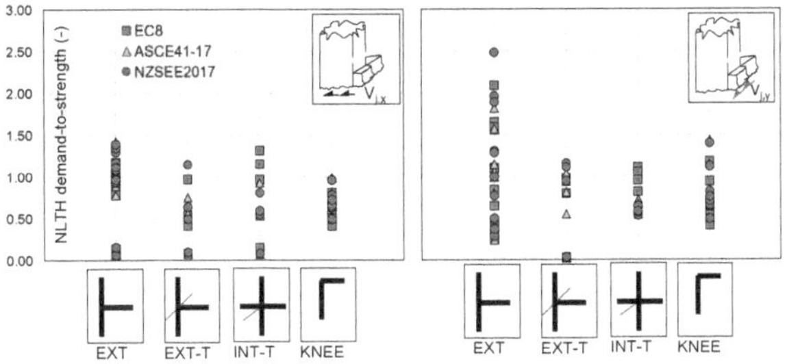

Figure 18. NLTH Demand-to-strength ratios for the beam–column joints of SPEAR frame. The reader is referred to the color version of this figure. (Note. NLTH demands are evaluated using extreme absolute values.)

Table 4. Shear strength and demand evaluated for beam–column joints of the SPEAR frame.

					Shear Strength					**Shear Demand**					
					$min(v_{j,x})$			$min(v_{j,y})$		$max(v_{j,x})$			$max(v_{j,y})$		
					All the Values Are Expressed in MPa $^{0.5}$										
COL (a)	**FL** (b)	**T-XZ** (c)	**T-YZ** (d)	ν (e)	**EC8** $v_{j,x-y}$	**ASCE41-17** $v_{j,x}$	$v_{j,y}$	**NZSEE2017** $v_{j,x}$	$v_{j,y}$	**MRSA** $v_{j,x}$	$v_{j,y}$	**NLSA** $v_{j,x}$	$v_{j,y}$	**NLTH** $v_{j,x}$	$v_{j,y}$
C8	1	EXT	EXT	0.03	0.471	0.498	0.498	0.397	0.397	0.531	0.517	0.703	0.700	0.515	0.778
C9	1	EXT-T	INT-T	0.10	0.607	0.664	0.996	0.511	1.043	0.337	0.588	0.925	0.900	0.049	0.672
C5	1	EXT	EXT	0.04	0.493	0.498	0.498	0.415	0.415	0.509	0.419	0.627	0.750	0.388	0.119
C1	1	INT-T	EXT-T	0.16	0.701	0.996	0.664	1.143	0.591	0.618	0.303	0.884	1.161	0.920	0.683
C2	1	EXT	EXT	0.11	0.626	0.498	0.498	0.528	0.528	0.408	0.364	1.074	0.728	0.709	0.525
C4	1	EXT	EXT	0.21	0.767	0.498	0.498	0.647	0.647	0.309	0.444	0.771	0.769	0.039	0.241
C7	1	EXT	EXT	0.08	0.571	0.498	0.498	0.481	0.481	0.378	0.562	0.768	0.556	0.075	0.241
C6	1	INT-T	EXT-T	0.04	0.504	0.996	0.664	0.940	0.425	0.391	0.404	0.784	0.714	0.075	0.014
C3	1	EXT-T	INT-T	0.31	0.884	0.664	0.996	0.747	1.352	0.231	0.673	1.521	0.886	0.369	0.718
C8	2	EXT	EXT	0.01	0.433	0.498	0.498	0.364	0.364	0.429	0.486	0.512	0.553	0.508	0.905
C9	2	EXT-T	INT-T	0.05	0.514	0.664	0.996	0.433	0.950	0.257	0.573	0.642	0.701	0.498	0.540
C5	2	EXT	EXT	0.02	0.447	0.498	0.498	0.377	0.377	0.370	0.402	0.535	0.662	0.441	0.493
C1	2	INT-T	EXT-T	0.08	0.574	0.996	0.664	1.009	0.484	0.462	0.304	0.684	0.838	0.553	0.536
C2	2	EXT	EXT	0.05	0.524	0.498	0.498	0.441	0.441	0.296	0.327	0.800	0.664	0.477	0.569
C4	2	EXT	EXT	0.11	0.614	0.498	0.498	0.518	0.518	0.241	0.393	0.563	0.695	0.500	0.397
C7	2	EXT	EXT	0.04	0.491	0.498	0.498	0.414	0.414	0.304	0.497	0.728	0.406	0.463	0.783
C6	2	INT-T	EXT-T	0.02	0.458	0.996	0.664	0.897	0.385	0.315	0.369	0.741	0.422	0.528	0.363
C3	2	EXT-T	INT-T	0.16	0.695	0.664	0.996	0.587	1.138	0.183	0.663	1.050	0.499	0.374	0.664
C8	3	KNEE	KNEE	0.00	0.404	0.332	0.332	0.340	0.340	0.173	0.257	0.498	0.234	0.195	0.217
C9	3	KNEE	KNEE	0.00	0.404	0.332	0.332	0.340	0.340	0.106	0.241	0.268	0.265	0.170	0.208
C5	3	KNEE	KNEE	0.00	0.404	0.332	0.332	0.340	0.340	0.140	0.188	0.249	0.306	0.358	0.228
C1	3	KNEE	KNEE	0.00	0.404	0.332	0.332	0.340	0.340	0.170	0.140	0.279	0.332	0.250	0.210
C2	3	KNEE	KNEE	0.00	0.404	0.332	0.332	0.340	0.340	0.118	0.151	0.332	0.263	0.143	0.123
C4	3	KNEE	KNEE	0.00	0.404	0.332	0.332	0.340	0.340	0.104	0.156	0.239	0.260	0.105	0.259
C7	3	KNEE	KNEE	0.00	0.404	0.332	0.332	0.340	0.340	0.133	0.230	0.562	0.166	0.168	0.235
C6	3	KNEE	KNEE	0.00	0.404	0.332	0.332	0.340	0.340	0.144	0.229	0.702	0.177	0.141	0.224
C3	3	KNEE	KNEE	0.00	0.404	0.332	0.332	0.340	0.340	0.080	0.323	0.384	0.174	0.224	0.201

Notes. (a) Column label according to Figure 7. (b) Floor number. (c,d) Joint type according to Figure 5. (e) Normalized axial load at top portion of the column. Shear demand at yielding of beam's longitudinal reinforcement (top layer) is equal to 0.89 MPa$^{0.5}$.

To summarize, beam–column joints of the SPEAR frame showed significant strength differences. Two major aspects impacted the most , i.e.: (i) the definition of the joint type in each flexural plane; and (ii) the use of a different building code. For the latter, a discussion on the strength accuracy (e.g., [4]) cannot be sustained conclusively because the compared shear demands represent a condition far from failure, as it was observed experimentally.

4.4. Comparison of Shear Demands

The envelope values of shear demand, computed for different structural analysis methods, are reported in Table 4 and Figure 17.

It is clearly recognized that NLSA: (i) gave larger values if compared with MRSA and NLTH; and (ii) generally exceeded the lowest predicted strength. The latter, which contradicts the experimental evidence, can be partly sustained by recalling what was claimed by Chopra for the estimation of plastic hinge rotations using the modal-pushover (MPA) [53]: *. . . All pushover analysis procedures considered do not seem to compute accurately local response quantities, such as hinge plastic rotations. Thus, the structural engineering profession should examine the present trend of comparing computed hinge plastic rotations against rotation limits established in FEMA-273 to judge structural performance. Perhaps structural performance evaluation should be based on storey drifts that are known to be closely related to damage and can be estimated to a higher degree of accuracy by pushover analyses.* In this context, element forces could also be assimilated to local response quantities, thus less accuracy should be expected when MPA is employed. As a proof of that, cases where the estimated shear demand overcame the yielding force were recognized. Specifically, according to Equation (4), a value of 0.88 $MPa^{0.5}$ was obtained for the hogging type moment, having considered an overstrength factor equal to γ_{Rd} 1.25. Although the sagging type moment gave half of the value, the comparison with envelopes of shear demand can be sustained only for the former case.

While the adoption of MPA has been considered as the state-of-the art for the inelastic analysis of non-symmetric structure (possibly torsionally flexible) under bi-axial seismic input, the authors are aware of other procedures such as the "modified" N2 method [30,31]. Although its computational effort is reduced with respect to MPA, the same lack of accuracy in predicting the element forces should be expected.

Differences were recognized by comparing NLTH values with those given by Stratan [36]. The deviation trend is not easily discernible to the point that, aside from general differences in modeling, both (i) a different definition of shear demand and (ii) use of mean extreme values rather than absolute might have an influence.

Finally, the shear demands are represented as a function joint type in Figure 19. The exterior joints with a transverse beam were characterized by the highest shear demand by using NLSA. Interior joints prevailed with MRSA. NLTH did not show a quite clear trend.

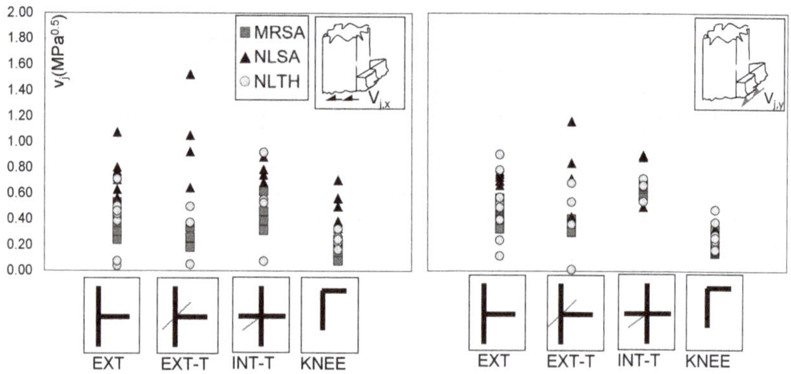

Figure 19. Shear demand of beam–column joints of SPEAR frame as a function of the joint type.

5. Conclusions

A numerical study was presented considering a notorious benchmark in the context of RC structures against earthquake loading, i.e., the SPEAR frame. This paper focused on shear demand at beam–column joints. Results were obtained using a three-dimensional numerical model validated against past studies addressing the same structure. Beam–column joints were modeled with rigid offsets. Flexural type plastic hinges were assumed to be inelastic sources for both beams and columns. Different methods for structural analysis were employed such as MRSA, NLSA and NLTH, assuming a comparable seismic demand. The obtained shear demands at joints were compared to shear strengths evaluated according to building codes targeted to existing structures. The following conclusions can be drawn:

1. The shear demand at a beam–column joint being a non-conventional output, a post-process based on nodal moments is needed. Two major hypotheses were made: (i) the internal lever arm at the column face cross section, of the beam, was assumed constant and (ii) the contribution to the horizontal equilibrium of the column's shear was neglected. The last assumption's results are conservative;
2. Shear strength evaluation should be extended to all the beam–column joints in a three-dimensional frame, distinguishing two flexural planes. Significant strength differences might be influenced by (i) the joint type attribute in each flexural plane and (ii) the use of different building codes;
3. The larger safety margin was recognized for corner joints by assigning to them the exterior joint type in both the flexural planes. Larger strength should be expected as a result of the confinement offered by the beams in both directions. Besides, elliptical strength interaction domain results were conservative when compared with NLTH shear demand orbits;
4. Among the reviewed building codes, EC8 produced larger strength predictions than ASCE41-17 and NZSEE2017. The latter produced the lowest ones. Discussion about the strength accuracy was not intended because the compared shear demands represent a condition far from failure;
5. NLSA was proven to estimate the larger shear demand with respect to MRSA and NLTH. Differences were explained as a consequence of the possible inaccuracy of NLSA, using modal combination, in the evaluation of elements' forces. Such evidence needs to be investigated further since NLSA is frequently adopted in the assessment of existing RC structures as a compromise between MRSA and NLTH.

Author Contributions: Conceptualization, A.M.; Methodology, A.M./G.M.; Software, A.M.; Validation, A.M.; Data Curation, A.M.; Writing—Original Draft and Reviews, A.M.; Writing—Reviews, G.M.; Supervision, G.M. All authors have read and agreed to the published version of the manuscript.

Funding: This research received no external funding.

Data Availability Statement: Data are available upon request.

Acknowledgments: Open access publication was financially supported by the Board of the PhD Programme of Structural, Seismic and Geotechnical Engineering at Politecnico di Milano.

Conflicts of Interest: The authors declare no conflict of interest.

Abbreviations

The following abbreviations are used in this manuscript:

CQC	Complete Quadratic Combination
dofs	degrees-of-freedom
MPA	Modal-pushover-Analysis
MRSA	Modal Response Spectrum
NLSA	Non-Linear Static Analysis

NLTH	Non-Linear Time History
PGA	Peak Ground Acceleration
PsD	Pseudo-dynamic
RC	Reinforced Concrete

References

1. Varum, H. Seismic Assessment, Strengthening and Repair of Existing Buildings. Ph.D. Thesis, Universidade de Aveiro, Aveiro, Portugal, 2003.
2. Kam, W.Y.; Pampanin, S.; Elwood, K. Seismic performance of reinforced concrete buildings in the 22 February Christchurch (Lyttelton) earthquake. *Bull. N. Z. Soc. Earthq. Eng.* **2011**, *44*, 239–278. [CrossRef]
3. Pantazopoulou, S.; Bonacci, J. On earthquake-resistant reinforced concrete frame connections. *Can. J. Civ. Eng.* **1994**, *21*, 307–324. [CrossRef]
4. Parate, K.; Kumar, R. Shear strength criteria for design of RC beam–column joints in building codes. *Bull. Earthq. Eng.* **2019**, *17*, 1407–1493. [CrossRef]
5. Celik, O.C.; Ellingwood, B.R. Modeling beam–column joints in fragility assessment of gravity load designed reinforced concrete frames. *J. Earthq. Eng.* **2008**, *12*, 357–381. [CrossRef]
6. Moehle, J.P. *Seismic Deisgn of RC Buildings*; McGraw-Hill Education: New York, NY, USA, 2015. [CrossRef]
7. Negro, P.; Fardis, M.N. *Seismic Performance Assessment and Rehabilitation of Existing Buildings (SPEAR) International Workshop Proceedings*; Office for Official Publication of the European Communities: Luxembourg, 2005.
8. Paulay, T.; Priestley, N. *Seismic Design for Concrete and Masonry Buildings*; Wiley: New York, NY, USA, 1992.
9. ACI. *ACI 318-19 Building Code Requirements for Structural Concrete and Commentary*; American Concrete Institute (ACI): Farmington Hills, MI, USA, 2019. [CrossRef]
10. *EN 1998-1:2004*; Eurocode 8: Design of Structures for Earthquake Resistance. Part1: General Rules, Seismic Actions and Rules for Buildings. European Committee for Standardization (CEN): Brussels, Belgium, 2004.
11. *ACI 352R-02*; Recommendations for Design of Beam-Column Connections in Monolithic Reinforced Concrete Structures (ACI-ASCE 352-02). American Concrete Institute (ACI): Farmington Hills, MI, USA, 2002.
12. *CEN/TC250/SC8*; prEN1998-1-3:2022. Eurocode 8: Design of Structures for Earthquake Resistance. Part 1–3: Assessment and Retrofitting of Buildings and Bridges. European Committee for Standardization (CEN): Brussels, Belgium, 2004.
13. NZSEE. *The Seismic Assessement of Existing Building*; Technical Report; New Zeland Society of Earthquake Engineering: Wellington, New Zealand, 2017.
14. ASCE. *ASCE/SEI, 41-17 Seismic Evaluation and Retrofit of Existing Buildings*; American Society of Civil Engineers: Reston, VA, USA, 2017; p. 623.
15. Oloukun, F.A. Prediction of concrete tensile strength from compressive strength: evaluation of existing relations for normal weight. *ACI Mater. J.* **1991**, *88*, 302–309.
16. Hakuto, S.; Park, R.; Tanaka, H. Seismic Load Test on Interior and Exterior Beam-Column Joints with Substandards Reinforcing Details. *ACI J.* **2000**, *97*, 11–25.
17. Kosmopoulos, A.; Fardis, M.N. Estimation of inelastic seismic deformations in asymmetric multistorey RC buildings. *Earthq. Eng. Struct. Dyn.* **2007**, *36*, 1209–1234. [CrossRef]
18. Di Ludovico, M. Seismic Behavior of a Full-Scale RC Structure Retrofitted Using GFRP Laminates. *J. Struct. Eng.* **2008**, *134*, 810–821. [CrossRef]
19. Negro, P.; Mola, E.; Molina, F.J.; Magonette, G.E. Full-scale PsD testing of a torsionally unbalanced three-storey non-seismic RC frame. In Proceedings of the 13th World Conference on Earthquake Engineering, Vancouver, BC, Canada, 1–6 August 2004; p. 968.
20. Molina, F.J.; Negro, P. Bidirectional pseudodynamic technique for testing a three-storey reinforced concrete building. In Proceedings of the 13th World Conference on Earthquake Engineering, Vancouver, BC, Canada, 1–6 August 2004; p. 75.
21. Reynouard, J.M.; Ile, N.; Brun, M. Inelastic seismic analysis of the SPEAR test building. *Eur. J. Environ. Civ. Eng.* **2010**, *14*, 855–867. [CrossRef]
22. Bento, R.; Bhatt, C.; Pinho, R. Using nonlinear static procedures for seismic assessment of the 3D irregular SPEAR building. *Earthq. Struct.* **2010**, *1*, 177–195. [CrossRef]
23. Freeman, S.A. Review of the Development of the Capacity Spectrum Method. *ISET J. Earthq. Technol.* **2004**, *41*, 113.
24. Bhatt, C.; Bento, R. Extension of the CSM-FEMA440 to plan-asymmetric real building structures. *Earthq. Eng. Struct. Dyn.* **2010**, *40*, 1263–1282. [CrossRef]
25. Combescure, A.; Gravouil, A. A numerical scheme to couple subdomains with different time-steps for predominantly linear transient analysis. *Comput. Methods Appl. Mech. Eng.* **2002**, *191*, 1129–1157. [CrossRef]
26. Brun, M.; Batti, A.; Limam, A.; Combescure, A. Implicit/explicit multi-time step co-computations for predicting reinforced concrete structure response under earthquake loading. *Soil Dyn. Earthq. Eng.* **2012**, *33*, 19–37. [CrossRef]
27. Di Ludovico, M. Comparative Assessment of Seismic Rehab Techniques on the Full Scale SPEAR Structure. Ph.D. Thesis, Università degli Studi di Napoli Federico II, Napoli, Italy, 2008.

28. Dolsek, M.; Fajfar, P. Simplified probabilistic seismic performance assessment of plan-asymmetric buildings. *Earthq. Eng. Struct. Dyn.* **2007**, 36, 2021–2041. [CrossRef]
29. Fajfar, P. A Nonlinear Analysis Method for Performance-Based Seismic Design. *Earthq. Spectra* **2000**, *16*, 573–592. [CrossRef]
30. Fajfar, P.; Marušić, D.; Peruš, I. Torsional effects in the pushover-based seismic analysis of buildings. *J. Earthq. Eng.* **2005**, *9*, 831–854. [CrossRef]
31. Fajfar, P.; Marušić, D.; Peruš, I. The Extension of the N2 method to asymmetric buildings. In Proceedings of the 4th European Workshop on the Seismic Behaviour of Irregular and Complex Structures, Thessaloniki, Greece, 26–27 August 2005; p. 41.
32. Mola, E.; Negro, P.; Pinto, A. Evaluation of current approaches for the analysis and design of multi-storey torsionally unbalanced frames. In Proceedings of the 13th World Conference on Earthquake Engineering, Vancouver, BC, Canada, 1–6 August 2004; p. 3304.
33. Pardalopoulos, S.I.; Pantazopoulou, S.; Manolis, G.D. On the Modeling and Analysis of Brittle Failure in Existing RC Structures Due to Seismic Loads. *Appl. Sci.* **2022**, *12*, 1602. [CrossRef]
34. Rozman, M.; Fajfar, P. Seismic response of a RC frame building designed according to old and modern practices. *Bull. Earthq. Eng.* **2009**, *7*, 779–799. [CrossRef]
35. Stratan, A.; Fajfar, P. *Influence of Modelling Assumptions and Analysis Procedure on the Seismic Evaluation of Reinforced Concrete GLD Frames;* Technical Report; University of Lubjana: Lubjana, Slovenia, 2002.
36. Stratan, A.; Fajfar, P. *Seismic Assessment of the SPEAR Test Structure;* Technical Report January; University of Ljubljana: Ljubljana, Slovenia, 2003.
37. CSI. *CSI Analysis Reference Manual. For SAP2000, ETABS, SAFE, CsiBridge;* [Version 2013]; CSI: Berkeley, CA, USA, 2013.
38. Giberson, F. The Response of Non-Linear Multi-Story Structures Subjected to Earthquake Excitation. Ph.D. Thesis, California Institute of Technology, Pasadena, CA, USA, 1967.
39. Bentz, E.C. Sectional Analysis of Reinforced Concrete Members. Ph.D Thesis, University of Toronto, Toronto, ON, Canada, 2000.
40. FEMA. *FEMA 356 Seismic Rehabilitation of Buildings;* FEMA: Washington, DC, USA, 2000.
41. Takeda, T. Reinforced Concrete Response to Simulated Earthquakes. *J. Struct. Div.* **1970**, *96*, 19–26. [CrossRef]
42. Pantazopoulou, S.; French, C. Slab Participation in Practical Design of R.C. Frames. *ACI Struct. J.* **2001**, *98*, 479–489.
43. Birely, A.C.; Lowes, L.N.; Dawn, E.L. Linear analysis of concrete frames considering joint flexibility. *ACI Struct. J.* **2012**, *109*, 381–391. [CrossRef]
44. Fajfar, P.; Dolsek, M.; Marušić, D.; Stratan, A. Pre-and post-test mathematical modelling of the SPEAR building. In Proceedings of the SPEAR International Workshop Proceedings, Ispra, Italy, 4–5 April 2005; pp. 173–188.
45. Papazafeiropoulos, G.; Plevris, V. OpenSeismoMatlab: A new open-source software for strong ground motion data processing. *Heliyon* **2018**, *4*, e00784. [CrossRef]
46. Chopra, A.K.; Goel, R.K. A modal pushover analysis procedure to estimate seismic demands for unsymmetric-plan buildings. *Earthq. Eng. Struct. Dyn.* **2004**, *33*, 903–927. [CrossRef]
47. Newmark, N.M. Method of Computation for Structural Dynamics. *J. Eng. Mech. Div.* **1959**, *2*, 1235–1264. [CrossRef]
48. Shayanfar, J.; Bengar, H.A.; Parvin, A. Analytical prediction of seismic behavior of RC joints and columns under varying axial load. *Eng. Struct.* **2018**, *174*, 792–813. [CrossRef]
49. Bathe, K.J. *Finite Element Procedures*, 2nd ed.; K.J. Bathe: Watertown, MA, USA, 2006
50. Menun, C. Envelopes for seismic response vectors, I: Applications. *J. Struct. Eng.* **2000**, *126*, 474–481. [CrossRef]
51. Clough, R.W.; Penzien, J. *Dynamics of Structures*, 3rd ed.; Computers and Structures: Berkeley, CA, USA, 2003.
52. Kurose, Y.; Guimaraes, G.; Zuhua, L.; Kreger, M.; Jirsa, J.O. Evaluation of slab-beam–column connections subjected to bidirectional loading. *ACI Spec. Publ.* **1991**, *123*, 39–67.
53. Chopra, A.K.; Goel, R.K. A modal pushover analysis procedure for estimating seismic demands for buildings. *Earthq. Eng. Struct. Dyn.* **2002**, *31*, 561–582. [CrossRef]

applied sciences

MDPI

Article

Seismic Resistance and Parametric Study of Building under Control of Impulsive Semi-Active Mass Damper

Ming-Hsiang Shih [1] and Wen-Pei Sung [2,*]

1 Department of Civil Engineering, National Chi-Nan University, Nantou 54561, Taiwan; mhshih@ncnu.edu.tw
2 Department of Landscape Architecture, National Chin-Yi University of Technology, Taichung 41170, Taiwan
* Correspondence: wps@ncut.edu.tw; Tel.: +886-963-179-668

Abstract: When high-rise buildings are shaken due to external forces, the facilities of the building can be damaged. A Tuned Mass Damper (TMD) can resolve this issue, but the seismic resistance of TMD is exhausted due to the detuning effect. The Impulsive Semi-Active Mass Damper (ISAMD) is proposed with fast coupling and decoupling at the active joint between the mass and structure to overcome the detuning effect. The seismic proof effects of a high-rise building with TMD and ISAMD were compared. The numerical analysis results indicate that: (1) the reduction ratio of the maximum roof displacement response and the mean square root of the displacement reduction ratio of the building with the ISAMD were higher than 30% and 60%, respectively; (2) the sensitivity of the efficiency index to the frequency ratio of the ISAMD was very low, and detuning did not occur in the building with the ISAMD; (3) to achieve stable seismic resistance of the ISAMD, its frequency ratio should be between 2 and 4; (4) the amount of displacement of the control mass block of the ISAMD can be reduced by enhancing the stiffness of the auxiliary spring of the ISAMD; and (5) the proposed ISAMD has a stable control effect, regardless of the earthquake distance.

Keywords: Impulsive Semi-Active Mass Damper; frequency ratio; mass ratio; structural displacement; maximum roof displacement

Citation: Shih, M.-H.; Sung, W.-P. Seismic Resistance and Parametric Study of Building under Control of Impulsive Semi-Active Mass Damper. *Appl. Sci.* **2021**, *11*, 2468. https://doi.org/10.3390/app11062468

Academic Editor: Maria Favvata

Received: 5 February 2021
Accepted: 5 March 2021
Published: 10 March 2021

Publisher's Note: MDPI stays neutral with regard to jurisdictional claims in published maps and institutional affiliations.

Copyright: © 2021 by the authors. Licensee MDPI, Basel, Switzerland. This article is an open access article distributed under the terms and conditions of the Creative Commons Attribution (CC BY) license (https://creativecommons.org/licenses/by/4.0/).

1. Introduction

As science and technology advance further, the demand for urban development continues, so high-rise buildings and skyscrapers are constantly being built. Many important cities think of such buildings as landmarks. However, if such buildings are shaken by wind or earthquake forces, the users may experience both physical and psychological discomfort. Such shaking can also damage the facilities in the building, such as elevators, water towers and pipelines. Although these effects are rarely considered as design control factors for structural security, long-term dynamic deformation can shorten the fatigue lifetime of structural materials. Thus, the shock absorption of high-rise buildings is a very important issue.

Unlike energy dissipation technology [1–16], a TMD needs to be installed on the higher floors of a building to provide a fine shock absorption effect. The weight of a mass damper is much less than that of the structure, and the available space for such an installation is limited. However, the TMD is a popular passive control technology for super high-rise buildings. Currently, TMDs have been installed in buildings worldwide, such as the Citigroup Center, USA [17]; CN Tower, Canada [18]; John Hancock Tower, Boston, USA [19]; and Taipei 101, Taiwan [20]. The main purpose of a mass damper is to reduce wind-generated building vibrations by 30–40%. Many empirical formulas of the optimal parameters for the frequency ratio and damper ratio of a TMD [21–27] have been proposed over the past 20 years. The optimization of the design of a TMD is not complicated if the basic natural frequency is calibrated well. However, changes to the service modes of a building, aging materials and other problems can contribute to changes in the structural frequency of a building, and. over time, the frequency ratio of the TMD will deviate from the effective frequency ratio interval. This deviation reduces the shock absorption effect of the TMD. Many studies have pointed out that, when the material or geometric conditions of a structure enter a nonlinear state under the action of a strong wind or seismic force, the shock absorption effect of a TMD is almost completely lost. The reason is that the basic natural frequency of the structure decreases and the frequency ratio of the TMD shifts. This phenomenon is called the "detuning effect" of a TMD. To mitigate or obviate this effect, many improvement strategies have been proposed by many scholars [28–44], including the Multi-Tuned Mass Damper (MTMD), Active Variable Interal-Semi-Active Mass Damper (AVI-SAMD) and Switched Semi-Active Mass Damper (Switch-type SAMD).

The natural frequency of the Semi-Active Mass Damper (SAMD) is similar to the natural frequency of the structure. Therefore, the natural frequency of the semi-active mass dampers with impulsive reaction [45] provides twice the natural frequency of the structure. Therefore, a kind damper was defined to distinguish the difference between SAMD and this new structural control mechanism, the Impulsive Semi-Active Mass Damper (ISAMD). The main purpose of ISAMD is to address the main defects of the TMD and thereby to improve the shock absorption effect. The ISAMD adopts the advantages of both the TMD and the Active Mass Damper (AMD) and requires only a limited power supply. The control method of the ISAMD involves simply locking and unlocking the connection between the structure and the control mass. These "Unlock" and "Lock" actions of the ISAMD are determined solely according to the structural reactions of the structure under excitation by external forces. Because this action changes the natural frequency of the entire structure, the structural reactions can be reduced. The research achievements of the impulse semi-active mass control mechanism [46] show that the optimal design frequency ratio for this ISMAD is around 4 with a mass ratio around 0.04–0.06 as well as suitable mass distribution at each floor. The frequency ratio of structure with ISAMD should be less than 4.0 to avoid enlarging the maximum acceleration responses. Regardless of the structural frequency misestimation ratio, the detuning phenomenon had little effect on structure under control of ISAMD. Nevertheless, a higher or lower misestimation ratio of the structure frequency caused a worse shock absorption effect of structure under control of TMD [47]. To overcome the defects of TMD, the control mechanism of the ISAMD and the Directional Active Joint of the ISAMD were developed in this research to derive the control law based on the control characteristics of the proposed ISAMD, and the Vector Form Instinct Finite Element method (VFIFE) was applied to perform a dynamic numerical simulation for a structure with TMD or ISAMD, respectively. Then, time histories of structural responses of high-rise buildings with a TMD and the ISAMD under excitation of various near-fault and far-field ground motion records were compared to determine the seismic proof capabilities and the maximum displacements of the control mass blocks of these two dampers under different parameters. In addition, the shock absorption effect and displacement of the control mass block of the ISAMD under various parameters with an auxiliary spring and damper are discussed in this paper, as is the seismic resistance of a building with a TMD or ISAMD affected by near-fault and far-field ground motion. The goal of this study was to develop this new ISAMD to reduce vibrations induced by wind and earthquake forces in high-rise buildings.

2. Concept of the ISAMD

The main idea of the ISAMD is that an active joint between a structure and a mass damper can be locked or unlocked to form a series, namely a structure–spring–active joint–mass damper, as shown in Figure 1.

The components of the proposed ISAMD are as follows: (A) control mass block; (B) active component unit (comprising (B1) switching spring and (B2) active joint); (C) vibration sensing unit (comprising (C1) structural vibration sensors and (C2) vibration sensors for the control mass block); (D) controller; and (E) fixed component unit (comprising (E1) fixed rebound device and (E2) energy dissipation device). The workflow of the proposed ISAMD is as follows:

(1) Motion of the structural floor and the control mass block is detected by motion sensors (accelerometers, C1 and C2). The signal is actively retrieved by the controller.
(2) The controller converts the acceleration signal to velocity and displacement signals.
(3) The controller determines whether to "Unlock" or "Lock" the active joint according to the control law and outputs the control signal to the active joint.
(4) The active joint (B2) releases or captures the switching spring (B1) according to the control signal from the controller.
(5) Steps (1)–(5) are repeated.

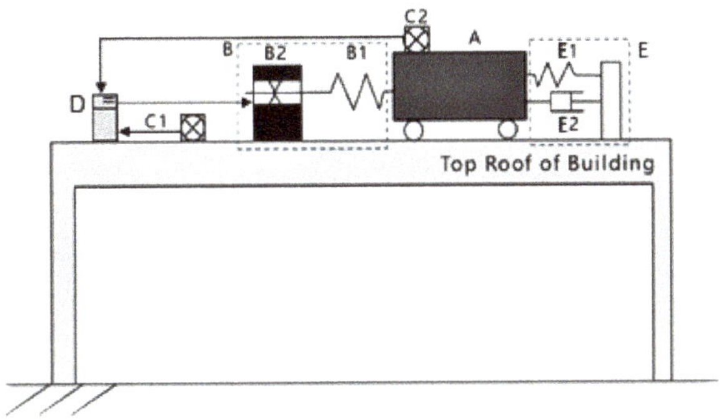

Figure 1. Composition of Impulsive Semi-Active Mass Damper.

The control can perform on/off switching by using a quick-reaction electromagnetic server or solenoid valve. In the "Unlock" status of the active joint, there may be no interaction between the control mass block and the structure, or a weak spring (E1) and energy dissipation device (E2) may maintain a weak interaction force to reduce the stroke of the control mass block. In the "Lock" status of the active joint, the control mass block and the main structure have a strong connection. The interaction force induced by the deformation of the strong spring is applied to change the movement behaviors of the main structure and the control mass block. The timing of the switching of the active joint can be achieved with positive/negative symbols of the work done by the control mass block on the main structure, as follows:

(1) Lock to Unlock: When the inner force of the switching spring (B1) begins to do positive work on the structure, the ISAMD switches to the "Unlock" status and the structure and the control mass become separated.
(2) Unlock to Lock: When the ISAMD switches to the "Lock" status, the switching spring (B1) performs negative work on the structure. Then, the controller (D) sends a "Lock" command to the active joint (B2) to reconnect the main structure and control mass with the switching spring.

3. Developing Directional Joint of the ISAMD

The process from Unlock to Re-Lock of structural control device can cause time delays due to equipment problems. The longer the time delay lasts, the less control it will have, which can seriously affect the control effect of control device. To reduce the defects of time delay problems and achieve the above Unlock/Lock switching process, a directional active joint is developed in this research. The main components are as follows: (A) bevel casting tube (blue); (B) locking steel ball (green); (C) switching tube (red); (D) push rod (yellow); (E) end cover plate (grey); (F) central slider (white); and (G) spring (black) (Figure 2). An exploded diagram of the directional active joint is provided in Figure 3. When a force on the central slider pushes it to the right, the steel ball is in contact with the central sliding rod and the double bevel casing. It is automatically locked and unable to move to the right. The steel balls on the left maintain contact with the central sliding rod and double bevel casing. Conversely, when the central slider slides to the left under force, the normal force of the steel balls on the left side disappears. Since there is no locking power, the sliding rods are free to move to the right. If the steel balls on the right (Figure 2) also maintain contact with the central sliding rod and the double bevel casing, the contact provides a restraint function opposite to that of the steel balls on the left side. Thus, the slider can slide to the right without moving to the left.

Therefore, this joint, which can optionally lock the movement in one direction, is installed in the ISAMD to accurately and automatically combine the control mass block and the main structure for a short impact and then separate them. In addition, this directional active joint can avoid delaying the re-lock action and allows easy simplification of the control law. It greatly increases the reliability of the control mechanism. The directional joint and the control mass block pass through the spring–damper to connect with the top of the structure at both ends of the sliding shaft, as shown in Figure 4. The main hardware of the ISAMD is shown in Figure 5.

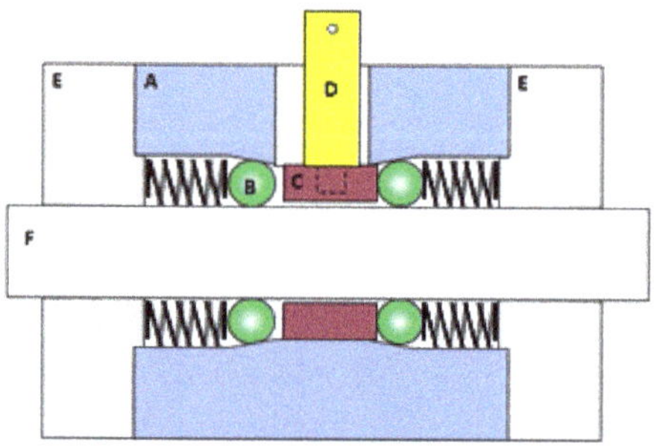

Figure 2. Section of directional active joint (schematic view).

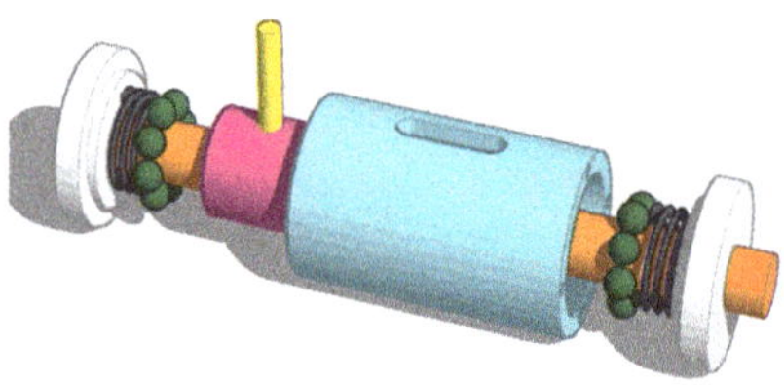

Figure 3. Exploded diagram of the directional active joint.

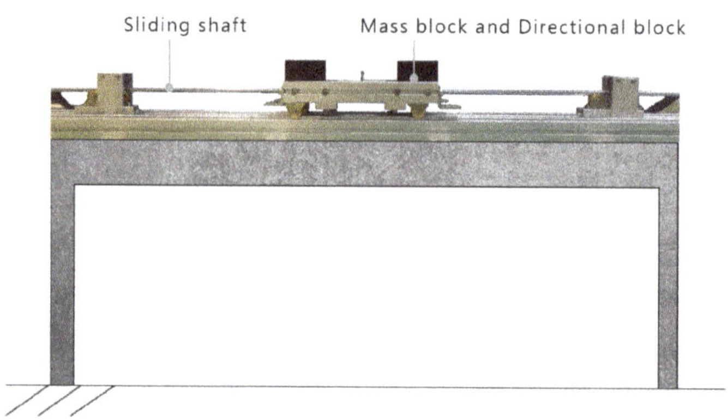

Figure 4. The main hardware of the ISAMD with the directional joint, sliding car and mass block.

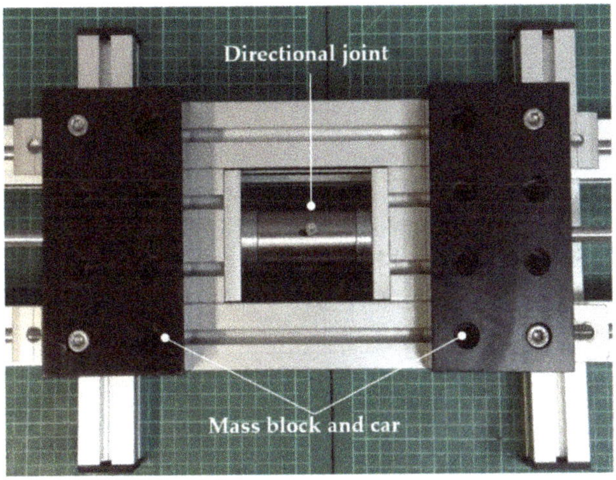

Figure 5. Combination of directional joint, sliding car and mass block.

4. Analysis Model

4.1. Control Law of the ISAMD

The control law of the ISAMD is based on the switching spring, which can capture the structure at the right time for the control mass block to absorb the kinetic energy of the structure. When the control mass block begins to do positive work on the structure, the switching spring is released to maximize the negative work to achieve the maximum energy dissipation effect. The basic condition for the timing is that the mass control block can do negative work on the structure. To maximize this negative work, it is also necessary to consider the structural velocity and acceleration responses. The proposed control law in this study is as follows:

$$\begin{array}{c}\text{Status of Joint is "Unlock", When } (W_V V_s - W_A A_s) \times V_{CS} \leq 0\text{, Switch to "Lock"} \\ \text{Status of Joint is "Lock", When } V_S \times D_{CS} > 0\text{, Switch to "Unlock"}\end{array} \quad (1)$$

where

V_S is the moving velocity of the controlled structure, where the rightward direction is positive;

V_{CS} is the moving velocity of the control mass block relative to that of the controlled structure, where the rightward direction is positive; and

D_{CS} is the displacement of the control mass block relative to that of the controlled structure when the active joint is locked. It is equal to the deformation of the switching spring.

When $V_S \times D_{CS} > 0$, the ISAMD does positive work on the structure. W_V is the weighting of velocity, which is greater than or equal to 0. W_A is the weighting of acceleration, which is greater than or equal to 0. The flowchart of the control law of the ISAMD is shown in Figure 6.

According to this control law, the lock/unlock time of the active joint must take into account the structural acceleration. In fact, the sensor detects the trend of velocity changes in its motion. Therefore, the acceleration sensor is not required as long as the first difference of the speed sensing value of the controller unit or the second difference of the displacement sensing value of the controller unit obtains acceleration responses.

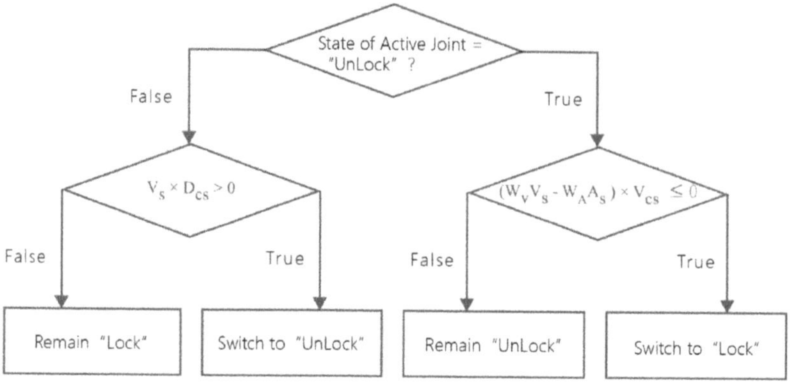

Figure 6. Control Low flowchart of the ISAMD.

4.2. One-Dimensional Vector Form Instinct Finite Element Method, VFIFE

During the action of the ISAMD, the directional joint of the ISAMD is applied to lock or unlock the connection between the ISAMD and the structure. Thus, the Vector Form Instinct Finite Element Method (VFIFE) is applied to analyze the structural responses of the structure with the ISAMD. The steps of the analysis are as follows:

Step 1: Discretization

The VFIFE method is used to discretize the joints of the nodes and elements, where the nodes are the vertices of the element. Nodes are assigned to the mass of the element and become the contact point between two separate elements. The element provides internal force based on the overall coordinates of its nodes and is assigned to the node. The equation is as follows:

$$\vec{F}^{k}_{in,j}(t_i) = \vec{f}\left(\vec{u}(t_i), \dot{\vec{u}}(t_i), \Omega^k\right) \tag{2}$$

where

$\vec{F}^{k}_{in,j}(t_i)$ is the internal force vectors of the k element at time point t_i to maintain this element at the time the node coordinate state is in the j_{th} node. This internal force is affected by the equation of motion, which includes mass matrix, damping matrix and stiffness matrix.

$\vec{u}(t_i)$ and $\dot{\vec{u}}(t_i)$ are the system displacement vector and velocity vector at time point t_i, respectively;

Ω^k represents the properties of the k_{th} element, including the material properties and geometric definitions.

Step 2: Force Equilibrium

At each time point t_i, the internal forces acting on each node associated with the element can be calculated by the node displacement and velocity vector at that time. The external and internal forces on the degrees of freedom of all nodes can be accumulated to achieve the unbalanced force vector on each node's degree of freedom. The equation can be expressed as follows:

$$\Delta\vec{F}_j(t_i) = \vec{F}_{ext,j}(t_i) - \sum_{k=1}^{n_e} \vec{F}^{k}_{in,j}(t_i) \tag{3}$$

where

$\Delta\vec{F}_j(t_i)$ is the balance force vector of the j_{th} node at time point t_i and

$\vec{F}_{ext,j}(t_i)$ is the external force vector of the j_{th} node at time point t_i.

Step 3: Predict displacement vector at the next time step t_{i+1}

The centered difference scheme is applied to explicitly estimate the displacement reaction at time point t_{i+1} under the condition of acceleration responses as a known condition at time point t_i:

$$\ddot{\vec{u}}_j(t_i) = \frac{\vec{u}_j(t_{i+1}) - 2\vec{u}_j(t_i) + \vec{u}_j(t_{i-1})}{\Delta t^2} \tag{4}$$

where

$\ddot{\vec{u}}_j(t_i)$ is the acceleration vector of the j_{th} node at time point t_i and

Δt is an analytical stride.

Then, the displacement reaction at time point t_{i+1} can be expressed as follows:

$$\vec{u}_j(t_{i+1}) = \boldsymbol{M}_j^{-1}\Delta\vec{F}_j(t_i)\Delta t^2 + 2\vec{u}_j(t_i) - \vec{u}_j(t_{i-1}) \tag{5}$$

where $\boldsymbol{M}_j$ is the mass matrix of the j_{th} node. It is a diagonal matrix.

Step 4: Repeat the above steps to complete the analysis.

5. Setting of the Analysis Model for the Shock Absorption Effect of a 10-Story Building with the ISAMD

To explore the reaction to seismic forces of a structure with multiple degrees of freedom and the ISAMD and define the effective interval of the control parameters of the ISAMD for control system design, a 10-story shearing building under excitation of various earthquake records was investigated. These numerical simulations compared the displacement reaction at the roof of the building and at the control mass block for a 10-story shearing building under control of a TMD and the ISAMD and under excitation of different earthquake records. Thus, the structural displacement and displacement reduction ratio of the control mass block under various combinations of parameters were compared to obtain the optimal design parameters. An analysis program with a VFIFE function was applied to perform numerical simulations. In those simulations, the element of the linear spring-damper in parallel was applied to simulate the elastic recovery force and damping provided by the column of the shear building as well as simulate the spring force and damping force between the TMD and the top floor of the main structure. The spring stiffness and weighting of the velocity and acceleration of the control law were set for the ISAMD. Time delay effects were not considered in this research.

5.1. Analysis Setting

The analyzed control subject matter in this study was the 10-story shearing building shown in Figure 7. To compare the control characteristics of the building under control of the ISAMD and TMD, this research analyzed the structural responses of the building with these two dampers under excitation of different earthquake records. The main parameters of the analysis included the parameters of the main structure and the control parameters of the ISAMD and TMD, described as follows.

5.1.1. Main Structure Parameters

The bare structure was the 10-story shearing building. Assuming a mass m_s of each floor of 500 tons, the story stiffness k_s and damper c_s were 883,645 kN·m and 2813 kN·s/m, respectively. The first modal frequency and damper ratio were 1.0 Hz and 0.01, respectively.

5.1.2. Control Parameters of the ISAMD

The control parameters of the ISAMD can be divided into two categories: (1) system hardware control parameters, including the control mass block ratio (μ), frequency ratio of the control mass block (γ_f) and the damping ratio of the control mass block (ξ_a); and (2) control law parameters, including the weight of velocity response (w_v) and acceleration response (w_a) of the control law.

The control mass block ratio can be defined by the rate of the mass of the control block and structure:

$$\mu = \frac{m_a}{\sum m_s} \tag{6}$$

Then, the frequency ratio of the control mass block can be defined as the rate of a natural vibration frequency and the first modal frequency of the main structures:

$$\gamma_f = \frac{f_a}{f_0} = \frac{\sqrt{k_a/m_a}}{2\pi f_0} \tag{7}$$

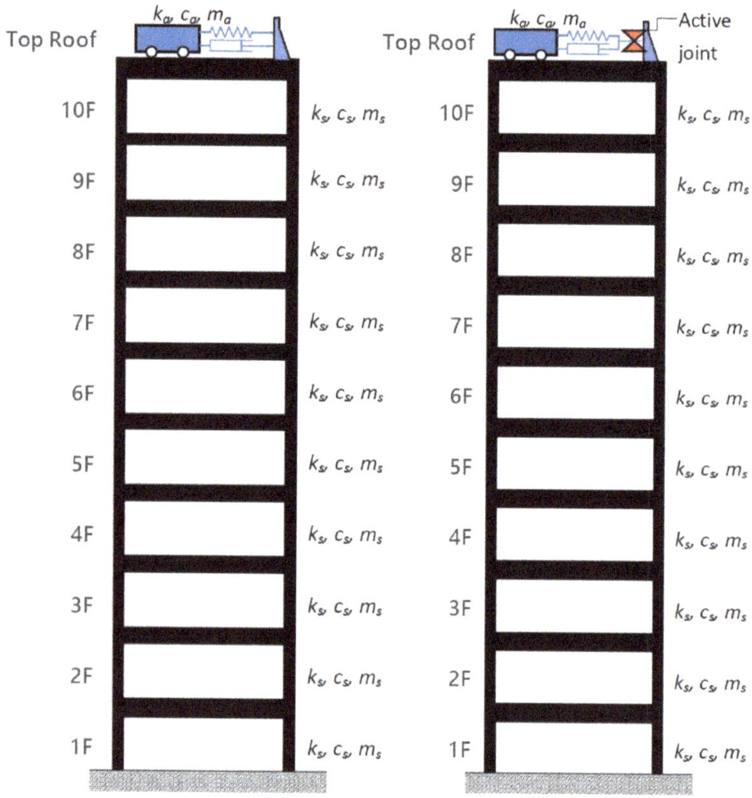

(**a**) Building under control of the TMD (**b**) Building under control of the ISAMD

Figure 7. Analysis model of Building with TMD and ISAMD.

The weight of velocity response (w_v) and acceleration response (w_a) of the ISAMD control law affect the timing of the "releasing" and "capturing" status of the control mass block and the main structure. If $w_v : w_a = 1 : 0$, the "releasing" (unlock) or "capturing" (lock) status is determined by the structural velocity responses at the installation position of the control mass block. The timing of the combination of the ISAMD and structure is at the point when the direction of the structural motion reverses, which is the maximum or minimum structural displacement, at which point the structural velocity is zero. In contrast, when $w_v : w_a = 0 : 1$, the "releasing" or "capturing" status is determined by the structural acceleration responses at the installation position of the control mass block. The timing of the combination of the ISAMD and the structure is at the point when the direction of the structural acceleration reverses, i.e. the structural displacement is zero and the structural acceleration reaches the maximum or minimum acceleration. The ratio between these two extreme ratios causes the switching time point to be between these two extreme time points, at which point the velocity and acceleration direction are the same. Five different proportions are compared to determine the seismic resistance in this study. The control parameters of the ISAMD are listed in detail in Table 1.

Table 1. Analysis setting value [47].

Analysis Parameter Category	Parameter Range
Parameters of main structure	Structure type: 10DOF shear building Story mass: m_a = 500 tn Inter-story stiffness: k_a = 883,645 kN/m Inter-story damping: c_a = 2813 kN·s/m First modal frequency: f_0 = 1.0 Hz First modal damping ratio: ξ_0 = 0.01
Control Parameters of the ISAMD	Mass ratio: μ = 0.02 Freq. ratio: γ_f = 0.8145~6.0 @rate = 0.95 Damping ratio: ξ_a = 0.01~0.10 Weight ratio: $w_v/w_a = 1/0, 4\pi f_0/0.414, 2\pi f_0/1, 2\pi f_0/2.414, 0/1$
Control Parameters of the TMD	Mass ratio: μ = 0.02 Freq. ratio: γ_f = 0.8145~1.2277@rate = 0.95 Detuning rate: Γ = 18.55%~22.77% Damping ratio: ξ_a = 0.071~0.10
Seismic waves	26 records of ground acceleration of earthquakes with epicentral distance = 1–339 km (see Table 2)

5.1.3. Control Parameters of the TMD

The control parameters of the TMD are as follows. The control mass block ratio (μ), frequency ratio (γ_f) and damper ratio (ξ_a) have the same definitions as those of the ISAMD. To compare the influence of the shock absorption ratio by the detuning effect, the frequency ratio (γ_f) of the control mass block of the TMD was changed. The range of variation of this frequency ratio was around 0.8145–1.2277. The detuning frequency (Γ) is the ratio of the difference of the frequency ratio of the control mass block and the optimal frequency ratio of the TMD to the optimal frequency ratio of the TMD:

$$\Gamma = \frac{\gamma_f - \gamma_{f,opt}}{\gamma_{f,opt}} \times 100\% \tag{8}$$

The control parameters of the TMD are listed in detail in Table 1 and were based on the formula of the optimal control parameters suggested by Lin et al. [48] to estimate the optimal frequency ratio and the damping ratio, 0.9694 and 0.07035, respectively. Therefore, the optimal frequency of the control mass block was 6.283 rad./s.

$$r_f = \frac{f_a}{f_p} = \left(\frac{a}{1+\mu}\right)^b;\ a = 1.0 - \frac{\xi_p}{4};\ b = 1.35e^{3.2\xi_p};\ \xi_a = 0.46\mu^{0.48} \tag{9}$$

5.1.4. Earthquake Records

The seismic action of a near fault is short in duration. Thus, the maximum structural displacement occurs under a 1.5–2.5 cycle action of the main shock wave. The shock absorption effect of the TMD comes from the kinetic energy of the control mass block. The TMD does not provide enough energy to resist the structural movement in the early stage of an earthquake. Therefore, there is no noteworthy shock absorption effect of the TMD on the maximum structural displacement reaction under the action of a near-fault ground motion. Conversely, the structural displacement reaction is controlled by the component of the resonant frequency under the action of a far-field ground motion. The maximum displacement reaction occurs at 3–5 reciprocating cycles because the control mass block has enough time to accumulate sufficient kinetic energy to confer seismic resistance. Therefore, the TMD and ISAMD have significant control effects on the maximum structural displacement reaction for far-field ground motions because the shock absorption effect of the structural control method is closely related to the actual time history of the earthquake force. To compare the control efficiencies of different control methods, diachronic analysis of a building under control of the TMD and ISAMD is necessary to analyze and compare the average ratio and standard deviation of the shock absorption effect. Records of 26 earthquakes with epicentral distances of 1–339 km were used to analyze the seismic resistance. Each earthquake's name, occurrence time, recording station, epicentral distance, seismic direction and original peak ground acceleration are listed in Table 2.

Table 2. Earthquake records for analytical use.

No.	Earthquake	Year	Station	Epi. Dist. (km)	Dir.	PGA (g)
1	Kobe, Japan	1995	KJMA	1	NS	8.21
2	Northridge, USA	1994	Tarzana	4	EW	17.45
3	Santa Barbara, USA	1978	UCSB Goleta	14	NS	3.40
4	Chi-Chi, Taiwan	1999	TCU075	18	NS	2.57
5	Northridge, USA	1994	Newhall	19	EW	5.72
6	Northridge, USA	1994	S_Monica	23	EW	8.66
7	Norcialtaly, Italy	2016	St_Angelo	28	-	2.06
8	Northridge, USA	1994	LA	38	EW	3.48
9	Northridge, USA	1994	SF bay	40	-	1.64
10	Norcialtaly, Italy	2016	Colfiolito	40	-	2.55
11	Loma Prieta, USA	1989	S_Cruz	48	NS	3.62
12	Calexico, USA	2010	Sierra	77	EW	4.91
13	Chile	2010	CCSP	109	EW	5.94
14	Sumatra, Indonesia	2007	PSKI	125	EW	1.24
15	Chile	2016	LL06	136	-	2.24
16	Chile	2010	VA03	168	-	2.68
17	Chile	2010	GO01	170	EW	2.32
18	Chile	2010	CUR	170	EW	4.66
19	Chile	2010	GO04	175	EW	2.34
20	Chile	1980	ANGO	209	NS	6.84
21	Chile	1980	MAT	230	NS	3.37
22	Alaska, USA	2016	HNE	254	EW	2.07
23	Chile	1980	LLO	274	NS	3.19
24	Chile	2010	S_Jose	333	NS	4.61
25	Chile	2010	S_Lucia	334	NS	2.39
26	Chile	2010	ColegioLasAmericas	339	NS	3.02

5.2. Indices of Control Performance

The shock absorption effects of these two dampers were compared in the building with TMD or ISAMD control under excitation of near-fault and far-field ground motion records. Six indices of control performance were defined to investigate the seismic resistance.

(1) Average ratio of the maximum roof displacement reaction (J_1):

$$J_1 = average\left(\frac{max(|Roof\, displacement\, with\, ISAMD\, or\, TMD|)}{max(|Roof\, displacement\, withoutcontrol|)}\right) \tag{10}$$

(2) Standard deviation of the maximum roof displacement reaction ratio (J_2):

$$J_2 = stdev\left(\frac{max(|Roof\, displacement\, with\, ISAMD\, or\, TMD|)}{max(|Roof\, displacement\, withoutcontrol|)}\right) \tag{11}$$

(3) Average Root Mean Square, RMS ratio of the roof displacement reaction (J_3):

$$J_3 = average\left(\frac{rms(Roof\, displacement\, with\, ISAMD\, or\, TMD)}{rms(Roof\, displacement\, withoutcontrol)}\right) \tag{12}$$

(4) Standard deviation of the average RMS ratio of the roof displacement reaction (J_4):

$$J_4 = stdev\left(\frac{rms(Roof\, displacement\, with\, ISAMD\, or\, TMD)}{rms(Roof\, displacement\, withoutcontrol)}\right) \tag{13}$$

(5) Average ratio of the roof maximum absolute acceleration reaction (J_5)

$$J_5 = average\left(\frac{max(Roof\, acceleration\, with\, ISAMD\, or\, TMD)}{max(Roof\, acceleration\, withoutcontrol)}\right) \tag{14}$$

(6) Average ratio of the maximum displacement of the control mass block (J_6):

$$J_6 = average\left(\frac{max(|Mass\, displacement\, with\, ISAMD|)}{max(|Mass\, displacement\, with\, TMD|)}\right) \tag{15}$$

All indices of control performance in this research follow the axiom of the Smaller the Better (STB).

6. Results and Discussion

6.1. Analysis Results

The analysis results of the control performance indices of the building with TMD or ISAMD control under excitation of 26 near-fault and far-field ground motion records are shown in Figures 8–13. The optimal control performance indices of the TMD and ISAMD are listed in Table 3. Table 3 reveals that the optimal control performance indices of the building under control of the TMD or ISAMD were relative to the frequency ratio of the control mass block γ_f. The ISAMD, without additional auxiliary dampers or springs and only one spring and active joint, switched the status of "Unlock" to "Lock".

Table 3. The optimal control performance indices for the building with TMD or ISAMD under excitation of 26 seismic records.

Parameter		J1	J3	J5	J6
TMD	Value	0.72	0.48	0.88	N.A.
	min. at *rf*	0.95	0.95	0.95	N.A.
ISAMD $W_V{:}W_A = 1{:}0$	Value	0.65	0.40	0.85	5.43
	min. at *rf*	1.59	1.43	1.67	2.19
ISAMD $W_V{:}W_A = \omega_0 : \sqrt{2}-1{:}$	Value	0.62	0.38	0.91	3.79
	min. at *rf*	2.1	2.1	1.45	5.78
ISAMD $W_V{:}W_A = \omega_0{:}1$	Value	0.65	0.38	0.91	3.87
	min. at *rf*	2.79	2.27	1	4.90
ISAMD $W_V{:}W_A = \omega_0 : \sqrt{2}+1{:}$	Value	0.67	0.38	0.93	3.98
	min. at *rf*	2.79	2.90	1.2	2.12
ISAMD $W_V{:}W_A = 0{:}1$	Value	0.68	0.4	0.94	3.71
	min. at *rf*	3.25	3.25	1.36	3.42

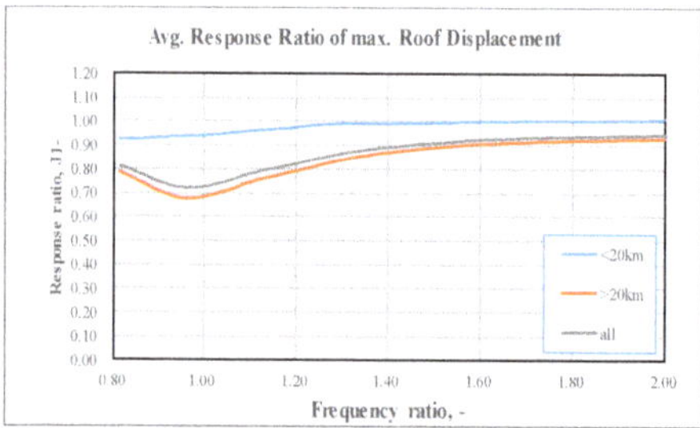

(**a**) Average response ratio of maximum roof displacement

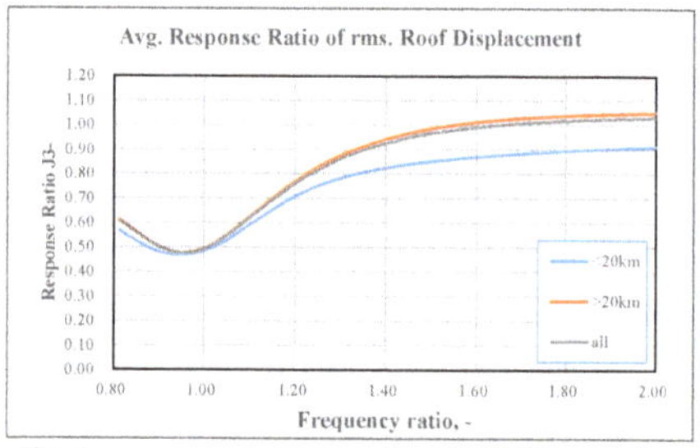

(**b**) Average response ratio of RMS roof displacement

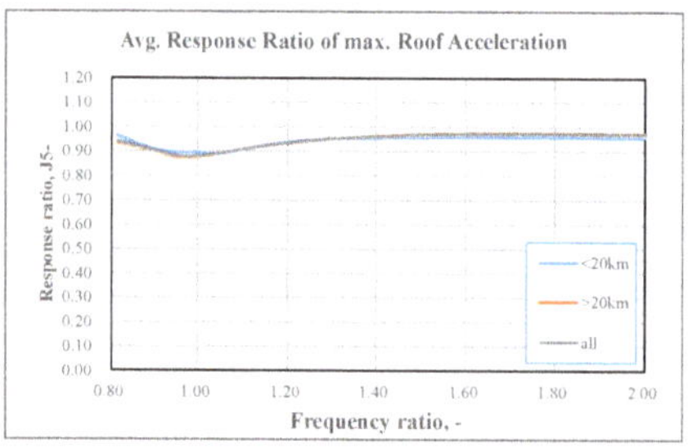

(**c**) Average response ratio of maximum roof acceleration

Figure 8. Response ratio of Structure with TMD under 26 seismic excitations [45].

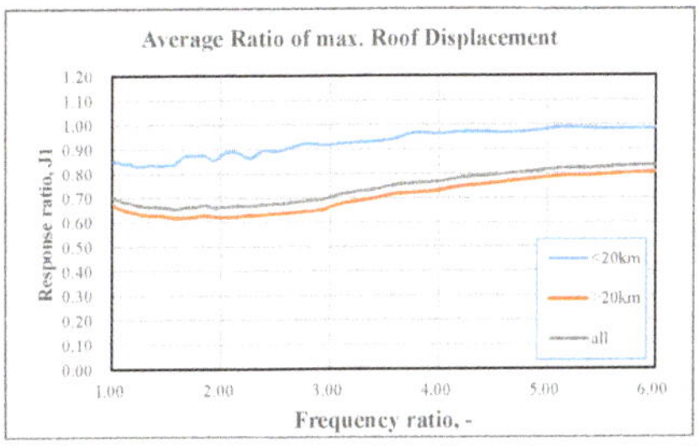

(**a**) Average response ratio of maximum roof displacement

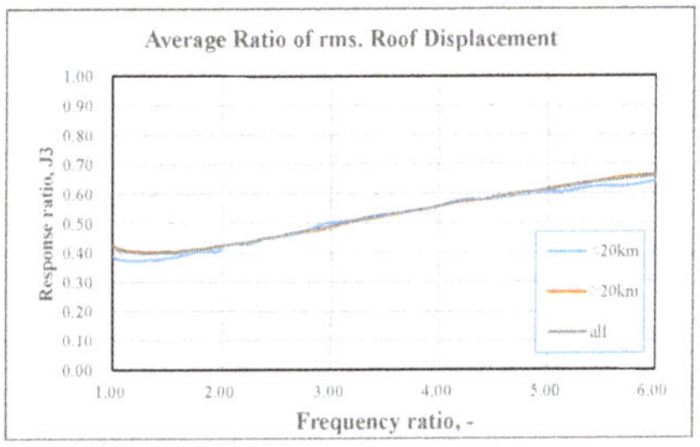

(**b**) Average response ratio of rms. roof displacement

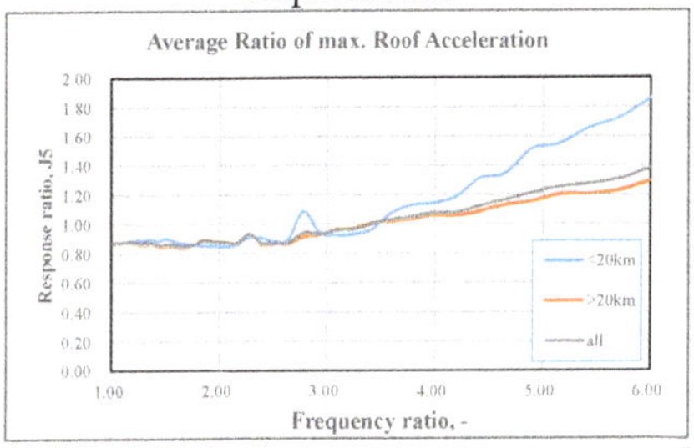

(**c**) Average response ratio of maximum roof acceleration

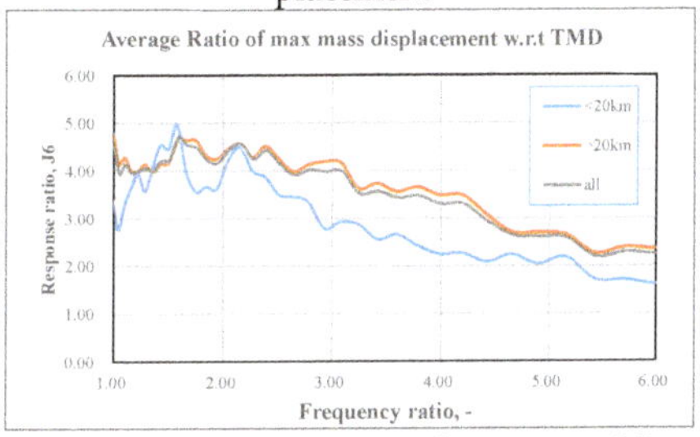

(**d**) Average response ratio of maximum Mass displacement with respect to TMD

Figure 9. Response ratio of Structure with the ISAMD with W_V:W_A = 1:0, without auxiliary damper and spring under 26 seismic excitations.

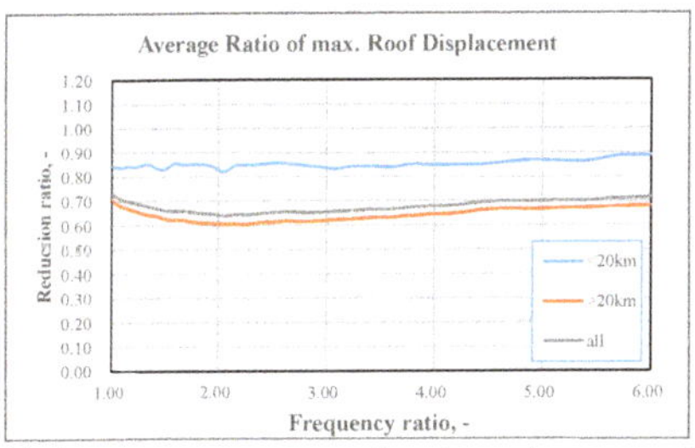

(**a**) Average response ratio of maximum roof displacement

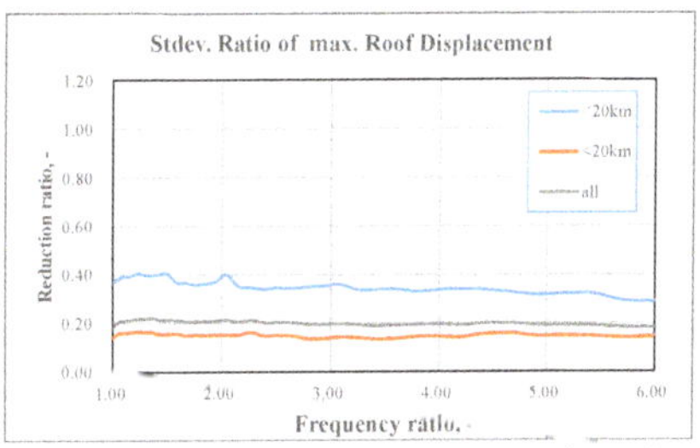

(**b**) Standard dev. response ratio of maximum roof displacement

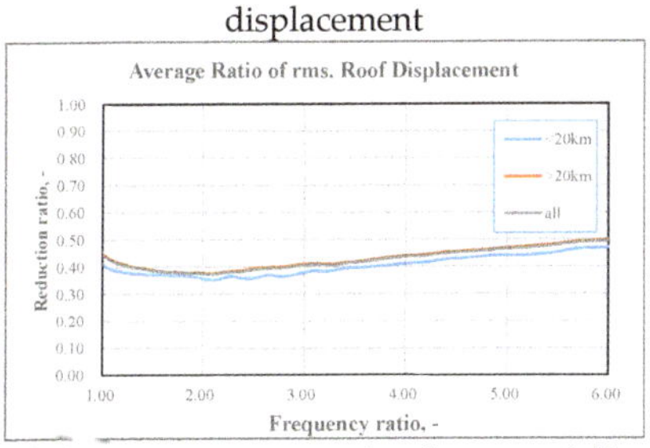

(**c**) Average response ratio of rms. roof displacement

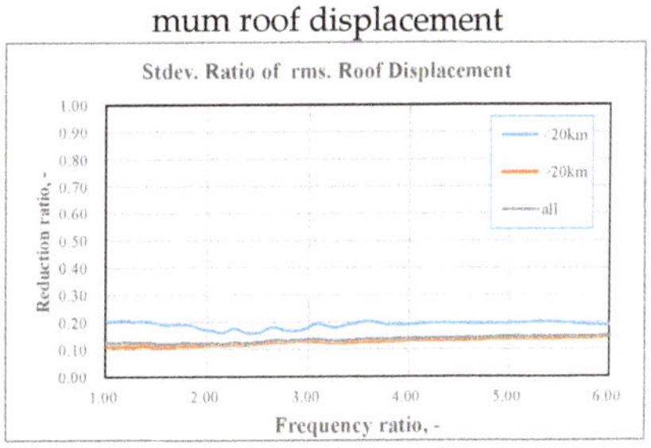

(**d**) Standard dev. response ratio of rms. roof displacement

Figure 10. *Cont.*

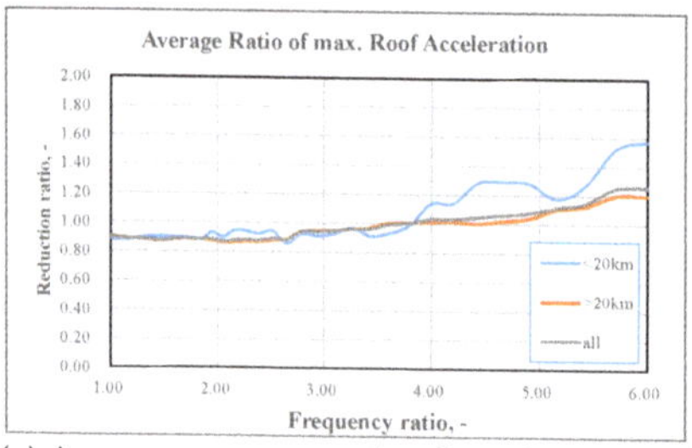

(**e**) Average response ratio of maximum roof acceleration

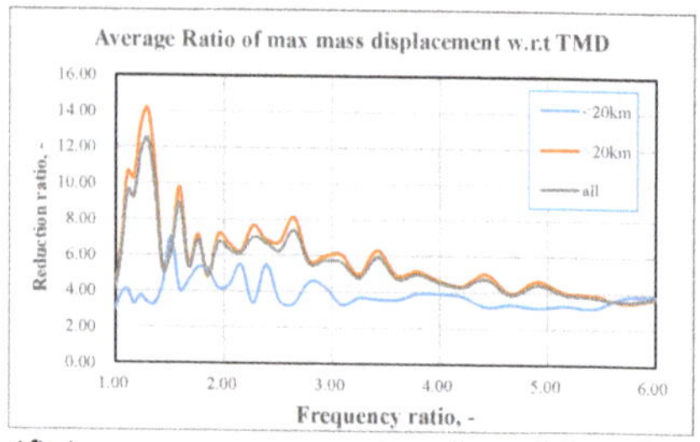

(**f**) Average response ratio of max Mass displacement with respect to TMD

Figure 10. Response ratio of Structure with the ISAMD with W_V:W_A = ω_0 : $\sqrt{2}-1$, without auxiliary damper and spring under 26 seismic excitations.

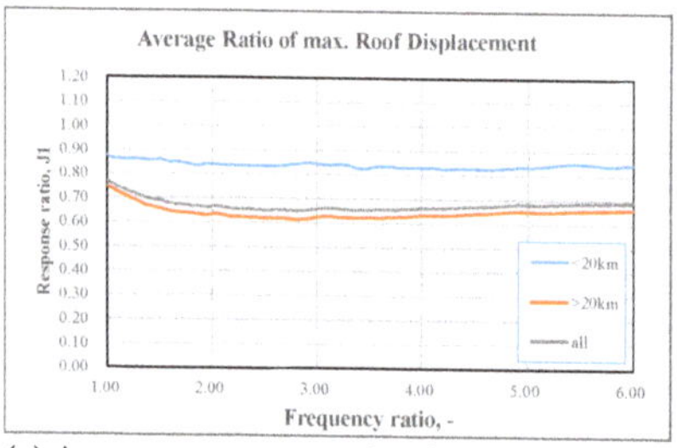

(**a**) Average response ratio of maximum roof displacement

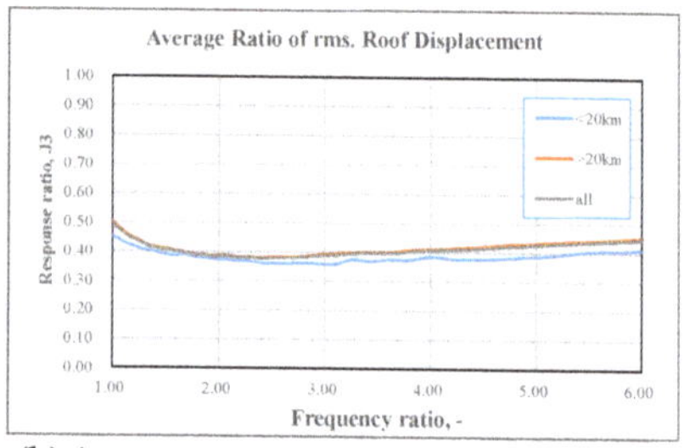

(**b**) Average response ratio of rms. roof displacement

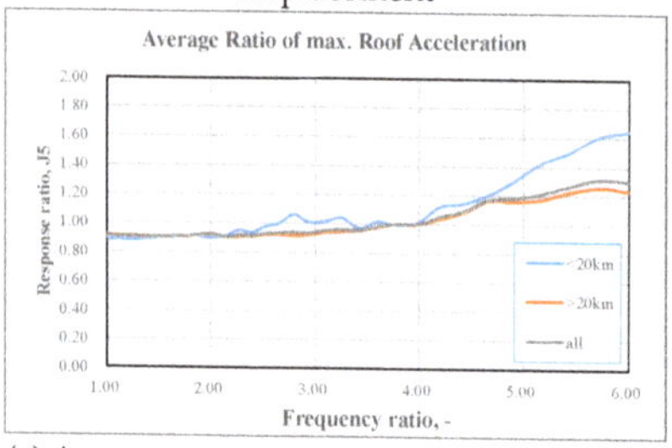

(**c**) Average response ratio of maximum roof acceleration

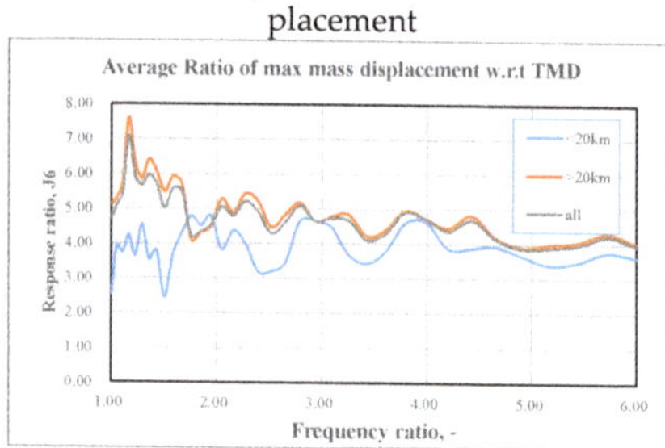

(**d**) Average response ratio of max Mass displacement with respect to TMD

Figure 11. Response ratio of Structure with the ISAMD with W_V:W_A = ω_0 : 1, without auxiliary damper and spring under 26 seismic excitations [45].

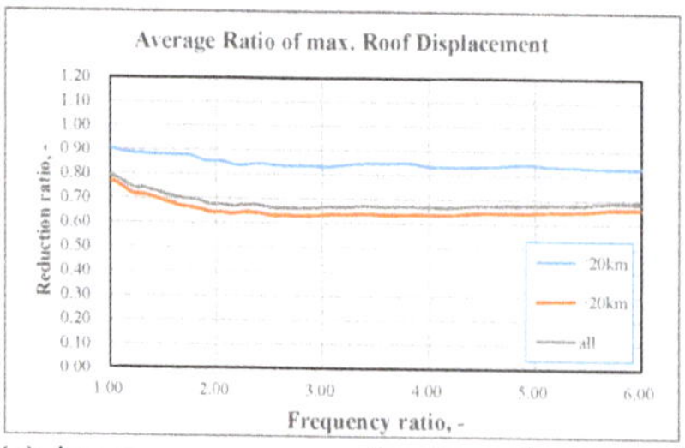

(**a**) Average response ratio of maximum roof displacement

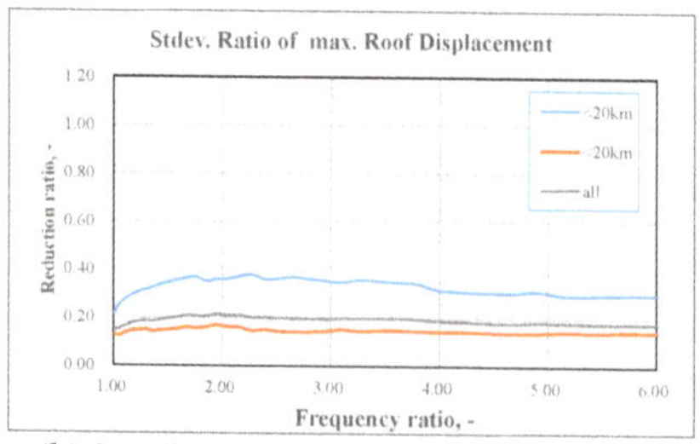

(**b**) Standard deviation response ratio of maximum roof displacement

Figure 12. *Cont.*

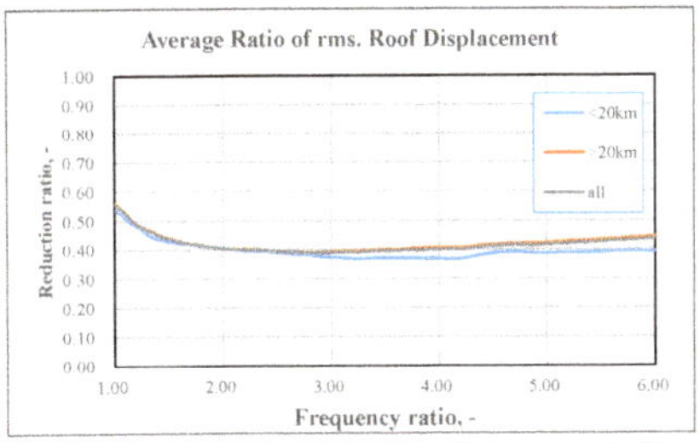

(**c**) Average response ratio of rms. roof displacement

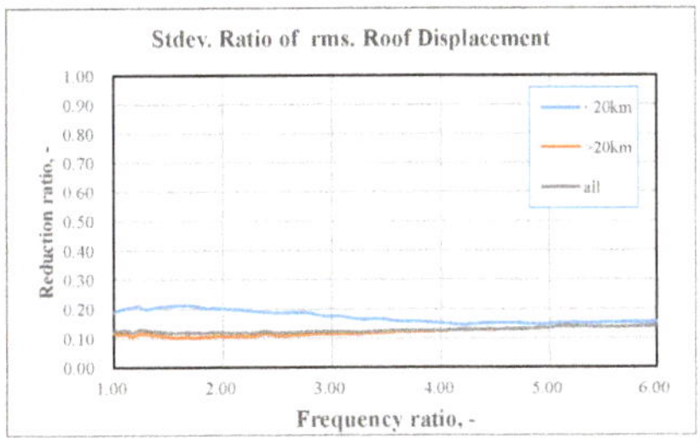

(**d**) Standard dev. response ratio of rms. roof displacement

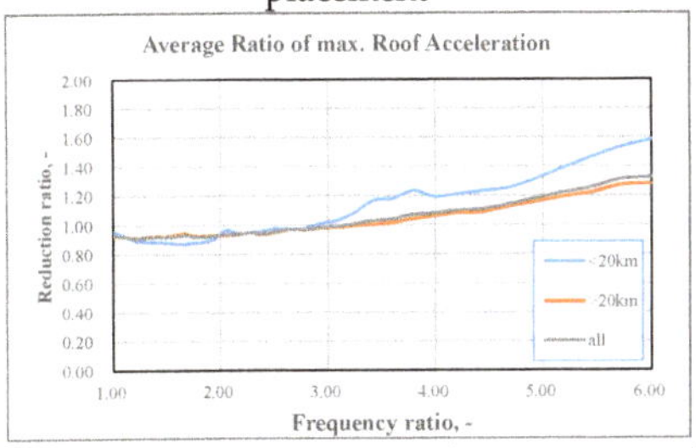

(**e**) Average response ratio of maximum roof acceleration

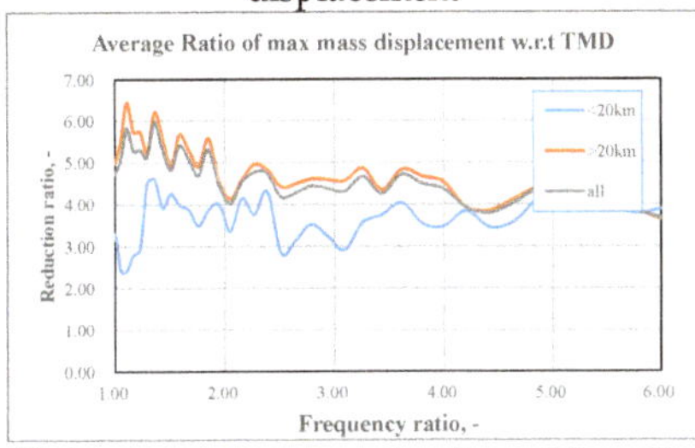

(**f**) Average response ratio of max Mass displacement with respect to TMD

Figure 12. Response ratio of Structure with the ISAMD with W_V:W_A = $\omega_0 : \sqrt{2}+1$, without auxiliary damper and spring under 26 seismic excitations.

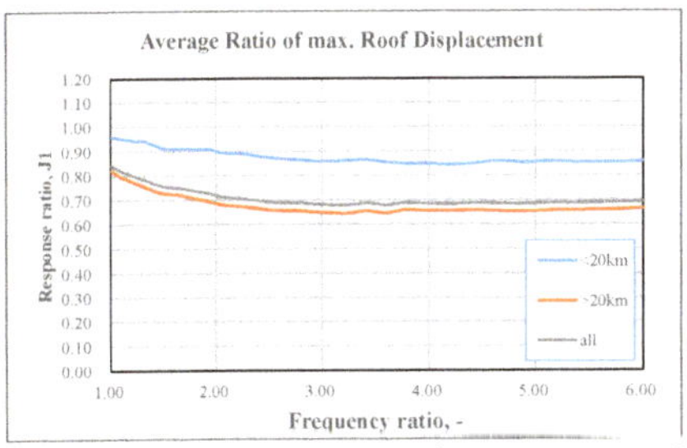

(**a**) Average response ratio of maximum roof displacement

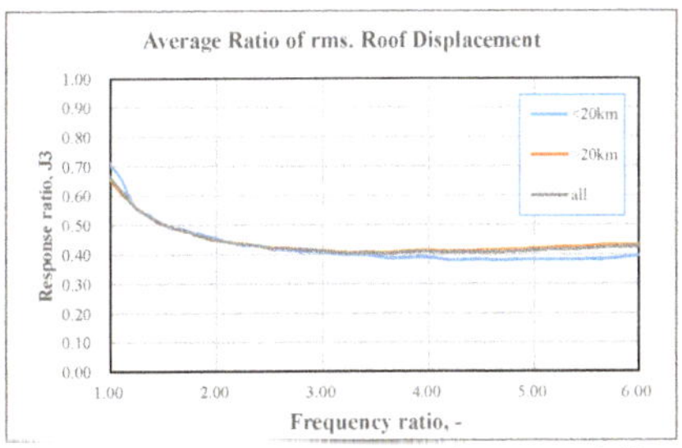

(**b**) Average response ratio of rms. roof displacement

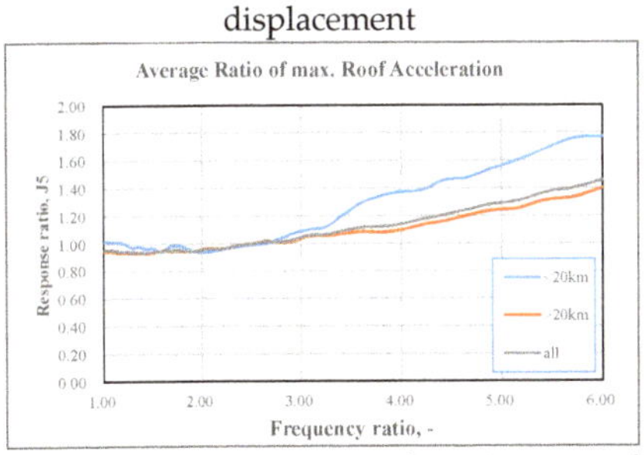

(**c**) Average response ratio of maximum roof acceleration

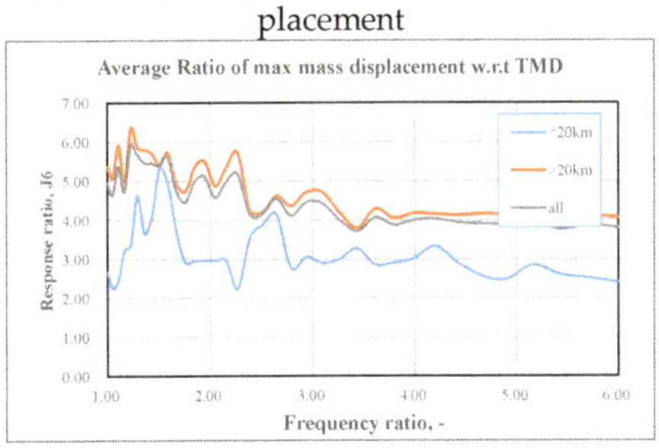

(**d**) Average response ratio of max Mass displacement w.r.t TMD

Figure 13. Response ratio of Structure with the ISAMD with W_V.W_A – 0:1, without auxiliary damper and spring under 26 seismic excitations.

6.2. Discussion

J1, as listed in Table 3, reveals that the maximum displacement reduction effect of the ISAMD was better than that of the TMD. The maximum displacement responses of the roof were further reduced by about 4–10%. The root mean square of the displacement reduction effect of the ISAMD was greater than that of the TMD, leading to a reduction of 8–10%. Conversely, the shock absorption effect of the structural acceleration responses of the building with the ISAMD was no better than that of the TMD. However, the maximum displacement response of the control mass block was significantly less than that of the TMD, so the required installation space of the ISAMD is 2–4 times that of the TMD. The comparison between the structural responses of the building with the TMD and the bare structure shown in Table 3 and Figure 6 indicates that the frequency ratio of the optimal control performance indices of the building with TMD was around 0.95. When the frequency ratio was slightly offset, the control performance indices J1, J3 and J5 increased. For example, when the frequency ratio was 1.2, J1, J3 and J5 were 0.83, 0.79 and 0.94, respectively. The shock absorption effect was very limited, manifesting the so-called detuning effect. In contrast, the relationship of the control performance indices to the frequency ratio, as presented in Table 2 and Figures 6–11, revealed that the minimum values of J1 and J3 occurred on a very flat curve. The frequency ratio was almost constant in this range. That is, the sensitivities of the control performance and frequency ratio were extremely low. Basically, this phenomenon could be expected because the maximum roof displacement reaction and the root mean square displacement of the building with the ISAMD had reduction rates of more than 30% and 60%, respectively. The control mass block displacement of the ISAMD was 2–4 times that of the TMD, as shown in Figures 9d, 10f, 11d, 12f and 13d. This suggests that the installation space of the ISAMD must be large. In fact, the higher ISAMD frequency of the control mass could provide the benefit of reducing the installation space, such as the simple pendulum TMD of the Taipei 101 building. It has a 6.8 s cycle, which requires a pendulum length of 11.48 m, occupying a space of four stories. Conversely, if the frequency ratio of the ISAMD is 3 with a 2.26 s cycle, the pendulum length only needs to be 1.27 m. It can be set up in one story. Therefore, the space requirement of the ISAMD may not be greater than that of the TMD.

6.2.1. The Maximum Roof Displacement and Root Mean Square Displacement of the Structure

The maximum roof displacement and root mean square displacement of the TMD under the condition of the optimal frequency ratio and those of the ISAMD with the weight ratio W_V:W_A = 1:0 without a parallel auxiliary spring or damper under excitation of 26 earthquake records are compared in Figures 14 and 15. These two figures show that the shock absorption ratio of the ISAMD was 10% higher than that of the TMD. The shock absorption effect of the TMD varied greatly with the seismic load, and the standard deviation was about 0.26 and 0.17. The standard deviations of the ISAMD were only 0.20 and 0.10. In other words, the reliability of the displacement reaction of the control structure with the ISAMD is better than that of the TMD.

6.2.2. Influence of W_V:W_A on the Control Performance of the ISAMD

The timing of the "Unlock" and "Lock" switching of the ISAMD is dependent on the control law of the ISAMD and is regulated by the velocity and acceleration weight. For noncyclic near-fault and far-field ground motion loads, the shock absorption effect of the ISAMD under various weight ratios should be investigated. The results of analyzing six control performance indices of a building under control of the ISAMD with five different weight ratios and without auxiliary stiffness or a spring under excitation of 26 seismic loads are listed in Table 2. The results for the control performance indices J1–J6 are shown in Figures 16–20.

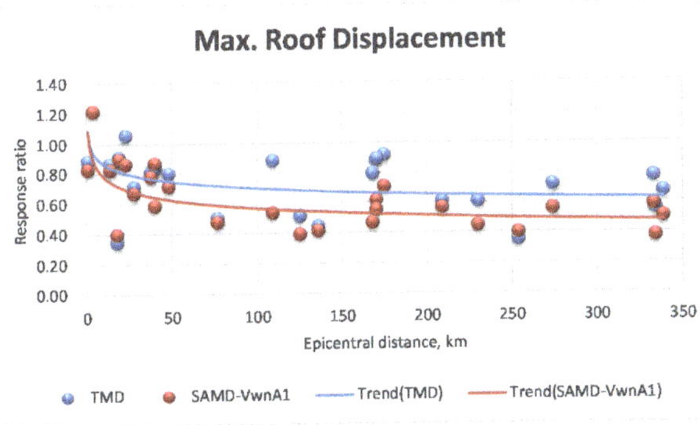

Figure 14. The maximal roof displacement ratio of the TMD with optimal *rf* and ISAMD W_V:W_A = ω_0 : 1, K = 0, C = 0.

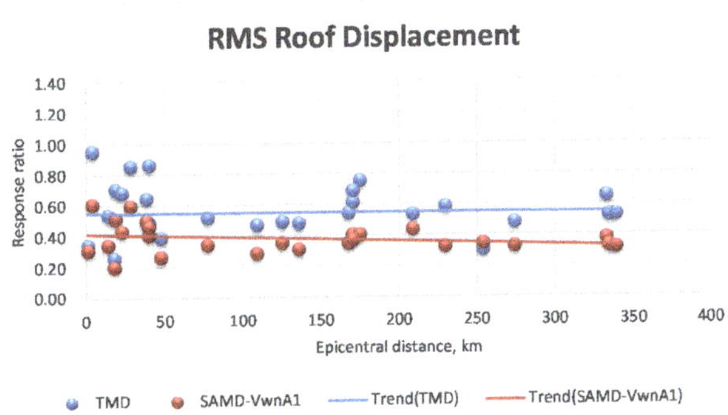

Figure 15. RMS roof displacement ratio of the TMD with optimal *rf* and ISAMD W_V:W_A = ω_0 : 1, K = 0, C = 0.

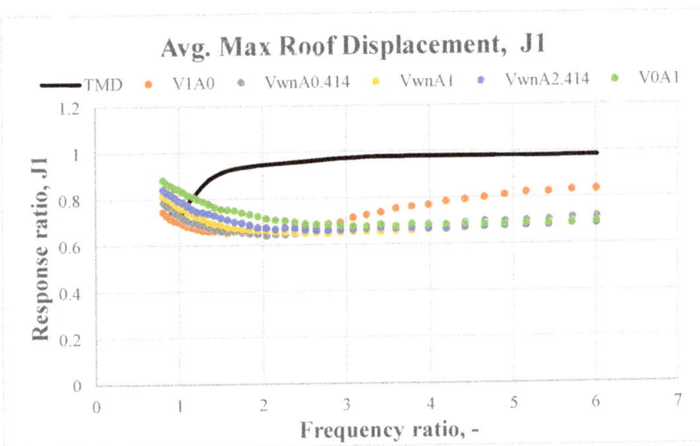

Figure 16. The average maximum roof displacement ratio of the building with the ISAMD, J1.

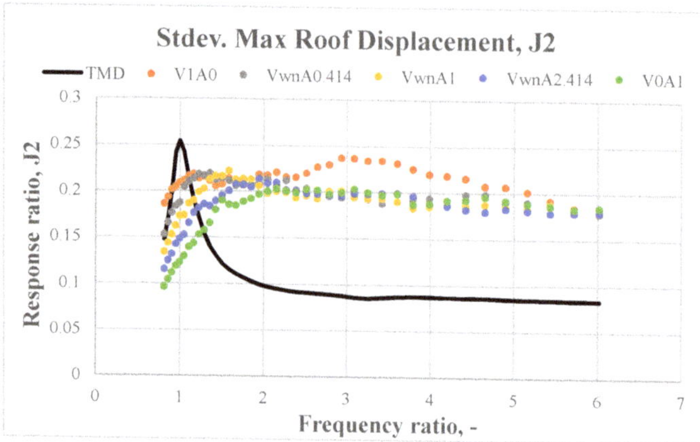

Figure 17. The standard deviation of the maximum roof displacement response ratios of the ISAMD, J2.

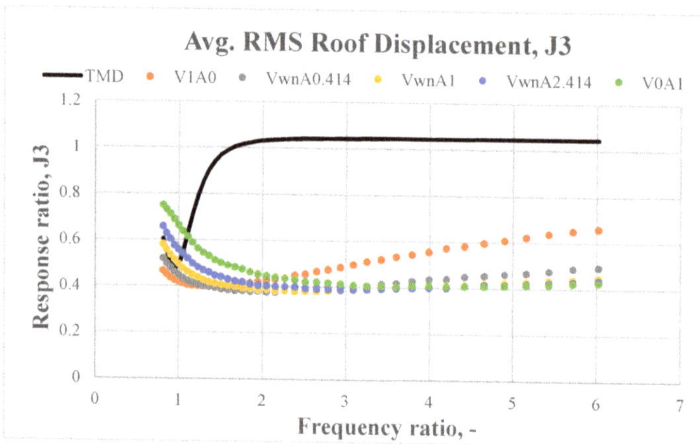

Figure 18. The average value of the root mean square roof displacement responses of the ISAMD, J3.

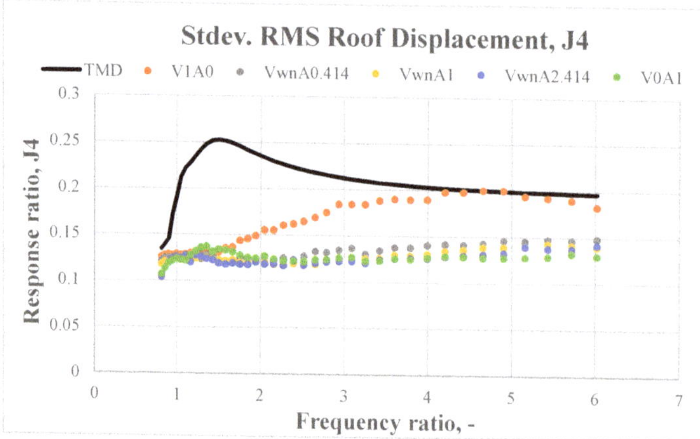

Figure 19. The standard deviation of the root mean square roof displacement response ratio of the ISAMD, J4.

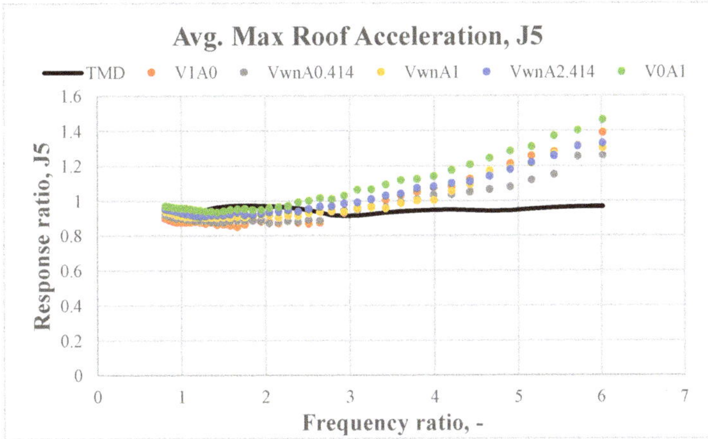

Figure 20. The average value of the maximum roof acceleration response ratio of the ISAMD, J5.

The minimum index of J1 was mostly within 0.65–0.67 for the ISAMD under all weight ratios, and the average seismic resistance ratios were 33–35% for the building with the ISAMD under excitation of 26 seismic records. The orange dotted line in Figure 16 shows that, when the ISAMD functions without consideration of acceleration responses (V0A1), the sensitivity of the control performance index J1 of the ISAMD to the frequency ratio (*rf*) is the highest weight ratio of the ISAMD. When the frequency ratio is greater than 2, the shock absorption effect is reduced. However, there is a large range of stiffness changes from the optimal frequency ratio, around 1.59–2. This is not a defect of design. Thus, if the weight of acceleration responses is only considered in J1, the value of J1 monotonically decreases with the frequency ratio. The frequency ratio should be very high to have the same shock absorption effect as other ratios. If it is too high, there will be other adverse reactions. Thus, adopting only the consideration of acceleration weight (V0A1) is not recommended. Figure 16 shows that the frequency ratio of 2–4 with a weight ratio of V1A0–V0A1 can achieve a stable shock absorption effect.

When the standard deviation of the maximum roof displacement reaction ratio is low, the sensitivity of the control performance to the seismic wave is low and the shock absorption control is stabler. Figure 17 shows that, when the frequency ratio is lower than 1, the value of J2 is the minimum. When the frequency ratio is large, it can steadily fall by 0.2. The shock absorption ratio is about 34% for $rf = 2.0$. The shock absorption effect of the ISAMD can be guaranteed.

Figures 18 and 19 reveal that the root mean square roof displacement shock absorption effect of the ISAMD is better than the roof displacement shock absorption effect of the ISAMD. The average shock absorption ratio is around 60%, and the standard deviation is also relatively small. Therefore, the ISAMD provides a fine damping effect. Figure 18 shows that, for the control effect of the ISAMD, the V1A0 weight ratio is slightly inferior to the other weight ratios.

The control performance index of the ISAMD of acceleration responses behaves poorly. The acceleration reaction of the ISAMD with a frequency ratio interval of 2–4 and a better displacement control effect is basically larger than those of the bare structure. The reason is that the control force of the control mass block is close to the impact force when the frequency ratio is large. Therefore, the acceleration responses are amplified.

Another disadvantage of the ISAMD is the displacement responses of the control mass block. Figure 21 shows that the maximum displacements of the ISAMD are several times larger than those of the TMD. For example, with a frequency ratio of 2–4, the maximum displacements of the control mass block are roughly 4–6 times those of the TMD. The large displacement responses of the control mass block restrict the application range of the ISAMD; it can only be installed in buildings with adequate space.

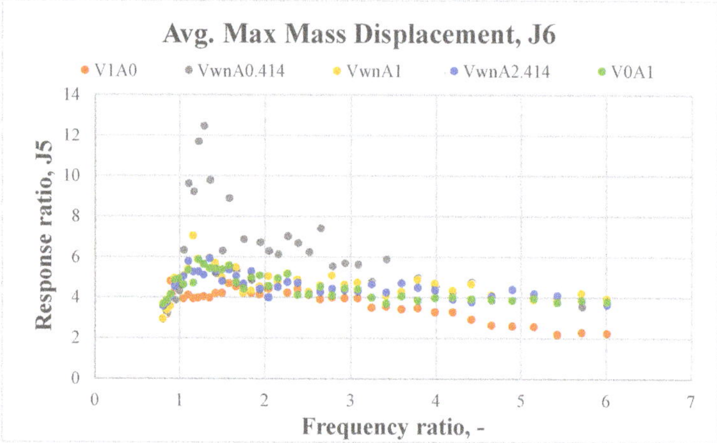

Figure 21. The average value of displacement responses of the control mass block of the ISAMD, J6.

6.2.3. Seismic Resistance and Mass Block Displacement Influence of the ISAMD with Auxiliary Damper and Spring

The large mass block displacement of the ISAMD is a major limitation of its application, as shown in Figure 21. Therefore, a feasibility study of the ISAMD with an auxiliary spring and damper installed to reduce the mass block displacement between the mass block and the structure was performed. The ISAMD in parallel with an auxiliary spring and damper is shown in Figure 22. The parameters of weight, added strength and damping coefficient interval of the ISAMD were W_V:W_A = ω_0 : 1, 0, 50, 100 and 200–3600 kN/m and c_a = 0, 20, 40 . . . 100 kN·s/m, respectively, to investigate the influence of seismic resistance and mass block displacement.

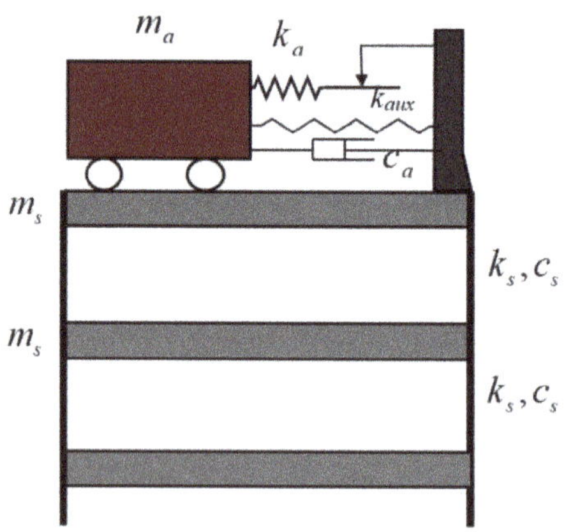

Figure 22. ISAMD Structure with auxiliary spring and damper.

The analysis of the control efficiency index J1 with the maximum displacement of the top floor showed that, when the stiffness of the auxiliary spring k_{aux} is much lower than the optimal stiffness of the TMD, J1 decreases as K_{aux} increases. When K_{aux} increases up to one-quarter of the optimal stiffness of the TMD, J1 increases as K_{aux} increases. Therefore, for the analysis example in this paper, the auxiliary spring stiffness of the ISAMD is 1/8–1/4 of the optimal stiffness of the TMD to effectively reduce the mass block displacement without reducing the shock absorption effect of the ISAMD displacement control. To facilitate the comparison of the influence of an auxiliary spring and damper on the control efficiency indices J1 and J5 and the mass displacement index J6, the average value of the average efficiency indices was defined as each index of the main spring frequency ratio of the ISAMD being between 2 and 3, W_V:W_A = $\omega_0 : 1$, the better parameter interval.

$$J_k^* = \frac{\sum_{i=1}^{NCase} J_{k,i}}{NCase} \quad (16)$$

where *NCase* is the number of analysis examples at which the frequency ratio of the ISAMD main spring is between 2 and 3.

Figures 23–25 show the influence of the auxiliary spring and damper on the top displacement control index, top floor acceleration control index and mass block displacement control index, respectively.

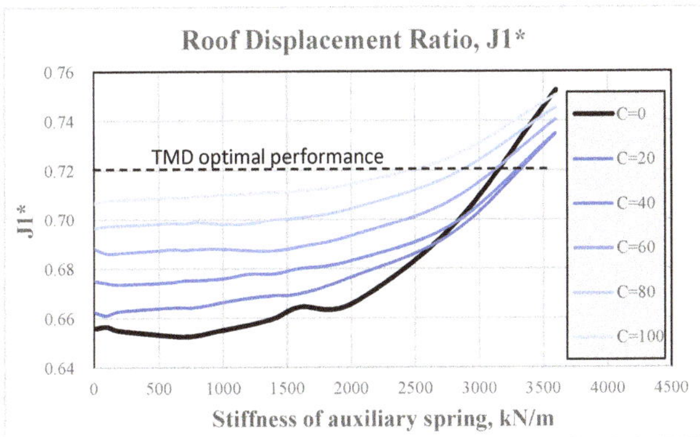

Figure 23. Influence of auxiliary spring and damper on the control index of maximal roof displacement.

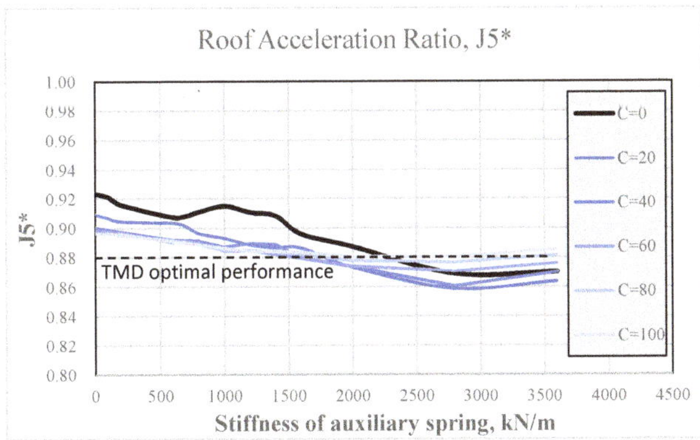

Figure 24. Influence of auxiliary spring and damper on control index of maximal roof acceleration.

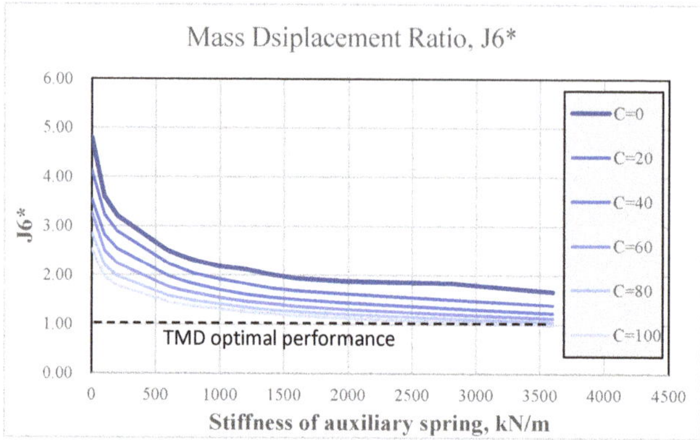

Figure 25. Influence of auxiliary spring and damper on control index of damper mass displacement.

The bold black lines in Figures 23–25 show the variation of the control indices of the ISAMD with only the auxiliary spring. The conclusions from the observations can be drawn as follows:

(1) Figure 25 shows that, when the spring stiffness is up to 1000 kN/m, equivalent to one-quarter of the optimal stiffness of the TMD, the mass block displacement is reduced to twice that of the TMD. The displacement of mass blocks is effectively controlled with a reduction rate of about 60% compared to the original displacement reaction.

(2) When the stiffness of the auxiliary spring is less than 1500 kN/m, the change in the displacement control effect for the maximum structural displacement reaction of the roof is less than 1%.

(3) The auxiliary spring reduces the mass block displacement without affecting the control of structural acceleration; structural acceleration at the roof is even slightly lower than that without the auxiliary spring.

(4) The ISAMD with an auxiliary damper has a more significant effect on reducing the mass block displacement than that of the ISAMD with the auxiliary spring. When the ISAMD is paired with auxiliary spring stiffness of 1000 kN/m, the damping coefficient is 80 kN·s/m, equivalent to the optimal damping coefficient of the TMD, and the mass block displacement is only 1.31 times that of the TMD. Nevertheless, the control effect index of the structural displacement reaction of the roof, J1*, is 0.70, indicating a 4% loss in seismic resistance.

(5) The ISAMD with the auxiliary damper obviously magnifies the structural displacement of the roof. The ISAMD with the auxiliary spring should be preferred to reduce the displacement of the mass block.

(6) The sensitivity of the ISAMD control indices to the frequency ratio of the main spring can be further reduced by the ISAMD with the auxiliary damper.

6.3. Influence of Near-Fault and Far-Field Ground Motion

The maximum structural displacement of the roof and the root mean square of the structural displacement reaction for the structure with the ISAMD under 26 seismic loads, as presented in Table 4 and Figure 26, were compared with those of the TMD. The optimal frequency ratio with the ISAMD was employed, with a weight ratio of W_V:W_A = 0:1 and without the auxiliary spring or damper. These ground motion records were classified into two categories by the epicentral distance of 20 km. The ground motion characteristics of these seismic records and the influence of the control effect on the structure with TMD and ISAMD were compared.

Table 4. Comparison of optimal TMD and ISAMD (W_V:W_A = 0:1, K = 0, C = 0).

Response	TMD (max)	ISAMD (max)	TMD (rms)	ISAMD (rms)
Average (overall)	0.77	0.63	0.57	0.38
Average (<20 km)	0.93	0.84	0.56	0.40
Average (>20 km)	0.73	0.58	0.58	0.38
Stdev. (overall)	0.26	0.20	0.17	0.10
Stdev. (<20 km)	0.46	0.29	0.28	0.16
Stdev. (>20 km)	0.18	0.15	0.14	0.08

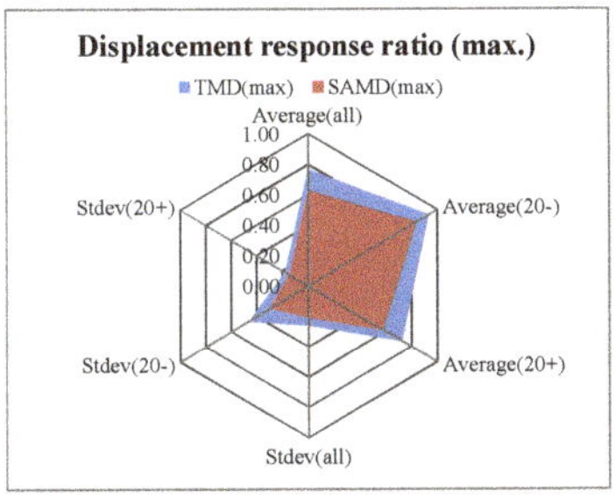

(**a**) The maximum displacement response ratio

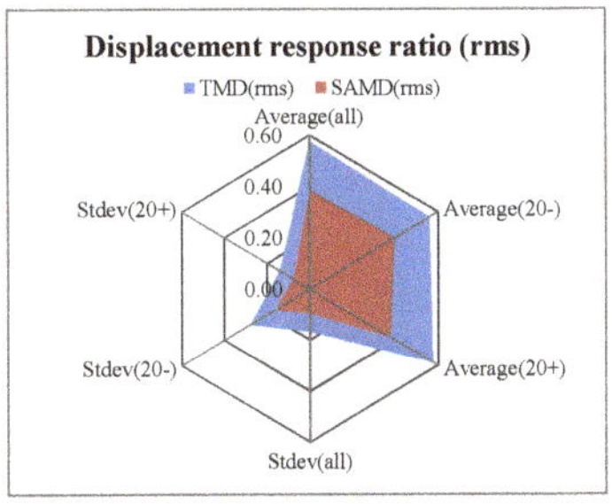

(**b**) Root mean square (RMS) displacement response ratio

Figure 26. Comparison of control effects of the TMD and ISAMD.

If the epicentral distance were less than 20 km, the control effect on the structure with TMD and ISAMD would be affected, as shown in Figure 24. The average response ratios and standard deviation were higher than the ensemble average. The average maximum reaction ratio and standard deviation of the structure with the TMD were 0.93 and 0.46, respectively, indicating that the probability of the maximal roof displacement response larger than that of a structure without control is nearly 39.64%. The average maximum reaction ratio and standard deviation of the structure with the ISAMD were 0.84 and 0.20, respectively, indicating that the probability of the maximal roof displacement response larger than that of a structure without control is nearly 11.20%.

7. Conclusions

An ISAMD is proposed in this paper, and a directional active joint is developed as the breaker to lock and unlock contact between the structure and damper. To achieve the maximum dissipation effect, the proposed ISAMD does negative work on the structure. To investigate the seismic resistance of the proposed ISAMD, VFIFE was used to derive a mathematical model based on the characteristics of the ISAMD. Then, a 10-story shearing building controlled by the TMD or ISAMD under excitation of 26 seismic wave records was used to investigate: (1) the structural displacement reactions of the top floor and the mass block displacement reactions of the TMD and ISAMD under various combinations of parameters; (2) the displacement reduction ratio of the mass control block; (3) the influence of different parameter combinations on the seismic dissipation effect; (4) the influence of the seismic resistance and mass block displacement of the ISAMD with an auxiliary damper and spring; and (5) the influence of near-fault and far-field ground motions.

From the above analysis and results, the following conclusions are presented:

1. The maximum displacement reduction effect of the ISAMD is only slightly better than that of the TMD, reducing the maximum displacement responses of the roof by 4–10%. The optimal control efficiency indices of the TMD control system happen at a frequency ratio of 0.95. When the frequency ratio is slightly offset, the seismic

resistance of the TMD is extremely limited due to the detuning effect. The sensitivity of the control indices of the ISAMD to the frequency ratio is very low, so detuning does not occur.

2. The maximum roof displacement reaction of the structure with the ISAMD has a reduction ratio above 30%, and the root mean square of displacement reaction indicates greater than 60% seismic resistance. The control mass block displacement of the ISAMD is 2–4 times greater than that of the TMD. The installation space of the ISAMD must therefore be large. However, when the control mass frequency ratio of the ISAMD is relatively high, the installation space does not need to be excessively large.
3. The frequency ratio should be around 2–4 for the structure with the ISAMD under different weight ratios to achieve a stable shock absorption effect. The average seismic resistance ratio is about 33–35%. The shock absorption effect of the root mean square of the roof structural displacement of the structure with the ISAMD is better than that of the maximum structural displacement reaction. The average seismic proof effect is consistently around 60%. Therefore, the ISAMD provides a very good damping effect. Nevertheless, the weight ratio of the ISAMD should avoid V1A0, as the control effect is slightly inferior to those of other weights.
4. The control mass block displacement of the ISAMD can be reduced effectively by increasing the stiffness of the auxiliary spring. The gradient of the control mass block displacement on the stiffness of the auxiliary spring has a decreasing tendency with lower stiffness.
5. The seismic resistance of the TMD changes greatly with the seismic load. However, the shock absorption effect of the ISAMD is stable, and the reliability of the structural displacement control effect of the structure with the ISAMD is better than that of the TMD, regardless of earthquake distance.
6. The design criteria of the proposed ISAMD should consider a frequency ratio and weight ratio of around 2–4 and less than natural frequency, respectively.

Author Contributions: Conceptualization, M.-H.S. and W.-P.S.; methodology, M.-H.S. and W.-P.S.; software, M.-H.S.; formal analysis, M.-H.S. and W.-P.S.; data curation, M.-H.S. and W.-P.S.; writing—original draft preparation, M.-H.S. and W.-P.S.; writing—review and editing, M.-H.S. and W.-P.S.; visualization, M.-H.S. and W.-P.S.; project administration, M.-H.S. and W.-P.S.; and funding acquisition, M.-H.S. and W.-P.S. Both authors have read and agreed to the published version of the manuscript.

Funding: This research was funded by Ministry of Science and Technology, Taiwan, grant number No. MOST-105-2221-E-260-003 and MOST-105-2221-M-167-001.

Informed Consent Statement: Not applicable for this study not involving humans.

Data Availability Statement: All data are available within the article and also from the corresponding author upon request.

Conflicts of Interest: The authors declare that there is no conflict of interests regarding the publication of this paper.

References

1. Basharkah, M.A.; Yao, J.T.P. *Reliability Aspects of Structural Control*; Structural Engineering (Technical Report); Purdue University, School of Civil Engineering: West Lafayette, IN, USA, 1982.
2. Choi, K.-M.; Jung, H.-J.; Lee, H.-J.; Cho, S.-W. Seismic protection of base-isolated building with nonlinear isolation system using smart passive control strategy. *Struct. Control Health Monit.* **2008**, *15*, 785–796. [CrossRef]
3. Chen, Y.; Cao, T.; Ma, L.; Luo, C. Structural vibration passive control and economic analysis of a high-rise building in Beijing. *Earthq. Eng. Eng. Vib.* **2009**, *8*, 561–568. [CrossRef]
4. Jung, H.-J.; Jang, D.-D.; Lee, H.-J.; Lee, I.-W.; Cho, S.-W. Feasibility Test of Adaptive Passive Control System Using MR Fluid Damper with Electromagnetic Induction Part. *J. Eng. Mech.* **2010**, *136*, 254–259. [CrossRef]
5. Murase, M.; Tsuji, M.; Takewaki, I. Smart passive control of buildings with higher redundancy and robustness using base-isolation and inter-connection. *Earthq. Struct.* **2013**, *4*, 649–670. [CrossRef]

6. Zhang, Z.; Balendra, T. Passive control of bilinear hysteretic structures by tuned mass damper for narrow band seismic motions. *Eng. Struct.* **2013**, *54*, 103–111. [CrossRef]
7. Cacciola, P.; Tombari, A. Vibrating barrier: A novel device for the passive control of structures under ground motion. *Proc. R. Soc. A Math. Phys. Eng. Sci.* **2015**, *471*, 20150075. [CrossRef]
8. Fisco, N.; Adeli, H. Smart structures: Part I—Active and semi-active control. *Sci. Iran.* **2011**, *18*, 275–284. [CrossRef]
9. Liu, K.; Chen, L.-X.; Cai, G.-P. Active control of a nonlinear and hysteretic building structure with time delay. *Struct. Eng. Mech.* **2011**, *40*, 431–451. [CrossRef]
10. Liu, K.; Chen, L.-X.; Cai, G.-P. Active control of buildings with bilinear hysteresis and time delay. *Int. J. Struct. Stab. Dyn.* **2013**, *13*. [CrossRef]
11. Xinfa, Z.; Zhenbin, P.; Lian-Guang, M.; Yu, S. Active Control Based on Prediction of Structural Vibration Feedback. In Proceedings of the 2014 Fifth International Conference on Intelligent Systems Design and Engineering Applications, Hunan, China, 15–16 June 2014; pp. 123–127.
12. Shih, M.-H.; Sung, W.-P.; Go, C.G. Development of accumulated semi-active hydraulic damper. *Exp. Tech.* **2002**, *26*, 29–32. [CrossRef]
13. Shih, M.-H.; Sung, W.-P. A design concept of autonomous controller for improving seismic proof capability of semi-active control device. *Exp. Tech.* **2009**, *34*, 20–26. [CrossRef]
14. Palacios-Quiñonero, F.; Rubió-Massegú, J.; Rossell, J.M.; Karimi, H.R. Semi-active-passive structural vibration control strategy for adjacent structures under seismic excitation. *J. Frankl. Inst.* **2012**, *349*, 3003–3026. [CrossRef]
15. Hiramoto, K.; Matsuoka, T.; Sunakoda, K. Simultaneous optimal design of the structural model for the semi-active control design and the model-based semi-active control. *Struct. Control Health Monit.* **2014**, *21*, 522–541. [CrossRef]
16. Shih, M.-H.; Sung, W.-P. Development of semi-active hydraulic damper as active interaction control device to withstand external excitation. *Sadhana* **2014**, *39*, 123–138. [CrossRef]
17. Available online: https://en.wikipedia.org/wiki/Citigroup_Center (accessed on 22 January 2021).
18. Available online: https://en.wikipedia.org/wiki/CN_Tower (accessed on 22 January 2021).
19. Available online: https://en.wikipedia.org/wiki/John_Hancock_Tower (accessed on 22 January 2021).
20. Available online: https://en.wikipedia.org/wiki/Taipei_101 (accessed on 22 January 2021).
21. Sadek, F.; Mohraz, B.; Taylor, A.W.; Chung, R.M. A method of estimating the parameters of tuned mass dampers for seismic applications. *Earthq. Eng. Struct. Dyn.* **1997**, *26*, 617–635. [CrossRef]
22. Lin, C.-C.; Ueng, J.-M.; Huang, T.-C. Seismic response reduction of irregular buildings using passive tuned mass dampers. *Eng. Struct.* **2000**, *22*, 513–524. [CrossRef]
23. Bakre, S.V.; Jangid, R.S. Optimum parameters of tuned mass damper for damped main system. *Struct. Control Health Monit.* **2007**, *14*, 448–470. [CrossRef]
24. Wang, J.-F.; Lin, C.-C.; Lian, C.-H. Two-stage optimum design of tuned mass dampers with consideration of stroke. *Struct. Control Health Monit.* **2009**, *16*, 55–72. [CrossRef]
25. Soong, T.; Spencer, B. Supplemental energy dissipation: State-of-the-art and state-of-the-practice. *Eng. Struct.* **2002**, *24*, 243–259. [CrossRef]
26. Lin, C.C.; Wang, J.F.; Chen, B.L. Train-Induced Vibration Control of High-Speed Railway Bridges Equipped with Multiple Tuned Mass Dampers. *J. Bridg. Eng.* **2005**, *10*, 398–414. [CrossRef]
27. Robinson, J.K.; Gamble, S.L.; Myslimaj, B.M. Supplemental damping and using tuned sloshing dampers. *Struct. Mag.* **2007**, *6*, 14–18.
28. Yamamoto, M.; Sone, T. Behavior of active mass damper (AMD) installed in high-rise building during 2011 earthquake off Pacific coast of Tohoku and verification of regenerating system of AMD based on monitoring. *Struct. Control Health Monit.* **2014**, *21*, 634–647. [CrossRef]
29. Alkhatib, R.; Golnaraghi, M.F. Active Structural Vibration Control: A Review. *Shock. Vib. Dig.* **2003**, *35*, 367–383. [CrossRef]
30. Abe, M. Semi-active tuned mass dampers for seismic protection of civil structures. *Earthq. Eng. Struct. Dyn.* **1996**, *25*, 743–749. [CrossRef]
31. Aldemir, U. Optimal control of structures with semi-active tuned mass dampers. *J. Sound Vib.* **2003**, *266*, 847–874. [CrossRef]
32. Cai, C.S.; Wu, W.J.; Araujo, M. Cable Vibration Control with a TMD-MR Damper System: Experimental Exploration. *J. Struct. Eng.* **2007**, *133*, 629–637. [CrossRef]
33. Lin, C.-C.; Lin, G.-L.; Wang, J.-F. Protection of seismic structures using semi-active friction TMD. *Earthq. Eng. Struct. Dyn.* **2009**, *39*, 635–659. [CrossRef]
34. Lin, G.-L.; Lin, C.-C.; Lu, L.-Y.; Ho, Y.-B. Experimental verification of seismic vibration control using a semi-active friction tuned mass damper. *Earthq. Eng. Struct. Dyn.* **2011**, *41*, 813–830. [CrossRef]
35. Nagarajaiah, S.; Sonmez, E. Structures with semi-active variable stiffness single/multiple tuned mass dampers. *J. Struct. Eng.* **2007**, *133*, 67–77. [CrossRef]
36. Nagarajaiah, S. Adaptive passive, semiactive, smart tuned mass dampers: Identification and control using empirical mode decomposition, hilbert transform, and short-term fourier transform. *Struct. Control Health Monit.* **2009**, *16*, 800–841. [CrossRef]
37. Chey, M.-H.; Chase, J.G.; Mander, J.B.; Carr, A.J. Semi-active tuned mass damper building systems: Design. *Earthq. Eng. Struct. Dyn.* **2009**, *39*, 119–139. [CrossRef]

38. Chey, M.-H.; Chase, J.G.; Mander, J.B.; Carr, A.J. Semi-active tuned mass damper building systems: Application. *Earthq. Eng. Struct. Dyn.* **2009**, *39*, 69–89. [CrossRef]
39. Spencer, B.F.; Nagarajaiah, S. State of the Art of Structural Control. *J. Struct. Eng.* **2003**, *129*, 845–856. [CrossRef]
40. Lin, C.C. *Semi-Active Resettable Variable Stiffness TMD for Seismic Protection*; Research Report; NSC-100-2221-E-005-054; National Science Council: Taipei, Taiwan, 2012.
41. Soto-Brito, R.; Ruiz, S.E. Influence of ground motion intensity on the effectiveness of tuned mass dampers. *Earthq. Eng. Struct. Dyn.* **1999**, *28*, 1255–1271. [CrossRef]
42. Lukkunaprasit, P.; Wanitkorkul, A. Inelastic buildings with tuned mass dampers under moderate ground motions from distant earthquakes. *Earthq. Eng. Struct. Dyn.* **2001**, *30*, 537–551. [CrossRef]
43. Chen, G.; Wu, J. Experimental study on multiple tuned mass dampers to reduceseismic responses of a three-storey building structure. *Earthq. Eng. Struct. Dyn.* **2003**, *32*, 793–810. [CrossRef]
44. Pinkaew, T.; Lukkunaprasit, P.; Chatupote, P. Seismic effectiveness of tuned mass dampers for damage reduction of structures. *Eng. Struct.* **2003**, *25*, 39–46. [CrossRef]
45. Shih, M.-H.; Sung, W.-P. Development of semi-active mass damper with impulsive reaction. *Sadhana* **2020**, *45*, 1–11. [CrossRef]
46. Shih, M.-H.; Sung, W.-P. Structural Control Effect and Performance of Structure Under Control of Impulse Semi-active Mass Control Mechanism. *Iran. J. Sci. Technol. Trans. Civ. Eng.* **2020**, 1–16. [CrossRef]
47. Shih, M.H.; Sung, W.P. Parametric Study of Impulse Semi-Active Mass Damper with Developing Directional Active Joint. *Arab. J. Sci. Eng.* **2021**, in press.
48. Lin, C.; Hu, C.; Wang, J.; Hu, R. Vibration control effectiveness of passive tuned mass dampers. *J. Chin. Inst. Eng.* **1994**, *17*, 367–376. [CrossRef]

applied sciences

Article

Seismic Response Study of L-Shaped Frame Structure with Magnetorheological Dampers

Jiangtao You [1], Yang Yang [2], Yongfeng Fan [3] and Xiangcheng Zhang [1,*]

1 Highway School, Henan College of Transportation, Zhengzhou 450006, China; youjiangtao2022@163.com
2 School of Mechanics and Safety Engineering, Zhengzhou University, Zhengzhou 450001, China; 13613788269@163.com
3 Henan Xinrong Expressway Construction Co., Ltd., Luoyang 471023, China; wyjd1224@163.com
* Correspondence: zxiangcheng_edu@yeah.net

Abstract: To analyze the multi-dimensional seismic mitigation performance and the torsional vibration characteristic of an L-shaped frame structure with a magnetorheological damper (MRD). Firstly, the mechanical property of the MRD is experimentally studied. Then, the calculation models of the L-shaped frame structure without and with MRDs are found through theoretical analysis, and the programs of the calculation models are compiled. Finally, the time history responses of the calculation models are calculated during the excitation of bi-directional seismic wave; the responses of displacement, acceleration, story drift ratio, shear force and bending moment and torsional vibration are then compared and analyzed. The results show that by adjusting the current, the damping force provided by the MRD can be adjusted continuously in the range of 3.1 to 120 kN. The compiled calculation model programs of the L-shaped frame structure without and with MRDs can effectively simulate the multi-dimensional seismic response of the structure. Reasonable arrangement of MR dampers can effectively reduce the displacement, acceleration, shear force, bending moment, multi-dimensional vibration, and torsional vibration response of L-shaped frame structures.

Keywords: magnetorheological damper; seismic response; story drift ratio; shear force; bending moment; torsional vibration

Citation: You, J.; Yang, Y.; Fan, Y.; Zhang, X. Seismic Response Study of L-Shaped Frame Structure with Magnetorheological Dampers. *Appl. Sci.* **2022**, *12*, 5976. https://doi.org/10.3390/app12125976

Academic Editor: Maria Favvata

Received: 12 February 2022
Accepted: 28 February 2022
Published: 12 June 2022

Publisher's Note: MDPI stays neutral with regard to jurisdictional claims in published maps and institutional affiliations.

Copyright: © 2022 by the authors. Licensee MDPI, Basel, Switzerland. This article is an open access article distributed under the terms and conditions of the Creative Commons Attribution (CC BY) license (https://creativecommons.org/licenses/by/4.0/).

1. Introduction

Magnetorheological fluid (MRF) is a new kind of smart fluid whose rheological characteristics are related to the strength of the magnetic field under the action of the magnetic field; MRFs can restore to the initial liquid state without the magnetic field [1]. Magnetorheological dampers (MRDs), semi-active control devices using MRFs, become effective damping control devices for controlling structural vibration because of advantages such as quick response, low energy consumption, large damping force, and being continuously adjustable. Therefore, MRDs have shown good application prospects in civil engineering earthquake mitigation and disaster prevention [2–8].

The mechanical performance of the MRD is an important factor affecting the efficiency of suppressing vibration control of structures [9]. Therefore, the performance of the MRD should be experimentally studied before the MRD is installed on the structure [10,11]. Since 1996, many important experimental studies on the performance of MRDs have been carried out. Spencer et al. [12] designed and tested an MRD that had a 200 kN damping force. In 2003, Fujitani et al. [13] developed a 400 kN MRD for a real base-isolated building, whose dynamic characteristics were verified through tests. Xu et al. [14] carried out tests on a multi-stage shear-valve mode MRD, and the test results showed that the energy dissipation and the damping force increased significantly with the current until magnetic saturation occurred. Tu et al. [15] presented a large-scale MRD that had a 500 kN damping force and carried out its performance experiments; the results showed that the damping force increased obviously with the increase of current, and when the current was 1.0 A, the maximum damping force provided by the MRD was 537 kN.

The trends of the irregular plane frame structure [16–19] have rapidly increased due to flexible column grid layouts and large practical spaces. However, the building structures with irregular planes are more susceptible to seismic forces than regular ones. The reason is that the center of mass has different positions from the center of rigidity in the irregular frame structure, and the torsional deformation of the structure caused by eccentricity is more likely to lead to structural failure. To reduce the torsional response of irregular structures, many experimental studies on the multi-dimensional seismic response of the irregular structure with MRDs have been carried out, and the results demonstrate that setting MRDs in the structure can efficaciously decrease the structural torsion response and the multi-dimensional dynamic response [20–26].

Dyke et al. [27] studied a three-story model structure with MRDs through tests and concluded that MRDs had a significant weakening effect on the peak and RMS responses caused by seismic excitations. Yoshida et al. [28] conducted experimental research on a two-story structure model with an irregular column distribution and concluded that the responses of the structure were greatly reduced after adopting the semi-active controller. Although the analysis of the structure with MRDs has the advantages of obtaining accurate results and high reliability, it also has the disadvantages of high costs and being time-consuming. Moreover, numerical simulation analysis of the structure with MRDs is an effective and economical method to study the effect of structural torsion and multi-dimensional dynamic responses.

Therefore, many scholars have used MATLAB programming to establish various models of the structures with MRDs. Zhang et al. [29] carried out the time history analysis on the elastic-plastic for the plane structure with MRDs, and the results showed that the amount of the yielding point and cracking point of the controlled structure obviously decreased and that MRDs can improve the seismic performance of the structure. Bhaiya et al. [30] established the elastic plane mode of a ten-story frame structure with MRDs and conducted the analysis and research under the action of the seismic force. Xu et al. [31] developed a time history analysis program on the elastic–plastic of the spatial frame structure with MRDs through using MATLAB programming.

However, the program of the spatial frame structure to MRDs in [29] ignored the function of the floor, and the other plane model cannot simulate the multi-dimensional vibration response and torsional vibration control effect of the L-shaped frame structure with MRDs. Compared with [32], this paper tests the mechanical properties of the self-made MRD and analyzes the shear force and bending moment of the structural columns. Additionally, the L-shaped frame structure with MRDs has not been studied up until now.

In this paper, the mechanical performance of the self-made MRD is experimentally studied at different displacement amplitudes and currents. The calculation models of the L-shaped frame structure with and without MRDs are established in view of the theories of spatial beam elements, shell elements, and semi-active control algorithms, and the programs of the calculation models are then compiled by MATLAB software. The dynamic time history analysis is conducted during the action time of bi-directional seismic waves, and the responses of displacement, acceleration, story drift ratio, shear force, bending moment, multi-dimensional vibration, and torsional vibration of the L-shaped frame structure without and with MRDs are then contrasted and analyzed.

The rest of the work includes Section 2, which provides details about the tests on the MRDs. Section 3 describes in detail the modeling theory of the L-shaped frame structure. Section 4 describes the model and explains the verification process. Additionally, Section 5 gives a detailed analysis and interpretation of results, and finally, Section 5 summarizes the research conducted and presents the future objectives that can extend it.

2. The Tests of the MRD

2.1. Structure of the MRD

The structure chart of the self-made four-coil shear valve mode MRD is shown in Figure 1. The MRD consists of a multi-stage piston, a piston rod, a cylinder, a guide sleeve, a cover plate, an O-ring, a ball-end piston rod, and coils, in which the multi-stage piston is made of electrical pure iron and the piston rod, cylinder, guide sleeve, cover plate, and cylinder are made of No. 45 steel. The parameters of each part of the MRD are shown in Table 1.

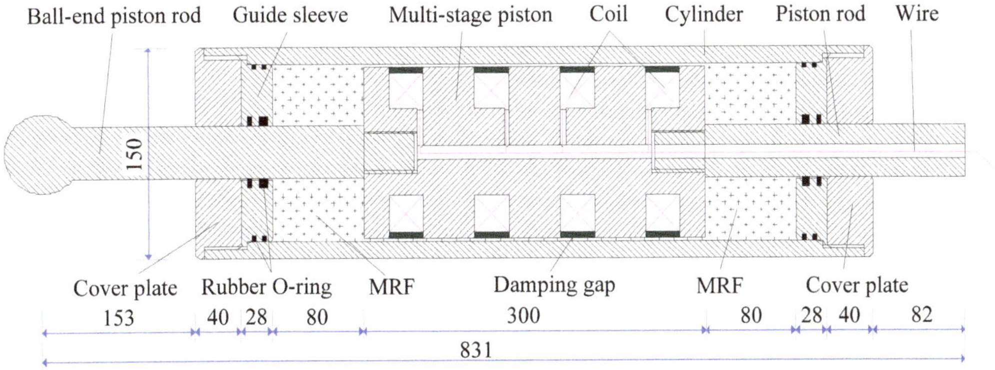

Figure 1. Structure schematic diagram of the MRD (mm).

Table 1. The structure parameters of the MRD.

Parameter	Size	Parameter	Size
Outer diameter of cylinder (mm)	150	Stroke (mm)	±80
Inner diameter of cylinder (mm)	120	Coil groove depth (mm)	25
Diameter of piston (mm)	118	Damping gap (mm)	1

The cavity of the MRD was filled with self-made MRF, whose main components were carbonyl iron power, sodium dodecyl sulfate, and methyl silicone oil. The masses of carbonyl iron powder and methyl silicone oil account for 78% and 20%, respectively. When the shear rate is 100 s^{-1}, the shear stresses of the MRF under different magnetic fields are given in Table 2.

Table 2. Shear stress of the MRF under different magnetic fields.

Magnetic Field Intensity (T)	0	0.05	0.10	0.15	0.20	0.25	0.30	0.35	0.40	0.45	0.50
Shear stress (Pa)	304.8	1340	4062.9	7928	10,921	15,016.7	18,346	21,312	23,550	25,498	26,732

2.2. Test Loading of the MRD

Tests on the MRD were conducted to research the characteristics of the MRD. During the test, the MRD was powered by a regulated DC power supply, and the current levels for the four coils were identical to the current levels from 0 A to 0.28 A at the interval of 0.14 A. For each current level, the MRD was held in a fixed position at one end, while the other end was subjected to a triangular displacement, as shown in Figure 2.

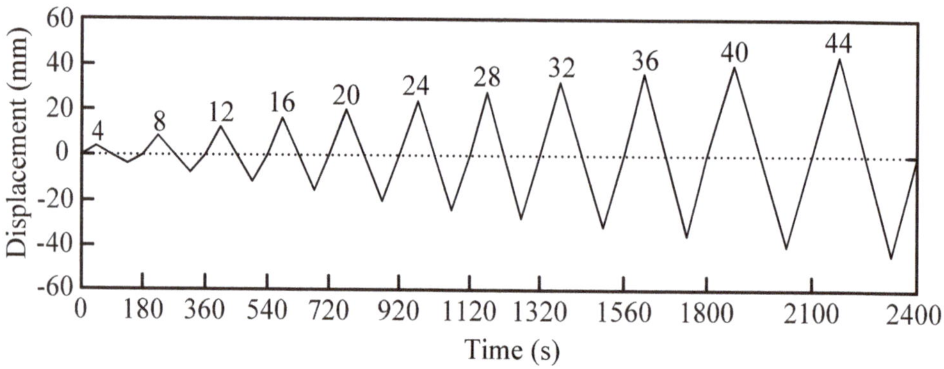

Figure 2. Loading protocol of the MRD.

2.3. Test Results Analysis

According to the test results, the variation of mechanical performance with current is plotted in Figure 3. As shown in Figure 3a, when the current is 0 A, the minimum damping force provided by the MRD can attain about 3.1 kN, while the maximum damping force can attain about 5.3 kN when the displacement amplitude is 40 mm. As shown in Figure 3c, when the current attains 0.28 A, the minimum damping force provided by the MRD can attain about 80 kN when the displacement amplitude is 4 mm, while the maximum damping force can attain 120 kN when the displacement amplitude is bigger than 4 mm. As seen in Figure 3b,c, in the case of the MRD power supply, the damping force can fluctuate within a certain range after it rises rapidly to the maximum value with the increase of displacement amplitude.

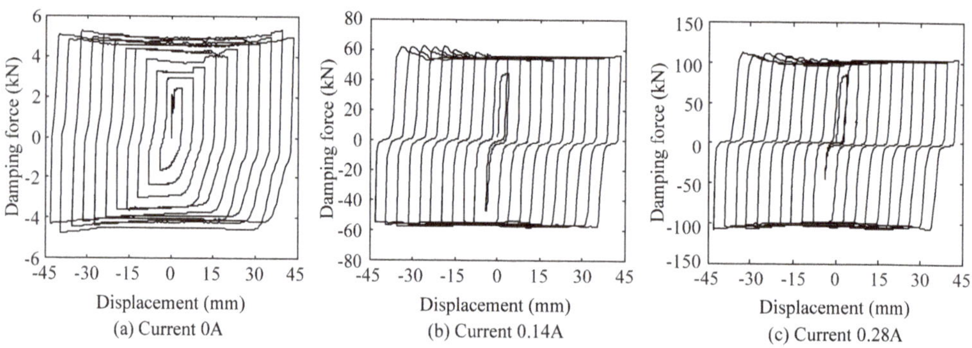

Figure 3. Damping force-displacement hysteresis curve of the MRD at different currents.

3. Modeling Theory of the L-Shaped Frame Structure

3.1. Theory of Spatial Beam Element

For the spatial beam element, as shown in Figure 4, every node has 6 degrees of freedom (DOFs), where u, v, and w are the axial displacements in x, y, and z directions, respectively; θ_x, θ_y, and θ_z are the angular displacements around x, y, and z directions, respectively. The numbers in brackets in Figure 4 denote the DOF number.

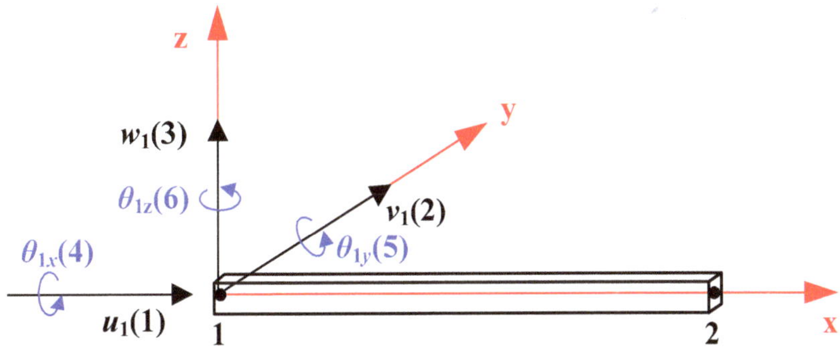

Figure 4. Node displacement and DOF number of beam element.

According to [33], the stiffness matrix of spatial beam element, considering shear deformation $\boldsymbol{k}_e$, is as follow:

$$\boldsymbol{k}_e = \begin{bmatrix} \boldsymbol{k}_{11} & \boldsymbol{k}_{12} \\ \boldsymbol{k}_{21} & \boldsymbol{k}_{22} \end{bmatrix} \quad (1)$$

$$\boldsymbol{k}_{11} = \begin{bmatrix} k_1 & 0 & 0 & 0 & 0 & 0 \\ 0 & k_2 & 0 & 0 & 0 & k_3 \\ 0 & 0 & k_4 & 0 & k_5 & 0 \\ 0 & 0 & 0 & k_6 & 0 & 0 \\ 0 & 0 & k_5 & 0 & k_7 & 0 \\ 0 & k_3 & 0 & 0 & 0 & k_8 \end{bmatrix} \quad (2)$$

$$\boldsymbol{k}_{22} = \begin{bmatrix} k_1 & 0 & 0 & 0 & 0 & 0 \\ 0 & k_2 & 0 & 0 & 0 & -k_3 \\ 0 & 0 & k_4 & 0 & -k_5 & 0 \\ 0 & 0 & 0 & k_6 & 0 & 0 \\ 0 & 0 & -k_5 & 0 & k_7 & 0 \\ 0 & -k_3 & 0 & 0 & 0 & k_8 \end{bmatrix} \quad (3)$$

$$\boldsymbol{k}_{12} = {\boldsymbol{k}_{21}}^{\mathrm{T}} = \begin{bmatrix} -k_1 & 0 & 0 & 0 & 0 & 0 \\ 0 & -k_2 & 0 & 0 & 0 & k_3 \\ 0 & 0 & -k_4 & 0 & k_5 & 0 \\ 0 & 0 & 0 & -k_6 & 0 & 0 \\ 0 & 0 & -k_5 & 0 & k_9 & 0 \\ 0 & -k_3 & 0 & 0 & 0 & k_{10} \end{bmatrix} \quad (4)$$

In Equations (2)–(4), $k_1 = \frac{EA}{L}$, $k_2 = \frac{12EI_z}{(1+\beta_z)L^3}$, $k_3 = \frac{6EI_z}{(1+\beta_z)L^2}$, $k_4 = \frac{12EI_y}{(1+\beta_y)L^3}$, $k_5 = -\frac{6EI_y}{(1+\beta_y)L^2}$, $k_6 = \frac{GJ}{L}$, $k_7 = \frac{(4+\beta_y)EI_y}{(1+\beta_y)L}$, $k_8 = \frac{(4+\beta_z)EI_z}{(1+\beta_z)L}$, $k_9 = \frac{(2-\beta_y)EI_y}{(1+\beta_y)L}$, $k_{10} = \frac{(2-\beta_z)EI_z}{(1+\beta_z)L}$, $\beta_y = \frac{12\mu EI_y}{GAL^2}$, $\beta_z = \frac{12\mu EI_z}{GAL^2}$.

Where E is the elastic modulus, A is the cross-sectional area, L is the length of the element, I is the cross-sectional inertia moment, G is the shear modulus, μ is Poisson's ratio, and J is the torsional moment of inertia. The stiffness matrix represents the system of linear equations that must be solved in order to ascertain an approximate solution to the differential equation.

The mass matrix of spatial beam element $\boldsymbol{m}_e$ is as follow [33]:

$$\boldsymbol{m}_e = \frac{\rho AL}{420} \begin{bmatrix} \boldsymbol{m}_{11} & \boldsymbol{m}_{12} \\ \boldsymbol{m}_{21} & \boldsymbol{m}_{22} \end{bmatrix} \quad (5)$$

$$\boldsymbol{m}_{11} = \begin{bmatrix} m_1 & 0 & 0 & 0 & 0 & 0 \\ 0 & m_3 & 0 & 0 & 0 & m_4 \\ 0 & 0 & m_3 & 0 & -m_4 & 0 \\ 0 & 0 & 0 & m_7 & 0 & 0 \\ 0 & 0 & -m_4 & 0 & m_9 & 0 \\ 0 & m_4 & 0 & 0 & 0 & m_9 \end{bmatrix} \quad (6)$$

$$\boldsymbol{m}_{22} = \begin{bmatrix} m_1 & 0 & 0 & 0 & 0 & 0 \\ 0 & m_3 & 0 & 0 & 0 & -m_4 \\ 0 & 0 & m_3 & 0 & m_4 & 0 \\ 0 & 0 & 0 & m_7 & 0 & 0 \\ 0 & 0 & m_4 & 0 & m_9 & 0 \\ 0 & -m_4 & 0 & 0 & 0 & m_9 \end{bmatrix} \quad (7)$$

$$\boldsymbol{m}_{12} = \boldsymbol{m}_{21}{}^{\mathrm{T}} = \begin{bmatrix} m_2 & 0 & 0 & 0 & 0 & 0 \\ 0 & m_5 & 0 & 0 & 0 & m_6 \\ 0 & 0 & m_5 & 0 & -m_6 & 0 \\ 0 & 0 & 0 & m_8 & 0 & 0 \\ 0 & 0 & m_6 & 0 & m_{10} & 0 \\ 0 & -m_6 & 0 & 0 & 0 & m_{10} \end{bmatrix} \quad (8)$$

In Equations (6)–(8), $m_1 = 140$, $m_2 = 70$, $m_3 = 156$, $m_4 = 22L$, $m_5 = 54$, $m_6 = -13L$, $m_7 = \frac{140J}{A}$, $m_8 = \frac{70J}{A}$, $m_9 = 4L^2$, $m_{10} = -3L^2$, and ρ is the density.

3.2. Theory of the Shell Element

For the shell element, as shown in Figure 5, each node also has 6 DOFs, and the DOFs of the shell element node are the same as those of the spatial beam element node. The numbers in brackets for Figure 5 denote the DOF number.

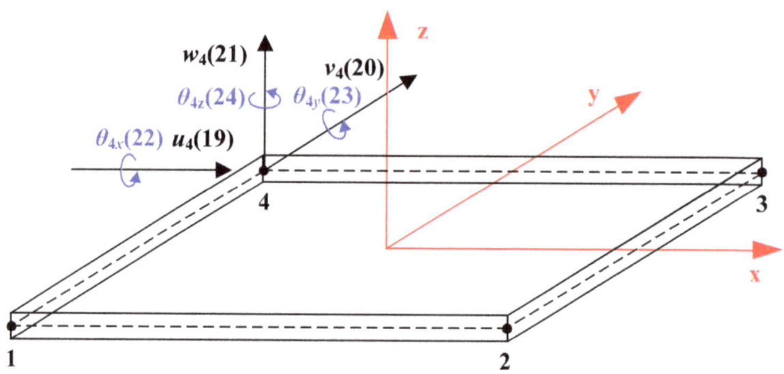

Figure 5. Node displacement and DOF number of the shell element.

The stiffness [32] of the shell element consists of the membrane stiffness $\boldsymbol{k}_e^m$ and the bending stiffness $\boldsymbol{k}_e^b$. $\boldsymbol{k}_e^m$, given in Equation (9), is used to deal with shell element membrane effects, whose corresponding DOFs are u, v and θ_z; $\boldsymbol{k}_e^b$ given in Equation (10) is used to deal with shell element bending effects, whose corresponding DOFs are w, θ_x, and θ_y; $\boldsymbol{k}_{ij}^m$ and $\boldsymbol{k}_{ij}^b$ are the submatrices with the dimension of 3×3. Then, the shell element stiffness matrix $\boldsymbol{k}_e$ is obtained by combining Equations (9) and (10), the dimension of which is 24×24, as shown in Equation (11).

$$\boldsymbol{k}_e^m = \begin{bmatrix} \boldsymbol{k}_{11}^m & \boldsymbol{k}_{12}^m & \boldsymbol{k}_{13}^m & \boldsymbol{k}_{14}^m \\ \boldsymbol{k}_{21}^m & \boldsymbol{k}_{22}^m & \boldsymbol{k}_{23}^m & \boldsymbol{k}_{24}^m \\ \boldsymbol{k}_{31}^m & \boldsymbol{k}_{32}^m & \boldsymbol{k}_{33}^m & \boldsymbol{k}_{34}^m \\ \boldsymbol{k}_{41}^m & \boldsymbol{k}_{42}^m & \boldsymbol{k}_{43}^m & \boldsymbol{k}_{44}^m \end{bmatrix} \quad (9)$$

$$k_e^b = \begin{bmatrix} k_{11}^b & k_{12}^b & k_{13}^b & k_{14}^b \\ k_{21}^b & k_{22}^b & k_{23}^b & k_{24}^b \\ k_{31}^b & k_{32}^b & k_{33}^b & k_{34}^b \\ k_{41}^b & k_{42}^b & k_{43}^b & k_{44}^b \end{bmatrix} \tag{10}$$

$$k_e = \begin{bmatrix} k_{11} & k_{12} & k_{13} & k_{14} \\ k_{21} & k_{22} & k_{23} & k_{24} \\ k_{31} & k_{32} & k_{33} & k_{34} \\ k_{41} & k_{42} & k_{43} & k_{44} \end{bmatrix} \tag{11}$$

In which

$$k_{ij} = \begin{bmatrix} k_{ij}^m(1,1) & k_{ij}^m(1,2) & 0 & k_{ij}^m(1,3) \\ k_{ij}^m(2,1) & k_{ij}^m(2,2) & 0 & k_{ij}^m(2,3) \\ 0 & 0 & k_{ij}^b & 0 \\ k_{ij}^m(3,1) & k_{ij}^m(3,2) & 0 & k_{ij}^m(3,3) \end{bmatrix} \quad i,j = 1,2,3,4 \tag{12}$$

Analogously, the shell element mass matrix [32] is derived in an identical method as the stiffness matrix. The membrane mass matrix m_e^m given in Equation (13) is used to deal with shell element membrane effects, whose corresponding DOFs are u, v and θ_z; the bending mass matrix m_e^b given in Equation (14) is used to deal with shell element bending effects, whose corresponding DOFs are w, θ_x, θ_y. m_{ij}^m and m_{ij}^b are the submatrices, with dimensions of 3×3. Then, the shell element mass matrix m_e is obtained by combining Equations (13) and (14), the dimension of which is 24×24, as shown in Equation (15).

$$m_e^m = \begin{bmatrix} m_{11}^m & m_{12}^m & m_{13}^m & m_{14}^m \\ m_{21}^m & m_{22}^m & m_{23}^m & m_{24}^m \\ m_{31}^m & m_{32}^m & m_{33}^m & m_{34}^m \\ m_{41}^m & m_{42}^m & m_{43}^m & m_{44}^m \end{bmatrix} \tag{13}$$

$$m_e^b = \begin{bmatrix} m_{11}^b & m_{12}^b & m_{13}^b & m_{14}^b \\ m_{21}^b & m_{22}^b & m_{23}^b & m_{24}^b \\ m_{31}^b & m_{32}^b & m_{33}^b & m_{34}^b \\ m_{41}^b & m_{42}^b & m_{43}^b & m_{44}^b \end{bmatrix} \tag{14}$$

$$m_e = \begin{bmatrix} m_{11} & m_{12} & m_{13} & m_{14} \\ m_{21} & m_{22} & m_{23} & m_{24} \\ m_{31} & m_{32} & m_{33} & m_{34} \\ m_{41} & m_{42} & m_{43} & m_{44} \end{bmatrix} \tag{15}$$

In which

$$m_{ij} = \begin{bmatrix} m_{ij}^m(1,1) & m_{ij}^m(1,2) & 0 & m_{ij}^m(1,3) \\ m_{ij}^m(2,1) & m_{ij}^m(2,2) & 0 & m_{ij}^m(2,3) \\ 0 & 0 & m_{ij}^b & 0 \\ m_{ij}^m(3,1) & m_{ij}^m(3,2) & 0 & m_{ij}^m(3,3) \end{bmatrix} \quad i,j = 1,2,3,4 \tag{16}$$

3.3. Coordinate Transformation

The k_e and m_e are element matrices in the local coordinate system (x, y, z), which need to be converted into the element matrices K_e and M_e in the global coordinate system (X, Y, Z) before the total matrix is assembled. K_e and M_e can be given as follow:

$$K_e = T^T k_e T, \qquad M_e = T^T m_e T \tag{17}$$

where the transformation matrix $\boldsymbol{T}$ can be given as follow:

$$\boldsymbol{T} = \begin{bmatrix} t & 0 & 0 \\ 0 & \ddots & 0 \\ 0 & 0 & t \end{bmatrix}_{a \times a} \quad (18)$$

in which a is equal to 4 and 8 for the spatial beam element and the shell element, respectively. $\boldsymbol{t}$ can be derived according to the direction cosine relationship between the global coordinate system and the local coordinate system.

$$\boldsymbol{t} = \begin{bmatrix} \cos(X, x) & \cos(X, y) & \cos(X, z) \\ \cos(Y, x) & \cos(Y, y) & \cos(Y, z) \\ \cos(Z, x) & \cos(Z, y) & \cos(Z, z) \end{bmatrix} \quad (19)$$

3.4. Equation of Motion

The equation of motion of the L-shaped structure with MRDs under seismic load can be given as follow:

$$\boldsymbol{M}\ddot{x}(t) + \boldsymbol{C}\dot{x}(t) + \boldsymbol{K}x(t) = -\boldsymbol{M}\boldsymbol{R}\ddot{x}_g(t) - \boldsymbol{H}\boldsymbol{F}(t) \quad (20)$$

where $\boldsymbol{M}$ is the mass matrix, $\boldsymbol{C}$ is the Rayleigh damping matrix, $\boldsymbol{K}$ is the stiffness matrix, $\boldsymbol{R}$ is the unit column vector, $\boldsymbol{H}$ is the location matrix of the MRD, $\ddot{x}$ is the acceleration vector of the structure, $\ddot{x}_g$ is the acceleration vector of seismic ground motion, and $\boldsymbol{F}$ is the control force matrix of the MRD.

3.5. Semi-Active Control Algorithm

At present, the classical linear quadratic control algorithm is mostly used to solve for the optimal control force [34]. In this algorithm, the optimal control force $\boldsymbol{U}$ is selected through the following performance indicators.

$$J = \frac{1}{2}\int_0^t \left[\boldsymbol{Z}^{\mathrm{T}}\boldsymbol{Q}\boldsymbol{Z} + \boldsymbol{U}^{\mathrm{T}}\boldsymbol{R}\boldsymbol{U}\right] dt + \frac{1}{2}\boldsymbol{Z}^{\mathrm{T}}\boldsymbol{Q}_0\boldsymbol{Z} \quad (21)$$

where the diagonal positive definite matrix $\boldsymbol{R}$, with the dimension of m × m, is the weighting matrix for the control matrix, where m is the number of MRDs, the diagonal positive semidefinite matrix $\boldsymbol{Q}_0$, and the diagonal positive definite matrix $\boldsymbol{Q}$ are the weighting matrix for the response of the structure, and $\boldsymbol{Z}$ is the state vector of the controlled system. The optimal control question of the system state is to find the optimal control vector in an infinite time interval; the optimal control force can be given as follows:

$$\boldsymbol{U} = \boldsymbol{R}^{-1}\boldsymbol{B}^{\mathrm{T}}\boldsymbol{P}\boldsymbol{Z} \quad (22)$$

in which the solution of the Riccati matrix algebraic equation $\boldsymbol{P}$ is as follows:

$$-\boldsymbol{P}\boldsymbol{A} - \boldsymbol{A}^{\mathrm{T}}\boldsymbol{P} + 1/2\boldsymbol{P}\boldsymbol{B}\boldsymbol{R}^{-1}\boldsymbol{B}^{\mathrm{T}}\boldsymbol{P} - 2\boldsymbol{Q} = 0 \quad (23)$$

In which, $\boldsymbol{A} = \begin{bmatrix} 0 & \boldsymbol{I} \\ -\boldsymbol{M}^{-1}\boldsymbol{K} & -\boldsymbol{M}^{-1}\boldsymbol{C} \end{bmatrix}$, $\boldsymbol{B} = \begin{bmatrix} 0 \\ -\boldsymbol{M}^{-1}\boldsymbol{H} \end{bmatrix}$, $\boldsymbol{Q} = \alpha\begin{bmatrix} \boldsymbol{K} & 0 \\ 0 & \boldsymbol{M} \end{bmatrix}$, $\boldsymbol{R} = \beta\boldsymbol{I}$.

In which α and β can be obtained through trial calculation, and $\boldsymbol{I}$ is the unit matrix with the dimension of m × m.

As can be seen in Section 2, the minimum and maximum damping forces of the MRD are F_{min} = 3.1 kN and F_{max} = 120 kN when the current is 0 A and 0.28 A, respectively; therefore, the damping force range of the MRD is 3.1 to 120 kN, as shown in Figure 4. The optimal control force should be regulated when it surpasses the adjustable range of the damping force developed by the MRD, or it is opposite to the relative displacement direction of the two ends of the MRD. In addition, some scholars have found that a real-time controlled MRD can trace the required damping force well [35–37]. Thus, the semi-active control strategy for adjusting the optimal control force is shown in Equation (24).

$$F_i = \begin{cases} \mathrm{sgn}(u_i)F_{\min} & \text{when } U_i \cdot u_i > 0 \text{ or } |U_i| < F_{\min} \\ U_i & \text{when } U_i \cdot u_i < 0 \text{ and } F_{\min} < |U_i| < F_{\max} \\ \mathrm{sgn}(u_i)F_{\max} & \text{when } U_i \cdot u_i < 0 \text{ and } |U_i| > F_{\max} \end{cases} \quad (24)$$

where U_i is the optimal control force of the ith MRD, u_i is the relative displacement at both ends of the ith MRD, and F_i is the damping force provided by a single MRD on the ith chevron support [38].

3.6. Determination of the MRD Location Matrix

Due to the fact that the damping force of the MRD is relatively small, two dampers are arranged on each chevron support, as seen in Figure 6. There are the following relationships between the damping force of MRD and the force on the corresponding nodes of the structure:

$$\begin{cases} F_{jx} = F_{kx} = F_i \\ F_{px} = F_{qx} = -F_i \\ F_{jz} = -F_{kz} = -2F_i h/d \end{cases} \quad (25)$$

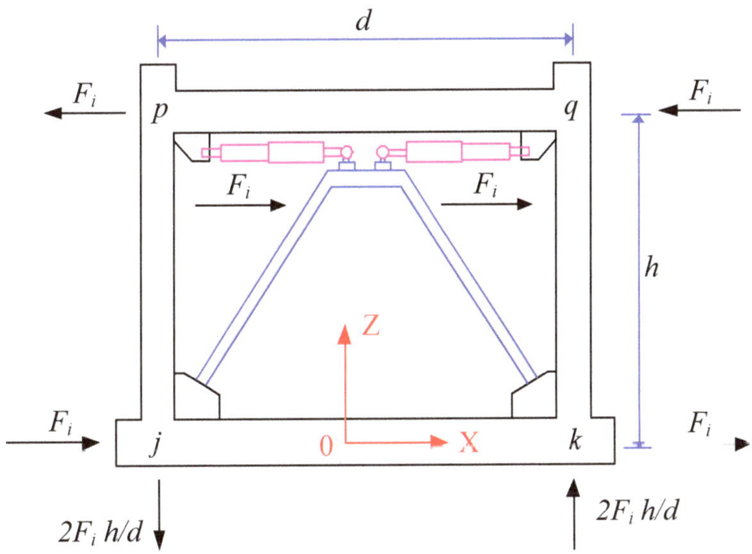

Figure 6. The schematic diagram of damping force distribution.

The location matrix $\boldsymbol{H}$ is used to distribute the control force vector $\boldsymbol{F}$ of the MRD to the corresponding nodes of the structure, whose dimension is n × m, in which n is the amount of the structural DOFs, and m is the number of MRDs [39,40]. For the MRDs on the ith chevron support, as shown in Figure 6, the elements corresponding to nodes j and k in location matrix $\boldsymbol{H}$ are: $\boldsymbol{H}(6j-5,i) = \boldsymbol{H}(6k-5,i) = 1$, $\boldsymbol{H}(6j-3,i) = -\boldsymbol{H}(6k-3,i) = -2h/d$, and the elements corresponding to nodes p and q in location matrix $\boldsymbol{H}$ are: $\boldsymbol{H}(6p-5,i) = \boldsymbol{H}(6q-5,i) = -1$, while the others without damping force are all 0.

4. Model Description and Verification

4.1. Model Description

The calculation model of the L-shaped frame structure with MRDs is presented in Figure 7. The L-shaped frame structure has ten floors with a total height of 34.5 m. The heights between the first and the third floors are 4.5 m, and the heights of the other floors are 3.0 m. The column spacing in the X direction is 6.3, 4.5, and 4.5 m, and column spacing in the Y direction is all 6.3 m. The cross-sectional sizes of X directional beams are 0.25 × 0.65, 0.25 × 0.6, and 0.25 × 0.6 m, and those of Y directional beams are all 0.3 × 0.7 m. The cross-sectional sizes of columns of 1 to 3 floors are 0.7 × 0.7 m, and those of other floors are 0.6 × 0.6 m. The thickness of all floors is 0.12 m; Poisson's ratio $\mu = 0.2$, elastic modulus $E = 3.25 \times 10^4$ N/mm^2, and density $\rho = 2500$ kg/m^3. Assuming the damping ratios of the first- to second-order vibration modes of the structure are set to 5%, the bottom columns are assumed to be consolidated. As the deformation of frame structure under seismic load is mainly concentrated in the bottom floors, the MRDs are, therefore, set on the first to sixth floors, as seen in Figure 7. Numbers in brackets of Figure 7 indicate the node number.

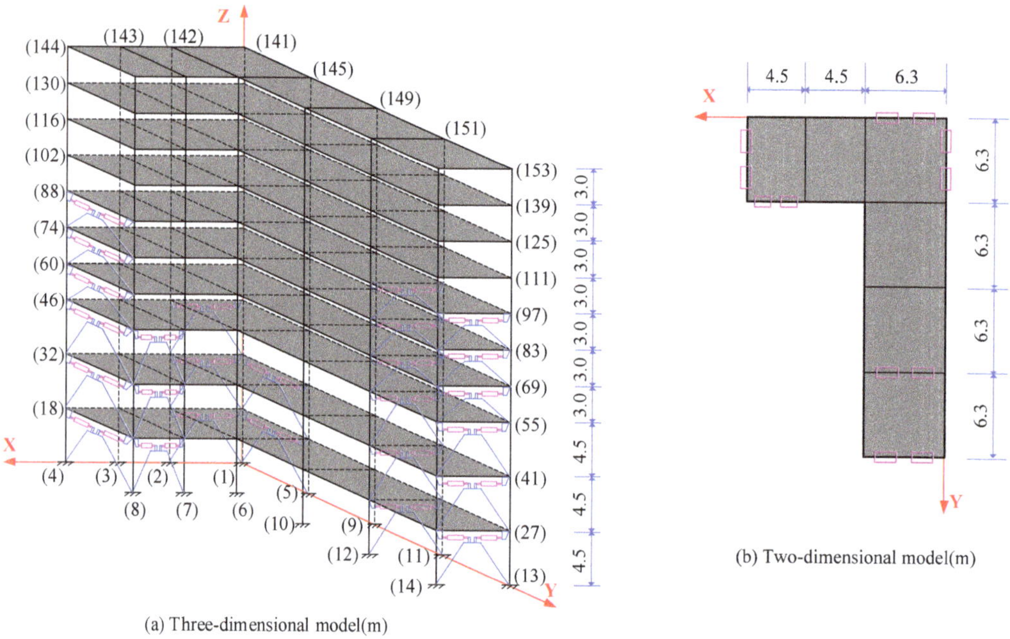

Figure 7. Calculation model of the L-shaped frame structure with MRDs.

4.2. Model Verification

According to the theories and methods described in Sections 3.1–3.4, the programs of the calculation model of the L-shaped frame structure with and without MRDs were compiled by MATLAB software. For verifying the validity of the programs, the model of the L-shaped frame structure is set up by ANSYS, which is the finite element software, in which the BEAM188 element is used for beams and columns and the SHELL 181 element is used for plates. The material parameters and geometry in the ANSYS model are the same as in the self-compiled program. For the ANSYS model and the self-compiled program, each beam and column is divided into 5 and 1 elements. The El-Centro wave, whose maximum acceleration is 0.7 m/s^2, is adopted in both the ANSYS model and the self-compiled program, and the peak acceleration ratio of the seismic wave in the horizontal direction is X:Y = 1:0.85. Contrasts of the dynamic response of Node 141 of the self-compiled program and the ANSYS model are shown in Figure 8.

As seen in Figure 8, throughout the duration of the El-Centro wave, the dynamic responses of the two models of Node 141 are basically consistent. Numerically, the relative errors of the maximum displacement in the horizontal bi-direction are 2.33% and 2.32%;

the relative errors of the maximum acceleration in the horizontal bi-direction are −0.95% and −0.52%. Moreover, the contrasts of the maximum displacement and acceleration of representative nodes between the ANSYS model and the self-compiled program can be seen in Tables 3 and 4.

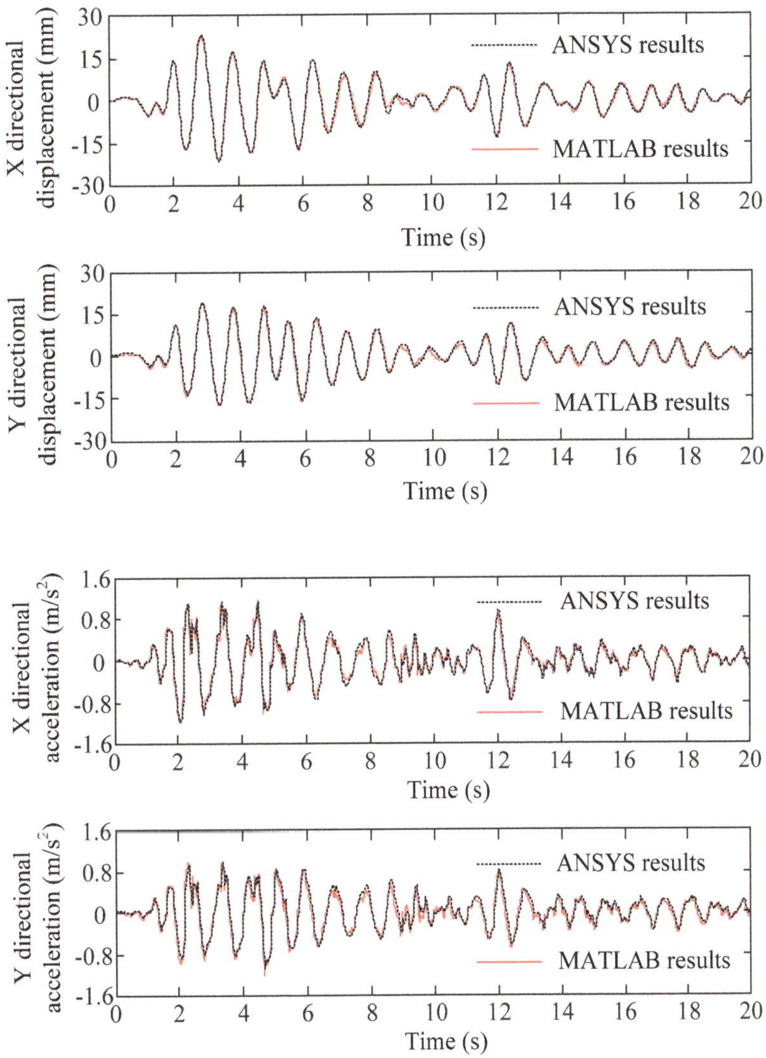

Figure 8. Contrasts of time history response of Node 141 between the self-compiled program and the ANSYS model.

Table 3. Contrasts of the maximum displacement of the nodes between the self-compiled program and the ANSYS model.

Floor (Node)		10 (154)	10 (148)	10 (141)	9 (140)	9 (134)	9 (127)	8 (126)	8 (120)	8 (113)
X direction	ANSYS (mm)	44.10	24.11	22.73	42.48	23.42	22.06	40.50	22.48	21.10
	Self-compiled (mm)	42.34	23.34	22.20	40.94	22.71	21.57	38.89	21.75	20.64
	Relative error (%)	3.97	3.20	2.33	3.63	3.05	2.22	3.96	3.26	2.18
Y direction	ANSYS (mm)	22.78	32.91	19.38	22.17	31.93	18.91	21.20	30.54	18.18
	Self-compiled (mm)	21.97	31.65	18.93	21.41	30.76	18.47	20.50	29.45	17.74
	Relative error (%)	3.54	3.84	2.32	3.44	3.65	2.34	3.31	3.56	2.45

Table 4. Contrasts of the maximum acceleration of the nodes between the self-compiled program and the ANSYS model.

Floor (Node)		10 (154)	10 (148)	10 (141)	9 (140)	9 (134)	9 (127)	8 (126)	8 (120)	8 (113)
X direction	ANSYS (m/s^2)	2.32	1.47	1.17	2.19	1.39	1.14	2.05	1.29	1.09
	Self-compiled (m/s^2)	2.22	1.43	1.18	2.11	1.37	1.135	1.98	1.27	1.088
	Relative error (%)	4.47	2.45	−0.95	3.52	1.52	0.43	3.41	1.89	0.12
Y direction	ANSYS (m/s^2)	1.53	2.01	1.19	1.49	1.96	1.17	1.43	1.87	1.13
	Self-compiled (m/s^2)	1.50	1.93	1.20	1.46	1.87	1.165	1.39	1.79	1.11
	Relative error (%)	2.16	3.96	−0.52	2.24	4.51	0.38	2.70	4.52	1.36

5. Analysis and Interpretation of Results

In this part, the compiled programs are used for calculating the time history response of the L-shaped frame structure with and without MRDs, and the seismic excitations are the Kobe wave and the Taft wave, whose peak accelerations are 2 m/s^2 and the peak acceleration ratio of the seismic wave in the horizontal bi-direction is X:Y = 1:0.85. Through the trial calculation [41], the value of weight matrix coefficients α, β of the LQR control algorithm is set to $\alpha = 100$ and $\beta = 3.5 \times 10^{-5}$.

5.1. Results Analysis on Multi-Dimensional Vibration Reduction of the Structure

Throughout the duration of seismic waves, contrasts on the dynamic displacement response of Node 154 of the L-shaped frame structure with and without MRDs are plotted in Figure 9. As seen in Figure 9, during the action time of the Taft wave, in contrast with the structure without MRDs, the three-directional displacement responses of Node 154 of the structure with MRDs are decreased significantly. The three-directional maximum displacements of Node 154 of the structure without MRDs are 71.79, 48.27, and 1.51 mm, respectively, and those with MRDs are 42.36, 38.42, and 1.25 mm, which are decreased by 41.01%, 20.41%, and 17.36%, respectively [42]. During the action time of the Kobe wave, the three-directional maximum displacements of Node 154 of the structure without MRDs are 89.56, 64.13, and 2.20 mm, and those with MRDs are 57.41, 53.87, and 1.68 mm, which are decreased by 35.90%, 16.01%, and 23.58%, respectively.

Under the excitation of seismic waves, contrasts on the dynamic acceleration response of Node 154 of the L-shaped frame structure with and without MRDs are plotted in Figure 10. As shown in Figure 10, during the action time of the Taft wave, in contrast with the structure without MRDs, the three-directional acceleration responses of Node 154 of the structure with MRDs are significantly decreased [43]. The three-directional maximum accelerations of Node 154 of the structure without MRDs are 3.29, 3.35, and 0.13 m/s^2, respectively, and those with MRDs are 2.77, 2.91, and 0.12 m/s^2, which are decreased by 15.87%, 13.03%, and 5.31%, respectively. During the action time of the Kobe wave, the three-directional maximum accelerations of Node 154 of the structure without MRDs are 3.56, 3.70, and 0.13 m/s^2, and those with MRDs are 2.87, 3.21, and 0.11 m/s^2, which are decreased by 19.35%, 13.29%, and 12.32%, respectively.

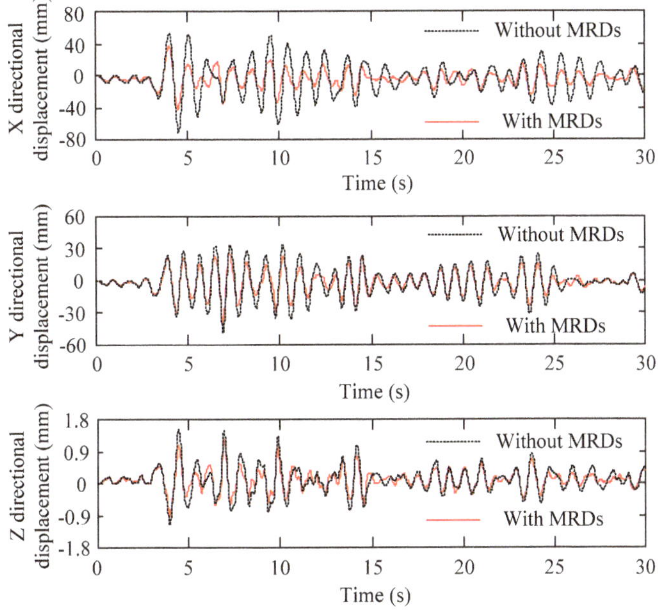

(a) During the excitation of Taft wave

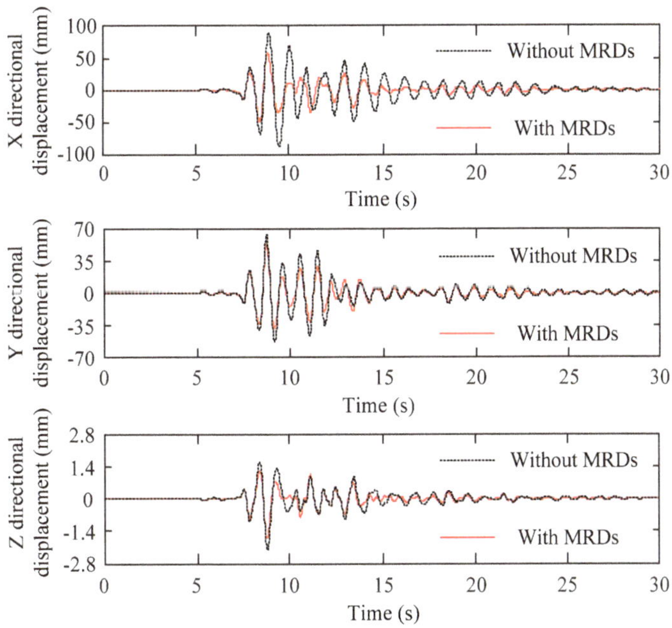

(b) During the excitation of Kobe wave

Figure 9. Contrasts of displacement response of Node 154 of the structure with and without MRDs.

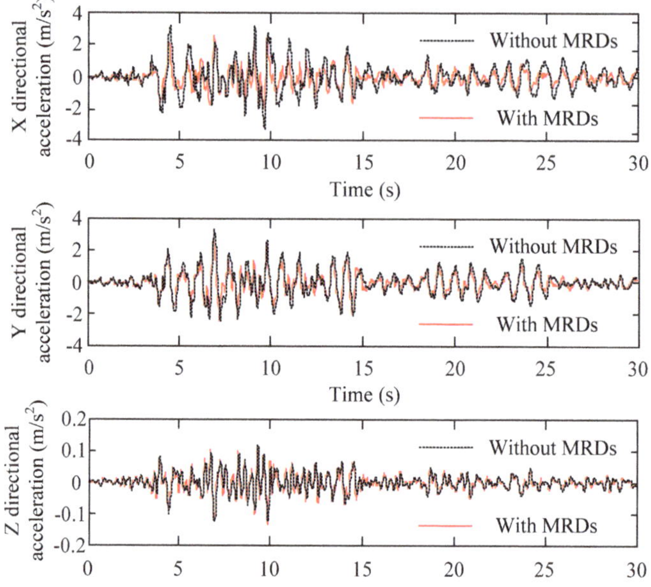

(a) During the excitation of Taft wave

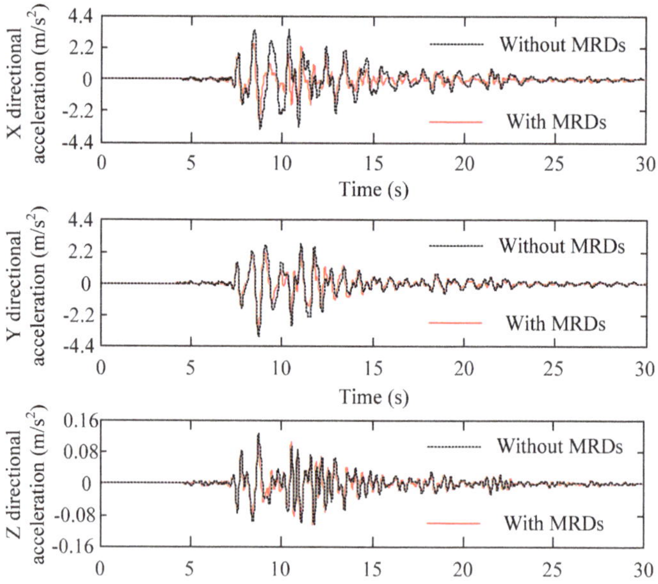

(b) During the excitation of Kobe wave

Figure 10. Contrasts of acceleration response of Node 154 of the structure with and without MRDs.

According to the differential equation of motion of the MRD damping structure [32], a control force matrix will be added because of the setting of MRDs in the structure; it is equivalent to increasing the stiffness and damping of the structure, and both of them can reduce the displacement response of the structure under the action of seismic waves; hence, by setting the MRD in the structure, the displacement control effect of the structure is very obvious, as shown in Figures 9 and 11. However, increasing the damping and stiffness of the structure has the opposite effect on the acceleration response of the structure under the action of seismic waves. Therefore, setting the MRD in the structure, the acceleration control effect of the structure is relatively general, as shown in Figures 10 and 11. Since there are no MRDs in the Z direction, the displacement and acceleration responses in the Z direction are relatively small.

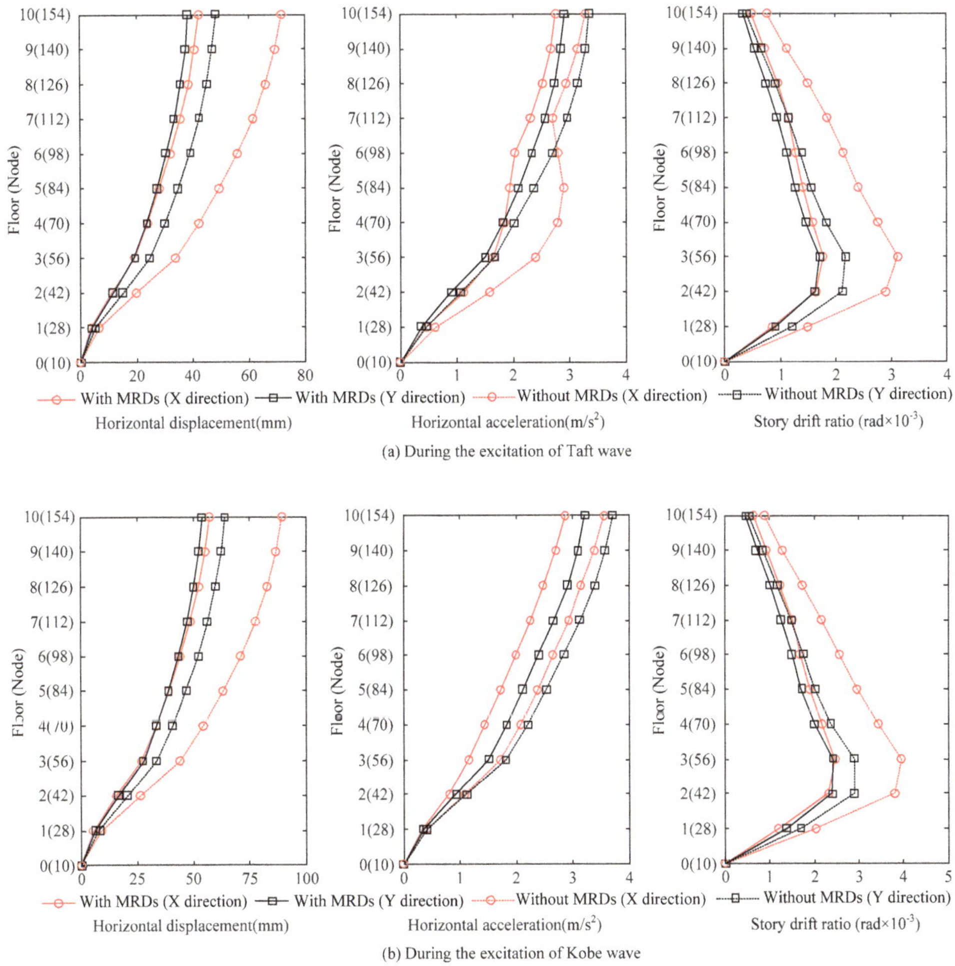

Figure 11. The max horizontal displacement, acceleration, and story drift ratio of every floor of the structure with and without MRDs.

5.2. Results Analysis on the Maximum Displacement, Acceleration, and Story Drift Ratio of the Structure

Under the excitation of seismic waves, the maximum horizontal displacement, acceleration, and story drift ratio of every floor of the L-shaped frame structure with and without MRDs are plotted in Figure 11. As shown in Figure 11a, during the action time of the Taft wave, the maximum horizontal displacement, acceleration, and story drift ratio of every floor of structure with MRDs are obviously lower than those without MRDs [44]. For Node 126 on the eighth floor, during the action time of the Taft wave, the horizontal bi-directional maximum displacements of the structure without MRDs are 66.05 and 45.07 mm, and those with MRDs are 38.62 and 35.77 mm, which are decreased by 41.53% and 20.64%, respectively. The horizontal bi-directional maximum accelerations of the structure without MRDs are 2.95 and 3.15 m/s^2, and those with MRDs are 2.54 and 2.74 m/s^2, which are decreased by 14.01% and 13.12%, respectively. The horizontal bi-directional story drift ratios of the structure without MRDs are 1.51×10^{-3} rad and 0.92×10^{-3} rad, and those with MRDs are 0.97×10^{-3} rad and 0.76×10^{-3} rad, which are decreased by 35.55% and 17.51%, respectively.

As shown in Figure 11b, for Node 126 on the eighth floor, during the action time of the Kobe wave, the horizontal bi-directional maximum displacements of the structure without MRDs are 82.99 and 59.99 mm, and those with MRDs are 52.67 and 50.32 mm, which are decreased by 36.53% and 16.14%, respectively. The horizontal bi-directional maximum accelerations of the structure without MRDs are 3.15 and 3.38 m/s^2, and those with MRDs are 2.48 and 2.91 m/s^2, which are decreased by 21.08% and 13.89%, respectively. The horizontal bi-directional story drift ratios of the structure without MRDs are 1.74×10^{-3} rad and 1.17×10^{-3} rad, and those with MRDs are 1.25×10^{-3} rad and 1.01×10^{-3} rad, which are decreased by 28.41% and 14.44%, respectively.

It can be seen from Figure 11 that the story drift ratios of the L-shaped frame structure with and without MRDs first increase and then decrease with the increase in the number of floors of the structure, and the maximum value appears on the third floor. This is because the bottom column of the structure is assumed to be consolidated, and the angle displacement of the bottom column is 0, resulting in a small story drift ratio between the bottom floors at not less than 1/2 of the second layer displacement angle; it shows that the envelope diagram is reasonable.

5.3. Results Analysis on Shear Force and Bending Moment of the Structural Columns

Under the excitation of seismic waves, the maximum shear force and bending moment of the column on every floor of the L-shaped frame structure with and without MRDs are plotted in Figure 12. As shown in Figure 12, the maximum shear force and the bending moment of the column on each floor of the structure with MRDs are obviously lower than those without MRDs, and the shear forces at both ends of the column are equal. Taking the column (Nodes 20 and 34) on the second floor as an example, as shown in Figure 12a, under the excitation of the Taft wave, the horizontal bi-directional maximum shear forces of the column of structure without MRDs are 184.90 and 220.93 kN, and those with MRDs are 128.35 and 167.67 kN, which are decreased by 30.58% and 24.11%, respectively. As shown in Figure 12b, throughout the duration of the Kobe wave, the horizontal bi-directional maximum shear forces of the column of structure without MRDs are 265.29 and 302.03 kN, and those with MRDs are 195.07 and 247.98 kN, which are decreased by 26.47% and 17.90%, respectively.

As shown in Figure 12, the inflection point of the column on the top floor appears in the lower middle of the column, the inflection point of the column on the ground floor appears in the upper middle of the column, and those on the other floors appear in the middle of the column. Throughout the duration of seismic waves, the maximum bending moments at both ends of columns on each floor of the structure with MRDs are obviously lower than those without MRDs. Taking the column (Nodes 20 and 34) on the second floor as an example, as shown in Figure 12a, during the action time of the Taft wave, the horizontal bi-directional maximum bending moments of the column bottom of structure without MRDs are 541.61 and 468.77 kN·m, and those with MRDs are 413.81 and 325.01 kN·m, which are decreased by 23.60% and 30.66%, respectively. As shown in Figure 12b, throughout the duration of the Kobe wave, the horizontal bi-directional maximum bending moments of the column bottom of structure without MRDs are 731.01 and

674.36 kN·m, and those with MRDs are 602.17 and 496.72 kN·m, which are decreased by 17.62% and 26.34%, respectively.

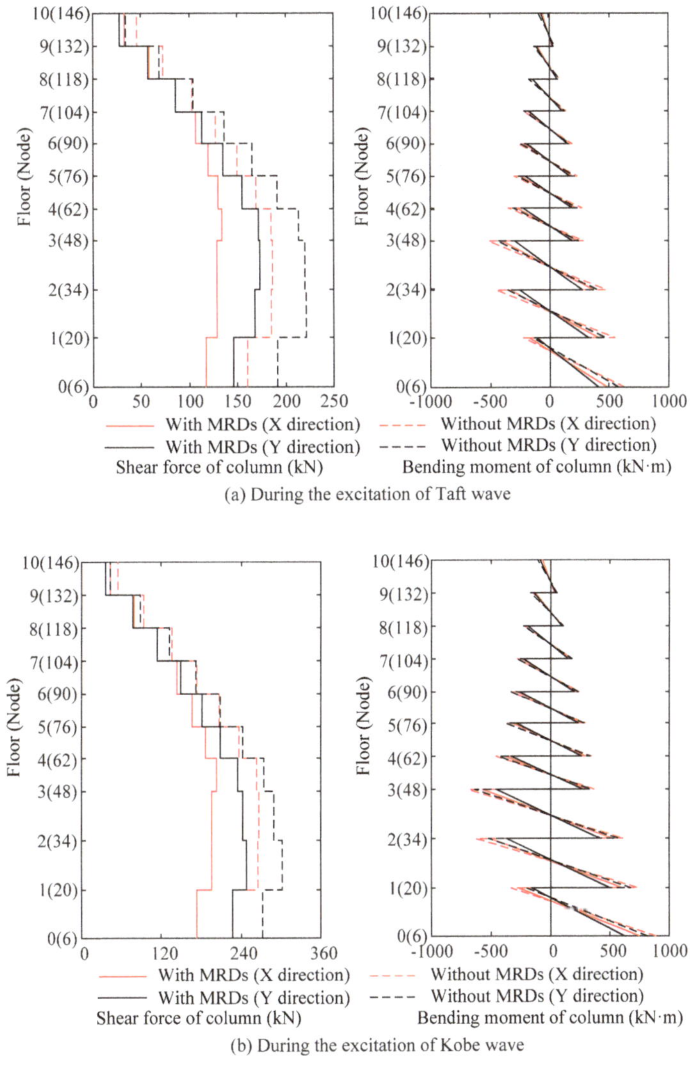

Figure 12. The max shear forces and bending moments of the columns of every floor of the structure with and without MRDs.

5.4. Results Analysis on Torsional Vibration Control of the Structure

Contrasts of the horizontal displacement response of the top node of the L-shaped frame structure without MRDs are plotted in Figure 13. As shown in Figure 13, the X directional displacement time history responses of Nodes 153 and 141 are significantly different, and the Y directional displacement time history responses of Nodes 144 and 141 are also different. As shown in Figure 13a, during the excitation of the Taft wave, the X directional maximum displacements of Nodes 153 and 141 are 71.79 and 48.20 mm, and the ratio is 1.489; the Y directional maximum displacements of Nodes 144 and 141 are 53.77 and 44.52 mm, and the ratio is 1.208. As shown in Figure 13b, during the excitation of the

Kobe wave, the X directional maximum displacements of Nodes 153 and 141 are 89.56 and 71.93 mm, and the ratio is 1.245; the Y directional maximum displacements of Nodes 144 and 141 are 69.62 and 62.01 mm, and the ratio is 1.123. The results show that the L-shaped structure without MRDs has obvious torsional vibration in the horizontal plane.

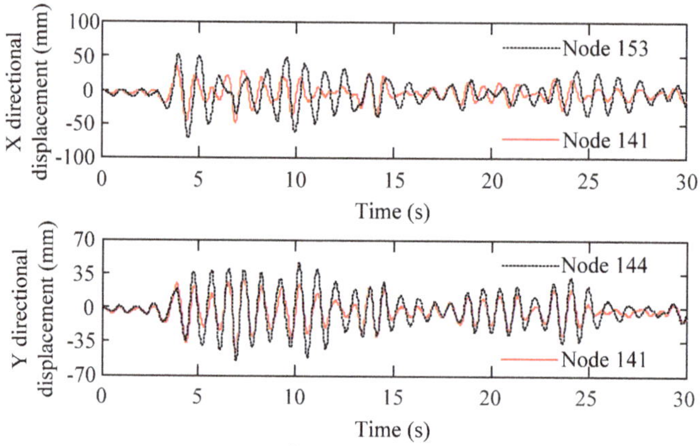

(a) During the excitation of Taft wave

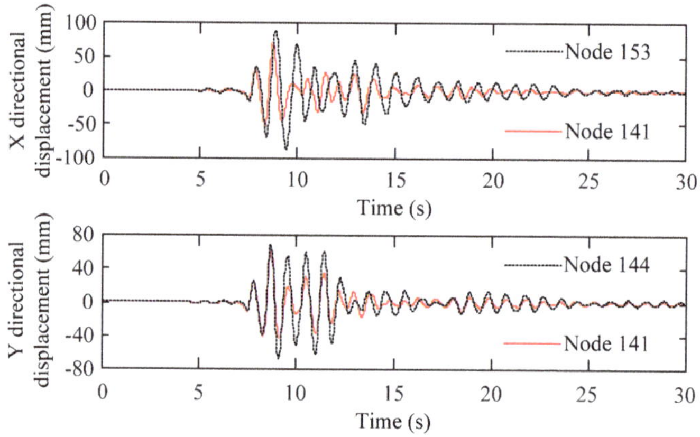

(b) During the excitation of Kobe wave

Figure 13. Contrasts of the horizontal displacement response of the top node of the structure without MRDs.

Contrasts of the horizontal displacement response of the top node of the L-shaped frame structure with MRDs are plotted in Figure 14. As shown in Figure 14, the X directional displacement time history responses Nodes 153 and 141 and Y directional displacement time history responses of Nodes 144 and 141 are almost identical. As shown in Figure 14a, during the excitation of the Taft wave, the X directional maximum displacements of Nodes 153 and 141 are 42.36 and 37.25 mm, and the ratio is 1.140; the Y directional maximum displacements of Nodes 144 and 141 are 40.26 and 37.25 mm, and the ratio is 1.081. As shown in Figure 14b, during the excitation of the Kobe wave, the X directional maximum displacements of Nodes 153 and 141 are 57.41 and 57.55 mm, and the ratio is 0.998; the Y directional maximum displacements of Nodes 144 and 141 are 57.79 and 51.92 mm, and the ratio is 1.113. The results show that, compared with the L-shaped frame structure without MRDs, the torsional vibration of the structure with MRDs is significantly reduced.

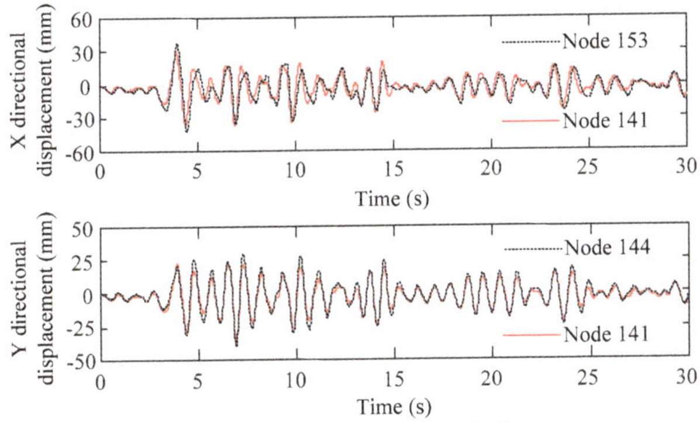

(a) During the excitation of Taft wave

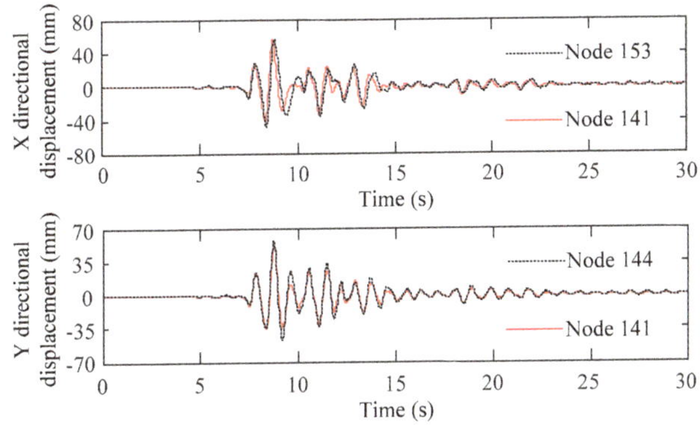

(b) During the excitation of Kobe wave

Figure 14. Contrasts of the horizontal displacement response at the top node of the structure with MRDs.

For the action time of seismic waves, the maximum horizontal displacement ratios of each floor of the L-shaped frame structure with and without MRDs can be seen in Tables 5 and 6. As shown in the tables, compared with the L-shaped frame structure without MRDs, the Y directional maximum displacement ratios of every floor of the structure with MRDs are getting closer to 1; the X directional maximum displacement ratios are obviously on the decrease. The results demonstrate that the L-shaped frame structural torsional vibration is significantly controlled.

Table 5. The maximum horizontal displacement ratios of every floor during the Taft wave excitation.

Floor (Node/Node)	X Direction		Floor (Node/Node)	Y Direction	
	Without MRDs	With MRDs		Without MRDs	With MRDs
1 (27/15)	1.395	1.046	1 (18/15)	1.253	1.103
2 (41/29)	1.445	1.087	2 (32/29)	1.261	1.112
3 (55/43)	1.480	1.109	3 (46/43)	1.249	1.101
4 (69/57)	1.503	1.116	4 (60/57)	1.243	1.103
5 (83/71)	1.506	1.122	5 (74/71)	1.235	1.102
6 (97/85)	1.499	1.127	6 (88/85)	1.226	1.097
7 (111/99)	1.494	1.135	7 (105/99)	1.218	1.092
8 (125/113)	1.491	1.141	8 (116/113)	1.212	1.086
9 (139/127)	1.489	1.140	9 (130/127)	1.208	1.081
10 (153/141)	1.489	1.140	10 (144/141)	1.208	1.081

Table 6. The maximum horizontal displacement ratios of every floor during the Kobe wave excitation.

Floor (Node/Node)	X Direction		Floor (Node/Node)	Y Direction	
	Without MRDs	With MRDs		Without MRDs	With MRDs
1 (27/15)	1.232	0.937	1 (18/15)	1.119	1.104
2 (41/29)	1.237	0.974	2 (32/29)	1.133	1.121
3 (55/43)	1.237	0.986	3 (46/43)	1.131	1.119
4 (69/57)	1.242	0.989	4 (60/57)	1.130	1.124
5 (83/71)	1.247	0.988	5 (74/71)	1.128	1.125
6 (97/85)	1.248	0.987	6 (88/85)	1.125	1.124
7 (111/99)	1.247	0.990	7 (105/99)	1.123	1.120
8 (125/113)	1.246	0.993	8 (116/113)	1.122	1.116
9 (139/127)	1.244	0.995	9 (130/127)	1.121	1.113
10 (153/141)	1.245	0.998	10 (144/141)	1.123	1.113

It can be seen from Figure 7 that the center of mass has different positions from the center of rigidity in the L-shaped frame structure; it directly leads to the obvious torsional vibration response under the action of seismic waves. Since the multi-floor L-shaped frame structure is a shear-type structure, many MRDs are arranged at the bottom floor of the structure, which can effectively increase the shear rigidity and torsional rigidity of the overall structure. This is the main reason why the multi-dimensional vibration and torsional vibration response of the L-shaped frame in this article are effectively controlled after setting the MRDs.

6. Conclusions

In the present study, extensive research about the mechanical performance of the self-made MRD is experimentally studied at different displacement amplitudes and currents. Finally, the main conclusions are summarized as follows:

1. The damping force provided by the MRD gradually increases with the current, and the adjustable scope of damping force is from 3.1 to 120 kN when the current level increases from 0 to 0.28 A.
2. The programs of the calculation model and time history analysis of the L-shaped frame structure without and with MRDs are proposed, which offers a new way for seismic response analysis of frame structures with MRDs.
3. In contrast with the L-shaped frame structure without MRDs, the three-directional displacement and acceleration, story drift ratio, shear force, and bending moment of each floor of the L-shaped frame structure with MRDs have been significantly decreased.
4. Throughout the duration of seismic waves, the L-shaped frame structure has the obvious response of torsional vibration, and the response of torsional vibration can be controlled by reasonably arranging MRDs in the structure.

A comparison of the most known machine learning algorithms will be considered in the same scenario for future applications. Moreover, a more comprehensive range of structural systems should be examined to incorporate them into the seismic response methodology.

Author Contributions: Conceptualization, Y.Y., J.Y., Y.F. and X.Z.; methodology, Y.Y., J.Y. and X.Z.; software, Y.Y.; validation, Y.Y., J.Y., Y.F. and X.Z.; formal analysis, Y.F. and X.Z.; investigation, Y.Y. and J.Y.; resources, Y.Y. and J.Y.; data curation, Y.Y., J.Y., Y.F. and X.Z.; writing—original draft preparation, Y.Y.; writing—review and editing, Y.Y., J.Y., Y.F. and X.Z.; visualization, Y.Y., Y.F. and X.Z.; supervision, X.Z.; project administration, X.Z.; funding acquisition, Y.Y., J.Y., Y.F. and X.Z. All authors have read and agreed to the published version of the manuscript.

Funding: This research is financially supported by the Key R & D and Promotion Projects of Henan Province (202102310239).

Institutional Review Board Statement: Not applicable.

Informed Consent Statement: Not applicable.

Data Availability Statement: Not applicable.

Acknowledgments: This research is financially supported by the Key R & D and Promotion Projects of Henan Province (202102310239). The support is gratefully acknowledged.

Conflicts of Interest: The authors declare that they have no known competing financial interests or personal relationships that could appear to influence the work reported in this paper.

References

1. Kolekar, S.; Venkateshv, K.; Oh, J.S.; Choi, S.B. Vibration controllability of sandwich structures with smart materials of electrorheological fluids and magnetorheological materials: A review. *J. Vib. Eng. Technol.* **2019**, *7*, 359–377. [CrossRef]
2. Rahman, M.; Ong, Z.C.; Julai, S.; Ferdaus, M.M.; Ahamed, R. A review of advances in magnetorheological dampers: Their design optimization and applications. *J. Zhejiang Univ.-Sci. A (Appl. Phys. Eng.)* **2017**, *18*, 991–1010. [CrossRef]
3. Bahman, F.A.; Hedayat, V.; Farzad, R.; Siamak, T. Control of the nonlinear building using an optimum inverse TSK model of MR damper based on modified grey wolf optimizer. *Eng. Struct.* **2020**, *214*, 110657.
4. Ding, Y.; Zhang, L.; Zhu, H.T.; Li, Z.X. A new magnetorheological damper for seismic control. *Smart Mater Struct.* **2013**, *22*, 115003-1–115003-12. [CrossRef]
5. Wang, W.; Hua, X.; Wang, X.; Wu, J.; Sun, H.; Song, G. Mechanical behavior of magnetorheological dampers after long-term operation in a cable vibration control system. *Struct. Control Health Monit.* **2019**, *26*, e22801-19. [CrossRef]
6. Xu, Z.D.; Shen, Y.P. Intelligent Bi-State Control for the Structure with Magnetorheological Dampers. *J. Intell. Mater. Syst. Struct.* **2003**, *14*, 35–42. [CrossRef]
7. Xu, Z.D.; Ge, T.; Liu, J. Experimental and theoretical study of high energy dissipation viscoelastic dampers based on acrylate rubber matrix. *J. Eng. Mech.* **2020**, *146*, 04020057. [CrossRef]
8. Xu, Z.D.; Xu, F.H.; Chen, X. Vibration suppression on a platform by using vibration isolation and mitigation devices. *Nonlinear Dynam.* **2016**, *83*, 1341–1353. [CrossRef]
9. Zhang, X.C.; Zhang, X.; Zhao, Y.; Zhao, J.; Xu, Z. Experimental and numerical studies on a composite MR damper considering magnetic saturation effect. *Eng. Struct.* **2017**, *132*, 576–585. [CrossRef]
10. Lv, H.Z.; Zhang, S.S.; Sun, Q.; Chen, R.; Zhang, W.J. The dynamic models, control strategies and applications for magnetorheological damping systems A Systematic Review. *J. Vib. Eng. Technol.* **2020**, *9*, 131–147. [CrossRef]
11. Parlak, Z.; Engin, T. Time-dependent CFD and quasi-static analysis of magnetorheological fluid dampers with experimental validation. *Int. J. Mech. Sci.* **2012**, *64*, 22–31. [CrossRef]
12. Carlson, J.D.; Spencer, B.F. Magnetorheological fluid dampers for semi-active seismic control. *Motion Vibr. Control* **1996**, *2*, 35–40.
13. Fujitani, H.; Sodeyama, H.; Tomura, T.; Hiwatashi, T.; Hata, K.; Sunakoda, K.; Morishita, S.; Soda, S. Development of 400 kN magnetorheological damper for a real base-isolated building. *Proc. SPIE-Int. Soc. Opt. Eng.* **2003**, *5052*, 265–276.
14. Xu, Z.D.; Sha, L.F.; Zhang, X.C.; Ye, H.H. Design, performance test and analysis on magnetorheological damper for earthquake mitigation. *Struct. Control Health Monit.* **2013**, *20*, 956–970. [CrossRef]
15. Tu, J.W.; Liu, J.; Qu, W.L.; Zhou, Q.; Cheng, H.B.; Cheng, X.D. Design and Fabrication of 500-kN Large-scale MR Damper. *J. Intell. Mater. Syst. Struct.* **2011**, *22*, 475–487. [CrossRef]
16. Khanal, B.; Chaulagain, H. Seismic elastic performance of L-shaped building frames through plan irregularities. *Structures* **2020**, *27*, 22–36. [CrossRef]
17. Abdel Raheem, S.E.; Ahmed, M.M.M.; Ahmed, M.M.; Abdel-shafy, A.G.A. Evaluation of plan configuration irregularity effects on seismic response demands of L-shaped MRF buildings. *Bull. Earthq. Eng.* **2018**, *16*, 3845–3869. [CrossRef]
18. Abdel Raheem, S.E.; Ahmed, M.M.M.; Ahmed, M.M.; Ahmed, M.M.; Abdel-shafy, A.G.A. Seismic performance of L-shaped multi-storey buildings with moment-resisting frames. *Proc. Inst. Civil Eng. Struct. Build.* **2018**, *171*, 395–408. [CrossRef]
19. Amarloo, N.; Emami, A.R. A 3-dimensional perspective for inter-storey drift, ductility and damage distributions in plan-irregular RC buildings considering seismic orientation effect. *Bull. Earthq. Eng.* **2019**, *17*, 3447–3474. [CrossRef]
20. Zafarani, M.M.; Halabian, A.M. Supervisory adaptive nonlinear control for seismic alleviation of inelastic asymmetric buildings equipped with MR dampers. *Eng. Struct.* **2018**, *176*, 849–858. [CrossRef]
21. Li, Z.X.; Lv, Y.; Xu, L.H.; Ding, Y.; Zhao, Q. Experimental studies on nonlinear seismic control of a steel-concrete hybrid structure using MR dampers. *Eng. Struct.* **2013**, *49*, 248–263. [CrossRef]
22. Bhatti, A.Q. Performance of viscoelastic dampers (VED) under various temperatures and application of magnetorheological dampers (MRD) for seismic control of structures. *Mech. Time-Depend. Mater.* **2012**, *17*, 275–284. [CrossRef]
23. Sakurai, T.; Morishita, S. Seismic response reduction of a three-story building by an MR grease damper. *Front. Mech. Eng.* **2017**, *12*, 224–233. [CrossRef]
24. Ali, S.F.; Ramaswamy, A. Hybrid structural control using magnetorheological dampers for base isolated structures. *Smart Mater. Struct.* **2009**, *18*, 055011. [CrossRef]
25. Dyke, S.J.; Spencer, B.F.; Sain, M.K.; Carlson, J.D. Modeling and control of magnetorheological dampers for seismic response reduction. *Smart Mater. Struct.* **1996**, *5*, 565–575. [CrossRef]
26. Xu, Z.D.; Zhao, Y.L.; Guo, Y.Q.; Yang, X.L.; Sarwar, W. Shaking table tests of magnetorheological damped frame to mitigate the response under real-time online control. *Smart Mater. Struct.* **2019**, *28*, 115021. [CrossRef]

27. Dyke, S.J.; Spencer, B.F.; Sain, M.K.; Carlson, J.D. An experimental study of MR dampers for seismic protection. *Smart Mater. Struct.* **1998**, *7*, 693–703. [CrossRef]
28. Yoshida, O.; Dyke, S.J.; Giacosa, L.M.; Truman, K.Z. Experimental verification of torsional response control of asymmetric buildings using MR dampers. *Earthq. Eng. Struct. Dyn.* **2003**, *32*, 2085–2105. [CrossRef]
29. Zhang, X.C.; Xu, Z.D. Testing and modeling of a CLEMR damper and its application in structural vibration reduction. *Nonlinear Dyn.* **2012**, *70*, 1575–1588. [CrossRef]
30. Bhaiya, V.; Shrimali, M.K.; Bharti, S.D.; Datta, T.K. Modified semi-active control with MR dampers for partially observed systems. *Eng. Struct.* **2019**, *191*, 129–147. [CrossRef]
31. Xu, F.H.; Xu, Z.D.; Zhang, X.C. Study on the space frame structures incorporated with magnetorheological dampers. *Smart Struct. Syst.* **2017**, *19*, 279–288. [CrossRef]
32. Zhao, J.; Li, K.; Zhang, X.C.; Sun, Y.P.; Xu, Z.D. Multidimensional vibration reduction control of the frame structure with magnetorheological damper. *Struct. Control Health Monit.* **2020**, *27*, e2572. [CrossRef]
33. Qian, R.J.; Yuan, X.F.; Lin, Z.B. *The Analysis Theory and Finite Element Method for Solid and Structure*; Southeast University Press: Nanjing, China, 2013; pp. 425–427+436–437. (In Chinese)
34. Motra, G.B.; Mallik, W.; Chandiramani, N.K. Semi-active vibration control of connected buildings using magnetorheological dampers. *J. Intell. Mater. Syst. Struct.* **2011**, *22*, 1811–1827. [CrossRef]
35. Ubaidillah, H.K.; Kadir, F.A.A. Modelling, characterisation and force tracking control of a magnetorheological damper under harmonic excitation. *Int. J. Model. Identif. Control* **2011**, *13*, 9–21. [CrossRef]
36. Russo, R.; Terzo, M. Modelling, parameter identification, and control of a shear mode magnetorheological device. *Proc. Inst. Mech. Eng. I* **2011**, *225*, 549–562. [CrossRef]
37. Weber, F. Robust force tracking control scheme for MR dampers. *Struct. Control Health Monit.* **2015**, *22*, 1373–1395. [CrossRef]
38. Ribeiro, T.; Bernardo, L.; Carrazedo, R.; Domenico, D. Seismic Design of Bolted Connections in Steel Structures—A Critical Assessment of Practice and Research. *Buildings* **2022**, *12*, 32. [CrossRef]
39. Nguyen, D.H.; Hong, W.-K. A novel erection technique of the l-shaped precast frames utilizing laminated metal plates. *J. Asian Archit. Build. Eng.* **2021**, *20*, 674–688. [CrossRef]
40. Chhatani, Y.; Pawade, P.Y.; Dabhekar, K.R.; Khedikar, I.P. Seismic performance of l-shaped building through plan irregularities. *IOP Conf. Ser. Mater. Sci. Eng.* **2021**, *1197*, 012050. [CrossRef]
41. Li, J.; Wang, Z.; Li, F.; Mou, B.; Wang, T. Experimental and numerical study on the seismic performance of an L-Shaped double-steel plate composite shear wall. *J. Build. Eng.* **2022**, *49*, 104015. [CrossRef]
42. Ping, C.Z.; Weiwei, S.; Yang, Y. Seismic behavior of reinforced concrete t-shaped columns under compression-bending-shear and torsion. *Earthq. Struct.* **2021**, *20*, 431–444.
43. Zhang, X.; Xia, Q.; Ye, B.; Yan, W.; Deng, Z.; Xiang, P. Seismic performances of src special-shaped columns and rc beam joints under double-direction low-cyclic reversed loading. *Front. Mater.* **2021**, *8*, 438. [CrossRef]
44. Zhang, M.J.; Xu, F.Y. Tuned mass damper for self-excited vibration control: Optimization involving nonlinear aeroelastic effect. *J. Wind Eng. Ind. Aerodyn.* **2022**, *220*, 104836. [CrossRef]

applied sciences

MDPI

Article

Comparison of Strengthening Solutions with Optimized Passive Energy Dissipation Systems in Symmetric Buildings

Charbel Mrad [1], Magdalini D. Titirla [2,*] and Walid Larbi [2]

1 Department of Civil Engineering, Institut Supérieur des Sciences Appliquées et Économiques (ISSAE), Conservatoire National des Arts et Métiers (CNAM LIBAN), Beirut 20239201, Lebanon; charbel.mrad@isae.edu.lb

2 Structural Mechanics and Coupled Systems Laboratory (LMSSC), Conservatoire National des Arts et Métiers (CNAM), 75003 Paris, France; walid.larbi@lecnam.net

* Correspondence: magdalini.titirla@lecnam.net

Abstract: The aim of this study is to compare the seismic response of reinforced concrete (RC) symmetric buildings, with a varied number of stories, strengthening with three types of passive energy dissipation systems, as tuned mass dampers, viscous dampers, and friction dampers. The paper presents an overview of design optimization with the object of minimizing certain functions: (i) the maximum displacement at the top of the structures, (ii) the base shear loads, and (iii) the maximum interstory drift. The objective functions were evaluated in three residents' buildings (a four-story building, a nine-story building, and a sixteen-story building) subjected to seven (real and artificial) seismic recorded accelerograms. For this purpose, 94 nonlinear dynamic analyses were carried out. The effects of each strengthening solution are presented, and from this innovative comparison (optimal design, three different passive energy systems, three different story numbers), further useful results were observed. The outcomes of the study show the effectiveness of a tuned mass damper (TMD) system, and how it might be better for tall and flexible structures than for stiffer structures. However, the response of the pendulum tuned mass damper (TMD) configuration is better than the conventional one because it acts in all directions. The viscous dampers (VDs) provide a significant reduction for mid-rise buildings, while friction dampers (FDs) boost the performance of all structures under seismic action, especially in terms of displacement, and they are more suitable for low-rise buildings.

Keywords: passive energy dissipation systems; tuned mass damper; viscous damper; friction damper; optimization; dynamic response

Citation: Mrad, C.; Titirla, M.D.; Larbi, W. Comparison of Strengthening Solutions with Optimized Passive Energy Dissipation Systems in Symmetric Buildings. *Appl. Sci.* **2021**, *11*, 10103. https://doi.org/10.3390/app112110103

Academic Editor: Maria Favvata

Received: 27 September 2021
Accepted: 23 October 2021
Published: 28 October 2021

Publisher's Note: MDPI stays neutral with regard to jurisdictional claims in published maps and institutional affiliations.

Copyright: © 2021 by the authors. Licensee MDPI, Basel, Switzerland. This article is an open access article distributed under the terms and conditions of the Creative Commons Attribution (CC BY) license (https://creativecommons.org/licenses/by/4.0/).

1. Introduction

During an earthquake, most structures have an inherent damping in them which results in some of the input seismic energy being dissipated, but a large amount of energy is absorbed by the structure, causing it to undergo several deformations and maybe even collapse. So, over the last year, there has been great interest in the creation of seismic energy dissipation devices that will absorb the majority of the seismic energy, but will not belong to the supporting structure of the construction (conventional braced frames). The main advantages of these are their easy replacement or repair. These devices belong to the passive energy dissipation systems, do not require external power to generate system control forces, and hence, are easy and cheap to implement in a structure [1–3]. Passive energy dissipation devices such as tuned mass dampers (TMD), viscous dampers (VD), and friction dampers (FD) have widely been used to reduce the dynamic response of civil engineering structures that are subjected to seismic loads. Their effectiveness for the seismic design of building structures is attributed to minimizing structural damages by absorbing the structural vibratory energy and by dissipating it through their inherent hysteresis behavior.

The passive TMD is undoubtedly a simple, inexpensive, and somewhat reliable means to suppress the undesired vibrations. The TMD concept was first applied by Frahm in 1909 [4] to reduce the rolling motion of ships and ship hull vibrations. A theory for the TMD was presented later by Ormondroyd and Den Hartog [5], followed by a detailed discussion of optimal tuning and damping parameters in Den Hartog's book on mechanical vibrations [6]. A number of TMDs have been installed in tall buildings, bridges, and towers. The first structure in which a TMD was installed is the Centrepoint Tower in Sydney Australia, which was conceived in 1968 [7]. There are many buildings in the United States, like the Citicorp Center in New York City [8] and the John Hancock Tower in Boston [9]; in Japan, there is the Chiba Port Tower [10] and others [11].

A VD damper generally consists of a piston within a damper housing filled with a compound of silicone or a similar type of oil, with the piston containing a number of small orifices through which the fluid may pass from one side of the piston to the other [12]. Viscoelastic materials are very popular in engineering [13,14]. As the damper piston rod and piston head are stroked, fluid is forced to flow through orifices either around or through the piston head. The first applications of VD dampers to structures were for reducing acceleration levels, or increasing human comfort, due to wind. In 1969, VD dampers were installed in the twin towers of the World Trade Center in New York as an integral part of the structural system. In 1982, VD dampers were incorporated into the 76-story Columbia Sea First Building in Seattle, Washington, to protect against wind-induced vibrations [15]. Applying the well-developed fluid damping technology to civil structures was relatively straightforward; within a short time, the first research projects were completed on the application of fluid dampers to a steel framed building [12] and an isolated bridge structure [16].

In a typical FD, the generated frictional force helps to dissipate the external energy and stabilize the structure under the dynamic excitation scenarios. The FDs are also not prone to thermal effects, and possess a stable hysteretic behavior for a considerable number of cycles under such dynamic excitations [17]. Based primarily upon an analogy to the automotive brake, Pall et al. [18] began the development of passive frictional dampers to improve the seismic response of structures. The objective is to slow the motion of buildings "by braking rather than breaking" [19]. After that, many researchers proposed friction dampers that focus on protection in the braced frames or in the joint connection [20–27]. Several of these devices have been selected for the seismic strengthening of existing or new buildings in the USA, Canada, and Japan [28–30].

This study compares the seismic response of three reinforced concrete (RC) symmetric buildings of varying stories and their strengthening with three types of passive energy dissipation systems, as tuned mass dampers, viscous dampers, and friction dampers. We focus on the optimal design of each building in minimizing (i) the maximum displacement at the top of the structures, (ii) the base shear loads, and (iii) the maximum inter-story drift. Three residents' buildings (a four-story building, a nine-story building, and a sixteen-story building) were subjected to seven (real and artificial) seismic recorded accelerograms. For this purpose, 94 nonlinear dynamic analyses were carried out. The effects of each strengthening solution are presented, and from this innovative comparison (optimal design, three different passive energy systems, and three different story numbers), further useful results were observed.

2. Description of Benchmark Investigated Buildings

Three symmetric, in plan, reinforced concrete residential buildings were studied in this paper. The three buildings were regular in plan according to EC8 [31], and they had the same external dimensions: 40.00 m in the longitudinal direction and 20.00 m in the transversal direction, as shown in Figure 1. The number of the stories was varied, with the constant height of each story equal to 3.50 m. The first building, mentioned from now on as « Low-rise », consisted of three stories; the second building, mentioned as « Mid-rise », consisted of eight stories; the third building, mentioned as « High-rise », consisted of 15 stories. More details of the construction elements are given in Table 1. The buildings had a structural system for resisting horizontal loads based to walls. Their distribution in plan was symmetric in both horizontal directions to avoid an additional torsional effect.

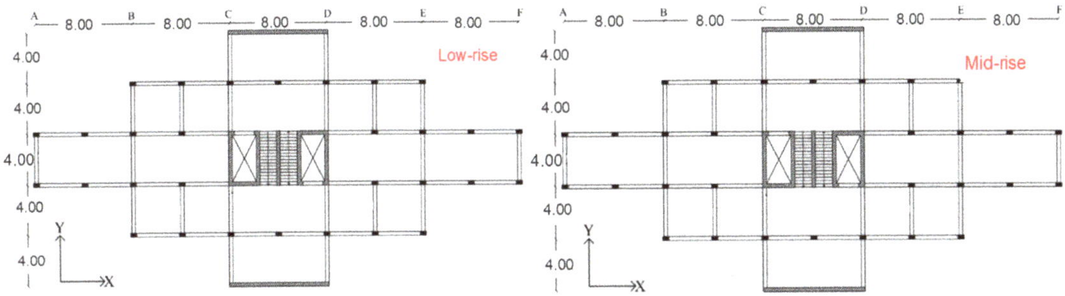

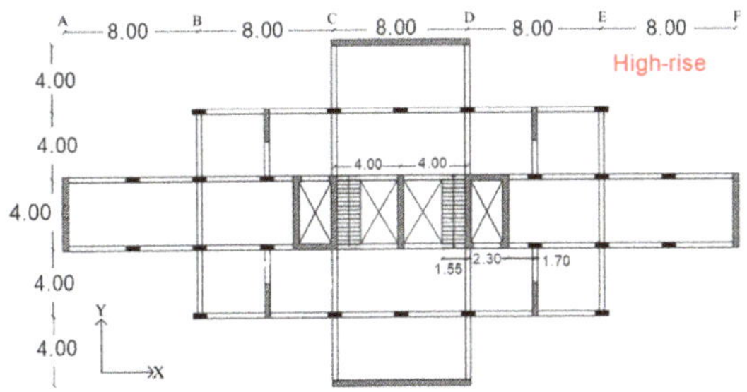

Notation:

1. Concrete C30/37 E=32.9 GPa
2. Steel rebar S500B E =210 GPa

Area	Dead load G [kN/m²]		Live load Q [kN/m²]
Residential	Self-weight +	6	1.5
Balcony		2	3.5
Stairs		2	3.5
Roof		0.12	0.8

Figure 1. Plan view of the buildings (units in m).

Table 1. Description of the investigated buildings.

Building Type	Low-Rise	Mid-Rise	High-Rise
Number of stories	G + 03	G + 08	G + 15
Story height		3.5 m	
Total height including roof level	14 m	31.5 m	56 m
Columns cross section	30 × 50 cm	30 × 80 cm	30 × 80 cm
Beams cross section		30 × 30 cm	
Walls thickness		30 cm	
Slab thickness		20 cm	

3. Building's Modeling

A finite element method (FEM) was constructed to model the structural system and mass distribution. Non-linear dynamic time history analyses were performed to account for geometrical and structural non-linearities. The beams and the columns were modeled as frame elements with rectangular cross sections (see Table 1), while the walls were modeled as shell elements. The rigid floor diaphragm assumption was used for the modeling of the stories, as the buildings are regular (EN 1998-1:2004, page 42, section 4.2.3.2) and in elevation (EN 1998-1:2004, page 43, section 4.2.3.3) [31]. For the walls and floors, a four-node shell elements was used in this study. The selected shells elements are homogeneous, with 6DOF in each node, and an appropriately selected mesh was used in order to have equilibrium between the accuracy of the results and the computation cost (103,200 DOF for the « low-rise » building, 272,400 for the « mid-rise », and 628,800 for the « high-rise »). In the three-dimensional structural model, elastic flexural stiffness and shear stiffness were taken into account, and equal to the one-half of the corresponding stiffness of the uncracked elements [31]. Material properties like concrete and steel rebars remain the same for all the stories, while the building is subjected to gravity and lateral loads (see Figure 1).

The optimal design of the strengthening solutions is presented in the Section 4 of this paper. A tuned-mass damper (TMD), also known as a pendulum damper, is not actually a damper, but rather a pendulum or another gravity-based oscillator that is attached to the structure in such a way that it counteracts the vibration of one or more fundamental modes, thereby reducing the wind and/or seismic response of those modes. A TMD was modeled using a spring-mass system with damping. A linear link element reproduced the spring properties, while the mass and weight was also assigned in the model. The details of the TMD mass (or PTMD) for each building are presented in the Section 4.1. The damping properties of nonlinear viscous dampers (VD) were based on the Maxwell model of viscoelasticity [32]. The nonlinear properties, as stiffness, damping coefficient, and damping exponent were specified, and modeled in series. A linear link object is most suitable unless nonlinear damping is assigned using a damping exponent other than 1.0. This enables the modeling of a linear dashpot parallel with linear stiffness for both linear and nonlinear analysis cases. The numerical modeling of friction dampers (FD) was very easy, since the hysteretic loop of the friction dampers is perfectly rectangular, similar to the perfectly elasto-plastic material. The friction dampers were modeled as a fictitious plasticity element having a yield force equal to the slip load. The FD and VD were positioned in steel diagonal brace elements. More details for the shape and the position of the steel diagonal braces are given in the Section 4 of this paper. The braces were modeled as a frame element.

All building models, i.e., the benchmark buildings and the alternative ones with the passive energy dissipation systems, were analyzed for seven different real and artificial accelerograms that were compatible to ground type B-dependent Eurocode 8 elastic spectra (seismic zone V according to the French national annex [33]). The selection of the accelerograms was based on the provisions of Eurocode 8 Part 1 [31]. The mass and stiffness proportional damping was chosen, and critical damping ratios equal to 5% and 4% were considered for the first and second period of the analyzed building systems, correspondingly. The strengthening solutions were carried out by maximizing structure performances as much as possible. This can be done by adopting an elastic linear behavior (behavior factor $q = 1$) to help prevent damages in structural elements that could compromise the durability of the structures. In order to take into account uncertainties linked to the location of the masses and the spatial variation of the seismic movement, EC8 requires an additional accidental eccentricity of at least 5% of the dimension of the building that is perpendicular to the direction of seismic action. This means that the center of gravity of each story must be offset in each direction of this eccentricity with respect to its nominal position. The adopted 5% eccentricity is considered constant, and repeated on each story in the same direction in the present study.

A nonlinear dynamic analysis was performed with seismic inputs described by bi-directional recorded accelerograms (Figure 2a), which have been applied at base level. Figure 2b shows the response spectra of the selected accelerograms compared to the EC8 elastic response spectrum Type 1, with peak ground acceleration equal to 0.3 g, ground type B, and 5% damping.

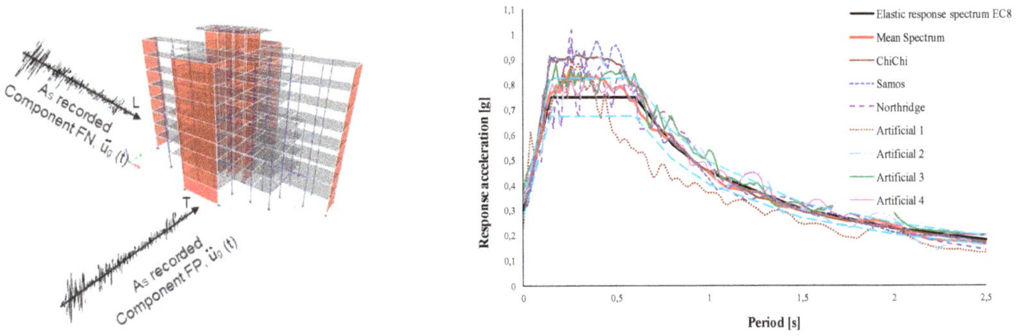

(**a**) (**b**)

Figure 2. (**a**) The bidirectional recorded accelerograms, (**b**) response spectra relative to the selected accelerograms compared to the EC8 elastic response spectrum Type 1 with peak ground acceleration equal to 0.3 g, ground type B, and 5% damping.

4. Optimal Design of Passive Energy Dissipation Systems

These days, there are numerous passive energy dissipation dampers, while the present study is focused on TMD, VD, and FD, which are described in the introduction. A design optimization technique for each system is presented in this section, covering damper characteristics and displacements. The design optimization sought to minimize (i) the maximum displacement at the top of the structures and (ii) the maximum inter-story drift.

4.1. Tuned Mass Damper

TMD is a motion-based passive system that consists of a mass m_d, a spring with spring stiffness k_d, and a dashpot with a damping coefficient c_d attached and typically tuned to the natural structural frequency [34]. During an earthquake, the damping system is stretched and compressed, reducing vibrations in the structure by increasing its effective damping [34]. A schematic representation of the 2 DOF (degree of freedom) system is shown in Figure 3, noting that m, k, and c represent, respectively, the main mass, stiffness, and inherent damping coefficient of the structural system. TMD is typically effective over a narrow frequency band. It is therefore important to be tuned to a particular natural frequency. The system efficiency decreases, with structures having several closely spaced natural frequencies [34].

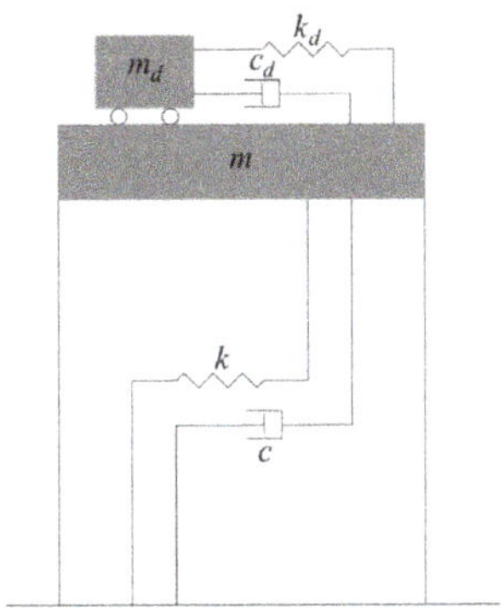

Figure 3. Schematic representation of the 2 DOF system.

An appreciation of TMD efficiency can be obtained by following the basic development of Den Hartog [6], which considers an undamped structural system subject to a sinusoidal excitation. Figure 4 shows that the dynamic amplification factor, R, which takes the damping effect of the TMD, is a function of the four essential variables: the mass ratio $\overline{m}$ ($\overline{m} = m_d/m$), the TMD damping ratio ξ_d ($\xi_d = \frac{c_d}{2 \cdot m_d \cdot \omega_d}$), the frequency ratio ν ($\nu = \omega_d/\omega$), and the forced frequency ratio λ ($\lambda = \overline{\omega}/\omega$), where $\omega = \sqrt{k/m}$ and $\omega_d = \sqrt{k_d/m_d}$ are the natural frequency of the structural system and TMD, respectively. The dynamic amplification factor R is expressed by the Equation (1):

$$R = \sqrt{\frac{(\nu^2 - \lambda^2)^2 + (2 \cdot \xi_d \cdot \nu \cdot \lambda)^2}{[(\nu^2 - \lambda^2)(1 - \lambda^2) - \nu^2 \cdot \lambda^2 \cdot \overline{m}]^2 + (2 \cdot \xi_d \cdot \nu \cdot \lambda)^2 (1 - \lambda^2 - \lambda^2 \cdot \overline{m})^2}} \quad (1)$$

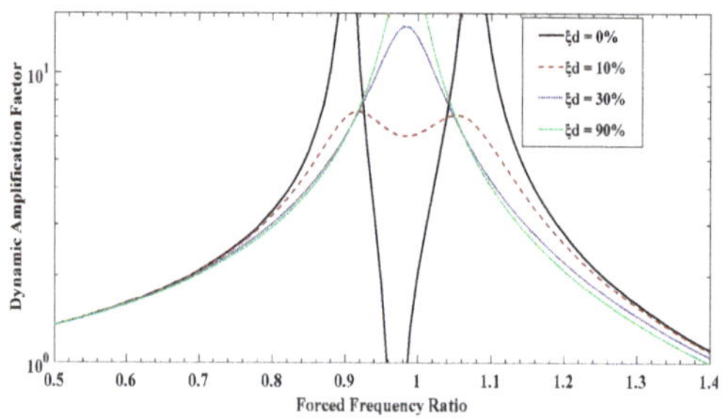

Figure 4. Dynamic amplification factor as a function of λ.

Figure 4 shows a plot of R as a function of λ for $\overline{m} = 0.05$ and $\nu = 1$. Without TMD damping, the response amplitude is infinite at two resonant frequencies of the 2 DOF systems. Furthermore, for an infinite TMD damping, the two masses are virtually fused to each other, leading the amplitude of resonant frequency to be infinite again [1]. Therefore, between these extremes, there is a value of ξ_d, for which the peak becomes a minimum.

An objective of installing TMD in structures is to bring the response amplitude down to its lowest possible value, 1; this is why the damping ratio of TMD must be carefully selected in such a way that small amplifications over a wider frequency bandwidth can be achieved. As can be seen in Figure 4, this can be achieved by taking a small value of ξ_d, like 30%. So, the effect of the TMD damping ratio is very essential. One observes that this parameter must exist but must not be high, because at this case, the amplifications are small and the frequency range in which the damper works is the biggest increasing damper efficiency. Outside of this range, the motion is not considerably influenced by the TMD system.

The conventional TMD described above requires a large mass and space for installation, thus creating architectural constraints [35]. An alternative approach is using a pendulum configuration PTMD. During ground motion, the pendulum produces a horizontal force which opposes the story motion [36]. This configuration type can be represented by an equivalent SDOF system attached to the story, as shown in Figure 5.

With the pendulum configuration [36], the equivalent stiffness is given by $k_{eq} = m_d \cdot g/L$, the natural frequency is expressed by $\omega_d = \sqrt{k_{eq}/m_d} = \sqrt{g/L}$ and the natural period is set by $T_d = 2 \cdot \pi \cdot \sqrt{L/g}$. The tuning parameters of PTMD are the mass m_d and the length L. This configuration is advantageous over the conventional TMD, especially for high-rise buildings, because its frequency can be retuned easily by modifying the cable length [37].

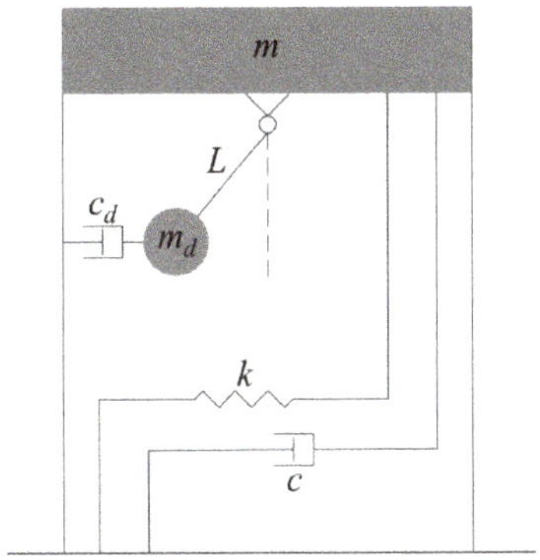

Figure 5. PTMD schematic representation.

The first important parameter in the optimal design process of the TMD system is the mass ratio $\overline{m}$. It is well known that the structural response decreases as $\overline{m}$ increases, but this ratio has a limit in practice which must not exceed 10% [34] based on geometrical and economical constraints. Then, optimized absorber parameters are calculated.

All real systems contain some inherent damping, meaning that an absorber is added to a lightly damped system. The effect of the inherent damping in the real system is an important design consideration on the optimum tuning parameters of TMD. Table 2 shows the equations used in the present study for ν and ξ_d, which includes the structural damping ratio of the primary structural system ξ, the derived rigidity k_d, and damping coefficient c_d of the TMD, respectively [38]. Also, Table 3 shows the structure and TMD optimal parameter values taken for the three investigated buildings, noting that T is the fundamental period of the structure in the transversal direction in which the TMD is applied.

Table 2. TMD optimized parameters expressions.

	Expressions	
Optimal tuning parameters of TMD given in [38]	$\nu = \frac{1}{1+\overline{m}} \cdot \left[1 - \xi \cdot \sqrt{\frac{\overline{m}}{1+\overline{m}}}\right]$	(2)
	$\xi_d = \frac{\xi}{1+\overline{m}} + \sqrt{\frac{\overline{m}}{1+\overline{m}}}$	(3)
Optimized absorber parameter	$k_d = \omega_d{}^2 \cdot m_d = \nu^2 \cdot \omega^2 \cdot \overline{m} \cdot m$	(4)
	$c_d = 2 \cdot \xi_d \cdot \omega_d \cdot m_d = 2 \cdot \xi_d \cdot \nu \cdot \omega \cdot \overline{m} \cdot m$	(5)

Table 3. Structure-TMD optimal parameter values.

Building Type	Structure		TMD		
	Parameter	Value	Parameter	Values	
Low-rise	T [s]	0.216	$\overline{m} = 0.5\%$	$\nu = 0.9915$	$\xi_d = 12.03\%$
	m [t]	1572.29	$\overline{m} = 1\%$	$\nu = 0.9852$	$\xi_d = 14.90\%$
Mid-rise	T [s]	0.985	$\overline{m} = 1\%$	$\nu = 0.9852$	$\xi_d = 14.90\%$
	m [t]	4626.92	$\overline{m} = 3\%$	$\nu = 0.9626$	$\xi_d = 21.92\%$
High-rise	T [s]	2.202	$\overline{m} = 2\%$	$\nu = 0.9735$	$\xi_d = 18.90\%$
	m [t]	8560.39	$\overline{m} = 3\%$	$\nu = 0.9626$	$\xi_d = 21.92\%$

In order to choose the most appropriate mass ratio, two analyses will be carried out for two mass ratios per building (Table 4). The structural response will be compared to the undamped case in Table 5.

Table 4. TMD design parameter values for the three investigated buildings.

Building Type	$\overline{m}$	m_d [t]	k_d [kN/m]	c_d [kN·s/m]
Low-rise	0.50%	7.86	6561.47	54.63
	1%	15.72	12,955.58	134.49
Mid-rise	1%	46.27	1829.10	86.70
	3%	138.8	5238.61	373.86
High-rise	2%	171.2	1321.15	179.82
	3%	256.8	1937.40	309.24

Table 5. Responses of the three investigated buildings for the two different mass ratio of TMD with margin from undamped case.

Building Type	Case	Direction	Fundamental Period		Top Roof Displacement		Base Shear	
			Value [s]	Margin	Value [cm]	Margin	Value [kN]	Margin
Low-rise	Undamped	Longitudinal	0.156		0.8556		9536.64	
		Transversal	0.216		0.4248		2772.68	
	Damped with $\overline{m} = 0.5\%$	Longitudinal	0.158	−1.28%	0.8243	3.66%	9530.28	0.07%
		Transversal	0.218	−0.93%	0.4183	1.53%	2751.67	0.76%
	Damped with $\overline{m} = 1\%$	Longitudinal	0.157	−0.64%	0.8237	3.73%	9287.71	2.61%
		Transversal	0.216	0.00%	0.4164	1.98%	2693.44	2.86%
Mid-rise	Undamped	Longitudinal	0.697		13.2068		24,870.8	
		Transversal	0.985		5.4141		4652.36	
	Damped with $\overline{m} = 1\%$	Longitudinal	0.71	−1.87%	12.6213	4.43%	23,726.16	4.60%
		Transversal	0.997	−1.22%	5.3094	1.93%	4567.02	1.83%
	Damped with $\overline{m} = 3\%$	Longitudinal	0.733	−5.16%	12.5036	5.32%	22,507.32	9.50%
		Transversal	1.031	−4.67%	4.7931	11.47%	4559.55	1.99%
High-rise	Undamped	Longitudinal	1.983		38.377		19,306.7	
		Transversal	2.202		12.5677		5222.92	
	Damped with $\overline{m} = 2\%$	Longitudinal	2.058	−3.78%	35.9845	6.23%	18,077.8	6.37%
		Transversal	2.284	−3.72%	10.2574	18.38%	5112.19	2.12%
	Damped with $\overline{m} = 3\%$	Longitudinal	2.091	−5.45%	36.3981	5.16%	17,476.06	9.48%
		Transversal	2.322	−5.45%	10.1373	19.34%	5331.14	−2.07%

Flexible buildings undergo larger horizontal displacements, which may result in significant damages. In this case, it is preferable to choose a relatively big mass ratio, unlike rigid buildings, for which it is advisable to adopt a relatively small ratio, because increasing its value does not provide any additional damping effect.

This is why the low-rise building is studied over the two following small mass ratio values 0.5% and 1%, the mid-rise building over 1% and 3%, while the high rise building will be studied for 2% and 3%.

As for the installation location of TMD, choosing the best location is the most important factor to consider in the optimal design process to show excellent control performance for the controlling dynamic response [39]. It is important to note that there are not enough studies carried on the installation of TMD in a spatial structure, as well as a lack of data on the optimal installation [40].

The TMD is commonly installed at the center plan to avoid creating torsional effects. In their study on the performance and placement of one or more TMDs in buildings, Almazan et al. [41] concluded that the optimum location is near the geometric center of the plan, whether for symmetric or asymmetric buildings. In addition to that, TMD is a motion-based system, which means that the TMD efficiency in reducing structural response is gained by applying it at the story that will experience the most motion. In symmetric buildings, it is usually on the upper story level [42].

The margins in Table 5 are calculated according to the undamped case, which means that a reduction is detected for positive values. The negative margins mean that there is an increase in parameter values after installing the TMD.

After focusing on the margin between the two proposed mass ratio for each of the three investigated buildings, it is clear that an increase in the mass ratio brings a relatively small additional damping for Artificial 4 ground motion. This is why for economic concerns, 0.5% is adopted for the low-rise building as a value of mass ratio, and 3% for the mid and high-rise building.

Optimized TMD design parameter values are listed in Table 6. However, the optimal position for irregular buildings is not necessarily on the upper story due to the different stiffness values for each story in elevation. Furthermore, the damper location in plan is considered a primary design variable; in this case, it depends essentially on the eccentricity between the center of mass and rigidity, but is always near the geometric center of the plan [41]. It is important to note that buildings optimal design is evaluated for Artificial 4 ground acceleration by applying a nonlinear time history analysis because its spectral response is close to the EC8 elastic response spectrum [31].

Table 6. Cable length required for the three investigated buildings.

Building Type	$\bar{m}$	ω_d [rad/s]	L [m]
Low-rise	0.50%	28.893	0.012
Mid-rise	3.00%	6.143	0.259
High-rise	3.00%	2.747	1.300

An alternative solution is using a PTMD configuration, which can provide additional damping, as it can act in all directions. Table 6 shows that the cable length required for the low and mid-rise building are too small, and thus, unrealistically small values. So, the PTMD configuration is only evaluated for the high-rise building. The cable that relates the additional mass to structure is composed of steel rods, with a circular section offering high axial rigidity. In this optimization section, a comparison between TMD and PTMD configuration is established. It is clear that the values obtained for the three parameters studied are relatively close. This can be explained by the fact that the building is not too flexible, which limits the performance of TMD and PTMD as well (Figure 6). However, the PTMD configuration is adopted in the high-rise building because it offers more reduction than the translational configuration, and its frequency can be easily retuned (Table 7).

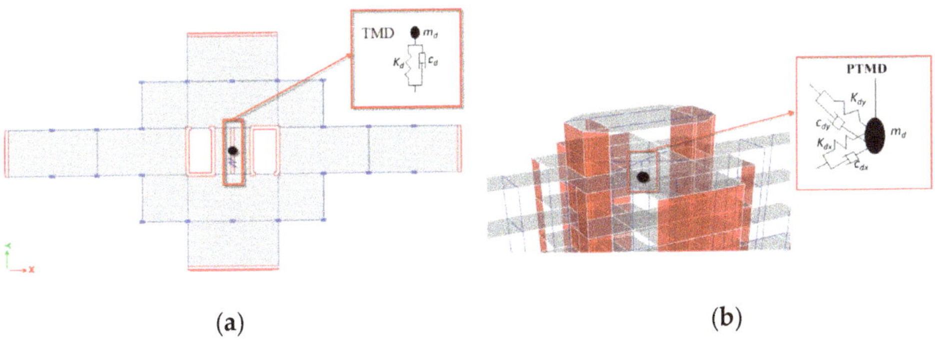

Figure 6. Modeling configuration of (**a**) TMD for low-rise and mid-rise buildings, (**b**) PTMD for high-rise building.

Table 7. Response of the high-rise building for the TMD and PTMD configuration.

Case	Longitudinal			Transversal		
	Undamped	With TMD	With PTMD	Undamped	With TMD	With PTMD
Fundamental period [s]	1.983	2.091	1.801	2.202	2.322	2.67
Top roof displacement [cm]	38.377	36.39	25.01	12.567	10.14	9.23
Base shear [kN]	19,306.7	17,476.06	16,974.11	5222.9	5331.14	4386.71

4.2. Viscous Dampers (VDs)

Viscous damping is the dissipation of energy that occurs when a particle in a vibrating system is resisted by a force, the magnitude of which is a constant independent of displacement and velocity, and the direction of which is opposite to the direction of the velocity of the particle. Uniaxial force is a result of a pressure across the piston head. Since the fluid is nearly incompressible, a reduction in fluid volume results in a restoring force, which is prevented by the use of a rod make-up accumulator [12]. Previous research show an increase in temperature can be significant, particularly in long-durations or large-amplitude seismic motions. This temperature is compensated by a relatively small effect through mechanisms [43,44].

An efficient mathematical model to describe VD behavior (linear or nonlinear) was proposed by Seleemah and Constatinou [45] based on experimental results. The force of the damper $P(t)$ is calculated by the following Equation (6):

$$P(t) = C_d|\dot{u}(t)|^{\alpha}\, sgn[\dot{u}(t)] \tag{6}$$

where C_d is the damping coefficient, $u(t)$ is the displacement across the damper, and α is a coefficient, depending on the piston head design and viscosity properties of fluid.

The coefficient α is the first important parameter to verify which could be less or equal to 1. Figure 7a describes the force-velocity relationship for linear and nonlinear behavior, while Figure 7b shows the force–displacement hysteretic loops. For earthquake resistance structures, the α coefficient has a value ranging from 0.3 to 1.0, in order to provide larger forces and to minimize shocks for high velocities with no degradation of performance [46]. In addition, the lowest value needed to maintain a high amount of energy absorbed per cycle of vibration is shown in Figure 7b, and minimizes at the same time the stress at adjacent structural members [47]. So, in our study, an α value equal to 0.3 has been selected.

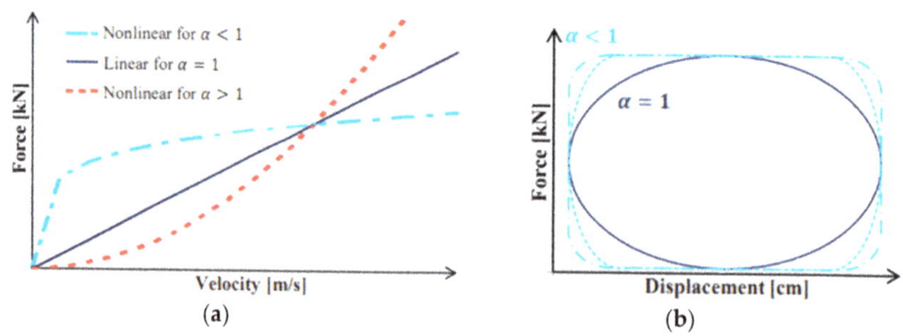

Figure 7. (**a**) Force-velocity behavior of VD and (**b**) Force-displacement hysteretic-loops.

Velocity is the second important parameter to fix. This is because the VD force varies with velocity, which is related to structural motion and depends on the structural fundamental period. The horizontal flexibility of the structure injects the full movement directly into the horizontal component of the damper, so VD is considered efficient for flexible rather than rigid structures [47]. In this study, a parametric study was done in order to select the correct velocity in accordance with previous studies [47].

The damping coefficient (Cd) is the third important parameter to define, related to the desired effective damping ξ_{eff}, and attributed to the structure. Design codes do not provide any substantial procedure for the distribution of the calculated damping coefficient over the whole building. These days, a large variety of methods have been proposed, classified between two categories: standard and advanced methods [48,49]. In the present study, the damping coefficient is distributed along the height of the building, based on the proportionality respective of the story shear force (Equation (7)); the effective damping ξ_{eff} is a sum of the structural inherent damping ratio (ξ_0) and the damping ratio of the viscous dampers (ξ_d), according to [50] recommendations (see Equation (8)).

$$C_{d,i} = \frac{V_i}{\sum V_i} \sum C_j \tag{7}$$

$$\xi_{eff} = \xi_0 + \xi_d = \xi_0 + \frac{\sum \lambda C_j {\phi_{rj}}^{1+\alpha} \cos^{1+\alpha} \theta_j}{2\pi A^{1-\alpha} \omega^{2-\alpha} \sum M_i {\phi_i}^2} \tag{8}$$

$$\lambda = 2^{2+\alpha} \frac{\Gamma^2\left(1+\frac{\alpha}{2}\right)}{\Gamma(1+\alpha)} \tag{9}$$

where A is the amplitude, ϕ_{rj} is the relative horizontal displacement of the damper, θ_j is the inclined angle of the damper j, ω is the loading frequency supposed equal to the natural structural frequency, M_i is the vibrating mass of the story i, ϕ_i is the modal displacement at story i, and λ is a parameter calculated by Equation (9) [51].

Del Gobbo [52] indicates that in order to establish the optimal effective damping, nonstructural elements must be taken into account. To have an essential damping ratio-repair cost relationship, the range of optimal effective damping is identified as 30 − 40% to minimize mean economic losses. However, the optimal damping amount also depends on the building's properties, such as the fundamental period of structure. Table 8 shows the selected, effective damping and velocities values, as well as the calculated damping coefficient.

Table 8. Effective damping and calculated damping coefficient for the three investigated buildings (α = 0.3).

Building Type	Direction	Fundamental Period [s]	Structural Rigidity Description	Suggested Velocity [m/s]	Suggestedeffective Damping ξ_{eff}	$\sum C_j$ [kN·(s/m)]
Low-rise	Longitudinal	0.156	Rigid	0.127	30%	71,537.07
	Transversal	0.216		0.127	30%	59,405.58
Mid-rise	Longitudinal	0.697	Semi-rigid	0.254	35%	56,981.16
	Transversal	0.985		0.254	35%	39,325.16
High-rise	Longitudinal	1.983	Flexible	0.381	40%	12,658.71
	Transversal	2.202		0.381	40%	12,095.19

The design optimization of VD is not limited only on the mechanical parameters of VDs, but also on the position of the dampers in the plan-view of the building. It is important to ensure that the dampers are located in a configuration that does not introduce eccentricity to the structure; this is why the most efficient placement would be equivalently about the building's center of mass, for example, along the perimeter of typical structures. The main reason is to be able to control any torsional motion of the building [47]. In our study, the VDs are positioned in steel diagonal braces, half of them working under compression, and the other under tension. Different configurations of VD's placement were studied, while two of them are illustrated in the Figure 8. At least two dampers were positioned in each direction and on each side of the building's center mass at every story, even though it is not required. It could be terminated before the top levels or alternated at different story levels. Moreover, to limit damper force output, more than two dampers per direction could be used, especially for buildings with large footprints.

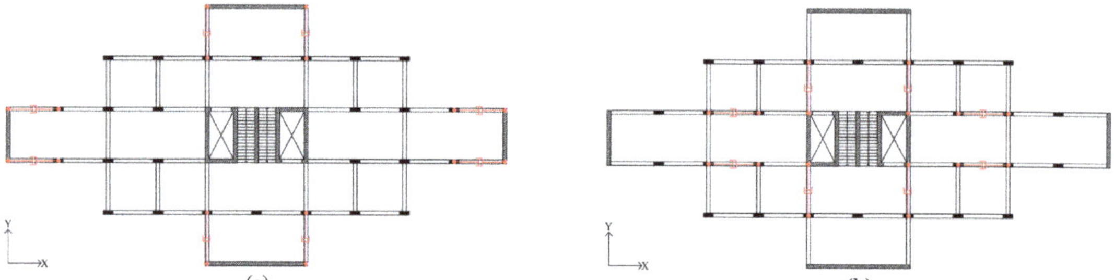

Figure 8. (**a**) Alternative 1 of dampers placement and (**b**) Alternative 2 of dampers placement.

To choose the best one for each building by evaluating the fundamental period, the top roof displacement and base shear obtained in the longitudinal and transversal directions is indicated in Table 9. It should be noted that dampers were installed at all levels with four systems per direction. It must be noticed that the dampers placement can affect a building's structural response by evaluating the structural response between the two alternatives of dampers placement shown in Table 9. In general, alternative 2 provides the best reduction, especially for the low and mid-rise building, and alternative 1 offers the smallest values of displacement for the high-rise building. So, alternative 2 is chosen for the low and mid-rise building, and alternative 1 for the high-rise building. Figure 9 shows the schematic configuration during the modeling of VDs.

Table 9. Responses of the three investigated buildings for the two alternatives of VDs placement.

Building Type	Direction	Fundamental Period [s]		Top Roof Displacement [cm]		Base Shear [kN]	
		Altern. 1	Altern. 2	Altern. 1	Altern. 2	Altern. 1	Altern. 2
Low-rise	Longitudinal	0.112	0.112	0.167	0.176	1973.2	1905.6
	Transversal	0.148	0.147	0.218	0.345	2308.6	2070.0
Mid-rise	Longitudinal	0.387	0.387	3.121	3.763	1888.1	7567.0
	Transversal	0.775	0.775	2.997	1.760	4991.5	83.3
High-rise	Longitudinal	0.543	0.543	5.812	6.821	10,828	322.5
	Transversal	0.989	0.989	4.604	4.803	178.6	14,737.6

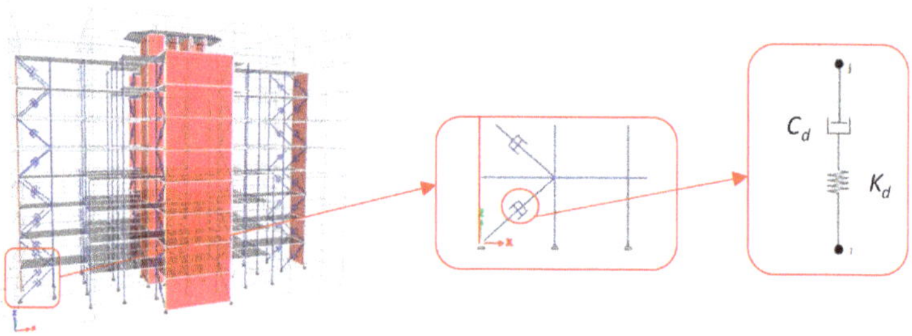

Figure 9. Modeling configuration of VDs.

4.3. Friction Dampers (FDs)

FD is a displacement-based system which dissipates energy through friction across the surfaces between two solid elements [1,3]. The dissipative mechanism generates heat through dry sliding friction with a stable hysteretic behavior [53]. A simple model for defining the behavior of the damper is given by the idealized Coulomb model of friction. The theory is based on the following hypotheses, which are experimentally validated [19]:

- Force independent of the apparent contact surface

- Force proportional to the total normal force acting through the interface
- Force independent of speed even with a slip at low speed

As a result, the force can be written using the following expression $F_t = \mu \cdot F_n$, where F_t and F_n represent the frictional and normal forces, respectively, and μ the coefficient of friction which depends on the selection of sliding materials and present conditions of the sliding interface. F_n and μ are maintained at constant values over extended durations of time, which is difficult to achieve in practice [3]. The damper hysteresis loop is rectangular, showing a great amount of energy dissipated per cycle of motion, and the cyclic behavior of FD is strongly nonlinear, as shown in Figure 10a. When the friction force is overcome, FD adds initial stiffness to the structural system. It is important to note that if no restoring force is provided, permanent structural deformation may exist after an earthquake [3]. As shown in Figure 10b, the response of the structure is highly affected by FD slip force, and a small variation of FD optimum slip load has a minimum effect on structure's response. The selected slip force must be high enough to prevent the damper from slipping under a small applied lateral loads value, and should be low enough to achieve slip before the yielding of the main structural elements [54].

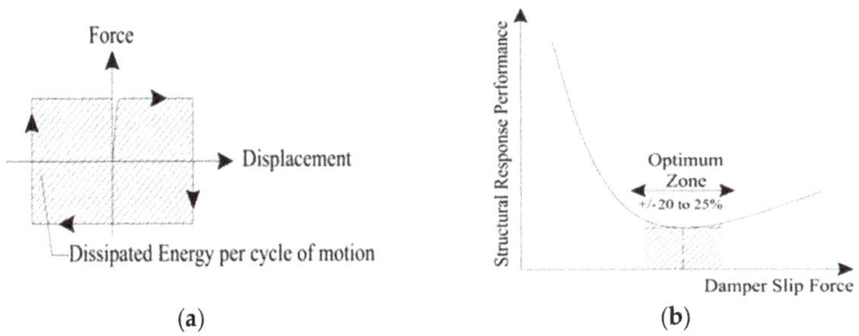

Figure 10. (**a**) Force-Displacement hysteresis loop of a friction damper, (**b**) optimal slip force effect on structural response.

A simple method used in the present study consists of taking a portion from the applied shear force, so the load at each story is estimated by the Equation (10)

$$F_{t,\ optimal} = \frac{1}{3}[V_i / n_i] \tag{10}$$

where $F_{t,opt}$ is the optimal slip force or frictional force, V_i is the shear load, and n_i is the number of dampers per direction in the story i. The shear load was calculated from the results of the Fast Nonlinear Analysis (FNA).

For the diagonal configuration to the damper-brace assembly, it is clear that the device and the brace are connected in series. The FD stiffness value is considered infinity, so the total stiffness value to integrate while modeling is equal to the brace system to avoid brace buckling, as explained in Equation (11) [55]:

$$k_{bd} = \frac{1}{\left(\frac{1}{k_b}\right) + \left(\frac{1}{k_d}\right)} \overset{k_d \to \infty}{\rightarrow} k_{bd} = k_b \tag{11}$$

The same two alternatives, as in the optimal design of VDs, were studied in the optimal design of the FDs (see Figure 8). Table 10 summarizes and compares the results. The authors selected configuration number 1 due to certain criteria: the significant reduction obtained in the longitudinal direction in terms of displacement and base shear. Although alternative 2 provides an important reduction in some buildings in the transversal direction, the alternative that is able to reduce top displacement and base shear values as much as possible was chosen. Figure 11 shows the schematic configuration during the modeling of FDs with the use of N-link plastic elements.

Table 10. Responses of the three investigated buildings for the two alternatives of FDs placement.

Building Type	Direction	Fundamental Period [s]		Top Roof Displacement [cm]		Base Shear [kN]	
		Altern. 1	Altern. 2	Altern. 1	Altern. 2	Altern. 1	Altern. 2
Low-rise	Longitudinal	0.158	0.158	0.228	0.290	6971.2	7296.0
	Transversal	0.218	0.218	0.126	0.057	1525.1	2000.0
Mid-rise	Longitudinal	0.707	0.708	1.682	1.775	14,030.2	14,639.4
	Transversal	0.998	0.999	1.124	0.298	3100.0	3500.0
High-rise	Longitudinal	2.016	2.016	6.915	7.787	12,373.0	13,950.1
	Transversal	2.239	2.239	4.349	2.672	3494.58	4000.0

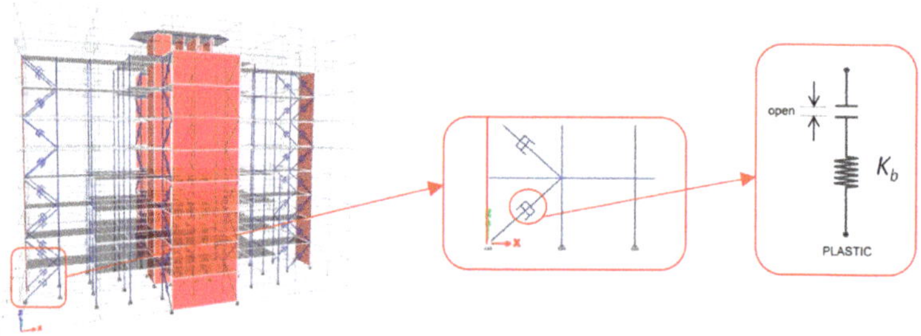

Figure 11. Modeling configuration of FDs.

5. Results and Discussion

The alternative design buildings with the three passive energy dissipation systems were redesigned. Focus was placed on the optimal design of the dissipated systems. The aim of the attempted redesign was to minimize (i) the maximum displacement at the top of the structures, (ii) the maximum inter-story drift, and (iii) the base shear loads. The results are presented in terms of two essential parameters: the maximum top roof displacement and base shear forces. In addition, the maximum inter-story drift is presented. A comparison of each parameter is established between the undamped and the damped cases with tuned mass damper, viscous, and friction dampers for the seven selected ground motions. An interpretation is established at the end in order to provide a conclusion on the comparative results, and to select the most suitable damper for each type of building.

5.1. Displacement at the Top of the Structures

Figure 12 illustrates the horizontal displacement at the top of each building in the longitudinal and transversal direction for the seven accelerograms. The percentage of reduction in the responses for the low-rise building (Figure 12a,b) equipped with friction dampers, in comparison with the structure without dampers, generally exceeds 70.29% in both directions, and reaches 86.10% with ChiChi earthquake excitation in the transversal direction. Although the reduction in the longitudinal direction with viscous dampers is bigger than those obtained with friction dampers, it is limited in the transversal direction, reaching a maximum of 39.49%. As for the damped case with TMD, the percentage of reduction does not exceed 12.94% for all earthquake records except Samos. By evaluating the mean value of percentage reduction in both directions, which is equal to 6.72% and 9.75% with a tuned mass damper, 91.11% and 30.82% with viscous dampers, and 71.44% and 76.87 % with friction dampers, it can be seen that friction dampers perform better than the two other types in the response reduction in the low-rise building.

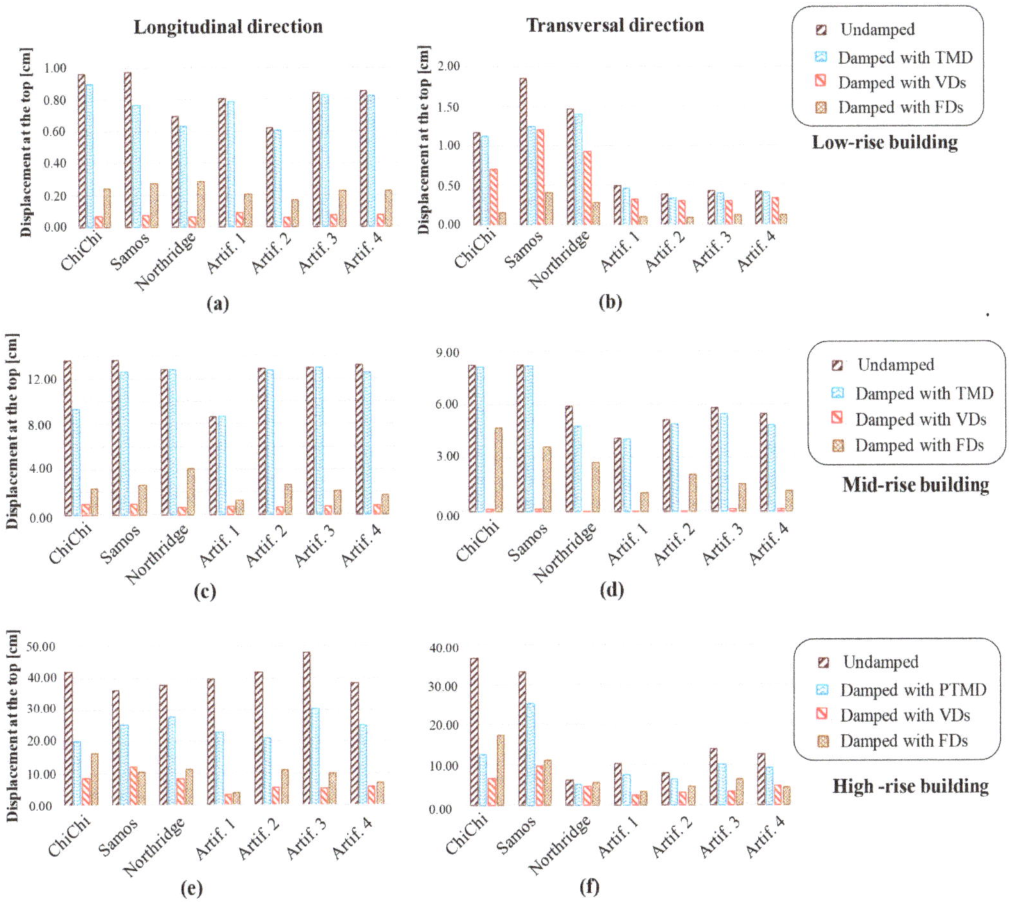

Figure 12. Horizontal displacement at the top of (**a**,**b**) the low-rise building, (**c**,**d**) the mid-rise building, and (**e**,**f**) the high-rise building.

The maximum roof displacement value of the mid-rise building for each history record is plotted in Figure 12 with and without dampers. According to the results, it can be seen that utilizing viscous dampers reduces the displacement the most in both horizontal directions, which goes beyond 91.46%. For comparison purposes, the maximum displacement values with friction dampers in an 87.26% and 79.25% reduction and with tuned mass damper in a 31.71% and 19.24% reduction, respectively, in longitudinal and transversal direction. Since the viscous and friction dampers have the greatest impact on the displacement by evaluating the mean value of percentage reduction in both directions, which is equal to 6.58% and 6.30% with a tuned mass damper, 93.70% and 98.96% with viscous dampers, and 81.14% and 63.29% with friction dampers, both systems seem to perform well under all earthquake records for the mid-rise building.

From the results of the high-rise building, all three types of dampers contribute to significant reduction in terms of displacement. One could observe that the percentage of reduction for the high-rise building equipped with friction dampers reaches a maximum of 90.36% in the longitudinal direction, and a maximum of 69.50% in the transversal direction, which is considered high. Viscous dampers also provide high values of reduction, reaching a maximum of 91.89% and 76.96% in both horizontal directions, respectively. Moreover, the pendulum configuration of PTMD offers a great reduction of 52.10% and 65.75% for the longitudinal and transversal direction, respectively. In terms of mean values, a reduction of 39% and 30% is detected with a tuned mass damper, 82.53% and 65.71% with viscous dampers, and 75.28% and 51.75% with friction dampers. Considering the mean values listed before, even though the PTMD system performs less than the two other damping systems, the reduction results are considered acceptable.

5.2. Base Shear Load

Figure 13 shows the results of the base shear load under the seven seismic ground records for the undamped and the three damped cases. For all damper systems, the ratio between the base shears of models with and without dampers for the low-rise building shows an important reduction (Figure 13a,b). Up to a 76.74% response reduction was achieved with viscous dampers in the longitudinal direction, and a maximum reduction attained by 36.61% in transversal direction. High values of reduction with friction dampers reach 39.93% and 59.14%, respectively, in both directions. With TMD, the percent of reductions are given as about a maximum of 15.74%. By evaluating the mean value of percentage reduction in both directions, which is equal to 2.54% and 5.32% with tuned mass damper, 79.96% and 31.06% with viscous dampers, and 27.38% and 41.66 % with friction dampers, both viscous and friction dampers appear to be effective for the low-rise building in base shear reduction.

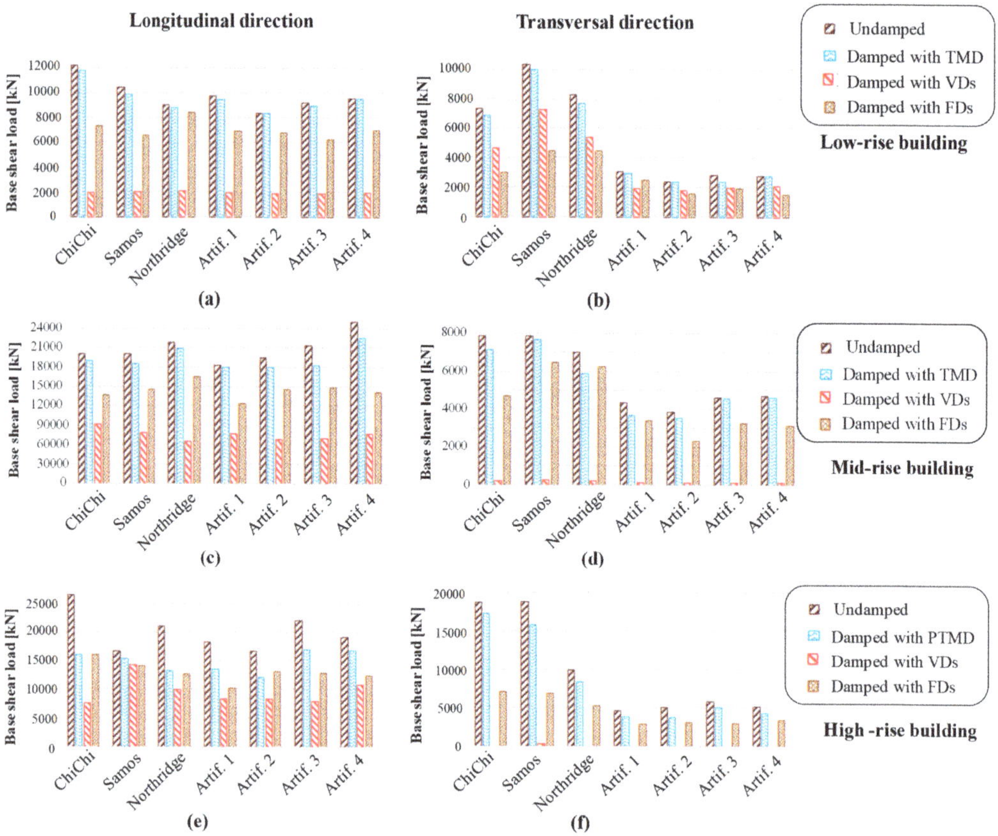

Figure 13. Base shear load for (**a**,**b**) the low-rise building, (**c**,**d**) the mid-rise building and (**e**,**f**) the high-rise building.

Tuning of the three studied dampers resulted in a 14.79%, 70.09%, and 43.59% maximum reduction of the base shear values with the tuned mass damper, viscous dampers, and friction dampers systems, respectively, in the longitudinal direction, and in a 16.14%, 98.22%, and 40.70% maximum reduction, respectively, in the transversal direction. Considering all the results obtained, and by evaluating the mean value of percentage reduction in both directions, which is equal to 7.12% and 7.72% with tuned mass damper, 63.91% and 97.62% with viscous dampers, and 30.97% and 27.57% with friction dampers, the mid-rise building equipped with viscous dampers have the greatest impact on the reduction of the base shear response in earthquake excitations compared to the two other damping systems.

Base shear results indicated a maximum reduction achieved with viscous dampers reaching 71.20% and 98.76% for ChiChi earthquake in longitudinal and transversal direction, respectively, and a maximum of 43.77% and 63.48% with friction dampers.

5.3. Interstory Drift

Due to a large number of diagrams, the authors have decided to present the diagrams of the interstory drift only for the accelerogram Samos (Figures 14 and 15). The interstory drift index is defined as interstory displacement, $\delta_{s,i}$, divided by story height, h_i. The relationship between the interstory drift index and the global drift index δ_t/h_t depends on the extent of inelasticity in the structure, the type of plastic hinge mechanism, and the importance of higher mode effects. This comparison validates the general conclusion of this study that is presented in the Section 6.

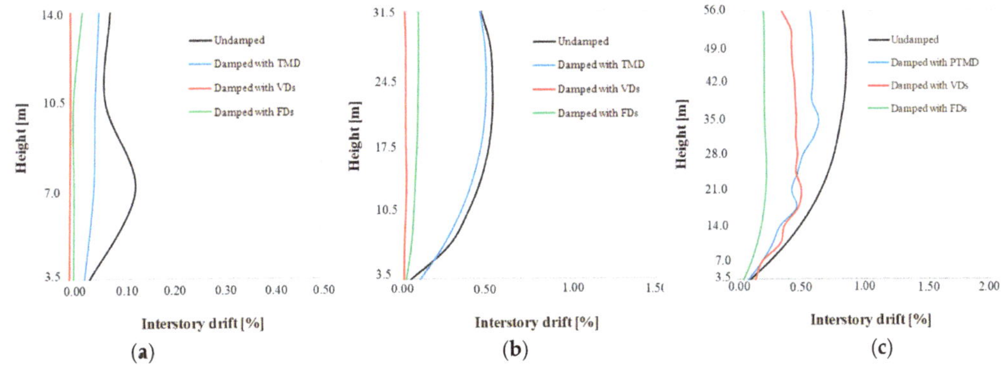

Figure 14. Maximum interstory drift for (**a**) low-rise, (**b**) mid-rise, and (**c**) high-rise building in the longitudinal direction.

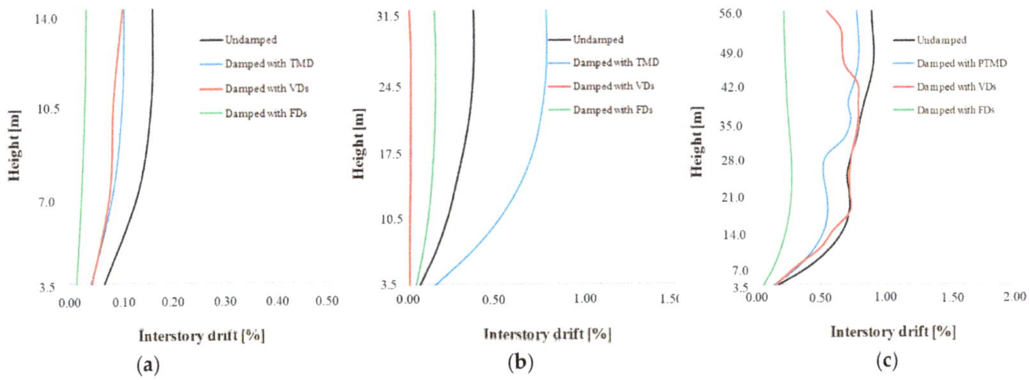

Figure 15. Maximum interstory drift for (**a**) low-rise, (**b**) mid-rise, and (**c**) high-rise building in the transversal direction.

5.4. Hysteretic Loops

In this section of our study, the hysteresis loops of the three dampers (PTMD, VD and FD) are presented in Figure 16. This figure presents the loops of the real accelerogram of Samos for the high-rise building. The shape of the loops is compared with the expected and well known shape of each damper based on previous studies [1–3,15–17,21–24,56,57] and the accuracy of this study is qualitative.

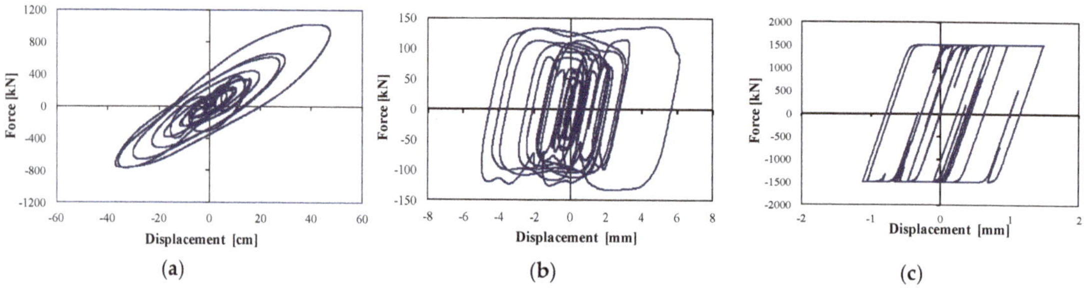

Figure 16. Hysteretic loops for (**a**) PTMD, (**b**) VD, (**c**) FD.

5.5. General Remarks

The comparative evaluation of the results obtained between undamped and damped case, discussed in terms of displacement and base shear, has led to the following interpretations:

- It is well known that the structural response reduction increases as the mass of TMD increases, but this mass has a limit in practice, due to geometrical and economical constraints. That is the reason why the mass ratio is not considered as an important value to optimize, and therefore, it is hard to achieve high reduction values practically. The results show that TMD systems are not effective for low and mid-rise buildings, because both the displacement and base shear values are barely affected, unlike high-rise building values. In fact, TMD are motion-based systems that demonstrate how their effectiveness is very limited for rigid buildings. As for the high-rise building, even though damped case with PTMD provides less reduction compared to the two other damped cases, it is considered acceptable and more suitable for this kind of structure.
- Structural strengthening with viscous damper systems is defined by the desired additional damping fixed in the preliminary design. From the results obtained, it has been observed that the structural response with the viscus dampers decreases well, showing better performance in terms of the displacement and base shear. In addition, viscous dampers are velocity-dependent systems, where its effectiveness increases with high velocities, usually for flexible buildings. Even though these systems are considered effective for the three studied buildings, they are considered more suitable for mid-rise buildings.
- Friction dampers' incorporation into the structures reduces considerably the building's response after optimizing dampers slip forces, their numbers, and locations under all earthquakes and types of buildings considered. It can be seen from the results obtained that the friction dampers are effective for both rigid and flexible buildings.

6. Conclusions

The present study compares the seismic response of three reinforced concrete (RC) symmetric buildings with varied number of stories strengthening with three types of passive energy dissipation systems, as tuned mass dampers, viscous dampers, and friction dampers. We focus the optimal design of each building on minimizing (i) the maximum displacement at the top of the structures, (ii) the base shear loads and (iii) the maximum interstory drift. Three residents' buildings (a four-story building, a nine-story building, and a sixteen-story building) were subjected to seven (real and artificial) seismic recorded accelerograms. The buildings were tested by considering a nonlinear dynamic analysis. The selected recorded time history functions fulfill the spectrum compatibility conditions required by the Eurocode.

The objective of the present paper was to optimize damper properties and placement in the selected buildings in order to maximize structural performance by providing high reduction, especially in terms of the displacement and base shear. A comparison was established between obtained results with the three types of passive dampers used, including tuned mass dampers (TMD), viscous dampers (VD), and friction dampers (FD) to choose the most suitable damping system for each type of structure, taking into account the damping–cost general relationship.

As a conclusion, the friction dampers were found suitable for the low-rise building, the viscous damping more preferable to incorporate in the mid-rise building, and the pendulum configuration of the tuned mass damper system more appropriate for the high-rise building. These results are valid for the previous symmetric structures under the earthquake considered in the present study. However, they provide a good insight into the effect of strengthening solutions with passive energy dissipation systems in symmetric reinforced concrete buildings. It is important to notice that for further investigation, buildings with other characteristics, such as irregularities in plan and elevation, should be also examined in order to study their effect on dampers design optimization, and choosing the most appropriate strengthening solutions for irregular buildings with different heights.

Author Contributions: All the authors contributed to the design and implementation of the research, to the analysis of the results and to the writing of the manuscript. All authors have read and agreed to the published version of the manuscript.

Funding: This research received no external funding.

Institutional Review Board Statement: Not applicable.

Informed Consent Statement: Not applicable.

Data Availability Statement: Not applicable.

Conflicts of Interest: The authors declare no conflict of interest.

References

1. Constantinou, M.C.; Soong, T.T.; Dargush, G.F. *Passive Energy Dissipation Systems for Structural Design and Retrofit*; MCEER Monograph Series, No; Multidisciplinary center for Earthquake Engineering Research: Buffalo, NY, USA, 1998.
2. Soong, T.T.; Spencer, B.F., Jr. Supplemental energy dissipation: State-of-the-art and state-of-the-practice. *Eng. Struct.* **2002**, *24*, 243–259. [CrossRef]
3. Symans, M.D.; Charney, F.A.; Whittaker, A.S.; Constantinou, M.C.; Kircher, C.A.; Johnson, M.W.; McNamara, R.J. Energy dissipation system for seismic applications: Current practice and recent developments. *J. Struct. Eng.* **2008**, *134*, 3–21. [CrossRef]
4. Frahm, H. Device for Damping Vibrations of Bodies. U.S. Patent No. 989,958, 30 October 1909.
5. Ormondroyd, J.; Den Hartog, J.P. The theory of the dynamic vibration absorber. *ASME J. Appl. Mech.* **1928**, *50*, 9–22.
6. Den Hartog, J.P. *Mechanical Vibrations*, 2nd ed.; McGraw-Hill: London, UK, 1940.
7. Kwok, K.C.S.; Macdonald, P.A. Full-scale measurements of wind-induced acceleration response of Sydney tower. *Eng. Struct.* **1990**, *12*, 153–162. [CrossRef]
8. McNamara, R.J. Tuned mass dampers for buildings. *J. Struct. Div.* **1977**, *103*, 1785–1798. [CrossRef]
9. Khan, F.R. 100 Storey John Hancock Center, Chicago: A case study of the design process. *Eng. Struct.* **1983**, *5*, 10–14. [CrossRef]
10. Kawabata, S.; Ohkuma, T.; Kanda, J.; Kitamura, H.; Ohtake, K. Chiba Port Tower. Full-scale measurement of wind actions Part 2. Basic properties of fluctuating wind pressures. *J. Wind Eng. Indust. Aerodyn.* **1990**, *33*, 253–262. [CrossRef]
11. Ueda, T.; Nakagaki, R.; Koshida, K. Suppression of Wind-Induced Vibration of Tower-Shaped Structures by Dynamic Dampers. *Struct. Eng. Inter.* **1993**, *3*, 50–53. [CrossRef]
12. Constantinou, M.C.; Symans, M.D. Experimental study of seismic response of buildings with supplemental fluid dampers. *Struct. Des. Tall Build.* **1993**, *2*, 93–132. [CrossRef]
13. Larbi, W. Numerical modeling of sound and vibration reduction using viscoelastic materials and shunted piezoelectric patches. *Comput. Struct.* **2020**, *232*, 105822. [CrossRef]
14. Larbi, W.; Deü, J.-F.; Ohayon, R. Vibroacoustic analysis of double-wall sandwich panels with viscoelastic core. *Comput. Struct.* **2016**, *174*, 92–103. [CrossRef]

15. Keel, C.J.; Mahmoodi, P. Designing of Viscoelastic Dampers for Columbia Center Building. In *Build Motion in Wind;* Isyumov, N., Tschanz, T., Eds.; ASCE: New York, NY, USA, 1986; pp. 66–82.
16. Tsopelas, P.; Constantinou, M.C. *Experimental and Analytical Study of a System Consisting of Sliding Bearings and Fluid Restoring Force/Damping Devices;* National Center for Earthquake Engineering Research: Taipei, China, 1994.
17. Moreschi, L.M.; Singh, M.P. Design of yielding metallic and friction dampers for optimal seismic performance. *Earthq. Eng. Struct. Dynam.* **2003**, *32*, 1291–1311. [CrossRef]
18. Pall, A.S.; Marsh, C.; Fazio, P. Friction joints for seismic control of large panel structures. *J. Prestress. Concr. Inst.* **1980**, *25*, 38–61. [CrossRef]
19. Pall, A.S.; Marsh, C. Response of friction damped braced frames. *J. Struct. Div. ASCE* **1982**, *108*, 1313–1323. [CrossRef]
20. Whittaker, A.S.; Bertero, V.V.; Thompson, C.L.; Alonso, J.L. Seismic testing of steel plate energy dissipating devices. *Earthq. Spectra* **1991**, *7*, 563–604. [CrossRef]
21. Grigorian, C.E.; Yang, T.S.; Popov, E.P. Slotted bolted connection energy dissipators. *Earth. Spectra* **1993**, *9*, 491–504. [CrossRef]
22. Ricles, J.M.; Sause, R.; Garlock, M.M.; Zhao, C. Post-tensioned seismic-resistant connections for steel frames. *J. Struct. Eng. ASCE* **2001**, *127*, 113–121. [CrossRef]
23. Christopoulos, C.; Filiatrault, A.; Uang, C.M.; Folz, B. Post tensioned energy dissipating connections for moment resisting steel frames. *J. Struct. Eng. ASCE* **2002**, *128*, 1111–1120. [CrossRef]
24. Kim, H.J.; Christopoulos, C. Friction damped post-tensioned self-centering steel moment-resisting frames. *J. Struct. Eng.* **2008**, *134*, 1768–1779. [CrossRef]
25. Titirla, M.; Papadopoulos, P.; Doudoumis, I. Finite Element modelling of an innovative passive energy dissipation device for seismic hazard mitigation. *Eng. Struct.* **2018**, *68*, 218–228. [CrossRef]
26. Fabbrocino, F.; Titirla, M.; Amendola, A.; Benzoni, G.; Fraternali, F. Innovative devices for the base isolation of existing buildings. In Proceedings of the COMPDYN 2017, 6th ECCOMAS Thematic Conference on Computational Methods in Structural Dynamics and Earthquake Engineering, Rhodes Island, Greece, 15–17 June 2017.
27. Titirla, M.; Katakalos, K.; Zuccaro, G.; Frabbrocino, F. On the mechanical response of an innovative energy dissipation device. *Ing. Sismic Int. J. Earthq. Eng.* **2017**, 126–138.
28. Pall, A.S.; Verganelakis, V.; March, C. Friction Dampers for seismic control of Concordia University Library Building. In Proceedings of the 5th Canadian Conference on Earthquake Engineering, Ottawa, ON, Canada, 6–8 July 1987.
29. Martinez-Romero, E. Experiences on the Use of Supplemental Energy Dissipators on Building Structures. *Earthq. Spectra* **1993**, *9*, 581–624. [CrossRef]
30. Pasquin, C.; Pall, A.; Pall, R. High-Tech Seismic Rehabilitation of Casino de Montreal. In Proceedings of the Structures Congress XII, Atlanta GA, USA, 24–28 April 1994; Volume 2, pp. 1292–1297.
31. CEN. *Eurocode 8: Design of Structures for Earthquake Resistance. Part 1: General Rules, Seismic Actions and Rules for Buildings;* EN 1998-1; European Committee for Standardization: Brussels, Belgium, 2004.
32. Malvern, L.E. *Introduction to the Mechanics of a Continuous Medium;* Prentice Hall: Hoboken, NY, USA, 1969.
33. *National Annex to NF EN 1998-1:2005;* The French Association for Standardization (AFNOR): Paris, France, 2005. (In French)
34. Farghaly, A.A.; Salem Ahmed, M. Optimum Design of TMD System for Tall Buildings. *Int. Sch. Res. Netw. Civ. Eng.* **2012**, *2012*, 716469. [CrossRef]
35. Nagase, T. Earthquake records observed in tall buildings with tuned pendulum mass damper. In Proceedings of the 12th World Conference on Earthquake Engineering, Auckland, New Zealand, 30 January–4 February 2000.
36. Connor, J.J. *Introduction to Structural Motion Control*, 1st ed.; Prentice Hall Pearson Education, Inc.: Upper Saddle River, NJ, USA, 2003.
37. Gutierrez Soto, M.; Adeli, H. Tuned mass dampers. *Arch. Comp. Methods Eng.* **2013**, *20*, 419–431. [CrossRef]
38. Sadek, F.; Mohraz, B.; Taylor, A.W.; Chung, R.M. A method of estimating the parameters of tuned mass dampers for seismic application. *Earthq. Eng. Struct. Dynam.* **1997**, *26*, 617–635. [CrossRef]
39. Kim, H.S.; Kang, J.W. Vibration control of smart TMD for retractable-roof spatial structure considering closed and open roof condition. *Int. J. Steel. Struct.* **2017**, *17*, 1537–1548. [CrossRef]
40. Lee, Y.Y.; Kim, H.S.; Kang, J.W. Seismic response control performance evaluation of tuned mass dampers for a retractable-roof spatial structure. *Int. J. Steel Struct.* **2021**, *21*, 213–224. [CrossRef]
41. Almazan, J.L.; Espinoza, G.; Aguirre, J.J. Torsional balance of asymmetric structures by means of tuned mass dampers. *Eng. Struct.* **2012**, *42*, 308–328. [CrossRef]
42. Elias, S.; Matsagar, V. Seismic response control of steel benchmark building with a tuned mass damper. *Asian. J. Civil. Eng.* **2020**, *21*, 267–280. [CrossRef]
43. Makris, N. Viscous heating of fluid dampers. I: Small-amplitude motions. *J. Eng. Mech.* **1998**, *124*, 1210–1216. [CrossRef]
44. Makris, N.; Roussos, Y.; Whittaker, A.S.; Kelly, J.M. Viscous heating of fluid dampers. II: Large-amplitude motions. *J. Eng. Mech.* **1998**, *124*, 1217–1223. [CrossRef]
45. Seleemah, A.; Constantinou, M.C. *Investigation of Seismic Response of Buildings with Linear and Nonlinear Fluid Viscous Dampers;* Report No. NCEER 97-0004; National Center for Earthquake Engineering Research, State University of New York at Buffalo: New York, NY, USA, 1997.

46. Hwang, J.S. Seismic design of structures with viscous dampers. In Proceedings of the International Training Programs for Seismic Design of Building Structures Hosted by National Centre for Research on Earthquake Engineering Sponsored by Department of International Programs, Taipei, China, 21–25 January 2002; National Science Council: Washington, DC, USA, 2002.
47. Fluid Viscous Dampers . In *General Guidelines for Engineers, Including a Brief History*, 3rd ed.; Taylor Devices Inc.: North Tonawanda, NY, USA, 2020.
48. Hwang, J.S.; Lin, W.C.; Wu, N.J. Comparison of distribution methods for viscous damping coefficients to buildings. *Struct. Infrastruct. Eng.* **2013**, *9*, 28–41. [CrossRef]
49. Whittle, J.K.; Williams, M.S.; Karavasilis, T.L.; Blakeborough, A. A comparison of viscous damper placement methods for improving seismic building design. *J. Earthq. Eng.* **2012**, *16*, 540–560. [CrossRef]
50. FEMA. *Commentary to NEHRP Guidelines for the Seismic Rehabilitation of Buildings*; FEMA 274; Building Seismic Safety Council, Federal Emergency Management Agency: Washington, DC, USA, 1997.
51. Taiyari, F.; Mazzolani, F.D.; Bagheri, S. Damaged-based optimal design of friction dampers in multistory chevron braced steel frames. *Soil Dyn. Earthq. Eng.* **2019**, *119*, 11–20. [CrossRef]
52. Del Gobbo, G.M. Placement of Fluid Viscous Dampers to Reduce Total-Building Seismic Damage. Ph.D. Thesis, University of Oxford, Oxford, UK, 2017.
53. Moreschi, L.M. Seismic Design of Energy Dissipation Systems for Optimal Structural Performance. Ph.D. Thesis, Virginia Polytechnic Institute and State University, Blacksburg, VA, USA, 2000.
54. Pall, A.; Pall, R.T. Performance-based design using Pall Friction Dampers—An economical design solution. In Proceedings of the 13th World Conference on Earthquake Engineering, Vancouver, CB, Canada, 1–6 August 2004.
55. Armali, M.; Damerji, H.; Hallal, J.; Fakih, M. Effectiveness of friction dampers on the seismic behavior of high rise building VS shear wall system. *Eng. Rep.* **2019**, *1*, e12075. [CrossRef]
56. Titirla, M.; Katakalos, K. Evaluation of an innovative passive mitigation device through experimental and numerical investigation. In Proceedings of the COMPDYN 2017, 6th ECCOMAS Thematic Conference on Computational Methods in Structural Dynamics and Earthquake Engineering, Rhodes Island, Greece, 15–17 June 2017.
57. Titirla, M.D.; Papadopoulos, P. Finite Element investigation of a new seismic energy absorption device through simultaneously yield and friction. In Proceedings of the COMPDYN 2015, 5th ECCOMAS Thematic Conference on Computational Methods in Structural Dynamics and Earthquake Engineering, Crete Island, Greece, 25–27 May 2015.

applied sciences

MDPI

Article

Reliability Analysis of Response-Controlled Buildings Using Fragility Curves

Ahmad Khalid Karimi, Edisson Alberto Moscoso Alcantara and Taiki Saito *

Department of Architecture and Civil Engineering, Toyohashi University of Technology, Toyohashi 441-8580, Japan; karimi.ahmad.khalid.wg@tut.jp (A.K.K.); moscoso.alcantara.edisson.alberto.eh@tut.jp (E.A.M.A.)
* Correspondence: saito.taiki.bv@tut.jp

Abstract: The number of buildings with passive control systems is steadily growing worldwide. For this reason, this study focuses on the reliability analysis of these systems employing fragility curves. The structural performance evaluation is obtained for a 10-story steel building with two different sections (trimmed and conventional). The trimmed section of the building was evaluated with hysteresis and oil dampers, while the conventional section of the building was evaluated without damper. The fragility curves were obtained from the incremental dynamic analysis using 20 ground motion records. Spectral acceleration response at the fundamental period of the building was considered and used as the intensity measure for the ground motion records. The maximum inter-story drift ratio of the building was employed as the damage measure. In addition, the seismic energy absorption rate was compared between hysteresis and oil dampers. As a result, hysteresis dampers were found to be more effective for high ground motion intensities. On the other hand, the oil damper dissipates energy immediately, even for low ground motion intensities. Furthermore, the combination of different types of dampers improved the seismic performance of the trimmed section of the building to almost the same level as the conventional section of the building. Eventually, a combination of hysteresis and oil dampers in a building is suggested to improve structural performance.

Keywords: response controlled systems; dampers' combination; fragility curves; incremental dynamic analysis

Citation: Karimi, A.K.; Moscoso Alcantara, E.A.; Saito, T. Reliability Analysis of Response-Controlled Buildings Using Fragility Curves. *Appl. Sci.* **2022**, *12*, 7717. https://doi.org/10.3390/app12157717

Academic Editor: Maria Favvata

Received: 20 June 2022
Accepted: 28 July 2022
Published: 31 July 2022

Publisher's Note: MDPI stays neutral with regard to jurisdictional claims in published maps and institutional affiliations.

Copyright: © 2022 by the authors. Licensee MDPI, Basel, Switzerland. This article is an open access article distributed under the terms and conditions of the Creative Commons Attribution (CC BY) license (https://creativecommons.org/licenses/by/4.0/).

1. Introduction

Earthquakes pose serious threats to life and infrastructure. The experience and knowledge gained through these events have improved our understanding of how to manage, mitigate and work towards the prevention of similar catastrophes. To reduce the impact of earthquakes on people and property, response-controlled systems are an advanced practice in managing the consequences of such disasters. In Japan, seismic response-controlled systems have been applied to almost all of the high-rise buildings constructed in the last several years in order to improve their structural safety and decrease damage sustained during seismic excitations [1]. Furthermore, these systems have been applied to rehabilitate the seismic resistance of existing structures. Different response-controlled techniques are presented, such as seismic isolation, dampers, and so on [2]. Moreover, researchers have combined different types of dampers to improve the seismic performance of structures. It is necessary to justify the adoption of a response-controlled system to building officials, owners, and users in terms of seismic performance during higher-intensity earthquakes.

There have been several studies on the performance of dampers and structures equipped with dampers. Hysteresis dampers, oil dampers, viscous dampers, and viscoelastic dampers are four major types of energy dissipation device [3]. The Japan Society of Seismic Isolation (JSSI) guidebook [4] for the design, fabrication, testing, quality control, and analytical modeling of various passive control systems was issued to cover the main concerns at all stages of design, manufacture, and construction of the above four major passive dampers. To improve the damping, stiffness, and strength characteresistics of the buildings, passive energy dissipation systems make use of a variety of materials and technologies. The dissipation may be achieved either by transferring energy into vibrating modes or by the conversion of kinetic energy into heat. The first mechanism consists of devices involving deformation of viscoelastic solids or fluids and those employing fluid orificing, while the latter group includes devices that operate on principles such as yielding of metals and frictional sliding. A third classification consists of re-centering devices that use either a preload generated by fluid pressurization or internal springs, or a phase transformation to produce a modified force–displacement response that includes a natural re-centering component [5].

Kam et al. [6] proposed a combination of various alternative energy dissipation elements (hysteretic, viscous, or visco-elastoplastic) in series and/or in parallel to self-centering elements and called it an advanced flag-shaped (AFS) system. They compared the seismic performance of AFS systems with that of the conventional system using a set of four single-degree of freedom systems under a suite of near-fault and far-field ground motions by performing nonlinear dynamic analyses. Chukka et al. [7] compared the seismic performance of an X-shaped metallic damper (XMD) and a fluid viscous damper (FVD) by analyzing five-, eight-, and ten-story reinforced concrete buildings without dampers, with XMD, and with FVD under eight different earthquake ground motions. They also discussed how the locations of the dampers affected the seismic response of the structure.

To examine the performance of a structure with passive dampers, seismic fragility analysis is commonly used, where the probability of reaching or exceeding a specified limit state of damage measure (DM) is calculated as a function of a specified intensity measure (IM) in a structure. For a specific limit state of damage, several IMs can be achieved under multiple earthquake excitations, and collectively they are referred to as the multi-recorded IM cluster. The IM cluster is a random function, and the average and standard deviation can be obtained by calculating the mathematical characteristic values of multiple IM values. According to [8], it is assumed that the conditional probability of DM to IM satisfies the lognormal distribution. This assumption is based on large amounts of data statistics experience with existing research results on engineering structures, and the advantage of this assumption is that the logarithm of DM, (ln(DM)), and the logarithm of IM, (ln(IM)), are linear in the logarithmic coordinate. Del Gaudio et al. [9] obtained the seismic fragility curves using lognormal distribution function to analyze 250 reinforced concrete structures in L'Aquila that were subjected to the 6 April 2009 L'Aquila earthquake and came up with a damage scenario that matched the real one.

Fragility curves are obtained using analytical methods such as Incremental Dynamic Analysis (IDA). The IDA is widely used to capture the overall seismic performance of structures. Alternative procedures for IDA include procedures such as Multiple-Stripe Analysis (MSA, [10]) and Cloud Analysis (CA, [11]), where MSA involves performing a series of nonlinear dynamic analyses at specific intensity levels, whereas CA aims at deriving IDA-based fragility curves by choosing ground motion intensities strategically in order to minimize the amount of scaling. CA requires linear regression predictions based on the results of the structural analysis of the un-scaled records to identify the range of intensity values near demand to capacity ratios equal to unity prior to performing actual CA. These nonlinear dynamic analyses are used to characterize the relationship between DMs and IMs and to perform fragility assessments based on recorded ground motions. Vamvatsikos et al. [12] simplified and standardized IDA's general procedure, creating a strong foundation for its future implementations. IDA is a way of subjecting analytical models of structures to a suite of Ground Motion Records (GMRs), and each GMR is scaled to several intensity levels designed to force the structure from the elastic range to the nonlinear range. IDA can analyze a structure's seismic performance, from elasticity through plasticity to collapse. IDA includes the selection of appropriate IM and DM, selection of a suitable and adequate number of ground motion intensities, and appropriate scaling of GMRs for higher intensities to cover the entire range of structural responses by a series of scale factors [12,13]. Then, it is possible to obtain the IDA curves that describe the relationship between the IM of the GMRs and the DM of the structure.

The selection of appropriate IM and DM is the first important step in performing IDA. A sufficient and efficient IM should be selected based on depicting a good correlation with the DM of choice, possessing low dispersion, and predicting a relatively better structural response using a relatively low number of GMRs [8]. The peaks of the ground acceleration, velocity, and displacement signals (PGA, PGV, and PGD), as well as the spectral acceleration value $Sa(T_1)$ corresponding to the fundamental period of a structure, are now used as typical ground motion IMs. Considering that a structure's displacement response, $Sa(T_1)$, is the most efficient IM, since it allows records to be selected regardless of magnitude, distance, or duration, as well as predicting the response with less uncertainty [14]. It is known that the spectral acceleration, $Sa(T_1)$, has a good correlation with the seismic damage measures in first-mode-dominated structures [15]. However, for taller or asymmetric structures where higher modes become important, improved IM alternatives should be sought [14]. On the other hand, the maximum inter-story drift ratio, θmax (hereafter referred to as story drift ratio) appears to be an effective damage measure index [14]. Mazza et al. [16] studied the predictive ability of nine spectral IMs for base-isolated structures subjected to near-fault earthquakes using three engineering demand parameters. Asgarian et al. [17] used IDA to study the Tehran communications tower with various types of DMs and IMs, concluding that $Sa(T_1)$ is more efficient than PGA. In addition, during IDA-based seismic fragility analysis, the story drift ratio is also used as the quantitative index to separate the performance levels. The performance-based guidelines established by the Japan Structural Consultant Association (JSCA) [18] can be used to identify structural damage levels based on the story drift ratio.

The second important step in performing IDA is the selection of appropriate GMRs possessing a minimum of scatter in their structural response. The selection of a proper suite of GMRs to reliably predict the limit-state capacity of buildings is still challenging due to the lack of a solid framework. Several methodologies for selecting suitable GMRs have been suggested. Most of these studies were looking for GMRs that were well-matched to the target spectrum, including in terms of design, uniform hazard, and conditional mean spectrums. Due to the uncertainty of earthquake excitation, a sufficient number of ground motions should be selected for IDA to accurately assess the seismic performance of the structure, and the selected earthquake records should cover the strongest earthquake action that the structure may suffer in the future. A set of ten to twenty records is usually enough to provide sufficient accuracy in the estimation of seismic demands, assuming a relatively efficient IM, like $Sa(T_1)$, for mid-rise buildings [19]. Each ground motion record needs to be appropriately scaled for higher intensities to cover the entire range of structural responses by a series of scale factors [13].

Since there is no efficient method to compare the structural performance of damper-equipped building, this paper employs a novel method and performs original work on the reliability analysis of response-controlled buildings using fragility curves. To evaluate and compare the seismic performance and reliability of damper-equipped buildings, this study analyzes the seismic performance of two major damper types (hysteresis and oil dampers) in a 10-story moment-resisting steel structure, and their fragility curves are then compared. Additionally, the efficiency of integrated systems was also investigated by combining the dampers in the building. The building models were analyzed using the frame analysis software STERA 3D [20], developed by one of the authors, and their dynamic responses were obtained. $Sa(T_1)$ was chosen as the IM since the building models used in this work are the first-mode-dominated structures. On the other hand, the story drift ratio was used as the DM. The story drift ratio is employed as a quantitative measure to distinguish structural performance levels during IDA-based seismic fragility assessments. To properly portray an earthquake scenario, a collection of 20 distinct Japanese GMRs that a building might face during its lifetime was selected. The IDA curves for each model were generated by analyzing the models for all 20 GMRs using the STERA 3D software. From the IDA curves, fragility curves for the story drift of the JSCA damage levels were obtained. Furthermore, energy responses were compared between the buildings with different types of dampers.

2. IDA Pre-Requirements (Target Building, Dampers, and Input Earthquakes)

2.1. Target Buildings Description and Configurations

A 10-story moment-resisting steel building was selected from the JSSI theme structures [21] to examine the performance of dampers. The building was designed with two different steel sections, namely the Trimmed Section (TS) and the Conventional Section (CS). The buildings have a rigid frame structure in both directions. A description of building models with different damper arrangements is given in Table 1. The columns of the buildings have a square box-shaped cross-section, while the beams are H-shaped. The dimensions of the structural columns listed in Table 2 represent the height, width, and thickness (H × B × t) of the column sections, while the dimensions listed in Table 3 represent the beam height, flange width, web thickness, and flange thickness (H × B × t_1 × t_2) of the structural beams, respectively. The TS building was deliberately designed to be weaker than the CS building to see the effect of the dampers. The steel was SN490B, which has yield and tensile strengths of 325 and 490 MPa, respectively [21]. The building models are first-mode-dominated structures, and the vibration effects of the other modes are minor, since more than 80% of mass participation is from the first mode of the buildings, as shown in Table 1. The STERA 3D software was used to model and analyze the target buildings. The beam element was modeled as a nonlinear bending spring, while the column element was modeled as a nonlinear vertical spring [22]. The analysis was performed in the longitudinal direction of the buildings only; hence, hysteresis and oil dampers were arranged in the longitudinal direction of the TS building, as shown in Figure 1. The top floor has dampers only in the center. The plan, elevation, and damper arrangement are shown in Figure 1a, while Figure 1b represents the 3D model of the building. The dimensions of structural columns and beams are summarized in Tables 2 and 3, respectively.

Table 1. Description of building models with different damper arrangements.

Concise Name	Model Description	Fundamental Period (T_1)	Mass Participation of First Mode
10F_TS_FM	10F TS Frame Model	2.018 s	82.8%
10F_TS_HM	10F TS Hysteresis Model	1.356 s	80%
10F_TS_OM	10F TS Oil Model	2.018 s	82.8%
10F_TS_HOHM	10F TS Hysteresis-Oil-Hysteresis Model	1.501 s	80.5%
10F_TS_OHOM	10F TS Oil-Hysteresis-Oil Model	1.685 s	81.7%
10F_CS_FM	10F CS Frame Model	1.323 s	84.2%

Table 2. Structural column dimensions (mm).

Story	Interior Column		Exterior Column		Corner Column	
	TS (H × B × t)	CS (H × B × t)	TS (H × B × t)	CS (H × B × t)	TS (H × B × t)	CS (H × B × t)
9–10	350 × 350 × 25	550 × 550 × 22	350 × 350 × 25	500 × 500 × 22	350 × 350 × 16	500 × 500 × 19
8	400 × 400 × 25	550 × 550 × 22	350 × 350 × 25	500 × 500 × 22	350 × 350 × 16	500 × 500 × 19
7	400 × 400 × 28	550 × 550 × 22	350 × 350 × 28	500 × 500 × 22	350 × 350 × 16	500 × 500 × 19
5–6	450 × 450 × 25	600 × 600 × 28	400 × 400 × 25	550 × 550 × 25	400 × 400 × 19	550 × 550 × 22
4	450 × 450 × 28	600 × 600 × 28	400 × 400 × 25	550 × 550 × 25	400 × 400 × 19	550 × 550 × 22
3	500 × 500 × 28	650 × 650 × 28	450 × 450 × 25	600 × 600 × 25	450 × 450 × 19	600 × 600 × 22
2	500 × 500 × 28	650 × 650 × 28	450 × 450 × 25	600 × 600 × 25	450 × 450 × 19	600 × 600 × 22
1	500 × 500 × 36	650 × 650 × 28	450 × 450 × 36	600 × 600 × 28	450 × 450 × 28	600 × 600 × 25

Table 3. Structural beam dimensions (mm).

Story	Trimmed Section (TS)			
	Longitudinal Direction (X)		Transverse Direction (Y)	
	Interior Beam ($H \times B \times t_1 \times t_2$)	Exterior Beam ($H \times B \times t_1 \times t_2$)	Short Span ($H \times B \times t_1 \times t_2$)	Long Span ($H \times B \times t_1 \times t_2$)
R	450 × 200 × 9 × 16	450 × 200 × 9 × 12	450 × 300 × 16 × 28	450 × 350 ×16 ×32
10	450 × 300 × 9 × 16	450 × 200 × 12 × 19	450 × 300 × 12 × 19	450 × 300 × 16 × 28
9	500 × 300 × 12 × 19	500 × 300 × 9 × 16	500 × 300 × 12 × 25	500 × 300 × 16 × 32
8	500 × 350 × 12 × 19	500 × 300 × 12 × 19	500 × 300 × 12 × 25	500 × 300 × 16 × 32
7	500 × 350 × 12 × 22	500 × 300 × 12 × 22	500 × 350 × 12 × 25	500 × 350 × 16 × 32
6	500 × 350 × 12 × 22	500 × 300 × 12 × 22	500 × 350 × 16 × 28	500 × 350 × 16 × 32
5	500 × 350 × 16 × 25	500 × 300 × 16 × 25	500 × 350 × 16 × 28	500 × 350 × 16 × 36
4	500 × 350 × 16 × 28	500 × 300 × 16 × 25	500 × 350 × 16 × 32	500 × 350 × 16 × 36
3	500 × 350 × 16 × 28	500 × 300 × 16 × 25	500 × 350 × 16 × 32	500 × 350 × 16 × 36
2	500 × 350 × 16 × 32	500 × 300 × 16 × 28	500 × 350 × 16 × 36	500 × 350 × 16 × 36
	Conventional Section (CS)			
	Longitudinal Direction (X)		Transverse Direction (Y)	
Story	Interior Beam ($H \times B \times t_1 \times t_2$)	Exterior Beam ($H \times B \times t_1 \times t_2$)	Short Span ($H \times B \times t_1 \times t_2$)	Long Span ($H \times B \times t_1 \times t_2$)
R	600 × 300 × 12 × 22	600 × 250 × 12 × 22	600 × 300 × 14 × 25	600 × 300 × 14 × 32
10	600 × 300 × 12 × 22	600 × 250 × 12 × 22	600 × 300 × 14 × 25	600 × 300 × 14 × 32
9	700 × 300 × 12 × 22	700 × 250 × 12 × 22	700 × 300 × 14 × 25	700 × 300 × 16 × 32
8	700 × 300 × 12 × 22	700 × 250 × 12 × 22	700 × 300 × 14 × 25	700 × 300 × 16 × 32
7	750 × 300 × 16 × 25	750 × 250 × 14 × 25	750 × 300 × 16 × 28	750 × 300 × 16 × 32
6	750 × 300 × 16 × 25	750 × 250 × 14 × 25	750 × 300 × 16× 28	750 × 300 × 16 × 32
5	750 × 300 ×16 × 28	750 × 250 × 16 × 28	750 × 350 × 16 × 28	750 × 350 × 16 × 32
4	750 × 300 × 16 × 28	750 × 250 × 16 × 28	750 × 350 × 16 × 28	750 × 350 × 16 × 32
3	750 × 300 × 16 × 28	750 ×250 × 16 × 28	750 × 350 × 16 × 28	750 × 350 × 16 × 32
2	800 × 300 × 16 × 32	800 × 300 × 16 × 28	800 × 300 × 16 × 32	800 × 300 × 16 × 32

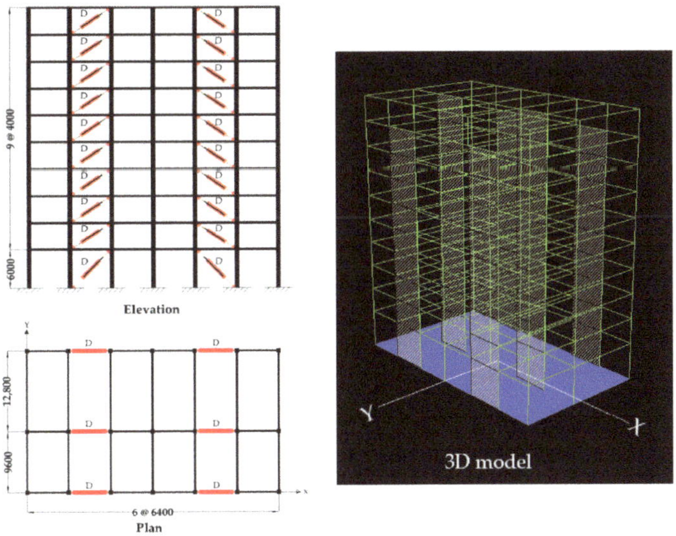

(a) (b)

Figure 1. Target building (dimensions in mm): (**a**) plan, elevation, and damper configuration, (**b**) 3D model.

2.2. Passive Dampers

The current study uses two types of dampers: hysteresis and oil dampers. A hysteresis damper is a deformation-dependent damper, such as the buckling restrained brace [22]. It consists of steel as a damping material to absorb vibration energy by means of its plastic deformation [23]. The hysteresis damper is modeled as a shear spring, as shown in Figure 2a, with a bi-linear force deformation relationship, as shown in Figure 2b [22]. An oil damper is a velocity-dependent damper that uses the orifice flow resistance mechanism. It consists of a low-viscosity oil in a cylindrical tube configuration. The force–velocity relationship of the oil damper normally appears as an ellipse hysteresis shape [24]. The bi-linear oil damper is modeled as a shear spring, including the Maxwell model with an elastic spring with stiffness, K_D, and a dashpot with bilinear type damping coefficient, C, connected in series [25]. The element model of the bi-linear oil damper is shown in Figure 3a. The force–velocity relationship of the dashpot is shown in Figure 3b. The technical parameters of both the hysteresis and oil dampers were calculated by Prof. Kasai [4] and are given in Table 4.

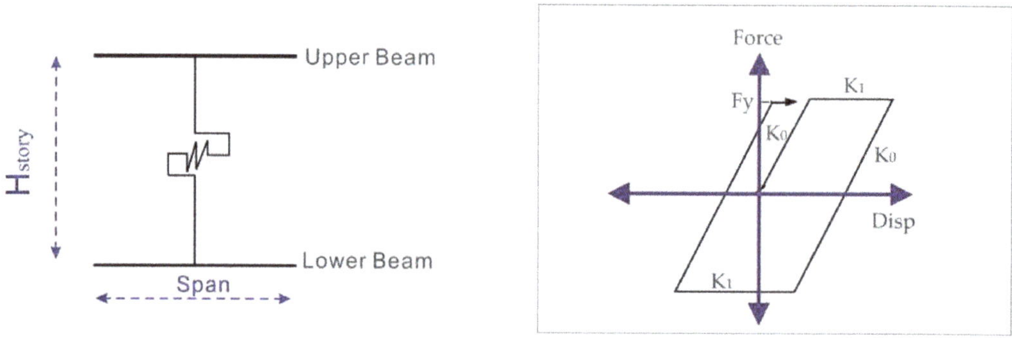

(a) (b)

Figure 2. Hysteresis damper: (**a**) element model, (**b**) bi-linear force–displacement relationship.

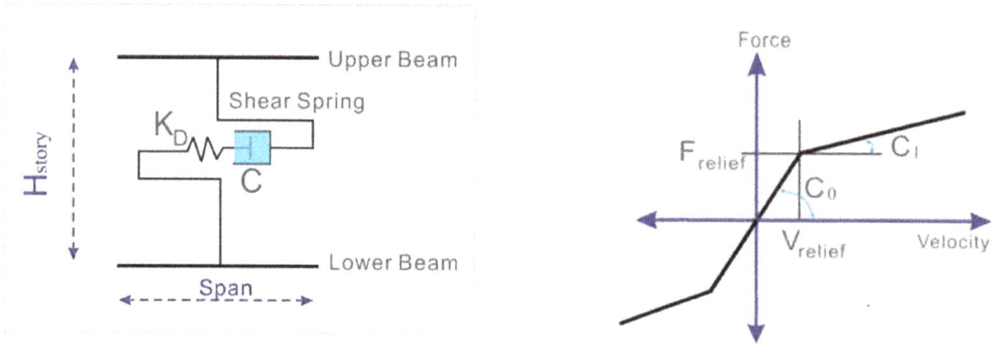

(a) (b)

Figure 3. Bi-linear type oil damper: (**a**) element model, (**b**) force–velocity relationship of oil damper.

Table 4. Technical parameters of hysteresis and oil dampers.

Story	Height (H)	Hysteresis Damper			Oil Damper			
		Damper Stiffness (K_D)	Force (Fy)	Stiffness Ratio (K_1/K_0)	Damper Stiffness (K_D)	Damping (C_0)	Relief Velocity (V_r)	Damping Ratio C_1/C_0
	m	kN/mm	kN		kN/mm	kN-s/mm	cm/s	
10	4.0	16.82	112.17	0.02	27.30	5.67	38.6	0.02
9	4.0	56.03	373.5	0.02	31.00	6.45	38.6	0.02
8	4.0	74.52	496.83	0.02	37.92	7.88	38.6	0.02
7	4.0	96.83	645.5	0.02	42.13	8.77	38.6	0.02
6	4.0	98.52	656.67	0.02	50.23	10.45	38.6	0.02
5	4.0	116.65	777.67	0.02	52.70	10.95	38.6	0.02
4	4.0	124.6	830.67	0.02	56.50	11.75	38.6	0.02
3	4.0	105.73	704.83	0.02	65.93	13.70	38.6	0.02
2	4.0	118.15	787.67	0.02	66.02	13.72	38.6	0.02
1	6.0	67.93	679.33	0.02	48.18	10.02	57.9	0.02

2.3. Ground Motion Record (GMR) Selection

The selection of a proper and adequate suite of GMRs is highly important when conducting the IDA of the building. Japan is located along the subduction zone, with numerous active faults. The target buildings are designed in accordance with Japanese standards, so the collection of 20 distinct Japanese GMRs presented in Table 5 were selected, with a moment magnitude scale ranging from 6 to 9 Mw, and an epicenteral depth ranging from 6 to 66 km, consisting of both near-field and far-field earthquakes. GMRs were collected from different sites to have a wide range of spectral intensity. The acceleration response spectra for all of the GMRs are given in Figure 4. They were used as the input ground acceleration in the longitudinal direction of the building. To obtain the IDA curves of the buildings, analyses were performed by gradually increasing the intensities of GMRs until the required damage levels were obtained.

Table 5. Summary of input earthquake ground motions.

	Record Name	Depth (km)	Magnitude (Mw)	Duration (s)	PGA (g)
1	JP_2000_Tottori	11	7.3	240	0.616
2	JP_2011_Tohoku	24	9	300	0.572
3	JP_2008_Iwate	8	7.2	300	0.701
4	JP_2007_Noto	11	6.9	300	0.864
5	JP_1997_Kagoshima	8	6.3	70	0.503
6	JP_2016_Kumamoto	12	7.3	300	0.630
7	JP_2003_Hokkaido	42	8	300	0.494
8	JP_2003_Miyagi	42	8	300	0.737
9	JP_2005_Miyagi	42	7.2	162	0.406
10	JP_2018_Osake	13	6.1	124	0.812
11	JP_2020_Hachinohe	35	6.3	156	0.326
12	JP_2009_Shizuoca	23	6.5	219	0.451
13	JP_1995_Kobe	17.9	6.9	50	0.630
14	JP_2021_Fukushima	55	7.3	300	0.528
15	JP_2004_Chuetsu	13	6.8	299	1.335
16	JP_2001_Geiyo	51	6.4	193	0.566
17	JP_2021_Miyagi	59	6.9	300	0.623
18	JP_2003_Hokkaido	42	8	300	0.414
19	JP_2011_Miyagi	142	7.1	193	0.428
20	JP_2011_Fukushima	39	7	194	0.415

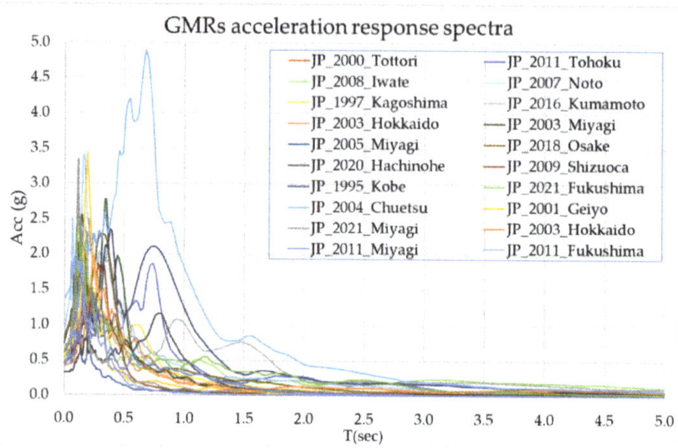

Figure 4. GMR acceleration response spectra (5% damping).

3. Fragility Curve Based on Incremental Dynamic Analysis

3.1. Intensity Measure (IM) and Damage Measure (DM) Selection

Since the building models employed in this study are first-mode-dominated structures, $Sa(T_1)$ is a better choice as the IM. On the other hand, the story drift ratio θmax was used as the DM index. The θmax is employed as a quantitative measure to distinguish performance levels during IDA-based seismic fragility assessments. The performance-based guidelines established by JSCA [18] were used as a reference to select the structural damage from three different levels as shown in Table 6. The selection of damage levels such as 1/200 shows minor damage, while 1/100 and 1/50 represent significant damage and collapse respectively.

Table 6. Structural damage levels based on the story drift ratio.

Structural Damage	No Damage	Minor Damage	Significant Damage	Severe Damage	Collapse
Story drift (θmax)	θmax ≤ 1/300	1/300 < θmax ≤ 1/150	1/150 < θmax ≤ 1/100	1/100 < θmax ≤ 1/75	θmax > 1/75

3.2. Scale Factors

The models were analyzed by gradually increasing the spectral acceleration $Sa(T_1)$ of the GMRs until the required story drift ratios were obtained. To achieve 1/50 (2%) θmax, the $Sa(T_1)$ increased from 0.01 to 1.5 (g) with an incremental step of 0.05 (g), where g is the gravity acceleration (9.8 m/s^2).

3.3. Applying IDA

IDA was applied to all six models using STERA 3D software [20]. The analysis was performed at each incremental step of ground motion intensity. A total number of 3600 analyses were performed for all six models using 20 GMRs commencing from 0.01 g with an incremental step of 0.05 g and stopping at 1.5 g. IDA graphs were obtained for each model, representing the θmax values as a function of increasing intensities of Sa (T_1). The IDA curves for the TS buildings are shown in Figure 5a–c, while the IDA curves for the CS building (10F_CS_FM) are shown in Figure 5d.

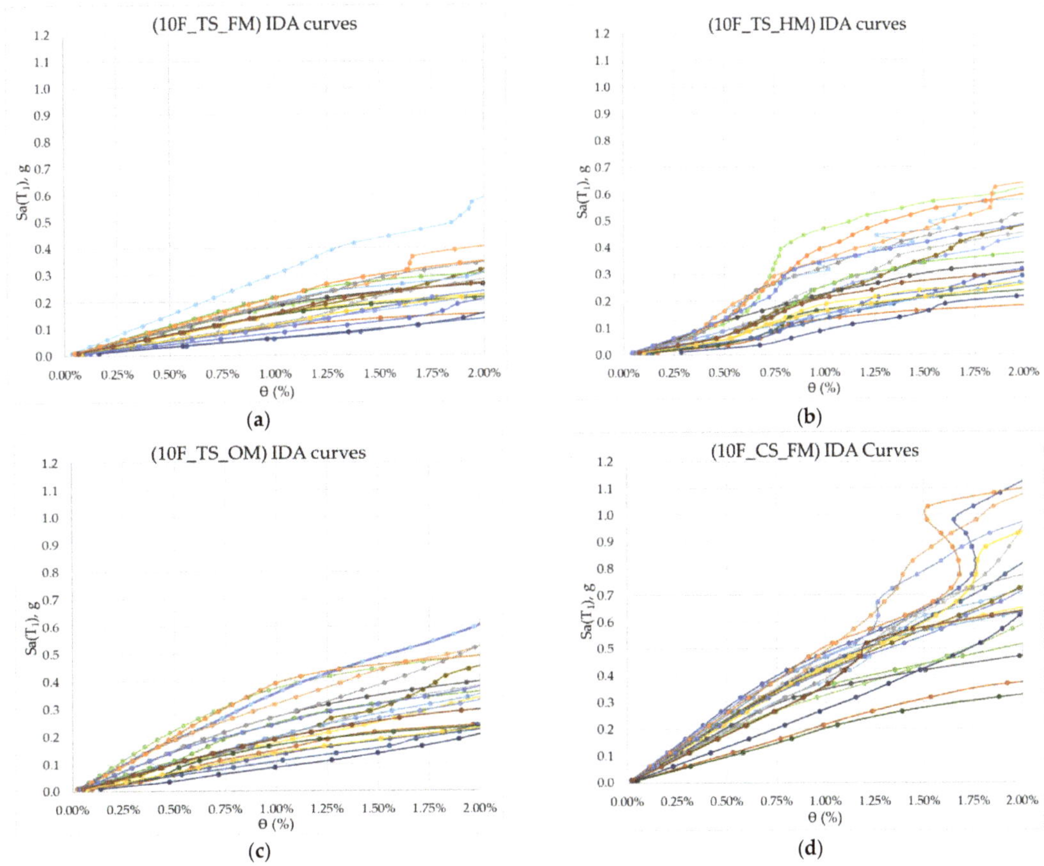

Figure 5. IDA curves for models: (**a**) 10F_TS_FM, (**b**) 10F_TS_HM, (**c**) 10F_TS_OM, (**d**) 10F_CS_FM.

3.4. Fragility Curves

Analyzing the models with 20 GMRs provides a bunch of discrete points on an IM_DM plot. The IM values corresponding to 0.5, 1.0, and 2.0% story drift were obtained using interpolation.

By using the estimated mean and standard deviation values of IM, the fragility curve was obtained as the lognormal cumulative distribution function expressed by Equation (1), as follows:

$$P(DM \geq DMco) = \varnothing\left(\frac{lnX - \mu lnX}{\sigma lnX}\right) \quad (1)$$

where Φ is the standard cumulative distribution function, lnX is the natural logarithm of the variable $X(\mathrm{Sa}(\mathrm{T}_1))$, and μlnX and σlnX are the mean and the standard deviation of the natural logarithm of X, respectively.

The lognormal distribution of IM ($\mathrm{Sa}(\mathrm{T}_1)$) corresponding to 0.5, 1.0, and 2.0% story drift are presented in Figure 6a for the TS building and in Figure 6b for the CS building. The mean (μ) and standard deviation (σ) of $\mathrm{Sa}(\mathrm{T}_1)$ for the required levels of damage are given in Table 7.

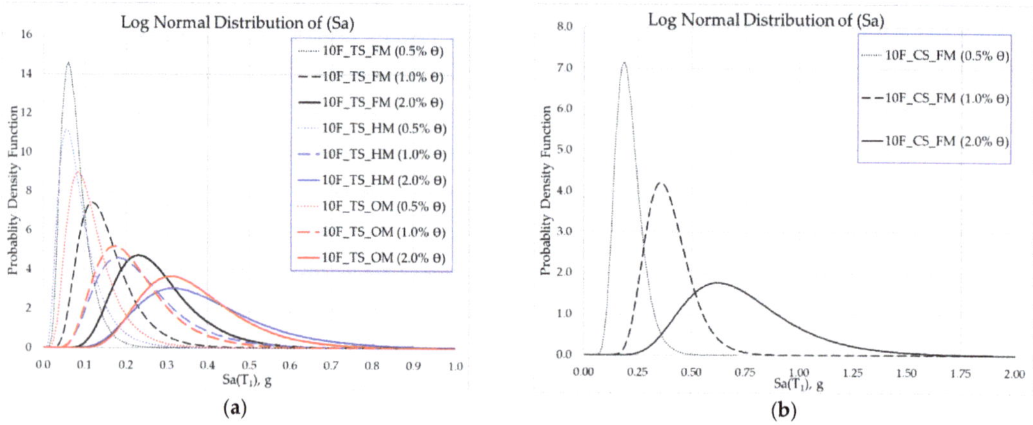

Figure 6. Lognormal distribution of Sa(T_1): (**a**) trimmed section (**b**) conventional section.

Table 7. Mean (μ) and standard deviation (σ) values.

θmax	10F_TS_FM		10F_TS_HM		10F_TS_OM		10F_CS_FM	
	μ	σ	μ	σ	μ	σ	μ	σ
0.5%	−2.645	0.421	−2.572	0.541	−2.242	0.466	−1.606	0.290
1.0%	−1.961	0.416	−1.517	0.434	−1.584	0.408	−0.965	0.257
2.0%	−1.348	0.346	−1.000	0.383	−1.056	0.332	−0.358	0.343

The fragility curves for the TS building are presented in Figure 7. The lower the fragility curve, the less the probability of damage. For instance, considering 0.5% story drift, the probability of damage is higher for the hysteresis damper model (10F_TS_HM (0.5%θ)) than for the oil damper model (10F_TS_OM (0.5%θ)) for any value of Sa(T_1). On the other hand, for 2% story drift, the probability of damage seems to be higher for the model equipped with oil dampers (10F_TS_OM (2.0%θ)) compared to that equipped with hysteresis dampers (10F_TS_HM (2.0%θ)).

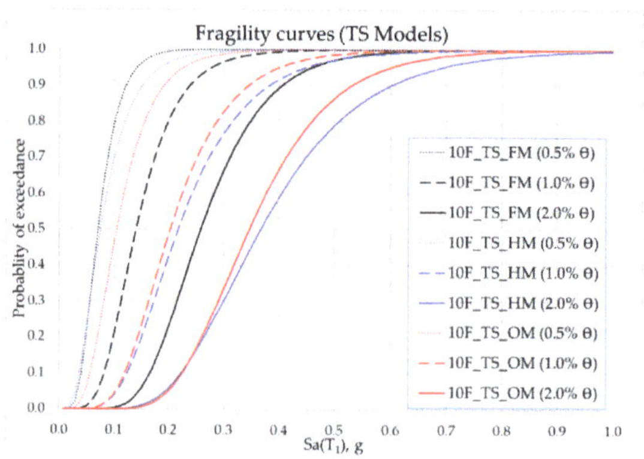

Figure 7. Fragility curves of trimmed section models.

3.5. Response Energy of the Dampers

Hysteretic dampers dissipate energy through the inelastic deformation of metallic substances. On the other hand, oil dampers use the orifice flow resistance mechanism as a response to energy. For nonlinear systems, the equation of motion can be expressed as [22]:

$$[M]\{\ddot{u}\} + [c]\{\dot{u}\} + Q(u,\dot{u}) = -[M][u]\begin{Bmatrix} \ddot{x}_0 \\ \ddot{Y}_0 \\ \ddot{Z}_0 \end{Bmatrix} = \{p\} \tag{2}$$

where $[u]$ is the displacement vector, $[M]$ and $[C]$ are the mass and damping matrices, $\{u\}$ and $\{p\}$ are the displacement and external force vectors, respectively, and $Q(u,\dot{u})$ is the nonlinear restoring force vector.

The equation of energy can be derived by multiplying the velocity vector, $\{\dot{u}\}^T$, and integrating by the time range [0–t]:

$$W_K + W_D + W_P = W_I \text{ Equation of energy} \tag{3}$$

$$W_K = \frac{\{\dot{u}\}^T[M]\{\dot{u}\}}{2} \text{ Kinematic energy} \tag{4}$$

$$W_D = \int_0^t \{\dot{u}\}^T[C]\{\dot{u}\}dt \text{ Damping energy} \tag{5}$$

$$W_P = \int_0^t \{\dot{u}\}^T Q(u,\dot{u})dt \text{ Potential energy} \tag{6}$$

$$W_I = \int_0^t \{\dot{u}\}^T\{P\}dt \text{ Input energy} \tag{7}$$

The STERA 3D software also [20] calculates the energy absorption rate of the structural members and the dampers as shown in Equation (3). The average rate of energy absorption by hysteresis and oil dampers with increasing values of ground motion intensity was calculated by the software as shown in Figure 8. It can be seen that the hysteresis damper's energy absorption rate gradually increases with increasing ground motion intensity and decreases after reaching its maximum absorption capacity. On the other hand, oil dampers store energy that is significant for smaller intensities of ground motion, but their capacity will decreases at higher intensities.

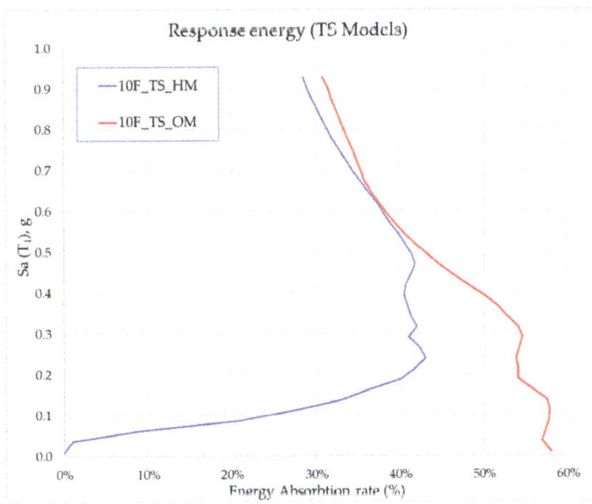

Figure 8. Energy response of hysteresis and oil dampers for TS models.

4. Combination of Different Types of Dampers

4.1. Combination of Hysteresis and Oil Dampers

To further improve the seismic performance of the damper-equipped models and to use the results of the fragility curves and the response energy of the hysteresis and oil dampers, two different combinations of hysteresis and oil dampers were considered. The first combination was dubbed the Hysteresis–Oil–Hysteresis Model (HOHM), because the hysteresis dampers in the center portion of the hysteresis model were replaced with the same number of oil dampers, while the hysteresis dampers in the two outer sections of the model remained the same as shown in Figure 9a. In contrast, the latter combination had the reverse arrangement, namely, the Oil–Hysteresis–Oil Model (OHOM), as shown in Figure 9b.

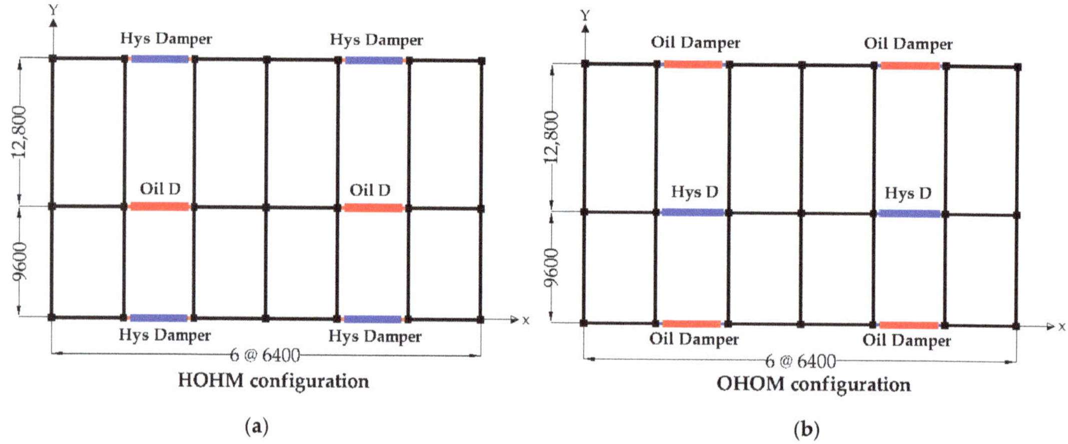

Figure 9. New combinations (dimensions in mm): (**a**) Hysteresis–Oil–Hysteresis Model; (**b**) Oil–Hysteresis–Oil Model.

4.2. Performance Evaluation of Combined Dampers Models

The same analysis was performed for the combined dampers models as for the individual damper-equipped models, and their fragility curves were compared with each individual damper-equipped model, as shown in Figure 10. Figure 10a–c show the probability of exceeding 0.5, 1.0, and 2.0% θmax, respectively, for all TS buildings. The 10F_TS_HOHM has the highest seismic performance for all performance levels, as seen in Figure 10a–c.

In addition, the TS building equipped with a Hysteresis–Oil–Hysteresis damper (10F_TS_HOHM) configuration has almost the same performance as the CS building (10F_CS_FM), as shown in Figure 10d. Therefore, this configuration of dampers in a TS building can significantly improve the seismic performance equivalent to the CS building.

Moreover, the average rate of energy absorption with increasing values of ground motion intensity was obtained for all of the models that included the combination models, and these values were then compared, as shown in Figure 11. It appears that the combined models use a mixture of the properties of hysteresis and oil dampers, and absorb energy more consistently for all levels of ground motion intensity.

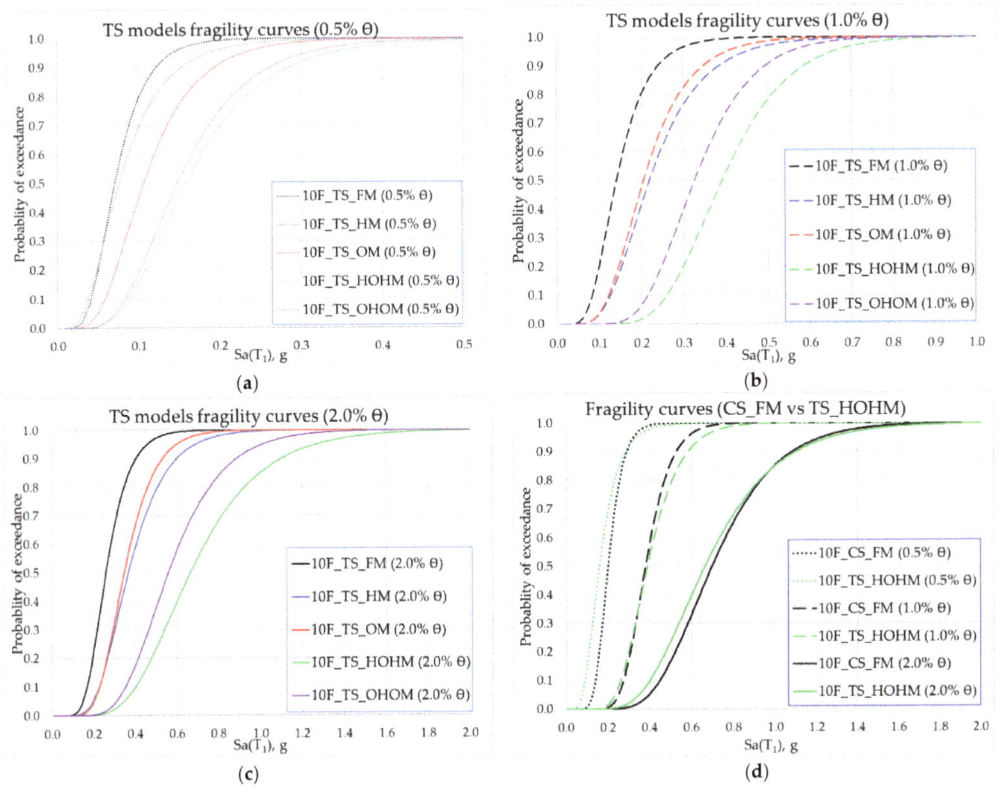

Figure 10. Fragility curves of all models: (**a**) TS (0.5%) θ, (**b**) TS (1.0%) θ, (**c**) TS (2.0%) θ, (**d**) 10F_CS_FM vs. 10F_TS_HOHM.

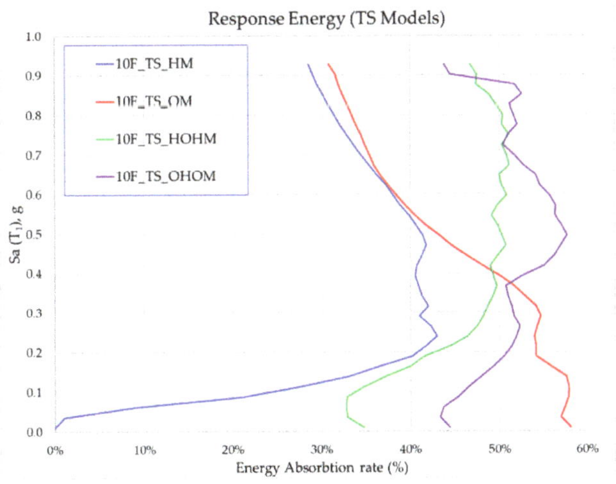

Figure 11. Energy response of all TS models.

5. Conclusions

Since there was no efficient method for comparing the performance of damper-equipped buildings, this paper employed a novel method and performed original work on the reliability analysis of response-controlled buildings using fragility curves. The building models were 10-story steel buildings consisting of two different sections (trimmed and conventional), and hysteresis and oil dampers were installed in the trimmed section of the building. The building models were created and analyzed with the use of a single type of damper and with a combination of different types of dampers. Moreover, the combination of hysteresis and oil dampers proposed in this paper significantly improved the seismic performce of the buildings. The following points are the main conclusions of this study:

- The IDA is an effective method that can accurately predict the fragility assessment and reliability of response-controlled buildings.
- $Sa(T_1)$ is an efficient IM, since it has a good correlation with the seismic damage measures of first-mode-dominated structures. On the other hand, the story drift ratio θmax appeared to be an effective DM index. The lognormal cumulative distribution function provides a good understanding of damage probability with respect to the intensity of an earthquake at the certain damage level.
- Hysteresis and oil dampers effectively improved the seismic performance of the TS building.
- On the basis of the fragility curves, hysteresis dampers are more effective for larger ground motion accelerations than oil dampers.
- Oil dampers dissipated energy before hysteresis dampers for small deformations.
- Combined hysteresis and oil dampers in the TS buildings exhibited better performance than the individual hysteresis and oil damper buildings.
- The TS building equipped with a Hysteresis–Oil–Hysteresis damper configuration has almost the same performance as the CS building.

Author Contributions: Data curation, E.A.M.A.; Investigation, A.K.K. and T.S.; Methodology, A.K.K., E.A.M.A. and T.S.; Writing—original draft, A.K.K.; Writing—review & editing, T.S. All authors have read and agreed to the published version of the manuscript.

Funding: This research received no external funding.

Informed Consent Statement: Not applicable.

Data Availability Statement: The data presented in this study are available on request from the corresponding author.

Conflicts of Interest: The authors declare no conflict of interests.

References

1. Ikeda, Y.; Yamamoto, M.; Furuhashi, T.; Kurino, H. Recent research and development of structural control in Japan. *Jpn. Archit. Rev.* **2019**, *2*, 219–225. [CrossRef]
2. Marko, J.; Thambiratnam, D.; Perera, N. Influence of damping systems on building structures subject to seismic effects. *Eng. Struct.* **2004**, *26*, 1939–1956. [CrossRef]
3. Kasai, K.; Kibayashi, M. JSSI manual for building passive control technology part 1. In Proceedings of the 13th World Conference on Earthquake Engineering Vancouver, Vancouver, BC, Canada, 1–6 August 2004; p. 2989.
4. JSSI. *Manual for Design and Construction of Passively-Controlled Buildings*, 3rd ed.; Japan Society of Seismic Isolation: Tokyo, Japan, 2013. (In Japanese)
5. Kubilay, K. Seismic Base Isolation And Energy Absorbing Devices. *Eur. Sci. J. ESJ* **2013**, 42–48. [CrossRef]
6. Kam, W.Y.; Pampanin, S.; Palermo, A.; Carr, A.J. Self-centering structural systems with combination of hysteretic and viscous energy dissipations. *Earthq. Eng. Struct. Dyn.* **2010**, *39*, 1083–1108. [CrossRef]
7. Chukka, N.D.K.R.; Krishnamurthy, M. Comparison of X-shaped metallic dampers with fluid viscous dampers and influence of their placement on seismic response of the building. *Asian J. Civ. Eng.* **2019**, *20*, 869–882. [CrossRef]
8. Nicolas Luco, C.A.C. Structure-Specific Scalar Intensity Measures for Near-Source and Ordinary Earthquake Ground Motions. *Earthq. Spectra* **2007**, *23*, 357–392. [CrossRef]
9. Del Gaudio, C.; Ricci, P.; Verderame, G.M.; Manfredi, G. Urban-scale seismic fragility assessment of RC buildings subjected to L'Aquila earthquake. *Soil Dyn. Earthq. Eng.* **2017**, *96*, 49–63. [CrossRef]

10. Jalayer, F.; Cornell, C.A. Alternative non-linear demand estimation methods for probability-based seismic assessments. *Earthq. Eng. Struct. Dyn.* **2009**, *38*, 951–972. [CrossRef]
11. Miano, A.; Jalayer, F.; Ebrahimian, H.; Prota, A. Cloud to IDA: Efficient fragility assessment with limited scaling. *Earthq. Eng. Struct. Dyn.* **2018**, *47*, 1124–1147. [CrossRef]
12. Vamvatsikos, D.A.C. Incremental dynamic analysis. *Earthq. Eng. Struct. Dyn.* **2002**, *31*, 491–514. [CrossRef]
13. Vamvatsikos, D.; Cornell, C.A. The incremental dynamic analysis and its application to performance-based earthquake engineering. In Proceedings of the 12th European Conference on Earthquake Engineering, London, UK, 9–13 September 2002; p. 479.
14. Iunio Iervolino, G.M. A review of ground motion record selection strategies for dynamic structural analysis. In *Modern Testing Techniques for Structural Systems*; Springer: New York, NY, USA, 2008; Volume 502, pp. 131–163.
15. Cheng, Y.; Dong, Y.-R.; Bai, G.-L.; Wang, Y.-Y. IDA-based seismic fragility of high-rise frame-core tube structure subjected to multi-dimensional long-period ground motions. *J. Build. Eng.* **2021**, *43*, 102917. [CrossRef]
16. Mazza, F.; Labernarda, R. Structural and non-structural intensity measures for the assessment of base-isolated structures subjected to pulse-like near-fault earthquakes. *Soil Dyn. Earthq. Eng.* **2017**, *96*, 115–127. [CrossRef]
17. Asgarian, B.; Yahyai, M.; Mirtaheri, M.; Samani, H.R.; Alanjari, P. Incremental dynamic analysis of high-rise towers. *Struct. Des. Tall Spec. Build.* **2010**, *19*, 922–934. [CrossRef]
18. JSCA. The guide to safe buildings. In *Performance-Based Seismic Design*; JASO: Tokyo, Japan; Available online: https://www.jaso.jp/ (accessed on 12 December 2021). (In Japanese)
19. Shome, N.C.C.A. *Probabilistic Seismic Demand Analysis of Nonlinear Structures, RMS-35*; Reliability of Marine Structures: Stanford, CA, USA, 1999.
20. Saito, T. Structural Earthquake Response Analysis STERA_3D. Available online: http://www.rc.ace.tut.ac.jp/saito/Software/STERA_3D/STERA3D_user_manual.pdf (accessed on 10 October 2021).
21. Sekiya, E.; Mori, H.; Ohbuchi, T.; Yoshie, K.; Hara, H.; Arima, F.; Takeuchi, Y.; Saito, Y.; Ishii, M.; Kasai, K. Details of 4-, 10-, & 20-Story Theme Structure Used for Passive Control Design Examples. In *The JSSI Response Control Committee Symposium on Passive Vibration Control*; JSSI: Tokyo, Japan, 2004. (In Japanese)
22. Saito, T. Structural Earthquake Response Analysis STERA_3D. Technical Manual. Available online: http://www.rc.ace.tut.ac.jp/saito/Software/STERA_3D/STERA3D_technical_manual.pdf (accessed on 10 October 2021).
23. Yasuhiro, N.M.H.; Takanori, S.; Tsugio, I. JSSI manual for building passive control technology part 6. In Proceedings of the 13th World Conference on Earthquake Engineering Vancouver, Vancouver, BC, Canada, 1–6 August 2004; p. 3451.
24. Yasuo, T.Y.G.; Fumiya, I.; Yuji, K. JSSI manual for building passive control technology part 3. In Proceedings of the 13th World Conference on Earthquake Engineering, Vancouver, BC, Canada, 1–6 August 2004; p. 2468.
25. Naqi, A.; Saito, T. Seismic Performance Evaluation of Steel Buildings with Oil Dampers Using Capacity Spectrum Method. *Appl. Sci.* **2021**, *11*, 2687. [CrossRef]

applied sciences

MDPI

Article

Fragility Curves and Probabilistic Seismic Demand Models on the Seismic Assessment of RC Frames Subjected to Structural Pounding

Maria G. Flenga and Maria J. Favvata *

Civil Engineering Department, University of Patras, 26504 Patras, Greece; mariaflenga@upnet.gr
* Correspondence: mfavvata@upatras.gr

Abstract: This study aims to evaluate five different methodologies reported in the literature for developing fragility curves to assess the seismic performance of RC structures subjected to structural pounding. In this context, displacement-based and curvature-based fragility curves are developed. The use of probabilistic seismic demand models (PSDMs) on the fragility assessment of the pounding risk is further estimated. Linear and bilinear PSDMs are developed, while the validity of the assumptions commonly used to produce a PSDM is examined. Finally, the influence of the PSDMs' assumptions on the derivation of fragilities for the structural pounding effect is identified. The examined pounding cases involve the interaction between adjacent RC structures that have equal story heights (floor-to-floor interaction). Results indicate that the fragility assessment of the RC structure that suffers the pounding effect is not affected by the examined methodologies when the performance level that controls the seismic behavior is exceeded at low levels of IM. Thus, the more vulnerable the structure is due to the pounding effect, the more likely that disparities among the fragility curves of the examined methods are eliminated. The use of a linear PSDM fails to properly describe the local inelastic demands of the structural RC member that suffers the impact effect. The PSDM's assumptions are not always satisfied for the examined engineering demand parameters of this study, and thus may induce errors when fragility curves are developed. Nevertheless, errors induced due to the power law model and the homoscedasticity assumptions of the PSDM can be reduced by using the bilinear regression model.

Keywords: RC structure; structural pounding; probabilistic seismic assessment; fragility curves; global and local EDPs; linear and bilinear PSDMs; PSDM's assumptions; nonlinear dynamic analyses; Eurocode 8

Citation: Flenga, M.G.; Favvata, M.J. Fragility Curves and Probabilistic Seismic Demand Models on the Seismic Assessment of RC Frames Subjected to Structural Pounding. *Appl. Sci.* **2021**, *11*, 8253. https://doi.org/10.3390/app11178253

Academic Editor: Amadeo Benavent-Climent

Received: 2 August 2021
Accepted: 1 September 2021
Published: 6 September 2021

Publisher's Note: MDPI stays neutral with regard to jurisdictional claims in published maps and institutional affiliations.

Copyright: © 2021 by the authors. Licensee MDPI, Basel, Switzerland. This article is an open access article distributed under the terms and conditions of the Creative Commons Attribution (CC BY) license (https://creativecommons.org/licenses/by/4.0/).

1. Introduction

In the last three decades, several researches have been undertaken to address the seismic performance of structures through probabilistic procedures. A key issue in these methods is the definition of accurate fragility curves that identifies the vulnerability of structures to seismic damage. Withman et al. [1] introduced the concept of damage probability matrices (DPM) to provide the probability that a particular level of damage will occur to a structural system, while the ground motion has been described by the modified Mercalli intensity (MMI) scale. Thereafter, in 1991, the Applied Technology Council (ATC) and the Federal Emergency Management Agency (FEMA) introduced the use of vulnerability functions for lifelines in the ATC-25 report [2]. For this purpose, the concept of a continuous relation between the seismic damage and ground motion intensity is established and a regression analysis of different damage probability matrices is performed. In 1997, the risk assessment software package Hazard United States (HAZUS) was first introduced by the Federal Emergency Management Agency (FEMA). The HAZUS earthquake model consisted of an inventory data base, a ground motion model, an earthquake model of buildings, and lifelines damage models in order to evaluate the probability of damage to buildings and infrastructures according to ground shaking data from the U.S. Geological Survey (USGS) shake map [3]. Nowadays, fragility curves have been evolved as an efficient tool for retrofit and design of infrastructures [4].

Different types of fragility curves have been reported in the literature, namely expert-based opinion, experimental, analytical, hybrid, and empirical curves. Expert opinion fragility curves are developed based on statistics estimations of the probable damage when different earthquake intensities occur. ATC-13 [5] developed DPMs for typical California infrastructures using MMIs based on opinion of 42 experts. In 2007, Kostov et al. [6] developed vulnerability curves for Sofia's buildings stock and lifelines according to European Macro Seismic Scale (EMS98). Nevertheless, the reliability of this type of fragility curves is questionable due to uncertainties contain the experts' judgements [7]. Empirical fragility curves are developed based on damage observations from past earthquakes. Basöz et al. [8] assessed the vulnerability of bridges based on the observed data of Loma Prietta 1989, and Northridge 1994 developed empirical fragility curves through logistic regression. In 2003, Rosseto and Elanshai [7] developed empirical fragility curves based on a database of 99 post-earthquake damage distributions observed in 19 earthquakes of 340,000 RC buildings. Other researchers [9–11] have also used post-earthquake damage data to develop empirical curves. On the other hand, the experimental fragilities are based on experimental results. Vosooghi and Saidi [12] developed experimental fragility curves for bridge piers based on shake table and cyclic load tests. They correlate the damage state and the damage index to develop fragility functions. In the case of absence of damage data from real earthquakes, analytical fragility curves can be developed using results of structural demands from analytical simulations of non-linear static analysis [13–15], non-linear dynamic analysis [16–25], and elastic spectral analysis [26,27]. Finally, the hybrid method uses a combination of different procedures (available damage statistics and nonlinear analysis) for the prediction of the damage levels. Kappos et al. [28,29] developed DPMs where part of intensity levels was constructed using available data from past earthquakes and results based on nonlinear dynamic analyses. In 2006, Kappos et al. [30] presented fragility curves for masonry and RC buildings combining statistical data derived from earthquake-damage Greek buildings and results of nonlinear static or dynamic analyses.

In order to develop fragility curves, the structural demands, the earthquake intensity measures, and uncertainties regarding demand and capacity have to be defined. Focused on the development of analytical fragility curves non-linear dynamic analysis procedures have to be performed. Thus, the influence of the record-to-record variability on structural seismic response is taken into account, as the structural model is subjected to a wide range of ground motions. The characteristics of ground motions are described through intensity measures (IMs) while the structural behavior through engineering demand parameters (EDPs). Nonlinear dynamic analysis procedures, such as cloud analysis (CA) [31–33], incremental dynamic analysis (IDA) [34], and multi-stripe analysis (MSA) [35,36], can be performed to generate samples of EDPs-IMs pairs.

Thereafter, the discrete probability data points for each level of earthquake intensity can be defined incorporating directly the results of the nonlinear analyses (e.g., IDA) in order to provide a form of empirical cumulative distribution function (CDF) of fragility curve. A key parameter for the development of a fragility curve is the definition of the median (μ) and the dispersion (β) of the EDP-IM values, and thus different approaches have been reported in the literature. Among them, the moment method, the method of maximum likelihood estimation, the IM percentiles method, and the concept of the probabilistic seismic demand model (PSDM) are widely used [37–40]. For the moment method (also known as Method A by Porter) [37], parameters of the fragility curves occur based on the earthquake intensities of the observed data. Baker [38] introduced the method of maximum likelihood estimation where the parameters of the fragility curves are deduced through maximizing the likelihood of producing the observed data. The IM percentiles [39] incorporate results of the IDA and are based on the summarized 16%, 50%, and 84% IDA curves, respectively. Cornel et al. [31] introduced the relationship between the IM and EDP that provides a closed form solution of fragility curves. In this method, the variation of the structural demand for a wide range of earthquake intensities can be described through regression models known as the probabilistic seismic demand model (PSDM). In this context, the linear regression model is commonly used. Nevertheless, when nonlinear seismic demands have to be described, the use of bilinear PSDM is suggested [40]. Ramammothy et al. [41] developed fragility curves based on linear and bilinear PSDMs for the vulnerability assessment of RC buildings and the evaluation of retrofitted strategy. Bai et al. [42] developed linear and bilinear PSDMs in order to achieve a better representation of the demand for the examined range of the IM. Freddi et al. [43] used bilinear PSDM to describe the local EDPs of low-ductility RC frames. Aljawhari et al. [44] used bilinear PSDM to study the influence of ground motion sequences on the fragility of RC moment-resisting frames. Other researchers developed fragility curves based on a Bayessian approach for the formulation of PSDMs [41,42,45,46]. Furthermore, Jalayer et al. [47] introduced a robust fragility assessment approach for the RC moment-resisting frame using Markov Chain Monte Carlo (MCMC) simulation.

Recently, fragility curves have also been used for the probabilistic evaluation of the pounding effect between adjacent buildings. Tubaldi et al. [40] developed fragility curves for the assessment of the seismic pounding risk of singe-degree-of-freedom (SDOF) and multi-degree-of-freedom (MDOF) systems, based on linear and bilinear PSDMs. Nazri et al. [48] developed fragility curves in terms of maximum interstory drift vs. PGA to study the structural pounding effect between the adjacent RC frames, taking into account the available separation distance and the structural irregularities. The results of the assessment were based on the IDA method. In 2020, Flegga and Favvata [49] investigated the floor-to-floor structural pounding effect between RC structures at different global and local performance levels as a function of PGA and the spectral acceleration, Sa. Kazemi et al. [50] studied the effect of the floor-to-floor structural pounding between RC and steel moment resisting frames (MRFs) at different global performance levels. Fragility curves based on IDA have been developed in terms of maximum interstory drift as a function of Sa. Finally, Flenga and Favvata [51] incorporated the local inelastic demands of a RC frame in the probabilistic assessment of the pounding risk. The floor-to-floor and the floor-to-column types of structural pounding were evaluated based on fragilities that have been developed through linear and bilinear PSDMs.

Based on this review, it can be stated that, although many different probabilistic methods have been proposed for the derivation of accurate fragility curves, a limited number of methodologies in the research area of the pounding risk between adjacent structures has been evaluated.

So, this study aims to evaluate five different methodologies reported in the literature for developing fragility curves to assess the seismic performance of real RC structures subjected to structural pounding. In this context, displacement-based and curvature-based fragility curves are developed. The validity of the assumptions commonly used to produce a PSDM is examined, and the influence of these assumptions on the derivation of the fragilities for the structural pounding effect is identified. EDPs of the RC structure that suffers the pounding effect have been estimated based on the results of IDAs [34].

2. Methodologies of Developing Fragility Curves

In modern seismic engineering, fragility curves are employed to assess the vulnerability of a structure due to seismic events. Fragility curves describe the probability an engineering demand parameter (EDP) to exceed the capacity level of $\hat{C}$ for a given IM, and can be calculated as:

$$P[\,EDP|IM \geq C|IM] = \Phi\left(\frac{\ln\mu - \ln\hat{C}}{\beta}\right) \quad (1)$$

The fragility assessment of the pounding effect is examined considering the following five methods:

(a) Empirical cumulative distribution function (CDF),
(b) Moment method (MM),
(c) Maximum likelihood estimation (MLE) method,
(d) 16%, 50%, 84% IM—percentiles, and
(e) Probabilistic seismic demand model (PSDM).

The empirical CDF method provides discrete probability data points at a given value of IM. For a continuous representation of the fragility curve, the median (μ) and the dispersion (β) of the responses have to be defined utilizing one of the other four methodologies. All methods are presented and discussed below. Fragilities based on the empirical CDF method are considered as the reference probability data.

2.1. Empirical Cumulative Distribution Function (CDF)

The implementation of IDA method results to EDPs values that correspond to each seismic excitation for a given value of IM. Therefore, stripes of discrete EDPs values are generated at increasing levels of IM. The probability an EDP to exceed a particular performance level for a given value IM is estimated as the sum of those events which exceed the capacity $\hat{C}$ over the total number of seismic excitations:

$$P\left[EDP \geq \hat{C}\right] = \frac{m}{n} \quad (2)$$

where m is the number of events which exceed a particular limit state (capacity level) $\hat{C}$ at the level of IM, and n is the total number of seismic excitations.

In Figure 1, the IDA curves for n ground motions and the corresponding fragility curve based on the empirical CDF method are presented. In Figure 1a, dots represent the structural response at a particular value of IM, while the vertical solid line depicts the capacity $\hat{C}$ at the performance level of the assessment. The red dots refer to the m events where the demand exceeds the capacity at the particular level of IM, while the blue dots refer to (n-m) events for which demand does not exceed capacity. The discrete probability data points (Figure 1b) at the performance level of $\hat{C}$ can be evaluated based on Equation (2). Therefore, the fragility curve based on the empirical CDF can be developed by connecting the discrete points with linear segments. Tubaldi et al. [40] utilized the discrete probability data points to validate the fragility curves developed through PSDM for different values of separation gap distance.

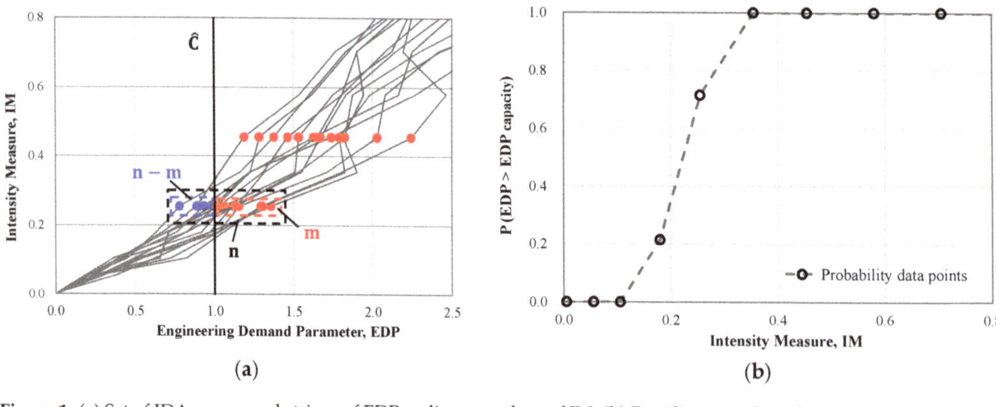

Figure 1. (**a**) Set of IDA curves and stripes of EDP at discrete values of IM, (**b**) Fragility curve based on empirical cumulative distribution function (CDF).

2.2. Moment Method

The moment method is used to define the median (μ) and the dispersion (β) in the case of continuous representation of the fragility curve. This method is also known as "Method A" by Porter et al. [37] and is based on the results of the IDA method. For the calculation of median (μ) and dispersion (β) values of IM at the onset of exceedance, a particular limit state is used considering lognormal distribution. For each seismic motion, the value of IM at the capacity level $\hat{C}$ is presented in Figure 2 with star point.

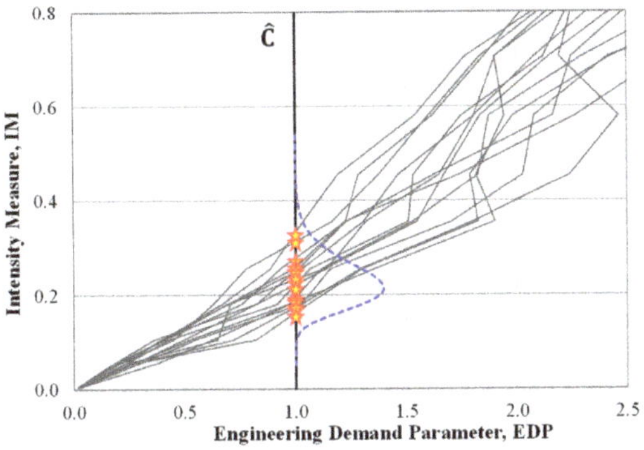

Figure 2. Set of IDAs curves and corresponding values of IM at the level of capacity $\hat{C}$.

The μ and β can be estimated as:

$$\mu = \frac{1}{n}\sum_{i=1}^{n} \ln IM_i \tag{3}$$

$$\beta = \sqrt{\frac{1}{n-1}\sum_{i=1}^{n}(\ln(IM_i) - \mu)^2} \tag{4}$$

where n is the number of seismic excitations and IM_i is the value of IM associated with the onset of exceedance a particular performance level. Aslani and Miranda [52] used this fitting approach to develop continuous fragility curves for different damage states in slab-column connections of non-ductile reinforced concrete buildings.

2.3. Maximum Likelihood Estimation (MLE) Method

The method of MLE [38] provides the median (μ) and dispersion (β) so that the fitted fragility function is more consistent with the probability data points obtained through Equation (2).

The likelihood function can be calculated as:

$$\text{Likelihood} = \prod_{j=1}^{m} \binom{n_j}{z_j} \Phi\left(\frac{\ln(IM_j) - \mu}{\beta}\right)^{z_j} \left(1 - \Phi\left(\frac{\ln(IM_j) - \mu}{\beta}\right)\right)^{n_j - z_j} \tag{5}$$

where m is the number of IM levels, z_j is the number of ground motions that exceed a performance level out of n_j seismic excitations, and Π denotes a product over all the considered IM levels.

Through maximizing the likelihood function the median (μ) and dispersion (β), are defined as:

$$\{\mu, \beta\} = \max_{\mu,\sigma} \sum_{j=1}^{m} \left\{ \ln\binom{n_j}{z_j} + z_j \ln \Phi\left(\frac{\ln(IM_j) - \mu}{\beta}\right) + (n_j - z_j) \ln\left(1 - \Phi\left(\frac{\ln(IM_j) - \mu}{\beta}\right)\right) \right\} \tag{6}$$

Baker [38] introduced the method of maximum likelihood estimation to develop fragility curves when incremental analysis was performed up to some level of IM for some ground motions.

2.4. 16%, 50%, 84%. IM—Percentiles

IDAs curves define the seismic demands of the structural model for a suite of ground motions, so a deterministic evaluation of structural behavior is provided. The summarization of IDAs curves into a central value (mean or median) and a measure of dispersion quantifies the randomness introduced by records [34].

Each EDP | IM point at the summarized IDAs curves of 16%, 50%, and 84% represents an upper value of EDP where 16%, 50%, and 84% of the records produce demand lower of this value for a given level of IM. So, in the context of fragility curves, the 16%, 50%, and 84% percentiles curves are also defined in this study.

The median value μ is defined as the intersection of the 50% summarized IDA curve with the level of capacity $\hat{C}$ ($\mu = IM_{c,50\%}$), while the dispersion β is calculated as [53]:

$$\beta = \frac{1}{2} \ln \frac{IM_{C,84\%}}{IM_{C,16\%}} \tag{7}$$

where, $IM_{C,84\%}$, $IM_{C,16\%}$ are the intersections of the corresponding summarized IDAs curves with the capacity level of $\hat{C}$ (see Figure 3). Freddi et al. [53] developed fragility curves for assessing the vulnerability of existing reinforced concrete buildings retrofitted by dissipated braces.

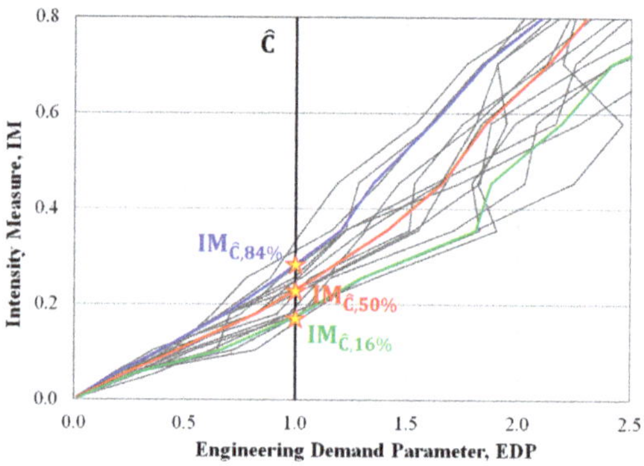

Figure 3. 16%, 50%, 84% summarized IDAs curves and corresponding values of IM at the level of capacity $\hat{C}$.

2.5. Probabilistic Seismic Demand Model (PSDM)

Another way of generating fragility curves is through the probabilistic seismic demand model [40–42,45–47,49,51]. A probabilistic seismic demand model is a mathematical relation between the structural response and the earthquake intensity measure (IM) [54]. The mathematical representation of the PSDM leads to a closed form solution that permits the definition of the fragility curves [31]. Therefore, Equation (1) is transformed to the following expression:

$$G_{EDP|IM}(C|IM) = P[\, EDP|IM \geq C|IM] = \Phi\left(\frac{\ln \hat{EDP}|IM - \ln \hat{C}}{\beta_{EDP|IM}}\right) \tag{8}$$

$\Phi(.)$ denotes the standard normal cumulative function, $\hat{EDP}|IM$ the median structural demand, $\hat{C}$ the median value of the capacity, and $\beta_{EDP|IM}$ the logarithm standard deviation.

The mathematical representation of median structural demand response $\hat{EDP}$ and the IM can be approximated by a two parameters power law model [31,55]:

$$\hat{EDP}|IM = a\, IM^{b} \quad (9)$$

The coefficients α and b are calculated through linear regression analysis of logarithm of IM and EDP, so the Equation (9) is transformed to the following expression:

$$\ln \hat{EDP}|IM = b\ln IM + \ln a + \varepsilon|IM \quad (10)$$

where, ε | IM is the random error with mean zero and variance σ^2.

The structural response demand is assumed to follow lognormal distribution [55] with logarithm standard deviation $\beta_{EDP|IM}$ which is calculated by the following equation:

$$\beta_{EDP|IM} = \sqrt{\frac{\sum_{i=1}^{n} \left(\ln EDP_i|IM - \ln \hat{EDP}|IM\right)^2}{n-2}} \quad (11)$$

Furthermore, a bilinear regression model is also considered to capture the nonlinear behavior of the structural members. The bilinear regression model can be described by the following expression:

$$\ln \hat{EDP}|IM = (a_1 + b_1 \ln IM)(1 - H_1) + \left[\ln EDP|IM^* + b_2(\ln IM - \ln IM^*)\right]H_1 + \varepsilon|IM \quad (12)$$

The coefficients α_1, b_1, and b_2 are defined through linear regression analysis and determine the slope of the two linear branches. H_1 is a dummy variable which is equal to $H_1 = 0$ for IM $\leq$ IM* and H1 = 1 for IM > IM*. Finally, the parameter IM* is the intersection of two linear branches, while $\ln EDP|IM^*$ is the value of the EDP at the intersection point.

3. Examined Case Study

3.1. Description

The probabilistic seismic assessment of an eight-story RC frame structure that suffers the impact effect from an adjacent three-story rigid structure ($T_{3\text{-story}} = \infty$) is evaluated. The story levels of the adjacent structures have the same height so that collisions may occur between the story diaphragms and consequently between the story masses (floor-to-floor interaction). The separation gap distance (d_g) between the adjacent structures is considered as: d_g = 0.0 cm (structures in contact from the beginning), d_g = 4.5 cm, and d_g = 9.0 cm (Eurocode's requirement). The seismic performance of the eight-story structure without the pounding effect is also included for comparison reasons. The fundamental period of the eight-story RC frame is 1.132 s and the yield displacement at the top story level is equal to 5 cm. The two adjacent buildings are considered as fixed-based structures, while the effect of the soil–structure interaction is not examined in this study.

3.2. Structural Design and Modelling Assumptions

The examined eight-story RC frame was designed according to Eurocodes 2 and 8, meeting the ductility capacity medium (DCM) criteria of the codes. The seismic behavior factor for the frame was q = 3.75. The mass M, was taken equal to

$$M = \frac{G + 0.3Q}{g} \quad (13)$$

where G is gravity loads, Q is live loads, and g is the acceleration of gravity.

The design base shear force V, was equal to

$$V = \frac{0.3g}{q} M \quad (14)$$

Reduced values of member moments of inertia (I_{ef}) were considered in the design to account for the cracking. The geometry of the eight-story RC frame and the reinforcement of columns at the top (up) and the bottom (dn) cross sections are presented in Figure 4a. Different reinforcement of the cross sections along the same column is due to the design code requirements.

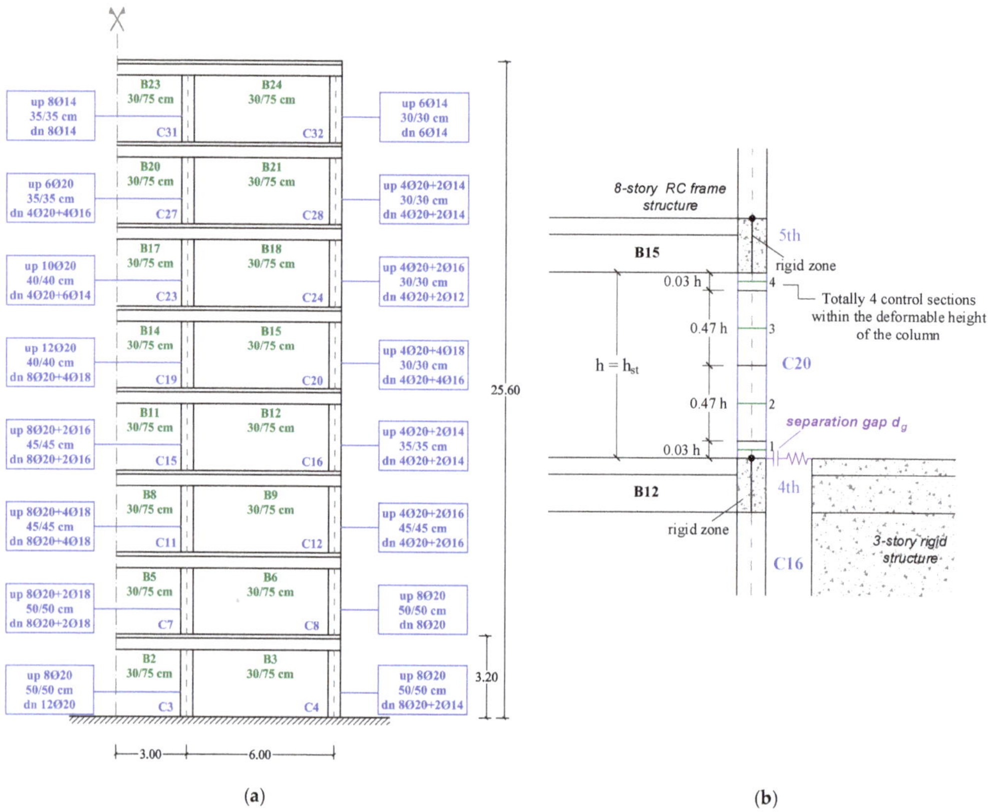

Figure 4. (**a**) Eight-story RC frame and (**b**) idealization of the contact area.

Using the computer program Drain-2dx [56] two types of one-dimensional beam-column element are employed for the finite mesh of the eight-story RC frame. The first type of element is used for the simulation of beams, and it is the common lumped plasticity model that considers the inelastic behavior concentrated in zero-length "plastic hinges" at the element's ends. The second element is the distributed plasticity element that accounts for the spread of inelastic behavior over the cross-section and along the deformable length of columns. The latter element performs numerical integration of the virtual work along the length of the member using data deduced from cross-section analysis at pre-selected locations (control sections, see Figure 4b). The cross-section analysis that is performed at the control sections is based on the fiber model, while the hysteretic behavior of columns is mainly based on the hysteretic rule of materials used in the analysis. More details about the fiber section model and the overall modelling assumptions can be found in Karayannis and Favvata [57].

Collisions are simulated using contact elements that become active when the corresponding nodes come into contact. The idealization of the contact area is presented in Figure 4b. The response of the contact element is described by [58,59]: (a) the negative direction of the X-axis that represents the condition that the buildings move away from each other; and (b) the positive direction of the X-axis that simulates the actual behavior of the structures in the case there is a small gap distance (d_g) between them.

4. Fragility Assessment of Structural Pounding

In this section, the research is focused on evaluating the above-mentioned methodologies (Section 2) to assess the seismic performance of the eight-story RC frame structure subjected to structural pounding. Displacement-based and curvature-based fragility curves are developed considering three different EDPs: (a) the maximum interstory drift, IDR_{max}; (b) the maximum top drift TDR_{max}; and c) the maximum curvature ductility demands $\mu_{\varphi,max}$ of the external columns at the pounding side of the eight-story RC frame.

The IDA method [34] was employed to define the seismic demands of the eight-story RC structure at discrete values of PGA due to the pounding effect. Two horizontal components of seven different seismic excitations (totally 14 records) extracted from PEER's database [60] are used with PGA to be scaled in the range of 0.005 g to 1.4 g, and 628 nonlinear dynamic analyses were performed. The seismic excitations cover a range of magnitude (M_w) between 6.2 to 7.9, and closet distance-to-rupture area (R) is between 9.6 km to 116.2 km. The soil average shear wave velocity in upper 30 m of soil ($v_{s,30}$) is up to 813.5m/s and corresponds to soil class A. The ground motions are selected to fit the Eurocode's 8 (EC8) elastic spectra based on the provisions of the Eurocode 8 (EN1998-1:2004, 3.2.3.1.2(4), [61]). Thus, taking into account the fundamental period T of the examined eight-story frame (T = 1.132 s), each record is scaled by a factor so the computed mean spectrum not to be less than 90% of the Eurocode's spectra in the range of periods between 0.2264 s (0.2 T) and 2.264 s (2.0 T). The main characteristics of the selected seismic excitations are presented in Table 1.

Table 1. Main characteristics of seismic excitations.

Seismic Excitations	Duration (s)	Maximum Acceleration α_{max} (m/s^2)		M_w [3]	R [4] (km)
		component FN [1]	component FP [2]		
Italy Arienzo, 1980 (EQ283)	24	0.268	0.405	6.9	52.9
Italy Auletta, 1980 (EQ284)	34	0.615	0.655	6.9	9.6
Chi-Chi Taiwan-06, 1999 (EQ3479)	42	0.073	0.070	6.3	83.4
Denali- Alaska, 2002 (EQ2107)	60	0.869	0.975	7.9	50.9
Loma Prieta, 1989 (EQ804)	25	1.090	0.509	6.9	63.1
Chi-Chi Taiwan-04, 1999 (EQ2805)	60	0.096	0.075	6.2	116.2
San Fernando, 1971(EQ59)	14	0.153	0.181	6.6	89.7

[1] Fault Normal component [2] Fault Parallel component [3] Moment Magnitude [4] Closest distance to rupture plane.

4.1. Displacement-Based Fragility Curves

Herein, the five different methodologies presented in Section 2 are used for the probabilistic evaluation of the pounding effect in terms of $IDR_{max} \mid PGA$ and $TDR_{max} \mid PGA$. The probability data that are directly deduced based on the empirical CDF method are considered as reference values to evaluate the accuracy and the limitations of the other four methods. Three different separation gap distances are examined: (a) d_g = 0.0 cm; (b) d_g = 4.5 cm; and (c) d_g = 9.0 cm, while results based on the free vibration mode (without pounding) of the eight-story RC structure are also presented.

Fragility curves are developed considering the damage states—capacity levels of [62]:

i. Immediate occupancy (IO) that corresponds to a maximum interstory drift (IDR_{max}) is equal to 1% of the story height (h_{st}), and

ii. 1% maximum top drift (TDR_{max}) as a function of the total height of the structure (H_{tot}).

Figure 5 shows the fragility curves in terms of IDR_{max} | PGA at the performance level of IO. As it can be observed, all the examined procedures are developing fragilities that are in a good agreement with the probability data points of the empirical CDF method. Nevertheless, in the case of analyzing the eight-story RC frame without the pounding effect, the PSDM-based fragility curve is shifted to greater values of PGA in comparison to the deduced fragilities based on the MLE, MM, and IM percentiles procedures.

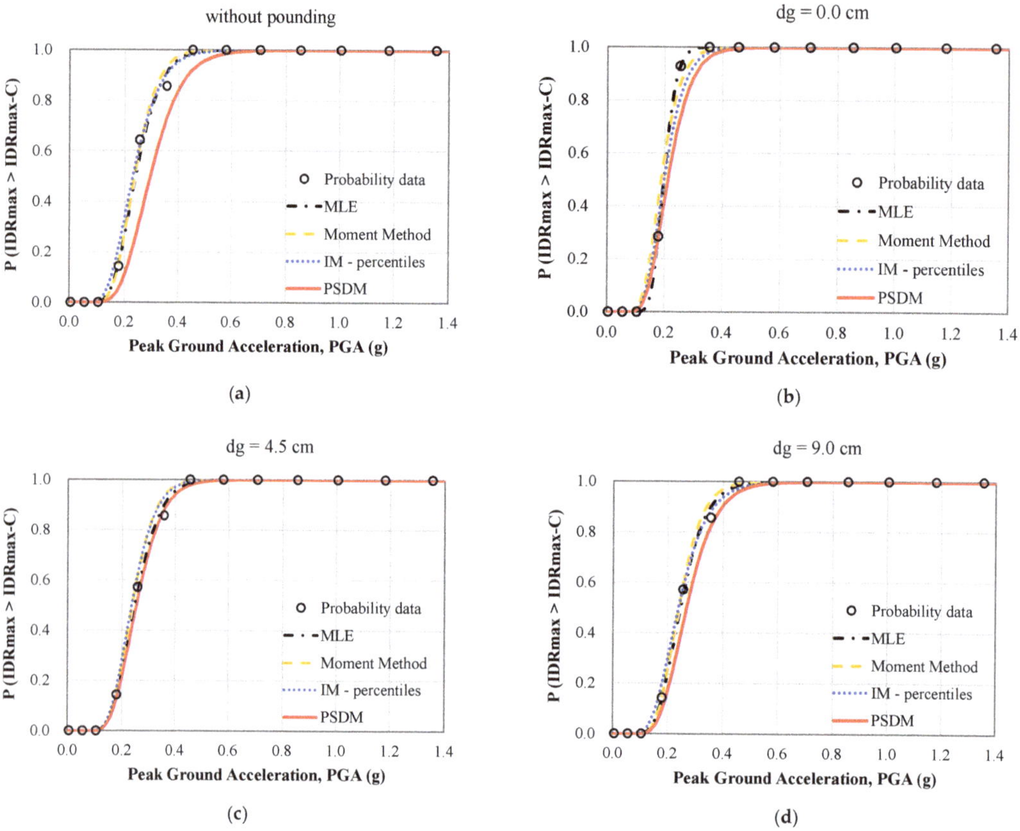

Figure 5. MLE, MM, IM percentiles, PSDM methodologies for the fragility assessment of the eight-story RC frame in terms of maximum interstory drift (IDR_{max} -%h_{st}) as a function of the PGA. Examined cases: (**a**) without the pounding effect, (**b**) d_g = 0.0 cm, (**c**) d_g = 4.5 cm, and (**d**) d_g = 9.0 cm.

In the case of developing fragilities in terms of TDRmax | PGA, the IM percentiles method gives more conservative results in comparison to the other methodologies (Figure 6). On the other hand, the PSDM-based fragility curves indicate that the corresponding seismic assessment of the RC frame against pounding is less vulnerable than the expected one (probability data of empirical method CDF) when d_g = 0.0 cm. Nevertheless, the MM fails to formulate fragility curves in cases of considering separation gap distance d_g = 0.0 cm and d_g = 4.5 cm. This is attributed to the fact that analyses fail to be carried up to values of PGA that exceed the examined performance level.

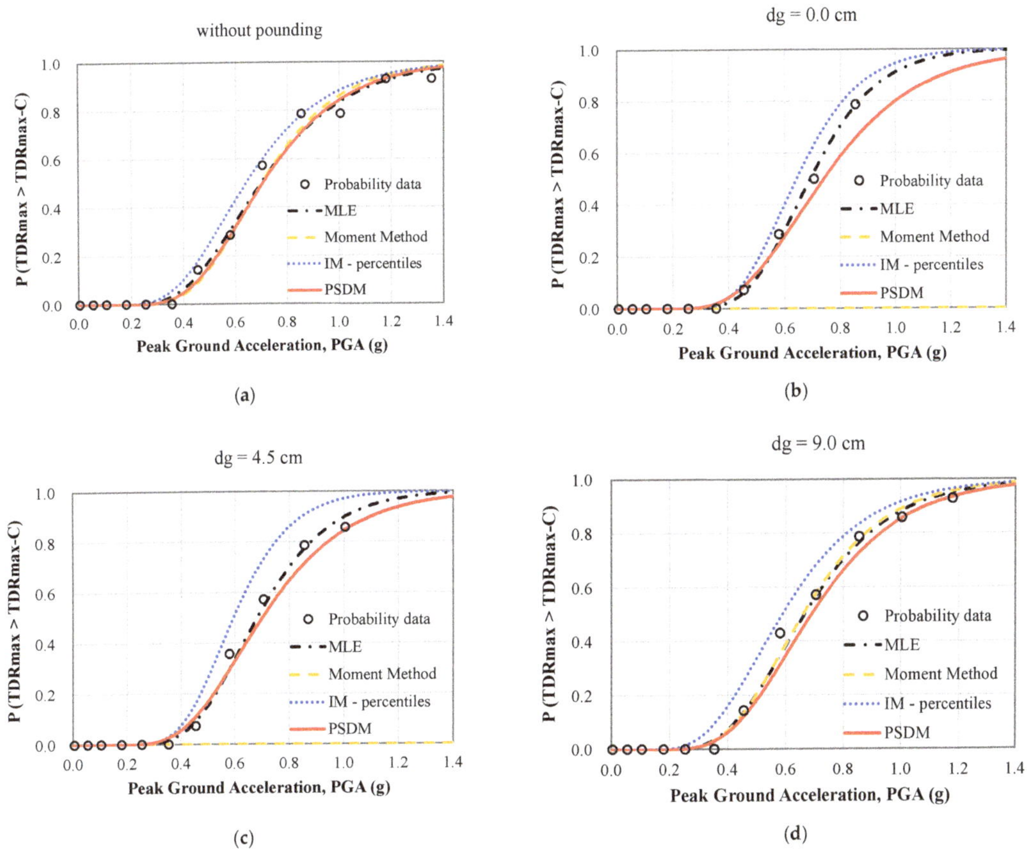

Figure 6. MLE, MM, IM percentiles, PSDM methodologies for the fragility assessment of the eight-story RC frame in terms of maximum top drift (TDR_{max} -%H_{tot}) as a function of the PGA. Examined cases: (**a**) without the pounding effect, (**b**) d_g = 0.0 cm, (**c**) d_g = 4.5 cm, and (**d**) d_g = 9.0 cm.

The difference among the fragility curves is owed to the estimated values of medians μ (Table 2). The estimated value of median μ in the case of PSDM-based fragility curve is greater than the other methods, which explains the shift of the particular curve. The values of median μ and dispersion β of all the examined cases of this study are presented in Table 2.

Regarding the fragility assessment of the pounding effect, results presented in Figures 5 and 6 indicate the following:

- Fragility curves that describe the pounding risk of the RC frame against IDR_{max} are shifted to lower values of intensity in comparison with the corresponding fragilities without pounding.
- The pounding risk is increased as the initial gap distance between the adjacent structures is decreased.
- The vulnerability of eight-story RC frame against TDR_{max} demands is almost identical either with or without considering the pounding effect.

Table 2. Median μ and logarithm standard deviation (dispersion) β.

		Methodology							
		MM		MLE		IM—Percentiles		PSDM	
EDPs	**Examined Case**	μ (g)	β	μ (g)	β	μ (g)	β	μ (g)	β1/β2 †
IDR_{max} ($\%h_{st}$)	without pounding	0.239	0.265	0.240	0.286	0.231	0.324	0.300	0.281
	d_g = 0.0 cm	0.198	0.170	0.192	0.267	0.201	0.274	0.215	0.270
	d_g = 4.5 cm	0.239	0.262	0.246	0.285	0.236	0.277	0.255	0.290
	d_g = 9.0 cm	0.243	0.266	0.246	0.285	0.240	0.330	0.270	0.287
TDR_{max} ($\%H_{tot}$)	without pounding	0.704	0.323	0.700	0.369	0.642	0.372	0.710	0.320
	d_g = 0.0 cm	- *	- *	0.691	0.273	0.644	0.272	0.740	0.352
	d_g = 4.5 cm	- *	- *	0.682	0.302	0.595	0.272	0.700	0.324
	d_g = 9.0 cm	0.661	0.338	0.665	0.347	0.589	0.384	0.691	0.330
C20 $\mu_{\varphi,max}$	without pounding	0.785	0.274	0.821	0.287	0.745	0.298	0.307	0.334
	d_g = 0.0 cm	0.364	0.373	0.368	0.315	0.403	0.453	0.308	0.289/0.676
	d_g = 4.5 cm	0.358	0.317	0.381	0.268	0.376	0.314	0.340	0.315/0.689
	d_g = 9.0 cm	0.536	0.238	0.546	0.231	0.530	0.279	0.445	0.300/0.768

† β1 depicts the β value at the first linear branch of bilinear PSDM or the β value of a linear PSDM β2 depicts the β value at the second branch of bilinear PSDM, * no available result.

4.2. *Curvature-Based Fragility Curves*

Fragility curves in terms of maximum curvature ductility $\mu_{\varphi,max}$ as a function of PGA have been developed based on the five examined procedures. The consistency of the fragility curves is examined for the most critical structural member against pounding. For the examined pounding cases, the local fragility curves of all the external columns at the pounding side of the eight-story RC frame are developed. These results indicate that the column at the fourth floor level is the most critically affected due to the pounding with the adjacent structure (column C20—Figure 1). All the fragility curves are developed at the yield capacity of the column in terms of $\mu_{\varphi,max}$ (i.e., damage level equal to $\mu_{\varphi,max}$ = 1).

So, in Figure 7 the local curvature-based fragilities of the column C20 are presented for the cases (i) d_g = 0.0 cm; (ii) d_g = 4.5 cm; (iii) d_g = 9.0 cm; and (iv) without the pounding effect. Results indicate that fragility curves based on the MLE and MM are very close to the probability data points of the empirical CDF. On the other hand, when the IM percentiles method is used, the fragility curve of the column against pounding effect is moved towards greater values of PGA in the case of d_g = 0.0 cm and in comparison to the corresponding curves based on the MLE, the MM, and the empirical CDF.

Considering the PSDM-based fragility curves, it can be observed that, for the examined cases, the curvature-based fragilities are shifted to lower values of the PGA in comparison to the other methodologies. However, an important issue here is the type of regression model that was used to produce the PSDM. As discussed in Section 2.5, linear and bilinear regression models are adopted for the PSDM. So, in Figure 7, fragility curves based on both linear (LPSDM) and bilinear PSDM (BPSDM) are presented. These results clearly indicate that the use of a linear PSDM can fail in properly describing the local inelastic demands of the structural member.

In the following sections, the assumptions of producing an accurate PSDM are further analyzed to give a better insight on the results of this study.

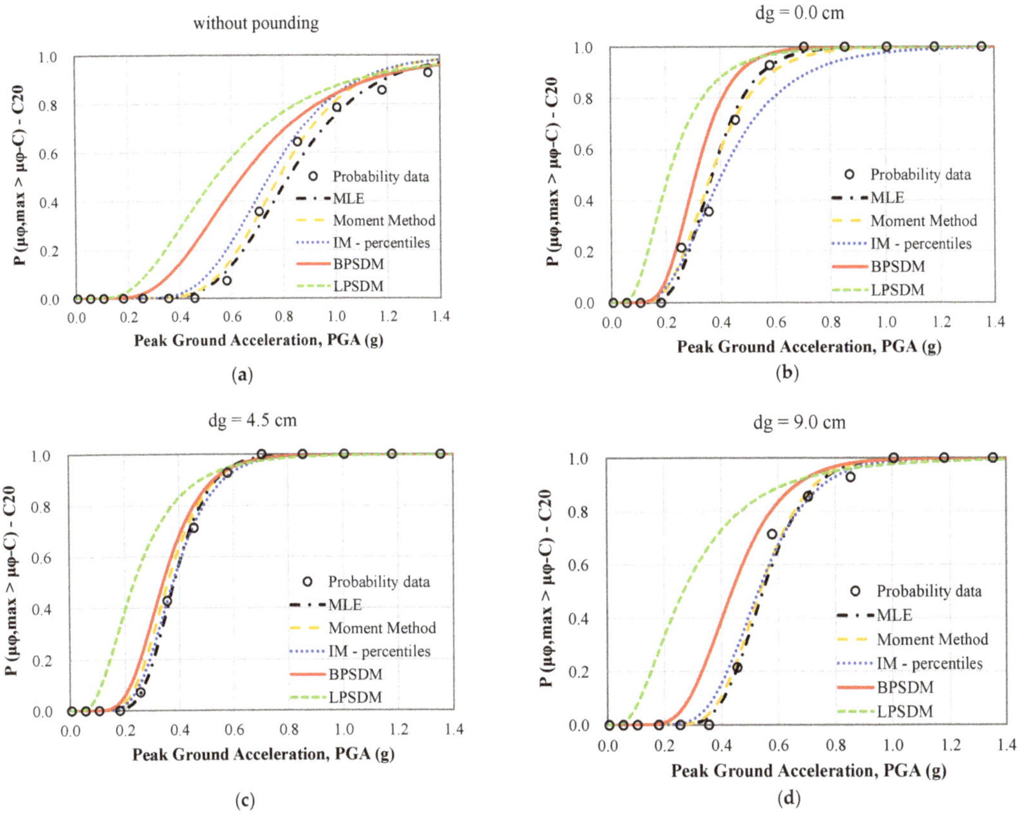

Figure 7. MLE, MM, IM percentiles, PSDM methodologies on the fragility assessment of the eight-story RC frame in terms of maximum curvature ductility $\mu_{\varphi,max}$ as a function of the PGA. Examined cases: (**a**) without the pounding effect, (**b**) d_g = 0.0 cm, (**c**) d_g = 4.5 cm, and (**d**) d_g = 9.0 cm.

5. Validity of PSDM's Assumptions

In this section, the validity of the assumptions that are commonly used to produce a PSDM is examined. The considered assumptions are:

i. lognormal distribution of the evaluated structural demands,
ii. power law model relationship between EDP and IM,
iii. constant logarithm standard deviation of structural demands over the examined range of IM (homoscedasticity assumption).

Lognormality assumption

The validity of the lognormality assumption is examined according to the Anderson–Darling (AD) test. In general, the Anderson–Darling test measures how well the data follow a particular distribution. The smaller the AD value, the better the distribution fits to the data. Through the AD test, the *p*-value is calculated in order to determine which distribution fits better to data set [63].

The AD test is implemented for each level of PGA considering a confidence level equal to 95%. For each level of PGA, the implemented hypothesis test is as follows: For H_o *(null hypothesis)*, the data follow lognormal distribution; and for H_1, the data do not follow the lognormal distribution. If the *p*-value for the AD test is lower than the significance level of 0.05, data do not follow lognormal distribution. On the other hand, the null hypothesis cannot be rejected. In Figure 8, indicative results of the probability plots at the level of PGA = 0.355 g are presented, while Table 3 shows the *p*-values of all the examined cases.

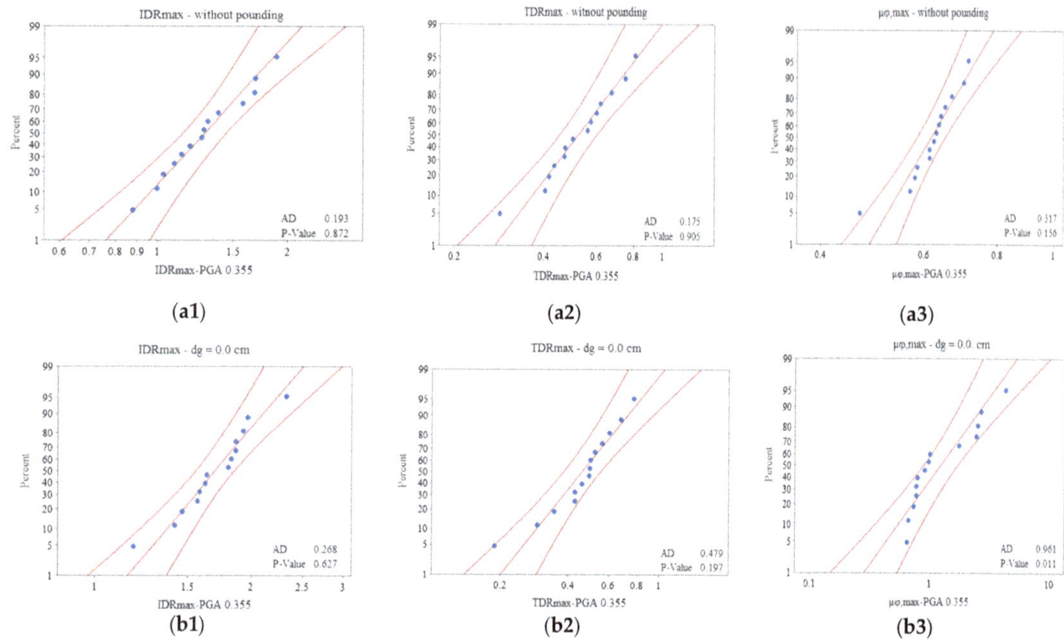

Figure 8. Lognormality assumption: Probability plot at PGA = 0.355 g in the case of (**a**) without the pounding effect, and (**b**) when d_g = 0.0 cm. Examined engineering demand parameters: (**1**) IDR_{max}, (**2**) TDR_{max}, and (**3**) $\mu_{\varphi,max}$.

It can be observed that, although the assumption of the lognormal distribution cannot be rejected in the case of considering the IDR_{max} and TDR_{max} seismic demands, in the case of $\mu_{\varphi,max}$, the distribution of the data does not match the lognormal distribution for all the examined levels of PGA (see Table 3).

Table 3. Check of the lognormality assumption (*p*-value cut off equal to 0.05).

EDPs	Examined Case	PGA (g)												
		0.005	**0.055**	**0.105**	**0.180**	**0.255**	**0.355**	**0.455**	**0.580**	**0.705**	**0.855**	**1.005**	**1.18**	**1.355**
IDR_{max} (%h_{st})	without pounding	0.391	0.288	0.349	0.516	0.605	0.872	0.851	0.150	0.559	0.617	0.800	0.61	0.34
	d_g = 0.0 cm	0.149	0.140	0.241	0.023	0.488	0.627	0.139	0.247	0.823	0.896	- *	- *	- *
	d_g = 4.5 cm	0.398	0.403	0.232	0.785	0.769	0.511	0.411	0.871	0.234	0.843	0.138	- *	- *
	d_g = 9.0 cm	0.398	0.403	0.232	0.773	0.952	0.660	0.760	0.316	0.079	0.483	0.109	0.93	- *
TDR_{max} (%H_{tot})	without pounding	0.848	0.958	0.948	0.723	0.759	0.905	0.662	0.241	0.314	0.841	0.904	0.31	0.40
	d_g = 0.0 cm	0.536	0.815	0.491	0.087	0.493	0.197	0.134	0.652	0.914	0.745	- *	- *	- *
	d_g = 4.5 cm	0.790	0.989	0.944	0.661	0.884	0.514	0.567	0.174	0.442	0.738	0.716	- *	- *
	d_g = 9.0 cm	0.790	0.989	0.944	0.674	0.759	0.844	0.584	0.264	0.388	0.516	0.829	- *	- *
C20 $\mu_{\varphi,max}$	without pounding	0.889	0.597	0.726	0.404	0.382	0.156	0.097	0.252	0.423	0.082	0.141	0.61	0.41
	d_g = 0.0 cm	0.196	0.771	0.047	0.902	0.035	0.011	0.096	0.301	0.187	0.099	- *	- *	- *
	d_g = 4.5 cm	0.794	0.536	0.535	0.967	0.309	0.178	0.388	0.777	0.130	0.210	0.100	- *	- *
	d_g = 9.0 cm	0.794	0.536	0.535	0.777	0.249	0.037	0.010	0.042	0.098	<0.005	<0.005	- *	- *

* no available result.

Power law model

The second assumption of the PSDM concerns the relationship between the median structural demand and the intensity measure (IM). It is assumed that the mathematical relationship between $\hat{EDP}$ and IM can be approximated by a power law model and, thus, the representation of EDP-IM pairs in log-log space (lognormality assumption) implies that the structural response variates linearly (Equation (10)) with the intensity measure. Therefore, the median structural demand lies on a line at any level of IM [64]. However, linear representation of the PSDM along the range of IM is not always accurate and the adoption of a bilinear model is required to describe the variation of the structural demand especially when the components exhibit significant nonlinear behavior. In Figures 9 and 10, the accuracy of using the power law model to describe the demands of IDR_{max} and TDR_{max} as a function of the PGA, is examined. In these figures, the blue dots show the structural demands retrieved based on IDAs, the black rhombus dots represent the median value of EDPs' stripes at each level of IM, and the red lines depict the produced PSDM. Figure 9 shows that the probabilistic seismic demand model in terms of IDR_{max} is in good agreement with the median demand at each level of IM for all the examined cases. The same results are also yielded in the case of TDR_{max}. As it can be observed in Figure 10, the median TDR_{max} demand at each level of PGA is very close with the probabilistic seismic demand model for all the examined cases. Therefore, the linear PSDM can be used to generate the seismic demands of the eight-story frame in terms of IDR_{max} and TDR_{max} in the context of the probabilistic assessment procedure.

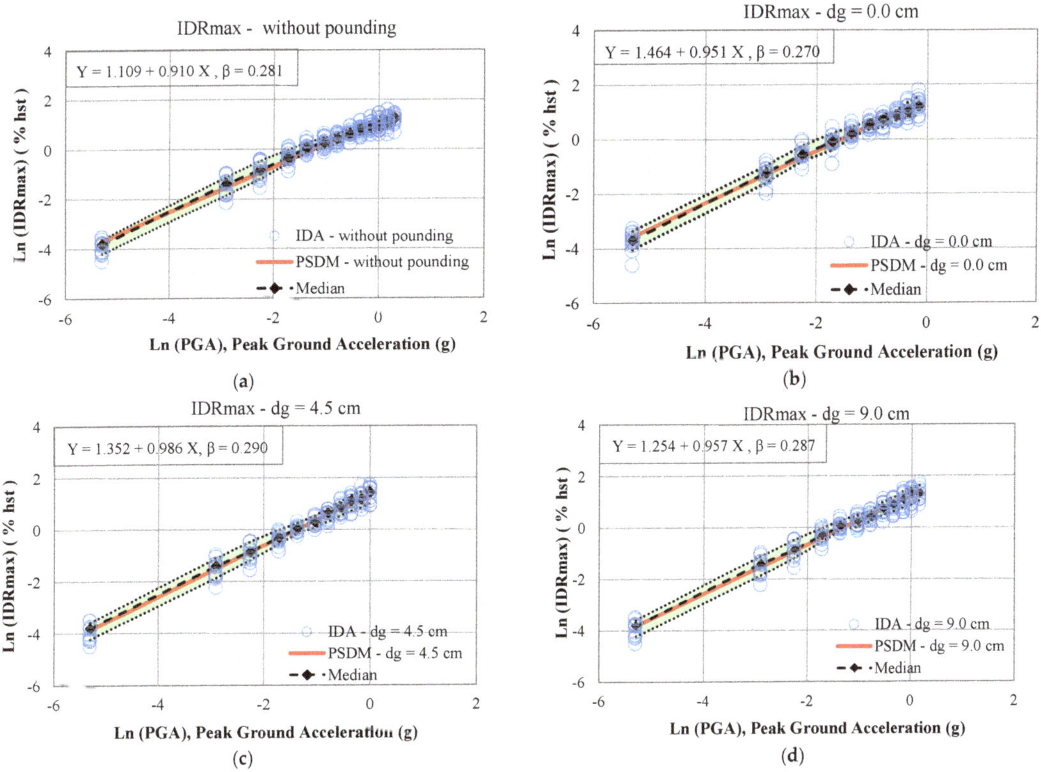

Figure 9. Comparative results in terms of IDR_{max} ($\%h_{st}$) at each level of PGA as deduced based on IDAs, PSDMs, and median values of demand. Examined cases: (**a**) without the pounding effect, (**b**) d_g = 0.0 cm, (**c**) d_g = 4.5 cm, and (**d**) d_g = 9.0 cm.

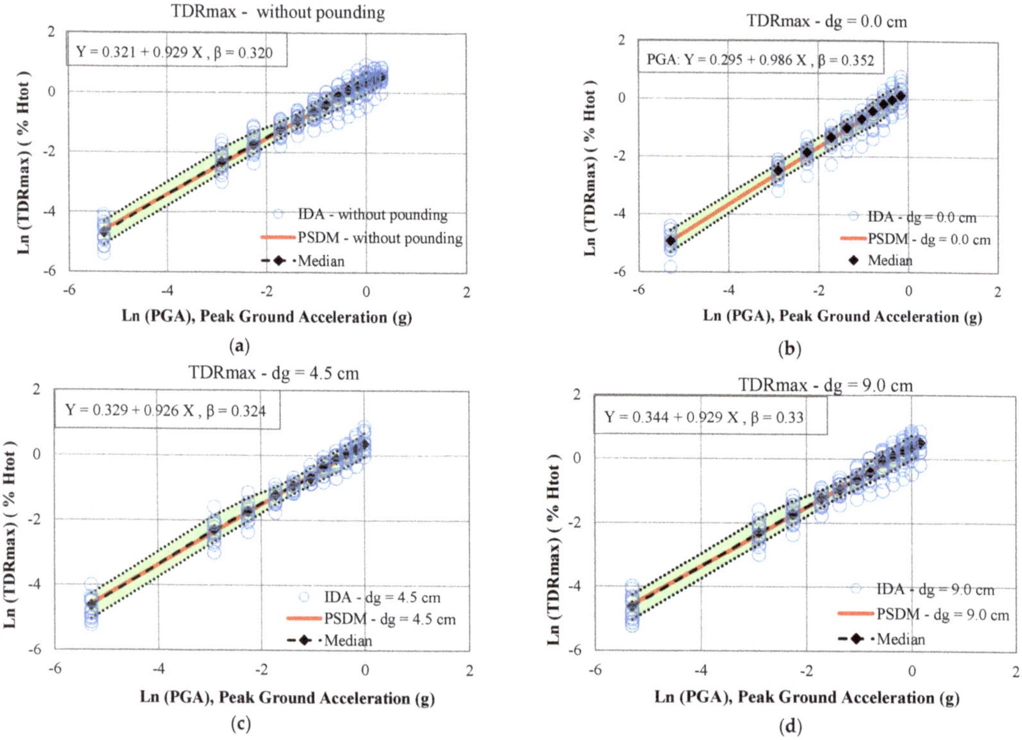

Figure 10. Comparative results in terms of TDR_{max} ($\%H_{tot}$) at each level of PGA as deduced based on IDAs, PSDMs, and median values of demand. Examined cases: (**a**) without the pounding effect, (**b**) d_g = 0.0 cm, (**c**) d_g = 4.5 cm, and (**d**) d_g = 9.0 cm.

In Figure 11, comparative results between linear and bilinear PSDMs of the column C20 in terms of $\mu_{\varphi,max}$ | PGA are presented. These results are compared with the corresponding median demands at each level of PGA.

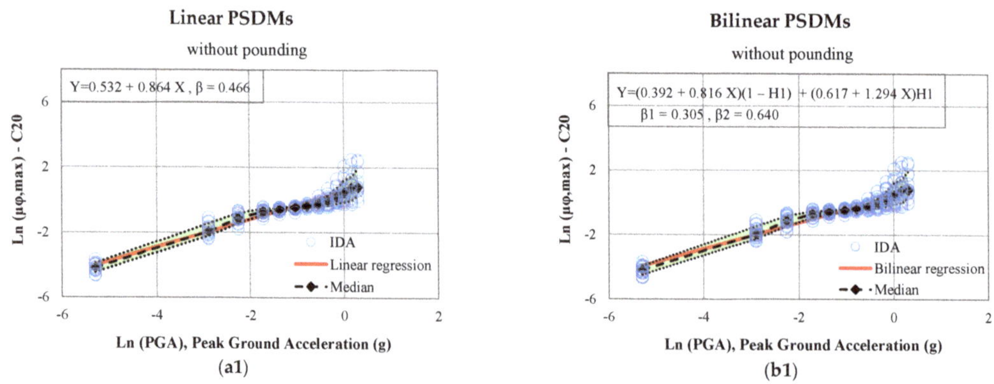

Figure 11. *Cont.*

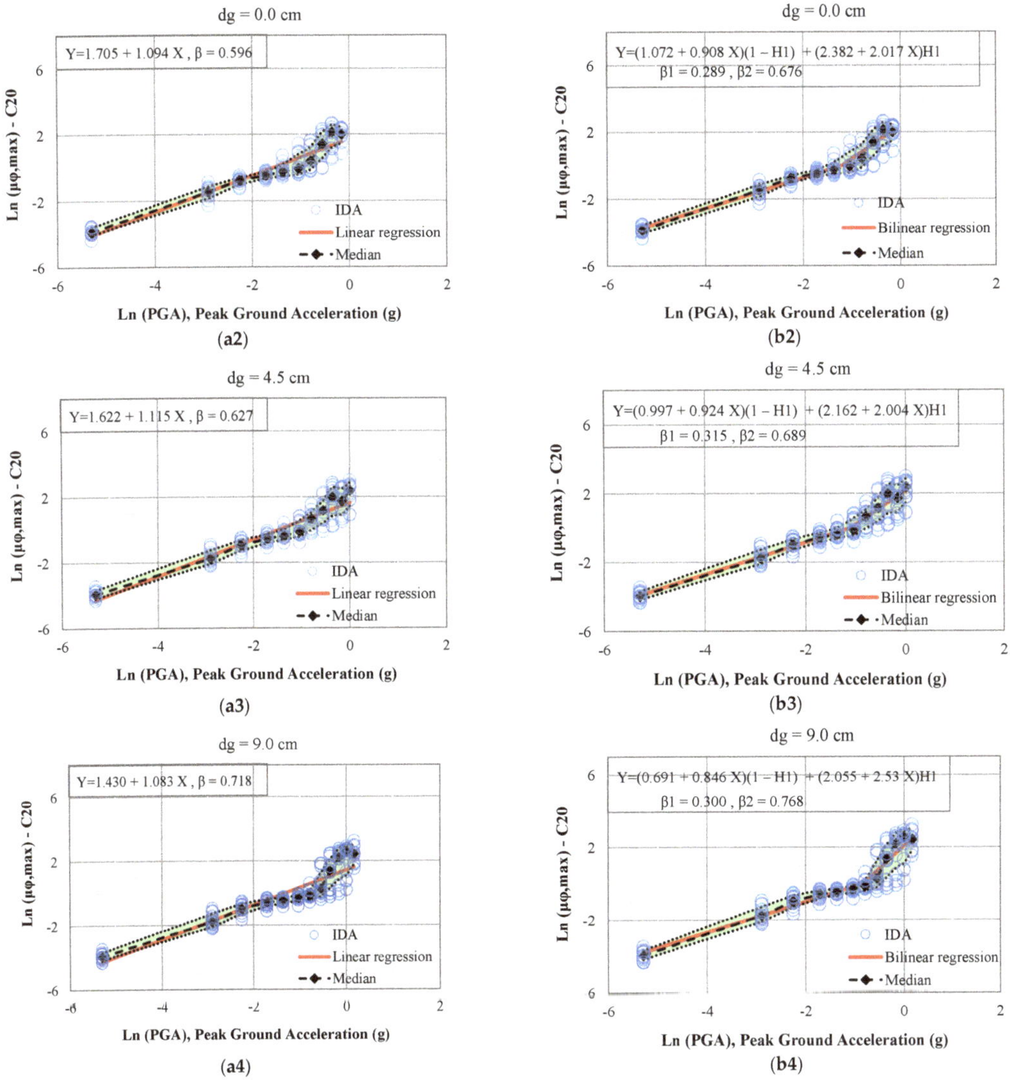

Figure 11. Comparative results in terms of $\mu_{\varphi,max}$ (column C20) at each level of PGA as deduced based on IDAs, PSDMs, and median values of demand. Regression models: (**a**) linear, and (**b**) bilinear. Examined cases: (**1**) without the pounding effect, (**2**) d_g = 0.0 cm, (**3**) d_g = 4.5 cm, and (**4**) d_g = 9.0 cm.

It can be observed that the median data are differentiated from the PSDMs in the case of using a linear regression model. These differences are more pronounced in the case where the pounding effect is evaluated, and especially when the structural members start to exhibit significant nonlinear behavior due to the interaction with the adjacent shorter and stiffer structure.

Homoscedasticity Assumption

The last assumption for the formulation of the PSDM concerns the logarithm standard deviation ($\beta_{EDP|IM}$) of the demand. In order to evaluate the validity of homoscedasticity assumption, the area limited between the mean plus and minus one standard deviation of the associated normal distribution at each level of IM is defined. So, in Figures 9–11, the green shaded area shows the variation of the demands along the examined range of PGA. Figure 9 shows that the variation of demands in terms of the IDR_{max} is not constant but is slightly reduced for values of PGA greater than 0.255 g. Similar results are also deduced in the case of TDR_{max}, as it can be observed in Figure 10.

Therefore, the homoscedasticity assumption of producing the PSDM of IDR_{max} and TDR_{max} is not satisfied within the overall range of PGA. Furthermore, the interaction between the adjacent structures has not altered the outcomes related to the homoscedasticity assumption of the PSDM in terms of displacement EDPs. On the contrary, the case of structural pounding between adjacent structures has altered the outcomes related to the homoscedasticity of the PSDM in terms of $\mu_{\varphi,max} \mid PGA$. As it is discussed below, this is attributed to the effect of pounding on the local inelastic demands of the column.

Results of Figure 11 clearly indicate that the linear PSDM cannot adequately describe the maximum local curvature ductility demands, i.e., $\mu_{\varphi,max}$ of the critical column, due to the pounding effect (Figure 11a). For the bilinear PSDMs, the validity of the homoscedasticity assumption is evaluated at each linear branch of the regression model. Small fluctuations in the dispersion along the first branch of the regression model are observed, and the response of the column is still an elastic one. On the other hand, the nonlinear behavior of the structural member results in a large variation of the demands that differs significantly for each ground motion. These variations are depicted at the second branch of the PSDM.

Thus, it could be pointed out that the nonlinear local demands of the structural member are not sufficiently reflected on the homoscedasticity assumption when only linear PSDM is adopted.

Interaction of Assumptions on Developing Fragility Curves

In this part of the study, the influence of the PSDM assumptions on the fragility curves is examined in order to provide a better insight of the bias induced. For this purpose, the following three different hypothesis cases are considered:

- Case 1 (*lognormality assumption*) In this case, only the lognormality assumption is considered for developing the fragility curves. So, the value of the probability is defined accounting the mean and the standard deviation of each distribution at a particular level of IM.
- Case 2 (*lognormality assumption and power law model*) The lognormality assumption is considered in combination with the power law model. The median of the structural demand at a particular level of IM is based on the PSDM, while the dispersion is calculated for each level of IM through Equation (11).
- Case 3 (*lognormality assumption, power law model, and homoscedasticity assumption*) The three basic assumptions of PSDM are considered for the development of the fragility curves.

The available separation gap distances d_g between the adjacent structures and the case of without pounding effect are also incorporated. Furthermore, the discrete probability data points (empirical CDF method) and the MLE-based fragility curve (as a reliable continuous fragility curve) are used for the quantification of the bias induced through assumptions.

In Figures 12 and 13, the fragility curves in terms of $IDR_{max} \mid PGA$ and $TDR_{max} \mid PGA$ based on the examined hypothesis cases (1–3) are presented. As observed, the examined fragility curves are in a good agreement with the assessment based on the MLE method. Therefore, the violation of the homoscedasticity assumption for both EDPs does not influence the fragility assessment of the pounding effect. So, PSDMs for both EDPs are sufficient estimators of the structural demand on the fragility assessment of the eight-story RC frame structure subjected to pounding.

In Figure 14, the corresponding fragility curves that are produced considering the three different hypothesis cases in terms of $\mu_{\varphi,max} \mid PGA$ at the performance level of DL, are presented. It is observed that the fragility curves are significantly influenced by the assumptions made on the formulation of PSDMs when the seismic performance of eight-story frame is evaluated without considering the pounding between the adjacent structure

(Figure 14a). The error induced due to the power law model and the homoscedasticity assumption in the case of $\mu_{\varphi,max}$ seems to be considerably high. However, this outcome is not so intense when the effect of structural pounding is assessed. This is attributed to the use of bilinear PSDM to describe the critical inelastic behavior of the column at low levels of IM (Figure 14b–d).

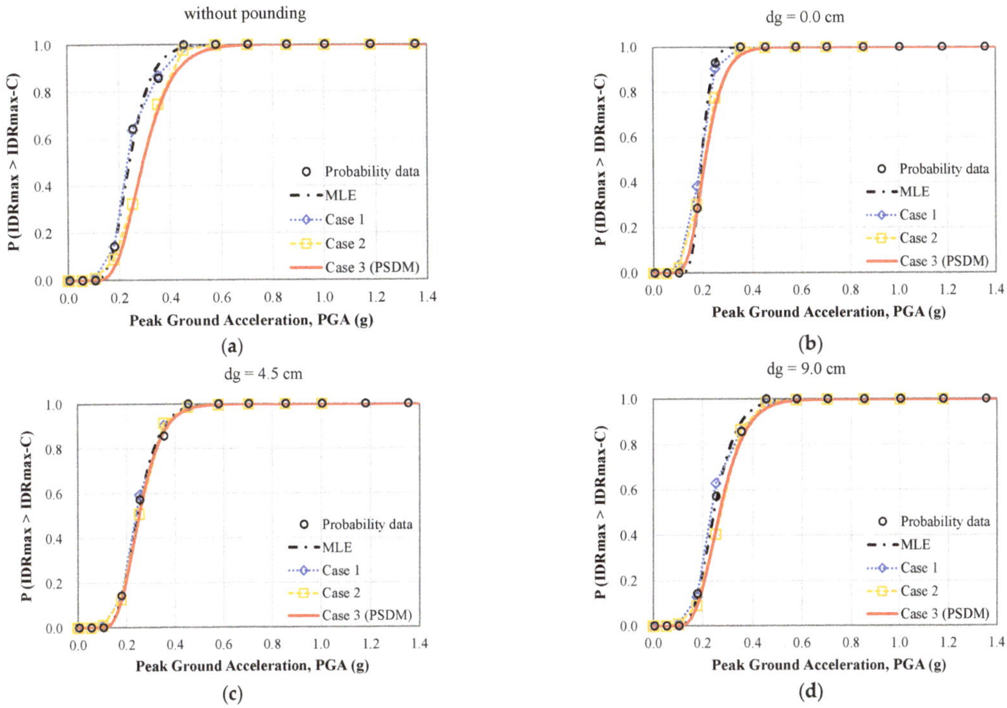

Figure 12. Influence of the PSDM assumptions on the fragility assessment of the eight-story RC frame in terms of IDR_{max} ($\%h_{st}$) as a function of the PGA. Examined assumptions: Case 1—lognormality, Case 2—lognormality and power law model, and Case 3—SDM's assumptions. Fragility curves (**a**) without the pounding effect, (**b**) d_g = 0.0 cm, (**c**) d_g = 4.5 cm, and (**d**) d_g = 9.0 cm.

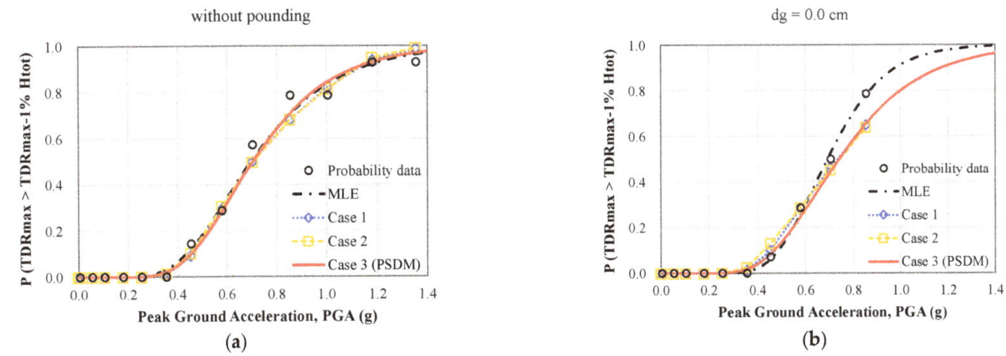

Figure 13. *Cont.*

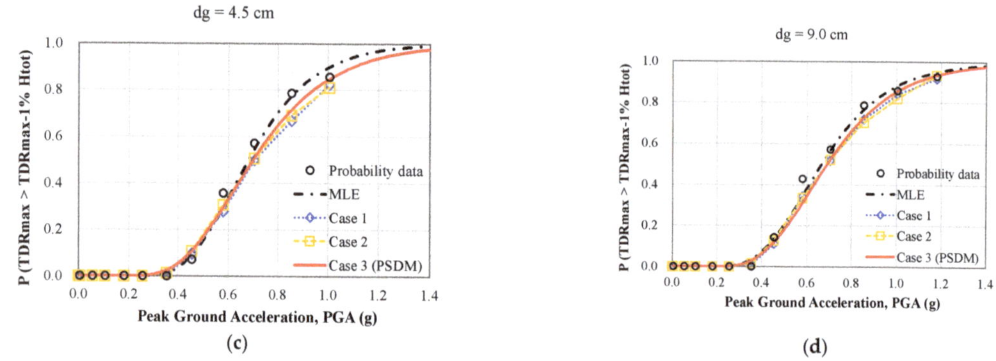

Figure 13. Influence of the PSDM assumptions on the fragility assessment of the eight-story RC frame in terms of TDR_{max} ($\%H_{tot}$) as a function of the PGA. Examined assumptions: Case 1—lognormality, Case 2—lognormality and power law model, Case 3—PSDM's assumptions. Fragility curves (**a**) without the pounding effect, (**b**) d_g = 0.0 cm, (**c**) d_g = 4.5 cm, and (**d**) d_g = 9.0 cm.

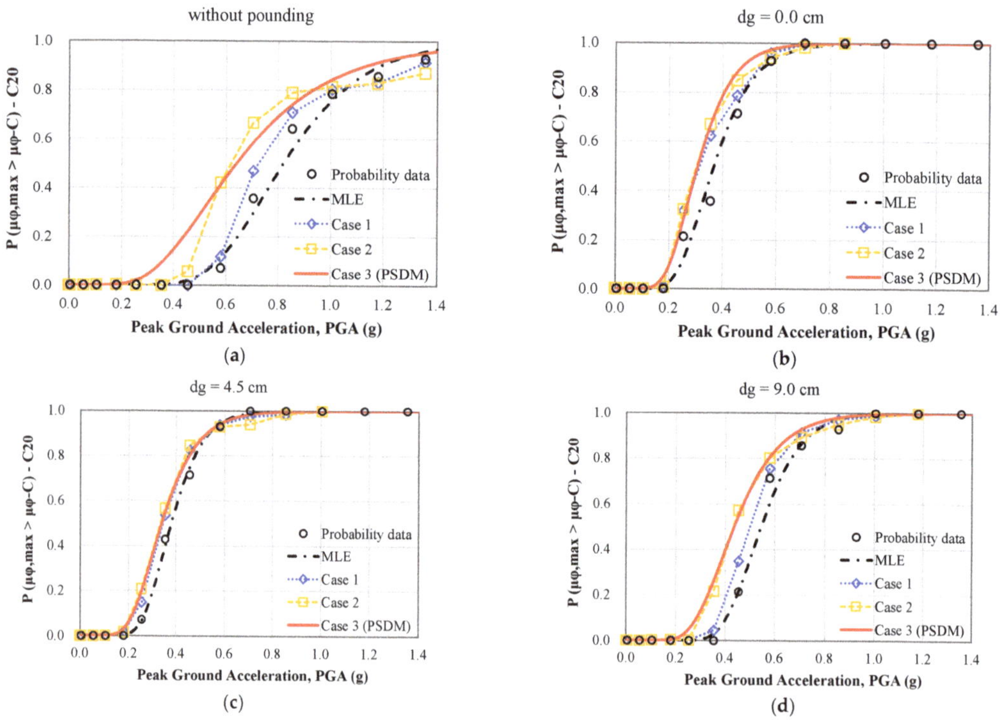

Figure 14. Influence of the PSDM assumptions on the fragility assessment of the eight-story RC frame in terms of $\mu_{\varphi,max}$ (C20) as a function of the PGA. Examined assumptions: Case 1—lognormality, Case 2—lognormality and power law model, Case 3—PSDM's assumptions. Fragility curves (**a**) without the pounding effect, (**b**) d_g = 0.0 cm, (**c**) d_g = 4.5 cm, and (**d**) d_g = 9.0 cm.

6. Conclusions

In this study, five different methodologies for developing fragility curves are evaluated to assess the seismic performance of real RC structures subjected to structural pounding. For this purpose, the discrete probability data points deduced based on the empirical

CDF method are utilized to validate the fragility curves developed through MLE, MM, IM percentiles, and PSDM procedures. Displacement-based and curvature-based fragility curves are developed. The use of probabilistic seismic demand models on the fragility assessment of the pounding risk is further estimated. Linear and bilinear PSDMs are developed, while the validity of the assumptions commonly used to produce a PSDM is examined. Finally, the influence of the PSDMs' assumptions on the derivation of the fragilities for the structural pounding effect is identified. The examined structural pounding cases are between an eight-story RC frame and a three-story rigid structure that have equal story heights (floor-to-floor interaction). The main outcomes of this study are as follows:

- The MLE, MM, and IM percentiles procedures are developing fragilities that are in a good agreement with the probability data points of the empirical CDF method.
- The IM percentiles method gives more conservative results in terms of $TDR_{max} \mid PGA$, in comparison to the other methodologies of this study. Nevertheless, in the case of $\mu_{\varphi,max} \mid PGA$, the fragility curve is moved towards greater values of PGA when $d_g = 0.0$ cm. This result indicates that the IM-percentiles-based local fragility curve cannot accurate capture the increased inelastic demands of the column due to the pounding effect, when $d_g = 0.0$ cm.
- The displacement-based fragilities that are developing from the PSDMs are shifted to greater values of PGA in comparison to the deduced fragilities based on the MLE, MM, and IM percentiles procedures.
- The curvature-based fragilities that are developing from the PSDMs are shifted to lower values of PGA in comparison to the deduced fragilities based on the MLE, MM, and IM percentiles procedures.
- Similar results regarding the fragility assessment of the RC structure between the examined methodologies are deduced when the performance level controls the seismic behavior of the eight-story RC frame structure at low levels of IM.
- The observed shift on the fragility curves is owed to the different values of medians μ that methodologies estimate.
- The lognormality assumption that is evaluated for each level of PGA showing that it is not always satisfied especially in the case of maximum curvature ductility.
- The homoscedasticity assumption of developing the PSDM of IDR_{max} and TDR_{max} is not satisfied within the overall range of PGA.
- The use of a linear PSDM fails to properly describe the local inelastic demands of the structural RC member.
- The nonlinear local demands of the structural member are not sufficiently reflected on the homoscedasticity assumption when only linear PSDM is adopted. The errors induced due to the power law model and the homoscedasticity assumptions of the PSDM can be reduced by using a bilinear regression model.

Author Contributions: Conceptualization, M.G.F. and M.J.F.; methodology, M.J.F.; investigation, M.G.F.; writing—original draft preparation, M.G.F.; writing—review and editing, M.J.F. All authors have read and agreed to the published version of the manuscript.

Funding: This research received no external funding.

Institutional Review Board Statement: Not applicable.

Conflicts of Interest: The authors declare no conflict of interest.

References

1. Whitman, R.V.; Biggs, J.M.; Brennan, J.E.; Cornell, C.A.; De Neufville, R.L.; Vanmarcke, E.H. Seismic Design Decision Analysis. *J. Struct. Div.* **1975**, *101*, 1067–1084. [CrossRef]
2. Applied Technology Council (ATC). *Seismic Vulnerability and Impact of Disruption of Lifelines in the Conterminous United States*; Report No. ATC-25; ATC: Redwood City, CA, USA, 1991.
3. Schneider, P.J.; Schauer, B.A. HAZUS—Its Development and Its Future. *Nat. Hazards Rev.* **2006**, *7*, 40–44. [CrossRef]
4. Mackie, K.; Stojadinovic, B. *Fragility Basis for California Highway Overpass Bridge Seismic Decision Making*; PEER 2005/12; University of California: Berkeley, CA, USA, 2005.

5. Applied Technology Council (ATC). *Earthquake Damage Evaluation Data for California*; Report No. ATC-13; ATC: Redwood City, CA, USA, 1985.
6. Kostov, M.; Kaneva, A.; Vaseva, M.; Stefanov, D.; Koleva, N. An Advanced spproach to earthquake risk scenarios of Sofia. In Proceedings of the 8th Pacific Conference on Earthquake Engineering, Singapore, Australia, 5–7 December 2007; pp. 1–9.
7. Rossetto, T.; Elnashai, A. Derivation of vulnerability functions for European-type RC structures based on observational data. *Eng. Struct.* **2003**, *25*, 1241–1263. [CrossRef]
8. Basöz, N.; Kiremidjian, A.S. *Evaluation of Bridge Damage Data from the Loma Prieta and Northridge CA Earthquakes*; Report No. MCEER-98-0004; University at Buffalo: Buffalo, NY, USA, 1997.
9. Sarabandi, P.; Pachakis, D.; King, S. Empirical fragility functions from recent earthquakes. In Proceedings of the 13th World Conference on Earthquake Engineering, Vancouver, BC, Canada, 1–6 August 2004; pp. 1–15.
10. Rota, M.; Penna, A.; Strobbia, C.; Magenes, G. Direct derivation of fragility curves from Italian post-earthquake survey data. In Proceedings of the 14th World Conference on Earthquake Engineering, Beijing, China, 12–17 October 2008; pp. 1–8.
11. Colombi, M.; Borzi, B.; Crowley, H.; Onida, M.; Meroni, F.; Pinho, R. Deriving vulnerability curves using Italian earthquake damage data. *Bull. Earthq. Eng.* **2008**, *6*, 485–504. [CrossRef]
12. Vosooghi, A.; Saiidi, M.S. Experimental Fragility Curves for Seismic Response of Reinforced Concrete Bridge Columns. *ACI Struct. J.* **2012**, *109*, 825–834. [CrossRef]
13. Mander, J.B.; Basöz, N. Seismic fragility curves theory for highway bridges. In Proceedings of the 5th US Conference on Lifeline Earthquake Engineering: Optimazing Post-Earthquake Lifeline System Reliability, Seattle, WA, USA, 12–14 August 1999; pp. 31–40.
14. Moschonas, I.F.; Kappos, A.J.; Panetsos, P.; Papadopoulos, V.; Makarios, T.; Thanopoulos, P. Seismic fragility curves for greek bridges: Methodology and case studies. *Bull. Earthq. Eng.* **2008**, *7*, 439–468. [CrossRef]
15. Shinozuka, M.; Feng, M.Q.; Kim, H.-K.; Kim, S.-H. Nonlinear Static Procedure for Fragility Curve Development. *J. Eng. Mech.* **2000**, *126*, 1287–1295. [CrossRef]
16. Karim, K.R.; Yamazaki, F. A simplified method of constructing fragility curves for highway bridges. *Earthq. Eng. Struct. Dyn.* **2003**, *32*, 1603–1626. [CrossRef]
17. Kwon, O.-S.; Elnashai, A.S. Fragility analysis of a highway over-crossing bridge with consideration of soil–structure interactions. *Struct. Infrastruct. Eng.* **2010**, *6*, 159–178. [CrossRef]
18. Nielson, B.G.; Desroches, R. Analytical Seismic Fragility Curves for Typical Bridges in the Central and Southeastern United States. *Earthq. Spectra* **2007**, *23*, 615–633. [CrossRef]
19. Nielson, B.G.; Desroches, R. Seismic fragility methodology for highway bridges using a component level approach. *Earthq. Eng. Struct. Dyn.* **2007**, *36*, 823–839. [CrossRef]
20. Padgett, J.E. Seismic Vulnerability Assessment of Retrofitted Bridges Using Probabilistic Methods. Ph.D. Thesis, Georgia Institute of Technology, Atlanta, GA, USA, 4 April 2007.
21. Mosleh, A.; Razzaghi, M.S.; Jara, J.; Varum, H. Seismic fragility analysis of typical pre-1990 bridges due to near- and far-field ground motions. *Int. J. Adv. Struct. Eng.* **2016**, *8*, 1–9. [CrossRef]
22. Mosleh, A.; Razzaghi, M.S.; Jara, J.; Varum, H. Development of fragility curves for RC bridges subjected to reverse and strike-slip seismic sources. *Earthquakes Struct.* **2016**, *11*, 517–538. [CrossRef]
23. Mosleh, A.; Jara, J.; Razzaghi, M.S.; Varum, H. Probabilistic Seismic Performance Analysis of RC Bridges. *J. Earthq. Eng.* **2018**, *24*, 1704–1728. [CrossRef]
24. Erberik, M.; Elnashai, A.S. Fragility analysis of flat-slab structures. *Eng. Struct.* **2004**, *26*, 937–948. [CrossRef]
25. Kirçil, M.S.; Polat, Z. Fragility analysis of mid-rise R/C frame buildings. *Eng. Struct.* **2006**, *28*, 1335–1345. [CrossRef]
26. Hwang, H.; Jernigan, J.B.; Lin, Y.-W. Evaluation of Seismic Damage to Memphis Bridges and Highway Systems. *J. Bridg. Eng.* **2000**, *5*, 322–330. [CrossRef]
27. Yu, O.; Allen, D.L.; Drnevich, V.P. Seismic vulnerability assessment of bridges on earthquake priority routes in Western Kentucky. Proceedings of 3rd US National Conference on Lifeline Earthquake Engineering, Los Angeles, CA, USA, 22–23 August 1991; pp. 817–826.
28. Kappos, A.J.; Pitilakis, K.; Stylianidis, K.C. Cost-Benefit analysis for the seismic rehabilitation of buildings in Thessaloniki, based on a hybrid method of vulnerability assessment. In Proceedings of the Fifth International Conference on Seismic Zonation, Nice, France, 17–19 October 1995; pp. 406–413.
29. Kappos, A.J.; Stylianidis, K.C.; Pitilakis, K. Development of Seismic Risk Scenarios Based on a Hybrid Method of Vulnerability Assessment. *Nat. Hazards* **1998**, *17*, 177–192. [CrossRef]
30. Kappos, A.J.; Panagopoulos, G.; Panagiotopoulos, C.; Penelis, G. A hybrid method for the vulnerability assessment of R/C and URM buildings. *Bull. Earthq. Eng.* **2006**, *4*, 391–413. [CrossRef]
31. Cornell, C.A.; Jalayer, F.; Hamburger, R.O.; Foutch, D.A. Probabilistic Basis for 2000 SAC Federal Emergency Management Agency Steel Moment Frame Guidelines. *J. Struct. Eng.* **2002**, *128*, 526–533. [CrossRef]
32. Bazzurro, P.; Cornell, C.A.; Shome, N.; Carballo, J.E. Three Proposals for Characterizing MDOF Nonlinear Seismic Response. *J. Struct. Eng.* **1998**, *124*, 1281–1289. [CrossRef]
33. Shome, N.; Cornell, C.A.; Bazzurro, P.; Carballo, J.E. Earthquakes, Records, and Nonlinear Responses. *Earthq. Spectra* **1998**, *14*, 469–500. [CrossRef]

34. Vamvatsikos, D.; Cornell, C.A. Incremental dynamic analysis. *Earthq. Eng. Struct. Dyn.* **2002**, *31*, 491–514. [CrossRef]
35. Jalayer, F.; Cornell, C.A. *A Technical Framework for Probability-Based Demand and Capacity Factor Design (DCFD) Seismic Formats*; PEER 2003/08; University of California: Berkeley, CA, USA, 2003.
36. Jalayer, F.; Cornell, C.A. Alternative non-linear demand estimation methods for probability-based seismic assessments. *Earthq. Eng. Struct. Dyn.* **2009**, *38*, 951–972. [CrossRef]
37. Porter, K.; Kennedy, R.; Bachman, R. Creating Fragility Functions for Performance-Based Earthquake Engineering. *Earthq. Spectra* **2007**, *23*, 471–489. [CrossRef]
38. Baker, J.W. *Fitting Fragility Functions to Structural Analysis Data Using Maximum Likelihood Estimation*. Working Paper–PEER. 2011. Available online: https://www.scirp.org/(S(vtj3fa45qm1ean45vvffcz55))/reference/ReferencesPapers.aspx?ReferenceID=1344212 (accessed on 1 August 2021).
39. Bakalis, K.; Vamvatsikos, D. Seismic Fragility Functions via Nonlinear Response History Analysis. *J. Struct. Eng.* **2018**, *144*, 04018181. [CrossRef]
40. Tubaldi, E.; Freddi, F.; Barbato, M. Probabilistic seismic demand model for pounding risk assessment. *Earthq. Eng. Struct. Dyn.* **2016**, *45*, 1743–1758. [CrossRef]
41. Ramamoorthy, S.K.; Gardoni, P.; Bracci, J.M. Probabilistic Demand Models and Fragility Curves for Reinforced Concrete Frames. *J. Struct. Eng.* **2006**, *132*, 1563–1572. [CrossRef]
42. Bai, J.-W.; Gardoni, P.; Hueste, M.B.D. Story-specific demand models and seismic fragility estimates for multi-story buildings. *Struct. Saf.* **2011**, *33*, 96–107. [CrossRef]
43. Freddi, F.; Padgett, J.E.; Dall'Asta, A. Probabilistic seismic demand modeling of local level response parameters of an RC frame. *Bull. Earthq. Eng.* **2016**, *15*, 1–23. [CrossRef]
44. Aljawhari, K.; Gentile, R.; Freddi, F.; Galasso, C. Effects of ground-motion sequences on fragility and vulnerability of case-study reinforced concrete frames. *Bull. Earthq. Eng.* **2020**, 1–31. [CrossRef]
45. Gardoni, P.; Der Kiureghian, A.; Mosalam, K.M. Probabilistic Capacity Models and Fragility Estimates for Reinforced Concrete Columns based on Experimental Observations. *J. Eng. Mech.* **2002**, *128*, 1024–1038. [CrossRef]
46. Gardoni, P.; Mosalam, K.M.; Der Kiureghian, A. Probabilistic seismic demand models and fragility estimates for rc bridges. *J. Earthq. Eng.* **2003**, *7*, 79–106. [CrossRef]
47. Jalayer, F.; Ebrahimian, H.; Miano, A.; Manfredi, G.; Sezen, H. Analytical fragility assessment using unscaled ground motion records. *Earthq. Eng. Struct. Dyn.* **2017**, *46*, 2639–2663. [CrossRef]
48. Nazri, F.M.; Miari, M.; Kassem, M.M.; Tan, C.-G.; Farsangi, E.N. Probabilistic Evaluation of Structural Pounding Between Adjacent Buildings Subjected to Repeated Seismic Excitations. *Arab. J. Sci. Eng.* **2018**, *44*, 4931–4945. [CrossRef]
49. Flegga, M.; Favvata, M. Global and local performance levels on the probabilistic evaluation of the structural pounding effect between adjacent rc structures. In Proceedings of the 11 International Conference on Structural Dynamics, EURODYN, Athens, Greece, 23–26 November 2020; pp. 3762–3779. [CrossRef]
50. Kazemi, F.; Miari, M.; Jankowski, R. Investigating the effects of structural pounding on the seismic performance of adjacent RC and steel MRFs. *Bull. Earthq. Eng.* **2020**, *19*, 317–343. [CrossRef]
51. Flenga, M.G.; Favvata, M.J. Probabilistic seismic assessment of the pounding risk based on the local demands of a multistory RC frame structure. *Eng. Struct.* **2021**, *245*, 112789. [CrossRef]
52. Aslani, H.; Miranda, E. Fragility assessment of slab-column connections in existing non-ductile reinforced concrete buildings. *J. Earthq. Eng.* **2005**, *9*, 777–804. [CrossRef]
53. Freddi, F.; Tubaldi, E.; Ragni, L.; Dall'Asta, A. Probabilistic performance assessment of low-ductility reinforced concrete frames retrofitted with dissipative braces. *Earthq. Eng. Struct. Dyn.* **2012**, *42*, 993–1011. [CrossRef]
54. Mackie, K.R.; Stojadinović, B. Comparison of Incremental Dynamic, Cloud, and Stripe Methods for Computing Probabilistic Seismic Demand Models. In Proceedings of the Structures Congress 2005: Metropolis and Beyond, New York, NY, USA, 20–24 April 2005; pp. 1–11. [CrossRef]
55. Shome, N. *Probabilistic Seismic Demand Analysis of Nonlinear Structures*; Report No. RMS-35; Department of Civil Engineering, Stanford University: Stanford, CA, USA, 1999.
56. Prakash, V.; Powell, G.H.; Gampbell, S. *DRAIN-2DX Base Program Description and User's Guide, UCB/SEMM*; Report No. 17/93; University of California: Orkland, CA, USA, 1993.
57. Karayannis, C.G.; Favvata, M.J. Inter-story pounding between multistory reinforced concrete structures. *Struct. Eng. Mech.* **2005**, *20*, 505–526. [CrossRef]
58. Karayannis, C.G.; Favvata, M.J. Earthquake-induced interaction between adjacent reinforced concrete structures with non-equal heights. *Earthq. Eng. Struct. Dyn.* **2004**, *34*, 1–20. [CrossRef]
59. Favvata, M.J. Minimum required separation gap for adjacent RC frames with potential inter-story seismic pounding. *Eng. Struct.* **2017**, *152*, 643–659. [CrossRef]
60. PEER Ground Motion Database. 2011. Available online: https://peer.berkeley.edu/peer-strong-ground-motion-databases (accessed on 10 February 2017).
61. Eurocode 8. *Design of Structures for Earthquake Resistance. Part 1: General Rules, Seismic Actions and Rules for Buildings*; EN 1998-1; European Committee for Standardization: Brussels, Belgium, 2004.

62. Applied Technology Council (ATC). *Seismic Evaluation and Retrofit of Concrete Buildings*; Report No. ATC-40; ATC: Redwood City, CA, USA, 1996; Volume 1.
63. Minitab. Available online: https://support.minitab.com (accessed on 12 May 2021).
64. Karamlou, A.; Bocchini, P.; Bochini, P. Computation of bridge seismic fragility by large-scale simulation for probabilistic resilience analysis. *Earthq. Eng. Struct. Dyn.* **2015**, *44*, 1959–1978. [CrossRef]

applied sciences

MDPI

Article

IMPA versus Cloud Analysis and IDA: Different Methods to Evaluate Structural Seismic Fragility

Carlotta Pia Contiguglia [1,*], Angelo Pelle [1], Bruno Briseghella [2] and Camillo Nuti [1]

[1] Department of Architecture, Roma Tre University, 00153 Rome, Italy; angelo.pelle@uniroma3.it (A.P.); camillo.nuti@uniroma3.it (C.N.)
[2] College of Civil Engineering, Fuzhou University, Fuzhou 350108, China; bruno@fzu.edu.cn
* Correspondence: carlottapiacontiguglia@uniroma3.it

Abstract: Well-known methods for seismic performance assessment, such as incremental dynamic analysis (IDA), multi-stripes analysis (MSA) and the cloud method, involve nonlinear response time-history analyses to characterize the relationship between the chosen damage measure versus intensity measure. Over the past two decades, many authors have proposed simplified procedures or nonlinear static approaches to develop fragility. In these procedures, the capacity of the system is evaluated by nonlinear static procedures (i.e., the capacity spectrum method (CSM), the N2 method, modal pushover analysis (MPA)) and the demand is derived by response spectra. In addition to the familiar ones, incremental modal pushover analysis (IMPA) is a novel nonlinear static procedure proposed in recent years, and it is used in this research to present an *IM*-based fragility estimation. The accuracy and effectiveness of different methods to assess vulnerability are investigated by comparing fragility curves derived by MPA-based cloud analysis, IMPA and cloud analysis against IDA. The comparison gives valuable insights on the influence of scaling on different sets of records; however, a more extended validation is needed to confirm the obtained results and draw more general conclusions. Results arise from two relatively small bins of record motions differing by ranges of Joyner-Boore distance and scattered in a range of magnitude are presented.

Keywords: IDA; the cloud method; IMPA; MPA; nonlinear static analysis; nonlinear dynamic analysis; fragility curve

Citation: Contiguglia, C.P.; Pelle, A.; Briseghella, B.; Nuti, C. IMPA versus Cloud Analysis and IDA: Different Methods to Evaluate Structural Seismic Fragility. *Appl. Sci.* **2022**, *12*, 3687. https://doi.org/10.3390/app12073687

Academic Editor: Maria Favvata

Received: 20 February 2022
Accepted: 4 April 2022
Published: 6 April 2022

Publisher's Note: MDPI stays neutral with regard to jurisdictional claims in published maps and institutional affiliations.

Copyright: © 2022 by the authors. Licensee MDPI, Basel, Switzerland. This article is an open access article distributed under the terms and conditions of the Creative Commons Attribution (CC BY) license (https://creativecommons.org/licenses/by/4.0/).

1. Introduction

Performance-based earthquake engineering (PBEE) procedures allow the prediction and evaluation of the probabilistic seismic performance of bridges and buildings in terms of system-level decision variables, such as loss of use, repair cost and casualties. In the United States, the first generation of PBEE assessment and design procedures for buildings (SEAOC Vision 2000, FEMA 273, ATC-40 [1–4]) took significant steps toward achieving performance-based earthquake engineering. Since then, the Pacific Earthquake Engineering Research Center (PEER) has been working on developing a more robust methodology that involves four stages: hazard analysis, structural analysis, damage analysis, and loss analysis [5]. In the third stage, damage analysis, fragility functions describe the conditional probability of component, element or system to be damaged for a given intensity measure. The first attempt to determine fragility curves can be dated back to 1975, when the Seismic Design Decision Analysis (SDDA) procedure was proposed in the US [6]. Further developments [7,8] were initially applied in the field of the nuclear industry to define a probabilistic relationship between an intensity measure representing seismic input and a damage measure representing the seismic failure of a component of a nuclear power plant. From then on, several methods to estimate fragility (expert-based, experimental, analytical, hybrid, empirical) have been developed by researchers worldwide, relying on different assumptions and restrictions to overcome prevalent intrinsic uncertainties. However, due to the extremely high subjectiveness, lack of data and other drawbacks typical of expert-based, empirical and experimental methods, the common practice has aroused its interest in analytical and hybrid methods during the last two decades.

Among analytical approaches to derivate fragility curves (probabilistic seismic demand model, elastic spectral analysis, nonlinear static analysis, linear or nonlinear time history analysis [9–17]), incremental dynamic analysis (IDA) is a parametric analysis method developed in 1998 and deeply discussed in 2002 [18,19]. IDA became a worldwide method used by engineers and researchers, and it is still widespread. However, the introduction of uncertainties due to an excessively coarse description of seismic input with varying intensity is inevitable. Eventually, many authors have pointed out that a simple amplitude scaling of ground motion records is one of the main shortcomings in IDA, together with its high computational demand [20–23]. In IDA, a certain number of inputs amplitude scaled to define $IM = im$, then NL-THA is performed and DCR_{LS} so determined are used to define the distribution of $DCR_{LS} \mid IM = im$. This process is repeated by varying the scale factor to define the seismic response in a whole range of seismic intensities. According to previous research [24–26], the median response of a structure subjected to scaled ground motion records is comparable to that of a structure subjected to unscaled earthquake ground motion records. However, because a single *IM* is a highly simplified description of ground-motion severity, the value of DCR_{LS} from different ground-motion with $IM = im$ may be different, and thus the relationship $DCR_{LS} \sim f(IM)$ is probabilistic. A common limitation in current databases is the lack of strong ground motion records covering high-intensity intervals at specific periods of the structure [27,28]. Thus, an excessive scaling to fit within high-intensity intervals may occur, biasing the structural response [29], generating a false correlation between *IM*s and EDPs, and increasing uncertainties in the structural response.

In contrast to IDA and MSA (multiple-stripe analysis), the cloud method [26,30,31] involves nonlinear analysis of the structure subjected to a sample of different distance/intensity combined unscaled as-recorded ground motion, which may reduce the number of analyses, uncertainties in seismic input with intensity, computational effort in defining a seismic fragility curve [32], and is based on a regression in the logarithmic space of structural response versus seismic intensity.

To respond to the need for simplified, faster and/or approximate methods, from the late 1990′s, many studies have been published regarding the use of pushover analysis procedures to assess seismic vulnerability, mostly on bridges. In this case, the capacity of the system is evaluated by using nonlinear static procedures (NSPs) (i.e., the capacity spectrum method [33–36], the N2 method [37–39], and modal pushover analysis [20]), while the demand is estimated by response spectra. To assess the reliability of these analytical procedures, different authors have compared developed fragility curves to those obtained by nonlinear time history analysis [34,36,38].

Among NSPs, a novel procedure called incremental modal pushover analysis (IMPA) has been proposed in recent years by Bergami and his co-workers [40]. IMPA requires the execution of modal pushover analysis (MPA) and the evaluation of structural performance within a range of different seismic intensity levels to develop a multimodal capacity curve in terms of base shear versus top displacement. This approach is suitable for performing a displacement-based design procedure and structural analysis of existing structures, yet authors have not suggested the analytical estimation of fragility.

This paper aims to evaluate the reliability of structural fragility derived by the methods mentioned, advancing an *IM*-based derivation of structural fragility, strikingly similar to IDA, based on IMPA. It is known that IDA has a small sensitivity to record-to-record variability compared to other methodologies. Nevertheless, the results indicate that, amidst its slightly higher sensitivity, IMPA has the advantage of requiring considerably smaller computational effort to perform the structural analysis.

The authors argue that scaling response spectra at a higher range of intensities might introduce less uncertainties than a simple amplitude scaling of ground motions. Further steps of this research will address how the uncertainties in the seismic input affect the reliability of IMPA versus IDA seismic fragility for strong ground motions.

In the following paragraph, analyses are carried out on a real RC frame belonging to a school building located in Norcia (Italy). Each nonlinear dynamic and static procedure is briefly presented, including a step-by-step computational procedure of IMPA. Finally, the reliability of pushover-based estimation of seismic vulnerability is assessed by comparing these curves to those obtained by IDA.

2. Methodology

2.1. Choice of Engineering Demand Parameter and Intensity Measure

The first-mode spectral acceleration $S_a(T_1,\xi = 5\%)$ is commonly used as an intensity measure (*IM*) parameter [20,21]. Shome et al. [24] stated that the nonlinear response of an MDOF structure dominated by the first mode of vibration depends on the "intensity" of the records at the first period of vibration, while magnitude and distance play a minor role in it. For these reasons, the 5% damped spectral acceleration at the structure's first-mode period $S_a(T_1,\xi = 5\%)$ or simply S_a is adopted as the *IM* in this work, since the structure selected as the case study is dominated by the first mode of vibration (structure's first-mode period of vibration $T_1 = 0.62$ s and the modal mass participation at first-mode is 82%, see Section 3.1 for more details).

In the literature, various engineering demand parameters (EDPs) have been proposed [24], somehow representative of the structure's local or global damaged state. In this study, the critical demand to the capacity ratio for the desired limit state (*LS*), denoted as DCR_{LS} [23,41], is assumed to be the EDP. It represents the demand-to-capacity ratio which brings the system closer to the onset of limit state (herein, the life-safety limit state). The weakest-link formulation is adopted to evaluate the DCR_{LS} (Equation (1)), which means that if the demand-to-capacity ratio D_{jl}/C_{jl} is equal to or higher than unity in just one element, then the structure attains the expected limit state for the *l*th mechanism.

$$DCR_{LS} = \max_l^{N_{mech}} \max_j^{N_e} \left(\frac{D_{jl}}{C_{jl}(LS)} \right) \quad (1)$$

where N_{mech} and N_e are the numbers of the considered potential mechanism of failure and the number of the elements taking part in the *l*th mechanism, respectively. D_{jl} and $C_{jl}(LS)$ are the demand and the limit state capacity, respectively, evaluated for the *j*th element of the *l*th mechanism.

In particular, in this work, only a ductile failure mechanism in columns and beams is considered as a potential failure mechanism ($N_{mech} = 1$). Therefore, in this deformation-based critical DCR_{LS}, the demand *D* is expressed in terms of maximum chord rotation in the *j*th component. Instead, the capacity C in terms of chord rotation is evaluated according to guidelines reported in the Commentary [42] of NTC 2018. Namely, the chord rotation for life-safety limit state is defined as $\frac{3}{4}$ of that corresponding to near-collapse limit state θ_u, evaluated according to Equation C8.7.2.5 of the Commentary (Equation (2)).

$$\theta_u = \left(\theta_y + (\phi_u - \phi_y) L_{pl} \left(1 - \frac{0.5 L_{pl}}{L_v} \right) \right) \quad (2)$$

where θ_y is the yield chord rotation, ϕ_u and ϕ_y are the ultimate and yield curvature, respectively, L_{pl} is the plastic hinge length, and L_v is the shear length.

2.2. Record Selection

The PEER Next Generation Attenuation (NGA)—West 2 Project [43] has been used to define the main database of 210 ground motions. In addition, two ground motions representing the 2016 Norcia earthquake, with epicentral distances of 4.6 km and 26.9 km, respectively, have been extracted from the Italian Accelerometric Archive [44] and included in the aforementioned database.

A set of 36 as-recorded ground motions listed in Table 1 has been defined with an average shear wave velocity to a depth of 30 m ($V_{s,30}$) falling between 213 m/s and 724 m/s, therefore corresponding to the types of mass B′ or C′ (according to Eurocode 8 [45]) which are mixed into the set. This latter comprehends highly scattered values of magnitude M_w, ranging from 5 to 7.5, and Joyner-Boore distance included between 0 km and 50 km. The set includes about 50% of near-fault (19 records), ranging from an epicentral distance of 0–10 km, and 50% of far-field records (17 records), or records with an epicentral distance greater than 10 km. The selection exhibits a prevalence of three fault mechanisms: normal, reverse, and strike-slip. Since the frame model used in this study is 2D (see Section 3), two orthogonal directions of the same seismic event are avoided. Record selection also comprehends a wide range of *IM* and distributed values of DCR_{LS}, with at least one-third of the values greater than 1 [23]. The original set of 36 records was split into two subsets depending on the epicentral distance as shown in Figures 1 and 2, and these were studied separately to investigate the different effects of near-fault versus far-field ground motions. It is well known that the proximity to the fault renders the same ground motions (NF) different from ordinary (FF) ground motions [46]. The near-fault records selection avoids including pulse-like ground motions. Impulsive signals have been identified using the open-source algorithm proposed by Shahi and Baker [47,48]. This can identify pulses at arbitrary orientations using continuous wavelet transforms of two horizontal orthogonal components of a ground motion to identify the orientation that may contain a pulse.

Table 1. Details of the two subsets of ground motion data deepened for the study based on the NGA—West 2 database.

File ID	Earthquake Name	RSN	Year	Mech.	M_w	R_{jb} [km]	$V_{s,30}$ [m/s]	DS-595 [s]	DS-575 [s]
1	"Oroville-01"	106	1975	Normal	5.89	7.79	680.37	3.4	1.5
2	"Oroville-03"	114	1975	Normal	4.7	7.35	418.97	4.4	1.3
3	"Santa Barbara"	136	1978	Reverse Oblique	5.92	0	514.99	7.5	4.3
4	"Tabas_Iran"	139	1978	Reverse	7.35	0	471.53	11.3	6.7
5	"Helena_Montana-01"	1	1935	Strike-slip	6	2.07	593.35	2.5	1.2
6	"Dursunbey_Turkey"	144	1979	Normal	5.34	5.57	585.04	2.5	1.4
7	"Coyote Lake"	145	1979	Strike-slip	5.74	5.3	561.43	8.5	2.7
8	"Norcia_Italy"	156	1979	Normal	5.9	1.41	585.04	5.7	2.7
9	"Livermore-02"	222	1980	Strike-slip	5.42	7.94	550.88	4.5	1.1
10	"Anza (Horse Canyon)-01"	226	1980	Strike-slip	5.19	5.85	617.78	2.4	1.1
11	"Mammoth Lakes-06"	249	1980	Strike-slip	5.94	6.45	373.18	5.1	2.5
12	"Izmir_Turkey"	134	1977	Normal	5.3	0.74	535.24	1.6	0.3
13	"Mammoth Lakes-07"	253	1980	Strike-slip	4.73	3.86	377.41	10.2	3.1
14	Imperial Valley-02	6	1940	Strike-Slip	6.95	6.09	213.44	24.2	17.7
15	Chalfant Valley-04	563	1986	Strike-Slip	5.44	8.88	316.19	17.1	7.7
16	Kalamata, Greece-01	564	1986	Normal	6.2	6.45	382.21	6.1	1.9
17	Kalamata, Greece-02	565	1986	Normal	5.4	4	382.21	4.2	1
18	Loma Prieta	752	1989	Reverse Oblique	6.93	8.65	288.62	13.2	5.6
19	Central Italy	n.a.	2016	Normal	6.5	4.6	498	n.a.	n.a.
20	"Kern County"	15	1952	Reverse	7.36	38.42	385.43	30.3	10.7
21	"Lytle Creek"	49	1970	Reverse Oblique	5.33	42.14	667.13	5.1	2.9
22	"Santa Barbara"	135	1978	Reverse Oblique	5.92	23.75	465.51	7	3.4
23	"San Fernando"	81	1971	Reverse	6.61	35.54	529.09	13.7	7.1
24	"Northern Calif-07"	101	1975	Strike-slip	5.2	28.73	567.78	5.7	4.3
25	"Oroville-02"	108	1975	Normal	4.79	12.07	377.25	7.1	3.3
26	"Friuli_ Italy-01"	125	1976	Reverse	6.5	14.97	505.23	4.9	2.5
27	"Coyote Lake"	152	1979	Strike-slip	5.74	20.44	362.98	8.2	3.9
28	"Norcia_ Italy"	157	1979	Normal	5.9	13.21	535.24	10.5	5.9
29	"Anza (Horse Canyon)-01"	225	1980	Strike-slip	5.19	12.24	724.89	2.1	0.7
30	"Victoria_ Mexico"	265	1980	Strike-slip	6.33	13.8	471.53	8.2	4.4
31	"Mammoth Lakes-04"	241	1980	Strike-slip	5.7	12.75	537.16	11.5	3.4
32	"Mammoth Lakes-09"	274	1980	Strike-slip	4.85	10.96	377.41	16	7.7
33	"Almiros_ Greece"	279	1980	Normal	5.2	13.25	412.68	10	4.6
34	"Coalinga-02"	370	1983	Reverse	5.09	24.23	467.03	13.7	8.6
35	"Borah Peak_ ID-02"	442	1983	Normal	5.1	16.31	468.44	5	2.3
36	Central Italy	n.a.	2016	Normal	6.5	26.9	n.a.	n.a.	n.a.

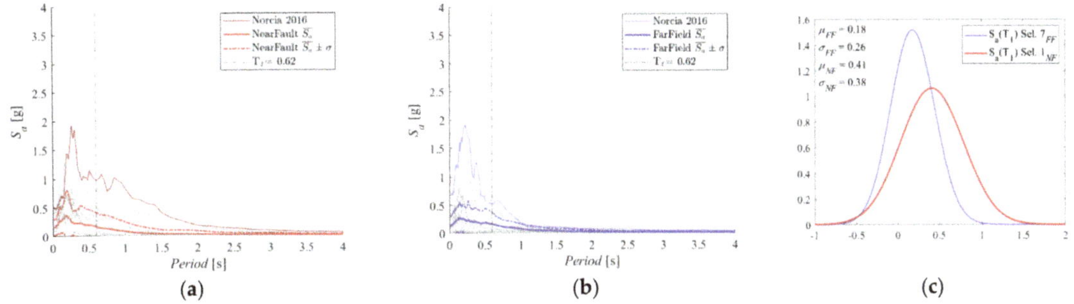

Figure 1. Elastic response spectra of the (**a**) near-fault and (**b**) far-field subset of records. The figures highlight the two records of the 2016 Norcia Earthquake, $\overline{S_a}$ is the average response spectra from the two set, and $\overline{S_a} \pm \sigma$ is the range of variance according to standard deviation. (**c**) Normal distribution of S_a for $T = T_1$.

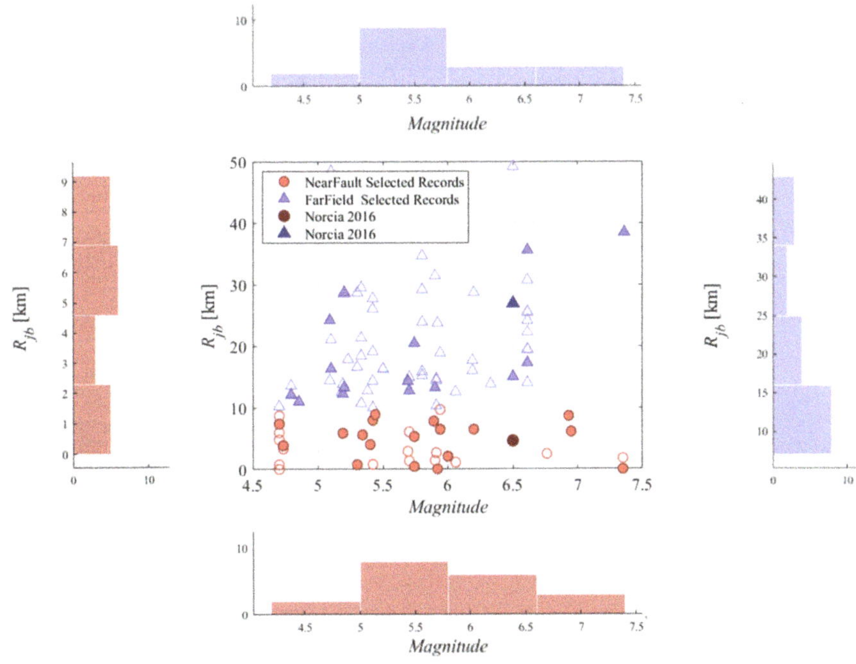

Figure 2. M_w magnitude–R_{jb} distance scatter diagrams of the two subsets, Sel. 1 $_{NF}$ and Sel. 7 $_{FF}$.

2.3. *Performed Nonlinear Analysis*

2.3.1. Cloud-Based Analysis

The cloud-based Analysis (CA) is particularly suitable to assess structural fragility both for the simplicity of its formulation and for the low required computational effort. Conversely, it is extremely sensitive to the record selections and based on a few simplifying assumptions, such as fixed standard error of the regression [23,31,32,39,41,49].

CA adopts a linear regression model in the logarithm scale to fit the pairs of demand to capacity ratio (DCR_{LS}) and IM, where DCR_{LS} are calculated through nonlinear analysis. The regression-based probability model describes the DCR_{LS} for a given IM level and can be evaluated by Equations (3) and (4):

$$\mathrm{E}[\ln DCR_{LS}|IM] = \ln \eta_{DCR_{LS}|IM} = \ln a + b \ln IM \tag{3}$$

$$\sigma_{\ln \beta_{DCR_{LS}|IM}} \cong \beta_{DCR_{LS}|IM} = \sqrt{\sum_{i=1}^{N} \left(\ln DCR_{LS,i} - \ln \eta_{DCR_{LS}|IM_i}\right)^2 / (N-2)} \tag{4}$$

where E[lnDCR_{LS} | IM] is the expected value for the natural logarithm of DCR_{LS} given IM, and $\eta_{DCRLS\,|\,IM}$ and $\sigma_{\ln DCRLS\,|\,IM}$ are the median and logarithmic standard deviation for DCR_{LS} given IM, respectively. The constants ln$a$ and b are the linear least square regression coefficients. Finally, the structural fragility obtained based on the CA is (Equation (5)):

$$\mathrm{P}(DCR_{LS} > 1|IM) = \mathrm{P}(\ln DCR_{LS} > 0|IM) = \Phi\left(\frac{\ln \eta_{DCR_{LS}|IM}}{\beta_{DCR_{LS}|IM}}\right) \tag{5}$$

where Φ is the standard Gaussian cumulative distribution function.

In this work, two different methodologies are adopted to find the relationships of IM versus DCR_{LS} for the structure under investigation. Namely, in one case, time-history analyses are employed to evaluate the demand D_{jl} (demand of the jth element of lth mechanism) at each time step. Within this paper, this approach is named dynamic cloud analysis, or shortly D-CA.

In the other case, the demand D_{jl} is computed by adopting the modal pushover analysis (MPA). This approach is referred to as MPA-CA. The modal pushover analysis [50,51] is a nonlinear static procedure based on static analysis of the structure subjected to lateral forces distributed over the building height according to nth modal shape. Chopra and Goel [51] showed that this procedure is accurate enough for practical application. The MPA procedure used in this work adopts the capacity spectrum method (CSM), a nonlinear static analysis procedure to assess the seismic vulnerability of buildings originally proposed by Freeman [52]. The procedure permits finding a correlation between earthquake ground motions and building performance [53] (ATC, 1982) comparing a response spectrum (representing structure demand) and a pushover curve (representing building capacity) by an iterative procedure. This latter aims to the definition of the performance point (PP), which represents the state of maximum inelastic displacement of a building for a given seismic event. To plot the two curves in the same chart, RS and pushover curves are transformed into an acceleration displacement response spectrum (ADRS). The whole procedure employed can be summarized in the following steps (see Figure 3):

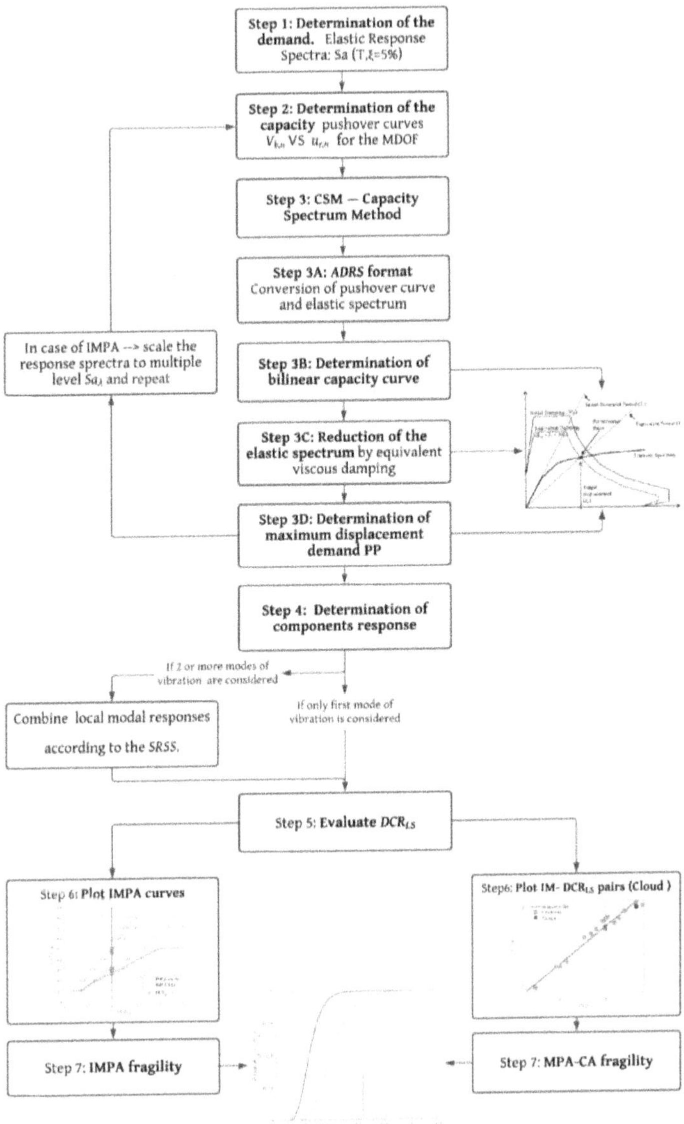

Figure 3. Flowchart of MPA-CA and IMPA procedures. $V_{b,n}$ and $u_{r,n}$ are the base shear and the top displacement respectively, PP is the performance point and DCR_{LS} is the critical demand to the capacity ratio for the desired limit state (LS).

1. Determine demand: elastic response spectrum (T, ξ = 5%);
2. Evaluate the capacity via pushover curves in terms of base shear $V_{b,n}$ versus top displacement $u_{r,n}$ for the MDOF structure subjected to lateral forces profile proportional to nth mode shape;
3. Determine maximum demand in terms of top displacement via the capacity spectrum method:

- Convert the pushover curve of the nth mode shape to a capacity curve in the ADRS format by (Equations (6)–(9)):

$$a_{C,n} = \frac{V_{b,n}}{M_{tot} \cdot \alpha_n} \tag{6}$$

$$d_{C,n} = \frac{u_{r,n}}{\Gamma_n \cdot \phi_{n,r}} \tag{7}$$

$$\Gamma_n = \frac{\phi_n^T M I}{\phi_n^T M \phi_n} \tag{8}$$

$$\alpha_n = \Gamma_n \frac{\phi_n^T M I}{M_{tot}} \tag{9}$$

where M_{tot} is the total mass of the structure, ϕ_n is the nth natural vibration mode, $\phi_{n,r}$ is the amplitude of the nth natural vibration mode at the roof of the structure, and Γ_n and α_n are the modal participation factor and modal mass of the nth mode, respectively;
- Convert 5% damped response spectrum from the standard pseudo-acceleration S_a versus the period of vibration T format to the ADRS format by (Equation (10)):

$$S_{De}(T) = S_a(T)\left(\frac{T}{2\pi}\right)^2 \tag{10}$$

where $S_D(T)$ is the displacement spectrum;
- Plot demand and capacity diagrams together in the ADRS space. Determine the bilinear capacity curve. Iteratively determine the displacement demand for the nth mode shape. In this step, the dynamic analyses of a sequence of equivalent linear systems with successively updated values of equivalent viscous damping are involved;
- Reduce the elastic spectrum by the equivalent viscous damping (Equation (11));

$$\eta = \sqrt{\frac{10}{5 + v_{eq}}} \tag{11}$$

- Determinate the performance point or the maximum expected demand in terms of top displacement;

4. Convert displacement demand found in step 3 to global top displacement and individual component of local deformation (i.e., interstory drift) for the nth mode shape;
5. Evaluate maximum demand to capacity ratio values according to Equation (1). If two or more modes of vibration are considered, combine the local modal responses according to the square-root-of-sum-of-squares (*SRSS*);
6. Estimate parameters of the linear regression model in the logarithm scale to fit the pairs of demand to capacity ratio (DCR_{LS}) and *IM*;
7. Draw structural fragility curve according to Equation (5).

2.3.2. Incremental Dynamic Analysis (IDA)

In IDA, a nonlinear structural model is subjected to a set of scaled ground motion records (accelerogram a_λ), each scaled to multiple levels of a monotonic scalable intensity measure such as S_a, PGA, PGV (herein $IM = S_a(T_1,\xi = 5\%)$)."As-recorded" unscaled time histories are scaled by using a non-negative scale factor (λ) to obtain a scaled accelerogram a_λ, in which amplitudes are scaled without changing the frequency content of signals. The output of the analysis is represented by a collection of IDA curves, which are a plot of the recorded DCR_{LS} (DM) against $S_a(T_1,\xi = 5\%)$ (*IM*), all parameterized on the same *IM*s and DM [19].

Among all the analytical methods to develop the fragility based on IDA, the following is one of the simplest proposed [54]:

$$P(LS|IM = x) = P(DCR_{LS} \geq 1|IM = x) = P\left(IM^{DCR=1} \leq x\right) \quad (12)$$

In an EDP-based interpretation of the fragility (Equation (12)), the conditional probability of exceeding a limit state given an *IM*, herein the spectral acceleration, is equal to the probability of the demand to capacity ratio of exceeding 1 for a given Sa.

However, it is possible to express the fragility also as the complementary cumulative distribution function or "*IM*-based fragility" (Equation (13)). Incremental dynamic analysis is well suited to be represented by *IM*-based derivation of fragility. This interpretation expresses the seismic fragility as the probability of spectral acceleration values—denoted as ${S_a}^{DCR=1}$ and defined by intercepting all the IDA curves with the DCR_{LS} = 1—to be smaller than a given value. DCR_{LS} = 1 represents the threshold of a limit state *LS* and the intersection provides the empirical distribution of the random variable (*IM*), to which a model such as the lognormal appearing in Equation (12) can be fitted.

$$P\left(IM^{DCR=1} \leq x\right) = \phi\left(\frac{\ln x - \ln \eta_{S_a|DCR=1}}{\beta_{S_a|DCR=1}}\right) \quad (13)$$

In this Equation (13), ϕ denotes the standard normal (Gaussian) cumulative distribution (CDF) of two-parameters (median or log of mean η and standard deviation β) estimated by the second-moment method or "METHOD A" (Equation (14)) proposed by Porter [54].

$$\ln \eta_{S_a|DRC=1} \cong \frac{\sum_{i=1}^{n} \ln S_a^{DCR=1}}{n} \qquad \beta_{S_a|DRC=1} \cong \sqrt{\frac{\sum_{i=1}^{n}\left(\ln S_{a,1}^{DCR=1} - \ln \eta_{S_a|DRC=1}\right)^2}{n-1}} \quad (14)$$

2.3.3. Incremental Modal Pushover Analysis (IMPA)

IMPA is a novel nonlinear static procedure proposed by Bergami and others first for buildings [40,55,56], and later adjusted also for bridges [57–59]. This procedure takes advantage of the simplicity of static analysis, but at the same time it grants the definition of the seismic demand for a certain range of intensity levels by scaling down response spectra. Conceptually, the procedure to find the maximum expected demand for the *j*th element of *l*th mechanism D_{jl} is the same as that describe for MPA-CA in sub-Section 2.3.1, yet in IMPA procedure response spectra are scaled to multiple levels of a chosen monotonic scalable intensity measure as S_a, PGA, PGV (herein $IM = S_u(T_1, \xi - 5\%)$) (Figure 4). For each intensity level, the performance point (P.P.) can be determined and the demand measure D_{jl} combined if two or more modes of vibration are considered, to define a multimodal DCR_{LS}. The output of the analysis can be represented as a collection of "IMPA" curves, which are a plot of the recorded DCR_{LS} against *IM*, wholly similar to IDA curves. The intersection of IMPA curves and the chosen threshold of the limit state (DCR_{LS} = 1) generates the empirical distribution of the random variable (*IM*) for the probabilistic model of the fragility aforementioned in Section 2.3.1 (Equations (12) and (13)).

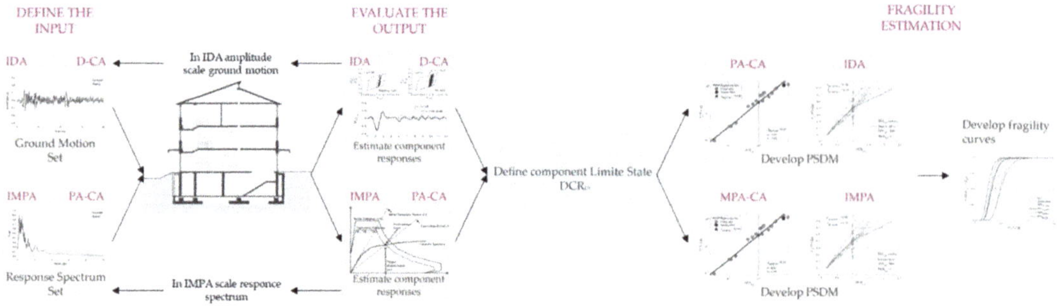

Figure 4. Flowchart schematically shows the main steps to develop seismic fragility curves for nonlinear static and dynamic procedures.

3. Numerical Application

3.1. Frame Description

The transverse frame modelled and analysed in this study comes from an actual school building in Norcia (Italy, 42.7941° North latitude, 13.0963° East longitude). The building, which originally consisted of three aligned blocks, was later joined into a single complex during the various refurbishment works. The building consists of a reinforced concrete (RC) frame structure with a footprint of 12.80 × 59.80 m and a maximum height (from the foundations), corresponding to the roof beams, of about 16 m. The building consists of a one-floor basement, a ground floor, three storeys and an attic above ground. The inter-story height is 3.50 m for the basement floor and ground floor, 3.30 m for the other three floors, and 2.5 m for the attic (Figure 5). Built in 1962, the school has survived a variety of seismic events before the 6.5 magnitude central Italy earthquake that caused severe damage to structural and especially non-structural parts in 2016. In accordance with the construction methods of the time of construction, the structure was designed using 2D models schematizing the reinforced concrete frames in the transverse direction of the building. Although the legislation of the time did not explicitly require it, the designer also took into account the seismic action by applying an acceleration of 0.07 g.

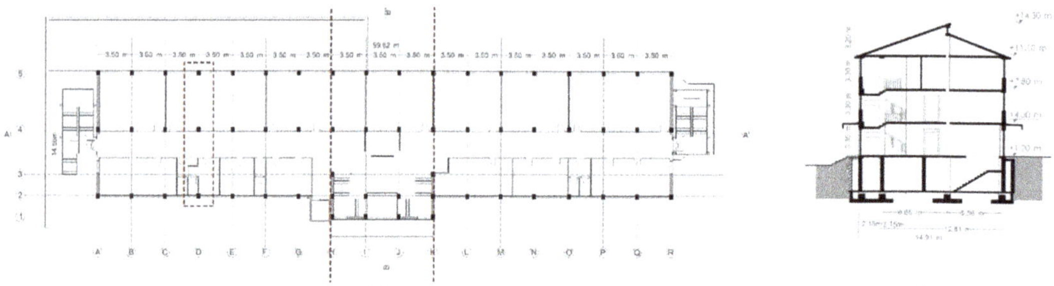

(a) (b)

Figure 5. *Cont.*

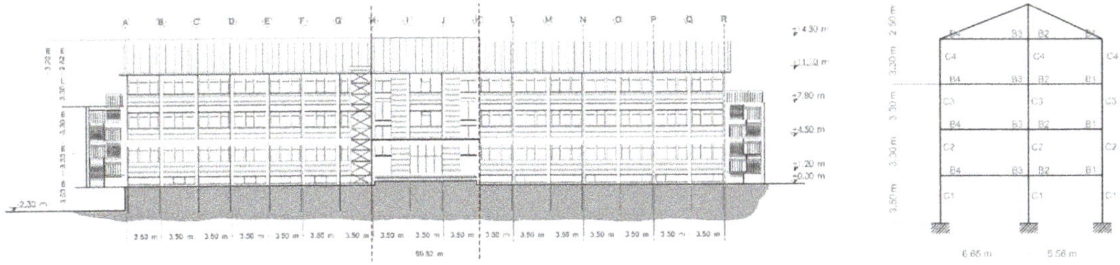

(c) (d)

Figure 5. (**a**) Plan; (**b**) transverse section; (**c**) elevation; (**d**) modelled frame. The structural joint originally designed for thermal deformations are highlighted by orange hidden lines.

The frame under study belongs to one of the two lateral blocks (Figure 5) and is a two-bay (5.65 m and 5.56 m span) regular cross frame. It is considered to be fixed at the base, and basement and soil-structure interaction has not been considered.

The geometry of the frame, the column and beam geometry, and the reinforcement details are shown in Figures 5 and 6. The permanent structural load (G1) and permanent non-structural load (G2) are calculated as G1 + G2 = 5.1 kN/m^2 (from ground floor to 2nd floor), G1 + G2 = 4.1 kN/m^2 (3rd floor), G1 + G2 = 4.22 kN/m^2 (roof beams), the live load is taken as Q1 = 3 kN/m^2 (from ground floor to 2nd floor), Q1 = 1 kN/m^2 (3rd floor), Q2 = 1.8 kN/m^2 (roof beams) and taken as concentrated gravity on the columns at the edge of each floor. Each floor was assigned a seismic mass equal to 1/6 of the total mass of one of the three original blocks of the building.

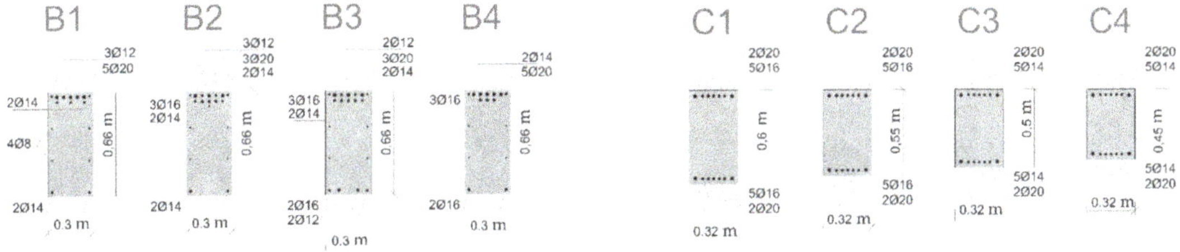

Figure 6. Columns (C) and beams (B) cross-sections and longitudinal reinforcement details.

Fundamental and second periods of the frame have been evaluated as T_1 = 0.62 s and T_2 = 0.21 s, respectively.

3.2. FE Model Description

The nonlinear FE model of the analysed frame was developed in the OpenSEES platform [60]. To account for the nonlinearity, "Beam With Hinges Element", already available in the OpenSEES library, was used to model columns and beams. This element adopts a lumped plasticity formulation with plastic hinges at the end of the element connected by an elastic link. This means that all nonlinearities are concentrated at the ends of the elements and can be only propagated along the length L_p of the plastic hinge, unlike the distributed plasticity formulation where they may spread along the whole element. Therefore, the length of the plastic hinge L_p plays an important role in avoiding the concentration of strain at the element ends. In this work, in good agreement with the L_p evaluated by the equation proposed by Priestley and Park [61], it is assumed to be equal to the cross-section height. The two-point Gauss integration was used on the element interior, while two-point Gauss-Radau integration was applied over lengths of $4L_p$ at the element ends. A total of six integration points were used [62,63]. To account for non-linearity, a four-point moment-curvature relationship was assigned to the element ends. The four-point law accounting for crack, yield, ultimate failure and collapse state (80% of ultimate failure) was evaluated through the software Response2000 [64], which can simulate nonlinear sectional behaviour by assuming a suitable law for the material. Due to the dependence of the sectional response on the applied axial load, it is assumed to be zero for beams, while it is estimated for the column considering their area of influence. Shear failure was not considered in the model. The Newton line search method was considered as the solution algorithm for the time-history analyses, which increased the effectiveness of the Newton–Raphson algorithm by introducing line search to solve the nonlinear residual equation. The tolerance and maximum number of iterations used were OpenSees default values [62]. Newmark integrator has been used and the convergence test was the normal displacement increment. The tolerance of the test is equal to 10^{-7} and the number of maximum iterations is 50. Rayleigh damping is adopted to account for energy dissipation.

In IDA, the scale factor λ was chosen to scale the spectral acceleration at the fundamental period $S_a(T_1,\xi = 5\%)$, which was scaled to $IM = a_\lambda \in [0.1$ g, 0.8 g$]$ with $\Delta_{a\lambda} = 0.1$ g. Similarly, in IMPA, the response spectra were scaled multiple times to obtain scaled spectral acceleration at the fundamental period $S_a(T_1,\xi = 5\%)$ equal to $IM = S_{a\lambda} \in [0.1$ g, 0.8 g$]$ with $\Delta S_{a\lambda} = 0.1$ g. The mass damping coefficient and the stiffness damping coefficient of the Rayleigh damping are evaluated by considering the first and the second natural frequency of the case study. The percentage of critical damping is equal to 5%.

4. Results

4.1. Nonlinear Static Analysis Results

The pushover analysis has been performed in displacement control to reach a target displacement of 350 mm. The number of steps to reach the target displacement is 350 calculation steps with an increment of 1 mm each step. Figure 7 shows the capacity curves obtained applying two load distributions proportional to the first and second modal shapes, respectively. The capacity curve for the first load distribution reaches a maximum base shear of 492 kN corresponding to a top displacement of about 220 mm. The sequence of the plastic hinges activation with the relative calculation step is shown in Figure 8. A plastic hinge is activated when the reinforcement of the section reaches and exceeds the yield point: the first plastic hinges activated in the columns rather than in the beams, particularly in the upper stories which, according to Eurocode 8, have weak column–strong beam connections.

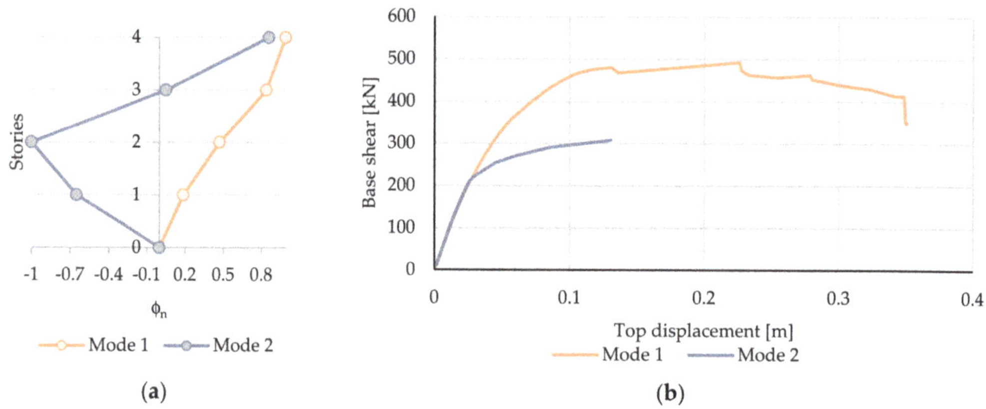

Figure 7. (**a**) First and second modal shapes and (**b**) capacity curves of the frame.

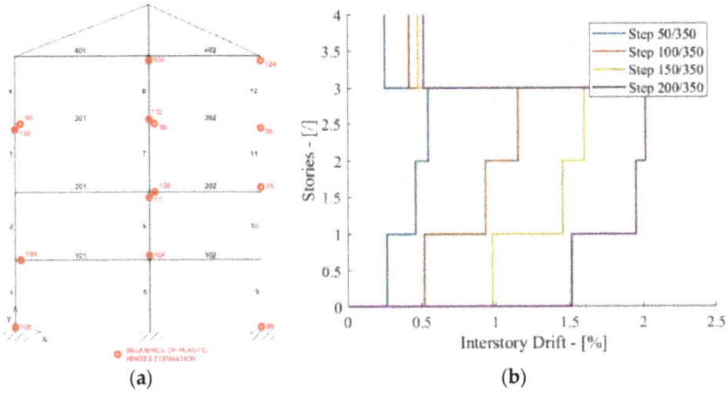

Figure 8. (**a**) Plastic hinges and (**b**) interstory drift at different steps of the pushover analysis.

Finally, Figure 8 shows the interstory drift at different steps of the analysis, with an interval of 50 steps. It shows a large concentration of interstory drift in the second and third stories, while the fourth one is moving almost rigidly.

4.2. Nonlinear Dynamic Analysis Results

A total number of six subsets of near-fault records and six subsets of far-field records were examined. The subsets were randomly generated from the main set of 210 records to comply with the general rules for record selection explained in Section 2.2. The mean and standard deviation of the normal distributions for magnitude, epicentral distance and spectral acceleration of each record selection were given in Table 2.

Table 2. Mean and standard deviation of each record selection magnitude, epicentral distance and spectral acceleration normal distributions.

	Sel. 1	Sel. 2	Sel. 3	Sel. 4	Sel. 5	Sel. 6	Sel. 7	Sel. 8	Sel. 9	Sel. 10	Sel. 11	Sel. 12
	NF	NF	NF	NF	NF	NF	FF	FF	FF	FF	FF	FF
μ_{Rjb}	4.84	4.31	4.73	4.63	4.56	4.28	20.80	19.55	19.32	20.89	19.66	18.84
σ_{Rjb}	2.95	3.04	3.19	3.08	2.97	3.04	10.09	8.97	6.68	9.84	9.19	6.44
μ_{Mw}	5.83	5.88	6.03	5.86	5.92	5.85	5.73	5.80	5.64	5.72	5.54	5.72
σ_{Mw}	0.72	0.77	0.71	0.82	0.82	0.82	0.75	0.73	0.59	0.60	0.56	0.60
μ_{Sa}	0.41	0.44	0.45	0.45	0.49	0.46	0.18	0.23	0.22	0.19	0.19	0.21
σ_{Sa}	0.39	0.40	0.37	0.44	0.40	0.42	0.26	0.26	0.29	0.27	0.27	0.27

As expected, the results show that IDA and IMPA are less dependent on record selection, with the mean values of fragility curves ranging from 0.463 g to 0.525 g and from 0.387 to 0.432, respectively. In contrast, D-CA and MPA-CA show greater dependence on record selection, with mean values between 0.479 g and 0.724 g and between 0.321 g and 0.621 g, respectively (Figure 9). It can be pointed out that IMPA provides the most conservative results for all the datasets studied, as shown in Table 3. Moreover, this methodology seems to be the more accurate with respect to IDA in estimating vulnerability for the far-field record selections.

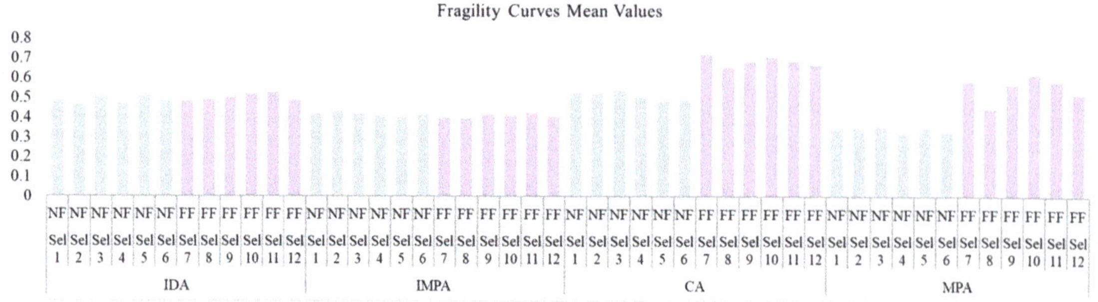

Figure 9. Histogram of mean values of fragility curves for different methodology and selection of records.

Table 3. Percentage variation of the 16th percentile, mean and 84th percentile of fragility curves for different methods with respect to IDA.

Methodology	**IDA**	**$IMPA_1$**	**D-CA**	**MPA_1-CA**	**IDA**	**$IMPA_1$**	**D-CA**	**MPA_1-CA**	**IDA**	**$IMPA_1$**	**D-CA**	**MPA_1-CA**	**IDA**	**$IMPA_1$**
Fractile	**0.16**	**0.16**	**0.16**	**0.16**	**0.5**	**0.5**	**0.5**	**0.5**	**0.84**	**0.84**	**0.84**	**0.84**	β	β
	[g]	%	%	%	[g]	%	%	%	[g]	%	%	%		
Sel. 1 $_{NF}$	0.388	−20%	1%	−34%	0.487	−15%	8%	−29%	0.602	15%	44%	−5%	0.23	0.29
Sel. 2 $_{NF}$	0.353	−13%	3%	−33%	0.464	−7%	12%	−25%	0.609	32%	62%	8%	0.27	0.35
Sel. 3 $_{NF}$	0.396	−21%	7%	−38%	0.500	−16%	8%	−29%	0.632	12%	38%	1%	0.24	0.29
Sel. 4 $_{NF}$	0.354	−15%	5%	−46%	0.466	−12%	8%	−32%	0.615	20%	45%	15%	0.28	0.31
Sel. 5 $_{NF}$	0.399	−30%	−11%	−42%	0.506	−21%	−5%	−31%	0.641	14%	28%	3%	0.24	0.37
Sel. 6 $_{NF}$	0.382	−19%	−4%	−40%	0.482	−14%	1%	−32%	0.608	15%	34%	−4%	0.23	0.29
Sel. 7 $_{FF}$	0.401	−27%	43%	8%	0.481	−18%	50%	21%	0.577	12%	90%	63%	0.18	0.30
Sel. 8 $_{FF}$	0.376	−26%	32%	−28%	0.488	−19%	35%	−9%	0.634	14%	78%	50%	0.26	0.35
Sel. 9 $_{FF}$	0.398	−23%	44%	3%	0.501	−17%	37%	14%	0.631	13%	64%	58%	0.23	0.31
Sel. 10 $_{FF}$	0.402	−30%	46%	−2%	0.515	−21%	37%	21%	0.659	16%	65%	91%	0.25	0.38
Sel. 11 $_{FF}$	0.429	−26%	34%	−3%	0.525	−19%	31%	12%	0.643	7%	57%	58%	0.20	0.29
Sel. 12 $_{FF}$	0.387	−22%	47%	−1%	0.485	−17%	38%	8%	0.607	11%	62%	49%	0.23	0.29
μ [g]	0.39	0.30	0.47	0.31	0.49	0.41	0.60	0.45	0.62	0.56	0.76	0.65		
σ	0.02	0.01	0.10	0.09	0.02	0.01	0.10	0.12	0.02	0.02	0.09	0.17		
CoV	0.05	0.05	0.21	0.29	0.04	0.03	0.16	0.27	0.03	0.04	0.12	0.27		

The two subsets of records studied in-depth (Sel. 1 $_{NF}$ and Sel. 7 $_{FF}$ (Table 1)) comprise two different records of the real seismic event to which the case study was exposed in 2016, the 6.5 magnitude Central Italy earthquake, which, as mentioned, caused severe damage to structural and especially non-structural component of the school. Figures 10 and 11 shows IDA curves and IMPA curves: the curves of IDA referring to the Norcia 2016 earthquake exceed the threshold of DCR_{LS} for the intensities of S_a equal to 0.5 g and 0.57 g for NF and FF selection, respectively. The results of D-CA show that, as expected, the DCR_{LS} threshold for the SLV limit state is exceeded for both the near-fault record and far-field record, with DCR_{LS} values of 2.57 and 1.01, respectively (Figure 12). Similarly, even higher DCR_{LS} are obtained from the cloud analysis based on nonlinear static analysis (Figure 13).

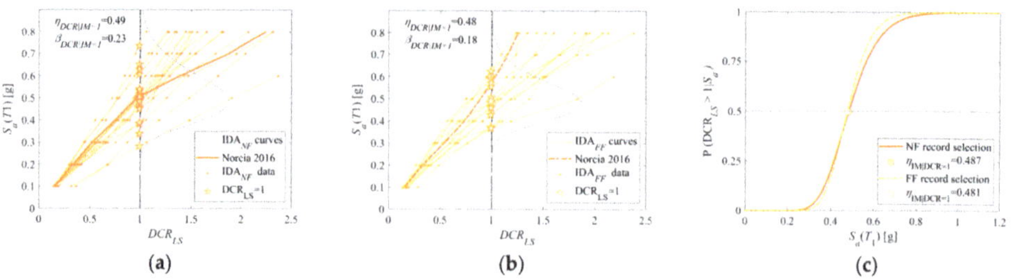

Figure 10. IDA: (**a**) comparison among IDA curves for near-fault record selection and (**b**) far-field record selection; (**c**) comparison between the two fragility curves obtained.

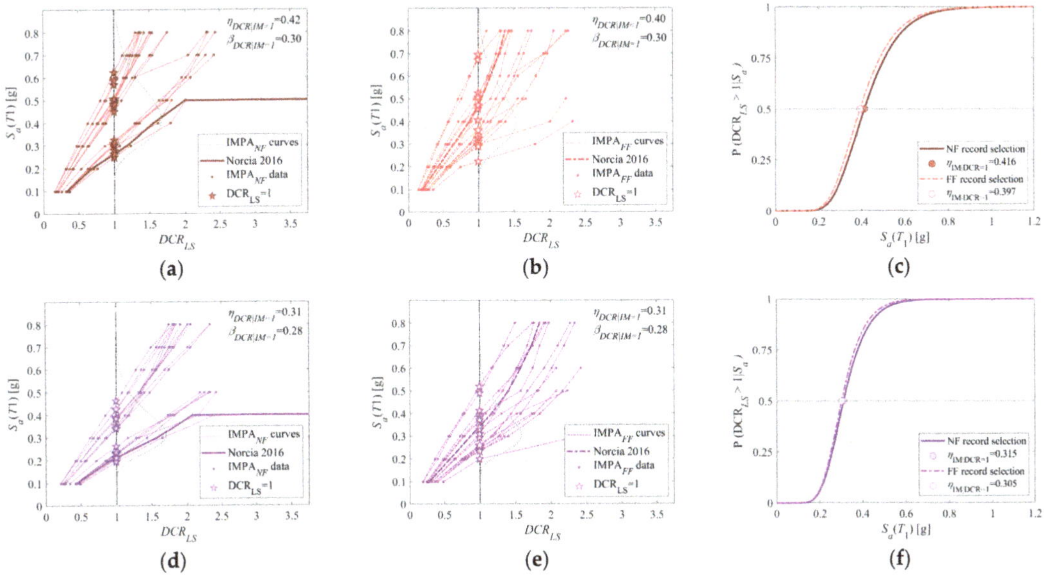

Figure 11. IMPA: (**a**) comparison among IMPA curves for near-fault record selection and (**b**) far-field record selection; (**c**) comparison between the two fragility curves obtained considering only the first mode; (**d**) comparison among IMPA curves for near-fault record selection and (**e**) far-field record selection; (**f**) comparison between the two fragility curves obtained considering the first mode and second mode.

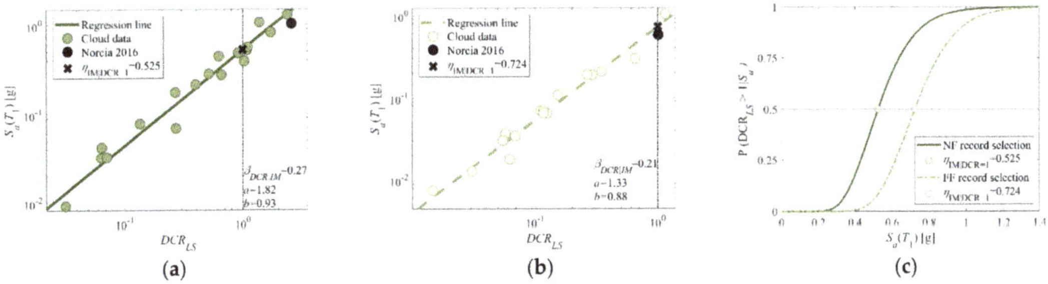

Figure 12. D-CA: (**a**) comparison among dynamic analysis cloud data regressions for near-fault record selection and (**b**) far-field record selection; (**c**) comparison between the two fragility curves obtained.

To check the consistency of modal and multimodal IMPA and pushover-based cloud, the developed fragility curves (Figure 14) are compared with those of IDA. The accuracy of the prediction of the different fragility models with respect to IDA is quantified by normalized root-mean-square deviation (*RMSD*). It is evaluated according to the following Equation (15), assuming that the values derived by Equation (12) (IDA) are the reference ones:

$$RMSD(\%) = \sum_{i=1}^{n} \sqrt{\frac{(\hat{y}_i - y_i)^2}{\hat{y}_i^2}} \quad (15)$$

where n is the number of points, and y_i and $\hat{y}_i$ are the predicted and reference probability of exceeding the considered limit state (LS), respectively.

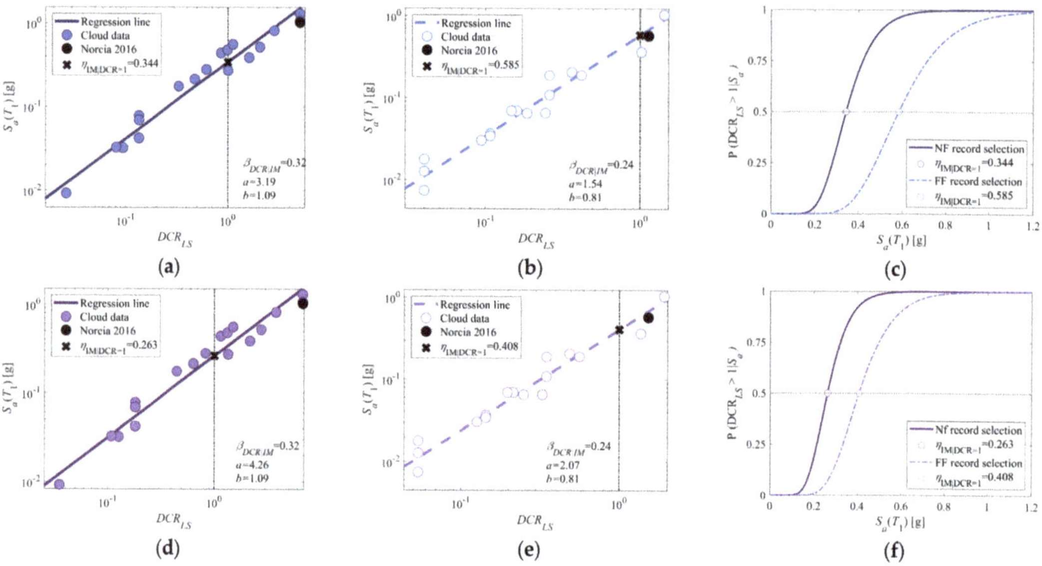

Figure 13. MPA-CA: (**a**) comparison among modal pushover analysis cloud data regressions for near-fault record selection and (**b**) far-field record selection; (**c**) comparison between the two fragility curves obtained considering only the first mode; (**d**) comparison among modal pushover analysis cloud data regressions for near-fault record selection and (**e**) far-field record selection, (**f**) comparison between the two fragility curves obtained considering the first and the second mode.

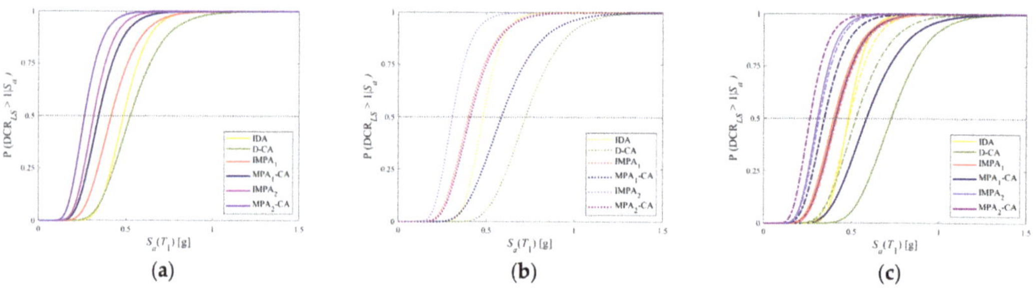

Figure 14. Comparison among fragility curves for all aforementioned methods: (**a**) comparison among fragility curves for near-fault (NF) selection (**b**) far-field (FF) selection, and (**c**) total.

The following Table 4 shows the comparison in terms of the percentage variation of the median, 16% and 84% fractiles of the fragility curves with respect to IDA and the absolute values of the standard deviation of each method. Regarding the selection of NF records, the cloud method appears solid in estimating the 50%, 16% and 84% fractiles compared to IDA with the smallest normalized root-mean-square deviation of 8%. However, when it comes to far-field records, IMPA appears to be the most accurate methodology for estimating fragility. The inclusion of two or more vibration modes in the assessment of the maximum multimodal DCR_{LS} does not seem to be essential, as it leads to very conservative results.

Table 4. Percentage values represent the percentage change with respect to the values observed for IDA.

Methodology	Near-Fault Record (Sel. 1)					Far-Field Record (Sel. 7)				
	$\eta_{16\%}$ [g]	$\eta_{50\%}$ [g]	$\eta_{84\%}$ [g]	β	*RMSD*	$\eta_{16\%}$ [g]	$\eta_{50\%}$ [g]	$\eta_{84\%}$ [g]	β	*RMSD*
IDA	0.388	0.487	0.612	0.22	-	0.481	0.401	0.577	0.18	-
D-CA	1%	8%	15%	0.26	8%	43%	50%	58%	0.21	41%
MPA_1-CA	−34%	−29%	−24%	0.33	11%	8%	21%	36%	0.24	20%
MPA_2-CA	−50%	−46%	−42%	0.33	15%	−24%	−15%	−5%	0.24	5%
$IMPA_1$	−20%	−15%	−9%	0.28	6%	−27%	−18%	−7%	0.30	6%
$IMPA_2$	−39%	−35%	−32%	0.27	13%	−42%	−37%	−30%	0.28	12%

5. Conclusions

This paper compares fragility curves obtained by various known static and dynamic nonlinear procedures. Incremental modal pushover analysis (IMPA) is proposed as an alternative to IDA, which is currently considered the most reliable method, to determinate IMPA curves and thus seismic fragility. Similarly, both MPA and NL-THA are used to determine capacity in the well-known cloud method. For this study, it was necessary to perform a relatively small number of nonlinear time histories using two different data sets. These differ in the range of Joyner-Boore distance (R_{jb}) and are scattered in a range of magnitude.

The following conclusions can be drawn from this study that is limited to a simplified 2D frame model and a small set of records:

- Comparison of fragility curves shows that, in the case of methodologies distinguished by scaling (in terms of accelerograms (IDA) or response spectra (IMPA)), near-fault records and far-field record selections have led to nearly equivalent results. In contrast, the results in terms of fragility when using records without scaling, i.e., in D-CA and MPA-CA, show clear differences in the whole range of intensities;
- Fragility curves that only consider the contribution of the first mode in determining DCR_{LS} have led to more accurate results in relation to IDA, so the inclusion of higher mode contributions does not seem to be essential for low to medium buildings (up to nine stories [20]);
- A total number of 12 subsets have been extracted from the main 210 set of records and exanimated, but results are not fully reported in this paper. The results have shown that D-CA leads to a smaller vulnerability than MPA-CA and IMPA in all selection and for the whole range of intensities. Methodologies based on the pushover analysis, on the contrary, have led to more conservative results, especially for 16% and 50% fractiles;
- IDA shows less sensitivity to record-to-record variability. It should be noted, however, that IMPA, despite its slightly greater sensitivity, has the advantage of a large reduction in the computational effort required to perform the structural analysis. In IMPA, the total time required relates mostly to the post-processing phase, which is no different for small 2D frames or more complex 3D buildings.

A more comprehensive validation is needed to confirm the obtained results and draw more general conclusions.

Author Contributions: Conceptualization, C.P.C., A.P., C.N. and B.B.; methodology, C.P.C., A.P. C.N. and B.B.; software, C.P.C. and A.P.; formal analysis, C.P.C. and A.P.; data curation, C.P.C. and A.P.; writing—original draft preparation, C.P.C. and A.P.; writing—review and editing, C.P.C., B.B., A.P. and C.N.; visualization, C.P.C. and A.P.; supervision, C.N. and B.B. All authors have read and agreed to the published version of the manuscript.

Funding: The authors gratefully acknowledge the funding received by The Laboratories University Network of Seismic Engineering (ReLUIS): research project ReLUIS/DPC 2019–2021 Reinforced Concrete Existing Structures.

Institutional Review Board Statement: Not applicable.

Informed Consent Statement: Not applicable.

Data Availability Statement: Publicly available datasets were analyzed in this study. This data can be found here: https://ngawest2.berkeley.edu/spectras/new?sourceDb_flag=1; https://www.seismosoc.org/Publications/BSSA_html/bssa_104-5/2013191-esupp/index.html (all accessed on 15 February 2022).

Conflicts of Interest: The authors declare no conflict of interest.

References

1. Krawinkler, H. Challenges and Progress in Performance-Based Earthquake Engineering. In Proceedings of the International Seminar on Seismic Engineering for Tomorrow—In Honor of Professor Hiroshi Akiyama, Tokyo, Japan, 26 November 1999.
2. Applied Technology Council. *ATC-20 Procedures for Postearthquake Safety Evaluation of Buildings*; Applied Technology Council: Redwood City, CA, USA, 1989.
3. FEMA. *NEHRP Guidelines for the Seismic Rehabilitation of Buildings*; Prepared by the Applied Technology Council; Federal Emergency Management Agency (FEMA): Washington, DC, USA, 1997; p. 273.
4. SEAOC Vision 2000 Committee. *Performance Based Seismic Engineering of Buildings*; Soulages, J., Ed.; Structural Engineers Association of California: Sacramento, CA, USA, 1995; Volume 2.
5. Porter, K.A. An overview of PEER's Performance-Based Earthquake Engineering Methodology. In Proceedings of the 9th International Conference on Applications of Statistics and Probability in Civil Engineering, San Francisco, CA, USA, 6–9 July 2003; pp. 1–8.
6. Whitman, R.V.; Biggs, J.M.; Brennan, J.E., III; Cornell, C.A.; De Neufville, R.L.; Vanmarcke, E.H. Seismic design decision analysis. *J. Struct. Div.* **1975**, *101*, 1067–1084. [CrossRef]
7. Kennedy, R.; Ravindra, M. Seismic fragilities for nuclear power plant risk studies. *Nucl. Eng. Des.* **1984**, *79*, 47–68. [CrossRef]
8. Kennedy, R.P.; Cornell, C.A.; Campbell, R.D.; Kaplan, S.; Perla, H.F. Probabilistic seismic safety study of an existing nucle-ar power plant. *Nucl. Eng. Des.* **1980**, *59*, 315–338. [CrossRef]
9. Muntasir Billah, A.H.M.; Shahria Alam, M. Seismic fragility assessment of highway bridges: A state-of-the-art review. *Struct. Infrastruct. Eng.* **2015**, *11*, 804–832. [CrossRef]
10. Nazri, F.M.; Miari, M.; Kassem, M.M.; Tan, C.-G.; Farsangi, E.N. Probabilistic Evaluation of Structural Pounding Between Adjacent Buildings Subjected to Repeated Seismic Excitations. *Arab. J. Sci. Eng.* **2018**, *44*, 4931–4945. [CrossRef]
11. Farsangi, E.N.; Yang, T.; Tasnimi, A. Influence of concurrent horizontal and vertical ground excitations on the collapse margins of non-ductile RC frame buildings. *Struct. Eng. Mech.* **2016**, *59*, 653–669. [CrossRef]
12. Kassem, M.; Nazri, F.M.; Farsangi, E.N. The seismic vulnerability assessment methodologies: A state-of-the-art review. *Ain Shams Eng. J.* **2020**, *11*, 849–864. [CrossRef]
13. Farsangi, E.N.; Tasnimi, A.A.; Mansouri, B. Fragility assessment of RC-MRFs under concurrent vertical-horizontal seismic action effects. *Comput. Concr.* **2015**, *16*, 99–123. [CrossRef]
14. Kassem, M.M.; Nazri, F.M.; Farsangi, E.N. The efficiency of an improved seismic vulnerability index under strong ground motions. *Structures* **2020**, *23*, 366–382. [CrossRef]
15. Farsangi, E.N.; Takewaki, I.; Yang, T.Y.; Astaneh-Asl, A.; Gardoni, P. *Resilient Structures and Infrastructure*; Springer: Berlin/Heidelberg, Germany, 2019.
16. Dehghani, S.; Fathizadeh, S.; Yang, T.; Farsangi, E.N.; Vosoughi, A.; Hajirasouliha, I.; Málaga-Chuquitaype, C.; Takewaki, I. Performance evaluation of curved damper truss moment frames designed using equivalent energy design procedure. *Eng. Struct.* **2020**, *226*, 111363. [CrossRef]
17. Rakicevic, Z.; Bogdanovic, A.; Farsangi, E.N.; Sivandi-Pour, A. A hybrid seismic isolation system toward more resilient structures: Shaking table experiment and fragility analysis. *J. Build. Eng.* **2021**, *38*, 102194. [CrossRef]
18. Luco, N.; Cornell, C.A. Effects of random connection fractures on the demands and reliability for a 3-story pre-Northridge SMRF structure. In Proceedings of the 6th US National Conference on Earthquake Engineering, El Cerrito, CA, USA, 31 May–4 June 1998; pp. 1–12.
19. Vamvatsikos, D.; Cornell, C. Applied Incremental dynamic analysis. *Earthq. Eng. Struct. Dyn.* **2002**, *31*, 491–514. [CrossRef]
20. Han, S.W.; Chopra, A.K. Approximate incremental dynamic analysis using the modal pushover analysis procedure. *Earthq. Eng. Struct. Dyn.* **2006**, *35*, 1853–1873. [CrossRef]
21. Mackie, K.R.; Stojadinović, B. Comparison of Incremental Dynamic, Cloud, And Stripe Methods for Computing Probabilistic Seismic Demand Models. In Proceedings of the 2005 Structures Congress: Metropolis and Beyond & 2005 Forensic Engineering Symposium, New York, NY, USA, 20–24 April 2005.
22. Kiani, J.; Khanmohammadi, M. New Approach for Selection of Real Input Ground Motion Records for Incremental Dynamic Analysis (IDA). *J. Earthq. Eng.* **2015**, *19*, 592–623. [CrossRef]
23. Jalayer, F.; Ebrahimian, H.; Miano, A.; Manfredi, G.; Sezen, H. Analytical fragility assessment using unscaled ground motion records. *Earthq. Eng. Struct. Dyn.* **2017**, *46*, 2639–2663. [CrossRef]

24. Shome, N.; Cornell, C.A.; Bazzurro, P.; Carballo, J.E. Earthquakes, records, and nonlinear responses. *Earthq. Spectra* **1998**, *14*, 469–500. [CrossRef]
25. Iervolino, I.; Cornell, C.A. Record Selection for Nonlinear Seismic Analysis of Structures. *Earthq. Spectra* **2005**, *21*, 685–713. [CrossRef]
26. Shome, N.; Cornell, C.A. *Probabilistic Seismic Demand Analysis of Nonlinear Structures*; RMS-35, RMS Program; Stanford University: Stanford, CA, USA, 1999; p. 320.
27. Vargas-Alzate, Y.F.; Hurtado, J.E. Efficiency of intensity measures considering near-and far-fault ground motion records. *Geosciences* **2021**, *11*, 234. [CrossRef]
28. Bradley, B.A. A critical examination of seismic response uncertainty analysis in earthquake engineering. *Earthq. Eng. Struct. Dyn.* **2013**, *42*, 1717–1729. [CrossRef]
29. Luco, N.; Bazzurro, P. Does amplitude scaling of ground motion records result in biased nonlinear structural drift responses? *Earthq. Eng. Struct. Dyn.* **2007**, *36*, 1813–1835. [CrossRef]
30. Bazzurro, P.; Cornell, C.A.; Shome, N.; Carballo, J.E. Three Proposals for Characterizing MDOF Nonlinear Seismic Response. *J. Struct. Eng.* **1998**, *124*, 1281–1289. [CrossRef]
31. Jalayer, F. Direct Probabilistic Seismic Analysis: Implementing Non-Linear Dynamic Assessments. Ph.D. Thesis, Stanford University, Stanford, CA, USA, 2003.
32. Miano, A.; Jalayer, F.; Ebrahimian, H.; Prota, A. Nonlinear Dynamic Analysis Procedure With A Limited Number of Analyses and Scaling. In Proceedings of the 7th ECCOMAS Thematic Conference on Computational Methods in Structural Dynamics and Earthquake Engineering (COMPDYN), Crete, Greece, 24–26 June 2019.
33. Mander, J.B. Fragility curve development for assessing the seismic vulnerability of highway bridges. *Res. Prog.* **1999**, *89*. Available online: https://www.semanticscholar.org/paper/Fragility-Curve-Development-for-Assessing-the-of-Mander/02d6253500a128cf1a9daf75172d965e6a43e7d1 (accessed on 15 February 2022).
34. Shinozuka, M.; Feng, M.Q.; Kim, H.-K.; Kim, S.-H. Nonlinear Static Procedure for Fragility Curve Development. *J. Eng. Mech.* **2000**, *126*, 1287–1295. [CrossRef]
35. Banerjee, S.; Shinozuka, M. Nonlinear Static Procedure for Seismic Vulnerability Assessment of Bridges. *Comput. Civ. Infrastruct. Eng.* **2007**, *22*, 293–305. [CrossRef]
36. Rossetto, T.; Gehl, P.; Minas, S.; Galasso, C.; Duffour, P.; Douglas, J.; Cook, O. FRACAS: A capacity spectrum approach for seismic fragility assessment including record-to-record variability. *Eng. Struct.* **2016**, *125*, 337–348. [CrossRef]
37. Dolšek, M.; Fajfar, P. IN2-A simple alternative for IDA. In Proceedings of the 13th World Conference on Earthquake Engineering, Vancouver, BC, Canada, 1–6 August 2004; pp. 1–6.
38. Faella, C.; Lima, C.; Martinelli, E. Non-linear Static Methods For Seismic Fragility Analysis And Reliability Evaluation of Existing Structures. In Proceedings of the 14th World Conference on Earthquake Engineering, Beijing, China, 12–17 October 2008; pp. 12–17.
39. Jalayer, F.; Ebrahimian, H.; Miano, A. *N2 with Cloud: A Non-Linear Dynamic Analysis Procedure for the Equivalent SDOF System*; Pisa University Press: Pisa, Italy, 2019; pp. 215–225.
40. Bergami, A.V.; Liu, X.; Nuti, C. Evaluation of a modal pushover based incremental analysis. In Proceedings of the ACE, Vietri sul Mare, Italy, 12–13 June 2015.
41. Jalayer, F.; De Risi, R.; Manfredi, G. Bayesian Cloud Analysis: Efficient structural fragility assessment using linear regression. *Bull. Earthq. Eng.* **2014**, *13*, 1183–1203. [CrossRef]
42. MIT Ministry of Infrastructures and Transportation. Istruzioni per l'Applicazione delle Nuove Norme Tecniche per le Costruzioni di Cui al Decreto Ministeriale 14 Gennaio 2008; Consiglio Superiore dei Lavori Pubblici. GU Serie Generale No. 47 del 26-02-2009—Suppl. Ordinario No. 27; Ordinario No. 27; No. 617 of 2/2/2009; Italy. 2009. Available online: http://www.cngeologi.it/wp-content/uploads/2016/06/Circ.-esplicativa-NTC-2008-n%C2%B0-617-del-02_02_2009.pdf (accessed on 15 February 2022).
43. Ancheta, T.D.; Darragh, R.B.; Stewart, J.P.; Seyhan, E.; Silva, W.J.; Chiou, B.S.J.; Donahue, J.L. NGA-West2 database. *Earthq. Spectra* **2014**, *30*, 989–1005. [CrossRef]
44. Luzi, L.; Pacor, F.; Puglia, R. *Italian Accelerometric Archive v 3.0—Istituto Nazionale di Geofisica e Vulcanologia*; Istituto Nazionale di Geofisica e Vulcanologia, Dipartimento della Protezione Civile Nazionale: Rome, Italy, 2019. [CrossRef]
45. EN 1998-1. *Eurocode 8: Design of Structures for Earthquake Resistance—Part 1: General Rules, Seismic Actions and Rules for Buildings*; European Committee for Standardization: Brussels, Belgium, 2005.
46. Somerville, P.G. Characterizing Near-Fault Ground Motion for the Design and Evaluation of Bridges. In Proceedings of the 3rd National Seismic Conference and Workshop on Bridges and Highways, Portland, OR, USA, 28 April–1 May 2002; Volume 28, pp. 1371–1448.
47. Shahi, S.K.; Baker, J.W. An efficient algorithm to identify strong velocity pulses in multi-component ground-motions. *Bull. Seismol. Soc. Am.* **2013**, *104*, 2456–2466. [CrossRef]
48. Ertuncay, D.; Malisan, P.; Costa, G.; Grimaz, S. Impulsive Signals Produced by Earthquakes in Italy and their Potential Relation with Site Effects and Structural Damage. *Geosciences* **2021**, *11*, 261. [CrossRef]
49. Miano, A.; Jalayer, F.; Ebrahimian, H.; Prota, A. Cloud to IDA: Efficient fragility assessment with limited scaling. *Earthq. Eng. Struct. Dyn.* **2017**, *47*, 1124–1147. [CrossRef]

50. Chopra, A.K.; Goel, R.K. *Capacity-Demand-Diagram Methods for Estimating Seismic Deformation of Inelastic Structures: SDF Systems*; Report No. PEER1999/02; Pacific Earthquake Engineering Research Center: Berkeley, CA, USA, 1999.
51. Chopra, A.K.; Goel, R.K. A modal pushover analysis procedure for estimating seismic demands for buildings. *Earthq. Eng. Struct. Dyn.* **2001**, *31*, 561–582. [CrossRef]
52. Freeman, S.A. The capacity spectrum method. In Proceedings of the 11th European Conference on Earthquake Engineering, Paris, France, 6–11 September 1998.
53. ATC-40. *Seismic Evaluation and Retrofit of Concrete Buildings*; Applied Technology Council: Redwood City, CA, USA, 1996.
54. Porter, K.; Kennedy, R.; Bachman, R. Creating Fragility Functions for Performance-Based Earthquake Engineering. *Earthq. Spectra* **2007**, *23*, 471–489. [CrossRef]
55. Bergami, A.V.; Liu, X.; Nuti, C. Proposal and Application of the Incremental Modal Pushover Analysis (IMPA). In Proceedings of the IABSE Conference Conference-Structural Engineering: Providing Solutions to Global Challenges, Geneva, Switzerland, 23–25 September 2015.
56. Bergami, A.V.; Forte, A.; Lavorato, D.; Nuti, C. The Incremental Modal Pushover Analysis (IMPA): Proposal and Application. In Proceedings of the 16WCEE, Santiago de, Chile, Chile, 9–13 January 2017.
57. Bergami, A.; Fiorentino, G.; Lavorato, D.; Briseghella, B.; Nuti, C. Application of the Incremental Modal Pushover Analysis to Bridges Subjected to Near-Fault Ground Motions. *Appl. Sci.* **2020**, *10*, 6738. [CrossRef]
58. Bergami, A.V.; Nuti, C.; Lavorato, D.; Fiorentino, G.; Briseghella, B. IMPAβ: Incremental Modal Pushover Analysis for Bridges. *Appl. Sci.* **2020**, *10*, 4287. [CrossRef]
59. Bergami, A.V.; Pelle, A.; Fiorentino, G.; Lavorato, D.; Giaccu, G.F.; Quaranta, G.; Briseghella, B.; Nuti, C. Seismic assessment of corroded concrete bridges using incremental modal pushover analysis. In *Proceedings of the Institution of Civil Engineers-Bridge Engineering*; Thomas Telford Ltd.: London, UK, 2021; pp. 1–29.
60. Mazzoni, S.; McKenna, F.; Scott, M.H.; Fenves, G.L. OpenSees Command Language Manual; Pacific Earthquake Engineering Research (PEER) Center. 2006, pp. 137–158. Available online: https://opensees.berkeley.edu/OpenSees/manuals/usermanual/OpenSeesCommandLanguageManualJune2006.pdf (accessed on 15 February 2022).
61. Priestley, M.J.N.; Park, R. Strength and ductility of concrete bridge columns under seismic loading. *Struct. J.* **1987**, *84*, 61–76.
62. Scott, M.H.; Fenves, G.L. Plastic Hinge Integration Methods for Force-Based Beam–Column Elements. *J. Struct. Eng.* **2006**, *132*, 244–252. [CrossRef]
63. Scott, M.H.; Ryan, K.L. Moment-Rotation Behavior of Force-Based Plastic Hinge Elements. *Earthq. Spectra* **2013**, *29*, 597–607. [CrossRef]
64. Bentz, E.C. *Sectional Analysis of Reinforced Concrete Members*; University of Toronto: Toronto, ON, Canada, 2000.

applied sciences

MDPI

Article

Irregularity of the Distribution of Masonry Infill Panels and Its Effect on the Seismic Collapse of Reinforced Concrete Buildings

Juan Carlos Vielma [1,*], Roberto Aguiar [2], Carlos Frau [3] and Abel Zambrano [4]

[1] Civil Engineering School, Pontificia Universidad Católica de Valparaíso, Valparaíso 2340000, Chile
[2] Department of Earth and Construction Sciences, Universidad de las Fuerzas Armadas, Sangolquí 171103, Ecuador; rraguiar@espe.edu.ec
[3] Regional Center for Technological Developments for Construction, Seismology and Seismic Engineering (CEREDETEC), Universidad Tecnológica Nacional, Mendoza 5500, Argentina; cdfrau@frm.utn.edu.ar
[4] Engineering Faculty, Universidad Laica Eloy Alfaro de Manabí, Manta 130203, Ecuador; angel.zambrano@uleam.edu.ec
* Correspondence: juan.vielma@pucv.cl; Tel.: +56-322-273-644

Abstract: On 16 April 2016, an earthquake of Mw 7.8 shook the coast of Ecuador, causing the destruction of buildings and a significant number of casualties. Following a visit by the authors to the city of Portoviejo during the debris removal and recovery stage, it was noted that several reinforced concrete buildings located on corners had collapsed in the central part of the city. These buildings were characterized by the presence of masonry at the edges of the buildings but not between the two mostly open-plan facades on the corner for practical reasons. This article reviews the effect of masonry infill panels on the seismic response of reinforced concrete structures. For this, a model that contains the geometric and mechanical characteristics typical of collapsed buildings was generated and subjected to nonlinear analysis, with both static and dynamic increments. The results show the clear influence of the masonry infill panels on the structural response through the torsional behavior that is reflected in the evolution of the floor rotations. Finally, dynamic incremental analysis is used to obtain the collapse fragility curve of the building, and a new damage measure based on floor rotations is proposed.

Keywords: masonry infill panels; torsional behavior; incremental dynamic analysis; collapse fragility curve; damage measure

Citation: Vielma, J.C.; Aguiar, R.; Frau, C.; Zambrano, A. Irregularity of the Distribution of Masonry Infill Panels and Its Effect on the Seismic Collapse of Reinforced Concrete Buildings. *Appl. Sci.* **2021**, *11*, 8691. https://doi.org/10.3390/app11188691

Academic Editor: Maria Favvata

Received: 28 July 2021
Accepted: 14 September 2021
Published: 17 September 2021

Publisher's Note: MDPI stays neutral with regard to jurisdictional claims in published maps and institutional affiliations.

Copyright: © 2021 by the authors. Licensee MDPI, Basel, Switzerland. This article is an open access article distributed under the terms and conditions of the Creative Commons Attribution (CC BY) license (https://creativecommons.org/licenses/by/4.0/).

1. Introduction

At present, design standards prescribe few or no recommendations aimed at considering the contribution of masonry to the response of a structure. This is especially important in regard to structures that are located in areas of high seismic hazard. On the other hand, in Latin America and in other regions of the world with developing countries, reinforced concrete buildings are frequently built with informal construction practices given the low technological level required [1–3]. Such buildings have a low ductility, which makes them vulnerable to damage [4]. The difference between masonry infilled frames and confined masonry lies in the construction methodology, resulting in completely different load transfer mechanisms [5].

Recent reports, produced after the inspection of areas affected by strong earthquakes, have concluded that informally constructed reinforced concrete buildings responded with a low level of damage or did not collapse due to the contribution of the masonry [6]. Notably, it has been determined that the contribution of masonry is beneficial when the construction process is carried out with confined masonry structural systems, which present an acceptable performance even for modified Mercalli intensities between VIII and IX [7].

This acceptable performance contrasts with the precarious performance of other informal construction typologies that include unreinforced masonry or adobe buildings, which have presented much more severe damage or even collapse [8].

Following the Wenchuan earthquake [9], various problems associated with the damage and even collapse of confined masonry buildings were detected, including floor diaphragm failure, excessive floor deformation failure, and local failure by debonding in beams and columns as a result of cyclical loads. The authors of this work have previously highlighted the adequate behavior of buildings designed in accordance with modern codes, without achieving a notable improvement in the global ductility. Similarly, based on observations of buildings damaged during the Jaya (Indonesia) earthquake, the following types of failures in confined masonry buildings were identified: loss of anchor, failure due to in-plane shear, failure due to out-of-plane action, and differential settlement [10].

In another post-earthquake building survey campaign, this time in Sumatra, the team identified the causes of failure of confined masonry structures to be shear column failure, longitudinal reinforcement buckling, and premature masonry collapse [11]. Likewise, the earthquake that affected Ludian Province in 2014 [12] provided an opportunity for the authors to corroborate that the confined masonry buildings designed, according to current standards demonstrated adequate performance, especially those with a uniform distribution of masonry.

However, the beneficial effect of masonry on the seismic response of a reinforced concrete structure can produce a detrimental effect if the contribution of its stiffness and resistance is not taken into consideration [6]. In the event that the distribution of the masonry in the structure does not follow the uniformity and symmetry criteria in both plan and elevation, it can produce an irregular structure, whose dynamic response is completely different from the dynamic response of the concrete structures analyzed alone, causing torsional effects that mainly affect the columns of the outer frames [13]. Additionally, an inadequate distribution of masonry can produce harmful effects, such as changes in the dissipated energy and the generation of local failure mechanisms, such as the effects of a short column [6].

This detrimental effect of masonry can be described by taking the Ecuador earthquake of 16 April 2016, as an example [14]. In the city of Portoviejo, several buildings collapsed, among which it was possible to identify buildings with reinforced concrete frames in which the presence of masonry presumably caused an irregularity in the plan, producing a predominant torsional seismic response, and consequently the collapse of buildings located on corners. In previous work [15], some of these corner buildings were observed to have undergone extensive damage or even collapse. Among the possible causes of extensive damage and collapse, the authors point out that the level of acceleration far exceeded the acceleration of normative design, the excessive flexibility of the structures, the construction of vertical extensions, the existence of ground floors with open plans (commercial space) and the effect of siege in some locations in the city. Notably, the capacity of buildings subjected to corrosive environments rapidly degrades [15], a situation that may have affected damaged buildings but is not addressed in this research. Figure 1 shows two buildings located on Chile Street in Portoviejo that are currently damaged; note the predominant structural typology based on columns and slabs with flat beams and open facades. Figure 2 shows pre-earthquake and post-earthquake images of the Marinero building, which is also located on Chile Street in Portoviejo and had to be demolished because it collapsed under the effect of torsion.

This article studies the influence of masonry on the response of a model that attempts to reproduce the typology and the mechanical and geometric characteristics of the corner buildings of Portoviejo, obtaining the nonlinear response through incremental push (pushover) analysis and incremental dynamic analysis (IDA), with some records corresponding to the Ecuador earthquake of 16 April 2016.

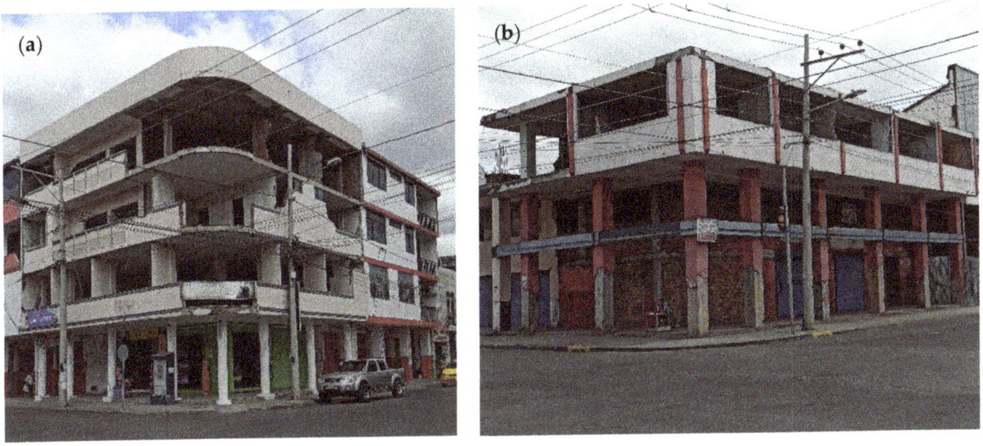

Figure 1. Current status of damaged buildings located on the corners of Chile Street, Portoviejo (**a**) building in use and (**b**) building closed.

Figure 2. Marinero building located on Chile Street in Portoviejo, (**a**) before the earthquake and (**b**) after the earthquake.

2. Numerical Model for Confined Masonry Buildings

It is well documented that today, fewer tools are available for the analysis of confined masonry structures than there are for the analysis of reinforced concrete or steel structures. In the formulation of calculation programs for confined masonry structures, a compromise between the precision and the complexity of the model is necessary [16].

The use of macroscale finite element models (macro-models) is common in the numerical study of structures based on confined masonry [1,5,17–20]. The authors developed a macro-model for simple masonry and confined masonry based on smeared-crack total strain, validated with experimental results [21]. In the study of [22], the dependence of the seismic behavior of the macro-models on the following parameters was described in detail:

- The strength of the masonry;
- The density of the masonry walls;
- The level of axial compression of the masonry walls;
- The characteristics of the confining elements.

It was detected in this work that the computational models reviewed to date could not be considered appropriate to carry out the conventional evaluation of the capacity of this type of structure, especially due to their numerical instability [22]. In [23], the aspects that influence the seismic response of masonry structures were studied and determined using pushover analysis, especially for cases that can be classified as exhibiting soft-floor behavior:

- Critical floor drift;
- The total number of floors in the building;
- The configuration of the applied lateral loads.

Among the models used for the analysis of masonry walls, whether simple, reinforced, or confined, the work of [19] incorporated the size of the masonry units (bricks) as a variable. The joint model was obtained by considering a connection with two nonlinear springs, one longitudinal and the other transverse, connected in parallel, and a contact element connected in series with the other springs. The study of [18] considered a constitutive model at the cohesion interface to simulate the behavior of cracks both in the concrete and in the masonry units and mortar joints, adopting an elastic-plastic model. During the numerical analysis of the cyclical response of masonry buildings, the influence of the following parameters has been determined: meshing, dilation angle, and boundary conditions [1].

The consideration of window and door openings in masonry walls introduces a number of additional modeling complications [5]. These openings have not been considered in this work. The presence of masonry has been included along only the perimeters of the studied building.

Most of the models used to numerically simulate the cyclical response of confined masonry buildings have been validated using laboratory tests in which the specimens were subjected to gravity loads and lateral loads in the masonry plane [1,24–26].

Inelastic Infill Panel Element

Among the finite element macro-models used to model inelastic infill masonry panels within frame structures, the one presented in the original work [27] and implemented by other authors [28], consisting of a four-node element, is highlighted here. In this macro-model, the structural elements of the frames act in association with the masonry infill panels, trying to reproduce what happens with these structural typologies when they are subjected to lateral displacements. In this macro-model, each masonry infill panel is represented by six struts that allows the interaction that occurs between the cracked masonry infill panels and the structural elements (beams and columns) that confine them to be captured (see Figure 3a). Each diagonal direction is represented by two parallel struts that receive the axial forces along the entire length of these diagonals, and a third diagonal that supports the shear that is transmitted from the top to the bottom of the masonry panel. This last strut acts in only the direction of the strut that works in compression, depending on the activation of the deformations reached in the panel (Figure 3c). The compression struts use a hysteretic strut model, and the shear strut uses a bilinear hysteretic model.

For the parallel struts, four internal nodes that allow the points of contact between the panel and the frame elements were set (Figure 3b), considering the dimensions of the cross-section of the columns and the beams that confine the panel. In addition, four dummy nodes were used to take into consideration the contact length between the panel and the structural frame. All the internal forces were transferred to the four outer nodes, which are defined as the common nodes where the beams and columns of the frame meet. By defining the internal nodes and the dummy nodes, the area of the compression or tension strut shown in Figure 3b was obtained. This area represents the cracking condition of the masonry infill panels, which reduces the contact between the panels and the frame elements that confine them.

The thickness of the masonry infill panels is defined from the width of the masonry units arranged in the analyzed structure, ignoring the contribution of the mortar used to bond the units. Finally, in this model, it is important to define the interstory drift that indicates the failure of the panel under the action of out-of-plane forces. This leads to the panels of the model being deactivated when this interstory drift is reached, so that the panels are unable to provide stiffness and resistance to the structure anymore, although the masonry continues to add weight to the structure.

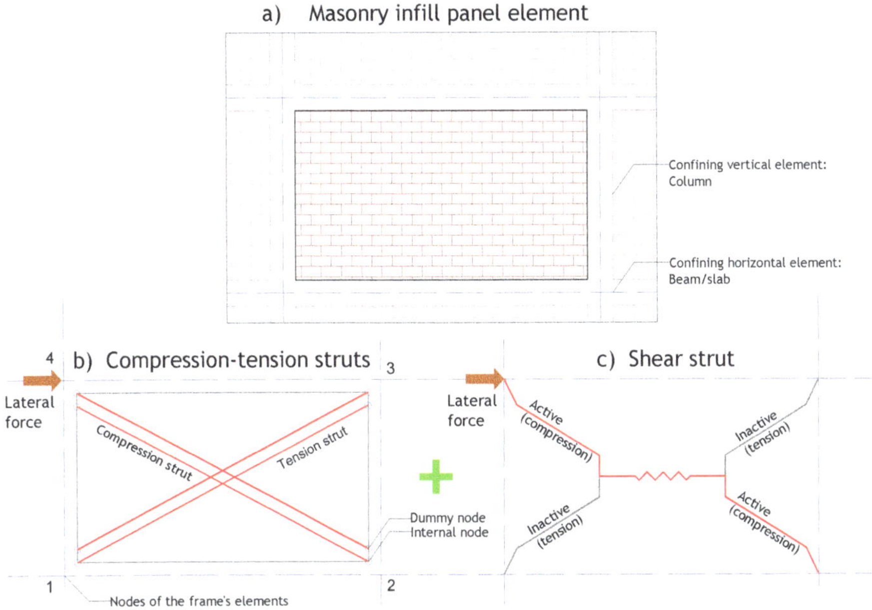

Figure 3. Confined masonry model (**a**) masonry infill panel and frame scheme, (**b**) compression-tension struts and (**c**) shear strut. Adapted from [27,28].

3. Damage in Structures with Masonry Infill Panels

It is important to characterize the failure of masonry infill panels. To do so, the results of post-earthquake surveys were consulted, where it was determined that global plasticization in confined masonry buildings occurs with drifts of approximately 0.5% and 0.9% [9]. In a numerical study [4], the authors reported the following drifts for confined masonry buildings: for formal buildings, drifts of 0.4% and 0.8% correspond to the maximum response and collapse, respectively, while for informal buildings, these drifts are 0.25% and 0.5%, respectively. On the other hand, the sudden loss of stiffness that affects the overall stiffness of buildings is achieved at very low floor drifts. The authors [15] have indicated that for a drift of 0.1%, 50% of the stiffness of the walls is lost, while for a drift of 0.5%, up to 84% of the stiffness of the walls is lost.

The type of failure of confined masonry buildings designed according to industry standards shows adequate resistance to strong seismic actions, as was demonstrated in [16], in which a value of up to 0.8 g was reached in shaking table tests, producing global collapse through the weak ground floor mechanism. In terms of damage to confined masonry buildings, two main types must be distinguished: out-of-plane stress failures and in-plane stress failures. For this last type of failure, numerous researchers have proposed certain displacements or drifts associated with certain behavioral states [6].

On the other hand, in [29], it was proposed that for interstory drifts between 0.4% and 1.5%, the masonry is expected to fail, while in [30], it was considered that cracking occurs in masonry walls at lower drifts of approximately 0.25%. An extreme state of damage in masonry walls has been reported in [5], indicating that the experimental results show that the bricks undergo crushing when an interstory drift of 1.4% occurs. This value is very close to that indicated in [6], where a value of 1.5% is set for the limit at which the masonry collapses. The improvement in the behavior of masonry walls can be achieved by increasing the presence of confining elements (especially columns) and placing steel meshes in the seat of the masonry walls [15]. Finally, in [9], the maximum lateral resistance of confined masonry structures subjected to the action of earthquakes was reached at values between 0.5% and 0.9%.

The following section provides a concise and precise description of the numerical results, their interpretation, and the conclusions that were drawn.

4. Case Study

The case study focuses on an archetype of buildings that suffer extensive damage during earthquakes due to being located on corners. The fragility of these buildings was due to not only the aspects indicated above but also the inadequate interaction of the masonry infill panels with a very fragile structure, which caused torsion that ultimately resulted in the collapse of the buildings. According to [17], the irregular distribution of masonry infill panels in reinforced concrete buildings contributes to their collapse during earthquakes; however, the irregular distribution studied did not induce torsion. In the generated model, the irregular distribution of the masonry panels that existed in the collapsed corner buildings of Portoviejo is reproduced. This masonry panels were on the sides that adjoined neighboring buildings, while the other sides presented openings in the case of windows of the residential floors and accesses in the case of the commercial first floor. Clearly, the regular distribution of the masonry panels can generate a completely different dynamic response; however, the study focuses on the response of the representative model of the collapse, suggesting that the response of buildings with a regular distribution of masonry panels should be studied in the future.

Description of the Corner Building

The frames of the structure are made up of reinforced concrete beams and columns. The slabs are ribbed in two directions with a thickness of 0.20 m and are filled with blocks of clay. The beams have inverted sections with dimensions of 0.50 × 0.20 m, and six longitudinal reinforcement bars 10 mm in diameter, and stirrups 6 mm in diameter separated every 0.15 m. According to these dimensions, the beams are embedded within the slabs. The columns have cross-sections with dimensions of 0.35 × 0.35 m, with six longitudinal reinforcement bars 16 mm in diameter and stirrups 8 mm in diameter separated every 0.20 m. Both the dimensions of the sections and the reinforcement considered for the elements correspond to conditions similar to those existing in the collapsed buildings in Portoviejo before the earthquake. Figure 4 shows the floor plan of the studied building, with four structural axes in each direction equally spaced every 3.5 m. The masonry infill panels are included toward the north and east ends of the plan. Figure 5 shows elevation views of the building's frames, where the height of the first floor is dedicated to commerce (3.5 m), while the rest of the floors have heights of 2.8 m since they are residences. The resulting model has similar geometric and mechanical characteristics in both the X and Y directions.

The characteristics of the materials used in the numerical model are summarized in Table 1. The resistance values of the concrete and steel used in the model are typical of the structures present in Portoviejo when the earthquake occurred. The behavior curve of Mander et al. is used for the model concrete [31], and the Menegotto and Pinto curve is used for the steel [32], which are available in the SeismoStruct program [33]. The resistance of the masonry units is estimated by taking into account that they are of the ceramic type and manufactured with low-quality control [4,17].

Traditionally, the analysis of this type of building is carried out neglecting the stiffness provided by the masonry, only considering its weight, or its mass in the dynamic analysis.

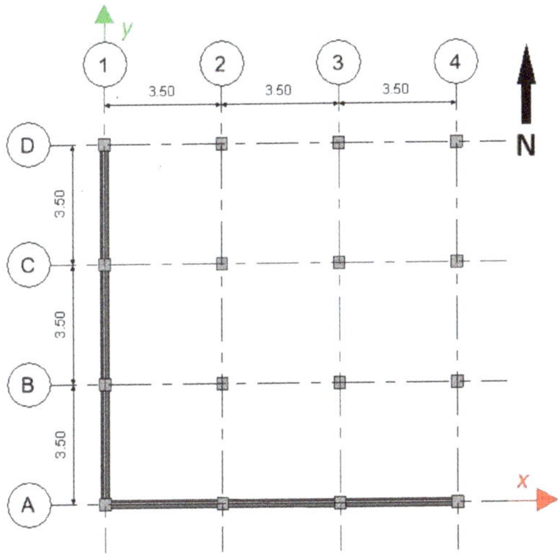

Figure 4. Typical plan of the studied building.

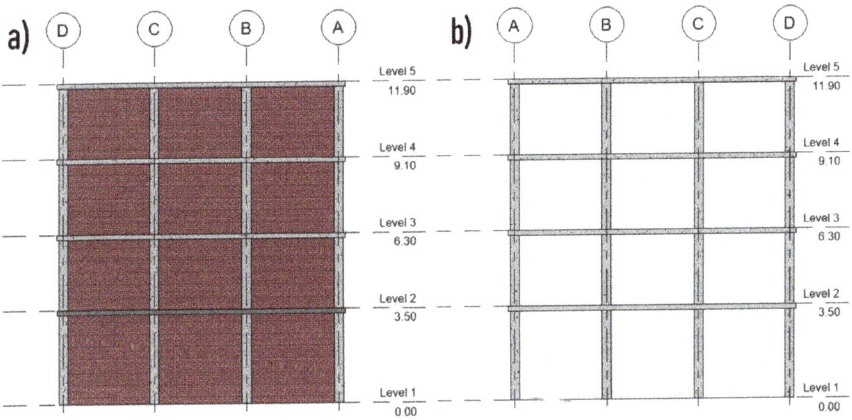

Figure 5. Elevation views: (**a**) west and (**b**) east.

Table 1. Strengths of the model materials.

Strength	Value
Concrete	20.6 (MPa)
Reinforcement	412 (MPa)
Masonry unit	14 (MPa)

To perform the nonlinear analysis, the characteristics of the macro-model struts are defined, and the compression strut parameters are summarized in Table 2. Of these parameters, the most relevant is the compressive strength, which corresponds to the strength of the masonry infill panel as a whole, not just the masonry units. This parameter shows high variability among the relevant studies consulted [1,2,4,9,15–19,21] and is determined based on the shear strength calculated under diagonal compression. The initial Young's modulus also exhibits high variability and has been determined to be 1000 times the value of the compressive strength, as recommended in previous works [34,35].

Table 2. Parameters of the compression strut implemented in the masonry model.

Curve Parameters	Assumed Value
Initial Young's modulus	3500 (MPa)
Compressive strength	3.5 (MPa)
Tensile strength	0.0 (MPa)
Strain at maximum stress	0.0012 (m/m)
Ultimate strain	0.024 (m/m)
Closing strain	0.004 (m/m)
Strut area reduction strain	0.0006 (m/m)
Residual strut area strain	0.001 (m/m)

The parameters of the shear strut are summarized in Table 3. The shear bond strength also tends to vary. Among those reported by different authors, a value of 0.3 MPa is selected for this work, which is the minimum reported in [36] and is within the range of values from 0.1 to 1.5 MPa suggested in [37]. The maximum shear strength depends on the masonry shear failure mode (shear friction failure, diagonal tensile failure, and compressive failure). It is estimated by adding an additional resistance to the already defined shear bond strength, reaching a value of 0.5 MPa.

Table 3. Parameters of the shear strut implemented in the masonry model.

Curve Parameters	Assumed Value
Shear bond strength	0.3 (MPa)
Friction coefficient	0.7
Maximum shear strength	0.5 (MPa)
Reduction shear factor	1.5

The model of the building studied is shown in Figure 6. In this figure, the masonry infill panels are located on the sides of the building that face other buildings, while the sides without masonry infill panels correspond to open facades due to the presence of windows (residential levels) and doors (commercial first level).

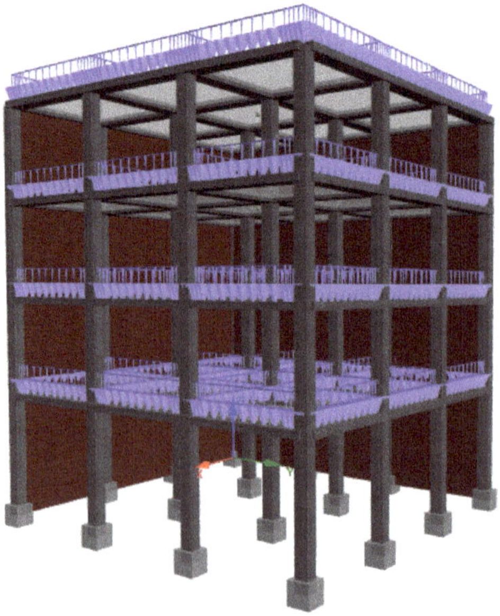

Figure 6. Perspective view of the model.

To obtain an approximation of the dynamic behavior of the model with masonry infill panels, modal analysis of the structure was performed, obtaining the values of the periods and the percentages of the participating masses of the first nine vibration modes, which are presented in Table 4. The second vibration mode corresponds to a translational mass with equal participatory masses in both directions, which gives rise to a similar behavior of the structure in both the X and Y directions due to the correspondence in the geometry in these directions. In turn, this allows the characterization of the seismic response by performing one-way analysis. The deformed shapes of the first three modes are shown in Figure 7. These first three modes of vibration concentrate the largest participatory mass in each direction of analysis.

Table 4. Dynamic parameters with masonry infill panels included in the model.

Mode	Period (s)	Ux	Uy	Rz
1	0.458	26.79%	26.79%	29.20%
2	0.458	42.16%	42.16%	0.00%
3	0.402	15.01%	15.01%	54.93%
4	0.132	3.48%	3.48%	3.79%
5	0.118	4.60%	4.60%	0.00%
6	0.065	1.41%	1.41%	5.40%
7	0.065	1.10%	1.10%	1.20%
8	0.059	1.30%	1.30%	0.00%
9	0.039	0.33%	0.33%	0.00%

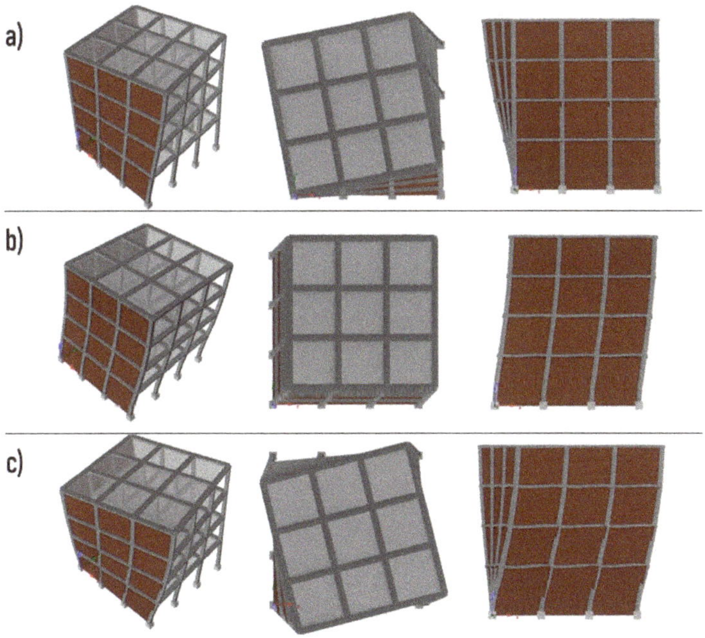

Figure 7. Perspective, plan, and front views of the deformed shapes corresponding to the (**a**) first mode, (**b**) second mode, and (**c**) third mode.

5. Determination of the Seismic Response of the Case Studied

With the structural model, two types of analyses are carried out in a nonlinear range that is the most appropriate to determine the seismic response of buildings [37]. The first is the analysis with incremental thrust, both conventional and adaptive, while the second is the IDA, using records of the seismic event that occurred on the Ecuadorian coast and that caused the failure of the buildings located in Portoviejo. The details of these analyses are provided below.

5.1. Nonlinear Static Analysis

Nonlinear static analysis (pushover) has been applied to the model considering the distribution of lateral loads corresponding to the distribution of the first mode of vibration [38], although it is evident that the structural axis with masonry infill panels (axis D) has a stiffness greater than that of a structural axis without masonry infill panels (axes A, B, and C) (Figure 4). In this way, the effect of the irregularity in the stiffness distribution on the torsional behavior of the structure can be captured. The nonlinear static analysis is carried out in both the conventional and adaptive modes, since the validity of the first type of analysis is limited by the irregular nature of the structure, conferred by the presence of unevenly distributed masonry infill panels. The advantage of adaptive nonlinear static analysis is that it allows the variations in the lateral stiffness of the model to be captured when it undergoes alterations in its elements. The capacity curves of the conventional nonlinear static analysis are shown in Figure 8, in which the curves corresponding to the X direction of analysis change in both the positive and negative directions (see Figure 8a,b, respectively). The capacity curves show the evolution of the normalized basal shear with respect to the seismic weight vs. the normalized ceiling displacement with respect to the total height of the building (global drift). Additionally, note that the nonlinear analysis is carried out in only the X direction since the Y direction is similar in both the geometric and mechanical characteristics adopted. These curves show how the failure of the masonry infill panels affects the overall response of the structure (for a global drift close to 0.8%); however, this failure is not well defined in the curves obtained from the conventional analysis.

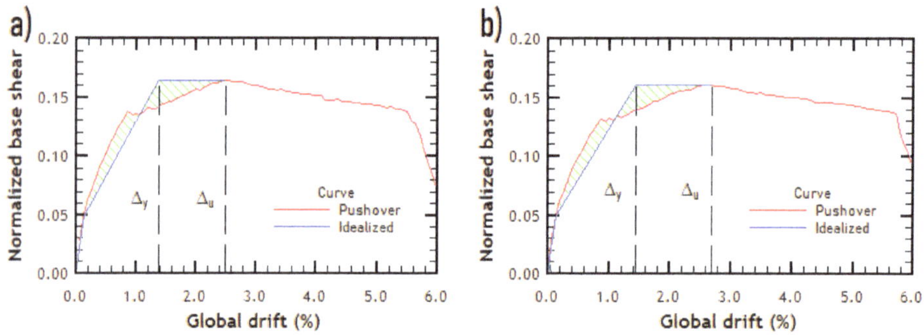

Figure 8. Resulting capacity and idealized curves obtained from conventional pushover: (**a**) lateral forces (+) and (**b**) lateral forces (−).

On the other hand, the capacity curves obtained when applying adaptive pushover analysis, which are presented as variants of the vector of lateral forces applied to the structure by virtue of the changes that take place in the modal shapes as different levels of element damage occur as the simulation progresses, are interesting. The capacity curves in Figure 9a,b are shown with the idealized curves, which are determined using the method proposed in [39,40]. With the ultimate displacement and the yield displacement of the idealized curve, the ductility (μ) of the model is determined according to Equation (1) with the ultimate displacement (Δ_u) and the yield displacement (Δ_y), and is summarized in Table 5. Note that, although the results of the ductility calculated with the conventional and adaptive curves are quite similar, the adaptive nonlinear static analysis has allowed the curves for which the global drift resulted in the failure of the masonry infill panels to be clearly identified.

$$\mu = \frac{\Delta_u}{\Delta_y} \tag{1}$$

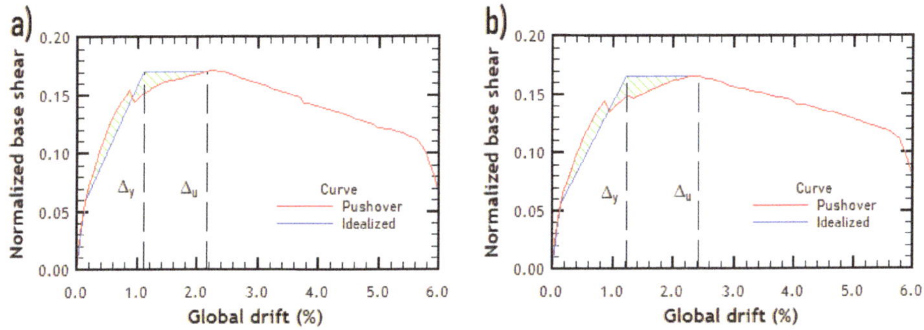

Figure 9. Resulting capacity and idealized curves obtained from adaptive pushover analysis: (**a**) lateral forces (+) and (**b**) lateral forces (−).

Table 5. Ductility values obtained from the idealized capacity curves.

Analysis Type	Δ_u [%]	Δ_y [%]	μ
Conventional (+)	2.4874	1.3896	1.7900
Conventional (−)	2.6891	1.4551	1.8481
Adaptive (+)	2.1512	1.1176	1.9248
Adaptive (−)	2.4201	1.2185	1.9861

Another of the characteristics of the response that is interesting to evaluate is the variation in the rotations with respect to the displacements at the center of gravity of the roof level during the pushover analysis. Figure 10a,b show the results of the rotations at the centers of gravity (CG) of each floor for both the conventional and adaptive incremental pushover analyses. Beyond the similarity of these curves, it can be seen that, at the beginning of the analysis, there is a linear increase in the rotations of the centers of gravity of each floor, but once a global drift of 0.8% is reached, there is a significant reduction in these rotations. However, residual rotations remain that cannot be recovered due to the displacement caused by the collapse of the building; thus, the failure mechanism clearly includes significant rotations in levels 3 and 4 of the building. This reduction in rotation is produced by the failure of the masonry infill panels, which cease to provide strength and stiffness to the structure, thereby reducing the eccentricity caused by torsion. Additionally, each floor reaches the maximum rotation at different displacements, although these displacements are very similar. On the other hand, the maximum rotation is reached at the roof level of the building (0.0074 rad). This implies that the columns farthest from the center of mass of this level reach global drifts approximately 0.3% higher than the previously mentioned global drift of 0.8%, which implies that the frames associated with these columns (perimeter frames) undergo greater damage as the lateral displacements increase.

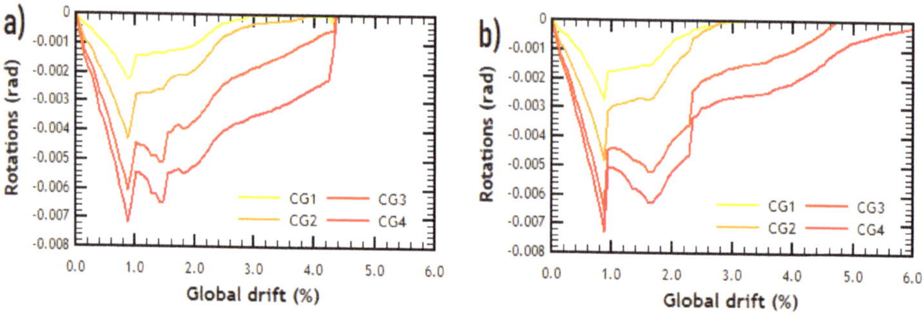

Figure 10. Variation in the rotations of the centers of gravity of each floor with respect to the lateral displacement according to different pushover analyses: (**a**) conventional pushover analysis and (**b**) adaptive pushover analysis.

5.2. Incremental Dynamic Analysis

IDA is an analysis that is carried out in the nonlinear range using a set of registers or synthetic accelerograms as dynamic excitation to obtain the performance of structures [41]. IDA is applied to the studied structure using a set of four records with a double component of the earthquake of 16 April 2016, obtained at different seismological stations in Ecuador [42]. Table 6 shows the main characteristics of these records, while Figures 11 and 12 show the seismic records and their response spectra, respectively. Note the different durations of the records, as well as the maximum acceleration values achieved [43]. IDA makes it possible to determine the seismic capacity of the structures considering the variability in the records used [44–46] and to objectively obtain the seismic performance factors of structural typologies [47].

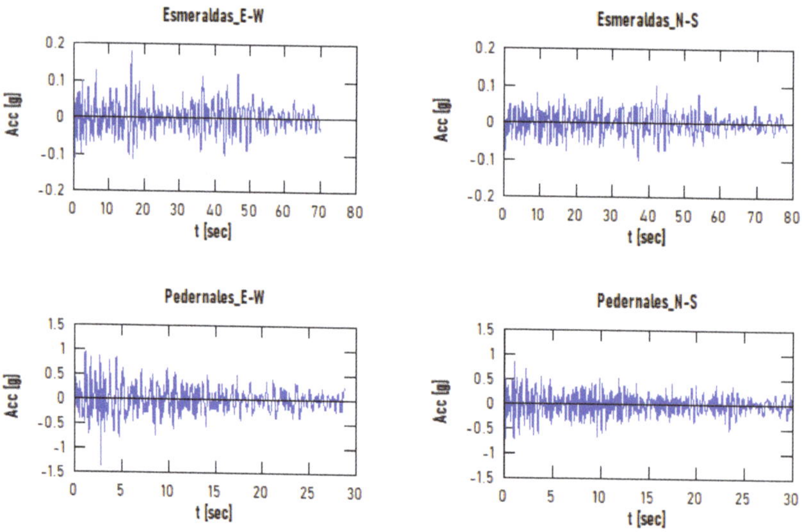

Figure 11. *Cont.*

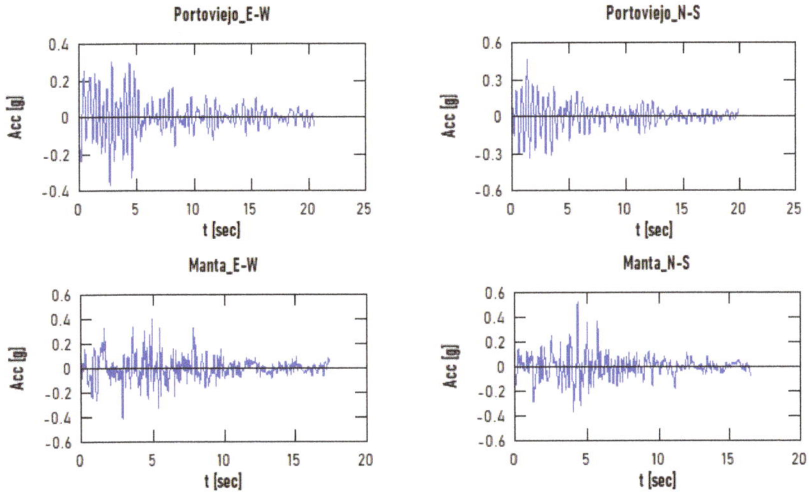

Figure 11. Acceleration records from the four stations reported.

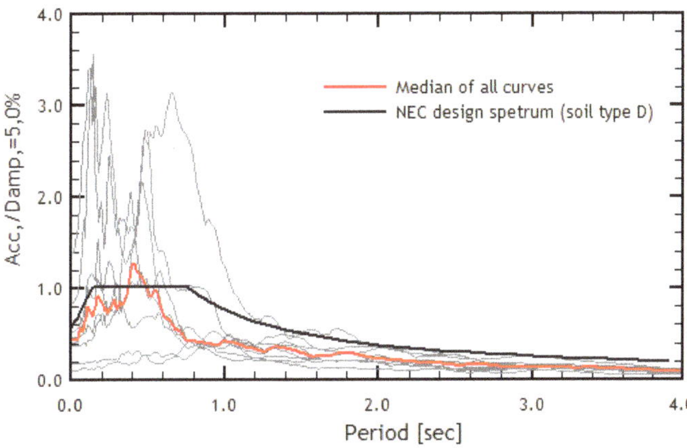

Figure 12. Response, mean, and elastic design spectra corresponding to the accelerograms used for the IDA.

Table 6. Information from the records used for IDA.

Earthquake	Date	Station ID	Location	Epicentral Distance [km]	Component	Peak Ground Acceleration [g]
Ecuador 7.8 Mw	16 April 2016	AES2	Esmeraldas	76	E–W	0.1706
					N–S	0.1122
		APED	Pedernales	36	E–W	1.3413
					N–S	0.8258
		APO1	Portoviejo	167	E–W	0.348
					N–S	0.4196
		AMNT	Manta	171	E–W	0.4226
					N–S	0.5111

Regarding the cyclical behavior of confined masonry structures, the energy dissipation capacity is determined by the throttling of the hysteresis loops [39,40]. This behavior is reflected in the IDA curves in Figure 13, obtained by plotting the values of the maximum interstory drifts (ID_{max}) obtained in the centers of gravity of the floors of the building vs. the spectral acceleration (S_a) corresponding to the predominant period in the direction of analysis. Note that the average curve has a collapse threshold value lower than 0.2 g. The collapse threshold S_{CT} is defined on the average curve at the point where it undergoes a reduction in stiffness, such that the current stiffness is less than 20% of the initial stiffness of the curve. This criterion for capturing the collapse threshold, based on stiffness reduction, corresponds to the criterion originally formulated in [41]. By including the interstory drift as a damage measure (DM) of the analysis, when an increase in the drift occurs, the reduction in the slope of the curve also occurs; therefore, the criterion adopted preserves a relationship with the overall damage of the structure.

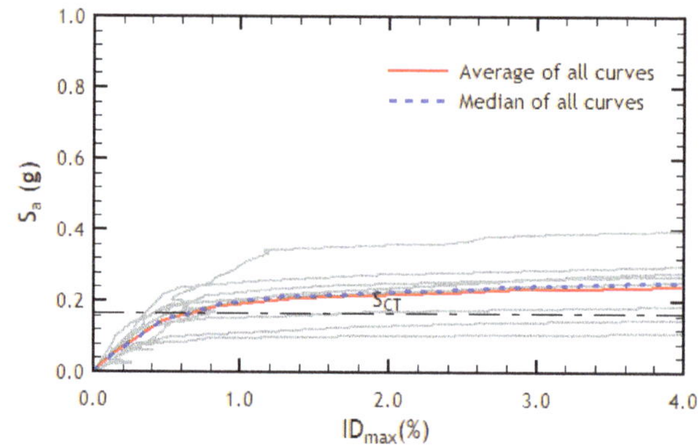

Figure 13. IDA curves obtained by applying the set of accelerograms.

Finally, with the S_{CT} value resulting from the set of IDA curves and with the total deviation calculated from the data recommended in FEMA P-695 [47], the collapse fragility curve that gives the probability values of exceedance for a horizontal acceleration of the ground is obtained. The total system collapse uncertainty β_{TOT} is computed with

$$\beta_{TOT} = \sqrt{\beta_{RTR}^2 + \beta_{DR}^2 + \beta_{TD}^2 + \beta_{MDL}^2} \leq 0.95 \quad (2)$$

where β_{RTR} is the record-to-record collapse uncertainty, β_{DR} is the design requirement-related collapse uncertainty, β_{TD} is the test data-related collapse uncertainty, and β_{MDL} is the modeling-related collapse uncertainty. The values of the different types of uncertainty recommended according to the specific conditions of the building studied are summarized in Table 7.

Table 7. Uncertainty values assumed for the determination of total system collapse uncertainty. Adapted from [47].

Uncertainty	Assumed Value
β_{RTR}	0.40
β_{DR}	0.50
β_{TD}	0.50
β_{MDL}	0.50

The collapse fragility curve is obtained using a lognormal distribution in which the mean value of the collapse spectral acceleration is $S_{CT} = 0.165$ g and the total system collapse uncertainty is $\beta_{TOT} = 0.95$ (Figure 14). The fragility curves are especially useful for determining the seismic vulnerability of structures [45,46] since they give the probability of exceeding a limit state of the specific behavior, in this particular case, the limit state of collapse. With the collapse fragility curve, it is possible to determine the exceedance probability corresponding to the acceleration demands introduced by the components of the Portoviejo record for the predominant period of the structure (1.01 g in the northeast direction and 2.13 g in the southwest direction). For these values, a probability of exceedance greater than 96% is reached, which explains the collapse of this type of structure during the earthquake of 16 April 2016.

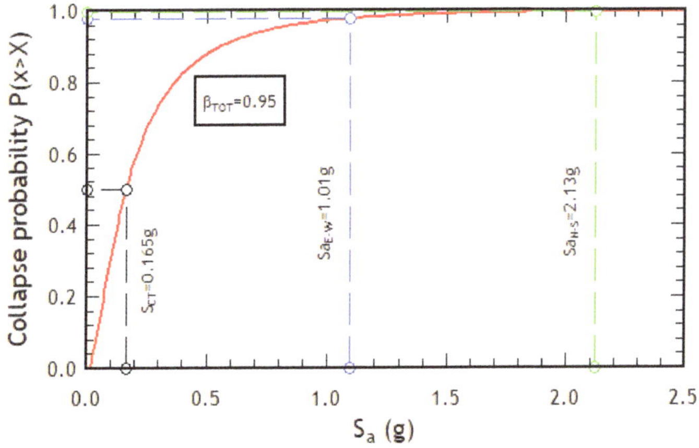

Figure 14. Collapse fragility curve obtained from the IDA analysis of the analyzed building.

5.3. Rotational Incremental Dynamic Analysis

IDA is usually performed by taking into consideration damage measures that have special significance from a performance point of view. This is how the use of global drift interstory drift has become widespread, which is related to the advance of the damage to the structures. However, given the special conditions exhibited by the building studied here, and similar buildings, it is necessary to ask whether the aforementioned parameters are the most appropriate to carry out an IDA that truly captures the effect of the torsion experienced by this type of building.

For this reason, as a result of this research, two candidates for damage measurement are explored and compared. First, the relative rotation of the center of gravity of each floor with respect to the foundation level is used as an engineering demand parameter. Thus, the curve of Figure 15a is obtained according to the evolution of the rotations of the centers of gravity as the value of the spectral acceleration is increased by 5% of the critical damping. The other candidate for the damage measure is the relative rotation of successive floors (see Figure 15b), which, although they result in curves similar to those of the floor rotations, do not show a trend as defined by the curves obtained using this last parameter as a damage measure.

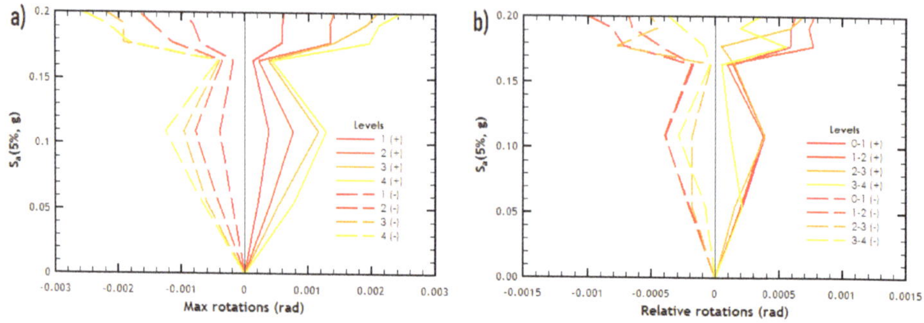

Figure 15. Variations in the (**a**) maximum rotations and (**b**) relative rotations of the center of gravity for each floor calculated for various values of spectral acceleration and 5% critical damping.

6. Discussion of the Results

Nonlinear static analysis provides a good characterization of the seismic response of regular structures. In the present research, this analysis is used, considering its limitations, to capture the torsional behavior of a complete structure and to use the results as a basis for characterization before applying IDA. The results of this analysis show that the structure presents a low displacement ductility, calculated from the capacity curves determined by applying conventional and adaptive procedures. On the other hand, it is possible to observe the influence of the stiffness and resistance of the masonry infill panels on the lateral deformations of the structure. Upon reaching a global drift of the center of gravity at the roof of approximately 1%, the masonry infill panels cease to provide stiffness, and there is a sudden loss in the overall stiffness of the structure, which affects the sharp reduction in rotation as the lateral loading increases, since the structure loses much of its eccentricity. In addition, upon reaching the maximum rotation of the roof's center of gravity, additional lateral displacements are imposed on the columns farthest from it, which implies an increase in the global drift of approximately 0.3%. This increase reflects a higher concentration of rotations in the frames than in the columns (perimeter frames) and is introduced due to the displacements with a component normal to the plane, which especially affects the masonry infill panels.

The IDA applied using accelerograms recorded during the Ecuador earthquake of 16 April 2016, results in curves that are not very dispersed, with very well defined points of collapse for the sections in which at least 20% of the initial stiffness has been lost.

Regarding the fragility curve obtained for this type of building, it is evident that, for moderate accelerations, for example, 0.2 g, the building would reach a high probability of exceeding the collapse limit state (approximately 60%). However, if the probability of exceedance is calculated for the basic acceleration of the current design for the coastal zone of Ecuador [48], which contains the city of Portoviejo (0.5 g), the probability of exceedance of the collapse limit state is approximately 90%. If the accelerations achieved in the response spectra of the two components of the record obtained in Portoviejo are taken into account, exceedance probabilities very close to 100% are obtained.

Based on the floor rotations obtained from the IDA, it can be clearly seen that, for both the negative and positive maximum rotations, the floors show an increase up to a certain value of spectral acceleration slightly higher than 0.10 g. At this threshold, the seismic action causes the collapse of the masonry infill panels, and without that contribution to the stiffness of the extreme frames, this change drastically reduces the torsional behavior of the building; all the rotations reduce to a minimum value corresponding to a spectral acceleration of approximately 0.17 g. After this, an increase in rotations is again experienced due to the increase in lateral displacements induced by the application of accelerograms with ordinates of greater amplitude and the presence of the weight of the masonry, which remains, although it no longer contributes to the lateral stiffness of the building. This behavior is similar to that experienced by the rotations of the centers of gravity when applying the different types of pushover analyses reported in Figure 8a,b.

On the other hand, the relative rotation of the centers of gravity between successive floors is considered applicable to measure the damage concentrated in a specific floor, considering a kind of local concentration of rotations similar to that observed when determining the interstory drifts based on the relative displacements. However, the curves shown in Figure 15b do not exhibit a clear trend in the evolution of these relative rotations, for which it is recommended to use the relative rotation of the center of gravity with respect to the foundation level as a damage measure (see Figure 15b), providing new possibilities for the study of the dynamic behavior of buildings that, due to their irregularity, are prone to a response dominated by torsion instead of translation.

7. Concluding Remarks

First, the need to incorporate the contribution of masonry infill panels to the strength and stiffness of the structure in design is evident, since at present, seismic codes take into account the contribution of only the seismic weight of buildings. The results presented above show how the response of the structure in the case study was altered by the presence of the masonry infill panels.

The model used here captured the failure mode of the corner buildings that failed in Portoviejo as a consequence of the earthquake that shook the Ecuadorian coast in 2016. The behavior of these buildings was clearly torsional due to the effect of the eccentricity introduced by the resistance and stiffness contributions of the masonry infill panels located on the boundaries of the buildings.

Although nonlinear static analysis (pushover) is not the most effective method of obtaining the nonlinear response of irregular structures, it provided certain indications that allow cataloging the seismic behavior of the case studied. On the one hand, it provided global ductility values lower than 2, which are characteristic of fragile buildings, and on the other hand, it provided evidence of the occurrence of the loss of lateral stiffness of corner buildings. Adaptive nonlinear static analysis was shown to be more suitable for capturing the masonry panel failure threshold than conventional analysis.

The incremental dynamic analysis of the model revealed the inadequate resistance of this type of building, which may result in an inadequate performance under dynamic action. The fragility curve obtained by applying a set of records obtained from the seismic event that triggered the failure of the Portoviejo buildings indicated a high probability that the limit state of collapse of this type of building was exceeded when considering the basic acceleration indicated by the Ecuadorian Construction Code for the site area or the values of the spectral acceleration corresponding to the response spectra of the components of the Ecuador earthquake of 16 April 2016, recorded in Portoviejo.

The results of the incremental dynamic analysis using the relative rotation of the centers of gravity with respect to the floor level as a damage measure show that the floor rotations, instead of the interstory drifts or the global drift, can provide useful indicators of the threshold at which significant changes in stiffness occur in buildings that are irregular in plan due to the presence of masonry infill panels. Further studies are required on different configurations of irregular buildings to validate the use of floor rotation as a damage measure. Based on the results obtained, it is suggested that studies are carried out on the effect of uniformly distributed masonry panels on the response of buildings.

Author Contributions: Conceptualization, J.C.V., R.A., C.F. and A.Z.; methodology, A.Z. and J.C.V.; software, J.C.V.; validation, R.A., C.F. and A.Z.; formal analysis, J.C.V.; investigation, J.C.V.; resources, A.Z.; data curation, A.Z. and C.F.; writing—original draft preparation, J.C.V.; writing—review and editing, J.C.V., R.A., C.F. and A.Z.; visualization, J.C.V.; supervision, R.A.; project administration, C.F.; funding acquisition, C.F., A.Z. and J.C.V. All authors have read and agreed to the published version of the manuscript.

Funding: This research was funded by the Pan-American Institute of Geography and History (grant number IPGH GEOF 01-2018), and the article processing charge (APC) was funded by Pontificia Universidad Católica de Valparaíso.

Data Availability Statement: The data is available in the google-drive of the corresponding author.

Acknowledgments: The authors are grateful to the destructive potential project work in response to the earthquake in Ecuador on 16 April 2016, presented to the Technical Assistance Program (PAT) of the Pan American Institute of Geography and History, IPGH GEOF 01-2018.

Conflicts of Interest: The authors declare no conflict of interest. The funders had no role in the design of the study; in the collection, analyses, or interpretation of data; in the writing of the manuscript; or in the decision to publish the results.

References

1. Borah, B.; Kaushik, H.B.; Singhal, V. Finite element modelling of confined masonry wall under in-plane cyclic load. *IOP Conf. Ser. Mater. Sci. Eng.* **2020**, *936*, 012020. [CrossRef]
2. Mebarki, A.; Boukri, M.; Laribi, A.; Farsi, M.; Belazougui, M.; Kharchi, F. Seismic vulnerability: Theory and application to Algerian buildings. *J. Seismol.* **2014**, *18*, 331–343. [CrossRef]
3. Piroglu, F.; Ozakgul, K. Site investigation of masonry buildings damaged during the 23 October and 9 November 2011 Van Earthquakes in Turkey. *Nat. Hazards Earth Syst. Sci.* **2013**, *13*, 689–708. [CrossRef]
4. Lovon, H.; Tarque, N.; Silva, V.; Yepez-Estrada, C. Development of fragility curves for confined masonry buildings in Lima, Peru. *Earthq. Spectra* **2018**, *34*, 1339–1361. [CrossRef]
5. Singhal, V.; Rai, D.C. Behavior of confined masonry walls with openings under in-plane and out-of-plane loads. *Earthq. Spectra* **2018**, *34*, 817–841. [CrossRef]
6. Vicente, R.S.; Rodrigues, H.; Varum, H.; Costa, A.; da Silva, J.A.R.M. Performance of masonry enclosure walls: Lessons learned from recent earthquakes. *Earthq. Eng. Eng. Vib.* **2012**, *11*, 23–34. [CrossRef]
7. Ademović, N.; Hadzima-Nyarko, M.; Zagora, N. Seismic vulnerability assessment of masonry buildings in Banja Luka and Sarajevo (Bosnia and Herzegovina) using the macroseismic model. *Bull. Earthq. Eng.* **2020**, *18*, 3897–3933. [CrossRef]
8. Preciado, A.; Ramirez-Gaytan, A.; Santos, J.C.; Rodriguez, O. Seismic vulnerability assessment and reduction at a territorial scale on masonry and adobe housing by rapid vulnerability indicators The case of Tlajomulco Mexico. *Int. J. Disaster Risk Reduct.* **2019**, *44*, 101425. [CrossRef]
9. Su, R.K.L. Typical Collapse Modes of Confined Masonry Buildings under Strong Earthquake Loads. Open Constr. Build. *Technol. J.* **2011**, *5*, 50–60. [CrossRef]
10. Idris, Y.; Cummins, P.; Rusydy, I.; Muskin, U.; Syamsidik, M.Y.H.; Meilianda, E. Post-Earthquake Damage Assessment after the 6.5 Mw Earthquake on December, 7th 2016 in Pidie Jaya, Indonesia. *J. Earthq. Eng.* **2019**, 1–17. [CrossRef]
11. Sanada, Y. Investigation and analysis of buildings damaged during the september 2007 Sumatra, Indonesia earthquakes. *J. Asian Archit. Build. Eng.* **2008**, *7*, 371–378. [CrossRef]
12. Chen, H.; Xie, Q.; Dai, B.; Zhang, H.; Chen, H. Seismic damage to structures in the M s6.5 Ludian earthquake. *Earthq. Eng. Eng. Vib.* **2016**, *15*, 173–186. [CrossRef]
13. Fardis, M.N. Seismic design issues for masonry- infilled rc frames. In Proceedings of the First European Conference on Earthquake Engineering and Seismology, Geneve, Switzerland, 3–8 September 2006; Paper 313.
14. Aguiar, R.; Mieles, Y. Analysis of buildings that collapsed of Portoviejo during the earthquake of 16 april 2016. *Rev. Int. Ing. Estruct.* **2016**, *21*, 257–282. [CrossRef]
15. Niu, L.H.; Zheng, S.S.; Li, L. Experimental study on the seismic behavior of masonry walls under offshore environment. *Structures* **2020**, *28*, 433–445. [CrossRef]
16. Marques, R.; Pereira, J.M.; Lourenço, P.B. Lateral in-plane seismic response of confined masonry walls: From numerical to backbone models. *Eng. Struct.* **2020**, *221*, 111098. [CrossRef]
17. Tomaževič, M.; Klemenc, I. Verification of seismic resistance of confined masonry buildings. *Earthq. Eng. Struct. Dyn.* **1997**, *26*, 1073–1088. [CrossRef]
18. Koutromanos, I.; Stavridis, A.; Shing, P.B.; Willam, K. Numerical modeling of masonry-infilled RC frames subjected to seismic loads. *Comput. Struct.* **2011**, *89*, 1026–1037. [CrossRef]
19. Campbell, J.; Durán, M. Numerical model for nonlinear analysis of masonry walls. *Rev. La Constr.* **2017**, *16*, 189–201. [CrossRef]
20. Angelillo, M.; Cardamone, L.; Fortunato, A. A numerical model for masonry-like structures. *J. Mech. Mater. Struct.* **2010**, *5*, 583–615. [CrossRef]
21. Medeiros, P.; Vasconcelos, G.; Lourenço, P.B.; Gouveia, J. Numerical modelling of non-confined and confined masonry walls. *Constr. Build. Mater.* **2013**, *41*, 968–976. [CrossRef]
22. Ahmed, A.; Shahzada, K. Seismic vulnerability assessment of confined masonry structures by macro-modeling approach. *Structures* **2020**, *27*, 639–649. [CrossRef]

23. Su, R.K.L.; Tang, T.O.; Lee, C.L. Evaluation of local and global ductility relationships for seismic assessment of regular masonry-infilled reinforced concrete frames using a coefficient-based method. *Earthq. Struct.* **2013**, *5*, 1–22. [CrossRef]
24. Chourasia, A.; Singhal, S.; Parashar, J. Seismic performance evaluation of full-scale confined masonry building using light weight cellular panels. *J. Build. Eng.* **2020**, *32*, 101473. [CrossRef]
25. Xu, H.; Wu, H.; Gentilini, C.; Su, Q.; Zhao, S. Experimental study on seismic failure modes of confined masonry structures with different enhancements. *Key Eng. Mater.* **2017**, *747*, 594–603. [CrossRef]
26. Nader, A.; Galal, K. Experimental investigation of axial load and detailing effects on the inelastic response of reinforced concrete masonry structural walls with boundary elements. *J. Struct.-Eng.* **2020**, *146*, 04020259. [CrossRef]
27. Crisafulli, F.J. Seismic Behaviour of Reinforced Concrete Structures with Masonry Infills. Ph.D. Thesis, University of Canterbury, Christchurch, New Zealand, 1997.
28. Smyrou, E.; Blandon, C.A.; Antoniou, S.; Pinho, R.; Crisafulli, F. Implementation and verification of a masonry panel model for nonlinear dynamic analysis of infilled RC frames. *Bull. Earthq. Eng.* **2011**, *9*, 1519–1534. [CrossRef]
29. Magenes, G.; Pampanin, S. Seismic response of gravity-load design frames with masonry infills. In Proceedings of the 13th World Conference on Earthquake Engineering, Vancouver, BC, Canada, 1–6 August 2004.
30. Bell, D.K.; Davidson, B.J. Evaluation of Earthquake Risk Buildings with Masonry Infill Panels. In Proceedings of the 2001 Technical Conference, Future Directions: A Vision for Earthquake Engineering in New Zealand, New Zealand Society for Earthquake Engineering Taupo, Taupo, New Zealand, 23–25 March 2001.
31. Menegotto, M.; Pinto, P.E. Method of analysis for cyclically loaded R.C. plane frames including changes in geometry and non-elastic behaviour of elements under combined normal force and bending. In *Symposium on the Resistance and Ultimate Deformability of Structures Acted on by Well Defined Repeated Loads*; International Association for Bridge and Structural Engineering: Zurich, Switzerland, 1973; pp. 15–22.
32. Mander, J.B.; Priestley, M.J.N.; Park, R. Theoretical stress-strain model for confined concrete. *J. Struct. Eng.* **1988**, *114*, 1804–1826. [CrossRef]
33. Seismosoft, SeismoStruct 2018—A Computer Program for Static and Dynamic Nonlinear Analysis of Framed Structures. 2018. Available online: http://www.seismosoft.com (accessed on 1 May 2018).
34. Sahlin, S. *Structural Masonry*; Prentice-Hall Inc.: Hoboken, NJ, USA, 1971.
35. Paulay, T.; Priestley, M.J.N. *Seismic Design of Reinforced Concrete and Masonry Buildings*; John Wiley & Sons Inc.: New York, NY, USA, 1992.
36. Hendry, A.W. *Structural Masonry*; Macmillan Education Ltd: London, UK, 1990.
37. Porcu, M.C.; Vielma, J.C.; Panu, F.; Aguilar, C.; Curreli, G. Seismic retrofit of existing buildings led by non-linear dynamic analyses. *Int. J. Saf. Secur.* **2019**, *9*, 201–212. [CrossRef]
38. Vielma, J.C.; Barbat, A.H.; Oller, S. Seismic performance of waffled-slab floor buildings. *Proc. Inst. Civ. Eng. Struct. Build.* **2009**, *162*, 169–182. [CrossRef]
39. Vielma, J.C.; Mulder, M. Improved procedure for determining the ductility of buildings under seismic loads. *Rev. Int. De Métodos Numéricos Para Cálculo y Diseño en Ing.* **2018**, *34*, 1–27.
40. Vielma, J.C.; Mulder, M. Seismic performance factors of buildings with irregularity in plan. In *Monograph CIMNE IS 76, Earthquake Engineering Monographs*; CIMNE: Barcelona, Spain, 2018.
41. Vamvatsikos, D.; Cornell, A. Incremental dynamic analysis. *Earthq. Eng. Struct. Dyn.* **2002**, *31*, 491–514. [CrossRef]
42. RENAC. *Accelerograms of the Earthquake of April 16 2016*; National Accelerometer Network of the Geophysical Institute of the National Polytechnic School: Quito, Ecuador, 2016.
43. Martineau, M.O.; Lopez, A.F.; Vielma, J.C. Effect of earthquake ground motion duration on the seismic response of a low rise RC building. *Adv. Civ. Eng.* **2020**, *2020*, 1–12. [CrossRef]
44. Vielma, J.C.; Barbat, A.H.; Oller, S. Seismic safety of low ductility structures used in Spain. *Bull. Earthq. Eng.* **2010**, *8*, 135–155. [CrossRef]
45. Vielma, J.C.; Porcu, M.C.; López, N. Intensity measure based on a smooth inelastic peak period for a more effective incremental dynamic analysis. *Appl. Sci.* **2020**, *10*, 8632. [CrossRef]
46. Vielma, J.C.; Cando, M. Calibration of the response reduction factors used in Ecuador for steel SMRF. *Bull. Int. Inst. Seismol. Earthq. Eng.* **2018**, *52*, 22–37.
47. FEMA P-695. *Quantification of Building Seismic Performance Factors*; FEMA: Washington, DC, USA, 2009.
48. NEC-15. Ecuadorian Construction Standard, SE-DS, Seismic Loads. In *Earthquake Resistant Design*; MIDUVI: Quito, Ecuador, 2015. (In Spanish)

applied sciences

MDPI

Article

Seismic Damage Probability Assessment of Isolated Girder Bridges Based on Performance under Near-Field Earthquakes

Qiang Liu * and Chunyan Yang

College of Harbour and Coastal Engineering, Jimei University, Xiamen 361021, China; ycy@jmu.edu.cn
* Correspondence: Liutanq007@aliyun.com; Tel.: +86-130-0391-0928

Abstract: This paper presents a copula technique for developing seismic fragility curves for an RC (reinforced concrete) isolated continuous girder bridge, by considering earthquake damage indicators such as bridge piers, isolated bearing components, and the main girder of collision damage. The results of this method are compared with those of the limit method of the first-order reliability theory. Meanwhile, the incremental dynamic analysis of the bridge structure under different failure conditions is carried out, and the randomness of the near-fault ground motion and the structural parameters are accounted. Based on the damage index of the isolated bridge under different damage conditions, the seismic fragility curves of each component and the whole isolated bridge are obtained. The research shows that the safety control of the isolated continuous girder bridge structure is mainly affected by the seismic fragility of the isolated bearing, the influence of bridge pier seismic fragility is relatively small, and the probability of beam collision in an isolated bridge is lower than that of a general bridge without isolation bearing. By applying the isolation scheme, the probability of different damage state of the bridge structure is greatly reduced, thus the seismic performance is improved. It also verifies the efficiency and superiority of copula technology. The results will provide a reference for future seismic damage prediction.

Keywords: RC isolated continuous girder bridge; the copula function; seismic fragility analysis; near-fault ground motions; damage index

Citation: Liu, Q.; Yang, C. Seismic Damage Probability Assessment of Isolated Girder Bridges Based on Performance under Near-Field Earthquakes. *Appl. Sci.* **2021**, *11*, 9595. https://doi.org/10.3390/app11209595

Academic Editor: Maria Favvata

Received: 23 September 2021
Accepted: 10 October 2021
Published: 14 October 2021

Publisher's Note: MDPI stays neutral with regard to jurisdictional claims in published maps and institutional affiliations.

Copyright: © 2021 by the authors. Licensee MDPI, Basel, Switzerland. This article is an open access article distributed under the terms and conditions of the Creative Commons Attribution (CC BY) license (https://creativecommons.org/licenses/by/4.0/).

1. Introduction

The bridge is an important hub of transportation systems, especially the reinforced concrete long-span continuous girder bridge, the most common type on highways in China. In recent years, it has been noticed that the long-period velocity and displacement pulse motion of the near-fault ground motion may adversely affect the seismic performance and design of the bridge [1–4]. All previous earthquake disaster statistics have shown that these bridges will sustain serious damage from earthquakes [5,6]. In order to effectively reduce the seismic damage to bridges, the isolation design of lead rubber bearing (LRB) is used in the actual engineering. The seismic responses of such isolated structures have already been widely investigated [7,8]. It was found that this kind of isolated bridge structure can not only meet the design requirements under various working conditions but can also effectively reduce the seismic effect on the bridge pier under the rare earthquake. However, near-fault ground motion with strong non-stationary characteristics adversely affects the seismic performance and design of the isolated bridge through support shedding, displacement of expansion joints, excessive relative displacement of pier beams, and falling beams caused by the collision of the main beams. What is more, the large bearing displacement and the isolation device failure can cause serious damage to the structure [9,10]. Therefore, evaluating the seismic performance of isolated bridges under near-fault ground motion is an important subject for study.

However, because of the uncertain structural and ground motion parameters, the probabilistic analysis method is generally used to study the seismic performance of bridges.

The probabilistic approach offers an efficient way to deal with various uncertainties such as aleatory or epistemic uncertainties in structures and ground motions [11]. Recently, the performance-based seismic design theory of bridges has drawn attention to the probabilistic seismic capacity of structures at multi-performance levels, and the seismic fragility analysis method is the most effective method to evaluate the seismic performance of structures [12,13]. At present, the seismic fragility analysis of bridges mainly includes empirical analysis and numerical simulation analysis; empirical fragility analysis establishes fragility curves based on the damage to bridge structures in real disasters. However, owing to the lack of disaster damage data for bridge structures, application of this method is extremely limited [14]. The numerical fragility analysis method is based on the numerical simulation and dynamic response analysis of the bridge structure to establish the fragility curve. This method can effectively solve the problems resulting from the empirical method, so it is widely used in the study of the seismic fragility of bridges [15].

In fact, the bridge is a high-order statically indeterminate structure, and the failure of one member of the whole structure does not mean the failure of the whole structure. Compared with those of the individual components, the seismic performance of the whole bridge system deserves more attention [16]. However, under the influence of an earthquake, the main beams, supports, columns, and piers of the bridge system are mutually affected. Therefore, accurate simulation of the correlation between the seismic responses of various components is one of the difficulties in seismic fragility research when shifting from a single component to the whole bridge system [17,18].

At present, in view of the correlation between the components in the overall bridge structure, it is assumed that the seismic responses of the components are either completely correlated or not, and the upper and lower boundary values of the fragility of the bridge system are obtained by numerical analysis. The upper bound value is conservatively used to describe the seismic fragility of structural systems [19,20]. However, when the bridge system contains a large number of components, the difference between the upper and lower limits of seismic fragility will be too wide [21,22]. The upper bound of the first-order limit method will obviously overestimate the seismic fragility of the structural system, which will seriously underestimate the safety and reliability evaluation of the entire bridge. In addition, the Monte Carlo sampling method can be used to establish the fragility curve of the bridge system, but the method needs to assume the edge and joint probability distribution types in advance when establishing the joint probability demand model [23,24]. However, there are still many problems in the above analysis methods, such as assuming a linear correlation between the seismic requirements of components, which involves a large amount of numerical sampling.

With the development of mathematical statistics, the copula theory, which deals with correlation between variables, has been gradually applied to the mechanical, civil, and hydraulic engineering fields [25]. At present, the copula function can describe not only the linear correlation between variables, but also the nonlinear correlation between variables, and is not limited by the edge distribution function [26]. The copula function model can accurately simulate the correlation characteristics between variables by separating the edge distribution of variables from the correlation between variables, thus simplifying the process of establishing the joint probability distribution model, which provides a new method for the seismic fragility assessment of the whole system of bridge structures.

Therefore, in this paper, we take an RC isolated continuous girder bridge as a case and use the performance-based probability analysis method to evaluate the seismic failure probability of the components and the bridge system. By considering the impact damage indexes of piers, bearings, and the collision of the main beams, based on the calculation of the seismic fragility of the piers, bearing components, and the collision of the main beams, a seismic fragility analysis of the isolation bridge system is carried out using the copula technique, and the results of this method and the first-order reliability theory method are compared to verify the accuracy and superiority of the copula method. In addition, in the seismic fragility analysis, to assess the uncertainty of the structural parameters, 100 near-fault ground motions are randomly selected from the PEER (Pacific Earthquake Engineering Research) database, and a large number of random ground motion structural samples are established. Thereafter, based on the damage indexes of bridges under different failure states, the seismic fragility curves of each component and of the isolated bridge system are obtained according to the capacity and demand of the structure. Finally, according to the actual seismic demand of similar as-built isolated bridges, the probabilities of damage states are obtained, and a basis is presented as a guide to isolated structural seismic design, reinforcement, and maintenance decision making, etc.

2. Calculation Methods

2.1. The Seismic Fragility Analysis Method

Structural seismic fragility is the probability that the structural damage will exceed a specified value when the ground motion intensity reaches a certain value [27–29]. The seismic fragility of bridges can be expressed as:

$$F_R(a) = P[EDP \geq LS|IM = a] = \int_V^{\infty} f_{R|IM}[r|a]dr \tag{1}$$

where, $F_R(a)$ is seismic fragility, P is the probability of the structure failure exceeding a specified value, EDP is the engineering demand parameter, LS is the state limit of structure, IM is the intensity coefficient of ground motion; $f_{R\,|\,IM}$ $[r\,|\,a]$ means that, when the ground motion intensity is IM, the structure (component) of a certain engineering demand parameter (EDP) has reached or exceeded the conditional probability density of a specified damage state.

Assuming that the relationship between EDP and IM follows the log normal distribution, the mean value of the structure (or component) is:

$$\overset{\wedge}{EDP} = b(IM)^c \tag{2}$$

where, c and b are correlation coefficients.

The seismic fragility of the structural limit state can be calculated. The smooth "seismic fragility curve" is obtained by statistical curve fitting:

$$P = [EDP \geq LS|IM] = 1 - \Phi(\frac{\ln(LS_m - \ln(bIM^C)}{\sqrt{\beta_{LS}^2 + \beta_{EDP}^2}}) \tag{3}$$

where, Φ is the cumulative density function of standard normal distribution, LS_m is log normal distribution under structural damage states, β_{LS} is log standard deviation of structural capacity, β_{EPD} is log standard deviation of structural requirements.

2.2. Seismic Fragility Analysis of Bridge Systems Based on the Copula Technique

The bridge structure is a series system composed of all the structural members of the bridge. Therefore, to evaluate the failure probability of the overall structure of the bridge, it is necessary to consider the seismic fragility evaluation of each of its components. The reliability theory is used to estimate the failure probability of the structure under each damage state:

$$P_{fs} = P[g_1(X) \leq 0, g_2(X) \leq 0, \ldots g_i(X) \leq 0] \tag{4}$$

where, P_{fs} is the failure probability of the bridge structural system, $g_1(X), g_2(X), g_i(X)$ and $g_i(X)$ are the functions of failure probability for each component of the bridge system. The probability of the simultaneous failure of all components is:

$$\begin{aligned} &P[g_1(X) \leq 0, g_2(X) \leq 0, \ldots g_i(X) \leq 0] \\ &= P\{F_1[g_1(X)] \leq F_1(0), F_2[g_2(X)] \leq F_2(0), \ldots, F_i[g_i(X)] \leq F_i(0)\} \\ &= C[F_1(0), F_2(0), \ldots, F_3(0)] = C\left(P_{f1}, P_{f2}, \ldots, P_{fi}\right) \end{aligned} \tag{5}$$

Then, the failure probability of the bridge system considering the correlation of components is obtained:

$$\begin{aligned} P_{fs} &= P[g_1(X) \leq 0 \cup g_2(X) \leq 0 \ldots \cup g_i(X) \leq 0] \\ &= P_{f1} + P_{f2} + \ldots + P_{fi} - C\left(P_{f1}, P_{f2}, \ldots P_{fi}\right) \end{aligned} \tag{6}$$

where, P_{f1}, P_{f2}, P_{fi} are the failure probabilities of each component of the bridge system; C is the copula function.

The seismic fragility of the bridge system can be obtained by substituting the seismic fragility of the bridge components into Equation (3). According to Equation (6), the copula function is the key to solving the fragility of the bridge system.

The above method can easily determine the seismic fragility of each component using IDA calculation results. However, it is difficult to calculate the seismic fragility of two or more components because of the interaction between components. This is because the joint probability distribution of multiple components cannot be directly established by the marginal distribution function of a single component. Therefore, this paper will use the copula technique to analyze the random seismic fragility of the bridge system.

The copula technique provides a common method for describing the nonlinear correlation between random variables. The copula function is defined as an n-dimensional probability distribution on $[0, 1]^n$, which is expressed as [30]:

$$C(u_1, u_2, \ldots u_n) = P(U_1 \leq u_1, U_2 \leq u_2, \ldots U_n \leq u_n) \tag{7}$$

where, $u_1, u_2, \ldots u_n$ is the sample set of uniform random variable U.

Given that $F_i(x_i)$ is an edge probability distribution, its joint probability distribution is represented by $F(x_1, x_2, \ldots x_n)$. Then, the relationship between $F_i(x_i)$ and $F(x_1, x_2, \ldots x_n)$ can be established by the copula function.

$$F(x_1, x_2, \ldots x_n) = C(F_1(x_1), F_2(x_2), \ldots F_n(x_n)) = C(u_1, u_2, \ldots u_n) \tag{8}$$

If $F_i(x_i)$ ($i = 1, \ldots, n$) are all continuous functions, then the copula function $C(u_1, u_2, \ldots u_n)$ is unique. It can be seen that the copula function connects the joint distribution of variables with their edge distribution. As a result of the copula function, the joint distribution function can be expressed explicitly with the edge distribution function. Although it is the connection function of the edge distribution, the form of the copula function is not limited by the edge distribution function.

$$P(X_1, X_2, \ldots X_n | IM) = C(P(x_1), P(x_2), \ldots P(x_n) | IM) \tag{9}$$

where, $P(X_i)(i = 1, \ldots, n)$ represents the seismic fragility of a single component, which can be obtained from Equation (3), and n represents the number of components damaged at the same time. By substituting Equation (9) into Equation (6), the seismic fragility of the bridge system can be expressed as:

$$\begin{aligned} P_{fs} = &\sum_{i=1}^{m} P(X_i|IM) - \sum_{1 \leq i \prec j \leq m} C(P(X_i), P(X_j)|IM) + \\ &\sum_{1 \leq i \prec j \prec k \leq m} C(P(X_i), P(X_j), P(X_k)|IM) + \ldots + \\ &(-1)^{m-1} C(P(X_1), P(X_2), \ldots, P(X_m)|IM) \end{aligned} \tag{10}$$

The seismic fragility of the bridge system can be accurately calculated by Equation (10). The key task of building the seismic fragility curve of the bridge system with the copula method is determining the edge distribution of each component of the bridge system and the best fitting copula function that accurately reflects the correlation of the seismic demand of each component [31].

In short, the key to the seismic fragility analysis of a bridge system is establishing the joint probability distribution function of the pier, support, and other components. However, it is difficult to directly establish the joint probability distribution function between components because of the correlation between the seismic requirements of the components. Therefore, the advantage of using the copula function method is that the correlation between the seismic requirements of components and the probability distribution function of each component edge can be separated, thus simplifying the modeling process of the joint distribution function. In addition, the copula function method not only considers the nonlinear correlation between the seismic requirements of components but also minimizes the numerical sampling, which greatly improves the computational efficiency.

2.3. Selection of the Best Fitting Copula Function

The empirical distribution function can be used to estimate the marginal distribution of the seismic requirements of a component since it is more flexible and has fewer limitations. Because the normal function and *t* copula function can describe the symmetry between component seismic requirements, an empirical distribution-based analysis method was used in this study to select the most suitable copula from the commonly used copula functions (normal function and t copula function). The normal copula function and *t* copula function with correlation coefficient ρ are expressed as:

$$C(X_1, X_2\cdots, X_n; \rho) = \Phi_\rho\left(\Phi^{-1}(u_1), \Phi^{-1}(u_2)\cdots, \Phi^{-1}(u_n)\right) \tag{11}$$

$$C(X_1, X_2\cdots, X_n; \rho, \nu) = \Phi_{\rho,\nu}\left(T_\nu{}^{-1}(u_1), T_\nu{}^{-1}(u_2)\cdots, T_\nu{}^{-1}(u_n)\right) \tag{12}$$

where, Φ_ρ is an n-dimensional standard normal distribution with ρ, Φ^{-1} is the inverse normal distribution, $\Phi_{\rho,\nu}$ is the n-dimensional *t* distribution with ρ and ν, and T^{-1} is the inverse *t* distribution.

Furthermore, the marginal distribution function based on the seismic demand of the component, and the empirical joint distribution function expressed by the empirical copula function are defined as:

$$\hat{C}(u_1, u_2, \ldots, u_n) = \frac{1}{n}\sum_{i=1}^{n} I_{[F_n(x_{1i})\le u_1]} \bullet I_{[F_n(x_{2i})\le u_2]} \cdots \bullet I_{[F_n(x_{mi})\le u_m]} \tag{13}$$

where $F_n(x_{ji})$ ($i = 1, \ldots, n$; $j = 1, \ldots, m$) denotes the marginal probability distribution function of components. $I(\bullet)$ denotes the indicator function, which is equal to 1 if the value of component marginal distribution is smaller than u_i, otherwise, it is equal to 0.

For the above two copula functions as alternative functions, the copula function with the smallest distance is the best fitting copula function describing the correlation between component seismic requirements. The squared Euclidean distance between the alternative function and the empirical copula function is defined as:

$$d^2 = \sum_{i=1}^{n}\left|\hat{C}(u_{1i}, u_{2i}, \cdots, u_{mi}) - C(u_{1i}, u_{2i}, \cdots, u_{mi})\right|^2 \tag{14}$$

where $C(\bullet)$ is the alternative copula function, $\hat{C}$ is the empirical copula function.

In short, according to the component seismic requirements obtained using IDA, the correlation parameters of two commonly used copula functions were estimated, and the squared Euclidean distance was calculated with Equation (14). Then the minimum distance criterion was used to select the best fitting copula function. Finally, the system fragility was obtained by substituting the fragility of each component and the best fitting copula function into Equation (10).

2.4. Basic Steps of Seismic Fragility Analysis

In this paper, the Latin hypercube sampling method (LHS) was applied to incorporate the uncertainty of the structure, and the randomness of ground motion was selected from the PEER (Pacific Earthquake Engineering Research) database for the seismic fragility analysis. The whole research process is summarized, and the technical route is given as follows (Figure 1).

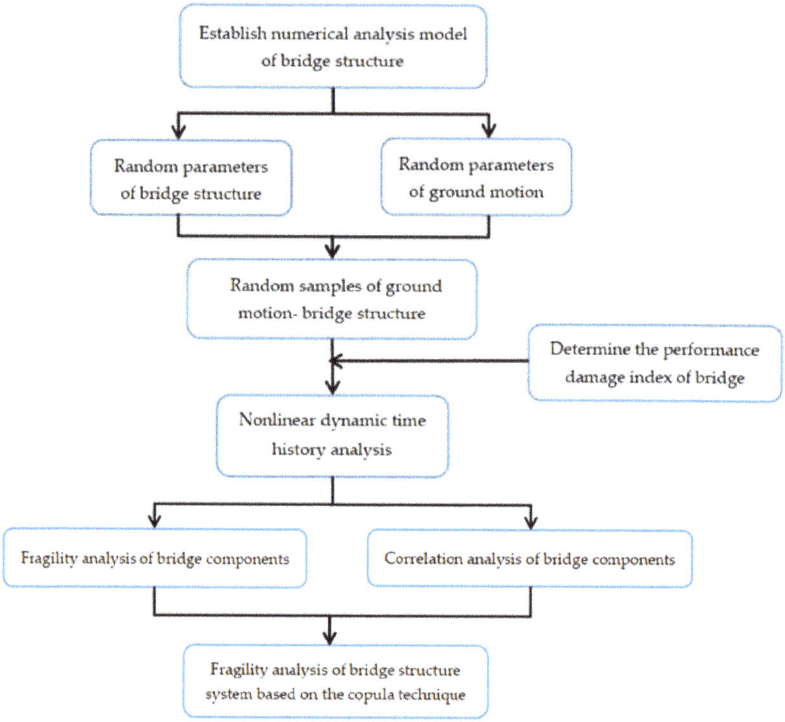

Figure 1. The technical route of the research process.

3. Application

3.1. The Isolated Bridge Geometry and Finite Element Model

The study object of this article is a six-span reinforced concrete continuous girder bridge with a total length of 120 m. The bridge arrangement is shown in Figure 2. The bridge superstructure is a 6 × 20 m cast-in-place C40 concrete box girder with a beam height of 1.3 m. The lower part of the structure is an RC circular pier column of C30 concrete. No. 1 and No. 5 piers are double-column piers 1.3 m in diameter and 8 m in height; No. 2, 3, and 4 piers are variable cross-section one-column circular bridge piers with a height of 8 m and diameter of 1.5 m. The main girder and the pier top are connected to an LRB500 isolation bearing by setting expansion joints on both sides of the stage. The bridge pier foundation material is C30 concrete. The seismic fortification intensity is 8 degrees in the area, and the site class is II.

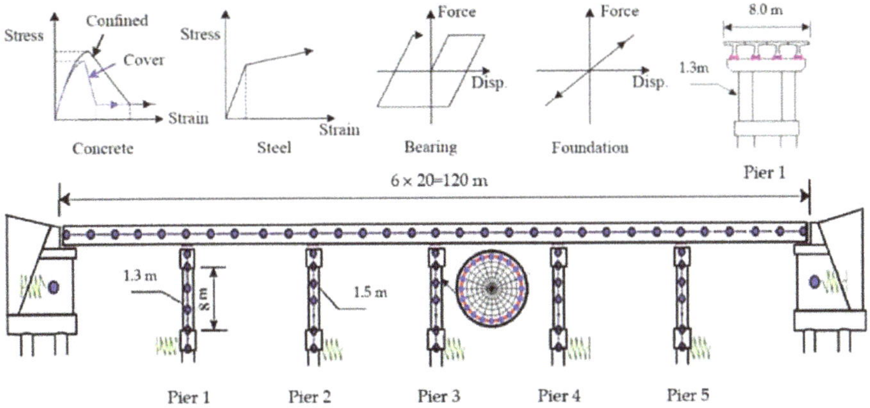

Figure 2. The isolated girder bridge structural configuration.

The finite element dynamic analysis model of the isolated bridge structure was established using OpenSees software [32], and the seismic response of the structure was measured by only the longitudinal seismic motion. The superstructure was modeled with elastic beam-column elements because they remain undamaged and in an elastic state in earthquakes. The damping was simulated with Rayleigh damping [23]. The bearings were modeled using zero-length elements [33], and the shear force–deformation relationship was assumed to be perfect elastic–plastic [34]. Regarding the pier, it often enters the plastic stage, and the plastic hinge appears. Therefore, in the time history analysis, the piers were simulated by nonlinear beam-column elements. The cross section of the element was defined by fibers with the corresponding material stress–strain relationship. The concrete behavior was modeled with Concrete01 in OpenSees. This material was defined with the Kent–Scott–Park model [35]. The reinforcing steel was modeled with Steel 01 in OpenSees, and the constitutive model is shown in Figure 1. The expansion joint between the main beam and the abutment was modeled with a nonlinear contact Gap unit, as shown in Figure 3. It is worth noting that when solving the contact collision problem, it is required that the two components in the collision should not have initial penetration, otherwise the wrong calculation results will be obtained. Therefore, when establishing the model, the spatial clearance at the contact should be maintained as far as possible, so as to conform to the actual spatial position of the bridge [36]. Therefore, this paper assumed that the value of the relative displacement of the abutment and the girder was set, and the collision would occur when the open value was reached.

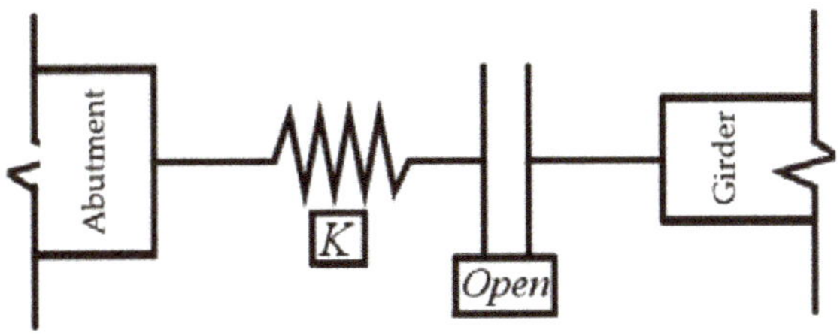

Figure 3. The element model for gap.

3.2. Selection of Near-Fault Ground Motion Records

The object of this paper was choosing near-fault ground motion. The accuracy and efficiency of the structural fragility analysis relies on selecting a reasonable classification and the appropriate number of seismic ground motions. The seismic ground motions should meet two requirements: the first is the ability to simulate the random process of an earthquake disaster itself, and the second is the selection of a large number of adequate seismic ground motions to reduce the impact of ground motion uncertainty on the results of the seismic fragility analysis. The general criterion for ground motion selection was to propagate uncertainty in the magnitude and epicentral distance. In the present study, a suite of ground motions typical of the bridge location were chosen from the Next Generation Attenuation database of the Pacific Earthquake Engineering Research Center (NGA-West2). Then, 100 near-field ground motions with a fault distance of less than 20 km and with significant velocity pulse effect were selected. The general properties of the selected ground motions were as follows: (1) the source-to-site distance (r) was $0 \leq r \leq 20$ km, (2) the moment magnitude of the selected near-field ground motions ranged from 6.5 to 7.6, (3) the average shear-wave velocity of the upper 30 m was accepted from 260 to 500 m/s, and (4) there was significant pulse velocity (a pulse-like waveform with a long period, rich medium and long-period components, and a large peak value). The selected near-fault ground motions covered a wide range of intensities to ensure that the selected suite represented both large and small earthquakes [23]. The distribution of PGAs in 100 near-fault ground motion records is shown in Figure 4.

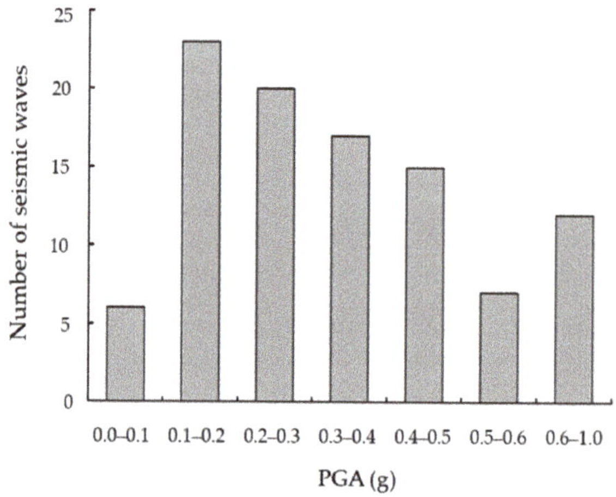

Figure 4. PGA distribution near-fault ground motion record.

3.3. *Stochastic Simulation of Near-Fault Ground Motion-Isolated Bridge Samples*

For isolated bridges, uncertainty parameters include yield strength of steel (Fy), quality of the upper structure (ρ is bulk density), the compressive strength of concrete (Fc), pre-yield stiffness of bearing (K1), and yield shear of bearing (Qy) [37,38]. The main uncertainty parameters and their probability distributions were processed with the Latin hypercube sampling (LHS) approach in the present study. In this paper, the curvature ductility of the pier bottom and the displacement of the isolation bearing were selected as the evaluation indexes, and the parameters of each group of structural samples were changed [39,40]. In order to eliminate the difference between each uncertain parameter in the structural analysis, the data above were normalized, and the sensitivity results of each structural parameter under the action of PGA (1.0 g) ground motion were obtained. The structure parameters of the sensitive variable probability distribution statistics are shown in Table 1. The analysis showed that the yield strength of steel, the quality of the upper structure, the compressive strength of concrete, the pre-yield stiffness of bearing (K1), and the yield shear of bearing (Qy) were sensitive to the seismic response of bridge structures. Thereafter, LHS (Latin hypercube sampling) was used to randomly combine five main parameters to form 10 isolated bridge structure samples, which ensured that these parameters (five main parameters) were taken equally from their probability distributions. Finally, combining the 10 isolated bridge structural samples with the selected 100 near-fault ground motion records, a group of 10 near-fault ground motion-isolated bridge samples were developed to determine the isolated bridge service life (seismic performance analysis).

Table 1. Statistics of the random variables.

Random Variable	Probability Distribution	Average Value	Standard Deviation	Unit
Fy	normal	385.42	28.32	Mpa
Fc	normal	27.68	5.23	Mpa
ρ	normal	28,930	2530	N/m^3
K1	uniform	12,357	—	kN/m
Qy	uniform	54.83	—	kN

3.4. Definition of Seismic Damage to Isolated Girder Bridges

The seismic performance level of structures is a finite state of damage, and the damage to structures should be related to the level of performance (FEMA 2008). Bridge structure damage from earthquake is mainly damage to the bridge pier, bearing, and main beam component, and especially to the bridge pier, which is the most fragile component. In this paper, the damage state of the bridge was defined in three aspects: the failure of the bridge piers, the failure of the isolated bearing, and the collision of the main beam (girder). Thus, the corresponding damage index was determined.

3.4.1. Bridge Pier

With the curvature of piers as engineering demand parameters (EDPs), and with the curvature ductility ratio μ as damage index, the damage index of bridge piers is determined by the moment–curvature relationship of the sections of different damage conditions [41,42]. Earthquake damages indicated that the bottom of the bridge pier was the first failure. Therefore, the damage state was defined by the curvature of the bottom section of the bridge pier, and the damage state and damage index were described at all levels [43], as shown in Table 2.

Table 2. Damage status and description of damage index for piers.

Damage State	Intact	Minor Damage	Medium Damage	Serious Damage	Collapse
Failure criterion	$\phi \leq \phi'_y$	$\phi'_y \leq \phi \leq \phi_y$	$\phi_y \leq \phi \leq \phi_d$	$\phi_d \leq \phi \leq \phi_u$	$\phi \geq \phi_u$

Note: ϕ'_y is the first reinforcement yielding, concrete cracking; ϕ_y is the appearance of the plastic hinge of the section; ϕ_d is the maximum value of the bending capacity; ϕ_u is the limit bending state.

3.4.2. Isolation Bearing

The failure state of the isolated bearing is mainly determined by the displacement and shear strain of the bearing. Five damage states, including basically intact, minor damage, moderate damage, severe damage, and collapse, were defined by Choi according to the displacement of the bearings [44]. Here, displacement of the isolated bearing based on bearing shear strain was adopted as the basis for evaluating earthquake damage. The definition of the deformation rate of the bearing under different damage states is given in Table 3.

Table 3. Description of isolated bearing damage indexes (unit: mm).

State	Intact	Minor Damage	Moderate Damage	Severe Damage	Collapse
Index	$d_\Delta \leq 100\%\gamma$	$d_\Delta \leq 150\%\gamma$	$d_\Delta \leq 200\%\gamma$	$d_\Delta \leq 250\%\gamma$	$d_\Delta > 250\%\gamma$

Note: d_Δ is relative horizontal displacement, γ is the total thickness of the rubber layer.

3.4.3. The Girder End and the Abutment Collision

Under the action of near-fault ground motion, a large relative displacement may occur at the girder end. When the relative displacement of the motion reaches or exceeds the width of the expansion joint between the girder end and the abutment, a collision may occur between the girder end and the abutment. Furthermore, this paper assumed that the expansion between the girder ends and the abutment joint width was 100 mm (gap = 100 mm). In the analysis, the change in temperature and the effect of material variation on the width of the expansion joint were considered. Under the action of a strong earthquake, when the relative displacement of the girder end reached or exceeded 100 ± 10 mm, the girder end was considered to collide with the abutment.

3.5. Probabilistic Seismic Demand Model and Correlation Analysis for Bridge Structure

The probabilistic seismic demand model describes the relationship between structural seismic demand parameters and the ground motion intensity index. Based on the time history analysis of 10 random samples of isolated bridge ground motion, the relationship between the maximum response of dimensionless components and the ratio of required capacity μ under each ultimate failure state and the seismic intensity PGA was obtained. Furthermore, the fitting result obtained by logarithmic regression analysis (that is, the least squares method used for regression analysis of seismic demand and ground motion parameters of structural components) was the probabilistic seismic demand model of the structure. The relationship between the demand capacity ratio μ of each component of the bridge structure under different failure states and the peak seismic acceleration PGA can be expressed as follows:

$$\ln(\mu) = A\ln(PGA) + B \tag{15}$$

where, μ is the requirement capacity ratio of the structure; A and B are regression coefficients.

According to Equation (15), the regression analysis of each component of the bridge system under various failure states could be carried out, and the correlation analysis of the seismic demands of each component could be carried out according to the analysis results of nonlinear dynamic time history, and the relevant parameters of the seismic demands of the structural components of the bridge were obtained, as shown in Table 4.

Table 4. Correlation coefficients of structure seismic demands.

The Correlation Coefficient	lnμ(φ)	lnμ(d_Δ)	lnμ(Gap)
lnμ(φ)	1	0.892	0.753
ln μ(d_Δ)	0.892	1	0.892
lnμ(Gap)	0.753	0.813	1

It can be seen from Table 4 that the correlation between pier and abutment collision was relatively weak, but the minimum correlation coefficient was still greater than 0.75. The results showed that the seismic requirements of each component of the bridge structure were significantly correlated.

4. Results and Discussion

4.1. The Seismic Fragility Analysis of Isolated Bridge Components

As an example, owing to limited space, only the No. 3 pier is provided, based on the damage index. The pier and isolation bearing of the seismic fragility curves under different damage conditions are plotted (obtained from Equation (3)) in Figures 5 and 6, respectively.

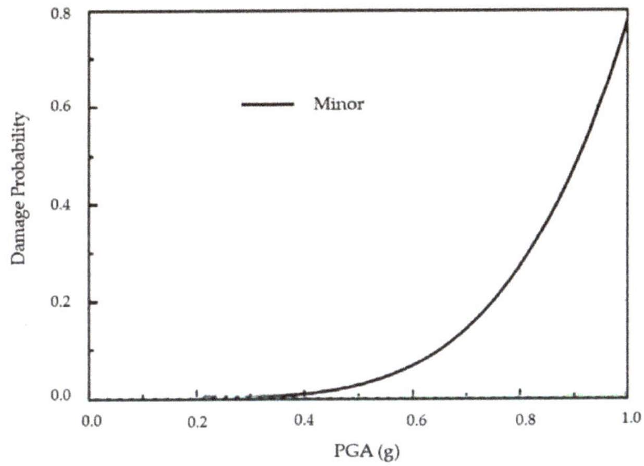

Figure 5. The seismic fragility curves of the pier.

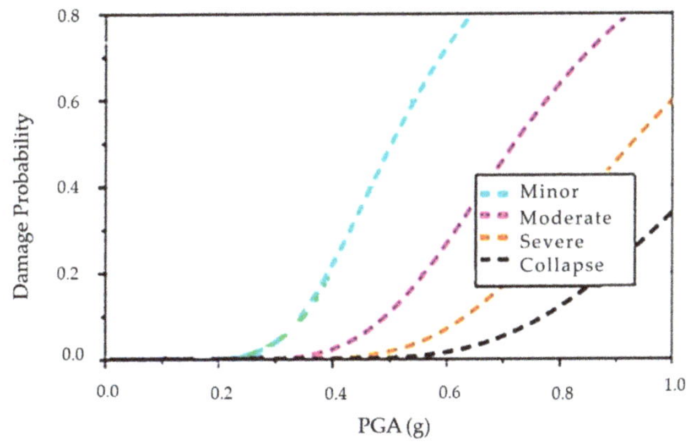

Figure 6. The seismic fragility curves of the isolated bearing.

In order to analyze the structure under the earthquake action, it was assumed that when the ground motion intensity was 0.11 g (PGA = 0.11 g), it was a frequent earthquake, and when the ground motion intensity was 0.51 g (PGA = 0.51 g), it was a rare earthquake.

As is shown in Figure 5, the No. 3 pier with isolated bridges remained intact under frequent earthquake action (PGA = 0.11 g). In the case of rare earthquake action (PGA = 0.51 g), the probability of minor damage to the bridge pier was 4%, but without moderate damage, severe damage, complete destruction, or other incidents. Minor damage did not occur in the frequent earthquake, and it had only a small probability of occurring from the rare earthquake. This was because the pier could bear a greater force, which reflected the requirements of ductility design of bridge components. It showed that the seismic demand on the bridge pier was greatly reduced in the isolation bearing structure.

According to the data shown in Figure 6, under frequent earthquake action (PGA = 0.11 g), the isolated bearing was not easily damaged. However, under rare earthquake action (PGA = 0.51 g), damage to the bearing occurred; the probability of minor, moderate, and severe damage was 52, 11, and 4%, respectively, but a collapse of the isolated bearing did not happen. Compared with the bridge pier, the seismic fragility of the isolated bearing played a key role in the safety control of the isolated continuous girder bridge. This was mainly because the isolated continuous girder bridge canceled the braking pier, the deformation of the isolated bearing was relatively large, so the seismic fragility of the isolated bearing was large.

In order to compare the failure probability of collision between the girder end and the abutment of the general bridge (without isolation device) and of the isolated bridge, the seismic fragility curves of the collisions between the girder and the abutment of two different bridges are given in Figure 7. Under the frequent earthquake action (PGA = 0.11 g), there was no collision between the girder end and abutment of these two types of bridges. However, in the case of rare earthquake action (PGA = 0.51 g), the probability of collision between the girder end and the abutment of the general bridge was about 78%, and the probability of collision between the girder end and the abutment of the isolated bridge was 63%. Therefore, for the bridge constructed using the isolation technology, the probability of collision was 15% lower than that of the general bridge. The main reason was that the relative displacement of the beam members from the earthquake was greatly reduced because of the isolation bearings.

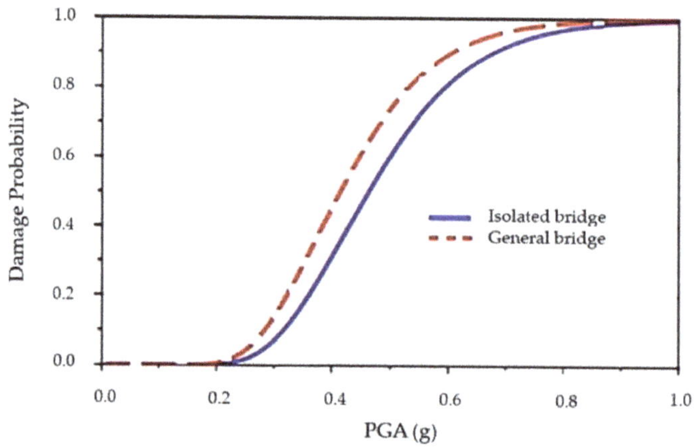

Figure 7. The seismic fragility curves of the girder collision (general bridge and isolated bridge).

4.2. Comparative Analysis of the Seismic Fragility of the Bridge Structure System Based on First-Order Reliability Theory and Copula Technique

In order to verify the accuracy of the copula function method, the upper and lower limits of the seismic fragility of bridge structures based on the first-order reliability theory were considered. According to the reliability theory, when the seismic demand on components is completely related, the maximum seismic fragility of components is the system seismic fragility, which constitutes the lower bound of the first-order limit method; when the seismic demand on components is completely unrelated, any component failure will lead to the failure of the series system, which constitutes the upper bound of the system seismic fragility. Therefore, the first-order boundary of the seismic fragility of the bridge system is:

$$\max_{i=1}^{m}[P(F_i)] \leq P_{sys} \leq 1 - \prod_{i=1}^{m}[1 - P(F_i)] \quad (16)$$

where, $P(F_i)$ indicates the probability of failure in the i-th damage state. P_{svs} is the failure probability of the bridge system in each damage state. m is the number of bridge system components.

By substituting the seismic fragility function of piers and bearings into Equation (16), the upper and lower boundaries of the seismic fragility of the system could be obtained. For comparison, the seismic fragility boundary of the bridge system based on the copula function and that based on the boundary method of first-order reliability theory are shown in Figure 8.

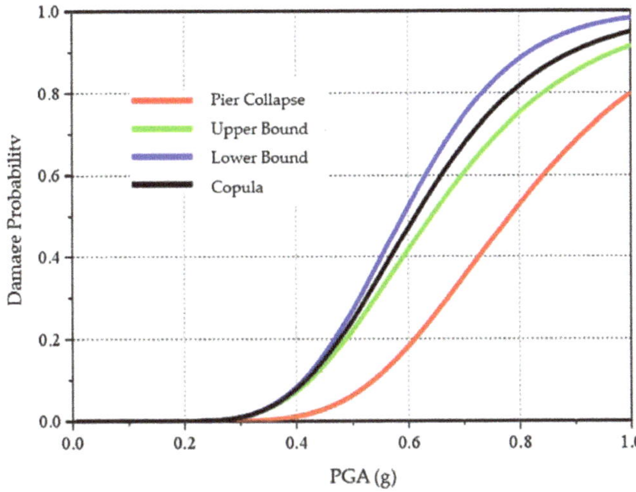

Figure 8. The seismic fragility curve of the overall bridge based on the copula function method.

As seen in Figure 8, the seismic fragility curve of the overall structure based on the first-order reliability theory could be used to evaluate the failure probability range of the overall structure of the bridge under earthquake action. Under the rare earthquake action (PGA = 0.51 g), the failure probability of the overall structure of the bridge was between 21 and 28%, and the upper boundary seismic fragility curve of the failure probability of the overall structure of the bridge and the seismic fragility curve of the bearing failure almost coincided. However, the seismic fragility curve of the bridge system based on the copula function was between the upper and lower bounds of the first-order boundary in the whole range of seismic intensity, and it was closer to the lower bound. If the lower bound described the seismic fragility of the bridge system, it would underestimate the fragility of the structure, while if the upper bound described the seismic fragility of the bridge system, it would obviously overestimate the fragility of the structure. With the increase in ground motion intensity, the difference between the upper and lower bounds obtained by the first-order reliability theory limit method increased, and the difference between the upper and lower bounds obtained by the copula function method and the seismic fragility of bridge system also increased. Therefore, when the intensity of ground motion is large, the limit method of the first-order reliability theory will fail to estimate the fragility of the seismic system of the structure and cannot reasonably evaluate the safety and reliability of the structure.

It can also be seen from Figure 8 that the seismic fragility curve of the bridge system based on the copula function method, under the rare earthquake action (PGA = 0.51 g), indicated that the failure probability of the whole bridge structure was 19%. However, no matter whether they were based on the first-order reliability theory or the copula function method for seismic fragility analysis, the pier and the bearing of the bridge structure system will not fail under frequent earthquake action (PGA = 0.11 g), and it is more likely to cause minor or medium damage. In the rare earthquake action (PGA = 0.51 g), the probability factors of complete failure of the pier and of the bearing were 8 and 21%, respectively. This embodied the structural requirements for multiple seismic designs.

In conclusion, compared to the first-order reliability theory, the failure probability of the whole bridge structure based on copula function analysis was smaller and the probability interval was safer and more reliable.

4.3. Comparative Analysis of Seismic Vulnerability between the Isolated Bridge System and General Bridge System

The comprehensive analysis from the previous section showed that there are significant dependences among component seismic demands since the components interact with each other under earthquake excitation. Therefore, the copula technique was applied to evaluate the seismic damage probability for the bridge structure system.

In order to compare the effect of the seismic isolation system, the seismic fragility curves of the general bridge and the isolated bridge were devised based on the copula function method, as shown in Figure 9.

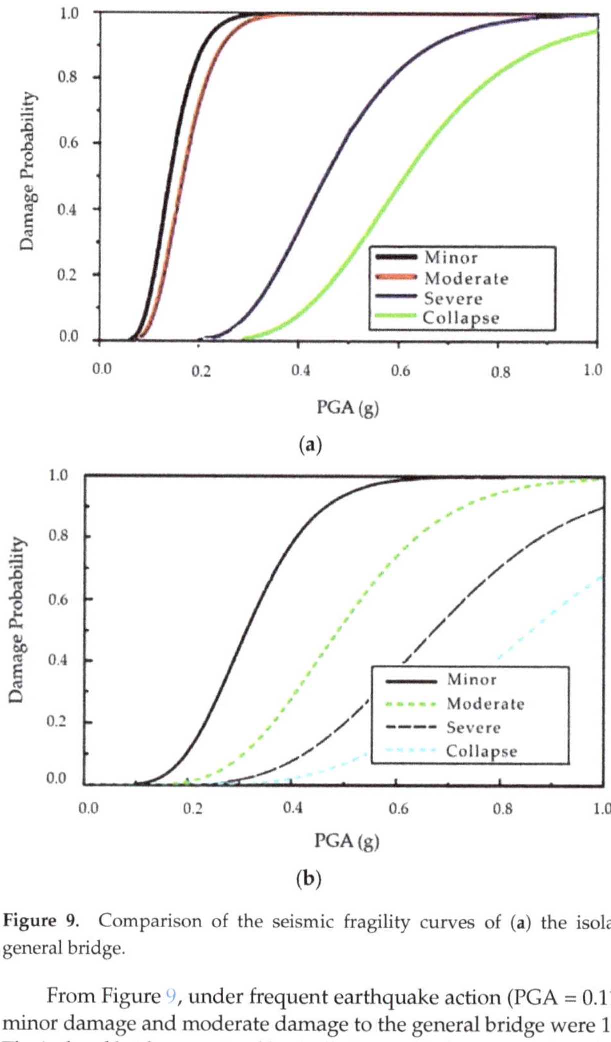

Figure 9. Comparison of the seismic fragility curves of (**a**) the isolated bridge and (**b**) the general bridge.

From Figure 9, under frequent earthquake action (PGA = 0.11 g), the probability of minor damage and moderate damage to the general bridge were 18 and 6%, respectively. The isolated bridge remained basically intact. Under rare earthquake action (PGA = 0.51 g), the probability of serious damage to the general bridge was 68%, and the probability of collapse 29%, but the probability of minor damage, moderate damage, severe damage, and collapse of isolated bridges were 92, 56, 20, and 10%, respectively. Therefore, the probability of different damage states of the bridge structure was greatly reduced with the seismic isolation scheme.

5. Conclusions

In this paper, an RC (reinforced concrete) isolated continuous girder bridge was selected as the research object. In order to reasonably evaluate the seismic performance of the isolated bridge, considering the randomness of the near-fault ground motion and structural parameters, an incremental dynamic analysis of the bridge structure under different failure states was carried out. With the copula function method, the joint distribution function of components was obtained, and the seismic fragility curve of the bridge system, considering the influence of multiple components and their demand correlation, was established, which was compared with the first-order limit reliability theory model. The main conclusions are as follows.

Compared with the traditional method, the copula function method has obvious advantages. It can not only overcome the limitation of the assumption that the seismic demands obey the lognormal distribution but can also describe the nonlinear correlation between the seismic demands on the bridge components, especially the tail correlation. This new method eliminates the limitation that only the Pearson correlation coefficient can describe linear correlation and thus simplifies the modeling process of the multivariate joint probability distribution function of the bridge structure. It provides a new model for the seismic vulnerability analysis of complex (multiple) bridge structures.

The fragility curve of bridge structures obtained with the copula method was located between the two limit values obtained with the first-order method, but there was still a large gap, which indicated that the limit error caused by the traditional fragility analysis method was relatively large. The calculation results showed that the failure probability of the whole bridge structure analyzed with the copula method had a smaller probability interval, and that the evaluation based on the copula method was more reasonable, safer, and more reliable, which also proved the calculation efficiency and superiority of the copula method.

Furthermore, in this study, the seismic fragility analysis of the isolated continuous girder bridge showed that the failure probability of the isolation bearing was relatively large and was more likely to occur as a limit state failure, while the failure probability of the pier was relatively small. Therefore, in the seismic design of this kind of isolated bridge structure, the seismic capacity of the isolation bearing should be the main consideration. At the same time, this study also showed that the probability of different failure states in the continuous bridge structure was greatly reduced with the seismic isolation scheme, which can greatly improve the seismic performance of the bridge.

Author Contributions: Conceptualization, investigation and writing, Q.L.; supervision, C.Y.; formal analysis, investigation and methodology, C.Y. All authors have read and agreed to the published version of the manuscript.

Funding: This research was supported by the Natural Scientific Research Foundation of Fujian, China (Grant No. 2021J01856).

Institutional Review Board Statement: Not applicable.

Informed Consent Statement: Not applicable.

Data Availability Statement: Not applicable.

Conflicts of Interest: The authors declare no conflict of interest.

References

1. Jangid, R.S.; Kelly, J.M. Base isolation for near-fault motions. *Earthq. Eng. Struct. Dyn.* **2001**, *30*, 691–707. [CrossRef]
2. Loh, C.H.; Wan, S.; Liao, W.I. Effects of hysteresis model on seismic demands: Consideration of near-fault ground motions. *Struct. Des. Tall Build.* **2002**, *11*, 155–169. [CrossRef]
3. Zhao, L.F.; Wei, D. Seismic analysis of RC frame structures subjected to near-fault pulse-like ground motions. *Eng. Mech.* **2005**, *25*, 150–193.
4. Yazdani, M.; Jahangiri, V.; Marefat, M.S. Seismic performance assessment of plain concrete arch bridges under near-field earthquakes using incremental dynamic analysis. *Eng. Fail. Anal.* **2019**, *106*, 104170. [CrossRef]

5. Guo, A.; Yuan, W.; Lang, C.; Guan, X.; Li, H. Time-dependent seismic demand and fragility of deteriorating bridges for their residual service life. *Bull. Earthq. Eng.* **2015**, *13*, 2389–2409. [CrossRef]
6. Crespi, P.; Zucca, M.; Longarini, N.; Giordano, N. Seismic assessment of six typologies of existing RC bridges. *Infrastructures* **2020**, *5*, 52. [CrossRef]
7. Hameed, A.; Koo, M.-S.; Do, T.D.; Jeong, J.-H. Effect of lead rubber bearing characteristics on the response of seismic isolated bridges. *KSCE J. Civ. Eng.* **2008**, *12*, 187–196. [CrossRef]
8. Han, Q.; Du, X.I.; Liu, W.G. Shaking table test of isolated continuous girder bridge model under different earthquake intensity excitation. *China J. Highw. Transp.* **2008**, *21*, 50–56.
9. Ge, S.J.; Xiong, Z.H.; Zhai, M.G.; Pan, C.P. Research on seismic fragility of medium and small sized concrete continuous girder bridge. *J. Highw. Transp. Res. Dev.* **2013**, *30*, 60–65.
10. Kang, Q.; Huang, X.; Chen, J.; Zhou, L. Seismic response analysis of multi-span continuous isolated bridge. *J. Earthq. Eng. Eng. Vib.* **2014**, *34*, 217–223.
11. Ghosh, J.; Sood, P. Consideration of time-evolving capacity distributions and improved degradation models for seismic fragility assessment of aging highway bridges. *Reliab. Eng. Syst. Saf.* **2016**, *154*, 197–218. [CrossRef]
12. Billah, A.M.; Alam, M.S. Seismic fragility assessment of highway bridges: A state-of-the-art review. *Struct. Infrastruct. Eng.* **2015**, *11*, 804–832. [CrossRef]
13. Yang, D.X.; Li, G.; Cheng, G.D. Seismic analysis of base-isolated structures subjected to near-fault pulse like ground motions. *J. Earthq. Eng. Eng. Vib.* **2012**, *3*, 68–77.
14. Gautam, D.; Rupakhety, R.; Adhikari, R. Empirical fragility functions for Nepali highway bridges affected by the 2015 Gorkha Earthquake. *Soil Dyn. Earthq. Eng.* **2019**, *126*, 105778. [CrossRef]
15. Mosleh, A.; Jara, J.; Razzaghi, M.S.; Varum, H. Probabilistic Seismic Performance Analysis of RC Bridges. *J. Earthq. Eng.* **2018**, *108*, 1477637. [CrossRef]
16. Zhang, J.; Bi, K.; Zheng, S.; Jia, H.; Zhang, D.Y. Seismic system reliability analysis of bridges using the multiplicative dimensional reduction method. *Struct. Infrastruct. Eng.* **2018**, *14*, 1455–1469. [CrossRef]
17. Wu, W.P.; Li, L.F. System seismic fragility analysis methods for bridge structures. *J. Vib. Shock* **2018**, *37*, 273–280.
18. Pan, Y.; Agrawak, A.K.; Ghosn, M. Seismic Fragility of continuous steel highway bridges in New York State. *J. Bridge Eng.* **2007**, *12*, 689–699. [CrossRef]
19. Simon, J.; Bracci, J.M.; Gardoni, P. Seismic response and fragility of deteriorated reinforced concrete bridges. *J. Struct. Eng.—ASCE* **2010**, *136*, 1273–1281. [CrossRef]
20. Kibboua, A.; Naili, M.; Benouar, D.; Kehila, F. Analytical fragility curves for typical Algerian reinforced concrete bridge piers. *Struct. Eng. Mech.* **2011**, *39*, 411–425. [CrossRef]
21. Choi, E.; DesRoches, R.; Nielson, B. Seismic fragility of typical bridges in moderate seismic zones. *Eng. Struct.* **2004**, *26*, 187–199. [CrossRef]
22. Tavares, D.H.; Suescun, J.R.; Paultre, P.; Padgett, J.E. Seismic fragility of a highway bridge in Quebec. *J. Bridge Eng.—ASCE* **2013**, *18*, 1131–1139. [CrossRef]
23. Nielson, B.G.; Des, R.R. Seismic fragility methodology for highway bridges using a component level approach. *Earthq. Eng. Struct. Dyn.* **2007**, *36*, 823–839. [CrossRef]
24. Wang, Q.A.; Wu, Z.Y.; Jia, Z.P. Multidimensional fragility analysis of bridge system under earthquake. *Eng. Mech.* **2013**, *30*, 192–198.
25. Shen, G.Y.; Yuan, W.C.; Pang, Y.T. Bridge seismic fragility analysis based on Nataf transformation. *Eng. Mech.* **2014**, *31*, 93–100.
26. Song, S.; Qian, Y.J.; Wu, G. Research on seismic fragility method of bridge system based on copula function. *Eng. Mech.* **2016**, *33*, 193–200.
27. Cornell, C.A.; Jalayer, F.; Hamburger, R.O.; Foutch, D.A. Probabilistic basis for 2000 SAC federal emergency management agency steel moment frame guideline. *J. Struct. Eng.* **2002**, *128*, 526–533. [CrossRef]
28. Vamvatsikos, D.; Cornell, C.A. Incremental dynamic analysis. *Earthq. Eng. Struct. Dyn.* **2002**, *31*, 491–514. [CrossRef]
29. Baker, J.W. Efficient analytical fragility function fitting using dynamic structural analysis. *Earthq. Spectra* **2015**, *31*, 579–599. [CrossRef]
30. Nelsen, R.B. *An Introduction to Copulas*; Springer Science and Business Media: New York, NY, USA, 2013.
31. Song, S.; Qian, Y.; Liu, J.; Xie, X.; Wu, G. Time-variant fragility analysis of the bridge system considering time-varying dependence among typical component seismic demands. *Earthq. Eng. Eng. Vib.* **2019**, *18*, 363–377. [CrossRef]
32. Mazzoni, S.; Mc, K.F.; Scott, M.; Fenves, G.L. *The Open Sees Command Language Manual, Version 2.0*; Pacific Earthquake Engineering Research Center, University of California at Berkeley: Berkeley, CA, USA, 2009.
33. Seo, J.; Linzell, D.G. Nonlinear seismic response and parametric examination of horizontally curved steel bridges using 3D computational models. *J. Bridge Eng.* **2013**, *18*, 220–231. [CrossRef]
34. Siqueira, G.H.; Sanda, A.S.; Paultre, P.; Padgett, J.E. Fragility curves for isolated bridges in Eastern Canada using experimental results. *Eng. Struct.* **2014**, *74*, 311–324. [CrossRef]
35. Mander, J.B.; Priestley, M.J.N.; Park, R. Observed stress-train behavior of confined concrete. *J. Struct. Eng.* **1988**, *114*, 1827–1849. [CrossRef]

36. Qi, X.J.; Shen, Y.G. Uneven distribution of pounding effect in a curved grider bridge unden earthquake. *J. Vib. Shock* **2012**, *31*, 72–78.
37. Ghosh, J.; Padgett, J.E. Impact of multiple component deterioration and exposure conditions on seismic vulnerability of concrete bridges. *Earthq. Struct.* **2012**, *3*, 649–673. [CrossRef]
38. Park, C.; Kim, N.; Haftka, R. The effect of ignoring dependence between failure modes on evaluating system reliability. *Struct. Multidiscip. Optim.* **2015**, *52*, 251–268. [CrossRef]
39. Monteiro, R.; Delgado, R.; Pinho, R. Probabilistic seismic assessment of RC bridges: Part I—Uncertainty models. *Structures* **2016**, *5*, 258–273. [CrossRef]
40. Monteiro, R.; Zelaschi, C.; Silva, A.; Pinho, R. Derivation of fragility functions for seismic assessment of RC bridge portfolios using different intensity measures. *J. Earthq. Eng.* **2019**, *23*, 1678–1694. [CrossRef]
41. Faria, R.; Oliver, J.; Cervera, M. A strain-based plastic viscous-damage model for massive concrete structures. *Int. J. Solids Struct.* **1998**, *35*, 1533–1558. [CrossRef]
42. Park, Y.J.; Ang, A.H.S. Mechanistic seismic damage model for reinforced concrete. *J. Struct. Eng.* **1985**, *111*, 722–739. [CrossRef]
43. Lu, B.Y.; Liu, B.Q.; Liu, M.; Xing, G.H.; Wu, T. Quantitative research on reinforced concrete performance index of reinforced concrete bridge column. *China J. Highw. Transp.* **2010**, *23*, 49–57.
44. FEMA 356. *Pre-Standard and Commentary for Seismic Rehabilitation of Buildings*; American Society of Civil Engineers: Washington, DC, USA, 2000.

applied sciences

MDPI

Article

Seismic Design and Performance Assessment of Frame Buildings Reinforced by Dual-Phase Steel

Jure Žižmond * and Matjaž Dolšek

Faculty of Civil and Geodetic Engineering, University of Ljubljana, 1000 Ljubljana, Slovenia; mdolsek@fgg.uni-lj.si

* Correspondence: jzizmond@fgg.uni-lj.si

Abstract: To improve the durability and serviceability of reinforced concrete structures, different variants of dual-phase reinforcing steel were developed within the research project NEWREBAR. The investigated variant of the new material, termed DPD2 steel, has a specific microstructure that increases the corrosion resistance, but its yielding strength is less than that of Tempcore steel B500B. DPD2 steel has no yielding plateau, which is characteristic of conventional reinforcing steel. Thus, it was investigated whether the current building codes can be used to design earthquake-resistant concrete structures reinforced by DPD2 steel bars. For this reason, three multi-story reinforced concrete frame buildings were designed according to Eurocode by considering DPD2 steel and, for comparison reasons, Tempcore steel B500B. Based on the nonlinear model, which was validated by cyclic test of columns, the seismic performance of DPD2 buildings was found to be improved compared to those designed with conventional B500B reinforcing steel. This can mainly be attributed to the substantial strain hardening of the DPD2 steel, which increases the overstrength factor of the structure by about 10%. However, for the improved seismic performance, the amount of steel in DPD2 buildings had to be increased in the design by approximately 20–25% due to the smaller yield strength of DPD2 steel. Nevertheless, it was demonstrated that Eurocode 8 could be used to design earthquake-resistant frame building reinforced with dual-phase reinforcing steel DPD2.

Keywords: dual-phase reinforcing steel; earthquake-resistant design; Tempcore reinforcing steel; reinforced concrete frames; seismic analysis; pushover analysis

Citation: Žižmond, J.; Dolšek, M. Seismic Design and Performance Assessment of Frame Buildings Reinforced by Dual-Phase Steel. *Appl. Sci.* **2021**, *11*, 4998. https://doi.org/10.3390/app11114998

Academic Editor: Maria Favvata

Received: 12 May 2021
Accepted: 25 May 2021
Published: 28 May 2021

Publisher's Note: MDPI stays neutral with regard to jurisdictional claims in published maps and institutional affiliations.

Copyright: © 2021 by the authors. Licensee MDPI, Basel, Switzerland. This article is an open access article distributed under the terms and conditions of the Creative Commons Attribution (CC BY) license (https://creativecommons.org/licenses/by/4.0/).

1. Introduction

The exposure of reinforced concrete structures to aggressive environmental conditions causes corrosion of steel reinforcing bars and deterioration of structural performance against gravity and seismic loads (e.g., [1–4]). The corrosion rate, however, also depends on the typology of the reinforcing steel. At present, the majority of new reinforced concrete structures in Europe are reinforced with Tempcore reinforcing bars. Tempcore steel is characterised by its highly suitable mechanical properties, excellent weldability, ductility, bendability and manageable production costs. Nevertheless, several authors [5–7] observed that the deformation and energy dissipation capacity of structural elements reinforced by Tempcore steel bars rapidly deteriorates if they are exposed to aggressive environmental conditions. Corrosion affects the bending and shear capacity of structural elements. Thus, the seismic performance of the structure can be reduced significantly during its lifetime (e.g., [8–10]). Celarec et al. [8] showed that the corrosion of stirrups could trigger the shear failure of structural elements, although the frame was designed and constructed according to the capacity design principles. Lavorato et al. [10], among others, reported that the corrosion might induce longitudinal bar buckling in sections that were originally designed to prevent such phenomena.

Therefore, it is sensible to investigate and improve the durability (i.e., resistance against corrosion) of reinforced concrete structures. The degradation of concrete structures is often improved by the utilisation of modified concrete mixture [11,12]. The most convenient approach is to increase the thickness of the concrete cover and improve the quality of concrete [13]. However, in recent years, it has become possible to enhance the properties of steel bars. For example, Lollini et al. [14] and Gu and Meng [15] presented the applications of stainless steel in construction. Maffei et al. [16] and Salvatore et al. [17] analysed the possibility of adopting dual-phase (DP) steels for civil construction. DP steels are widely used in the automotive sector due to their excellent ductile properties and improved durability performance. However, the use of DP steel in construction is limited because the technology for mass production of reinforcing bars has yet to be fully developed. One attempt was made by Lorusso et al. [18], who presented the application of DP steels in wires for the reinforcement of concrete structures. Authors found that the DP steels had a greater capacity for energy absorption, ultimate elongation and hardening exponent than ATR500N, although they showed a slight decrease in resistance. However, their product did not reach the levels of elongation at maximum strength or at rupture which are typical for traditional Tempcore steel.

The production of dual-phase reinforcing steel, which has similar characteristics in terms of strength and deformation capacity to traditional Tempcore steel (e.g., B500B, B450C), was addressed within the European research project 'NEW dual-phase steel REinforcing BARs for enhancing capacity and durability of antiseismic moment-resisting frames' (NEWREBAR). The project was funded by the Research Fund for Coal and Steel (RFCS). The technology for producing DP reinforcing bars is one of the results of this project. In particular, two DP steel grades (i.e., DPD2 and DPF2) were developed and tested [19], and the industrial feasibility of DP rebars using existing plants was also evaluated [20]. It was shown that the existing plants have to be upgraded in order to allow for industrial production of the new steel.

The two grades of DP steel (i.e., DPD2 and DPF2) were produced using the same production process, but they have different chemical compositions of produced coils. Consequently, grade DPD2 has higher strength but a lower deformation capacity in comparison to DPF2. Howeve, details about the production of steels are beyond the scope of this paper because it focuses on the usability of DP steel for the construction of reinforced concrete buildings in seismic prone areas [19,21]. The corrosion resistance of innovative DP reinforcing steels is improved against Tempcore steel [21] as a consequence of their specific microstructure, characterised by the direct embedment of martensite into the ferrite matrix. In the case of Tempcore steel, the martensite is formed on the outer surface of the bars, which increases their hardness, while the remaining ferritic–perlitic core maintains the typical ductility of hot-rolled bars [22].

Besides the corrosion resistance, the stress–strain curve of DP steel is different from that of conventional Tempcore steel. The stress–strain curve of DP steel does not indicate a typical yielding plateau, and it has a different hardening ratio and ultimate stress in comparison to conventional reinforcing steels. As a consequence, the use of DP steel bars in reinforced concrete (RC) structures can alter the maximum strain demand in reinforcing bars. Additionally, the use of DP steel bars affects the ductility of the structural element and the global ductility of the structure, both of which are essential for achieving the life safety requirements of the current codes for earthquake-resistant structure design. As DP steel is a new type of reinforcing steel, it is not yet understood whether the current building codes [13,23] can be used for the earthquake-resistant design and the seismic performance assessment of concrete structures reinforced by DP steel bars. For this purpose, this research was conceived and performed in order to provide an insight into the seismic performance of concrete structures reinforced by DP steel. The research involved the earthquake-resistant design and seismic performance assessment of three multi-storey frame buildings reinforced with B500B and DPD2 bars. The impact of new DP steel on design and seismic performance results is commented on and evaluated.

In the first part of the paper, the mechanical properties of the traditional Tempcore B500B reinforcing steel and dual-phase DPD2 reinforcing steel are presented and compared. Then, the simplified nonlinear models are described, and the cyclic response of a column reinforced by B500B or DPD2 steel bars is validated by the results of the cyclic tests [24,25]. Finally, the findings of the earthquake-resistant design of the frame buildings are presented, and the difference between the seismic performance of frame buildings reinforced with DPD2 and B500B reinforcing bars are discussed.

2. Mechanical Properties of Tempcore Steel and Dual-Phase Steel DPD2

Tempcore steel grade B500B and innovative DP steel grade D2 (i.e., DPD2 steel grade), which was developed within the European research project NEWREBAR [19], are produced using different procedures. Reinforcing steel B500B is produced using the Tempcore process, which is characterised by a quenching phase following the self-tempering phase. This process provides good strength and ductility and moderate production costs. However, several researchers [6,26] have recently observed various durability problems associated with Tempcore steel that reduce both the deformation and energy dissipation capacities of structural elements if exposed to aggressive environmental conditions. However, the resistance of DP reinforcing steel against corrosion is better, due to its specific microstructure, which causes also differences in the stress–strain relationship in comparison to that of the frequently used Tempcore steel (Figure 1). The stress–strain relationship of DP steel corresponds to DPD2 steel grade manufactured within the European research project NEWREBAR [21], whereas the stress–strain curve of Tempcore B500B was obtained from the results of tensile tests performed within a research project that focused on the effects of corrosion on the mechanical behaviour of steel reinforcing bars [27] and a research project sponsored by the Slovenian Research Agency [24]. From Figure 1, it can be observed that the stress–strain relationship of B500B has a yielding plateau, which is typical for heat-treated steel. This phenomenon is not seen in the case of the DPD2 steel grade, which is characterised by a continuous yielding, similar to that presented in the literature for DP steel grades [28]. Therefore, the yield strength of reinforcing steel DPD2 has to be determined based on a permanent offset of 0.2% of the gage length under load.

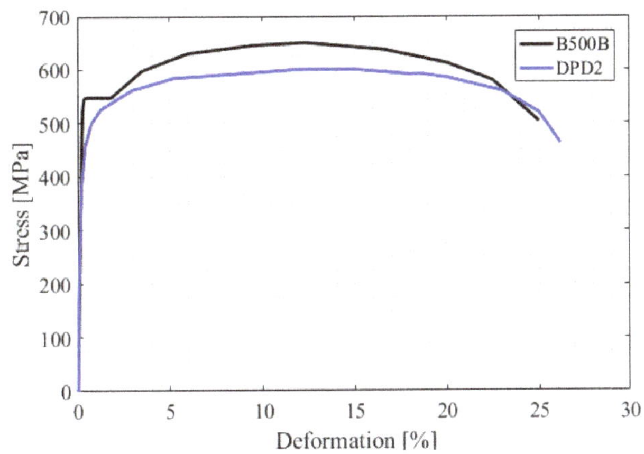

Figure 1. Stress-strain relationships of B500B and DPD2 steel grades.

The mechanical properties of the steel grades presented in Figure 1 are elaborated in Table 1, where $f_{y,k}$ is the characteristic yield strength (i.e., fifth percentile of the measured yield strengths), $f_{y,m}$ is the mean yield strength, f_u is the mean value of the maximum tensile strength, A_{gt} is the deformation at maximum strength, A_5 is the deformation of rupture of the bar, and k is the hardening ratio. It can be concluded that the DPD2 steel grade has a higher value of ultimate deformation (A_5; 26.2% vs. 25.0%) and higher hardening ratio (k; 1.28 vs. 1.19). On the other hand, the B500B steel grade has a higher yield and tensile strength. The characteristic yield strength of B500B is higher than that of DPD2 by a factor of 1.25.

Table 1. Mechanical properties of B500B and DPD2 steel grades taken from the experimental results [21,24,27].

Steel Grade	$f_{y,k}$ [MPa]	$f_{y,m}$ [MPa]	f_u [MPa]	A_{gt}	A_5	k
B500B	500	548	650	12.3%	25.0%	1.19
DPD2	400	461	590	11.9%	26.2%	1.28

The results of low-cycle fatigue tests [29] also revealed that DPD2 reinforcing steel has almost the same energy dissipation capacity as traditional Tempcore steel if the energy dissipation capacity was measured in terms of the maximum cycle numbers at the designated imposed deformation and the total dissipative energy.

3. Mathematical Modelling of Cyclic Response of Columns Reinforced with B500B and DPD2

3.1. Description of Code-Based Mathematical Modelling of Seismic Response of RC Columns

The code-based mathematical model of an RC column comprises an elastic element with nonlinear flexural hinges at both ends of the column. If the model is used for simulation of the cyclic test of a column, then the nonlinear flexural hinge is used only at the base of the column (see Figure 2a). Such a model is presented in the following and is used to simulate the cyclic response of columns reinforced with both B500B and DPD2 steel bars. All simulations were performed using OpenSees [30]; therefore, some features of the modelling refer to the OpenSees command language.

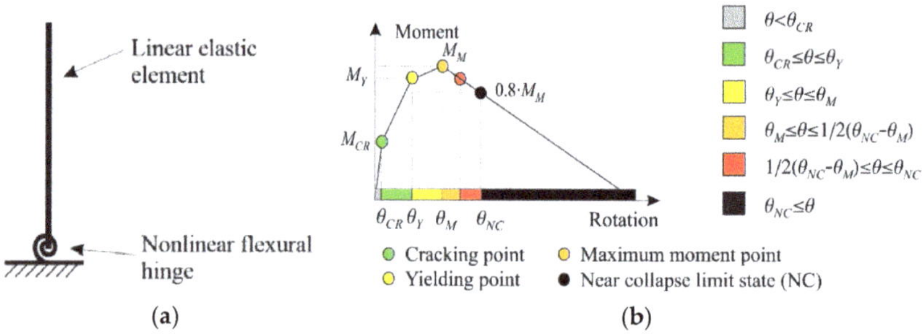

Figure 2. (**a**) mathematical model of column for simulation of cyclic response and (**b**) four-linear moment-rotation relationship of plastic hinge.

The linear elastic element was modelled using the 'elasticBeamColumn' element available in OpenSees. The element is defined based on the height of column used in experiments and has the characteristics of a concrete cross-section (i.e., elastic modulus of concrete, and area and moments of inertia of cross-section).

The four-linear moment–rotation relationship of the nonlinear flexural hinge, similar to that proposed by Ferreira et al. [31], was defined. The cracking of the concrete cross-section is represented by the first characteristic point (CR) from the moment–rotation relationship, whereas other points are related to the reinforcement yielding (Y), the maximum moment (M) and the near-collapse limit (NC) state (see Figure 2b). Note also that an additional point (see a red dot in Figure 2b) is presented on the moment–rotation relationship. This point was added to improve the presentation of the damage of the structural elements.

The data required for the calculation of the moment–rotation relationship (see Figure 2b) are elastic proprieties of structural elements, the geometry of cross-section of the structural element, longitudinal reinforcement, transverse reinforcement, the material strength, level of normalised axial force and assumed ratio r_{CM} (see below). The cracking M_{CR}, yielding M_Y, and the maximum M_M moments were calculated by performing a moment–curvature analysis of the cross-section, which was divided into many fibres. Each fibre was defined with an appropriate uniaxial material, available in OpenSees. The confined and unconfined concrete were modelled using 'Popovics Concrete Material' (or 'Concrete04 Material') [32]. In the case of unconfined cover concrete, the parameters of 'Popovics Concrete Material' were assessed according to the requirements of Eurocode 2 [13], whereas the concrete compressive strength was obtained from the compressive test performed on cubes taken from the concrete used to cast the column. The parameters of the confined concrete were estimated according to the draft of the new Eurocode 8-3 [33]. In this case, the confined concrete maximum compressive strength f_{cc} and the corresponding strain $\varepsilon_{c0,c}$ can be determined using the following expressions

$$f_{cc} = f_c(1 + K) \tag{1}$$

$$\varepsilon_{c0,c} = \varepsilon_{c0}(1 + 5K) \tag{2}$$

where f_c is the concrete compressive strength of unconfined concrete, whereas ε_{c0} is the compressive strain in the concrete at maximum strength, which is assumed to be 0.002. The parameter K is calculated with the following equation

$$K = 3.5\left(\frac{\alpha \cdot \rho_{sw} \cdot f_{yw}}{f_c}\right)^{\frac{3}{4}} \tag{3}$$

where f_{yw} is the mean yield strength of transverse steel, α is the confinement effectiveness factor, and ρ_{sw} is the ratio of transverse reinforcement. From Equation (3), it can be seen that the response of the confined concrete does not only depend on the amount of transverse reinforcement but also on the strength of the steel used for reinforcement, as was also found by Wang et al. [34]. For rectangular sections, the ρ_{sw} is defined as

$$\rho_{sw} = \frac{A_{sw}}{b_w \cdot s_w} \tag{4}$$

where A_{sw} is the area of the transverse reinforcement bars of the cross-section, b_w is the width of cross-section, and s_w is the spacing of the stirrups measured from the centreline. The confinement effectiveness factor α is defined as follows

$$\alpha = \left(1 - \frac{s_w}{2b_o}\right)\left(1 - \frac{s_w}{2h_o}\right)\left(1 - \frac{\sum\limits_{i=1}^{n} b_i^2}{6b_o h_o}\right) \tag{5}$$

where b_o and h_o are the dimensions of the rectangular of the confined concrete measured to the centreline of the perimeter bar, b_i is the i-th centreline spacing between longitudinal bars, laterally restrained by a stirrup corner or hook, and n is the number of longitudinal bars, laterally restrained along the perimeter of the cross-section.

The reinforcing steels B500B and DPD2 were modelled using 'MultiLinear uniaxialMaterial' [35], which can take the softening branch after the maximum strength into account. 'MultiLinear uniaxialMaterial' is defined by specifying points of the stress–strain (force-displacement) curve.

The cracking moment, $M_{CR,}$ of the moment–rotation relationship of the nonlinear flexural hinge corresponded to the cracking of concrete in the first fibre in tension. The yielding moment M_Y was reached when the strain in the first reinforcing bar was equal to the yield strain of the steel ε_{sy}. The maximum moment M_M was determined from the results of the moment–curvature analysis of the cross-section of the column. The moment corresponding to near-collapse limit state M_{NC} was defined at 80% of the maximum moment in the softening branch of the moment–rotation relationship.

The rotation in the nonlinear flexural hinge at the occurrence of concrete cracking (M_{CR}) was determined by assuming a linear curvature along the distance L_V

$$\theta_{CR} = \frac{M_{CR}L_V}{3EI} \tag{6}$$

where EI is the product of elastic modulus of concrete and moment of inertia of the cross-section, and $L_V = M/V$ is the length from the plastic hinge to the point of zero moment, which, in this case, is equal to the height of the column. The yield rotation θ_Y was calculated following the draft of the new Eurocode 8-3 [33]

$$\theta_Y = \varphi_y \frac{L_V + a_V z}{3} + 0.0019\left(1 + \frac{h}{1.6L_V}\right) + \frac{\varphi_y d_{bl} f_y}{8\sqrt{f_c}} \tag{7}$$

where h is the depth of the rectangular column in the direction of loading, f_y is the yield strength of the steel, f_c is the compressive strength of concrete, d_{bl} is the (mean) diameter of the tension reinforcement, $a_V z$ is the tension shift of the bending moment diagram, which was calculated according to Eurocode 2 [13] 9.2.1.3(2), and φ_y is the yield curvature of a cross-section, which was obtained by performing moment–curvature analyses in OpenSees.

The rotation at the near-collapse limit state was also estimated in accordance with the draft of the new Eurocode 8-3 [33]

$$\theta_{NC} = \theta_Y + (\varphi_u + \varphi_y)L_{pl}\left(1 - \left(\frac{0.5L_{pl}}{L_V}\right)\right) + \Delta\theta_{u,slip} \tag{8}$$

where φ_u is the ultimate curvature of a cross-section and was obtained by performing moment–curvature analyses in OpenSees. The L_{pl} is the length of the plastic hinge, whereas $\Delta\theta_{u,slip}$ is the post-yield fixed-end rotation due to yield penetration in the anchorage zone beyond the yielding end of the element, and should be taken as

$$\Delta\theta_{u,slip} = 9.5d_b \frac{\varphi_u + \varphi_y}{2} \tag{9}$$

where d_b is the bar diameter. The ultimate curvature of the cross-section was also calculated with consideration of the draft of the new Eurocode 8-3, which prescribes that the ultimate curvature φ_u in a critical zone can occur before (condition a) or after (condition b) spalling of the concrete cover. Both ultimate curvature models foresee attainment of the ultimate curvature when the tension strain in the bar equals the ultimate tension strain (ε_{su}) or the compressive strain in the concrete equals the ultimate compressive strain of unconfined (ε_{cu}) or confined ($\varepsilon_{cu,c}$) concrete. However, the criteria for the ultimate tension strain of the steel and ultimate compressive strain of the concrete differ for conditions (a) and (b). The ultimate strains in steel and concrete for condition (a) are as follows:

- For the steel bar in tension

$$\varepsilon_{su} = 0.4\varepsilon_{su,nom} \tag{10}$$

- For the (unconfined) concrete

$$0.0035 \leq \varepsilon_{cu} = (18.5/h[mm])^2 \leq 0.01 \tag{11}$$

where $\varepsilon_{su,nom}$ is the uniform elongation at tensile strength in a standard steel coupon test, and h is the depth of the unspalled section. For condition (b), which refers to the situation when the ultimate curvature is reached after spalling of the concrete cover, the ultimate strains in steel and concrete are defined as the minimum of the following strains:

- For the steel bar in tension

$$\varepsilon_{su} = \frac{4}{15}\varepsilon_{su,nom}\left(1+3\frac{d_{bL}}{s_w}\right)\left(1-0.75e^{-0.4N_{b,compr}}\right) \quad (12)$$

- For the confined concrete core inside the steel ties

$$\varepsilon_{cu,c} = \varepsilon_{cu} + 0.04\sqrt{\frac{\alpha\rho_{sv}f_{yw}}{f_c}} \quad (13)$$

where d_{bL} is the diameter of the longitudinal bars, $N_{b,compr}$ is the number of bars near the extreme compression fibres (those in the outermost layer of compression bars for a rectangular compression zone), and ε_{cu} is obtained from Equation (11), replacing h with the depth of the confined core (h_o). The ρ_{sv} is the volumetric ratio of transverse reinforcement defined as the volume of confining hoops over the volume of a concrete core, and α is the confinement effectiveness (see Equation (5)).

According to Eurocode 8-3, condition (b) applies when the calculated flexural resistance of the confined concrete section exceeds 80% of the resistance of the full unspalled concrete section at the instant when the strain, either in steel or in the concrete, becomes equal to the ultimate strain. In all other cases, the ultimate curvature of the section is defined by condition (a).

The length of the plastic hinge has an important influence on the near-collapse limit-state rotation. According to the draft of the new Eurocode 8-3, it is determined as follows

$$L_{pl} = \left(1-\frac{1}{3}\sqrt{\min\left(2.5;\max\left(0.05;\frac{b_w}{h}\right)\right)}\right)\cdot\left(1+0.4\min\left(9;\frac{L_V}{h}\right)\right)\cdot(1-0.45\min(0.7;\nu))\cdot(0.3h) \quad (14)$$

where b_w is the width of the section parallel to the shear force, and $\nu = N/A_cf_c$ is the normalised axial force, where A_c is the cross-sectional area, and the axial force N is considered positive for compression.

Based on the definitions from the Eurocode and the adopted four-linear moment-rotation relationship (see Figure 2b), it is possible to calculate the rotation at maximum moment θ_M using rules of similar triangles (i.e., the ratio between the "distances" M_M-M_C and θ_M-θ_C is equal to the ratio between the "distances" M_{NC}-M_C and θ_{NC}-θ_C (see Figure 2b))

$$\begin{aligned} &\frac{M_M - M_C}{\theta_C - \theta_M} = \frac{M_{NC} - M_C}{\theta_C - \theta_{NC}} \rightarrow \frac{M_M - M_C}{\theta_M \cdot r_{CM} - \theta_M} = \frac{0.8M_M - M_C}{\theta_M \cdot r_{CM} - \theta_{NC}} \\ &\rightarrow \quad \theta_M = \theta_{NC}\frac{(M_M - M_C)}{(M_M - M_C + (r_{CM} - 1)(1-0.8)M_M)} \end{aligned} \quad (15)$$

where rotation at the near-collapse limit state θ_{NC} is estimated according to Equation (8), the moments at the characteristic points C and M are calculated as discussed above, and r_{CM} is the ratio between the rotation at zero moment θ_C and the rotation at maximum moment θ_M. Note that, in the derivation, it was also taken into account that the moment corresponding to near-collapse limit state M_{NC} was defined at 80% of the maximum moment in the softening branch of the moment–rotation relationship. As the four-linear moment–rotation relationship of the plastic hinge of the column is simplistic, the rotation at zero moment θ_C is not realistic. Thus, it makes sense to define the r_{CM} based on experimental results in order to calibrate the moment–rotation relationship at least to the near-collapse limit state. The r_{CM} is not prescribed by the codes (e.g., [36]). However, in the example of the simulation of the seismic response of the four-storey reinforced-concrete frame building [37], it was shown that the match between the results of numerical simulation and the experimental test was quite good if r_{CM} = 3.5 was used for both columns and beams.

The moments at characteristic points of the four-linear moment–rotation relationship of the plastic hinge of the column have a rigorous physics background, whereas the formulas for corresponding rotations are based on empirical regression models. Therefore, it is expected that the characteristic rotations obtained from the particular experiment may differ from the values for the yield or near-collapse rotation from the empirically based regression models (Equations (7) and (8)). In this study, it was observed that the yield rotation, estimated according to Equation (7), was notably different from the yield rotation from the cyclic test of examined specimens. To calibrate the mathematical model with the results of the particular cyclic tests of the columns reinforced by B500B and DPD2 steel bars, the yield rotation according to Equation (7) was multiplied by a factor γ, as discussed in Sections 3.2 and 3.3.

The described mathematical model of the plastic hinge of the RC column was realised in OpenSees by two zero-length elements, connected in parallel. The three-linear moment-rotation relationship (i.e., uniaxial material 'Hysteretic') was assigned to each zero-length element, aiming to at simulate flexural behaviour, whereas the axial, shear and torsional characteristics were defined using 'Elastic uniaxialMaterial'. The characteristics of each three-linear moment–rotation relationship were defined in such a manner that the moment–rotation relationship of the coupled zero-length elements was equal to the four-linear moment–rotation relationship described above. The rules for determining the parameters of each of the two zero-length elements with the three-linear moment–rotation relationship are schematically presented in Figure 3. Note that the rotations θ_{CR} and θ_Y define the rotation at the first characteristic point (CP1) of the three-linear backbone 1 and 2, respectively. The rotations of the second (CP2) and third (CP3) characteristic points are equal to θ_M and θ_C and are the same for both backbones. The moment corresponding to the third characteristic point (CP3) is equal to zero for both backbones. In contrast, the moments of the first two points (CP1 and CP2) of backbones are assessed based on the assumption that the increase in the moment of the moment–rotation relationship between rotations corresponding to θ_Y and θ_M is the same for both backbones.

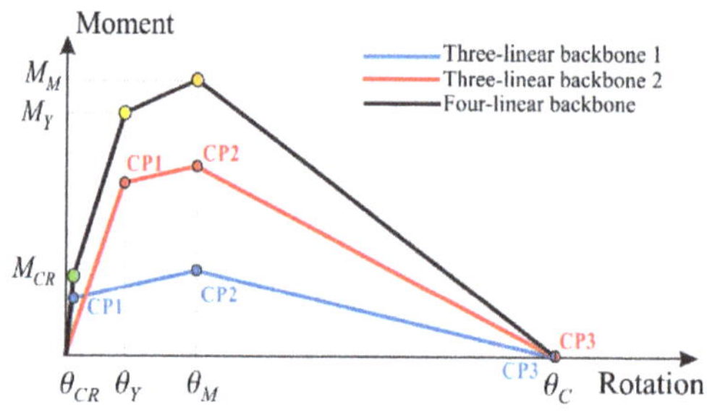

Figure 3. Schematic presentation of coupling two three-linear moment–rotation relationships in parallel in order to obtain a four-linear moment–rotation relationship in a flexural plastic hinge.

For the reader who is not familiar with OpenSees, it may be interesting to describe the uniaxial material 'Hysteretic'. The moment–rotation envelope of this material is defined by three characteristic points in a positive direction and three characteristic points in a negative direction. The cyclic behaviour is then controlled by two parameters defining pinching, the energy-dependant damage parameter, the ductility-dependant damage parameter and the ductility-dependant parameter β, which affects the unloading stiffness. As the ductilities of 'Hysteretic uniaxialMaterial' materials at a certain rotation are not the same, the ductility-dependant damage parameter and β have to be appropriately calibrated in order to achieve a similar hysteretic response of 'Hysteretic uniaxialMaterial' materials. For reinforced concrete elements, the values of parameter β are typically set to approximately 0.75 for a three-linear backbone with smaller initial stiffens (i.e., backbone 2) and 0.69 for another three-linear backbone (i.e., backbone 1). Note also that the damage and pinching parameters are usually set to zero in the simulation of a complex multi-degree-of-freedom structure, because non-zero values can produce convergence problems.

3.2. Numerical Simulation of Cyclic Test of Column Reinforced with B500B

The proposed model was used to simulate the cyclic response of a column reinforced with reinforcing steel B500B and tested within a research project sponsored by the Slovenian Research Agency [24]. The specimen height was 1.70 m with a cross-section b/h = 30/30 cm. The square column was reinforced with 8 bars of diameter ϕ 16 mm in the longitudinal direction and ϕ8 mm/7.5 cm $n = 2 + \sqrt{2}$ in the transverse direction (see Figure 4). The mean compressive strength of concrete was obtained from the experimental test on the cylinder, and it was equal to 41.2 MPa, whereas for the reinforcing steel, the nominal strength of the B500B was assumed. The column was loaded with an axial force that corresponded to the normalised axial force 0.15. The distance between the column base and the centre of the horizontal hydraulic jack, which was used to impose the displacements to the column, was 1.50 m.

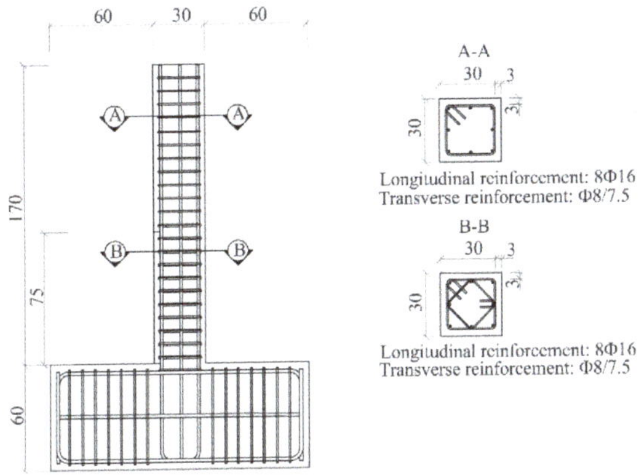

Figure 4. Elevation view of specimen with reinforcement, and cross-sections of column of specimen reinforced with B500B.

The simulation of the response of the column was performed by imposing on the model the displacements that were used during the cyclic test. The model of the column was developed according to the rules defined in Section 3.1. The elastic part of the model was based on the concrete cross-section dimensions and the elastic modulus of concrete (E_c = 36,000 MPa, see Table 2). The moment–rotation relationship was assessed using the characteristics of the reinforced concrete cross-section at the base of the column. For the reinforcing steel, the stress–strain curve for B500B, presented in Figure 1, was used. The material properties of confined and unconfined concrete are presented in Table 2. Note that the axial force (N) was not applied to the columns, but its effect was taken into account in the calculation of the moment–rotation relationship (see Section 3.1).

Table 2. Compressive strength (f_c), corresponding compressive strain at f_c (ε_{c0}), ultimate strain (ε_{cu}), axial tensile strength (f_{ct}), tensile strain at f_{ct} (ε_t) and elastic modulus (E_c) of confined and unconfined concrete.

	Unconfined Concrete	**Confined Concrete**
f_c [MPa]	41.2	56.8
ε_{c0} [‰]	2.22	6.41
ε_{cu} [‰]	3.50	19.66
f_{ct} [MPa]	3.1	3.1
ε_t [‰]	0.086	0.086
E_c [MPa]	$3.6 \cdot 10^4$	$3.6 \cdot 10^4$

The characteristic points of the moment–rotation relationship were calculated as defined in Section 3.1 by assuming r_{CM} = 3.5. The parameter β was considered, as defined in Section 3.1 (i.e., 0.75 and 0.69 for, respectively, uniaxial material *Hysteretic* corresponding to zero-length element with smaller and higher initial stiffness (see zero-length element 2 in Figure 3)). The pinching and damage parameters were set as 0.

The model for the moment–rotation envelope of the column was further calibrated to improve the match between the simulated and measured cyclic response of the column. For this purpose, only the yielding rotation from Equation (7) was multiplied by the correction factor γ = 0.60, and the r_{CM} was set to 7. The parameter β and the damage parameter (*damage1*) of the column were also estimated based on the experimental results. For the zero-length element with smaller initial stiffness, the parameter β was set to 0.75, and *damage1* was taken to equal 0.009, whereas, for another zero-length element (Figure 3), the corresponding values were considered to equal 0.55 and 0.001.

The cyclic test results and the corresponding simulations using the basic and the calibrated model are presented in Figure 5. From Figure 5a, it can be observed that the basic model overestimates the yield rotation and rotation at the maximum moment. In addition, it underestimates the amount of dissipated hysteretic energy. The results of the simulation with the calibrated model are significantly improved. Note also that the near-collapse rotation of the plastic hinge, calculated according to Equation (8), is sufficiently accurate for both models.

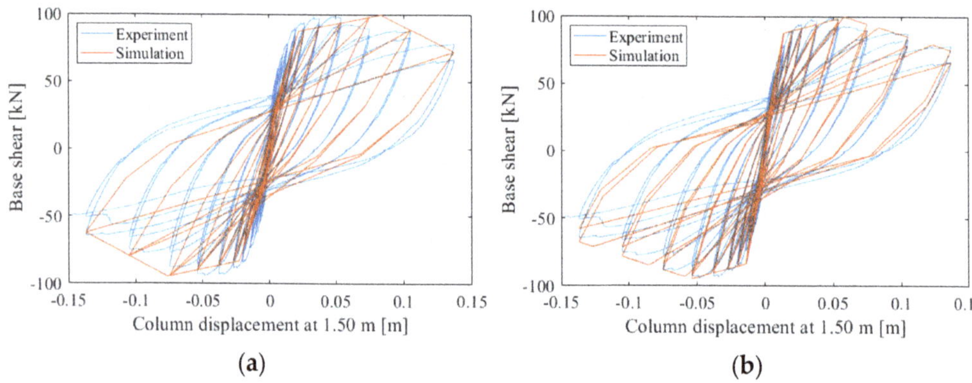

Figure 5. Base shear—displacement at height 1.50 m relationship observed during cyclic test and that obtained from simulation of cyclic response of a column reinforced with B500B using (**a**) basic and (**b**) calibrated model.

3.3. Numerical Simulation of Cyclic Test of Column Reinforced with DPD2

The proposed model was also used to simulate the cyclic response of the column, which was tested within the NEWREBAR research project [25]. The column height was 1.75 m, with a cross-section b/h = 35/35 cm. In this case, the column was reinforced with DPD2 steel bars. Eight bars of diameter ϕ16 mm were placed in the longitudinal direction and ϕ 10/7.5 cm $n = 2 + \sqrt{2}$ in the transverse direction (see Figure 6). The mean compressive strength of concrete, which was obtained from the experimental test on the cylinder, was relatively low, and amounted to 21.1 MPa. The axial force corresponded to the normalised axial force 0.30. The distance between the column base and the centre of the horizontal hydraulic jack was 1.53 m.

The response of the column was simulated by imposing the displacements that were measured during the cyclic test. The basic model was developed according to the rules defined in Section 3.1. The elastic part of the numerical model was modelled based on the dimensions of the concrete cross-section and the elastic modulus of concrete (E_c = 30,000 MPa, see Table 3). The moment–rotation relationship was assessed using the characteristics of the reinforced concrete cross-section and the results of moment–curvature analysis. For such an analysis, the confined and unconfined concrete, the mechanical properties of which are shown in Table 3, were used to define the properties of the uniaxial material 'Concrete04'. The stress–strain relationship of the DPD2 reinforcing steel, which is presented in Figure 1, was used to determine the properties of the uniaxial material 'MultiLinear'.

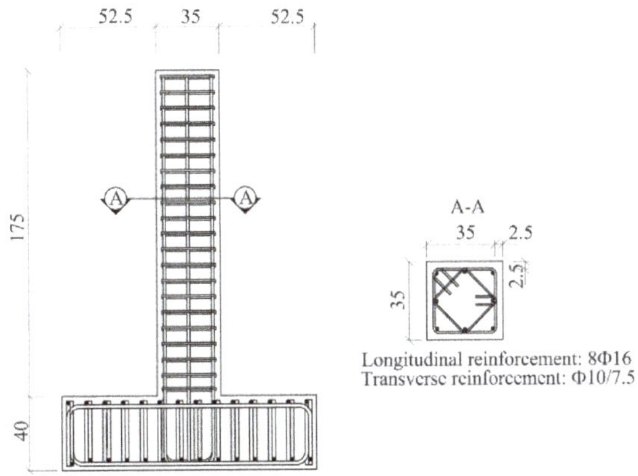

Figure 6. Elevation view of specimen with reinforcement and cross-section of column of specimen reinforced with DPD2.

Table 3. Compressive strength (f_c), corresponding compressive strain at f_c (ε_{c0}), ultimate strain (ε_{cu}), axial tensile strength (f_{ct}), tensile strain at f_{ct} (ε_t) and elastic modulus (E_c) of confined and unconfined concrete.

	Unconfined Concrete	Confined Concrete
f_c [MPa]	21.1	35.7
ε_{c0} [‰]	1.80	8.04
ε_{cu} [‰]	3.50	23.28
f_{ct} [MPa]	1.7	3.1
ε_t [‰]	0.056	0.086
E_c [MPa]	$3.0 \cdot 10^4$	$3.0 \cdot 10^4$

The basic model was defined as described in Section 3.1. The values of r_{CM}, β, the pinching and damage parameters, were assumed to be equal to those used in the case of the basic model of a column reinforced with B500B. The basic model was then calibrated to improve the simulation of the envelope of the force–displacement relationship. In this case, the yielding rotation based on Equation (7) was multiplied with the correction factor $\gamma = 0.75$. The ratio between the rotation at zero moment θ_C and the rotation at maximum moment θ_M (r_{CM}) was set to 7, which is equal to the value used in the case-calibrated model of the B500B column. The ductility-dependant parameters β and *damage1* of the zero-length element with smaller initial stiffness were assumed to be equal to 0.60 and 0.011, respectively. For another zero-length element (Figure 3), the corresponding values were considered equal to 0.64 and 0.0005, respectively.

The results of the simulations of the cyclic test of the column using the basic and the calibrated model are presented in Figure 7. The conclusions are similar to those given for column reinforced with B500B. In this case, the basic model also overestimates the yield rotation and rotation at the maximum moment. The amount of dissipated hysteretic energy is also significantly underestimated. From the results of the cyclic test, it can also be observed that the strength was different in the positive and negative direction. This phenomenon, which cannot be modelled by simplified nonlinear models, is partly a consequence of the issues during the cyclic test, as explained in [25]. The results of the numerical simulation match particularly well with the experimental results for a positive direction of displacement, even in the near-collapse range. This match indicates that the near-collapse rotation, calculated according to Equation (8), is sufficiently accurate, at least for this particular column, reinforced with DPD2 steel bars. However, in another direction, the strength and the near-collapse rotation capacity seem to be slightly overestimated by the numerical simulation.

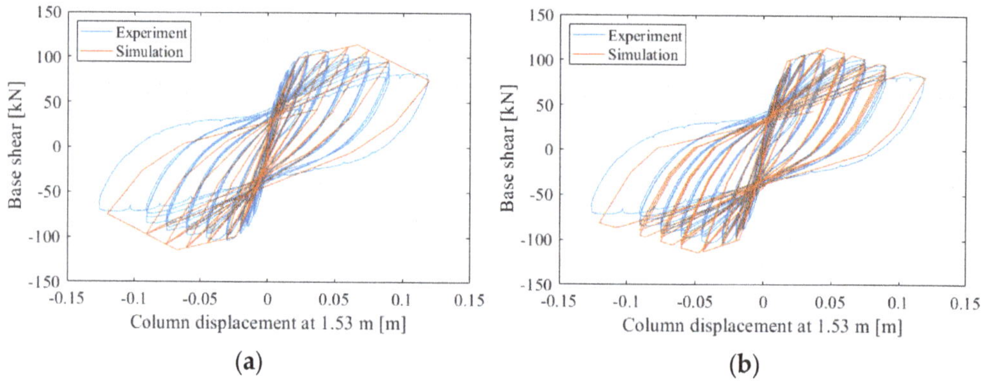

Figure 7. Base shear—displacement at height 1.53 m relationship observed during cyclic test and that obtained from simulation of cyclic response of a column reinforced with DPD2 using (**a**) basic and (**b**) calibrated model.

4. Design and Seismic Performance Assessment of DPD2 and B500B Buildings

The parametric study was conceived and performed, aiming to investigate the usability of the current earthquake-resistant design procedures for concrete frames reinforced by DPD2 steel. Three multi-storey RC frame buildings were designed according to Eurocode 8-1 and assessed using pushover and response history analyses. The seismic design parameters and the investigated buildings are first introduced. This is followed by a description of the mathematical model of the RC frame building, which is based on the model of the calibrated columns. The results of the seismic performance of DPD2 RC frame buildings are presented in Section 4.3, along with the results obtained for B500B buildings. For these cases, B500B and DPD2 steel were taken into account in the design and the seismic performance assessment. The second variant of DPD2 buildings was also defined to investigate the impact of the lower strength of DPD2 steel. In this case, the reinforcement in the B500B building was replaced by the same amount of DPD2 reinforcement. These buildings are termed DPD2 buildings, designed as B500B.

4.1. Description of Analysed Buildings

The 4-, 8- and 11-storey reinforced-concrete frame buildings (Figure 8) were designed and analysed. The building structures were designed for the gravity and seismic load combinations prescribed by Eurocode 0 [29]. The design peak ground accelerations amounted to 1.2·0.30 g = 0.36 g (soil type B), 1.15·0.25 g = 0.29 g (soil type C) and 1.2·0.25 g = 0.30 g (soil type B), respectively, for the 4-, 8- and 11-storey buildings. The behaviour factor was assumed to equal 3.9, which is defined in Eurocode 8 for multi-storey multi-bay frames designed for ductility class medium (DCM). Concrete C30/37 was prescribed in the design of the 4- and 8-storey buildings, whereas C35/45 was selected in the case of the 11-storey building. Selected properties of the structures and their design parameters are presented in Table 4. The largest design base shear–weight ratio was observed for the 4-storey building (15.5%), whereas the smallest Fb/W ratio was observed for the 11-storey building (5.2%).

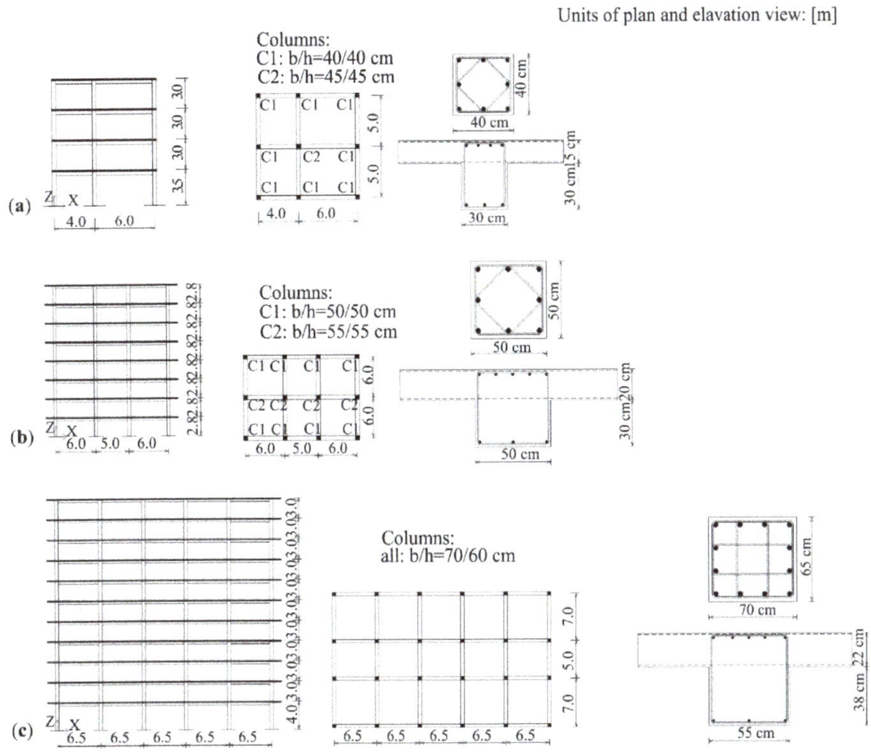

Figure 8. Elevations, plan views and reinforcement in typical columns and beams of (**a**) 4-, (**b**) 8- and (**c**) 11-storey buildings.

Table 4. Total mass, first vibration period, reference peak ground acceleration, soil factor, spectral acceleration corresponding to first vibration mode from the elastic spectrum and design base shear–weight ratio of 4-, 8- and 11-storey buildings.

Building	Total Mass (t)	Period T_1 [s]	$a_{gD,R}$ [g]	Soil Factor	$S_{e,D}(T_1)$ [g]	F_b/W
4-storey	339	0.68	0.30	1.2	0.66	15.5%
8-storey	2338	1.21	0.25	1.15	0.36	7.7%
11-storey	10,221	1.67	0.25	1.2	0.23	5.2%

The average longitudinal reinforcing ratios of columns and beams in the critical zone (zone of the plastic hinge) of each storey are shown in Figure 9. As the design yield strength of the DPD2 reinforcing steel ($f_{yd,DPD2}$ = 348 MPa) is lower than that of the B500B reinforcing steel ($f_{yd,B500B}$ = 435 MPa), the required amount of DPD2 reinforcement is significantly increased for most of the structural elements. The highest longitudinal reinforcing ratios of the columns can be observed for 4-storey buildings, followed by the 8- and 11- storey buildings. The longitudinal reinforcing ratios of the columns of the 4-storey building varied between 1.5% and 2.0%, and were equal to approximately 1% in the case of the 11-storey building. In the critical zones of the columns of the 4-storey building, the amount of longitudinal reinforcement of DPD2 building is, on average, higher by a factor of 1.24 than the amount of longitudinal reinforcement of the B500B building. This result is practically equal to the ratio between the design yield strengths of the two types of steel (i.e., $f_{yd,B500B}/f_{yd,DPD2}$ = 435 MPa/348 MPa = 1.25). For an 8-storey building, the factor is slightly smaller and amounts to 1.20. On the other hand, the longitudinal reinforcement of the columns of the 11-storey DPD2 and B500B building are practically the same, which is a consequence of the fact that the reinforcement in the columns of both variants of buildings is governed by the minimum requirements of Eurocode 8 [23] (i.e., the amount of longitudinal reinforcement should be higher than 1% of the area of the concrete cross-section).

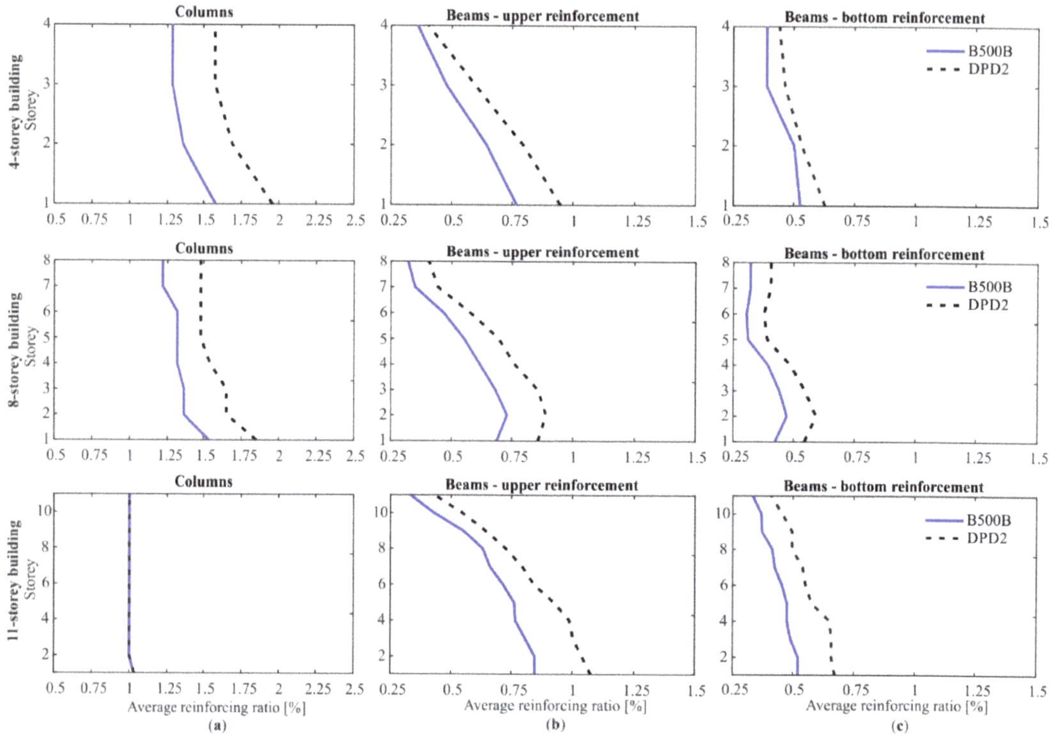

Figure 9. Average longitudinal reinforcing ratios in critical regions of (**a**) columns and (**b**) and (**c**) beams at storey level of 4-, 8- and 11-storey building.

The highest longitudinal reinforcing ratios of beams can be observed for the 11-storey building, followed by the 4- and 8- storey buildings. By comparing the amount of longitudinal reinforcement of the beams, it can be concluded that, for the 4-storey DPD2 building, the top and bottom longitudinal reinforcement of the beams are, on average, higher, respectively, by a factor of 1.21 and 1.14, compared to the reinforcement in the B500B building. In the case of 8- and 11-storey buildings, these factors are slightly increased, and amount to between 1.23 and 1.28.

4.2. Description of Mathematical Model for Nonlinear Seismic Analysis of RC Frame Building

Simplified nonlinear three-dimensional frame models with concentrated plasticity were developed for nonlinear static and dynamic analyses of RC frame buildings (Figure 10). The model of the building structure includes the calibrated model of the column (Section 3). Other modelling features are consistent with Eurocode 8 [23,36] requirements. Thus, the beam and column flexural behaviour were based on one-component lumped plasticity elements, comprising an elastic element and two inelastic rotational hinges (defined by a moment–rotation relationship) on each side of the elastic element. The element formulation assumed an inflexion point at the midpoint of the element. For beams, the plastic hinge was used for major axis bending only. For columns, two independent plastic hinges for bending about the two principal axes were taken into account. The floor diaphragms were assumed to be rigid in their own planes, and the masses and moments of inertia of each floor were lumped at the corresponding centre of gravity. Beams were modelled using an effective beam width defined according to Eurocode 2 [13]. The moment–rotation relationship was modelled by a four-linear relationship (see Figure 2b in Section 3.1).

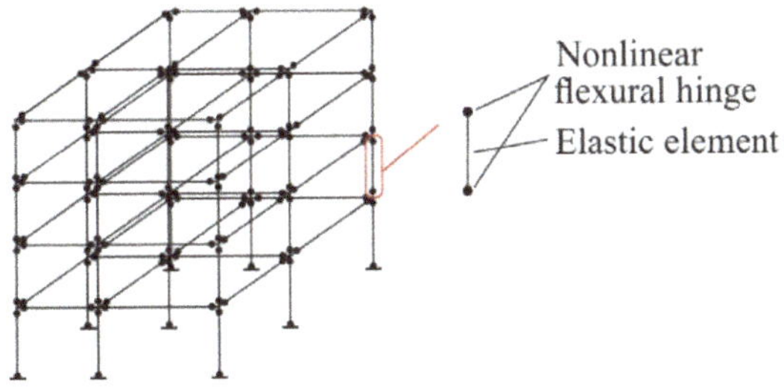

Figure 10. Simplified nonlinear model for seismic performance assessment of frame buildings.

The moment–rotation envelope of plastic hinges was calculated based on the procedure used in the case of the calibrated models of the columns, as presented in Section 3. However, for elements reinforced with traditional Tempcore steel, it may be interesting to use a different procedure (e.g., [38]) for the definition of the moment–rotation relationship. Nevertheless, the moment–curvature analysis was performed for each beam and column by taking the geometry of the section, confined and unconfined concrete and longitudinal reinforcement into account. However, the moment–rotation relationship of the plastic hinge of columns was calculated based on rectangular cross-sections and actual reinforcement, whereas the properties of plastic hinges in beams were calculated based on a T-shaped beam cross-section and three layers of longitudinal reinforcement (i.e., the reinforcement layer at the top and bottom reinforcement of the T cross-section, and reinforcement layer at the bottom level of the flange). Zero axial force and the axial load due to gravity loads were taken into account when determining the moment–rotation relationship for beams and columns, respectively. The distance from the element end to the point of zero moment L_V was assumed to be to a half-length of the element. For determination of the ratio r_{CM}, a model which is dependent on the axial force was proposed. The model was developed based on the results of the calibration of the numerical simulation of cyclic response of the columns (see Sections 3.2 and 3.3) and the results of the various experimental tests (e.g., [39]). For the cross-sections with a normalised axial force higher than 0.15, the r_{CM} was assumed to equal 7 (Section 3), but for normalised axial forces between 0 and 0.15, a linear function with a value r_{CM} between 2.5 and 7 was taken into account. Note that the r_{CM} was introduced based on the results of the simulation of cyclic response of the columns (see Section 3.2) and the results of the various experimental tests (e.g., [39]). The parameter β, which controls unloading stiffness, was defined as in the case of calibrated models of the columns (Section 3), whereas the *damage1* parameter was set to 0 to avoid convergence issues. However, in the process of calibration of the models of columns, it was found that the final results were not significantly affected by adopting a *damage1* close to 0.

The moment–rotation relationships in the plastic hinges of beams and columns were calculated using the mean values of material characteristics, as prescribed in Eurocode 8-3 [36]. Therefore, the mean concrete compressive strength was assumed to be 8 MPa higher than the characteristic value of compressive cylinder strength (i.e., 38 MPa and 43 MPa for C30/37 and C35/45, respectively) [13], whereas the stress–strain relationships for the B500B and DPD2 were the same as those presented in Figure 1.

Gravity load was represented by the uniformly distributed load on the beams and/or by concentrated loads at the top of the columns. A total of 5% critical damping was considered proportional to the mass. Models of all building variants were generated by the modified version of the PBEE toolbox [37], whereas the analyses were performed with OpenSees [30]. The PBEE toolbox is a simple yet effective tool for the seismic performance assessment of reinforced concrete frames, using simplified nonlinear models. This includes different functions for calculation of the moment–rotation relationship of the plastic hinges in the columns and beams, functions for the generation of the Tcl input code for OpenSees, functions for the post-processing of the analysis results and functions for structural performance assessment.

The lumped plasticity models are simplistic. Therefore, some phenomena observed in reinforced concrete structures (e.g., buckling of rebar bars in compression, bond-slip [40]) cannot be directly simulated. These phenomena are taken into account only indirectly through the empirical-based regression equations for the estimation of the limit-state deformation capacity, which is usual practice in the assessment. The model does not account for simultaneous nonlinear effects due to biaxial bending. The nonlinear element's flexural behaviour is modelled independently in two perpendicular directions. Furthermore, the moment–rotation relationship of plastic hinges is based on the constant value of the axial force during the analysis. The potential shear failure is also not simulated by the model used in this study because the shear failure is not critical [41] for code-compliant (i.e., [23]) structures that are not subjected to aggressive environmental conditions. However, although the lumped plasticity models are simplistic, it was found several times that they can produce sufficiently good results (e.g., [37,42]). As they are not computationally demanding, they are attractive for use in the simulation of the seismic response of entire structures. The fibre elements model is a possible alternative to the seismic analysis of building structures [40,43]. Such models provide more information on the local level, can simulate biaxial bending and can consider the axial-flexure load. However, it may be too complex for multi-storey structures in terms of computational cost, especially in the case of an iterative earthquake-resistant design based on nonlinear models [44]. The detailed finite element models could be an alternative, but they are not yet used for the calculation of the global seismic response of a structure because they are not computationally robust and are extremely computationally demanding (e.g., [45]).

4.3. Pushover Analyses

The capacity of the buildings was investigated using pushover analyses, which were performed by utilising the modal pattern of lateral forces. The resulting pushover curves are presented in Figure 11. The pushover curves are provided for the X direction only because the pushover curves for the Y direction are similar. Thus, the presentation of more results does not affect the conclusion of the study. The maximum strength of DPD2 buildings is higher than that of the B500B buildings, and it is observed at a higher roof displacement. The DPD2 buildings become more flexible at the initiation of the nonlinear behaviour, which begins with the cracking of concrete. After the yielding of columns at the base, the pushover curves of B500B buildings are practically horizontal, whereas hardening is more pronounced in the case of DPD2 buildings. All these differences in the pushover curves of DPD2 and B500B buildings are primarily the consequences of variations in the stress–strain relationship of the two types of steel. DPD2 steel bars are more deformable than the B500B bars, but they are not characterised by the yielding plateau, which is typical of the conventional reinforcing steel (i.e., for Tempcore steel B500B). However, to achieve the appropriate strength of DPD2 buildings, the quantity of steel had to be increased in the design. If the amount of steel in DPD2 buildings was equal to that designed for B500B buildings, then the strength would be significantly lower (see the pushover curve of DPD2 buildings designed as B500B buildings in Figure 11).

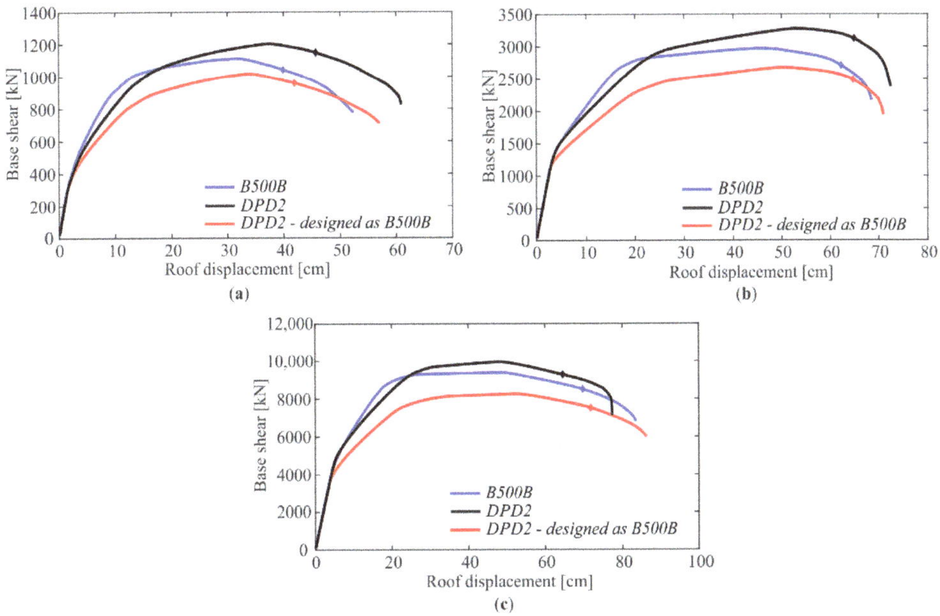

Figure 11. Pushover curves of variants of (**a**) 4-, (**b**) 8- and (**c**) 11-storey buildings. Diamond-shaped points on each of the pushover curves correspond to near-collapse displacement.

The maximum base shear (F_{max}) and the roof displacement at the near-collapse (NC) limit state (d_{NC}) are listed in Table 5 and serve as a precise analysis of the described observations. The maximum strength of DPD2 buildings is from 6 to 10% higher than that of B500B buildings. On the other hand, the maximum strength of B500B buildings is from 9 to 14% higher than that observed for DPD2 buildings designed as B500B buildings.

Table 5. Maximum base shear (F_{max}), near-collapse roof displacement (d_{NC}) and median spectral accelerations at first vibration period, causing a near-collapse limit state ($S_{e,NC}$).

Building	Designed According to	F_{max} [kN]	d_{NC} [cm]	$S_{e,NC}$ [g]
4-storey	B500B	1113	40.0	2.84
	DPD2	1203	45.7	3.20
	DPD2—designed as B500B	1018	42.0	2.89
8-storey	B500B	2974	62.4	1.48
	DPD2	3273	65.0	1.49
	DPD2—designed as B500B	2672	64.9	1.42
11-storey	B500B	9402	69.7	0.91
	DPD2	9968	64.5	0.83
	DPD2—designed as B500B	8272	71.8	0.87

The near-collapse roof displacements d_{NC} were estimated by assuming that the near-collapse limit state at the structural level is attained when the near-collapse is observed in the first column. From Table 5, it is clear that the d_{NC} increases with the number of stories (Table 5). However, the trend regarding the difference in the d_{NC} of DPD2 and B500B buildings is not clear. A significantly and slightly higher d_{NC} can be observed, respectively, for 4- (45.7 cm) and 8-storey (65.0 cm) DPD2 buildings, whereas in the case of the 11-storey building, the d_{NC} was observed to be slightly higher for the B500B building. Note that the latter observation is not the consequence of the mechanical characteristics (i.e., stress-strain curve) of the DPD2 bars. The increase in the deformation capacity of the 11-storey B500B building was triggered by the higher column-to-beam strength ratio value, which was observed because the required quantity of longitudinal reinforcement in the columns of the B500B building was less than that of the DPD2 building. However, in both cases, the required quantity of longitudinal reinforcement was less than 1%, which is the minimum requirement according to Eurocode 8.

The only noticeable trend in deformation capacity is that the deformation capacity of the DPD2—designed as B500B buildings is always approximately 4% higher than that of B500B buildings. However, DPD2 buildings designed as B500B buildings are only addressed here for comparative reasons.

By analysing the capacity of the DPD2 buildings, it can be concluded that Eurocode standards can be used for the design of DPD2 buildings. In most cases, the lateral strength and deformation capacity were increased in comparison to those observed in buildings reinforced with B500B bars. As the design procedure of DPD2 buildings and B500B buildings were the same, the increase in the lateral strength of DPD2 structures can be attributed to the substantial hardening of DPD2 steel after yielding, which increases the overstrength factor of the DPD2 buildings by about 10%. However, special attention should be paid when the minimum requirements of the standard dictate the column reinforcements. In these cases, the new, stricter minimum requirement should be defined if the same capacity as the buildings designed with conventional B500B reinforcing steel is required.

4.4. Results of Seismic Performance Assessment

The observations from the pushover analyses were verified by means of nonlinear response history analyses. For this purpose, incremental dynamic analysis (IDA) [46] was performed, aiming to estimate the spectral acceleration causing the near-collapse limit state, which was attained when a near-collapse rotation was observed in the first column, as defined in the previous section.

The spectral acceleration at the first vibration period was adopted for the intensity measure. The ground motions for IDA were selected based on a conditional spectrum approach [47], using a SHARE seismic hazard model [48]. The mean magnitude and mean distance were obtained from seismic hazard disaggregation for Ljubljana and for spectral acceleration, which corresponded to a return period of 2475 years ($S_{e,2475}$) and the first vibration period of the structure. Consequently, the selected sets of ground motion are building-specific. For each building, 30 ground motions (see Figure 12) were selected from the combined NGA [49] and RESORCE [50] strong ground motion databases, which contain 9188 ground motions. All of 30 ground motions correspond to events with magnitudes between 4.5 and 7, and source-to-site distances between 5 and 50 km. For the analysis of the 4- and 11- storey buildings, the ground motions were recorded on soil with a shear-wave velocity $v_{s,30}$ between 360 and 800 m/s, whereas, in the case of an 8-storey building, the $v_{s,30}$ was between 180–360 m/s.

The resulting median spectral accelerations causing the near-collapse limit state $S_{e,NC}$ are presented in Table 5, whereas the median IDA curves are shown in Figure 13. The median IDA curves of DPD2 and B500B buildings are highly similar. Slight differences can be observed in a range close to dynamic instability. The median spectral acceleration causing the near-collapse limit state $S_{e,NC}$ of the 4-storey DPD2 building was observed to be higher than that of the DPD2 building, whereas the opposite can be observed for the 11-storey building. This trend was expected because, in the case of frame buildings, the deformation capacity controls the ground motion intensity, causing the near-collapse limit state. Therefore, the IDA curves of 8-storey DPD2 and B500B buildings are practically equal, because this was also the case for the near-collapse roof displacement of buildings observed in the pushover analysis. It is interesting to note that the seismic performance, in terms of the median IDA curves of DPD2 designed as B500B buildings, is not significantly reduced, although the maximum strength of these buildings is significantly smaller than for the other buildings. Namely, the seismic performance of the medium-to-long period building is controlled by the deformation capacity, which was quite similar for variants of the investigated buildings (Figure 11).

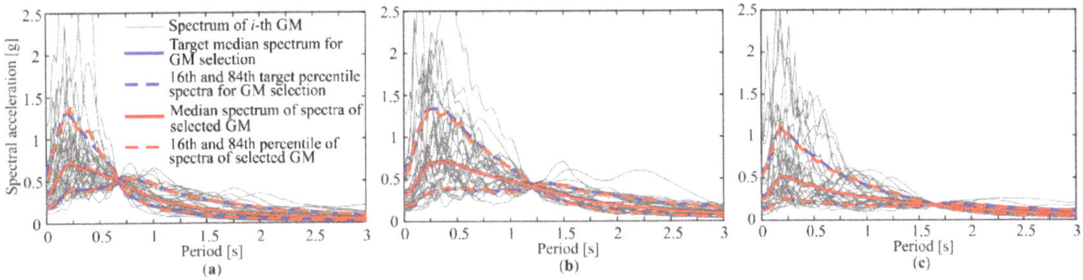

Figure 12. Target conditional acceleration spectra (median, 16th and 84th percentile), corresponding spectra of selected ground motions for an earthquake scenario based on $S_{e,2475}$, and acceleration spectra of each ground motion used for assessment of (**a**) 4-, (**b**) 8- and (**c**) 11-storey buildings.

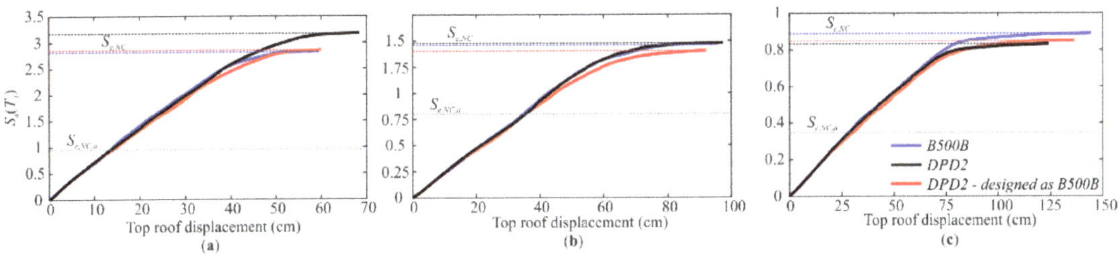

Figure 13. Median IDA curve of (**a**) 4-, (**b**) 8- and (**c**) 11-storey buildings.

To determine whether the life safety objective was fulfilled, the $S_{e,NC}$ were compared to the risk-targeted spectral acceleration causing the NC of structure $S_{e,NC,a}$, which was calculated according to [51], and represents the target median value of spectral acceleration causing the NC limit state. To calculate $S_{e,NC,a}$ the target (acceptable) annual probability of collapse $P_{C,a}$ was set to 10^{-4} (0.5%/50 years). This is a rather high value, but it is close to the values adopted in some other codes [52]. The seismic hazard curves for spectral acceleration at the first vibration period of the building, which are also required to calculate $S_{e,NC,a}$, were obtained utilising the SHARE seismic hazard model [48]. The dispersion of the spectral acceleration causing the collapse $\beta_{Se,C}$ was assumed to be 0.40 [53], and the limit-state reduction factor γ_{ls}, which transforms $S_{e,C}$ to $S_{e,NC}$, was assumed to equal 1.15 [53]. Based on these definitions and assumptions, the target values of $S_{e,NC,a}$ amounted to 0.95 g, 0.80 g and 0.35 g, respectively, for 4-, 8- and 11-storey buildings (see grey dashed lines in Figure 13). From Figure 13, it can be observed that the actual median values of spectral acceleration which caused the NC limit state are significantly higher than $S_{e,NC,a}$. Therefore, it can be claimed that all buildings are overdesigned if the life safety objective is defined by a high value of the annual probability of collapse, $P_{C,a} = 10^{-4}$.

The results of the analyses showed that the seismic performance of DPD2 buildings designed by Eurocode 8 are improved in comparison to the seismic performance of B500B buildings. The advantage of DPD2 building is in their higher resistance to corrosion [21]. However, the amount of reinforcement is increased.

5. Conclusions

The seismic performance of frame buildings designed by Eurocode 8 and reinforced by recently developed dual-phase reinforcing steel DPD2 and by conventional Tempcore-reinforcing steel B500B was investigated. The seismic performance of buildings was assessed using the improved numerical model of columns with concentrated plasticity. The developed model of columns, which account for four linear moment–rotation relationships, was validated by the results of the cyclic test. The conclusions are as follows:

1. Only the yield rotation of Eurocode empirical regression models had to be calibrated to adequately simulate the moment–rotation relationship from the cyclic tests of particular columns. However, it is suggested that the modelling uncertainty is investigated further at the level of DPD2 buildings, for example, by performing a pseudo-dynamic test using the full-scale building;
2. For the majority of DPD2 buildings, the maximum lateral strength (6–10%), deformation capacity (up to 15%) and seismic intensity causing a near-collapse limit state (up to 13%) were observed to be improved in comparison to B500B buildings;
3. Because the seismic performance of DPD2 buildings is better than that of B500B buildings if they are designed by Eurocode 8, it can be concluded that the DPD2 reinforcing steel can be used for the construction of DPD2 buildings designed by Eurocode 8;
4. The reinforcement quantity of DPD2 buildings is increased in the design primarily because of the smaller yield strength of DPD2 steel. However, this is not a prevailing source of the improved lateral strength of DPD2 buildings;
5. The improved performance of the DPD2 building in terms of strength is attributed to substantial strain hardening of the DPD2 steel after the yielding point, which is not the case for the Tempcore steel with a typical yielding plateau;
6. DPD2 reinforcing steel improves seismic resistance and resistance against corrosion, but also increases the amount of reinforcement required. However, some modifications to the existing design rules may also be introduced in the future, aiming to decrease the required amount of reinforcement and to modify the minimum required amount of reinforcement.

Author Contributions: Conceptualization, J.Ž. and M.D.; methodology, J.Ž.; software, J.Ž. and M.D.; formal analysis, J.Ž.; investigation, J.Ž.; writing—original draft preparation, J.Ž. and M.D.; writing—review and editing, J.Ž. and M.D.; supervision, M.D.; project administration, J.Ž. and M.D.; funding acquisition, M.D. Both authors have read and agreed to the published version of the manuscript.

Funding: This research was funded by Research Fund for Coal and Steel, grant number RFSR-CT-2015-00023 and by the research program Earthquake Engineering (P2-0185), funded by the Slovenian Research Agency". The APC was funded by the research program Earthquake Engineering, funded by the Slovenian Research Agency.

Institutional Review Board Statement: Not applicable.

Informed Consent Statement: Not applicable.

Data Availability Statement: Not applicable.

Conflicts of Interest: The authors declare no conflict of interest.

Nomenclature

B500B	Tempcore reinforcing steel with B500B characteristics
DPD2	Dual phase reinforcing steel type D2
DPF2	Dual phase reinforcing steel type F2
DPD2	Dual phase reinforcing steel type D2
CP1-CP3	three characteristic points of three-linear backbone
CR	characteristic point of the moment-rotation relationship associated with cracking of cross-section
M	characteristic point of the moment-rotation relationship associated with a maximum moment of cross-section

NC	near collapse limit state/characteristic point of the moment-rotation relationship associated with the near-collapse limit state of an element
RC	reinforced concrete
Y	characteristic point of the moment-rotation relationship associated with yielding of cross-section
A_c	cross-sectional area
A_{gt}	deformation at a maximum strength of reinforcing steel
A_{sw}	area of the transverse reinforcement bars of cross-section
A_5	deformation of rupture of the bar
E_c	elastic modulus of concrete
EI	product of elastic modulus of concrete and moment of inertia of the cross-section
F_b	design base shear
F_{max}	maximum base shear
K	confinement factor
L_{pl}	length of the plastic hinge
L_V	length between the plastic hinge and the point of zero moment
M_{CR}	moment at cracking of cross-section
M_M	maximum moment of cross-section
M_{NC}	moment corresponding to near-collapse limit state, which was defined at 80% of M_M
M_Y	moment at yielding od cross-section
N	axial force, which is considered to be positive for compression
$N_{b,compr}$	number of bars near the extreme compression fibres
$P_{C,a}$	target (acceptable) annual probability of collapse
$S_{e,D}(T_1)$	spectral acceleration corresponding to first vibration mode from the elastic spectrum
$S_{e,C}$	median spectral accelerations at the first vibration period causing the collapse of the structure
$S_{e,NC}$	median spectral accelerations at first vibration period causing a near-collapse limit state
$S_{e,NC,a}$	risk-targeted spectral acceleration causing NC of structure
$S_{e,2475}$	spectral acceleration for a return period of 2475 years
T_1	first fundamental period
W	weight of the structure
$a_{gD,R}$	design peak ground acceleration
a_{VZ}	tension shift of the bending moment diagram
d_b	bar diameter
d_{bl}	(mean) diameter of the tension reinforcement
b_i	*i*-th centreline spacing between longitudinal bars laterally restrained by a stirrup corner or hook
b_w	width of the cross-section
b_o	width of the rectangular of the confined concrete measured to the centreline of the perimeter bar
d_{NC}	roof displacement at near-collapse limit state
f_c	concrete compressive strength of (unconfined) concrete
f_{cc}	confined concrete maximum compressive strength
f_{ct}	axial tensile strength of concrete
f_u	mean value of the maximum tensile strength of reinforcing steel
$f_{yd,DPD2}$	design yield strength of the DPD2 reinforcing steel
$f_{yd,B500B}$	design yield strength of the B500B reinforcing steel
$f_{y,k}$	characteristic yield strength of reinforcing steel
$f_{y,m}$	mean yield strength
f_y	yield strength of the steel
f_{yw}	mean yield strength of transverse steel
h	depth of the rectangular cross-section/the depth of the unspalled section
h_o	depth of the rectangular of the confined concrete measured to the centreline of the perimeter bar
k	hardening ratio of reinforcing steel
n	number of longitudinal bars laterally restrained along the perimeter of the cross-section
r_{CM}	ratio between θ_C and θ_M
s_w	spacing of the stirrups measured from the centreline.

$v_{s,30}$	shear-wave velocity at last 30 m
$\Delta\theta_{u,slip}$	post-yield fixed-end rotation due to yield penetration
θ_C	rotation at zero moment
θ_M	rotation at maximum moment
θ_{NC}	rotation at near-collapse limit state
θ_Y	yield rotation
ϕ	diameter of bar
α	confinement effectiveness factor
$\beta_{Se,C}$	dispersion of the spectral acceleration causing collapse
γ_{ls}	limit-state reduction factor
ε_{cu}	ultimate compressive strain of unconfined concrete.
$\varepsilon_{cu,c}$	ultimate compressive strain of confined concrete
ε_{c0}	compressive strain in the concrete at maximum strength
$\varepsilon_{c0,c}$	strain at the confined concrete maximum compressive strength
ε_{su}	ultimate tension strain of bar
$\varepsilon_{su,nom}$	uniform elongation at tensile strength in a standard steel coupon test
ε_{sy}	yield strain of the steel
ε_t	tensile strain at f_{ct}
ν	normalised axial force
ρ_{sv}	volumetric ratio of transverse reinforcement defined as the volume of confining hoops over the volume of a concrete core
ρ_{sw}	ratio of transverse reinforcement
φ_y	yield curvature of a cross-section
φ_u	ultimate curvature of a cross-section

References

1. Berto, L.; Caprili, S.; Saetta, A.; Salvatore, W.; Talledo, D. Corrosion effects on the seismic response of existing rc frames designed according to different building codes. *Eng. Struct.* **2020**, *216*, 110397. [CrossRef]
2. Val, D.V.; Stewart, M.G. Reliability Assessment of ageing reinforced concrete structures–Current situation and future challenges. *Struct. Eng. Int.* **2009**, *19*, 211–219. [CrossRef]
3. Anoop, M.B.; Rao, K.B. Seismic damage estimation of reinforced concrete framed structures affected by chloride-induced corrosion. *Earthq. Struct.* **2015**, *9*, 851–873. [CrossRef]
4. Yuksel, I.; Coskan, S. Earthquake response of reinforced concrete frame structures subjected to rebar corrosion. *Earthq. Struct.* **2013**, *5*, 321–341. [CrossRef]
5. Apostolopoulos, C.A. Mechanical behavior of corroded reinforcing steel bars S500s tempcore under low cycle fatigue. *Constr. Build. Mater.* **2007**, *21*, 1447–1456. [CrossRef]
6. Caprili, S.; Salvatore, W. Cyclic behaviour of uncorroded and corroded steel reinforcing bars. *Constr. Build. Mater.* **2015**, *76*, 168–186. [CrossRef]
7. Meda, A.; Mostosi, S.; Rinaldi, Z.; Riva, P. Experimental evaluation of the corrosion influence on the cyclic behaviour of RC columns. *Eng. Struct.* **2014**, *76*, 112–123. [CrossRef]
8. Celarec, D.; Vamvatsikos, D.; Dolšek, M. Simplified estimation of seismic risk for reinforced concrete buildings with consideration of corrosion over time. *Bull. Earthq. Eng.* **2011**, *9*, 1137–1155. [CrossRef]
9. Xiong, C.; Deng, X.; Liang, Y.; Li, Q.; Huang, J.; Lin, X.; Li, D. Regional seismic damage simulation of corroded RC frame structures: A case study of Shenzhen City. *Appl. Sci.* **2020**, *10*, 4818. [CrossRef]
10. Lavorato, D.; Fiorentino, G.; Pelle, A.; Rasulo, A.; Bergami, A.V.; Briseghella, B.; Nuti, C. A corrosion model for the interpretation of cyclic behavior of reinforced concrete sections. *Struct. Concr.* **2020**, *21*, 1732–1746. [CrossRef]
11. Maraveas, C. Durability issues and corrosion of structural materials and systems in farm environment. *Appl. Sci.* **2020**, *10*, 990. [CrossRef]
12. Al-Obaidi, S.; Bamonte, P.; Ferrara, L.; Luchini, M.; Mazzantini, I. Durability-based design of structures made with ultra-high-performance/ultra-high-durability concrete in extremely aggressive scenarios: Application to a geothermal water basin case study. *Infrastructures* **2020**, *5*, 102. [CrossRef]
13. CEN. *EN1992-1-1, Eurocode 2: Design of Concrete Structures–Part 1-1: General Rules and Rules for Buildings*; European Committee for Standardisation: Brussels, Belgium, 2004.
14. Lollini, F.; Carsana, M.; Gastaldi, M.; Redaelli, E. Corrosion behaviour of stainless steel reinforcement in concrete. *Corros. Rev.* **2019**, *37*, 3–19. [CrossRef]
15. Gu, L.; Meng, X.H. Review on research and application of stainless steel reinforced concrete. In Proceedings of the International Conference on Mechatronics, Manufacturing and Materials Engineering (MMME 2016), Hong Kong, China, 11–12 June 2016.
16. Maffei, B.; Salvatore, W.; Valentini, R. Dual-phase steel rebars for high-ductile r.c. structures, Part 1: Microstructural and mechanical characterization of steel rebars. *Eng. Struct.* **2007**, *29*, 3325–3332. [CrossRef]

17. Salvatore, W.; Buratti, G.; Maffei, B.; Valentini, R. Dual-phase steel re-bars for high-ductile r.c. structures, Part 2: Rotational capacity of beams. *Eng. Struct.* **2007**, *29*, 3333–3341. [CrossRef]
18. Lorusso, H.; Burgueño, A.; Egidi, D.; Svoboda, H. Application of dual phase steels in wires for reinforcement of concrete structures. *Procedia Mater. Sci.* **2012**, *1*, 118–125. [CrossRef]
19. Caprili, S.; Salvatore, W.; Valentini, R.; Ascanio, C.; Luvarà, G. A new generation of high-ductile dual-phase steel reinforcing bars. *Constr. Build. Mater.* **2018**, *179*, 66–79. [CrossRef]
20. Caprili, S.; Mattei, F.; Salvatore, W.; Ascanio, C.; Luvarà, G. Industrial and techno-economic feasibility of concrete structures reinforced with DP rebars. *Constr. Build. Mater.* **2021**, *283*, 122793. [CrossRef]
21. Caprili, S.; Salvatore, W.; Valentini, R.; Ascanio, C.; Luvarà, G. Dual-phase steel reinforcing bars in uncorroded and corroded conditions. *Constr. Build. Mater.* **2019**, *218*, 162–175. [CrossRef]
22. Bautista, A.; Pomares, J.C.; González, M.N.; Velasco, F. Influence of the microstructure of TMT reinforcing bars on their corrosion behavior in concrete with chlorides. *Constr. Build. Mater.* **2019**, *229*, 116899. [CrossRef]
23. CEN. *EN1998-1, Eurocode 8: Design of Structures for Earthquake Resistance–Part 1: General Rules, Seismic Actions and Rules for Buildings*; European Committee for Standardisation: Brussels, Belgium, 2004.
24. Dolšek, M.; Kramar, M.; Gams, M.; Bohinc, U.; Anžlin, A.; Lajkovič, G.; Mihelčič, F.; Požonec, V. *Poročilo o preizkusih armiranobetonskih stebrov za ARRS projekt J2-5461*; Slovenian National Building and Civil Engineering Institute: Ljubljana, Slovenija, 2016. (In Slovenian)
25. Caprili, S.; Chellini, G.; Mattei, F.; Romis, F.; Salvatore, W. *NEWREBAR Research Project: Deliverable, D.4.1: Report on Design, Execution and Results of Tests on Full-Scale Prototypes*; University of Pisa, Department of Civil and Industrial Engineering: Pisa, Italy, 2019.
26. Apostolopoulos, C.A.; Papadakis, V.G. Consequences of steel corrosion on the ductility properties of reinforcement bar. *Constr. Build. Mater.* **2008**, *22*, 2316–2324. [CrossRef]
27. Salvatore, W.; Caprili, S.; Braconi, A.; Finetto, M.; Bianco, L.; Ascanio, C.; Moersch, J.; Apostolopoulos, C.; Ferreira Pimenta, G. *Effects of Corrosion on Low-Cycle Fatigue (Seismic) Behaviour of High-Strength Steel Reinforcing Bars (RUSTEEL)*; Publications Office of the European Union: Luxembourg, 2014.
28. Colla, V.; De Sanctis, M.; Dimatteo, A.; Lovicu, G.; Solina, A.; Valentini, R. Strain hardening behavior of dual-phase steels. *Metall. Mater. Trans. A Phys. Metall. Mater. Sci.* **2009**, *40*, 2557–2567. [CrossRef]
29. Caprili, S.; Salvatore, W.; Valentini, R.; Apostolopoulos, C.; Drakakaki, I.; Cabral, A.M.; Reis, S. *NEWREBAR research project: Deliverable, D.2.1: Report on Mechanical, Micro-Structural and Corrosion Properties of Reinforcing Selected DP Steel Rebars*; University of Pisa, Department of Civil and Industrial Engineering: Pisa, Italy, 2019.
30. McKenna, F.; Fenves, G.L. Open System for Earthquake Engineering Simulation (OpenSees). Available online: http://opensees.berkeley.edu (accessed on 14 March 2018).
31. Ferreira, T.M.; Costa, A.A.; Vicente, R.; Varum, H. A simplified four-branch model for the analytical study of the out-of-plane performance of regular stone URM walls. *Eng. Struct.* **2015**, *83*, 140–153. [CrossRef]
32. OpenSees Concrete04 Material. Available online: http://opensees.berkeley.edu/OpenSees/manuals/usermanual/1626.htm (accessed on 20 January 2021).
33. CEN. *wdEN1998-3 NEN SC8 PT3, Eurocode 8: Design of Structures for Earthquake Resistance–Part 3: Assessment and Retrofitting of Buildings and Bridges*; CEN European Committee for Standardisation: Brussels, Belgium, 2018.
34. Wang, P.; Shi, Q.; Wang, F.; Wang, Q. Seismic behaviour of concrete columns with high-strength stirrups. *Earthq. Struct.* **2020**, *18*, 15–25. [CrossRef]
35. OpenSees MultiLinear Material. Available online: http://opensees.berkeley.edu/wiki/index.php/MultiLinear_Material (accessed on 20 January 2021).
36. CEN. *EN1998-3, Eurocode 8: Design of Structures for Earthquake Resistance–Part 3: Assessment and Retrofitting of Buildings*; European Committee for Standardisation: Brussels, Belgium, 2005.
37. Dolsek, M. Development of computing environment for the seismic performance assessment of reinforced concrete frames by using simplified nonlinear models. *Bull. Earthq. Eng.* **2010**, *8*, 1309–1329. [CrossRef]
38. Liu, Z.; Li, S. Development of an ANN-based lumped plasticity model of RC columns using historical pseudo-static cyclic test data. *Appl. Sci.* **2019**, *9*, 4263. [CrossRef]
39. Taylor, A.W.; Kuo, C.; Wellenius, K.; Chung, D. *A Summary of Cyclic Lateral Load Tests on Rectangular Reinforced Concrete Columns*; Building and Fire Research Laboratory, National Institute of Standards and Technology: Gaithersburg, MA, USA, 1997.
40. Rasulo, A.; Pelle, A.; Lavorato, D.; Fiorentino, G.; Nuti, C.; Briseghella, B. Finite element analysis of reinforced concrete bridge piers including a flexure-shear interaction model. *Appl. Sci.* **2020**, *10*, 2209. [CrossRef]
41. Žižmond, J.; Dolšek, M. Evaluation of factors influencing the earthquake-resistant design of reinforced concrete frames according to Eurocode 8. *Struct. Infrastruct. Eng.* **2016**, *12*, 1323–1341. [CrossRef]
42. Žižmond, J.; Dolšek, M. Modeliranje efektivne širine pasnice grede za nelinearno analizo armiranobetonske okvirne stavbe (Modelling of effective flange width of beam for nonlinear analysis of reinforced concrete frame building). *Gradb. Vestn.* **2014**, *63*, 26–39. (In Slovenian)

43. Rasulo, A.; Pelle, A.; Lavorato, D.; Fiorentino, G.; Nuti, C.; Briseghella, B. Seismic assessment of reinforced concrete frames: Influence of shear-flexure interaction and rebar corrosion. In *International Conference on Computational Science and Its Applications*; Gervasi, O., Murgante, B., Misra, S., Garau, C., Blečić, I., Taniar, D., Apduhan, B.O., Rocha, A.M.A.C., Tarantino, E., Torre, C.M., et al., Eds.; Springer: Cham, Switzerland, 2020; pp. 463–478.
44. Lazar Sinković, N.; Brozovič, M.; Dolšek, M. Risk-based seismic design for collapse safety. *Earthq. Eng. Struct. Dyn.* **2016**, *45*, 1451–1471. [CrossRef]
45. Zoubek, B.; Isakovic, T.; Fahjan, Y.; Fischinger, M. Cyclic failure analysis of the beam-to-column dowel connections in precast industrial buildings. *Eng. Struct.* **2013**, *52*, 179–191. [CrossRef]
46. Vamvatsikos, D.; Cornell, C.A. Incremental dynamic analysis. *Earthq. Eng. Struct. Dyn.* **2002**, *31*, 491–514. [CrossRef]
47. Baker, J.W. Conditional mean spectrum: Tool for ground-motion selection. *J. Struct. Eng.* **2011**, *137*, 322–331. [CrossRef]
48. Woessner, J.; Laurentiu, D.; Giardini, D.; Crowley, H.; Cotton, F.; Grünthal, G.; Valensise, G.; Arvidsson, R.; Basili, R.; Demircioglu, M.B.; et al. The 2013 European seismic hazard model: Key components and results. *Bull. Earthq. Eng.* **2015**, *13*, 3553–3596. [CrossRef]
49. Chiou, B.; Darragh, R.; Gregor, N.; Silva, W. NGA project strong-motion database. *Earthq. Spectra* **2008**, *24*, 23–44. [CrossRef]
50. Akkar, S.; Sandikkaya, M.A.; Şenyurt, M.; Azari Sisi, A.; Ay, B.O.; Traversa, P.; Douglas, J.; Cotton, F.; Luzi, L.; Hernandez, B.; et al. Reference database for seismic ground-motion in Europe (RESORCE). *Bull. Earthq. Eng.* **2014**, *12*, 311–339. [CrossRef]
51. Žižmond, J.; Dolšek, M. Formulation of risk-targeted seismic action for the force-based seismic design of structures. *Earthq. Eng. Struct. Dyn.* **2019**, *48*, 1406–1428. [CrossRef]
52. ATC. *FEMA P695, Quantification of Building Seismic Performance Factors*; Federal Emergency Management Agency: Washington, DC, USA, 2009.
53. Dolšek, M.; Lazar Sinković, N.; Žižmond, J. IM-based and EDP-based decision models for the verification of the seismic collapse safety of buildings. *Earthq. Eng. Struct. Dyn.* **2017**, *46*, 1–18. [CrossRef]

applied sciences

MDPI

Article

Numerical Analysis Exterior RC Beam-Column Joints with CFRP Bars as Beam's Tensional Reinforcement under Cyclic Reversal Deformations

Violetta K. Kytinou, Parthena-Maria K. Kosmidou and Constantin E. Chalioris *

Laboratory of Reinforced Concrete and Seismic Design of Structures, Civil Engineering Department, School of Engineering, Democritus University of Thrace, 67100 Xanthi, Greece; vkytinou@civil.duth.gr (V.K.K.); pkosmido@civil.duth.gr (P.-M.K.K.)
* Correspondence: chaliori@civil.duth.gr; Tel.: +30-25-4107-9632

Abstract: In this paper the cyclic lateral response of reinforced concrete (RC) beam-column joints with composite carbon fiber-reinforced polymer (CFRP) bars as a longitudinal reinforcement in the beam is simulated with finite element (FE) modeling using software Abaqus. An experimental project of two full-scale joint specimens subjected to cyclic loading with supplementary accompanying pull-out tests of CFRP bars is also included in this study. These test results are used to calibrate the developed FE model, the constitutive laws of the materials and the bond response between CFRP bars and concrete. Comparisons between test data and numerical results indicate that the calibrated model accurately predicts the cyclic response of RC beam-column joint specimens with CFRP longitudinal bars as the beam's tensional reinforcement. A parametric analysis is also performed to provide useful concluding remarks concerning the design of concrete joints with composite bars and the ability of CFRP bars to substitute for conventional steel bars in RC structural members under seismic excitations.

Keywords: reinforced concrete; beam-column joints; carbon fiber-reinforced polymer longitudinal bars; seismic performance; finite element analysis; tests; bond; cyclic loading

Citation: Kytinou, V.K.; Kosmidou, P.-M.K.; Chalioris, C.E. Numerical Analysis Exterior RC Beam-Column Joints with CFRP Bars as Beam's Tensional Reinforcement under Cyclic Reversal Deformations. *Appl. Sci.* **2022**, *12*, 7419. https://doi.org/10.3390/app12157419

Academic Editor: Maria Favvata

Received: 27 June 2022
Accepted: 21 July 2022
Published: 24 July 2022

Publisher's Note: MDPI stays neutral with regard to jurisdictional claims in published maps and institutional affiliations.

Copyright: © 2022 by the authors. Licensee MDPI, Basel, Switzerland. This article is an open access article distributed under the terms and conditions of the Creative Commons Attribution (CC BY) license (https://creativecommons.org/licenses/by/4.0/).

1. Introduction

Fiber-reinforced polymer (FRP) materials are widely accepted as an important aspect of modern civil infrastructure. The advantages of FRP materials over traditional building materials such as conventional steel and steel-reinforced concrete (RC) rest in their enhanced structural performance in regard of stability, stiffness, strength and durability [1–3]. FRP materials are available in a variety of forms, including cables, sheets, plates and more. The use of FRP materials is now widespread throughout the world. As a result, the importance of cost-effective and reliable FRP structures increases. The majority of early research efforts [4,5] concerning FRP reinforcement (strips, sheets and other textiles) focused on rehabilitation and strengthening applications of masonry and concrete structures [6–9]. Extensive experimental and theoretical research has also been conducted on the strengthening and retrofitting of deficient or/and damaged concrete structures [10–13]. Nevertheless, the application of FRP rebars in concrete structures continues to be a popular topic of research, as all the design aspects and implementation drawbacks have not been thoroughly investigated and discussed, especially concerning RC joints [14]. In particular, the joint behavior and the framework of design standards have not been thoroughly examined. It is evident from a review of the prior studies that considerable research has been performed in the field of FRP beam-column joint wrappings for repairing and rehabilitation purposes [15–18], whereas research on using nonmetallic rebars for beam-column joint implementations is limited. In recent years, nonmetallic reinforcements have gained prominence in international structural applications [19–22]. Even though they have been used successfully in construction, nonmetallic reinforcements will only be widely used in engineering if design specifications and guidelines are developed promptly [23].

The typical method for designing an earthquake-resistant RC structure relies on the plastic deformation (ductility) of the components, which is primarily caused by the yielding of reinforcing steel [24,25]. However, the FRP materials display a linear elastic stress–strain relationship up to failure (no yielding), which has prompted some concerns over their use in prone to earthquakes areas. However, fundamental research in this field reveals that concrete structures reinforced with FRP may sustain substantial lateral displacements without collapsing [26–29]. This was attributable to the high tensile strength and low modulus of elasticity of FRPs, which permits the structures to tolerate significant deformations before exceeding their ultimate bearing capacity [30].

The characteristics impacting the performance of RC with FRP reinforcement (FRP-RC) moment-resisting frames must be thoroughly investigated in light of the encouraging findings of early research on the potential of utilizing FRPs in RC structures located in seismic zones. To attain this objective, it is vital to comprehend the behavior of beam-column joints as a critical factor in the lateral stability of frames [31,32]. Several researchers have examined the seismic performance of externally applied glass FRP reinforcement in existing RC beam-column joints, focusing on aspects such as joint shear stress, reinforcement details, the existence of lateral beams and the compressive strength of concrete [33–39]. However, several known characteristics that impact the seismic performance of FRP-RC beam-column joints have not yet been addressed. Current regulations and design recommendations for FRP-RC structures lack complete seismic protections due to a lack of research and data. The implementation of FRP-RC frames in seismic zones is not feasible without a thorough understanding of the properties influencing the behavior of beam-column joints.

Recent research has demonstrated that carbon FRP (CFRP) bars may efficiently replace steel reinforcing bars in concrete members. The bonding strength of FRP reinforcing bars (rebars) in concrete is significantly lower than that of standard steel rebars, despite the evident benefits of FRPs [40–43]. The bond between concrete and reinforcing bars affects the transmission of stress among them, and thus debonding has become one of the most challenging aspects of the investigation of concrete structures [44]. The bond between FRP reinforcing bars and the concrete matrix is complex, and a number of variables can affect the bonding properties of FRP reinforcements to concrete [45–49]. The geometry and surface characteristics of FRP rebars, concrete compressive strength, confinement pressure, rebar diameter and location in the cast and specimen, embedment length, temperature variations and environmental factors all impact bonding between FRP rebars and concrete [50,51].

Numerous experimental investigations have been performed to examine the bond strength of FRP rebars in concrete and the effects of factors such as fiber type, surface treatment, bar diameter and temperature on the bond properties of FRP rebars [52–56]. In addition, these studies demonstrated that the needed embedment lengths of these rebars must be carefully considered. In contrast to steel rebars, there is little information in the literature about the bonding behavior of CFRP rebars in normal and high-strength concrete members, particularly under cyclic loading circumstances [57].

To explore the seismic behavior of RC external beam-column joints with CFRP longitudinal bars in the beam, experimental testing followed by finite element (FE) model analysis were performed in this study. The influence of CFRP bar bond slip on joint hysteretic performance was also examined. An FE model was developed to simulate the behavior of the CFRP-RC joints. The accuracy of the FE simulation was accomplished by accounting for the nonlinear behavior of concrete, the decrease in concrete's compressive strength caused by crack initiation, the confinement given by the internal reinforcement and the bond slip relation between the reinforcement and concrete. For the purpose of accurately simulating the bond strength of the CFRP bars, further pull-out tests were conducted to determine the bond stress versus slip behavior. Finally, further FE analyses were performed to compare the performance of the CFRP-RC joints to conventionally steel-reinforced RC joints and to evaluate the potential of replacing conventional reinforcement with CFRP bars.

2. Experimental Test Program

The experimental program of this study was carried out to investigate the ability of FRP bars to substitute conventional steel bars in RC structural members during seismic excitations. Two (2) external T-shaped beam-column RC joints at full size (scale 1:1) were subjected to cyclic reversal loading with increasing imposed displacement. The beams were reinforced with CFRP longitudinal bars, whereas the columns were commonly reinforced with deformed steel bars.

2.1. Geometry and Reinforcement Characteristics of the Specimens

The specimens shared the same geometrical characteristics. The column's overall length was 2950 mm with a cross-section of 350/250 mm, and the beam's overall length (net span) was 1875 mm with a cross-section of 250/350 mm. The clear concrete cover was 20 mm. Figures 1 and 2 depict the geometry, cross-sectional dimensions and reinforcing features of the tested beam-column joint specimens. The columns of both specimens contained four longitudinal bars with a 14 mm diameter, placed at each angle, and two longitudinal bars with a 12 mm diameter positioned in the middle of the sides. As transverse reinforcement in the column area, stirrups with an 8 mm diameter were arranged per 100 mm. No shear reinforcement was installed in the joint area in either of the specimens as shown in Figure 2.

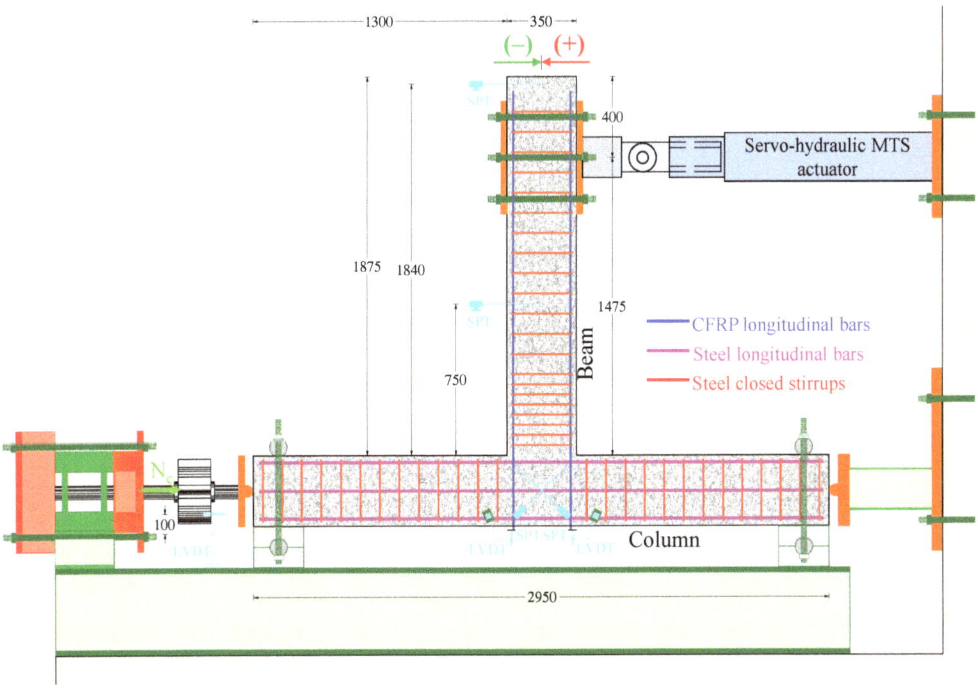

Figure 1. Schematic drawing of the test setup, geometry and reinforcement of the beam-column joint specimens subjected to cyclic reversal deformations (dimensions in mm).

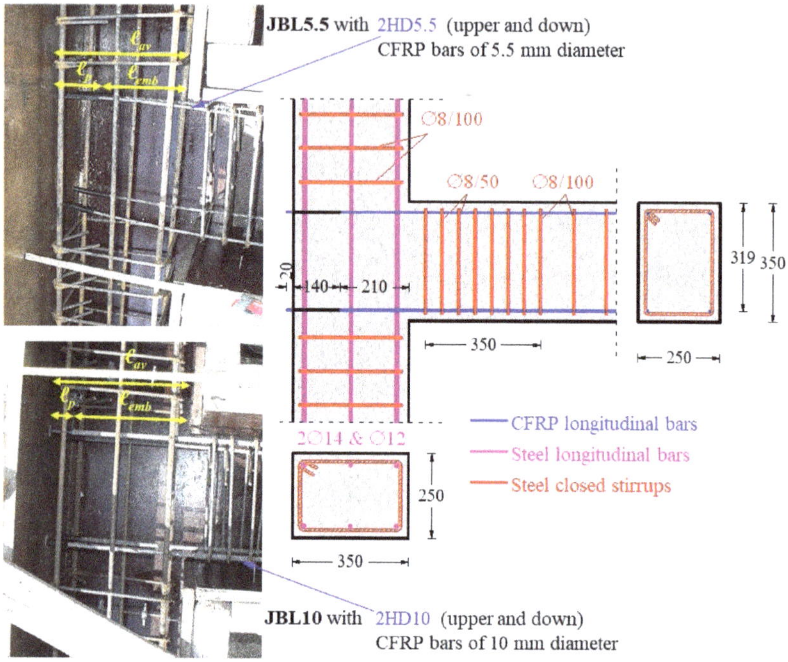

Figure 2. Geometry and reinforcement details of the joint region in specimens JBL5.5 and JBL10 (dimensions in mm).

The specimens differed in the longitudinal reinforcement of the beam. JBL5.5 contained four CFRP longitudinal bars of a 5.5 mm diameter positioned on the top and bottom sides of the beam, and JBL10 contained four CFRP bars of a 10 mm diameter as top (2 bars) and bottom (2 bars) longitudinal reinforcing bars, respectively. The beams were designed as under-reinforced in accordance with ACI 440.1R-15 [58] design criteria. In particular, the beams of the beam-column joint specimens were designed in order for their longitudinal CFRP reinforcing bars to fail under pull-out bond failure between the bar and concrete instead of fiber rupture at CFRP bar ultimate tension capacity. The reinforcement ratio of the longitudinal CFRP bars was 0.06% and 0.20% for specimens JBL5.5 and JBL10, respectively. Furthermore, the anchorage length of the longitudinal CFRP bars utilized in the beams was specified in accordance with the design requirements of ACI 440.1R-15 [58]. Based on these provisions, or the specimens, straight anchoring lengths of $\ell_{\mathrm{emb}} = 38\varnothing_f$ and $30\varnothing_f$ were calculated and adopted for the specimens JBL5.5 and JBL10, respectively. This way, pull-out bond failure of the longitudinal CFRP bars during the cyclic loading procedure was anticipated. To ensure that the tested specimen had the desired anchorage length ℓ_{emb}, a stiff polyvinyl chloride (PVC) pipe with a diameter of 10 mm was placed around the remaining part of the bars near the outside end of the column to form noncontact (de-bonded) zones between the bars and the surrounding concrete, as shown in Figure 2. The overall length of this PVC pipe was $\ell_p = \ell_{av} - \ell_{emb} = 350 - 210 = 140$ mm, where ℓ_{av} is the available anchorage length for the longitudinal bars of the beams, which is equal to the column's width.

Both specimens' shear transverse reinforcement comprised of closed deformed steel stirrups of an 8 mm diameter spaced uniformly at 50 mm (8/50 mm) along the critical region of the beams that equals to 350 mm, and at 100 mm (8/100 mm) outside this area.

For the JBL10 specimen, the anchorage length of the CFRP bars was determined to be $\ell_{emb} = 30\varnothing_f = 300$ mm. For the HD10 CFRP bars, a rigid PVC pipe with an external diameter of 16 mm and a total length of $\ell_p = \ell_{av} - \ell_{emb} = 350 - 300 = 50$ mm was selected and applied around the remaining portion of the bars near the exterior end of the columns.

2.2. Mechanical Properties of Materials

2.2.1. Concrete

Supplementary compression tests of six standard 150/300 mm cylinders were also performed to measure the concrete's compressive strength with grade C20/25. On the day that the tests were conducted, three cylinders were subjected to axial compression, while three others were tested under splitting tension. The average compressive and splitting tensile strengths of the employed concrete were f_c = 28.0 Mpa and $f_{ct,spl}$ = 2.05 Mpa, respectively, for all specimens.

2.2.2. Steel and CFRP Reinforcement

The HD5.5 and HD10 longitudinal CFRP bars of the beams were supplied by Sintecno Company. They were produced using the pultrusion method and were comprised of continuous longitudinal carbon fibers with a high tensile strength joined together with an epoxy resin. The HD5.5 and HD10 bars contained approximately 57% and 60% carbon fiber, respectively. These carbon fibers have nominal tensile strengths and elastic moduli in excess of 4 Gpa and 230 Gpa, respectively. Due to a particular surface treatment with quartz sand, the outer layer of the utilized CFRP bars had a rough external surface. The nominal ultimate tensile strength and elastic modulus of the used CFRP bars were f_{fu} = 1.8 Gpa and E_f = 130 Gpa, respectively.

The experimentally determined yield tensile strength, f_y, of the deformed steel bars was 550 Mpa for the Ø14 and Ø12 longitudinal bars of the columns, respectively, and 550 Mpa for the Ø8 utilized stirrups.

2.3. Experimental Setup and Instrumentation

The experimental setup and instrumentation are presented in Figure 3a,b. After performing a 90-degree counterclockwise rotation, the specimens of each beam-column joint were positioned so that the column was in the horizontal position and the beam was in the vertical position. In order to replicate the inflection points of the columns in the center, the specimen was supported with equipment that allowed rotation.

During the testing, a constant $N_c = 0.05A_c f_c$ column compressive axial load was applied (where A_c is the area of the column cross-section). All of the specimens were subjected to full cycle deformations that were applied by a swivel connector that was attached to the actuator near the free end of the beam. Throughout the whole loading procedure, the column axial load was maintained at a constant amount of 122.5 kN ($=N_c = 0.05A_c f_c = 0.05 \times 250 \times 350 \times 28.0 \times 10^{-3}$) for all specimens.

The cyclic displacement-controlled loading history at the beam end began after applying the constant axial load to the column. The specimens were subjected to increasing cyclic reversal deformation across an eight-step loading history, with maximum displacement, $d = \pm 8.50$ mm, ± 12.75 mm, ± 17.00 mm, ± 25.50 mm, ± 34.00 mm, ± 51.00 mm, ± 68.00 mm and ± 85.00 mm.

As indicated in Figure 4, each loading step contained three complete loading cycles for each story drift (SD) level. It is noted that d = 17 mm corresponds to SD = 1%. The loading pace was 0.05 mm/s for the initial four loading phases and then raised to 1 mm/s for the last four loading steps.

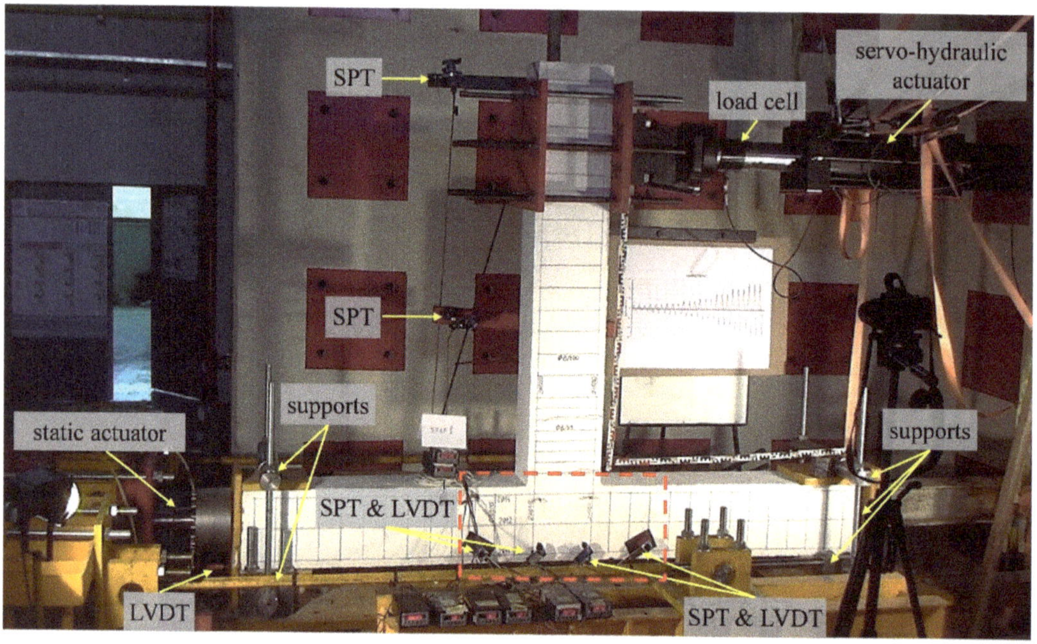

(**a**)

(**b**)

Figure 3. (**a**) Experimental setup and instrumentation of the full-scale beam-column joint specimens under lateral imposed cyclic reversal deformations; and (**b**) LVDT set for measuring the slip of the beams' longitudinal CFRP bars (two LVDTs with 0.01 mm accuracy).

The load was measured with an accuracy of 0.05 kN by a load cell. The displacements of the beams were measured using two string position transducers placed 0.75 m and 1.84 m from the end of the column (see also Figure 1), while the slip of the longitudinal FRP bars of the beams was measured using two linear variable differential transducers (LVDT) with 0.01 mm accuracy (see Figure 3b).

In addition, two SPTs were inserted on the front side of the joint core to assess the shear deformation until specimen failure. Moreover, one LVDT was installed at the left end of the specimen column to monitor the support during the test. The photographs of Figure 3a,b provide additional information on test rig instrumentation.

Step	SD (%)	*d* (mm)
1°	0.50	8.50
2°	0.75	12.75
3°	1.00	17.00
4°	1.50	25.50
5°	2.00	34.00
6°	3.00	51.00
7°	4.00	68.00
8°	5.00	85.00

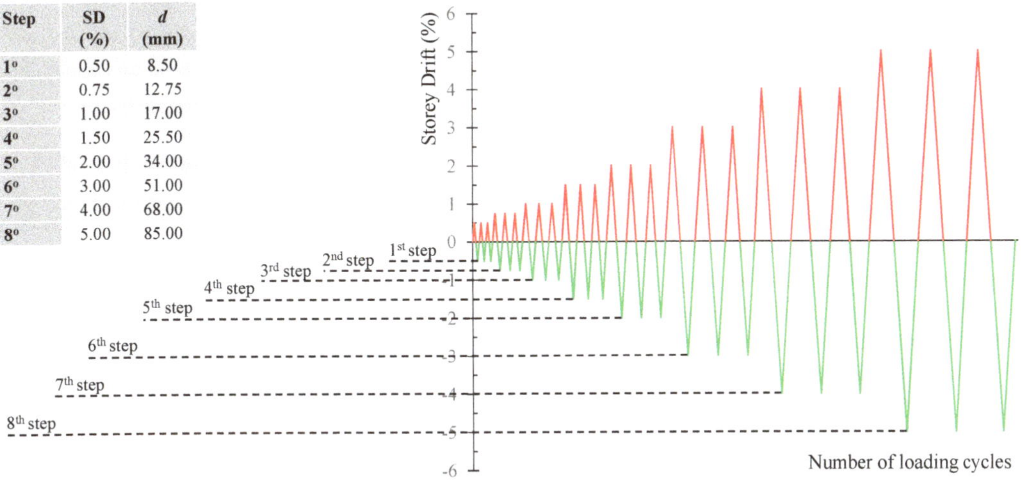

Figure 4. Cyclic reversal loading sequence: eight (8) loading steps consisting of three (3) full loading cycles at each loading step.

2.4. Bond Strength Evaluation—Pull-Out Tests

Pull-out tests were conducted to determine the bond slip behavior of the bars embedded in concrete in order to accurately simulate the contact law in the subsequent FE analysis. Pull-out specimens with a central rebar arrangement were used.

Each bar was embedded in a 300 × 300 × 200 mm concrete cube. Test specimens comprised of 1.20 m length bars embedded in concrete cubes over a length of five times their diameter (5Ø in Figure 5). In order to accomplish the specified embedment length, contact between the concrete and the bar was avoided using PVC tubing during the casting of concrete, as depicted in Figure 5. A steel tube was bonded to the loading end of the FRP bar using adhesive to form the loading end grip for the adjustment to the test setup.

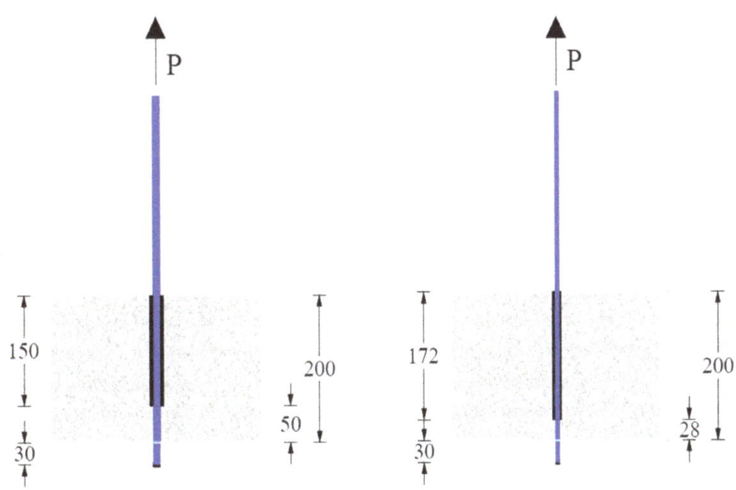

(**a**) (**b**)

Figure 5. Geometric characteristics of pull-out specimens: (**a**) 10 mm diameter CFRP bar and (**b**) 5.5 mm diameter CFRP bar.

Figure 6 shows a schematic of the pull-out specimen. For each rebar, four identical specimens were tested to ensure the test setup's reliability and the test findings' scatter (Figure 7).

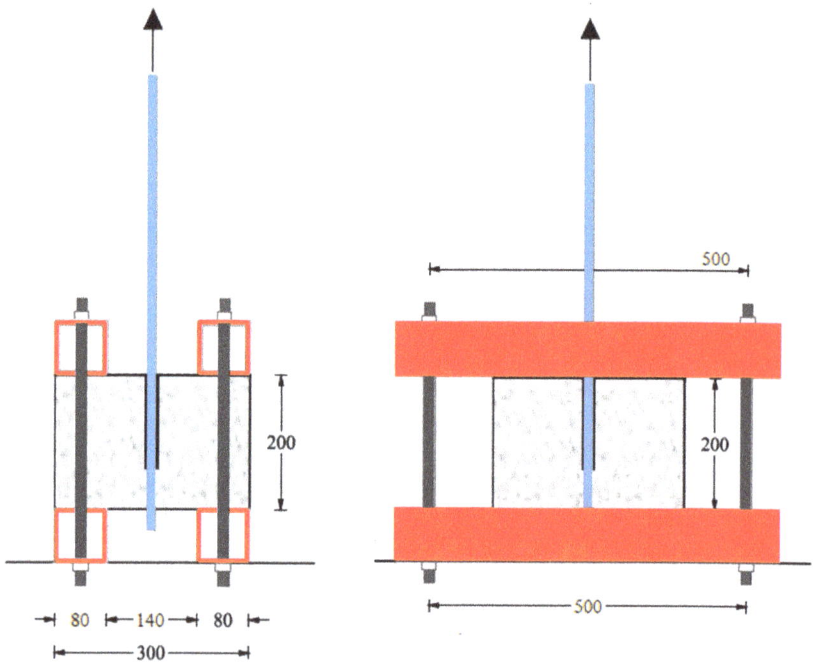

Figure 6. Test setup of pull-out specimens.

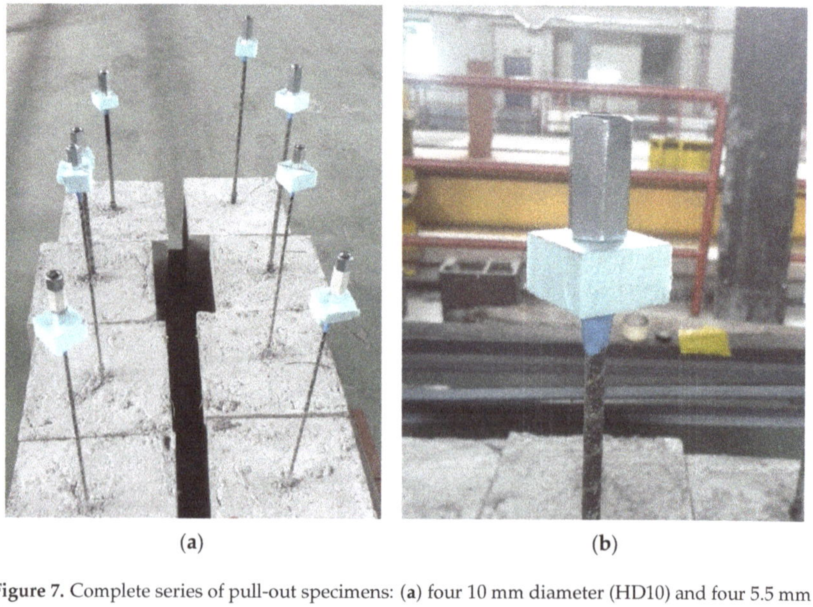

(**a**) (**b**)

Figure 7. Complete series of pull-out specimens: (**a**) four 10 mm diameter (HD10) and four 5.5 mm diameter (HD5.5) CFRP bars and (**b**) bonded steel tube to the loading end.

The test setup of the pull-out experiments is illustrated in Figure 8. The load was applied at a rate of 0.5 mm/min to the CFRP reinforcing bar. In order to determine the post peak behavior, all tests were conducted in displacement control mode. During the test, three laser sensors, as depicted in Figure 8, measured the slip of the loaded end of the bar. Typical test results of the performed pull-out tests are presented in Figure 8 in terms of average bond stress versus slip behavioral curves.

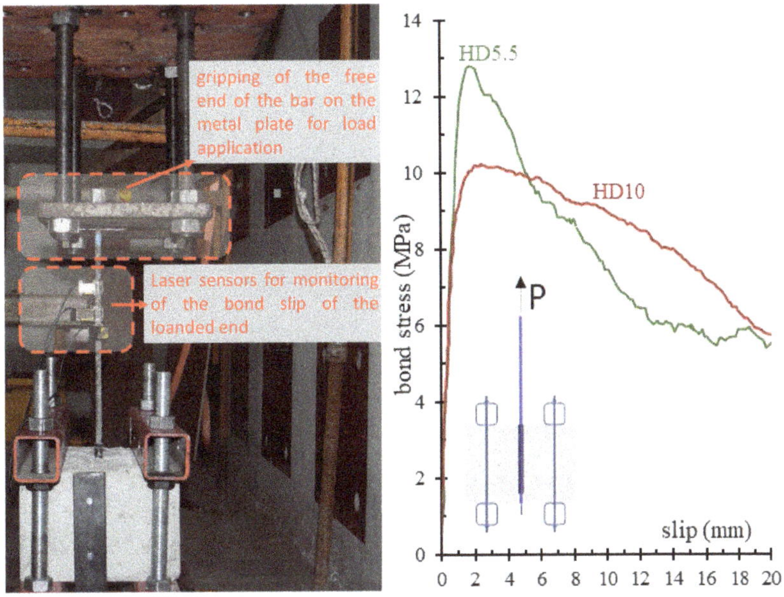

Figure 8. Test rig, instrumentation and results of the pull-out tests.

3. FE Model

In this section, the FE software, Abaqus, was used to study the behavior of full-scale beam-column joints under lateral imposed cyclic reversal deformations [59]. A 3D FE model was created to account for both the material and geometric nonlinearity, as well as the bond interaction of the longitudinal CFRP reinforcement and the surrounding concrete.

3.1. Meshing

All parts of the beam-column joints were simulated using 3D elements; the concrete and the CFRP bars were modelled using eight-node linear brick elements (C3D8R), while the longitudinal and transversal reinforcement was simulated with two-node linear 3D truss elements (T3D2) in the developed FE model. The mesh arrangement that was adopted in the numerical model was nonuniform to reduce simulation time and avoid convergence issues at the expected regions of substantial inelastic deformations, while a finer mesh was applied in areas of importance and at regions where stress concentration was expected to appear (Figure 9). In the areas of a fine mesh, the maximum mesh element size was set to 25 mm, while it was 50 mm in the rest.

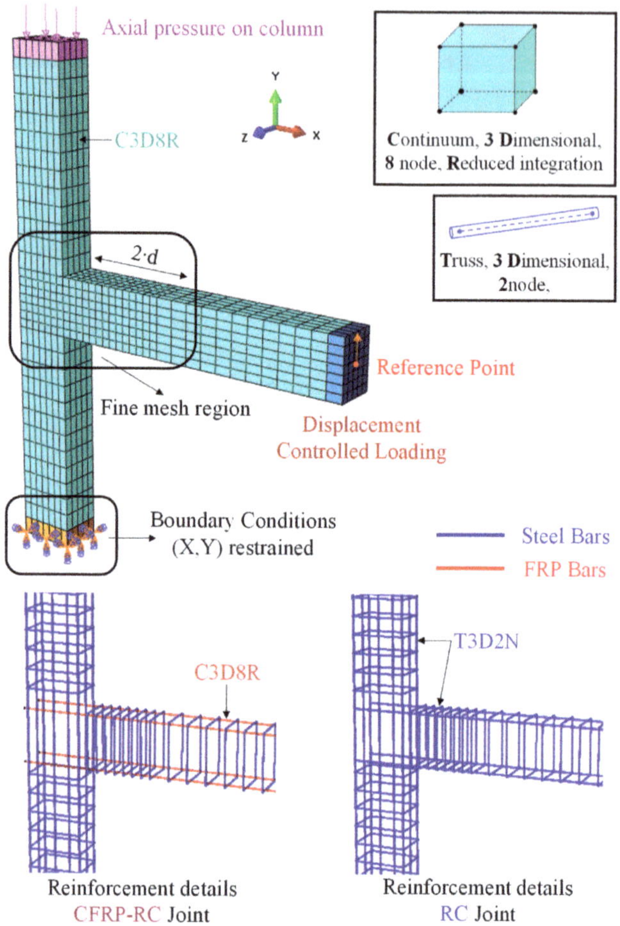

Figure 9. Details of the 3D FE simulation used to simulate the CFRP-RC and RC beam-column joints.

3.2. *Material Constitutive Relationship*

In order to appropriately simulate the CFRP-RC and RC joints' behavior, the FE simulation took into account the nonlinearity of concrete under compression and tensile cracking, the elastic–plastic behavior of steel reinforcement, and the elastic–brittle performance of the CFRP reinforcement.

3.2.1. Concrete

In the current FE model, the concrete damaged plasticity (CDP) model available in Abaqus [53] is used to simulate concrete behavior. The CDP model is designed to analyze RC structures imposed to monotonic, cyclic or dynamic stresses. The CDP model enables the definition of both the elastic and plastic properties of concrete. In previous studies, the authors provide further information about the constitutive laws and parameters used [60–62].

3.2.2. Steel

The conventional steel reinforcement (longitudinal bars and stirrups) was described as an elastic–plastic material with a strain-hardening behavior. The material demonstrates an elastic linear behavior, following the modulus of elasticity, up to yielding point. After the yielding point is reached, the strain hardening effect begins with an incline of $0.01E_s$ [63] and reaches its maximum at a strain value of 0.05. Furthermore, the FE analysis required the steel's poison's ratio, which was assumed to be 0.3.

Concerning the interaction of steel reinforcement with concrete (column longitudinal bars and stirrups of the beam and column), a full bond was adopted by the use of the embedded region method.

3.2.3. CFRP Reinforcing Bars

The stress–strain curve for CFRP rebars was described as a linear elastic material up to the point of ultimate tensile strength and, afterwards, the brittle failure of the material was considered. The characteristics of CFRP bars varied in the transverse direction because of their fibrous nature, and it would be more appropriate to simulate them as anisotropic materials. However, for the sake of computational simplicity, an isotropic material assumption is used. The model's accuracy in the current case is unaffected by this supposition as the beam-column joints investigated in the present study are designed not to fail in shear; the contribution of the CFRP reinforcement's shear dowel action is presumed to be minimal. Consequently, the shear failure of the CFRP bars is not included in this work. The longitudinal beam reinforcement mostly contributes by transmitting axial tensile forces, and, hence, CFRP bars contribute in the fiber direction. The modulus of elasticity and ultimate strength are the mechanical properties necessary to define the CFRP material within the FE model.

3.3. Bond Slip Interaction—Cohesive Method

In this model, the surface-based cohesive behavior was employed to represent the bond stress versus slip behavior between CFRP bars and concrete, utilizing the Abaqus cohesive approach. This approach, referred to as the surface-to-surface method, can be used to determine the contact parameters of two surfaces in contact. In comparison to the node-to-surface approach, this method was selected because it generates more trustworthy results [64]. Moreover, the surface-to-surface technique applies contact criteria in an average sense over the regions that surround the slave nodes, suggesting that each restriction will not only consider one slave node but also nearby slave nodes [65].

To assign surface-to-surface contact, the user must identify two surfaces that are in touch. The surface that has a coarse mesh is defined as the master surface, while the surface that contains a fine mesh is defined as the slave surface.

Abaqus calculates the bond behavior for this contact model using the linear elastic traction–separation model, as illustrated in Figure 10. This model begins by making the assumption of elastic behavior, then proceeds on to consider the initiation and progression of damage. The elastic behavior is expressed in terms of an elastic constitutive matrix that connects normal and shear stresses to normal and shear interface separations. Considering uncoupled normal and tangential stiffness components, the traction–separation behavior matrix is as follows

$$T = \begin{Bmatrix} t_n \\ t_s \\ t_t \end{Bmatrix} = \begin{bmatrix} k_{nn} & 0 & 0 \\ 0 & k_{ss} & 0 \\ 0 & 0 & k_{tt} \end{bmatrix} \begin{Bmatrix} \delta_n \\ \delta_s \\ \delta_t \end{Bmatrix} = K\delta, \tag{1}$$

where t_n = nominal traction in the normal direction; t_s and t_t = nominal stresses in two local shear directions; δ_n, δ_s and δ_t = corresponding displacements. k_{ss}, k_{tt}, and k_{nn} are derived as follows [66]

$$k_{ss} = k_{tt} = \frac{\tau_m}{s_m}, \tag{2}$$

$$k_{nn} = 100 \cdot k_{ss} = 100 \cdot k_{tt}, \tag{3}$$

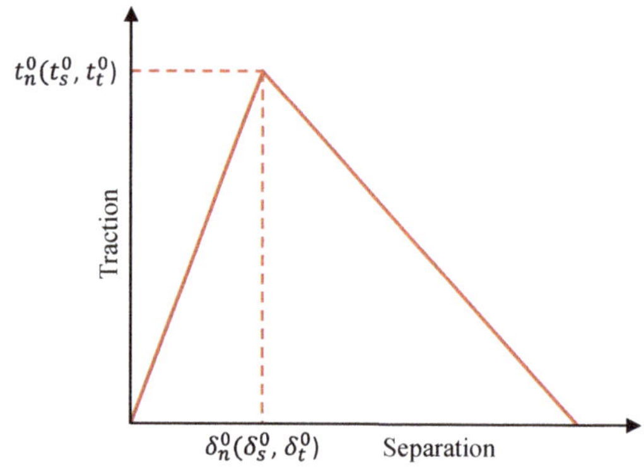

Figure 10. Traction versus separation model adopted in the performed FE analysis.

The pull-out tests that were conducted in the current research provided the specifications for maximum bond strength (τ_m) and slip at maximum bond strength (s_m) (see also Figure 8).

3.4. Boundary Conditions and Load Application

To achieve convergence, the boundary conditions were modelled equivalently to the experimental test's supports. To simulate the boundary conditions for the beam-column joints in the experimental testing, the top face of the column was restricted in the X-direction resulting in a roller support, while at the bottom of the column, a hinge support was formed by constraining against the X and Y movement.

As indicated in Figure 9, the column's axial load was applied to the surface of the column, while the displacement control load was applied to the reference point on the beam profile. The load was applied steadily and gradually in accordance with the loading sequence of the experimental procedure.

4. Results and Discussion

4.1. Verification of the FE Model

To evaluate the ability of the developed FE simulation to estimate the lateral response of beam-column joints using CFRP bars as a longitudinal reinforcement in the beam, a comparison was made between the predicted and experimental CFRP-RC beam-column joint specimens. The next section provides specific comparative results. Figure 11 shows a comparison between the applied load versus SD curves predicted using the FE analysis (noted as "FEA") and the experimental one (noted as "Test"). In both investigated joints the model predicts the maximum load at a satisfactory level as well as the post cracking behavior.

Figure 12 presents a comparison between the predicted area of damage occurrence and the experimental results. Clearly, the model was capable of accurately predicting the failure of the specimens. For the specimen JBL10 a large crack opened in the beam area close to the joint connection, while no cracking occurred in the joint body. The corresponding mode of failure predicted by the FE analysis also illustrates damage close to this area, while in the joint core no damage is represented as in the experimental specimen. Only low strain concentration around the CFRP bars is illustrated, which is attributed to the slippage of the CFRP bars.

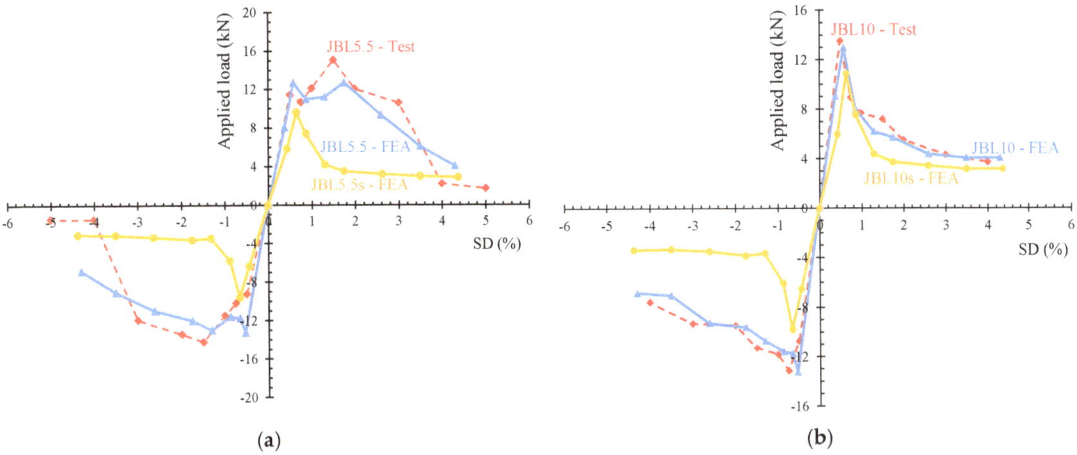

Figure 11. Comparative lateral load versus SD diagrams of the corresponding specimens: (**a**) JBL5.5—Test, JBL5.5—FEA, JBL5.5s—FEA and (**b**) JBL10—Test, JBL10—FEA, JBL10s—FEA.

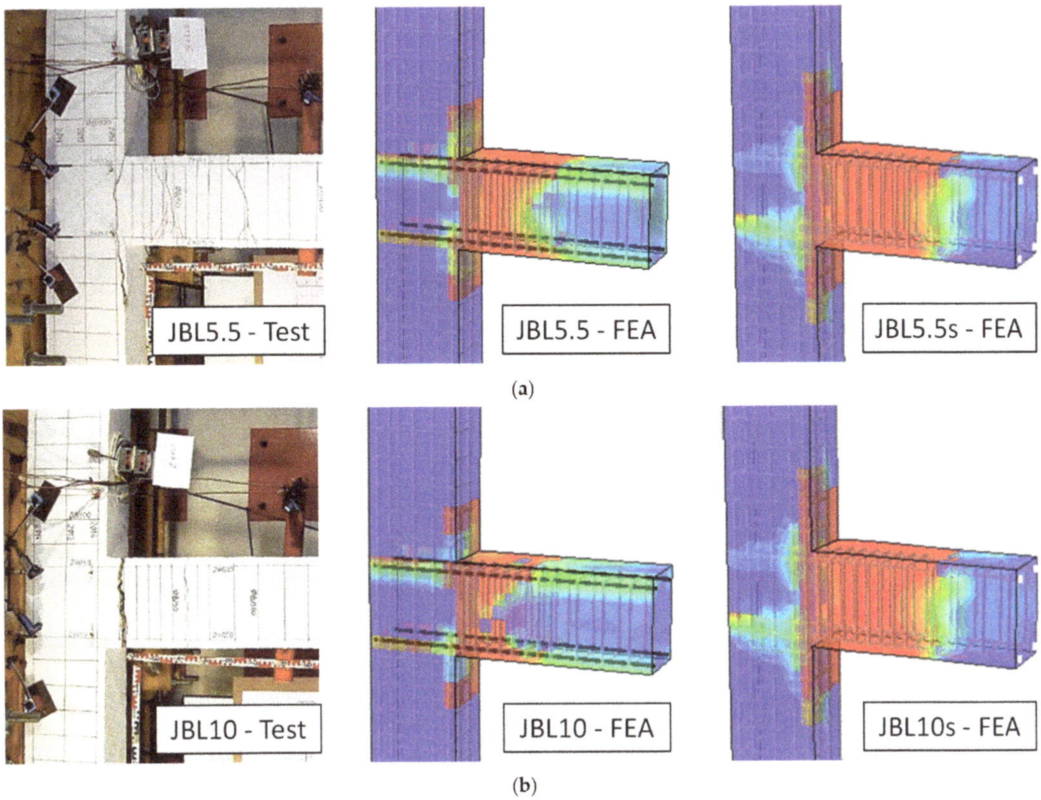

Figure 12. Verification of FE model prediction of damage area: (**a**) JBL5.5—Test, JBL5.5—FEA, JBL5.5s—FEA and (**b**) JBL10—Test, JBL10—FEA, JBL10s—FEA.

Figure 13 illustrates that the developed model can also successfully capture the CFRP bar slippage. In particular, for the specimen JBL10, the predicted slip is almost identical to the corresponding slip calculated in the experimental test. In the case of the JBL5.5 specimen, the FE analysis predicted a smaller slip than that experimentally measured but this still remains within acceptable limits. The observed differences between the experimental results and FE model predictions of slip of JBL5.5 could be attributed to the protentional discrepancies and uncertainties of the CFRP bar and concrete bond slip simulation adopted in the cohesive approach described in Section 3.3.

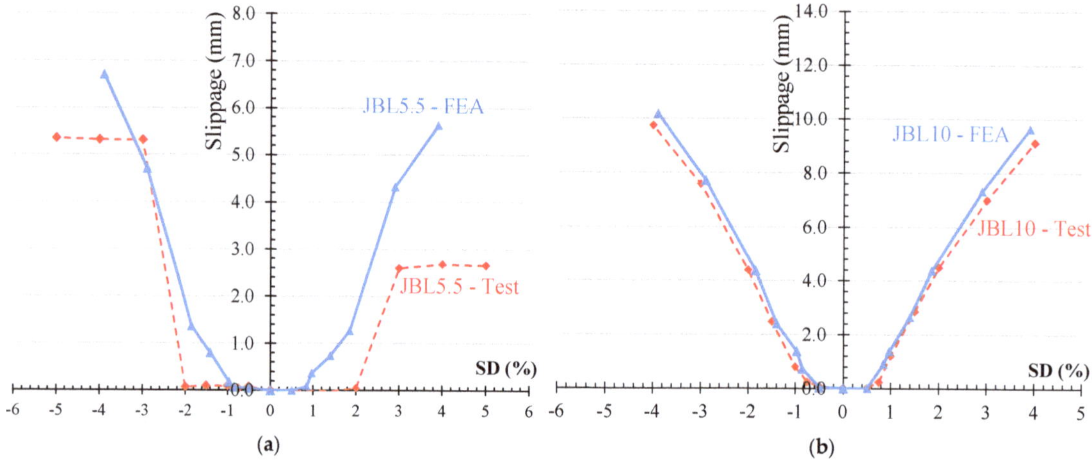

Figure 13. Verification of slip prediction of FE model in terms of slippage versus SD diagrams of the corresponding specimens: (**a**) JBL5.5—Test, JBL5.5—FEA and (**b**) JBL10—Test, JBL10—FEA.

Figure 14 represents the energy dissipation calculated from the experimental results simply summing up the amount of energy that was lost over the course of several load-displacement cycles and the energy dissipation extracted from the FE analysis. In terms of energy dissipation, the FE simulation is satisfyingly close to the experimental results.

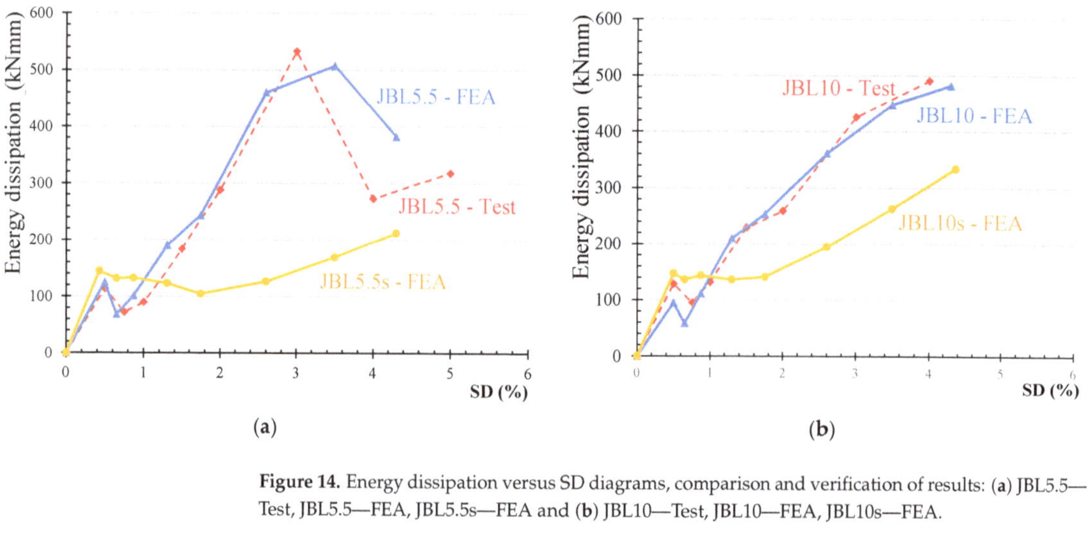

Figure 14. Energy dissipation versus SD diagrams, comparison and verification of results: (**a**) JBL5.5—Test, JBL5.5—FEA, JBL5.5s—FEA and (**b**) JBL10—Test, JBL10—FEA, JBL10s—FEA.

According to the validation results, the developed FE simulation can accurately estimate the lateral response of CFRC beam-column joints.

4.2. Use of CFRP Bars versus Conventional Steel Rebars as Beam Longitudinal Reinforcement

Since the experimental program did not contain steel RC joints with conventional steel rebars in the beam, the authors were not able to validate the developed FE simulation against such experimental results of RC joints. Nonetheless, the previously developed FE models by the authors [52,53], which were validated for the monotonic and cyclic response of RC beams against shear and flexure, are equally applicable for predicting the behavior of RC joints. Notably, no significant alterations to the established FE analysis for the CFRP-RC joints were performed aside from employing longitudinal steel reinforcing bars in the beam region shown in Figure 9.

Compared to the CFRP-RC specimens (JBL10, JBL5.5), the steel-reinforced specimens (JBL10s, JBL5.5s) demonstrated a quicker decline in load bearing capacity (Figure 11). The steel RC specimens exhibited linear behavior up to an SD = 0.6%. Due to the yielding of the steel reinforcement, the rate of growth in the specimens' lateral capacity reduced and stayed practically constant up to an SD = 4%.

At greater SD ratios, the CFRP-RC maintained a higher load. This suggests that CFRP-RC beam-column joints may withstand a seismic event with minimal residual damage. This function decreases the post event cost of repairing the framed structure and permits the complete restoration of the original CFRP-RC section performance by replacing the damaged concrete. However, this is not applicable for steel RC sections, which are intended to yield at such a point of ultimate loading. Even by repairing the damaged concrete, the original section's strength and stiffness (prior to yielding) cannot be recovered.

After attaining their design capacity, specimens JBL10 and JBL5.5 revealed a decrease in lateral load resistance at an SD = 1%. This decrease is attributable to the slip of the beam longitudinal bars that were anchored in the joint as a result of bar anchoring failure. As seen in Figure 11a,b, despite the anchorage's failure, both specimens were still able to withstand about 35.0% of their initial capacity through their final loading cycles until the completion of the test at an SD = 4%. Despite the fact that the slip of the bar is a brittle form of failure, it must be emphasized that specimens JBL10 and JBL5.5 were initially intended to attain the desired shear stress in the joint body when the beam achieves its flexural capacity. The failure due to slip of the bars occurred after the joint achieved the intended load level, as illustrated in Figure 13a,b, and that has no effect on the study's analysis or findings given the study's scope. In addition, earthquake occurrences are regarded as extreme–ultimate limit state situations; hence, the performance of both specimens after the failure occurrence is significant since it exhibited a residual strength greater than 35% of the maximum load after the slip of the bars.

The envelope of the specimens' total dissipated energy at each loading versus SD is depicted in Figure 14. It is evident that both CFRP-RC specimens had a comparable degree of dissipated energy up to an SD = 3%, but the energy dissipated at failure slightly differed for each specimen. In addition, the figure indicates a substantial association between the energy lost due to hysteretic activity and the extent of damage detected during the test. As the specimens fail, the inclination of the curve of cumulative energy dissipation versus SD rises, indicating the concrete deterioration rate owing to the development of plastic hinges. Nevertheless, the increase in energy dissipation during the final steps of loading for specimens JBL10 and JBL5.5 is mostly attributable to the slippage of the longitudinal reinforcement of the beam, as previously noted. Figure 14 also demonstrates that specimens with conventional reinforcement dissipated less energy than those with CFRP. While reduced energy dissipation is regarded as a disadvantage, it also implies that the joint recovers its previous form when the stresses are eliminated, needing only a minimal degree of restoration after undergoing such loading action.

5. Conclusions

This paper deals with the application of CFRP bars as a beam's longitudinal reinforcement in RC beam-column joint full-scale specimens under cyclic lateral deformations. An experimental investigation that involved two full-scale CFRP-RC joints was conducted and accompanied with eight CFRP bar pull-out tests in order to obtain the bond stress versus slip behavior for the bars. Furthermore, a numerical study was performed. An FE model was developed that could successfully predict the behavior of CFRP-RC joints. Additionally, a comparative study of steel RC joints and CFRP-RC joints was conducted to investigate the potential replacement of the conventional steel reinforcement with CFRP bars. On the basis of the study's findings and discussion, the following conclusions may be drawn:

1. The experimentally evaluated specimens' load-carrying capacity and load-deflection response could be accurately predicted using the nonlinear FE model provided in this work. The failure mechanism, bond slip of CFRP bars and energy dissipation comparison data demonstrated that the model can properly predict the behavior of CFRP-RC joints.
2. The slippage of the CFRP bars occurs after the reach of the ultimate load and thus does not significantly affect the post failure behavior and the hysteretic performance of the joint. The investigated CFRP-RC joints exhibited a residual strength higher than 35% of the flexural capacity after the occurrence of slippage. However, more investigation must be conducted in order to ensure the post failure mode and overall response of joints with CFRP longitudinal reinforcement bars.
3. The dissipated energy of the CFRP-RC joints increased gradually following the specimen's mode of failure. However, after a certain point the continuous increase up to failure is possibly attributed to the bar slippage. After the reach of an approximate SD = 0.5%, the dissipated energy of the steel-reinforced joints was significantly lower than that of the CFRP-RC joints.

Overall, the behavior of RC joints with beams' CFRP longitudinal reinforcements indicated promising results and the favorable contribution of the CFRP bars in the seismic performance of RC joints. CFRP bars have great potential of partially or even fully replacing conventional steel bars in RC structural elements. More research must be conducted though in order to form regulations and design specifications so that CFRP bars become widely applicable in the construction industry.

Author Contributions: Conceptualization, C.E.C.; data curation, V.K.K. and P.-M.K.K.; formal analysis, V.K.K. and P.-M.K.K.; investigation, V.K.K., P.-M.K.K. and C.E.C.; methodology, V.K.K. and P.-M.K.K.; project administration, C.E.C.; software, V.K.K.; supervision, C.E.C.; validation, V.K.K.; visualization, V.K.K. and P.-M.K.K.; writing—original draft, V.K.K.; writing—review and editing, C.E.C. All authors have read and agreed to the published version of the manuscript.

Funding: This research received no external funding.

Institutional Review Board Statement: Not applicable.

Informed Consent Statement: Not applicable.

Data Availability Statement: The data presented in this study are available on request from the corresponding author.

Acknowledgments: The contribution of the personnel of the Laboratory of Reinforced Concrete and Seismic Design of Structures in Democritus University of Thrace on the experimental procedure is sincerely appreciated.

Conflicts of Interest: The authors declare no conflict of interest.

References

1. ACI 440R-96. Abstract of: State-of-the-art-report on fiber reinforced plastic (frp) for concrete structures. *ACI Struct. J.* **1995**, *92*, 5. [CrossRef]
2. Bakis, C.E.; Nanni, A.; Terosky, J.A.; Koehler, S.W. Self-monitoring, pseudo-ductile, hybrid FRP reinforcement rods for concrete applications. *Compos. Sci. Technol.* **2001**, *61*, 815–823. [CrossRef]
3. Bank, L.; Gentry, T.; Barkatt, A.; Prian, L.; Wang, F.; Mangla, S. Accelerated aging of pultruded glass/vinylester rods. In Proceedings of the Second International Conference on Composites in Infrastructure National Science Foundation, Tucson, AZ, USA, 5–7 January 1998; Volume 2.
4. Rizkalla, S.H.; Mufti, A.A.; ISIS Canada (Organization). *Reinforcing Concrete Structures with Fibre Reinforced Polymers*; ISIS Canada: Winnipeg, MB, Canada, 2001; ISBN 978-0-9689006-6-6.
5. Qureshi, J. A review of fibre reinforced polymer structures. *Fibers* **2022**, *10*, 27. [CrossRef]
6. Juozapaitis, A.; Sandovič, G.; Jakubovskis, R.; Gribniak, V. Effects of flexural stiffness on deformation behaviour of steel and frp stress-ribbon bridges. *Appl. Sci.* **2021**, *11*, 2585. [CrossRef]
7. Golias, E.; Vougioukas, E.A.; Wittemann, K.; Kalogeropoulos, G.I.; Karayannis, C.G. Cyclic response of RC beam-column joints strengthened with transverse steel bars and with C-FRP diagonal ties. *Acta Polytech.* **2022**, *62*, 274–282. [CrossRef]
8. Karayannis, C.G.; Golias, E.; Kalogeropoulos, G.I. Influence of carbon fiber-reinforced ropes applied as external diagonal reinforcement on the shear deformation of RC joints. *Fibers* **2022**, *10*, 28. [CrossRef]
9. Karayannis, C.G.; Golias, E. Full-scale experimental testing of RC beam-column joints strengthened using CFRP ropes as external reinforcement. *Eng. Struct.* **2022**, *250*, 113305. [CrossRef]
10. Salleh, N.; Hamid, N.A.; Majid, M.A. Finite element modelling of concrete beams reinforced with hybrid fiber reinforced bars. *IOP Conf. Ser. Mater. Sci. Eng.* **2017**, *271*, 012093. [CrossRef]
11. Fahmy Mohamed, F.M.; Abd-ElShafy Zainab, E.; Wu, Z. Experimental and numerical evaluation of the shear behavior of reinforced concrete T-beams with hybrid steel-FRP stirrups. *J. Compos. Constr.* **2017**, *21*, 04017007. [CrossRef]
12. Kalogeropoulos, G.I.; Tsonos, A.-D.G.; Konstandinidis, D.; Tsetines, S. Pre-earthquake and post-earthquake retrofitting of poorly detailed exterior RC beam-to-column joints. *Eng. Struct.* **2016**, *109*, 1–15. [CrossRef]
13. Zia, A.; Pu, Z.; Holly, I.; Umar, T.; Tariq, M.A.U.R. Development of an analytical model for the FRP retrofitted deficient interior reinforced concrete beam-column joints. *Appl. Sci.* **2022**, *12*, 2339. [CrossRef]
14. Ibrahim, H.A.; Fahmy, M.F.M.; Wu, Z. Numerical study of steel-to-FRP reinforcement ratio as a design-tool controlling the lateral response of SFRC beam-column joints. *Eng. Struct.* **2018**, *172*, 253–274. [CrossRef]
15. Golias, E.; Zapris, A.G.; Kytinou, V.K.; Osman, M.; Koumtzis, M.; Siapera, D.; Chalioris, C.E.; Karayannis, C.G. Application of X-shaped CFRP ropes for structural upgrading of reinforced concrete beam-column joints under cyclic loading-experimental study. *Fibers* **2021**, *9*, 42. [CrossRef]
16. Golias, E.; Zapris, A.G.; Kytinou, V.K.; Kalogeropoulos, G.I.; Chalioris, C.E.; Karayannis, C.G. Effectiveness of the novel rehabilitation method of seismically damaged RC joints using C-FRP ropes and comparison with widely applied method using C-FRP sheets—Experimental investigation. *Sustainability* **2021**, *13*, 6454. [CrossRef]
17. Karayannis, C.G.; Golias, E. Strengthening of deficient RC joints with diagonally placed external C-FRP ropes. *Earthq. Struct.* **2021**, *20*, 123–132.
18. Karayannis, C.G.; Golias, E. Full scale tests of RC joints with minor to moderate seismic damage repaired using C-FRP sheets. *Earthq. Struct.* **2018**, *15*, 617–627.
19. Ahmed, E.A.; Benmokrane, B.; Sansfaçon, M. Case study: Design, construction, and performance of the la chancelière parking garage's concrete flat slabs reinforced with GFRP bars. *J. Compos. Constr.* **2017**, *21*, 05016001. [CrossRef]
20. Karayannis, C.G.; Kosmidou, P.-M.K.; Chalioris, C.E. Reinforced concrete beams with carbon-fiber-reinforced polymer bars—Experimental study. *Fibers* **2018**, *6*, 99. [CrossRef]
21. Murad, Y.; Tarawneh, A.; Arar, F.; Al-Zu'bi, A.; Al-Ghwairi, A.; Al-Jaafreh, A.; Tarawneh, M. Flexural strength prediction for concrete beams reinforced with FRP bars using gene expression programming. *Structures* **2021**, *33*, 3163–3172. [CrossRef]
22. Jeong, Y.; Kim, W.; Gribniak, V.; Hui, D. Fatigue behavior of concrete beams prestressed with partially bonded CFRP bars subjected to cyclic loads. *Materials* **2019**, *12*, 3352. [CrossRef]
23. Hadhood, A.; Mohamed, H.M.; Benmokrane, B.; Nanni, A.; Shield, C.K. Assessment of design guidelines of concrete columns reinforced with glass fiber-reinforced polymer bars. *ACI Struct. J.* **2019**, *116*, 193–207. [CrossRef]
24. ACI-ASCE Committee 352. Recommendations for Design of Beam-Column Joints in Monolithic Reinforced Concrete Structures. *ACI J. Proc.* **1976**, *73*, 375–393. [CrossRef]
25. Karayannis, C.G.; Chalioris, C.E. Capacity of RC joints subjected to early-age cyclic loading. *J. Earth. Eng.* **2000**, *04*, 479–509. [CrossRef]
26. Said, A.M.; Nehdi, M.L. Use of FRP for RC frames in seismic zones: Part II. performance of steel-free GFRP-reinforced beam-column joints. *Appl. Compos. Mater.* **2004**, *11*, 227–245. [CrossRef]
27. Hasaballa, M.; Amr, E.; El-Salakawy, E. Seismic behavior of beam-column joints reinforced with GFRP bars and stirrups. *J. Compos. Constr.* **2011**, *15*, 875–886. [CrossRef]
28. Hasaballa, M.; El-Salakawy, E. Anchorage performance of GFRP headed and bent bars in beam-column joints subjected to seismic loading. *J. Compos. Constr.* **2018**, *22*, 04018060. [CrossRef]

29. Ha, G.-J.; Cho, C.-G.; Kang, H.-W.; Feo, L. Seismic improvement of RC beam-column joints using hexagonal CFRP bars combined with CFRP sheets. *Compos. Struct.* **2013**, *95*, 464–470. [CrossRef]
30. El-Mandouh, M.A.; Omar, M.S.; Elnaggar, M.A.; Abd El-Maula, A.S. Cyclic behavior of high-strength lightweight concrete exterior beam-column connections reinforced with GFRP. *Buildings* **2022**, *12*, 179. [CrossRef]
31. Flenga, M.G.; Favvata, M.J. Fragility curves and probabilistic seismic demand models on the seismic assessment of RC frames subjected to structural pounding. *Appl. Sci.* **2021**, *11*, 8253. [CrossRef]
32. Flenga, M.G.; Favvata, M.J. Probabilistic seismic assessment of the pounding risk based on the local demands of a multistory RC frame structure. *Eng. Struct.* **2021**, *245*, 112789. [CrossRef]
33. Ghomi, S.K.; El-Salakawy, E. Seismic performance of GFRP-RC exterior beam-column joints with lateral beams. *J. Compos. Constr.* **2016**, *20*, 04015019. [CrossRef]
34. Hasaballa, M.; El-Salakawy, E. Shear capacity of exterior beam-column joints reinforced with GFRP bars and stirrups. *J. Compos. Constr.* **2016**, *20*, 04015047. [CrossRef]
35. Murad, Y.Z. Retrofitting interior RC beam-to-column joints subjected to quasi-static loading using NSM CFRP ropes. *Structures* **2021**, *34*, 4158–4168. [CrossRef]
36. Murad, Y.Z.; RHunifat, R.; Wassel, A.L.B. Interior reinforced concrete beam-to-column joints subjected to cyclic loading: Shear strength prediction using gene expression programming. *Case Stud. Constr. Mater.* **2020**, *13*, e00432. [CrossRef]
37. Balamuralikrishnan, R.; Saravanan, J. Finite element analysis of beam-column joints reinforced with GFRP reinforcements. *Civil Eng. J.* **2019**, *5*, 2708–2726. [CrossRef]
38. Ghomi, S.K.; El-Salakawy, E. Seismic behavior of exterior GFRP-RC beam-column connections: Analytical study. *J. Compos. Constr.* **2018**, *22*, 04018022. [CrossRef]
39. Tiwary, A.K.; Singh, S.; Chohan, J.S.; Kumar, R.; Sharma, S.; Chattopadhyaya, S.; Abed, F.; Stepinac, M. Behavior of RC beam–column joints strengthened with modified reinforcement techniques. *Sustainability* **2022**, *14*, 1918. [CrossRef]
40. Balendran, R.V.; Rana, T.M.; Maqsood, T.; Tang, W.C. Application of FRP bars as reinforcement in civil engineering structures. *Struct. Surv.* **2002**, *20*, 62–72. [CrossRef]
41. El-Nemr, A.; Ahmed, E.A.; Barris, C.; Benmokrane, B. Bond-dependent coefficient of glass- and carbon-FRP bars in normal- and high-strength concretes. *Constr. Build. Mater.* **2016**, *113*, 77–89. [CrossRef]
42. Islam, S.; Afefy, H.M.; Sennah, K.; Azimi, H. Bond characteristics of straight- and headed-end, ribbed-surface, GFRP bars embedded in high-strength concrete. *Constr. Build. Mater.* **2015**, *83*, 283–298. [CrossRef]
43. Lin, X.; Zhang, Y.X. Bond-slip behaviour of FRP-reinforced concrete beams. *Constr. Build. Mater.* **2013**, *44*, 110–117. [CrossRef]
44. Zhou, Y.; Wu, G.; Li, L.; Guan, Z.; Guo, M.; Yang, L.; Li, Z. Experimental investigations on bond behavior between FRP bars and advanced sustainable concrete. *Polymers* **2022**, *14*, 1132. [CrossRef] [PubMed]
45. Rolland, A.; Quiertant, M.; Khadour, A.; Chataigner, S.; Benzarti, K.; Argoul, P. Experimental investigations on the bond behavior between concrete and FRP reinforcing bars. *Constr. Build. Mater.* **2018**, *173*, 136–148. [CrossRef]
46. Xue, W.; Zheng, Q.; Yang, Y.; Fang, Z. Bond behavior of sand-coated deformed glass fiber reinforced polymer rebars. *J. Reinf. Plast. Compos.* **2014**, *33*, 895–910. [CrossRef]
47. Hao, Q.; Wang, Y.; He, Z.; Ou, J. Bond strength of glass fiber reinforced polymer ribbed rebars in normal strength concrete. *Constr. Build. Mater.* **2009**, *23*, 865–871. [CrossRef]
48. Baena, M.; Torres, L.; Turon, A.; Barris, C. Experimental study of bond behaviour between concrete and FRP bars using a pull-out test. *Compos. Part B Eng.* **2009**, *40*, 784–797. [CrossRef]
49. Sun, Y.; Liu, Y.; Wu, T.; Liu, X.; Lu, H. Numerical analysis on flexural behavior of steel fiber-reinforced LWAC beams reinforced with GFRP bars. *Appl. Sci.* **2019**, *9*, 5128. [CrossRef]
50. Cosenza, E.; Manfredi, G.; Realfonzo, R. Behavior and modeling of bond of FRP rebars to concrete. *J. Compos. Constr.* **1997**, *1*, 40–51. [CrossRef]
51. Gao, J.; Xu, P.; Fan, L.; Terrasi, G.P. Study on Bond-Slip Behavior between Seawater Sea-Sand Concrete and Carbon Fiber-Reinforced Polymer (CFRP) Bars with Different Surface Shapes. *Polymers* **2022**, *14*, 2689. [CrossRef]
52. Achillides, Z.; Pilakoutas, K. Bond behavior of fiber reinforced polymer bars under direct pullout conditions. *J. Compos. Constr.* **2004**, *8*, 173–181. [CrossRef]
53. Aiello, M.A.; Leone, M.; Pecce, M. Bond performances of FRP rebars-reinforced concrete. *J. Mater. Civil Eng.* **2007**, *19*, 205–213. [CrossRef]
54. Roman, O.; Yuan Robert, L. Bond strength of fiber reinforced polymer rebars in normal strength concrete. *J. Compos. Constr.* **2005**, *9*, 203–213. [CrossRef]
55. Tighiouart, B.; Benmokrane, B.; Gao, D. Investigation of bond in concrete member with fibre reinforced polymer (FRP) bars. *Constr. Build. Mater.* **1998**, *12*, 453–462. [CrossRef]
56. Malvar, L.J.; Cox, J.V.; Cochran, K. Bergeron bond between carbon fiber reinforced polymer bars and concrete. I: Experimental study. *J. Compos. Constr.* **2003**, *7*, 154–163. [CrossRef]
57. Akbas, T.T.; Celik, O.C.; Yalcin, C.; Ilki, A. Monotonic and cyclic bond behavior of deformed CFRP bars in high strength concrete. *Polymers* **2016**, *8*, 211. [CrossRef] [PubMed]
58. ACI Committee 440. *Guide for the Design and Construction of Concrete Reinforced with FRP Bars (ACI 440.1R-15)*; American Concrete Institute (ACI): Farmington Hills, MI, USA, 2015; p. 88.

59. Dassault Systèmes Simulia System Information. *Abaqus 2017 User's Manual*; Version 6.12.1; SIMULIA: Providence, RI, USA, 2017.
60. Chalioris, C.E.; Kytinou, V.K.; Voutetaki, M.E.; Karayannis, C.G. Flexural damage diagnosis in reinforced concrete beams using a wireless admittance monitoring system—Tests and finite element analysis. *Sensors* **2021**, *21*, 679. [CrossRef]
61. Kytinou, V.K.; Chalioris, C.E.; Karayannis, C.G.; Elenas, A. Effect of steel fibers on the hysteretic performance of concrete beams with steel reinforcement—Tests and analysis. *Materials* **2020**, *13*, 2923. [CrossRef]
62. Kytinou, V.K.; Chalioris, C.E.; Karayannis, C.G. Analysis of residual flexural stiffness of steel fiber-reinforced concrete beams with steel reinforcement. *Materials* **2020**, *13*, 2698. [CrossRef]
63. Brahim, A.M.A.; Fahmy, M.F.M.; Wu, Z. 3D finite element modeling of bond-controlled behavior of steel and basalt FRP-reinforced concrete square bridge columns under lateral loading. *Compos. Struct.* **2016**, *143*, 33–52. [CrossRef]
64. Mahini, S.S.; Ronagh, H.R. Numerical modelling of FRP strengthened RC beam-column joints. *Struct. Eng. Mech.* **2009**, *32*, 649–665. [CrossRef]
65. Dabiri, H.; Kheyroddin, A.; Kaviani, A. A Numerical study on the seismic response of RC wide column-beam joints. *Int. J. Civil Eng.* **2019**, *17*, 377–395. [CrossRef]
66. Henriques, J.; Simões da Silva, L.; Valente, I.B. Numerical modeling of composite beam to reinforced concrete wall joints: Part I: Calibration of joint components. *Eng. Struct.* **2013**, *52*, 747–761. [CrossRef]

applied sciences

MDPI

Article

Evaluation of Structural Performance of Post-Installed Anchors Embedded in Cracked Concrete in Power Plant Facilities

Sangmoon Lee and Wooyoung Jung *

Department of Civil Engineering, Gangneung-Wonju National University, Gangneung 210-702, Korea; idealmoon@gwnu.ac.kr
* Correspondence: woojung@gwnu.ac.kr

Abstract: In this study, a field survey was conducted on the fixed anchorages of the operation and power generation facilities installed in domestic power plants. A static/dynamic performance evaluation was conducted to present safety evaluation guidelines that meet the domestic seismic performance requirements. Seismic performance tests were performed on the post-installed set anchors M10 and M12, which are mainly used for anchorages in accordance with the US and European seismic performance standards. The dynamic shear test results showed that the M12 anchor met the seismic performance verification criterion, whereas the M10 anchor did not because its dynamic performance was reduced, owing to the cyclic loading. In the results of the dynamic pull-out test, M12 also met the seismic performance verification criterion, whereas M10 was safe only in a non-cracked state. In summary, the seismic performance of M12 in both cracks and non-cracks was satisfied, but, in the case of M10, the results were not satisfied in cracks. This was an experimental study; it will be necessary to conduct additional analytical research in the future to verify the reliability and parameters of the experiment.

Keywords: post-installed anchor; structural performance evaluation; dynamic load protocol; shear; pull-out; cracked concrete; seismic

Citation: Lee, S.; Jung, W. Evaluation of Structural Performance of Post-Installed Anchors Embedded in Cracked Concrete in Power Plant Facilities. *Appl. Sci.* **2021**, *11*, 3488. https://doi.org/10.3390/app11083488

Academic Editor: Maria Favvata

Received: 15 March 2021
Accepted: 6 April 2021
Published: 13 April 2021

Publisher's Note: MDPI stays neutral with regard to jurisdictional claims in published maps and institutional affiliations.

Copyright: © 2021 by the authors. Licensee MDPI, Basel, Switzerland. This article is an open access article distributed under the terms and conditions of the Creative Commons Attribution (CC BY) license (https://creativecommons.org/licenses/by/4.0/).

1. Introduction

1.1. Background

Most operation and power generation facilities that are used in power plants have important functional purposes and are independently constructed on concrete slabs applying fixation systems. Post-installed anchors are mainly used as a general fixation system and are widely used for attaching or fixing structures, owing to their flexibility and ease of construction [1]. However, such anchors may affect the safety of the entire structure when exposed to direct events, such as an earthquake, or indirect events, such as concrete cracking due to aging [2].

According to a Federal Emergency Management Agency (FEMA) report, the frequency of occurrence of earthquakes with a magnitude of 5.0 or higher has been continuously increasing over the last 10 years, and there has been an increasing amount of damage to non-structural elements, such as operation and power generation facilities in power plants [3]. Most of this damage has involved an overturning or sliding of the facility from damaged anchors embedded in cracked concrete. The occurrence of such damage may lead to secondary damage, causing more serious social chaos from the functional loss and malfunction of the facilities [4].

Despite the increasing importance of anchors used in fixation units, such as in operation and power generation facilities, there are no seismic evaluation or verification criteria for such units in South Korea. In this study, static and dynamic experiments on the anchorages of non-structural elements were conducted using the seismic performance evaluation criteria of anchorages in the United States and Europe.

1.2. Fixation Unit Anchor System

1.2.1. Types of Post-Installed Anchor Bolts

Anchor bolts are installed in concrete by drilling holes using hammer or core drills after the concrete has been poured and hardened. They can be divided into mechanical post-installed anchor bolts, which produce the main bearing capacity by frictional force and mechanical locking devices, and adhesive post-installed anchor bolts, which generate the main bearing capacity by the adhesive force of the attached compounds [5]. Figure 1 shows post-installation anchors commonly used for field installation in Korea, and, in this study, set anchors were targeted.

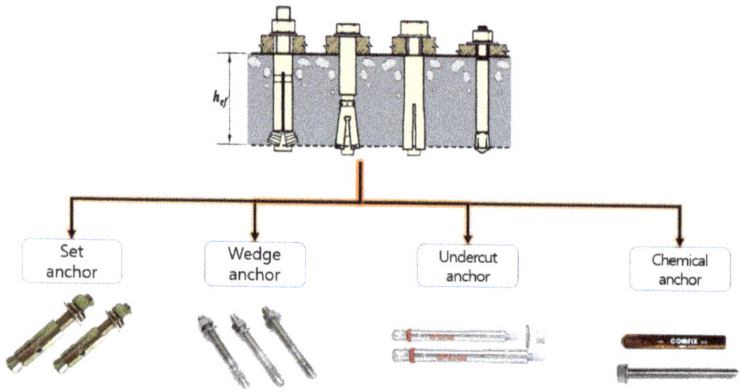

Figure 1. Typical post-installed anchors.

1.2.2. Design Standards and Failure Modes for Concrete-Embedded Anchor Bolts

In this study, the design load was calculated by referring to the domestic "Concrete Structure Standard (2012) Appendix II Concrete Anchors" design standards, and the level of safety was examined through a comparison with the results of static tests (pull-out and shear tests) [6]. The basic concepts of the anchor bolts subjected to tensile and shear loads are shown in Equations (1) and (2) below. Regarding the nominal tensile strength and shear strength, the smallest values among the possible anchor failure modes were used, as presented in Table 1. Anchor bolts were installed.

$$\varnothing N_n \geq N_{ua} \quad (1)$$

where N_{ua} is the required tensile strength and N_n is the nominal tensile strength.

$$\varnothing V_n \geq V_{ua} \quad (2)$$

where V_{ua} is the required shear strength and V_n is the nominal shear strength.

Table 1. Failure mode of anchor bolt subjected to strength.

Division	Tensile Strength	Shear Strength
Steel Failure	N_{sa}	V_{sa}
Pullout	N_{pn}	
Side-Face Blowout	N_{sb}	
Concrete Breakout	N_{cb}	V_{cb}
Concrete Splitting		
Concrete Pryout		V_{cp}

2. Overseas Seismic Performance Evaluation Criteria

Concrete with embedded anchors generally exhibits cracks when its durability is significantly reduced in the event of an earthquake. Additionally, a stress distribution occurs along the anchor installation locations, as shown in Figure 2, and the bearing capacity of the anchors is rapidly weakened [7]. In the evaluation of the seismic performance of concrete anchors, these behavioral characteristics are considered, and the effect of the cracks in the fixation unit of the concrete on the adhesive force of the anchors is evaluated. The average maximum loads are calculated by evaluating the static performance of two concrete types (non-cracked and cracked concrete), and a dynamic load protocol is presented based on such calculations. Finally, the seismic safety of the concrete-embedded anchor bolts is evaluated by assessing the performance under these dynamic loads [8].

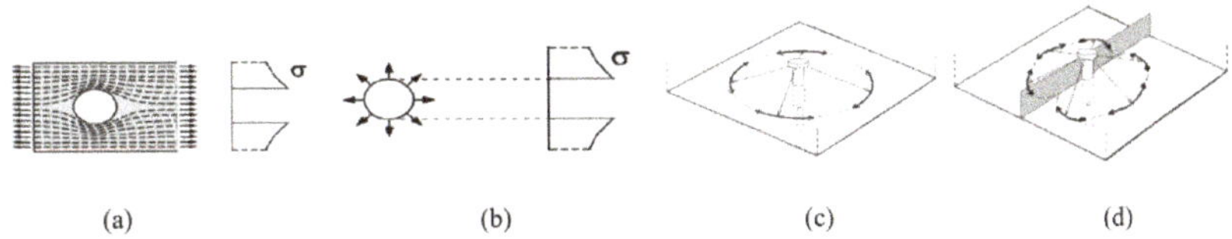

Figure 2. Condition of concrete under cracking: (**a**) stress concentration around the anchor hole; (**b**) stress in concrete induced by the anchor expansion or by the load acting on the anchor; (**c**) stress distribution in non-cracked concrete; (**d**) stress distribution in cracked concrete.

In the United States, the seismic performance of post-installed anchor bolts embedded in cracked concrete is evaluated in accordance with the American Concrete Institute (ACI) 355.2 standards [9] because they may cause concrete cracks or be pulled out in the event of an earthquake [10]. According to the ACI 355.2 standards, a simulated seismic load test is conducted by simulating an earthquake using cyclic loading, and it is assumed that post-installed anchor bolts are embedded in concrete with a 0.5-mm-wide crack. The presented cyclic loading is shown in Figure 3. For the cyclic loading pattern, a 50% load with average strength, determined through a static performance test, is applied 10 times; a 37.5% load is applied 30 times, and a 25% load is applied 100 times. Therefore, cyclic loading is applied 140 times until failure. In this instance, the loading rate must be between 0.1 and 2 Hz. The seismic performance is considered verified if the dynamic performance through cycling loading meets 80% of the static performance.

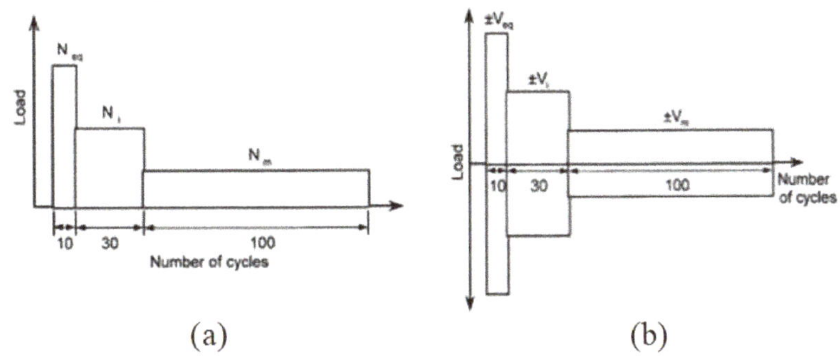

Figure 3. Number of cycles in loading pattern according to American Concrete Institute (ACI) 355.2 and TR 049: (**a**) pull-out force; (**b**) shear force. (Unit: mm)

TR 049, which are the design standards published by the European Organization for Technical Approval (EOTA), include the design standards for mechanical post-installed anchors [11]. For the C1 classification for non-structural elements, it is assumed that they are embedded in concrete with a 0.5-mm-wide crack. In addition, a 50% load with average strength, as determined through a static performance test, is applied 10 times; a 37.5% load is applied 30 times, and a 25% load is applied 100 times, in the same way as with the ACI 355.2 standards. The seismic performance is also considered to be verified if 80% of the static performance is met during the static loading test.

3. Experiment Overview

3.1. Selection of Test Targets through a Field Survey

In this study, a field survey was conducted to evaluate the seismic performance of the post-installed anchor bolts embedded in an actual site. The Daecheong hydroelectric power plant located in Daejeon Metropolitan City was selected as the test bed. As shown in Figure 4, most of the fixation unit anchors constructed at the site were found to be cast-in-place anchors. In the case of anchor damage or equipment without anchors, post-installed anchors were applied through a seismic reinforcement added in the year 2000.

(a) (b) (c)

Figure 4. Field survey for an anchorage in power plant: (**a**) Gas Circuit Breaker (GCB) cabinet; (**b**) governor cabinet; (**c**) main control cabinet.

For the post-installed anchors applied on-site, M10 and M12 set anchors were used. These anchors were selected as the test targets in this study. Figure 5 shows the M10 and M12 set anchors selected as test targets, and Table 2 shows their specifications.

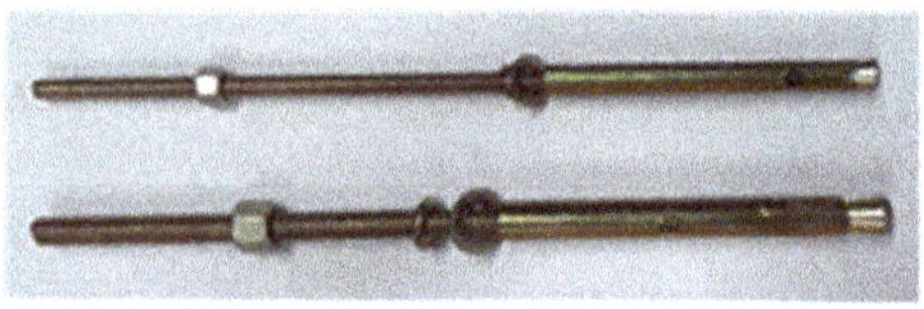

Figure 5. Post-installed anchors applied to experiments.

Table 2. Post-installed anchor specifications.

Division	Diameter (mm)	Length (mm)	Embedded Depth (mm)
Anchor	Φ10	250	100
	Φ12	250	100

3.2. Specimen Design and Fabrication

At the time of the field survey, concrete in the fixation units of power generation facilities in the test bed exhibited considerable concrete degradation from aging as a few decades had passed since the pouring of the concrete. This means that the operation facilities are highly likely to overturn and slide in the event of an earthquake, owing to a reduction in the anchor–concrete adhesive force caused by cracks in the concrete with a fixation unit. To evaluate the level of safety against this problem, in this study, specimens were fabricated, and tests were conducted by dividing the concrete with a fixation unit into two states (non-cracked and cracked concrete).

3.2.1. Non-Cracked Concrete

For the fabrication of the specimens, the design strength of 21 MPa measured in the fixation units of the power generation facilities in the test bed using a Schmidt hammer, a non-destructive test tool, was also considered as the design strength of the concrete. Specimens were fabricated with a size of 500 mm (width) × 500 mm (depth) × 200 mm (height) considering the fracture radius of an anchor bolt in accordance with ACI 318, as shown in Figure 6. Five specimens were fabricated for each of the pull-out and shear tests for each set anchor diameter. Thus, a total of 20 specimens were fabricated. The set anchors used were M10 and M12, and a 250 mm length and a 100 mm embedded depth were selected as their specifications, considering the fixation unit anchor bolts installed at the actual site and the experimental equipment set.

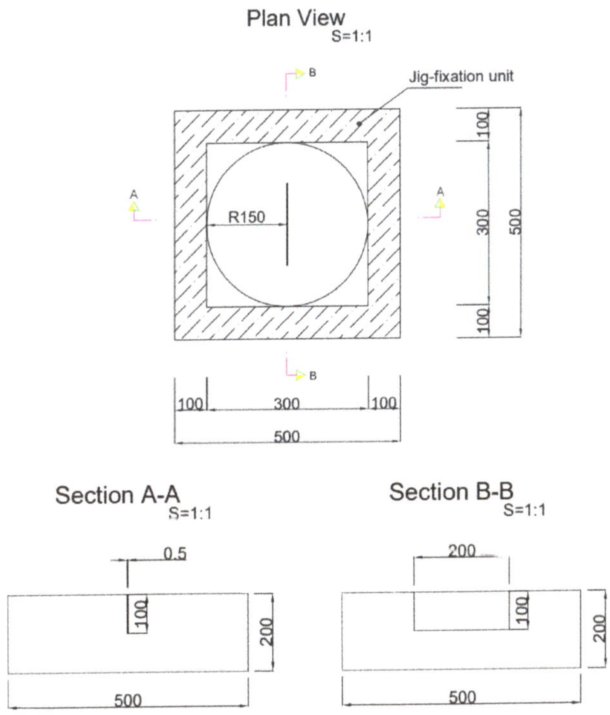

Figure 6. Schematics of concrete specimen (Unit: mm).

3.2.2. Cracked Concrete

Cracked concrete specimens were fabricated in the same way as non-cracked concrete specimens with a specific size. An artificial crack was simulated in the concrete by referring to the seismic performance evaluation criteria of the United States and Europe. In the case of artificial cracks [12], the experiments were conducted by fabricating the test specimen by inserting a 0.5 mm panel, not by using a general method, in order to conduct the experiment under the same conditions. A 0.5-mm-wide artificial crack was generated in the specimens by embedding a stainless-steel plate, at a depth of up to 100 mm, during concrete pouring and then removing it after curing. Five specimens were fabricated for each of the pull-out and shear tests for each set anchor type. Thus, a total of 20 specimens were fabricated. Figure 7 shows the fabrication process for the cracked concrete specimens.

Figure 7. Process of making concrete specimens.

4. Performance Evaluation of Post-Installed Anchors

A static performance evaluation was conducted by installing M10 and M12 set anchors, among various types of post-installed anchors, into two types of concrete specimens (non-cracked and cracked concrete). The main purpose of the experiment was the development of a dynamic load protocol for seismic performance evaluation. The load protocol applied to the dynamic tests was proposed by investigating the structural behavior under a pull-out and shear load and using the average maximum load from the results.

4.1. Static Loading Test

4.1.1. Experimental Method

The static performance of the post-installed anchors was evaluated by conducting shear and pull-out tests using a dynamic 1000 kN universal testing machine (UTM) in a domestic certification test research center in accordance with American Society for Testing and Materials (ASTM) E 488 [13]. For the applied loading condition, a continuous loading method (T-01), in which the load is slowly applied until failure, was used to obtain the load–displacement data for the full behavior until the failure of the anchor bolt. The loading rate was set to 0.4 mm/min, allowing the tests to be slowly conducted while observing the concrete–anchor bond behavior.

In the case of the shear tests, each concrete specimen with anchor bolts was erected vertically and fixed at the UTM, as shown in Figure 8a, and the equipment set for the shear tests was then fastened with the anchor. In the case of the pull-out tests, by contrast, each of the concrete specimens with the anchor bolts was fixed to the reaction floor of the UTM, as shown in Figure 8b, and the hydraulic tension grip of the UTM load cell was then fastened with the anchor. The tests were stopped when damage occurred or when the load resistance was lost from such damage. From the five tests conducted, three test results with an error rate of less than 20% were selected.

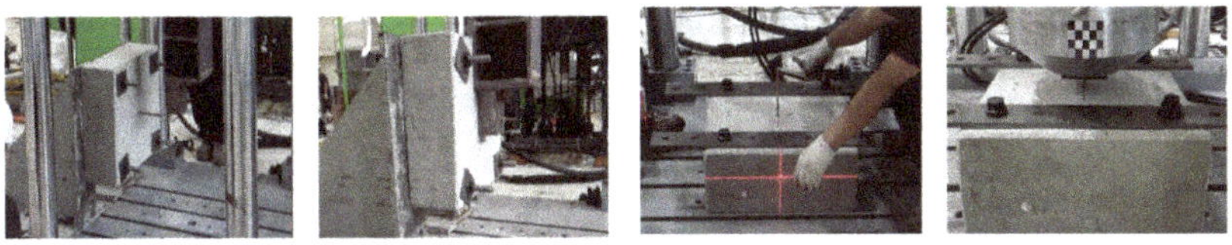

(**a**) (**b**)

Figure 8. Test setup for performance evaluation of static loading: (**a**) shear tests; (**b**) pull-out tests.

Based on the selected test results, the average maximum loads were calculated, and the pull-out performance was evaluated. The result was used to propose a load protocol for use in dynamic performance evaluation. Table 3 provides a list of the specimen names.

Table 3. Test specimens and parameters for static loading test.

Specimen Name	Test Method	Remark (Specimen Test Classification)
SSN1001–03	T-01	Test No. 01 to 03 (static_shear_non-cracked_10Ø)
SSC1001–03		Test No. 01 to 03 (static_shear_cracked_10Ø)
SSN1201–03		Test No. 01 to 03 (static_shear_non-cracked_12Ø)
SSC1201–03		Test No. 01 to 03 (static_shear_cracked_12Ø)
SPN1001–03		Test No. 01 to 03 (static_pull-out_non-cracked_10Ø)
SPC1001–03		Test No. 01 to 03 (static_pull-out_cracked_10Ø)
SPN1201–03		Test No. 01 to 03 (static_pull-out_non-cracked_12Ø)
SPC1201–03		Test No. 01 to 03 (static_pull-out_cracked_12Ø)

4.1.2. Shear Performance Evaluation

In the shear test results of the M10 and M12 set anchors, both anchors exhibited a steel failure instead of a concrete–anchor bond failure, as shown in Figure 9. It appears that the steel failure occurred before the load affected the concrete, owing to the sufficient bearing capacity of the concrete for the anchor installed at the center of the specimen and because of the distance of the concrete–anchor generated from the way in which the anchor was set. The differences in performance depending on the presence of a concrete crack were found to be insignificant, as shown in Figure 10. Table 4 summarizes the average maximum load for each test.

Table 4. Respective results of the shear tests.

Specimen Test Classification			Max. Average Load
Shear tests	Φ10	Non-cracked concrete	16.08 kN
		Cracked concrete	15.85 kN
	Φ12	Non-cracked concrete	25.44 kN
		Cracked concrete	24.52 kN

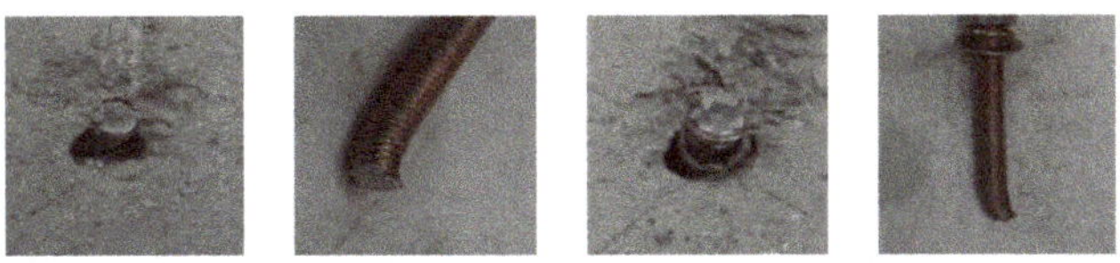

(a) (b)

Figure 9. Failure mode for static shear tests: (**a**) M10 set anchor; (**b**) M12 set anchor.

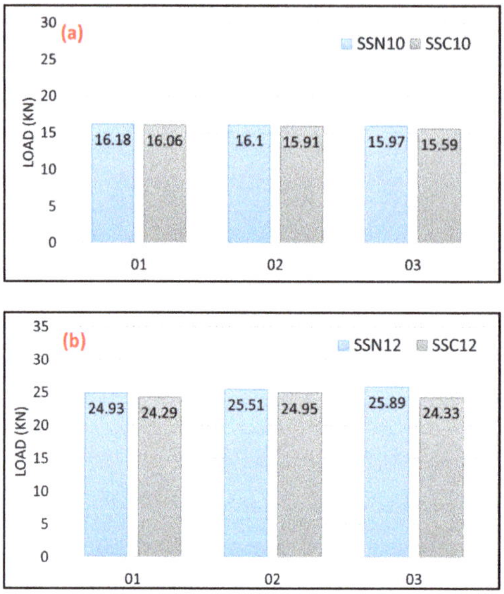

Figure 10. Comparison of results of shear tests: (**a**) M10 set anchor; (**b**) M12 set anchor.

4.1.3. Pull-Out Performance Evaluation

For the pull-out tests on the M10 set anchor, a steel anchor failure finally occurred in both the cracked and non-cracked concrete, as shown in Figure 11a. The maximum pull-out load in the cracked concrete specimens was smaller than that in the non-cracked concrete specimens, owing to a reduction in adhesive force between the concrete and anchor. For the M12 set anchor, a concrete cone failure by the adhesive force occurred only under the slightly increased anchor diameter, as shown in Figure 11b. As with M10, the maximum pull-out load in the cracked concrete specimens was smaller than that in the non-cracked concrete specimens, owing to a reduction in the adhesive force. Figure 12 shows the differences in performance depending on the presence of a concrete crack, and Table 5 shows the average maximum load for each test.

Table 5. Respective esults of the pull-out tests.

Specimen Test Classification			Max. Average Load
Pull-out tests	Φ10	Non-cracked concrete	26.03 kN
		Cracked concrete	23.46 kN
	Φ12	Non-cracked concrete	31.89 kN
		Cracked concrete	25.93 kN

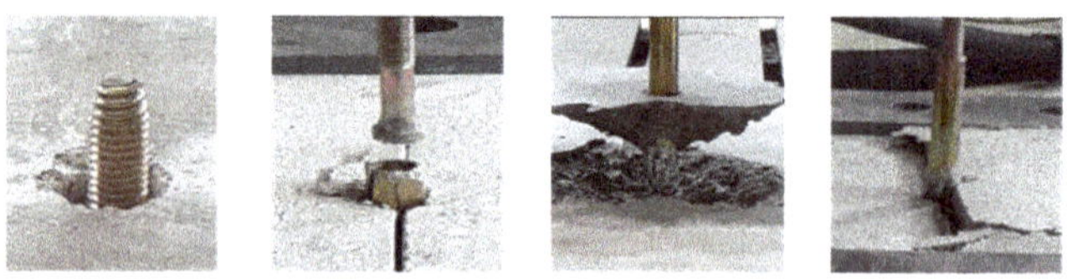

(a) (b)

Figure 11. Failure mode for static pull-out tests: (**a**) M10 set anchor; (**b**) M12 set anchor.

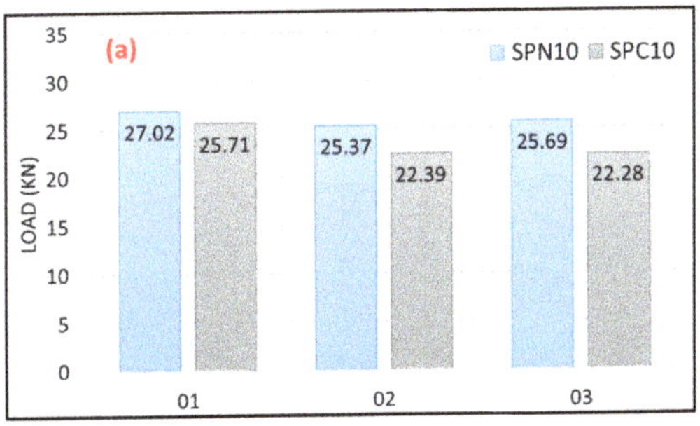

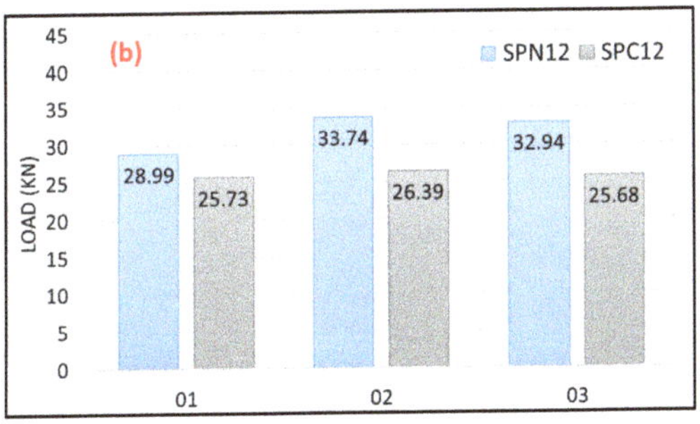

Figure 12. Comparison of results of pull-out tests: (**a**) M10 set anchor; (**b**) M12 set anchor.

4.2. Dynamic Loading Test

4.2.1. Presentation of Dynamic Load Protocol

In this study, a load protocol for a seismic safety evaluation under a dynamic load was presented by referring to the ACI 355.2 standards of the United States for verification of the seismic performance of power generation facilities in power plants and the TR 049 Category C1 standards of Europe, which consider non-structural elements. Based on the average maximum loads derived from the static performance evaluation, a dynamic load protocol was presented, as shown in Table 6. The protocol was used to evaluate the dynamic performance of the M10 and M12 set anchors. Figure 13 shows the dynamic load protocol of the M10 set anchor, and Figure 14 shows that of the M12 set anchor. The seismic performance of the concrete anchors installed in power generation facilities was considered verified if the maximum pull-out and shear load values of the dynamic tests derived under the presented cyclic loading met 80% of the performance presented in the static tests.

Table 6. Load protocol of post-installed anchors.

Anchor	Test	Applied Load (kN) 50% (10 Times)	37.5% (30 Times)	25% (100 Times)	Static Max. Average Load (kN) (Refer to Tables 4 and 5)
Φ10	Shear	8	6	4	16.08
	Pull-out	13	9.5	6.4	26.03
Φ12	Shear	13	9.8	6.5	25.44
	Pull-out	16	12	8	31.89

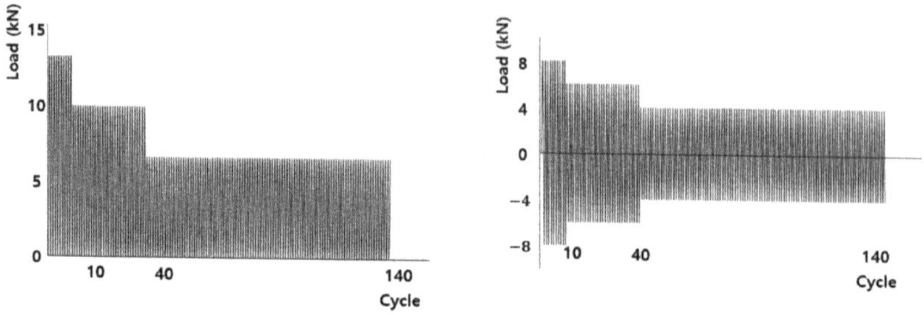

Figure 13. Load protocol shape of M10 set anchor: pull-out and shear dynamic force.

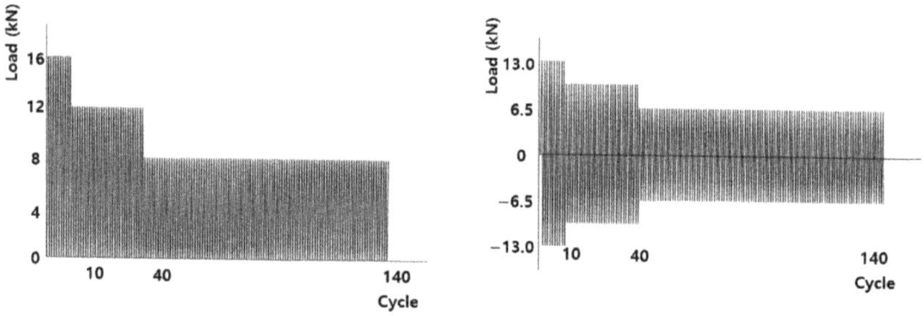

Figure 14. Load protocol shape of M12 set anchor: pull-out and shear dynamic force.

4.2.2. Experimental Methods

During the dynamic tests, there was clearance between the concrete and each anchor owing to the sleeve of the set anchors. As such, clearance may cause errors from a slip of the equipment set and the vibration of the anchor under the actual cyclic loading, and an equipment set that can deliver cyclic loading more accurately was fabricated. In addition, during the tests, the rate of the cyclic loading was reduced from 1.0 to 0.2 Hz, as determined through trial and error, to deliver the load to the actual anchor more accurately. Experimental methods other than the loading rate and equipment set were conducted in the same way as the static performance tests. The tests were stopped when the load resistance was lost from damage to the anchor or concrete. From the five tests conducted, the results of three tests with an error rate of less than 20% were selected. Table 7 provides a specimen list with the specimen names.

Table 7. Test specimens and parameters for dynamic loading test.

Specimen Name	Test Method	Remark (Specimen Test Classification)
DSN1001–03	T-01	Test No. 01 to 03 (dynamic_shear_non-cracked_10Ø)
DSC1001–03		Test No. 01 to 03 (dynamic_shear_cracked_10Ø)
DSN1201–03		Test No. 01 to 03 (dynamic_shear_non-cracked_12Ø)
DSC1201–03		Test No. 01 to 03 (dynamic_shear_cracked_12Ø)
DPN1001–03		Test No. 01 to 03 (dynamic_pull-out_non-cracked_10Ø)
DPC1001–03		Test No. 01 to 03 (dynamic_pull-out_cracked_10Ø)
DPN1201–03		Test No. 01 to 03 (dynamic_pull-out_non-cracked_12Ø)
DPC1201–03		Test No. 01 to 03 (dynamic_pull-out_cracked_12Ø)

4.2.3. Shear Performance Evaluation

In the dynamic shear test results of the M10 and M12 set anchors, both anchors exhibited steel failure. In the case of the M10 anchors, an early anchor failure occurred owing to the cyclic loading in the tests compared to the M12 anchors. This appears to be due to the difference in diameter. The differences in dynamic shear performance depending on the presence of a concrete crack were found to be insignificant, as in the static shear tests.

Table 8 summarizes whether the shear maximum loads generated during the tests met 80% of the static shear performance (seismic performance verification criterion).

Table 8. Dynamic shear test results of M10 and M12 maximum loads.

Anchor	Concrete	Test No.01 (kN)	Test No.02 (kN)	Test No.03 (kN)	Criteria 80% (Refer to Table 6)
Φ10	Non-crack	8.39	8.64	8.10	12.86 kN
	Crack	8.31	8.17	8.36	
Φ12	Non-crack	25.43	25.80	25.86	20.82 kN
	Crack	25.32	24.72	25.01	

- DSN10 and DSC10

Figure 15 shows the load–displacement curves, and Figure 16 shows an average summary of the values for each case-specific dynamic shear test. As the figure shows, the dynamic shear performance of the M10 anchors does not meet the seismic performance verification criteria (horizontal lines) for both cracked and non-cracked concrete. Based on this, the anchors of M10 are not considered suitable for anchors of fixed parts.

- DSN12 and DSC12

Figure 17 shows the load–displacement curves, and Figure 18 shows an average summary of the values for each case-specific dynamic shear test. As the figure shows, in the case of M12 anchors, unlike M10 anchors, both cracked and non-cracked concrete met the seismic performance verification criteria.

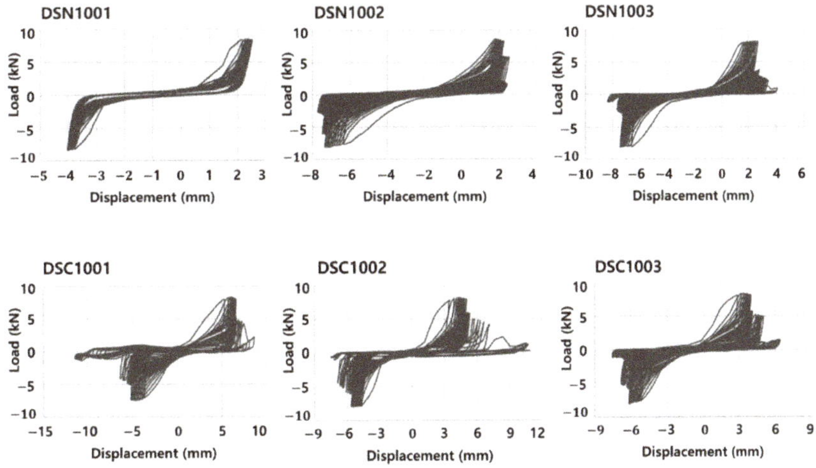

Figure 15. Load–displacement curve of dynamic shear tests for M10 anchors.

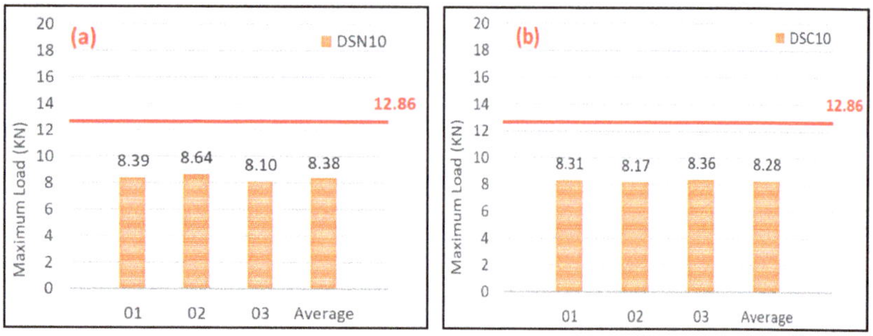

Figure 16. Comparison results of the dynamic shear tests for M10 anchors: (**a**) non-cracked concrete; (**b**) cracked concrete.

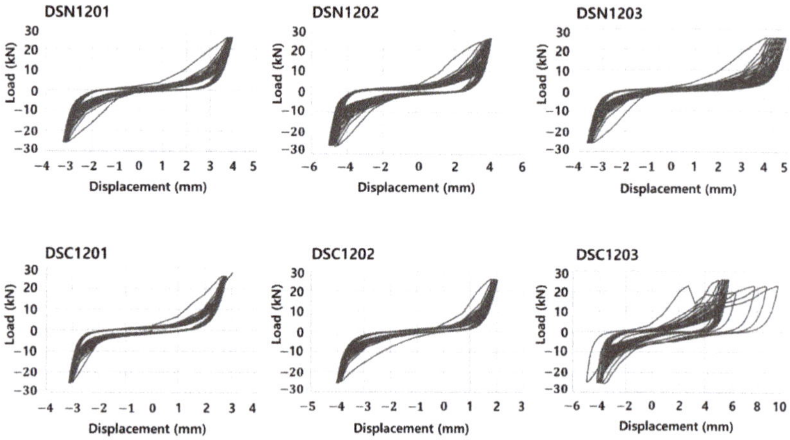

Figure 17. Load–displacement curve of dynamic shear tests for M12 anchors.

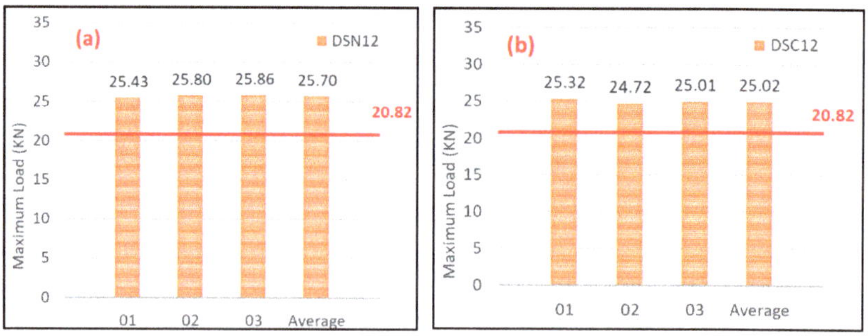

Figure 18. Comparison results of the dynamic shear tests for M12 anchors: (**a**) non-cracked concrete; (**b**) cracked concrete.

4.2.4. Pull-Out Performance Evaluation

In the case of the M10 set anchor, a pull-out failure of the steel anchor occurred. The pull-out loads in the cracked concrete specimens were smaller than those in the non-cracked concrete specimens. In the case of the M12 set anchor, a concrete cone failure occurred as in the static pull-out tests. As with M10, the pull-out loads in the cracked concrete specimens were smaller than those in the non-cracked concrete specimens. It was found that the M10 set anchor exhibited a failure of the anchor itself before the concrete–anchor bond failure, owing to the relatively smaller diameter. When the maximum loads of the dynamic pull-out tests were compared with those of the static tests, the maximum pull-out loads under dynamic loads were found to be higher. This appears to be because the clearance between the concrete and the set anchor, which existed at the time of the set anchor installation, contributed to the concrete–anchor frictional force that was generated by vibration under the application of cyclic loading.

Table 9 summarizes whether the pull-out maximum loads generated during the tests met 80% of the static pull-out performance (seismic performance verification criterion).

Table 9. Dynamic pull-out test results of M10 and M12 maximum loads.

Anchor	Concrete	Test No.01 (kN)	Test No.02 (kN)	Test No.03 (kN)	Criteria 80% (Refer to Table 6)
Φ10	Non-crack	20.52	20.30	22.88	20.35 kN
	Crack	14.43	17.90	15.22	
Φ12	Non-crack	42.47	41.25	40.39	25.51 kN
	Crack	39.15	38.95	40.70	

- DPN10 and DPC10

Figure 19 shows the load–displacement curves, and Figure 20 shows an average summary of the values for each case-specific dynamic pull-out test. As the figure shows, the dynamic pull-out performance of the M10 anchors does not met the seismic performance verification criteria (horizontal lines) for cracked concrete. Based on this, the set anchors of M10 are not considered suitable for anchors of fixed parts.

- DPN12 and DPC12

Figure 21 shows the load–displacement curves, and Figure 22 shows an average summary of the values for each case-specific dynamic pull-out test. As shown in the figure, both cracked and non-cracked concrete met the seismic performance verification criteria for the M12 anchor, as shown in the M10 anchor.

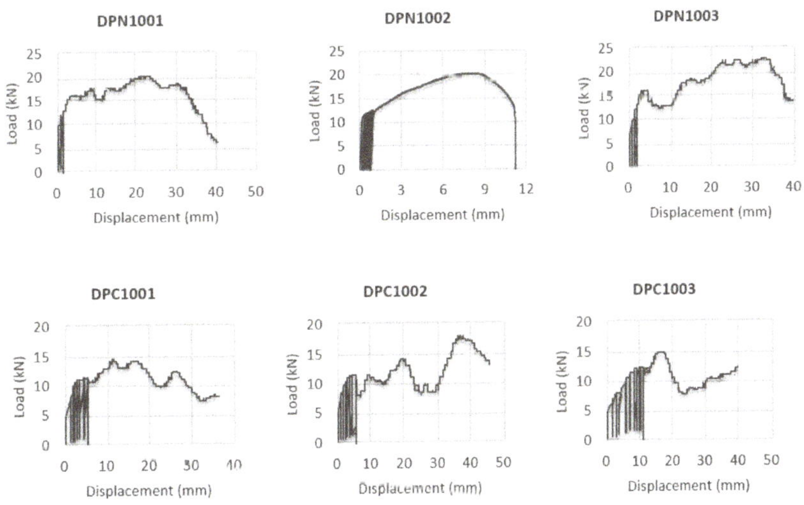

Figure 19. Load–displacement curve of dynamic pull-out tests for M10 anchors.

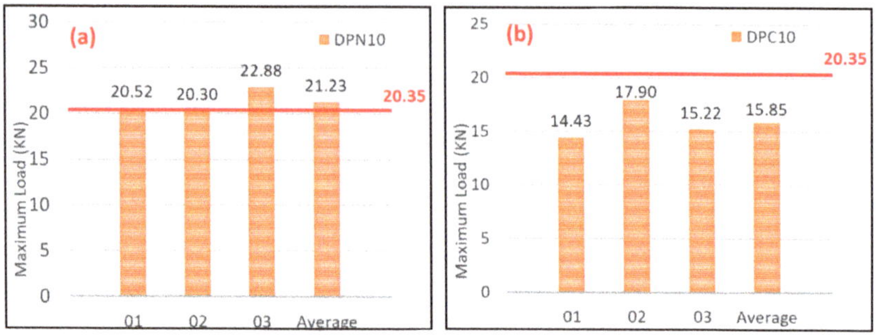

Figure 20. Comparison results of the dynamic pull-out tests for M10 anchors: (**a**) non-cracked concrete; (**b**) cracked concrete.

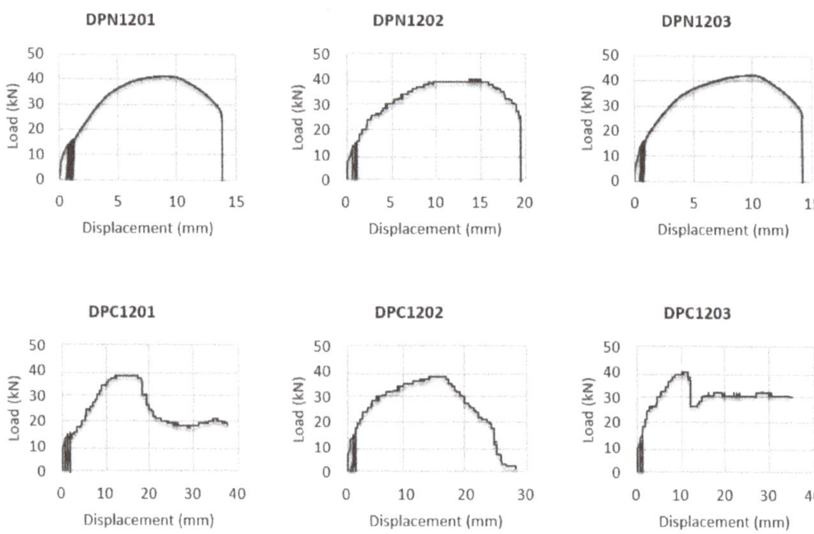

Figure 21. Load–displacement curve of dynamic pull-out tests for M12 anchors.

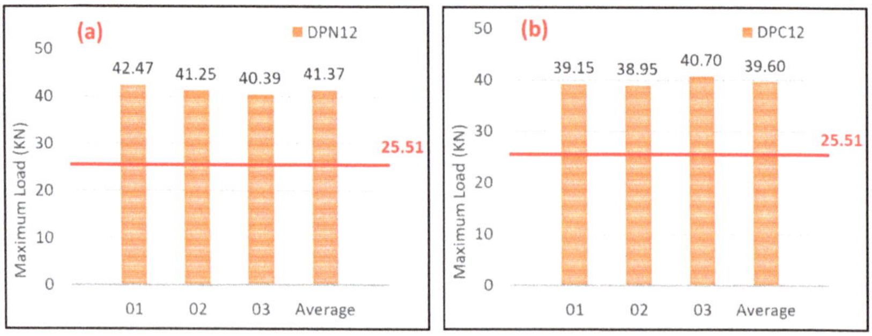

Figure 22. Comparison results of the dynamic pull-out tests for M12 anchors: (**a**) non-cracked concrete; (**b**) cracked concrete.

5. Conclusions

In this research, a field study was conducted on the fixation unit of post-installed anchors used in operation and power generation facilities installed in domestic power plants. Based on the results, a static/dynamic performance evaluation was conducted to determine the safety evaluation guidelines that meet the domestic seismic performance requirements.

- In this study, the average maximum loads of post-installed anchors were calculated by evaluating the static performance of two concrete types (non-cracked and cracked concrete), based on which a dynamic load protocol was presented. A dynamic performance evaluation was conducted using this protocol to evaluate the seismic safety of concrete-embedded anchor bolts.
- The static shear test results of the M10 and M12 set anchors showed that steel failure occurred in both the cracked and non-cracked concrete. It appears that the anchor failure occurred before the load affected the concrete, owing to the sufficient bearing capacity of the concrete for the anchor bolt installed at the center of the concrete specimen and because of the concrete–anchor distance generated during the anchor installation.
- In the static pull-out test results, the M10 set anchor exhibited an anchor steel failure in both the cracked and non-cracked concrete. The M12 set anchor showed a concrete cone failure by the concrete–anchor adhesive force, owing to its slightly increased diameter. As with the M10 set anchor, the maximum pull-out load in the cracked concrete specimens was smaller than that in the non-cracked concrete specimens, owing to a reduction in the concrete–anchor adhesive force.
- In the dynamic shear test results of the M10 and M12 set anchors, both anchors exhibited steel failure. The M12 set anchor met the seismic performance verification criterion (80% of the static performance), but the M10 set anchor could not because its dynamic performance was reduced, owing to the cyclic loading.
- The results of the dynamic pull-out test showed that the M10 anchor met the seismic performance verification criterion in a non-cracked state, i.e., at the time of installation, but could not meet the criterion when cracks occurred, owing to aging and other loads. In the case of the M12 set anchor, the maximum pull-out loads under dynamic loads were found to be higher than those of the static tests. This appears to be because the clearance between the concrete and the set anchor, which existed at the time of the set anchor installation, contributed to the concrete–anchor frictional force generated by vibrations under the application of cyclic loading.

Because this was an experimental study, it will be necessary in the future to additionally conduct analytical research to verify the reliability and applied parameters of the experiment.

Author Contributions: Conceptualization, S.L. and W.J.; methodology, S.L.; software, S.L.; validation, S.L. and W.J.; formal analysis, S.L.; investigation, S.L.; resources, S.L.; data curation, S.L.; writing—original draft preparation, S.L.; writing—review and editing, S.L., and W.J. All authors have read and agreed to the published version of the manuscript.

Funding: This research was funded by the power plant research program (21IFIP-B128598-05) funded by the Ministry of Land, Infrastructure, and Transport.

Institutional Review Board Statement: Not applicable.

Informed Consent Statement: Not applicable.

Data Availability Statement: Not applicable.

Conflicts of Interest: The authors declare no conflict of interest.

References

1. Jung, W.; Kwon, M.; Kim, J.; Ju, B. Performance Evaluation of the Post-Installed Anchor for Sign Structure. *Int. J. Civ. Environ. Eng.* **2012**, *6*, 1097–1102.
2. Lee, S.M.; Jung, W.Y. Evaluation of anchorage performance of the switchboard cabinet under seismic loading condition. *Adv. Mech. Eng.* **2020**, *12*, 1687814020926309. [CrossRef]
3. *Reducing the Risks of Nonstructural Earthquake Damage—A Practical Guide*; FEMA E-74; FEMA: Washington, DC, USA, 2012.
4. Oh, S.H.; Park, H.Y.; Choi, K.K. Seismic Damage Status and Characteristics of Non-structural Elements. *Korean Soc. Noise Vib. Eng.* **2018**, *28*, 71–77.
5. Park, Y.M.; Jeon, M.; Lee, K.J.; Kim, C.H. A Study on the Concrete Breakout Capacity Evaluation of Medium-to-Large size CIP Anchor Bolts under Tension Loading. *J. Korean Soc. Steel Constr.* **2011**, *23*, 493–501.
6. *Building Code Requirements for Structural Concrete (ACI 318-11) and Commentary*; ACI Committee 318; ACI: Farmington Hills, MI, USA, 2011; 503p.
7. Jang, J.B.; Suh, Y.P.; Lee, J.R. A Study on the Evaluation of Concrete Breakout Strength for Cast-In-Place Anchor with Crack. *J. Korean Soc. Civ. Eng.* **2004**, *24*, 647–652.
8. Park, Y.M.; Ju, H.J.; Kim, D.H.; Kang, M.K.; Lee, J.H. Static Shear Resistance of Cast-In-Place Anchors in Cracked Concrete. *J. Korean Soc. Steel Constr.* **2015**, *27*, 87–97. [CrossRef]
9. *Evaluating the Performance of Post-Installed Mechanical Anchors in Concrete, Reported by ACI Committee 355*; ACI 355.2-01; ACI: Farmington Hills, MI, USA, 2002.
10. Mahrenholtz, C.; Eligehausen, R.; Hutchinson, T.C.; Hoehler, M.S. Behavior of post-installed anchors tested by stepwise increasing cyclic crack protocols. *ACI Struct. J.* **2017**, *114*, 623. [CrossRef]
11. *Post-Installed Fasteners in Concrete under Seismic Action*; EOTA TR 049; EOTA: Brussels, Belgium, 2016.
12. Kim, S.-Y.; Yu, C.-S.; Yoon, Y.-S. Sleeve-type expansion anchor behavior in cracked and uncracked concrete. *Nucl. Eng. Des.* **2004**, *228*, 273–281. [CrossRef]
13. *Standard Test Methods for Strength of Anchors in Concrete and Masonry Elements*; ASTM E 488-96; ASTM: West Conshohocken, PA, USA, 2003.

applied sciences

MDPI

Article

Shaking Table Tests of Lead Inserted Small-Sized Laminated Rubber Bearing for Nuclear Component Seismic Isolation

Gyeong-Hoi Koo [1,*], Tae-Myung Shin [2] and Sang-Jin Ma [2]

1 Korea Atomic Energy Research Institute, Daejeon 305-353, Korea
2 School of Mechanical, Automotive and Aeronautical Engineering, Korea National University of Transportation, Chungju 27469, Korea; tmshin@ut.ac.kr (T.-M.S.); fjjang82@naver.com (S.-J.M.)
* Correspondence: ghkoo@kaeri.re.kr

Abstract: To assure seismic isolation performance against design and beyond design basis earthquakes in the nuclear facility components, the lead inserted small-sized laminated rubber bearings (LRB), which has a 10 kN vertical design load, have been designed and quasi-statically tested to validate their design mechanical properties in previous studies. Following this study, the seismic shaking tests of these full-scale LRBs are performed and discussed in this paper with the dummy mass system to investigate actual seismic isolation performance, dynamic characteristics of LRBs, consistency of the LRB's quality, and so on. To study the seismic isolation performance, three beam structures (S1–S3) with different natural frequencies were installed both on the shaking table and the dummy mass supported by four LRBs: (1) S1: structure close to seismic isolation frequency; (2) S2: structure close to peak input spectral frequency; (3) S3: structure in the high-frequency region. The test results are described in various seismic levels of OBE (Operating Basis Earthquake), SSE (Safe Shutdown Earthquake), and BDBE (Beyond Design Basis Earthquake), and are compared with the analysis results to assure the seismic isolation performance and the LRB's design parameters. From the results of the shaking table tests, it is confirmed that the lead inserted small-sized LRBs reveal an adequate seismic isolation performance and their dynamic characteristics as intended in the LRB design.

Keywords: seismic isolation; laminated rubber bearing; lead rubber bearing; shaking table test; nuclear facility component; beyond design basis earthquake; LRB design; seismic performance; seismic isolation frequency

Citation: Koo, G.-H.; Shin, T.-M.; Ma, S.-J. Shaking Table Tests of Lead Inserted Small-Sized Laminated Rubber Bearing for Nuclear Component Seismic Isolation. *Appl. Sci.* **2021**, *11*, 4431. https://doi.org/10.3390/app11104431

Academic Editor: Maria Favvata

Received: 19 April 2021
Accepted: 10 May 2021
Published: 13 May 2021

Publisher's Note: MDPI stays neutral with regard to jurisdictional claims in published maps and institutional affiliations.

Copyright: © 2021 by the authors. Licensee MDPI, Basel, Switzerland. This article is an open access article distributed under the terms and conditions of the Creative Commons Attribution (CC BY) license (https://creativecommons.org/licenses/by/4.0/).

1. Introduction

After Fukushima nuclear power plant accident, many efforts have been made to enhance the seismic capacity of nuclear power plants, especially for old nuclear power plants in operation, and to accommodate the beyond design basis earthquakes.

In general, the seismic capacity of the nuclear power plants has been handled at the plant level, but recently the individual seismic capacity of the safety-related facility components such as control cabinet, emergency diesel generator, remote shutdown console, battery pack, spent fuel rack, and so on has become important in determining the seismic capacity of nuclear power plants. Especially enhancing the seismic capacity of the operating nuclear power plants, which have experienced various aging environments for a long service lifetime, is a critical issue in points of safety against potential beyond design basis earthquakes.

There may be many ways to increase the seismic capacity of individual nuclear facility components [1–3], but applying the customized seismic isolators to individual nuclear components can be one of the effective approaches [4,5].

There are many types of seismic isolators such as friction pendulum system, rigid sliding bearing with damper, elastic sliding bearing with damper, laminated rubber bearing, etc. [5–9]. Among them, the laminated rubber bearings (LRB) have been popularly developed in many countries for nuclear applications because they can support very heavy-weight and are very flexible in shear deformation against horizontal earthquakes. Due to these characteristics, LRBs have been generally developed for seismic isolation of buildings [10–16]. There are many studies on shaking table tests of LRBs for a building structure seismic isolation, but most of them are performed with the reduced scale LRBs [17–20]. For this reason, there are very few studies of the shaking table tests for the full-scale LRBs.

In previous studies, the small-sized LRBs with 10 kN vertical design load and 2.3 Hz seismic isolation frequency have been developed for individual nuclear component seismic isolation instead of entire building isolation. The developed LRB uses natural rubber and a lead plug is inserted in the center for damping. The design mechanical properties of LRB have been identified by the quasi-static tests [21].

In this study, the shaking table tests of the full-scale LRBs developed in previous studies are carried out and their seismic isolation performance and dynamic characteristics are investigated through a well-established test matrix. The test conditions included in the test matrix are established to investigate (1) three levels of the seismic load, i.e., OBE (Operating Basis Earthquake), SSE (Safe Shutdown Earthquake), BDBE (Beyond Design Basis Earthquake), (2) damping function of the lead plugin continuous aftershock conditions, (3) consistent fabrication quality of small-sized LRBs, (4) extreme behavior of LRB, and (5) confirm the LRB design parameters. The used input motions in shaking table tests correspond to the typical floor response spectrum at 137 ft elevation of the nuclear power plant building structure, where the seismic isolation target components are expected to be located. To confirm the seismic isolation performance, three beam structures with different natural frequencies are installed both on the shaking table and the dummy mass supported by the LRBs, and their seismic responses of non-seismic isolation and seismic isolation cases are measured for comparison. Seismic analyses with two kinds of LRB models of the Bi-linear and the equivalent stiffness-damping are performed with the identified mechanical properties in previous studies and their results are compared with those of the tests to confirm the dynamic design characteristics.

2. Review of LRB Design Characteristics

In previous studies, the lead inserted small-sized LRBs have been designed with 100 mm outer diameter and a design vertical inertia load of 10 kN. In this design, the lead plug with a 21.5 mm diameter is inserted into the LRB center position to take the role of energy dissipation. Ultimately this is expected to reduce the maximum shear displacement of LRB during the seismic events.

The total rubber height is 20 mm with appropriate shape factors of $S_1 = 9.9$ and $S_2 = 5.0$, where S_1 represents the ratio of the area of the rubber plate where the surface pressure applies and the free surface area where the rubber plate; S_2 represents the ratio of the diameter and the total rubber height.

The design shear displacement of LRB is 35 mm corresponding to 175% shear strain of LRB based on the total rubber height of 20 mm. To accommodate the beyond design basis earthquake, the 70 mm corresponding to 350% shear strain of LRB is set to increase the seismic fragility capacity for the peak ground motions.

Table 1 presents the summary of the design parameters of the lead inserted small-sized LRB to be carried out by the shaking table tests to identify its seismic isolation performance characteristics.

Table 1. Summary of LRB design parameters [21].

Design Parameters	Design Value
Outer Diameter (D_o)	100 mm
Lead Plug Diameter (d)	21.5 mm
Total Rubber Height (t_R)	20 mm
Total LRB Height (t)	34 mm
Shape Factor (S_1)	9.9
Shape Factor (S_2)	5.0
Design Vertical Load (P_d)	10 kN
Design Shear Displacement (D_d)	35 mm (175%)
Beyond Design Shear Displacement (D_b)	80 mm (400%)

The LRB dynamic characteristics corresponding to the design parameters are presented in Table 2. As shown in the table, the horizontal seismic isolation frequency at the design shear displacement level is 2.3 Hz for a design and the identified frequency by the quasi-static tests is 2.25 Hz [21]. The critical damping ratio is 25.5% for a design and the identified value by the quasi-static tests is 24.1%. For equivalent bi-linear characteristics of LRB, the designed tangential stiffness and the characteristic strength are 123 kN/m and 3 kN, respectively, and the identified values by the quasi-static tests are 122 kN/m and 2.72 kN, respectively.

Table 2. LRB dynamic characteristics at design level [21].

Parameters	Design	Quasi-Static Tests
Horizontal Isolation Frequency (F_{iso})	2.3 Hz	2.25 Hz
Equivalent Horizontal Stiffness (K_{eff})	209 kN/m	199.3 kN/m
Critical Damping Ratio (ξ)	25.4%	24.1%
Tangential Stiffness (K_t)	123 kN/m	122 kN/m
Characteristic Strength (Q_d)	3 kN	2.72 kN

3. Seismic Shaking Table Tests

3.1. Performance of Shaking Table Test Facility

Table 3 presents the shaking table information used in this study. As shown in the table, the table size is 4 m × 4 m and the maximum loading capacity is 300 kN. The information related to the control software of the MTS 469D can be found in the MTS, 2021 [22].

Table 3. Shaking table performance.

Item	Performance
Max. Loading (kN)	300
Table Size (m)	4 × 4
Control Axes	6 DOF (Translation 3 axes, Rotational 3 Axes)
Max. Displacement (mm)	Hor. (X, Y) = ±300, Ver. (Z) = ±150
Max. Velocity (m/s)	Hor. (X, Y) = 1.5, Ver.(Z) = 1.0
Max. Acceleration (g)	Hor. (X, Y) = 3.0, Ver.(Z) = 5.0
Frequency Range (Hz)	0.1 ~ 60
Excitation Mechanism	Electro-hydraulic Servo, 3 Variable Control
Control Software	MTS 469D
Feedback Data Acquisition	63 Channels (Sampling Rate = 512 Hz)

3.2. Description of Seismically Isolated Test Mockup

As shown in Table 1, the designed LRB has a vertical load of 10 kN ($\simeq$ 1 ton) to perform the seismic isolation frequency of 2.3 Hz. To validate their seismic isolation characteristics during the actual seismic loading conditions, the total dummy mass of 4 tons is fabricated with a rectangular shape (1300 mm × 1300 mm × 320 mm) and four LRBs are installed between the shaking table and the bottom of the dummy mass at each corner with bolted joint connection, as shown in Figure 1.

Figure 1. Photos of test mockup with four LRBs and four tons dummy mass system.

To investigate the seismic isolation performance induced by LRBs, three beam structures (S1–S3) are installed both on the shaking table and on the dummy mass, which will represent the actual superstructures to be seismically isolated. The target natural frequencies for beam structures are 5 Hz for S1, 9 Hz for S2, and 20 Hz for S3. When considering the design seismic isolation frequency of 2.3 Hz, the S1 structure is set to investigate the seismic responses of the structure near the seismic isolation frequency. The S2 structure is aimed to investigate the seismic isolation performance for the strong input motion having a spectral peak around 10 Hz. The S3 structure is prepared to validate the seismic isolation responses of superstructures having high-frequency dynamic characteristics.

For the measurement of seismic responses, the accelerometers are installed on the shaking table for the input seismic motion and on the center of the dummy mass for the seismic isolation response. To obtain the seismic responses of the beam structures both on the shaking table and on the dummy mass, the accelerometers are attached at top of each beam structure. To measure the LRB shear displacement response, which will be the relative displacement between the shaking table and the dummy mass, the LVDT (Linear Variable Differential Transformer) is installed between the sensor supports (Figure 1) and the dummy mass.

3.3. Seismic Input Motions

The seismic input motion used for the shaking table test is obtained from the typical floor response spectrum at an elevation of 137 ft in nuclear power plant building structure, which is the location where the target facility components can be seismically isolated by using LRBs to increase the seismic capacity or to accommodate the beyond design basis earthquakes.

Figure 2 presents the design response spectrum (DRS) considered in this shaking table test. This spectrum is obtained from the system (nuclear building structures) seismic analysis using the seismic time history response analysis method with the design response spectrum conformed to the US NRC RG 1.60 [23]. As shown in the figure, the zero period acceleration value is 1.22 g, amplified about four times. The broadened spectral peak ranges from 9 Hz to 12 Hz. Figure 3a shows the generated artificial acceleration time history enveloped the DRS, which is the required shaking table input motion. Figure 3b is the actual shaking table motion measured at the shaking table.

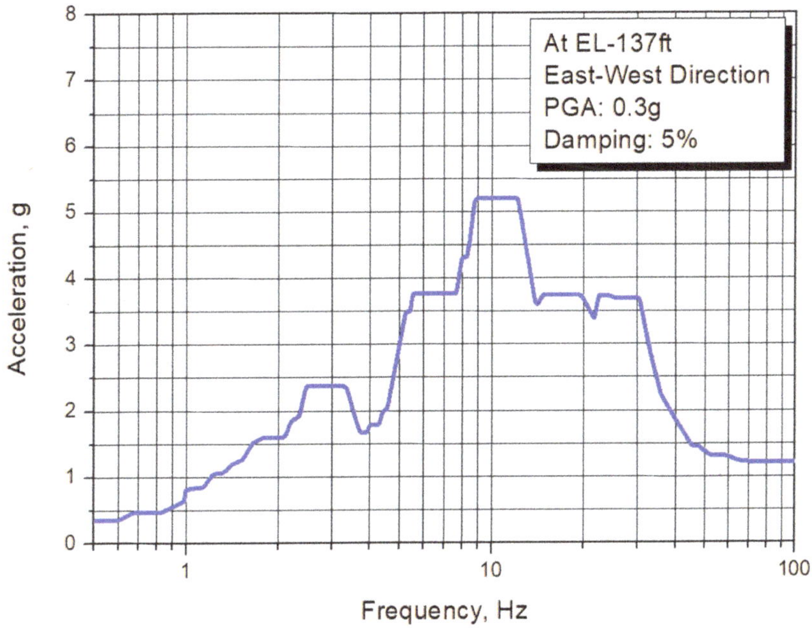

Figure 2. Design response spectrum used in tests (PGA = 0.3 g).

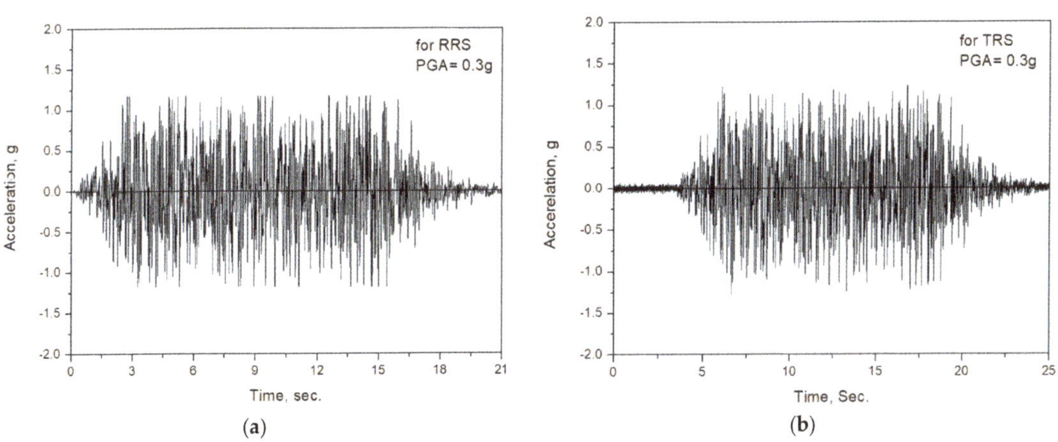

Figure 3. Acceleration time histories for shaking table input motion (PGA = 0.3 g): (**a**) Corresponding to RRS; (**b**) Corresponding to TRS.

Figure 4 presents the comparison of the DRS, the required response spectrum (RRS) corresponding to Figure 3, and the test response spectrum (TRS) corresponding to Figure 4. As shown in the figure, we can see that the RRS and the TRS are almost the same spectral shape and these appropriately envelope the DRS.

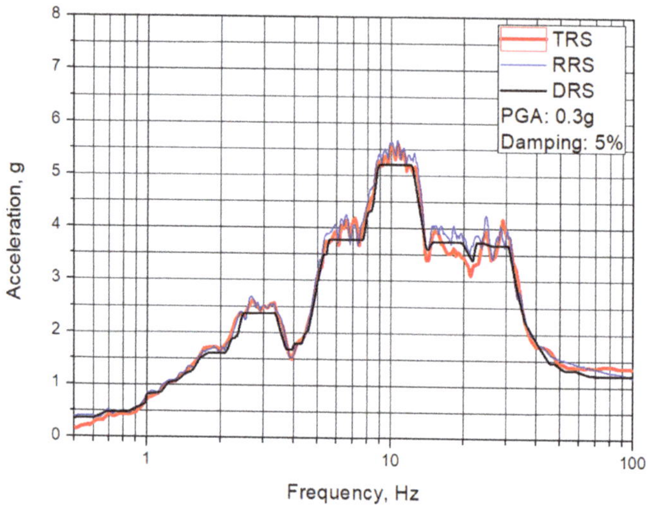

Figure 4. Design response spectrum used in tests (PGA = 0.3 g).

The original required sampling rate for the RRS is 200 Hz (0.005 s), which is a typical value used for the nuclear seismic design. In this study, a faster sampling rate of 512 Hz was used for the shaking table tests with consideration of the test facility performance.

3.4. Test Matrix

Table 4 presents the summary of the test matrix carried out in this shaking table test. As shown in the table, three sets of the test mockup (OPT-2A, OPT-2B, and OPT-2C), which can validate the reliability for the consistency of LRB fabrication quality, are prepared to investigate the seismic isolation characteristics and performance. Each set is composed of four LRBs with the same dummy mass.

Table 4. Test Matrix.

Test Mockup ID	Test ID	PGA (g)	PGA with IEAF (F_μ) [1] (g)	Remarks
OPT-2A	OPT-2A-1	0.5	0.625	Seismic Fragility Capacity
	OPT-2A-2	0.3	0.3	SSE
	OPT-2A-3	0.1	0.1	OBE
OPT-2B	OPT-2B-1	0.5	0.625	Seismic Fragility Capacity
	OPT-2B-2	0.3	0.3	SSE
	OPT-2B-3	0.1	0.1	OBE
OBT-2C	OPT-2C-1	0.3	0.3	SSE
	OPT-2C-2	0.6	0.75	For Ultimate Behavior

[1] F_μ = 1.25: Inelastic energy absorption factor (IEAF) for building structure [24].

As shown in the test matrix, four PGA levels of a seismic input motion such as 0.1 g, 0.3 g, 0.5 g, and 0.6 g are considered to be tested. In these load levels, 0.1 g and 0.3 g are the levels of the operating basis earthquake (OBE) and the safe shutdown earthquake (SSE) respectively. Moreover, 0.5 g is the level of the beyond design basis earthquake (BDBE) corresponding to the required seismic fragility capacity for 0.3 g SSE. The level of PGA = 0.6 g is the seismic fragility capacity corresponding to 0.36 g SSE level and is set to investigate the ultimate behavior of LRBs, which may occur in the seismically isolated nuclear facility components located at the 137 ft elevation level of the nuclear building structures.

In the test mockups of OPT-2A and OPT-2B, the shaking table tests are sequentially carried out from 0.5 g level to 0.1 g. The purpose of this test plan is to investigate whether the seismic isolation performance can be achieved even when the smaller earthquakes occur continuously in the deformed state of the inserted lead plug after experiencing a large earthquake history. This test plan is also reasonable when considering that the primary purpose of developing the lead inserted small-sized LRB is to accommodate the beyond design basis earthquakes in the nuclear facility components. The tests for 0.3 g and 0.1 g after the 0.5 g test are set to investigate the LRB performance for aftershocks.

All input motions corresponding to each PGA level were obtained by linearly scaling for the input motion of PGA = 0.3 g presented in Figure 3. When considering an actual earthquake event, the PGA levels above 0.3 g of SSE in the test matrix will be greater than linearly scaled levels because the design floor response spectrum used for the shaking table tests will be practically reduced due to the plastic behavior of the building structures, which can be represented by the inelastic energy absorption factor (IEAF). When taking the IEAF to be 1.25 as recommended in EPRI/NP-6041-SL [24], the PGA levels for 0.5 g and 0.6 g can be treated as 0.625 g and 0.75 g levels, respectively.

All tests are carried out with an interim time of about 10 min after each test listed in test ID to minimize the variation of the rubber characteristics due to the temperature increase that may occur in the rubber in the test performed just before.

3.5. Results of Shaking Table Tests

3.5.1. Seismic Isolation Characteristics and Performance

To investigate the seismic isolation characteristics of the lead inserted small-sized LRB developed for an individual nuclear facility component seismic isolation, the test results for the representative design basis earthquake level of PGA = 0.3 g are presented in Figure 5. From the spectrum responses of Figure 5b corresponding to the test results of acceleration time history responses shown in Figure 5a, we can see that the peak spectral frequency band around 10 Hz in the input shaking table motion is significantly shifted to the 2.6 Hz in the dummy mass response, which is slightly higher than the 2.3 Hz of the design seismic isolation frequency in Table 2. From this result, it can be confirmed that the LRB on development in this study has one of the typical seismic isolation characteristics that can assure the seismic isolation performance for the nuclear facility components generally having around 10 Hz dynamic frequencies.

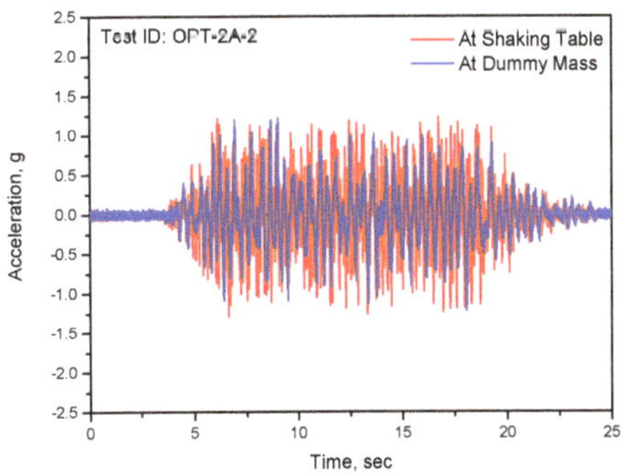

(a)

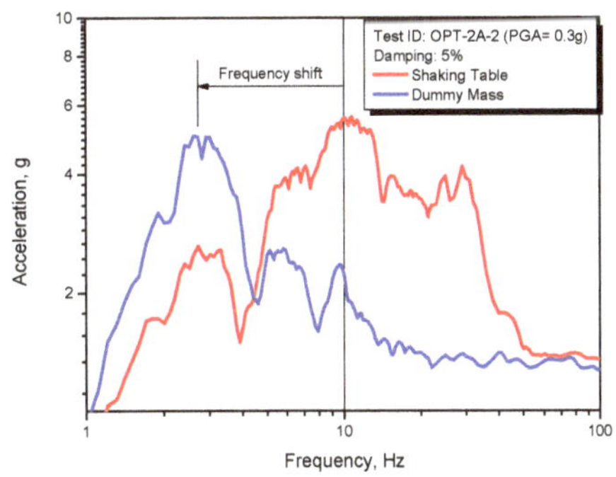

(b)

Figure 5. Test results for safe shutdown earthquake (PGA = 0.3 g): (**a**) Acceleration time histories; (**b**) Corresponding response spectrum.

Figures 6–13 show the test results of the acceleration time history responses for the individual test ID with a comparison between the isolated structures, i.e., installed on the dummy mass, and the non-isolated structures, i.e., installed on the shaking table.

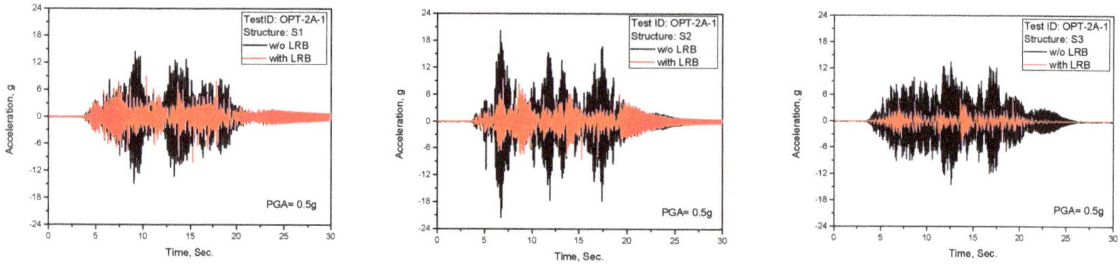

(a) (b) (c)

Figure 6. Test results of acceleration time history response (TEST ID: OPT-2A-1, PGA = 0.5 g): (**a**) S1; (**b**) S2; (**c**) S3.

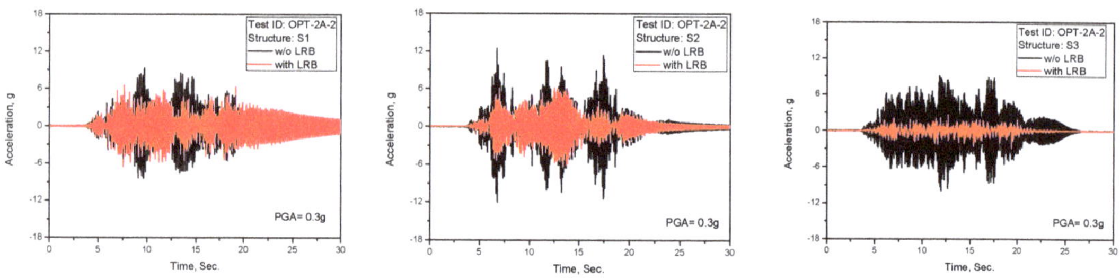

(a) (b) (c)

Figure 7. Test results of acceleration time history response (TEST ID: OPT-2A-2, PGA = 0.3 g): (**a**) S1; (**b**) S2; (**c**) S3.

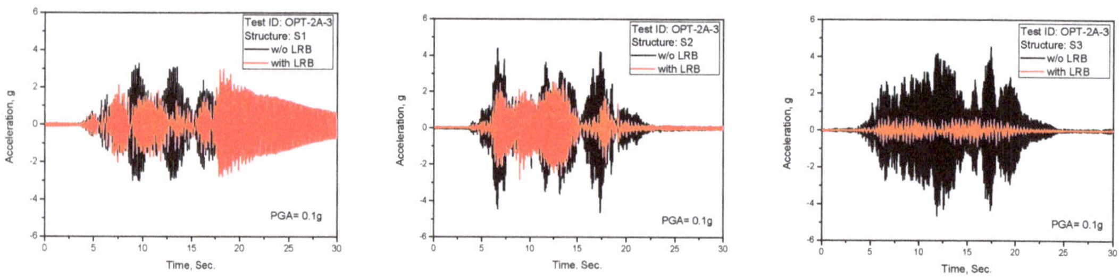

(a) (b) (c)

Figure 8. Test results of acceleration time history response (TEST ID: OPT-2A-3, PGA = 0.1 g): (**a**) S1; (**b**) S2; (**c**) S3.

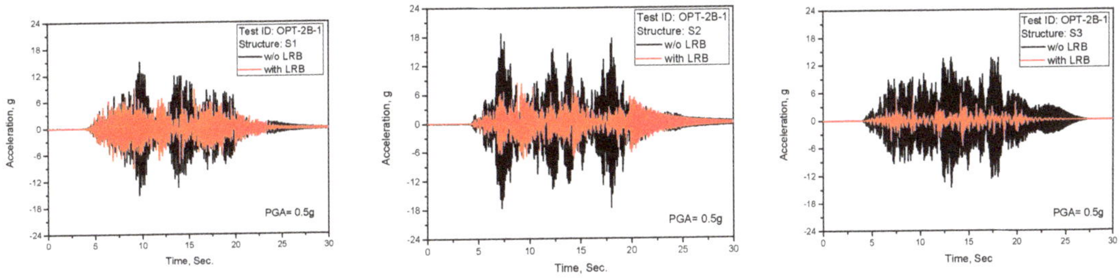

(**a**) (**b**) (**c**)

Figure 9. Test results of acceleration time history response (TEST ID: OPT-2B-1, PGA = 0.5 g): (**a**) S1; (**b**) S2; (**c**) S3.

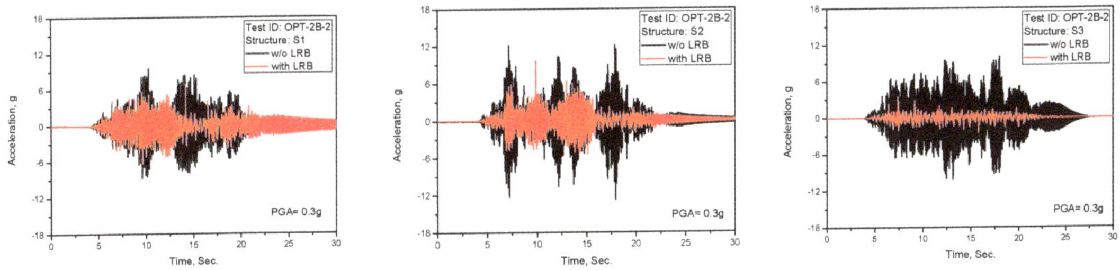

(**a**) (**b**) (**c**)

Figure 10. Test results of acceleration time history response (TEST ID: OPT-2B-2, PGA = 0.3 g): (**a**) S1; (**b**) S2; (**c**) S3.

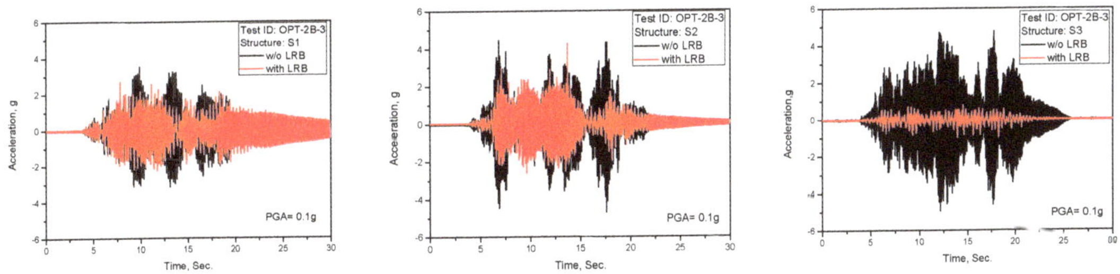

(**a**) (**b**) (**c**)

Figure 11. Test results of acceleration time history response (TEST ID: OPT-2B-3, PGA = 0.1 g): (**a**) S1; (**b**) S2; (**c**) S3.

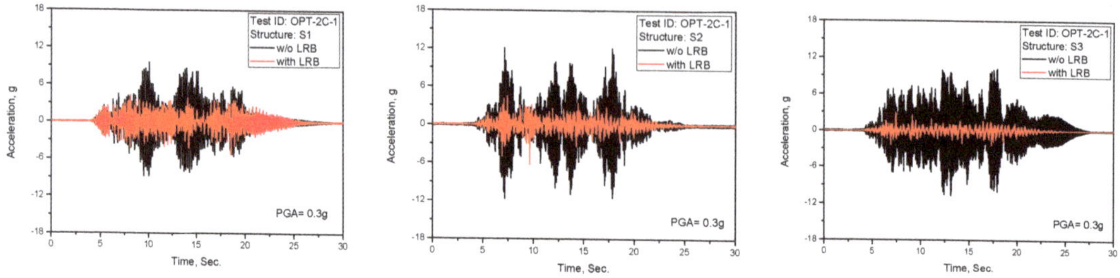

(a) (b) (c)

Figure 12. Test results of acceleration time history response (TEST ID: OPT-2C-1, PGA = 0.3 g): (**a**) S1; (**b**) S2; (**c**) S3.

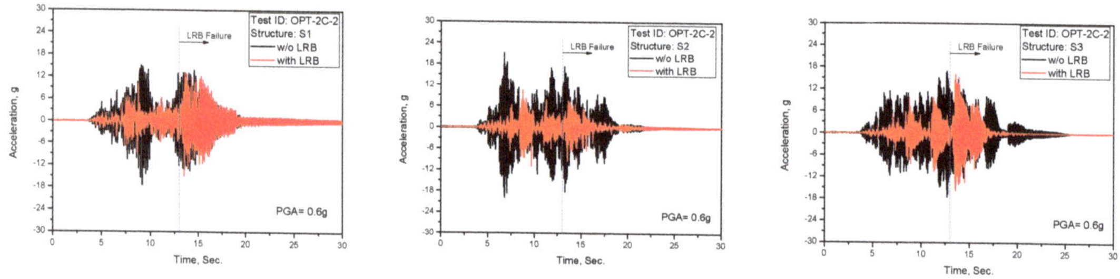

(a) (b) (c)

Figure 13. Test results of acceleration time history response (TEST ID: OPT-2C-2, PGA = 0.6 g): (**a**) S1; (**b**) S2; (**c**) S3.

As clearly shown in the above test results of the acceleration time history responses, the seismically-isolated structures reveal the seismic isolation performance as expected from the frequency shift effect presented in Figure 5b of results of the response spectrum, which are for the shaking table and the dummy mass.

Figure 14 presents the comparison of the spectral peak values between seismically-isolated structures and the non-isolated structures calculated from the above acceleration time history responses. As shown in the figure, we can see that the LRB gives better seismic isolation performance as the input earthquake levels increase.

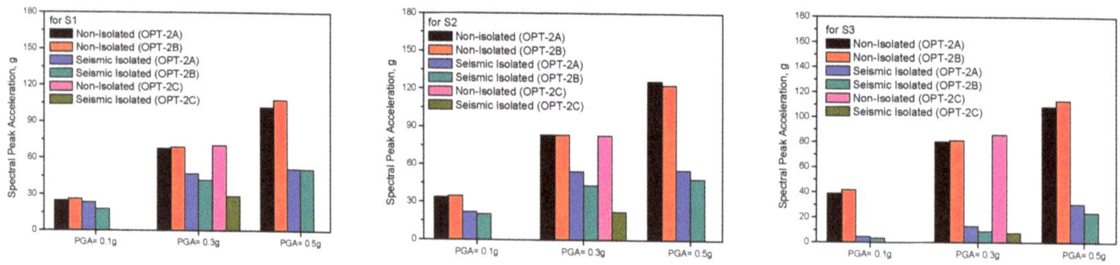

(a) (b) (c)

Figure 14. Comparison of spectral peak values calculated from the test results: (**a**) S1; (**b**) S2; (**c**) S3.

From the above acceleration seismic responses, the results of the shaking table tests of the lead inserted small-sized LRB are summarized as follows;

1. It is confirmed that the fabricated lead inserted small-sized LRB gives a consistent seismic isolation performance through the repeated shaking table tests with different sets of the fabricated LRB as intended in the test matrix to check the reproducibility of LRB;
2. Contrary to concerns about the lead plug after experiencing larger earthquakes, the lead inserted small-sized LRB reveals adequate seismic isolation performance in subsequent shaking table tests;
3. In particular, the seismic isolation performance in a frequency range of 9 to 20 Hz, which corresponds to the resonance frequency band of most nuclear facility components, is confirmed as expected in the LRB design;
4. When the input earthquake level is small, such as PGA = 0.1 g, the amplitude of the cyclic shear displacement of LRB occurs is also small and the initial stiffness effects of the lead plug increase the seismic isolation frequency. As a result, the seismic isolation performance is relatively reduced.
5. In the test results of the PGA = 0.6 g in Figure 13, LRB failure occurred after 13 s due to exceeding the LRB shear strain limits of 600% (120 mm), which has been confirmed in previous quasi-static tests [21].

3.5.2. LRB Shear Displacement Responses

In general, one of the main issues in the seismic isolation design, regardless of isolation device types, is how to control the appropriate relative displacement between the seismic excitation base and the seismically isolated superstructures in point of the interface design. Figure 15 presents the maximum shear displacements of LRB measured at the center and the edge of dummy mass for all test IDs. As shown in the figures, the LRB reveals well-controlled shear displacement behavior up to the beyond design basis earthquake of PGA = 0.5 g, which is 0.625 g in a seismic fragility point of view.

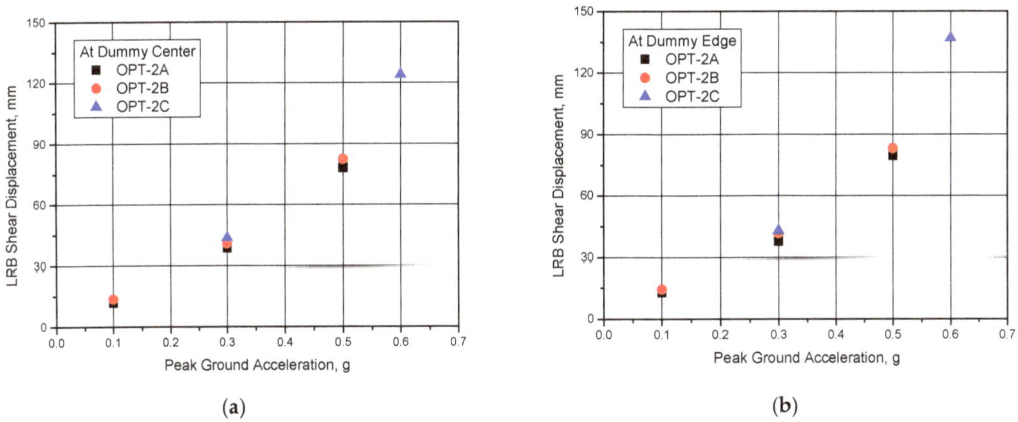

Figure 15. Measured maximum shear displacement of LRB: (**a**) At the center of dummy mass; (**b**) At the edge of dummy mass.

The averaged maximum shear displacements of LRB corresponding to the design basis earthquake of SSE (PGA = 0.3 g) and the beyond design basis earthquake, PGA = 0.5 g were measured to be 41 mm and 80 mm respectively, which were very close to the design target value of 35 mm and 80 mm corresponding to the shear strain of 175% and 400% respectively as presented in Table 1.

Figure 16 represents the measured shear displacement time history response of LRB for PGA = 0.5 g (Test ID: OPT-2A-1). As shown in the figure, the LRB shear displacement time history responses measured at the center and the edge of the dummy mass is almost the same, with less than the 1.5 mm difference. This result indicated that no rotational behavior occurred in the dummy mass and that all LRBs used in the test had the same mechanical characteristics.

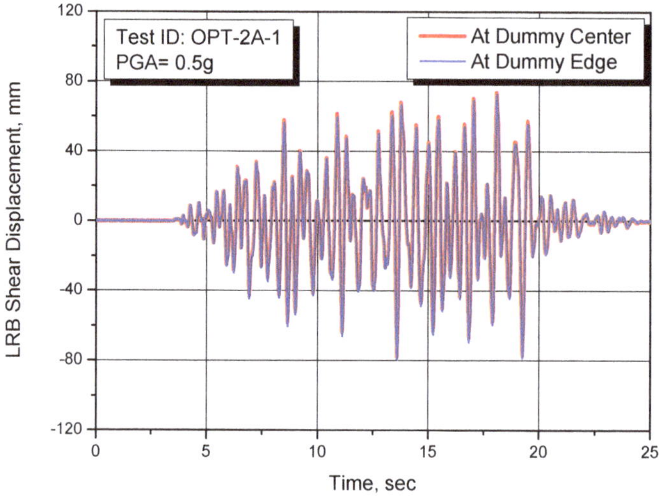

Figure 16. Measured LRB shear displacement time history response for Test ID: OPT-2A-1, PGA = 0.5 g.

In the case of the ultimate test with PGA = 0.6 g (Test ID: OPT-2C-2), the failure of LRB occurred after 13 s during the test. At this time, the maximum shear displacement was measured to be 137 mm (685%), significantly exceeding the beyond design shear displacement of 80 mm. Figure 17 presents the photos showing the failed LRB after tests.

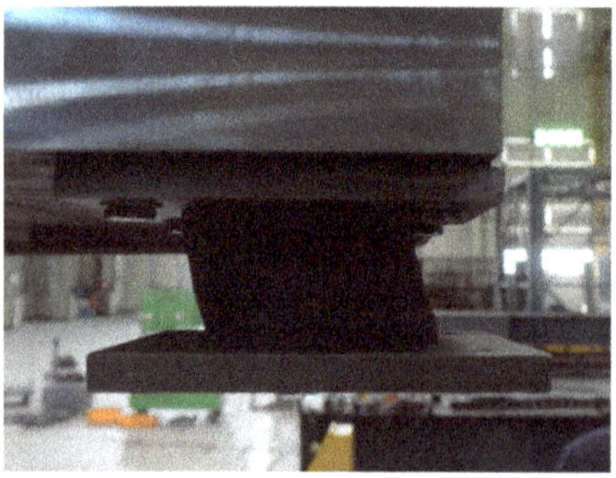

(**a**)

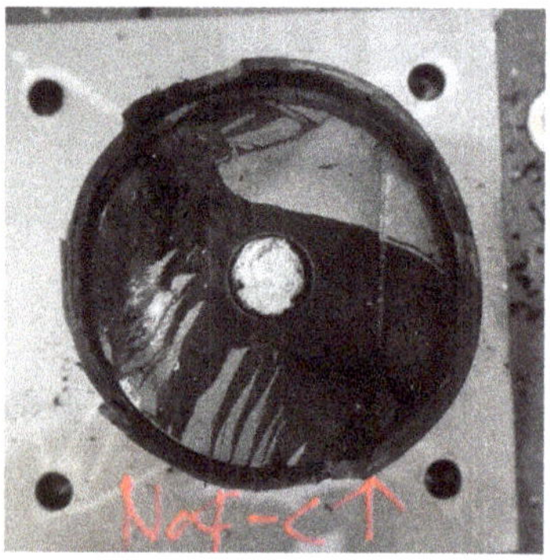

(**b**)

Figure 17. Photos of failed LRB after ultimate test with PGA = 0.6 g (Test ID, OPT-2C-2): (**a**) Front view with permanent deformation; (**b**) Top view with failed section.

4. Numerical Seismic Analysis and Discussions

To confirm the seismic isolation characteristics of LRB, the numerical analyses are carried out for the representative Test ID, OPT-2A-2 (PGA = 0.3 g), which is for the design basis earthquake of SSE (PGA = 0.3 g). The used analysis seismic input motion is shown in Figure 5a, which is measured at the shaking table.

For the numerical seismic analysis, the ANSYS commercial finite element program [25] is used.

4.1. Seismic Analysis Modeling

Figure 18 presents the schematic concept of the seismically isolated test mockup performed in this study and the corresponding finite element model used for the seismic analysis. As shown in Figure 18a, there are three different beam structures both on the shaking table and on the dummy mass, which is intended to express a seismically isolated structure and the non-isolated structure. Figure 18b is the corresponding seismic analysis model used in this study.

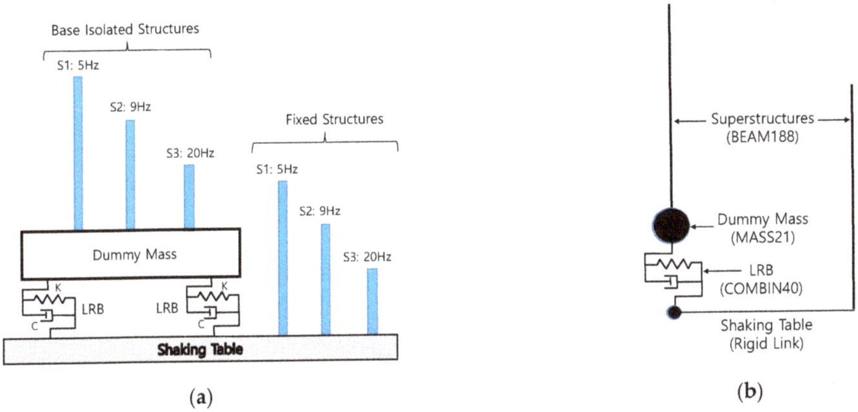

Figure 18. Seismic analysis model: (**a**) Schematics of the test mockup; (**b**) Finite element model.

Table 5 presents the input parameters for the finite element modeling used in the seismic analysis. As shown in the table, two types of LRB analysis modes are used: the Bilinear model and equivalent stiffness-damping model. For the LRB model, the element type of COMBIN40 in the ANSYS is used with input parameters of elastic stiffness (K_e), tangential stiffness (K_t), and characteristic strength (Q_d) for the bilinear model, and of equivalent stiffness (K_{eq}) and critical damping ratio (ξ) for the equivalent spring-damper model. And for the beam structure model, the element type of BEAM188 is used with total height (L), area moment of inertia (I_{zz}), and Rayleigh damping parameters of *ALPHA* and *BETA*.

Table 5. Analysis input parameters for finite element modeling.

Structure ID	Beam Model				LRB Models				
					Bi-Linear			Equiv. Stiffness-Damping	
	L (mm)	I_{zz} (m^4)	*ALPHA*	*BETA*	K_e (kN/m)	K_t (kN/m)	Q_d (kN)	K_{eq} (kN/m)	ξ (%)
S1	47.0	0.84E-11	0.863	1.349E-4	187.6	122	2.7	199.3	30
S2	42.2	2.00E-11	3.247	4.569E-5	187.6	122	2.7	199.3	30
S3	34.6	6.75E-11	1.626	2.059E-4	187.8	122	2.7	199.3	30

4.2. Comparison with Test Results

Figures 19–21 present the comparison results of the response spectrum obtained from tests and analyses for individual beam structures on the shaking table and the dummy mass. As shown in the figures, the calculated response spectrum from the seismic analyses is in good agreement with those of the tests. Overall, it can be seen that the used two analysis models of LRB give very similar results but in the case of the equivalent stiffness-damping model, the value of critical damping ratio used in the analysis is 30%, which is slightly higher than the 24.1% obtained from the quasi-static tests in Table 1.

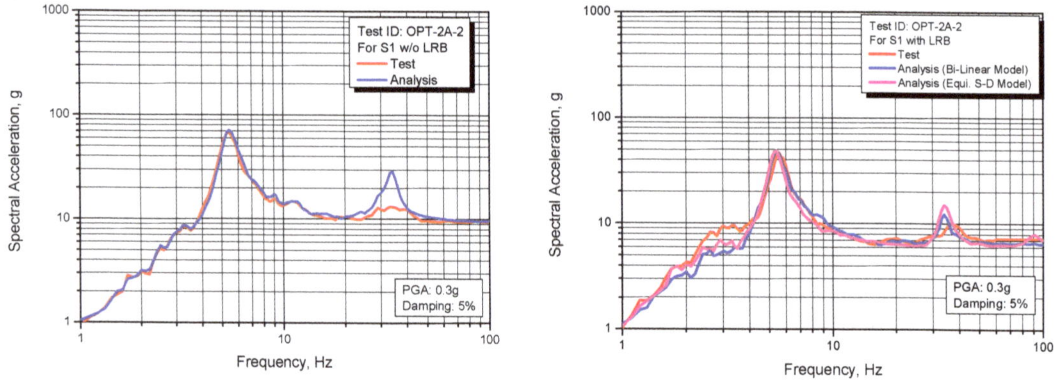

(a) (b)

Figure 19. Comparison of response spectrum between test and analysis for S1 structure: (**a**) Non-isolated; (**b**) Seismically isolated.

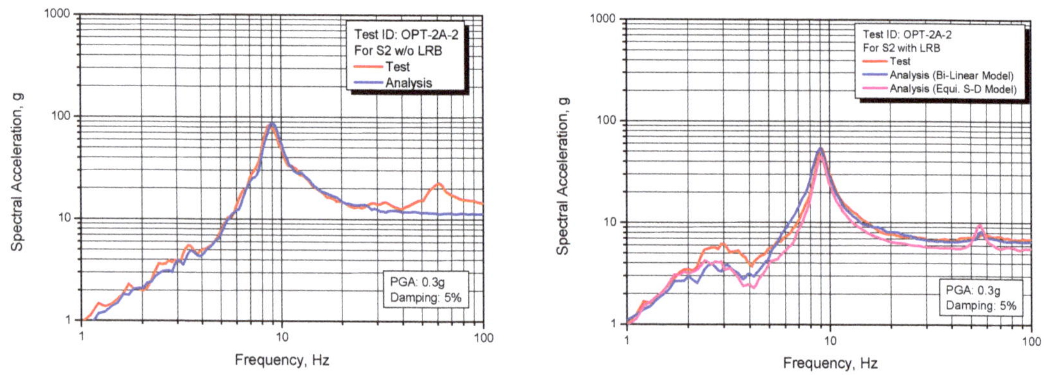

(a) (b)

Figure 20. Comparison of response spectrum between test and analysis for S2 structure: (**a**) Non-isolated; (**b**) Seismically isolated.

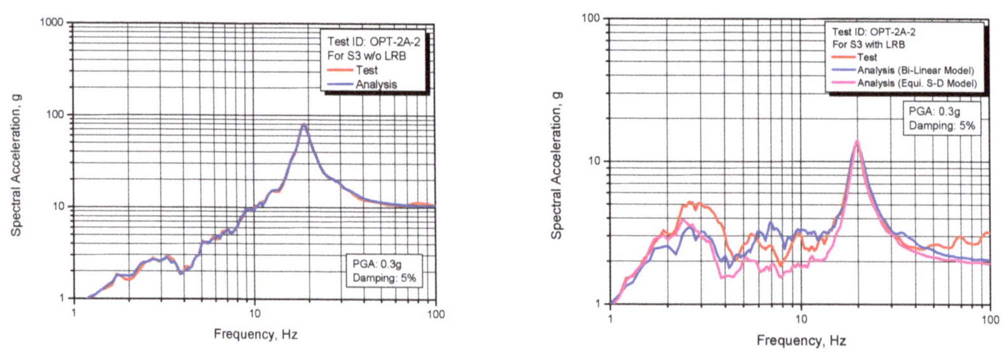

(a) (b)

Figure 21. Comparison of response spectrum between test and analysis for S3 structure: (**a**) Non-Isolated; (**b**) Seismically Isolated.

Figure 22 presents the results for the LRB shear displacement responses. In this figure, Figure 22a shows comparison results of the maximum values between test and analysis. As shown in the figure, the maximum shear displacements of LRB in the analysis are 35.6 mm for the Bi-linear model and 37.3 mm for the equivalent stiffness-damping model. These results were slightly less than the 38.6 mm of the test result but were in good agreement with the 35 mm of the design target value presented in Table 1. Figure 22b shows the analysis result of the hysteretic behavior of LRB with the shear displacements vs. restoring forces. From the results, it could be confirmed that the Bi-linear characteristics of LRB identified from previous quasi-static tests can be used for the seismic analysis.

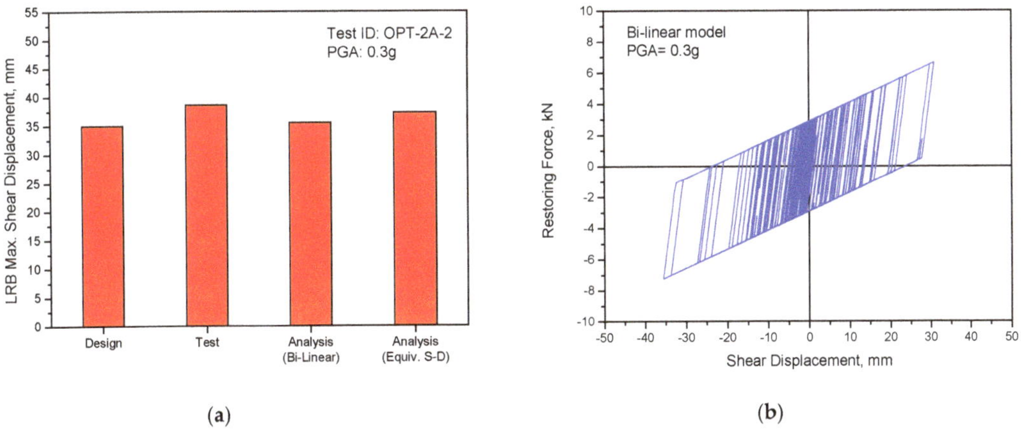

Figure 22. LRB shear displacement responses: (**a**) Comparison of maximum values; (**b**) Hysteretic response with Bi-linear model.

5. Conclusions

In this study, the shaking table tests of the lead inserted small-sized LRBs were performed with the dummy mass system and the applicability of the horizontal seismic isolation to the nuclear facility components was investigated in detail.

Through all shaking table tests specified in the test matrix and the seismic verification analyses, some valuable conclusions are derived as follows;

1. The tests of full-scaled LRBs can assure the reliability of the seismic isolation performance by removing the uncertainties potentially contained in the reduced-scale tests.
2. It was confirmed that the mechanical characteristics of LRB identified by the quasi-static tests were very similar to the results of the dynamic seismic shaking table tests.
3. The lead plug inserted in the center of the LRB reveals a good damping function by adequately restricting the shear displacement as targeted in the LRB design.
4. As intended in the test matrix starting tests sequentially from large to small level earthquakes, it was confirmed that LRB maintains seismic isolation performance against aftershocks even after experiencing the level of beyond design basis earthquakes.
5. The smaller the seismic input load level, the less the seismic isolation effect. This is because in the region where the shear displacement of LRB is small, the initial stiffness effect of the lead plug is large, and the frequency shift effect of the seismic isolation system is reduced.
6. As a result of performing the same repeated shaking table tests for each different LRB set of mockups, almost the same test results were obtained, and from this, the reliability of manufacturing the lead inserted small-sized LRB was confirmed.
7. It is confirmed that the developed LRB in this study has a seismic capacity accommodating PGA = 0.5 g (seismic fragility capacity, 0.625 g) at the 137 ft elevation of the nuclear power plant building structures.
8. It is recommended that the allowable seismic shear strain of LRB developed in this study is 400% for application to the nuclear facility components. When applied to

input seismic load exceeding this stain level, more additional shaking table tests are required to confirm higher shear strain limit.

9. From the comparison results between the shaking table tests and the seismic analyses, it is found that the Bi-linear model or the equivalent stiffness-damping model of LRB can be used in the seismic analysis as a simple method when using well-identified stiffness and damping value of LRB.

Author Contributions: Conceptualization, G.-H.K., T.-M.S., and S.-J.M.; methodology, G.-H.K. and T.-M.S.; validation, G.-H.K.; formal analysis, G.-H.K.; investigation, T.-M.S. and S.-J.M.; writing—original draft preparation, G.-H.K.; writing—review and editing, T.-M.S. and S.-J.M.; funding acquisition, G.-H.K., T.-M.S. All authors have read and agreed to the published version of the manuscript.

Funding: This study was funded by the Ministry of Trade, Industry, and Energy through KETEP (Korea Institute of Energy Technology Evaluation Planning) (No. 20181510102380).

Institutional Review Board Statement: The study was conducted according to the guidelines of the Declaration of Helsinki, and approved by the Institutional Review Board.

Informed Consent Statement: Informed consent was obtained from all subjects involved in the study.

Data Availability Statement: Not Applicable.

Acknowledgments: This study was supported by the Ministry of Trade, Industry, and Energy through KETEP (Korea Institute of Energy Technology Evaluation Planning) (No. 20181510102380).

Conflicts of Interest: The authors declare no conflict of interest.

References

1. Kwag, S.Y.; Kwag, J.S.; Lee, H.H.; Oh, J.H.; Koo, G.H. Enhancement in the Seismic Performance of a Nuclear Piping System using Multiple Tuned Mass Dampers. *Energies* **2019**, *12*, 2077. [CrossRef]
2. Cho, S.G.; Chang, S.K.; Sung, D.Y. Application of tuned mass damper to mitigation of the seismic responses of electrical equipment in nuclear power plants. *Energies* **2020**, *13*, 427. [CrossRef]
3. Kostarev, V.; Petrenko, A.; Vasilev, P.A. New Method for Essential Reduction of Seismic and External Loads on NPP's Structures, Systems and Components. In Proceedings of the 17th International Conference on SMiRT, Prague, Czech Republic, 17–22 August 2003; Volume K13-1.
4. Carlos, M.V.; Tianjian, J. Seismic protection technology for nuclear power plants: A systematic review. *J. Nucl. Sci. Technol.* **2015**, *52*, 607–632.
5. *Seismic Isolation Systems for Nuclear Installations*; IAEA-TECDOC-1905; IAEA: Vienna, Austria, 2020.
6. Qin, C.; Liu, W.; He, W. Seismic Response Analysis of Isolated Nuclear Power Plants with Friction Damper Isolation System. *AASRI Procedia* **2014**, *7*, 26–31. [CrossRef]
7. *Seismic Isolation of Nuclear Power Plants Using Sliding Bearing*; NUREG/CR-7254; NRC: North Bethesda, MD, USA, 2015.
8. Kamrava, A. Seismic isolators and their types. *Curr. World Environ.* **2015**, *10*, 27–32. [CrossRef]
9. Koo, G.H.; Lee, J.H.; Yoo, B.; Ohtori, Y. Evaluation of Laminated Rubber Bearing for Seismic Isolation Using Modified Macro-Model with Parameter Equations of Instantaneous Apparent Shear Modulus. *Eng. Struct.* **1999**, *21*, 594–602. [CrossRef]
10. Eidinger, J.K.; Tajirian, F.F.; Kircher, C.A.; Vaidya, N.; Constantinou, M.; Kelly, J.M.; Ovadia, D.; Seidensticker, R. Base Isolation for Nuclear Power and Nuclear Material Facilities. In Proceedings of the Second DOE Natural Phenomena Hazards Mitigation Conference, Lawrence Livermore National Lab, kNoxville, TN, USA, 3–5 October 1989; pp. 46–57.
11. Fujita, T.; Fujita, S.; Suzuki, S.; Yoshizawa, T. Experimental Study of Laminated Rubber Bearings for Earthquake Isolation of Buildings. *Trans. Jpn. Soc. Mech. Eng.* **1987**, *53*, 71–76. [CrossRef]
12. Koshida, H.; Yasaka, A. Vibration tests and earthquake observation results of base-isolated building. *Seism. Shock Vib. Isol.* **1989**, *181*, 135–140.
13. Tsai, H.C.; Kelly, J.M. Seismic response of heavily damped base isolation systems. *Earthq. Eng. Struct. Dyn.* **1993**, *22*, 633–645. [CrossRef]
14. Alhan, C.; Oncu, S. Performance limits of seismically isolated buildings under near-field earthquakes. *Eng. Struct.* **2016**, *116*, 83–94. [CrossRef]
15. Kubo, T.; Yamamoto, T.; Saro, K.; Jimbo, M.; Imaoka, T.; Umeki, Y. A Seismic Design of Nuclear Reactor Building Structures Applying Seismic Isolation System in A High Seismicity Region–A Feasibility Case Study In JAPAN. *Nucl. Eng. Technol.* **2014**, *46*, 716. [CrossRef]
16. Zhai, C.; Zheng, Z.; Li, S.; Pan, X. Damage Accumulation of a Base-isolated RCC Building under Main shock-aftershock Seismic Sequences. *KSCE J. Civ. Eng.* **2017**, *21*, 364–377. [CrossRef]

17. Sato, N.; Watanabe, Y.; Kato, A.; Ohba, M.; Fukushima, Y.; Iizuka, M.; Yoshikawa, K.; Umeki, K.; Suhara, J.; Murazumi, Y.; et al. Shaking Table Test and Analysis on Ultimate Characteristics of the Lead Rubber Bearing for Base Isolated FBR Plant. In Proceedings of the 12th World Conference on Earthquake Engineering, Auckland, New Zealand, 30 January–4 February 2000.
18. Ohtori, Y. Experimental Study on Mechanical Characteristics of High Damping Rubber Bearings of Various Shapes. *J. Struct. Eng. (Trans. Archit. Inst. Jpn.)* **1997**, *43B*, 125–133.
19. Moteki, M.; Kawai, N.; Ishida, K.; Yabana, S.; Nojima, O. Shaking table test on ultimate behavior of seismic isolation system, Part 1: Outline of the test and response of superstructure. In *Proceedings of the 10th World Conference on Earthquake Engineering*; Madrid, Spain, 19–24 July 1992, pp. 2271–2276.
20. Sato, N.; Kato, A.; Fukushima, Y.; Iizuka, M. Shaking table tests on failure characteristics of base isolation system for a DFBR plant. *Nucl. Eng. Des.* **2002**, *212*, 293–305. [CrossRef]
21. Koo, G.H.; Jung, J.Y.; Lee, J.H.; Shin, T.M. Development of Small-Sized Lead Inserted Laminated Rubber Bearing for Nuclear Component Seismic Isolation. *Energies* **2020**, *13*, 3193. [CrossRef]
22. MTS. 2021. Available online: https://www.mts.com/en/forceandmotion/geociviltesting/index.htm (accessed on 14 April 2021).
23. *Design Response Spectra for Seismic Design of Nuclear Power Plants*; Regulatory Guide 1.60; Nuclear Regulatory Commission: Rockville, MD, USA, 1973.
24. *A Methodology for Assessment of Nuclear Power Plant Seismic Margin (Revision 1)*; EPRI/NP-6041-SL; Electric Power Research Institute (EPRI): Washington, DC, USA, 1991.
25. ANSYS. *Mechanical APDL Release 15.0*; ANSYS, Inc.: Canonsburg, PA, USA, 2013.

applied sciences

MDPI

Article

Shaking Table Tests to Validate Inelastic Seismic Analysis Method Applicable to Nuclear Metal Components

Gyeong-Hoi Koo [1,*], Sang-Won Ahn [2], Jong-Keun Hwang [2] and Jong-Sung Kim [3]

[1] Korea Atomic Energy Research Institute, Daejeon 34057, Korea
[2] SMARTEK E&C, Daejeon 34013, Korea; smartekenc@naver.com (S.-W.A.); jkhwang0@hanmail.net (J.-K.H.)
[3] Department of Nuclear Engineering, Sejong University, Seoul 05006, Korea; kimjsbat@sejong.ac.kr
* Correspondence: ghkoo@kaeri.re.kr; Tel.: +82-42-868-2950

Abstract: The main purpose of this study is to perform shaking table tests to validate the inelastic seismic analysis method applicable to pressure-retaining metal components in nuclear power plants (NPPs). To do this, the test mockup was designed and fabricated to be able to describe the hot leg surge line nozzle with a piping system, which is known to be one of the seismically fragile components in nuclear steam supply systems (NSSS). The used input motions are the displacement time histories corresponding to the design floor response spectrum at an elevation of 136 ft in the in-structure building in NPPs. Two earthquake levels are used in this study. One is the design-basis safe shutdown earthquake level (SSE, PGA = 0.3 g) and the other is the beyond-design-basis earthquake level (BDBE, PGA = 0.6 g), which is linearly scaled from the SSE level. To measure the inelastic strain responses, five strain gauges were attached at the expected critical locations in the target nozzle, and three accelerometers were installed at the shaking table and piping system to measure the dynamic responses. From the results of the shaking table tests, it was found that the plastic strain response at the target nozzle and the acceleration response at the piping system were not amplified by as much as two times the input earthquake level because the plastic behavior in the piping system significantly contributed to energy dissipation during the seismic events. To simulate the test results, elastoplastic seismic analyses with the well-known Chaboche kinematic hardening model and the Voce isotropic hardening model for Type 316 stainless steel were carried out, and the results of the principal strain and the acceleration responses were compared with the test results. From the comparison, it was found that the inelastic seismic analysis method can give very reasonable results when the earthquake level is large enough to invoke plastic behavior in nuclear metal components.

Keywords: inelastic seismic analysis; shaking table test; nuclear metal component; design-basis earthquake; beyond-design-basis earthquake; plastic strain; strain gauge; Chaboche kinematic hardening model; Voce isotropic hardening model; strain gauge rosettes; principal strain

Citation: Koo, G.-H.; Ahn, S.-W.; Hwang, J.-K.; Kim, J.-S. Shaking Table Tests to Validate Inelastic Seismic Analysis Method Applicable to Nuclear Metal Components. *Appl. Sci.* **2021**, *11*, 9264. https://doi.org/10.3390/app11199264

Academic Editor: Marco Vona

Received: 2 September 2021
Accepted: 29 September 2021
Published: 6 October 2021

Publisher's Note: MDPI stays neutral with regard to jurisdictional claims in published maps and institutional affiliations.

Copyright: © 2021 by the authors. Licensee MDPI, Basel, Switzerland. This article is an open access article distributed under the terms and conditions of the Creative Commons Attribution (CC BY) license (https://creativecommons.org/licenses/by/4.0/).

1. Introduction

In the current seismic design of nuclear power plants (NPPs), the peak ground acceleration (PGA) level of safe shutdown earthquake (SSE) is generally taken to be 0.3 g as a design-basis earthquake. Recently, there have been attempts to increase the PGA level above 0.3 g to enhance the seismic safety of NPPs. Furthermore, after the Fukushima NPP accident, the consideration of the beyond-design-basis earthquake (BDBE) became one of the big issues at the design stage, especially for safety-related nuclear facility components having small seismic margins. In fact, it is true that increasing the value of the SSE level is a big burden for the seismic design of NPPs. There have been many research and development efforts to resolve BDBE issues for the seismic design of NPPs. As one of the hardware approaches to accommodate large earthquakes, seismic isolation design has been studied by many countries for a whole nuclear island building isolation [1,2] or an individual facility component isolation [3,4]. Additionally, seismic energy absorbers such as tuned mass and dampers (TMDs) have been studied for application to seismic fragile piping systems or any other safety-related facilities [5].

Recently, as one of the approaches against large earthquakes, studies on the new seismic design acceptance criteria for large earthquakes have been actively conducted for the application of nuclear metal components [6–9]. The main concept of this design criteria provides strain-based design limits to protect the pressure-retaining nuclear metal components against an inelastic strain-induced failure mode. Actually, current design criteria in nuclear codes and standards such as ASME BPVC III [10] and RCC-MR [11] provide the stress-based design limits for the elastically calculated stress values. Therefore, they may not provide adequate limits for actual seismic failure modes in cases when the reversing dynamic plastic responses are significant.

In the previous studies on the new strain-based seismic design criteria, the feasibility of this approach was investigated for the nuclear pipe and nozzle. The most important thing when applying this approach is to calculate accurate inelastic strain values using inelastic seismic analysis [12,13]. Then, seismic shaking table tests are inevitably required in order to validate the inelastic seismic analysis method. There has been research on seismic shaking table tests for the piping system [14,15].

In this study, the test mockup for simulating the hot leg surge line nozzle, which is known as one of the fragile components in nuclear steam supply systems (NSSS), is designed and tested on a shaking table for two earthquake levels of PGA = 0.3 and 0.6 g. For validation of the inelastic seismic analysis method, the finite element model using the ANSYS program [16] is established with material constitutive equations of the Chaboche kinematic hardening model and the Voce isotropic hardening model for Type 316 stainless steel. Through the investigation of the test results and the comparison of seismic time history responses such as the principal strain and accelerations between the measured data and the inelastic seismic analysis results, the inelastic seismic analysis method for large earthquakes was validated and confirmed to be useful for the seismic design of nuclear metal components.

2. Design of Seismic Test Mockup

2.1. Configuration and Dimensions

In this study, the seismic test mockup is designed to be able to simulate the actual nuclear component installed in the in-structure building. This will be consistent with the purpose of validating the strain-based seismic design method developed in previous studies [9,13]. To do this, the metal component nozzle, which is known as one of the seismically fragile components in nuclear steam supply systems (NSSS), was selected as the reference for the test mockup. Specifically, the hot leg surge line nozzle connecting the hot-leg pipe and the surge line piping system, which is installed at the elevation of 136 ft in the in-structure building, is referenced in this test mockup design.

The main design concept of the test mockup is to configure the piping layout, resulting in the maximum accumulated plastic strain at the safe-end region connecting the nozzle and the piping system in the same condition as the actual hot-leg surge line nozzle in NSSS. Figure 1 presents the designed overall configuration and seismic directional axis of the test mockup. As shown in the figure, the test mockup consists of a target nozzle, horizontal and vertical pipes, one elbow, and two added masses. The total height of the test mockup is 1132 mm, and the total mass is about 158 kg. Table 1 reveals the summary of the design parameter values of the test mockup.

The used material of the test mockup is Type 316 stainless steel, which has an elastic modulus of 190 GPa, a density of 7970 kg/m^3, and a Poison's ratio of 0.27 at room temperature.

2.2. Analysis Model and Dynamic Characteristics of Test Mockup

To derive an appropriate plastic strain in the nozzle part, enough to validate the inelastic seismic analysis method, the test mockup is designed to have dominant dynamic characteristics resonant in the peak spectral frequency band of the seismic input motions.

Figure 2 presents the target required response spectrum, representing those at the elevation of 136 ft in the in-structure building (ISB) of NSSS. As shown in the figure, the peak spectral frequency band is 9 to 12 Hz for the EW (east–west) direction, 5 to 20 Hz for the NS (north–south) direction, and 15 to 21 Hz for the V (vertical) direction.

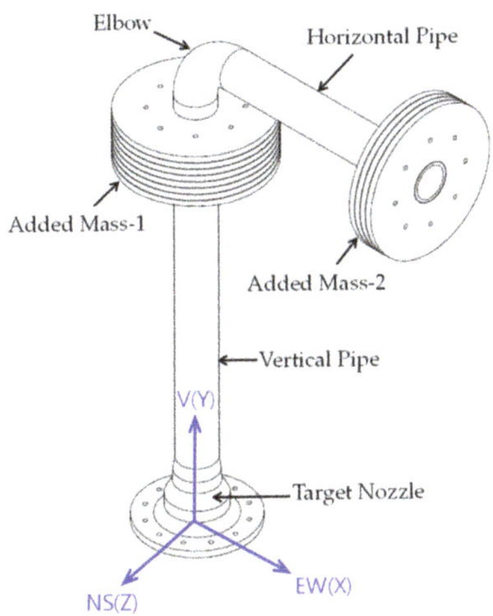

Figure 1. Overall configuration of the test mockup and seismic directional axis.

Table 1. Summary of test mockup design parameters.

Design Parameters	Design Value
Pipe Outer Diameter (mm)	89.1
Pipe Thickness (mm)	7.6
Vertical Pipe Length (mm)	850
Horizontal Pipe Length (mm)	550
Elbow Curvature (mm)	114.3
Added Mass-1 (kg)	80
Added Mass-2 (kg)	40
Total Height of Nozzle (mm)	167.35
Nozzle Outer Diameter (mm)	133.34

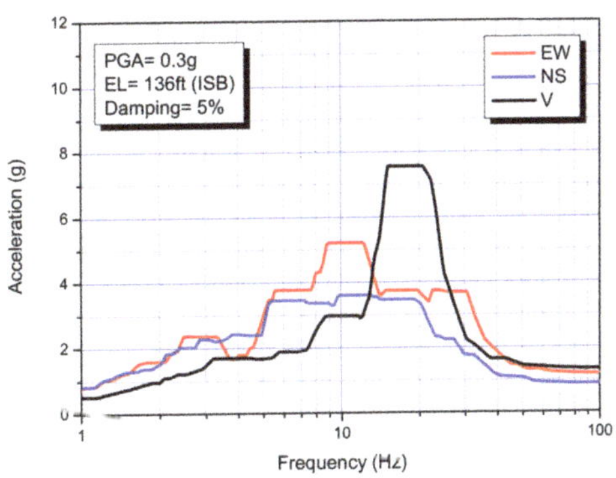

Figure 2. Target required response spectrum corresponding to PGA = 0.3 g.

The dynamic characteristics of the test mockup are investigated by the modal analysis by using the ANSYS [16] commercial finite element program. Figure 3 shows the finite element seismic analysis model of the test mockup. As shown in the model, the element type of the nozzle is a three-dimensional solid element of SOLID185. For the piping system, the three-dimensional pipe elements of PIPE289 and ELBOW290 are used for the straight run section and for the elbow, respectively. To rigidly couple the solid element of the nozzle and the line element of the pipe, the element type of MPC184 is used. The added masses in the test mockup are modeled with a structural mass element of MASS21, including the effects of the rotary moments of inertia.

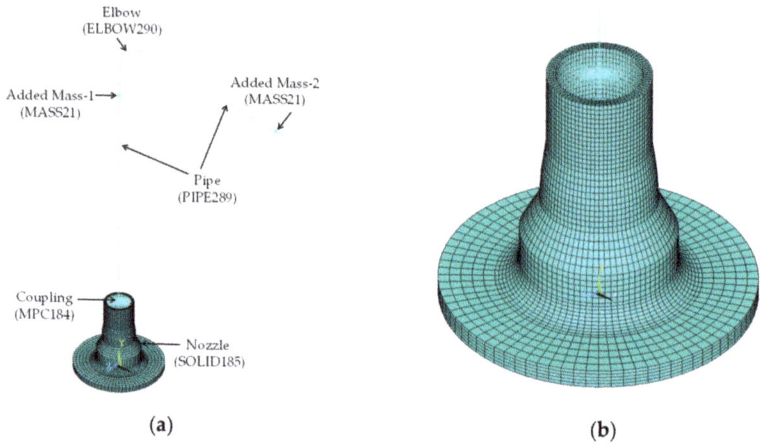

Figure 3. Finite element analysis model: (**a**) total model; (**b**) zoomed nozzle model.

Related to the boundary conditions, it is assumed that the seismic input motions are applied to the nozzle bottom surface as the design response spectrum is specified at the hot-leg piping system. Then, the modal analysis is carried out with the fixed condition of the nozzle bottom surface.

Table 2 presents the summary results of the modal analysis. For the horizontal directions, the dominant fundamental frequency is 11.74 Hz for the NS direction and 11.99 Hz for the EW direction. Additionally, the dominant natural frequency in the vertical direction is 27.63 Hz, slightly apart from the vertical target peak response frequency band. From the modal analysis, it is confirmed that the dynamic characteristics of the test mockup will meet the intended design goal, deriving the sufficiently inelastic strains at the nozzle.

Table 2. Summaries of modal analysis.

Mode No.	Frequencies (Hz)	Modal Participation Factors			Effective Mass		
		EW	NS	V	EW	NS	V
1	11.74	0.00	10.07	0.00	0.00	101.46	0.00
2	11.99	−8.74	0.00	4.35	76.30	0.00	18.96
3	25.89	0.00	5.47	0.00	0.00	29.94	0.00
4	27.63	7.74	0.00	5.05	59.84	0.00	25.54
5	101.61	0.00	0.07	0.00	0.00	5.81	0.00

Figure 4 presents the results of the mode shapes for dominant natural frequencies. From the results, it is expected that Mode 1 and Mode 3 will contribute to the torsional seismic response at the nozzle, and Mode 2, Mode 4, and Mode 5 will contribute to the bending seismic responses at the nozzle.

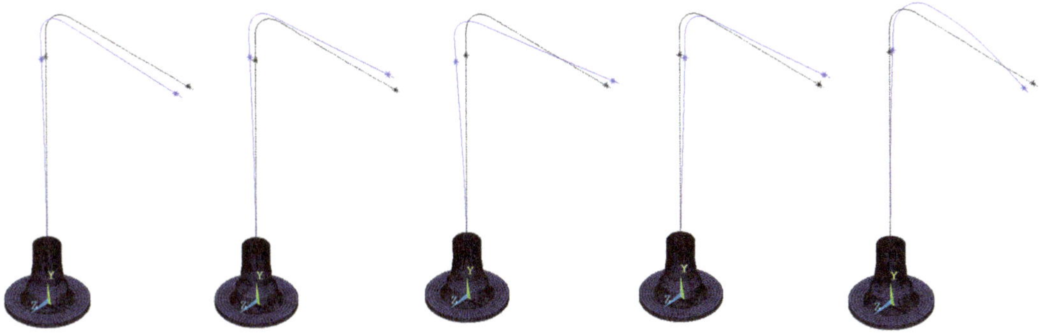

(a) (b) (c) (d) (e)

Figure 4. Mode shapes for test mockup: (**a**) Mode 1 (11.74 Hz); (**b**) Mode 2 (11.99 Hz); (**c**) Mode 3 (25.89 Hz); (**d**) Mode 4 (27.63 Hz); (**e**) Mode 5 (101.61 Hz).

The deadweight effect of the added masses attached to the pipe was examined and included in the modal analysis, but it was found that it has minimal effects on the dynamic characteristics of the test mockup.

3. Seismic Shaking Table Tests

3.1. Performance of Shaking Table Test Facility

Table 3 shows the performance information of the MTS shaking table (2209 × 2173 mm) used in tests. As shown in the table, the test facility can perform the maximum excitation level of 4.9 g for the horizontal direction and 6.0 g for the vertical direction with the loaded maximum of 2000 kg on the table. The frequency range for excitation is 0.8 to 100 Hz.

Table 3. Shaking table performance.

Item	Performance
Max. Loading (kg)	2000
Table Size (mm)	2209 × 2173
Excitation Axes	6 DOF (Translation 3 axes, Rotational 3 Axes)
Max. Displacement (mm)	Hor. (X, Z) = (±125, ±140), Ver. (Y) = ±110
Max. Accel. for bare table (g)	Hor. (X, Z) = 17, Ver. (Y) = 21
Max. Accel. with max load (g)	Hor. (X, Z) = 4.9, Ver. (Y) = 6.0
Frequency Range (Hz)	0.8~100
Excitation Mechanism	Electro-hydraulic Servo, 3 Variable Control
Control Software	354.20/MTS

3.2. Description of Test System and Sensors

As shown in Figure 5, the nozzle bottom flange is connected to the adapter flange by bolts, and then the adapter flange is installed on the shaking table by the bolted connection. Then, it can be assumed that the bolted connection of the test mockup with the shaking table is strong enough to assure the fixed boundary condition, as intended. In order to apply appropriate seismic inertia load on the nozzle, two added masses are attached to the pipes. These masses, Added Mass-1 and Added Mass-2, are composed of eight and four plates combined by bolts, respectively.

For the measurement of seismic strain responses, five strain gauges are attached to the nozzle safe end region where the maximum plastic strain responses are expected. They are almost equally spaced in the circumferential direction at the location of the safe-end region, as shown in Figure 6a. Figure 6b shows the pattern of the general-purpose foil strain gauge type used in this test. Table 4 presents the specification of the used strain gauge.

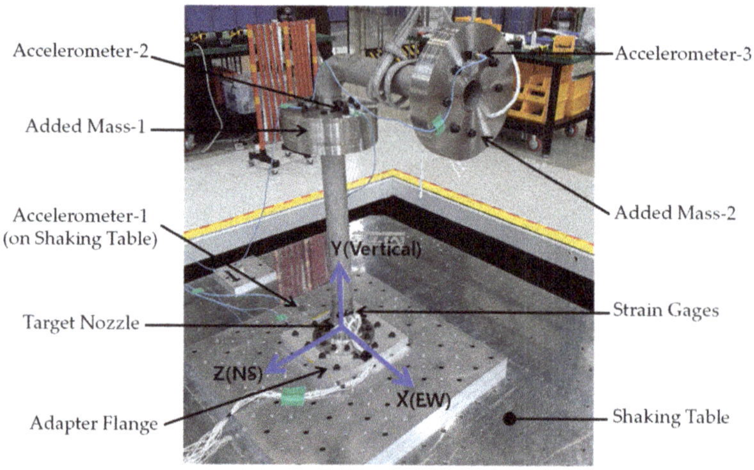

Figure 5. Photo of test mockup with sensors on the shaking table.

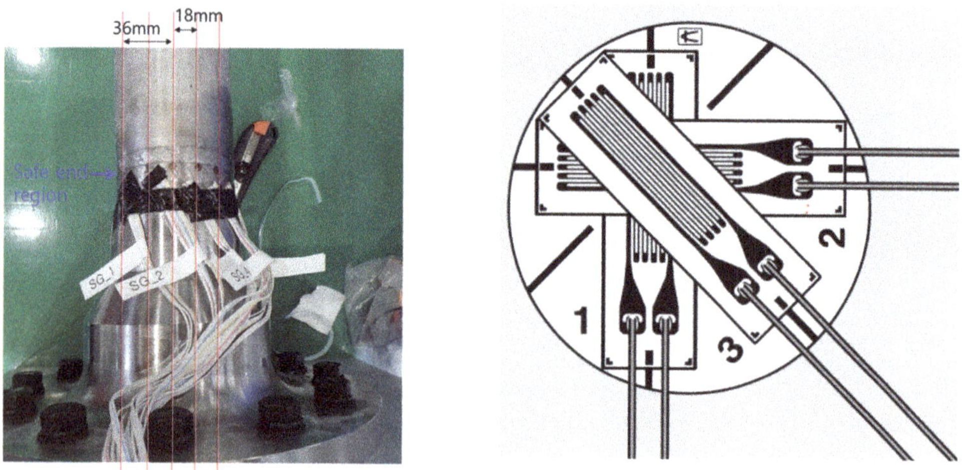

(**a**) (**b**)

Figure 6. Location and pattern of the strain gauges attached on the test mockup: (**a**) photo of attached strain gauge; (**b**) strain gauge rosette pattern.

Table 4. Specifications of the used strain gauge.

Item	Specifications
Model Name	KFGS-1-120-D17-11L15M3S (KYOWA)
Gauge Length (mm)	1
Gauge Resistance (Ω)	120 ± 0.7%
Gauge Pattern	Triaxial 0°–45°–90° Rosette, Round base
Adoptable Thermal Expansion	11.7×10^{-6}/°C
Applicable Adhesive	CC-33A, EP-340

As shown in Figure 5, three accelerometers are installed at the test system (one on the shaking table and two on the piping system) to measure the seismic acceleration responses. Since the seismic inelastic strain responses at the nozzle depend on what the seismic responses are at the piping system, it is important to investigate the acceleration responses. The accelerometer installed on the shaking table is to measure the input seismic motions needed for the inelastic seismic analyses. Table 5 presents the specifications of the used accelerometers.

Table 5. Specifications of the used accelerometers.

Item	Specifications
Model Name	Type 8396A (KISTLER)
Sensing Type	MEMS Variable Capacitance, Silicon Sensing Element
Measuring Range (g)	50
Measuring Axis	Triaxial
Measuring Freq. Range (Hz)	0.5~5000
Operating Temperature (°C)	−55~125

3.3. Results of Shaking Table Tests

3.3.1. Strain Time History Seismic Responses

From the measured strain data (ε_a, ε_b, and ε_c) obtained from the triaxial 0°–45°–90° rosette strain gauges, as conceptually described in Figure 7, the maximum principal strain (ε_{max}), the minimum principal strain (ε_{min}), the maximum shear strain (τ_{max}), and the directional angle of the principal strain (θ) can be calculated using the equations, as follows [17]:

$$\varepsilon_{max} = \frac{1}{2}\left[\varepsilon_a + \varepsilon_c + \sqrt{2\left\{(\varepsilon_a - \varepsilon_b)^2 + (\varepsilon_b - \varepsilon_c)^2\right\}}\right] \quad (1)$$

$$\varepsilon_{min} = \frac{1}{2}\left[\varepsilon_a + \varepsilon_c - \sqrt{2\left\{(\varepsilon_a - \varepsilon_b)^2 + (\varepsilon_b - \varepsilon_c)^2\right\}}\right] \quad (2)$$

$$\tau_{max} = \sqrt{2\left\{(\varepsilon_a - \varepsilon_b)^2 + (\varepsilon_b - \varepsilon_c)^2\right\}} \quad (3)$$

$$\theta = \frac{1}{2}\tan^{-1}\left(\frac{2\varepsilon_b - \varepsilon_a - \varepsilon_c}{\varepsilon_a - \varepsilon_c}\right) \quad (4)$$

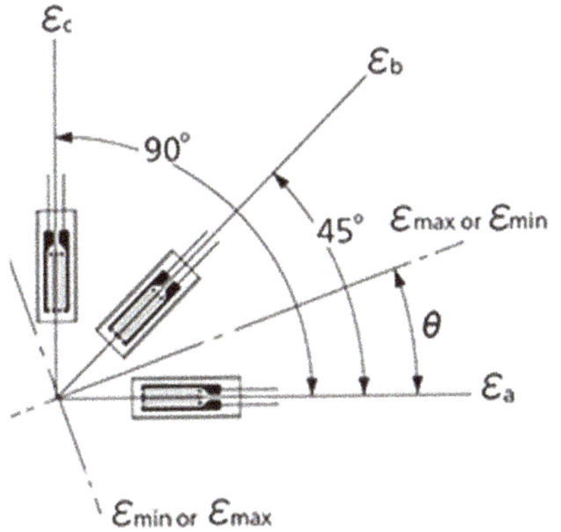

Figure 7. Concept of principal strains from the triaxial 0°–45°–90° rosette gauge pattern.

In the above equation, the angle, θ, is the angle of the maximum principal strain to the ε_a axis when $\varepsilon_a > \varepsilon_c$ or the angle of the minimum principal strain to the ε_a axis when $\varepsilon_a < \varepsilon_c$.

The accurate coordinates of the five strain gauges attached at the nozzle region and their corresponding node numbers in the seismic analysis model of Figure 3 are listed in Table 6.

Table 6. Coordinates of the attached strain gauges.

Strain Gauge ID	X (EW) (mm)	Y (V) (mm)	Z (NS) (mm)	Corresponding Node [(1)]
SG-1	26.2	156.0	36.0	19118
SG-2	40.7	156.0	18.1	19125
SG-3	44.5	156.0	0.0	137
SG-4	40.7	156.0	−18.1	851
SG-5	26.2	156.0	−36.0	858

The origin global coordinates (X = 0, Y = 0, Z = 0) are at the center of the nozzle bottom surface. [(1)] The detailed node locations are described in Figure 17(b) below.

Figures 8 and 9 present the test results of the principal strain time history responses for PGA = 0.3 and 0.6 g, respectively. These are calculated from the measured strains of ε_a, ε_b, and ε_c using Equations (1) and (2). As shown in the figures, the strain response time histories of the maximum principal strain and the minimum principal strain are almost symmetric. This means that the seismic strain responses at the nozzle have fully reversing characteristics with almost the same tension and compression behavior. Among the strain gauges, we can see that SG-5 shows the most significant seismic strain responses.

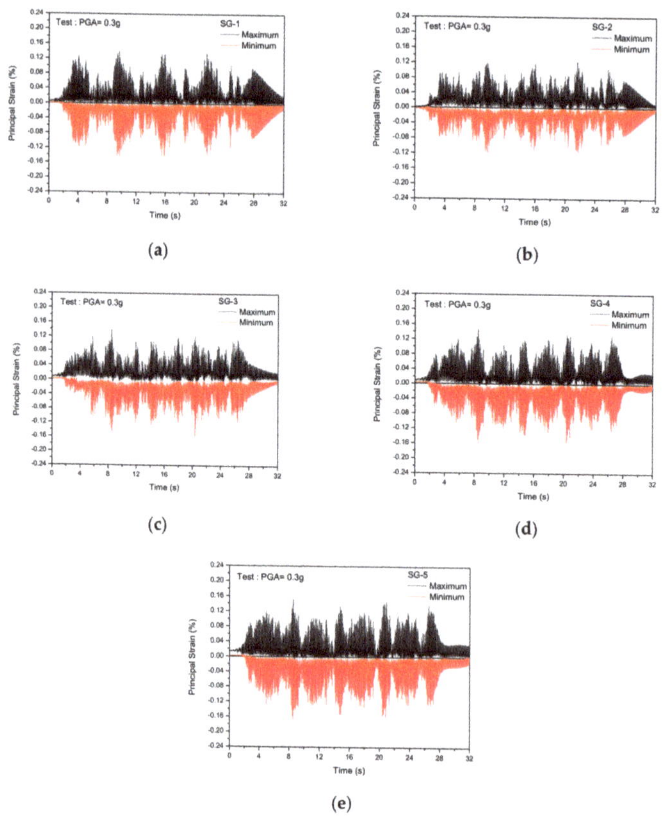

Figure 8. Test results of principal strain time history responses for PGA = 0.3 g: (**a**) SG-1; (**b**) SG-2; (**c**) SG-3; (**d**) SG-4; (**e**) SG-5.

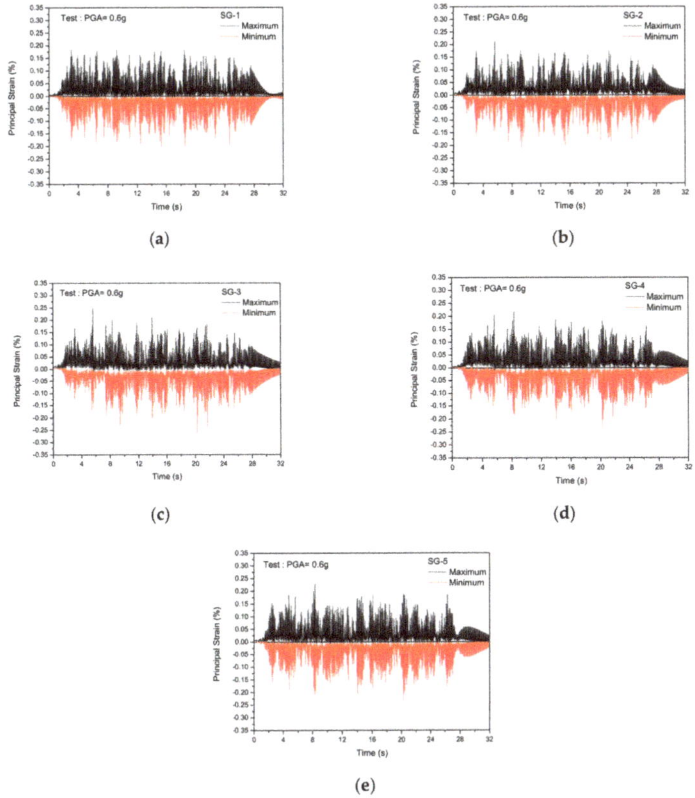

Figure 9. Test results of principal strain time history responses for PGA = 0.6 g: (**a**) SG-1; (**b**) SG-2; (**c**) SG-3; (**d**) SG-4; (**e**) SG-5.

3.3.2. Acceleration Time History Seismic Responses

Figures 10–12 present the measured acceleration time history responses for PGA = 0.3 g. As shown in Figure 10, measured at the shaking table, the maximum values are 1.74 g for EW, 1.52 g for NS, and 1.92 g for the vertical direction. These measured data will be used for the validation of the inelastic seismic analysis as input motions.

Table 7 presents the test results of the zero-period acceleration (ZPA) obtained from the shaking table tests of two scale input motions (PGA = 0.3 g and PGA = 0.6 g) and the ZPA ratio of the two cases for each exciting direction. As shown in the table, the ZPA values of target input motions for PGA = 0.6 g are linearly set to be two times those of PGA = 0.3 g. Then, the ZPA ratio measured at the shaking table is almost 2.0, the same as the target ZPA ratio. However, the ZPA ratio values measured at the end of the pipe are much less than 2.0. These results mean that significant plastic behavior might be occurring in the piping system during the PGA = 0.6 g scale seismic shaking table test. Since this plastic behavior in the piping system has a role in energy dissipation in the seismic responses, the acceleration responses do not increase linearly two times even if the input motions increase two times. From these results, it can be seen that in the case of an earthquake level large enough to exceed the design level, a more accurate seismic response can be obtained using the inelastic seismic analysis method, and the effect of reducing the seismic response can be obtained. This will be discussed in the results of the inelastic analysis in the section below.

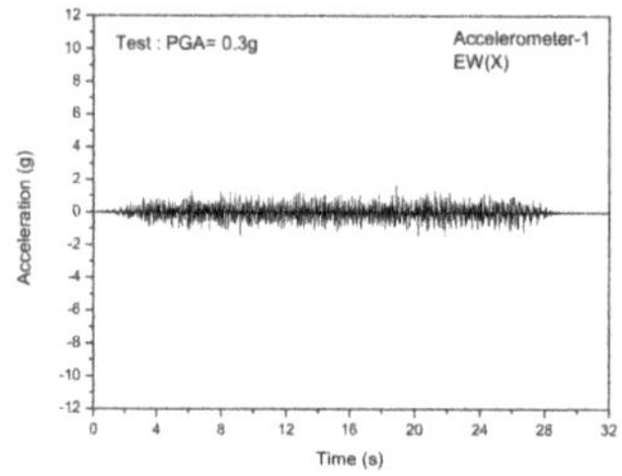

(**a**)

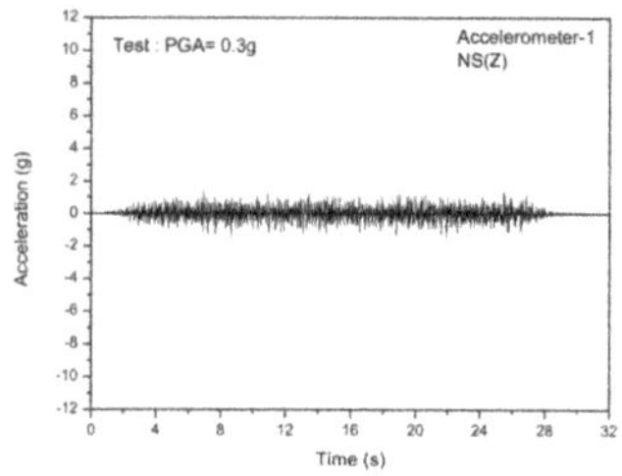

(**b**)

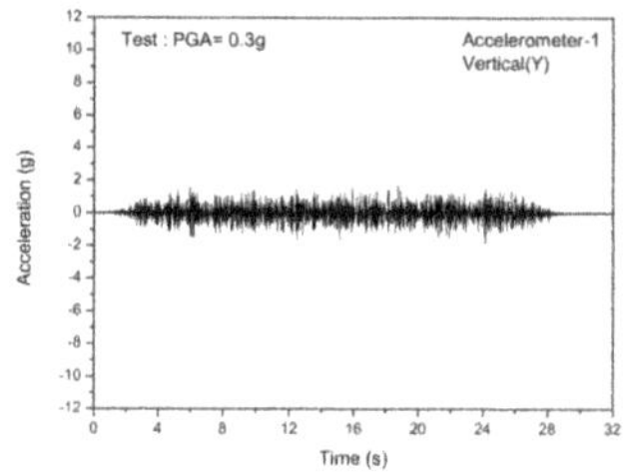

(**c**)

Figure 10. Test results of acceleration responses at the shaking table for PGA = 0.3 g: (**a**) EW (X); (**b**) NS (Z); (**c**) V (Y).

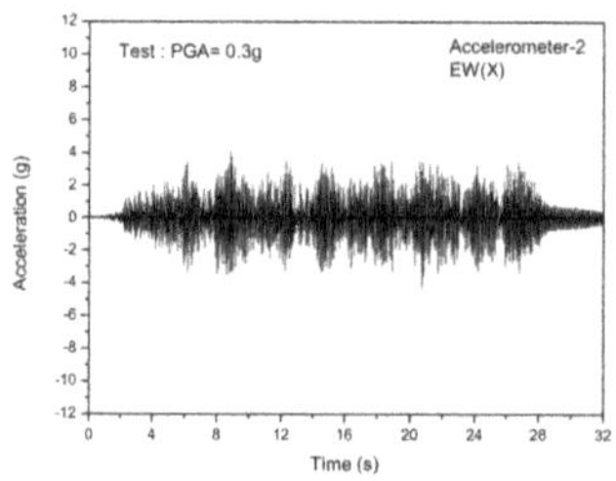

(**a**)

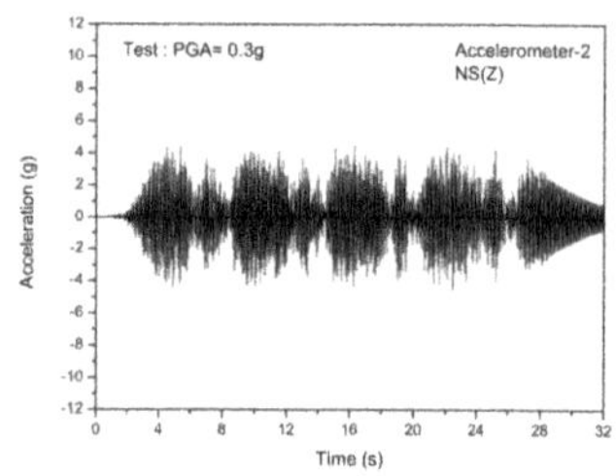

(**b**)

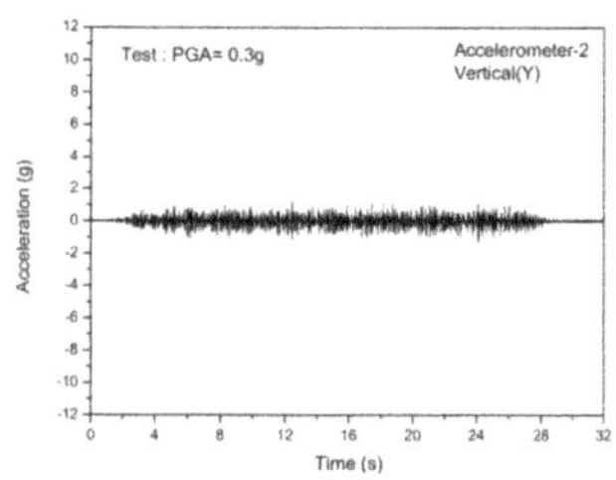

(**c**)

Figure 11. Test results of acceleration responses at Accelerometer-2 for PGA = 0.3 g: (**a**) EW (X); (**b**) NS (Z); (**c**) V (Y).

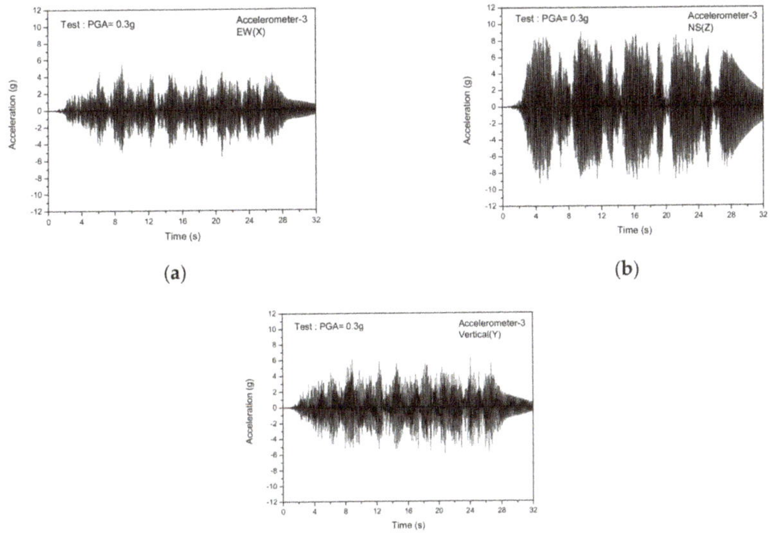

(a) (b) (c)

Figure 12. Test results of acceleration responses at Accelerometer-3 for PGA = 0.3 g: (**a**) EW (X); (**b**) NS (Z); (**c**) V (Y).

Table 7. Test results of ZPA responses.

Directions	PGA	Target Input Motions		Shaking Table (Accelerometer-1)		End of Pipe (Accelerometer-3)	
		ZPA (g)	ZPA Ratio *	ZPA (g)	ZPA Ratio	ZPA (g)	ZPA Ratio
EW (X)	0.3 g 0.6 g	1.18 2.35	2.0	1.71 3.33	1.95	5.55 8.76	1.57
NS (Z)	0.3 g 0.6 g	0.86 1.75	2.0	1.47 3.13	2.12	9.26 12.70	1.37
V (Y)	0.3 g 0.6 g	1.33 2.65	2.0	1.86 3.62	1.95	6.23 9.88	1.58

* The ZPA ratio represents the ratio of ZPA responses corresponding to PGA = 0.6 g and PGA = 0.3 g.

4. Validation of Inelastic Seismic Analysis

To validate the inelastic seismic analysis method, which can be used for strain-based seismic design, inelastic seismic time history analyses were performed, and their results were compared with those of the tests.

4.1. Analysis Modeling

4.1.1. Dynamic Characteristics

The used finite element model for the test mockup is shown in Figure 3. To confirm the dynamic characteristics of the analysis model, resonance searching tests were carried out, with the random input motions having the frequency range of 1.0 to 100.0 Hz. Figure 13 presents the frequency response functions measured by Accelerometer-3 (shown in Figure 5).

As shown in Figure 13, the measured dominant resonance frequencies are about 11.75 Hz for the 1st mode and 28 Hz for the 2nd mode. Table 8 presents the comparison of the resonance frequencies between tests and analyses.

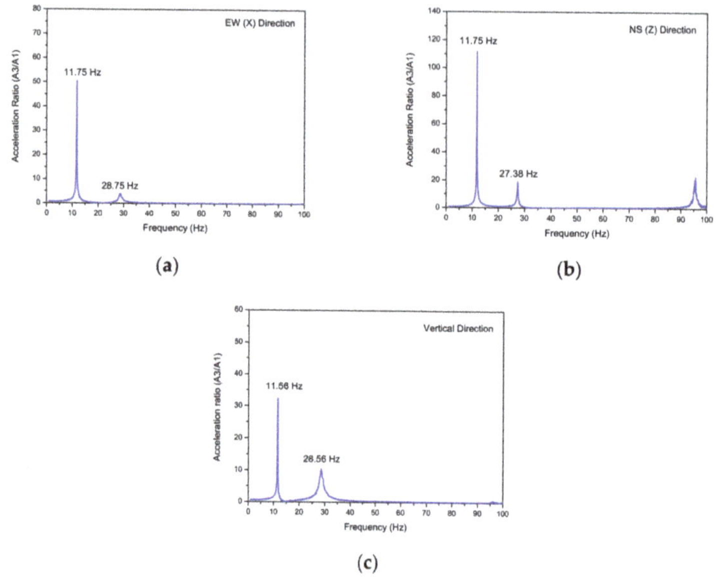

Figure 13. Resonance test results for each direction: (**a**) EW (X); (**b**) NS (Z); (**c**) vertical (Y).

Table 8. Comparison of resonance frequencies between tests and analyses.

Directions		Mode 1 (Hz)	Mode 2 (Hz)
EW (X)	Test	11.75	28.75
	Analysis	11.99	27.63
NS (Z)	Test	11.75	27.38
	Analysis	11.74	25.89
V (Y)	Test	11.56	28.56
	Analysis	11.99	27.63

As shown in Table 8, the dominant modes for EW and NS directions are almost the same in the tests but slightly different in the modal analysis results. In the test results, the frequency of Mode 1 is the same in both directions because it is not easy to separate such a closed mode in the test. However, it is confirmed that the used seismic analysis model describes the dynamic characteristics of the test mockup in good agreement.

4.1.2. Structural Damping Value

To be used for inelastic seismic analyses, the actual structural damping value corresponding to the test mockup is required. As shown in Figure 13, the test mockup is considered to be a very light damped system with dominantly well-separated modes. Therefore, the well-known half-power bandwidth method [18] can be used to identify the structural damping ratio. Figure 14 illustrates the concept for the dominant first natural frequency of the EW direction. In this method, it is assumed that half the total power of dissipation in this mode occurs in the frequency band between f_1 and f_2, where f_1 and f_2 are the frequencies corresponding to an amplitude of $f_c/\sqrt{2}$.

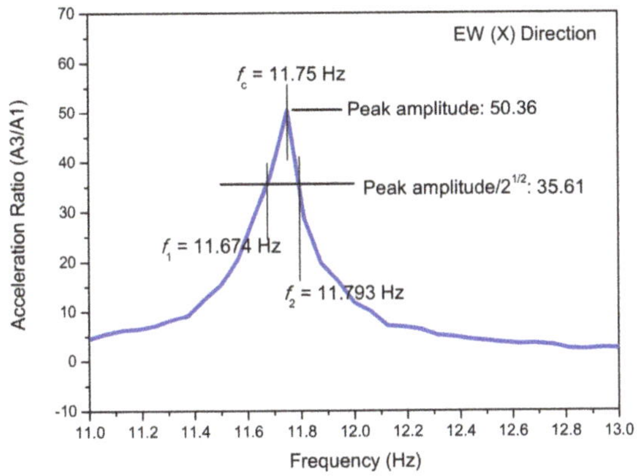

Figure 14. Spectrum illustrating the half-power bandwidth method for damping estimation (1st EW mode).

As illustrated in Figure 14, the critical damping ratio, ξ can be approximately determined by the following relationship;

$$\xi = \frac{f_1 - f_2}{2f_c} \tag{5}$$

From Equation (5), the obtained critical damping values based on the first natural frequency mode for each direction are 0.5% for EW, 0.4% for NS, and 0.8% for the vertical direction. Then, the averaged value of 0.57% for the three directions is used in this study.

4.1.3. Inelastic Material Model

For inelastic seismic analyses, the well-known constitutive equations of Chaboche's kinematic hardening model [19,20] is used as follows:

$$\dot{\alpha}_{ij} = \sum_{k=1}^{3}\left[\frac{2}{3}C_k\dot{\varepsilon}_{ij}^{p} - \gamma_k\left(\alpha_{ij}\right)_k\dot{p}\right] \tag{6}$$

where $\dot{\alpha}_{ij}$ and $\dot{p}$ indicate the revolution of back stress and an accumulated plastic strain, respectively. Additionally, C_k and γ_k (k = 1~3) are material constants to be used in the ANSYS program.

For the isotropic hardening model, the inelastic Voce model [21] is used as follows:

$$\dot{R} = b\left[Q - R\right]\dot{p} \tag{7}$$

where $\dot{R}$ indicates the revolution of drag stress. Additionally, b and Q are material constants.

Table 9 presents the used material constants required in Equations (6) and (7) for Type 316 stainless steel [22].

Table 9. Material constants for inelastic material models.

Material	σ_{yo} $\times 10^6$ (Pa)	E $\times 10^9$ (Pa)	C_1 $\times 10^9$	C_2 $\times 10^9$	C_3 $\times 10^9$	γ_1 $\times 10^3$	γ_2 $\times 10^3$	γ_3	b	Q $\times 10^6$
Type 316SS	135	190	120	20.2	10.67	1.0	1.0	1.0	45.0	85

4.1.4. Seismic Input Motions

Figure 15 presents the displacement time history input motions used for the inelastic seismic analyses, which are measured at the shaking table. The correlation coefficients are 0.015 between EW and NS, 0.0015 between NS and V and 0.0003 between V and EW, which are much less than the criteria value of 0.16 required for the seismic input motions [23]. Therefore, since the independence of the input motions is guaranteed, the inelastic seismic analyses can be performed by applying them simultaneously.

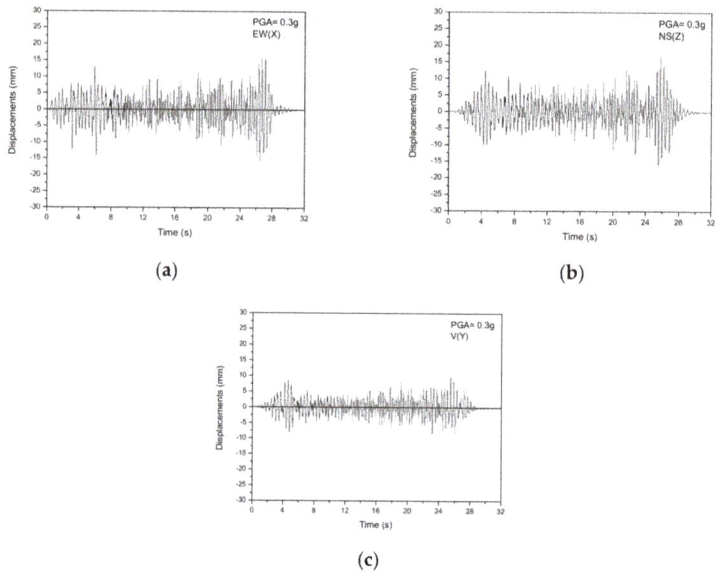

Figure 15. Displacement input motions used for analyses for PGA = 0.3 g: (**a**) EW (X); (**b**) NS (Z); (**c**) V (Y).

The time interval (Δt) used in the inelastic seismic analysis is 3.91 ms; then, the cut-off frequency ($1/2\Delta t$), 128 Hz, can sufficiently cover the seismic cut-off frequency of 33 Hz in NPP seismic design. The total analysis duration is 32 s. Figure 16 presents the test response spectrum (TRS) calculated from the measured shaking table motions of Figure 15. Compared with the target required response spectrum in Figure 2, it is confirmed that the TRS envelops the target response spectrum.

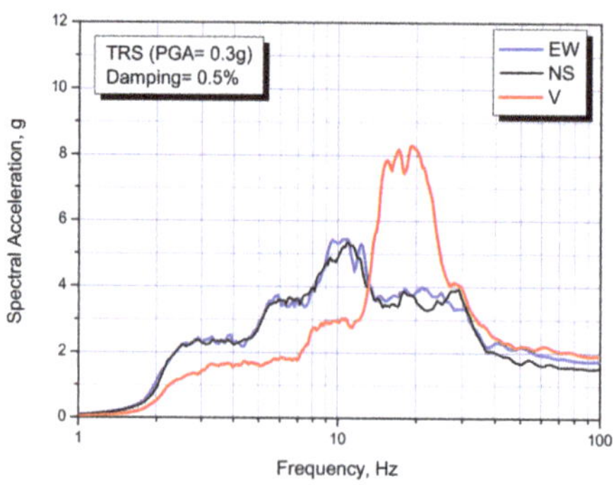

Figure 16. Test response spectrum of shaking table motions.

4.2. Results of Inelastic Seismic Analyses

4.2.1. Validation of Strain Time History Responses

Figure 17 presents the inelastic seismic time history analysis results, representing the accumulated equivalent plastic strain distribution at the end time of PGA = 0.3 g in the nozzle. As shown in the results, we can see that the maximum seismic strain responses occur in the region of the nozzle's safe end, as expected in the test mockup design. This result is considered to be dominantly caused by the bending motions of the piping system. In addition, the frequencies of Mode 1 (11.74 Hz, torsional) and Mode 2 (11.99, bending) are in a closed mode. It is judged that these closed bending and torsional modes strongly influence the determination of the location where the maximum strain response occurs.

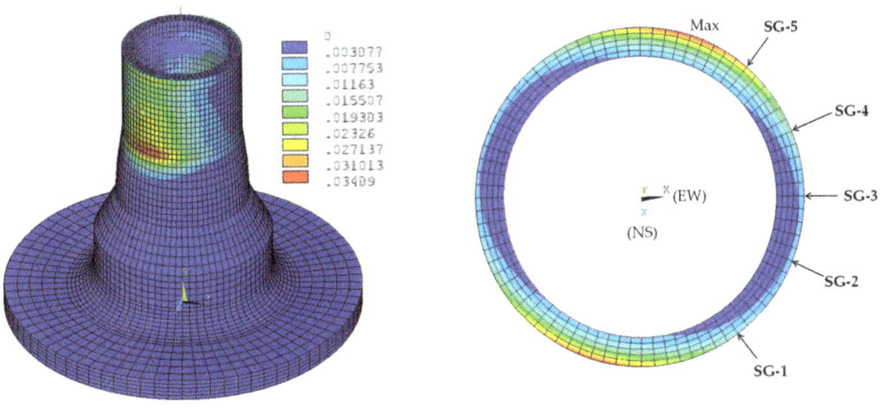

(a) (b)

Figure 17. Inelastic seismic analysis results of accumulated equivalent plastic strain distributions at the nozzle (PGA = 0.3 g): (**a**) isometric view; (**b**) sectional view at the maximum region and node points corresponding to locations of strain gauges attached in the tests.

If we look more closely in Figure 17b, which shows the inelastic strain distribution in the cross-section where the maximum strain occurs, the largest inelastic strain occurs along the (EW, NS) direction. From the sectional view, it is expected that the largest seismic responses will occur at the SG-5 location among the strain gauges in tests.

Figure 18 presents the material hysteretic responses at the location of the maximum accumulated equivalent plastic strain in Figure 17.

To compare the inelastic seismic analysis results with the test results, the maximum and minimum principal strains are calculated at each node, corresponding to the locations of the strain gauges, as shown in Figure 17b. Figures 19 and 20 present the analysis results of the principal strains for PGA = 0.3 and 0.6 g, respectively. As expected from the inelastic seismic analysis, the largest strain responses occur at SG-5.

When compared, the seismic strain time history responses of Figures 19 and 20 with the test results of Figures 8 and 9, the overall strain wave shapes and amplitudes are similar. Specifically, we can see that the largest strain responses occur at the same location of SG-5 in the tests and analyses, as expected in Figure 17. In general, the maximum accumulated equivalent plastic strain, which can be a failure mode in strain-based seismic design criteria, occurs at the location where the maximum value of the time history response occurs, but this is not always the case.

Table 10 presents the comparison results of the maximum principal strain values of the time history responses between the tests and inelastic seismic analyses for the case of PGA = 0.3 g and PGA = 0.6 g. As shown in the table, the results of the inelastic strain responses are in good agreement with the test results.

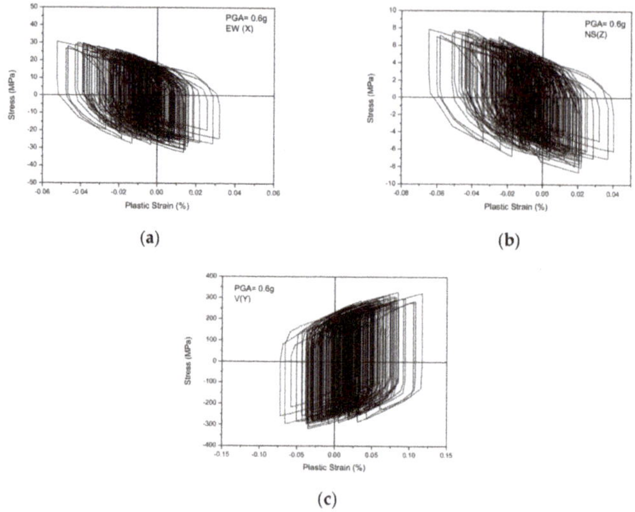

Figure 18. Hysteretic behavior of the material at the location of the maximum accumulated equivalent plastic strain for PGA = 0.6 g: (**a**) EW (X); (**b**) NS (Z); (**c**) V (Y).

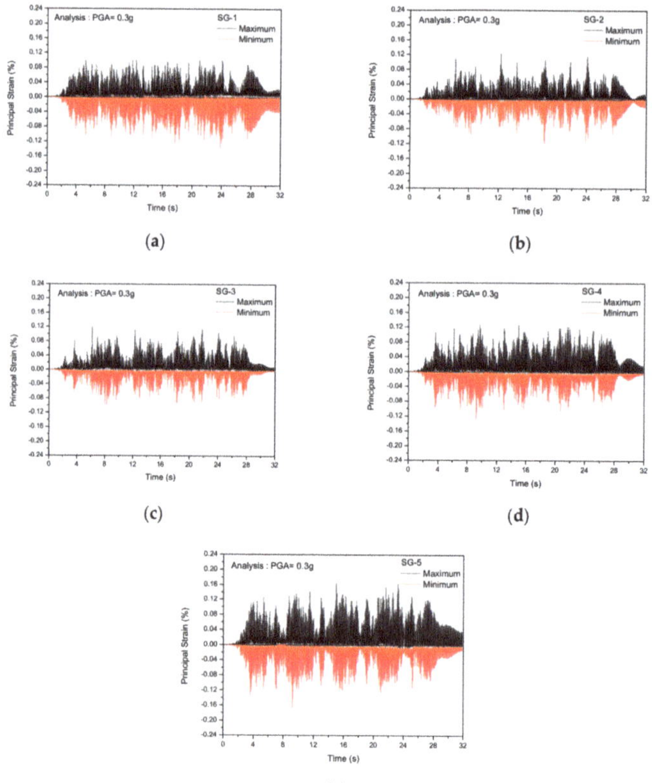

Figure 19. Inelastic seismic analysis results of principal strain time history responses for PGA = 0.3 g: (**a**) SG-1; (**b**) SG-2; (**c**) SG-3; (**d**) SG-4; (**e**) SG-5.

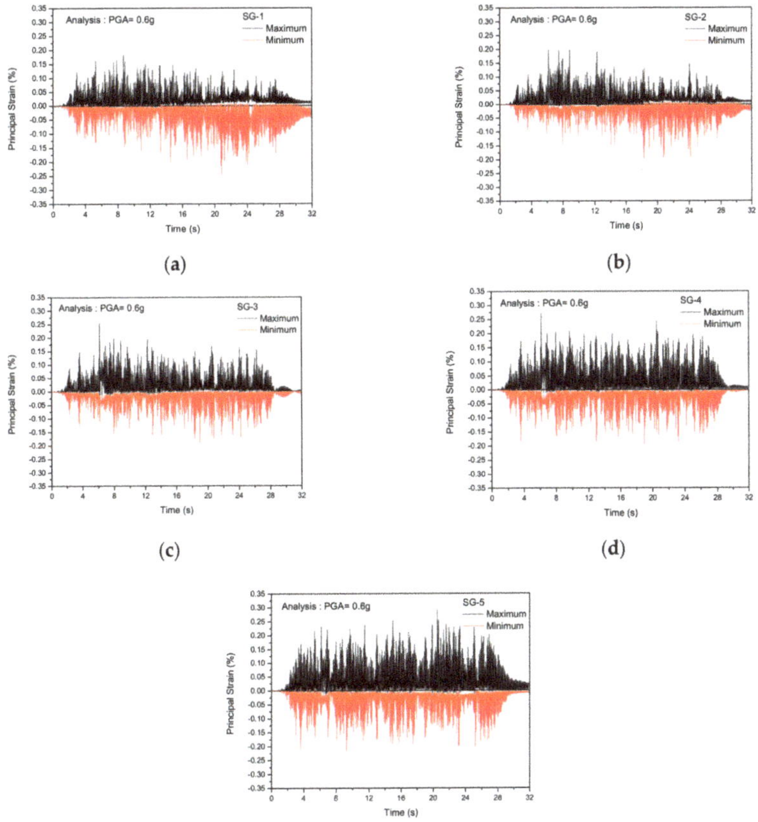

(e)

Figure 20. Inelastic seismic analysis results of principal strain time history responses for PGA = 0.6 g: (**a**) SG-1; (**b**) SG-2; (**c**) SG-3; (**d**) SG-4; (**e**) SG-5.

Table 10. Comparison results of principal strain responses.

Strain Gauge ID	Tests (%)				Inelastic Seismic Analysis (%)			
	PGA = 0.3 g		PGA = 0.6 g		PGA = 0.3 g		PGA = 0.6 g	
	Max(ε_{max})	Min(ε_{min})	Max(ε_{max})	Min(ε_{min})	Max(ε_{max})	Min(ε_{min})	Max(ε_{max})	Min(ε_{min})
SG-1	0.139	−0.144	0.185	−0.202	0.101	−0.139	0.184	−0.247
SG-2	0.125	−0.119	0.212	−0.211	0.123	−0.120	0.199	−0.197
SG-3	0.139	−0.158	0.249	−0.262	0.119	−0.099	0.255	−0.189
SG-4	0.148	−0.161	0.217	−0.239	0.127	-0.128	0.273	−0.194
SG-5	0.152	−0.168	0.228	−0.230	0.165	−0.167	0.290	−0.216

4.2.2. Validation of Acceleration Time History Responses

The investigation of the acceleration time history responses at the pipe is important in pointing out that the seismic strain responses in the nozzle depend on the piping's seismic behavior. Figures 21 and 22 present the acceleration time histories at the locations of Accelerometer-2 and Accelerometer-3 (see Figure 5). As shown in the figures, the overall response waveforms are very similar to the test results of Figures 11 and 12. The largest acceleration response occurs in the NS (Z) direction. This will dominantly excite Mode 1 and Mode 3, invoking torsional responses at the nozzle.

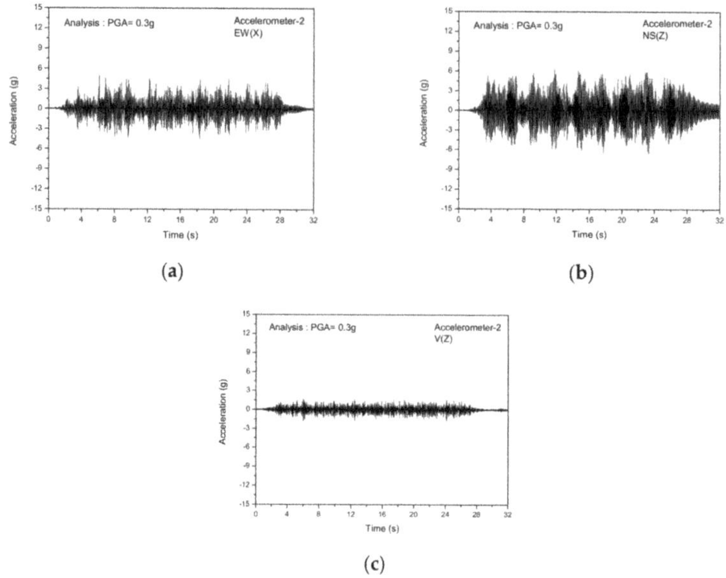

Figure 21. Inelastic seismic analysis results of acceleration responses at Accelerometer-2 for PGA = 0.3 g: (**a**) EW (X); (**b**) NS (Z); (**c**) V (Y).

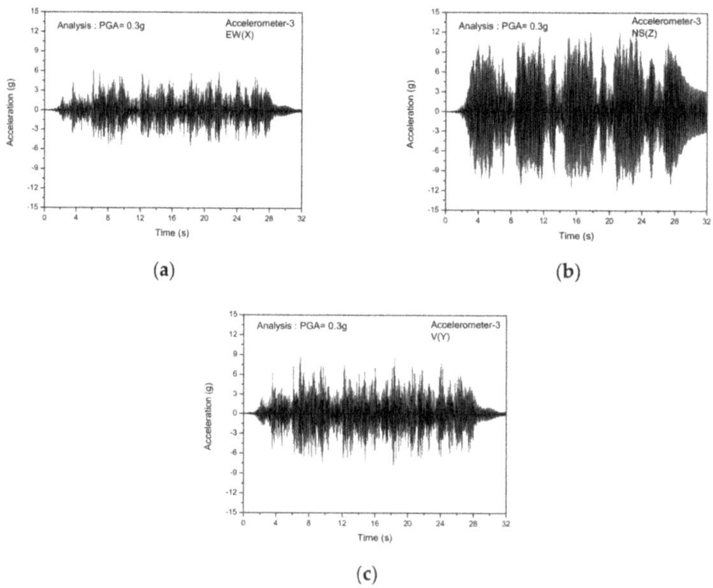

Figure 22. Inelastic seismic analysis results of acceleration responses at Accelerometer-3 for PGA = 0.3 g: (**a**) EW (X); (**b**) NS (Z); (**c**) V (Y).

Table 11 presents the comparison of ZPA values between the tests and analyses at the pipe end (Acceleration-3). The ZPA value corresponds to the maximum amplitude in the acceleration time history responses. As shown in the table, the analysis results reveal slightly larger values compared with the test results, but we can see that overall response characteristics are in good agreement.

Table 11. Comparison results of ZPA responses at the pipe end (Accelerometer-3).

Directions	PGA (g)	Tests (g)	Inelastic Seismic Analysis (g)
EW	0.3 0.6	5.55 8.76	6.17 9.90
NS	0.3 0.6	9.26 12.70	12.68 16.80
V	0.3 0.6	6.23 9.88	8.71 12.60

Figure 23 presents the comparison results of the response spectrum, calculated from the acceleration time history responses at the pipe end (Accelerometer-3). As shown in the figure, the results obtained from the inelastic seismic analysis reveal good agreement with those of the tests.

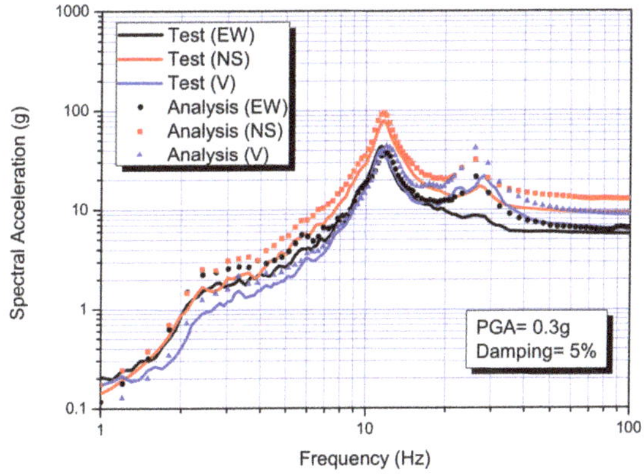

Figure 23. Comparison results of the response spectrum at the pipe end (Acceleromter-3) for PGA = 0.3 g.

5. Conclusions

In this study, shaking table tests are performed to validate the inelastic seismic analysis method applicable to the nuclear metal components. To do this, a test mockup, which can simulate the hot leg surge line nozzle, known as one of the nuclear seismic fragile components in NSSS, was designed and tested on the shaking table with seismic input motions corresponding to design-basis earthquake (PGA = 0.3 g) and beyond-design-basis earthquake (PGA = 0.6 g) levels. To validate the inelastic seismic analysis method, detailed comparisons of seismic responses between the tests and inelastic seismic analyses were carried out, especially for the seismic responses of the principal strains at the nozzle and the accelerations at the pipe. From this study, some meaningful conclusions have been derived, as follows:

1. The seismic responses obtained from the inelastic seismic time history analyses with an accurate inelastic material model using Chaboche's kinematic hardening model and the Voce isotropic hardening model for Type 316 stainless steel are in good agreement with those of the seismic shaking table tests.
2. From the comparison of the seismic strain time history responses at the nozzle between the tests and inelastic analyses, the location of the maximum strain responses from the inelastic analyses was found to be almost the same as the locations in the test results.
3. The structure damping value for the piping systems, recommended in the US NRC RG 1.61 [24], is 4% for the SSE level. However, in this study, the piping system made of Type 316 stainless steel revealed a much lighter damping value of about 0.57%.

4. For earthquakes large enough to result in inelastic behavior at the nuclear metal components, the inelastic seismic analysis is useful for reducing the seismic responses by energy dissipation due to hysteretic damping.
5. From the validation results obtained from the test mockup simulating the actual nuclear component in this study, it is assured that the inelastic seismic analysis method can be used for the seismic design of nuclear metal components in large earthquake scenarios such as the beyond-design-basis earthquake, which can cause significant plastic deformations.

Author Contributions: Conceptualization, G.-H.K., S.-W.A., J.-K.H. and J.-S.K.; methodology, G.-H.K., S.-W.A. and J.-S.K.; validation, G.-H.K.; formal analysis, G.-H.K.; investigation, S.-W.A. and J.-S.K.; data curation, S.-W.A.; writing—original draft preparation, G.-H.K.; writing—review and editing, S.-W.A., J.-K.H. and J.-S.K.; funding acquisition, G.-H.K., J.-K.H. and J.-S.K. All authors have read and agreed to the published version of the manuscript.

Funding: This study was funded by the Ministry of Trade, Industry, and Energy through KETEP (Korea Institute of Energy Technology Evaluation Planning; No. 20181510102380).

Institutional Review Board Statement: Not applicable.

Informed Consent Statement: Not applicable.

Data Availability Statement: Not applicable.

Acknowledgments: This study was supported by the Ministry of Trade, Industry, and Energy through KETEP (Korea Institute of Energy Technology Evaluation Planning; No. 20181510102380).

Conflicts of Interest: The authors declare no conflict of interest.

References

1. *Seismic Isolation Systems for Nuclear Installations*; IAEA-TECDOC-1905; IAEA: Vienna, Austria, 2020.
2. Koshida, H.; Yasaka, A. Vibration tests and earthquake observation results of base-isolated building. *Seism. Shock. Vib. Isol.* **1989**, *181*, 135–140.
3. Koo, G.H.; Jung, J.Y.; Lee, J.H.; Shin, T.M. Development of Small-Sized Lead Inserted Laminated Rubber Bearing for Nuclear Component Seismic Isolation. *Energies* **2020**, *13*, 3193. [CrossRef]
4. Kostarev, V.; Petrenko, A.; Vasilev, P. A New Method for Essential Reduction of Seismic and External Loads on NPP's Structures, Systems and Components. In Proceedings of the 17th International Conference on Structural Mechanics in Reactor Technology, Prague, Czech Republic, 17–22 August 2003; p. 3216.
5. Kwag, S.Y.; Kwag, J.S.; Lee, H.H.; Oh, J.H.; Koo, G.H. Enhancement in the Seismic Performance of a Nuclear Piping System using Multiple Tuned Mass Dampers. *Energies* **2019**, *12*, 2077. [CrossRef]
6. *Alternative Rules for Level D Service Limits of Class 1, 2, and 3 Piping Systems, Section III, Division 1, ASME Code Case (Record No. 13-1438)*; ASME: New York, NY, USA, 2019.
7. Masaki, M.; Akihiro, O.; Tomoyoshi, W.; Izumi, N.; Tadahiro, S.; Masaki, S. Seismic Qualification of Piping Systems by Detailed Inelastic Response Analysis: Part 1—A Code Case for Piping Seismic Evaluation Based on Detailed Inelastic Response Analyses. In Proceedings of the ASME 2017 PVP Conference, Waikoloa, HI, USA, 16–20 July 2017; Volume 8.
8. Akihiro, O.; Tadahiro, S.; Masaki, M.; Izumi, N.; Tomoyoshi, W.; Masaki, S. Seismic Qualification of Piping Systems by Detailed Inelastic Response Analysis: Part 2—A Guideline for Piping Seismic Inelastic Response Analysis. In Proceedings of the ASME 2017 PVP Conference, Waikoloa, HI, USA, 16–20 July 2017; Volume 8.
9. Koo, G.H.; Kim, J.S.; Kim, Y.J. Feasibility Study on Strain-Based Seismic Design Criteria for Nuclear Components. *Energies* **2020**, *13*, 4435. [CrossRef]
10. ASME. *BPVC Section III, Rules for Construction of Nuclear Facility Components, Mandatory Appendix XIII, Design Based on Stress Analysis*; ASME: New York, NY, USA, 2021.
11. AFCEN. *RCC-M: Design and Construction Rules for Mechanical Components of PWR Nuclear Islands*; AFCEN: Lyon, France, 2017.
12. Kim, J.Y.; Lee, J.M.; Park, J.G.; Kim, J.S.; Cho, M.K.; Ahn, S.W.; Koo, G.H.; Lee, B.H.; Huh, N.S.; Kim, Y.J.; et al. Round robin analysis to investigate sensitivity of analysis results to finite element elastic-plastic analysis variables for nuclear safety class 1 components under severe seismic load. *Nucl. Eng. Technol.* **2021**. [CrossRef]
13. Koo, G.H.; Kwag, S.; Nam, H.S. Study on Inelastic Strain-Based Seismic Fragility Analysis for Nuclear Metal Components. *Energies* **2021**, *14*, 3269. [CrossRef]
14. Ravikiran, A.; Dubey, P.N.; Agrawal, M.K.; Reddy, G.R.; Singh, R.K.; Vaze, K.K. Experimental and Numerical Studies of Ratcheting in a Pressurized Piping System under Seismic Load. *J. Press. Vessel. Technol.* **2015**, *137*, 3. [CrossRef]

15. Shirai, E.; Eto, K.; Umemoto, A.; Yoshii, T.; Kondo, M.; Shimizu, H.; Tai, K. Inelastic Seismic Test of the Small Bore Piping and Support System: Part 2—Static Failure Test for Piping Support Equipment. In Proceedings of the ASME 2008 PVP Conference, Chicago, IL, USA, 27–31 July 2008; Volume 8.
16. ANSYS. *ANSYS Mechanical APDL Release 15.0*; ANSYS, Inc.: Canonsburg, PA, USA, 2014.
17. Young, W.C.; Budynas, R.G.; Sadegh, A.M. *Roark's Formulas for Stress and Strain*, 8th ed.; McGraw Hill: New York, NY, USA, 2012.
18. Harris, C.M. *Shock and Vibration Handbook*; McGraw-Hill Book Company: New York, NY, USA, 1988.
19. Chaboche, J.L.; Rousselier, G. On the plastic and viscoplastic constitutive equations—Part II: Application of internal variable concepts to the 316 stainless steel. *J. Press. Vessel. Technol.* **1983**, *105*, 159–164. [CrossRef]
20. Chaboche, J.L. Constitutive equations for cyclic plasticity and cyclic viscoplasticity. *Int. J. Plast.* **1989**, *5*, 247–302. [CrossRef]
21. Voce, E. A Practical Strain hardening Function. *Metallurgia* **1955**, *51*, 219–226.
22. Koo, G.H.; Yoon, J.H. Inelastic Material Models of Type 316H for Elevated Temperature Design of Advanced High Temperature Reactors. *Energies* **2020**, *13*, 4548. [CrossRef]
23. *Standard Review Plan, 3.7.1 Seismic Design Parameters*; NUREG-0800; U.S. Nuclear Regulatory Commission: Washington, DC, USA, 2012.
24. *Damping Values for Seismic Design of Nuclear Power Plants; Regulatory Guide 1.61*; U.S. Nuclear Regulatory Commission: Washington, DC, USA, 2007.

applied sciences

MDPI

Article

Experimental and Finite Element Analysis of External ALC Panel Steel Frames with New Semi-Rigid Connector

Kewei Ding [1,*], Da Zong [1], Yunlin Liu [1], Shulin He [1] and Wanyu Shen [2]

1 School of Civil Engineering, Anhui Jianzhu University, Hefei 230601, China; zongda327@gmail.com (D.Z.); kzdy@ahjzu.edu.cn (Y.L.); heshulin4@gmail.com (S.H.)
2 Anhui Fuhuang Steel Structure Co., Ltd., Hefei 238076, China; shenwy@fuhuang.com
* Correspondence: dingkw@ahjzu.edu.cn

Abstract: In this paper, a new ALC panel connector was proposed. It has a good engineering economy and high fault tolerance. A quasistatic loading experiment was carried out to verify the feasibility of the external ALC panel steel frame under seismic loading. The test phenomena, hysteretic curve, skeleton curve, stiffness degradation, and energy dissipation of two sets of full-scale specimens were analyzed and discussed. Moreover, the simulation of pendulous Z-panel connectors with different thicknesses was carried out using ABAQUS software. The comparison reveals that the semi-rigid connection has a full hysteresis curve, good energy dissipation capacity, and a 15% increase in peak load capacity. Finally, similar results for different thicknesses in the use of pendulous Z-panel connectors reveal that using the 6 mm connector may be the most economical solution for engineering.

Keywords: steel structures; autoclaved lightweight aerated concrete (ALC) panel; seismic behavior; finite element analysis; parametric study

Citation: Ding, K.; Zong, D.; Liu, Y.; He, S.; Shen, W. Experimental and Finite Element Analysis of External ALC Panel Steel Frames with New Semi-Rigid Connector. *Appl. Sci.* **2021**, *11*, 10990. https://doi.org/10.3390/app112210990

Academic Editor: Maria Favvata

Received: 18 October 2021
Accepted: 17 November 2021
Published: 19 November 2021

Publisher's Note: MDPI stays neutral with regard to jurisdictional claims in published maps and institutional affiliations.

Copyright: © 2021 by the authors. Licensee MDPI, Basel, Switzerland. This article is an open access article distributed under the terms and conditions of the Creative Commons Attribution (CC BY) license (https://creativecommons.org/licenses/by/4.0/).

1. Introduction

With urbanization accelerating, traditional construction techniques cannot meet the current increasing demand for housing. Prefabricated building is considered a major trend for future development because of its good seismic performance, easy construction, low greenhouse gas emissions, and high utilization of building materials [1–3]. As a type of enclosure system of prefabricated buildings, autoclaved lightweight concrete (ALC or AAC) panels are made of several fine aggregates (cement, sand, gypsum, aluminum powder, etc.) by high temperature and pressure [4,5]. Many scholars conducted in-depth studies. Pehlivanlı [6] et al. studied the strength of AAC blocks with different fiber additions. The result shows that the flexural and compressive strengths of the test blocks are increased, especially for the carbon fiber group. Seddighi et al. [7] investigated the mechanical properties of AAC blocks incorporating graphene. They demonstrate that the addition of graphene brings a significant increase to the compressive and tensile strength and impact resistance of AAC. In addition, other researchers have also attempted to improve the seismic performance of structures by changing material parameters and effective out-of-plane reinforcement techniques. Rousakis et al. [8] investigated RC-framed infilled wall structures reinforced by polyurethane joints (PUFJ) or polyurethane-impregnated fiber grids (FRPU). The result shows that its application improves the initial stiffness, base shear, and maintains a high horizontal drift. By using in-plane cyclic shear-compression tests, Penna et al. [9] studied AAC masonry walls with bed-joint (BJR) reinforcement specimens. They show a significant increase in maximum deformation capacity, shear strength, and seismic performance compared to an unreinforced wall. Kałuża et al. [10] studied the deformation process of in-plane AAC walls with different joint treatments. They find that the specimens reinforced with GFRP show some improvement in deformation, stiffness, and load-carrying capacity. Binici et al. [11] proposed a new innovative fencing system. The results show that its application achieves the required 2% in-plane deformation without cracking and maintains its out-of-plane stability. Deng et al. [12] studied the seismic performance of URM walls reinforced by HDC, which improves the ultimate displacement, energy dissipation, and bearing capacity. De Paula Salgado et al. [13] studied the seismic performance of AAC infill walls with the addition of curauá fiber–cement composites. The result showes satisfactory post-peak ductility because the energy is not abruptly lost but gradually releases throughout its deflection-softening behavior. To improve the strength and ductility of concrete in seismicloading, antiseismic fabric, especially double-sided reinforcement of expansive glass particle plastering is used by Arslan et al. [14]. Erdem et al. [15] studied the seismic performance of three different planar RC frames with flexible connections, including U-slot, T-slot, and female and male slots. The test shows that flexible connections exhibit a bare frame-like behavior, protecting the infill wall and optimizing the rigid wall force pattern. The connection between the prefabricated structure and the main structure is also critical. There are differences in the basic mechanical properties of the different connections (such as beam-column joints, laminated slabs, etc.) [16–21]. However, little research has been performed on the connection of the external ALC panel. The main connection methods are hook head bolt and ADR connectors (as shown in Figure 1). The external hook head bolt is rigidly connected to the ALC panel. The hole will be damaged under seismic loading and increase the risk of overall instability. The ADR connector ensures a flexible connection, but it increases the construction cost. The tests showed that the bolted connector and the swing connector have good mechanical properties, but they ignore the innovation of the connector [22,23].

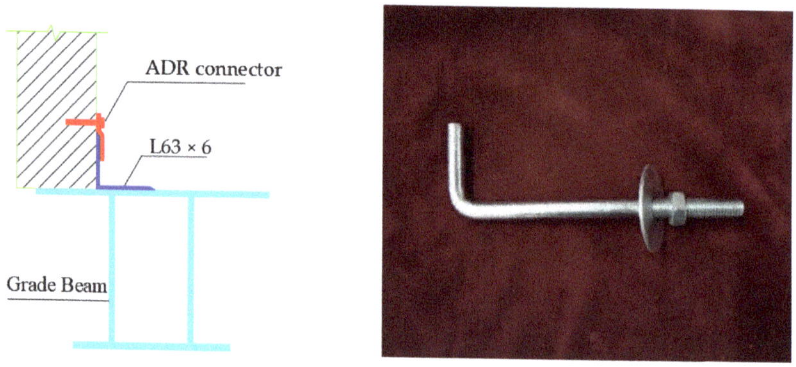

(a) (b)

Figure 1. Connector schematic: (**a**) schematic of ADR connector; (**b**) schematic of external hook head bolt.

Based on the above works, this paper proposed a new connector called the pendulous ALC Z-panel connector. The connector significantly improves the deformation adaptability of the structure. It also reduces construction error and engineering costs. In order to verify the feasibility of the new connector, two sets of full-scale external ALC-frames were tested using horizontal low cyclic loads. Moreover, finite element software was used to model and analyze the connector. The failure modes, hysteresis curves, skeleton curves, and stiffness degradation curves were compared in detail. Finally, the analysis of several groups of connectors of different thicknesses gave rationalization suggestions for practical application.

The rest of the paper is described as follows: In Section 2, we present the specimen design details, material properties, and test details. Section 3 analyzes the experimental results (experimental phenomena, hysteresis curves, skeleton curves, and stiffness degradation curves). In Section 4, finite element software verifies the simulation accuracy, and the connectors with different thicknesses are analyzed. Finally, Section 5 presents some conclusions.

2. Experimental Work

2.1. Design of Specimen

A new pendulous ALC Z-panel connector was proposed, which reduces the effect of seismic load on the external ALC panel steel and improves the deformation resistance of the overall structure. Figures 2 and 3 show the installation diagram of the connector, which consists of the ALC panel, high-strength bolts, and upper and lower parts. The upper connector is divided into an open-hole Z-connector and open-slot L-connector, and the lower connector is composed of a bearing plate and limit hole (as shown in Figures 4–6). The specific installation process is as follows. Firstly, the steel frame is installed, then the upper and lower parts are positioned and installed. The ALC panel is lifted into the appropriate position. Finally, the upper Z-connector protrusion is stuck together with the L-connector and connected to the ALC panel with the high-tensile bolt. The ALC panel is placed on the lower support plate and connected with high-strength bolts through the limit holes. Caulking mortar is applied after the ALC panel installation (as shown in Figure 7).

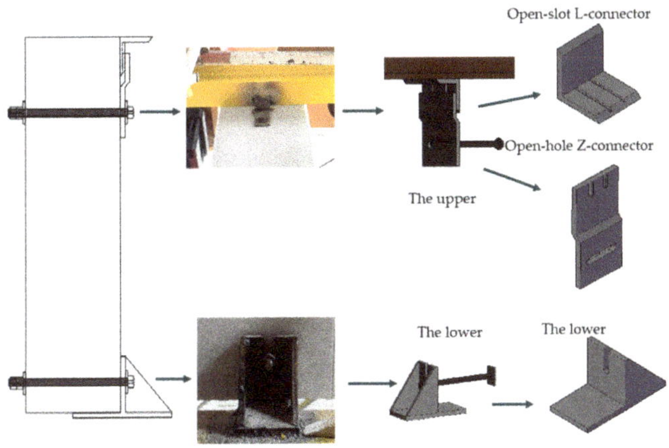

Figure 2. Connector installation diagram.

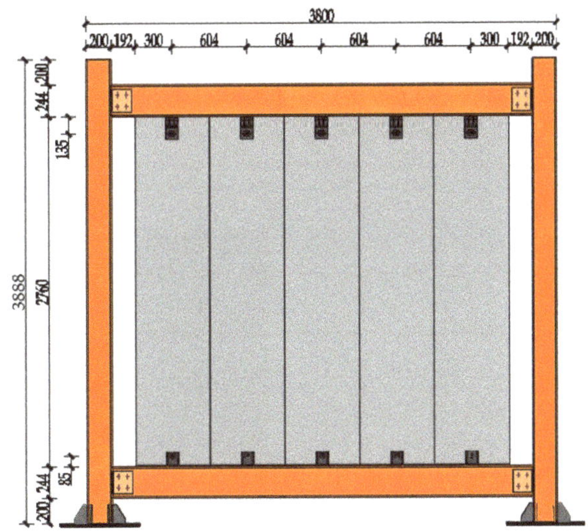

Figure 3. Connector installation dimension schematic (units: mm).

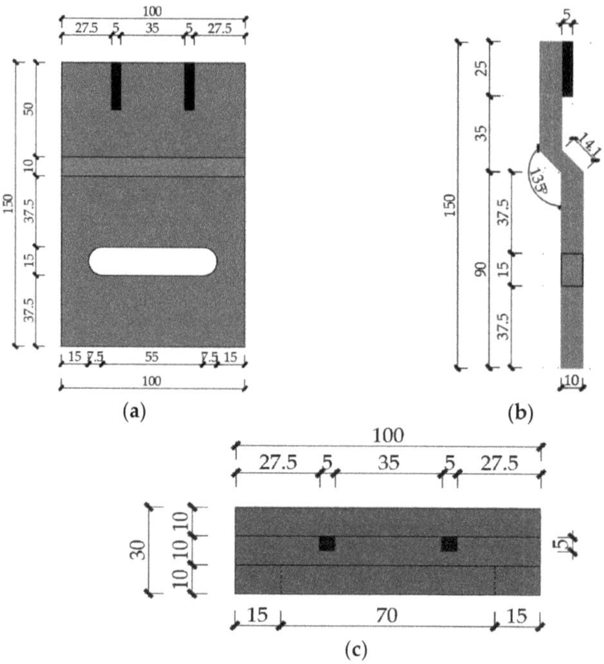

Figure 4. Geometric dimension and reinforcement distribution of the Z-connector: (**a**) front view; (**b**) side view; (**c**) top view (units: mm).

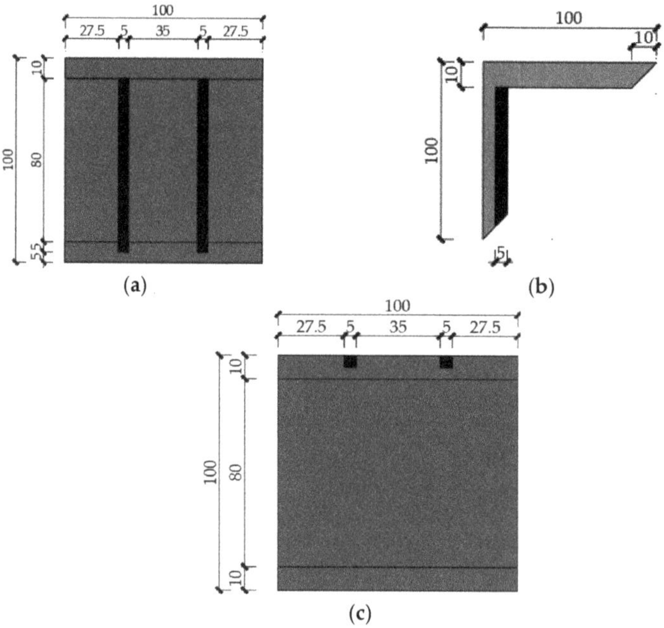

Figure 5. Geometric dimension and reinforcement distribution of the L-connector: (**a**) front view; (**b**) side view; (**c**) top view (units: mm).

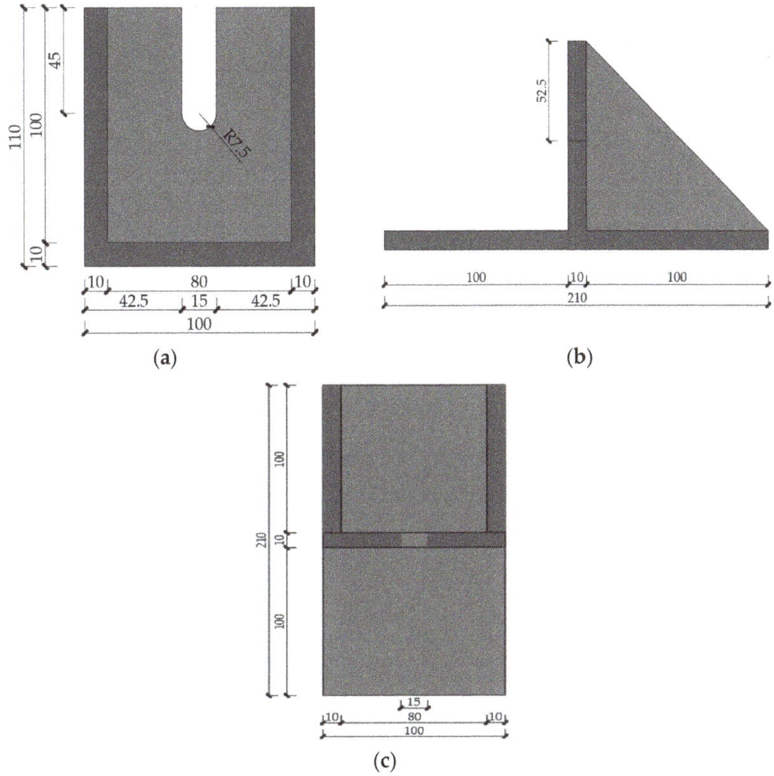

Figure 6. Geometric dimension and reinforcement distribution of the lower connector: (**a**) front view; (**b**) side view; (**c**) top view (units: mm).

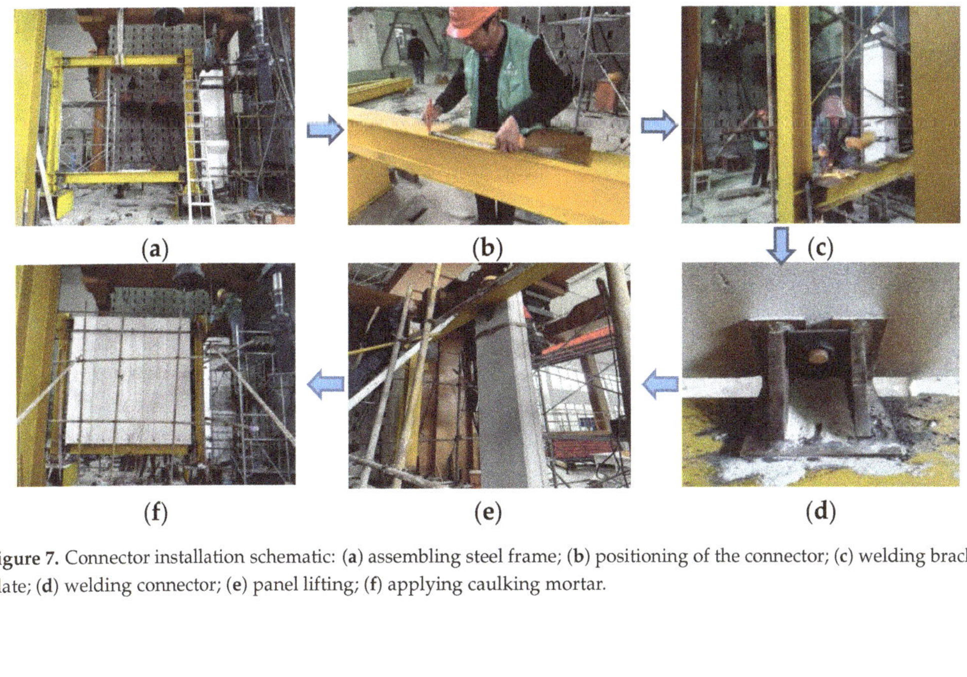

Figure 7. Connector installation schematic: (**a**) assembling steel frame; (**b**) positioning of the connector; (**c**) welding bracket plate; (**d**) welding connector; (**e**) panel lifting; (**f**) applying caulking mortar.

2.2. Material Properties

According to the standard GB/T11969-2020 "Test methods of autoclaved aerated concrete" [24], 9 ALC blocks were prepared to test the compressive strength and E. Their sizes were 100 mm × 100 mm × 100 mm and 100 mm × 100 mm × 300 mm (as shown in Figure 8). Tensile tests were performed on the same batches of steel to determine their yield strength, ultimate strength, modulus of elasticity, and elongation (as shown in Figure 9). The design requirements can be found in GB/T 228.1-2010 "Metallic Materials-Tensile testing" [25]. Tables 1 and 2 show the autoclaved lightweight concrete and the steel mechanical performance index.

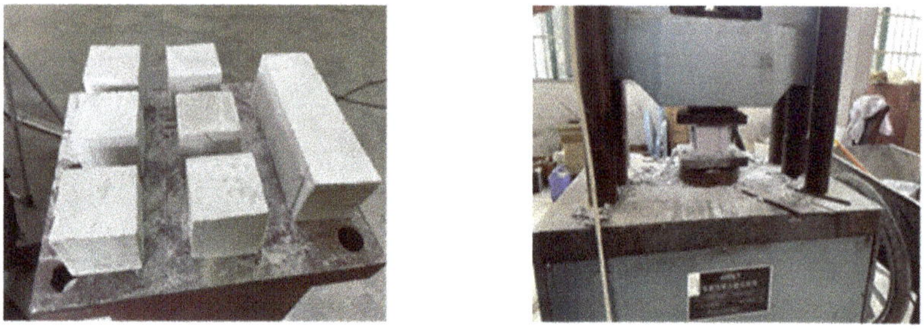

(**a**) (**b**)

Figure 8. ALC compression test: (**a**) ALC specimens; (**b**) electric servo universal testing machine.

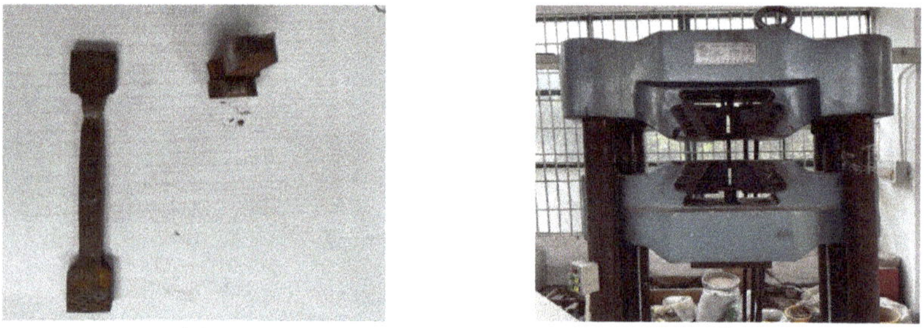

(**a**) (**b**)

Figure 9. Steel compression test: (**a**) steel specimen; (**b**) electric servo universal testing machine.

Table 1. Material properties of autoclaved lightweight concrete.

Specimen	Specimen Dimension (mm)	Cube Compressive Strength (MPa)	Elastic Modulus (GPa)
Sac1	100 × 100 × 100	3.89	
Sac2	100 × 100 × 100	2.97	
Sac3	100 × 100 × 100	3.26	
Sac4	100 × 100 × 100	3.78	
Sac5	100 × 100 × 100	3.96	
Sac6	100 × 100 × 100	3.49	
Average		3.56	
Sae1	100 × 100 × 300		1640
Sae2	100 × 100 × 300		1880
Sae3	100 × 100 × 300		1790
Average			1770

Table 2. Mechanical properties of steel.

Specimen	Sectional Dimension (mm)	Thickness (mm)	Yield Stress (N/mm^2)	Ultimate Stress (N/mm^2)	Elongation
Steel beam flange	HM 244 × 175 × 7 × 11	11	263.4	401.6	25.2%
Steel beam web	HM 244 × 175 × 7 × 11	7	275.3	411.3	22.3%
Steel column flange	HW 200 × 200 × 8 × 12	12	289.5	435.4	24.7%
Steel column web	HW 200 × 200 × 8 × 12	8	278.2	409.8	20.8%
Q345		10	376.6	510.1	19.6%

2.3. Experimental Device and Loading System

The two sets of test specimens were ultimately determined to be single-span planar steel frames with a single floor (3888 mm height and 3800 mm span), based on the site dimensions and practical application requirements of the key laboratory hall of structure and underground space at Anhui Jianzhu University. The FW-1 and FW-2 connector groups denote the hooked bolt and pendulous Z-panel connector groups, respectively. This test utilized an American MTS servo loading system with a stroke of 250 mm. A quasistatic loading system was used for this test. A horizontal, low-circumferential reciprocating load was given to the column's highest point. The loading plate served as a link between the actuator and the frame column (as shown in Figure 10). The steel frames were constructed of hot-rolled Q235 H-beams. More precisely, the beam section measured HM244 mm × 175 mm × 7 mm × 11 mm, and the column section measured HW200 mm × 200 mm × 8 mm × 12 mm, which were all supplied by Anhui Fu Huang Steel Structure Co., Ltd. (Hefei, China).

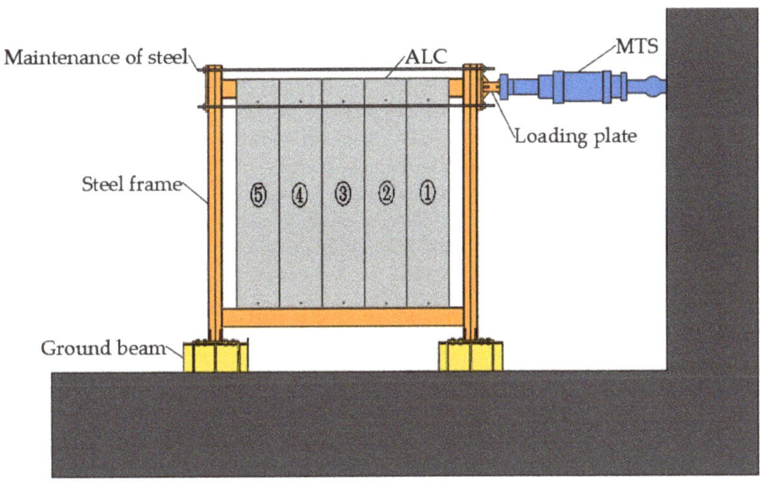

Figure 10. Schematic of test working conditions.

According to the interstory displacement angle index, 5 mm, 6 mm, 7.5 mm, 8.6 mm, 10 mm, 12 mm, and 15 mm cycles of 3-turn loading and 30 mm, 40 mm, 60 mm, 75 mm, 90 mm, and 105 mm cycles of 2-turn loading were taken into consideration, and they can be found in GB50011-2010 "Code for Seismic Design of Buildings" (as shown in Figure 11) [26]. The end of the test was marked by (1 or 2):

1. When the core components (beam, column, connectors) showed apparent damage.
2. The bearing capacity of the member was reduced to 85% of the ultimate load.

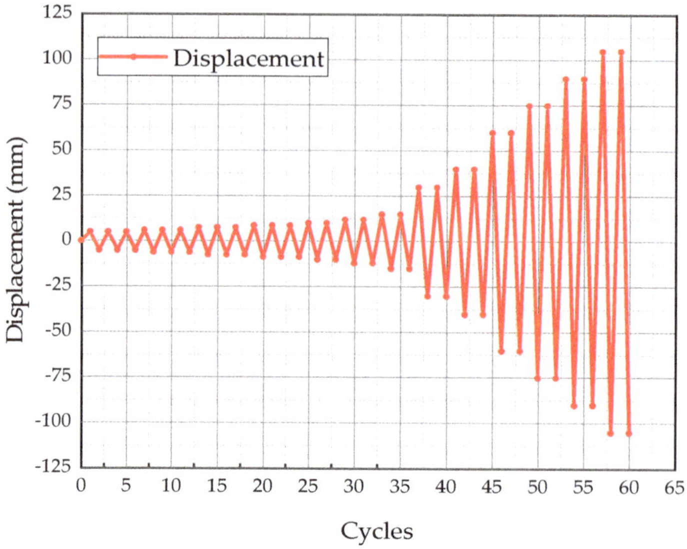

Figure 11. Loading history.

3. Experiment Results and Analysis

3.1. Specimen FW-1

The schematic diagram of the tested phenomena in the FW-1 group is shown in Figure 12. At the early stage of specimen loading, when the beam end was loaded at 1/600 (Δ = 5 mm), there was no noticeable change in deformation and strain of the specimen, and the panels did not produce cracks. When the displacement angle reached 1/500 (Δ = 6 mm), cracks appeared at the bottom of the mortar of panel joints of No. 1 and No. 2 (Figure 12a). When the displacement angle reached 1/400 (Δ = 7.5 mm), the sound of the panel corner rubbing and the crack of inter-slab splicing mortar continued to increase (Figure 12b). When the displacement angle reached 1/350 (Δ = 8.6 mm), the sound of mutual extrusion between panels could be clearly heard, and vertical cracks were produced in the splicing of the No. 1 and No. 2 panels (Figure 12c). Some small pieces of ALC fell off during the loading of 1/300 (Δ = 10 mm) and 1/250 (Δ = 12 mm) (Figure 12d). When the displacement angle reached 1/200 (Δ = 15 mm), there was evident slippage between the enclosure reinforcement and the end panel, and there was a noticeable misalignment between the panel where the splices fell off. When the displacement angle reached 1/50 (Δ = 60 mm), the crack tended to gradually become wider. At displacement angle 1/40 (Δ = 75 mm), the angle weld at the upper part of panel No. 2 broke off, and a crack appeared at the hook headbolt hole at the lower part of panel No. 3. When the displacement angle reached 3/10 (Δ = 90 mm), there was serious breakage at the welded joint of the left upper beam connector, and the cracks of the panels continued to expand (Figure 12e). At displacement angle 7/20 (Δ = 105 mm), each panel corner was damaged on a large scale, and the No. 3 panel presented oblique cracks (Figure 12f).

3.2. Specimen FW-2

The schematic diagram of the tested phenomena in the FW-2 group is shown in Figure 13. When the displacement angles reached 1/600 (Δ = 5 mm), 1/500 (Δ = 6 mm), and 1/400 (Δ = 7.5 mm), there was no obvious phenomenon in the overall specimen. The change of strain was within a reasonable range, the panel did not produce cracks, and a grating appeared during the experiment when the upper connector bolt and the long circular hole slid. When the displacement angles reached 1/350 (Δ = 8.6 mm), 1/300 (Δ = 10 mm), and 1/250 (Δ = 12 mm), there were no obvious phenomena. Only the two kinds of friction noise could be heard (the friction noise among the panel, bolt, and connector, and the friction noise between the enclosure reinforcement and the loading plate). When the displacement angle reached 1/200 (Δ = 15 mm), vertical cracks appeared at the bottom of panel No. 2 (Figure 13a). When the displacement angle reached 1/100 (Δ = 30 mm), the cracks at the bottom of the panels were extended. When the displacement angle reached 1/75 (Δ = 40 mm), a number of small fragments came off, and the bonding mortar cracked between panel No. 1 and panel No. 2, and panel No. 4 and panel No. 5. (Figure 13b). When the displacement angle reached 1/50 (Δ = 60 mm), a corner of the back of panel No. 4 was broken (Figure 13c). When the displacement angle reached 3/10 (Δ = 90 mm), new cracks appeared at the bottom of the panel, and vertical cracks appeared near the bolt holes at the connector on panel No. 4. When the displacement angle reached 7/20 (Δ = 105 mm), the ALC panel corner was broken severely, and the relative sliding between the bolt and the bolt hole could be clearly observed during the loading process (Figure 13d).

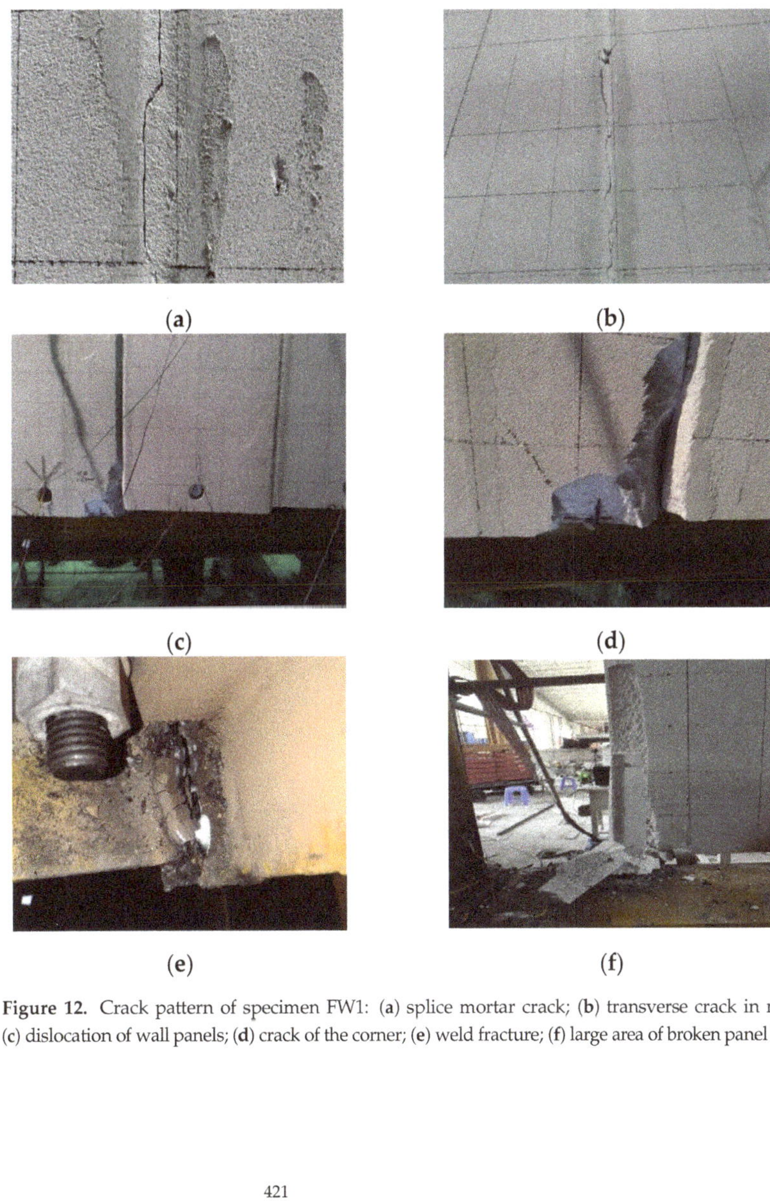

Figure 12. Crack pattern of specimen FW1: (**a**) splice mortar crack; (**b**) transverse crack in mortar; (**c**) dislocation of wall panels; (**d**) crack of the corner; (**e**) weld fracture; (**f**) large area of broken panel corner.

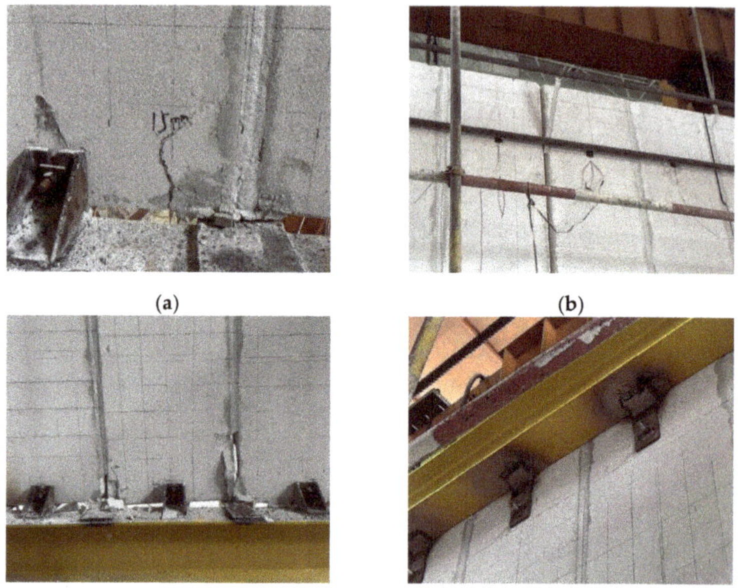

(a) (b) (c) (d)

Figure 13. Crack pattern of specimen FW-2: (**a**) crack in the bottom of the plane; (**b**) vertical cracks are produced; (**c**) ALC panel corner shattering; (**d**) slippage of connectors was obvious.

3.3. Hysteresis Curve

Figure 14 shows a comparison of the two sets of hysteresis curves. The hysteresis curve of FW-2 is fuller and encloses a larger area than FW-1. The bearing capacity of FW-2 is significantly greater than that of FW-1 at the same inter-story displacement angle, which indicates that FW-2 consumes more energy than FW-1 and absorbs seismic loads more effectively during earthquakes. The two sets of hysteresis curves behave asymmetrically, and the negative bearing capacity values are greater than the positive bearing capacity value. With the increase in loading displacement, FW-2 gradually approaches the inverse S-shape and "pinche", which indicates slippage in the upper part of the pendulous Z-panel connector and some residual deformation of the connection. The characteristic values on hysteresis curves are shown in Table 3.

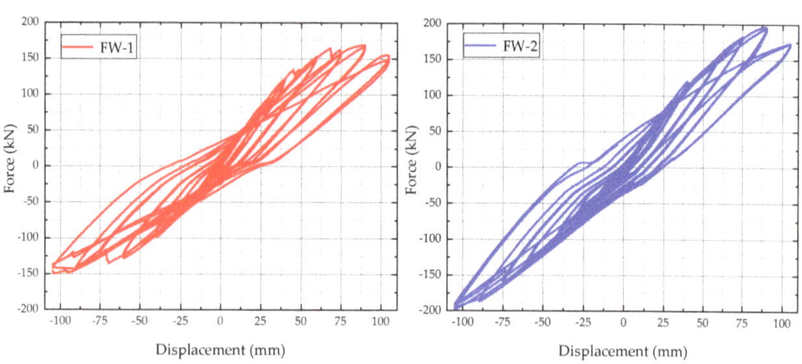

(a) (b)

Figure 14. Hysteresis curves of FW-1 (**a**) and FW-2 (**b**).

Table 3. Characteristic values on hysteresis curves.

Specimen	Yielding Point		Ultimate Point	
	$P_{y,t}$ (kN)	$\Delta_{y,t}$ (mm)	$P_{m,t}$ (kN)	$\Delta_{m,t}$ (mm)
FW-1	137.65	51.75	169.19	68.58
FW-2	152.31	58.32	195.15	89.13

3.4. Skeleton Curve

The skeleton curve is obtained by connecting the peak points in the same direction on the hysteresis curve in sequence. It reflects the characteristics of the specimens at different stages of stress and deformation. The trend of skeleton curves of FW-1 and FW-2 are similar. However, the positive and negative displacements are asymmetric. The value of positive displacement is greater than negative displacement, resulting from the different fine cracks and ground beam slippage caused by the positive and negative displacements during the test (as shown in Figure 15). Before the displacement reaches 60 mm, FW-2 is lower than FW-1 due to the pendulous Z-panel connector setting of a long circular hole, which reduces the impact of load on the panel and increases overall stability. The drop of FW-2 is slower after passing the peak point, and the bearing capacity of FW-1 decreases rapidly after passing the peak point. It indicates that the carrying capacity of the FW-2 group is higher than the FW-1 group, which reduces the risk of overall instability.

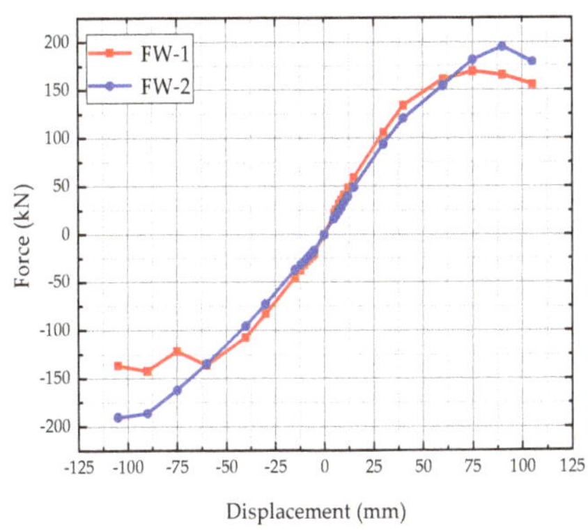

Figure 15. Skeleton curves of FW-1 and FW-2.

3.5. Stiffness Degradation

Figure 16 uses the secant stiffness to reflect the accumulation of structural damage at the two groups of connectors under external loads expressed as follows.

$$K = \frac{|+F_i| + |-F_i|}{|+X_i| + |-X_i|} \tag{1}$$

where F_i and X_i are the peak load and peak displacement in the i-th loading regime, respectively. The initial stiffness of specimen FW-2 is smaller than that of specimen FW-1, and the stiffness degradation rate is slower in the middle and later stages of the test. It is due to the bolt sliding in the long circular hole of the ALC pendulous Z-panel connector, which creates a buffer zone between the frame and the ALC panel. At the beginning of the test, the provision of the long circular hole reduces the lateral stiffness of the overall structure and the effect of seismic loading on the ALC panel, while increasing the stability of the ALC panel. In the later part of the test, the bolt contacts the edge of the long circular hole of the connector and provides load-bearing capacity for the whole structure.

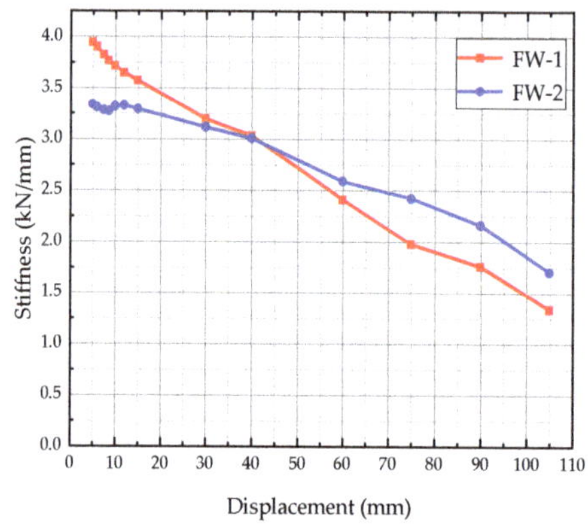

Figure 16. Stiffness degradation of FW-1 and FW-2.

3.6. Energy Dissipation

The area enclosed by the load–displacement hysteresis loop and the energy dissipation factor E was used to investigate the energy dissipation capacity of the two sets of specimens, which can be found in JGJ/T 101-2015 "Specification for seismic test of buildings" [27].

Figure 17 shows the total amount of total energy dissipation in both groups, and it can be observed that the trend of the FW-1 and FW-2 curves is similar. However, the values of specimen FW-2 are larger than that in FW-1.

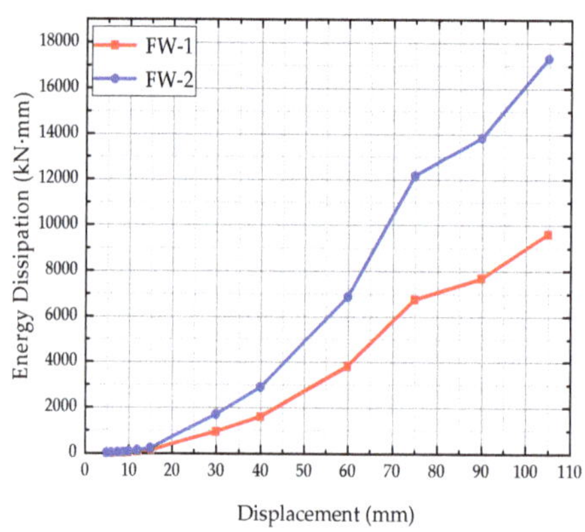

Figure 17. Energy dissipation of FW-1 and FW-2.

The energy dissipation coefficient (E) is shown in Figure 18 and defined as

$$E = \frac{S_{ABC} + S_{CDA}}{S_{OBE} + S_{ODF}} \tag{2}$$

where S_{ABC} and S_{CDA} in the numerator are the areas of the hysteresis curves ABC and CDA, and S_{OBE} and S_{ODF} in the denominator are the areas of OBE and ODF (as shown in Figure 19). Specimen groups show approximately the same growth trend. In general, the values of FW-2 are larger and faster than those from FW-1. The pendulous Z-panel connector group has better participation in energy dissipation and improves stiffness than the hooked bolt connector.

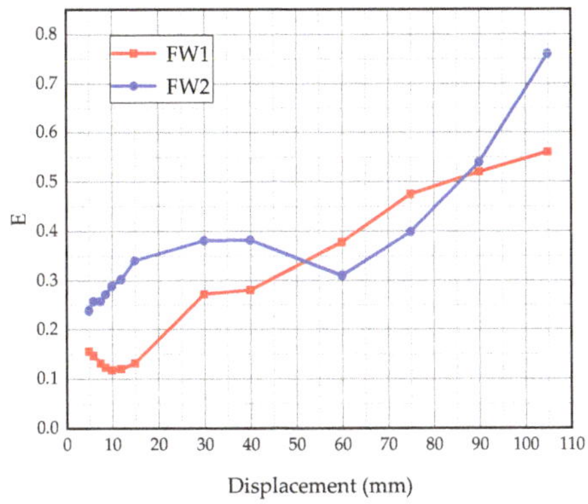

Figure 18. Energy dissipation coefficient of FW-1 and FW-2.

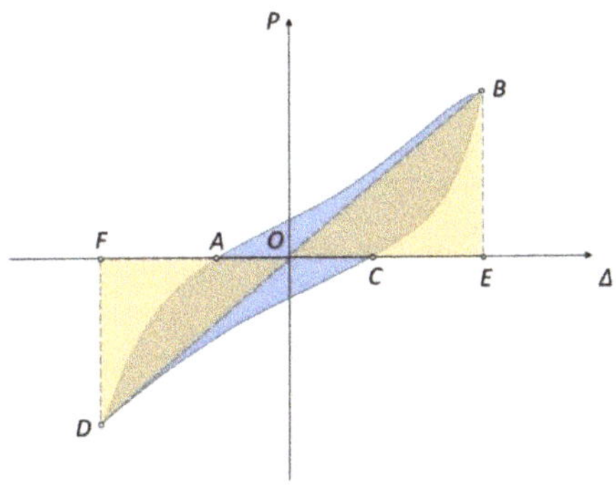

Figure 19. Idealized hysteretic relationship.

4. Finite Element Analysis

In order to fully reflect the force mechanism of the two groups of ALC panel steel frame systems, a series of validation and parametric analyses are performed on the pendulous Z-panel connector and the external hooked head bolt connector.

4.1. Finite Element Models

The data simulation and analysis were performed by using ABAQUS finite element software [28,29]. The two models were identical in size, reinforcement distribution diameter, and ALC panel grade. A hexahedral linear reduction integral solid element (C3D8R) was chosen to simulate the column footing, steel beam-column, connector, and bolt to simplify the model. The ALC panel internal reinforcement used linear truss elements (T3D2) to simplify the model. In order to improve the calculation efficiency, the ALC panel bolt holes were refined, and other parts used a thicker mesh (as shown in Figure 20).

The "Tie" constraint replaced all welding operations (between beams and columns, between connectors and beams, etc.). The steel bar was assembled into the ALC panel by using the embed command. Due to the slip effect, the contact surfaces between the connector, bolt, and panel were set to hard contact.

The steel beam and column used Q235B, and the connector used Q345B. A 10.9-grade high-strength bolt (24 diameters) was used for the beam and column connections, and a 5.6-grade bolt (14 diameters) was used for the connector with the ALC panel. The concrete damage model was used to simulate the damage of the ALC panel, which could represent the tensile and compressive performance of the ALC panel in detail. More material properties are shown in Table 4.

(a) (b)

Figure 20. Finite element model with (**a**) FW-1 and (**b**) FW-2.

Table 4. Material properties.

Material Type		Density (t/mm^3)	Elastic Modulus (MPa)	Poisson's Ratio
Steel	Q235B	7.89×10^{-9}	200,000	0.3
	Q345B	7.89×10^{-9}	206,000	0.3
ALC		5×10^{-10}	1770	0.2

According to the real conditions of the test, the H-shaped column's bottom X, Y, and Z directions were constrained by the translation and rotation (UX = UY = UZ = URX = URY = URZ = 0), and the coupling point (UY = UZ = URX = URY = URZ = 0, UX = 1) of the beam end simulate the same low-cycle reciprocating load.

4.2. Comparison of Simulation Results and Experimental Results

Figure 21 shows the stress cloud diagram of the test models. The stress concentration points in the models are consistent with the damage phenomena of the specimens in the experiments, which mainly appear around the beam-column joints. Figure 22 shows the comparison of an experimental hysteresis curve and a simulated one. The overall trends of the hysteresis curves are approximately the same. The hysteresis loop area of the simulation is larger than that of the experiment. The peak value of the simulation is slightly lower than the experimental value, and the "pinch" effect of the experiment is more obvious than the simulation results (as shown in Figure 22). The trends of simulated energy dissipation and energy dissipation coefficient are approximately the same as the experiment. The simulated values are larger than the experimental values, because the area of the simulated hysteresis loop is larger than the experimental one (as shown in Figure 23). The finite element model simplifies the complex boundary conditions in the experiment (slippage of the ground beam, small gaps between specimens, etc.). Moreover, the steel mesh slippage on the ALC panel was not considered. However, all the above differences are reasonable explanations, and the results of the ABAQUS simulation are basically consistent with the experimental results.

4.3. Parametric Study

The damage of the specimen and stress concentration points are mainly found at the bolt hole and the beam-column joint. It indicates that the connector influences the force performance of the ALC panel steel frame. Based on the hooked bolt connector of the general conditions, the pendulous Z-panel connectors of 6 mm, 8 mm, and 10 mm are designed to provide references for practical engineering applications (as shown in Table 5).

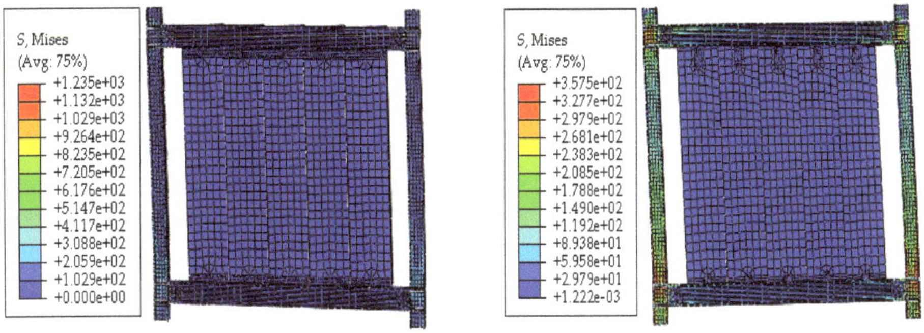

(**a**) (**b**)

Figure 21. Stress cloud diagram of the test models: (**a**) FW-1; (**b**) FW-2.

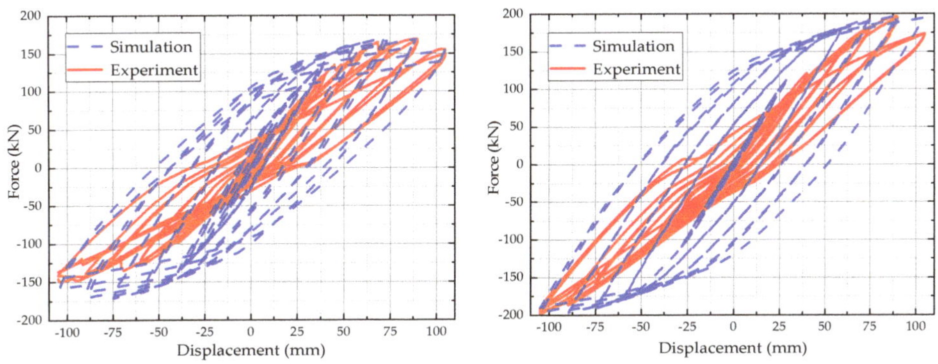

(**a**) (**b**)

Figure 22. Comparison of (**a**) FW-1 and (**b**) FW-2 experiment and simulation hysteresis curve.

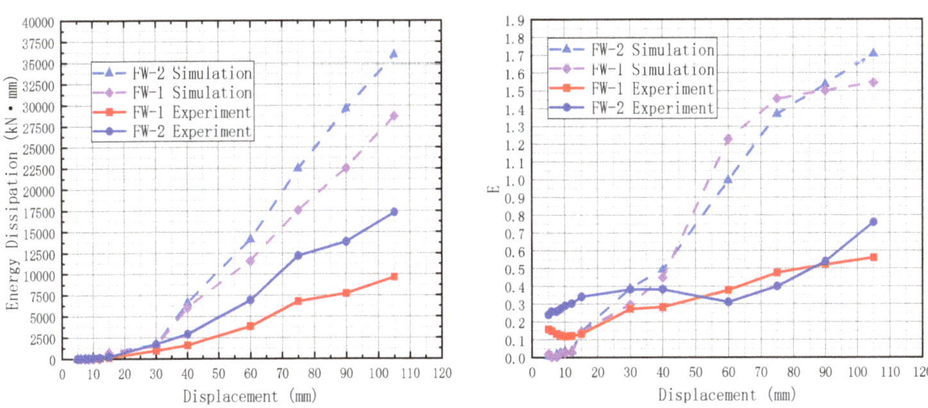

(**a**) (**b**)

Figure 23. Energy dissipation comparison of FW-1 and FW-2: (**a**) Energy dissipation; (**b**) Energy dissipation coefficient.

Table 5. Simulation connector type.

Component Number	Type	Features
GT	Traditional connectors	6 mm
TK-6	Fabricated new connectors	6 mm
TK-8	Fabricated new connectors	8 mm
TK-10	Fabricated new connectors	10 mm
TK-12	Fabricated new connectors	12 mm

4.4. Stress Cloud, Hysteresis Curve, and Skeleton Curve

Figure 24 shows the stress clouds for connectors with different thicknesses. The damage phenomenon of the ALC panel is similar for all four groups of specimens, which all appear around the beam-column joints. This is very different from the hook head bolt group, which appears around weld between bolt and angle steel. In terms of the ALC panel, the maximum stress value of 6 mm is 2.527 MPa, which is more than 5% lower than the other groups. When the displacement reaches the limit value of the elastic–plastic displacement angle 1/50 (Δ = 60 mm), the stress value of TK-6 is 1.986 MPa. Its stress is the smallest among the five groups (as shown in Table 6). It indicates that the thinner connector has more elastic deformation under seismic loading. It reduces the damage around the bolt hole and optimizes the force performance of the ALC panel.

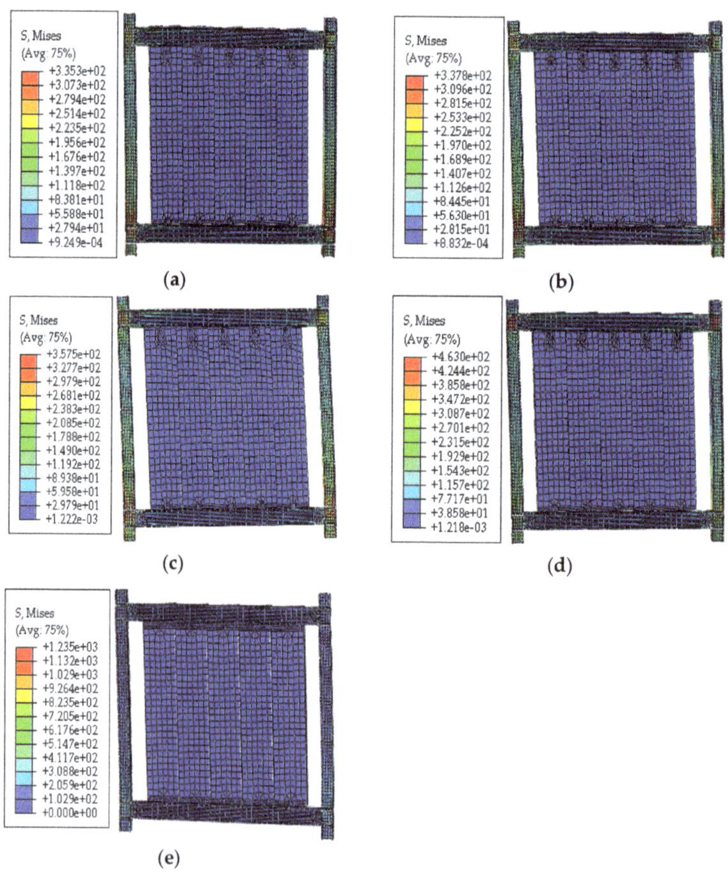

Figure 24. Stress clouds images: (**a**) TK-6; (**b**) TK-8; (**c**) TK-10; (**d**)TK-12; and (**e**) GT.

Table 6. ALC panel bolt hole stress.

Specimen	Characteristic Point		Maximum Point	
	Displacement (mm)	Mises Stress (MPa)	Displacement (mm)	Mises Stress (MPa)
TK-6	60	1.986	105.00	2.527
TK-8	60	2.076	105.00	2.785
TK-10	60	2.042	105.00	2.612
TK-12	60	2.174	105.00	2.659
GT	60	4.018	−105.00	5.133

The hysteresis curves images for GT, TK-6, TK-8, TK-10, and TK-12 are presented in Figure 25. Although their trends are basically the same, they have some differences in peak displacement and peak load. Specifically, the hysteresis curve is fuller as the thickness of the connector increases. At small displacements, the five sets of images appear to largely coincide with the thin hysteresis loops where the specimens are in the linear elastic phase. With the displacement increases, the tendency of specimen TK-6 decreases earlier than TK-8, TK-10, and TK-12, but this situation is much better than the GT group.

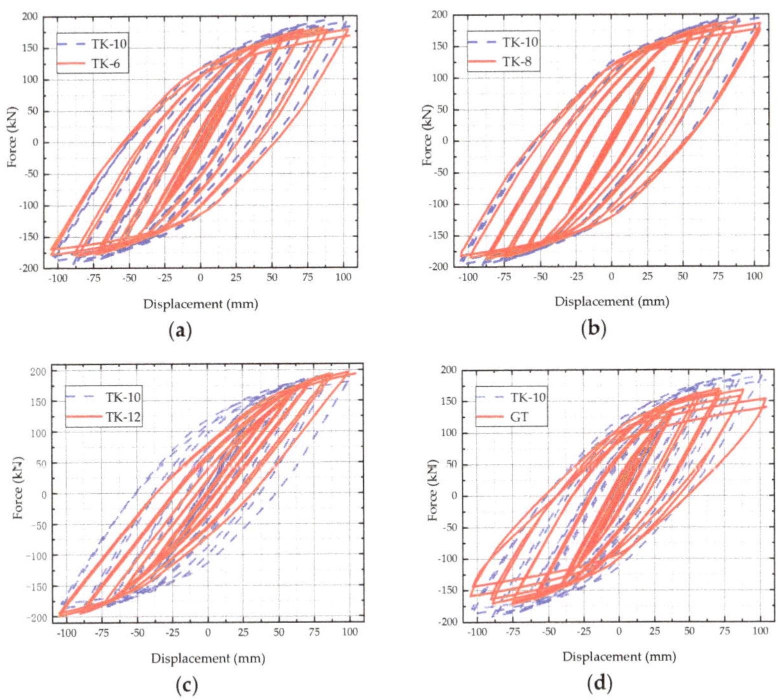

Figure 25. Hysteresis curve of FEM: (**a**) TK-6 and TK-10; (**b**) TK-8 and TK-10; (**c**) TK-10 and TK-12; (**d**) TK-10 and GT.

Figure 26 shows the relationship of the load–displacement skeleton curves for GT, TK-6, TK-8, TK-10, and TK-12. The five curves have the same variation. The peak of the image becomes more prominent as the thickness of the connector increases. TK-6 has a slight difference in its peak compared to the other groups, and its descent is relatively flat. It indicates that the different thickness connectors improve the bearing capacity and stiffness of the ALC panel steel frame, although the effect is slight.

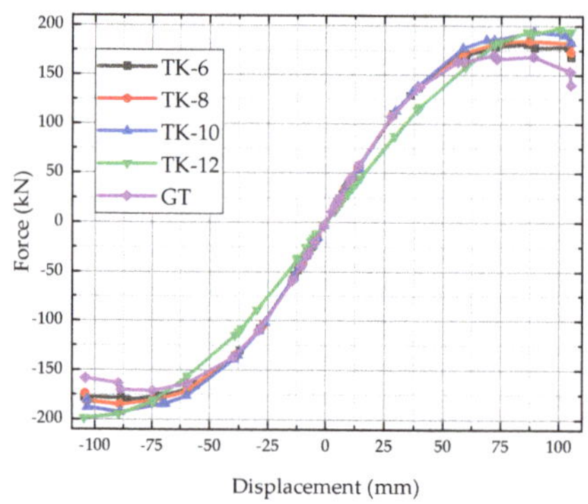

Figure 26. Skeleton curve of TK-6, TK-8, TK-10, TK-12, and GT.

5. Conclusions

This paper compares the external hooked bolt connector and the pendulous Z-panel connector. The following conclusions can be drawn from the tested and numerical simulations.

1. The damaged parts of the two groups of the external ALC panel tests are substantially different. For the FW-1 group, the damage first appears at the joints of the panels with clear penetration cracks when the displacement angle reaches 1/500 (Δ = 6 mm). For the FW-2 group, the damage first appears at the bottom span of the panel with slight cracks when the displacement angle reaches 1/200 (Δ = 15 mm). It was proved that the pendulous Z-panel connector optimizes force performance of the ALC panel steel frame, enhances energy dissipation capacity, ensures the integrity of the ALC panel, and avoids the risk of secondary instability.
2. The pendulous Z-panel (FW-2) connector method is similar to the semi-rigid connector in terms of the skeleton curve and the stiffness degradation curve in terms of experimental phenomenon and data. The overall damage phenomenon of the pendulous Z-panel connector group is significantly delayed compared to the external hooked bolt connector group because of the bolt sliding in the top long circular hole. This setting effectively releases the lateral force at the initial loading stage, then it reduces the damage to the ALC panel by the earthquake load under unfavorable working conditions and enhances the main structure's stability.
3. The connector is one of the significant influencing factors to ensure the stability of the external panel-framesynergistic system. Specifically, compared with FW-1, the yield load and peak load of FW-2 increase by 10.6% and 15.3%, respectively, and the yield displacement and peak displacement are increased by 12.6% and 29.9%, respectively. After reaching the peak load, the load capacity of FW-1 decreases rapidly, while the load capacity of FW-2 decreases slowly, although the initial stiffness of FW-2 is smaller. The analysis proves that this connector has good load carrying capacity and ductility.
4. The ABAQUS simulation revealed that the pendulous Z-panel connector of 6 mm has a larger load capacity and energy dissipation capacity than the external hooked bolt connector, which shows a better performance index. The TK-6 has a 7.6% decrease in load capacity compared to TK-10, and the decreasing segment is advanced. However, it has a similar change trend, and the stress around the ALC bolt hole is the smallest among the five groups. Having similar results for different thicknesses in the use of pendulous Z-panel connectors indicated that it would be more reasonable to use the 6 mm connector, which is the most economical solution for engineering.

Author Contributions: Conceptualization, K.D. and Y.L.; methodology, K.D.; software, K.D. and D.Z.; validation, Y.L.; formal analysis, D.Z.; investigation, K.D. and D.Z.; resources, K.D.; data curation, D.Z. and S.H.; writing—original draft preparation, K.D. and D.Z.; writing—review and editing, K.D. and Y.L.; visualization, D.Z.; supervision, K.D., Y.L. and W.S.; project administration,

K.D. and W.S.; funding acquisition, K.D., Y.L. and W.S. All authors have read and agreed to the published version of the manuscript.

Funding: This research was funded by the University Synergy Innovation Program of Anhui Province (GXXT-2019-005).

Institutional Review Board Statement: Not applicable.

Informed Consent Statement: Not applicable.

Data Availability Statement: Not applicable.

Acknowledgments: This research was financially supported by the University Synergy Innovation Program of Anhui Province (GXXT-2019-005), the University-Industry Collaborative Education Program of the Ministry of Education (202002129042), and the Scientific Research Project of Anhui Education Department- Key Project (KJ2020A0490).

Conflicts of Interest: The authors declare no conflict of interest.

References

1. Luo, T.; Xue, X.; Wang, Y.; Xue, W.; Tan, Y. A Systematic Overview of Prefabricated Construction Policies in China. *J. Clean. Prod.* **2021**, *280*, 124371. [CrossRef]
2. Tam, V.W.Y.; Tam, C.M.; Zeng, S.X.; Ng, W.C.Y. Towards Adoption of Prefabrication in Construction. *Build. Environ.* **2007**, *42*, 3642–3654. [CrossRef]
3. Jurczak, R.; Szmatuła, F. Evaluation of the Possibility of Replacing Fly Ash with Glass Powder in Lower-Strength Concrete Mixes. *Appl. Sci.* **2021**, *11*, 396. [CrossRef]
4. Narayanan, N.; Ramamurthy, K. Structure and Properties of Aerated Concrete: A Review. *Cem. Concr. Compos.* **2000**, *22*, 321–329. [CrossRef]
5. Karakurt, C.; Kurama, H.; Topçu, İ.B. Utilization of Natural Zeolite in Aerated Concrete Production. *Cem. Concr. Compos.* **2010**, *32*, 1–8. [CrossRef]
6. Pehlivanlı, Z.O.; Uzun, İ.; Demir, İ. Mechanical and Microstructural Features of Autoclaved Aerated Concrete Reinforced with Autoclaved Polypropylene, Carbon, Basalt and Glass Fiber. *Constr. Build. Mater.* **2015**, *96*, 428–433. [CrossRef]
7. Seddighi, F.; Pachideh, G.; Salimbahrami, S.B. A Study of Mechanical and Microstructures Properties of Autoclaved Aerated Concrete Containing Nano-Graphene. *J. Build. Eng.* **2021**, *43*, 103106. [CrossRef]
8. Rousakis, T.; Vanian, V.; Fanaradelli, T.; Anagnostou, E. 3D FEA of Infilled RC Framed Structures Protected by Seismic Joints and FRP Jackets. *Appl. Sci.* **2021**, *11*, 6403. [CrossRef]
9. Penna, A.; Mandirola, M.; Rota, M.; Magenes, G. Experimental Assessment of the In-Plane Lateral Capacity of Autoclaved Aerated Concrete (AAC) Masonry Walls with Flat-Truss Bed-Joint Reinforcement. *Constr. Build. Mater.* **2015**, *82*, 155–166. [CrossRef]
10. Kałuża, M. Analysis of In-Plane Deformation of Walls Made Using AAC Blocks Strengthened by GFRP Mesh. *Procedia Eng.* **2017**, *193*, 393–400. [CrossRef]
11. Binici, B.; Canbay, E.; Aldemir, A.; Demirel, I.O.; Uzgan, U.; Eryurtlu, Z.; Bulbul, K.; Yakut, A. Seismic Behavior and Improvement of Autoclaved Aerated Concrete Infill Walls. *Eng. Struct.* **2019**, *193*, 68–81. [CrossRef]
12. Deng, M.; Zhang, W.; Yang, S. In-Plane Seismic Behavior of Autoclaved Aerated Concrete Block Masonry Walls Retrofitted with High Ductile Fiber-Reinforced Concrete. *Eng. Struct.* **2020**, *219*, 110854. [CrossRef]
13. de Paula Salgado, I.; de Andrade Silva, F. Flexural Behavior of Sandwich Panels Combining Curauá Fiber-Reinforced Composite Layers and Autoclaved Aerated Concrete Core. *Constr. Build. Mater.* **2021**, *286*, 122890. [CrossRef]
14. Arslan, M.E.; Celebi, E. An Experimental Study on Cyclic Behavior of Aerated Concrete Block Masonry Walls Retrofitted with Different Methods. *Constr. Build. Mater.* **2019**, *200*, 226–239. [CrossRef]
15. Erdem, M.M.; Emsen, E.; Bikçe, M. Experimental and Numerical Investigation of New Flexible Connection Elements between Infill Walls-RC Frames. *Constr. Build. Mater.* **2021**, *296*, 123605. [CrossRef]
16. Ding, K.; Liu, J.; Ren, J.; Ma, W. Dynamic Responses of Cellular Metal-Filled Steel Beam-Column Joint Under Impact Loading. *J. Shanghai Jiaotong Univ. Sci.* **2020**, *25*, 384–393. [CrossRef]
17. Ding, K.; Han, M.; Li, X. Experimental research on seismic performance of semi-rigid bolted beam-column joints. *J. Hefei Univ. Technol.* **2020**, *43*, 805–810. (In Chinese)
18. Ding, K.; Chen, D.; Liu, Y.; Xia, S. Theoretical and experimental study on mechanical, behavior of laminated slabs with new type joints. *China Civ. Eng. J.* **2015**, *48*, 64–69. (In Chinese)
19. Ding, K.; Zhang, Y. Experimental study on seismic performance of fabricated bolted joint under low-cycle reciprocating loads. *Results Eng.* **2021**, *9*, 100208. [CrossRef]
20. Ding, K.; Liu, J.; Ma, W.; Liu, Y. Experimental study on seismic performances of a new type of fabricated semi-rigid beam-to-column connection. *China Civ. Eng. J.* **2021**, *54*, 1–7, 56.
21. Ding, K.; Ye, Y.; Ma, W. Seismic performance of precast concrete beam-column joint based on the bolt connection. *Eng. Struct.* **2021**, *232*, 111884. [CrossRef]

22. Wang, J.; Li, B. Cyclic Testing of Square CFST Frames with ALC Panel or Block Walls. *J. Constr. Steel Res.* **2017**, *130*, 264–279. [CrossRef]
23. Miao, Q.; Lu, Q.; Liu, H.; Yang, Y.; Zhang, Q. Research on key technology of autoclaved aerated concrete exterior wall panel system. *Build. Struct.* **2019**, *49*, 645–649. (In Chinese)
24. PRC. *Test Methods of Autoclaved Aerated Concrete GB/T 11969-2020*; China Building Industry Press: Beijing, China, 2021.
25. PRC. *Metallic Materials-Tensile Testing GB/T 228.1-2010*; China Building Industry Press: Beijing, China, 2011.
26. PRC. *Code for Seismic Design of Buildings GB 50011-2010*; China Building Industry Press: Beijing, China, 2011.
27. PRC. *Specification for Seismic Test of Buildings JGJ/T 101-2015*; China Building Industry Press: Beijing, China, 2015.
28. Zhao, X.; Wu, M.; Kong, D.D.; Chen, S.W. Study on Finite Element Model of Infilled Walls of Steel Frames. *Adv. Mater. Res.* **2011**, *250*, 2424–2427. [CrossRef]
29. Preti, M.; Bettini, N.; Plizzari, G. Infill Walls with Sliding Joints to Limit Infill-Frame Seismic Interaction: Large-Scale Experimental Test. *J. Earthq. Eng.* **2012**, *16*, 125–141. [CrossRef]

applied sciences

MDPI

Article

Seismic Performance of Dam Piers Retrofitted with Reinforced Polymer Cement Mortar

Yoshimi Sonoda [1,*], Hiroki Tamai [1] and Hirotsugu Ikeda [2]

[1] Department of Civil Engineering, Kyushu University, Fukuoka 819-0395, Japan; tamai@doc.kyushu-u.ac.jp
[2] Research Institute Technical Solution Headquarters, Kyushu Electric Power Co., Inc., Fukuoka 815-8520, Japan; Hirotsugu_Ikeda@kyuden.co.jp
* Correspondence: sonoda@doc.kyushu-u.ac.jp

Abstract: The deterioration of existing reinforced concrete (RC) structures is regarded as a problem worldwide. In Japan especially, the deterioration of RC dam structures has become severe. Many such dams meet design standards that were in place at the time of construction but do not meet the current seismic design standards, and appropriate seismic retrofitting is required. If the dam pier, which is an important part of the dam related to water storage, is damaged by an earthquake, the gate cannot be opened or closed, and the amount of water stored cannot be controlled. Therefore, the seismic retrofitting of dam piers is a top priority. However, various construction restrictions exist for dam piers, such as only the cross-section on the downstream side can be reinforced, and not on the upstream side where water is stored. Thus, it is difficult to apply the same reinforcement method that is applied to the piers of general road bridges. Therefore, in this study, we confirm the effectiveness of the SRS method (seismic retrofitting using cement mortar for shotcrete), which is suitable for partial reinforcement. Specifically, the dam piers of four types of existing dams were modeled using the three-dimensional finite element method, and a seismic response analysis was performed by inputting two types of seismic waveforms having different characteristics. As a result, the reinforcement effect of this method was verified according to structural characteristics. Furthermore, the effect of the reinforcement range on the reinforcement effect was clarified.

Keywords: dam pier; reinforced concrete (RC); reinforced polymer cement mortar (RPCM); SRS method; seismic response analysis; FEM; reinforcement effect

Citation: Sonoda, Y.; Tamai, H.; Ikeda, H. Seismic Performance of Dam Piers Retrofitted with Reinforced Polymer Cement Mortar. *Appl. Sci.* **2021**, *11*, 7255. https://doi.org/10.3390/app11167255

Academic Editor: Maria Favvata

Received: 7 June 2021
Accepted: 3 August 2021
Published: 6 August 2021

Publisher's Note: MDPI stays neutral with regard to jurisdictional claims in published maps and institutional affiliations.

Copyright: © 2021 by the authors. Licensee MDPI, Basel, Switzerland. This article is an open access article distributed under the terms and conditions of the Creative Commons Attribution (CC BY) license (https://creativecommons.org/licenses/by/4.0/).

1. Introduction

Dams are constructed mainly for flood control and water use. A dam pier is a tower-shaped reinforced concrete (RC) structure that opens and closes gates, such as spillways installed at the top of a gravity dam, as shown in Figure 1. It supports the drain gates attached to the dam, and thus crucial to water storage. If the dam pier breaks down during an earthquake, the water storage function of the dam is lost and there is a significant impact on the surrounding environment; therefore, the seismic resistance of the dam pier must be carefully considered. However, some dams in Japan have severely deteriorated due to aging. To improve their earthquake resistance, various reinforcement methods that are used for bridge piers [1] have been considered for dam pier reinforcement, such as retrofitting with concrete, steel jacketing and fiber sheets [2–6]. The reinforcing effect of these methods has been confirmed by some studies, but it is difficult to use these methods to strengthen dam piers. Owing to the structural characteristics of dam piers such as the one shown in Figure 1, there are limitations involved in their reinforcement: (i) it is difficult to install a formwork on the dam body; (ii) to reduce the percentage inhibition of the cross-sectional area of a river, the cross-section after reinforcement should not be as large as possible, (iii) if the weight increase due to the reinforcement is large, it will affect the dam body; (iv) as shown in Figure 2, only partial reinforcement is possible owing to the influence of gates and water storage. In light of these restrictions, the application of the SRS (seismic retrofitting using cement mortar for shotcrete) method is considered to be optimal for reinforcing dam piers. This method involves the placement of rebar on the existing parts of the dam and then spraying polymer cement mortar (PCM) after scraping their surface. In this study, the reinforced part on which the PCM is sprayed after the reinforcing bar is attached is called reinforced polymer cement mortar (RPCM). As shown in Figure 3, this method is suitable for partial reinforcement. It is possible to reinforce the existing part by installing the reinforcing bar on the existing concrete part after chipping and spraying PCM on it. In addition, this method has the advantages of a lesser reinforcement thickness than the conventional RC winding method, no need to install formwork and suitability to partial reinforcement. This method is easy to apply partially and can reduce the reinforcement thickness and additional mass because reinforcement is performed after scraping the surface. This means that the SRS construction method can reduce the risk that an increase in the structure's weight will increase the seismic load. Furthermore, by using PCM for the surface layer, which has a dense internal structure, high permeability resistance can be ensured, so it can be said that this method is suitable for dam piers.

Figure 1. Dam pier.

Many past studies on the seismic resistance of dams have examined the seismic resistance of the entire dam structure [7–10]. However, there are few studies on the seismic resistance of dam piers that are relevant to the present study. In addition, many studies have been conducted on the bending reinforcement effect of reinforcement methods such as RC winding reinforcement, steel plate winding reinforcement and fiber reinforcement as a method of reinforcing similar structural members such as RC piers. However, few studies [11,12] have been conducted on the reinforcement effect of the SRS method. Nakamura et al. [11] confirmed the seismic retrofitting effect of this method by reinforcing RC column specimens that modeled existing bridge piers by the SRS method and by conducting a peak-to-peak alternative load test that assumed earthquake conditions. As a result, it was confirmed that the specimen to which the method was applied showed an increase in horizontal physical strength, sufficient toughness and energy absorption capacity, and it had almost the same seismic retrofitting effect as the design. In addition, by applying this method to existing RC pier type specimens with a cut-off part, the bending strength of the cut-off part was improved, and the seismic reinforcement effect was confirmed. A review of previous studies revealed that there are no studies on the seismic resistance of dam piers to which the SRS method was applied, which demonstrates the novelty of this study.

Based on the above, to confirm the effectiveness of the SRS method as a seismic retrofitting method, four types of existing dam piers were modeled by FEM in this study and the reinforcement effect of the SRS method was examined by performing seismic response analysis. Furthermore, the effect of the cross-sectional shape and reinforcement range on the reinforcement effect was quantitatively analyzed.

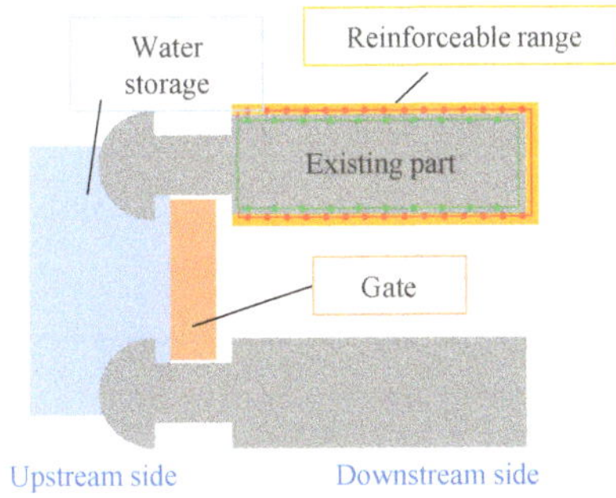

Figure 2. Reinforceable range.

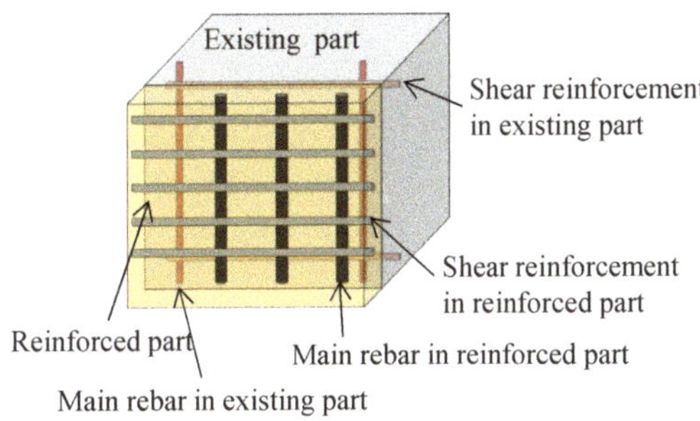

Figure 3. Modeling of reinforcement part in bird's-eye view.

2. Application of SRS Method to RC Beam Members (and Validation of FE Model)

To clarify the static load-bearing performance when the SRS method is applied to a bending fracture type RC beam, which is a primary structural member, and to verify the accuracy of the finite element (FE) analysis for the RC beam reinforced by the SRS method, a flat beam specimen that modeled the dam pier reinforced by the SRS method was produced, and a static loading experiment was conducted. Subsequently, a static test simulation using FEM was performed, and the analysis results were compared with the experimental results to confirm their validity.

2.1. Outline of Static Loading Experiment

Figures 4 and 5 show the cross-sectional dimensions of the beam specimens. In this study, the RC beam that was not reinforced by the SRS method is referred to as the RC specimen, and the RC beam reinforced with the SRS method is referred to as the RPCM specimen. D10 reinforcing bars were placed on the tensile side of the RC specimen, D6 reinforcing bars were placed on the compression side, and D10 reinforcing bars were placed at 100 mm intervals for shear reinforcement. For the RPCM specimen, a mortar layer with D6 reinforcement in the longitudinal direction was added around the cross-section of the RC specimen. In addition, D6 rebars with a thickness of 25 mm were also arranged as shear reinforcements at 100 mm intervals in the mortar layer. In this study, a simple reinforcement which was expected to adhere between the reinforcing bars and PCM was used for the ordinary purpose of seismic retrofitting of dam piers, and it has been confirmed that unexpected peeling or delamination do not occur in the bending test of the RPCM beam. However, additional studies will be required for other applications in which a large shear force or a tensile force that causes peeling or delamination in the PCM layer.

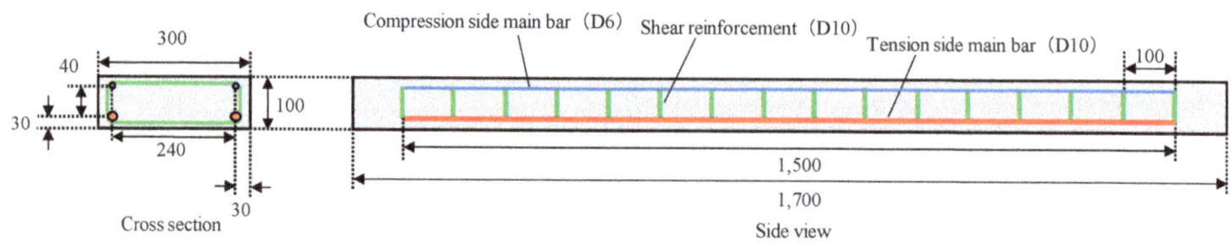

Figure 4. Dimensions of unreinforced specimen.

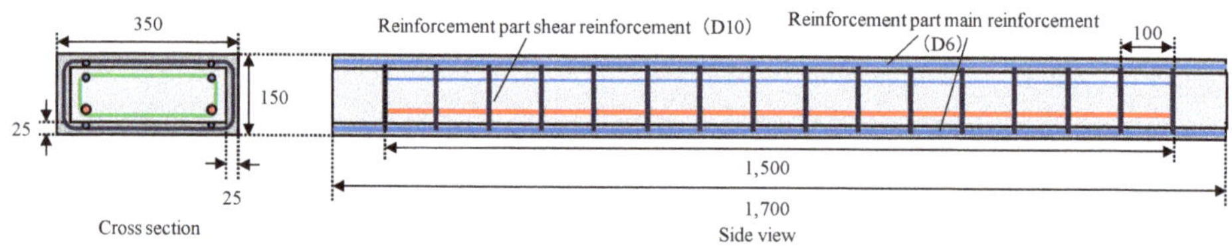

Figure 5. Dimensions of the RPCM reinforced specimen.

Table 1 shows the average values of the material properties obtained using the material test. In this study, three specimens with the same cross-section specifications were prepared as both RC and RPCM specimens, and three static loading tests were conducted on each specimen. Two-point static bending tests were performed with a pure bending section length of 500 mm, as shown in Figure 6, and Figure 7 shows a photograph of the experimental equipment.

Table 1. Average value of the material property (Unit: MPa).

Material	Compressive Strength	Tensile Strength	Static Modulus
Concrete	41.2	3.07	27,389
PCM	56.5	5.73	28,373

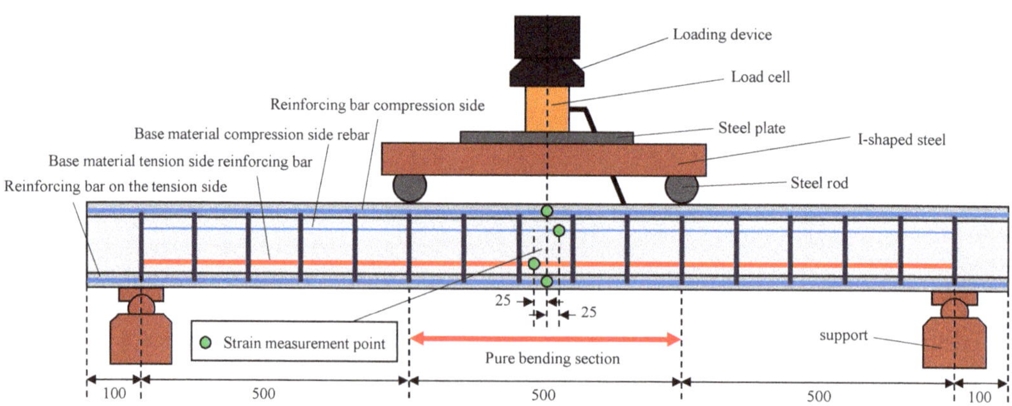

Figure 6. Installation of a specimen.

Figure 7. Set-up of a specimen.

2.2. Outline of FE Analysis

In this analysis, to confirm whether the reinforcement effect of the SRS method and the elasto-plastic behavior until the final failure could be accurately predicted by FEM, we attempted to reproduce the experimental results for both the RC and the RPCM specimens. The analysis models are presented in Figure 8. The concrete was modeled using a three-dimensional solid element, and the reinforcing bar was modeled using a truss element. Complete adhesion between the concrete and rebar elements was assumed. Regarding the boundary conditions of the specimen, an analytical model that discretized only half of the domain was used, as shown in Figure 8, considering the symmetry at midspan for both the RC and RPCM specimens. Complete adhesion between the concrete and the RPCM was also assumed at their interface because there was no significant separation during the experiment. For the material properties of both the RC and RPCM specimens, the values of compressive strength, tensile strength and static elastic modulus obtained by the material test shown in Table 1 were used.

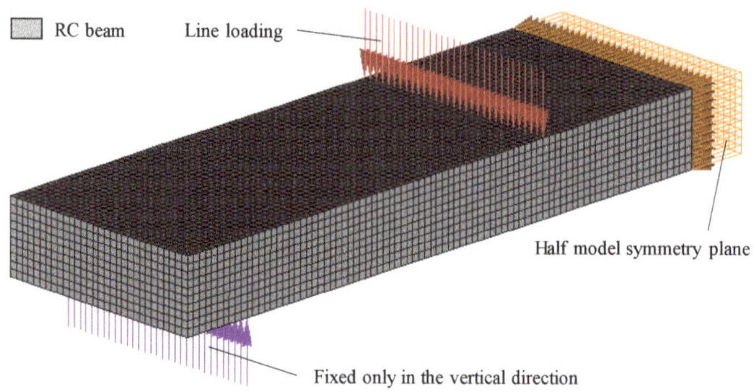

(a)

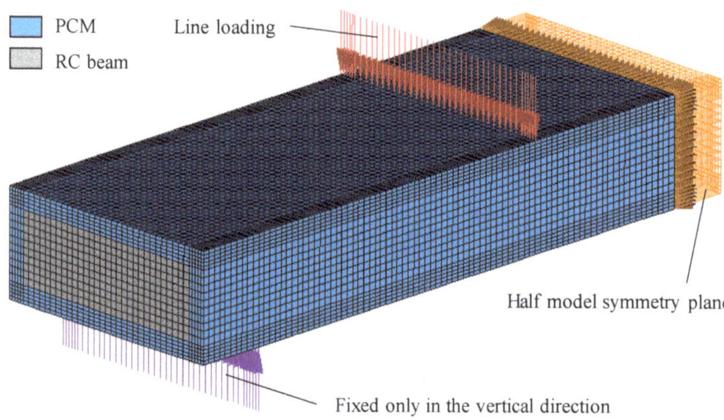

(b)

Figure 8. Analysis model: (**a**) RC specimen; (**b**) RPCM specimen model.

2.3. Results and Discussion

Figure 9 shows a comparison of the load-displacement relationship between the RC and RPCM specimens. The load of the analysis result was obtained by doubling the vertical reaction force at the fixed support of the 1/2 model, considering symmetry.

Figure 10 shows a comparison of the load-strain relationship of the RPCM specimen. In the analysis, although the strain on the tension side of the rebar in the existing concrete domain decreased immediately after the occurrence of concrete cracks from bending, the experimental results could be reproduced quite well on both the compression and tension sides. On the one hand, regarding the strain of the reinforcing bars in the PCM layer, it was confirmed that the reinforcing bars on the compression side could be accurately reproduced only up to the yield load level, but the tensile reinforcing bars could be reproduced up to the final failure level. Therefore, the validity of the analysis method and FE model used in this study was confirmed.

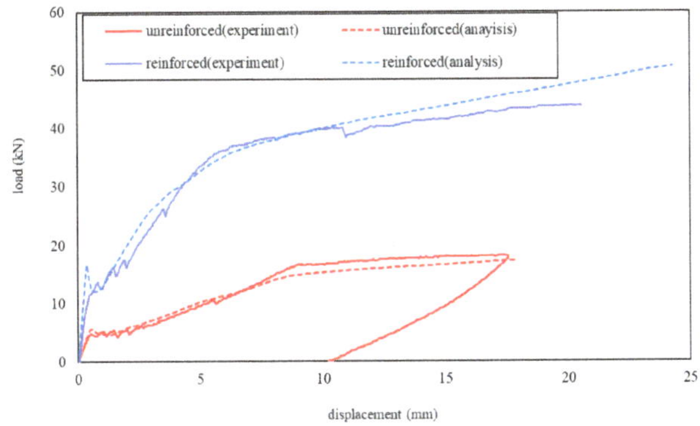

Figure 9. Comparison of the load-displacement relationship.

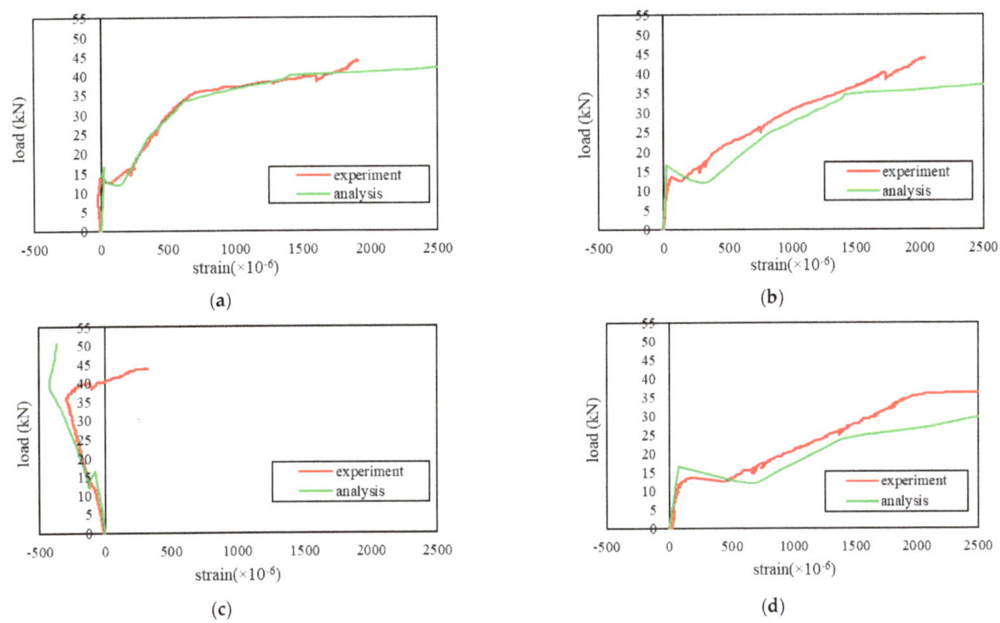

Figure 10. Load−strain relationship of each of the reinforcing bars: (**a**) compression strain of the rebar in the concrete; (**b**) tensile strain of the rebar in the concrete; (**c**) compression strain of the rebar in the PCM layer; (**d**) tensile strain of the rebar in the PCM layer.

3. Outline of Seismic Response Analysis for Dam Piers of Existing Dams

3.1. Shape and Modeling of the Target Dam Piers

To investigate the influence of the structural characteristics of dam piers on the reinforcing effect of the SRS method, we targeted the dam piers of four existing dams. In this study, they are called TYPE-A, -B, -C and -D, respectively. Figure 11 shows their shapes and dimensions, as well as the FEM model. As shown in Figure 11a, TYPE-A has dimensions such as a height of approximately 17.5 m, a length in the upstream and downstream directions of approximately 10 m, and a thickness in the dam axis direction (this becomes a weak axis direction) of approximately 2 m. For reinforcement, D16 was used as the main reinforcement, and D13 was used as the shear reinforcement. The bar arrangement interval was 600 mm for the main bar and 600 mm for the shear reinforcing bar. Next, as shown in Figure 11b, TYPE-B has a height of approximately 25 m, a length in the upstream and downstream directions of approximately 20 m, and a thickness in the dam axis direction of approximately 3 m. It is a dam pier that is larger than the TYPE-A model in both thicknesses. Regarding the reinforcement, D22 and D19 were used as the main reinforcements, and D13 was used as the shear reinforcement in the upper part. The bar arrangement interval was 158 mm for the main bar and 2000 mm for the shear reinforcing bar. At the bottom, D25 was arranged as the main reinforcement, and D16 was arranged as the shear reinforcement. The bar arrangement interval was 181 mm for the main bar and 500 mm for the shear reinforcing bar.

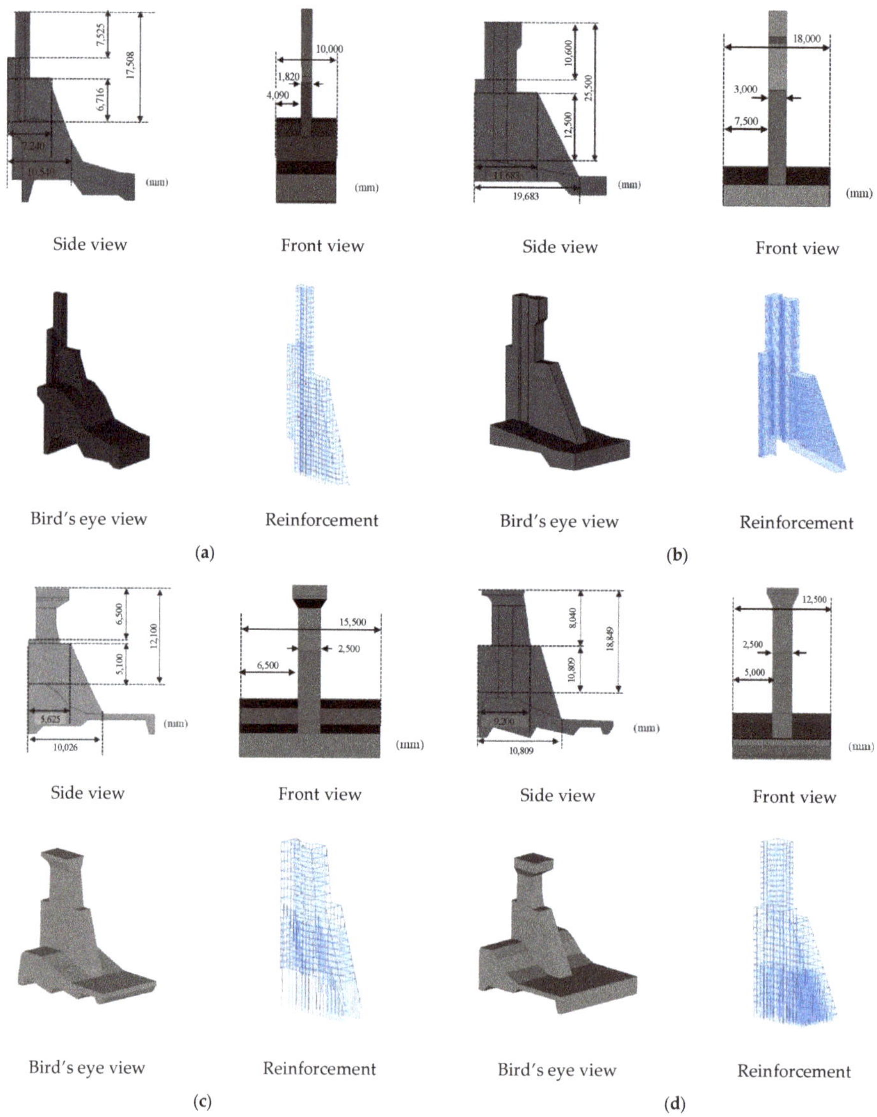

Figure 11. FEM analysis models of dam pier: (**a**) TYPE-A dam pier; (**b**) TYPE-B dam pier; (**c**) TYPE-C dam pier; (**d**) TYPE-D dam pier.

The major difference between the four types of dam piers is the shape and amount of the reinforcement. TYPE-A has a low and thick shape, and TYPE-B has a high and thin shape. In addition, TYPE-A has sparse reinforcement, and TYPE-B has dense reinforcement. As shown in Figure 11c, TYPE-C has dimensions such as a height of approximately 12.1 m, a length in the upstream and downstream directions of approximately 10 m and a thickness in the dam axis direction of approximately 2.5 m. For reinforcement, D25 is used as the main reinforcement, and D16 is used as the shear reinforcement. The bar arrangement interval was 450 mm for the main bar and 800 mm for the shear reinforcing bar. TYPE-C is lower and thicker than the TYPE-A and -B models, so it has a chunky shape overall. As shown in Figure 11d, TYPE-D has dimensions such as a height of approximately 15.4 m, a length in the upstream and downstream directions of approximately 10 m and a thickness in the dam axis direction of approximately 2.5 m. For reinforcement, D16 is used as the main reinforcement, and D12 is used as the shear reinforcement. The bar arrangement interval was 500 mm for the main bar and 1000 mm for the shear reinforcing bar. The TYPE-D model has the sparsest reinforcement among the four dam types.

The differences in the structural characteristics of these four dam piers are shown for the slenderness ratio and natural frequency. Figure 12 shows the slenderness ratio of the four dams and the amount of reinforcing bar in the existing part. The slenderness ratio is defined by Equation (1). Here, i is the radius of gyration of area, A is the cross-sectional area, B is the horizontal width of the cross-section and H is the vertical width of the cross-section. In addition, the amount of reinforcing bars in the existing part is defined by the total cross-sectional area of the reinforcing bars included in the unit length.

$$\lambda = \frac{L}{i} = L\sqrt{\frac{A}{I}} = L\sqrt{\frac{BH}{\frac{1}{12}BH^3}} = L\sqrt{\frac{12}{H^2}} = \frac{2\sqrt{3}L}{H} \tag{1}$$

The red dots in Figure 12 represent the four target dam piers, and the blue dots represent other dam piers owned by the same power company. From this figure, it can be seen that the four dam piers targeted in this study were evenly extracted based on the slenderness ratio of approximately 10 to 30 general dam piers, and the overall tendency could be determined by confirming these reinforcing effects. In addition, there is a positive correlation between the slenderness ratio and the amounts of reinforcing bar, and it is recognized that these four dams had the required amounts of reinforcing bar for their slenderness ratio. On the one hand, the five dam piers at the lower right are the groups that are expected to be reinforced by PCM because the amount of existing reinforcing bar is insufficient for their slenderness ratio. The effect of PCM reinforcement on dam piers in this group is discussed in the last section.

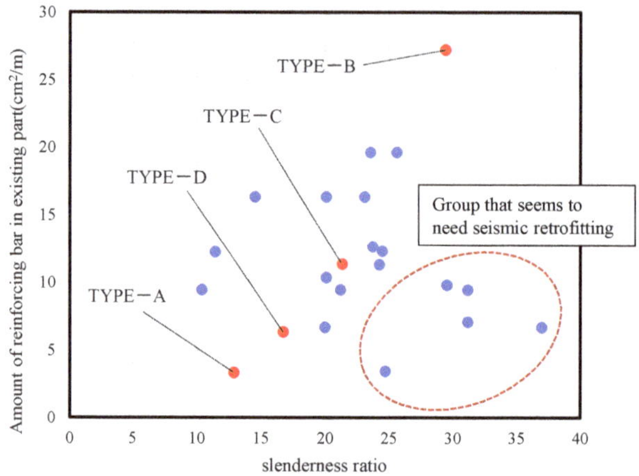

Figure 12. Classification of dams by slenderness ratio and amount of reinforcing bar.

Table 2 shows the eigenvalue analysis results for the four dam piers without reinforcement. From the results, it can be seen that the natural frequencies of the four dams vary from 2 Hz to 6 Hz in terms of the primary mode, and the vibration characteristics of the dams are covered as widely as possible. However, we decided to study the above four dams from TYPE-A to -D.

Table 2. Results of eigenvalue analysis.

Natural Frequency (Hz)	TYPE-A	TYPE-B	TYPE-C	TYPE-D
Primary mode	2.75	2.08	5.73	4.32
Secondary mode	6.52	7.22	9.84	13.2
3rd mode	10.3	8.33	23.0	16.7
4th mode	18.9	9.69	27.8	19.5

For the discretization of these four dam piers using the FE method, the concrete was modeled with 8-node solid elements and the reinforcing bars with 2-node truss elements. The nodes of the concrete and reinforcing bars were the same nodes, and complete adhesion was assumed.

3.2. Modeling of the Reinforcement Part

The FE model reinforced using the SRS method was modeled assuming that the existing deteriorated concrete elements were replaced with PCM elements, and new reinforcing bars were placed based on the actual reinforcing bar placement, as shown in Figure 13.

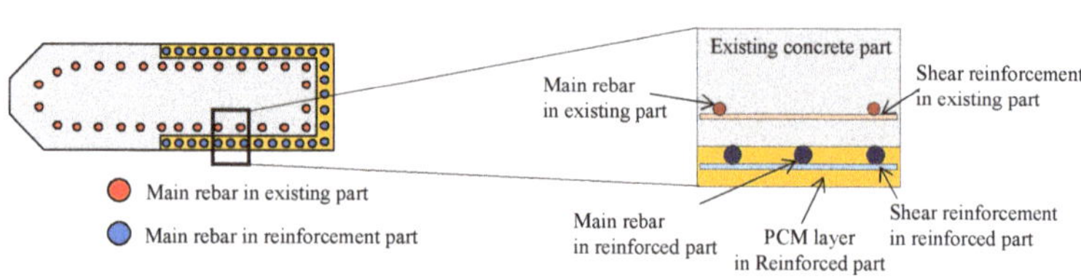

Figure 13. Modeling of the reinforcement part.

The PCM cover was 69 mm, and the main reinforcing bar diameter was D25 with a spacing of 250 mm for both types.

3.3. Material Model

In this study, the elastic-plastic material properties of the existing parts and reinforcing parts of dam piers are considered. Table 3 lists the material properties of the concrete and PCM.

Table 3. Physical characteristics of concrete.

Material Properties		Existing Part	Reinforcement Part (PCM)
Density	kg/mm^3	2290	2300
Young's modulus	N/mm^2	17,500	27,000
Poisson's ratio	-	0.17	0.2
Compressive strength	N/mm^2	17.2	64.5
Tensile strength	N/mm^2	1.8	4.12
Maximum aggregate diameter	mm	60	0.1

The stress-strain relationship on the compression side is shown in Equations (2)–(5) and conforms to the equation in the "Concrete Standard Specification" of the JSCE [13]. The strain at the compression strength was defined by the inverse calculation of Equation (3). Figure 14 shows the stress-strain curves on the compression side.

$$\sigma_c = E_0 K(\varepsilon_c - \varepsilon_p) \tag{2}$$

$$E_0 = 2f_c/\varepsilon_{peak} \tag{3}$$

$$K = \exp\left\{-0.73\frac{\varepsilon_c}{\varepsilon_{peak}}\left(1 - \exp\left(-1.25\frac{\varepsilon_c}{\varepsilon_{peak}}\right)\right)\right\} \tag{4}$$

$$\varepsilon_p = \varepsilon_c - 2.86\cdot\varepsilon_{peak}\left\{1 - \exp\left(-0.35\frac{\varepsilon_c}{\varepsilon_{peak}}\right)\right\} \tag{5}$$

where σ_c is the compressive stress (MPa), E_0 is the initial stiffness (MPa), K is the elastic rigidity residual rate, ε_c is the compressive strain, ε_p is the plastic strain, f_c is the compressive strength (MPa) and ε_{peak} is the strain at compression strength. As shown in Figure 15, the stress-strain relationship on the tensile side was derived from the tensile softening curve based on the tensile stress-crack width relations from the "Concrete Standard Specification" of the JSCE [13], as shown in Equations (6) and (7).

$$w_t = \frac{5G_f}{f_t} \tag{6}$$

$$G_f = 10(d_{max})^{\frac{1}{3}}\cdot f_c^{\frac{1}{3}} \tag{7}$$

where w_t is the crack width (mm), G_f is the tensile fracture energy (N/mm), f_t is the tensile strength (MPa) and d_{max} is the maximum dimension of the coarse aggregate (mm). In addition, the von Mises yield condition with no hardening (complete elastic-plastic characteristics) was applied to the steel material, as shown in Figure 16. The post-yield handling of the rebar is thought to have some effect on the elasto-plastic response of the member, but since the main purpose of this study is a relative comparison of the four existing dam piers, this modeling will not be mentioned in depth. In addition, kinematic hardening was assumed for the steel material. Table 4 lists the material properties of the steel bars for the concrete and PCM parts.

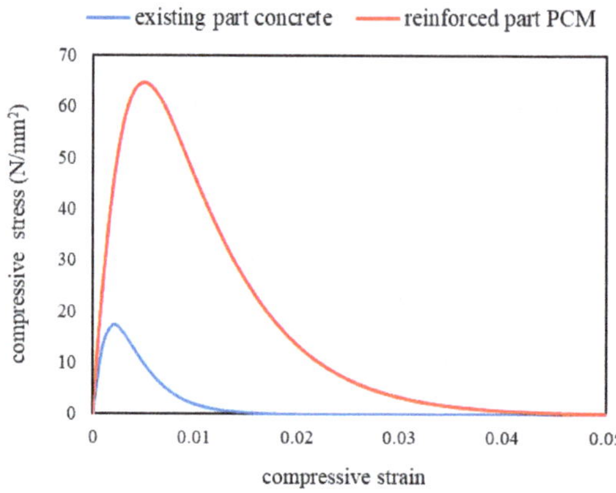

Figure 14. Stress-strain relationship of concrete (compression side).

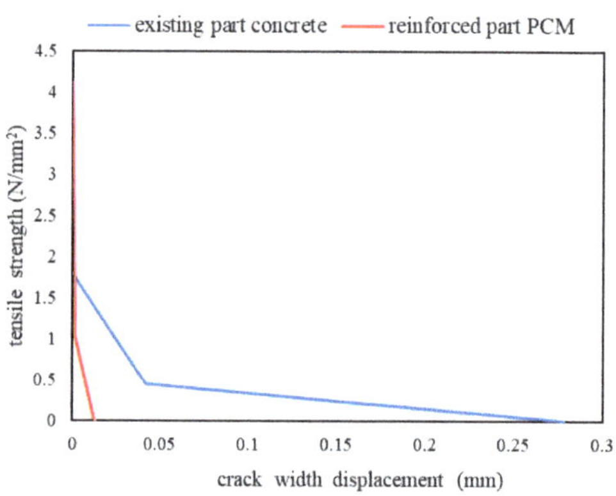

Figure 15. Concrete stress–crack displacement relationship (tension side).

Table 4. Physical characteristics of steel materials.

Material Properties	Unit	Existing Part		Reinforcement Part	
		Main Rebar	Shear Reinforcement	Main Rebar	Shear Reinforcement
Material		SR235		SD345	
Mass density	kg/m^3	7850			
Young's modulus	N/mm^2	200,000			
Yield modulus	N/mm^2	235		345	
Poisson's ratio	-	0.3			

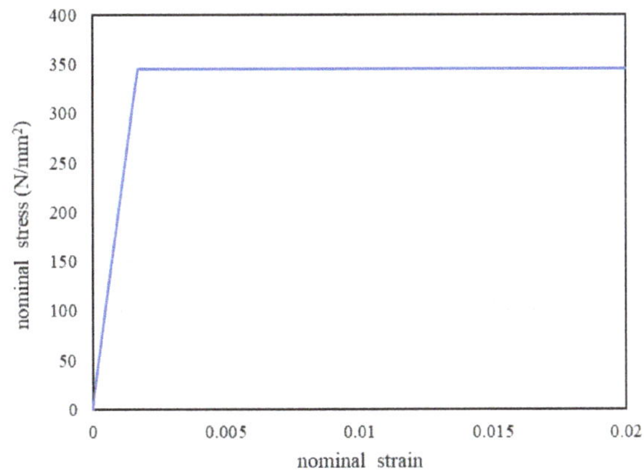

Figure 16. Stress—strain relationship of steel, converted to a single axis.

3.4. Boundary Condition

As shown in Figure 17, the boundary conditions of the dam pier were set based on actual mechanical conditions. Specifically, for both dams, the dam axis direction and the upstream and downstream directions were fixed at the bottom surface of the model. In this analysis, we focused on the dynamic behavior of only one dam pier and modeled only the levee body. The ground acceleration, as shown in Figure 18, was input in the dam axis direction of the full analysis model. In this study, seismic wave A and seismic wave B were applied to TYPE-A, -B, -C and -D. These acceleration waveforms were created from the waveforms used to verify the seismic performance of the dams and assumed a level 2 seismic motion. Seismic wave A had a maximum acceleration of 300 gal (3 m/s^2) and a duration of approximately 20 s, while seismic wave B had a maximum acceleration of 441 gal (4.41 m/s^2) and a duration of 24 s.

For all dams, an additional mass imitating the gate was provided to the gate support of the dam pier. The external force applied to the dam pier due to the vibration of the gate was considered. In addition, a switchgear was installed at the top of the pier, and an additional mass that was equivalent to the mass of the switchgear was applied to the top of the dam pier, as shown in Figure 17. Furthermore, for the TYPE-A dam, hydrostatic pressure and hydrodynamic pressure were applied to the dam pier, considering the amount of water stored. In this analysis, the hydrodynamic pressure was calculated using the Westergaard [14] equation shown in Equation (8). Here, m_d is the additional mass that represents hydrodynamic pressure; γ_w is the unit volume weight of water; g is the gravitational acceleration; b is the length in the upstream and downstream directions of the part where the dam pier and the water storage are in contact; H is the water storage level; h_1 and h_2 indicate the water depth at the representative points.

$$m_d = \int_{h_1}^{h_2} \frac{7}{8} \frac{\gamma_w}{g} b\sqrt{Hh}dh = \frac{7}{12} \frac{\gamma_w}{g} b\left(\sqrt{Hh_2{}^3} - \sqrt{Hh_1{}^3}\right) \quad (8)$$

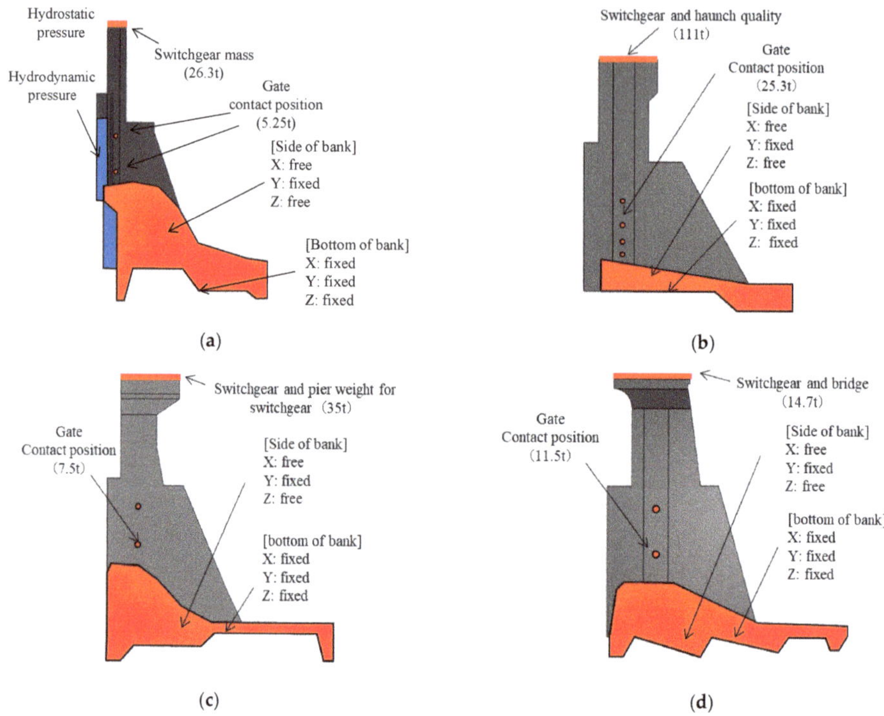

Figure 17. Boundary conditions: (**a**) TYPE-A dam model; (**b**) TYPE-B dam model; (**c**) TYPE-C dam model; (**d**) TYPE-D dam model.

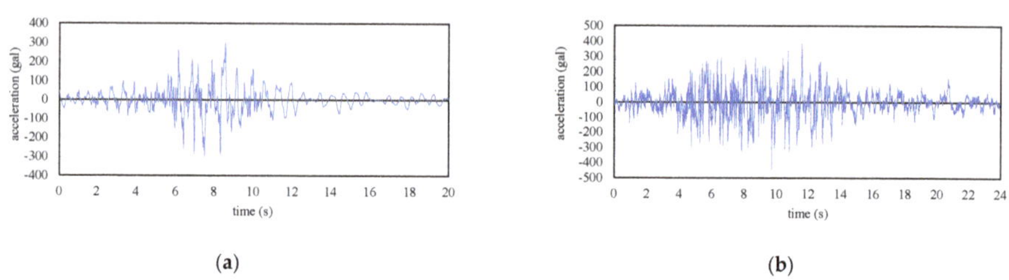

Figure 18. Input acceleration waveforms: (**a**) wave A; (**b**) wave B.

3.5. Analysis Method and Cases

The FE analysis software DIANA 10.3 was used for this analysis. The Newmark β method ($\beta = 0.25$) was used for the time integration method, and the time step was set to 0.01 s. This dynamic analysis assumed Rayleigh attenuation. Equations (9) and (10) show the equations for the Rayleigh attenuation parameters α and β, where ω_1 and ω_2 are the primary and secondary frequencies, respectively, and h_1 and h_2 are the critical attenuations (=5%) corresponding to each mode.

$$\beta = 2(\omega_1 h_1 - \omega_2 h_2) / \left(\omega_1^2 - \omega_2^2\right) \tag{9}$$

$$\alpha = 2\omega_1 h_1 - \beta\omega_1^2 \tag{10}$$

There are two main types of dam gates: roller gates and radial gates. In this study, a U-shaped reinforcement range was assumed for the roller gate, as shown in Figure 19.

Full (universal) reinforcement was not adopted, but full reinforcement (100%) was also calculated as a comparative example to understand the effect of partial reinforcement. The main rebar used was D25, and D22 was used as the shear rebar in the PCM layer, as shown in Figure 20.

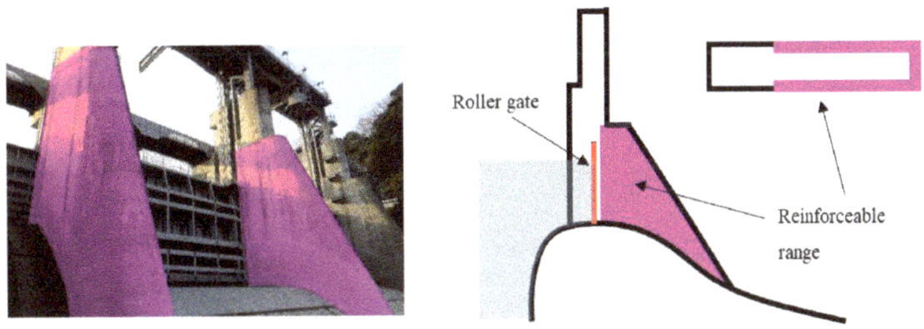

Figure 19. Roller gate and reinforceable range.

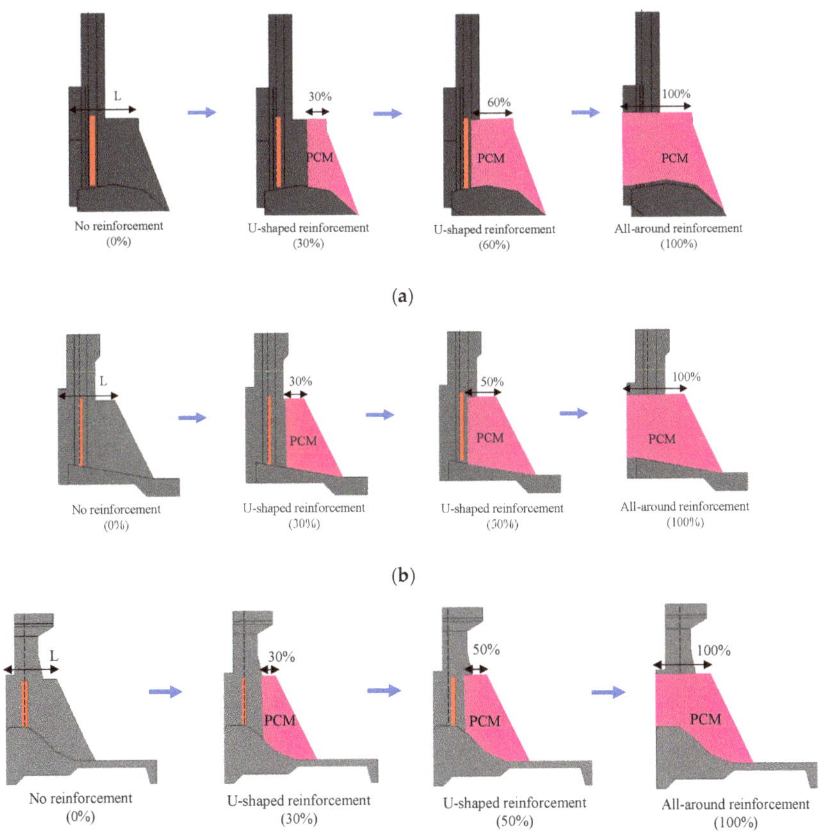

(a)

(b)

(c)

Figure 20. *Cont.*

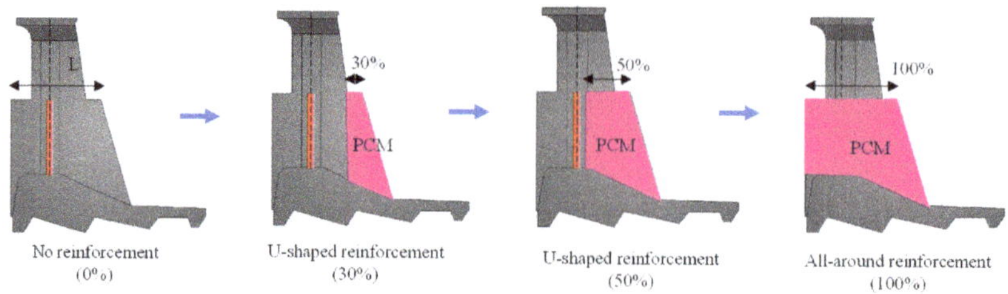

(d)

Figure 20. Reinforcement range: (**a**) TYPE-A dam model; (**b**) TYPE-B dam model; (**c**) TYPE-C dam model; (**d**) TYPE-D dam model.

4. Results and Discussion

4.1. Seismic Resistance Evaluation Method for Dam Pier by FE Analysis

In this study, the reinforcement effect of the SRS method on the seismic resistance performance of dam piers during and after an earthquake was determined from the damaged state. In general, dam piers need to function even if they are damaged after an earthquake. The highest priority function is, of course, the water storage function; however, to maintain the water storage function, it is also important that the gate can be opened and closed smoothly and that water leakage due to damage to the skeleton is prevented [15]. To ensure that these functions were not impaired, we decided to confirm that the residual displacement of the dam pier at the gate position was within the allowable value. However, the residual displacement depends not only on the maximum acceleration of the seismic waves but also on their historical characteristics. In addition, the damage distribution in the structure and the seismic wave amplitude characteristics in the final stage seem to have a significant impact. Therefore, it is difficult to predict the residual displacement quantitatively and accurately using numerical analysis. Therefore, we decided to make a relative comparison of the reinforcement effect on the dam, focusing not only on the residual displacement but also on the maximum displacement. In addition, the strength of the dam pier itself was evaluated by checking the crack distribution in the concrete in the non-reinforced part and the reinforced part, and the strain of the main reinforcing bar.

The results and discussions are given below after Section 4.2.

4.2. Seismic Response of Dam Piers

Figures 21–24 show the results of the displacement-time history of the four dam types using wave A and wave B, respectively. From these results, it can be seen that the maximum displacement of each type is reduced owing to the reinforcement effect.

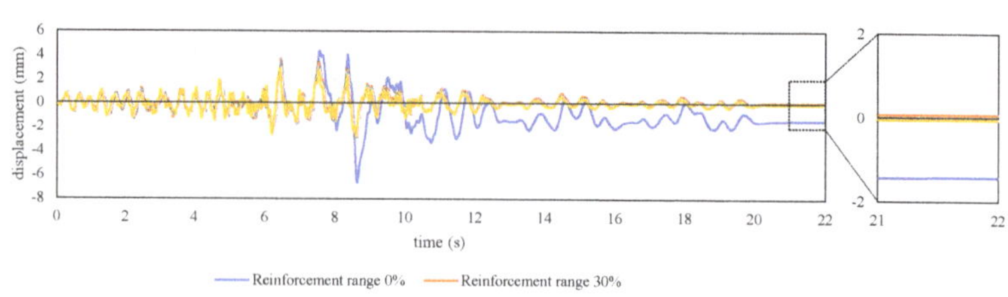

(a)

Figure 21. *Cont.*

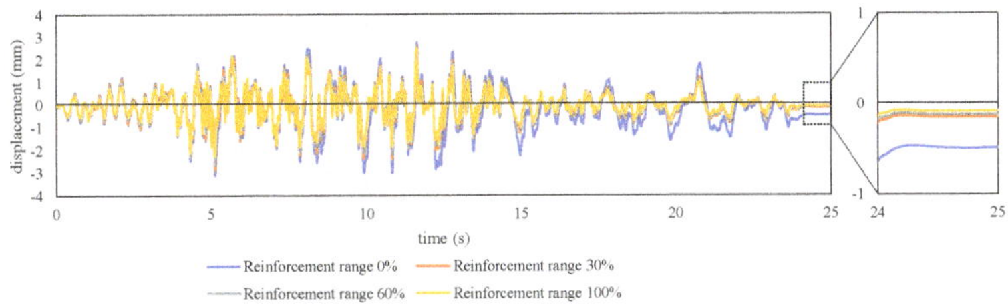

(**b**)

Figure 21. Comparison of displacement–time history at the top of the pier in TYPE-A: (**a**) under wave A; (**b**) under wave B.

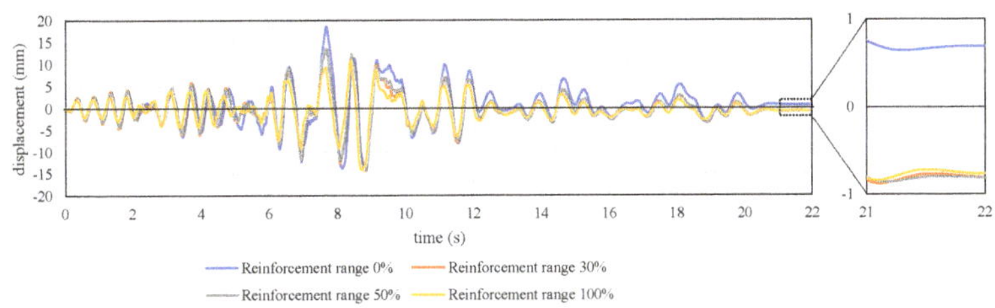

(**a**)

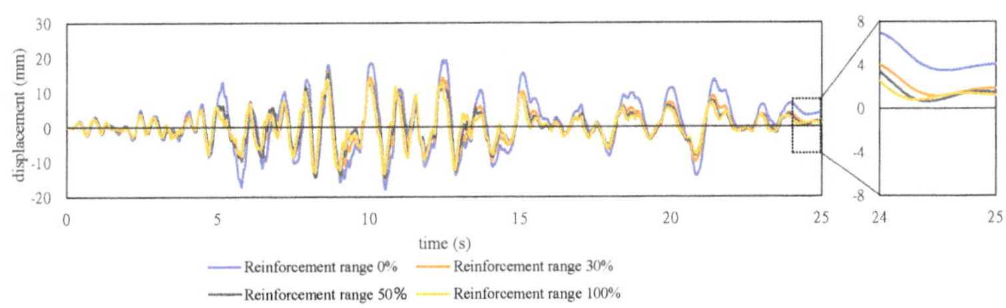

(**b**)

Figure 22. Comparison of displacement–time history at the top of the pier in TYPE-B: (**a**) under wave A, (**b**) under wave B.

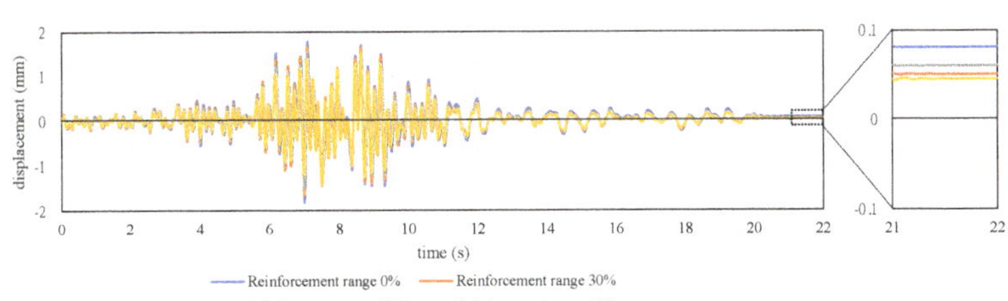

(**a**)

Figure 23. *Cont.*

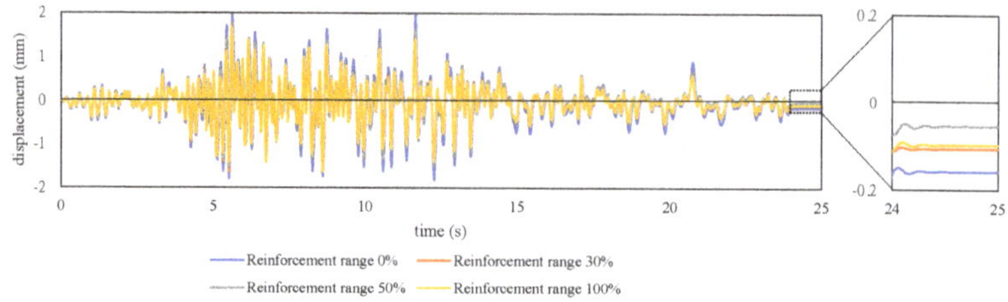

(b)

Figure 23. Comparison of displacement−time history at the top of the pier in TYPE-C: (**a**) under wave A; (**b**) under wave B.

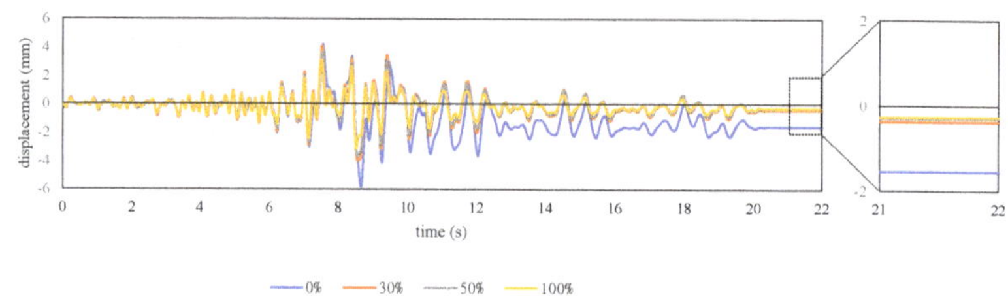

(a)

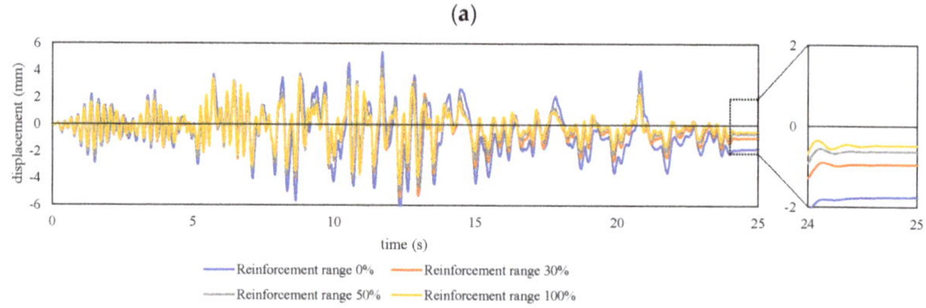

(b)

Figure 24. Comparison of displacement−time history at the top of the pier in TYPE-D: (**a**) under wave A; (**b**) under wave B.

In Figure 21a, by comparing the results of the RC and RPCM models (reinforcement range of 60%) in TYPE-A, it can be confirmed that the displacement response of the RPCM model becomes smaller than that of the RC model after 7 s under wave A. In the case of Figure 21b, it can also be confirmed that the displacement response of the RPCM model becomes smaller than that of the RC model after 12 s under wave B. This is because the maximum principal strain of the RC model throughout the base of the dam pier reached a strain level that caused bending cracks, as shown in Figure 25a, which caused a decrease in rigidity. However, in the case of the RPCM model, the bending cracks in the concrete at the base of the pier are limited to one part; thus, their maximum displacement decreased.

Next, the strain of the existing reinforcing bars was also compared between the RC and RPCM models at the time of maximum displacement, as shown in Figure 26. It was confirmed that many of the reinforcing bars reached the yield strain in the RC model. In contrast, in the RPCM model, the strain of all the existing reinforcing bars was less than the yield strain.

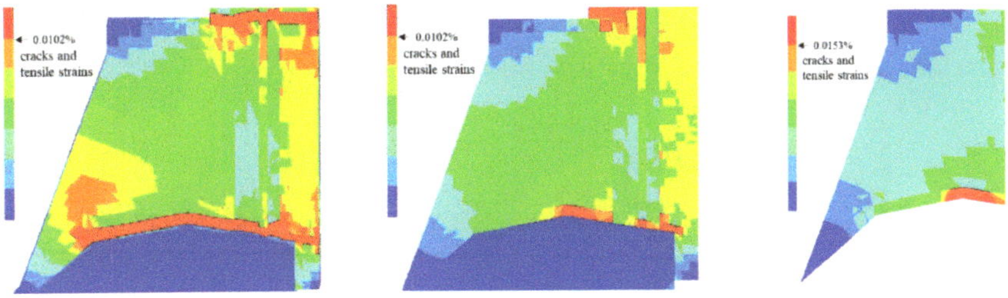

(**a**) RC model (**b**) RPCM model (60% reinforced case): (**left**) Existing part; (**right**) Reinforcement part

Figure 25. Maximum principal strain distribution of TYPE-A at 8.65 s.

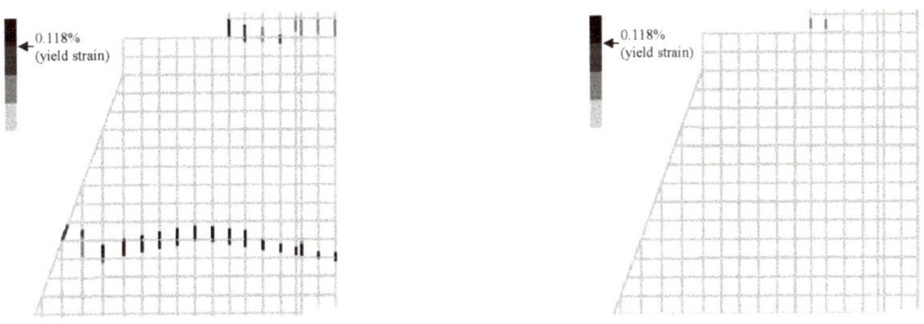

(**a**) (**b**)

Figure 26. Comparison of reinforcing bar strain in the vertical direction of TYPE-A: (**a**) RC model; (**b**) RPCM model.

4.3. Effect of Reinforcement Range on Reinforcement Effect

To discuss the effect of the reinforcement range on the reinforcement effect, we focused on the axial strain distribution in both the existing and reinforcement parts. Here, the results of two different reinforcement ranges (30% and 60%) for the TYPE-A dam under wave A are compared, as shown in Figures 27 and 28. First, it was confirmed that the axial strain of the existing reinforcing bar did not exceed the yield strain in either case and that they had a sufficiently reinforcing effect. In addition, the tendencies of both are similar, and it can be said that the effect of the reinforcement range on the vibration characteristics of the entire structure is small. As shown in Figures 27b and 28b, it was confirmed that relatively large strains were concentrated at the root of and upstream of the reinforcing bar at the reinforcement part. When the seismic motion that exceeds the seismic load is applied, it is thought that the weakest reinforcing bars will yield first, so it is necessary to take measures such as using large reinforcing bars for these parts.

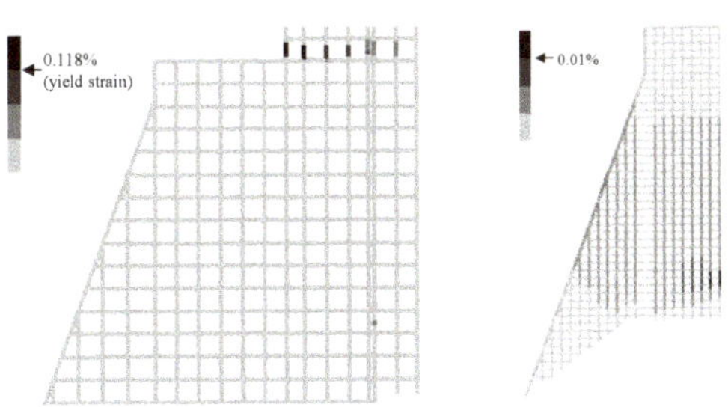

(**a**) (**b**)

Figure 27. Strain distribution of the reinforcing bars (TYPE-A, wave A, reinforcing range 30%): (**a**) existing part: (**b**) reinforcement part.

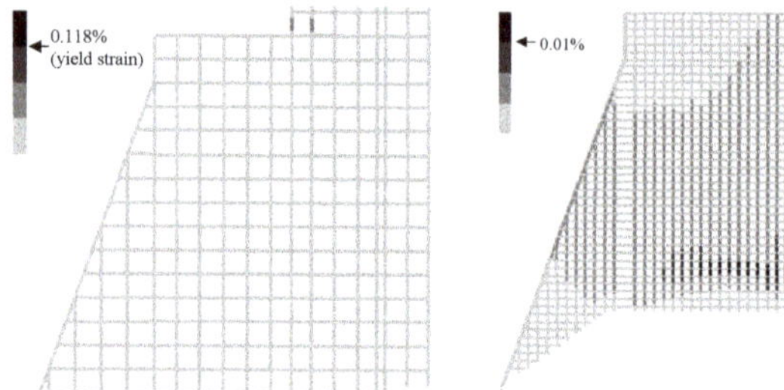

(a) (b)

Figure 28. Strain distribution of the reinforcing bars (TYPE-A, wave A, reinforcing range 60%): (**a**) existing part: (**b**) reinforcement part.

4.4. Comparison of Reinforcement Effect of Each Dam

Figures 29–32 show a comparison of each dam type's maximum displacement and residual displacement for four different reinforcement ranges by seismic wave A and seismic wave B. In the TYPE-A model, the reinforcing effect was recognized for both types of seismic waves. In particular, for seismic wave A, it was found that by reinforcing the cross-sectional area of the column by 30%, a remarkable reinforcement effect was observed, such that the maximum displacement was reduced by 45%, and the residual displacement was almost zero. Next, in the TYPE-B model, the reinforcing effect was relatively smaller than that in the TYPE-A model, but a certain reinforcing effect was confirmed for both seismic waves A and B. In particular, a clear reinforcement effect was confirmed for the residual displacement due to seismic wave B, such as the 58% reduction in residual displacement from reinforcing 30% of the column cross-section. In the TYPE-D model, similar to the TYPE-A model, a clear reinforcing effect was observed for both types of seismic waves. In particular, it was found that the residual displacement was significantly reduced by simply reinforcing 50% of the column cross-section. However, for the TYPE-C model, the reinforcement effect was found to be significantly smaller than that of the other three dams. One of the reasons is that the TYPE-C model has high rigidity; thus, the maximum displacement and residual displacement due to the two types of seismic waves are both small, so the need for reinforcement by the SRS method was initially minimal. The next section examines the causes of the different reinforcement effects for each dam and seismic wave. In some cases (for example, in the case of Figure 31b), the residual displacement is slightly large even though the reinforcement range is large. It can be said that this is because the final residual displacement of the element that repeats compression and tension changes sensitively depending on the mesh size and handling after cracking occurs. However, the reinforcing effect of the RPCM was observed in all cases.

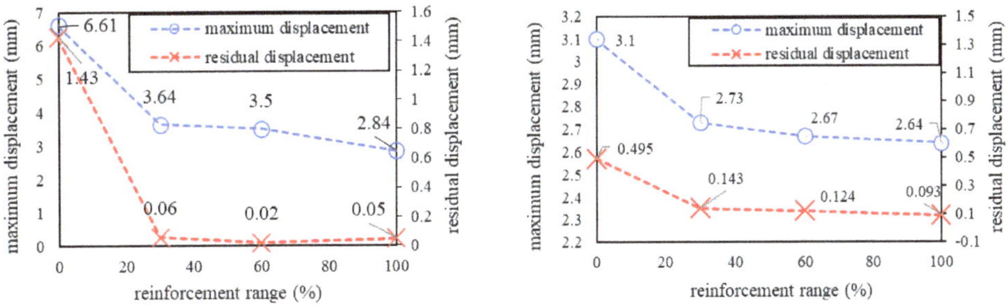

(a) (b)

Figure 29. Comparison of maximum displacement and residual displacement of TYPE-A: (a) wave A; (b) wave B.

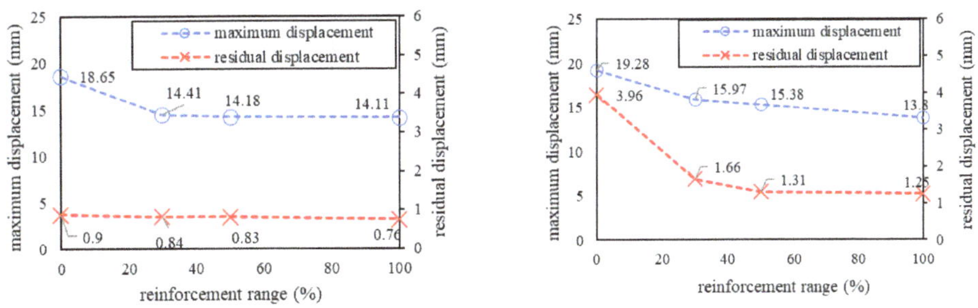

(a) (b)

Figure 30. Comparison of maximum displacement and residual displacement of TYPE-B: (a) wave A; (b) wave B.

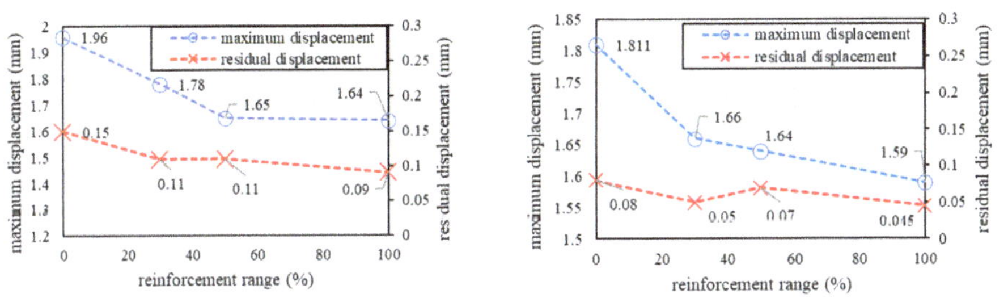

(a) (b)

Figure 31. Comparison of maximum displacement and residual displacement of TYPE-C: (a) wave A; (b) wave B.

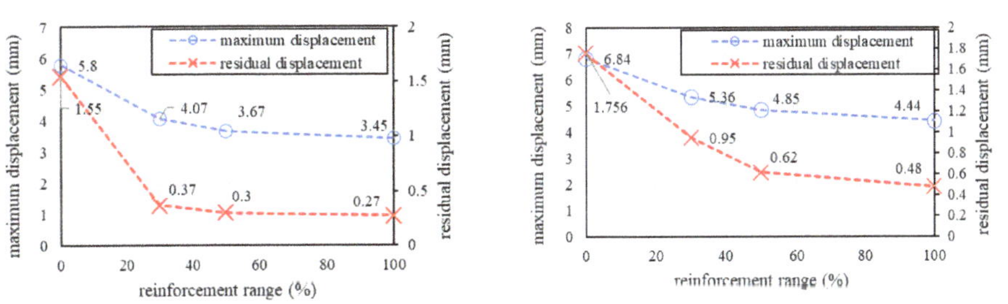

(a) (b)

Figure 32. Comparison of maximum displacement and residual displacement of TYPE-D: (a) wave A; (b) wave B.

4.5. Strain Distribution of Existing Reinforcing Bars: Comparison of Unreinforced and Reinforced Case

As shown in Section 4.2, there was no significant difference in displacement between the unreinforced and RPCM-reinforced cases until bending cracks occurred in the elastic region. However, it was confirmed that when the displacement increased, the reinforcing bars of the existing part yielded in the unreinforced case, so there was a difference between the unreinforced and RPCM-reinforced cases. Figures 33 and 34 show the axial strain distribution of the existing reinforcing bars at the peak displacement of each of the four dams under seismic wave A and seismic wave B, respectively. The TYPE-A model showed a tendency for the maximum displacement reduction effect to differ significantly between the two types of seismic waves, and the ratio of existing reinforcing bars that yielded in this type was larger for seismic wave A.

Next, the reinforcing effects of TYPE-B, TYPE-C and TYPE-D were compared using the strain distribution of the existing reinforcing bars, as shown in Figures 33 and 34. These figures show that the number of existing reinforcing bars that reached yield strain before and after reinforcement decreased in TYPE-B and TYPE-D dams. However, existing reinforcing bars that reached yield strain were not observed in the TYPE-C. Thus, there was no significant difference in displacement between the unreinforced and RPCM-reinforced TYPE-C cases. Therefore, it is concluded that a large reinforcing effect can be obtained in the case of a dam pier having a small amount of existing reinforcing bar and a large mass of attached structures such as switchgear, pier for switchgear and door body. In addition, it was confirmed that when a large number of existing reinforcing bars reach yield strain, as in TYPE-A and TYPE-D shown in Figures 29a and 32a, even partial reinforcement has almost the same effect as 100% reinforcement.

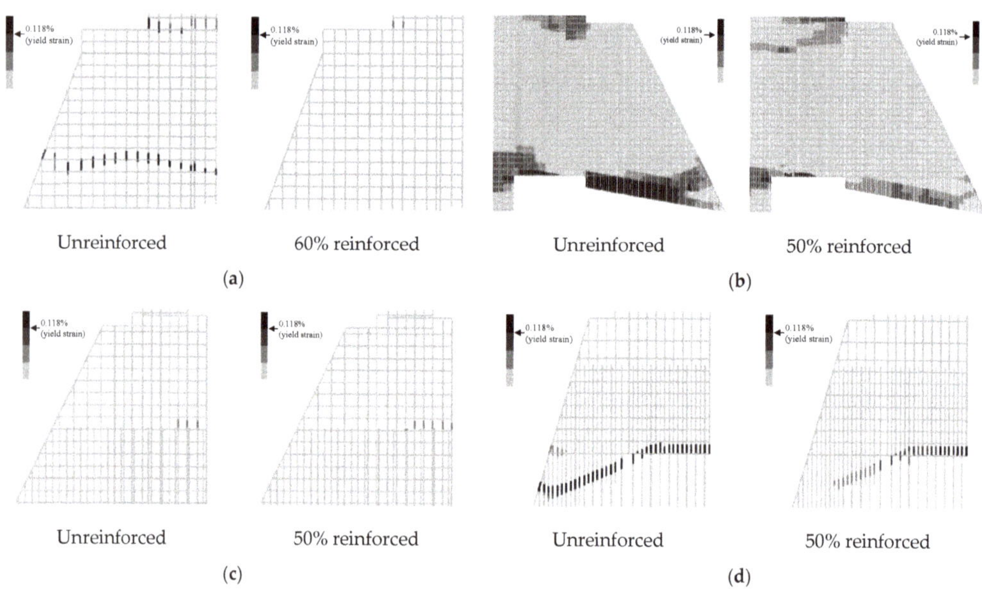

Figure 33. Strain distribution of existing reinforcing bars at peak displacement in wave A: (**a**) Type-A; (**b**) Type-B; (**c**) Type-C; (**d**) Type-D.

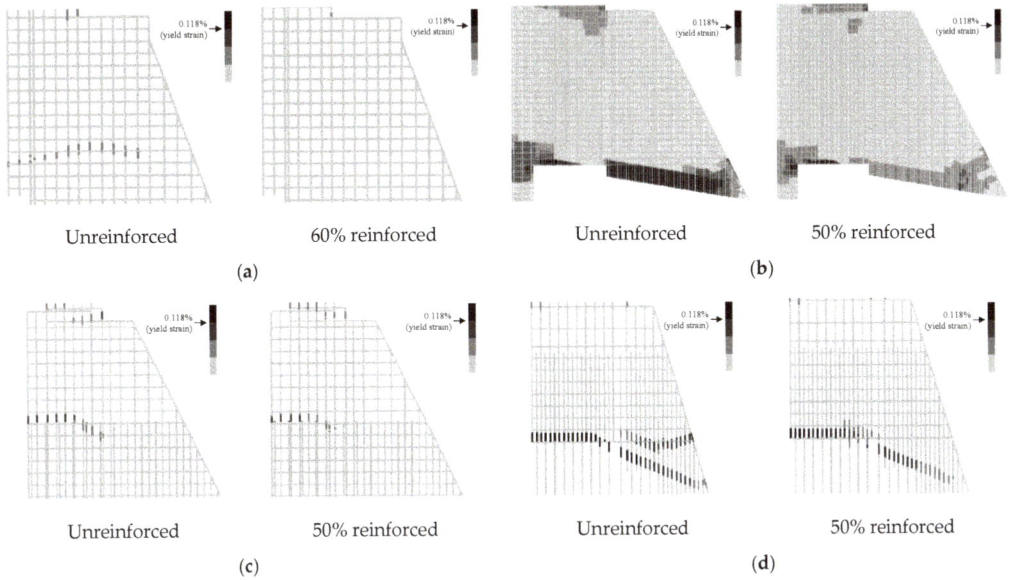

Figure 34. Strain distribution of existing reinforcing bars at peak displacement in wave B: (**a**) TYPE-A; (**b**) TYPE-B; (**c**) TYPE-C; (**d**) TYPE-D.

4.6. Reinforcement Effect by Structural Characteristics of Dam Pier

Figure 35 shows the classification of dams in the area of the graph near the four dams discussed in the previous section, using the amount of reinforcing bars in the existing cross-section and the slenderness ratio of the column shape, and Table 5 shows the natural period of each dam. It was found that the TYPE-C dam shows the least damage and has the shortest natural period; thus, the TYPE-C dam has higher rigidity than the other dams. As shown by arrow (A) in Figure 35, TYPE-A and TYPE-D have a smaller slenderness ratio than other dams, but the amount of existing reinforcing bars is small, and the load-bearing performance is also lower than that of other dams. Therefore, the reinforcing effect of the PCM method is clearly shown for TYPE-A and TYPE-D. On the other hand, although the TYPE-B dam has a larger amount of reinforcing bars than the TYPE-C dam, the slenderness ratio of the column shape is large, and the additional mass is also larger than that of the TYPE-C. Therefore, the reinforcement effect of TYPE-B is higher than that of TYPE-C. In summary, the reinforcement effect of the SRS method is likely to be obtained in dams with a long natural period.

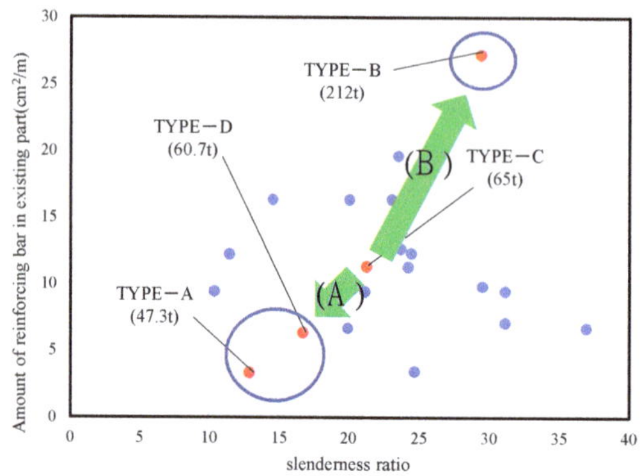

Figure 35. Classification of gravity dams when viewed in terms of the amount of existing reinforcing bars and slenderness ratio of dams near the area where the four dams exist.

Table 5. Natural period of each dam.

Natural Period (s)	TYPE-A	TYPE-B	TYPE-C	TYPE-D
Primary mode	0.353	0.482	0.175	0.243
Secondary mode	0.153	0.139	0.102	0.076
3rd mode	0.0973	0.120	0.0435	0.059
4th mode	0.053	0.103	0.036	0.051

Figure 36 shows that the five dams in the red dashed circle are the groups that are most likely to be reinforced by RPCM because of their insufficient amount of reinforcing bars. To confirm the effect of the RPCM reinforcement, TYPE-B′ having a lower amount of existing reinforcing bars than TYPE-B (shown in Figure 36) was modeled, and its elastic-plastic behavior under earthquake conditions was calculated. Specifically, the cross-sectional area of the rebar elements in the TYPE-B FE model was uniformly reduced, and the total number and spacing of reinforcing bars did not change in TYPE-B′.

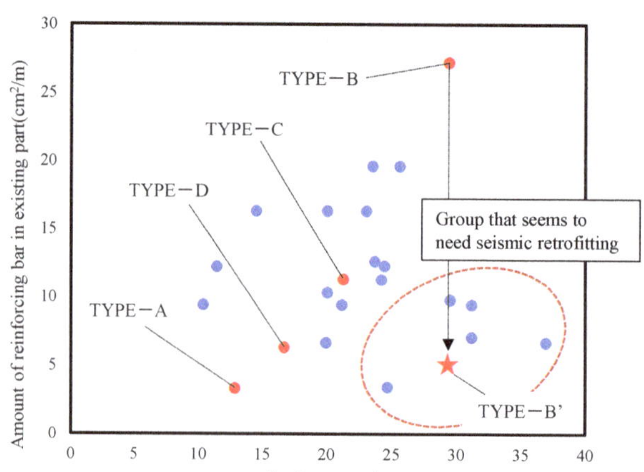

Figure 36. Classification of dams by slenderness ratio and amount of reinforcing bars (Added TYPE-B′).

Figure 37 shows the displacement-time history of TYPE-B′ under seismic waves A and B, respectively. From these results, it can be seen that the maximum displacement and residual displacement of TYPE-B′ are significantly reduced owing to the reinforcement effect of the RPCM. In addition, it was confirmed that reinforcement by RPCM is sufficiently effective only if an area of approximately 30% of the entire cross-section of the column is reinforced to reduce the residual displacement.

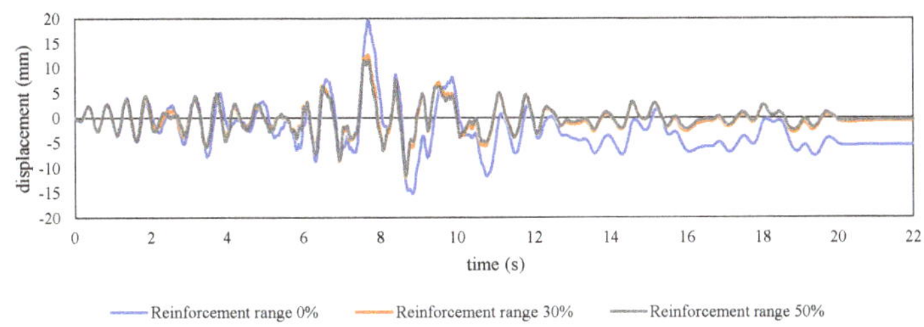

(**a**)

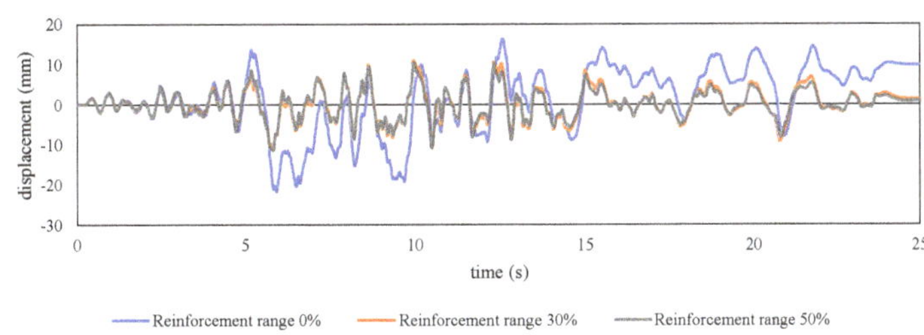

(**b**)

Figure 37. Comparison of displacement-time history at the top of the pier in TYPE-B′: (**a**) wave A; (**b**) wave B.

As a result, it was concluded that the RPCM reinforcement method is very effective for existing aged dam piers with insufficient rigidity.

5. Conclusions

The following were confirmed by performing seismic response analysis using FEM.

1. It was confirmed that the displacement response of the dam pier could be reduced by using the SRS method, and the effect of the reinforcement range by the SRS method on the maximum displacement and residual displacement could be quantitatively determined.
2. In the case of the dam pier, it was difficult to reinforce the entire cross-section owing to the characteristics of the structure. In this study, it was confirmed that a sufficient reinforcement effect was obtained even with partial reinforcement.
3. It was determined that the maximum stress occurred between the existing RC cross-section and the reinforced PCM layer, especially on the upstream side. Therefore, when a dam pier was reinforced using this method, using a large-diameter reinforcing bar at the boundary between the existing RC cross-section and the PCM layer is conceivable.
4. The reinforcing effect was remarkable in the case where the existing reinforcing bar yielded.

5. It was found that the SRS method had a considerable reinforcing effect on dams with a long natural period, which have low rigidity relative to their mass.
6. In fact, according to the TYPE-B′ calculation that simulates a group with a small amount of existing reinforcing bar relative to their slenderness ratio, it was confirmed that if a column cross-sectional area greater than 30% is reinforced using RPCM, the residual displacement after an earthquake can be significantly reduced.

Author Contributions: Conceptualization, Y.S., H.T. and H.I.; methodology, Y.S. and H.T.; software, H.T.; validation, Y.S. and H.T.; investigation, H.I.; resources, H.I.; writing—original draft preparation, Y.S. and H.T.; writing—review and editing, Y.S. and H.T.; visualization, H.T.; supervision, Y.S.; project administration, H.I. All authors have read and agreed to the published version of the manuscript.

Funding: This research received no external funding.

Institutional Review Board Statement: Not applicable.

Informed Consent Statement: Not applicable.

Data Availability Statement: Not applicable.

Acknowledgments: This paper is based on the research conducted by Kazuki Yamazaki and Junji Cho for their master's theses at Kyushu University. We would like to express our gratitude for the research activities.

Conflicts of Interest: The authors declare no conflict of interest.

References

1. Nobuyuki, M.; Tsutomu, S. Current Status of Seismic Methods for Railway Reinforced Concrete Viaducts. *JCI Concr. J.* **1997**, *35*, 9–17.
2. Gkournelos, P.D.; Triantafillou, T.C.; Bournas, D.A. Seismic Upgrading of Existing Reinforced Concrete Buildings: A State-of-the-Art Review. *Eng. Struct.* **2021**, *240*, 112273. [CrossRef]
3. Priestley, M.N.; Seible, F.; Xiao, Y.; Verma, R. Steel Jacket Retrofitting of Reinforced Concrete Bridge Columns for Enhanced Shear Strength. Part 1: Theoretical Considerations and Test Design. *ACI Mater. J.* **1994**, *91*, 394–405. [CrossRef]
4. Nigel Priestley, M.J.; Seible, F.; Xiao, Y.; Verma, R. Steel Jacket Retrofitting of Reinforced Concrete Bridge Columns for Enhanced Shear Strength. Part 2: Test Results and Comparison with Theory. *ACI Mater. J.* **1994**, *91*, 537–551. [CrossRef]
5. Balsamo, A.; Colombo, A.; Manfredi, G.; Negro, P.; Prota, A. Seismic Behavior of a Full-Scale RC Frame Repaired using CFRP Laminates. *Eng. Struct.* **2005**, *27*, 769–780. [CrossRef]
6. Thermou, G.E.; Papanikolaou, V.K.; Kappos, A.J. Flexural Behaviour of Reinforced Concrete Jacketed Columns under Reversed Cyclic Loading. *Eng. Struct.* **2014**, *76*, 270–282. [CrossRef]
7. Léger, P.; Leclerc, M. Evaluation of Earthquake Ground Motions to Predict Cracking Response of Gravity Dams. *Eng. Struct.* **1996**, *18*, 227–239. [CrossRef]
8. Mridha, S.; Maity, D. Experimental Investigation on Nonlinear Dynamic Response of Concrete Gravity Dam-Reservoir System. *Eng. Struct.* **2014**, *80*, 289–297. [CrossRef]
9. Zhang, S.; Wang, G. Effects of Near-Fault and Far-Fault Ground Motions on Nonlinear Dynamic Response and Seismic Damage of Concrete Gravity Dams. *Soil Dyn. Earthq. Eng.* **2013**, *53*, 217–229. [CrossRef]
10. Wang, G.; Wang, Y.; Lu, W.; Yu, M.; Wang, C. Deterministic 3D Seismic Damage Analysis of Guandi Concrete Gravity Dam: A Case Study. *Eng. Struct.* **2017**, *148*, 263–276. [CrossRef]
11. Satoru, N.; Shinichi, H. Seismic Retrofitting for Existing RC Bridge Pier by PCM Shotcrete Spray. *JCI Concr. J.* **2007**, *29*, 1219–1224.
12. Tomoaki, S.; Yasuhiro, M. Structural Performance of RC Column with Sidewalls Strengthened by Polymer-Cement Mortar. *JCI Concr. J.* **2007**, *29*, 1117–1122.
13. JSCE. *Standard Specifications for Concrete Structures-2012, Structural Performance Evaluation*; Japan Society of Civil Engineering: Tokyo, Japan, 2012.
14. Westergaard, H.M. Water Pressures on Dams during Earthquakes. *Trans. ASCE* **1933**, *98*. [CrossRef]
15. The River Bureau of MLIT. *Guidelines for Seismic Performance Evaluation of Dams During Large Earthquakes (Draft)*; The River Bureau of the Ministry of Land, Infrastructure and Transport (MLIT): Tokyo, Japan, 2005.

applied sciences

MDPI

Article

Reinforced Concrete Beams Retrofitted with External CFRP Strips towards Enhancing the Shear Capacity

George C. Manos * and Konstantinos B. Katakalos

Laboratory of Strength of Materials and Structures, Aristotle University, 54006 Thessaloniki, Greece; kkatakal@civil.auth.gr
* Correspondence: gcmanos@civil.auth.gr

Abstract: The practical difficulties in upgrading the structural performance of existing reinforced concrete (RC) structures is discussed, when retrofitting structural members by conventional RC jacketing. The use of retrofitting schemes employing externally applied fiber reinforcing polymer (FRP) strips attracted considerable research attention as a preferable alternative. Such retrofitting FRP schemes aiming to upgrade the shear capacity of existing RC beams have been examined in many published works employing such externally applied FRP shear reinforcing schemes without confronting the practical difficulties arising from the presence of the RC slab. Anchoring external CFRP strips aiming to shear upgrade, which is the focus here, overrides this difficulty. It is shown that effective anchoring, using either mechanical anchors such as the ones devised by the authors or CFRP anchor ropes produced by the industry, can effectively upgrade the shear capacity of an RC T-beam under-designed in shear to the desired level. A novel laboratory test set-up, devised by the authors, can be utilized to quantify the tensile capacity of CFRP stirrups with or without anchors, that can be of practical use. The predicted, according to design guidelines, upgraded shear capacity of the tested prototype RC T-beam, employing the used shear retrofitting schemes, under-estimates the measured shear capacity by 58%. This conservatism can counter-balance uncertainties arising from in situ conditions in constructing the various parts of such a shear retrofitting scheme.

Keywords: upgrading old RC structures; shear retrofit; T-beams; CFRP strips; anchoring devices

Citation: Manos, G.C.; Katakalos, K.B. Reinforced Concrete Beams Retrofitted with External CFRP Strips towards Enhancing the Shear Capacity. *Appl. Sci.* **2021**, *11*, 7952. https://doi.org/10.3390/app11177952

Academic Editor: Maria Favvata

Received: 21 July 2021
Accepted: 23 August 2021
Published: 28 August 2021

Publisher's Note: MDPI stays neutral with regard to jurisdictional claims in published maps and institutional affiliations.

Copyright: © 2021 by the authors. Licensee MDPI, Basel, Switzerland. This article is an open access article distributed under the terms and conditions of the Creative Commons Attribution (CC BY) license (https://creativecommons.org/licenses/by/4.0/).

1. Introduction

In many European countries, a large proportion of the building stock constructed after the end of WWII is multi-story reinforced concrete (RC) structural formations composed of slabs, beams, columns, and shear walls cast in place. The ease offered by this type of construction and the relatively low cost compared to alternative structural forms led to RC building becoming dominant in many countries. Unfortunately, a large number of these buildings were designed and constructed to resist low levels of seismic actions as shown by the development of serious structural damage when subjected to strong earthquake excitations [1]. During the last thirty years, a large volume of research on the seismic response of RC structural components and structural systems has provided a strong basis for upgrading the seismic provisions of RC structural design. Therefore, RC buildings that are currently designed and constructed according to such upgraded seismic code provisions can meet future earthquake excitations, as defined from probabilistic studies for each country, developing controlled structural damage that ensures that such RC buildings will not collapse. Moreover, the structural damage which will develop for the "design" earthquake is repairable. In this way, upgraded design seismic codes safeguard against the loss of life as well as against excessive repair costs for contemporary RC buildings. In contrast, old existing buildings are quite vulnerable to severe structural damage for such "design" earthquakes that could lead to collapse and the loss of life, as was the case during many strong past earthquakes around the world. Confronted with such a seismic risk, research efforts have focused on devising retrofitting techniques capable of upgrading the seismic resistance of such relatively "weak" RC structural formations. This was a difficult task because, whereas reinforced concrete construction offers very substantial flexibility when cast in place, it offers no such flexibility when it is hardened. Therefore, it is quite difficult to introduce any alterations or additions to the structural elements themselves or to the structural system as a whole. A typical retrofitting technique devised for strengthening such "weak" RC structures is to upgrade the capacity of selected critical structural elements of the structural system (e.g., weak columns) with RC jackets, providing in this way the required upgrade in strength and ductility (Figure 1a,b).

A well-known construction technique for strengthening existing reinforced concrete (RC) structural elements is to apply such RC jacketing surrounding the old structural elements with these new RC jackets. RC jackets are constructed using concrete of high bond strength and low shrinkage characteristics together with additional steel reinforcement and occasionally fiber-reinforced polymer composites. A variety of techniques have been investigated in the past to strengthening under-designed RC beams, RC columns, and RC joints employing such RC jacketing techniques. Published research [2–11] has shown that such external RC jacketing improves both the flexural and shear capacity, increases stiffness and ductility together, and provides for higher axial load capacity to under-designed structural members than prior to such RC jacketing. Code provisions [12] include guidelines for the design of RC jackets. RC jacketing was utilized in the past either by strengthening structural members being damaged after a strong intensity earthquake sequence or as a preventing measure for "weak" structures due to the outcome of a relevant inspection prior to an earthquake. Such RC jacketing during the last decades has been the most favorable choice for structural engineers in seismic-prone areas.

Figure 1. (**a**) RC jacket of columns employing shotcrete (gunite). (**b**) RC jacket of column and beam at the region of their joints employing shotcrete (gunite) [13].

However, there are also certain difficulties in employing such RC jacketing techniques. It requires specialized in situ labor and equipment; it also requires perforating the floor RC slab, by partially breaking structural elements in order to place the needed additional steel reinforcement, either longitudinal or transverse steel reinforcement. Moreover, employing such RC jacketing techniques increases the dimensions of the upgraded structural members, resulting in an increase in their stiffness and their dead load. This may not be always desirable. At the same time, such RC jackets result in a decrease in the available internal free space or applying RC jackets may face prohibitions in the external space of the structural system from adjacent buildings. The increase in mass and stiffness, resulting from RC jacketing, may change the dynamic characteristics of the whole structural system and may also cause, in some cases, undesirably increased demands at specific structural elements.

Such RC jacketing of critical structural elements can be combined, when possible, with additions to a structural system of extra shear walls along its height in proper locations.

This combination can reduce excessive displacement response in a way that the demands to the critical elements are met with the corresponding capacities [13]. This retrofit concept is quite effective; however, the construction of such jackets represents a degree of practical difficulties and response issues, as mentioned before.

During the last decades, alternative retrofitting techniques were developed aiming to provide practical solutions to the in situ RC jacketing difficulties for the various RC structural elements [14–21]. Obviously, apart from the structural elements themselves, the more difficult regions for effectively retrofitting an existing RC structure are the critical areas of the structural member connections (joints) as well as the connections with the foundation. Both, the structural connections and the foundation are very critical areas that require special consideration in both identifying the nature of the probable structural damage as well as proposing countermeasures. The main flexural structural damage in slabs and beams develops in the areas of maximum bending moments. For the beams, this usually develops near the joints with the columns and shear walls where large bending moments are expected to arise due to the seismic loads. Similarly, at the ends of the beams are areas of large shear forces from the combination of earthquake forces with the dead and live loads; these will cause the appearance of shear damage in the form of diagonal cracks. The presence of large bending moments mainly from the seismic loads together with large axial forces will cause the formation of flexural damage at the top and the toe of columns, whereas the presence of shear forces from the seismic loads together with axial forces will lead to the formation of shear damage at the columns, a very dangerous form of damage (see Figure 2a,b). The presence of large shear forces from seismic loads, together with relatively low-level axial forces, will lead to the development of another dangerous form of shear damage in the shear walls (see Figure 3b), whereas the presence of short columns will lead to the development of large shear forces from seismic loads and the development again of shear damage, as shown in Figure 3a.

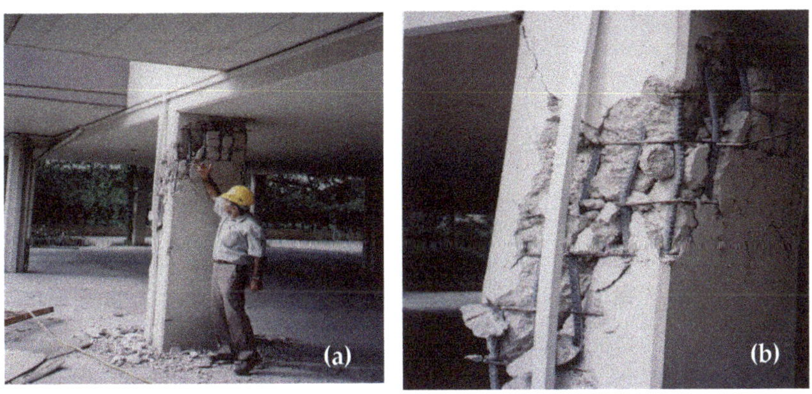

Figure 2. (**a**) Shear failure of an RC column at the ground floor of a four-story RC building. Athens, Greece 1999 earthquake sequence. (**b**) Detail of the same shear damage [1].

Alternative retrofitting techniques to RC jacketing have been developed, which employ fiber-reinforced polymers (FRP) attached externally with special organic resins or inorganic matrices to weak structural elements. This external application has in some cases specific practical advantages when the resulting structural performance is upgraded in the desired way [22–34]. It can be applied to various structural elements of an existing RC structure with a varying degree of effectiveness and difficulty. Thus, such a retrofit is able to easily upgrade the flexural capacity of slabs as the demands are mainly flexural. The same also applies to the flexural upgrade of RC beams for gravitational loads. The shear upgrade presents certain difficulties that are discussed within the rest of the sections of this study. Moreover, the flexural or shear upgrade of beams or columns for seismic loading is more difficult as the regions to be retrofitted are near and include the RC column-to-beam joints. For the RC columns, the flexural upgrade is again quite difficult as the regions to be retrofitted are the ones where the columns are joined with the foundation or with the beam-to-column joints. Alternatively, the shear and the compression capacity of RC columns can be effectively upgraded by externally applied CFRP close hoop strips. A large volume of research has been published utilizing strips made by FRP systems based on epoxy resins as well as high-strength steel fibers (known as SRG or SRP) with either cementitious grouts (SRG) or with organic resins (SFRP) by several researchers [14–43]. A typical mode of failure is reported to be the delamination of these FRP strips [35–37]. Using FRP strips based on steel wires is relatively limited when compared to the strips based on carbon fibers (CFRP).

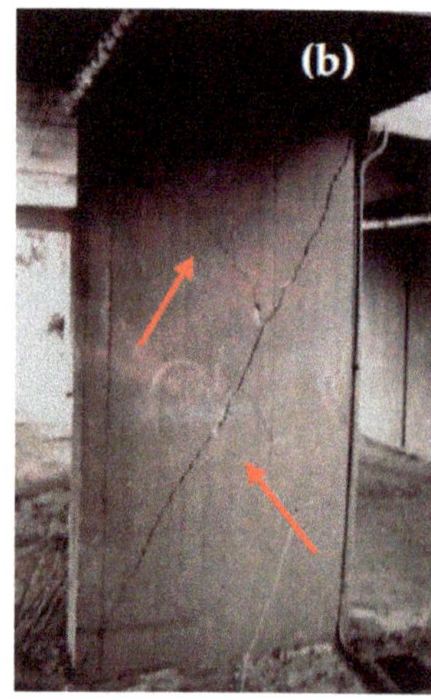

Figure 3. (**a**) Shear failure of an RC short column at the ground floor of a two-story RC building. Aigio, Greece 1995 earthquake sequence. (**b**) Detail of shear damage of a shear wall at the ground floor of a four-story RC building. Pyrgos, Greece 1993 earthquake sequence [1].

This study focuses on the specific problem of RC T-beams which are under-designed in shear. This is a relatively common problem that is due to outdated seismic code provisions with relatively low seismic force levels used in the past, which resulted in relatively low shear force demands for such RC beams in multi-story RC buildings. According to such an outdated seismic design, the shear demands for T-beams were resulting from the combinations including dead and live load and they were met by relatively light transverse reinforcing with open hoop steel stirrups. Instead, current seismic codes result in much larger shear force demands than before at regions where such beams join the columns. In addition, they require that the transverse reinforcement consists of closed hoop steel stirrups. Therefore, it is necessary to strengthen such under-designed RC T-beams deficient in shear capacity. Figure 1b depicts a retrofitting scheme whereby such shear capacity upgrade is carried out by constructing an RC jacket employing extra closed hoop stirrups narrowly spaced. Such a scheme is relatively difficult because it requires the partial breaking of the slab in order to place the closed hoop stirrups and then cast the RC jacket. A relatively simple alternative is to try to attach externally, on the web of such RC T-beam, strips of fiber-reinforced polymers (FRP) in the form of stirrups bonding them to the concrete surface with special epoxy resins. However, even in this case, the main difficulty is to try to add closed hoop FRP stirrups. A large number of experimental investigations were performed with rectangular beam specimens which did not include the slab. In this case, the absence of the slab allows the externally attached FRP strips to have the form of closed hoop stirrups. However, in the vast majority of multistory RC buildings, the slabs are cast in unison with the supporting beams, thus forming typical T-beam structural elements; this does not allow easy passing of such FRP strips in order to form the closed hoop stirrups that are required in order to bring the shear capacity to meet the demands posed by the upgraded seismic design. It was demonstrated that applying open hoop FRP strips, thus avoiding confrontation with breaking the slab, leads to limited shear upgrade because these open hoop FRP stirrups suffer from premature debonding, thus being completely neutralized [35–37]. This is presented and discussed in Sections 2–4. In Figure 4 an RC T-beam is shown schematically provided with external FRP reinforcement in order to increase its flexural and shear capacity. The positive bending moment capacity increase, as shown in Figure 4, is enhanced by FRP strips bonded at the bottom side of this T-beam with an FRP strip of one or multiple layers (indicated with red color) having as width the width of the bottom side of this T-beam. For negative bending moment capacity increase, the FRP strip should be bonded alternatively at the top side of this T-beam. In this case, this becomes relatively easy because the FRP strip width can be much wider than that bonded at the bottom side. This externally bonded CFRP strip will function in the same way as the internal longitudinal steel reinforcement developing tensile axial forces. The shear capacity is similarly enhanced by the transverse shear FRP open hoop strips (indicated with blue color) bonded externally; they will also develop axial forces resisting in this way the shear force demand after the formation of the diagonal shear cracks through the concrete volume. They will function in the same way as steel shear reinforcement placed internally in the form of steel stirrups.

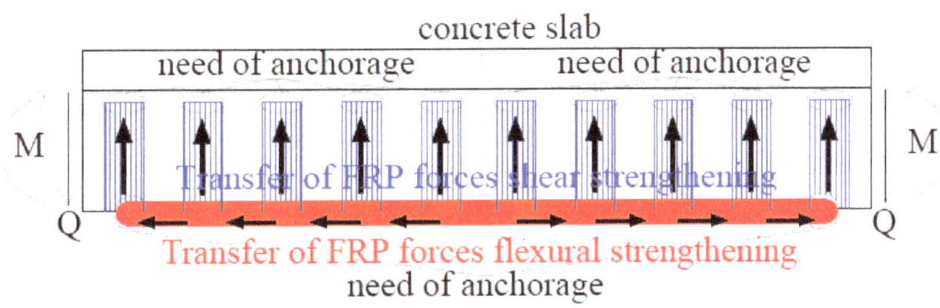

Figure 4. Force transfer mechanisms for externally bonded FRP strips either for flexural or shear retrofit.

For either flexural or shear retrofitting utilizing externally bonded FRP strips, it is critical to investigate the capacity of the relevant bond surfaces which will be called to transfer the axial forces that develop in these FRP strips in each case. When the level of these axial tensile forces exceeds the corresponding bond capacity the debonding mode of failure appears rendering the relevant FRP reinforcement completely ineffective. This represents a serious limitation in applying such external FRP retrofitting either for flexural or shear upgrading. One effective way towards this objective has been to provide various practical schemes of anchoring that combined with the FRP strips can transfer the desired level of axial force beyond the limit posed by the relevant bonding surface as indicated in Figure 4. This has been the focus of the research reported here.

In order to confront the unsatisfactory performance due to the debonding mode of failure various anchoring schemes have been devised [44–57]. Tanarslan et al. [44] reported on the performance of RC shear deficient T-beams strengthened with carbon fiber reinforced polymers (CFRP) systems with and without mechanical anchoring in several different configurations tested under cyclic loading conditions. Manos et al. [45–53], as well as Katakalos et al. [54], used either CFRP or SFRP for shear strengthening. The authors of this work have devised and tested such anchoring schemes [48] which combined with externally applied FRP strips can provide the required substantial shear upgrade of RC T-beams, deficient in shear, as is presented in what follows. Moreover, a number of alternative anchoring devices to be used in shear strengthening schemes together with externally applied FRP strips are also presented and discussed. All these anchoring schemes together with external FRP strips can form equivalent closed hoop stirrups which can provide the required upgrade in shear strength for such deficient RC T-beams. In the following sections, the force transfer mechanism that is mobilized for open hoop CFRP strips with or without anchoring devices, used as transverse shear reinforcement, is presented and discussed. This is carried out initially utilizing a novel experimental set-up employing small-length "unit T-beam" specimens hosting such CFRP strips (Sections 2 and 3). By comparing the obtained experimental results, the capacity of such CFRP strips, with or without anchoring devices, can be demonstrated. Moreover, the various failure modes involved in these transfer force mechanisms and their corresponding bearing capacity can be studied in some depth. These tested novel anchoring schemes have been devised by the authors (Patent No: EP2336455-(A1), 2011 [48]) and by the industry [52]. Both of these anchoring schemes try to tackle in an efficient and practical way the difficulty posed by the presence of the RC slab as part of the RC T-beam cross-section, as previously outlined. After studying the force transfer mechanism through the relatively small-length "unit T-beams", the capability and effectiveness of such CFRP strips as transverse shear reinforcement, with or without anchoring devices, is demonstrated by applying the same anchoring schemes to an RC T-beam of prototype dimensions being subjected to appropriate laboratory testing as is described in Sections 2 and 4. This "unit T-beam" and prototype T-beam experimental sequence is presented in a combined way within the present manuscript. In this way, the effectiveness of such a shear upgrade of under-designed RC T-beams is clearly demonstrated in a stepwise combined and documented way. Such a procedure has an additional practical significance because it can be also utilized for similar alternative retrofitting schemes.

2. Materials and Methods

In order to check the ability of such equivalent closed hoop stirrup schemes to upgrade the shear capacity RC T-beam structural elements, specimens of prototype dimensions are constructed and tested in the laboratory. These specimens are deliberately designed to be deficient in shear from the beginning, as is shown by the reinforcing details in Figure 5a,b. Figure 5c depicts the way such a specimen is supported and loaded with a loading arrangement known as four-point flexure, whereby the load is gradually increased monotonically until each specimen reaches its limit state. As can be seen in Figure 5a–c, such a "prototype RC T-beam" specimen is provided with sufficient top and bottom longitudinal steel reinforcement (three reinforcing steel bars of 20 mm diameter each) to provide considerable flexural capacity. In terms of geometry, support conditions, loading and longitudinal, or transverse reinforcing this prototype T-beam is symmetric along a vertical axis going through its mid-span. Three distinct regions can be seen along the specimen's length with respect to its shear reinforcement. At the very far left and right ends, the transverse reinforcement consists of narrow spaced (every 15 mm) steel close hoop stirrups of 8mm diameter. This is done in order to prohibit any shear failure at these regions hosting the supports as well as to provide sufficient confinement in order to enhance the bond strength of the longitudinal reinforcement in this region. Next, there is a 900 mm long central region where the steel close hoop stirrups of 8 mm diameter are spaced every 70 mm. This central region of the specimen is not required to resist any shear because of the employed loading; however, the used stirrups can enhance the flexural capacity by providing confinement to the compressive zone as well as prohibiting, up to a degree, of the buckling of the longitudinal reinforcement at the compressive tope zone because of the development of large bending moments at this region.

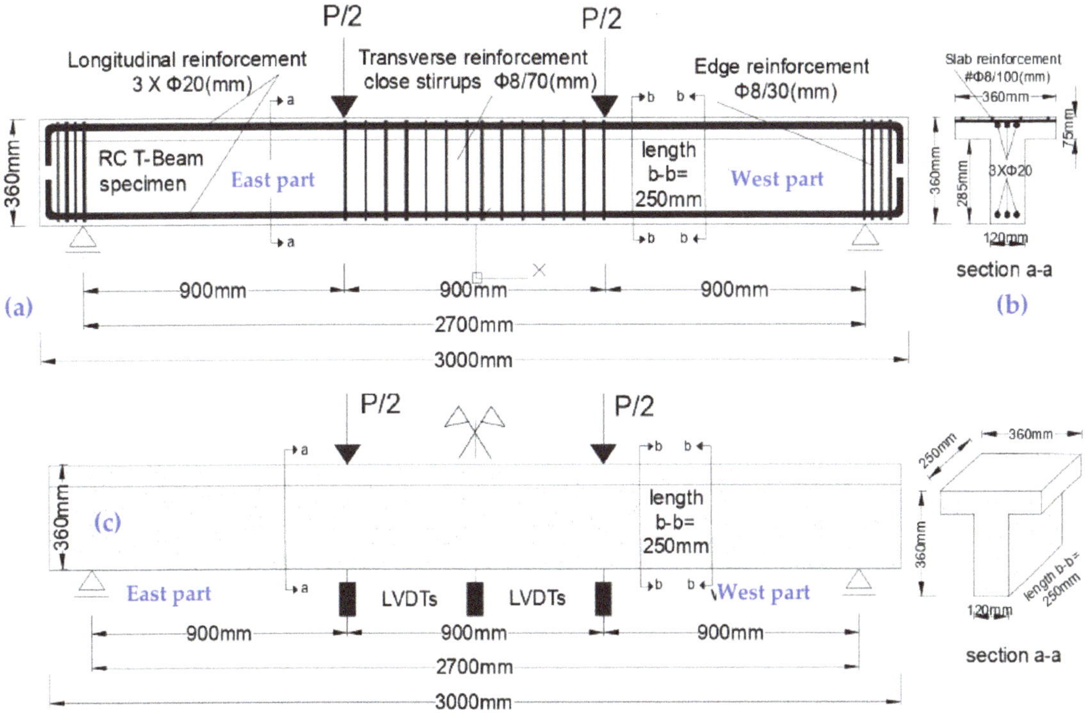

Figure 5. (**a**) Reinforcing details of RC T-beam specimen. (**b**) Cross-section RC T-beam specimen. (**c**) Four-point loading arrangement used in all tests together with the concept of forming the "unit T-beam" specimens (see also Figure 7a,b).

Finally, there are two regions between the supports and the central region that are left without any internal shear reinforcement in terms of steel stirrups (east and west parts). These two regions develop under the employed loading large shear demands. Therefore, it is expected that because of this intentional shear deficiency the prototype RC T-beam in its virgin state will fail in shear in these regions. In this way, in the initial loading stage, this represents the initial control stage whereby the shear resistance is provided only by the concrete volume. Next, during three distinct loading stages corresponding to three distinct retrofitting schemes (see Section 2.2), these regions are provided with an externally applied specific shear reinforcing scheme. Each of these three distinct retrofitting schemes employs either open hoop or equivalent closed hoop FRP strips; the measured shear capacity during each of these three distinct retrofitting stages is compared to the corresponding capacity of the control virgin stage. In this way, the ability of each retrofit to upgrade the shear resistance is demonstrated. The results of the observed performance of this prototype RC T-beam, without or with a shear strength upgrade, are presented in Section 4. The prototype RC T-beam as well as the "unit T-beams", described in Section 2.1, were built with the same concrete (the concrete compressive strength is listed in Table 1), reinforcing bars (typical tensile test shown in Figure 6a), and carbon FRP strips (typical tensile test shown in Figure 6b). From a series of laboratory tests, the compressive strength of the concrete was found equal to f_{ck} = 19.94 MPa, the yield and ultimate strength of the longitudinal reinforcing bars were found equal to 524.6 MPa and 683 MPa, whereas the ultimate strain of the CFRP material was found equal to 0.015 for Young's modulus approximately equal to 260 GPa which is in agreement with the value given by the manufacturers.

Table 1. Results of concrete cylinders tested in uni-axial compression. The specimens' diameter = 150 mm, height = 300 mm.

Specimen No	Maximum Compressive Axial Load (kN)	Compressive Strength (MPa)
Specimen 1	420	23.77
Specimen 2	330	18.67
Specimen 3	329	18.62
Specimen 4	331	18.73
Specimen 5	352	19.92
Average concrete cylinder compressive strength f_{ck} = 19.94 MPa (STDEV = 2.21 MPa)		

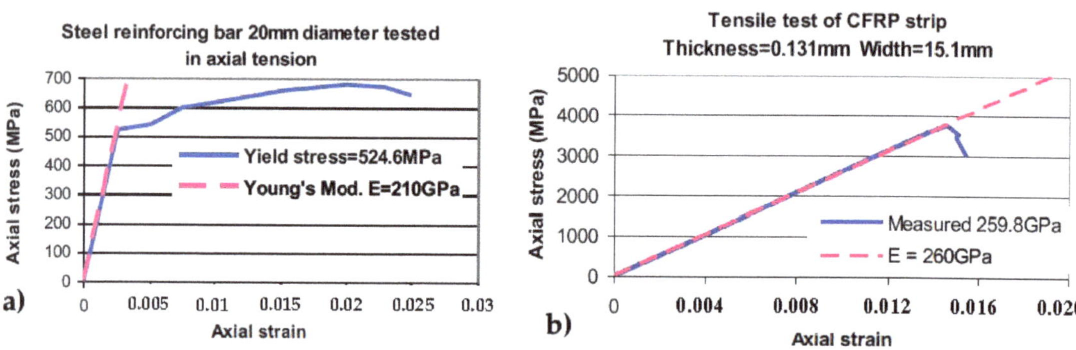

Figure 6. (**a**) Typical tensile test results of longitudinal reinforcement. (**b**) Typical tensile test results of CFRP strip.

2.1. A Novel Laboratory Set-Up to Test the Force Transfer Mechanism for Open Hoop CFRP Strips with or without Specific Mechanical Anchoring Devices

Before testing the simple-supported prototype RC T-beam depicted in Figure 5a–c without or with CFRP external transverse shear reinforcement, a simple experimental procedure was conceived in order to study the force transfer mechanisms that develop in this external CFRP strip transverse shear reinforcement and the concrete volume. For this purpose, T-beam specimens were constructed of relatively small length having the same cross-section as the prototype T-beam of full length (Figure 5b). The length of these specimens, denoted as "unit T-beams", was equal to 250 mm, which was sufficient to provide the necessary bond surface to house a CFRP external transverse shear reinforcing strip having a width of approximately 100 mm, as shown in Figure 7a,b (see also Figure 10a–c).

Figure 8a–d depicts the corresponding cross-sections of such "unit T-beam" specimens hosting a typical CFRP shear strip (100 mm wide) bonded to such a "unit T-beam" specimen with resin provided by the manufacturers, following the typical for the CFRP construction technique. A simple experimental loading procedure is used, aiming to quantify the axial tensile capacity of FRP shear strips without or with an anchoring device, which is its main forcing transfer mechanism, as indicated in Figure 9a–c, and the interaction with the concrete volume. As can be seen in Figure 8a–d, these specimens have the same cross-section as the prototype full-length RC T-beam (Figure 5a–c). However, as explained before, the length of these specimens, named "unit T-beams", is equal to 250 mm which is a portion of the full-length prototype RC T-beam, as indicated in Figures 5a–c, 7a,b and 10a–c, which is sufficient to host the width of one CFRP shear strip. In order to study the force transfer behavior of the CFRP stirrups, each "unit T-beam" is subjected to a tensile stress field by applying an external vertical load, as shown in Figures 7b and 9a–c with an arrow. In this way, the force transfer mechanism under investigation was replicated, as will be explained in detail in what will follow. The experimental set-up for testing these "unit T-beam" specimens is shown in Figure 9a–c. Each of these specimens, after the CFRP strip was set approximately seven days after being bonded at both sides of the "unit T-beam" in all cases, were loaded axially as indicated in these figures. In this way, each side of the tested CFRP strip is subjected to tension with an axial tensile force resultant assumed to be $\frac{1}{2}$ of the externally applied load that needs to be transferred in a way similar to the force transfer mechanism shown in Figure 7a,b. As described before, the same force transfer mechanism develops in the corresponding full-length prototype RC T-beam (Figures 5a–c and 7a) in order to resist the shear, after the development of the diagonal shear cracks. The CFRP strips are bonded in both cases, with or without the presence of an anchoring device. Instrumentation was provided to monitor the variation of the externally applied tensile axial load, as well as the relative slip displacement of the top end of the attached CFRP strip and the concrete surface of the web. Four strain gauges (s.g.1 to s.g.4, shown in Figure 9a–c) were bonded in place, two at each side of the CFRP strip. These strain gauges were placed at the vertical axis of symmetry of each CFRP strip/specimen at two heights along the bonded surface as shown in Figure 9a–c. The applied axial loading in this way reproduced the state of stress that develops after the formation of diagonal shear cracks at open hoop FRP strips in prototype T-beams (Figure 7a).

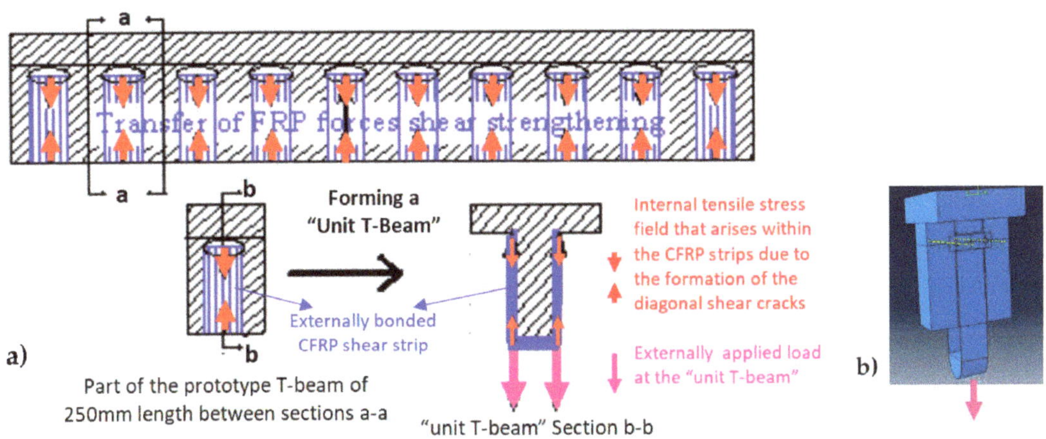

Figure 7. (**a**) Force transfer mechanisms for externally bonded CFRP strips for shear retrofit and the concept of forming a "unit T-beam" specimen. (**b**) Three-dimensional drawing of a limited length unit "T-beam" specimen including a single CFRP strip external transverse shear reinforcement with a steel anchoring device together with the externally applied load.

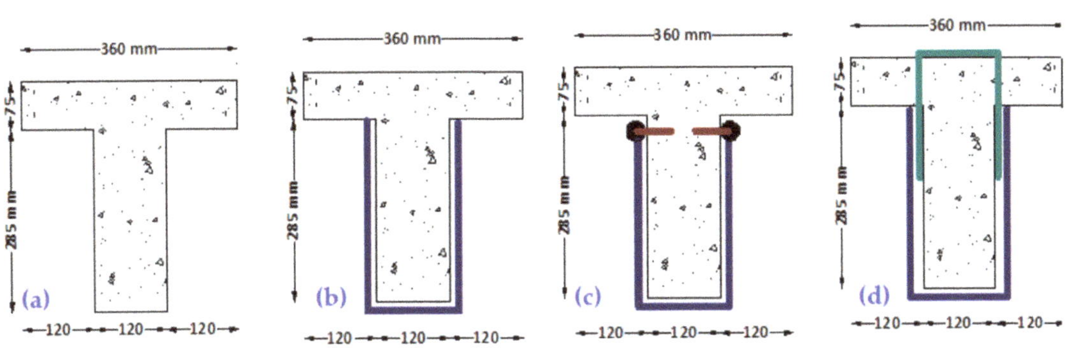

Figure 8. (**a**) R/C T-beam without an FRP strip (**b**) R/C T-beam with an open hoop FRP strip simply attached. (**c**) Open hoop FRP strip anchored with a mechanical anchor. (**d**) Open hoop FRP strip anchored with a rope FRP anchor.

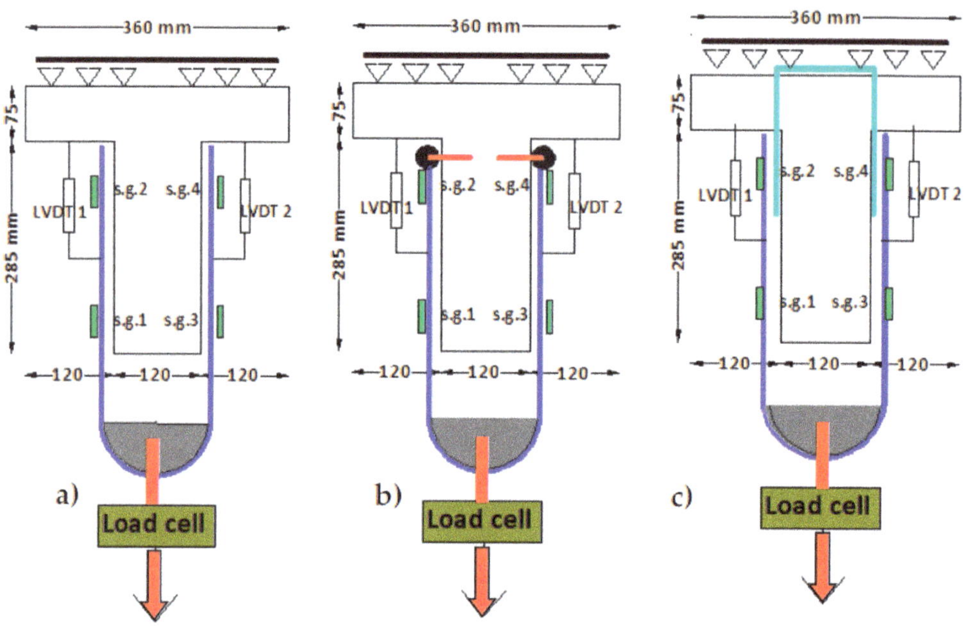

Figure 9. Testing three different open hoop CFRP strips employing a "unit T-beam" loading arrangement. (**a**) R/C "unit T-beam" with an open hoop FRP strip simply attached. (**b**) R/C "unit T-beam" with an open hoop FRP strip anchored with a mechanical anchor. (**c**) R/C "unit T-beam" with an open hoop FRP strip anchored with a rope FRP anchor.

The following limit states were expected to occur during the "unit T-beam" loading. (a) Debonding of the CFRP strip of the concrete bond surface; (b) tensile rupture of the CFRP strip; (c) failure of the anchoring device. In all these "unit T-beams" shear strengthening schemes, shown in Figures 8a–d and 9a–c, open hoop CFRP strips were employed in an effort to avoid breaking the RC slab of the "unit T-beam", apart from drilling relatively small diameter holes. This technique is designed to be applied in the same way to the full-length prototype RC T-beams, providing in this way a construction retrofitting technique that has a significant practical advantage. In the first scheme, the open hoop CFRP strip was simply attached at the sides (webs) and bottom of the "unit T-beam", as shown in Figures 8b and 9a, leaving the R/C slab undisturbed (Lu et al., 2005 [35], Wu et al., 2010 [36], Manos et al., 2013 [37]). Alternatively, in the second scheme, the open hoop CFRP strip was again attached at the sides of the "unit T-beam" employing this time side mechanical anchors devised by the authors [48], as shown in Figures 8c and 9b. Finally, in the last shear retrofitting scheme (Figure 8d), before attaching the open hoop CFRP strip at the sides and bottom of the R/C T-beam, as was carried out before, a CFRP anchor rope, which was specially provided by the FRP industrial suppliers [52], was inserted from the top of the slab through 16 mm diameter holes that were drilled for this purpose, as shown in Figures 8d and 9c. After this, CFRP anchor rope is placed in position through these holes and its fibers are spread out at the sides of the "unit T-beam" in such a way that this rope becomes flat and obtains a considerable width in order to be attached to the open hoop CFRP strip placed from the bottom of the T-beam. Epoxy resin is used to both fill the fibers of this CFRP anchor rope as well as to attach these spread rope fibers to the fibers of the open hoop CFRP strip. The same retrofitting process was also applied when upgrading in shear the prototype RC T-beam, as described in Section 2.2.

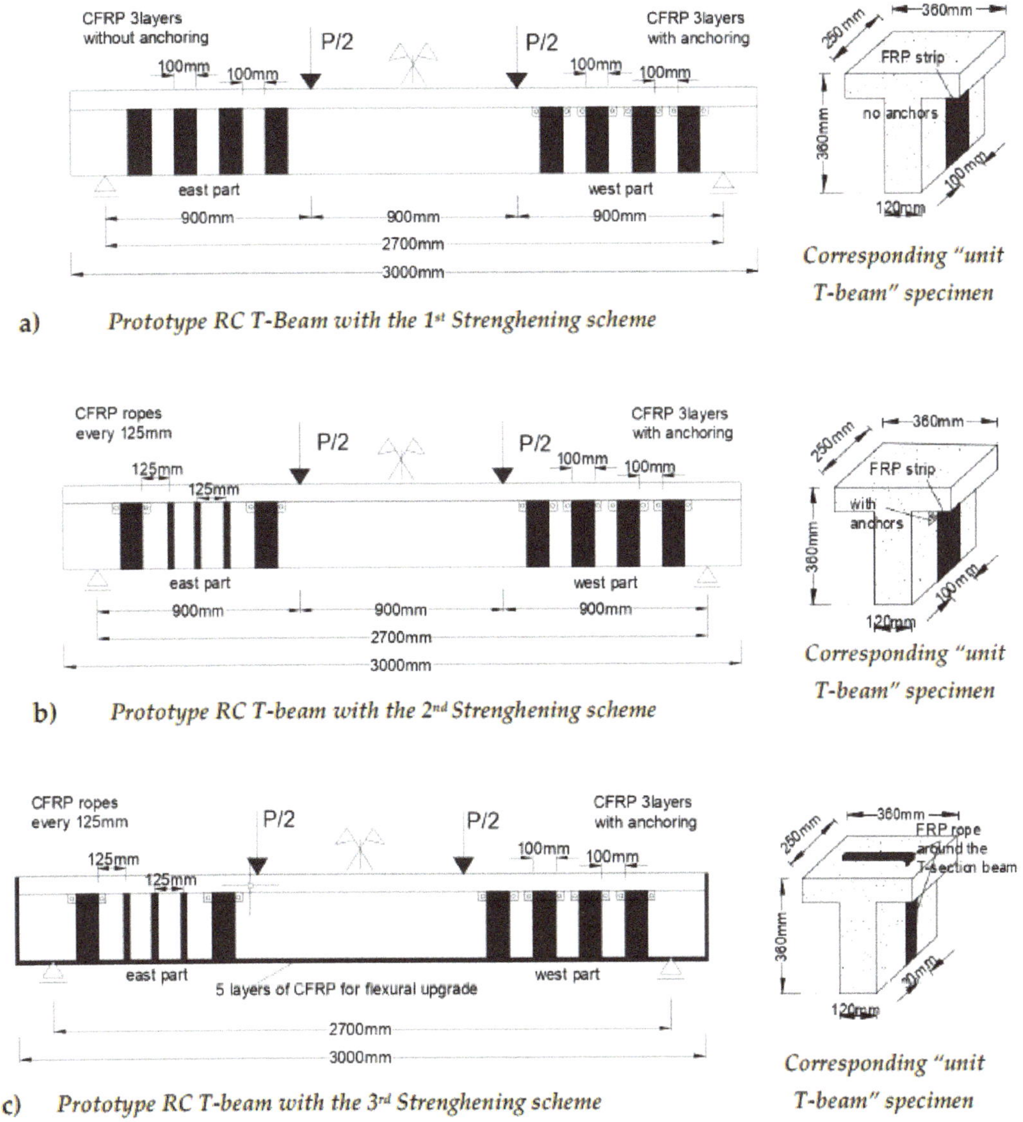

Figure 10. (**a**) 1st strengthening scheme of R/C T-beam (see also Figures 8b and 9a). (**b**) 2nd strengthening scheme of R/C T-beam (see also Figures 8c and 9b). (**c**) 3rd strengthening scheme of R/C T-beam (see also Figures 8d and 9c).

2.2. Four-Point Bending of a Prototype RC T-Beam with Various Shear Retrofirring Schemes

The full-length prototype RC T-beam described earlier (Figure 5a–c) was tested under four-point bending loading in order to quantify its shear capacity under the following four distinct stages. This prototype RC T-beam had, when tested during all four stages described in detail from (a) to (d) in what follows (Figures 5 and 10a–d), a clear span equal to 2700 mm. The central vertical load was monotonic and was applied through a stiff steel girder supported at the upper side of the T-beam at two points located 900 mm from the two end vertical supports. This vertical load was measured by a load cell located at mid-span, whereas the vertical deflections were recorded by displacement transducers at mid-span as well as at the two locations coinciding with the two points of applied load as indicated in Figure 5c.

(a) At this virgin stage, the prototype RC T-beam, having its east and west parts intentionally deficient in shear, was loaded prior to any shear retrofitting (Figure 5c) until the appearance of diagonal cracks, indicating that the shear capacity limit state was reached.
(b) At this 2nd stage, the 1st shear strengthening scheme was applied by employing the external application of open hoop CFRP strips. At the West part, four (4) 3-layer open hoop CFRP strips were employed (Figure 10a) having 0.131 mm thickness, 100 mm width, and spaced at 200 mm intervals measured from their centerline. These West part CFRP strips employed the anchor scheme of Figures 8c and 9b. At the East part, four (4) 3-layer open hoop CFRP strips were employed again, the same as at the west part of the T-beam, however, without any anchors (Figures 8b and 9a). Each CFRP strip layer was 0.131 mm thick and 100 mm wide and were spaced at 200 mm intervals measured from their centerline, as were the CFRP strips at the East part. This was carried out in order to study the debonding mode of failure for the CFRP strips attached to this part. As will be discussed in Section 4, the limit state, in this case, was the debonding of the unanchored CFRP strips of the East part.
(c) At this 3rd stage, the 2nd shear strengthening scheme was applied. The two debonded at the previous stage unanchored CFRP strips, located at the East part, were replaced by three closed hoop CFRP anchor rope stirrups each with a cross-section equal to 28.0 mm^2 (see Section 3.3). Holes were drilled in the slab of the specimens for these closed hoop ropes to go through, whereas these ropes took the shape of a CFRP strip along the webs and the bottom side of the specimen. The CFRP strips anchored with steel anchored devices at the West part were left without any modification because they did not exhibit signs of any distress during the previous stage (Figure 10b).
(d) Finally, at the 4th stage, the 3rd strengthening scheme was applied. It involved flexural strengthening consisting of five (5) CFRP layers (each layer being 0.131 mm thick and 120 mm wide) attached at the bottom side of the T-beam specimen, as is depicted in Figure 10c.

The obtained response during all these four distinct stages for the examined prototype RC T-beam during the above four distinct stages are presented and discussed in Section 4.

3. Measured Response from Testing the Capacity and Force Transfer Mechanism of External CFRP Shear Stirrups without or with an Anchoring Device Utilizing "Unit T-Beam" Specimens

In what follows, the measured response obtained from two distinct sequences of tests employing "unit T-beam" specimens with various types of CFRP stirrups, with or without anchoring, is presented and discussed. Figure 11a,b depict the measured response in terms of applied total tensile load versus the strain readings that developed at the locations of the four strain gauges shown in Figure 9a–c. Figure 11a depicts a typical plot of the measured response for the CFRP strips without any anchoring (Figures 8b and 9a), whereas, Figure 11b is the corresponding typical response for the CFRP strips that are provided with steel anchors in addition to bonding at the sides of the "unit T-beam" specimens (Figures 8c and 9b).

It can be seen in Figure 11a, although the strain readings exhibit quite a different variation versus the applied load at the initial stages of the loading sequence, the strain value recorded by all four strain gauges becomes almost identical when the load reaches its maximum value prior to the debonding failure. The strain readings at locations s.g.1 and s.g.3 increase in a more gradual trend with the load increase. This should be attributed to the CFRP—concrete volume interaction at the bonding surface, which is not uniform from the beginning of the loading sequence. From the strain recordings, it can be seen that the CFRP strips are more stressed at the lower side (near the location where the load is applied) rather than the upper side. This process depends on the amplitude of the load and how the bond surface interacts with the CFRP strip and the host concrete surface. During a previous investigation performed by the authors, it was shown that the proper treatment of this host concrete surface prior to bonding of the CFRP strips could result in a substantial increase in the corresponding force transfer capacity [37]. As expected, the presence of the anchoring devices diminishes the importance of this bonding mechanism because the force transfer mechanism and the corresponding capacity depend primarily on the effective performance of the anchoring device. The steel anchoring device employed here has been developed and patented by the authors [48] to be effective after being through numerous laboratory tests.

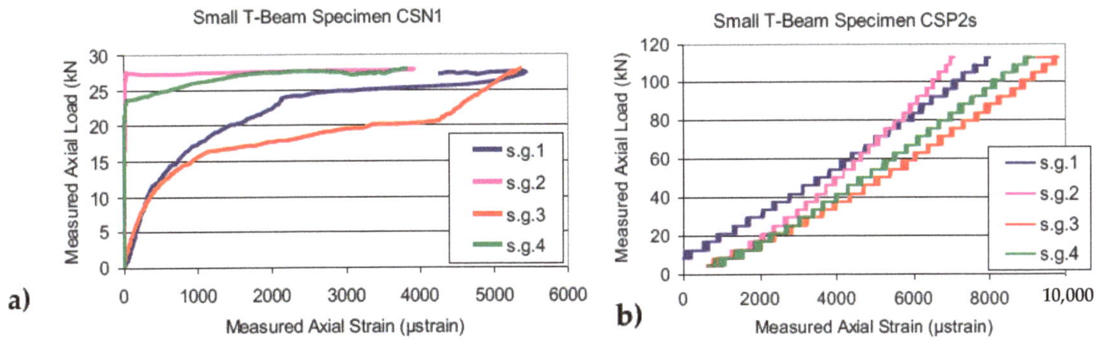

Figure 11. Typical strain gauge readings during the loading process of the CFRP strips bonded to the "unit T-beam" specimens (**a**) CFRP strips without any anchors (Figures 8b and 9a) (**b**) CFRP strips with steel anchors (Figures 8c and 9b).

As can be seen in Figure 11b, the variation of the strain readings against the load at the four strain gauge locations, when the anchor devices are present, are not very different. At the maximum load for this case, the readings at locations s.g.1 and s.g.3 are almost the same and larger than the corresponding readings at locations s.g.2 and s.g.4. This must be attributed to the fact that the presence of the anchoring device disturbs the uniformity of the axial stress field at close proximity to these anchoring devices, whereas this axial stress field is certainly less disturbed at the level of s.g.1 and s.g.3. Following this rationale, the maximum average axial strain value measured at locations s.g.1 and s.g.3 is utilized in Table 2, which includes a summary of the obtained results

Table 2. Results of "unit T-beam" specimens with open hoop CFRP strips with and without the use of mechanical anchors (Figures 8a–d and 9a–c).

"Unit T-beam" Specimen Code Name	Maximum Average CFRP Strip Axial Strain Values Measured at Locations s.g.1–3 (μstrain)	Total Maximum Measured Axial Load (kN)	Total Axial Load (kN) Obtained from the Max Average CFRP Axial Strain Values Measured at Locations s.g.1–3/ Failure Mode
(1)	(2)	(3)	(4)
CSN1 * single CFRP layer without anchor, Figures 8b and 9a	5670	27.94	34.17/ Debonding
CRN1 ** single CFRP layer without anchor, Figures 8b and 9a	7114	42.67	42.87/ Debonding
CSP2s * CFRP with two layers and mechanical anchoring of Figures 8c and 9b	9518	113.0	114.71/ Fracture of FRP
CRP2s ** CFRP with two layers and mechanical anchoring of Figures 8c and 9b	8689	102.7	104.72/ Fracture of FRP

* No special treatment of the bond surface apart from careful cleaning. ** The bond surface was made rough with a special hammer.

3.1. "Unit T-Beams" with Open Hoop CFRP Strips Employing Specific Mechanical Anchoring Devices

The tested "unit T-beam" specimens in this sequence are listed in column (1) of Table 2 with their code name stating the presence (or not) of an anchoring device for the CFRP strips. Moreover, the preparation of the concrete surface where the CFRP is attached is also indicated. Column 2 of Table 2 lists the average maximum value of the axial strain that was measured at locations s.g.1 and s.g.3 [53]. Column (3) of Table 2 lists the maximum total applied load value during each test. Column (4) lists the value of the total axial load based on the measured average maximum axial strain value (column 2) and the measured Young's modulus value of the used CFRP strips equal to 234 GPa. It is assumed that the CFRP strips of the tested specimen develop this maximum strain value for the whole width of its cross-section at both sides of the "unit T-beam". Each CFRP layer had a thickness of 0.131 mm and a width of 100mm. A reasonably good agreement can be seen by comparing the measured total maximum load values (Table 2 column 3) with the corresponding values calculated in the way described earlier (Table 2 column 4). It can be concluded that, for design purposes, it is very important to be able to ascertain with confidence the axial strain level that can be assumed to arise in these FRP strips. Towards this objective, it is important to approximate the effectiveness of the bonding surface or of a specific anchoring device and on this basis to adopt the appropriate axial strain value for the FRP strips. The following summarizes the most important observations of the behavior exhibited by either the used CFRP strips or the specific mechanical anchoring devices used in these tests.

(a1) The debonding of the CFRP strip from the concrete surface was observed for strain/stress levels well below the strain limits given by the manufacturers of the FRP materials. The strain/stress levels accompanying this debonding mode of failure continually decrease when one increases the layers of the FRP strip, and consequently its thickness and cross-sectional area, rendering such layer increase totally ineffective unless it is combined with some type of anchoring. This type of failure, which is expected to occur in similar practical applications, is depicted in Figure 12a as observed during the current investigation.

Figure 12. (**a**) Debonding mode of failure. (**b**) Failure of the anchoring scheme accompanied with debonding. (**c**) Tensile failure of the FRP strip.

(a2) From the preceding discussion, it becomes obvious that the debonding mode of failure prevails in almost all cases where an open hoop FRP strip is simply attached without any anchoring. However, the effective anchoring of such an open hoop FRP strip is not easy. Thus, the second category of modes of failure includes limit states in which the final debonding and failure of the FRP strip is a result of the interaction between the FRP strip and the used anchoring scheme. In many cases, the employed anchoring scheme is insufficient to withstand the level of axial force that the FRP strip can withstand by itself in ideal axial tension conditions leading to either local failure of parts of the anchoring scheme or local failure of the FRP strip in areas neighboring the anchor or both. Again, the increase of the layers of the FRP strip, and consequently of its thickness and cross-sectional area, results in a corresponding increase in the demands on the various parts of the anchoring scheme to withstand this increased CFRP strip capacity leading to partial successive failure of the anchoring device. This type of failure is depicted in Figure 12b as was observed during the current investigation for an anchoring scheme that proved ineffective and is not reported further in this paper.

(a3) The final mode of failure is in a form of tensile failure of the FRP strip. The closer this tensile failure resembles an ideal symmetric axial tensile failure of the FRP strip the higher the axial strain/stress levels that would develop, thus, resulting in higher exploitation of the capabilities of the FRP material. This desirable FRP strip performance is observed when the used anchoring scheme is effective in inhibiting any asymmetric local deformation patterns for the axial tensile force levels that correspond to such relatively high strain/stress levels of the FRP strip. The final limit state condition is that of the fracture of the FRP strip that is obviously preceded by its debonding. This type of failure is depicted in Figure 12c, as observed during the current investigation. For a given effective anchoring scheme linked with an FRP strip having a given number of layers, a successive increase in the number of layers will eventually lead to the failure of the anchoring scheme, unless it is properly redesigned.

3.2. "Unit T-Beams" with Open Hoop CFRP Strips Employing CFRP Anchor Ropes

This section presents the measured response obtained from an additional loading sequence investigating a different type of anchoring scheme (Figure 13a) utilizing again the "unit T-beam" loading process. This time, before attaching the open hoop CFRP strip at the sides and bottom of the R/C beam, a CFRP anchor rope is inserted from the top through 16mm diameter holes that are drilled in the R/C slab of the T-beam for this purpose. The effective cross-sectional area of this CFRP rope is equal to 33.1 mm^2 and Young's modulus is equal to 240 GPa. After this CFRP anchor rope has been placed in position through these holes, its fibers are spread at the sides of the beam in a way that this rope becomes flat and obtains a considerable width in order to be attached to the single-layer open hoop CFRP strip, which is put in place from the bottom of the T-beam. This anchoring scheme was studied in two different ways. First, one anchor rope was used with its axis located at the mid-axis of the width of the open hoop CFRP strip.

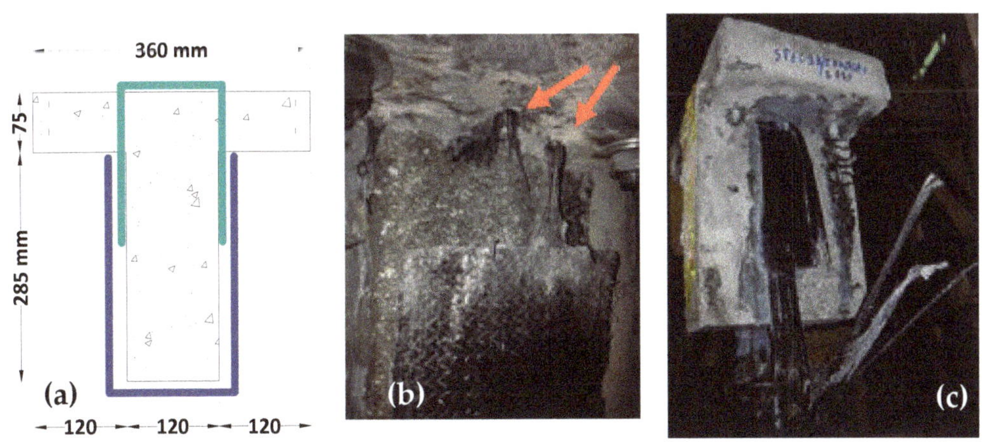

Figure 13. (**a**) R/C T-beam with an attached open hoop FRP strip anchored with an FRP anchor. (**b**) Mode of failure of specimen SW600C/1 No 1. Fracture of the CFRP rope. (**c**) Mode of failure of specimen SW600C/2 No 1. Fracture of the CFRP strip.

First, there are the three specimens, using a single anchor rope, with the code name SW600C/1 No1, No2, and No3 (column1, Table 3). In addition, a double anchor rope was placed side-by-side along the width of the open hoop CFRP strip. There are four specimens utilizing this double anchor rope with the code name SW600C/2 No1, No2, No3, and No4 (coulmn1, Table 3). In column 2 of Table 3, the total maximum measured load for each specimen is listed, whereas, column 3 lists the average values of the axial strains measured at both sides of the CFRP strip at the time instant that the specimen was subjected to the maximum load value of column 2. Column 4 of Table 3 lists the axial load valued obtained on the basis of the strain values of column 3 (with the value of Young's Modulus equal to 234 GPa) and the cross-sectional area of the CFRP strip. Because this anchoring scheme used in all these cases is not expected to disturb the distribution of the axial stress field at the location of the four strain gauges, the average value of all four strain gauges during the occurrence of the maximum load is utilized in Table 3. The following are the main observations.

Table 3. Measured tensile capacity of open hoop CFRP strips anchored with CFRP ropes.

Code Name of Specimen	Maximum Value of the Total Measured Axial Load (KN)	Average Strain from Both Sides of the Strip at Maximum Load (μstrain)	Total Axial Load (kN) Obtained from the Average CFRP Axial Strains Measured at Maximum Load	Mode of Failure
(1)	(2)	(3)	(4)	(5)
SW600C/1 No 1 1 layer CFRP *, single anchor rope **	60.88	3900	60.42	Fracture of anchor rope at upper corner
SW600C/1 No 2 1 layer CFRP *, single anchor rope **	68.76	4400	68.16	Delamination of FRP strips from anchor
SW600C/1 No 3 1 layer CFRP *, single anchor rope **	68.72	4400	68.16	Fracture of anchor rope at upper corner
SW600C/2 No 1 1 layer CFRP *, double anchor rope **	79.46	5200	80.55	Fracture of FRP strip
SW600C/2 No 2 1 layer CFRP *, double anchor rope **	97.18	6400	99.14	Fracture of FRP strip
SW600C/2 No 3 1 layer CFRP*, double anchor rope **	61.86	4200	65.06	Fracture of FRP strip
SW600C/2 No 4 1 layer CFRP *, double anchor rope**	105.98	5300	82.10	Fracture of FRP strip

* 1 layer CFRP strip Area A1 = 33.1 mm^2. ** CFRP Anchor Rope Area A2 = 28.0 mm^2.

(a2) When a single CFRP rope was used in the anchoring scheme of the 1-layer open hoop CFRP strip, the observed failure was mainly at this anchor rope (see Figure 13b). On the contrary, when double CFRP anchor ropes were used to anchor the open hoop CFRP strips the obtained tensile capacity resulted in an effective anchoring scheme leading to the tensile fracture of the single-layer CFRP strip (Figure 13c).

(b2) As can be seen from the obtained maximum axial load values listed in Table 3 (column 2), when a single CFRP anchor rope is used the average maximum axial load value is equal to 66.12 kN (SDEV = 4.54 kN, 6.9% of the average maximum value). In comparison, when double CFRP anchor ropes are used then the average maximum axial load value is equal to 86.12 kN (SDEV = 19.58 kN, 22.7% of the average maximum value), representing a substantial increase. The use of double anchor ropes also succeeded in changing the mode of failure from the anchor rope to the CFRP strip in all cases, which should be considered as a preferable performance.

(c2) A relatively large SDEV value in the axial tensile load capacity results can be seen when double anchor ropes are used. This indicates a degree of uncertainty in achieving desired high values of tensile bearing capacity when applying a relatively large number of anchor ropes. The largest measured tensile capacity when using double anchor ropes is equal to 105.98 kN, which represents a 60% increase from the measured average capacity when using a single anchor rope. At the same time, the smallest bearing capacity when using double anchor ropes is equal to 61.86 kN, which represents a 7% decrease from the measured average capacity when using a single anchor rope. This should be attributed to the interaction between the anchor ropes and the CFRP strip at the common bond surface and should be investigated further.

(d2) Reasonably good agreement can be seen by comparing the measured total maximum load values (column 2 of Table 3) with the corresponding values calculated in the way described earlier (column 4 of Table 3), with the exception of the specimen listed in the last row of Table 3. It can be again concluded that for design purposes it is very important to be able to ascertain with confidence the axial strain level that can be assumed to be able to be sustained in these CFRP strips. Towards this objective, it is important to approximate the effectiveness of the bonding surface or of a specific anchoring device and on this basis to adopt the appropriate axial strain value for the CFRP strips.

3.3. "Unit T-Beams" with Either Closed Hoop CFRP Strips or Closed Hoop Single CFRP Anchor Rope

In order to have a direct measurement of the tensile capacity of either the CFRP strips themselves or the CFRP anchor ropes when in position, extra "unit T-beam" specimens were constructed whereby the CFRP strip (specimens ref-1 and ref-2, Figure 14a,b) and the CFRP rope (specimens SWFX No1, No2 and No3, Figure 14c,d) were accommodated in a closed hoop formation and were subjected to the same loading arrangement depicted in Figure 9c. The obtained results are listed in Table 4.

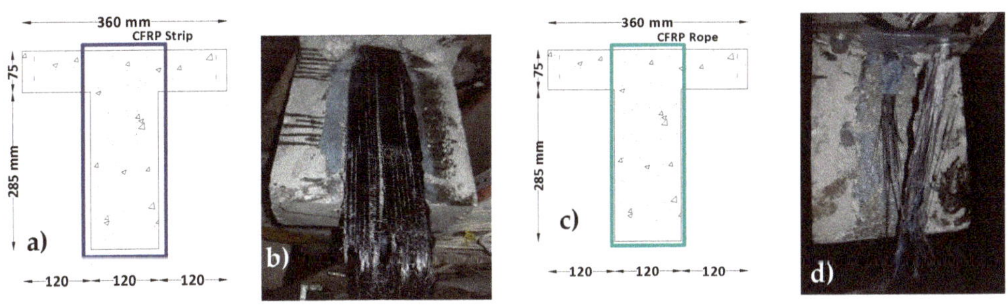

Figure 14. (**a**) "Unit T-beam" specimens CFRP strip Ref-1 and Ref-2. (**b**) Failure mode of specimen CFRP strip Ref-1. (**c**) "Unit T-beam" specimens CFRP Rope SWFX No1, No2 and No3. (**d**) Failure mode of specimen CFRP Rope SWFX No 2.

As can be seen from the obtained maximum axial load values listed in Table 4, when a closed hoop single CFRP anchor rope is used the average maximum axial load value is equal to 70.79 kN (SDEV = 4.21 kN, 5.9% of the average maximum value). When a closed hoop single CFRP strip is used, the average maximum axial load value is equal to 85.63 kN (SDEV = 18.43 kN, 21.52% of the average maximum value). From these maximum load values, it can be concluded that when the used single layer CFRP strip is anchored with a single CFRP anchor rope the tensile failure is expected to occur at the rope (70.79 kN < 85.63 kN). On the contrary, when the used single layer CFRP strip is anchored with a double CFRP anchor rope the tensile mode of failure is expected to occur on the CFRP strip (85.63 kN < δ * 70.79 kN, with the value of δ being larger than 1.25 signifying the degree of effectiveness of the double anchor rope when compared to that of a single anchor rope, given the uncertainty of the performance of multi-anchor ropes described in Section 3.2 and Table 3.

Table 4. Measured tensile capacity of either closed hoop CFRP or closed hoop CFRP anchor ropes (Figure 14a,d).

Code Name of Specimen	Maximum Total Measured Axial Load (KN)	Average Strain from Both Sides of the Strip at Maximum Load (μstrain)	Total Axial Load (kN) Obtained from the FRP Axial Strains Measured at Maximum Load	Mode of Failure
CFRP Strip Ref-1 Closed hoop strip *	98.66	6600	102.23	Fracture of FRP strip
CFRP Strip Ref-2 Closed hoop strip *	72.60	5100	79.00	Fracture of FRP strip
CFRP Rope SWFX No 1 Closed hoop anchor rope **	69.08	-	-	Fracture of anchor rope
CFRP Rope SWFX No 2 Closed hoop anchor rope **	75.58	-	-	Fracture of anchor rope
CFRP Rope SWFX No 3 Closed hoop anchor rope **	67.70	-	-	Fracture of anchor rope

* 1 layer CFRP strip Area A1 = 33.1 mm^2. ** CFRP Anchor Rope Area A2 = 28.0 mm^2.

4. Results of the Measured Behaviour of a Prototype R/C T-Beam in Its Virgin Condition as Well as Being Upgraded in Shear with Various External CFRP Strips

In what follows, the measured response of the tested prototype RC T-beam during the four distinct stages of sequentially applying is described in Section 2.2. CFRP strip shear retrofitting schemes are presented and discussed.

4.1. Virgin Prototype RC T-Beam without Any External CFRP Strip Shear Reinforcement

Initially, this prototype RC T-beam was loaded at its virgin stage until the shear limit-state was reached with the appearance of diagonal shear cracking patterns at the East and West parts for a maximum shear force value equal to 57.39 kN. This is depicted in Figure 15.

Figure 15. Virgin prototype RC T-beam that reached a shear limit state under four-point bending.

4.2. Prototype RC T-Beam witht External CFRP Strip Shear Reiforcement Which Includes at the East Part Unancored CFRP Strips (1st Shear Strengthening Scheme)

The limit state this time (1st shear strengthening scheme, Figure 16) resulted, as expected, in the debonding mode of failure of the East side unanchored CFRP strips as shown in Figure 17b for a shear force equal to 166.77 kN. The corresponding maximum bending moment value is equal to 150.09 kNm. This shear force value is more than three times larger than the shear capacity measured for the un-strengthened virgin T-beam. The variation of the applied shear force versus the vertical deflection of the virgin and the strengthening with this 1st shear strengthening scheme T-beam is depicted in Figure 18a, whereas, Figure 18b depicts the comparison of the measured bending moment response between the Virgin T-beam and the T-beam having been retrofitted with the 1st strengthening scheme (a-Shear force versus the deflection at mid-span, b-Bending moment versus the deflection at mid-span). A comparison of the modes of failure after the maximum load was reached is depicted in Figure 17a for the virgin T-beam without any external shear CFRP strips (development of diagonal shear cracks), and in Figure 17b for the T-beam retrofitted according to the 1st shear strengthening scheme. As can be seen in Figure 17b, the limit state, in this case, was the debonding of the open hoop CFRP strips without anchoring. It is important to underline that the West part of this T-beam, although subjected to the same shear force level as the East part, did not show signs of any distress. This is due to the presence of the effective anchors that accompanied the open hoop CFRP strips at this location. The design of this CFRP anchoring scheme was facilitated by specially designed software [47] as well as valid numerical simulations [51].

Figure 16. The East and the West parts of the prototype RC T-beam with the *1st shear strengthening* scheme under four-point bending.

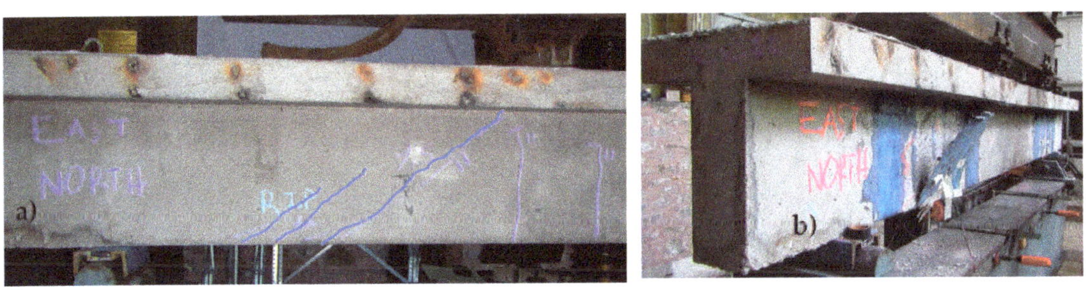

Figure 17. Comparison of failure modes between the prototype RC Virgin T-beam and this T-beam having been retrofitted with the 1st strengthening scheme. (**a**) Diagonal shear cracks at the East part of the Virgin T-beam. (**b**) The debonding of two open hoop CFRP strips without anchoring at the East part of this T-beam retrofitted with the *1st strengthening scheme*.

4.3. Prototype RC T-Beam witht External CFRP Shear Reiforcement Having all the CFRP Stirups Either Anchored at the West Part or Closed CFRP Rope Both at the East Part (2nd Shear Strengthening Scheme)

At this stage (2nd shear strengthening scheme), the two debonded at the previous stage unanchored CFRP strips, located at the East part, were replaced by three closed hoop CFRP anchor rope stirrups similar to the one tested in Section 3.3 (see Section 2.2 and Figure 10b). These closed hoop CFRP ropes can be seen in Figure 19a, whereas, Figure 19b depicts the traces of the diagonal cracks. Further development of these diagonal cracks was successfully prohibited by these closed hoop anchor ropes.

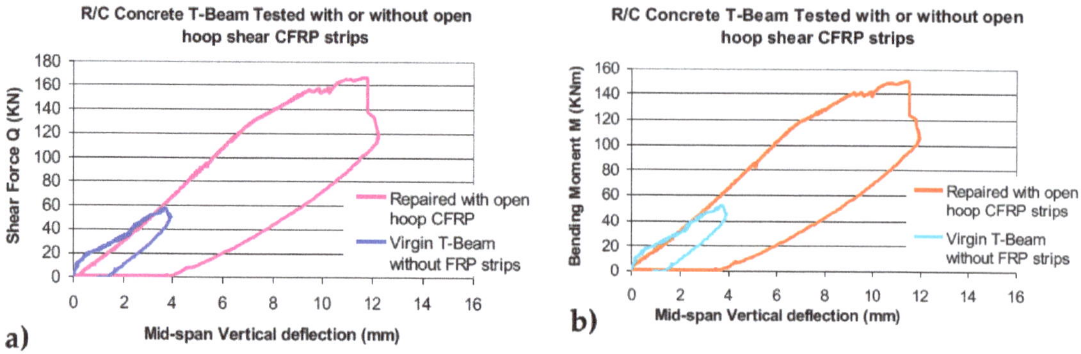

Figure 18. Comparison of the measured response between the Virgin RC prototype T-beam and this T-beam having been retrofitted with the *1st strengthening scheme*. (**a**) Shear force versus the deflection at mid-span. (**b**) Bending moment versus the deflection at mid-span.

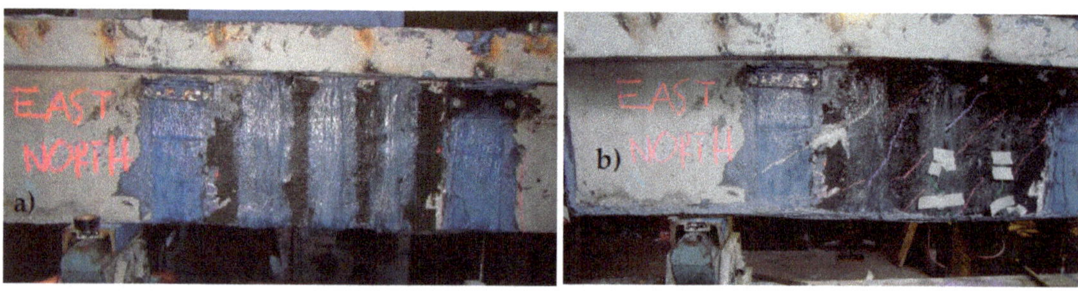

Figure 19. (**a**) The *2nd strengthening scheme* (East part). (**b**) Diagonal shear cracks at the East part of the prototype RC T-beam, which were checked by the applied CFRP anchor poles.

Figure 20a,b depicts the comparison of the measured response between the Virgin T-beam and the T-beam having been retrofitted with either the *1st* or *2nd strengthening scheme* (a-Shear-force versus the deflection at mid-span, b-Bending-moment versus the deflection at mid-span). This *2nd strengthening scheme* was capable of upgrading the shear resistance of the specimen in such a way that the flexural mode of failure developed this time. This is evident from the ductile nature of the variation of either the shear force (Q, Figure 20a) or the bending moment (M, Figure 20b) response versus the deflection at mid-span. The maximum shear force recorded for the 2nd shear strengthening scheme was equal to 197.43 kN and the corresponding maximum bending moment value was equal to 177.68 k Nm.

4.4. Prototype RC T-Beam witht External CFRP Shear Reiforcement Having All the CFRP Stirups Either Anchored at the West Part or Closed CFRP Rope Both at the East Part Together with a Flexural Upgrade (3rd Strengthening Scheme)

Finally, for the *3rd strengthening scheme 3*, whereas all types of external shear CFRP stirrups were left unchanged from the previous stage, a flexural strengthening was added consisting of five (5) CFRP layers (see Section 2.2). This is also depicted in Figure 21 (see also Figure 10c). This is evident from the measured shear force (Q) or bending moment (M) response versus the deflection at mid-span depicted in Figure 22a,b, respectively. The maximum shear force recorded for the 3rd shear strengthening scheme was equal to 229.55 kN and the corresponding maximum bending moment value was equal to 206.60 kNm. The used flexural strengthening scheme prohibited the development of any flexural mode of failure, as was the case for the 2nd strengthening scheme, and allowed the development of the shear limit state. This was of course a research objective, whereas the design objective is the ductile flexural response to prevail. In Figure 23a the West part of this 3rd strengthening scheme of the prototype RC T-beam prior to testing is shown; the externally bonded CFRP strips anchored with mechanical anchor [48] can be seen in this figure. The same West part of this T-beam is shown in Figure 23b at the end of testing for this 3rd strengthening scheme. The amplitude of the stress levels that developed at the CFRP strips and their anchors during this loading sequence led to the partial failure of the used anchoring devices when the applied load reached its maximum value as sown in Figure 23b. The widening of the shear diagonal cracks at this West part and the consequebt partial crushing of the concrete volume in the anchoring region reduced considerably the effectiveness of this anchoring scheme and led to the observed considerable decrease in the shear capacity of the tested T-beam (Figure 22a).

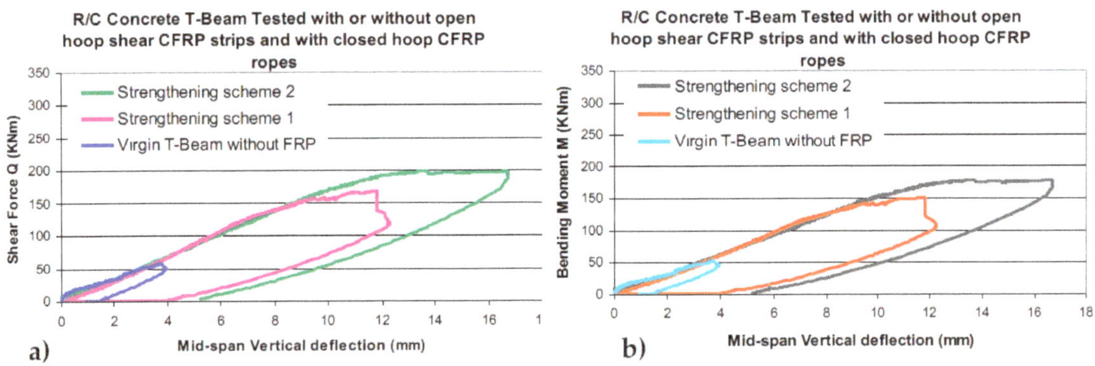

Figure 20. Comparison of the measured response between the prototype RC Virgin T-beam and this T-beam having been retrofitted with the *1st and 2nd strengthening schemes*. (**a**) Shear force versus the deflection at mid-span. (**b**) Bending moment versus the deflection at mid-span.

Figure 21. The *3rd strengthening scheme*. Open hoop CFRP strips with mechanical anchors at the West part and closed hoop anchor ropes at the East part (The same as in *Strengthening scheme 2*). The flexural upgrade of the 3rd strengthening scheme consisted of five (5) CFRP layers (each 0.131 mm thick and 120 mm wide) attached at the bottom side of the T-beam specimen.

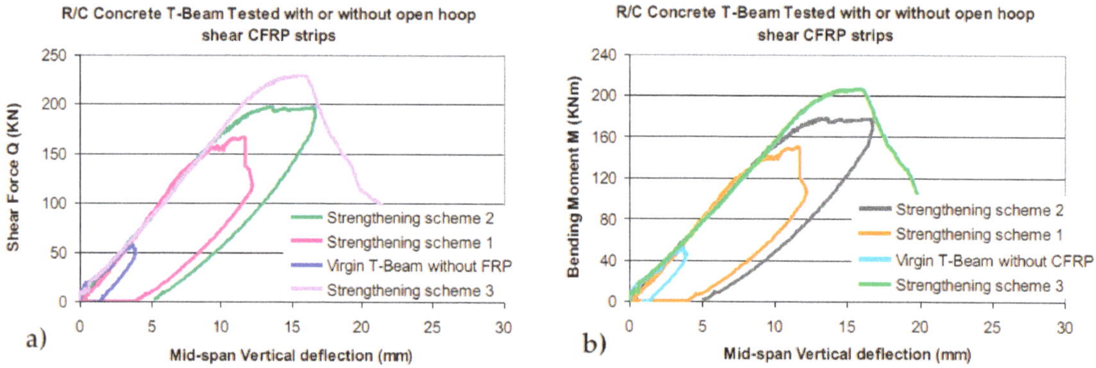

Figure 22. Comparison of the measured response between the prototype RC Virgin T-beam, this T-beam having been retrofitted with the *1st, 2nd, and 3rd strengthening schemes*. (**a**) Shear force versus the deflection at mid-span. (**b**) Bending moment versus the deflection at mid-span.

Figure 23. (**a**) The *3rd strengthening scheme* (West part). (**b**) The widening of the shear diagonal cracks and the partial crushing of the concrete volume in the anchoring region reduced considerably the effectiveness of this anchoring scheme.

5. Discussion of the Shear Performance of the Prototype R/C T-Beam without and with Shear Strengthening

This section summarizes the observed performance of the tested prototype RC T-beam when virgin and compared it with the corresponding performance of the same T-beam being retrofitted with the described in Section 4 strengthening schemes 1st, 2nd, and 3rd.

Table 5 lists the measured response in terms of the maximum measured values of shear force (Q_m) and bending moment (M_m) for the virgin T-beam, and the corresponding values for the T-beam being retrofitted in three distinct stages as described in Sections 2.2 and 4. The same table also lists predicted maximum shear force and bending moment values, as will be described in what follows. In columns (1) and (2) of this table, the description of this T-beam and the relevant retrofitting scheme is described in brief. Columns (3) and (4) of Table 5 list the measured maximum values of the shear force and the bending moment, respectively. In column (3), the measured increase in the maximum shear force (Q_m) that could be resisted during the relevant test, as a result of the applied strengthening scheme, is also listed. These increased values (%) were obtained using as a basis the maximum shear force measured for the virgin T-beam. It must be underlined again here that this specimen was intentionally under-designed in shear. Because of this, the increase in shear capacity that was finally achieved reached 300%. However, in practical cases whereby the structural member will possess a certain level of initial shear capacity from the existing internal shear steel reinforcement, the shear capacity increase by such retrofitting is expected to reach a rather modest value. The following are the important points that must be considered here from the presented results.

Table 5. Measured maximum shear force and bending moment values for the virgin and retrofitted specimen and corresponding predicted values.

Description of Tested T-Beam	Retrofitting Description	Maximum Measured Shear Force Q_m (kN)/increase %	Maximum Measured Bending Moment M_m (kNm)	Observed Mode of Failure	Predictions Shear Capacity V_p (kN)/ Flexural Capacity M_p (kNm)
(1)	(2)	(3)	(4)	(5)	(6)
Virgin	—	Q_m = 57.39/0%	M_m = 51.65	Diagonal Shear cracks	V_p =51.0 [58] /M_p =173.07
1st shear strengthening scheme	Unanchored single layer shear CFRP strips	Q_m = 166.77/190.6%	M_m = 150.09	Depending on unanchored CFRP open hoop strips (East). Anchored CFRP strips (West) performed OK	strip debV_p = 115.11 /M_p = 173.07
2nd shear strengthening scheme	Anchored single layer shear CFRP strips (West) or anchored ropes (East)	Q_m = 197.43/244.0%	M_m = 177.68	Anchored CFRP ropes (East) and anchored CFRP strips (West) performed OK. Flexural mode of failure	strip rupV_p =151.11 rope rupV_p =146.14 /M_p =173.07
3rd flexural strengthening scheme	Five (5) CFRP layers attached at the bottom side of the T-beam for flexural upgrading	Q_m = 229.55/300.0%	M_m = 206.60	Anchors of CFRP strips were damaged (West). Anchored CFRP ropes (East) performed OK	strip rupV_p = 151.11 rope rupV_p = 146.14 /M_p =206.26

- The 1st Strengthening scheme has the limitation of employing unanchored CFRP strips. This limitation should be considered in design as it means that the full tensile potential of the applied CFRP strip layers is not fully exploited.
- This limitation was dealt with by the 2nd strengthening scheme where all the parts of the external shear CFRP reinforcement were either CFRP strips provided with mechanical anchors (East) or closed hoop CFRP anchor ropes (West). This resulted in such a considerable upgrade of the resulting shear capacity that this prototype RC T-beam could not be forced enough in order to reach its shear capacity at this retrofitting stage. It instead reached its flexural limit state.
- Because of this, the 3rd strengthening scheme was applied in order to upgrade the flexural capacity of the prototype RC T-beam at this stage in order to prohibit the flexural failure and to lead towards reaching again a shear limit state. This time, the shear limit state appeared in the form of partial failure of the employed mechanical anchoring devices, which became ineffective, as described. This fact points to the importance of designing the anchoring system to be able to sustain the full potential capacity of the CFRP strips that are supported by it. As was underlined with the first comment, the full tensile potential of the applied CFRP strip layers is not exploited either because of the premature debonding or of the failing of the employed anchoring devices.
- All the above require a careful step-by-step design of all these partial aspects of a shear retrofitting scheme, as will be also discussed below.

In column 6 of Table 5, the predicted shear force (V_p) and bending moment (M_p) capacity values are listed based on the detailing of the original virgin T-beam together with all the additional CFRP detailing of each retrofitting scheme. Towards this objective, a software built for this purpose was utilized, that calculated both the shear and flexural capacity of such a T-beam [49] at the various stages. The calculations to obtain the flexural capacity are based on a well-established theory of the RC cross-sections assuming as limit states either the yield and fracture of the longitudinal tensile reinforcement or the limit tensile strain (0.7%) of the longitudinal CFRP strip attached at the bottom side of such a T-beam together with the compressive limit-state of the top fiber of the cross-section (0.3% limit compressive strain). All the measured geometric details of the T-beam cross-section and mechanical properties of all the used materials (see Section 2), are utilized by this software as input data. Towards obtaining the shear capacity predictions (V_p) for the tested T-beam, use is made of the contribution by the concrete part (V_c) adopting the formula proposed by Zsutty [58]. The contribution of the various schemes of CFRP external shear reinforcement (V_{CFRP}) is added on the concrete contribution, as shown by the following Equations (1) and (2), where the cross-section of a single CFRP strip is equal to A_{CFRP} the CFRP Young's modulus is equal to E_{CFRP} and ε_{CFRP} is the maximum tensile strain level assumed to develop at the CFRP strip being crossed by the diagonal shear cracks. Coefficient 2 at the front of the right part of Equation (2) accounts for the fact that the development of the diagonal shear crack is resisted by both sides of the hoop-shaped shear reinforcement surrounding the T-beam cross-section from both sides.

$$V_p = V_c + V_{CFRP} \tag{1}$$

$$V_{CFRP} = 2 \cdot A_{CFRP} \cdot E_{CFRP} \cdot \varepsilon_{CFRP} \tag{2}$$

The provisions of the Greek guidelines [12] for retrofitting RC structural elements with externally attached FRP strips are utilized to calculate the CFRP external shear contributions. These guidelines provide upper acceptable limits for the tensile strain (ε_{CFRP}) that such external shear reinforcement is allowed to develop. This tensile strain limit has a relatively low value for unanchored FRP strips, thus taking into account the debonding mode of failure. As indicated by Equation (1), the shear CFRP strip contribution is added to the shear contribution of the concrete resulting in the shear force predictions assuming *debonding* limit-state which is denoted as strip debV_p in Table 5 column 6. The presence of an anchoring system or closed hoop strips leads to accepting relatively higher tensile strain values (ε_{CFRP}) than before, according to the same provisions. In this case, the presence of anchoring or the nature of closed hoops is assumed to lead to the tensile *rupture* of the FRP material as limit-state. Again, applying Equation (2) and assuming as limit-state the rupture of the CFRP (either strip or rope) the corresponding shear capacity predictions are obtained which are denoted as either strip rupV_p or rope rupV_p (Table 5 column 6). In predicting the shear force or bending moment capacities, in the way described, the safety coefficients adopted in relevant design guidelines were set to be equal to 1. This way of obtaining the shear force capacity of RC beams strengthened by external FRP shear reinforcement is in line with the rationale followed by many relevant design guidelines [12,59–61] with a varying degree of complexity. D'Antino and Triantafillou [62] present an extensive review of a wide range of such design guidelines utilizing an extensive database of experimental results of RC beams strengthened with externally applied shear FRP retrofitting schemes. The following summarizes the main observations from comparing these predictions with the corresponding measurements.

- The shear force capacity of the concrete, as found by the virgin beam results, is very well predicted. The concrete shear force resistance is also kept when predicting the total shear force resistance for the 2nd, 3rd, and 4th strengthening scheme. This is due to the fact that the formation of diagonal cracking during the loading of the virgin T-beam at all subsequent stages did not reduce this shear resistance because the loading stopped before any significant widening of these diagonal cracks could take place. All the subsequent shear retrofitting schemes managed to prohibit any detrimental widening of these diagonal cracks till the last stage (last phase of loading during the 4th strengthening scheme). This can be seen in Figure 23a,b.
- For the 2nd strengthening scheme, which exhibited flexural limit state, its measured flexural capacity of the tested RC T-beam is also very well predicted Therefore, the discrepancies between measured and predicted capacity values are limited to the CFRP shear contributions.
- For the 1st shear strengthening scheme the predicted shear capacity value, based on the debonding limit state, is equal to 69% of the measured value.

- The 2nd shear strengthening scheme cannot be used for this purpose because during this testing stage the shear capacity was not reached due to exceeding the flexural limit state.
- The 3rd shear strengthening scheme can be used with certain reservations because it is based on the assumption of the rupture of either the CFRP shear strips with anchors or the CFRP closed hoop anchor ropes. However, the actual limit state, as described, was that of partial damage of the used mechanical anchors. This is a topic of a more detailed analysis beyond the current presentation. Despite this limitation, a comparison of the predicted shear capacity values (based on the assumed strip rupture limit state) with the measured value (actually resulting from the anchor damage) reveals that the predicted shear capacity value corresponds to approximately 65% of the measured value.
- The above discrepancies would be larger if the comparison would be made excluding the concrete shear force contribution. In this case for the case of debonding limit-state, the predicted CFRP contribution is equal to 58.6% of the measured value which is almost the same as for the case of the rupture limit-state whereby the predicted CFRP contribution is equal to 58% of the measured value.
- A possible explanation for these discrepancies is the inherent conservativism in the design guidelines reluctant to adopt either higher values of axial tensile strains for these externally applied CFRP shear reinforcement than the ones assumed or the participation of a larger number of CFRP strips than the ones assumed. This conservativism is justified, up to a point, because during practical applications in situ conditions, such as the preparation of the bond surface, the rounding of the corners, the proper attachment of the CFRP strips or ropes were found to have an important influence on the final effectiveness of such shear retrofitting schemes. In addition, it was also shown that employing wider CFRP strips does not lead to the expected increase in the shear capacity. Moreover, it must be underlined that the reported results were obtained for monotonic slow-rate loading. Finally, as was shown by the 3rd shear strengthening scheme of the current study, the performance of each of the adopted anchoring scheme components is another additional critical step for the effectiveness of a retrofitting scheme for practical applications. Various anchoring schemes are proposed in the literature combined with specific externally applied CFRP retrofitting schemes towards upgrading the flexural or the shear capacity of RC beams [45–57]. The importance of effective anchoring has been emphasized throughout this study.

6. Conclusions

- Using a simple laboratory test setup, devised by the authors, the tensile capacity of CFRP strips without any anchors or with mechanical anchors or anchor ropes can be found together with the corresponding tensile capacity and mode of failure of the assembly (CFRP strip and anchor). This may be of practical use when testing the effectiveness of such an external shear retrofitting scheme.
- An effective anchoring, using either a mechanical anchor such as the one devised by the authors or a CFRP anchor rope produced by the industry, can upgrade substantially the shear capacity of a RC T-beam under-designed in shear. The examined anchors in this study resolve the retrofitting difficulty created by the presence of RC slabs, thus having an advantage in practical applications.
- The predicted, according to design guidelines, upgraded shear capacity of the tested prototype RC T-beam with the used shear retrofitting schemes, under-estimate the measured shear capacity by 58%. This conservatism can counter-balance uncertainties arising from in situ conditions in constructing the various parts of such a shear retrofitting scheme. It must be also underlined that the results presented here were derived for monotonic and not for cyclic loading.
- The emphasis in this work was given to externally applied CFRP shear retrofitting schemes by examining ways to counteract one of their basic disadvantages which

is premature debonding. It must be underlined that this is not always possible. Therefore, applying traditional RC jacketing schemes remains a valid alternative, despite practical difficulties.

7. Patents

G. C. Manos, K. Katakalos, and V. Kourtides, "Construction System for Strengthening an Existing Structure with Tension Sheets and a Respective Anchoring Device and Method" Patent No: EP2336455-(A1), 2011.

Author Contributions: G.C.M., was responsible for the concept and the design/execution of all the experimental sequences, the supervision of all the tests, the recording and analysis of all the experimental results, the writing of the manuscript and the observations and the conclusions drawn. K.B.K., was responsible for the design and construction of all specimens, the execution of all tests and the recording and analysis of all the experimental results as well as of the administration of the contacts with the industry providing all the retrofitting materials. He also designed and edited all included figures and edited the final text. All authors have read and agreed to the published version of the manuscript.

Funding: This research received no external funding. The facilities of the Laboratory of Strength of Materials and Structure of Aristotle University were utilized exclusively during all experimental sequences.

Institutional Review Board Statement: Not required.

Informed Consent Statement: Not applicable.

Data Availability Statement: Any additional information can be requested by the corresponding author.

Acknowledgments: The material for the CFRP strips and the CFRP anchor ropes were provided by Sika Hellas. The epoxy resins were also provided by Sika Hellas under the code name SikaDur330.

Conflicts of Interest: The authors declare no conflict of interest.

References

1. Manos, G.C. Consequences on the Urban Environment in Greece Related to the Recent Intense Earthquake Activity. *J. Civ. Eng. Arch.* **2011**, *5*, 1065–1090. [CrossRef]
2. Alcocer, S.A.; Jirsa, J.O. Strength of reinforced concrete frame connections rehabilitated by jacketing. *ACI Struct. J.* **1993**, *90*, 249–261.
3. Rodriguez, M.; Park, R. Seismic Load Tests on Reinforced Concrete Columns Strengthened by Jacketing. *ACI Struct. J.* **1994**, *91*, 150–159. [CrossRef]
4. Bracci, J.M.; Reinhorn, A.M.; Mander, J.B. Seismic resistance of reinforced concrete frame designed for gravity loads: Performance of structural systems. *ACI Struct. J.* **1995**, *92*, 597–609.
5. Tsonos, A.G. Lateral load response of strengthened reinforced concrete beam-to-column joints. *ACI Struct. J.* **1999**, *96*, 46–56.
6. Fardis, M.N.; Biskinis, D.E. Performance-based engineering for earthquake resistant reinforced concrete structures: A volume Honoring Shunsuke Otani. In *Deformation Capacity of RC Members, as Controlled by Flexure or Shear*; Kabeyasawa, T., Shiohara, H., Eds.; University of Tokyo: Tokyo, Japan, 2003; pp. 511–530.
7. Engindeniz, M.; Kahn, L.F.; Zureick, A.-H. Repair and strengthening of reinforced concrete beam–column joints: State of the art. *ACI Struct. J.* **2005**, *102*, 1–14.
8. Vandoros, K.G.; Dritsos, S.E. Concrete jacket construction detail effectiveness when strengthening RC columns. *Constr. Build. Mater.* **2008**, *22*, 264–276. [CrossRef]
9. Karayannis, C.G.; Chalioris, C.E.; Sirkelis, G.M. Local retrofit of exterior RC beam–column joints using thin RC jackets—An experimental study. *Earthq. Eng. Struct. Dyn.* **2008**, *37*, 727–746. [CrossRef]
10. Calvi, G.M. Choices and criteria for seismic strengthening. *J. Earthq. Eng.* **2013**, *17*, 769–802. [CrossRef]
11. Vitiello, U.; Asprone, D.; Di Ludovico, M.; Prota, A. Life-cycle cost optimization of the seismic retrofit of existing RC structures. *Bull. Earthq. Eng.* **2017**, *15*, 2245–2271. [CrossRef]
12. KANEPE. *Guidelines for Retrofitting RC Structures*; Greek Organization of Antiseismic Planning and Protection: Athens, Greece, 2013.

13. Manos, G.C.; Papanaoum, E. Assessment of the earthquake behavior of Hotel Ermionio in Kozani, Greece constructed in 1933 before and after its recent retrofit. In *Earthquake Engineering Retrofitting of Heritage Structures, Design and Evaluation of Strengthening Techniques*; Syngellakis, S., Ed.; Wessex Institute of Technology: Southampton, UK, 2013; pp. 25–40. ISBN 978-1-84564-754-4. eISBN 978-1-84564-755-1.
14. Mosallam, A.S. Evaluation and construction of composite strengthening systems for the Sauvie Island Bridge. In *FRP Composites in Civil Engineering—CICE 2004*; CRC Press: Boca Raton, FL, USA, 2004; pp. 715–723.
15. Mosallam, A.S.; Bayraktar, A.; Elmikawi, M.; Pul, S.; Adanur, S. Polymer composites in construction: An overview. *SOJ Mater. Sci. Eng.* **2015**. Available online: https://escholarship.org/content/qt5xf7s8nj/qt5xf7s8nj.pdf (accessed on 20 July 2021).
16. Saadatmanesh, H.; Malek, A.M. Design guidelines for flexural strengthening of RC beams with FRP plates. *J. Compos. Constr.* **1998**, *2*, 158–164. [CrossRef]
17. Nanni, A. Flexural Behavior and Design of RC Members Using FRP Reinforcement. *J. Struct. Eng.* **1993**, *119*, 3344–3359. [CrossRef]
18. Shahawy, M.; Beitelman, T.E. Static and Fatigue Performance of RC Beams Strengthened with CFRP Laminates. *J. Struct. Eng.* **1999**, *125*, 613–621. [CrossRef]
19. Bakis, C.E.; Bank, L.C.; Brown, V.L.; Cosenza, E. Fiber-Reinforced Polymer Composites for Construction—State of the Art Review. *J. Compos. Constr.* **2002**, *6*, 73–87. [CrossRef]
20. Minnaugh, P.L.; Harries, K.A. Fatigue behavior of externally bonded steel fiber reinforced polymer (SFRP) for retrofit of reinforced concrete. *Mater. Struct.* **2008**, *42*, 271–278. [CrossRef]
21. Manos, G.C.; Katakalos, K. The Use of Fiber Rein-forced Plastic for the Repair and Strengthening of Existing Reinforced Concrete Structural Elements Damaged by Earthquakes. In *Fiber Reinforced Polymer—The Technology Applied for Concrete Repair*; Masuelli, M.A., Ed.; Intech Open: London, UK, 2013; ISBN 978–953–51–0938–9.
22. Huang, X.; Birman, V.; Nanni, A.; Tunis, G. Properties and potential for application of steel reinforced polymer and steel reinforced grout composites. *Compos. Part B Eng.* **2005**, *36*, 73–82. [CrossRef]
23. Katakalos, K.; Papakonstantinou, C. Fatigue of Reinforced Concrete Beams Strengthened with Steel-Reinforced Inorganic Polymers. *J. Compos. Constr.* **2009**, *13*, 103–112. [CrossRef]
24. Mossallam, A.S. (Ed.) *Innovative Systems for Seismic Repair and Rehabilitation of Structures, Design and Applications*; Sage Publishing Company: Thousand Oaks, CA, USA, 2000; ISBN 9781566769648. ISBN 1566769647.
25. Khalifa, A.; Nanni, A. Rehabilitation of rectangular simply supported RC beams with shear deficiencies using CFRP composites. *Constr. Build. Mater.* **2002**, *16*, 135–146. [CrossRef]
26. Anil, O. Strengthening of RC T-section beams with low strength concrete using CFRP composites subjected to cyclic load. *Constr. Build. Mater.* **2008**, *22*, 2355–2368. [CrossRef]
27. Casadei, P.; Nanni, A.; Alkhrdaji, T.; Thomas, J. Performance of Double-T Prestressed Concrete Beams Strengthened with Steel Reinforcement Polymer. *Adv. Struct. Eng.* **2005**, *8*, 427–442. [CrossRef]
28. Barton, B.; Wobbe, E.; Dharani, L.; Silva, P.; Birman, V.; Nanni, A.; Alkhrdaji, T.; Thomas, J.; Tunis, G. Characterization of reinforced concrete beams strengthened by steel reinforced polymer and grout (SRP and SRG) composites. *Mater. Sci. Eng. A* **2005**, *412*, 129–136. [CrossRef]
29. Papakonstantinou, C.G.; Katakalos, K. Flexural behavior of reinforced concrete beams strengthened with a hybrid inorganic matrix—steel fiber retrofit system. *Struct. Eng. Mech.* **2009**, *31*, 567–585. [CrossRef]
30. Thermou, G.E.; Katakalos, K.; Manos, G. Experimental investigation of substandard RC columns confined with SRG jackets under compression. *Compos. Struct.* **2018**, *184*, 56–65. [CrossRef]
31. Thermou, G.E.; Katakalos, K.; Manos, G. Influence of the cross section shape on the behaviour of SRG-confined prismatic concrete specimens. *Mater. Struct.* **2015**, *49*, 869–887. [CrossRef]
32. Thermou, G.E.; Katakalos, K.; Manos, G. Concrete confinement with steel-reinforced grout jackets. *Mater. Struct.* **2014**, *48*, 1355–1376. [CrossRef]
33. Arduini, M.; Di Tommaso, A.; Nanni, A. Brittle failure in FRP plate and sheet bonded beams. *ACI Struct. J.* **1997**, *94*, 363–370.
34. Prota, A.; Yong, T.K.; Nanni, A.; Pecce, M.; Manfredi, G. Performance of shallow reinforced concrete beams with externally bonded steel-reinforced polymer. *ACI Struct. J.* **2006**, *103*, 163–170.
35. Lu, X.Z.; Teng, J.G.; Yea, L.P.; Jiang, J.J. Bond–slip models for FRP sheets/plates bonded to concrete. *J. Eng. Struct.* **2005**, *27*, 920–937. [CrossRef]
36. Wu, Y.; Zhou, Z.; Yang, Q.; Chen, W. On shear bond strength of FRP-concrete structures. *Eng. Struct.* **2010**, *32*, 897–905. [CrossRef]
37. Manos, G.; Katakalos, K.; Kourtides, V. The Influence of Concrete Surface Preparation when Fiber Reinforced Polymers with Different Anchoring Devices are Being Applied for Strengthening R/C Structural Members. *Appl. Mech. Mater.* **2011**, *82*, 600–605. [CrossRef]
38. Chen, J.F.; Teng, J.G. Shear Capacity of Fiber-Reinforced Polymer-Strengthened Reinforced Concrete Beams: Fiber Reinforced Polymer Rupture. *J. Struct. Eng.* **2003**, *129*, 615–625. [CrossRef]
39. Rousakis, T.C.; Saridaki, M.E.; Mavrothalassitou, S.A.; Hui, D. Utilization of hybrid approach towards advanced database of concrete beams strengthened in shear with FRPs. *Compos. Part B Eng.* **2015**, *85*, 315–335. [CrossRef]
40. Rousakis, T.C.; Saridaki, M.E. Advanced database of concrete beams strengthened in shear with FRPs. In Proceedings of the Twenty-Second Annual International Conference on Composites/Nano-Engineering (ICCE-22), Saint Julian's, Malta, 13–19 July 2014; pp. 13–19.

41. Mitolidis, G.J.; Salonikios, T.N.; Kappos, A.J. Test results and strength estimation of R/C beams strengthened against flexural or shear failure by the use of SRP and CFRP. *Compos. Part B Eng.* **2012**, *43*, 1117–1129. [CrossRef]
42. Wobbe, E.; Silva, P.; Barton, B.L.; Dharani, L.R.; Birman, V.; Nanni, A.; Alkhrdaji, T.; Thomas, J.; Tunis, G. *Flexural Capacity of RC Beams Externally Bonded with SRP and SRG*; Society for the Advancement of Material and Process Engineering: Covina, CA, USA, 2004; pp. 2995–3002.
43. Gao, P.; Gu, X.; Mosallam, A.S. Flexural behavior of preloaded reinforced concrete beams strengthened by prestressed CFRP laminates. *Compos. Struct.* **2016**, *157*, 33–50. [CrossRef]
44. Tanarslan, H.M.; Altin, S. Behavior of RC T-section beams strengthened with CFRP strips, subjected to cyclic load. *Mater. Struct.* **2010**, *43*, 529–542. [CrossRef]
45. Manos, G.C.; Katakalos, K.; Kourtides, V. Study of the Anchorage of Carbon Fiber Plastics (CFRP) Utilized to Upgrade the Flexural Capacity of Vertical R/C Members. In Proceedings of the 14th World Conference on Earthquake Engineering (14WCEE), Beijing, China, 12–17 October 2008; pp. 1079–1095.
46. Manos, G.C.; Katakalos, K.; Kourtides, V. Cyclic Behavior of a Hybrid Anchoring Device Enhancing the Flexural Capacity and Ductility of an R/C Bridge-Type Pier Strengthened with CFRP Sheets. *J. Civil Eng. Res.* **2013**, *3*, 52–64.
47. Manos, G.C.; Katakalos, K.; Koidis, G.; Papakonstantinou, C.G. Shear Strengthening of R/C Beams with FRP Strips and Novel Anchoring Devices. *J. Civil Eng. Res.* **2012**, *2*, 73–83. [CrossRef]
48. Manos, G.C.; Katakalos, K.; Kourtides, V. Construction Structure with Strengthening Device and Method. European Patent Office Patent Number WO2011073696 (A1), 23 June 2011.
49. Manos, G.C.; Katakalos, K.; Papakonstantinou, C.; Koidis, G. Enhanced Repair and Strengthening of Reinforced Concrete Beams Utilizing External Fiber Reinforced Polymer Sheets and Novel Anchoring Devices. In Proceedings of the 15th World Conference on Earthquake Engineering, Lisbon, Portugal, 24–28 September 2012.
50. Manos, G.C.; Katakalos, K. Investigation of the Force Transfer Mechanisms for Open Hoop FRP Strips Bonded on R/C Beams with or without Anchoring Devices. *Open J. Civ. Eng.* **2013**, *3*, 143–153. [CrossRef]
51. Manos, G.C.; Theofanous, M.; Katakalos, K. Numerical simulation of the shear behaviour of reinforced concrete rectangular beam specimens with or without FRP-strip shear reinforcement. *J. Adv. Eng. Softw.* **2014**, *67*, 47–56. [CrossRef]
52. Manos, G.C.; Katakalos, K. *Experimental Investigation of Concrete Prismatic Specimens, Strengthened with CFRP Strips and FRP Anchorage, under Tensile Loading*; Unpublished Internal Technical Report to Sika Hellas; Aristotle University: Thessaloniki, Greece, 2015.
53. Manos, G.C.; Katakalos, K.; Mpalaskas, K. Experimental and Numerical Study of a Novel Anchoring Devices for Open Hoop FRP Strips for under Shear Forces. In Proceedings of the 16th European Conference on Earthquake Engineering, 16th World Conference on Earthquake Engineering, Thessaloniki, Greece, 18–21 June 2018.
54. Katakalos, K.; Manos, G.; Papakonstantinou, C. Seismic Retrofit of R/C T-Beams with Steel Fiber Polymers under Cyclic Loading Conditions. *Buildings* **2019**, *9*, 101. [CrossRef]
55. Ababneh, A.N.; Al-Rousan, R.Z.; Ghaith, I.M. Experimental study on anchoring of FRP-strengthened concrete beams. *Structures* **2019**, *23*, 26–33. [CrossRef]
56. Eslami, A.; Moghavem, A.; Shayegh, H.R.; Ronagh, H.R. Effect of FRP stitching anchors on ductile performance of shear-deficient RC beams retrofitted using FRP U-wraps. *Structures* **2019**, *23*, 407–414. [CrossRef]
57. Chalioris, C.E.; Zapris, A.G.; Karayannis, C.G. U-Jacketing Applications of Fiber-Reinforced Polymers in Reinforced Concrete T-Beams against Shear—Tests and Design. *Fibers* **2020**, *8*, 13. [CrossRef]
58. Zsutty, T.C. Beam Shear Strength Prediction by Analysis of Existing Data. *J. ACI* **1968**, *65*, 943–951.
59. *Eurocode 8—Design of Structures for Earthquake Resistance—Part 3: Assessment and Retrofitting of Buildings 1998–3:2005*; CEN: Brussels, Belgium, 2005; pp. 35–55.
60. ACI. *Guide for the Design and Construction of Externally Bonded FRP Systems for Strengthening Concrete Structures*; ACI 440.2R-08; American Concrete Institute: Farmington Hills, MI, USA, 2008; p. 45.
61. FIB. *Externally Bonded FRP Reinforcement for RC Structures*; Bulletin No. 14; FIB: Lausanne, Switzerland, 2001.
62. D'Antino, T.; Triantafillou, T.C.; D'antino, T. Accuracy of design-oriented formulations for evaluating the flexural and shear capacities of FRP-strengthened RC beams. *Struct. Concr.* **2016**, *17*, 425–442. [CrossRef]

applied sciences

MDPI

Article

On the Modeling and Analysis of Brittle Failure in Existing R/C Structures Due to Seismic Loads

Stylianos I. Pardalopoulos [1], Stavroula J. Pantazopoulou [2] and George D. Manolis [1,*]

[1] Department of Civil Engineering, Faculty of Engineering, Aristotle University, GR 54124 Thessaloniki, Greece; stylpard@civil.auth.gr
[2] Department of Civil Engineering, The Lassonde Faculty of Engineering, York University, Toronto, ON M3J 1P3, Canada; pantazo@yorku.ca
* Correspondence: gdm@civil.auth.gr

Abstract: Brittle failure is often observed in older reinforced concrete (R/C) buildings that have been designed prior to the 1980s following an earthquake event. Since this has ramifications on any subsequent repair protocol, it is important to quantify the remaining strength capacity for this class of building to determine a retrofit strategy. Following along these lines, an analytical-numerical methodology is presented for use as a third-tier seismic assessment which is based on a previously developed second-tier strength assessment criterion coming from a procedure known as Rapid Seismic Assessment (RSA). The assessment framework is performance-based, aiming to determine whether estimated local drift demands can be tolerated without failure developing along the load resistance path of substandard R/C buildings. This enables the development of guidelines for modeling all possible strength mechanisms that occur in the structural system of substandard R/C buildings. An application example using data provided by a benchmark experiment involving a full-scale R/C building helps to illustrate and then validate the proposed modeling procedures and establish their accuracy and efficiency for use by practicing engineers.

Keywords: reinforced concrete; substandard buildings; brittle failure; earthquake loads; rapid seismic assessment

Citation: Pardalopoulos, S.I.; Pantazopoulou, S.J.; Manolis, G.D. On the Modeling and Analysis of Brittle Failure in Existing R/C Structures Due to Seismic Loads. *Appl. Sci.* **2022**, *12*, 1602. https://doi.org/10.3390/app12031602

Academic Editor: Maria Favvata

Received: 29 December 2021
Accepted: 31 January 2022
Published: 2 February 2022

Publisher's Note: MDPI stays neutral with regard to jurisdictional claims in published maps and institutional affiliations.

cc BY

Copyright: © 2022 by the authors. Licensee MDPI, Basel, Switzerland. This article is an open access article distributed under the terms and conditions of the Creative Commons Attribution (CC BY) license (https://creativecommons.org/licenses/by/4.0/).

1. Introduction

Existing reinforced concrete (R/C) buildings encompass a variety of structures, ranging from cases that were designed and constructed according to modern forms of detailing (see ASCE/SEI 41-17 [1], EN1998-1 [2]) to structures that were built before the early 1980s under design code frameworks (summarized in the FIB Bulletin 24 [3]) that are now often characterized as substandard. The seismic vulnerability of existing R/C buildings, particularly that of substandard construction, is continuously demonstrated when strong earthquakes strike urban areas (e.g., Loma Prieta 1989, Northridge 1994, Kobe 1995, Kocaeli 1999, Athens 1999, Bhuj 2001, Port-au-Prince 2010, Gorkha 2015, and Mexico City 2017). Field observations have repeatedly suggested that most of the buildings that collapsed did not show signs of flexural yielding (such as densely spaced flexural cracks normal to the member axis), nor the development of any kind of global ductility prior to collapse. Rather, it appears that most of the collapsed structures had formed a brittle mechanism upon failure, marked by severe damage localization in a few areas with high shear demand, such as the disintegration of exterior frame joints, shear failure of severely unconfined captive columns, and soft-story damage in sway-frames.

Given the need for identifying among the vast inventory of existing R/C buildings those cases that are potentially vulnerable to collapse, thus representing a serious hazard to human safety in the event of a moderate or a strong earthquake, extensive research has been conducted over the past three decades and different assessment procedures have been developed. Analytical assessment procedures, also known as third-tier evaluation procedures (ASCE/SEI 41-17 [1], EN1998-3 [4], Greek Code of Structural Interventions [5], FEMA-356 [6], Vamvatsikos and Cornell [7]) include the mathematical simulation of the examined building using advanced numerical tools such as 3D finite element (FE) models combined with elastic and/or inelastic analysis methods. This category of assessment procedures enables the simulation of complex mechanisms throughout the structural system of buildings and evaluation of demand and supply in R/C structures in terms of deformation (either drift or ductility ratios at milestone performance points). Since they originate from design procedures that were developed for modern R/C structures, most of the analytical assessment procedures simulate the response of R/C members using nonlinear moment-rotation envelope curves associated with zones of inelasticity-spread, demonstrating a strain hardening, post-yielding plateau. However, in poorly detailed R/C members, strength and deformation capacity seems to be controlled by localization of the prevailing mode of failure which depends on the rate of degradation of the alternative strength mechanisms with increasing ductility demand and load history [8–27]. In order for the analysis results to be realistic, member resistance skeleton curves must necessarily reflect the brittle modes of failure that typically prevail in poorly reinforced members, a feature that leads to an unstable solution and lack of convergence with most Newton-Raphson type schemes used to perform nonlinear structural analysis. This also affects the post-processing of the numerical analysis results [28,29].

A new methodology for modeling member response of substandard R/C buildings for seismic evaluation is presented here. This is motivated by the need for analytical modeling tools that can effectively capture the seismic behavior of substandard R/C structural elements. The proposed methodology can be used for a third-tier seismic evaluation procedure, using commercial software that is easily accessible to practitioners. Concepts from the strength assessment criterion in the Rapid Seismic Assessment (RSA) procedure are adopted here (Pardalopoulos and Pantazopoulou [30], Pardalopoulos et al. [31,32]). A core objective of the method is to determine whether or not the local drift demands can be tolerated without failure along the load resistance path of substandard R/C buildings. Building on the aforementioned concepts, we provide guidelines for modeling all possible strength mechanisms that may develop in the various components of the structural system of substandard R/C buildings. In order to validate the introduced modeling procedures and to establish their accuracy and efficiency, an application example is presented, illustrating in practical terms the steps needed for evaluating the seismic response of a substandard, full-scale R/C test building.

2. Procedures for Simulating Strength Mechanisms in Substandard R/C Buildings

Contrary to analysis procedures used in the design of new R/C buildings that focus on the numerical simulation of ductile resistance mechanisms in structural members (under the proviso that all undesirable brittle failure modes will be capacity-designed and therefore suppressed) in the analytical seismic assessment of poorly reinforced R/C buildings, the potential prevalence of brittle-type mechanisms of resistance across the structural system must be examined. The typical layout of the structural system of R/C buildings constructed up to the early 1980s (at a time when construction details were still relatively primitive since their role in the seismic response was not yet fully understood) included poorly reinforced frames comprising columns of section sizes ranging between 250 mm and 500 mm, beams of a 150 mm to 250 mm width and 600 mm to 700 mm height (including the slab thickness), and slabs with a thickness ranging between 120 mm and 160 mm. Longitudinal reinforcement usually comprised relative low amounts of StI (f_y = 220 MPa, f_u = 340 MPa) to StIII (f_y = 420 MPa, f_u = 500 MPa) smooth bars, lap-spliced to arbitrary lengths and under poor confinement conditions. Column and beam stirrups were usually 6 mm to 8 mm, smooth, rectangular, StI bars (usually mild steel), with 90° hooks in the ends, spaced at 200 mm to 300 mm on center (o.c.). Beam-column joints were usually left without stirrups for the convenience of construction. Concrete quality was usually

Bn150 to Bn200 as per DIN 1045 [33], corresponding to C12/15 and C16/20 concrete grades according to EN1992-1-1 [34].

To address the need for an effective simulation of the seismic response of substandard R/C buildings, the present research introduces new procedures for modeling the mechanisms of resistance developing in old-type R/C structural elements. Based on the principles of the RSA System [30–32], emphasis is placed on numerical simulation procedures that are applicable to column lines of existing R/C buildings, non-conforming to modern standards of seismic design and detailing. These lines usually collapse in a brittle manner, forming a mechanism characterized by pronounced localization of damage in a few locations with a high shear demand before there is a chance for the development of any form of redistribution and ductility (see Lang and Marshall [35], Augenti and Parisi [36], Mehrabian and Haldar [37], Dogangun [38], Joeng and Elnashai [39], and Varum [40]).

The modeling procedures introduced here include the simulation of the following mechanisms: M1: non-linear flexural response of columns and beams; M2: shear failure of the column web; M3: anchorage failure of longitudinal reinforcement of columns; M4: attainment of the development capacity in column lap splices; and M5: beam-column joint shear failure. The proposed procedures for simulating the seismic response of substandard R/C buildings are summarized in Figure 1 and discussed in detail in the following sections. Note that simulation of mechanisms {M2–M5} along column lines of existing R/C building is meaningful only if they prevail over mechanism M1. According to the RSA system, this occurs when the corresponding resistance ratios (r) that represent the shear strength associated with the examined failure mode, normalized by the shearing force required in order to support flexural yielding in the column, are equal to or greater than 1.0.

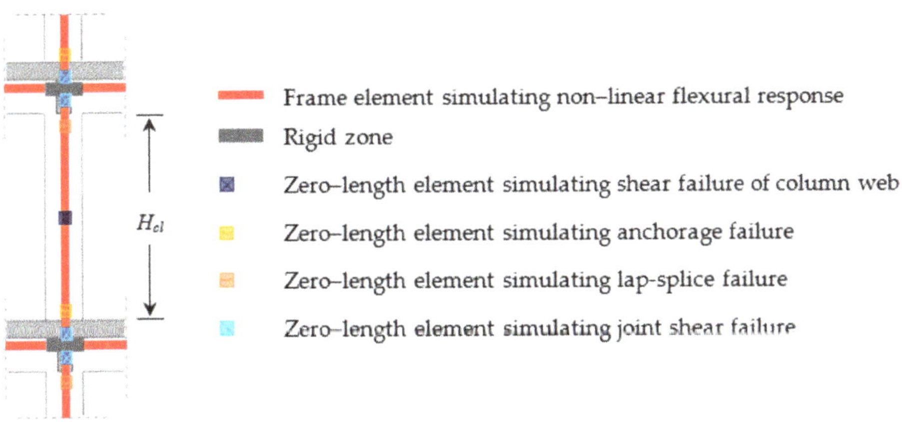

Figure 1. Simulation of strength mechanisms that develop within the structural system of substandard R/C buildings. Hatched regions represent cross-sections of members extending normal to the plane of view (e.g., slab and transverse beams).

2.1. *Simulation of InelasticFlexural Response in Columns and Beams*

Flexural behavior of R/C structural elements is associated with ductile member response and materializes only if it can be supported by all other resistance mechanisms developing along the element length. The inelastic flexural response is linked with the formation of extensive cracking, perpendicular to the axis of bending of R/C elements, in regions where the longitudinal reinforcement yields in tension. In columns and beams subjected to earthquake loading, flexural cracking usually forms in both ends, in a length approximately equal to twice the elements' width. By way of contrast, in R/C beams subjected to excessive gravity loading, the yielding of longitudinal reinforcement occurs at locations of maximum values for the developing bending moment.

Simulation of the inelastic flexural response of R/C elements may be performed using different, alternative approaches that are documented in the international literature. The most preferable approach in simulating inelastic flexural response in R/C elements subjected to biaxial bending is the use of linear elements encompassing fiber discretization of their cross-section [41]. According to this approach, the reinforcing bars plus the confined and the unconfined concrete of an R/C structural element are all modeled as separate fibers, possessing stress-strain relationships that characterize their response under uniaxial loading (see Figure 2). Based on the assumption that plane sections remain plane during member deformation, member stresses and deformations in both section directions are calculated through the integration of the response of the fibers over the element section. For R/C beams under seismic excitation that are primarily subjected to uniaxial bending, an alternative simulation approach is through the use of a lumped plasticity approach which combines a linear element with non-linear hinges at the locations where beams connect with other structural members (columns and/or beams). Inelastic response in each hinge is then defined by appropriate force-deformation relations, derived from sectional analysis of the corresponding element and classical mechanics.

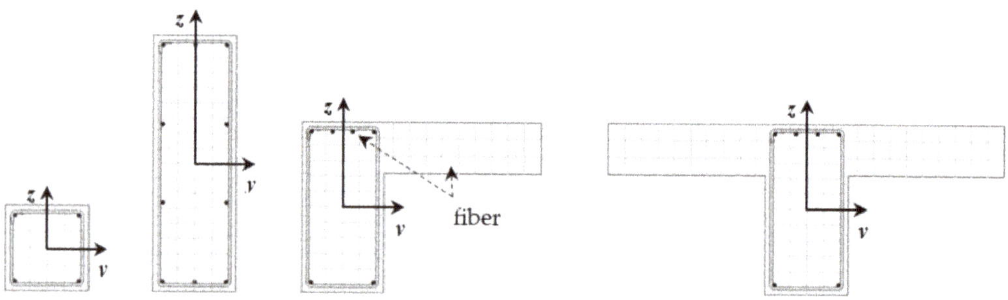

Figure 2. Simulation of the inelastic flexural response of R/C columns and beams using fiber discretization of cross-sections.

2.2. Simulation of Shear Failure of the Column Web

Shear failure of an R/C element web develops when the transverse reinforcement cannot provide adequate supporting shear resistance to displacement reversals beyond flexural yielding. In a third-tier seismic evaluation of existing R/C buildings, simulation of shear failure of the column web is essential, as this mechanism is linked to brittle member failure that can compromise the load-carrying capacity of columns and potentially lead to the collapse of the building.

The best approach to simulate the brittle response of R/C columns to shear failure of the web is to use a non-linear, zero-length element (the two end element nodes are distinct from each other but placed at the same location) at mid-height of the column's clear (deformable) length, H_{cl} (see Figure 1). Element response in each section direction is defined by the multilinear Moment–Rotation (M–θ) relationship illustrated in Figure 3, where $M_v = V_v \cdot (H_{cl}/2)$ is the bending moment associated with the shear strength V_v, of the column web in the examined direction. This V_v is calculated according to the RSA system from Equation (A4) in the Appendix A. Coefficient r_v in Figure 3 is the resistance ratio to shear failure of the column web, $r_v = V_v/V_{flex}$, where V_{flex} is calculated according to Equation (A3) in the Appendix A. Note that the column normalized axial load appearing in the expressions given in the Appendix A (denoted as ν_d) is obtained from static analysis of the examined R/C building to gravity loading and for the vertical component of the seismic load combination $G_k + \psi_E \cdot Q_k$, as perEN1998-1 [2].

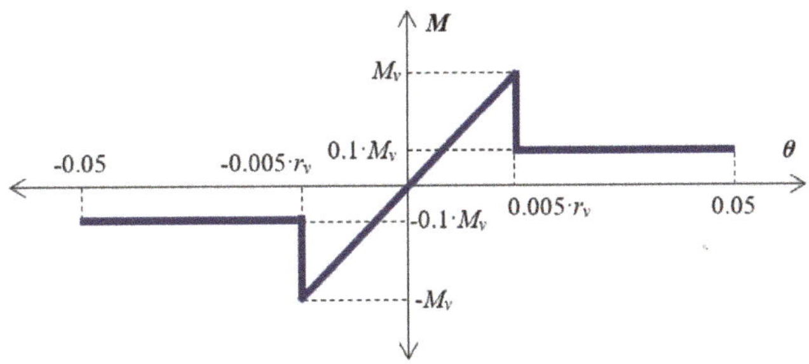

Figure 3. Moment–Rotation (M–θ) relationship used in simulating the response of R/C columns to shear failure in the web.

2.3. Simulation of Anchorage and Lap-Splice Failure of Longitudinal Reinforcement in Columns

Failure of anchorage and lap-splice of longitudinal bars limits the force developing in the bars to a value lower than their true axial strength and strain capacity. Both anchorage and lap-splice failures result from insufficient bond strength and are manifested with the formation of a single crack, perpendicular to the axis of the member. For columns, this occurs below the beam-column joint and above the base of the column in cases of anchorage and lap-splice failures, respectively.

In the framework of a third-tier seismic assessment of existing R/C buildings, anchorage and lap-splice failure of longitudinal reinforcement can be simulated with the use of two different zero-length elements, located at the top and the bottom of the clear length of an R/C column, as illustrated in Figure 1. The response of the zero-length elements is defined in each of the two-section directions by the M–θ trilinear relationships of Figure 4, where $M_a = V_v \cdot (H_{cl}/2)$, $M_{lap} = V_{lap} \cdot (H_{cl}/2)$. The column shear strengths against anchorage and lap-splice failure of the longitudinal reinforcement, V_a and V_{lap}, respectively, are calculated using the closed-form expressions given in the Appendix A and their respective resistance ratios, $r_a = V_a/V_{flex}$ and $r_{lap} = V_{lap}/V_{flex}$.

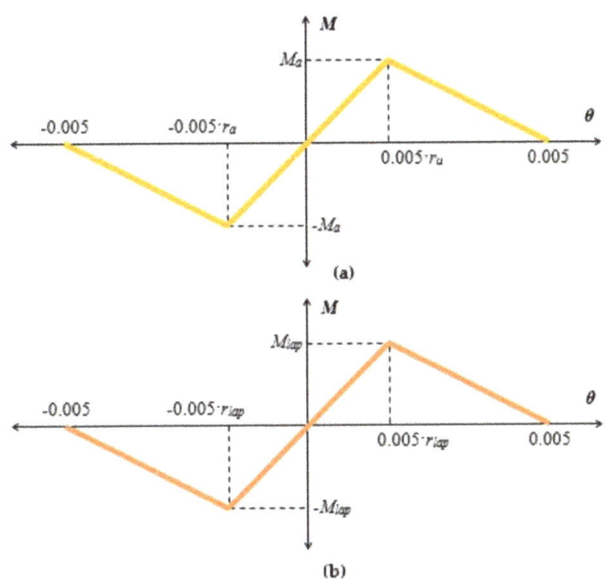

Figure 4. Moment–Rotation (M-θ) relationship for simulating (**a**) anchorage and (**b**) lap-splice behavior of longitudinal reinforcement in substandard R/C columns.

2.4. Simulation of Beam-Column Joint Behavior

Shear failure of beam-column joints in R/C buildings during earthquakes is a pronounced brittle mode of failure that may cause excessive flexibility of the overall frame and a consequent loss of vertical load-carrying capacity [3]. The integrity of R/C beam-column joints is secured through the compressive stress field acting within the joint which enables the concrete to participate in the joint shear action and allows (through the development of high bond stresses) for steep force gradients along with the beam and column primary reinforcements. In R/C buildings designed and constructed under contemporary standards (ASCE/SEI 41-17 [1], EN1998-1 [2], etc.), beam-column joints are dimensioned so as to sustain the development of the flexural strengths of the adjacent frame elements at the joint faces without significant degradation of the bond along with beam and column primary reinforcement. In substandard R/C buildings, the absence of a limited number of stirrups within the beam-column joints does not provide sufficient confinement to support the formation of the diagonal strut mechanism that is essential for force transfer between adjacent members during earthquake excitations.

Simulation of R/C beam-column joint behavior in FE models of existing R/C buildings can be achieved through the use of zero-length elements placed above and below the intersection of the beam and column elements used for simulating the mechanical behavior within the joint (see Figure 1). The joint response in each of the building principal plan dimensions is defined by the multilinear M–θ relationship of Figure 5. In this figure, $M_j = V_j \cdot (H_{cl}/2)$ is the bending moment associated with the shear capacity V_j of an R/C beam-column joint according to the RSA (see Equations (A11) and (A12) in the Appendix A, for unreinforced and well-reinforced joints, respectively), whereas r_j (= V_j/V_{flex}) is the resistance ratio to shear failure in the beam-column joints.

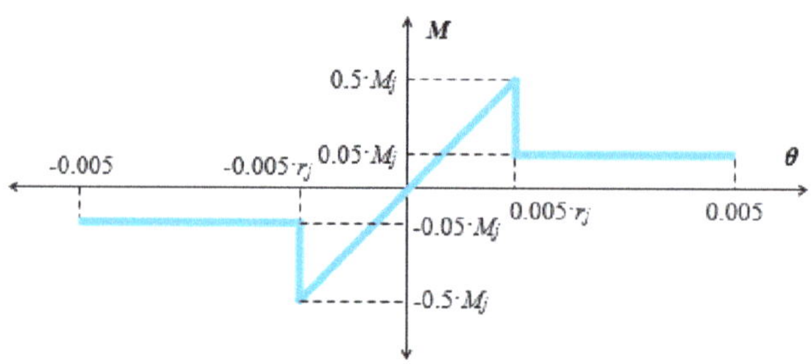

Figure 5. Moment–Rotation (M–θ) relationship representing the shear response of R/C beam-column joints in an intermediate floor of a building (the 0.5 factor for the moment is replaced by 1.0 for a roof joint in the building).

3. Example Application of the Proposed Numerical Simulation Procedures

The accuracy of the present numerical simulation methodology is investigated through the analytical evaluation of the seismic response of the SPEAR test building that was constructed and tested in the European Laboratory of Seismic Assessment (ELSA) at the Joint Research Center (JRC) in Ispra, Italy. This testing was conducted within the framework of the Seismic Performance Assessment and Rehabilitation (SPEAR) European research program (Fardis [42]; Negro et al. [43], Jeong and Elnashai [44]).

3.1. Geometric and Loading Characteristics of the SPEAR Building

The SPEAR building was a full-scale, three-story, 2 × 2 bay, torsionally sensitive R/C building which is considered representative of the structures that were designed and constructed throughout southern Europe from the 1950s until the mid-1980s (Figure 6). The building was designed for gravity loads alone, in addition to the self-weight of the R/C structural elements, as well as for 0.5 kN/m^2 and 2.0 kN/m^2 gravity loads on slabs accounting for finishing and live loads, respectively. All stories had an identical plan configuration with external dimensions of 12.825 m × 10.75 m, comprising eight columns of a 250 mm rectangular cross-section and one strong column of 250 mm × 750 mm cross-section. The columns were connected to beams having 250 mm × 500 mm web cross-sectional dimensions. Slab thickness was 150 mm, while story height in all cases was 3.00 m o.c. The columns had longitudinal reinforcement ∅12 mm smooth bars (see Figure 2), which were lap-spliced over 400 mm at the base for all three levels with a hook formation at the end of the bars. All columns had ∅8 mm perimeter stirrups, spaced at 250 mm o.c. which did not continue in the beam-column joints. Longitudinal reinforcement at both ends of all beams was 4 ∅12 mm bars at the top and 2 ∅12 mm bars at the bottom. Exceptions were: (i) beam B4 with 4 ∅20 mm plus 2 ∅12 mm bars at the top and 3 ∅20 mm bars at the bottom of both ends; (ii) beam B8 that was reinforced with 2 ∅20 mm top bars plus 4 ∅12 mm top bars at the connection with column C4, 4 ∅12 mm bars at the connection with column C7 and 2 ∅12 mm bars at both ends; (iii) beam B10 at the top had 4 ∅20 mm plus 2 ∅12 mm bars at the connection with column C5, 2 ∅20 mm plus 2 ∅12 mm bars at the connection with column C8 and 2 ∅20 mm bars at both ends; and (iv) beam B12 had 3 ∅20 mm plus 2 ∅12 mm bars at the top and 2 ∅20 mm bars at the bottom of both ends. The mean concrete compressive strengths f_{cm} obtained from tests on cylindrical core specimen taken from the building after completion of the experiment was found equal to 24.73 MPa, 26.70 MPa, and 25.32 MPa for the first, second, and third story columns, respectively. Finally, uniaxial tensile tests on steel bar coupons yielded stress of reinforcement f_y that was equal to 479.45 MPa, 474.11 MPa, and 396.87 MPa in the cases of the ∅8 mm, the ∅12 mm and the ∅20 mm bars, respectively.

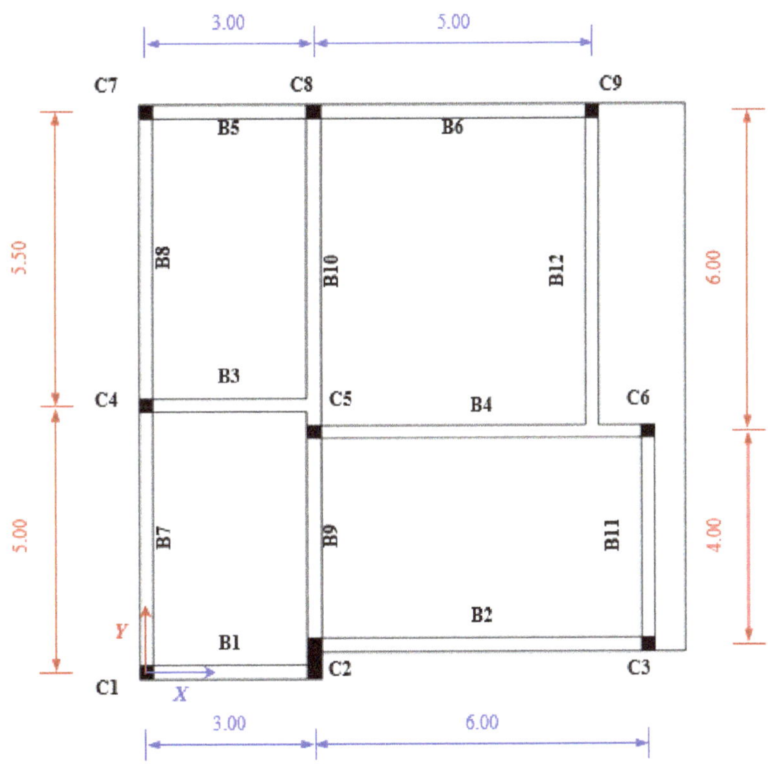

Figure 6. Typical story plan of the SPEAR test building at ELSA-JRC, Ispra, Italy (units in m).

The testing protocol at JRC consisted of a series of simulated ground motions of increasing intensity which were applied to the R/C structure through a pseudo-dynamic procedure. The basic accelerograms used were the two components of the Montenegro 1979 (Herceg Novi) ground motion, sequentially scaled upwards to 0.02 g, 0.15 g, and 0.20 g, labeled as the s10, s11, and s12 pseudo-dynamic tests. These accelerograms were applied in both the *X* and *Y* plan directions and were normalized with respect to the peak ground acceleration (*PGA*), as shown in Figure 7.

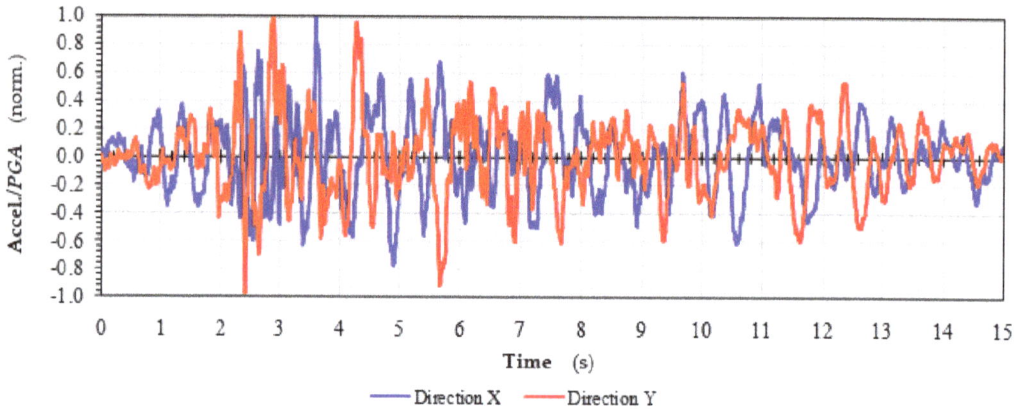

Figure 7. Time-histories of the two components of ground acceleration used in the testing of the SPEAR building at ELSA-JRC.

3.2. *Numerical Simulation and Analysis of the SPEAR Building*

In order to evaluate the seismic response of the SPEAR test building, the structural system was simulated as a 3D-FE model comprising linear and zero-length elements as described in Section 2, using the OpenSees platform [41].

At first, the flexural response of the columns was modeled using two equal-length nonlinear Beam-Column element objects with a longitudinal fiber section discretization along the column clear length H_{cl} (see Figure 1). The reinforcing steel material properties were represented by the uniaxial, bilinear stress-strain relationship with kinematic hardening (Steel01 Material) with an initial elastic modulus equal to 200 GPa and a strain-hardening ratio of 3.3‰ and 5.3‰ in the cases of the ∅12 mm and the ∅20 mm bars, respectively. The inelastic concrete behavior was modeled using the uniaxial Kent-Scott-Park stress-strain envelope with degrading unloading/reloading hysteresis loops and no tensile strength (Concrete01 Material). Specifically, the concrete had an initial tangent modulus of $2 \cdot f_{cm}/\varepsilon_{cy}$ (ε_{cy} = 0.0022) and an ultimate strength and deformation capacity of $f_{cu} = 0.2 \cdot f_{cm}$ and ε_{cu}= 0.0035 according to Park, Priestley, and Gill [45,46]. Details of the fiber discretization of the column cross-sections are depicted in Figure 2.

Next, the modeling of the flexural response of beams was similar to that of columns, using the nonlinear Beam-Column element objects with longitudinal fiber section discretization. The cross-section of beam elements was a T-shape for all beams with effective width equal to $b_w + 2 \cdot d$ (b_w: beam web width; d: beam static height), except for beams {B1, B2, B5, B6, B7, and B8} which had an L-shape cross-section with effective width equal to $b_w + d$. Beam finite elements spanning areas within the beam-column joints were modeled as rigid zones (see Figure 1).

Finally, the brittle response mechanisms of column failure (shear failure of the column web, of the anchorage, and lap-splice failure of column longitudinal reinforcements) were all simulated using Zero-Length element objects in the locations shown in Figure 1. Mechanical properties of the Zero-Length elements were defined according to the discussion in Sections 2.2–2.4, further considering column shear strengths corresponding to the different response mechanisms as calculated from Equations (A1)–(A6). The normalized average axial load ν in the columns that were used in Equations (A1)–(A6) was calculated from a static analysis of the FE building model under the self-weight of the structural elements and the additional gravity loads applied to the SPEAR building during the tests [42–44]. It is noted that the true axial load is variable, fluctuating about an average value due to overturning effects in the ground motion, with maximum values recorded in the columns that are furthest from the center of mass. Table 1 presents the normalized average axial load and the shear strengths of the SPEAR building columns corresponding to the mechanisms considered in the RSA system [30–32], whereas Table 2 presents the parameters defining the multilinear M–θ relationships used for simulating the responses of the zero-length elements parallel to the two principal plan directions.

Regarding the building mass needed for conducting a dynamic analysis, the numerical simulations used lumped masses defined at the centroid of each structural element (i.e., the slabs, the clear length of columns and beams, and the beam-column joints). Lumped masses were also added at individual nodes which were inserted at locations on the building floors where water tanks were placed during the tests for simulating the operational loads. For more information on the location of these tanks in the SPEAR building see References [42–44].

Diaphragm action at the floor levels was simulated by slaving the translational and rotational degrees of freedom acting parallel to the floor plane of all nodes located at the mid thickness of the building's slabs. This was done by employing the rigid-Diaphragm multi-point constraint object at each floor level, in order to connect all intersected column nodes, all nodes located at the centroid of the corresponding R/C slabs, and all nodes simulating the aforementioned water tanks representing the SPEAR building operational loads. Furthermore, gravity loads associated with the SPEAR building and used for static analysis of the FE model were simulated as uniformly distributed loads along with the frame elements of the model.

The first step was to investigate the accuracy of the numerical simulation procedures proposed in this research. To this end, the SPEAR building FE model was subjected to time-history dynamic analyses for the cases of the s10 (PGA = 0.02 g), the s11 (PGA = 0.15 g), and the s12 (PGA = 0.20 g) earthquake records. Damping in the FE model during the dynamic analyses was represented by Rayleigh damping coefficients using the principal translational mode with the greatest period of vibration and the first torsional period of vibration in the X–Y plane. These were computed from a modal analysis of the FE building model, to which the standard 5% viscous damping was applied.

3.3. Evaluation of the Analyses Results

In order to evaluate the accuracy of the results obtained from a third-tier seismic assessment of substandard R/C buildings simulated according to the proposed modeling procedures, the computed response of the FE building model of the SPEAR building is compared to the actual response recorded at ELSA–JRC during the testing. To demonstrate the significance of simulating the brittle mechanisms of failure in substandard R/C buildings, these comparisons considered modifications in the structural response as follows: (i) a first FE building model of the SPEAR building accounted only for the inelastic flexural response of the R/C structural elements, while (ii) in a second FE building model, zero-length elements simulating mechanism {M2–M5} along column lines according to Section 2 were also included.

Figures 8–13 present a comparison between the computed time histories of the horizontal displacements along with the X and Y plan directions, U_X and U_Y, versus those recorded at ELSA–JRC during the s10, s11, and s12 pseudo-dynamic tests (black lines). Specifically, the numerical responses obtained from the time-history dynamic analyses of the FE building model using the proposed simulation procedure (dark grey lines) and those of the FE building model, accounting only for an inelastic flexural response, are plotted concurrently (light gray lines). These displacements correspond to the vibration of the center of mass

(*CM*) of the three floors of the SPEAR building. Note that the overestimation of flexural stiffness leads to very low estimates of displacements and a stiffer overall structure, i.e., a lower natural period leading to unconservative results for the displacement demands and the anticipated damage. In the s11 earthquake case, the displacement demands estimated from analysis of the FE building model accounting only for inelastic flexural response are downgraded by 24% to 58% in the *X* plan direction and by 16% to 37% in the *Y* plan direction as compared to the response of the test building. In the case of the s12 earthquake case, the estimated displacement demands according to the analysis results of the same FE building model are downgraded by 32% to 61% and by 51% to 57% in the *X* and *Y* plan directions, respectively, as compared to the actual building response. On the other hand, the addition of the zero-length elements seem to have lengthened the natural period, leading to convergence of the displacement response and damage as compared to the response of the test building. Decisions made based on these two bounding estimates respectively would lead to little or no retrofit in the former case, and excessive and invasive retrofit in the latter case. It is also noted that as the intensity of shaking and damage accumulation from previous shaking increases in the structure, the waveforms of the more compliant model where all mechanism formations have been activated converge to the experimental response. This primarily occurs during the duration of the response pulses, which suggests convergence in the fundamental period computation.

Table 1. Normalized average axial load and shear strengths of the SPEAR building columns calculated according to the RSA system [30–32].

Story	Column	v	Plan Direction *X*					Plan Direction *Y*				
			$V_{flex,X}$ (kN)	$V_{v,X}$ (kN)	$V_{a,X}$ (kN)	$V_{lap,X}$ (kN)	$V_{j,X}$ (kN)	$V_{flex,Y}$ (kN)	$V_{v,Y}$ (kN)	$V_{a,Y}$ (kN)	$V_{lap,Y}$ (kN)	$V_{j,Y}$ (kN)
	C1,1	0.06	24.50	39.36	24.50	18.17	32.64	24.50	39.36	24.50	18.17	32.64
	C2,1	0.06	65.75	39.46	65.75	40.51	65.74	208.03	128.81	208.03	113.55	105.58
	C3,1	0.10	28.80	50.58	28.80	22.60	36.77	28.80	50.58	28.80	22.60	36.77
	C4,1	0.14	32.44	59.60	32.44	26.36	40.27	32.44	59.60	32.44	26.36	40.27
1	C5,1	0.29	42.69	87.76	42.69	37.06	51.07	42.69	87.76	42.69	37.06	51.07
	C6,1	0.23	38.60	79.50	38.60	32.79	46.65	38.60	79.50	38.60	32.79	46.65
	C7,1	0.08	26.33	38.99	26.33	20.06	34.46	26.33	38.99	26.33	20.06	34.46
	C8,1	0.16	34.14	64.27	34.14	28.12	41.91	34.14	64.27	34.14	28.12	41.91
	C9,1	0.18	35.14	67.43	35.14	29.18	43.04	35.14	67.43	35.14	29.18	43.04
	C1,2	0.03	22.38	40.06	22.38	16.23	31.36	22.38	40.06	22.38	16.23	31.36
	C2,2	0.04	59.49	39.95	59.49	34.52	63.58	189.69	130.38	189.69	95.49	102.52
	C3,2	0.06	25.51	39.48	25.51	19.45	34.47	25.51	39.48	25.51	19.45	34.47
	C4,2	0.09	28.18	38.90	28.18	22.21	37.11	28.18	38.90	28.18	22.21	37.11
2	C5,2	0.18	37.40	70.81	37.40	31.70	45.75	37.40	70.81	37.40	31.70	45.75
	C6,2	0.14	33.50	60.33	33.50	27.68	42.09	33.50	60.33	33.50	27.68	42.09
	C7,2	0.05	23.71	39.67	23.71	17.63	32.79	23.71	39.67	23.71	17.63	32.79
	C8,2	0.10	29.70	51.38	29.70	23.76	38.47	29.70	51.38	29.70	23.76	38.47
	C9,2	0.11	30.28	52.74	30.28	24.38	39.10	30.28	52.74	30.28	24.38	39.10
	C1,3	0.01	19.79	40.35	19.79	13.40	28.06	19.79	40.35	19.79	13.40	28.06
	C2,3	0.02	52.10	40.25	52.10	26.53	57.29	167.86	131.36	167.86	71.95	92.84
	C3,3	0.03	21.36	39.97	21.36	15.03	29.76	21.36	39.97	21.36	15.03	29.76
	C4,3	0.04	22.76	39.78	22.76	16.46	31.15	22.76	39.78	22.76	16.46	31.15
3	C5,3	0.11	30.04	53.06	30.04	23.94	38.21	30.04	53.06	30.04	23.94	38.21
	C6,3	0.08	26.50	39.02	26.50	20.31	34.90	26.50	39.02	26.50	20.31	34.90
	C7,3	0.02	20.41	40.16	20.41	14.04	28.76	20.41	40.16	20.41	14.04	28.76
	C8,3	0.06	24.32	39.40	24.32	18.07	32.78	24.32	39.40	24.32	18.07	32.78
	C9,3	0.05	24.17	39.59	24.17	17.89	32.52	24.17	39.59	24.17	17.89	32.52

Table 2. Moment-Rotation relationships used in the zero-length elements of the SPEAR FE building model. Note: Symbols (-) indicate that brittle fracture was not considered and (;) separate values between the X and Y directions.

Story	Column	$M_{v,X}$ $M_{v,Y}$ (kN-m)	$\theta_{0.005r\_v,X}$ $\theta_{0.005r\_v,Y}$ (rad)	$M_{a,X}$ $M_{a,Y}$ (kN-m)	$\theta_{0.005r\_a,X}$ $\theta_{0.005r\_a,Y}$ (rad)	$M_{lap,X}$ $M_{lap,Y}$ (kN-m)	$\theta_{0.005r\_lap,X}$ $\theta_{0.005r\_lap,Y}$ (rad)	$M_{j,X}$ $M_{j,Y}$ (kN-m)	$\theta_{0.005r\_j,X}$ $\theta_{0.005r\_j,Y}$ (rad)
1	C1,1	-	-	-	-	22.71	0.0037	-	-
	C2,1	49.33 ; 161.01	0.0030 ; 0.0031	- ; -	- ; -	50.64 ; 141.94	0.0031 ; 0.0027	- ; 131.98	- ; 0.0025
	C3,1	-	-	-	-	28.25	0.0039	-	-
	C4,1	-	-	-	-	32.95	0.0041	-	-
	C5,1	-	-	-	-	46.33	0.0043	-	-
	C6,1	-	-	-	-	40.99	0.0042	-	-
	C7,1	-	-	-	-	25.07	0.0038	-	-
	C8,1	-	-	-	-	35.15	0.0041	-	-
	C9,1	-	-	-	-	36.47	0.0042	-	-
2	C1,2	-	-	-	-	20.29	0.0036	-	-
	C2,2	49.94 ; 162.98	0.0034 ; 0.0034	- ; -	- ; -	43.15 ; 119.36	0.0029 ; 0.0025	- ; 128.15	- ; 0.0027
	C3,2	-	-	-	-	24.31	0.0038	-	-
	C4,2	-	-	-	-	27.77	0.0039	-	-
	C5,2	-	-	-	-	39.63	0.0042	-	-
	C6,2	-	-	-	-	34.60	0.0041	-	-
	C7,2	-	-	-	-	22.04	0.0037	-	-
	C8,2	-	-	-	-	29.70	0.0040	-	-
	C9,2	-	-	-	-	30.47	0.0040	-	-
3	C1,3	-	-	-	-	16.75	0.0034	-	-
	C2,3	50.32 ; 164.20	0.0039 ; 0.0039	- ; -	- ; -	33.16 ; 89.94	0.0025 ; 0.0021	- ; 116.04	- ; 0.0028
	C3,3	-	-	-	-	18.78	0.0035	-	-
	C4,3	-	-	-	-	20.57	0.0036	-	-
	C5,3	-	-	-	-	29.93	0.0040	-	-
	C6,3	-	-	-	-	25.39	0.0038	-	-
	C7,3	-	-	-	-	17.55	0.0034	-	-
	C8,3	-	-	-	-	22.59	0.0037	-	-
	C9,3	-	-	-	-	22.36	0.0037	-	-

Table 3 summarizes the peak displacement values obtained from the sequence of analyses for the center of mass of the three floors to the different intensity ground motions. Relative lateral drift ratios were obtained by dividing the relative floor displacement values by the clear floor height. Peak values, as well as the time instant of their occurrence, are also given in Table 3.

It is observed that the results of the flexural analysis (Fiber Model), where all the other premature local failure mechanisms were omitted in the model, fail to reproduce the salient characteristics of the response including the intensity of the drift demands by systematically underestimating them. In the s11 case, the analysis results of the fiber model lead to an underestimation of the actual drift demands by 24% in the first and by 69% in the second and third story with regards to the vibration in the *X* plan direction. These numbers respectively are 16%, 40%, and 53% in the first, second, and third stories in reference to the vibrations in the *Y* plan direction. In the s12 case, the drift demands obtained from analysis of the fiber FE model in the first, second, and third stories were underestimated by 32%, 73%, and 75% in the *X* plan direction and by 51%, 63%, and 64% in the *Y* plan direction as compared to the drift demands recorded during the testing of the SPEAR building. On the contrary, the complete model where local mechanism formation was enabled showed an increased rate of convergence with the experimental values as the intensity of the shaking and damage accumulation increased. In the s11 earthquake case, the results obtained from analysis of the complete model deviate by {+17%; −5%; +9%} in the X plan direction and by {+37%; +25%; +43%} in the Y plan direction with regards to the drift demands recorded at the first, the second, and the third story of the SPEAR building. Furthermore, in the case of the s12 pseudo-dynamic testing, the corresponding deviations of the analytical and experimental results were {−18%; −46%; +27%} in the X plan direction and {−43%; −32%; +35%} in the Y plan direction.

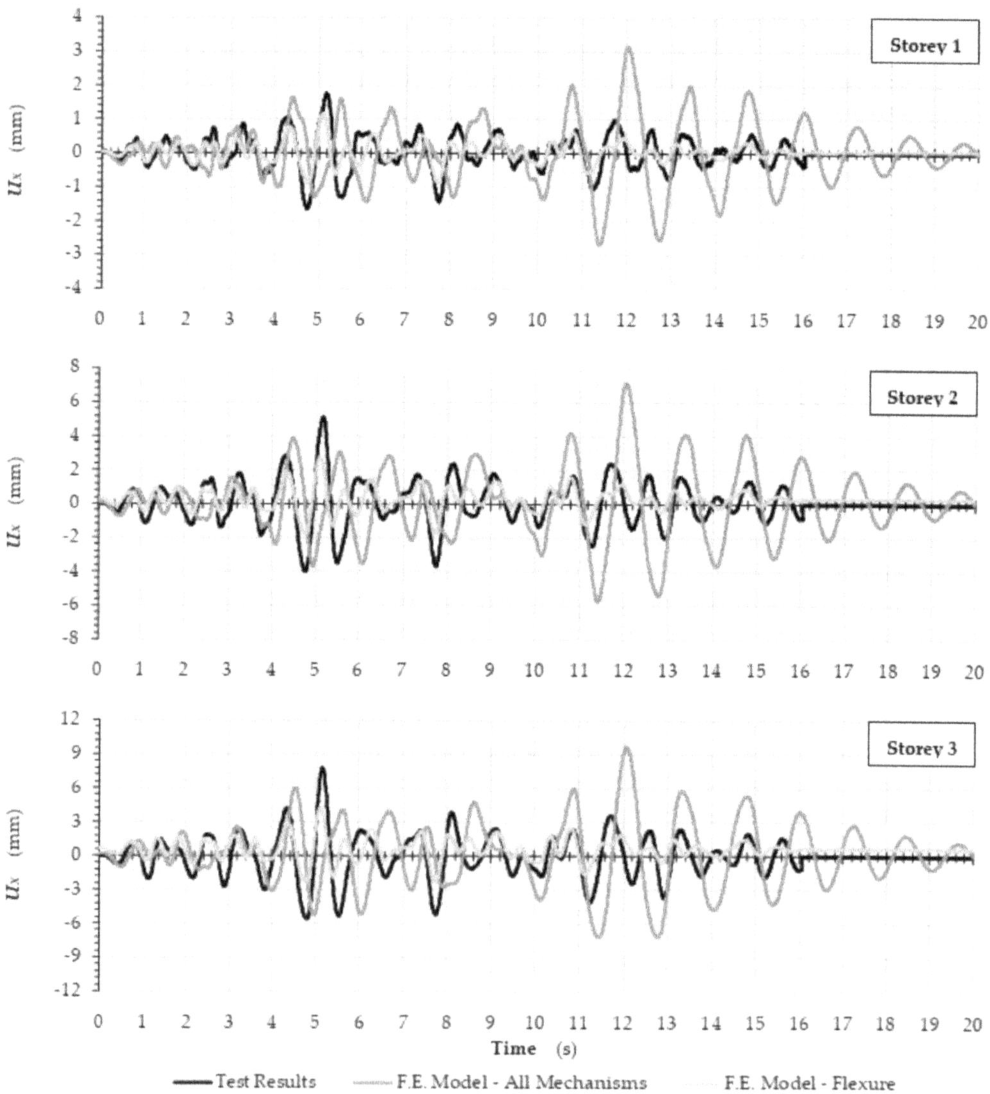

Figure 8. Comparison of the horizontal displacements in the X-plan direction measured at the SPEAR building during the s10 earthquake case and the corresponding response estimated from a dynamic analysis using two different FE building models.

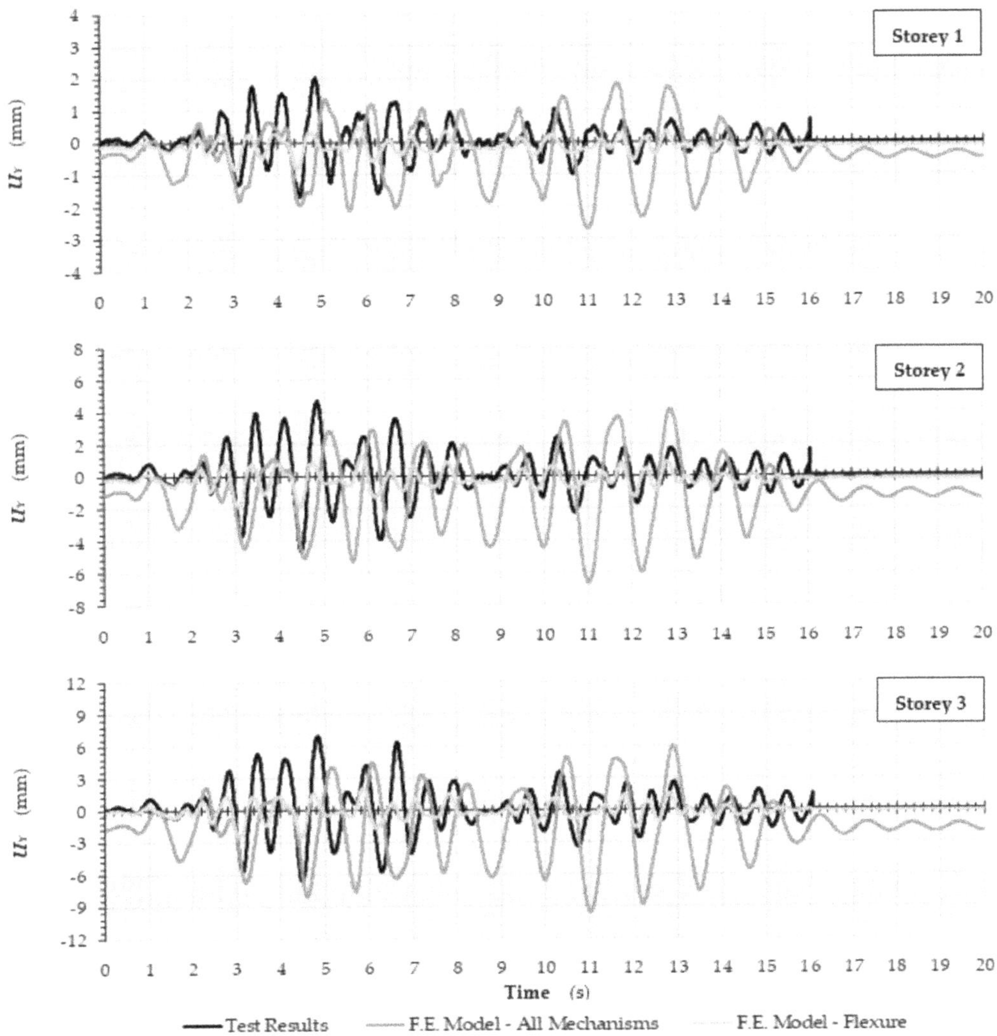

Figure 9. Comparison of the horizontal displacements in the Y-plan direction measured at the SPEAR building during the s10 earthquake case and the corresponding response estimated from a dynamic analysis using two different FE building models.

In reference to the instant of maximum horizontal translation of the centers of mass of the three floors, the analysis results converged to the experimental values in the longitudinal (X) direction (both in terms of the instant of time of occurrence as well as in terms of the magnitude). There was, however, a deviation in the transverse (Y) direction, particularly in the higher intensity earthquake motions s11 and s12. This deviation in the convergence of the estimated responses of both FE building models with the responses recorded at the SPEAR building in the Y direction is mainly due to the degree of accuracy prescribed in simulating the flexural response of the columns of the SPEAR building through fiber discretization. A discretization of the core of the building columns that is finer than the one used in the two FE building models (Figure 2), combined with the use of more advanced stress-strain relationships from the OpenSees library for simulating the inelastic response of steel and concrete materials, would most probably result in a better convergence between the analytically and experimentally obtained responses. However, such fine-tuning of the flexural behavior of the FE building models would have not affected the degree of influence that the simulated brittle mechanisms of response along the column lines of the SPEAR building have on the overall seismic response of the FE building models.

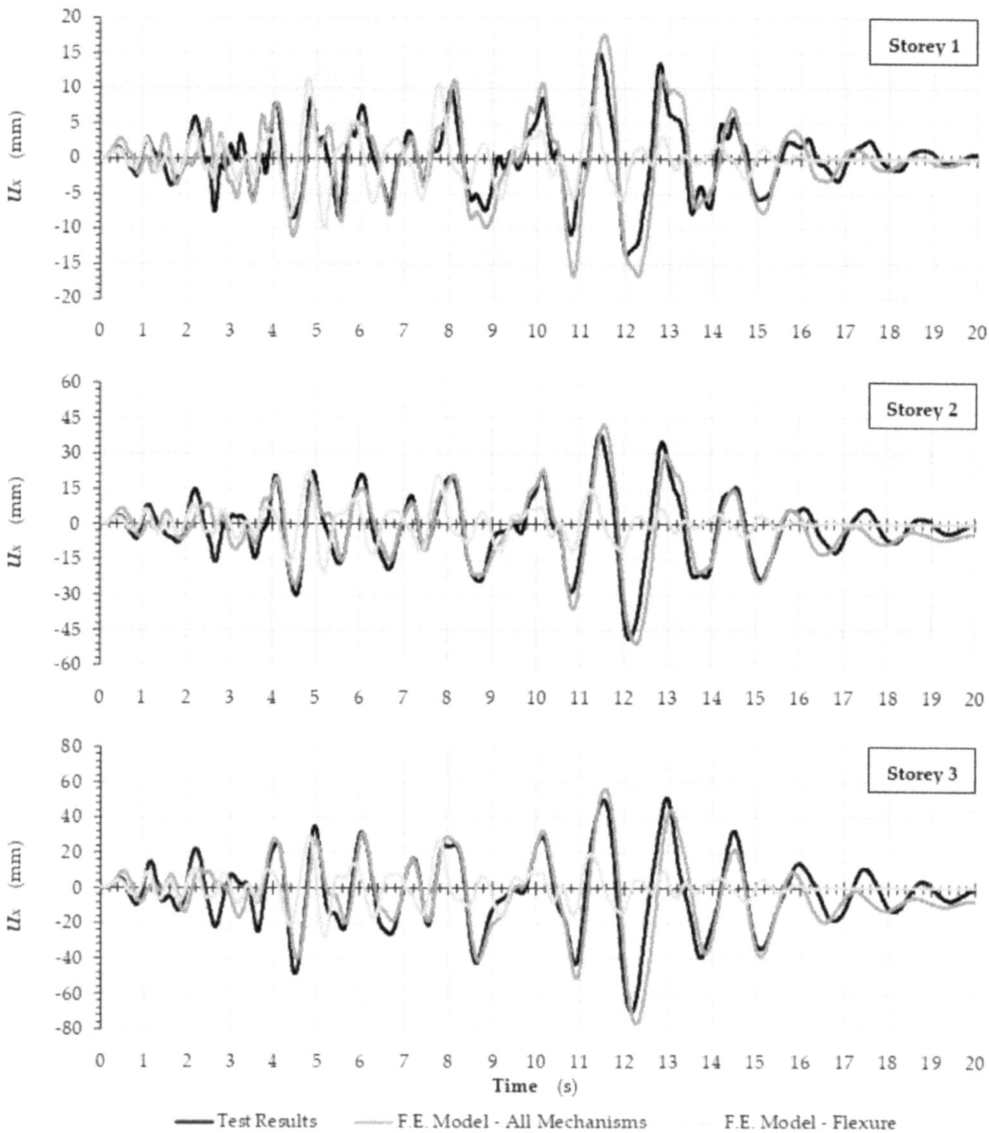

Figure 10. Comparison of the horizontal displacements in the *X*-plan direction measured at the SPEAR building during the s11 earthquake case and the corresponding response estimated from a dynamic analysis using two different FE building models.

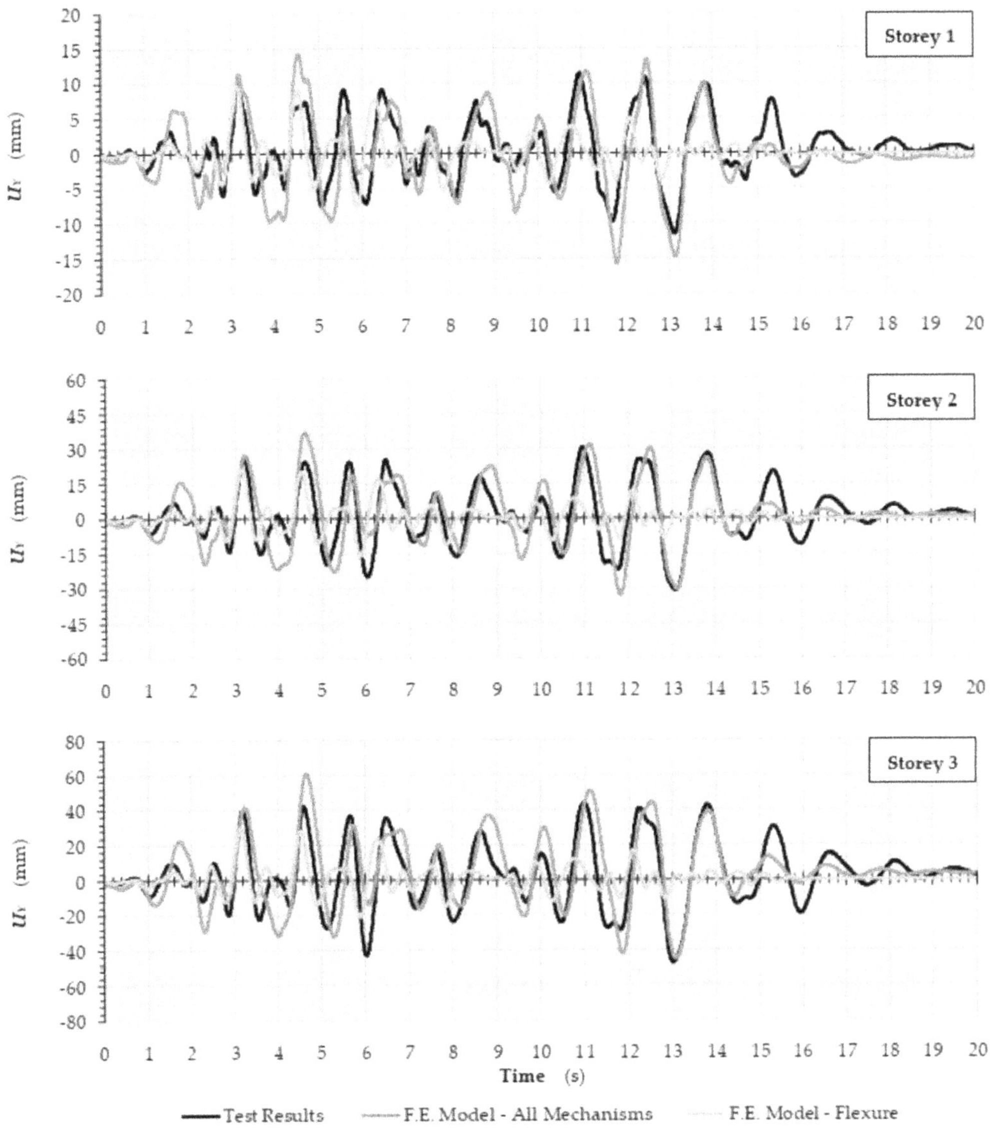

Figure 11. Comparison of the horizontal displacements in the *Y*-plan direction measured at the SPEAR building during the s11 earthquake case and the corresponding response estimated from a dynamic analysis using two different FE building models.

Finally, with regards to the additional computational cost incurred from implementing the present analytical-numerical methodology for simulating the brittle mechanisms of response along the column lines of sub-standard R/C buildings, the conclusion is as follows: No significant difference was observed in terms of convergence rate and time requirements by either simulating the brittle mechanisms of failure in the FE building models according to the proposed methodology or by simulating only the nonlinear flexural response of the same buildings.

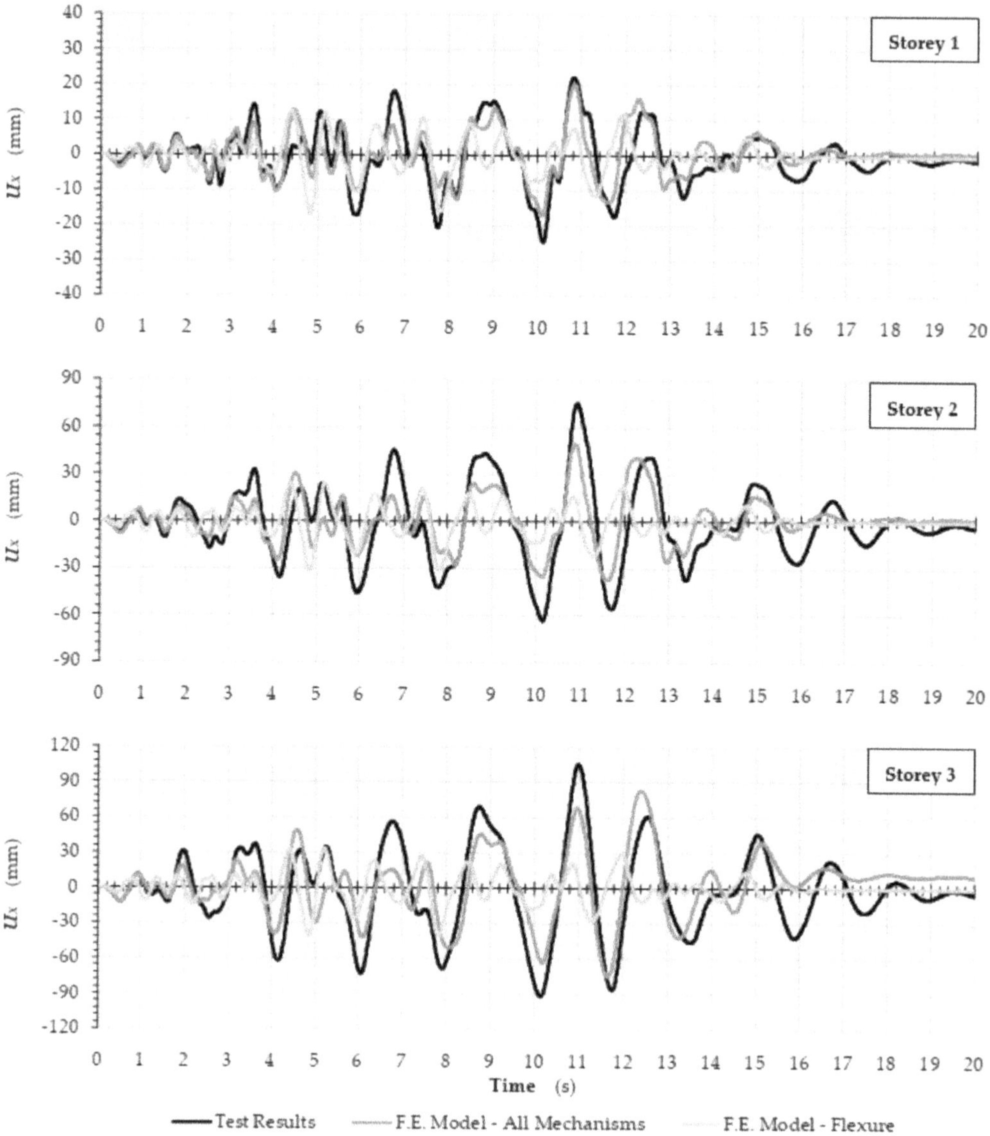

Figure 12. Comparison of the horizontal displacements in the *X*-plan direction measured at the SPEAR building during the s12 earthquake case and the corresponding response estimated from a dynamic analysis using two different FE building models.

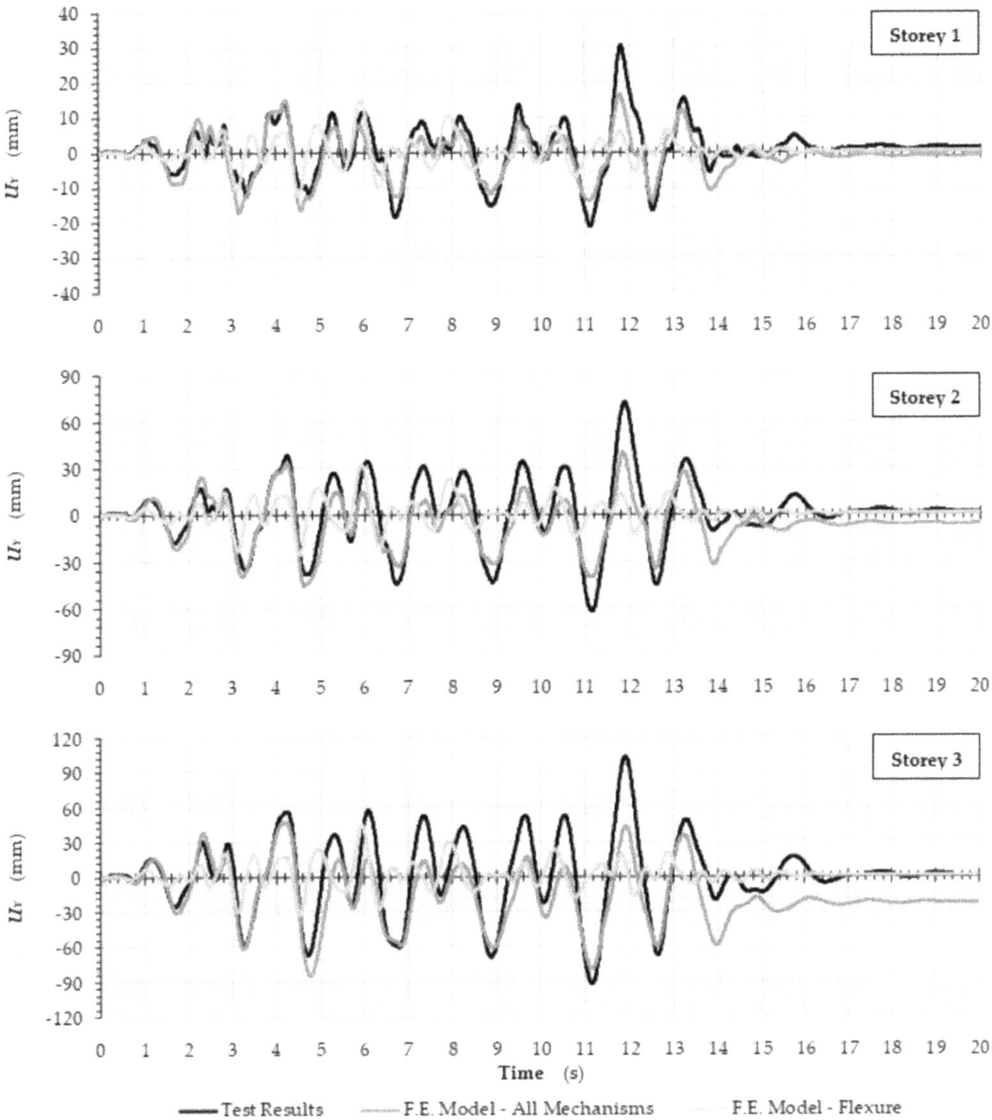

Figure 13. Comparison of the horizontal displacements in the *Y*-plan direction measured at the SPEAR building during the s12 earthquake case and the corresponding responses estimated from a dynamic analysis using two different FE building models.

Table 3. Peak floor displacements and inter-story drifts in the two plan directions of the SPEAR building were obtained from the test and the analyses (ID = Inter-story drift; FD = Floor Displacement).

Analysis Case	PlanStory	Story No.	Test Results			FE Model–All Mechanisms			FE Model–Flexure Only		
			Time (s) at Peak FD	Peak ID × 10³ rad	Time (s) at Peak ID	Time (s) at Peak FD	Peak ID × 10³ rad	Time (s) at Peak ID	Time (s) at Peak FD	Peak ID × 10³ rad	Time (s) at Peak ID
s10	X	1	5.17	0.591	5.17	5.10	0.385	5.10	12.03	1.049	12.03
		2	5.17	1.127	5.16	5.10	0.544	5.11	12.05	1.335	12.06
		3	5.15	0.943	5.14	5.11	0.487	5.11	12.07	0.937	12.13
	Y	1	4.86	0.674	4.86	3.07	0.279	3.07	11.01	0.899	11.01
		2	4.49	1.028	4.49	3.07	0.358	4.47	11.01	1.324	11.01
		3	4.82	0.952	6.63	3.07	0.267	5.87	11.01	1.054	4.61
s11	X	1	11.41	5.026	11.41	11.52	5.904	11.52	4.81	3.812	4.81
		2	12.12	12.052	12.13	12.28	11.487	12.27	4.82	3.731	5.19
		3	12.16	8.054	13.02	12.27	8.762	12.25	4.83	2.511	5.19
	Y	1	13.11	3.877	13.11	11.79	5.294	11.79	3.13	3.252	3.13
		2	13.09	6.648	13.04	4.59	8.319	4.63	3.14	3.975	3.15
		3	13.00	6.076	6.02	4.65	8.716	4.69	3.14	2.849	3.16
s12	X	1	10.13	8.197	10.13	10.84	6.723	10.84	4.83	5.595	4.83
		2	10.92	19.022	10.95	10.87	10.308	10.91	4.84	5.140	7.82
		3	10.97	11.931	11.04	12.39	15.164	12.46	4.85	3.031	7.81
	Y	1	11.80	10.185	11.80	3.17	5.760	3.17	5.91	4.985	5.91
		2	11.88	15.737	11.92	4.63	10.750	4.66	5.92	5.832	5.92
		3	11.90	10.857	11.96	4.75	14.613	4.78	5.92	3.895	5.94

4. Conclusions

This work investigates the validity of a proposed methodology for the performance assessment of older, substandard moment-resisting R/C frame buildings following an earthquake event. The classification 'sub-standard' comes from the fact that R/C construction prior to the mid-1980s did not benefit from the findings of subsequent research regarding pertinent seismic detailing of R/C structures which has now filtered into contemporary building codes. The proposed methodology is based on the earlier work by two of the authors leading to the development of a procedure known as Rapid Seismic Assessment (RSA) which hinges on whether or not the local drift demands can be tolerated without failure along the load resistance path. Guidelines are provided here for modeling all brittle localized strength mechanisms that may occur in the load path of frame structures under lateral loading which is manifested in substandard R/C buildings subjected to seismic loads. An application example using test data from a well-documented experimental program helps to clarify the importance of these mechanisms in a realistic evaluation of the seismic response for this category of buildings.

Author Contributions: Conceptualization, S.I.P., S.J.P. and G.D.M.; methodology, S.I.P. and S.J.P.; software, S.I.P.; validation, S.I.P.; formal analysis, S.I.P., S.J.P. and G.D.M.; investigation, S.J.P.; resources, G.D.M.; data curation, S.I.P.; writing—original draft preparation, S.I.P.; writing—review and editing, G.D.M. and S.J.P.; critical review, G.D.M.; project administration, G.D.M. All authors have read and agreed to the published version of the manuscript.

Funding: This research received no external funding.

Institutional Review Board Statement: Not applicable.

Informed Consent Statement: Not applicable.

Data Availability Statement: Data generated by the FE numerical simulations of the SPEAR testing building carried out for the purposes of this work are available in the form of text files by directly contacting the first author (SIP).

Acknowledgments: The Civil Engineering department of Aristotle University is acknowledged for providing a post-doctoral appointment to the first author (SIP) for the period 2017–2021.

Conflicts of Interest: The authors declare no conflict of interest.

Appendix A

The individual strength terms of an R/C column for an RSA are calculated using closed form expressions which lend themselves to easy spreadsheet calculations. These expressions represent the current state-of-the-art information in the field and may be subject to revisions as the knowledge base in R/C leads to improved models for the individual resistance mechanisms.

The first step towards the calculation of the individual strength terms of an R/C column is the determination of the normalized depth of the compression zone, $\xi = x/d$, at the ultimate limit state, which is estimated for a specified axial load ratio $\nu = N/(A_c \cdot f_c)$ from the following interpolations:

$$\text{if } v_{bal} < v < v_{max},\ \xi = \xi_{bal,u} + (1-\xi_{bal,u}) \cdot \frac{v - v_{bal}}{v_{max} - v_{bal}} \tag{A1}$$

$$\text{if } v_{min} < v \leq v_{bal},\ \xi = \delta_2 + (\xi_{bal,u} - \delta_2) \cdot \frac{v - v_{min}}{v_{bal} - v_{min}} \tag{A2}$$

The values of ν_{min}, ν_{bal}, and ν_{max} correspond to characteristic values of the depth of compression zone x, for $\varepsilon_{cu} = 0.005$, and are given as follows:

$$\text{for } x = d_2 \Rightarrow \xi_2 = \delta_2,\ \frac{v_{min}}{A_c \cdot f_c} = 0.72 \cdot \delta_2 - \frac{f_y}{(1-\alpha) \cdot f_c} \cdot [\rho_{s1} + \rho_v \cdot (1 - 2 \cdot \delta_2)] \tag{A3}$$

$$\text{for } x = x_{bal} \Rightarrow \xi_{bal,u} = 0.64,\ \frac{v_{bal}}{A_c \cdot f_c} = (\rho_{s2} - \rho_{s1}) \cdot \frac{f_y}{f_c} + 0.462 + 0.275 \cdot \rho_v \cdot \frac{f_y}{f_c} \tag{A4}$$

$$\text{for } x = d \Rightarrow \xi_d = 1,\ \frac{v_{max}}{A_c \cdot f_c} = 0.85^2 + \frac{f_y}{f_c} \cdot (\rho_{s2} + 0.8 \cdot \rho_v) \tag{A5}$$

With the interpolated value of ξ, the following strength terms are now determined:

Flexural shear demand:

$$V_{flex} = \left[\rho_{\ell,tot} \cdot \frac{f_y}{f_c} \cdot (1 - 0.4 \cdot \xi) + v \cdot \left(\frac{h}{d} - 0.8 \cdot \xi\right)\right] \cdot \frac{b \cdot d^2 \cdot f_c}{H_{cl}} \tag{A6}$$

Exhaustion of shear strength:

$$\text{If } v < 0.10: V_v = A_{tr} \cdot f_{st} \cdot \frac{d \cdot (1 - 0.4 \cdot \xi)}{s} \cdot \cot \theta_v \tag{A7}$$

$$\text{If } v \geq 0.10: V_v = v \cdot b \cdot d \cdot f_c \cdot \tan \alpha + A_{tr} \cdot f_{st} \cdot \frac{d \cdot (1 - 0.4 \cdot \xi)}{s} \cdot \cot \theta_v \tag{A8}$$

Anchorage failure of longitudinal reinforcement:

$$V_a = \left[\rho_{\ell,tot} \cdot \frac{\min\left\{ \frac{4 \cdot L_a \cdot f_b}{D_b} + \alpha_{hook} \cdot 50 \cdot f_b \ ;\ f_y \right\}}{f_c} \cdot (1 - 0.4 \cdot \xi) + v \cdot \left(\frac{h}{d} - 0.8 \cdot \xi\right)\right] \cdot \frac{b \cdot d^2 \cdot f_c}{H_{cl}} \tag{A9}$$

Lap failure of longitudinal reinforcement:

$$V_{lap} = \frac{\left[\min\left\{ \left(\begin{array}{c} \mu_{fr} \cdot L_{lap} \cdot \left[\frac{A_{tr}}{s} \cdot f_{st} + \alpha_b \cdot (b - N_b \cdot D_b) \cdot f_t\right] + \\ + \alpha_{hook} \cdot 50 \cdot N_b \cdot A_b \cdot f_b \end{array} \right) ;\ N_b \cdot A_b \cdot f_y \right\} \cdot d \cdot (1 - 0.4 \cdot \xi) + \atop + v \cdot b \cdot d^2 \cdot f_c \cdot (0.5 \cdot h/d - 0.4 \cdot \xi) \right]}{H_{cl}/2} \tag{A10}$$

Shear capacity of joints:

Unreinforced or lightly reinforced joints :$V_j = \gamma_j \cdot 0.5 \cdot \sqrt{f_c} \cdot \sqrt{1 + \frac{v_j \cdot f_c}{0.5 \cdot \sqrt{f_c}} \cdot \frac{b_j \cdot d \cdot d_{beam}}{H_{cl}}}$ (A11)

Well reinforced joints :$V_j = \left[\gamma_j \cdot 0.5 \cdot \sqrt{f_c} \cdot \sqrt{1 + \frac{v_j \cdot f_c}{0.5 \cdot \sqrt{f_c}} \cdot \frac{b_j \cdot d \cdot d_{beam}}{H_{cl}}} \right] \cdot \sqrt{1 + \rho_{j,horiz} \cdot \frac{f_{st}}{f_t}}$ (A12)

In the above expressions, the following terms appear:

- $\rho_{\ell,tot} = A_{s,tot}/(b \cdot d)$ is the total longitudinal reinforcement ratio of a column with external dimensions $h \times b$,
- $A_{s,tot}$ is the total area of the longitudinal reinforcement at the column's critical section,
- d is the column effective depth,
- f_y is the longitudinal reinforcement yield stress,
- f_c is the concrete compressive strength,
- ξ $(= x/d)$ is the normalized depth of compression zone,
- v is the axial load ratio acting on the cross-section $(N_{g+0.3q}/(b \cdot d \cdot f_c))$,
- H_{cl} is the column's deformable length,
- $\tan\alpha = (h/d - 0.8 \cdot \xi) \cdot d/H_{cl}$, where a $(\leq \theta_v)$ is the angle of inclination of the diagonal strut created between the centroids of the compression zones at the top and bottom column cross-sections of the column. This represents the strut forming by the axial load acting on the column according to Priestley et al. [44],
- θ_v = {45° when $v < 0.10$, 30° when $v \geq 0.25$, whereas for $0.10 \leq v < 0.25$ θ_v is calculated from linear interpolation} is the angle of sliding plane. Specifically, θ_v is the angle forming between the longitudinal member axis and a major inclined crack developing in the plastic hinge region of the column. It determines the number of stirrup legs that are intersected by the inclined sliding plane,
- h_{st} is the height of the stirrup legs,
- A_{tr} is the total area of stirrup legs in a single stirrup pattern, which are intersected by the inclined sliding plane,
- s is the stirrup spacing,
- f_{st} is the stirrup yield stress,
- L_α is the anchorage length of the longitudinal reinforcement,
- D_b is the diameter of longitudinal reinforcing bars,
- α_{hook} is a binary index (1/0) to account for hooked anchorages ($\alpha_{hook} = 0 \Rightarrow$ no hooks),
- $f_b = 2 \cdot f_{b,o}$ is the concrete bond stress, where $f_{b,o} = n_1 \cdot (f_c/20)^{0.5}$, n_1 = {1.80 for ribbed bars; 0.90 for smooth bars},
- μ_{fr} is the friction coefficient {$0.2 \leq \mu_{fr} \leq 0.3$ for smooth bars; $1.0 \leq \mu_{fr} \leq 1.5$ for ribbed bars},
- L_{lap} is the lap-splice length,
- α_b is a binary index {1 or 0} depending on whether ribbed or smooth reinforcement has been used,
- N_b is the number of longitudinal bars in tension,
- A_b is the area of a single tension bar,
- $f_t = 0.3 \cdot f_c^{2/3}$ is the concrete tensile strength,
- γ_j = {1.40 for interior joints; 1.00 for all other cases, whereas, for joints without stirrups these values are reduced to 0.4 and 0.3 respectively},
- v_j is the (service) axial load acting on the bottom of the column adjusted at the top of the joint (compression is positive),
- $b_j = (b + b_{beam})/2$ is the joint width, where b_{beam} is the web width of the adjacent beam,
- d_{beam} is the beam depth, and
- $\rho_{j,horiz} = A_{tr}/(s \cdot b_j)$.

References

1. *ASCE Standard, ASCE/SEI, 41–17;* Seismic Evaluation and Retrofit of Existing Buildings. American Society of Civil Engineers: Reston, VA, USA, 2017.
2. *EN 1998-1;* Eurocode 8: Design of Structures for Earthquake Resistance—Part 1: General Rules, Seismic Actions and Rules for Buildings. European Committee for Standardization (CEN): Brussels, Belgium, 2004.
3. *FIB Bulletin 24;* Seismic Assessment and Retrofit of Reinforced Concrete Buildings, State-of-Art Report prepared by Task Group 7.1. Federation of Structural Concrete (FIB): Lausanne, Switzerland, 2003.
4. *EN 1998-3;* Eurocode 8: Design of Structures for Earthquake Resistance—Part 3: Assessment and Retrofitting of Buildings. European Committee for Standardization (CEN): Brussels, Belgium, 2005.
5. EPPO. *Greek Code of Structural Interventions;* Earthquake Planning and Protection Organization of Greece (EPPO): Athens, Greece, 2012.
6. *FEMA-356 Pre-Standard;* Commentary for the Seismic Rehabilitation of Buildings. Federal Emergency Management Agency (FEMA): Washington, DC, USA, 2000.
7. Vamvatsikos, D.; Cornell, C.A. Incremental dynamic analysis. *Earthq. Eng. Struct. Dyn.* **2002**, *31*, 491–514. [CrossRef]
8. Syntzirma, D.V.; Pardalopoulos, S.I.; Pantazopoulou, S.J. Experimental evaluation of strength assessment procedures for R/C elements with sub-standard details. *Eng. Struct.* **2020**, *224*, 111191. [CrossRef]
9. Henkhaus, K.; Pujol, S.; Ramirez, J. Axial failure of reinforced concrete columns damaged by shear reversals. *J. Struct. Eng.* **2013**, *139*, 1172–1180. [CrossRef]
10. Sokoli, D.; Ghannoum, W.M. High-strength reinforcement in columns under high shear stresses. *ACI Struct. J.* **2016**, *113*, 605–614. [CrossRef]
11. Panagiotakos, T.; Fardis, M.N. Deformation of R/C members at yielding and ultimate. *ACI Struct. J.* **2001**, *98*, 135–148.
12. Pujol, S.; Ramirez, J.A.; Sozen, M.A. Drift capacity of reinforced concrete columns subjected to cyclic shear reversals. *ACI Spec. Publ.* **1999**, *187*, 255–274.
13. Di Ludovico, M.; Verderame, G.M.; Prota, A.; Manfredi, G.; Cosenza, E. Cyclic behavior of nonconforming full-scale RC columns. *J. Struct. Eng.* **2014**, *140*, 04013107. [CrossRef]
14. Lynn, A.; Moehle, J.P.; Mahin, S.; Holmes, W. Seismic evaluation of existing RC building columns. *Earthq. Spectra* **1996**, *12*, 715–739. [CrossRef]
15. Aboutaha, R.S.; Engelhardt, M.D.; Jirsa, J.O.; Kreger, M.E. Experimental investigation of seismic repair of lap splice failures in damaged concrete columns. *ACI Struct. J.* **1999**, *96*, 297–306.
16. Tastani, S.P.; Pantazopoulou, S.J. Yield penetration in seismically loaded anchorages: Effects on member deformation capacity. *Earthq. Struct.* **2013**, *5*, 527–552. [CrossRef]
17. Thanh Ngoc Tran, C.; Li, B. Ultimate displacement of reinforced concrete columns with light transverse reinforcement. *J. Earthq. Eng.* **2013**, *17*, 282–300. [CrossRef]
18. Elwood, K.J.; Moehle, J.P. Drift capacity of reinforced concrete columns with light transverse reinforcement. *Earthq. Spectra* **2005**, *21*, 71–89. [CrossRef]
19. Inel, M.; Aschheim, M.A.; Pantazopoulou, S.J. Seismic deformation capacity estimates for concrete columns. *Mag. Concr. Res.* **2007**, *59*, 297–310. [CrossRef]
20. Tastani, S.P.; Thermou, G.E.; Pantazopoulou, S.J. Yield penetration in bar anchorages and the effect on rotation capacity. In *Special Volume Compiled by the Institut fur Werkstofeim Bauwesen in Honor of Prof. R. Eligehausen;* Universitat Stuttgart: Stuttgart, Germany, 2012.
21. Pantazopoulou, S.J. Detailing for reinforcement stability in RC members. *J. Struct. Eng.* **1998**, *124*, 623–632. [CrossRef]
22. Pujol, S.; Sozen, M.; Ramirez, J. Displacement history effects of drift capacity of reinforced concrete columns. *ACI Struct. J.* **2006**, *103*, 253–262.
23. Ranf, R.T.; Eberhard, M.O.; Stanton, J.F. Effects of displacement history on lightly confined, reinforced concrete bridge columns. *ACI Spec. Publ.* **2006**, *236*, 23–42.
24. Wood, S.L.; Sittipunt, C. Cyclic response of reinforced concrete structural walls. *ACI Spec. Publ.* **1996**, *162*, 399–430.
25. Zhu, L.; Elwood, K.J.; Haukaas, T. Classification and seismic safety evaluation of existing reinforced concrete columns. *J. Struct. Eng.* **2007**, *133*, 1316–1330. [CrossRef]
26. *ACI 318;* Building Code Requirements for Structural Concrete and Commentary. American Concrete Institute (ACI): Farmington Hills, MI, USA, 2019.
27. Berry, M.P.; Eberhard, M.O. Practical performance model for bar buckling. *J. Struct. Eng.* **2005**, *131*, 1060–1070. [CrossRef]
28. Ruggieri, S.; Porco, F.; Uva, G.; Vamvatsikos, D. Two frugal options to assess class fragility and seismic safety for low-rise reinforced concrete school buildings in Southern Italy. *Bull. Earthq. Eng.* **2021**, *19*, 1415–1439. [CrossRef]
29. Ruggieri, S.; Porco, F.; Uva, G. A practical approach for estimating the floor deformability in existing RC buildings: Evaluation of the effects in the structural response and seismic fragility. *Bull. Earthq. Eng.* **2020**, *18*, 2083–2113. [CrossRef]
30. Pardalopoulos, S.I.; Pantazopoulou, S.J. Rapid seismic assessment of two four-storey R/C test buildings. *Bull. Earthq. Eng.* **2019**, *17*, 1379–1406. [CrossRef]
31. Pardalopoulos, S.I.; Pantazopoulou, S.J.; Lekidis, V.A. Simplified method for rapid seismic assessment of older R/C buildings. *Eng. Struct.* **2018**, *154*, 10–22. [CrossRef]

32. Pardalopoulos, S.; Thermou, G.E.; Pantazopoulou, S.J. Screening criteria to identify brittle R/C structural failures in earthquakes. *Bull. Earthq. Eng.* **2013**, *11*, 607–636. [CrossRef]
33. *DIN 1045*; Beton und Stahlbetonbau: Bemessung und Ausführung. Deutsches Institut für Normung (DIN): Berlin, Germany, 1972.
34. *EN1992-1-1*; Eurocode 2: Design of Concrete Structures—Part 1-1: General Rules and Rules for Buildings. European Committee for Standardization (CEN): Brussels, Belgium, 2004.
35. Lang, A.F.; Marshall, J.D. Devil in the details: Success and failure of Haiti's non-engineered structures. *Earthq. Spectra* **2011**, *27*, S345–S372. [CrossRef]
36. Augenti, N.; Parisi, F. Learning from construction failures due to the 2009 L'Aquila, Italy, earthquake. *ASCE J. Perform. Constr. Facil.* **2010**, *24*, 536–555. [CrossRef]
37. Mehrabian, A.; Haldar, A. Some lessons learned from post-earthquake damage survey of structures in Bam, Iran earthquake of 2003. *Struct. Surv.* **2005**, *23*, 180–192. [CrossRef]
38. Dogangun, A. Performance of reinforced concrete buildings during the May 1, 2003 Bingol Earthquake in Turkey. *Eng. Struct.* **2004**, *26*, 841–856. [CrossRef]
39. Jeong, S.H.; Elnashai, A.S. Analytical and experimental seismic assessment of irregular RC buildings. In Proceedings of the 13th World Conference on Earthquake Engineering (13th WCEE), Vancouver, BC, Canada, 1–6 August 2004.
40. Varum, H. Seismic Assessment, Strengthening and Repair of Existing Buildings. Ph.D. Thesis, Universidade de Aveiro, Aveiro, Portugal, 2003.
41. *OpenSees—Open System for Earthquake Simulation*; Pacific Earthquake Engineering Research Center (PEER): Berkeley, CA, USA, 2020; Available online: https://opensees.berkeley.edu/ (accessed on 1 December 2021).
42. Fardis, M.N. *Design of an Irregular Building for the SPEAR Project—Description of the 3-Storey Structure*; University of Patras Publication: Patras, Greece, 2002.
43. Negro, P.; Mola, E.; Molina, F.J.; Magonette, G.E. Full-scale PSD testing of a torsionally unbalanced three-storey non-seismic RC frame. In Proceedings of the 13th World Conference on Earthquake Engineering (13th WCEE), Vancouver, BC, Canada, 1–6 August 2004.
44. Jeong, S.-H.; Elnashai, A.S. *Analytical Assessment of an Irregular RC Full Scale 3D Test Structure*; Mid-America Earthquake Center (MAE) Publication: Urbana, IL, USA, 2004.
45. Park, R.; Priestley, M.J.N.; Gill, W.D. Ductility of square confined concrete columns. *ASCE J. Struct. Eng.* **1982**, *108*, 929–950. [CrossRef]
46. Priestley, M.N.; Seible, F.; Calvi, G.M. *Seismic Design and Retrofit of Bridges*; Wiley: Hoboken, NJ, USA, 1996.

applied sciences

MDPI

Article

An Analytical Study on the Pull-Out Strength of Anchor Bolts Embedded in Concrete Members by SPH Method

Chi Lu * and Yoshimi Sonoda

Department of Civil Engineering, Kyushu University, Motooka, Nishi-ku, Fukuoka 819-0395, Japan; sonoda@doc.kyushu-u.ac.jp
* Correspondence: luchi@doc.kyushu-u.ac.jp

Abstract: As an important method for connecting structural members, anchor bolts have been installed in many situations. Therefore, accurate evaluation of the pull-out strength of anchor bolts has always been an important issue, considering the complicated actual installation conditions and the problem of aging deterioration of the structural members. In general, the patterns of pull-out failure of anchor bolts can be classified into three types: adhesion failure, cone failure, and bolt break. However, it sometimes shows a mixed fracture pattern, and it is not always easy to predict the accurate pull-out strength. In this study, we attempted to evaluate the pull-out strength of anchor bolts under various installation conditions using SPH, which can analyze the crack growth process in the concrete. In particular, the anchor bolt-concrete interface model was introduced to SPH analysis in order to consider the bond failure, and it was confirmed that various failure patterns and the load capacity could be predicted by proposed SPH method. After that, the influence of several parameters, such as bond stress limit, anchor bolt diameter, and the anchor bolt embedment depth on the failure patterns and the load capacity, were investigated by numerical calculation. Furthermore, several useful suggestions on the pull-out strength of anchor bolts under improper installation conditions, such as the ends of members for the purpose of seismic retrofitting, are presented.

Keywords: concrete; anchor bolt; pull-out strength; SPH method

Citation: Lu, C.; Sonoda, Y. An Analytical Study on the Pull-Out Strength of Anchor Bolts Embedded in Concrete Members by SPH Method. *Appl. Sci.* **2021**, *11*, 8526. https://doi.org/10.3390/app11188526

Academic Editor: Maria Favvata

Received: 14 August 2021
Accepted: 10 September 2021
Published: 14 September 2021

Publisher's Note: MDPI stays neutral with regard to jurisdictional claims in published maps and institutional affiliations.

Copyright: © 2021 by the authors. Licensee MDPI, Basel, Switzerland. This article is an open access article distributed under the terms and conditions of the Creative Commons Attribution (CC BY) license (https://creativecommons.org/licenses/by/4.0/).

1. Introduction

When constructing a structure in the civil engineering field, a joining method that has structural continuity between concrete members and other members (reliable stress transmission) is very important, and joining with anchor bolts is one of the typical methods not only in new structures but also in existing structures. In addition, anchor bolts are used to connect the seismic retrofitting members, such as bridge restrainers [1] and shear walls, to existing concrete structures. Anchor bolts are also applied in railway sleepers [2], modular wall constructions [3], tunnel roofs [4], and nuclear-related facilities [5], and they are expected to continue to be used in the future. However, only a simple evaluation has been made for the ultimate strength when a pull-out load is applied to the anchor bolt. Therefore, there is a concern that the pull-out strength of bolts will be significantly reduced due to improper installation conditions and the aging deterioration of concrete materials.

In general, fracture modes of anchor bolts subjected to a pull-out load are mainly classified into bolt fracture, concrete body fracture, and bond fracture. Table 1 shows the typical fracture modes assumed when designing anchor bolts and shows the equations for calculating the pull-out strength of each fracture type. Basically, it is designed using the pull-out strength of the fracture mode, which has the lowest strength among the assumed these fracture modes shown in Table 1.

However, when installing new anchor bolts in an existing concrete structure, it may be difficult to secure sufficient anchor embedding depth and spacing between adjacent anchor bolts, depending on the arrangement of existing reinforcing bars inside. Furthermore, it is well known that the actual fracture mode of anchor bolt joint cannot be clearly classified into three types, and that a combined fracture mode in which cone fracture and adhesive fracture might occur depending on the installation conditions. For example, Figure 1a [6] shows the fracture of an anchor bolt in the pull-out test conducted by the authors, and a clear combined fracture mode in which cone fracture and adhesion fracture are mixed is obtained. Thus, a simple evaluation based on the strength comparison of each mode shown in Table 1 may not be applicable. Therefore, it is very important to establish a method that can predict the accurate fracture mode and pull-out strength of anchor bolts with various material properties, structural specifications, and loading conditions.

Table 1. Typical fracture patterns of anchor bolts and corresponding strength.

Anchor Bolt Fracture	Cone Fracture	Bond Fracture
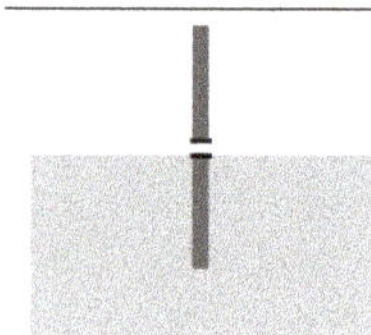	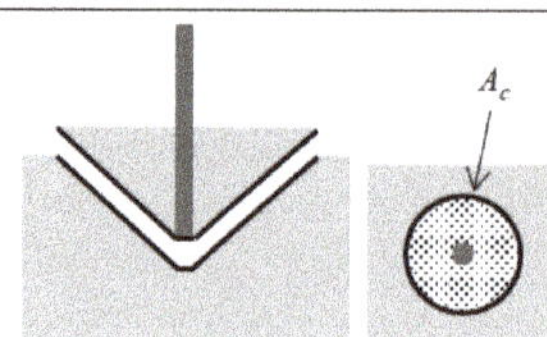	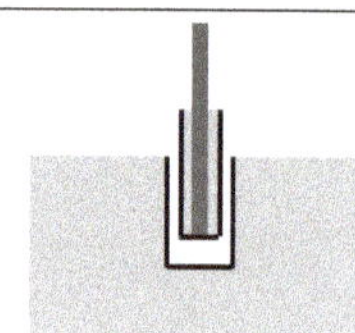
$T = \sigma_y a_0$	$T = 0.23\sqrt{\sigma_B}A_c$	$T = \tau\pi d_a l_e$

T: Maximum pull-out force, σ_y: Steel yield strength; a_0: Anchor bolt nominal cross-section area, σ_B: Concrete compressive strength; τ: Maximum bond stress, d_a: Anchor bolt nominal diameter, l_e: embedment depth.

In addition, when installing anchor bolts in an existing concrete structure where complex rebars arrangements are already embedded inside, it is often difficult to secure sufficient embedding depth, anchor bolt spacing, and distance from the edge of the structural member depending on the position of the existing internal reinforcing bars. Figure 1b shows an example of damage to an anchor bolt joint with a bridge collapse prevention device attached by the earthquake force. This accident is a typical fracture due to the insufficient distance from the edge, and this installation situation is not included in Table 1 either. Thus, in order to design and apply anchor bolts to structure members with sufficient structural safety, the installation situations and the damage patterns which are not included in the existing specifications are also needed to be taken into consideration. In this study, minimum allowable embedding depth, minimum bolt spacing, and minimum distance from the edge of structural member were evaluated, respectively, and proper reinforcement for the lack of distance from the edge of structural member was considered.

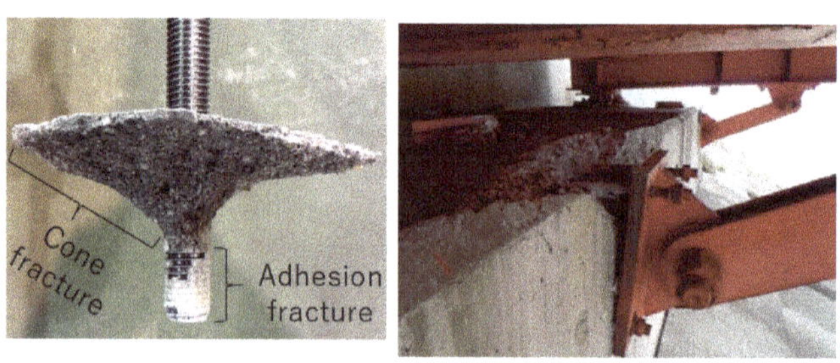

(**a**) (**b**)

Figure 1. Actual fracture patterns in experiment and structural members. (**a**) Complex fracture pattern shown in anchor bolt pull-out experiment [6]. (**b**) Actual anchor bolt fracture due to the insufficient distance from the edge of the concrete abutment.

Under these backgrounds, many researchers have already conducted the research on the pull-out fracture of anchor bolts. For example, there are experimental studies [5,7–9], some of them focusing on the damage of the base concrete [10,11], and others focusing on the bond fracture between concrete and anchor bolt [12,13], and studies focusing on the long-term strength against aging performance [14].

On the other hand, there are many analytical studies using numerical method and not only general FE method [15,16] but also new mesh-free methods, such as Galerkin method [17], and peridynamic theory [18]. Furthermore, there are studies that use neural networks to predict the strength of anchor bolts from installation conditions [19].

In this study, we will analyze the effect of basic factors (embedded depth, bolt diameter, etc.) on the load bearing performance of anchor bolts using SPH (Smoothed Particle Hydrodynamics) [20], referring to the existing research results. The advantage of using the SPH method, which is one of the mesh-free methods, is that, unlike the ordinary finite element method, the particles are not constrained by the conditions of displacement field, and fracture between particles is easily reproduced [21,22].

With reference to these research results, we will propose an evaluation method that can accurately predict the ultimate strength of anchor bolts and simulate the fracture process of pulling out anchor bolts. Specifically, we will introduce an analysis model that can accurately produce the crack development during the pull-out process and evaluate concrete fracture, bolt fracture, and steel-concrete bond fracture in the SPH method.

The fracture mode obtained by the analysis was compared with the experimental results to reproduce the pull-out fracture of the anchor bolt under various installation conditions. Considering the actual installation of the anchor bolts on an existing structure where rebars are already embedded inside, there may be certain restriction on the anchor bolt embedment depth, the interval between anchor bolts, and the distance from the free-end of the structure members. Thus, these are chosen as the input parameter in the analysis to evaluate the performance of anchor bolts under inappropriate condition. Then, we presented several useful suggestions on the pull-out strength of anchor bolts under improper installation conditions. Finally, considering the installation conditions shown in the Figure 1b, the attempt of applying PCM material [23,24] as a reinforcement is performed, and several suggestions are made when using such kind of reinforcement.

2. Analysis Method in This Study

In this study, SPH method, which is one of the most popular mesh-free methods, is adopted in order to analyze local fracture phenomena of concrete which are difficult to simulate by FEM. The physical quantity of each particle is evaluated by the weighted average of the adjacent particles using the kernel function, as shown in Figure 2, and the basic equation of the SPH method is written as Equation (1).

$$< f(x_i) >= \int_{\Omega} f(x_i) W(x_i - x_j, h)\, dx_j, \quad (1)$$

where W is the kernel function, h is the smoothing length, terms inside angel bracket are SPH approximations, Ω is the integral area within the support domain, and x_i, x_j are the position of particle i and j.

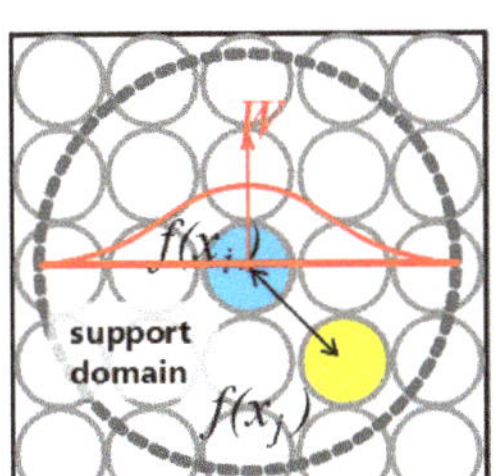

Figure 2. Influence domain of SPH particles (kernel function).

The smoothing length h is very important in SPH analysis. If it is too large, the accuracy of the analysis would be affected due to the smoothed-out details and local properties; on the other hand, if it is too small, the number of particles within smoothing domain may not be enough to calculate accurate response of particles [25]. Thus, in this study, an adaptive smooth length h_a is utilized. Since SPH method is to calculate weighted average values within the smoothing length, an adaptive smoothing length is better for the calculation of local deformation and stress concentration. The adaptive smoothing length h_a is changed during the analysis with the density of the particle, which is widely used in SPH analyses, as shown in Equation (2).

$$h_a = h_0 \left(\frac{\rho_0}{\rho_{current}} \right)^{\frac{1}{3}}, \tag{2}$$

where h_0 is the initial smoothing length, ρ_0 is the initial density of the particle, and $\rho_{current}$ is the density of particle calculated by SPH process, shown in Equation (3).

$$\rho_{current} = \sum_{j=1}^{N} m_j W_{ij}. \tag{3}$$

By the above calculation, the smoothing length h_a is decreased when the influence domain is compressed with the increase of density.

When adaptive smoothing length h_a is adopted in SPH analysis, to obey Newton's third law and conservation of momentum, the following changes are applied to the calculation of kernel function, shown in Equation (4).

$$< f(x_i) >= \int_{\Omega} f(x_i) W \left(x_i - x_j, \frac{h_i + h_j}{2} \right) dx_j. \tag{4}$$

By the above calculation, the interaction force between pairing particles by SPH analysis maintains the same.

If two particles are positioned with smaller distance than h_0 at the beginning of the analysis, we defined that the two particles are "SPH-linked", and the SPH-link between them will break when the distance between them is larger than h_a, or the particles enter the crushing state, which is explained in the next section. Damage in solids is generally unrecoverable, and solid material, such as concrete, is not able to bond together due to the decrease of distance. Thus, if two particles are not SPH-linked at the beginning of the analysis, or SPH-link breaks by concrete fracture, no SPH link will be generated, even if the distance between particles become smaller than h_a.

To calculate the bond stress between concrete and steel anchor bar, the normal direction of the contact surface $\boldsymbol{n}$ should be defined at the beginning of the numerical analysis. When the steel anchor is being pulled out, displacement $\boldsymbol{dr}$ would occur within the pair of neighboring concrete particle and steel particle, as shown in the Figure 3a. Using the normal direction and the vector of the two particles, the slip along the tangential direction $\boldsymbol{t}$ can be calculated as in Equation (5). Finally, by using the relationship between bond stress τ and slip distance $|\boldsymbol{t}|$, the bond stress can be obtained, as shown in Figure 3b and Equation (6).

$$\boldsymbol{t} = -(\boldsymbol{dr} - \boldsymbol{dr} \cdot \boldsymbol{n}), \tag{5}$$

$$\tau_{ij} = f_b(|\boldsymbol{t}|). \tag{6}$$

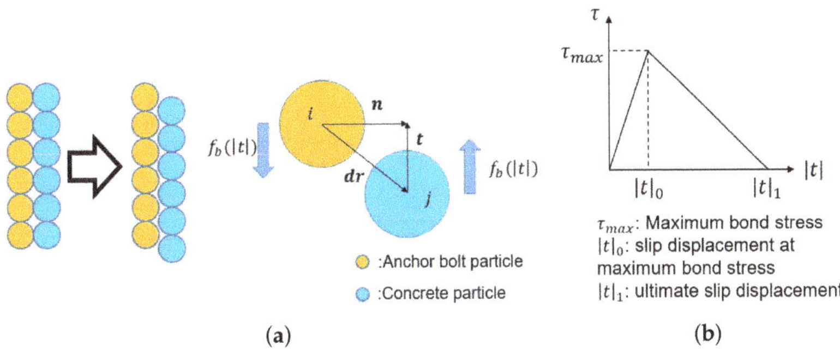

Figure 3. Calculation of bond stress. (**a**) Slip displacement between steel particle and concrete particle. (**b**) Relationship between slip displacement and bond stress abutment.

3. Overview of Crack Growth Analysis

To analyze the elasto-plastic behavior of concrete, a pressure-dependent yield function, such as Drucker-Prager's equation, is often used for the yield criteria. However, it is well known that linear Drucker-Prager's yield surface overestimates compressive strength of concrete under high hydrostatic pressure condition. Therefore, non-linear Drucker-Prager's yield function which improves linear Drucker-Prager's equation is applied in this study. Non-linear Drucker-Prager yield function applied in this research is represented by the following Equation (7).

$$f(I_1, J_2) = \sqrt{J_2} - \sqrt{\frac{\gamma^2 - \beta I_1}{3}} = 0, \quad (7)$$

where γ is $\sqrt{(f_c f_t)}$, β is $(f_c - f_t)$, I_1 is the first invariant of stress, J_2 is the second invariant of deviatoric stress, f_c is the uniaxial compressive strength, and f_t is the uniaxial tensile strength. Figure 4 shows non-linear Drucker-Prager yield surface in the $\sqrt{J_2} - I_1$ plane.

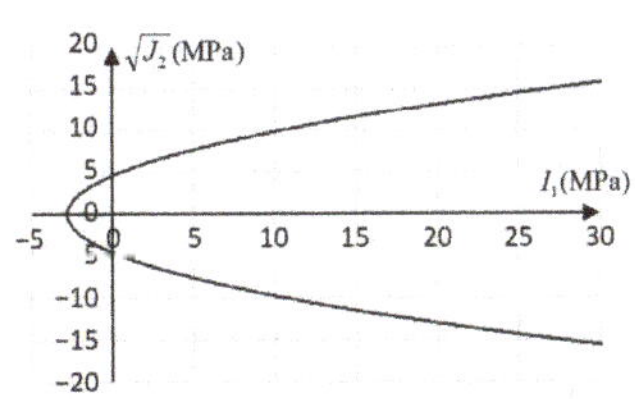

Figure 4. Nonlinear Drucker-Prager yield criterion.

Furthermore, bilinear softening of concrete is considered on the tensile stress side according to the Japanese specification [26], as shown in Figure 5, where σ_t is the tensile stress, f_t is the tensile strength of concrete, and G_F is the fracture energy of concrete determined from Japanese specifications [26].

In addition, considering the degree of damage in which the load transmission capacity in the concrete cross section decreases due to the accumulation of plastic strain, the relationship represented by the sigmoid function, as shown in Equation (8) and Figure 6, was assumed between the degree of damage and the plastic strain.

$$D_{pr} = \frac{D_{lim}}{1 + \exp[-k(\varepsilon^p_{pr} - \frac{\varepsilon^p_{max}}{2})]}, (0 \leq D_{pr} \leq 0.4), \quad (8)$$

where D_{pr} is the damage in the principal direction; k is the gradient adjustment constant, and $k = 300$ is applied in this study; ε^p_{pr} is the accumulated plastic strain along the principal direction; ε^p_{max} is the maximum limit of equivalent plastic strain, and $\varepsilon^p_{max} = 0.03$ is used.

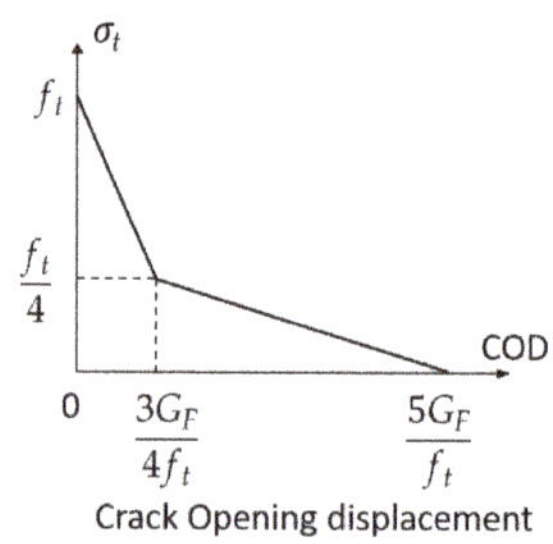

Figure 5. Tensile softening model of the concrete.

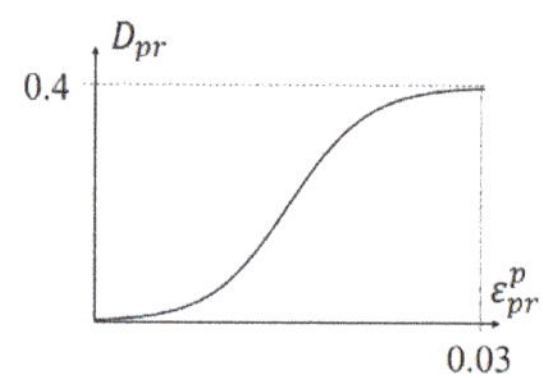

Figure 6. Damage index and principal strain relationship in the tensile side.

Compressive softening is also considered according to the Popovic's equation [27], as shown in Equations (9)–(11).

$$\sigma_{pr} = f_c \cdot \frac{n(\frac{\varepsilon_{pr}}{\varepsilon_{co}})}{(n-1) + (\frac{\varepsilon_{pr}}{\varepsilon_{co}})^n}, \tag{9}$$

$$\varepsilon_{co} = \frac{f_c}{E_c(1 - \frac{1}{n})}, \tag{10}$$

$$n = \exp(0.0256 \cdot f_c), \tag{11}$$

where f_c is the uniaxial compressive strength, ε_{pr} is the principal strain, and ε_{co}, n are parameters determined from material test results. Hardening is considered on the compression stress side, as shown in Equation (12).

$$H = \frac{d\sigma}{d\varepsilon^p}. \tag{12}$$

To obtain the the loss of elastic stiffness due to the increase of damage, the damage index in the global coordinate is calculated first with D_{pr} using Equation (13).

$$D_i = \sum_{i=1}^{3} |D_{pr} e_{i,pr}|, (i = 1,2,3), \tag{13}$$

where D_i is the damage index in the global coordinate, and $e_{i,pr}$ is unit vector of principal direction. This calculation process is shown in Figure 7.

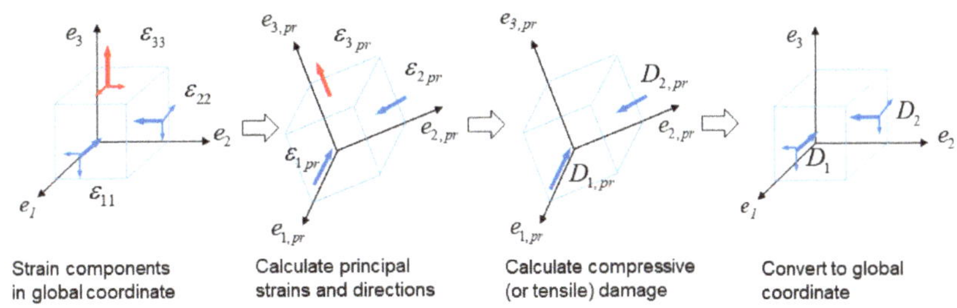

Figure 7. Damage index calculation process.

Then, the decrease ratio of elastic stiffness d_{ij} is calculated by Equation (14).

$$d_{ij} = \sqrt{(1-D_i)(1-D_j)}. \tag{14}$$

Finally, the elastic stiffness matrix with damage can be written as Equation (15):

$$E^e = \begin{pmatrix} (\lambda+2\mu)d_{11} & \lambda d_{12} & \lambda d_{13} & 0 & 0 & 0 \\ \lambda d_{12} & (\lambda+2\mu)d_{22} & \lambda d_{23} & 0 & 0 & 0 \\ \lambda d_{13} & \lambda d_{23} & (\lambda+2\mu)d_{33} & 0 & 0 & 0 \\ 0 & 0 & 0 & 2\mu d_{12} & 0 & 0 \\ 0 & 0 & 0 & 0 & 2\mu d_{23} & 0 \\ 0 & 0 & 0 & 0 & 0 & 2\mu d_{13} \end{pmatrix}, \tag{15}$$

where $\lambda = \mu E/[(1+\mu)(1-2\mu)]$, $\mu = E/[2(1+\nu)]$, and E, ν are the Young's modulus and the Poisson's ratio, respectively.

In this study, to avoid the unrealistic volume overlap between particles, when the volumetric strain of concrete particle ε_v reaches a certain threshold value (we call it erosion limit) ε_{v_lim}, the particles are regarded as a crushing state, and stress and stiffness of the particle are assumed to be zero; however, the particle itself is not erased in order to maintain the mass conservation, and the kinetic energy and the momentum of the particle is also preserved. Based on the assumptions mentioned above, the stress-strain relationship of concrete is schematically shown in Figure 8a,b.

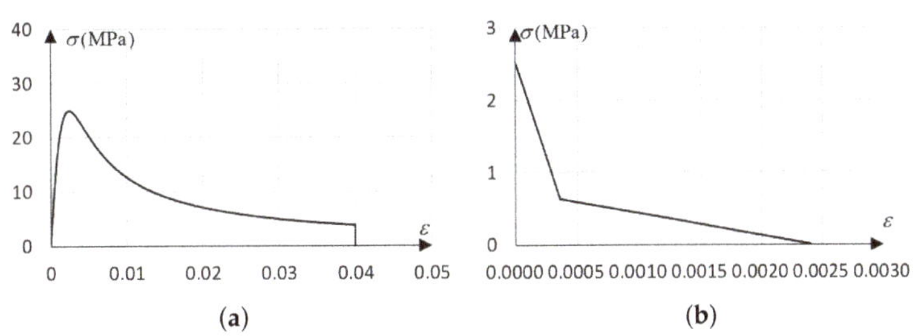

Figure 8. Uniaxial stress-strain relationship of concrete. (**a**) Compressive stress-strain relations. (**b**) Tensile stress-strain relations.

For the steel material, von Mises yield criterion is used, and strain hardening exponent is introduced to describe the hardening process. The stress-strain relationship of steel is shown in Figure 9.

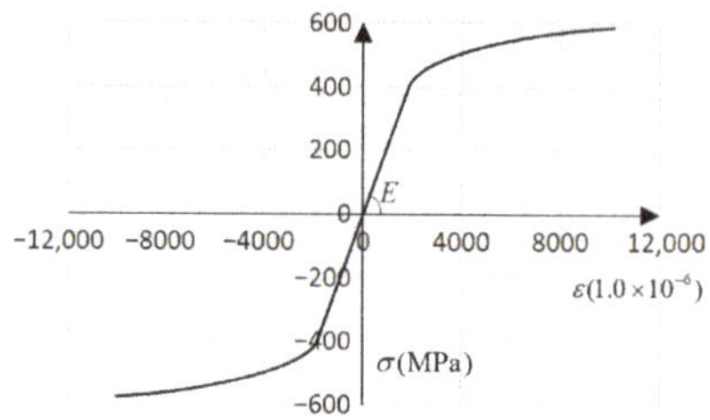

Figure 9. Stress-strain relationship of steel.

4. Anchor Bolt Pull-Out Analysis

4.1. Verification of the Proposed Method

In general, the fracture pattern when pulling out an anchor bolt embedded in a concrete member is complicated, and, as shown in experiment [6], the actual fracture pattern is often a combination of three fracture patterns (bond fracture, corn fracture, and bolt fracture). Therefore, in this study, we conducted a simulation analysis of the anchor bolt pull-out test and confirmed whether the fracture process and the pull-out strength in various fracture cases can be reproduced. The analytical model is shown in Figure 10a. Each particle in the model is 3 mm in diameter, and the total number of particles is about 200,000. An anchor bolt is embedded in the center of the concrete block. As a boundary condition, the vertical degrees of freedom of the particles (black area in the figure) at the four corners of upper surface were constrained. Table 2 shows the material constants used in the analysis.

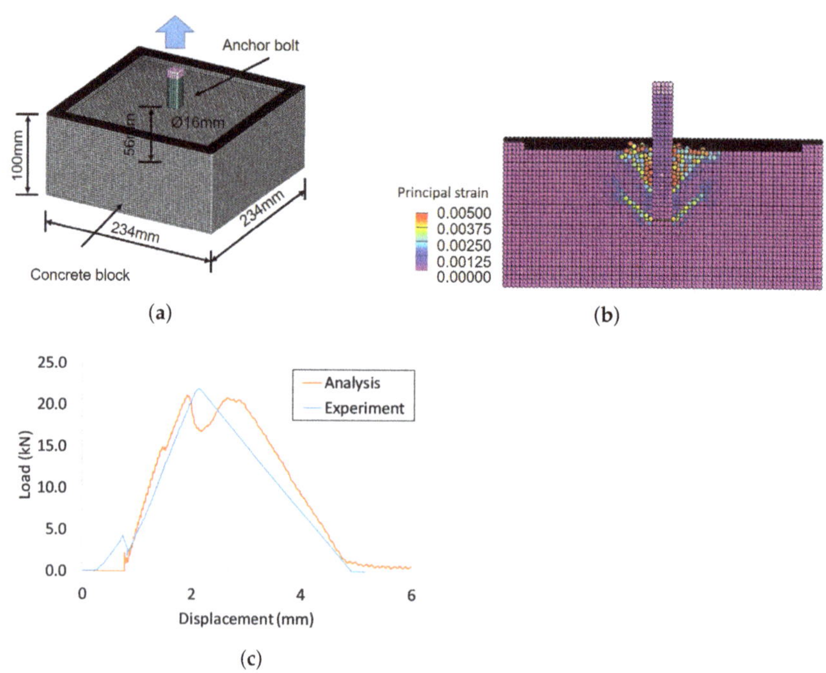

Figure 10. SPH analysis for the pull-out process in experiment [6]. (**a**) Analysis model. (**b**) Complex fracture pattern by numerical analysis. (**c**) Comparison of load-displacement relationship.

Figure 10b,c shows the final fracture pattern and the load-displacement relationship obtained from the proposed numerical analysis method. It can be found that the maximum load of the experiment is well reproduced by the numerical analysis, and the fracture pattern shown in the experiment is also reproduced by the numerical analysis. Bond fracture occurs at the lower part of the anchor, and cone fracture occurs at the upper part of the anchor. From these results, it is shown that the pull-out strength and fracture type of anchor bolts can be grasped by the proposed method.

Table 2. Material constants used in the analysis.

	Steel	Concrete
Compressive strength (MPa)	408.2	48.0
Tensile strength (MPa)	408.2	3.1
Poisson's ratio	0.30	0.20
Density (kg/m^3)	79,000	2350.0
Young's modulus (MPa)	214,000	21,430

4.2. Effect of Installation Conditions on the Fracture Mode

Next, in order to make a basic consideration of the effects of anchor bolt installation conditions (bolt diameter, embedding depth, joint strength between bolt and concrete) on pull-out strength and fracture pattern, an analysis model with the same dimensions as the specimens used in the previous study [6] was examined. An analysis model was created in which anchor bolts were embedded in the center of a concrete block, and analysis was performed by constraining the vertical displacement of the particles at the four corners, as shown in Figure 10. Here, assuming anchor bolts used to attach seismic retrofitting members to concrete blocks, we examined load-bearing performance when the embedding depth is not sufficient or when chemical adhesives with different adhesive strengths are used. Table 3 shows the material parameters used in the analysis, and Table 4 shows the condition parameters for anchor installation. In this study, 0.5 m/s was selected as the pull-out speed given to the anchor bolts according to the maximum speed level assumed for ordinary seismic wave.

Table 3. Material constants used in the analysis with different installation conditions.

	Steel	Concrete
Compressive strength (MPa)	408.2	25.0
Tensile strength (MPa)	408.2	2.5
Poisson's ratio	0.30	0.20
Density (kg/m^3)	79,000	2350.0
Young's modulus (MPa)	214,000	21,430

Table 4. Analysis cases.

Case	Maximum Bond Stress τ (MPa)	Anchor Bolt Diameter d (mm)	Anchor Bolt Embedment Depth l_e (mm)
1	8	16	48
2	12	16	48
3	16	16	48
4	12	8	48
5	12	24	48
6	12	16	32
7	12	16	64

First, the crack growth process of concrete with the increase of pull-out displacement of anchor bolts was investigated, and the adhesive stress distribution that changed, at the same time, was analytically considered. Figure 11 show the final crack pattern of Case 1, 2, and 3, respectively. With the increase of adhesive stress limit, the fracture pattern changed from bond fracture, bond-cone combined fracture, to cone fracture.

As shown in Figure 11, there are many small cracks around the anchor bolt, and a major crack can be found at the bottom of the anchor bolt. In the case of adhesive stress limit is 16 MPa, the major crack developed to the surface of the concrete, forming a cone shaped fracture pattern.

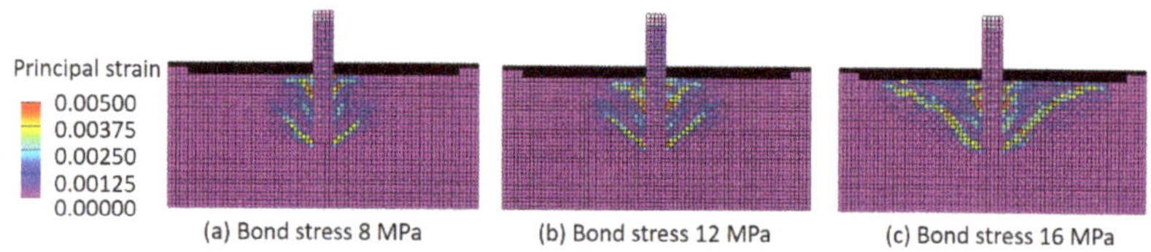

Figure 11. Effect of adhesive stress on the fracture pattern.

The bond stress development are shown in Figure 12. Due to the mechanical symmetry of the loading condition, the resulting figure shows only zoomed half of the analysis area. Focusing on the result of Figure 12a,b with the adhesive stress limit set to 8 MPa, when the pull-out displacement reaches $x = 75$ μm, fine cracks are seen near the anchor bolt, and the bond stress rises to about 5 MPa in the entire circumference of the anchor bolt. When pull-out displacement $x = 225$ μm, a major crack that could lead to cone fracture can be found at the bottom of the anchor bolt. In addition, adhesive fracture was confirmed in the concrete area with a depth of 10 to 15 mm around the anchor bolt. Regarding the adhesive stress distribution, it was found that the stress decreased to zero at the bottom and at about 10–15 mm depth of the concrete. This phenomenon indicate that the bond stress can be influenced by cracks. At the upper part, the constraint from the surrounding concrete is relatively low. Micro cracks and major radial cracks can both be found in this area. Thus, the bond stress in this area decreases at an earlier stage [12,15].

In Figure 12c,d, where the adhesive stress limit is 12 MPa, crack development and bond stress distributions are similar to the previous case. With the increase of adhesive stress limit, the cone area at the upper part of the anchor bolt when $x = 225$ μm is larger than that in Case 1. The final fracture pattern changed from bond fracture to bond-cone complex fracture. In Figure 12e,f, where the bond stress is 16 MPa, the crack development and the bond stress distribution of $x = 75$ μm is similar to previous cases. When $x = 300$ μm, the crack reaches the surface of the concrete block, and the bond stress on the lower part of the anchor bolt decreases, forming a cone shaped fracture pattern.

The load-displacement relationship of Case 1, 2, and 3 is shown in Figure 13, where it shows that, with the increase of the maximum bond stress, the maximum load is increasing. For the case with bond stress of 8 MPa where the fracture pattern is bond fracture, the shape of the load-displacement curve is similar to the bond-stress curve mentioned in Figure 3b. For the case with bond stress of 12 MPa where the fracture pattern is compound fracture, it can be found that the area of the load-displacement curve is larger than the previous curve, indicating that the energy consumed in compound fracture is larger than in bond fracture. For the case with bond stress of 16 MPa, the maximum load is larger than the previous cases, and the area of the load-displacement curve, or the energy consumed, is even larger. Because of concrete cracks generated in this process, more noise can be observed in the curve.

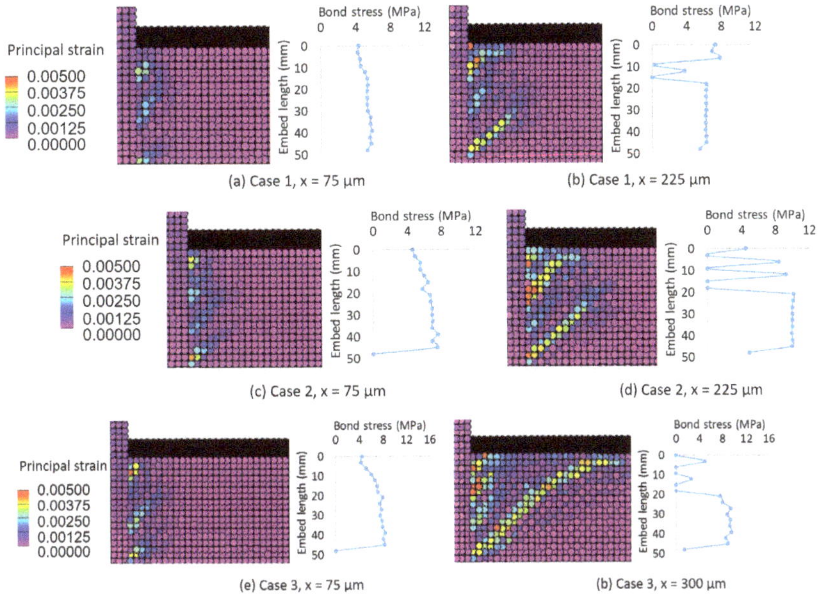

Figure 12. Crack development and bond stress distribution. (**a**,**b**) Case 1. (**c**,**d**) Case 2. (**e**,**f**) Case 3.

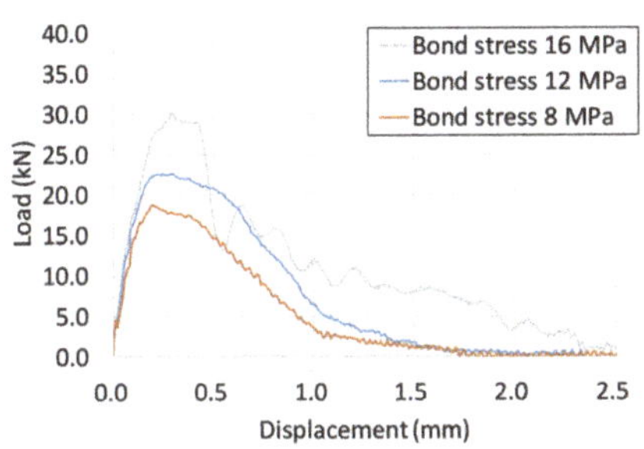

Figure 13. Load-displacement relationships with different bond stress.

For case 2, 6, and 7, when increasing the diameter of the anchor bolt, the damage pattern changed from bond damage to cone damage, shown in Figure 14, and the load-displacement relationship is shown in Figure 15. When the diameter of the anchor bolt increases, more bond force is provided by the increased interface area between anchor bolt and concrete. This bond force is subjected to the same amount of concrete in a cone-shaped area; thus, more concrete is damaged with more bond force by the pull-out process.

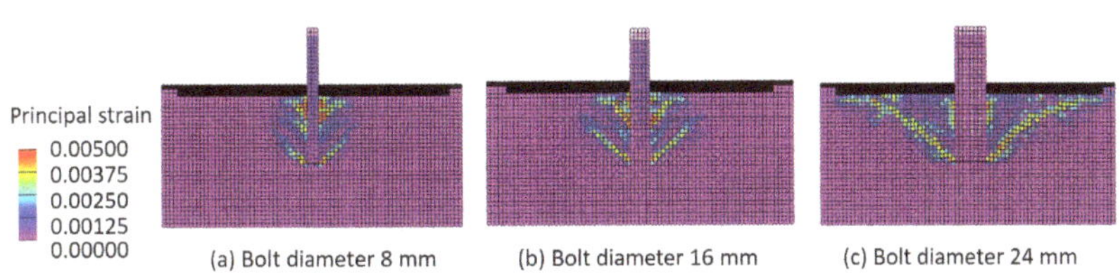

Figure 14. Effect of anchor bolt diameter on the fracture pattern.

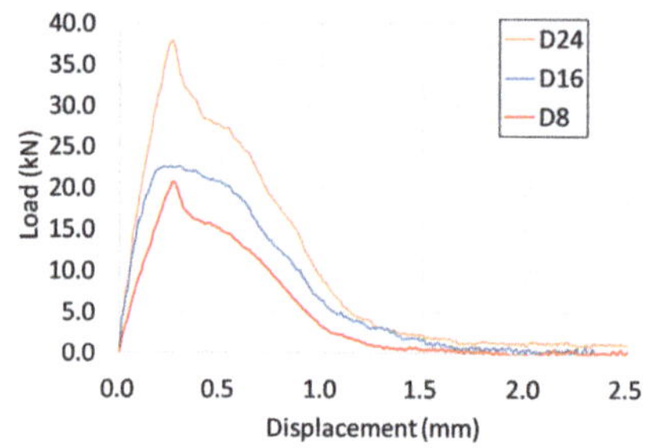

Figure 15. Load-displacement relationships with different anchor bolt diameters.

For case 2, 4, and 5, when changing the embedment depth of the anchor bolt from shallow to deep, the damage pattern changed from cone damage to bond damage, shown in Figure 16, and the load-displacement relationship is shown in Figure 17. When the embedment depth of the anchor bolt increases, the volume of the cone-shaped area also increases. Although more bond force is provided by the increased interface area, the fracture pattern changes from cone fracture to bond fracture due to more concrete participating in resisting the bond force.

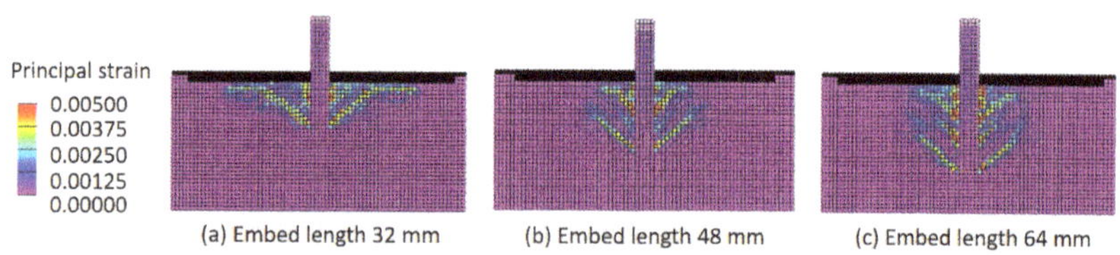

Figure 16. Effect of embedment depths on the fracture pattern.

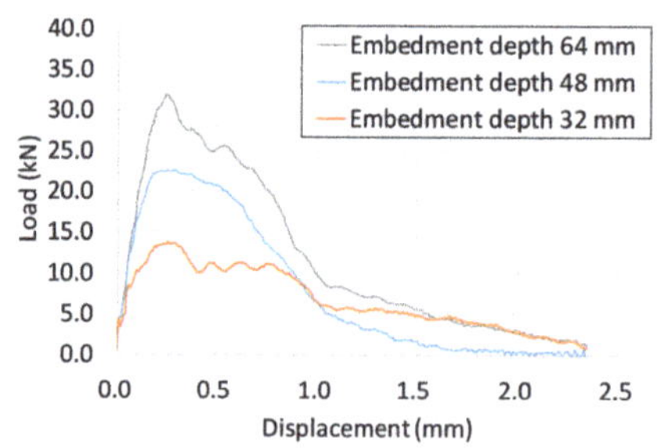

Figure 17. Load-displacement relationships with different embed depths.

To summarize, the analysis cases with bond stress of 12 MPa and their final crack pattern are listed in Table 5 and Figure 18. In this figure, the horizontal axis is the ratio between embedment depth l_e and the anchor bolt diameter d, while the vertical axis is the maximum load P in the pull-out process. When $l_e/d < 3$, the fracture pattern is cone fracture; when $l_e/d > 3$, the fracture pattern is bond fracture; and, when $l_e/d = 3$, the fracture pattern is cone-bond complex fracture. This figure shows the tendency that, in this study, with the increase of the value of l_e/d, the fracture pattern changes from cone fracture to bond fracture.

Table 5. Analysis cases and final crack pattern.

Case	Maximum Bond Stress (MPa)	Anchor Bolt Diameter (mm)	Anchor Bolt Embedment Depth (mm)	Final Crack Pattern
2	12	16	48	Bond-cone complex fracture
4	12	8	48	Bond fracture
5	12	24	48	Cone fracture
6	12	16	32	Cone fracture
7	12	16	64	Bond fracture

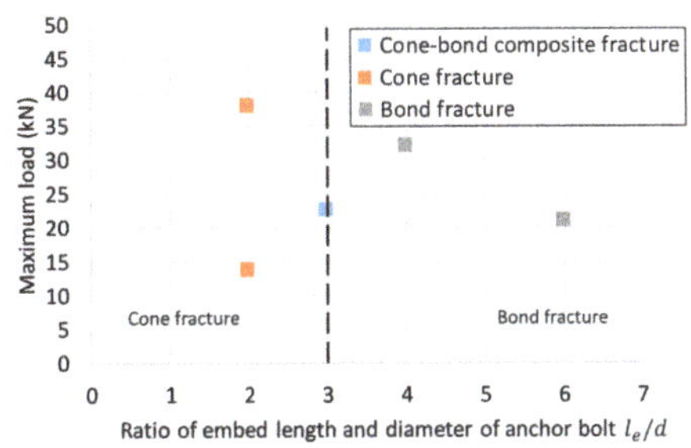

Figure 18. Classification of the pull-out fracture pattern of an anchor bolt.

4.3. Influence of the Spacing between Anchor Bolts

Then, the influence of spacing between anchor bolts to the load capacity of the anchor bolt is investigated. The analysis model is shown in Figure 19, with different spacing between anchor bolts *s*, from 234 mm to 25 mm. The sides of the upper surface of the concrete block are fixed, and the material parameters are the same with previous analysis. The development process of the cracks during the pull-out process is shown in Figures 20–22.

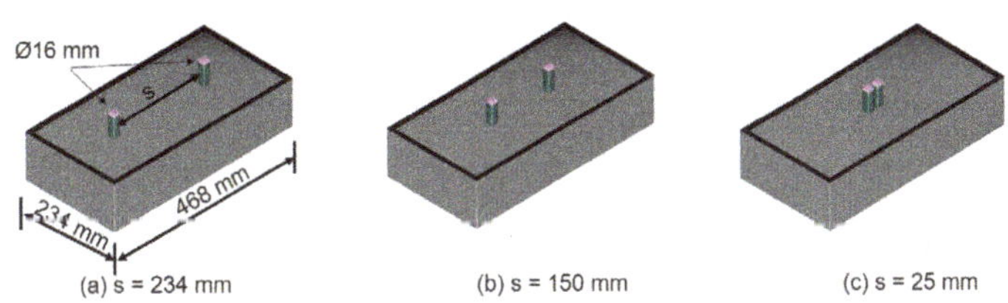

Figure 19. Analysis model with different spacing.

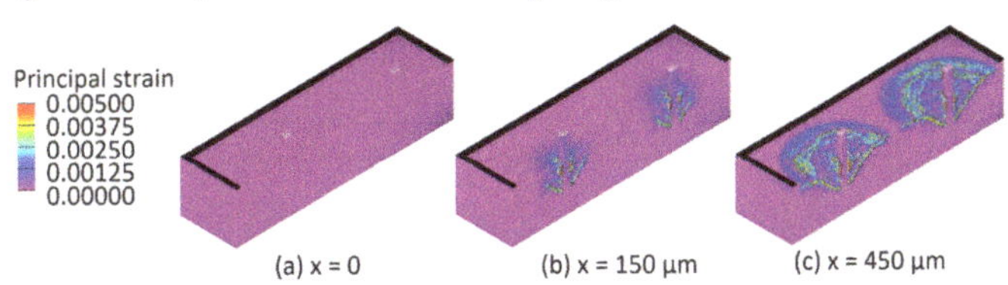

Figure 20. Crack development when the spacing *s* is 234 mm.

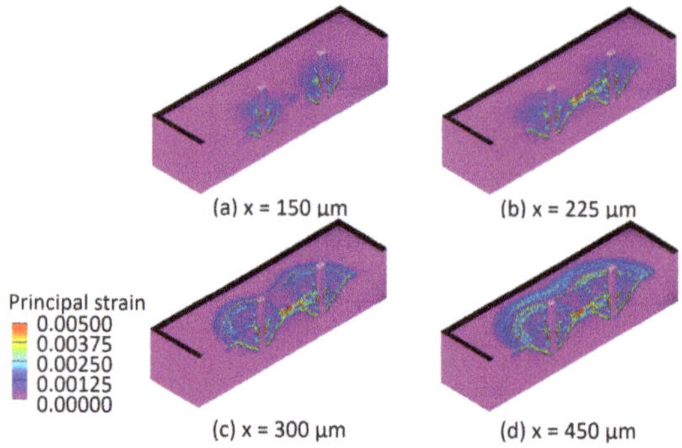

Figure 21. Crack development when the spacing s is 150 mm.

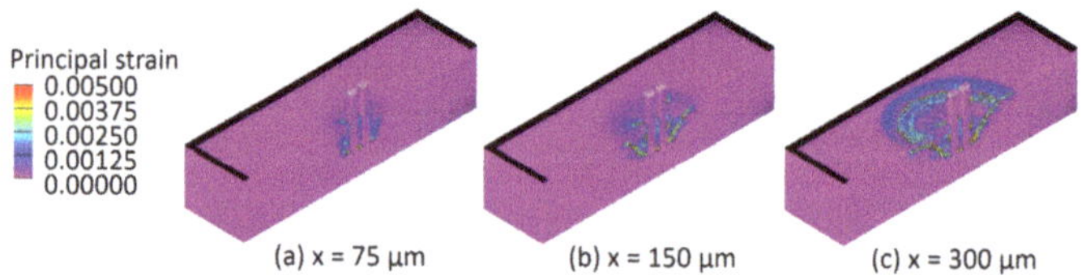

Figure 22. Crack development when the spacing s is 25 mm.

In the case where the spacing is 234 mm, the development process is similar to the case with a single anchor bolt. Inclined cracks generate from the bottom of the anchor bolt and, finally, reach the surface of the concrete block. Two separate cone areas are formed. When the spacing s is 150 mm, the inclined cracks also generate at the bottom of the anchor bolt. When the pull-out displacement $x = 225$ µm, the cracks located between the two anchor bolts joint with each other, forming a long crack connecting the two anchor bolts. When $x = 300$ µm and $x = 450$ µm, the cracks reach the surface of the concrete, and a complicated double-cone fracture surface is obtained. When the spacing s is 25 mm, cracks generate from the bottom of the anchor bolt. The cracks outside the anchor bolts are long and inclined, and the cracks between the anchor bolts are short and horizontal. When $x = 225$ µm and $x = 300$ µm, the inclined cracks reach the surface of the concrete block. The shape and the area of the emerged cone shape becomes more like the cone shape of a single anchor bolt.

The load-displacement curves are shown in Figure 23. In the case where the spacing s is 150 mm, the total area of the emerged cone shape changed little compared to the two separate cone shapes, and the maximum load remains same for spacing of 200 mm and 150 mm. When the spacing s is 25 mm, the two anchor bolts just behave as a single anchor bolt, and the maximum load is about half of the previous cases. To make the tendency more clear, relationship between the ratio between the total maximum load and the maximum load of a single anchor bolt P_{total}/P_{single} and the spacing s is shown in Figure 24a, and the relationship between the ratio between the total maximum load and the maximum load of a single anchor bolt P_{total}/P_{single} and the ratio between spacing and embedment depth s/l_e is shown in Figure 24b. In Figure 24a, it is found that the maximum load remains unchanged when the spacing s is larger than a certain value in different case, which is also described in Reference [28]. In Figure 24b, it is clear that the load decrease is not noticeable until s/l_e reaches 2.0. Compared to existing studies that considered the effect of anchor bolt diameter on sufficient embedment depth [28], this study investigated the effect on pull strength of adjacent bolt spacing at relatively shallow embedding depths. As a result, it is confirmed that an interval of 2.0 times of the embedment depth is needed as a minimum distance of adjacent bolt spacing to keep the pull-out strength. This result indicates that a minimum spacing of 2.0 times of the embedment depth should be ensured for anchor bolts to provide sufficient resistance against pull-out load.

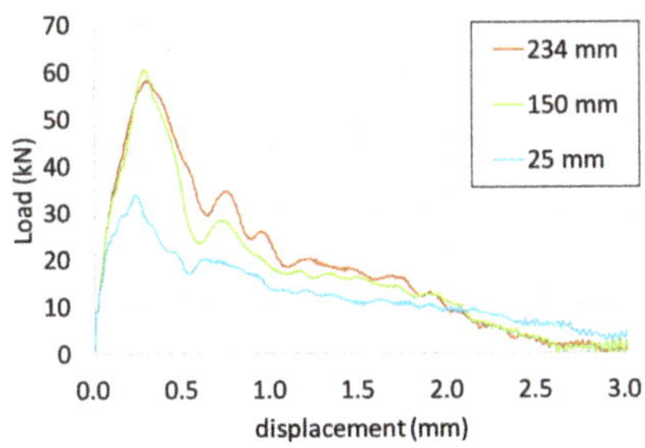

Figure 23. Load-displacement relationships with different spacing.

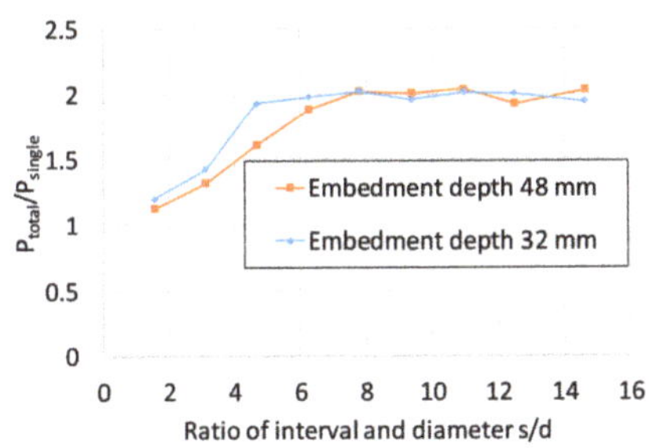

(**a**)

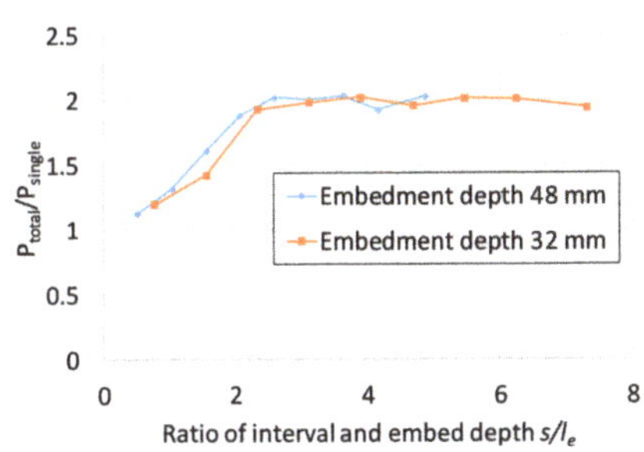

(**b**)

Figure 24. Pull-out strength influenced by input parameters. (**a**) Effect of bolt spacing on pull-out strength (compared to the strength of a single bolt). (**b**) Effect of embedded depth on pull-out strength (compared to the strength of a single bolt).

4.4. Influence of the the Distance from Free Edge

The influence of distance between the edge of the concrete block and the anchor bolt is investigated. The analysis model is shown in Figure 25. The embedment depth l_e is 48 mm, and distance s between the edge of the concrete and the anchor bolts varies from 100 mm to 17 mm. As a boundary condition, the vertical displacement of the black color area (along 3 sides) in the upper surface are fixed, and the other side is not fixed. Figure 26 show the principal strain distribution in half the region of the analysis model in consideration of mechanical symmetry under the 0.45 mm pull-out displacement level.

From the final crack pattern, it was found that, when there is enough distance from the edge of concrete block, a cone-shaped crack pattern can be observed. However, the cone-shaped crack becomes incomplete with the decrease of distance s. In particular, the crack at the bottom of the anchor bolt extended to the lateral surface of the concrete without changing direction in the case of $s/l_e = 0.4$, while, in other cases, the cracks turned. From the load-displacement relationship shown in Figure 27 and the relationship between the maximum load and the ratio of distance to free edge and embedment depth s/l_e shown in Figure 28, it can be recognized that the maximum pull-out strength decreased when s/l_e is lower than 1.0, and the maximum pull-out strength in case of $s/l_e = 0.4$ decreased around 40% compared with other cases. As a reference, about 50% of the maximum load drop can be found in the experiment [8,16]. Compared to the existing studies [8] that showed the influence of the anchor bolt diameter under sufficient embedment depth, we investigated the minimum distance from the free end to keep the pull-out strength of the anchor bolt, and it has been found that maintaining proper pull-out strength requires a distance from the free end that is greater than 1.0 times the anchor bolt embedding depth. This is also shown by the tendency of the experiments conducted in Reference [16]. Thus, it indicates that $s/l_e = 1.0$ should be guaranteed when installing anchor bolts at the edges of the concrete structures. In addition, compared to the analysis results in the previous section, it can be found that the analysis model with insufficient distance between the anchor bolt and the free edge is similar to a half model of cases with insufficient intervals between anchor bolts considering symmetricity. When sufficient valid base concrete around a single anchor bolt is not guaranteed, the pull-out performance of this single anchor bolt is limited. When installing anchor bolts under these conditions, it is necessary to avoid densely placed rebar areas to ensure sufficient embedding depth for the bolts and the surrounding concrete area. In the next section, we will consider an example of countermeasures when a sufficient concrete area cannot be secured around the anchor bolt.

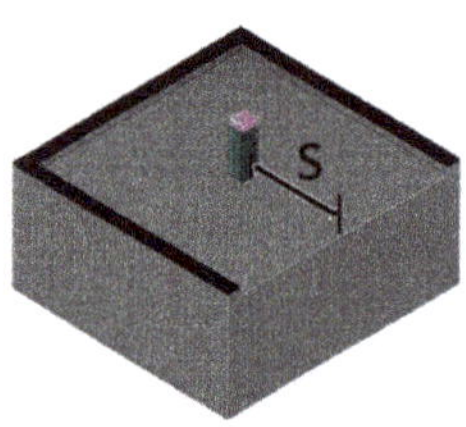

Figure 25. Analysis model with one side free.

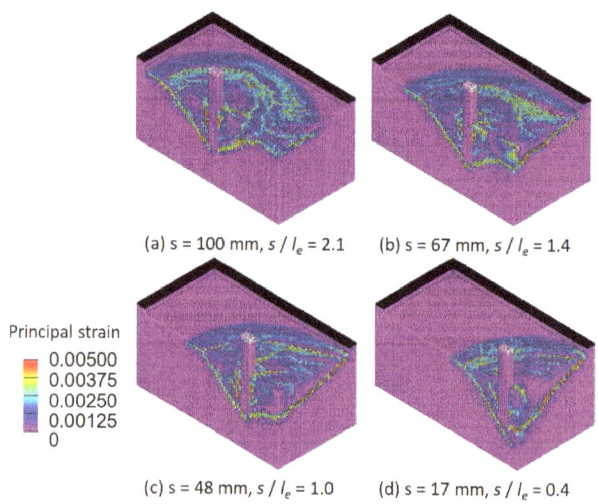

Figure 26. Crack patterns with different distance between the free edge and the anchor bolt.

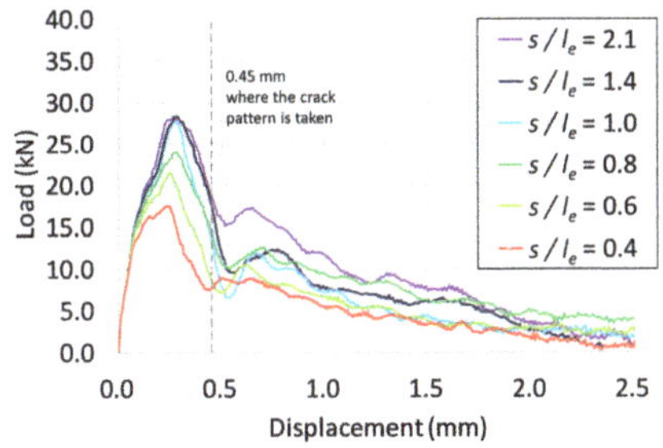

Figure 27. Load-displacement relationships with different distance to free edge.

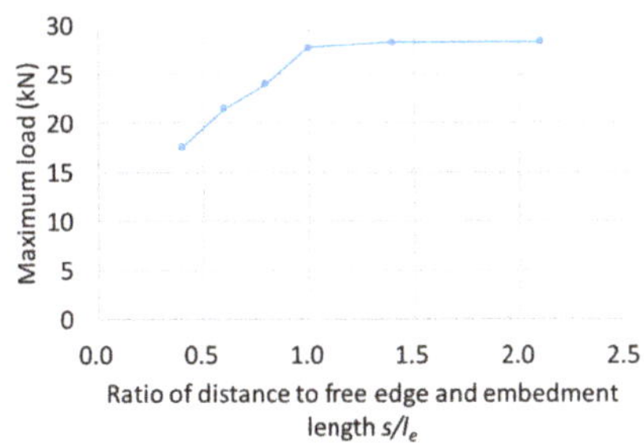

Figure 28. Relationship between maximum load and ratio of distance free edge and embedment depth s/l_e.

4.5. Measures against Insufficient Distance from the Free End with PCM

As clarified in the previous section, if the anchor bolt installation position is close to the free end, the pull-out strength will decrease. We investigated a method that does not reduce the pull-out strength as much as possible, even when anchor bolts are unavoidably attached near the free end by using PCM (Polymer Cement Mortar).

PCM is a simple method of spraying reinforced mortar onto the existing concrete of an aged RC structure, and it has already been used in many ways [23,24]. In this study, we calculated how much the anchor bolt pull-out strength can be prevented from decreasing by PCM method when the anchor installation position is close to the free end.

The analysis model of this simulation is shown in Figure 29, where a PCM reinforcement layer with 3-mm or 6-mm thickness is added to the existing concrete surface. In Figure 29a, the PCM area is smaller, but it is able to cover the projection of the anchor bolt on the lateral surface of the concrete block, while, in Figure 29b, the PCM area is sufficient to cover sufficient the corn failure domain. The material parameters of PCM is shown in Table 6. The boundary conditions are the same as the previous section, where the three sides of the upper surface displayed in black in the figure are fixed.

Figure 30 shows the crack pattern obtained in SPH analysis, and Figure 31 shows the load-displacement relationship. It is recognized from Figure 30a,b that, since the strength of the PCM material is higher than that of existing concrete, the cracks can be seen growing under the PCM area. Thus, this change in the crack growth path causes a slight increase in the pull-out strength of the anchor bolt.

On the other hand, in the case of Figure 30c,d, when the PCM reinforcement area is wide, a crack occurred in the PCM layer, which has a high tensile strength than that of existing concrete, and a clear effect was seen, such as an improvement in pull-out strength of about 40%. To summarize these results, it is possible to improve the pull-out strength of anchor bolts with insufficient distance from free end by using the PCM method. To achieve a better performance of the PCM reinforcement layer, the reinforced area by PCM should be large enough to cover the assumed cone failure domain of the concrete.

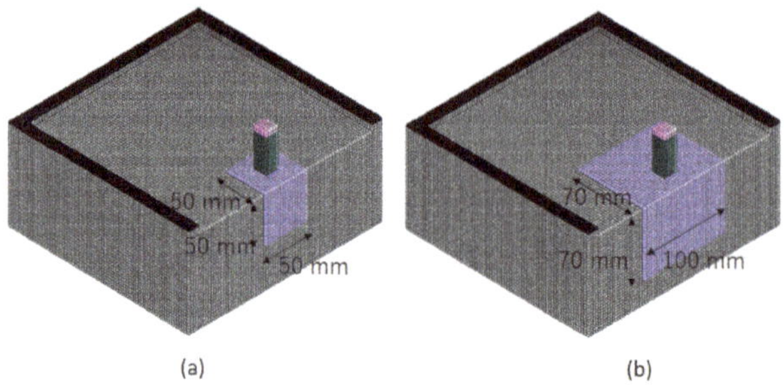

Figure 29. Analysis model with PCM reinforcement layer. (**a**) Model with PCM layer of small area. (**b**) Model with PCM layer of large area.

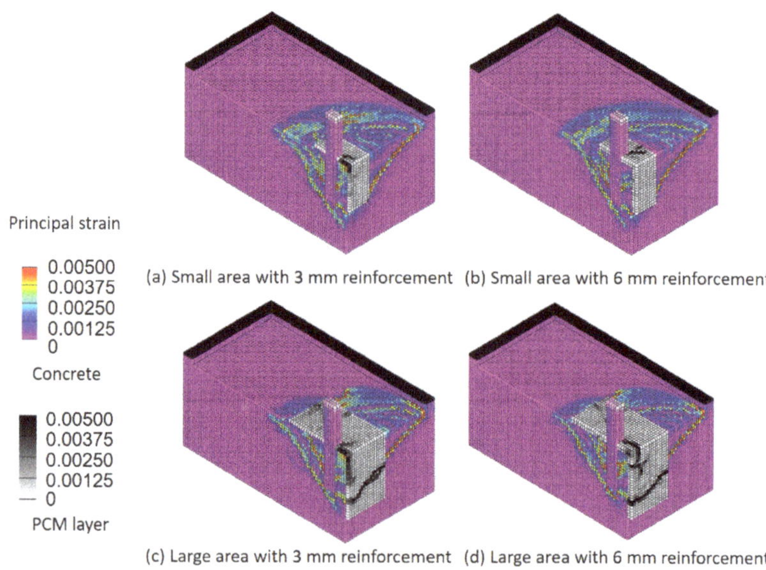

Figure 30. Crack patterns with PCM reinforcement layer.

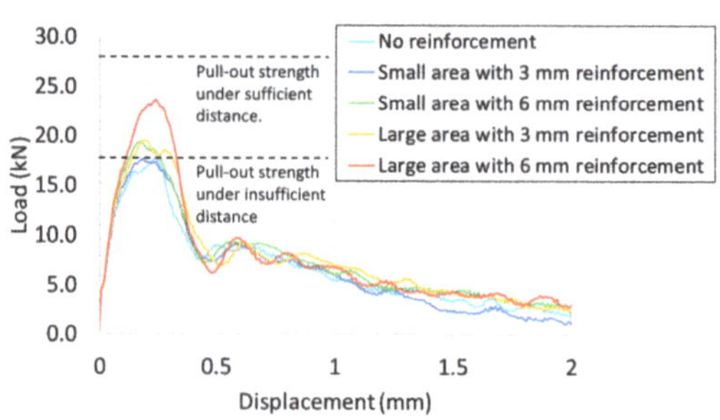

Figure 31. Load-displacement relationships with PCM reinforcement layer.

Table 6. Material parameter of PCM.

Compressive Strength (MPa)	Tensile Strength (MPa)	Poisson's Ratio	Density (kg/m^3)	Young's Modulus (MPa)
60.0	6.0	0.2	2300	27,000

5. Conclusions

In this study, an analysis method that can accurately evaluate the pull-out strength of anchor bolts embedded in concrete using the SPH method is presented and considered the ultimate pull-out strength under various conditions (bolt diameter, embedded depth, adjacent bolt spacing, etc.). The results obtained in this study are summarized as follows.

1. As a result of comparison with the existing experimental values, it was confirmed that the crack growth analysis using the SPH method proposed in this study was very effective on the accurate estimation of ultimate pull-out strength of anchor bolt.
2. When changing the parameters of the analysis cases, concrete base fractures are likely to occur with lower embedding depth and higher bond stress, and bond fractures are likely to occur with deeper embedding depth and lower bond stress.
3. It was found that, if the distance between two adjacent bolts was shortened under the installation conditions where cone fracture occurred, the cone fracture regions of the individual bolts overlapped, and the total pull-out strength of the two bolts decreased. According to the analysis results, if the distance between two adjacent bolts was smaller than embedded depth, the total pull-out strength decreases around 50% compared to the sufficient distance condition.
4. It was also recognized that, when the distance from the edge was larger than the embedment depth, the pull-out strength could maintain the certain level. However, if the distance from the edge becomes smaller than embedded depth, the pull-out strength decreases with the distance.
5. This research simulation shows that, when the distance from the anchor bolt to the edge is about half the embedding depth, the pull-out strength drops to about 50% of the original strength. However, even in such a case, it was confirmed from the analysis results that it is possible to prevent a decrease in the pull-out strength of the anchor bolt by using a PCM material.

In this study, it was shown that the pull-out strength of anchor bolts embedded in concrete could be accurately evaluated by the proposed SPH method. By using this method, it is possible not only to predict the pull-out strength and fracture mode under various installing conditions of anchor bolts in the structure but also to quantify the effect of the reinforcement method, such as the PCM method.

Author Contributions: Conceptualization, Y.S.; methodology, C.L. and Y.S.; software, C.L. and Y.S.; validation, C.L.; formal analysis, C.L.; investigation, C.L. and Y.S.; resources, C.L. and Y.S.; data curation, C.L.; writing original draft preparation, C.L.; writing review and editing, Y.S.; visualization, C.L.; supervision, Y.S.; project administration, Y.S. Both authors have read and agreed to the published version of the manuscript.

Funding: This research received no external funding.

Institutional Review Board Statement: Not applicable.

Informed Consent Statement: Not applicable.

Data Availability Statement: The data that support the findings of this study are available from the corresponding author, upon reasonable request.

Conflicts of Interest: The authors declare no conflict of interest.

References

1. Xiang, N.; Alam, M.S. Comparative seismic fragility assessment of an existing isolated continuous bridge retrofitted with different energy dissipation devices. *J. Bridge Eng.* **2019**, *24*, 04019070. [CrossRef]
2. Yu, P.; Manalo, A.; Ferdous, W.; Salih, C.; Abousnina, R.; Heyer, T.; Schubel, P. Failure analysis and the effect of material properties on the screw pull-out behaviour of polymer composite sleeper materials. *Eng. Fail. Anal.* **2021**, *128*, 105577. [CrossRef]
3. Sharda, A.; Manalo, A.; Ferdous, W.; Bai, Y.; Nicol, L.; Mohammed, A.; Benmokrane, B. Axial compression behaviour of all-composite modular wall system. *Compos. Struct.* **2021**, *268*, 113986. [CrossRef]
4. Yuan, C.; Fan, L.; Cui, J.F.; Wang, W.J. Numerical simulation of the supporting effect of anchor rods on layered and nonlayered roof rocks. *Adv. Civ. Eng.* **2020**, *2020*, 4841658. [CrossRef]
5. Hashimoto, J.; Takiguchi, K. Experimental study on pullout strength of anchor bolt with an embedment depth of 30 mm in concrete under high temperature. *Nucl. Eng. Des.* **2004**, *229*, 151–163. [CrossRef]
6. Munemoto, S.; Sonoda, Y. Experimental analysis of anchor bolt in concrete under the pull-out loading. *Procedia Eng.* **2017**, *171*, 926–933.
7. Richardson, A.; Dawson, S.; Campbell, L.; Moore, G.; Mc Kenzie, C. Temperature related pull-out performance of chemical anchor bolts in fibre concrete. *Constr. Build. Mater.* **2019**, *196*, 478–484. [CrossRef]
8. Yilmaz, S.; Özen, M.A.; Yardim, Y. Tensile behavior of post-installed chemical anchors embedded to low strength concrete. *Constr. Build. Mater.* **2013**, *47*, 861–866. [CrossRef]
9. Epackachi, S.; Esmaili, O.; Mirghaderi, S.R.; Behbahani, A.A.T. Behavior of adhesive bonded anchors under tension and shear loads. *J. Constr. Steel Res.* **2015**, *114*, 269–280. [CrossRef]
10. Nilforoush, R.; Nilsson, M.; Elfgren, L. Experimental evaluation of tensile behaviour of single cast-in-place anchor bolts in plain and steel fibre-reinforced normal-and high-strength concrete. *Eng. Struct.* **2017**, *147*, 195–206. [CrossRef]
11. Eligehausen, R.; Bouska, P.; Cervenka, V.; Pukl, R. Size effect of the concrete cone failure load of anchor bolts. In *Fracture Mechanics of Concrete Structures*; Bažant, Z.P., Ed.; Elsevier Science: London, UK, 1992; pp. 517–525.
12. Bajer, M.; Barnat, J. The glue—Concrete interface of bonded anchors. *Constr. Build. Mater.* **2012**, *34*, 267–274. [CrossRef]
13. Liu, Q.; Chai, J.; Chen, S.; Zhang, D.; Yuan, Q.; Wang, S. Monitoring and correction of the stress in an anchor bolt based on Pulse Pre-Pumped Brillouin Optical Time Domain Analysis. *Energy Sci. Eng.* **2020**, *8*, 2011–2023. [CrossRef]
14. Delhomme, F.; Debicki, G.; Chaib, Z. Experimental behaviour of anchor bolts under pullout and relaxation tests. *Constr. Build. Mater.* **2010**, *24*, 266–274. [CrossRef]
15. Satoh, A.; Takeda, K.; Murakami, K. FEM analysis on combined bond-cone fracture of a post-installed adhesive anchor filled with UHPFRC. *Theor. Appl. Fract. Mech.* **2019**, *100*, 46–54. [CrossRef]
16. Obata, M.; Inoue, M.; Goto, Y. The failure mechanism and the pull-out strength of a bond-type anchor near a free edge. *Mech. Mater.* **1998**, *28*, 113–122. [CrossRef]
17. Soparat, P.; Nanakorn, P. Analysis of anchor bolt pullout in concrete by the element-free Galerkin method. *Eng. Struct.* **2008**, *30*, 3574–3586. [CrossRef]
18. Lu, J.; Zhang, Y.; Muhammad, H.; Chen, Z.; Xiao, Y.; Ye, B. 3D analysis of anchor bolt pullout in concrete materials using the non-ordinary state-based peridynamics. *Eng. Fract. Mech.* **2019**, *207*, 68–85. [CrossRef]
19. Saleem, M. Assessing the load carrying capacity of concrete anchor bolts using non-destructive tests and artificial multilayer neural network. *J. Build. Eng.* **2020**, *30*, 101260. [CrossRef]
20. Lucy, L.B. A numerical approach to the testing of the fission hypothesis. *Astron. J.* **1977**, *82*, 1013–1024. [CrossRef]
21. Libersky, L.D.; Petschek, A.G. Smooth particle hydrodynamics with strength of materials. In *Advances in the Free-Lagrange Method Including Contributions on Adaptive Gridding and the Smooth Particle Hydrodynamics Method*; Springer: Berlin/Heidelberg, Germany, 1991; pp. 248–257.
22. Libersky, L.D.; Petschek, A.G.; Carney, T.C.; Hipp, J.R.; Allahdadi, F.A. High strain Lagrangian hydrodynamics: A three-dimensional SPH code for dynamic material response. *J. Comput. Phys.* **1993**, *109*, 67–75. [CrossRef]
23. Monier, A.; Zhe, X.; Huang, H.; Zhishen, W. External flexural strengthening of rc beams using BFRP grids and PCM. *J. Jpn. Soc. Civ. Eng. Ser. A2 (Appl. Mech. (AM))* **2017**, *73*, I_417–I_427. [CrossRef]
24. Higashi, Y.; Li, B.; Jiang, Y. Reinforcement effect of PCM shotcrete method using FRP grid for tunnel maintenance. In Proceedings of the 2012 International Conference on Renewable Energy Research and Applications (ICRERA), Nagasaki, Japan, 11–14 November 2012; IEEE: Piscataway, NJ, USA, 2012; pp. 1–5.
25. Liu, G.R.; Liu, M.B. *Smoothed Particle Hydrodynamics: A Meshfree Particle Method*; World Scientific: Singapore, 2003.
26. JSCE (Japan Society of Civil Engineers). *Standard Specifications for Concrete Structures: Design*; JSCE: Tokyo, Japan, 2012; pp. 37–38.
27. Popovics, S. A numerical approach to the complete stress-strain curve of concrete. *Cem. Concr. Res.* **1973**, *3*, 583–599. [CrossRef]
28. Eligehausen, R.; Cook, R.A.; Appl, J. Behavior and design of adhesive bonded anchors. *ACI Struct. J.* **2006**, *103*, 822.

applied sciences

MDPI

Article

Modal-Based Ground Motion Selection Method for the Nonlinear Response Time History Analysis of Reinforced Concrete Shear Wall Structures

Yang Liu [1,2,3]

[1] Shock and Vibration of Engineering Materials and Structures Key Laboratory of Sichuan Province, Mianyang 621010, China; yliubp@hqu.edu.cn
[2] Key Laboratory of Structural Engineering and Disaster Prevention of Fujian Province, Xiamen 361021, China
[3] College of Civil Engineering, Huaqiao University, Xiamen 361021, China

Abstract: This paper presents a modification of the modal-based ground motion selection (MGMS) method for improving the reliability of the nonlinear response time history analysis (NLRHA) of reinforced concrete (RC) shear wall structures. The original MGMS procedure quantified the impact of frequency content combinations in the time domain (FCCTD) of input ground motions (IGMs) on the seismic response of building structures using the level of interaction of the first three modes induced by IGMs. However, previous research found that the first two modes have far larger modal mass coefficients than those of higher modes and dominate the vibration of the RC shear wall structures with a symmetric plan. Therefore, the MGMS procedure should be modified by employing the interaction of the first two modes induced by IGMs to properly account for the effect of the FCCTD of IGMs on the seismic response of structures. In the MGMS procedure for RC shear wall structures, seven IGMs that caused the most significant interactions of the first two modes were selected from a suite of twenty seed IGMs, which were chosen with a conventional spectra-matching-based IGMs selection procedure for the NLRHA of the structure. A comprehensive case study involving three RC shear walls with different heights was conducted to investigate the capability of the MGMS in selecting suitable IGMs for the NLRHA of RC shear wall structures. Sets of seed IGMs were selected, adopting conditional mean spectra and design spectra as the target spectra. It was found that the seismic demands computed using MGMS selected IGMs can ensure a more reliable and reasonable computation of seismic demands compared with conventional spectra-matching-based IGMs selection methods.

Keywords: nonlinear response time history analysis; reinforced concrete shear wall structure; modal-based ground motion selection; frequency contents combination in the time domain; dominated mode

Citation: Liu, Y. Modal-Based Ground Motion Selection Method for the Nonlinear Response Time History Analysis of Reinforced Concrete Shear Wall Structures. *Appl. Sci.* **2021**, *11*, 8230. https://doi.org/10.3390/app11178230

Academic Editor: Maria Favvata

Received: 11 August 2021
Accepted: 31 August 2021
Published: 5 September 2021

Publisher's Note: MDPI stays neutral with regard to jurisdictional claims in published maps and institutional affiliations.

Copyright: © 2021 by the author. Licensee MDPI, Basel, Switzerland. This article is an open access article distributed under the terms and conditions of the Creative Commons Attribution (CC BY) license (https://creativecommons.org/licenses/by/4.0/).

1. Introduction

Tall buildings in moderate and high seismicity regions extensively adopt reinforced concrete (RC) shear wall structures to resist lateral loads because of their remarkable seismic performance. Owing to the increased seismic risk, accurate computation of the seismic demand is crucial to the seismic design of RC shear wall structures. The most precise method in computing seismic demands of building structures is the nonlinear response time history analysis (NLRHA) method. Still, the reliability and reasonability of the NLRHA are limited by the representative input ground motions (IGMs) because of the extreme uncertainty and contingency of earthquake events [1–5]. The available IGM selection procedures [6–9] firstly evaluate local geological features and basic properties of the IGMs [10–15] to ensure these features of IGMs fulfil the requirement of the site. Then, the selection procedures adopt various spectra matching techniques to modify IGMs to provide the response spectra IGMs that match closely with the target spectrum [16].

In the past two decades, plenty of research efforts have been made in developing spectra-matching-based IGM selection procedures for selecting and scaling proper IGMs for the NLRHA of building structures [17–28]. However, the structural seismic demand is affected by the peak modal responses and the modal response combinations in the time domain, which is determined by the frequency contents combinations in the frequency domain (FCCFD) and time-domain (FCCTD) of IGMs, respectively [29]. For high-rise buildings, the peak modal responses and modal response combinations in the time domain are of the same importance to the seismic demand of the building structure. The response spectrum of IGMs quantifies the effect of FCCFD of IGMs. Thus, most of the available IGMs selection procedures that employ spectra matching techniques can only consider the impact of FCCFD of IGMs. The research of Wang [20] and Anajafi [24] showed that even if the individual records are tightly matched to the target spectrum, the record to record variability in the inelastic structural responses might be still noticeable. Moreover, Anajafi and Medina [24] also illustrated that the record-to-record variability in higher-mode dominated responses is very sensitive to the value of the target damping. These findings highlighted the shortcoming of the spectral matching technique.

Recently, a modal-based ground motion selection (MGMS) procedure was developed by Liu et al. [30] to consider the effect of FCCTD of IGMs in IGM selection. In the MGMS procedure, the impact of FCCTD of IGMs on the seismic demand of structures is quantified with the interaction of the first three modes. The MGMS procedure's capability to select more reasonable IGMs for the NLRHA of structures has been proved for frame structures. At the same time, the deflection profile of the wall and frame structures is flexural and shear profiles, respectively, when the structures are subjected to the action of lateral load. Different deflection profiles of the wall and frame structures indicate different structural behavior [31,32]. The contribution of the seismic response of dominated modes to the final seismic demand shall also be different. Therefore, the capability of the MGMS procedure in improving the reliability of the NLHRA of RC shear wall structures is not guaranteed.

In this paper, the MGMS procedure is modified to take the modal response characteristics of RC shear wall structures into account in selecting IGMs for the NLRHA of the structure. In the MGMS procedure for wall structures, selected IGMs can induce the most significant interaction of the first two dominated modes, of which the modal mass coefficients are larger than the sum of modal mass coefficients of all the higher-order modes. A comprehensive case study was conducted on three RC shear wall structures with different heights. Seed IGMs were selected and scaled, with target spectra being conditional mean spectra [33] and ASCE 7 design spectra [6] to investigate the reliability and reasonability of the MGMS procedure when the target spectra vary. The results show that the MGMS procedure can effectively and consistently select suitable IGMs for the NLRHA of RC shear wall structures.

This paper has six sections, and the first section presents the background introduction, research significance and structure of the paper. Section 2 first provides the basic information on the MGMS procedure and discusses why modifications should be made to the selection procedure to ensure a more reasonable selection of IGMs for the NLRHA of RC shear wall structures. Then, Section 2 presents details about the proposed MGMS procedure for RC shear structures. Section 3 shows detailed information on a comprehensive case study, including structural prototypes, seed IGMs and IGM selection for the NLRHA of RC shear wall structures. Section 4 presents the NLRHA results adopting IGMs of different selection methods and discussions. Section 5 summarizes the findings of the paper and draws conclusions based on the findings. Appendix A provides details of the RC shear wall structures and seed IGMs.

2. Modal-Based Ground Motion Selection Procedure for RC Shear Wall Structures

2.1. Original Modal-Based Ground Motion Selection Procedure

The structural dynamics theory is firstly adopted to demonstrate the effect of IGMs on the peak modal responses and modal response combinations in the time domain. The equation of motion of a multi-degree-of-freedom system under the IGM is expressed as [29]

$$\boldsymbol{m}\ddot{\boldsymbol{u}} + \boldsymbol{c}\dot{\boldsymbol{u}} + \boldsymbol{k}\boldsymbol{u} = -\boldsymbol{m}\boldsymbol{i}\ddot{u}_g(t) \tag{1}$$

where $\boldsymbol{u}$ is story displacements vector, $\boldsymbol{k}$, $\boldsymbol{c}$ and $\boldsymbol{m}$ stand for the stiffness, damping and mass matrices, respectively, and $\boldsymbol{i}$ is the influence vector. $\ddot{u}_g(t)$ is the acceleration history of the input ground motion. As the loading history does not affect the independence of the lateral force, the solution of Equation (1) can be expressed as:

$$u(t) = \sum_{n=1}^{N} u_n(t) = \sum_{n=1}^{N} \Gamma_n \boldsymbol{\phi}_n D_n(t) = \sum_{n=1}^{N} \Gamma_n \boldsymbol{\phi}_n D_n a_n(t) \tag{2}$$

where

$$\Gamma_n = \frac{L_n}{M_n},\ L_n = \boldsymbol{\phi}_n^T \boldsymbol{m}\boldsymbol{i} \text{ and } M_n = \boldsymbol{\phi}_n^T \boldsymbol{m} \boldsymbol{\phi}_n \tag{3}$$

and $u_n(t)$ is the displacement history of nth mode; $\boldsymbol{\phi}_n$ is the mode shape of nth vibration mode, and N is the total number of mode; Γ_n is the modal participating factor; M_n is the generalised mass of mode n; L_n is the modal excitation factor of mode n; $D_n(t)$ is the top displacement time history the ESDOF of nth mode; D_n is the peak displacement of the ESDOF of nth mode; $a_n(t) = \frac{D_n(t)}{D_n}$ is the normalized top displacement time history of nth mode.

It is clear from Equation (2) that the seismic demand of tall buildings is determined by both peak modal response D_n, and modal response combination in the time domain $a_n(t)$. The D_n and $a_n(t)$ are of the same importance when the higher-mode effect is significant. Spectrum matching-based IGM selecting and scaling procedures only ensure that the value of D_n fulfils specific requirements but ignores the impact of $a_n(t)$ on the structural seismic demand, resulting in significant variations in the computed seismic demand [20,21,24].

Driven by the need to take the effect of FCCTD of IGMs on the structural seismic demand into account in the selection of IGMs and improve the reliability of NLRHA, the modal-based ground motion selection (MGMS) procedure [30] was developed. The MGMS procedure selects seven IGMs that lead to the most severe interaction of the first three modes from a group of twenty seed IGMs, primarily obtained through the common two-step IGMs selection method. A previous study has shown that with reasonable consideration of the effect of FCCTD of IGMs, the MGMS procedure can select more reasonable IGMs for the NLRHA of frame structures.

2.2. Proposed Modal-Based Ground Motion Selection Procedure for RC Shear Wall Structures

The modal mass coefficient α_n, which can be calculated using Equation (4), is the primary factor that affects the number of modes considered in the MGMS procedure.

$$\alpha_n = \frac{m_{eff}}{\sum_{i=1}^{N} m_j} = \frac{1}{\sum_{i=1}^{N} m_j} \frac{L_n^2}{M_n} \tag{4}$$

For most structures with symmetric plane arrangements, the sum of modal mass coefficients of the first three modes is higher than 90% [29]. Thus, the MGMS included the first three modes to account for the effect of IGMs on the multi-mode interaction. At the same time, it is found that for RC shear wall structures with a symmetric plan, the modal mass coefficients of the first two modes are larger than the sum of the modal mass coefficients of the higher-order modes [31,34]. At the same time, the modal mass coefficients of other higher modes are smaller than the sum of the modal mass coefficients of the higher-order modes. The relations of the modal mass coefficients of RC shear wall structures with a symmetric plan can be expressed as:

$$\alpha_i \geq \sum_{i+1}^{N} \alpha_n \text{ for } i = 1,\ 2 \quad (5)$$

$$\alpha_i < \sum_{i+1}^{N} \alpha_n \text{ for } i = 3,\ 4, \ldots,\ N-2 \quad (6)$$

The large modal mass coefficient values of the first two modes show the overwhelming domination of the first two modes in the vibration of the RC shear wall structures. Thus, considering the effect of IGMs on the interaction of the first two dominated modes is more reasonable in selecting IGMs for the NLRHA of RC shear walls with a symmetric plan. Otherwise, the higher-order mode effect will be exaggerated in the IGM selection, and the reliability of the NLRHA of shear wall structures will be affected.

Accounting for the domination of the first two modes in the vibration of the RC shear wall structures, the MGMS procedure was modified by considering the first two modes' interaction in the IGMs selection. After selecting twenty seed IGMs through a spectrum-matching procedure, the NL-ESDOFs of the first two modes were developed through the modal pushover analysis procedure:

(1). Apply lateral force whose distribution is $m\boldsymbol{\phi}_n$, where $n = 1, 2$, to the structure separately, and generate two sets of curves of base shear-roof displacement ($V_{bn} - u_{rn}$). Convert the ($V_{bn} - u_{rn}$) curves to bilinear curves.

(2). Transfer the ($V_{bn} - u_{rn}$) bilinear curves to the ($F_{sn}/L_n - D_n$) pushover curves of the NL-ESDOFs using Equation (7):

$$\frac{F_{sn}}{L_n} = \frac{V_{bn}}{M_n^*} \text{ and } D_n = \frac{u_{rn}}{\Gamma_n \phi_{rn}} \quad (7)$$

(3). The vibration period of NL-ESDOFs can be calculated as:

$$T_n^* = 2\pi\sqrt{\frac{L_n D_{ny}}{F_{sny}}} \text{ and } D_{ny} = \frac{u_{rny}}{\Gamma_n \phi_{rn}} \quad (8)$$

(4). Develop the NL-ESDOFs model using the base shear-top displacement relations of ($F_{sn}/L_n - D_n$) curves and T_n^* as the vibration period of NL-ESDOFs.

After obtaining the model of the NL-ESDOFs, NLRHA of NL-ESDOFs are conducted with twenty seed IGMs to compute the roof displacement time histories and the peak absolute roof displacement of the NL-ESDOFs $u_{jn}(t_k)$ and $\hat{u}_{jn} = \max_{t_k \in T_0} |u_{jn}(t_k)|$. Where $n = 1, 2$, is the mode ID; $j = 1, 2, \ldots, 20$, is the ID of the seed IGMs; t_k is the time point within the duration of the IGM j, T_{j0}. The level of multi-mode interaction caused by IGM j was then quantified by the combination factor β_j, which is the largest value of the factor β_{jk} that establishes the following inequalities simultaneously for all $t_k \in T_{j0}$:

$$\left(|u_{j1}(t_k)| \geq |\Gamma_1| \beta_{jk} \hat{u}_{j1}\right) \quad (9)$$

$$\left(|u_{j2}(t_k)| \geq \beta_{jk} \hat{u}_{j2}\right) \quad (10)$$

The β_j shall range from 0 to $1/|\Gamma_1|$. If β_j equals to 0, it means that under the action of motion j, at least one dominated mode does not contribute to the seismic response of the wall structures for all $t_k \in T_{j0}$. On the other hand, if β_j equals $1/|\Gamma_1|$, the first mode reaches the largest roof displacement, and the second mode has roof displacement just below the largest value simultaneously for at least one time point within the duration of IGM j. These two cases are sporadic when the wall structures are subjected to natural earthquakes. The MGMS procedure then selects seven IGMs with the largest value of β_j to carry out NLRHA of the full RC wall structure model. The average demands of NLRHA adopting the 7 MGMS-selected IGMs are the seismic demand of the RC wall structure. The procedure of selecting IGMs with the MGMS procedure for RC shear wall structures with a symmetric building plan is shown in Figure 1.

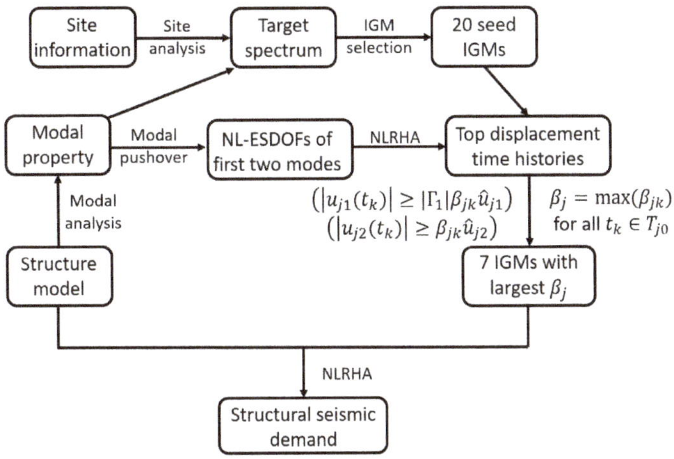

Figure 1. Flowchart of MGMS procedure of wall structures.

3. Case Study

3.1. Structure Prototype

In the case study, three RC shear walls, namely W1, W2 and W3, with story numbers of 8, 20 and 30, respectively, were studied. These three wall structures were all modified from the 8-story RC shear walls in the FEMA-440 [34]. Detailed information of design procedure of the RC shear walls can be found in [34,35]. The elevation and the reinforcement details of the three RC shear wall structure models can be found in Appendix A. The assumed gravity load for each story was 1468 kN for the 8-story wall structure and 2348.48 kN for the 20-story and 30-story shear wall structures.

The SAP 2000 Software nonlinear version [36] was employed to model the RC shear wall structures. The direct integration method adopting Wilson-θ algorithm was used to conduct NLRHA of wall structures, and the value of θ was set to be 1.4 to ensure the convergence of the integration algorithm. Grade 60 steel rebar was adopted for the W1 and W2, and Grade 75 steel rebar was adopted for W3. The yield strength of the Grade 60 and Grade 75 steel rebar is 413.7 MPa and 517.1 MPa, respectively. The Park parametric stress-strain relation was employed to model the nonlinear property of the steel rebar. The concrete property was modeled adopting the Mander concrete stress–strain curve. The unconfined concrete has a specified compressive strength of 27.6 MPa for W1 and W2 and 34.4 MPa for W3. The steel and concrete hysteresis behavior was modeled using the kinematic and concrete hysteresis models [36], respectively.

A multi-layer shell element was used to model the shear walls. In all the wall models, only the flexural inelasticity of shear walls was modeled, and the shear deformation of walls was set to be elastic since it was assumed that walls have sufficient shear strength. Table 1 summarizes the properties of the first five modes of the wall structures, where it is found that the modal mass coefficients of all the RC shear wall structures fulfill the relations of Equations (5) and (6) simultaneously, and the first two modes dominate the structural response of all the wall structures.

Table 1. Modal properties of shear wall structures.

NO.	Mode									
	1		2		3		4		5	
	T (s)	α	T (s)	α	T (s)	α	T (s)	α	T (s)	α
W1	0.625	0.680	0.127	0.200	0.061	0.045	0.047	0.009	0.041	0.016
W2	1.830	0.642	0.341	0.205	0.145	0.060	0.090	0.027	0.066	0.015
W3	2.392	0.630	0.436	0.200	0.182	0.064	0.110	0.031	0.079	0.020

In this study, 5% Rayleigh damping was assigned to the first and third modes to model the inherent damping of the structure. A few studies in the past [37–39] illustrated that this approach could result in fictitiously large damping forces, which underestimated the first-mode and higher-mode dominated responses. These studies proposed solutions to mitigate this shortcoming. Future studies should also consider alternative approaches proposed in the literature for modelling the viscous damping of the structure and investigate the effect of modelling damping on record-to-record variabilities.

3.2. Characteristics of the Seed IGMs

The target spectra for selecting and scaling the seed IGMs are conditional mean spectra (CMSs) and the ASCE 7 design spectra (DSs). Two DSs were constructed using two suites of parameters to consider different intensity levels. Table 2 summarizes the design parameters for the construction of the DSs.

Table 2. Design parameters of the ASCE design spectra.

Spectrum No.	S_{ds} (g)	S_{d1} (g)	T_L (s)
DS 1	1.00	0.75	12
DS 2	1.20	0.85	10

Note: S_{ds} and S_{d1} are spectral accelerations at 0.2 s and 1 s, respectively. T_L is the transition period of the long-period.

For wall structure Wn (n = 1, 2, 3), two CMSs, namely Wn CMS 1 and Wn CMS 2, was constructed adopting the fundamental mode of the structure T1 of Wn as the conditional period. Thus, overall six CMSs have been built. The Abrahamson–Silva-Kamai model, Boore–Stewart–Seyhan–Atkinson, Campbell–Bozorgnia, and Chiou–Youngs models were adopted to construct CMSs. Each target CMS is the average spectrum of CMSs built with the four ground motion models mentioned above using the same construction conditions. The moment magnitude was 8.5 and 9.0 for Wn CMS 1 and Wn CMS 2, respectively. The fault type is a strike-slip fault, and the distance from the site to the rupture plane is not less than 12 km. The Vs_{30} of site soil is the average shear velocity of top 30 m site soil and is assumed to be 400 m/s. Therefore, the site is classified as Class C of NEHRP. Detailed information for constructing the CMS are summarizes in Table 3, and the target CMSs and DSs are shown in Figure 2. It is seen from Figure 2 that CMSs built with the exact moment magnitude, but different conditioning periods are identical.

Table 3. Conditions for the construction of the CMSs.

Conditional Mean Spectrum	Conditioning Period	Moment Magnitude	Distance to Rupture Plane	V_{s30}
W1 CMS 1	0.625 s	8.5	≥12 km	400 m/s
W2 CMS 1	1.830 s			
W3 CMS 1	2.391 s			
W1 CMS 2	0.625 s	9.0		
W2 CMS 2	1.830 s			
W3 CMS 2	2.391 s			

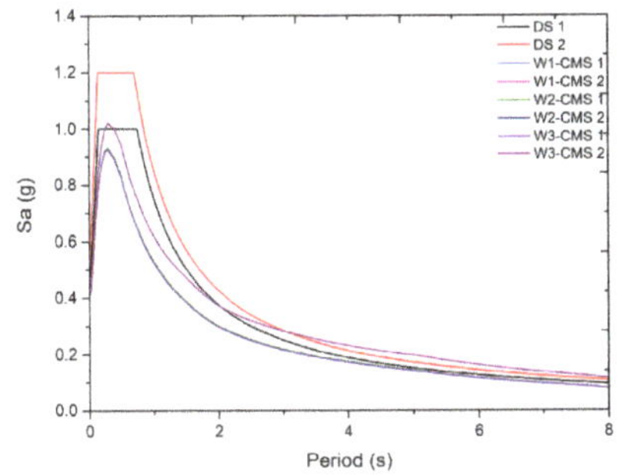

Figure 2. Target spectra for the IGM selection and scaling.

A set of twenty seed IGMs were selected and scaled from the strong ground motion database of the Pacific Earthquake Engineering Research Centre (PEER) [40], concerning each target spectrum. Therefore, eight sets of seed IGMs were selected. The criteria for the IGMs selection were: (1) the soil type of the site is class C of NEHRP for all the IGMs; (2) the distance from the recording stations to the epicentre should be not less than 12 km; (3) the range of moment magnitudes of earthquake events for all the IGMs is 6.5 to 9.0.

To ensure the response spectra of selected IGMs fit well with the target spectra, the average computed weighted mean squared error (MSE) between the response spectrum of each IGMs and target spectrum was minimized. When selecting and scaling seed IGMs for motion sets 1 and 2 with respect to the DSs, the scale factor has a value range of 1.0 to 5.0, and the weight value of 1.0 is set for periods ranging from 0.1 s to 5.0 s in the computation of MSE. The scale factor value ranges from 3.0 to 5.0 for motion sets 3–8, whose IGMs were selected and scaled with respect to CMS [41]. When minimizing the MSE for motion sets 3–8, the weight value is set to be 1.0 for from 0.2 T1 to 2 T1, where T1 is the conditioning period for the target CMS. Detailed criteria for the scaling of IGMs are listed in Table 4. Detailed information on the seed IGMs is provided in Appendix A.

Table 4. IGMs scaling criteria.

Motion Set	Target Spectrum	Scaling Factor	The Period Range for Minimizing the MSE	Structures Adopted for
1	DS 1	1.0–5.0	1.0 s–5.0 s	W1, W2, W3
2	DS 2	1.0–5.0	1.0 s–5.0 s	W1, W2, W3
3	W1 CMS 1	3.0–5.0	0.125 s–1.250 s	W1
4	W1 CMS 2	3.0–5.0	0.125 s–1.250 s	W1
5	W2 CMS 1	3.0–5.0	0.366 s–3.660 s	W2
6	W2 CMS 2	3.0–5.0	0.366 s–3.660 s	W2
7	W3 CMS 1	3.0–5.0	0.478 s–4.781 s	W3
8	W3 CMS 2	3.0–5.0	0.478 s–4.781 s	W3

3.3. IGM Selection for the NLRHA of RC Shear Wall Structures

To investigate the capability of the MGMS on selecting more proper IGMs for the NLRHA of RC shear wall structures, two spectra-matching-based IGM selection methods were employed to select IGMs from the seed motion sets. These two IGM selection procedures are uniformly weighted closest spectra matching procedure (UW-CM), which is widely used in the research and engineering practice, and variably weighted closest spectra matching procedure (VW-CM), which was recently developed in [28]. The UW-CM selects seven IGMs that have the lowest MSE, for which the weight value is 1.0 for all the period points within the period range of 0.2 Tl and 2 Tl, between the response spectra of the IGMs and target spectrum. The VW-CM method also selects seven IGMs with the lowest MSE. Still, the weight value for different periods varies to take into account the contribution of the different modes to the structural seismic responses. The VW-CM method calculates the MSE of motion j with the following equation:

$$MSE_j^{VW} = \sum_{i=1}^{n}\left[\frac{\alpha_i \sum_{T=0.5(T_{i+1}+T_i)}^{0.5(T_{i+1}+T_i) or 1.5T_1}\left(S_{a_j}(T) - S_a^t(T)\right)^2}{\sum_{i=1}^{n}\alpha_i}\right] \quad (11)$$

where $S_{aj}(T)$ and $S_a^t(T)$ are the spectrum acceleration value of motion j and target spectrum, respectively; T_i is period of ith mode; α_i is the modal mass coefficient of ith mode; n is the number of modes that ensure the sum of modal mass coefficients is not less than 90%.

Since the whole set of seed IGMs have similar response spectra and the broadest range of FCCTD, the mean seismic demand of NLRHA with the entire set of seed IGMs is the most reliable seismic demand. Thus, in this study, the mean seismic demand computed with all the seed IGMs of the set was employed as the reference seismic demand, and the average spectrum of IGMs of the whole motion set was adopted as the target spectrum, when adopting CM procedures to select IGMs from each seed motion set. The mean seismic demands from NLRHA with IGMs selected with different IGM selection methods were compared to investigate the rationality of the selected IGMs. NLRHA with motion sets 1 and 2 were conducted for all the shear wall structures. IGMs of sets 3–8 were adopted for the NLRHA of wall structures whose fundamental mode period is the same as the conditioning period of the target CMS of the set. Table 4 summarizes the motion sets for the NLHRA of different wall structures.

It is known from the selection process that the MGMS procedure ensures that the response spectra of the selected IGMs match well with the mean spectrum of the set, and the chosen IGMs can cause the most significant interaction of the first two dominated modes. Although both UW-CM and VW-CM procedures ensure that the response spectra of the selected IGMs match best with the mean spectrum of the set, the VW-CM considers the different contributions of modes to the seismic response of structures. The mean spectra of the twenty seed IGMs of the set (symbolized as Mean-20), mean spectra of MGMS selected IGMs (symbolized as MGMS-Wn), mean spectra of UW-CM-selected IGMs (symbolized as UW-CM-Wn) and mean spectra of VW-CM-selected IGMs (symbolized as VW-CM-Wn) are presented in Appendix A.

4. Results and Discussions

4.1. Comparison of the Results of MGMS Procedures

Figure 3 presents the seismic demands of wall structure W2 under motion set 6 adopting IGMs selected by MGMS procedures with and without considering the modal response characteristics of the RC shear wall structures, which is denoted as 2-mode and 3-mode, respectively. In Figure 3, the mean demands of the NLRHA with the whole set of IGMs are plotted and marked as "Mean". The presented MGMS procedure for RC shear wall structures, considering modal response characteristics of shear wall structures, selected seven IGMs that induced the most significant interaction of the first two modes from the seed IGMs set. The original MGMS procedure that does not take modal response characteristics of the RC shear wall structures into account selected seven IGMs that caused the most significant interaction of the first three modes for the NLRHA of the shear wall structures following the procedure in Reference [30].

It is found that no matter whether the modal response characteristics of the shear walls were considered, the MGMS procedures can select proper IGMs for the NLRHA of the structure and seismic demands by both MGMS procedures are close to the benchmark demands. It is also noticed that IGMs selected by the MGMS procedure considers the modal response characteristics of the RC shear walls led to a more accurate computation of most of the seismic demands since the seismic demands approximated the benchmark demand better. The more reliable computed seismic demand shows the necessity and rationality of considering the modal response characteristics of shear wall structures in selecting IGMs with the MGMS procedure.

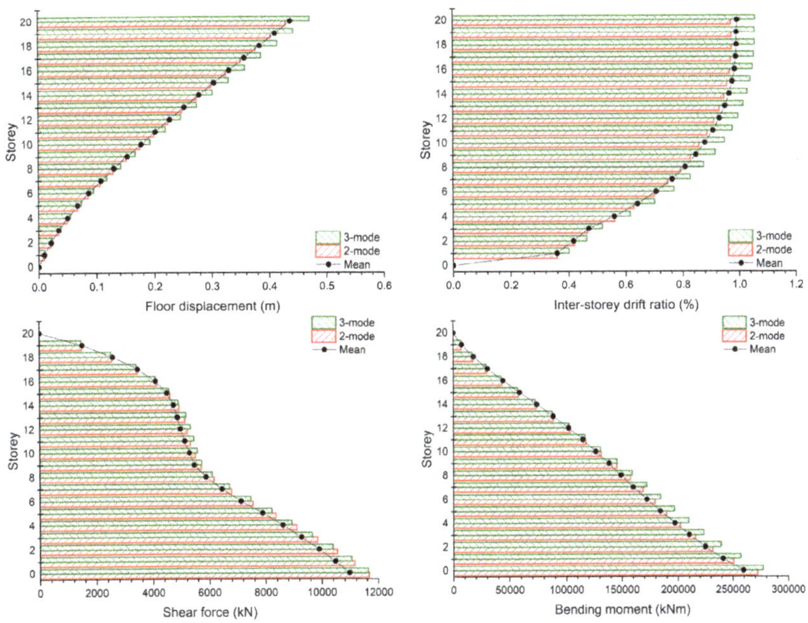

Figure 3. Comparison of the benchmark seismic demands and seismic demands computed by IGMs selected by MGMS procedures for motion set 6.

4.2. Comparison of Seismic Demands by Different IGM Selection Procedures

Figures 4–7 present the floor displacements, inter-story drift ratio, shear force, and bending moment computed by the IGMs from motion sets 2. The mean seismic demands of the set plus and minus one standard deviation of the demands, denoted as Mean + σ and Mean − σ, are also plotted. It is clear from Figures 4–7 that the ratio of one standard deviation of seismic deformation to the mean deformation is much larger than that of seismically induced force. As a result, the seismic deformation is more sensitive to the variation of IGMs, compared with seismically induced force.

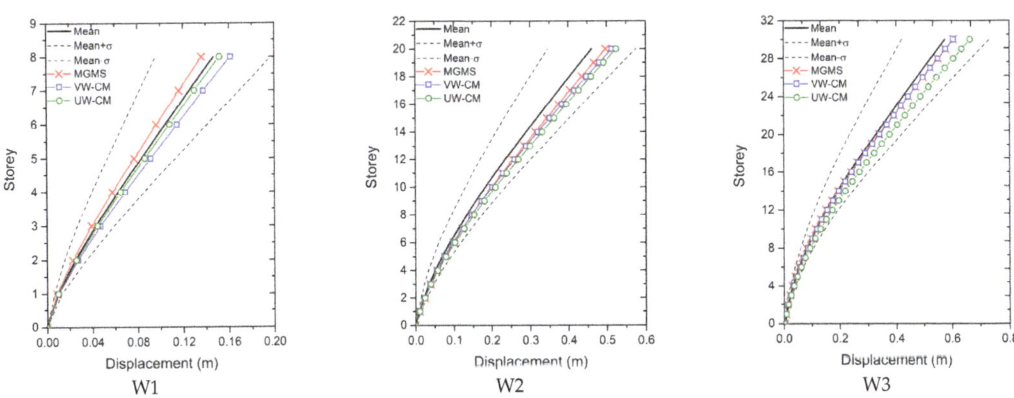

Figure 4. Comparison of the floor displacements of wall structures under motion Set 2.

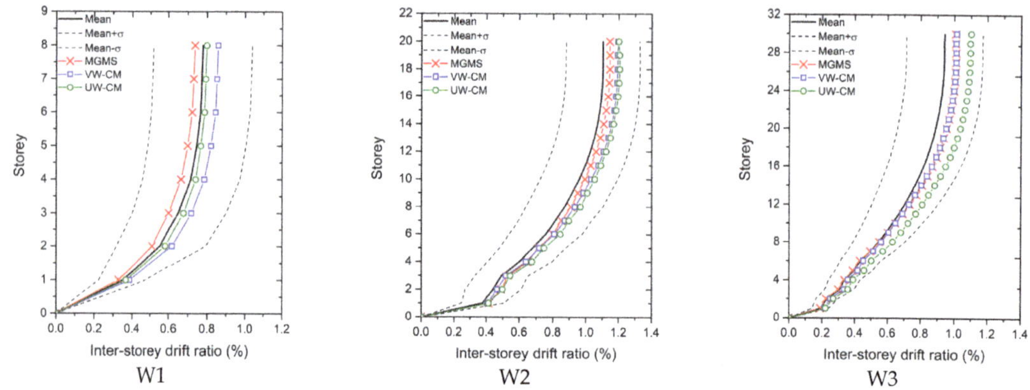

Figure 5. Comparison of the inter-story drift ratio of wall structures under motion Set 2.

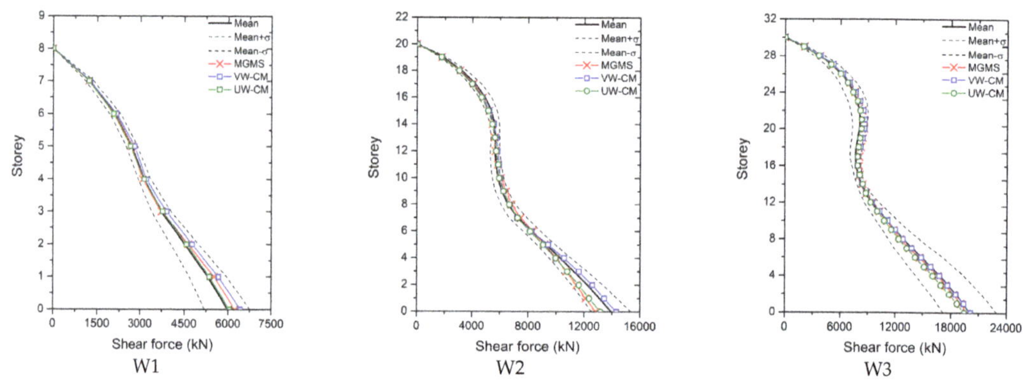

Figure 6. Comparison of the shear force of wall structures under motion Set 2.

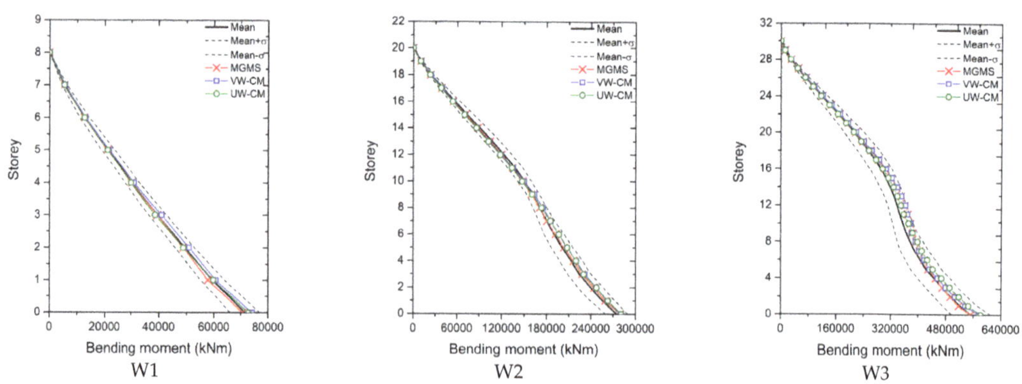

Figure 7. Comparison of the bending moment.

As shown in Figure 4, the IGMs selected by the MGMS procedure and CM procedures caused a less and more conservative prediction of the floor displacement for all the stories of W1, respectively. Displacement calculated with UW-CM selected IGMs matched the best with the mean displacement of the set. When calculating the floor displacement of W2, NLRHA with IGMs chosen by all the selection methods led to an overestimated floor displacement. Still, the displacement by the MGMS-IGMs approximates the reference displacement the best. Floor displacement of W3 under motion set 2 calculated by IGMs selected by all the selection procedures is conservative. The displacements of MGMS-IGMs and VW-CM-IGMs are similar and close to the benchmark displacement.

Considering the inter-story drift ratio of W1, the UW-CM selected IGMs led to the most reliable computation of the drift ratio, although the drift ratio computed with UW-CM-IGMs is conservation. The MGMS selected IGMs and VW-CM selected IGMs induced an under- and overestimation of the inter-story drift ratio of W1, respectively. The results of the MGMS-IGMs are closer to the mean drift ratio of the set. As for W2 and W3, inter-story drift ratios were overestimated by IGMs selected by all the IGM selection methods, but those computed with the MGMS-IGMs matched the best with the reference drift ratio. Comparing the inter-story drift ratio of W2 and W3 calculated by the CM procedures, the results from NLRHA with VW-CM selected IGMs have better approximations to the benchmark drift ratio.

It can be found in Figures 7 and 8 that IGMs of MGMS and both CM procedures could ensure shear force and bending moment of all the RC shear wall structures approximate the reference forces well in the upper floors. For shear force at lower stories, the UW-CM selected IGMs, VW-CM selected IGMs and MGMS-selected IGMs led to the most reliable computation of shear force of W1, W2 and W3, respectively. The bending moment of W1 at lower stories computed with UW-CM-IGMs is the closest to the reference bending moment, and the bending moment of W2 and W3 at lower floors from NLRHA with MGMS-IGMs has the best approximation to the mean moment of the set.

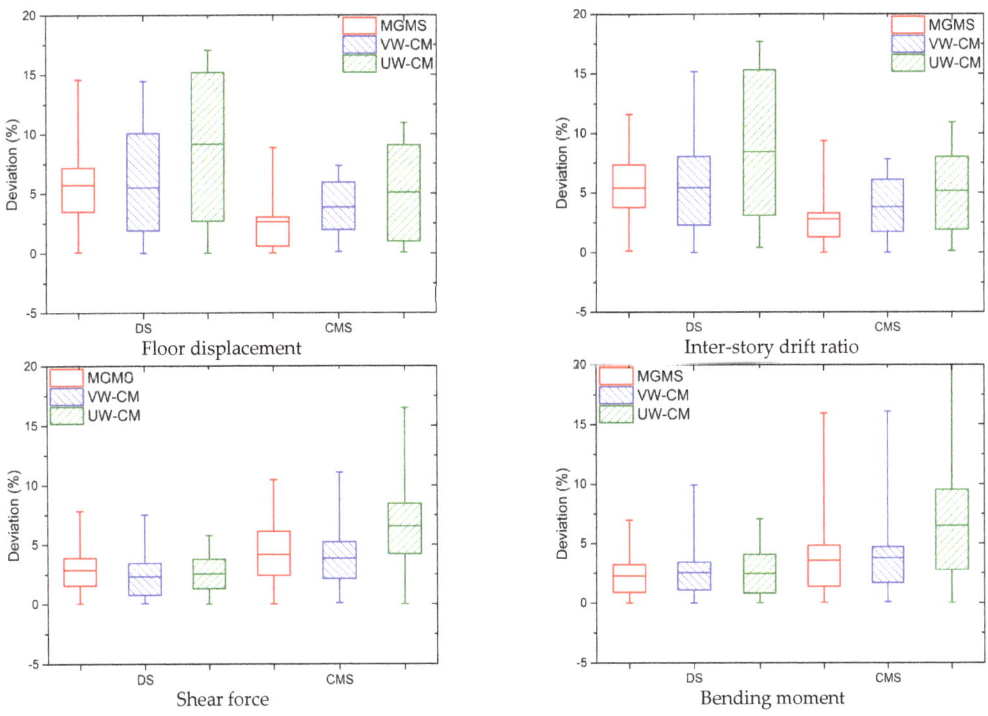

Figure 8. Deviations between the seismic demands from different IGM selection procedures and benchmark demands.

By comparing the seismically induced deformation and force of the wall structures under motion set 2, it is found that both MGMS procedure and CM procedures can select IGMs that ensure the seismic demands close to the mean demands of NLRHA with IGMs of the whole motion set. Additionally, MGMS-IGMs tend to provide a more reliable estimation of the seismic demands than CM procedures selected IGMs.

4.3. Comparison of the Deviation of the Computed Demands

The deviation between the seismic demands by different IGM selection procedures and the benchmark demands was calculated to quantitatively compare the reliability and reasonability of the seismic demands of NLRHA with a diverse selection of IGMs. The deviation δ_N^{ij} shows the relative difference between the benchmark demands of motion set j and demands computed using IGMs from IGM selection method N at the ith story, respectively, which are calculated as:

$$\delta_N^{ij} = \left| \frac{d_N^{ij} - \overline{d^{ij}}}{\overline{d^{ij}}} \right| \times 100\% \tag{12}$$

where N is the IGM selection methods, namely the MGMS method, UW-CM method and VW-CM method; d_N^{ij} is the i-th story's seismic demands calculated using IGMs selected by method N for motion set j; $\overline{d^{ij}}$ is the benchmark demands ith story for motion set j, which are the means demand of NLRHA with all IGMs of the set.

The statistics of the deviations are presented using box plots in Figure 8. Deviations of MGMS procedure and CM procedures are compared for motion sets containing seed IGMs selected and scaled concerning design spectra and conditional mean spectra. For each plot, the highest and lowest edges of the cirrus stand for the largest and lowest deviation. The horizontal axial value of the line inside the box is the mean deviation. The horizontal axial values of the upper and lower box edges are the deviation value of 75th and 25th percentiles, respectively. The mean deviation value shows the reliability of the IGM selection procedure. Additionally, as the smaller the difference between the 75th percentile deviation and 25th percentile deviation (75th-25th deviation difference) is, the trend of distribution of computed seismic demands along the height of the structure is closer to that of the benchmark seismic demands. Therefore, the 75th-25th deviation difference reveals the reasonability of the IGMs selected by the different selection methods.

Comparing the mean deviation of the seismic demands, it is evident that the MGMS-IGMs and VW-CM-IGMs can achieve a much lower mean deviation than that of the UW-CM selected IGMs, except for computing shear forces with IGMs selected with DSs being the target spectra. Therefore, the reliability of NLRHA with UW-CM-IGMs is lower than that of NLRHA with IGMs chosen by the other two IGM selection methods. The mean deviation of seismic deformation, including inter-story drift ratio and floor displacement, by MGMS-IGMs and VW-CM-IGMs is similar when using the DS as the target spectra. Meanwhile, a more than 27% lower mean deviation of seismic deformation by IGMs selected by MGMS procedure is noticed when selecting and scaling IGMs referring to CMSs. Considering the seismically induced forces, the MGMS selected IGMs, and the VW-CM selected IGMs have a higher mean deviation of shear force and bending moment, respectively, but the difference is insignificant.

The 75th-25th deviation difference of most seismic demands of the UW-CM method is much higher than that of MGMS and VW-CM methods, indicating lower reasonability of the UW-CM selected IGMs. Comparing the 75th-25th deviation difference of seismic deformation of the VW-CM and MGMS methods, it is found that the deviation difference of the MGMS procedure is over 38% and 54% lower than that of the VW-CM method when the target spectrum is DSs and CMSs, respectively. Compared with the VW-CM method, the MGMS procedure tends to have a slightly lower and higher 75th-25th deviation difference of seismically induced forces, when the target spectra for selecting and scaling the IGMs are DSs and CMs, respectively.

Based on the comparison of deviation of the computed demands, it is clear that considering the modal contribution in the IGM selection procedure, the VW-CM procedure can select more reliable and reasonable IGMs, compared with the UW-CM procedure.

Compared with the VW-CM procedure, the MGMS procedure for RC shear wall structures can achieve a noticeable lower mean deviation and 75th-25th deviation difference of the seismic deformation while having a similar mean deviation and deviation difference of the seismically induced forces. Meanwhile, it is seen in Figure 8 that the seismically induced force is less sensitive to the variation of the IGMs and the seismic deformation is the control seismic demand in the seismic design and analysis of building structures. Therefore, selecting IGMs with MGMS procedure could considerably improve the reliability and reasonability of the NLRHA of RC shear wall structures.

5. Conclusions

This paper presents a modification on the modal-based ground motion selection (MGMS) procedure for more proper consideration of the modal response characteristics of the RC shear wall structures in the input ground motions (IGMs) selection procedure. The first two modes dominated the dynamic behaviour of most shear wall structures with a symmetric plan. Therefore, the proposed MGMS procedure for the wall structures considers the impact of frequency contents combinations in the time domain (FCCTD) of IGMs on the seismic demand of building structures with the interaction of the first two modes. The capability of the MGMS in selecting more proper IGMs for the NLRHA of RC shear wall structures was verified with a comprehensive case study, where three reinforced concrete shear wall structures with different heights are studied. The conditional mean spectra (CMSs) and ASCE 7-16 design spectra (DSs) were adopted to select eight overall sets of twenty seed IGMs. Mean seismic demands of the RC shear walls computed using the whole set of seed IGMs were employed as the reference demands for the comparison. Two spectrum-matching based IGM selection methods that select IGMs whose response spectra have the closest matching concerning the mean spectrum of the set were adopted to investigate the MGMS procedure's capability to improve the reliability and reasonability of the NLRHA. These two IGM selection procedures are uniformly weighted closest spectra matching procedure (UW-CM) and variably weighted closest spectra matching procedure (VW-CM). From the comparison of the results, the following conclusions can be drawn:

1. Considering the modal response characteristics of RC shear wall structures, the MGMS procedure for shear wall structures presented led to a more reliable computation of seismic demands than the original version of the MGMS procedure.
2. Compared with the CM procedure with uniformly weight value for all the period points in computing difference between the response spectra of IGMs and mean spectra of the set (UW-CM), the CM procedure adopting variable weight value (VW-CM) procedure can ensure a more reliable and reasonable computation of seismic demands of RC shear wall structures.
3. Compared with the VW-CM procedure, the presented MGMS procedure could noticeably improve the reliability and reasonability of the computed seismic deformation, including the floor displacement and inter-story drift ratio, and achieve similar reliability and reasonability in calculating the seismically induced force.
4. Since the MGMS procedure just requires conducting NLRHA of equivalent single-degree-of-freedom systems, the computational consumption is minor. Taking advantage of high efficiency and great effectiveness in improving the reliability and reasonability of the NLRHA, the MGMS is an excellent supplement to the seismic design codes of practice's IGM selection procedure for the NLRHA of wall structures.

Funding: This research was funded by Open Fund of Shock and Vibration of Engineering Materials and Structures Key Laboratory of Sichuan Province under grand No. 19kfgk06, and the National Natural Science Foundation of China under Grant No. 52108454, and Research Start-up Fund for High-level Talents of Huaqiao University under grand No. 20191XD042. The author is grateful to Professor Yang Liu from Huaqiao University for his valuable advice and help during the study.

Institutional Review Board Statement: Not applicable.

Informed Consent Statement: Not applicable.

Conflicts of Interest: The author declares no conflict of interest.

Nomenclature

The following acronyms and symbols are used in this paper.

CMS	Conditional mean spectrum	$\boldsymbol{i}$	Influence vector
CM-UW	Uniformly weighted closest spectra matching procedure	L_n	Modal excitation factor of mode n
CM-VW	Variably weighted closest spectra matching procedure	m, c and k	Mass, damping, and stiffness matrices
DS	design spectrum	M_n	Generalized mass of mode n
FCCFD	Frequency contents combination in the frequency domain	$S_{aj}(T)$	Spectrum acceleration value of motion j at period T
FCCTD	Frequency contents combination in the time domain	$S_a^t(T)$	Spectrum acceleration value of target spectra at period T
IDR	Inter-story drift ratio	S_{d1}	Design spectrum acceleration at 1.0 s
IGMs	Input ground motions	S_{ds}	Design spectrum acceleration at 0.2 s
MGMS	Modal-based ground motion selection	t_k	Any time point within the duration of IGM
MSE	Mean computed weighted mean squared error	T_{j0}	duration of motion j
NL-ESDOF	Nonlinear equivalent single-degree-of-freedom system	T_L	Long-period transition period
NLRHA	nonlinear response time history analysis	T_n^*	Vibration period of NL-ESDOF of mode n
SDOF	Single-degree-of-freedom system	$\hat{u_{jn}}$	The maximum displacement of NL-ESDOF of mode n under motion j
$\boldsymbol{u}$	Displacement vector of floor	$u_{jn}(t_k)$	Displacement of NL-ESDOF of mode n under motion j at time tk
$a_n(\mathrm{t})$	Normalized top displacement time history of mode n	u_{rn}	Roof displacement of mode n
$\overline{d_{ij}}$	Benchmark demands ith story for motion set j	V_{bn}	The base shear force of mode n
d_M^{ij}	Seismic demands at ith story calculated using IGMs from selection method M for motion set j	α_i	Modal mass coefficient of mode i
D_n	Peak displacement of mode n	β_j	Maximum modal combination factor of motion j
d_n	Displacement of ESDOFs of mode ^{n}	β_{jk}	Maximum modal combination factor of motion j at time tk
$D_n(t)$	Top displacement time history of the mode n	Γ_n	the modal participating factor of mode n
d_{ny}	Yield displacement of ESDOFs of mode n	δ_M^{ij}	The relative difference between the benchmark demands of motion set j and demands computed using IGMs from selection method M at the ith story
F_{sn}	Restoring force of ESDOFs of mode n	ϕ_n	Mode shape of mode n
F_{sny}	Yield force of ESDOFs of mode n	Φ_{rn}	Mode shape value at the roof of mode n

Appendix A

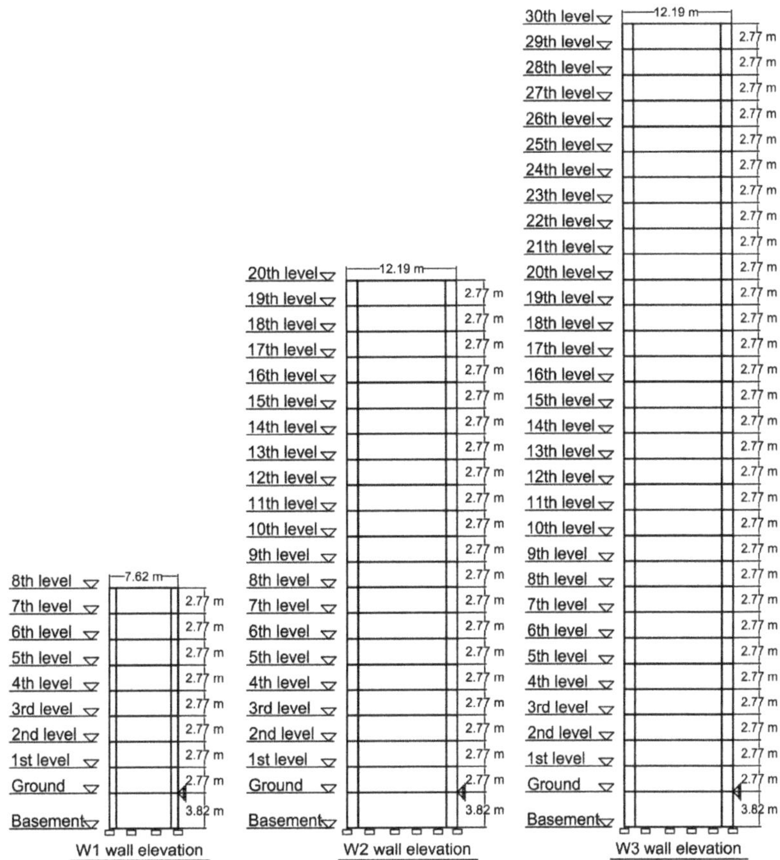

Figure A1. Elevation and plan views of wall structures.

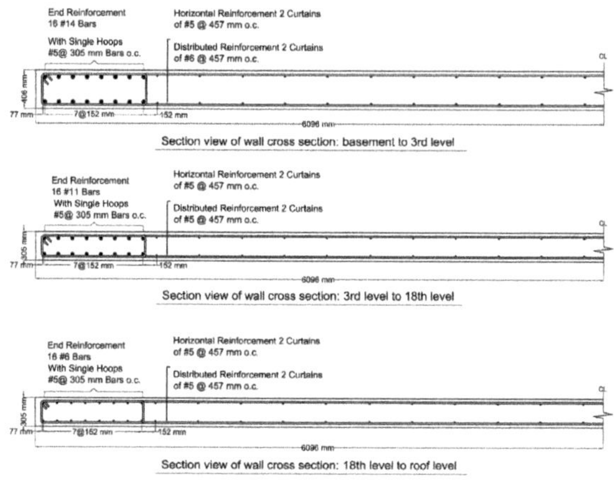

W1

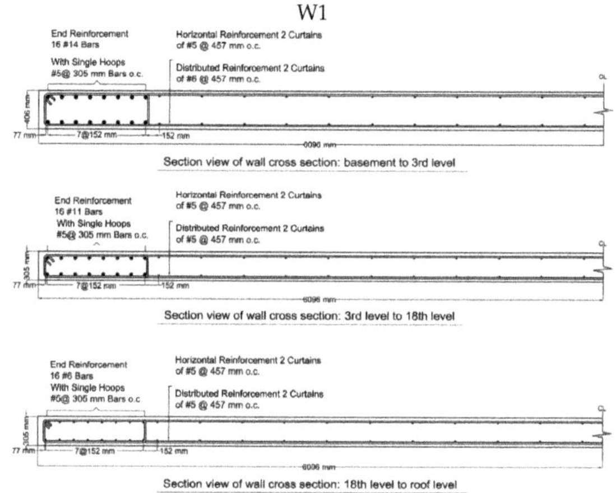

W2

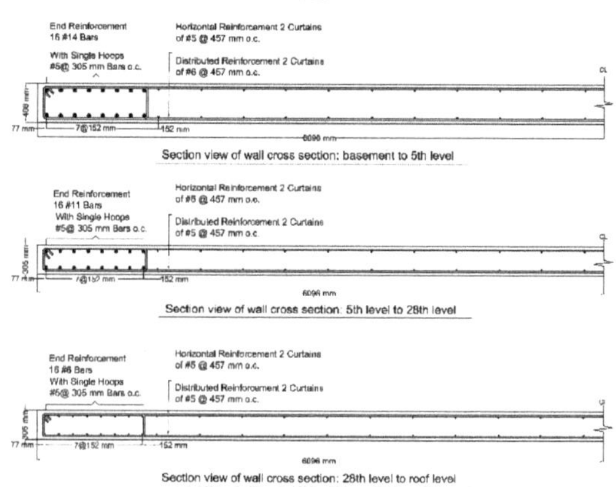

W3

Figure A2. Section views of wall structures.

Table A1. Details of IGMs of motion sets 1 and 2.

Record ID	Motion Set 1			Motion Set 2		
	Scale Factor	RSN No.	Component (deg)	Scale Factor	RSN No.	Component (deg)
1	4.764	RSN762	0	4.106	RSN15	21
2	4.582	RSN769	0	3.828	RSN289	0
3	2.303	RSN787	270	4.319	RSN735	0
4	3.548	RSN827	0	3.975	RSN755	195
5	4.805	RSN1261	E	4.168	RSN827	0
6	4.790	RSN1263	E	4.620	RSN1005	90
7	4.415	RSN1277	E	3.888	RSN1282	E
8	4.663	RSN127	E	4.179	RSN1297	N
9	4.332	RSN1300	N	3.610	RSN1471	E
10	2.098	RSN1484	E	2.247	RSN1541	E
11	2.348	RSN1500	E	2.619	RSN1762	90
12	4.396	RSN1522	E	4.536	RSN1794	90
13	2.326	RSN1762	90	1.404	RSN3748	270
14	3.832	RSN1794	90	2.127	RSN3750	270
15	1.133	RSN3748	70	3.357	RSN3751	270
16	3.021	RSN3751	270	4.136	RSN3757	90
17	3.642	RSN3757	90	2.624	RSN4865	NS
18	4.190	RSN4844	NS	4.146	RSN4872	NS
19	3.803	RSN4872	NS	2.729	RSN5778	NS
20	3.756	RSN6980	E	2.862	RSN5806	NS

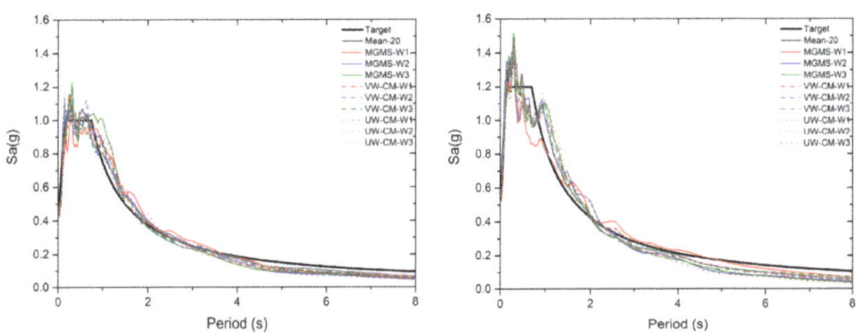

Set 1 Set 2

Figure A3. Mean spectra of IGMs selected from motion sets 1 and 2 with different selection procedures.

Table A2. Details of IGMs of motion sets 3 and 4.

Record ID	Motion Set 3			Motion Set 4		
	Scale Factor	RSN No.	Component (deg)	Scale Factor	RSN No.	Component (deg)
1	4.572	RSN28	0	3.666	RSN796	0
2	3.839	RSN736	137	3.656	RSN827	0
3	3.266	RSN827	0	3.281	RSN832	0
4	3.544	RSN838	0	3.976	RSN1019	0
5	3.552	RSN1019	0	3.009	RSN1083	170
6	4.657	RSN1029	0	4.225	RSN1166	180
7	3.774	RSN1166	180	3.127	RSN1208	E
8	3.615	RSN1277	E	4.047	RSN1277	E
9	4.153	RSN1293	N	4.080	RSN1280	E
10	4.606	RSN1346	N	4.649	RSN1293	N

Table A2. *Cont.*

Record ID	Motion Set 3			Motion Set 4		
	Scale Factor	RSN No.	Component (deg)	Scale Factor	RSN No.	Component (deg)
11	4.136	RSN1349	N	4.629	RSN1349	N
12	4.613	RSN1436	E	3.281	RSN1471	E
13	3.340	RSN1488	E	3.738	RSN1488	E
14	2.997	RSN1548	E	3.178	RSN1794	90
15	2.964	RSN3757	90	3.318	RSN3757	90
16	3.370	RSN4844	NS	3.772	RSN4844	NS
17	3.741	RSN4892	NS	4.187	RSN4892	NS
18	4.702	RSN5681	NS	3.154	RSN5284	NS
19	3.061	RSN6948	E	3.426	RSN6948	E
20	3.277	RSN6949	W	3.085	RSN6980	E

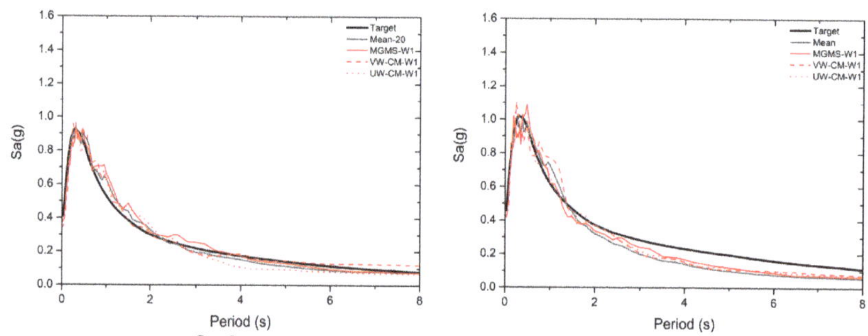

Figure A4. Mean spectra of IGMs selected from motion sets 3 and 4 with different selection procedures.

Table A3. Details of IGMs of motion sets 5 and 6.

Record ID	Motion Set 5			Motion Set 6		
	Scale Factor	RSN No.	Component (deg)	Scale Factor	RSN No.	Component (deg)
1	3.2018	RSN731	0	3.7418	RSN731	0
2	2.9568	RSN832	0	3.7123	RSN736	137
3	3.4254	RSN838	0	4.8197	RSN812	0
4	3.9506	RSN1261	E	3.4185	RSN827	0
5	3.5022	RSN1277	E	3.4556	RSN832	0
6	4.7352	RSN1279	E	3.0436	RSN1208	E
7	4.5283	RSN1285	E	4.617	RSN1261	E
8	4.6728	RSN1335	E	4.093	RSN1277	E
9	4.0228	RSN1339	E	4.7014	RSN1339	E
10	4.3041	RSN1436	E	4.8635	RSN1431	E
11	4.5664	RSN1470	E	4.2951	RSN1466	E
12	3.3515	RSN1475	E	3.0868	RSN1471	E
13	2.9285	RSN3757	90	4.3791	RSN1522	E
14	4.796	RSN3994	90	3.4225	RSN3757	90
15	3.3816	RSN4844	NS	3.952	RSN4844	NS
16	2.9919	RSN4872	NS	3.4966	RSN4872	NS
17	3.065	RSN5284	NS	3.582	RSN5284	NS
18	4.4291	RSN5681	NS	3.8479	RSN5776	NS
19	4.4052	RSN5796	NS	3.7347	RSN6901	W
20	3.1957	RSN6901	W	3.3329	RSN6980	E

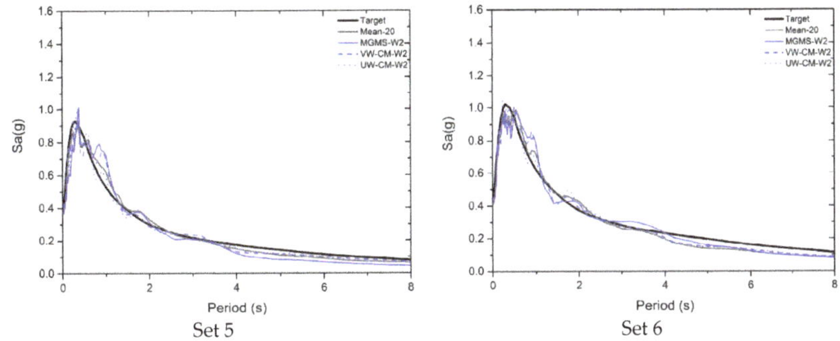

Figure A5. Mean spectra of IGMs selected from motion sets 5 and 6 with different selection procedures.

Table A4. Details of IGMs of motion sets 7 and 8.

Record ID	Motion Set 7			Motion Set 8		
	Scale Factor	RSN No.	Component (deg)	Scale Factor	RSN No.	Component (deg)
1	3.233	RSN731	0	4.0338	RSN731	0
2	3.4122	RSN812	0	4.2574	RSN812	0
3	4.8609	RSN1232	E	3.2824	RSN827	0
4	3.9131	RSN1261	E	3.6365	RSN832	0
5	3.2159	RSN1263	E	3.1732	RSN1208	E
6	4.0754	RSN1285	E	4.0125	RSN1263	E
7	3.7487	RSN1339	E	4.6773	RSN1339	E
8	4.3745	RSN1358	E	3.9016	RSN1464	E
9	4.6811	RSN1436	E	3.9419	RSN1465	E
10	3.1593	RSN1465	E	3.3486	RSN1467	E
11	3.2727	RSN1469	E	4.0834	RSN1469	E
12	3.9213	RSN1470	E	4.1845	RSN1473	E
13	3.3015	RSN1522	E	4.1193	RSN1522	E
14	4.1946	RSN1575	E	3.682	RSN1523	E
15	4.2539	RSN1588	N	3.8839	RSN1525	E
16	4.1117	RSN3994	90	2.9612	RSN3747	270
17	3.1024	RSN4882	NS	3.2803	RSN4848	NS
18	3.8213	RSN5472	NS	3.8709	RSN4882	NS
19	4.228	RSN5681	NS	4.7679	RSN5472	NS
20	4.1963	RSN5804	NS	4.7607	RSN5783	NS

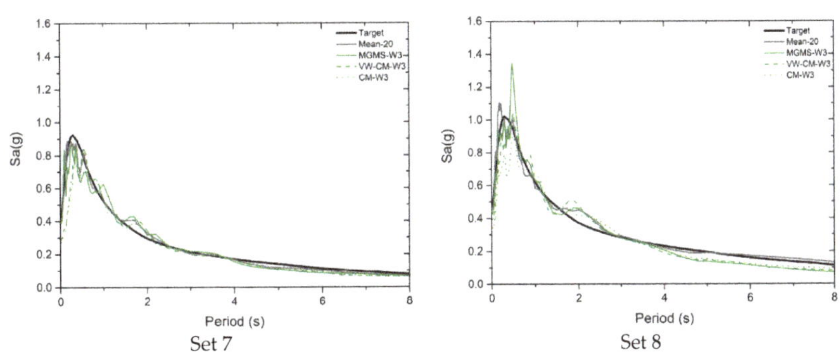

Figure A6. Mean spectra of IGMs selected from motion sets 7 and 8 with different selection procedures.

References

1. Katsanos, E.; Sextos, A.; Manolis, G.D. Selection of earthquake ground motion records: A state-of-the-art review from a structural engineering perspective. *Soil Dyn. Earthq. Eng.* **2010**, *30*, 157–169. [CrossRef]
2. Chieffo, N.; Clementi, F.; Formisano, A.; Lenci, S. Comparative fragility methods for seismic assessment of masonry buildings located in Muccia (Italy). *J. Build. Eng.* **2019**, *25*, 100813. [CrossRef]
3. Forcellini, D. Numerical simulations of liquefaction on an ordinary building during Italian (20 May 2012) earthquake. *Bull. Earthq. Eng.* **2019**, *17*, 4797–4823. [CrossRef]
4. Forcellini, D. Soil-structure interaction analyses of shallow-founded structures on a potential-liquefiable soil deposit. *Soil Dyn. Earthq. Eng.* **2020**, *133*, 106108. [CrossRef]
5. Petridis, C.; Pitilakis, D. Fragility curve modifiers for RC dual buildings to include nonlinear site effects and SSI. *Earthq. Spectra* **2020**, *36*, 1930–1951. [CrossRef]
6. ASCE. *Minimum Design Loads for Buildings and Other Structures (ASCE/SEI 7-17)*; American Society of Civil Engineering: Reston, VA, USA, 2017.
7. CEN. *ENV 1998-1 Eurocode 8: Design of Structures for Earthquake Resistance—Part 1: General Rules, Seismic Actions and Rules for Buildings*; European Committee for Standardisation: Brussels, Belgium, 2004.
8. Chinese Standard. *Code for Seismic Design of Buildings, GB 50011-2010*; Chinese Building Press: Beijing, China, 2008.
9. NZS. *Structural Design Actions (NZS 1170.5)*; Standards New Zealand Technical Committee: Wellington, New Zealand, 2004.
10. Baker, J.W.; Cornell, C.A. A vector-valued ground motion intensity measure consisting of spectral acceleration and epsilon. *Earthq. Eng. Struct. Dyn.* **2005**, *34*, 1193–1217. [CrossRef]
11. Tarbali, K.; Bradley, B.A. The effect of causal parameter bounds in PSHA-based ground motion selection. *Earthq. Eng. Struct. Dyn.* **2016**, *45*, 1515–1535. [CrossRef]
12. Barbosa, A.R.; Ribeiro, F.L.A.; Neves, L. Influence of earthquake ground-motion duration on damage estimation: Application to steel moment resisting frames. *Earthq. Eng. Struct. Dyn.* **2016**, *46*, 27–49. [CrossRef]
13. Pan, Y.; Ventura, C.E.; Finn, W.L.; Xiong, H. Effects of ground motion duration on the seismic damage to and collapse capacity of a mid-rise woodframe building. *Eng. Struct.* **2019**, *197*, 109451. [CrossRef]
14. Tao, D.; Ma, Q.; Li, S.; Xie, Z.; Lin, D.; Li, S. Support Vector Regression for the Relationships between Ground Motion Parameters and Macroseismic Intensity in the Sichuan–Yunnan Region. *Appl. Sci.* **2020**, *10*, 3086. [CrossRef]
15. Nguyen, V.-Q.; Aaqib, M.; Nguyen, D.-D.; Luat, N.-V.; Park, D. A Site-Specific Response Analysis: A Case Study in Hanoi, Vietnam. *Appl. Sci.* **2020**, *10*, 3972. [CrossRef]
16. O'Donnell, A.P.; Kurama, Y.C.; Kalkan, E.; Taflanidis, A. Experimental evaluation of four ground-motion scaling methods for dynamic response-history analysis of nonlinear structures. *Bull. Earthq. Eng.* **2017**, *15*, 1899–1924. [CrossRef]
17. Naeim, F.; Alimoradi, A.; Pezeshk, S. Selection and Scaling of Ground Motion Time Histories for Structural Design Using Genetic Algorithms. *Earthq. Spectra* **2004**, *20*, 413–426. [CrossRef]
18. Al Atik, L.; Abrahamson, N. An Improved Method for Nonstationary Spectral Matching. *Earthq. Spectra* **2010**, *26*, 601–617. [CrossRef]
19. Kayhan, A.H.; Korkmaz, K.A.; Irfanoglu, A. Selecting and scaling real ground motion records using harmony search algorithm. *Soil Dyn. Earthq. Eng.* **2011**, *31*, 941–953. [CrossRef]
20. Wang, G. A ground motion selection and modification method capturing response spectrum characteristics and variability of scenario earthquakes. *Soil Dyn. Earthq. Eng.* **2011**, *31*, 611–625. [CrossRef]
21. Reyes, J.C.; Riaño, A.C.; Kalkan, E.; Quintero, O.A.; Arango, C.M. Assessment of spectrum matching procedure for nonlinear analysis of symmetric- and asymmetric-plan buildings. *Eng. Struct.* **2014**, *72*, 171–181. [CrossRef]
22. Jiang, W.; Li, B.; Xie, W.-C.; Pandey, M.D. Generate floor response spectra: Part 1. Direct spectra-to-spectra method. *Nucl. Eng. Des.* **2015**, *293*, 525–546. [CrossRef]
23. Han, S.W.; Ha, S.J. Assessment of ground motion selection criteria specified in current seismic provisions with an accurate selection algorithm. *Bull. Earthq. Eng.* **2017**, *15*, 4113–4132. [CrossRef]
24. Anajafi, H.; Medina, R.A. Uncertainties in using the spectrum matching technique for generating synthetic ground motions. In Proceedings of the 11th National Conference in Earthquake Engineering, Los Angeles, CA, USA, 25–29 June 2018.
25. Anajafimarzijarani, H. Improved Seismic Design of Non-Structural Components (NSCs) and Development of Innovative Control Approaches to Enhance the Seismic Performance of Buildings and NSCs. Master's Thesis, University of New Hampshire, Durham, UK, 2018.
26. Mergos, P.E.; Sextos, A.G. Selection of earthquake ground motions for multiple objectives using genetic algorithms. *Eng. Struct.* **2019**, *187*, 414–427. [CrossRef]
27. Baker, J.W.; Lee, C. An Improved Algorithm for Selecting Ground Motions to Match a Conditional Spectrum. *J. Earthq. Eng.* **2018**, *22*, 708–723. [CrossRef]
28. Zhang, R.; Wang, D.S.; Chen, X.Y.; Li, H.N. Weighted scaling and selecting method of ground motions in time-history analysis considering influence of higher modes. *China Civil. Eng. J.* **2019**, *52*, 53–68.
29. Chopra, A. *Dynamics of Structures: Theory and Applications to Earthquake Engineering*, 5th ed.; Pearson Prentice Hall: Upper Saddle River, NJ, USA, 2017.

30. Liu, Y.; Kuang, J.; Yuen, T.Y. Modal-based ground motion selection procedure for nonlinear response time history analysis of high-rise buildings. *Earthq. Eng. Struct. Dyn.* **2019**, *49*, 95–110. [CrossRef]
31. Stafford, B.; Coull, A. *Tall Building Structures: Analysis and Design*; John Wily: New York, NY, USA, 1991.
32. Moradi, M.J.; Hariri-Ardebili, M.A. Developing a Library of Shear Walls Database and the Neural Network Based Predictive Meta-Model. *Appl. Sci.* **2019**, *9*, 2562. [CrossRef]
33. Baker, J.W. Conditional Mean Spectrum: Tool for Ground-Motion Selection. *J. Struct. Eng.* **2011**, *137*, 322–331. [CrossRef]
34. FEMA. *Improvement of Nonlinear Static Seismic Analysis Procedures: FEMA-440*; Federal Emergency Management Agency: Redwood, CA, USA, 2005.
35. Huang, K. Continuum MDOF Model for Seismic Analysis of Wall-Frame Structures. Master's Thesis, Hong Kong University of Science and Technology, Hong Kong, China, 2014.
36. CSI SV. *18: Integrated Finite Element Analysis and Design of Structures Basic Analysis Reference Manual*; Computers and Structures Inc.: Berkeley, CA, USA, 2018.
37. Ryan, K.L.; Polanco, J. Problems with Rayleigh Damping in Base-Isolated Buildings. *J. Struct. Eng.* **2008**, *134*, 1780–1784. [CrossRef]
38. Anajafi, H.; Medina, R.A.; Santini-Bell, E. Effects of the improper modeling of viscous damping on the first-mode and higher-mode dominated responses of base-isolated buildings. *Earthq. Eng. Struct. Dyn.* **2019**, *49*, 51–73. [CrossRef]
39. Kitayama, S.; Constantinou, M.C. Effect of superstructure modeling assumptions on the seismic performance of seismically isolated buildings. *Earthq. Eng. Struct. Dyn.* **2021**, *50*, 1805–1823. [CrossRef]
40. Chiou, B.S.J.; Darragh, R.; Gregor, N.; Silva, W.J. NGA Project Strong-Motion Database. *Earthq. Spectra* **2008**, *24*, 23–44. [CrossRef]
41. Du, W.; Ning, C.; Wang, G. The effect of amplitude scaling limits on conditional spectrum-based ground motion selection. *Earthq. Eng. Struct. Dyn.* **2019**, *48*, 1030–1044. [CrossRef]

applied sciences

MDPI

Article

Seismic Analysis Method for Underground Structure in Loess Area Based on the Modified Displacement-Based Method

Ruijie Zhang [1], Dan Ye [1,2,*], Jianting Zhou [1] and Dengzhou Quan [3]

1 State Key Laboratory of Mountain Bridge and Tunnel Engineering, Chongqing Jiaotong University, Chongqing 400074, China; xjdzhangruijie@163.com (R.Z.); jtzhou@cqjtu.edu.cn (J.Z.)
2 School of Tourism Management and Services, Chongqing University of Education, Chongqing 400067, China
3 School of Civil Engineering, Chang'an University, Xi'an 710061, China; qdz0809@chd.edu.cn
* Correspondence: yedan@cque.edu.cn

Abstract: At present, the seismic design research of underground structures in loess areas is lagging behind compared with practical engineering requirements. The selection of seismic calculation methods and parameters does not consider the influences of the special geological conditions in various regions, so their usefulness is limited. Based on the above problems, a modified displacement-based method (DBM) was proposed and its application was compared with the most commonly used methods of analysis (force-based design method, displacement-based design method, detailed equivalent static analysis numerical method, and the full dynamic time-history method). The results were also validated by considering data from shaking table tests conducted on a case study involving the underground Feitian Road subway station in Xi'an. The results show that compared with DBM, the average accuracy of the modified DBM technique is improved by 41.65%. The modified DBM offers good accuracy, simplicity in its model, a rapid analysis time, and easy convergence.

Keywords: traffic engineering; seismic calculation; loess area; DBM; DESANM

Citation: Zhang, R.; Ye, D.; Zhou, J.; Quan, D. Seismic Analysis Method for Underground Structure in Loess Area Based on the Modified Displacement-Based Method. *Appl. Sci.* **2021**, *11*, 11245. https://doi.org/10.3390/app112311245

Academic Editor: Maria Favvata

Received: 9 October 2021
Accepted: 18 November 2021
Published: 26 November 2021

Publisher's Note: MDPI stays neutral with regard to jurisdictional claims in published maps and institutional affiliations.

Copyright: © 2021 by the authors. Licensee MDPI, Basel, Switzerland. This article is an open access article distributed under the terms and conditions of the Creative Commons Attribution (CC BY) license (https://creativecommons.org/licenses/by/4.0/).

1. Introduction

With the growth of population and industrial activities, the shortage of available space directly restricts the rapid development of big cities. To solve this problem, many countries have begun to increase their development and utilization of urban underground space. Earthquake damage to underground structures is thus mitigated, as the confining pressure exerted by the surrounding soil can improve the level of structural safety in the event of an earthquake. Hence, seismic calculations and seismic measures are not applied to the underground structures associated with subway systems. However, earthquake disasters in recent decades have affected thinking around this traditional concept, especially the Hanshen earthquake which damaged the 3-km long subway tunnel and five subway stations, indicating the possibility of subway underground structural damage and secondary disasters remains significant [1–3]. Subway stations are a service-oriented public facility and host a concentrated population (many service facilities), and require a long time to expedite emergency evacuation.

At present, the seismic design research of underground structures in loess areas is lagging compared with the engineering requirements, for example, the selection of design parameters related to the characteristics of loess soil lacks a clear set of rules, thus hindering the safe and efficient seismic design of underground structures in loess area.

Currently, the relevant codes for the design of underground structures include: the *Code for Seismic Design of Buildings* (GB 50011-2010) [4] and the *Code for Design of Civil Air Defense Basements* (GB 50038-2005) [5]. However, these do not provide relevant seismic calculation methods. The inertial force method for underground structures in soft soil has been given in the *Code for Seismic Calculation of Subway Building Structures* (DG/TJ08-2008) [6], but it is not necessarily suitable for those loess areas. The response acceleration method and response displacement method provided in the *Code for Seismic Calculation of Urban Rail Transit Structures* (GB 500909-2014) [7] have no specific provisions on the horizontal relative displacement of the stratum and the foundation spring stiffness parameters. The differences of ground motion characteristics among different soil bodies are not considered in the *Code for Seismic Design of Underground Structures* (GB/T 51336-2018) and other international codes [8–17].

In recent years, many studies about seismic design and analysis of underground structures have been published, but most of them do not consider the dynamic characteristics of soil or the methods are too complex to be suitable for engineering design [18–22]. Seismic design and analysis methods mainly include coupled and decoupled approaches, numerical dynamic analyses, and quasi-static calculation methods. A soil-structure interaction analysis of underground tunnels was performed by Kisiridis in1983 [23]. The magnitude and distribution of static normal soil stresses against underground structural cylinders were studied by Penzien and Wu in 1998 [24]. New analytical solutions for a deep tunnel in a saturated poro-elastic ground were explored by Bobet A [25]. The analytical solutions for the thrust and moment in the lining of a circular tunnel due to seismic-induced ovaling deformation were studied by Park et al. [26]. An analytical solution for a rectangular opening in an infinite elastic medium subjected to far-field shear stresses was proposed for drained and undrained loading conditions [27]. The main limitations of the decoupled approaches were investigated and discussed through a large set of numerical simulations [28]. The main results of several numerical dynamic analyses of propped embedded retaining structures in the time domain were demonstrated by Soccodato FM and Tropeano G [29]. The seismic behavior of a multi-propped retaining structure was evaluated considering soil-structure interaction effects [30]. The quasi-static methods mainly include the force-based method (FBM), displacement-based method (DBM), and detailed equivalent static analysis numerical method (DESANM) [7–9,18,19,31–39]. Coupled and decoupled approaches and numerical dynamic analyses are too complex to be suitable for engineering design, and the quasi-static methods applicable to engineering design do not consider the effects of different soil properties

In conclusion, the selection of seismic calculation methods and parameters does not consider the influences of the special geological conditions in various regions, so their usefulness is limited. The seismic analysis of subway station structures in loess areas is yet to be codified. In the existing research results, the seismic performance of underground subway station structures in loess is rarely researched. For the practical engineering of subway underground structures in a loess area, there is no reference seismic experiment, reliable quasi-static calculation method, or seismic parameter-calculation method yet available.

Based on the above problems, herein, a modified DBM was proposed and its application was compared with the most commonly used methods of analysis (FDM, DBM, DESANM, and full dynamic-time history analysis (FDTHA)). The results were validated using data from a shaking table test conducted on the basis of a case study (the Feitian Road underground subway station in Xi'an). The modified DBM offers good accuracy, a simple model, rapid modeling, and easy convergence. It provides engineering designers with a seismic design and analysis method for underground structures with convenient application and high precision. Therefore, the results obtained in this study can be considered useful to designers who are required to address the seismic design of underground structures.

In Section 2, the main seismic calculation methods of underground structures were analyzed, and the methods of calculation of foundation reaction spring stiffness parameters and formation horizontal relative displacement parameters were explored. In Section 3, the shaking-table test scheme design and the FDTHA simulation modeling of a subway station in a loess area were conducted. In Section 4, the results of shaking-table testing and the FDTHA numerical simulation were studied, and the seismic calculation methods for the main underground structures were compared. In Section 5, the DBM was modified according to the experimental results and simulation results, and the accuracy of the modified MDB was verified.

2. Seismic Calculation Method and Important Calculation Parameters of Subway Underground Structures

2.1. FBM

When the FBM is used in the seismic calculation of underground stations, the static forces such as the load exerted by the soil on the roof, the inertial force of each component, and the increment of active lateral earth pressure at each point of the external wall are often used to replace the seismic force [18,19]. The calculation model under lateral seismic action is shown in Figure 1 [31–33], where P_i represents the inertial force representing the weight of the floor, F_1 is the inertial force of the side wall, F_2 denotes the sum of ground and roof overburden forces, and subscripts 1 to 3 represent the top, middle, and bottom plates, respectively; K is the foundation spring stiffness, V and H denote the structural bottom plate and side wall, and Δe refers to the increment of lateral earth pressure caused by the earthquake. The horizontal inertial force can be calculated according to the *Code for Seismic Calculations in Railway Engineering* (GB 50111-2006) [9].

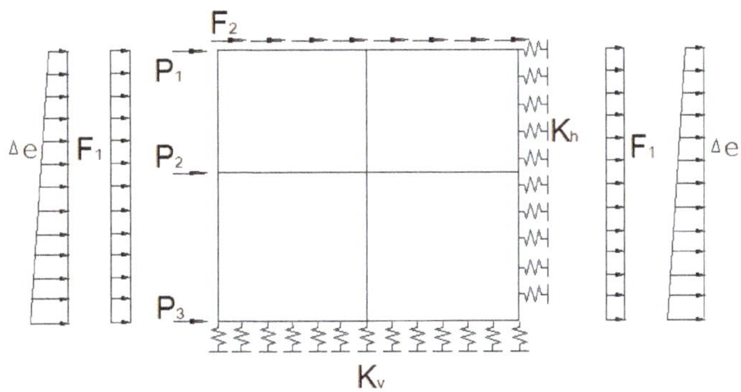

Figure 1. FBM calculation model.

2.2. DBM

The DBM [7,8] emphasizes the interaction between the soil and the underground structure. In this method, the foundation reaction spring was adopted to simulate the surrounding soil, and the horizontal relative displacement of the soil layer was applied at the end of the foundation reaction spring. The calculation model of the DBM under transverse earthquake action is illustrated in Figure 2, where 1, 2, 3, and 4 represent the ground, bedrock, soil displacement, and subway station acceleration, respectively, K_{sv} is the tangential shear foundation spring stiffness of the top and bottom plates of the structure, K_{sh} represents the tangential shear foundation spring stiffness of the side wall of the structure, k_v is the normal compression foundation spring stiffness of the top and bottom plate of the structure, K_h denotes the normal compression foundation spring stiffness of the side wall of the structure, τ_B is the friction shear force per unit area produced by the soil on the structural floor, and τ_U is the friction shear force per unit area of the soil acting on the roof of the structure.

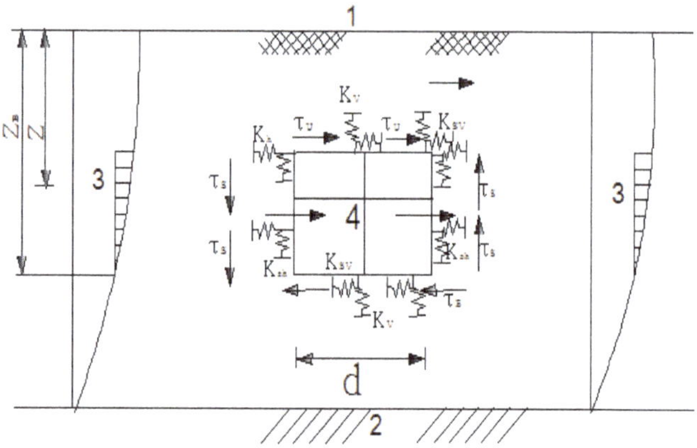

Figure 2. DBM calculation model.

2.3. DESANM

The DESANM [7,8,31,36–38] requires the engineer to establish soil and structure models at the same time in order to better reflect the interaction between soil and surrounding geotechnical media. This method reflects the phenomenon that the deformation difference between underground structure and surrounding soil changes in an irregular manner during an earthquake. The basic equation of DESANM is as shown in Equation (1).

$$[K]\{u\} = -[M](\{\ddot{u}T\} + [R]\{\ddot{u}\mathrm{gT}\}) = -[M]\{\ddot{u}T\} \quad (1)$$

The DESANM can be used in the automatic calculation of the interactive force between the soil medium and an underground structure through finite element analysis software, thus avoiding the error caused by improper selection of foundation reaction spring parameters when establishing spring stiffness values. The calculation model of DESANM under lateral earthquake action is illustrated in Figure 3, where 1 and 2 are equivalent lateral inertial accelerations of the soil and structure, respectively.

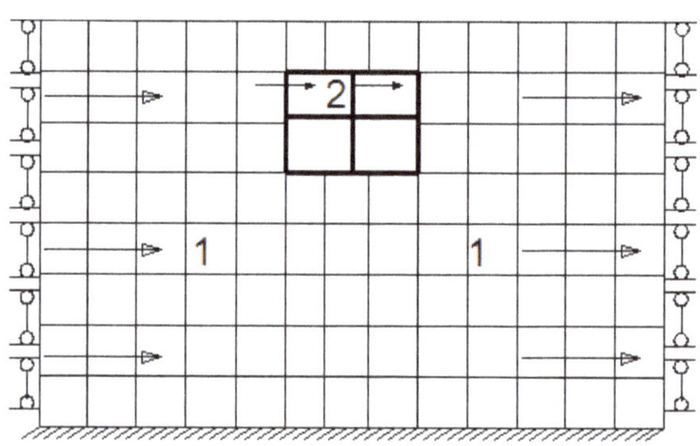

Figure 3. DESANM calculation model.

2.4. FDTHA

Although the three-dimensional model has the advantages of high calculation accuracy, the modeling is complex and the solution does not readily converge, posing a difficulty for engineering designers. The complete soil-structure system can be modeled and analyzed using 2-d numerical models. FDTHA is considered among the most sophisticated and accurate methods for the seismic analysis of underground structures [7,8,31,36–38].

The method can efficiently describe the kinematic and inertial aspects of the soil-structure interaction and the complex geometry of the soil deposit. FDTHA can be used to investigate the seismic behavior of an underground structure through numerical simulation [29,31].

2.5. Important Calculation Parameters

(1) Foundation Spring Stiffness

As an important parameter of the FBM and the DBM, the foundation reaction spring stiffness parameter plays an important role in FBM and the DBM. As the key parameter of the DBM and the DESANM, the horizontal relative displacement parameter of the stratum under seismic action has a significant influence on the seismic calculation results. For the value of foundation reaction spring stiffness parameter, the relevant norms make no clear provisions, and the relevant research has reached no unified conclusion.

The finite element calculation model has been established, as shown in Figure 4. In Figure 4, the width of the soil model is six to seven times that of the underground structure, and the depth of the soil model extends to the bedrock surface. The influences of soil properties and structural shape characteristics on foundation spring stiffness in loess area were evaluated, as shown in Figure 5. The proposed equation of foundation spring stiffness was fitted using MATLAB, as shown in Equations (2)–(7). The proposed equation can be used for the seismic calculation of underground station structures with various rectangular section sizes in loess sites.

$$\begin{aligned} K1 = &-0.1041E\left(\tfrac{1}{h^2} - 3.6\tfrac{1}{h} - 0.05\right)\left(d^3 - 17.682d^2 + 144.992d + 1459.612\right) \\ &\left(\tfrac{1}{b^3} - 0.15\tfrac{1}{b^2} - 0.003\tfrac{1}{b} + 0.003\right)\tfrac{1}{4.417-v} - 9.125 \times 10^4 \end{aligned} \quad (2)$$

$$K2 = 2.696 \times 10^{-6}E\left(\frac{1}{h^2} - 9.16\frac{1}{h} - 0.18\right)\left(d^3 - 6.912d^2 - 450.422d - 9408.926\right)(0.83 + \nu) + 1540 \quad (3)$$

$$K_3 = -13.92E\frac{1}{H^{1.1}}\left(\frac{1}{b^2} - 1.794\frac{1}{b} - 0.087\right)\left(\frac{1}{1.128 - \nu}\right) + 8.076 \times 10^6 \quad (4)$$

$$K_4 = -28.15E\frac{1}{H^{0.3}}\left(\frac{1}{b^2} - 0.4\frac{1}{b} + 0.009\right)\left(\frac{1}{5.8 - \nu}\right) + 1.122 \times 10^7 \quad (5)$$

$$\begin{aligned} K_5 = &\ 0.002413E(d^3 - 7.373d^2 - 223.914d + 136.272) \\ &\left[\left(\tfrac{1}{b}\right)^3 - 0.329\left(\tfrac{1}{b}\right)^2 + 0.007\left(\tfrac{1}{b}\right) - 4.9 \times 10^{-5}\right]\left(\tfrac{1}{0.7-\nu} + 6.579\right) + 805.6 \end{aligned} \quad (6)$$

$$\begin{aligned} K_6 = &\ 0.0006815E(d^3 - 18.856d^2 + 149.856d + 314.46) \\ &\left[\left(\tfrac{1}{b}\right)^3 - 0.299\left(\tfrac{1}{b}\right)^2 + 0.049\left(\tfrac{1}{b}\right) - 0.00026\right]\left(\tfrac{1}{0.7-\nu} + 17.158\right) - 8769 \end{aligned} \quad (7)$$

When $B \leq 30$ m, $H \leq 30$ m, and 3 m $\leq D \leq 3.5$ m, the following relationship holds:

$$K_4 = (0.5 \sim 0.6)K_3 \quad (8)$$

$$K_2 = (0.85 \sim 0.9)K_1 \quad (9)$$

where K_1 and K_2 represent the normal and tangential stiffness of the side, respectively (Pa); K_3 and K_4 are the normal and tangential stiffness of the bottom surface, respectively (Pa); K_5 and K_6 denote the normal and tangential stiffness of the top surface (Pa); b and h are the width and height of the structural section, respectively (m); d refers to the burial depth of the roof (m); H is the distance between the station floor and bedrock (m); and E and v represent the elastic modulus (Pa) and Poisson's ratio, respectively.

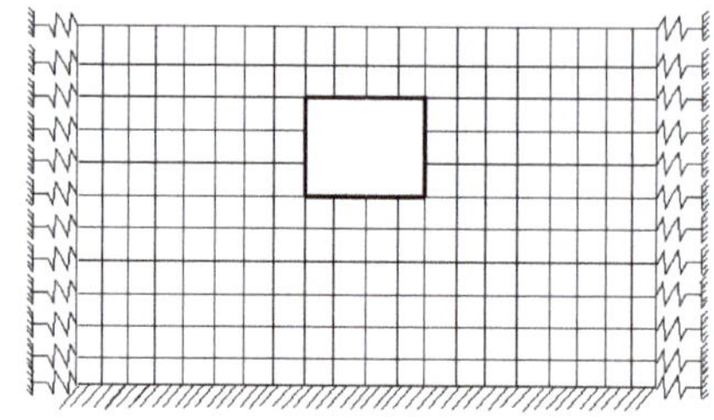

(**a**)

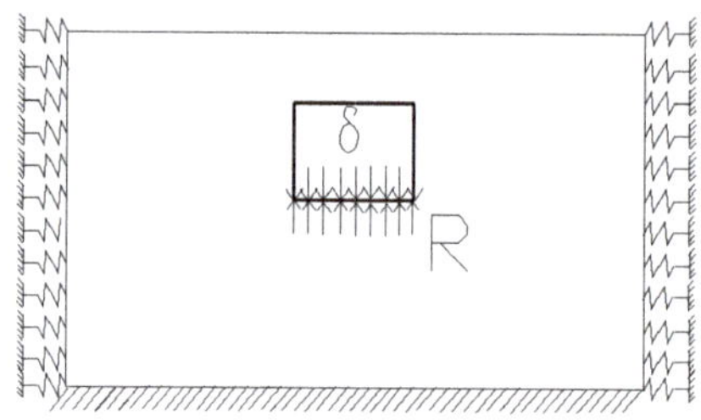

(**b**)

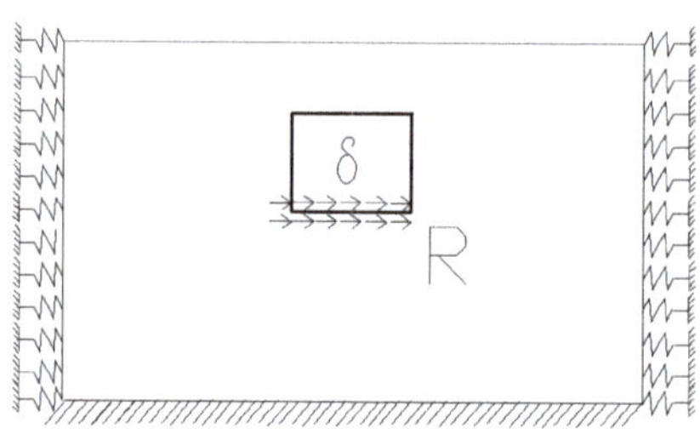

(**c**)

Figure 4. Finite element calculation model of reaction spring parameters of underground structure foundation. (**a**) Meshed model, (**b**) normal spring stiffness calculation model, (**c**) tangential spring stiffness calculation model.

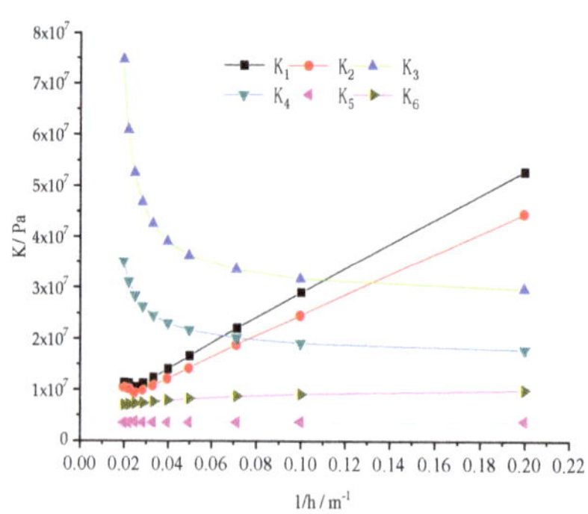

(**a**)

Figure 5. *Cont.*

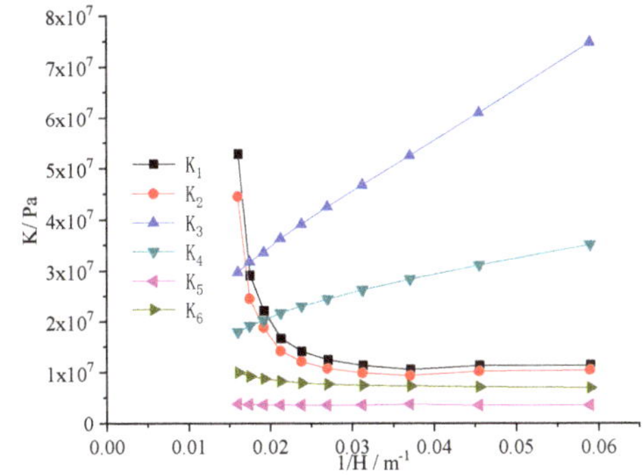

(**b**)

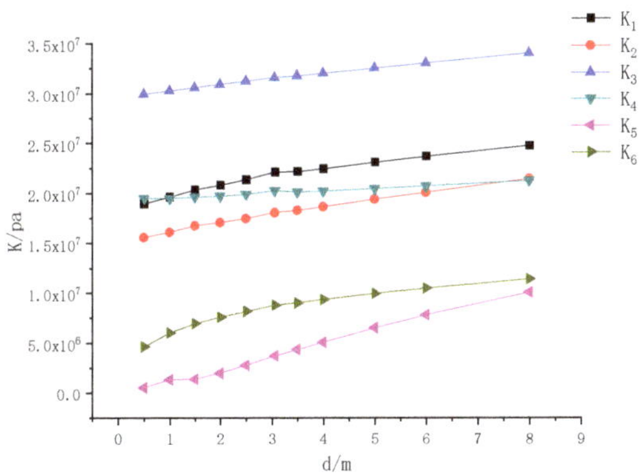

(**c**)

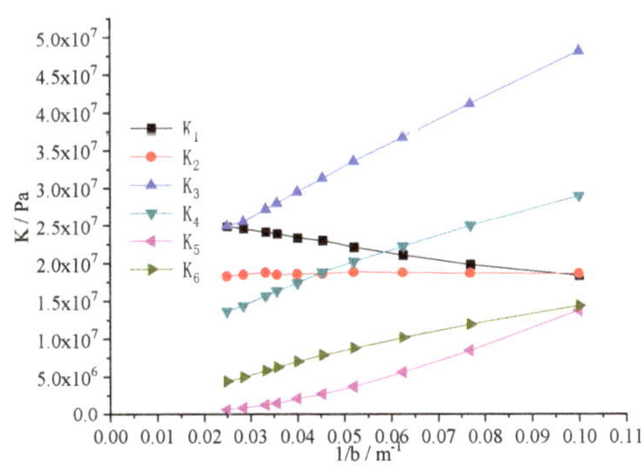

(**d**)

Figure 5. *Cont.*

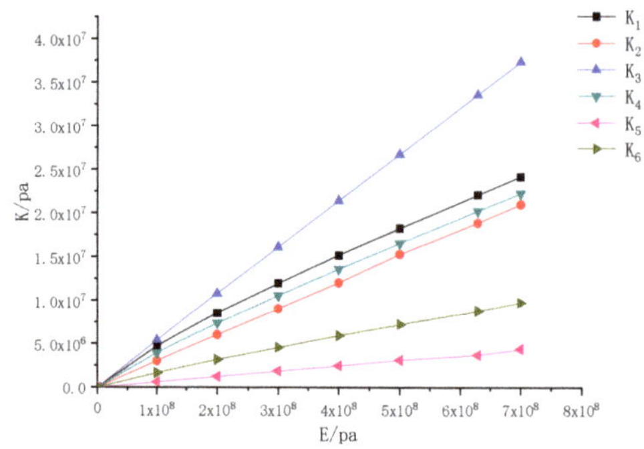

(e)

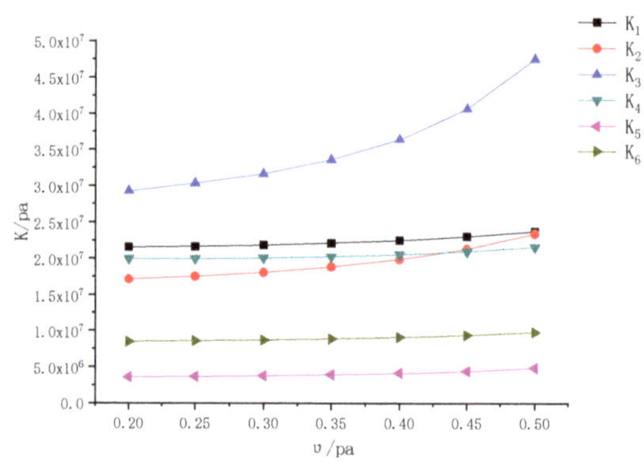

(f)

Figure 5. Influences of various factors on foundation spring stiffness: (**a**) K *versus* $1/h$, (**b**) K *versus* $1/H$, (**c**) K *versus* d, (**d**) K *versus* $1/b$, (**e**) K *versus* E, (**f**) K *versus* v.

(2) Horizontal Relative Displacement of Strata in a Loess Area

A numerical model of free field dynamic response considering seismic intensity, ground motion characteristics and loess soil characteristics was established, as shown in Figure 6. Where 1 is an infinite element boundary, 2 denotes a fixed boundary, and 3 is bedrock. The maximum response horizontal displacement of free field was obtained by inputting the Xi'an artificial wave, Taft wave, and Songpan wave in turn (Table 1 and Figure 7).

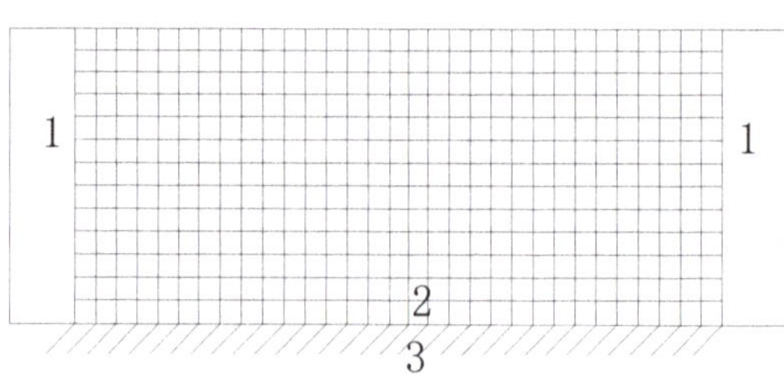

Figure 6. A numerical model of free field vibration.

Table 1. Maximum free-field horizontal displacement.

Seismic Fortification Intensity/Degree	7	7.5	8	8.5	9
Peak acceleration of seismic wave/g	0.1	0.15	0.2	0.3	0.4
Peak value of horizontal displacement/m	0.251	0.502	0.528	0.847	1.141

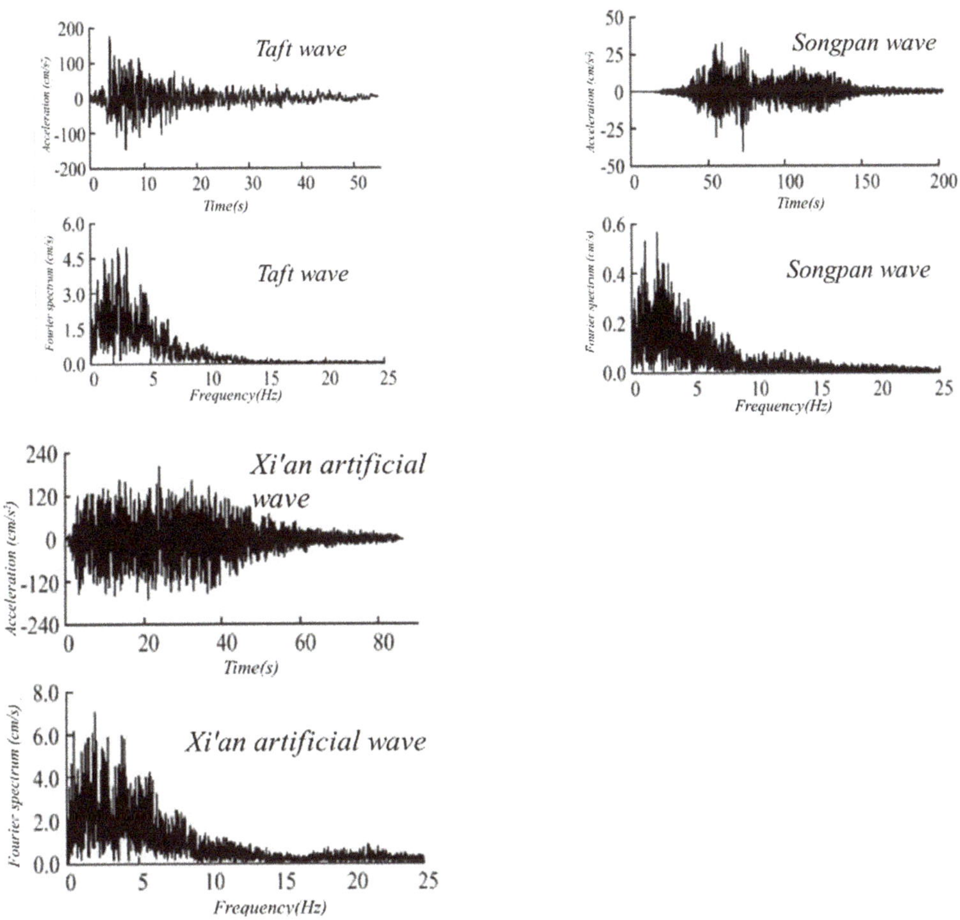

Figure 7. Acceleration and Fourier spectra of input ground motion.

3. Experiment and Simulation

3.1. Background to the Experiment

The geomorphic unit of Xi'an Feitian Road Station belongs to the second and third grade loess tableland. The lithologic characteristics of the site strata are summarized in Tables A1 and A2. The table is arranged from top to bottom according to the order of soil layers from shallow to deep. The underground subway station is a reinforced concrete structure with a total height of 14.01 m and a total width of 19.2 m. The longitudinal spacing of the center pillar is 9 m. The cross section of the center pillar measures 0.8 m × 1.2 m. The depth of soil above the roof is 3.459 m. The density of concrete ρ is 2.5 g/cm^3, its modulus of elasticity E is 35 GPa, and Poisson's ratio ν is 0.15. A typical cross section is shown in Figure 8.

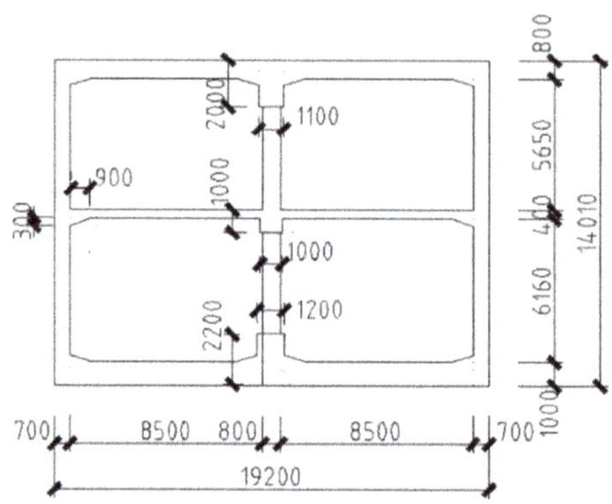

Figure 8. Schematic diagram of a typical cross section (dim.: mm).

The station is located in Chang'an district of Xi'an city. The site type in this area is class II, the basic seismic intensity is 8 degrees, and the characteristic period of seismic response spectrum is 0.4 s. The seismic parameters of station engineering are displayed in Table 2.

Table 2. Ground motion parameters.

Position	Parameter	50-Year Exceedance Probability			100-Year Exceedance Probability	
		63%	10%	2%	10%	2%
Ground	Tg(s)	0.38	0.43	0.68	0.59	0.83
	Ag(g)	0.079	0.235	0.457	0.340	0.575
Floor	Tg(s)	0.60	0.64	0.90	0.75	1.00
	Ag(g)	0.051	0.151	0.358	0.247	0.441

3.2. Shaking-Table TEST Scheme

Shaking-table tests on loess subway station models are mainly used to assess the seismic mechanism and soil structure dynamic interaction mechanism of subway station structures in loess sites. Due to the small size of the subway station model used here, it is difficult to eliminate the gravitational distortion effect by the artificial mass model with full counterweight, so an added-mass model was adopted. Based on the Buckingham π theorem, the length, elastic modulus, and acceleration were selected as basic physical quantities, and the table size, dynamic performance, bearing tonnage, and other supporting equipment performance of the test system were fully considered to ascertain the similarity relationship of the model system (Table A3). In the test, the method of sticking lead blocks into the structure was used to realize the additional artificial mass. Taking Feitian Road Station of Xi'an Metro Line 4 as the prototype structure, the subway station model was established by using particulate concrete and galvanized steel wire. The loess was taken from the foundation pit of Feitian Road Station of Xi'an Metro Line 4. When preparing the model foundation, the loess was layered into the model soil box, and the water content and density of the model foundation were controlled according to the natural water content and density of the prototype site soil. The excitation system used in this test was a horizontal one-way high-performance seismic simulation shaking table produced by MTS Company, Eden Prairie, MN, USA. The table measures 3.36 m × 4.86 m, the maximum load is 25 t, and the maximum acceleration is 1.0 g. The Taft wave, Songpan wave and Xi'an artificial wave were selected as input ground motions, respectively. Since the subway station was not situated in an active fault zone, only the transverse ground motion was input. The sensor arrangement is shown in Figure 9.

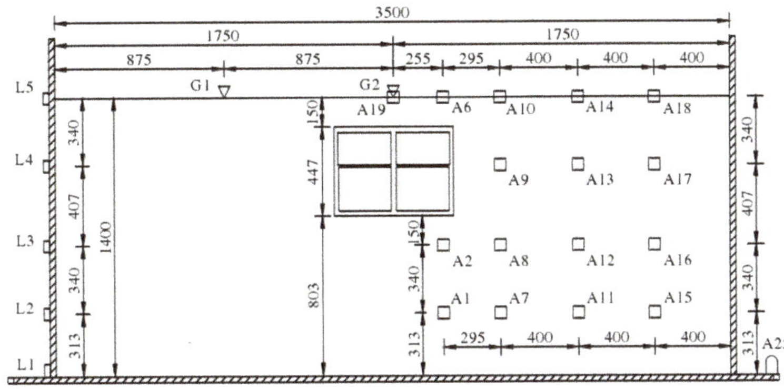

Figure 9. Sensor layout (dimensions: mm).

3.3. FDTHA Modeling

Although the three-dimensional model has the advantages of high calculation accuracy, the modeling is complex and the solution does not readily converge, posing a difficulty for engineering designers. In addition, considering the long length of the flying station model, the size and structure of each section are similar; the three-dimensional dynamic interaction system of loess subway underground structure was considered as a two-dimensional plane strain problem in the seismic analysis by a time-history analysis method.

The typical section of loess site and subway underground structure was numerically simulated, and the model was established as shown in Figure 10 [39], where 1 represents an infinite element boundary, 2 is a fixed boundary, and 3 denotes the bedrock.

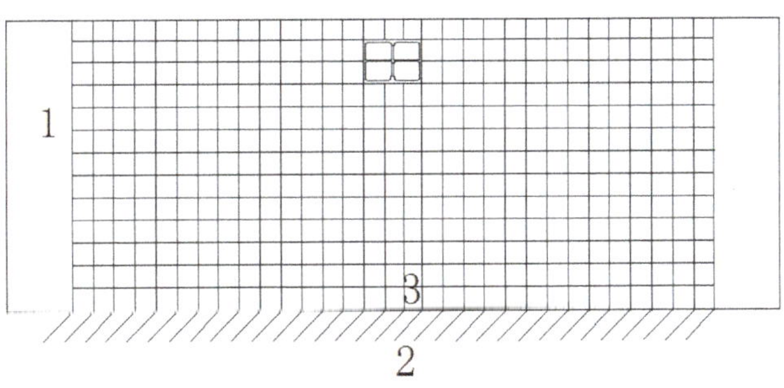

Figure 10. Schematic representation of the time-history analysis calculation model.

The finite element mesh was used to simulate the near-field region, and the infinite element boundary was utilized to simulate the far-field region far away from the structure. The size of the finite element mesh (Figure 10) is 150 m × 70 m (width × height), the height of the infinite element mesh shall be consistent with that of the finite element mesh, and the width is unlimited. The Taft wave, Songpan wave, and Xi'an synthetic wave were selected. The plastic damage model of concrete in ABAQUS finite element analysis software was employed to simulate the mechanical behavior of the prototype subway station concrete. The parameters used in the plastic damage model are listed in Table 3. The CPE4R (Four node bilinear plane strain quadrilateral element, reduced integration) element was used for on-site soil and the CPE4 element was used to model the subway station structure. The initial stress on the soil was calculated using the geostatic module in ABAQUS. To simulate the in-situ stress on the soil and how it affects the adjacent underground structure, the model states of excavation, support, construction of underground structure, and backfilling were established to simulate the construction process of this subway station, and the stress state in the soil after construction was taken as the initial stress field for the later dynamic analysis.

Table 3. Parameters used in the plastic damage model.

Parameter	Value	Parameter	Value
Density/kg/m^3	2500	Angle of dilation ψ/°	30
Elastic modulus E/MPa	0.66×10^4	Coefficient of viscosity μ	0.0005
Poisson's ratio ν	0.2	Tensile variable ω_t	0
Ultimate compressive stress/MPa	5.39	Compression variable ω_c	1
Invariant stress ratio K_c	0.667	Damping ratio ξ	0.1

4. Results

4.1. Earthquake Damage

The seismic damage to this structure was observed by the shaking-table test (Figures 11 and 12). The column is the most severely damaged component in the subway station model. There are vertical cracks in the upper middle column, and the longitudinal reinforcement of the lower middle column is exposed with significant spalling (a typical shear-compression failure). The most severe damage to the side wall and the center column occurs at the joint with the top and bottom plates, and the seismic damage entails significant concrete spalling and the armpit angle reinforcement is pulled out. The roof and floor remain in good condition after the earthquake. The widest crack is 15 mm across and the maximum vertical differential settlement is 32 mm.

(a) (b) (c)

Figure 11. Earthquake damage phenomenon in the shaking-table test: (**a**) failure mode of the join between central column and slab, (**b**) failure mode of the joint between side wall and slab, (**c**) failure mode of the joint between side wall and slab.

(a) (b) (c)

Figure 12. Final seismic failure phenomenon of the foundation in the model test: (**a**) macroscopic earthquake damage, (**b**) fracture distribution, (**c**) crack and settlement measurement.

4.2. Reliability Analysis of FDTHA

Under the action of seismic waves with different peak accelerations, when the maximum shear deformation occurs on the surface and bottom of the model foundation, the comparison of the horizontal relative displacement of each measuring point relative to the bottom of the model foundation is shown in Figure 13. The distribution of the horizontal relative displacement of the model foundation along the soil depth in the numerical simulation and shaking-table test is consistent. Acceleration sensors were, respectively, arranged along the side wall of the model structure from bottom to top. Under the action of seismic wave, the comparisons of acceleration response time history and corresponding Fourier spectrum of each measuring point in the model structure between numerical simulation and shaking-table test are demonstrated in Figure 14. The time-history waveform, amplitude, and Fourier spectrum of the acceleration response in the model structure recorded by the numerical simulation and shaking-table test are similar. Therefore, the numerical simulation and the numerical model of the dynamic interaction between the loess and a subway station established in this paper can be deemed to have simulated the acceleration response of a subway station structure reliably.

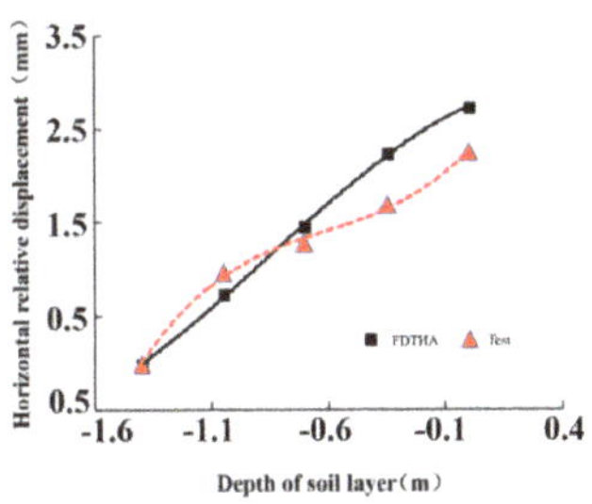

Figure 13. Time-history analysis and horizontal relative displacement of model foundation in the shaking-table test.

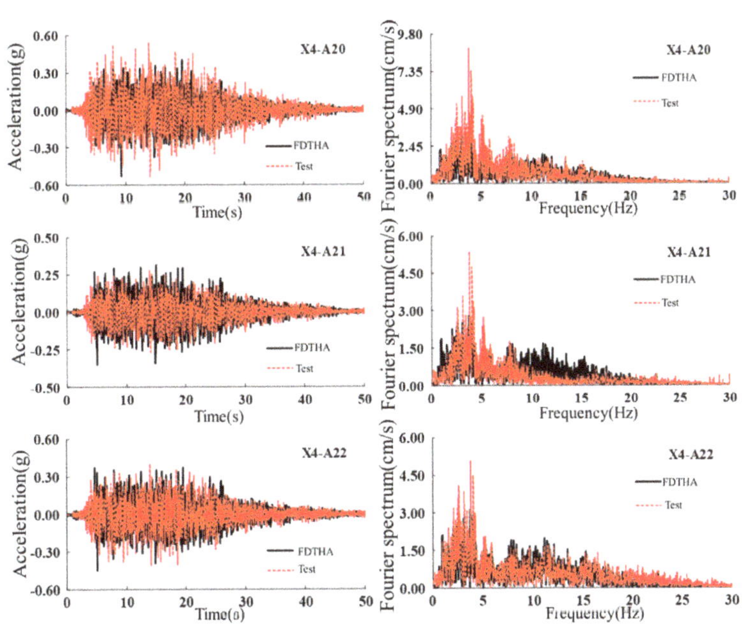

Figure 14. Time-history analysis and acceleration response of model structure in the shaking table test.

4.3. Reliability Analysis of Seismic Calculation Methods for Subway Underground Structure

The accuracy of the other three quasi-static seismic calculation methods was compared and evaluated based on the structural internal force output results of the time-history analysis method (Table A4). Observation points a and b mark the top and bottom of the upper column, respectively; c and d mark the top and bottom of the lower column, respectively; e and f mark the top and bottom of the upper wall, respectively; and g and h mark the top and bottom of the lower wall, respectively. According to the data in Table 4 and the shaking-table damage phenomenon, the difference is presented as follows:

Table 4. Comparison of structural internal forces in normal use.

Observation Point	Bending Moment/kN·m		Shear Force/kN		Axial Force/kN	
	Normal Use Stage	Under Lateral Seismic Action	Normal Use Stage	Under Lateral Seismic Action	Normal Use Stage	Under Lateral Seismic Action
a	−0.014	215.6	0.005	−201	−1180	−1031.7
b	78.2	−153.9	−132.000	−284	−1260	−1154.5
c	0.005	131.6	0.003	−419.6	−1560	−1421.9
d	90.5	−197.5	−134	−517.6	−428	−1558.6
e	−248	−658.2	−976	−118.2	−547	−506.9
f	−550	49.2	1050	−180.9	−727	−794.8
g	−347	−224.9	−183	−272.5	−610	−854.3
h	−418	−1352.4	568	−358.2	−1080	−1395

(1) After selecting reasonable seismic design parameters (the practical data tabulated herein were used for the horizontal relative displacement of the stratum in the loess area, and the comprehensive recommended equation proposed in this paper was used for the foundation spring stiffness), the results of the structural internal force of the DESANM are more consistent with the time-history analysis method. The FBM that neglects the seismic shear force leads to underestimation of the internal forces. The bending moment and shear force output by DBM are closest to those predicted by use of the time-history analysis method, but the accuracy of the axial force output remains insufficient.

(2) From the shaking-table test, the top of the upper column is the most severely damaged, and the bottom of the lower column follows. The results of time-history analysis and DESANM are consistent with the damage seen in the shaking-table test; the maximum bending moment on the middle column is located at the top of the upper column, followed by the bottom of the lower column. Therefore, the comprehensive recommended equation and practical tabulated data better reflect the effect of soil on the structure.

4.4. Seismic Analysis

(1) The internal forces on an underground structure, as calculated by FBM, are small, which is related to the fact that the friction and shear stresses caused by soil deformation are not considered in the FBM. The DBM is simple to establish, and the principle is to simulate the interaction between soil and structure by establishing springs and applying horizontal displacement to the key stratum.

(2) The DBM simulates the interaction between soil and structure by establishing springs and applying horizontal displacement to a stratum, and the model is simple. The correct values of spring stiffness and ground horizontal displacement make the output of structural shear force and bending moment approach those calculated using the time-history method of analysis. The output accuracy of shear force and bending moment values is higher than that of DESANM, but the accuracy of axial force output is poorer than that of DESANM.

(3) The loess stratum is established in the calculation model of the DESANM, so the calculation relies on fewer parameters. When the value of horizontal relative displacement is accurate, the results of structural internal force of DESANM are consistent with that of the time-history method of analysis. However, the calculation is more onerous, ranking second only in complexity compared to that in the time-history analysis.

5. Modification of DBM

To provide the designers of such an underground subway station with good seismic calculation accuracy and a simple calculation model, the modified DBM was used to make the axial force output more accurate.

5.1. Influences of Transverse Seismic Force on Subway Underground Structure

Only dead load and live load are input into the time-history analysis model to obtain the structural internal forces acting on the station in its normal service conditions, therefore, the influences of transverse seismic forces thereon can be determined.

It can be seen from Table 4 that, differing from the trend in shear force results and bending moment output, the axial forces are similar to those predicted using FDTHA. Therefore, the axial forces on such an underground structure are greatly affected by vertical load and less affected by transverse seismic load. According to the earthquake damage phenomenon affecting such a system in loess deposits, the soil still undergoes vertical displacement under the action of transverse earthquake excitation, and then generates vertical seismic earth pressures on the underground structure. Therefore, the neglect of the vertical seismic earth pressure of soil on the structure is the reason why the axial force predicted by DBM is too small. For this reason, the vertical seismic earth pressure of soil on the structure and the correct calculation of vertical earth pressure were incorporated in the original DBM calculation model.

5.2. Vertical Seismic EARTH Pressure

The tendency of the soil mass to undergo vertical relative displacement with horizontal displacement was revealed through shaking-table testing and analysis of the numerical model of the dynamic response to a transverse seismic wave. The results of the vertical relative displacement are listed in Table 5. The origin is set at the surface, and the change of vertical relative displacement along the burial depth (z) is assumed to be a cosine function $\mu_{a1(z)}$, as shown in Equation (10).

$$u_{a1}(z) = \frac{1}{2}u_{\max 2} \bullet \cos\frac{\pi z}{2H_b} \tag{10}$$

where the $\mu_{\max 2}$ is the peak vertical relative displacement of the stratum, H_b is the burial depth to the bedrock surface (dimension: m).

Table 5. Peak vertical displacements of the soil mass.

Seismic Fortification Intensity/Degree	7	7.5	8	8.5	9
Peak acceleration of seismic wave/g	0.1	0.15	0.2	0.3	0.4
Peak value of surface displacement/mm	10.5	13.3	16.2	19.7	24.1

The earth pressure generated by the soil on the top plate of the structure under earthquake was calculated by using Equation (11).

$$N_1 = K \bullet \mu_2(z) \tag{11}$$

where K is the stiffness parameter of the normal foundation reaction spring of the structural roof (Pa), N_1 denotes the vertical seismic earth pressure at the roof (N), and z is the burial depth of the roof (m).

5.3. Earth Pressure Acting on the Floor

The base plate exerts pressure on the foundation, and its reaction force is the support force of the foundation on the base plate. The support force is in line with the base pressure in the opposite direction. The supporting force is called the bottom plate pressure, and the calculation method is consistent with the base pressure, as shown in Equation (12).

$$N_2 = \gamma d \tag{12}$$

where N_2 represents the bottom plate pressure (N), d denotes the burial depth of the bottom plate (m), γ is the bulk unit weight of the soil mass (N/m^3).

5.4. Reliability Analysis: Modified DBM

The comparison of structural axial force values after correction by DBM is embodied by Table 6. The accuracy of the axial force predicted by the modified DBM is greatly improved, and the discrepancy is within 16%, and the average error is 6.65%, which agrees with the axial force results of the FDTHA method. The error in DBM axial force result is as high as 86.47%, and the average error is 48.30%, which is related to the problem whereby the DBM technique ignores the soil to estimate the vertical seismic earth pressure and vertical earth pressure on the structure. Compared with DBM, the accuracy of our modified DBM is improved by 80.17%, and the average error is reduced by 41.65%. The modified DBM overcomes the problem whereby the DBM ignores the soil to estimate the vertical seismic earth pressure and vertical earth pressure of the structure, which leads to the underestimation of the structural axial force. This method improves the calculation accuracy of the DBM and provides a simple and fast calculation method with good seismic calculation accuracy for designers of underground subway stations.

Table 6. Comparison of axial force predicted by the modified DBM.

Observation Point		Modified DBM/kN	DBM/kN	FDTHA/kN	Error in Modified DBM	Error in DBM
Central column	a	−1172.70	−510.95	−1031.70	13.70%	−50.47%
	b	−1184.20	−522.45	−1154.50	2.60%	−54.75%
	c	−1399.20	−712.10	−1421.90	−1.60%	−49.92%
	d	−1411.90	−724.80	−1558.60	−9.40%	−53.50%
Flank	e	−509.40	−388.26	−506.90	0.50%	−23.41%
	f	−671.30	−739.00	−794.80	−15.50%	−7.02%
	g	−823.80	−330.18	−854.30	−3.60%	−61.35%
	h	−1307.50	−188.70	−1395.00	−6.30%	−86.47%

6. Conclusions

(1) A new analysis method useful for the evaluation of the seismic behavior of underground structures in loess area was proposed based on the DBM and it is named the modified DBM. The DBM was modified according to the results obtained from a shaking-table test and numerical simulations. The results show that the modified displacement-based method improves the accuracy of structural axial force output and compensates for the defects of the DBM. The modified DBM is applicable to any typology of underground structure and it can be applied to the seismic design of underground structure.

(2) The modified DBM was compared with the most commonly used methods of analysis (FBM, DESANM, DBM, and FDTHM). The results were validated considering using data from a shaking table test based on the Feitian Road underground subway station in Xi'an. The comparison of output results shows that the modified DBM offers good accuracy, a simple model, rapid modeling, and easy convergence.

(3) Based on the finite element analysis method, a practical table of horizontal relative displacements of strata in loess areas, and the comprehensive recommended equation

of foundation reaction spring stiffness, were provided. The practical table and the comprehensive recommended equation provide a method of estimation of those seismic calculation parameters required by designers of underground subway stations in loess areas.

(4) Due to the special properties of loess, the modified DBM is suitable for the seismic response analysis of underground structures in loess area. However, whether it is applicable to other soils remains to be studied in future research.

Author Contributions: Conceptualization, R.Z. and D.Q.; methodology, D.Y.; Project administration, J.Z. All authors have read and agreed to the published version of the manuscript.

Funding: This study was supported by Chongqing Postdoctoral Special Fund (020919014) and The Natural Science Foundation of Chongqing, China (cstc2020jcyj-bshX0118) and National Natural Science Foundation of China (U20A20314).

Institutional Review Board Statement: Not applicable.

Informed Consent Statement: Not applicable.

Data Availability Statement: All data are available from the author.

Conflicts of Interest: The authors declare no conflict of interest.

Appendix A

Table A1. Physico-mechanical properties of soil layer.

Number	Soil	Weight Density/kN/m^3	Elastic Modulus/MPa	Cohesion/kPa	Internal Friction Angle/°	Liquidity Index
3-1-1	New loess	16.2	7	27	24.5	0.17
3-2-1	Paleosol	17.4	10	45	24	0.15
4-1-1-1	Old loess	16.1	9	35	23	0.3
4-2-1	Paleosol 2	17.6	11	44	23	0.31
4-1-1-2	Old loess 2	16.9	10	36	22.5	0.44

Table A2. Soil properties with increasing depth.

Number	Soil	Thickness/m	Equivalent Shear Wave Velocity/m/s	Density/kg/m^3
3-1-1	New loess	7.3	223.8	1620
3-2-1	Paleosol	3	335.5	1740
3-1-1	New loess	8.8	348.4	1620
3-2-1	Paleosol	4.2	354	1740
3-1-1	New loess	3.3	360	1620
3-2-1	Paleosol	3.6	361.7	1740
3-1-1	New loess	3.6	371.9	1620
4-2-1	Paleosol 2	5.1	384.9	1760
3-1-1	New loess	2.6	403.8	1620
4-2-1	Paleosol 2	2.3	422.8	1760
4-1-1-1	Old loess	4	422.8	1610
4-2-1	Paleosol 2	2.1	458.5	1760
4-1-1-1	Old loess	3.2	465	1610
4-2-1	Paleosol 2	2.5	465.3	1760
4-1-1-1	Old loess	8	475.7	1610
4-2-1	Paleosol 2	1.6	483.5	1760
4-1-1-2	Old loess 2	4.8	488.6	1690

Table A3. Similar constants in the test model.

Physical Property	Physical Quantity	Similarity Relationship	Similarity Constant
Geometric characteristics	Length l	S_l	1/30
	Area A	$S_A = S_l^2$	1/900
	Linear displacement l	S_l	1/30
Material characteristics	Elastic modulus E	S_E	1/5
	Stress σ	$S_\sigma = S_E$	1/5
	Density ρ	$S_\rho = S_E/(S_l S_\alpha)$	3.0
	Quality m	$S_m = S_\sigma S_l^2/S_\alpha$	1.11×10^{-4}
Load performance	Point force F	$S_F = S_\sigma S_l^2$	2.22×10^{-4}
	Line load q	$S_q = S_\sigma S_l$	6.67×10^{-3}
	Moment M	$S_M = S_\sigma S_l^3$	7.41×10^{-6}
Dynamic characteristics	Time t	$S_t = S_l^{0.5} S_\alpha^{-0.5}$	0.13
	Velocity v	$S_t = S_l^{0.5} S_\alpha^{0.5}$	0.26
	Acceleration α	S_α	2.0

Table A4. Comparison of internal forces at control points of structures with different calculation methods.

		Bending Moment/N·m				Shear Force/N				Axial Force/N			
Observation Point		FDTHA	FBM	DBM	FBM	FDTHA	FBM	DBM	FBM	FDTHA	FBM	DBM	FBM
Central column	a	215,555	10,423.3	238,803	279,008	−49,184	−4141.44	−44,832	−69,461.7	−1,031,720	−4141.44	−510,953	−1,045,410
	b	−153,906	−6832.71	−181,235	−198,130	−67,739	−4141.44	−44,832	−178,655	−1,154,460	−4141.44	−522,453	−1,140,410
	c	131,605	−2701.63	121,486	363,861	−43,857	741.297	−109,800	−92,933.5	−1,421,870	741.297	−712,100	−1,297,840
	d	−197,523	930.728	−214,180	−615,633	−50,276	741.297	−109,800	−261,956	−1,558,590	741.297	−724,804	−1,408,240
Flank	e	−658,209	−66,093.9	−677,207	−435,546	769,365	−99,877.4	659,369	−532,570	−506,949	−99,877.4	−388,257	−536,758
	f	49,262	45,317.8	43,402	502,996	−143,782	29,138.5	−724,622	−178,655	−794,779	29,138.5	−738,997	−772,260
	g	−224,909	−30,203.7	−251,916	−18,654.4	231,968	−40,416.1	661,587	−248,115	−854,289	−40,416.1	−330,183	−855,921
	h	−1,352,410	−147,243	−1,768,540	363,861	−1,995,360	100,578	−1,435,260	−2,190,510	−1,394,980	100,578	−1,887,020	−1,411,240

References

1. An, X.; Shawky, A.; Maekawa, K. The collapse mechanism of a subway station during the great Hanshin earthquake. *Cem. Concr. Compos.* **1997**, *19*, 241–257. [CrossRef]
2. Samata, S.; Ohuchi, H.; Matsuda, T. A study of the damage of subway structures during the 1995 Hanshin—Awaj i earthquake. *Cem. Concr. Compos.* **1997**, *19*, 223–239. [CrossRef]
3. St. John, C.M.; Zahrah, T.F. Aseismic design of underground structures. *Tunn. Undergr. Space Technol.* **1987**, *2*, 165–197. [CrossRef]
4. MOHURD; General Administration of Quality Supervision, Inspection and Quarantine of the People's Republic of China. *Code for Seismic Design of Building*; China Architecture & Building Press: Beijing, China, 2010; pp. 6–43.
5. MOHURD; General Administration of Quality Supervision, Inspection and Quarantine of the People's Republic of China. *Code for Design of Civil Defence Basement*; China Architecture & Building Press: Beijing, China, 2005; pp. 10–35.
6. MOHURD; General Administration of Quality Supervision, Inspection and Quarantine of the People's Republic of China. *Code for Seismic Design of Subway Structures*; China Architecture & Building Press: Shanghai, China, 2009; pp. 43–67.
7. MOHURD; General Administration of Quality Supervision, Inspection and Quarantine of the People's Republic of China. *Code for Seismic Design of Urban Rail Transit Structures*; China Planning Press: Beijing, China, 2014; pp. 34–40.
8. MOHURD; General Administration of Quality Supervision, Inspection and Quarantine of the People's Republic of China. *Standard for Seismic Design of Underground Structures*; China Architecture & Building Press: Beijing, China, 2018; pp. 6–50.
9. Ministry of Railways of the People's Republic of China. *Code for Seismic Design of Railway Engineering*; China Planning Press: Beijing, China, 2006; pp. 40–43.
10. Ministry of Transport of the people's Republic of China. *Specifications for Seismic Design of Highway Tunnels*; People's communications Publishing House Co., Ltd.: Beijing, China, 2019; pp. 10–13.
11. The State Bureau of Quality and Technical Supervision; MOHURD. *Code on Geotechncal Investigations for Metro and Light Rail Transit*; China Planning Press: Beijing, China, 1999; pp. 13–20.
12. MOHURD; General Administration of Quality Supervision, Inspection and Quarantine of the People's Republic of China. *Code for Geotechnical Investigations of Urban Rail Transit*; China Planning Press: Beijing, China, 2012; pp. 6–10.
13. AFPS/AFTES. *Guidelines on Earthquake Design and Protection of Underground Structures*; Version 1; Working Group of the Frech Association for Seismic Engineering (AFPS); French Tunneling Association (AFTES): Paris, France, 2001.
14. ALA. *Seismic Fragility Equationtions for Water Systems, Part1-Guideline*; ASCE-FEMA: Reston, VA, USA, 2001.
15. ASCE Committee in Gas and Liquid Fuel Lifelines. *Guidelines for the Seismic Design of Oil and Gas Pipeline Systems*; Technical Council on Lifeline Earthquake Engineering; ASCE: New York, NY, USA, 1984.
16. Kuesel, T.R. Earthquake design criteria for subways. *J. Struct. Div.* **1969**, *95*, 1213–1231. [CrossRef]

17. Japan Road Associaton. *Guide Specifications of Design and Construction of Underground Parking Lots*; Japan Road Association: Tokyo, Japan, 1992.
18. Wang, J.N. *Seismic Design of Tunnels*; Parson Brinckerhoff Inc.: New York, NY, USA, 1993.
19. Penzien, J. Seismically induced racking of tunnels lining. *Earthq. Eng. Struct. Dyn.* **2000**, *27*, 683–691. [CrossRef]
20. Kawashima, K. Seismic analysis of underground structures. *J. Disaster Res.* **2006**, *1*, 378–389. [CrossRef]
21. Kiyomiya, O. Earthquake-resistant design features of immersed tunnels in Japan. *Tunn. Undergr. Space Technol.* **1995**, *10*, 463–475. [CrossRef]
22. Hashash, Y.M.A.; Hook, J.J.; Schmidit, B.; Yao, J.C. Seismic design and analysis of underground structures. *Tunn. Undergr. Space Technol.* **2001**, *16*, 247–293. [CrossRef]
23. Kisiridis, A.E. Seismic analysis of underground structures. *Int. J. Soil Dyn. Earthq. Eng.* **1983**. [CrossRef]
24. Penzien, J.; Wu, C. Stresses in linings of bored tunnels. *Earthq. Eng. Struct. Dyn.* **1998**, *27*, 283–300. [CrossRef]
25. Bobet, A. Effect of pore water pressure on tunnel support during static and seismic loading. *Tunn. Undergr. Space Technol.* **2003**, *18*, 377–393. [CrossRef]
26. Park, K.H.; Tantayopin, K.; Tontavanich, B.; Owatsiriwong, A. Analytical solution for seismic-induced ovaling of circular tunnel lining under no-slip interface conditions: A revisit. *Tunn. Undergr. Space Technol.* **2009**, *24*, 231–235. [CrossRef]
27. Bobet, A. Drained and undrained response of deep tunnels subjected to far-fieldshear loading. *Tunn. Undergr. Space Technol.* **2010**, *25*, 21–31. [CrossRef]
28. Zucca, M.; Valente, M. On the limitations of coupled and decoupled approaches for the seismic behaviour evaluation of shallow multi-propped underground structures embedded in granular soils. *Eng. Struct.* **2020**, *211*, 110497. [CrossRef]
29. Soccodato, F.M.; Tropeano, G. The role of ground motion characters on the dynamic performance of propped retaining structures. In Proceedings of the 6ICEGE, 6th International Conference on Earthquake Geotechnical Engineering, Christchurch, New Zealand, 1–4 November 2015.
30. Zucca, M.; Tropeano, G.; Crespi, P.; Erbì, E. 2D linear equivalent analysis for the seismic vulnerability evaluation of multi-propped retaining structures. In Proceedings of the XVII ECSMGE—European Conference on Soil Mechanics and Geotechnical Engineering, Reykjavik, Iceland, 1–6 September 2019.
31. ISO 23469. *Bases for Design of Structures-Seismic Actions for Designing Geotechnical Works*; ISO TC 98/SC3/WG10; ISO International Standard: Geneva, Switzerland, 2005.
32. Liu, R.S.; Hu, S.Q.; Shi, H.B. Study on seismic loading of pseudo-static approach used in the seismic design of underground structure. *Chin. J. Geotech. Eng.* **2007**, *29*, 237–242. (In Chinese) [CrossRef]
33. Liu, J.B.; Wang, W.H.; Zhao, D.D. Comparison of the pseudo-static method for seismic analysis of the underground structures. *Eng. Mech.* **2003**, *30*, 105–111. (In Chinese) [CrossRef]
34. Gingery, J. A simplified method for estimating shear strains for ovaling and racking analysis of tunnels. In Proceedings of the 4th International Conference on Earthquake Geotechnical Engineering, Thessalonoki, Greece, 25–28 June 2007.
35. Newmark, N.M. Problems in wave propagation in soil and rock. In Proceedings of the International Symposium on Wave Propagation and Dymamic Properties of Earth Materials, Albuquerque, NM, USA, 1–3 August 1968.
36. FWHA. *Technical Manual for Design and Construction of Road Tunnels-Civil Element*; Puplication No. FHWA-NHI-10-034; U.S. Department of Transportation, Federal Highway Administration: Washington, DC, USA, 2009; p. 702.
37. Liu, J.B.; Wang, W.H.; Zhao, D.D.; Zhang, X.B. Integral response deformation method in seismic analysis of complex section underground structures. *China Civ. Eng. J.* **2014**, *47*, 134–142. (In Chinese) [CrossRef]
38. Liu, J.B.; Wang, D.Y.; Tan, H. Response displacement methods for longitudinal seismic response analysis of tunnel structures. *J. Vib. Shock.* **2019**, *38*, 104–111. (In Chinese)
39. Yin, S.Z.; Ye, D.; Liang, Y.J. Application of 2D arc consistent viscous-spring artifical boundary element in the ABAGUS. *World Earthq. Eng.* **2017**, *33*, 69–74. (In Chinese)

applied sciences

MDPI

Article

Time-Dependent Damage Estimation of a High-Rise Steel Building Equipped with Buckling-Restrained Brace under a Series of Earthquakes and Winds

Ahmad Naqi, Tathagata Roy and Taiki Saito *

Department of Architecture and Civil Engineering, Toyohashi University of Technology, Aichi 441-8580, Japan; ahmad.naqi.sl@tut.jp (A.N.); tathagataroy1989@gmail.com (T.R.)
* Correspondence: tsaito@ace.tut.ac.jp

Abstract: This study investigates the cumulative damage of a 20-story high-rise steel building equipped with buckling-restrained braces (BRB) under the likely occurrence of earthquake and wind events in the design life of the building. The objective of this research is to introduce a method for evaluating the cumulative damage of BRBs under multi-hazard events that are expected to occur during the service life of a high-rise building in order to achieve a safer building. A methodology is proposed using a Poisson point process to estimate the timeline of earthquake and wind events, wherein the events are assumed to be independent in nature. The 20-story high-rise steel building with BRBs is designed according to the Japanese standard and analyzed using the finite element approach, considering nonlinearities in the structural elements and BRBs. The building is analyzed consecutively using the timeline of earthquakes and winds, and the results are compared with those under individual earthquakes and winds. In addition to the responses of the frame such as the floor displacement and acceleration, the damage of BRBs in terms of the damage index, the energy absorption, the plastic strain energy, and the maximum and cumulative ductility factor are evaluated. It is observed that the BRB's fatigue life under multi-hazard scenarios is a multi-criteria issue that requires more precise investigation. Moreover, the overall building's performance and BRB's cumulative damage induced by the sequence of events in the design life of the building is significantly larger than that under an individual event.

Keywords: steel structure; buckling-restrained brace; cumulative damage; multi-hazard; wind

Citation: Naqi, A.; Roy, T.; Saito, T. Time-Dependent Damage Estimation of a High-Rise Steel Building Equipped with Buckling-Restrained Brace under a Series of Earthquakes and Winds. *Appl. Sci.* **2021**, *11*, 9253. https://doi.org/10.3390/app11199253

Academic Editor: Maria Favvata

Received: 17 September 2021
Accepted: 30 September 2021
Published: 5 October 2021

Publisher's Note: MDPI stays neutral with regard to jurisdictional claims in published maps and institutional affiliations.

Copyright: © 2021 by the authors. Licensee MDPI, Basel, Switzerland. This article is an open access article distributed under the terms and conditions of the Creative Commons Attribution (CC BY) license (https://creativecommons.org/licenses/by/4.0/).

1. Introduction

The buckling-restrained brace (BRB) is an energy-absorbing device that consists of a detached steel core brace inside a steel restrainer infilled with concrete mortar. This unique combination enables the BRB to exhibit stable hysteresis behavior both in compression and tension. The BRB was originally invented in Japan, and the first practical application was reported by Fujimoto et al. in 1988 [1]. Since the first successful application of BRB for 10- and 15-story steel frame buildings in 1989, the device has gained popularity in seismically active regions around the world [2]. The excellent performance of BRB as an energy-absorbing device has been proved by many experimental tests [3–10]. The use of BRB has become one of the efficient measures to minimize the effect of seismic loadings in low- to high-rise buildings [11,12]. Moreover, the BRB is extensively employed to retrofit the existing buildings to increase the building strength and energy dissipation capacity [13–17].

The use of BRBs is common in Japan for high-rise buildings [18] to mitigate their performance since the high-rise buildings experience a large amplitude shaking due to the effect of resonance with the long period component of earthquake ground motions. Since Japan is a country that experiences frequent earthquakes and typhoons [19], the performance of the high-rise building with BRBs must be examined under the multi-hazard scenario of earthquakes and long-duration winds. Although the performance of BRBs in high-rise buildings under either earthquake and wind loads is extensively studied in Japan and worldwide, limited studies had been conducted to evaluate the cumulative damage of BRBs in the successful application of wind and earthquakes. It is because most of the seismic provisions and design guidelines specify the design method for the single event of earthquakes and winds [20].

Under the multi-hazard scenario, high-rise buildings are exposed to a series of high-cycle low-strength dynamic wind loads and low-cycle high-strength earthquake loads. The literature review of BRB's fatigue life under the low-cycle excitation [21–25] indicates that the devices are capable of withstanding the design load and enduring the design threshold successfully. The effect of low-cycle fatigue of BRBs for three consecutive Level-2 earthquakes is investigated by Usami et al. [21] and the BRBs' fatigue performance is found to be in an acceptable range. On the other hand, although the failure mechanism of wind-induced fatigue for steel buildings has been examined by researchers [26–29], there are few studies to discuss the BRBs under the combination of low- and high-cycle excitations in high-rise buildings.

The objective of this research is to evaluate the cumulative deformation capacity of BRBs installed along the height of a 20-story steel frame building under multi-hazard scenarios. The target building is designed to satisfy the design criteria recommend in the Japanese regulation. Then, the performance of the building and the cumulative deformation of BRBs are evaluated under successive application of multi-hazards in the building lifetime, consisting of the low-cycle earthquake and high-cycle wind loadings. The damage assessment method presented in this study can help designers to achieve safer buildings evaluating the cumulative damage of BRBs under multi-hazard events.

2. Target Steel Building with BRB Devices

2.1. Building Description

The target building is a 20-story steel moment-resisting frame designed according to Japanese standards [30–32]. It is assumed that the target building is located in the Aichi Prefecture, Japan, with a 100-year service period. The elevation of the frame and the configuration of BRBs along the height of the frame are presented in Figure 1. The total height of the frame is 71 m, while the ground floor height is 4.5 m, and the remaining floor height is 3.5 m. There are five spans, with three spans of 6.4 m and two spans of 4 m. The hollow square-box and wide flange of SN490B steel type are selected for the column and beam elements, with the yield strength of 325 MPa and tensile strength of 490 MPa. The member size and thickness of the elements in each story are summarized in Table 1.

Table 1. Structural details of beam and column elements (mm).

Steel Beam				Steel Box Column		
Story	H × B	t_W	t_F	Story	H × B	t
19–20	450 × 250	9	16	-	-	-
17–18	450 × 250	9	16	17–20	350 × 350	19
14–16	500 × 250	12	22	13–16	400 × 400	19
11–13	500 × 300	12	16	9–12	450 × 450	19
8–10	500 × 300	12	25	6–8	450 × 450	22
5–7	550 × 300	12	16	4–5	500 × 500	22
2–4	550 × 300	12	25	2–3	500 × 500	25
1	600 × 300	12	19	1	550 × 550	35

In the table, t_w and t_F are the web and flange thickness of the I-beam, where t is the thickness of the box column.

The design guideline of buildings with passive control devices, JSSI [32], is used to determine the size and number of BRBs. The JSSI guideline adopts the procedure developed by Kasai et al. (1998) [33] to obtain the necessary stiffness and damping of additional BRBs. The method idealizes the building as an equivalent single degree of freedom (SDOF) system and estimates the amount of additional stiffness and damping of passive control devices according to the target performance level under design earthquake loads. For the current example, the target performance levels are the maximum inter-story drift ratio of less than 1% and the ductility factor of BRBs of less than 7, under the design earthquake and wind loads. Table 2 presents the estimated parameters of BRBs, while the device configuration is shown in Figure 1.

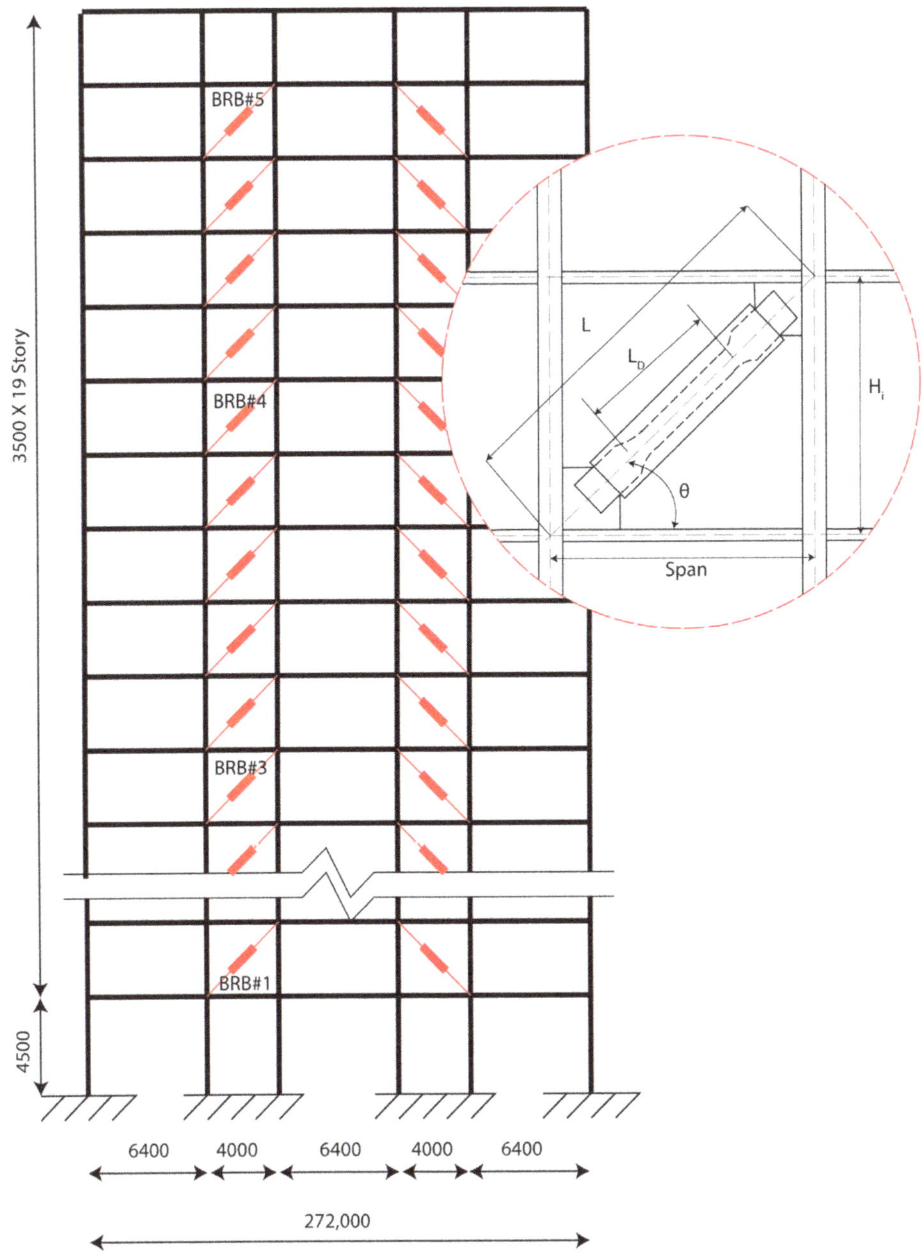

Figure 1. Twenty-story steel prototype building elevation and BRB configuration (dimensions in mm).

Table 2. Technical parameters of design BRBs.

Story	Yielding Capacity	Initial Stiffness	K_1/K_{ai1} [1]
	F_{iy} (*kN*)	K_{ai} (*kN/mm*)	
20	-	-	-
17–19	250	100	0.02
14–16	345	140	0.02
11–13	435	175	0.02
8–10	430	172	0.02
5–7	460	185	0.02
2–4	385	155	0.02
1	-	-	-

[1] Ratio of post-yield to pre-yield stiffness.

2.2. *Numerical Model of the Target Building*

The 20-story steel frame is modeled by finite element software, STERA_3D (STructural Earthquake Response Analysis 3D), which was developed by one of the authors [34]. The beam elements are presented by two nonlinear flexural springs at both ends. The column elements have nonlinear axial springs distributed in the sections of both ends. The steel strength is modified to be 1.1 times more than the nominal strength, and the ratio of post-yield stiffness to the initial stiffness is 0.001. Figure 2 represents the hysteresis behavior of the nonlinear bending spring, which is defined as the moment–rotation relationship under cyclic loadings. The beam–column connection is assumed to be rigid, where the rigid zone length for the beam element is set to be half of the column width. The BRB element, as presented in Figure 3, is defined as a shear spring in a frame with the bilinear hysteresis and initial stiffness, K_1, and secondary stiffness, K_2.

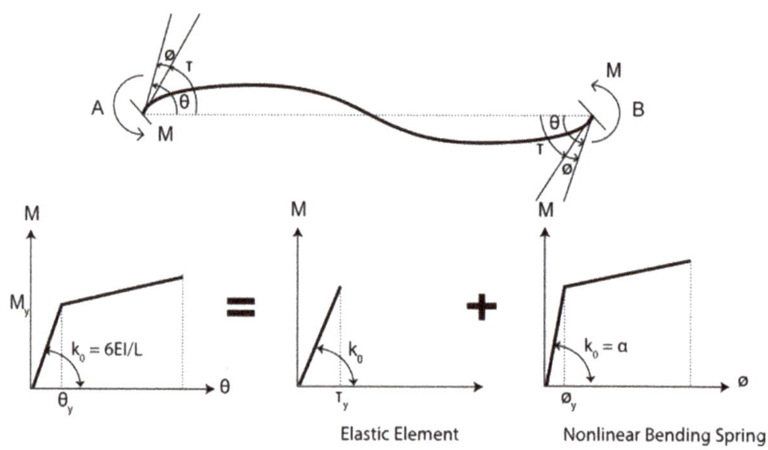

Figure 2. Hysteresis model of nonlinear bending spring of steel column and beam.

2.3. *Performance Evaluation of Target Building*

2.3.1. Pushover Analysis

The strength capacity and story drift distribution of the target building are evaluated using the nonlinear static pushover analysis. The model is pushed horizontally until the drift ratio at the equivalent height of the model reached 1/50 using the lateral load distribution given by the Japanese standard [30,31]. The relationship between the base shear coefficient and the drift ratio at the equivalent height of the model, for both cases, without and with supplemented BRBs, is illustrated in Figure 4. The result indicates that the lateral strength of the target frame is increased for the frame with supplemented BRBs by 135%, 115%, and 120% when the building pushover reaches 0.5%, 1.0%, and 2.0% of the drift ratio, respectively.

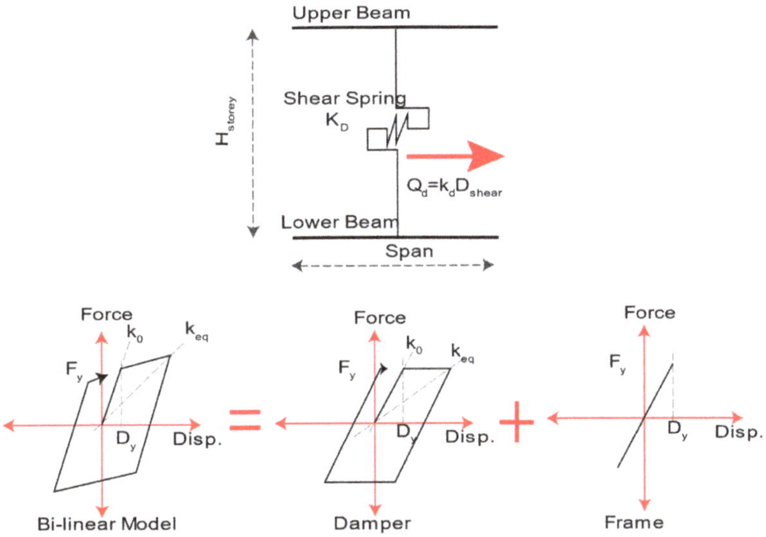

Figure 3. Configuration scheme and bilinear hysteresis model of nonlinear shear springs for BRBs.

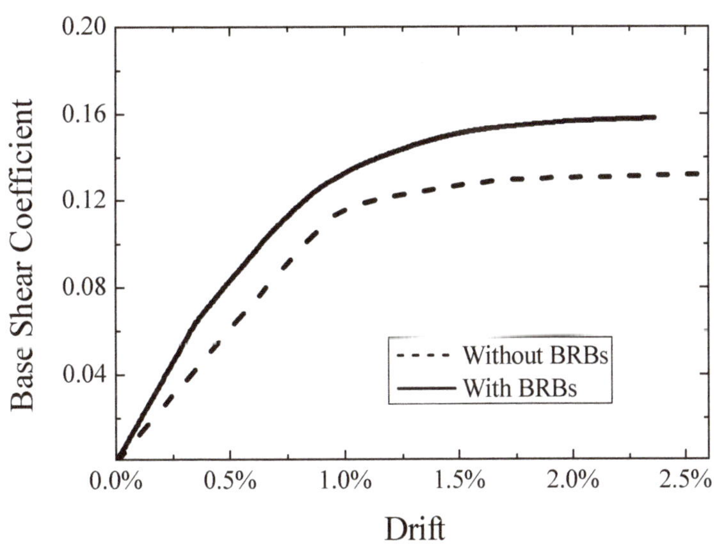

Figure 4. Base shear coefficient and drift ratio at the equivalent height of the target model.

2.3.2. Natural Period and Mode Shape

The dynamic characteristics of the target frame corresponding to the 1st to 3rd modes of vibration are obtained by the eigenvalue analysis. As shown in Figure 5, the natural period becomes shorter after installing BRBs, while the mode shape is not changed significantly.

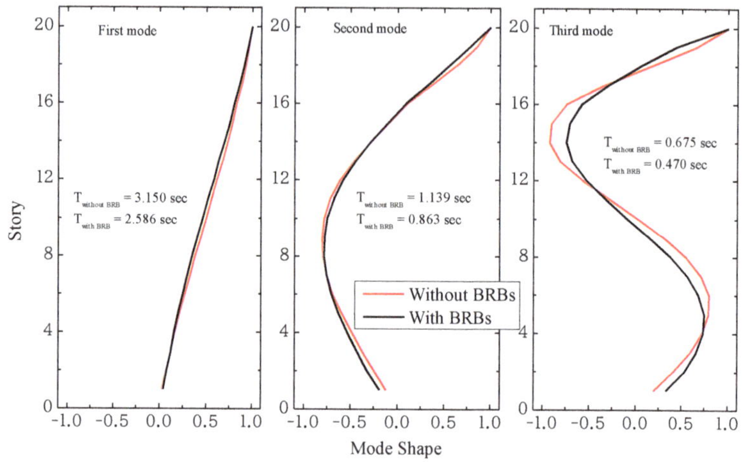

Figure 5. Normalized mode shape of the target high-rise building with and without BRBs devices.

2.3.3. Time History Analysis under Design Earthquakes

The target frame is analyzed under the design earthquakes and wind loads prescribed in the Japanese standard [30] to validate the design procedure and evaluate the contribution of BRBs. Five sets of earthquake ground motions are considered as listed in Table 3. Three of them are the observed earthquake ground motions scaled to Level 2 to have the maximum velocity of 50 cm/s. The other two earthquakes are artificial earthquake ground motions to have the response spectra to be compatible with the Level 2 design spectrum of the Japanese standard [31] and the phase spectra of historical earthquakes. The acceleration response spectra of selected earthquakes are illustrated in Figure 6. As for the wind load, four sets of design wind load of the different average speeds of 17, 20, 25, and 31 m/s are generated according to the AIJ recommendations [30]. Since the shape of the building is regular, the wind loads are applied at the center of gravity on each floor, while the distribution of the wind loads along the height of the building is decided according to the Japanese standard [31].

Table 3. List of selected earthquake ground motions.

Categories	No.	Event	Year	Station
Scaled earthquake to be compatible of 50 cm/s	1	Imperial Valley	1940	El Centro
	2	Kern County	1952	Taft
	3	Kobe	1995	JMA
Artificially generated earthquake to be compatible with Level 2	4	Tohoku	1978	Tohoku Univ.
	5	Tokachi Oki	1968	Hachinohe

The maximum story drift ratio of the target frame with and without BRBs under the design earthquake and wind loads is presented in Figure 7. For the design earthquake loads, the maximum story drift is controlled to satisfy the 1% design drift, while the average result indicates a 50% decrease in the critical stories compared to the target frame without BRBs. On the other hand, the maximum story drift under the wind loads also indicated an improvement of 50% in contrast to the model without BRBs, which reaches the 1.2% story drift in critical floors, in the case of a design wind load of 31 m/s. Hence, the design procedure is a straightforward tool that can determine the amount of additional stiffness to be provided by BRBs without any iteration to satisfy the design targets.

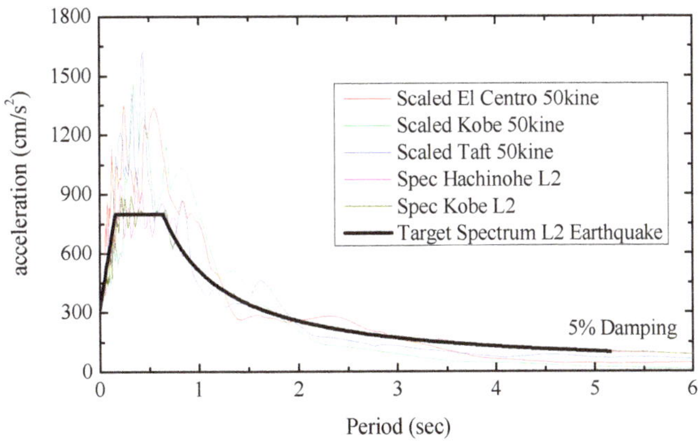

Figure 6. Response acceleration spectrum of selected ground motions (5% damping).

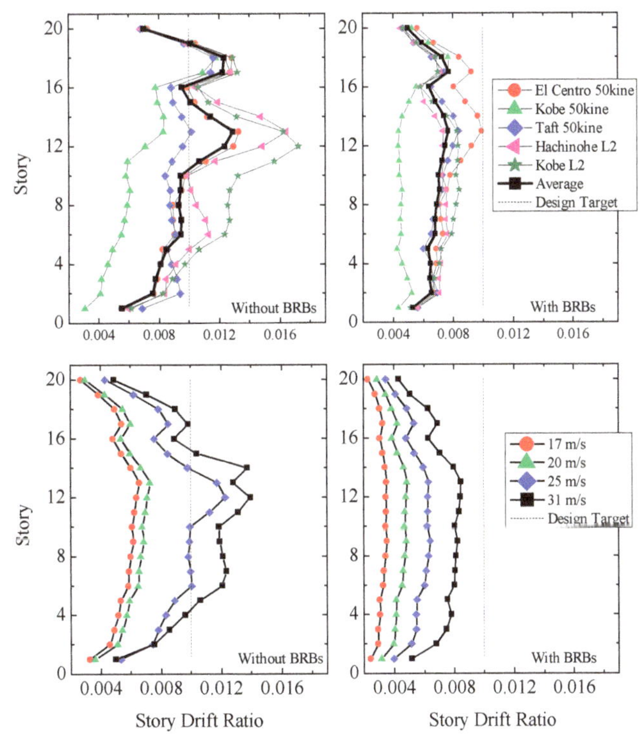

Figure 7. Maximum story drift ratio of high-rise steel building with and without BRBs under design earthquake and wind loads.

Furthermore, in Figure 8, for the earthquake and wind loads, the maximum ductility factor and the amount of dissipated energy of each BRB along the height of the building are evaluated. For most of the earthquake loads, the ductility factor of BRBs is in the range of the pre-defined design drift, which is considered to be $\mu = 7$, except for the lower stories for which it is exceeded. In the case of design wind loads, the ductility factor is decreasing significantly in upper stories. As for the amount of dissipated energy, in both cases, the BRBs in the upper floors are less effective at dissipating the earthquake or wind loads.

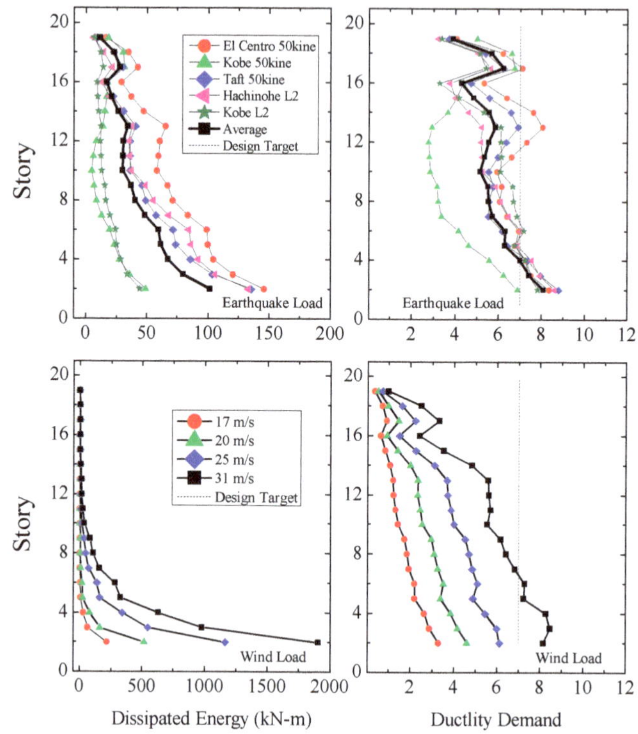

Figure 8. Ductility demand and energy dissipation rate of BRBs along the height of a high-rise steel building under earthquake and wind loads.

3. Procedure to Generate Probable Multi-Hazard Scenarios in Building Lifetime

The likely recurrence of earthquake and wind events during the building lifetime is crucial to be addressed, as it enables the decision-makers to evaluate the building performance in any specific service period. Thus, a practical procedure to estimate the likelihood of the occurrence of multi-hazard scenarios in the structural design life is developed and presented by Roy et al. [35]. The procedure is based on the Poisson process, which uses the anticipated return period and intensity to generate the earthquake and wind events.

From the Gutenberg–Richter equation, the annual occurrence rate, $N(I)$, of earthquake loads with an intensity more than I is given as,

$$\ln N(I) = a - bI. \tag{1}$$

The average recurrence period, $T(I)$, satisfies the following relationship,

$$N(I) = \frac{1}{T(I)}. \tag{2}$$

If we consider two earthquake events having the intensity of I_{100} and I_{500}, with the correspondent return periods of 100 and 500 years, respectively, the constants, a and b, of the Gutenberg–Richter equation can be obtained as given in Equations (4) and (5) and Table 4.

$$\left\{ \begin{array}{l} \ln(0.01) = a - bI_{100} \\ \ln(0.002) = a - bI_{500} \end{array} \right\} \tag{3}$$

$$b = \frac{\ln(0.01) - \ln(0.002)}{I_{500} - I_{100}} \tag{4}$$

$$a = \ln(0.01) + bI_{100} \tag{5}$$

Table 4. Parameters of the probabilistic distribution of earthquake and wind intensity for Aichi Prefecture Japan.

Event	Intensity I [1,2]	Unit	Return Period (Years)	Recurrence Rate	Parameters	
					a	b
Earthquake	600 943	gal gal	100 500	0.01 0.002	−1.7898	0.0047
Wind	16 36	m/s m/s	1 500	1 0.002	4.9717	0.3107

[1] in case of an earthquake: "I" denotes zero-period (PGA) acceleration from the response spectrum. [2] in case of wind: "I" denotes the mean wind speed.

Now, if $N(I_0)$ is an earthquake event with intensity more than the I_0, the cumulative distribution function (CDF) of the intensity I is given as

$$F(i) = P(I \leq i) = 1 - \frac{N(i)}{N_0}, \tag{6}$$

where $N_0 = N(I_0)$.

Once the probability of earthquake intensity is determined, the probability of earthquake occurrence, $P(t)$, assuming that the earthquake events occur individually with an average occurrence rate $N(I)$, can be estimated. The probability of n earthquake events of intensity more than I in the t-years is expressed as the Poisson process as

$$P_n(t) = \{N(I)t\}^n \, \frac{e^{-N(I)t}}{n!}. \tag{7}$$

In Equation (7), setting $n = 0$, the probability of no earthquake events of intensity more than I in the t-years can be determined as

$$P_0(t) = e^{-N(I)t}. \tag{8}$$

Similarly, the probability of occurrence of more than one earthquake event exceeding intensity I in the t-years could be given as

$$P(t) = 1 - P_0(t) = 1 - e^{-N(I)t}. \tag{9}$$

The probabilistic model of the occurrence of wind events is created in the same way.

In order to generate the events in the expected lifetime, the proposed procedure initially consists of estimation of the probable earthquake and wind intensity, and subsequently, for the estimated intensity, calculating the probable occurrence with the generation of earthquake and wind events. In the first step, the sample of wind or earthquake intensity, I_i (i = 1, 2, .., L), is generated from the CDF, Equation (6), using the uniform random variable F_i, where $F_i \in [0, 1]$. In conjunction, the sample of the returned period of intensity, ${}_iT_j$ (j = 1, 2, .., M), is generated from the CDF, Equation (8), using the uniform random variable P_i, where $P_i \in [0, 1]$. Lastly, the L events with different intensity, I_i (i = 1, 2, .., L), is arranged corresponding to the M return period, ${}_iT_j$ (j = 1, 2, .., M). For the multi-hazard scenarios, the earthquake and wind events are assumed to occur independently, where the final timeline is added together, as further illustrated in the proposed procedure's flowchart; see Figure 9.

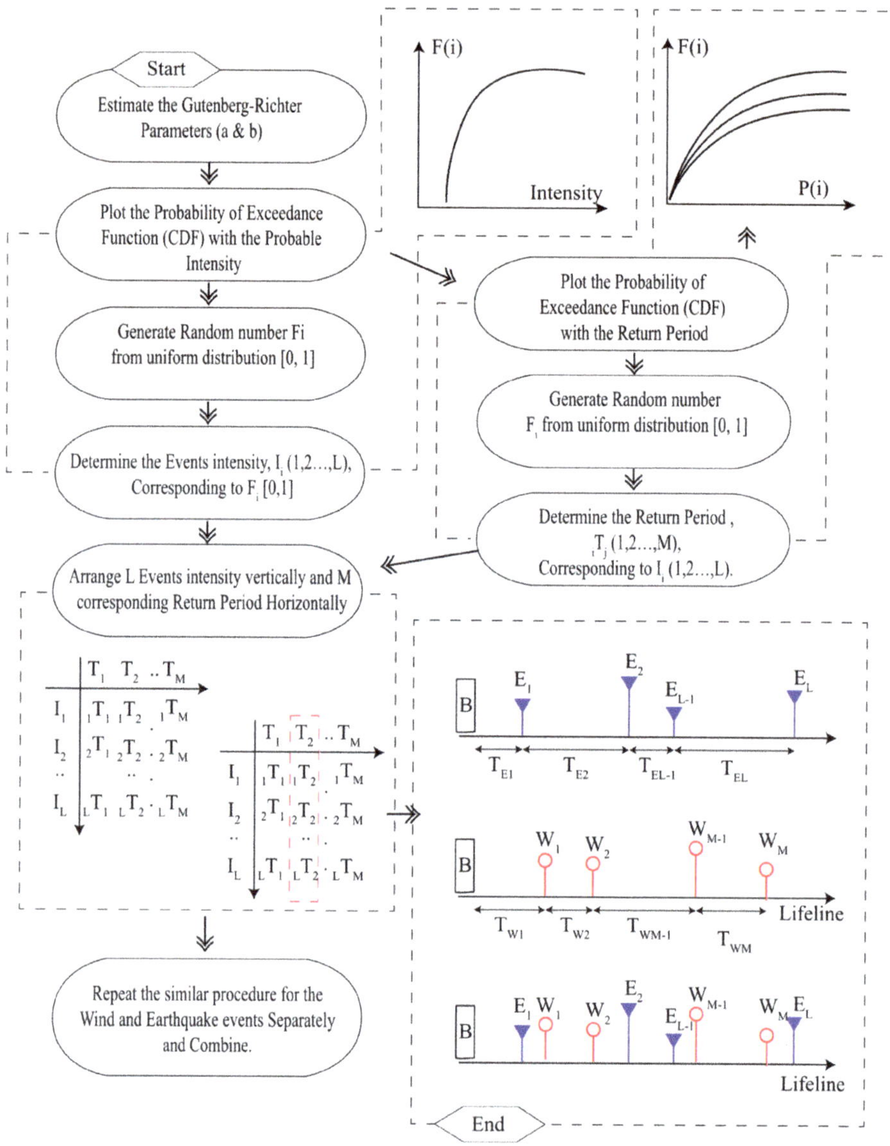

Figure 9. Flowchart of probable earthquake and wind events generation in the project lifetime.

4. Multi-Hazard Events

4.1. Earthquake Loads

Once the probable earthquake events are sampled for the project lifetime, the next task is to generate the ground motion time history data to be compatible with sampled earthquake intensities. In this regard, the Uniform Hazard Spectrum (UHS) provided in the AIJ (2019) for Nagoya City is used, as shown in Figure 10.

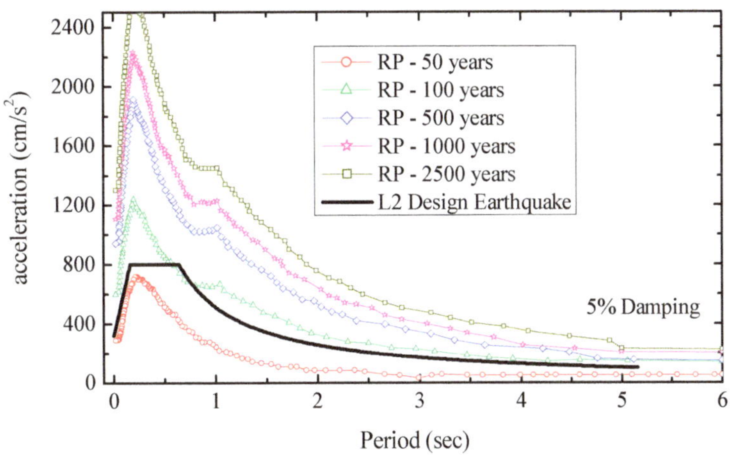

Figure 10. Uniform hazard spectrum provided in AIJ (2019) for Nagoya city, Aichi, Japan.

The recommended UHS is the acceleration response spectrum of 5% damping in the engineering bedrock (V_s = 292 m/s) level, with the probability of exceedance of 39%, 10%, 5%, and 2% over the 50 years. In case the intensity (PGA) of the probable earthquake is not matching the intensity of the recommended UHSs, the interpolation technique is used to generate the appropriate response spectra. Figure 11 illustrates the steps required to generate spectrum-compatible ground motions using the Fourier and phase spectra, following the algorithm developed by one of the authors, STERA_WAVE [36]. For the phase spectrum, corresponding to the PGA intensity of the estimated probable earthquake in the project lifetime, the historical earthquake has been carefully selected from Table 5, which has a similar PGA. The sequence of the selected phase spectrum to generate the sampled earthquake ground motion in each case is presented in Figure 12. In the case that the probably estimated intensity is smaller than 250 gal, the uniform random phase angle is adopted, and the ground motion is generated by multiplying the envelope function proposed by Jennings et al. [37].

Table 5. List of the earthquake used for a phase spectrum to scale the probable earthquake events.

No.	Phase Spectrum	Component	Date	Station	Recorded PGA (gal)
1	Random Phase	-	-	-	>250.000
2	Kumamoto	NS	3 January 2019	Wadamachi Eda	261.697
3	Tottori	NS	6 October 2000	Yonago City	280.233
4	Tottori	EW	21 October 2016	Kurayoshi City	285.811
5	Toho-Oki	NS	4 October 1994	Honamachi	454.975
6	Noto Hanto	NS	25 March 2007	Wajima City	463.544
7	Miyagi	EW	26 May 2003	Izumimachi	655.161
8	Iburi	NS	6 September 2018	Atsuma Kananuma	662.241
9	Osaka	NS	18 June 2018	Hirakata City	690.169
10	Tokachi-Oki	NS	26 September 2003	Makuhetsu-cho	754.200
11	Niigata-Chuetsu	NS	23 October 2004	Ojiya City Castle	779.244
12	Kushiro Oki	NS	15 January 1993	Kushiro	814.906
13	Kobe	NS	17 January 1995	JMA	818.066
14	Tohoku	EW	11 March 2011	Ofunato	944.072

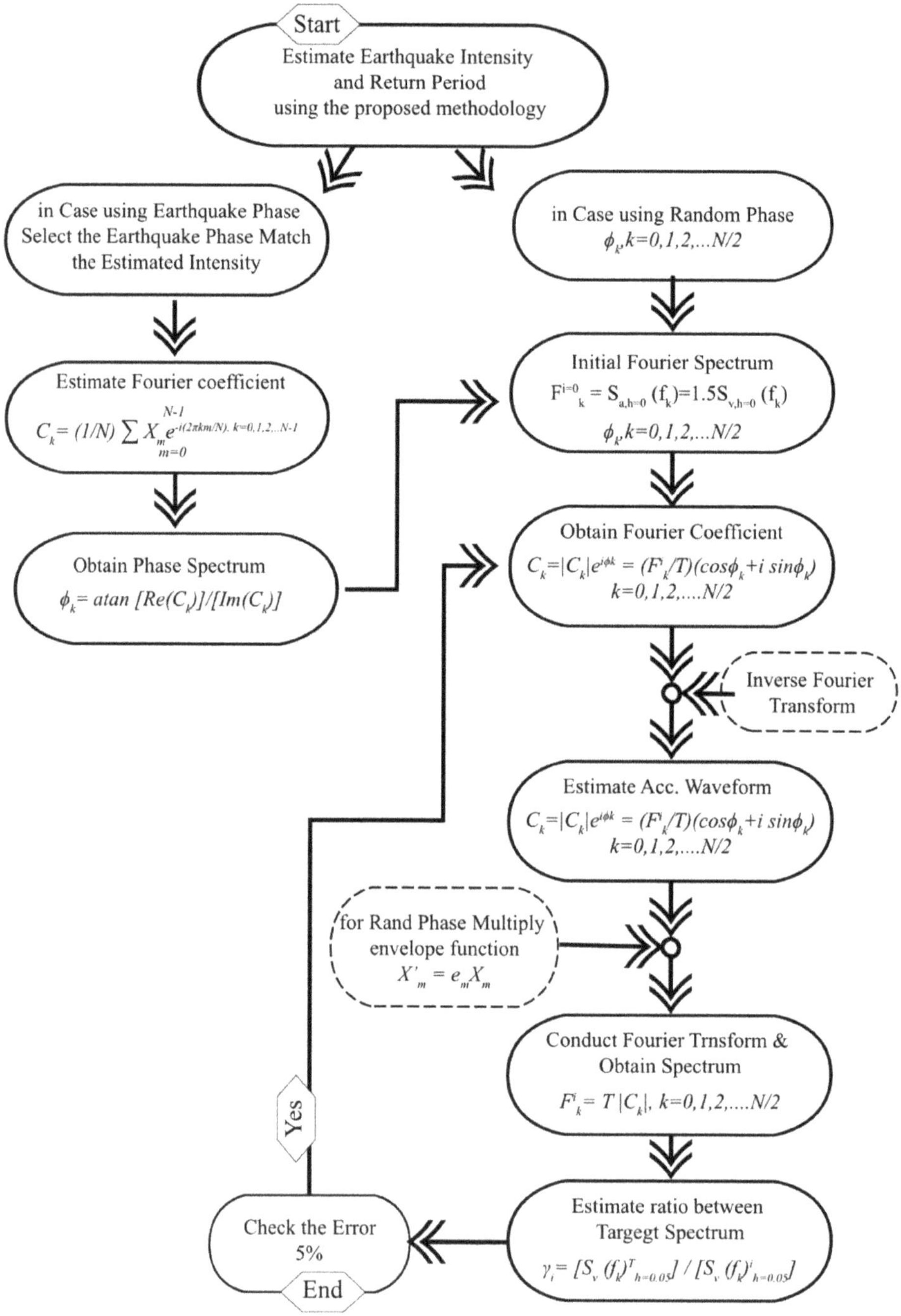

Figure 11. Flowchart of the STERA_Wave algorithm to generate spectrum-compatible earthquakes.

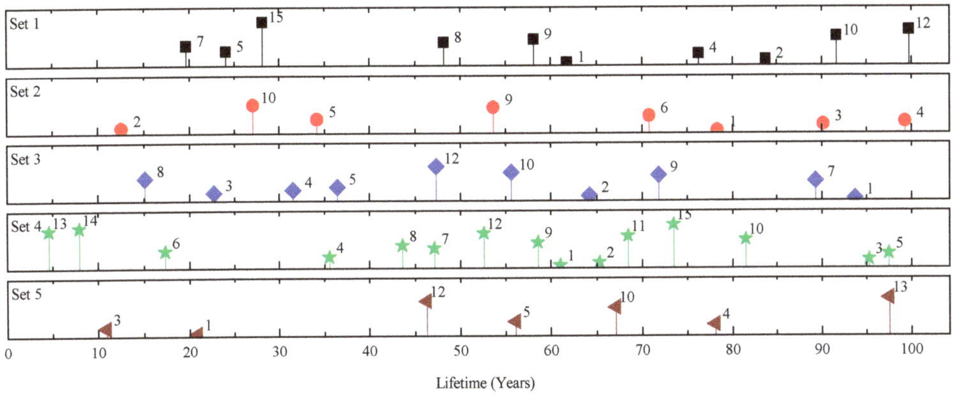

Figure 12. The sequence of the phase spectrum earthquakes used to generate compatible earthquakes in multi-hazard scenarios (the numbers are assigned according to Table 5).

4.2. Long-Duration Wind Loads

Similar to the earthquake, after the wind events in the project lifetime are sampled, the wind load time-history data are generated following the AIJ (2015) [30] recommendations. The natural wind speed is consisting of the mean wind speed component ($\overline{U}(z)$) and the fluctuating component ($u(t)$) in the longitudinal direction, which is changing with time. The AIJ (2015) [30] adopted the von Karman power spectrum to present the power spectral density of the fluctuating component of wind events, as given,

$$S_u(f) = \frac{4\sigma_u^2\{L_z/\overline{U}(z)\}}{[1+70.8\{fL_z/\overline{U}(z)\}^2]^{5/6}} \tag{10}$$

where in the equation, f is the frequency spectrum of wind loads, σ_u is the standard deviation of the fluctuating component, $\overline{U}(z)$ is the mean wind speed at 10 m of ground level, and L_Z is the turbulence scale.

The co-coherence of wind turbulence along with the structure height, which defines the frequency-dependent spatial correlation of wind speed fluctuation, is also considered as given below,

$$\gamma(f,d_j) = \frac{\text{Re}\,\lfloor S_{i_1,i_2}(f,d_j)\rfloor}{\sqrt{S_{i_1}(f)\cdot S_{i_2}(f)}} = exp\left(-C_{ij}\frac{fd_j}{\overline{U}(z)}\right),\ i = u,v,w;\ j = v,w \tag{11}$$

$$\gamma(f,d_j) = exp\left(-\frac{\sqrt{(C_{uy}\cdot fd_y)^2+(C_{uz}\cdot fd_z)^2}}{\overline{U}(z)}\right) \tag{12}$$

where d_j is the distance of two referenced nodes of Euclidian distance, S_u is the power spectral density of turbulence in the longitudinal direction of a single point, and C_{ij} is the decay coefficients. The current work employed the decay coefficients from the Norwegian Public Roads Administration (NPRA) Handbook N400 [38], and the power law is employed as well to estimate the wind speed force at story levels. Furthermore, the conventional spectral method introduced by Deodatis and Shinozuka (1991) [39] and Deodatis (1996) [40] is adopted to estimate the fluctuating component of wind load by simulating the multi-variate random process in the time domain. The entire process discussed in this section is further illustrated in the flowchart, see Figure 13, and the corresponding parameters are summarized in Table 6.

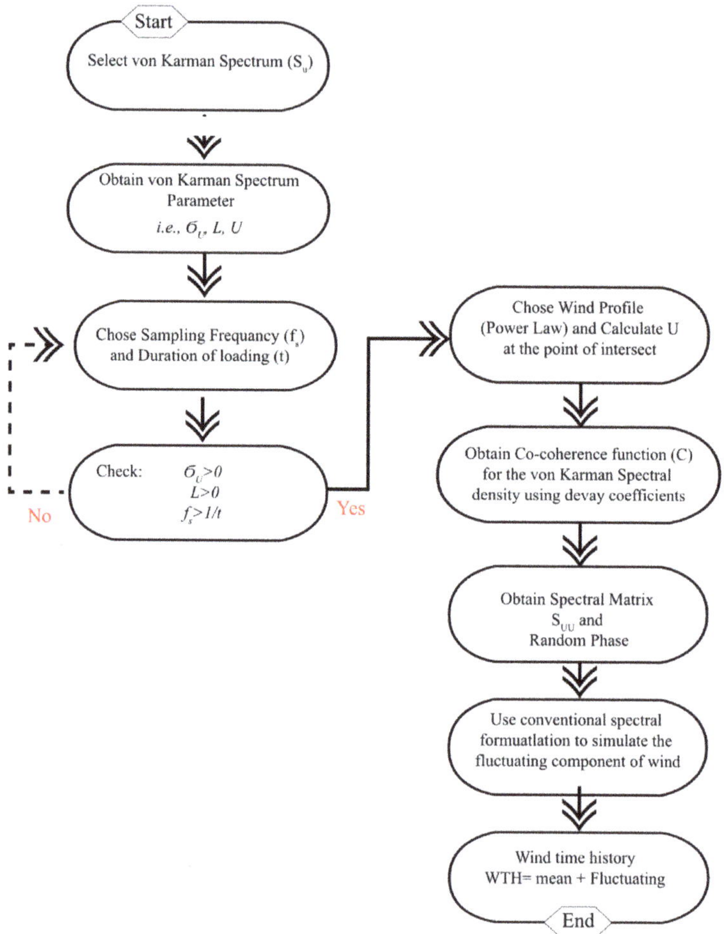

Figure 13. Flowchart for generating wind loads time history.

Table 6. Corresponding parameters of long-duration wind loads.

No.	Description	Parameter	Unit
1	Turbulence Length Scale	75.27	m
2	Standard Deviation of Fluctuating Component	6.73	m/s
3	Cutoff Frequency	5.00	Hz
4	Mean of Duration	7500.00	s
5	Coefficient of Variation	10.00	%

4.3. Multi-Hazard Timeline for the Target Building

Five random sets of timelines are developed from the earthquake and wind events of different return periods, as discussed in previous sub-sections, and presented in Figure 14. In each set of timelines, an average of ten earthquakes of different PGA (range from 250 to 1100 gal) is anticipated, where the earthquake events' details and sequence are summarized in Table 5. Similarly, in each set of timelines, an average of 43 wind events (range from 16 to 31 m/s) is expected to occur.

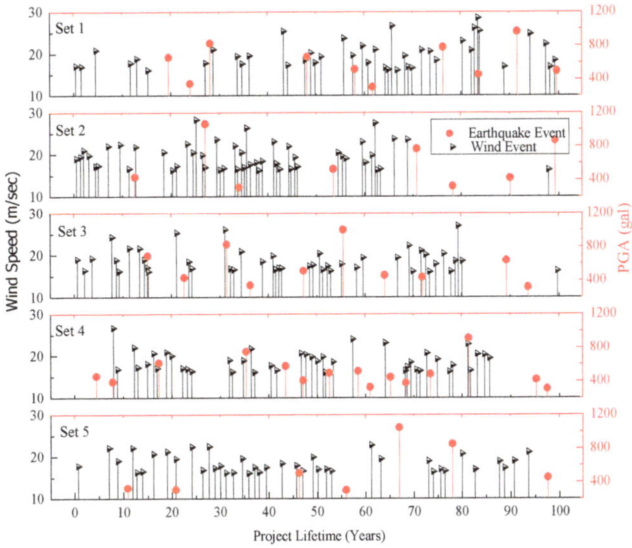

Figure 14. Probable multi-hazard scenarios in the target building lifetime.

5. Cumulative Damage Model

5.1. Fatigue Evaluation of BRBs Device

The fatigue life of BRBs is originated with the initiation of fatigue by the first yielding, and it grows slowly and gradually under each cyclic loadings until it reaches the failure thresholds. Thus, the fatigue life can be defined from the number of cycles N_f of the specific intensity that a BRB can sustain before the failure happens. For the earthquake and wind loads, which induce cyclic loads with variable intensities, the strain-based estimation of fatigue life is suitable. For this purpose, first, the random cyclic loads are sorted in bins of similar strain; then, for each of the bins, the cumulative damage is computed from the *S-N* curves, and finally, using the Miner's rule, the fatigue failure is evaluated.

5.2. Variable-Load Cycle Counting

The BRBs under earthquake and wind events experience variable-load cycles over the time of excitation. Therefore, it requires a proper procedure to estimate the number of cycles and classify the time-varying load history to a constant and equivalent load of similar amplitude. The rainflow counting method, which was developed by Matsuishi and Endo (1968), has demonstrated good agreement with the observed fatigue damage and was widely adopted by researchers [41]. Thus, the current work practiced the rainflow counting method to estimate the number of cycles corresponding to the strain amplitude of BRBs under wind and earthquake excitations.

5.3. Strain–Cycle Relationship of BRBs

Coffin (1962) and Manson (1966) defined the relationship between the number of cycles until the failure, N_f, and the strain amplitude, $\Delta\varepsilon_t$, which leads to fatigue failure, as given in Equation (13). Usually, this relationship is derived from experimental results or empirical formulas based on statistical data. The fatigue life of BRBs, manufactured in Japan, has been studied extensively under different loading scenarios such as the shaking table test [42], the constant amplitude fatigue test [43], the BRB frame shaking table test under random amplitude [44], and the truss frame test under gradually increasing amplitude [45]. Takeuchi et al. [46] observed that the fatigue properties of BRBs are not considerably changed for different types of steel material, wherein the approximate strain-failure cycle relation is given as below,

$$\begin{aligned}\Delta\varepsilon_t[\%] &= 0.5N_f^{-0.14}\ (\Delta\varepsilon_t < 0.1\%)\\ \Delta\varepsilon_t[\%] &= 20.48N_f^{-0.49}(0.1\% \le \Delta\varepsilon_t < 2.2\%)\\ \Delta\varepsilon_t[\%] &= 54.0N_f^{-0.71}(\Delta\varepsilon_t \ge 2.2\%).\end{aligned} \tag{13}$$

5.4. Miner's Damage Rule

According to the Miner rule, if there are j number of strains of different intensities in a strain profile, the cumulative damage index (CDI) is defined as,

$$CDI = \sum_{i=1}^{j} \frac{n_i}{N_i} \le 1 \tag{14}$$

where n_i is the number of cycles of the ith strain, and N_i is the number of cycles to the failure point under the ith strain. The failure occurs when the CDI reaches one.

Now, by substituting the BRBs strain–cycle relation, Equation (13), into the Miner rule, Equation (14), the cumulative damage of BRBs subjected to random cyclic loads of earthquake and wind excitations can be presented as follows,

$$\begin{aligned}CDI &= \sum_{i=1}^{j} \frac{n_i}{\left(\frac{\Delta\varepsilon_t}{0.50}\right)^{-\frac{1}{0.14}}}\ (\Delta\varepsilon_t < 0.1\%)\\ CDI &= \sum_{i=1}^{j} \frac{n_i}{\left(\frac{\Delta\varepsilon_t}{20.48}\right)^{-\frac{1}{0.49}}}(0.1\% \le \Delta\varepsilon_t < 2.2\%)\\ CDI &= \sum_{i=1}^{j} \frac{n_i}{\left(\frac{\Delta\varepsilon_t}{54.0}\right)^{-\frac{1}{0.71}}}(\Delta\varepsilon_t \ge 2.2\%)\end{aligned}\ . \tag{15}$$

5.5. Plastic Strain Energy (PSE)

The study conducted by Chung and Lee (1994) [47] introduces the plastic strain energy (PSE) as a damage criterion, because in each cycle of excitation, an irrecoverable amount of PSE is inserted into the steel elements, which in response initiates and propagates the fatigue crack. Therefore, the cumulative damage of BRBs is further evaluated in terms of plastic strain energy (PSE), which is defined as the ratio of the area of plastic strain energy (E_t) by the triangular area (E_y) shaped by the yield and deformation strengths of RBR, as shown in Figure 15a. The experimental study conducted by Iwata and Murai (2006) [48] introduces the PSE as a suitable indicator for the performance evaluation of BRBs in moment frames. The study recommends Equation (16) as a threshold to evaluate the safety of BRB in terms of cumulative plastic strain energy [48].

$$PSE = 150\ R \tag{16}$$

$$R = \frac{P_E}{P_y} \tag{17}$$

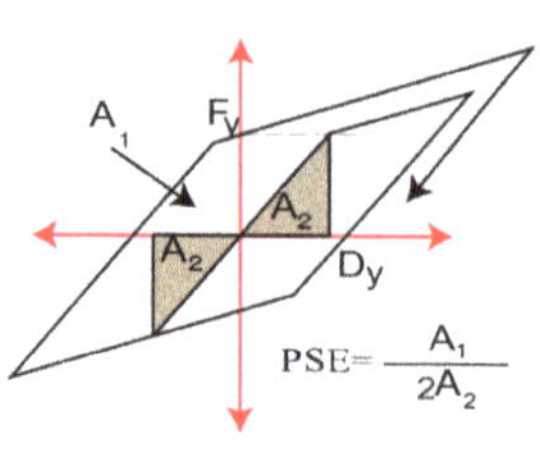

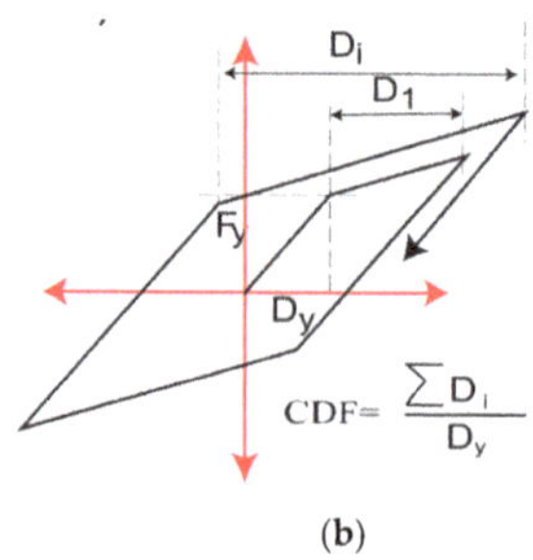

Figure 15. (**a**) Plastic strain energy (PSE); (**b**) cumulative ductility factor (CDF).

In the equation, P_E is the buckling strength of the restrainer, and P_y is the yield strength of the steel core plate. The R factor is an indicator of the stable energy absorption capacity of BRBs.

5.6. Cumulative Ductility Factor (CDF)

Similarly, the accumulative damage of BRBs in the high-rise steel building under successive application of multi-hazard scenarios is investigated in terms of cumulative ductility factor (CDF). The CDF is an important indicator that describes the plastic deformation capability of the BRB devices before resulting in structural destruction. As presented in Figure 15b, the CDF is the normalized summation of total plastic deformation to the yield strength of BRB, as defined in Equation (18).

$$CDF = \frac{\Sigma\,(D_i)}{D_y} \quad (18)$$

$$CDF = 75\,R \quad (19)$$

Researchers extensively examined the low-cycle fatigue capacity of BRBs in terms of cumulative ductility demands [49–54] and recommend the minimum criteria. FEMA-450 [51] and ANSI/AISC [54] recommend that the CDF demand should be at least 140 and 200, respectively, to ensure the cumulative ductility capability of the device. In the current research study, the result of the experimental study conducted by Iwata and Murai (2006) [48], as given in Equation (19), is utilized to evaluate the BRBs performance in the building lifetime.

6. Result and Discussion

6.1. Outline of Analysis and Results

For the target building, the overall performance is investigated in terms of story drift and shear force. The cumulative damage index, plastic strain energy, cumulative and maximum ductility demands are used to investigate the progressive damage of BRBs. For this purpose, five BRBs are selected along with the height of the target building to examine the results. The selected BRBs that are located in the 2nd, 5th, 10th, 15th, and 19th stories are denoted as BRB#1–#5, respectively, as shown in Figure 1. For the selected dampers, Table 7 summarizes the absorbed energy, the number of cycles, and strain amplitude obtained from the rainflow counting method after successive analysis of the five sets of lifetime multi-hazard scenarios. Since the average strain amplitude is higher for the damper installed in lower stories, the BRBs' contribution to absorb the MH energy becomes larger in response the accumulative damage increase. Table 7 also indicates that the energy absorption rate in upper stories is significantly smaller in comparison to the BRBs in lower stories.

Table 7. Cycle counting, strain amplitude, and CDI of selected BRBs under multi-hazard timelines.

Selected Damper	MH Scenarios	CDI	Strain Amplitude (%)		Cycle	Energy
			Ave.	Max	No.	kN-m
BRB#1	Set 1	0.643	0.048	1.210	129,094	25,400
	Set 2	0.573	0.046	1.440	141,647	22,310
	Set 3	0.501	0.042	1.018	138,566	21,310
	Set 4	0.434	0.040	1.362	138,988	16,240
	Set 5	0.363	0.043	1.092	126,927	13,120
BRB#2	Set 1	0.105	0.031	0.800	128,968	3838
	Set 2	0.074	0.029	0.783	140,602	3074
	Set 3	0.087	0.028	1.067	136,815	3564
	Set 4	0.060	0.026	1.214	138,187	2038
	Set 5	0.043	0.027	0.844	126,173	1356
BRB#3	Set 1	0.032	0.025	0.762	129,226	930
	Set 2	0.026	0.024	0.758	139,249	751
	Set 3	0.027	0.023	0.751	135,340	932
	Set 4	0.024	0.021	0.701	137,201	654
	Set 5	0.019	0.022	0.780	126,453	461

Table 7. *Cont.*

Selected Damper	MH Scenarios	CDI	Strain Amplitude (%)		Cycle	Energy
			Ave.	Max	No.	kN-m
BRB#4	Set 1	0.015	0.019	0.654	129,787	385
	Set 2	0.014	0.018	0.708	140,681	307
	Set 3	0.015	0.018	0.664	134,786	365
	Set 4	0.011	0.016	0.582	137,618	321
	Set 5	0.011	0.016	0.698	126,165	234
BRB#5	Set 1	0.005	0.009	1.210	142,085	99
	Set 2	0.004	0.008	0.339	152,142	94
	Set 3	0.005	0.008	0.406	148,793	88
	Set 4	0.004	0.008	0.367	151,671	83
	Set 5	0.004	0.007	0.422	135,930	69

6.2. Story Drift Ratio of the Target Building

Figure 16 presents the envelope of maximum story drift and shear force of the target building under the five sets of multi-hazard scenarios. Comparing the maximum story drift ratio with the design target of 1%, the maximum story drift obtained from multi-hazard timelines is 2.0–5.5 times larger than the design target. Although the story drift amplification for the lower story in contrast to the upper story is three times larger, the upper stories are not able to endure the design drift as well. In addition, the distribution of the maximum story shear force is approximately equal to the distribution of the story shear force (dotted line) when the seismic base shear is $C_S = 0.2\,W$, where W is the weight of the building.

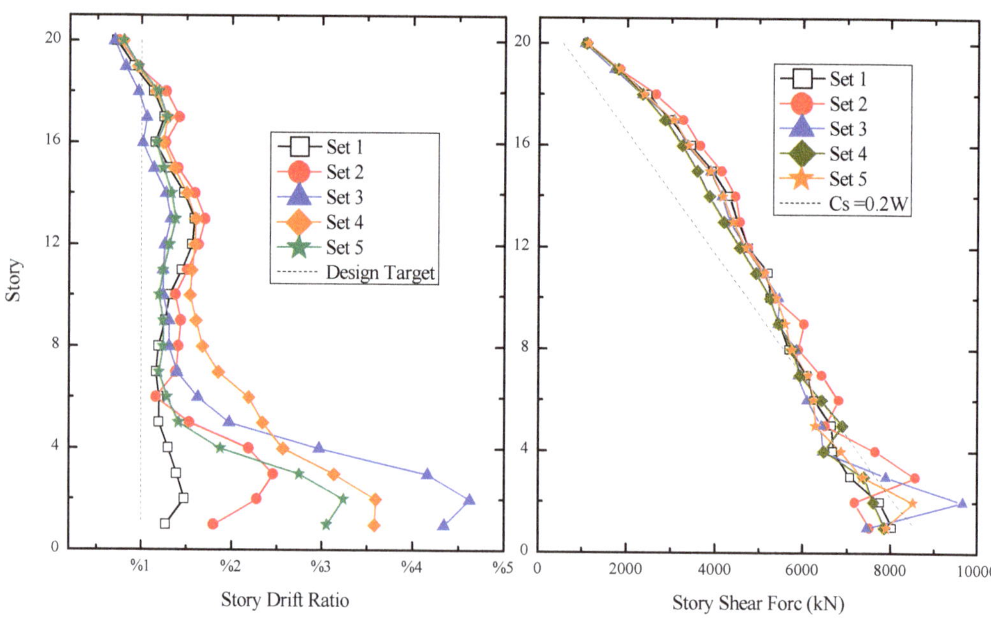

Figure 16. Maximum story drift ratio and shear force under multi-hazard scenarios.

6.3. Cumulative Damage Index (CDI)

Figure 17 demonstrates the increase in cumulative damage index (CDI) for the selected BRBs under the application of five sets of multi-hazard scenarios. As indicated in Tables 1 and 7, the estimated CDI for BRB#1–5 with an average strain amplitude of 0.05, 0.03, 0.025, 0.02, and 0.01% is about 0.64, 0.1, 0.03, 0.015, and 0.005, respectively, which indicates that CDI is decreasing in upper stories. Accordingly, none of the selected BRBs are reaching the failure status (CDI = 1.0) according to Miner's rule.

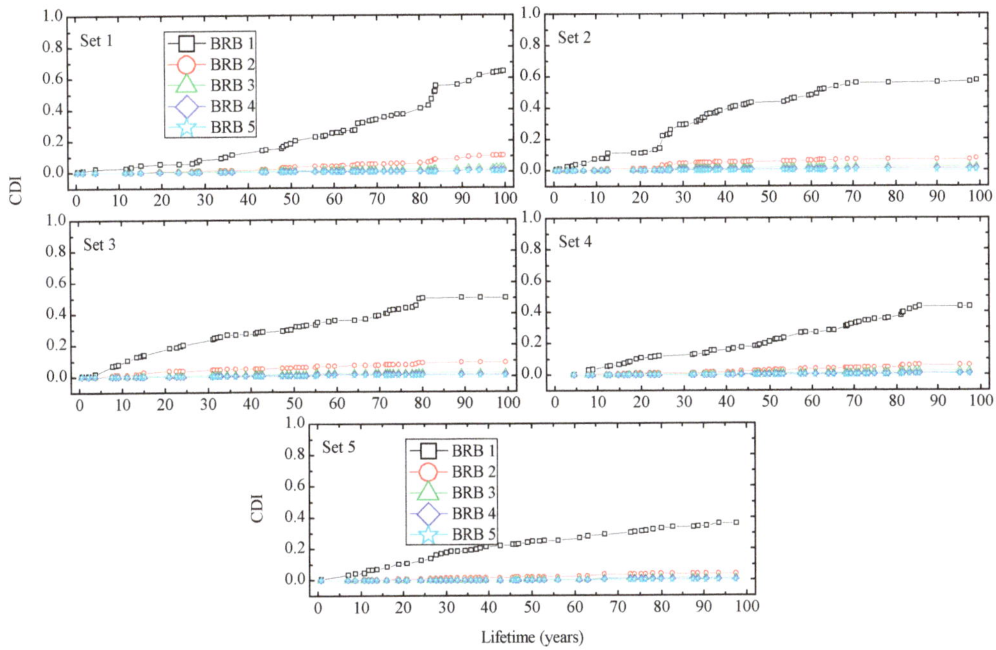

Figure 17. Cumulative damage index (CDI) of selected BRBs under multi-hazard scenarios.

6.4. Plastic Strain Energy (PSE)

Figure 18 presents the increase in estimated plastic strain energy (PSE) for the five selected BRBs under the successive analysis of multi-hazard scenarios. Since the PSE of BRB is associated with the R factor as $PSE = 150\ R$ [48], for a certain (R = 1, 2, 3, and 4), the PSE is marked to illustrate the BRB's progress damage. It is observed that the fatigue life of BRB#1, in terms of plastic strain energy, is about 25 years and 80 years for R = 1 and 4, respectively. Meanwhile, the estimated plastic strain for the BRB#5 (CDI = 0.005, R = 1) is below 150 after a service period of 100 years. In contrast to the CDI, the PSE reveals that even the BRBs with an average strain amplitude of less than >0.03% are prone to progressive damage under multi-hazard scenarios.

6.5. Cumulative Ductility Factor (CDF)

Figure 19 shows the increase in cumulative ductility factor (CDF) for the selected BRBs corresponding to the five sets of the multi-hazard scenarios. In the figure, the horizontal lines mark the CDF criteria according to Equation (18) for the certain R factors (R = 1, 2, 3, and 4, CDF = 75, 150, 225, and 300) to measure the progressive damage of selected BRB in terms of CDF. In contrast to the CDI and PSE, it is observed that all the selected BRBs are reaching their maximum capacity in terms of cumulative ductility factor. The effective age of the BRBs (with R = 4) is about 40–75 years corresponding to the BRB's location in the high-rise building and average strain amplitude. On the other hand, the CDF capacity of the BRBs (with R = 1) is about 20 years, regardless of the average strain amplitude and damper location. Comparing the results of CDI and PSE with the cumulative ductility factor, it is important to point out that the BRBs that satisfy the CDI and PSE indexes are severely prone to progressive damage in terms of CDF. Therefore, the current study recommends precisely evaluating the fatigue life BRBs under different multi-hazard scenarios in terms of different fatigue criteria.

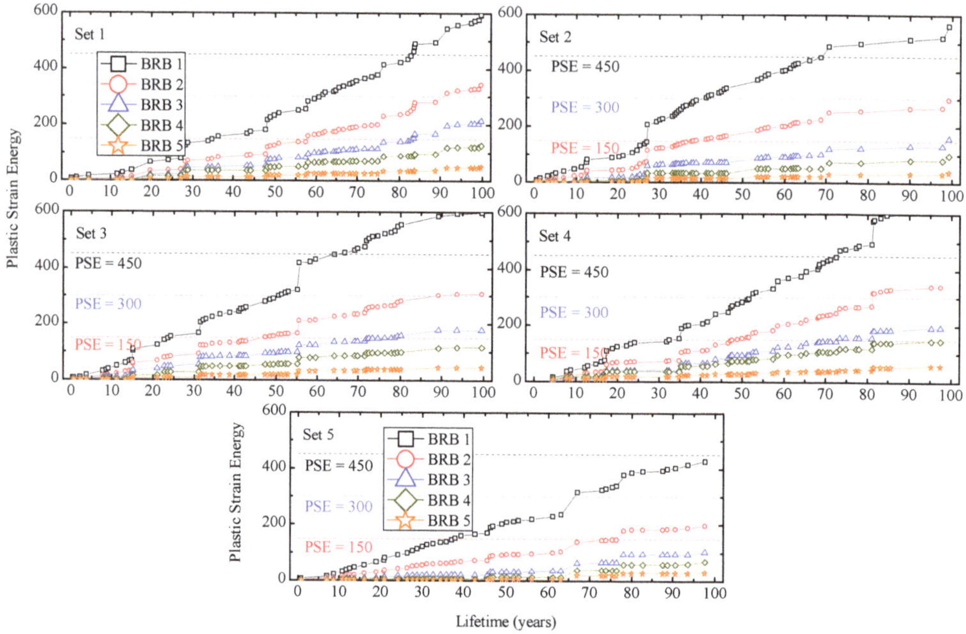

Figure 18. Plastic strain energy (PSE) of selected BRBs under multi-hazard scenarios.

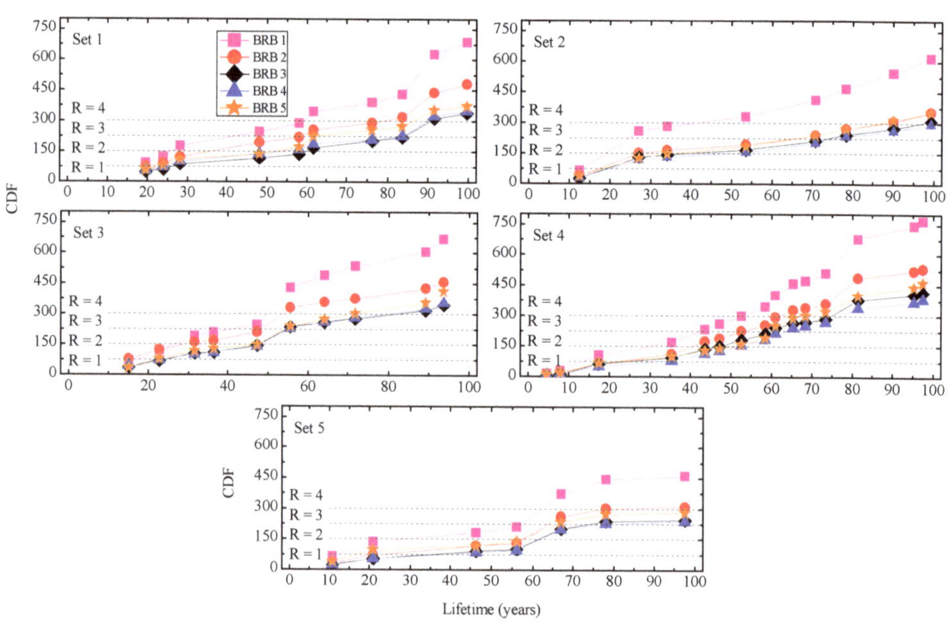

Figure 19. Cumulative ductility factor (CDF) of selected BRBs under multi-hazard scenarios.

6.6. Maximum Ductility Factor of BRBs

For the selected BRBs, the envelope of maximum ductility factor under successive application of multi-hazard scenarios is presented in Figure 20. According to the literature review, for the common types of BRB in Japan [47–49], the maximum ductility demand for BRBs with different configuration and cyclic loading protocols are reported to have a range of 20–24 before exceeding the ultimate capacity. Comparing the maximum ductility factor under Level-2 earthquake (μ = 3–8) and multi-hazard scenarios (μ = 10–25), it is seen that the ductility demand is three times larger in comparison to a Level-2 earthquake, as recommend in Japanese design provision.

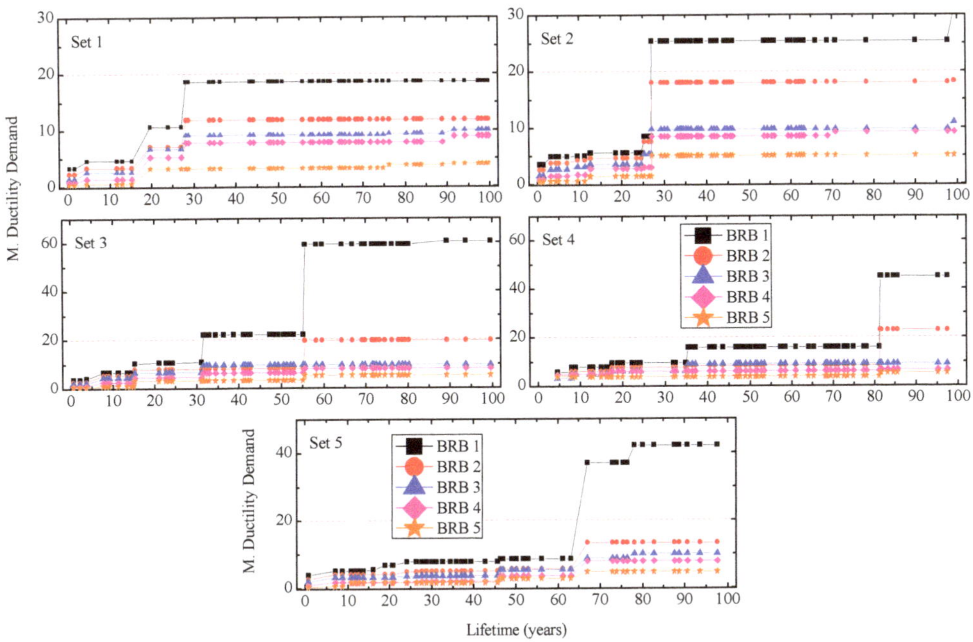

Figure 20. Maximum ductility demand of selected BRBs under multi-hazard scenarios.

7. Conclusions

The current research investigated the fatigue life of BRBs elements and their effect on the overall performance of high-rise steel buildings by predicting the sequence of multi-hazard scenarios in the building lifetime. Toward this objective, first, a 20-story moment resisting frame is designed to resist the gravitational loads; then, the simplified method presented in the JSSI manual is applied to design the required BRB elements to satisfy the 1% story drift ratio (design target) imposed by the lateral excitation. In the next step, the Poisson process-based procedure is used to randomly approximate the probable multi-hazard scenarios during the project's lifetime. The proposed concept is applied to generate five random scenarios, and the target building performance under multi-hazard scenarios is assessed. The main conclusions of the study are drawn as follows.

- Although for the design earthquake and wind loads, as recommend by AIJ (2019), the target building satisfies the 1% story drift ratio, but the building performance under multi-hazard scenarios is about 2.0–5.0 times larger. It is because, first, the estimated intensities of multi-hazard events from the proposed procedure are slightly larger in comparison to the design earthquake; second, the contribution of multi-hazard scenarios to the progressive damage of BRBs amplifies the overall building's performance.

- Under the multi-hazard scenarios, the progressive damage of BRBs that are in upper stories is considerably smaller than that of the BRBs located in lower stories. This is indicating that the BRBs in lower stories are severely prone to low-cycle fatigue damage in the building service period.
- The progressive damage of BRBs under multi-hazard scenarios, which is studied in terms of cumulative damage index (CDI), cumulative ductility factor (CDF), plastic strain energy (PSE), and maximum ductility, reveal that the fatigue life of a BRB is a multi-criteria issue. It is found that although the BRBs can satisfy one or two of the thresholds, under the successive application of wind and earthquake excitation, the devices fail to endure all the aforementioned criteria. Corresponding to the designated criteria, the fatigue life is variable as well. Therefore, the current study recommends the multi-criteria be incorporated in the design phase of the high-rise building with BRBs.
- Considering the multi-criteria investigated in this study, this study observed that the effective service life of BRBs is much smaller than the design service life.
- Under the successive analysis of multi-hazard scenarios, the ductility demand of BRBs is three times larger than that during Level-2 earthquakes.

Author Contributions: Conceptualization, T.S. and A.N.; methodology, A.N. and T.R.; software, T.S.; writing—original draft preparation, A.N.; writing—review and editing, T.S.; supervision, T.S. All authors have read and agreed to the published version of the manuscript.

Funding: This research received no external funding.

Institutional Review Board Statement: Not applicable.

Informed Consent Statement: Not applicable.

Data Availability Statement: The data presented in this study are available on request from the corresponding author.

Conflicts of Interest: The authors declare no conflict of interest.

References

1. Fujimoto, M.; Wada, A.; Saeki, E.; Watanabe, A.; Hitomi, Y. A study on the unbounded brace encased in buckling-restraining concrete and steel tube. *J. Struct. Eng.* **1988**, *34*, 249–258. (In Japanese)
2. Takeuchi, T. Buckling-Restrained Brace: History, Design and Applications. *Key Eng. Mater.* **2018**, *763*, 50–60. [CrossRef]
3. Black, C.J.; Makris, N.; Aiken, I.D. Component Testing, Seismic Evaluation and Characterization of Buckling-Restrained Braces. *J. Struct. Eng.* **2004**, *130*, 880–894. [CrossRef]
4. Tremblay, R.; Bolduc, P.; Neville, R.; Devall, R. Seismic testing and performance of buckling-restrained bracing systems. *Can. J. Civ. Eng.* **2006**, *33*, 183–198. [CrossRef]
5. Ju, Y.K.; Kim, M.-H.; Kim, J.; Kim, S.-D. Component tests of buckling-restrained braces with unconstrained length. *Eng. Struct.* **2009**, *31*, 507–516. [CrossRef]
6. Fahnestock, L.A.; Ricles, J.M.; Sause, R. Experimental Evaluation of a Large-Scale Buckling-Restrained Braced Frame. *J. Struct. Eng.* **2007**, *133*, 1205–1214. [CrossRef]
7. Tsai, C.S.; Chen, W.S.; Lin, Y.C.; Yang, C.T.; Tsou, C.P. Seismic Responses of a Full-Scale Steel Structure Using Multi-Curved Buckling Restrained Braces. In Proceedings of the ASME 2007 Pressure Vessels and Piping Conference. Volume 8: Seismic Engineering, San Antonio, TX, USA, 22–26 July 2007; pp. 55–61. [CrossRef]
8. Kaneki, Y.; Hikone, S.; Yamashita, T.; Iwata, M. Seismic Strengthening by buckling restrained braces arranged diagonally. *J. Struct. Constr. Eng.* **2008**, *73*, 2215–2222. [CrossRef]
9. Shima, Y.; Ding, Y.; Zhao, J. Proposal for the structural design method of a sustainable building structure system. *J. Struct. Constr. Eng.* **2009**, *74*, 1179–1185. (In Japanese) [CrossRef]
10. Alemayehu, R.W.; Kim, Y.; Bae, J.; Ju, Y.K. Cyclic Load Test and Finite Element Analysis of NOVEL Buckling-Restrained Brace. *Materials* **2020**, *13*, 5103. [CrossRef] [PubMed]
11. Guerrero, H.; Ji, T.; Teran-Gilmore, A.; Escobar, J.A. A method for preliminary seismic design and assessment of low-rise structures protected with buckling-restrained braces. *Eng. Struct.* **2016**, *123*, 141–154. [CrossRef]
12. Watanabe, A. Design and application of buckling-restrained braces. *Intern. J. High-Rise Build.* **2018**, *7*, 215–221.
13. Çelebi, M.; Kashima, T.; Ghahari, F.; Koyama, S.; Taciroğlu, E.; Okawa, I. Before and after retrofit behavior and performance of a 55-storey tall building inferred from distant earthquake and ambient vibration data. *Earthq. Spectra.* **2017**, *34*, 1599–1626. [CrossRef]

14. Di Sarno, L.; Manfredi, G. Seismic retrofitting with buckling restrained braces: Application to an existing non-ductile RC framed building. *Soil Dyn. Earthq. Eng.* **2010**, *30*, 1279–1297. [CrossRef]
15. Almeida, A.; Ferreira, R.; Proença, J.M.; Gago, A.S. Seismic retrofit of RC building structures with Buckling Restrained Braces. *Eng. Struct.* **2017**, *130*, 14–22. [CrossRef]
16. Saingam, P.; Sutcu, F.; Terazawa, Y.; Fujishita, K.; Lin, P.; Celik, O.C.; Takeuchi, T. Composite behavior in RC buildings ret-rofitted using buckling-restrained braces with elastic steel frames. *Eng. Struct.* **2020**, *219*, 110896. [CrossRef]
17. Castaldo, P.; Tubaldi, E.; Selvi, F.; Gioiella, L. Seismic performance of an existing RC structure retrofitted with buckling restrained braces. *J. Build. Eng.* **2021**, *33*, 101688. [CrossRef]
18. Takeuchi, T.; Wada, A. Review of buckling-restrained brace design and application to tall buildings. *Key Eng. Mater.* **2018**, *763*, 50–56. [CrossRef]
19. RSMC Tokyo-Typhoon Center: Best Track Data (1951–2020), Japan Meteorological Agency (JMA). Available online: https://www.jma.go.jp/jma/jma-eng/jma-center/rsmc-hp-pub-eg/besttrack.html (accessed on 3 February 2021).
20. Duthinh, D.; Simiu, E. Safety of Structures in Strong Winds and Earthquakes: Multihazard Considerations. *J. Struct. Eng.* **2010**, *136*, 330–333. [CrossRef]
21. Usami, T.; Wang, C.; Funayama, J. Low-Cycle Fatigue Tests of a Type of Buckling Restrained Braces. *Procedia Eng.* **2011**, *14*, 956–964. [CrossRef]
22. Wang, C.-L.; Usami, T.; Funayama, J. Improving Low-Cycle Fatigue Performance of High-Performance Buckling-Restrained Braces by Toe-Finished Method. *J. Earthq. Eng.* **2012**, *16*, 1248–1268. [CrossRef]
23. Liu, Y.S.; Chen, K.P.; Li, G.Q.; Sun, F.F. Low-Cycle Fatigue Performance of Buckling Restrained Braces and Assessment of Cumulative Damage under Severe Earthquakes. *Key Eng. Mater.* **2018**, *763*, 867–874. [CrossRef]
24. Xu, Z.-D.; Dai, J.; Jiang, Q.-W. Study on fatigue life and mechanical properties of BRBs with viscoelastic filler. *Steel Compos. Struct.* **2018**, *26*, 139–150.
25. Tong, C.; Wu, J.; Hua, K.; Xie, L. Low-Cycle Fatigue Life Estimation Curve for Buckling-Restrained Braces Based on Cumulative Plastic Deformation. *J. Earthq. Eng.* **2020**, 1–29. [CrossRef]
26. Repetto, M.P.; Solari, G. Wind-induced fatigue collapse of real slender structures. *Eng. Struct.* **2010**, *32*, 3888–3898. [CrossRef]
27. Jia, J. Wind and structural modelling for an accurate fatigue life assessment of tubular structures. *Eng. Struct.* **2011**, *33*, 477–491. [CrossRef]
28. Fang, Z.; Li, A.; Li, W.; Shen, S. Wind-Induced Fatigue Analysis of High-Rise Steel Structures Using Equivalent Structural Stress Method. *Appl. Sci.* **2017**, *7*, 71. [CrossRef]
29. Hoveidae, N. Ultra-low cycle fatigue fracture life of a type of buckling restrained brace. *J. Rehabil. Civ. Eng.* **2018**, *6*, 29–42.
30. AIJ. *Recommendations for Loads on Buildings*; Architectural Institute of Japan: Tokyo, Japan, 2019.
31. BSL. *The Building Standard Law of Japan*; The Building Center of Japan: Tokyo, Japan, 2004.
32. JSSI. *Report of Investigation Committee on Response Control Buildings*; The Japan Society of Seismic Isolation: Tokyo, Japan, 2012.
33. Kasai, K.; Fu, Y.; Watanabe, A. Passive Control Systems for Seismic Damage Mitigation. *J. Struct. Eng.* **1998**, *124*, 501–512. [CrossRef]
34. Saito, T. Structural Earthquake Response Analysis 3D, Version 10.5 (STERA_3D v10.5). 2020. Available online: http://www.rc.ace.tut.ac.jp/saito/software-e.html (accessed on 27 February 2021).
35. Roy, T.; Saito, T.; Matsagar, V. Multihazard framework for investigating high-rise base-isolated buildings under earthquakes and long-duration winds. *Earthq. Eng. Struct. Dyn.* **2020**, *50*, 1334–1357. [CrossRef]
36. Saito, T. Structural Earthquake Response Analysis WAVE Version 1.0 (STERA_WAVE v1.0). 2019. Available online: http://www.rc.ace.tut.ac.jp/saito/software-e.html. (accessed on 27 February 2021).
37. Jennings, P.C.; Housner, G.W.; Tsai, N.C. Simulated earthquake motions for design purpose. In Proceedings of the 4th World Conference on Earthquake Engineering, Santiago, Chile, 13–18 January 1969; pp. 145–160.
38. Wang, J.; Cheynet, E.; Snæbjörnsson, J.; Jakobsen, J.B. Coupled aerodynamic and hydrodynamic response of a long span bridge suspended from floating towers. *J. Wind. Eng. Ind. Aerodyn.* **2018**, *177*, 19–31. [CrossRef]
39. Deodatis, G.; Shinozuka, M. Weighted Integral Method. II: Response Variability and Reliability. *J. Eng. Mech.* **1991**, *117*, 1865–1877. [CrossRef]
40. Deodatis, G. Simulation of Ergodic Multivariate Stochastic Processes. *J. Eng. Mech.* **1996**, *122*, 778–787. [CrossRef]
41. Marsh, G.; Wignall, C.; Thies, P.R.; Barltrop, N.; Incecik, A.; Venugopal, V.; Johanning, L. Review and application of Rain-flow residue processing techniques for accurate fatigue damage estimation. *Int. J. Fatigue* **2016**, *82*, 757–765. [CrossRef]
42. Hasegawa, H.; Takeuchi, T.; Iwata, M.; Yamada, S.; Akiyama, H. Dynamic performances of unbonded braces. *AIJ Tech. Rep.* **1999**, *9*, 103–106. (In Japanese) [CrossRef]
43. Nakamura, H.; Takeuchi, T.; Maeda, Y.; Nakata, Y.; Sasaki, T.; Iwata, M.; Wada, A. Fatigue properties of practical-scale unbonded braces. *Nippon. Steel Tech. Rep.* **2000**, *82*, 51–57.
44. Yamaguchi, M.; Yamada, S.; Takeuchi, T.; Wada, A. Seismic performance of buckling resistant brace within a steel frame in the case of ultimate earthquake. *J. Constr. Steel.* **2004**, *12*, 207–210. (In Japanese)
45. Takeuchi, T.; Uchiyama, T.; Suzuki, K.; Ookouchi, Y.; Ogawa, T.; Kato, S. Seismic retrofit of truss tower structures using buckling restrained braces. *J. Struct. Constr. Eng.* **2005**, *589*, 129–136. (In Japanese) [CrossRef]

46. Takeuchi, T.; Ida, M.; Yamada, S.; Suzuki, K. Estimation of Cumulative Deformation Capacity of Buckling Restrained Braces. *J. Struct. Eng.* **2008**, *134*, 822–831. [CrossRef]
47. Chung, Y.W.; Lee, W.J. Cyclic plastic strain energy as a damage criterion and environmental effect in Nb-bearing high strength, low alloy steel. *Mater. Sci. Eng. A* **1994**, *186*, 121–128. [CrossRef]
48. Iwata, M.; Murai, M. Buckling-restrained brace using steel mortar planks; performance evaluation as a hysteretic damper. *Earthq. Eng. Struct. Dyn.* **2006**, *35*, 1807–1826. [CrossRef]
49. Takeuchi, T.; Ohyama, T.; Ishihara, T. Cumulative cyclic deformation capacity of high-strength steel frames with energy dissipation braces—Seismic performance of high-strength steel frames with energy braces part 1. *J. Struct. Constr. Eng.* **2010**, *75*, 1671–1679. [CrossRef]
50. Takeuchi, T.; Hajjar, J.F.; Matsui, R.; Nishimoto, K.; Aiken, I.D. Effect of local buckling core plate restraint in buckling restrained braces. *Eng. Struct.* **2012**, *44*, 304–311. [CrossRef]
51. FEMA-450. *NEHRP Recommended Provisions for Seismic Regulations for New Buildings and Other Structures*; Federal Emergency Management Agency: Washington, DC, USA, 2003.
52. Sabelli, R. *Research on Improving the Design and Analysis of Earthquake-Resistant Steel Braced Frames*; EERI: Oakland, CA, USA, 2001.
53. Usami, T.; Kasai, A.; Kato, M. Behavior of buckling restrained brace members. In Proceedings of the 4th International Con-ference on STESSA 2003—Behavior of Steel Structures in Seismic Areas, Naples, Italy, 9–12 June 2003; pp. 211–216.
54. ANSI/AISC 360-10. *Seismic Provisions of Structural Steel Buildings*; American Institute of Steel Construction: Chicago, IL, USA, 2010.

applied sciences

MDPI

Article

Evolution of Seismic Site Classification According to the Criteria in Chilean Design Codes

Edgar Giovanny Diaz-Segura

Faculty of Engineering, Escuela de Ingeniería Civil, Pontificia Universidad Católica de Valparaíso, Av. Brasil 2147, 3tr Floor, Valparaíso 2340000, Chile; edgar.diaz@pucv.cl

Abstract: Design codes establish seismic site classifications to determine the seismic demand of a structure according to the response of the soil foundation under the action of earthquake ground motions; the site classification can even condition the feasibility of a project. The occurrence of great earthquakes in Chile has tested its design codes, generating much information and experience regarding the seismic design of structures that have allowed researchers to identify variations in seismic demands according to the kind of ground foundation and to propose seismic site classification methods in Chilean regulations since the 1930s; countries in the vanguard of seismic design, such as the USA, Japan, and New Zealand, proposed methods even earlier. In this document, the evolution of methodologies for seismic site classification according to the criteria in Chilean codes is analysed from their implementation in the 1930s to the most recently proposed design code NCh 433, 2018–2021. Although the distinctive features of each country shape the criteria in their design codes, clear knowledge of the evolution of established criteria from their origins is considered an important tool that contributes to the better understanding, interpretation and application of the seismic site classification methodologies contained in a design code with better criteria. Likewise, the review indicates a distinct need to conduct a continuous evaluation of the classification criteria supported by records of new earthquakes, as well as by physical and numerical models that allow incorporating variables which condition the response of the terrain such as topography, lateral heterogeneities, and basic effects.

Keywords: seismic site classification; seismic design codes; earthquake resistance design; design criteria; soil response

Citation: Diaz-Segura, E.G. Evolution of Seismic Site Classification According to the Criteria in Chilean Design Codes. *Appl. Sci.* **2021**, *11*, 10754. https://doi.org/10.3390/app112210754

Academic Editor: Maria Favvata

Received: 3 August 2021
Accepted: 4 November 2021
Published: 15 November 2021

Publisher's Note: MDPI stays neutral with regard to jurisdictional claims in published maps and institutional affiliations.

Copyright: © 2021 by the author. Licensee MDPI, Basel, Switzerland. This article is an open access article distributed under the terms and conditions of the Creative Commons Attribution (CC BY) license (https://creativecommons.org/licenses/by/4.0/).

1. Introduction

The current main objective of seismic site classification is to assign design response spectra or to define parameters for its establishment according to the non-linear response of the soil foundation to shear waves generated from the bedrock upward to the ground surface by the action of earthquake ground motions.

Parameters to carry out seismic site classification remain objects of study and analysis. Despite the discrepancies generated by the geological, geomorphological and laws of each country, some consensus has been achieved at an international level, sometimes indirectly, with respect to parameters that must be applied for such seismic classification. An example of the latter is the known parameter Vs30 proposed by Borcherdt and Glassmoyer [1] and Borcherdt [2], which incorporates many design codes as the main classification parameter [3]. Although it has been the subject of some debate [4–6], the Vs30 parameter is accepted internationally. Note that some questions also arise due to the complexity and nature of the variables that intervene in the non-linear seismic response of the ground, which has hindered the definition of absolute or unified criteria at the normative level. Given the uncertainties generated in the analysis, perhaps in the future it could be convenient to incorporate complementary data analysis tools based on the neuro-fuzzy logic technique into design code committees [7].

On the other hand, the high seismic demand to which the Chilean infrastructure has been subjected because of great earthquakes during the last 100 years, especially recent large-magnitude earthquakes (Mw 8.8, 8.4, 8.3), has demanded that Chilean engineering be able to define ad hoc characterization parameters. This situation has allowed the response of the ground foundation for a structure to be categorized, which is required to establish the seismic demand for its seismic resistance design. Currently, the Chilean seismic resistance design code, named NCh 433, in its most recent modified proposal (version 2018–2021), contains important advances with respect to the methodology of seismic site characterization and would be a pioneering proposal at the international design code level. This position is not unprecedented for Chile, as the country has developed normative regulations for construction according to its investigations and experience and had integrated qualifying factors for almost 90 years before similar developments by other countries, such as the USA, Japan, and New Zealand. The objective has been to make a difference in seismic requirements according to ground conditions and has produced the seismic classification criteria recorded in different proposed design codes.

In this document, a review and an analysis of the evolution of methodologies for seismic site classification according to Chilean regulations are performed, from their implementation in the 1930s to the most recent proposal in 2018–2021, which considers some limitations of the use of Vs30 as the only dynamic parameter of soil for classification [3–5]. These methodologies incorporate the fundamental period measured by the H/V spectral ratio method (HVR), T_g, as a complementary parameter of seismic classification. Having distinct knowledge of the evolution of ruling criteria from their origins is considered an important tool that contributes to understanding, interpreting and applying better criteria for ground classification methodologies contained in a design code.

2. Brief Historical Context of Chilean Construction Regulations

For the Chilean territory, from colonial times, urban planning regulations, such as the Discovery Ordinances, New Populations and Pacifications of 1573, were issued to regulate the conformation, growth or reforms of cities, mainly motivated by demographic growth and for reasons of health [8].

In Chile as a republic, the first approach to regulations related to urban subjects was decreed in 1854, which assigned to city halls the responsibility for defining urban development guidelines about construction or urbanism. Between 1874 and 1912, urbanism laws were enacted by different city halls in Chile; these laws were collected during the reform of the city halls law in 1915 [8]. Although Chile has historically recorded more than 10 earthquakes with great magnitude since achieving independence in 1810 (today, earthquakes are catalogued with magnitudes higher than Mw 8.0 [9]), laws or regulations issued prior to 1927 did not include any type of seismic consideration for construction. As an international context for this decade, the first regulation of seismic design in the world was decreed in 1924 in Japan.

In 1928, the Mw 7.6 earthquake in Talca greatly damaged structures, which emphasized the urgent need to regulate building construction while considering the seismic demands of the country. From the previous background, in 1929, a law project was decreed that authorized developing a general ordinance to establish regulations to which building construction must conform in the different districts of Chile. These regulations would consider criteria to prevent collapse, for the first time indicating that construction must *"avoid, as much as possible, the risks coming from an earthquake"*. Thus, between 1930 and 1931, the First Law and Ordinance of Construction and Urbanization (OGCU), began seismic resistance regulation for construction in Chile [10], joining countries with normative design criteria, such as Japan, 1924 [11], the USA, 1927 [12], and New Zealand, 1935 [13]. Note that criteria and definitions with respect to seismic design are contained mainly in the version of the OGCU established in 1936, OGCU-1936 [14].

Between 1936 and 1971, the main regulation with seismic design criteria was the OGCU, which underwent different changes motivated by the large seismic events that occurred in the country. In 1972, the first Chilean seismic design code, referred to as NCh 433, was promulgated, the most recent version of which corresponds to the proposal evaluated between 2018 and 2020, which is in the process of being approved.

3. Seismic Site Classification Methodologies According to Chilean Design Regulations or Seismic Codes

3.1. Construction and Urbanism General Ordinance (1930–1936)

As indicated above, the first seismic regulations were established in OGCU-1936 [14], which set criteria for seismic design by referring to a variation in seismic solicitation according to ground conditions. This regulation shows the first approximation to seismic site classification, which is an aspect of design codes that was not considered during this period.

OGCU-1936, similar to the Japanese model [11], considered seismic action an equivalent force applied to the centre of gravity of a structure; the magnitude of the horizontal component was equal to the building weight multiplied by a seismic coefficient, while the vertical component was considered to be 50% of the horizontal component. This standard was based on a criterion proposed in the Japanese code of 1924, with a seismic coefficient equal to 0.1, i.e., 10% of the structure's weight. However, the OGCU defined seismic coefficients between 0.05 and 0.1 according to the seismic and geologic characteristics of the zone where the construction was to be located and the quality of the foundation ground. These coefficients were categorized with the expression *"allowable resistance of ground"*, as shown in Table 1. Considering this criterion, the *"allowable resistance of ground"* would be one of the first parameters for seismic site classification; starting from this point, the expected seismic solicitation magnitude for design was defined. In addition to that indicated in Table 1, Art. 152 of the OGCU-1936 also established conditions for "seismic action" based on the kind of ground, assigning for its consideration the factor shown in Table 2. However, the ambiguous statement of Art. 152 does not allow us to identify the use or application of such a factor in a settled way; whether it involved a balance of seismic coefficients (del Canto, et al., 1940) is also not indicated, but if it did consider a balance, this method could generate a solicitation for more than 10% of building weight for low-quality ground.

Table 1. Selection criteria for seismic coefficients according to soil resistance from OGCU-1936 (Ordinance of Construction and Urbanization).

Allowable Resistance of Ground Foundation	Horizontal Seismic Coefficient, k_h	Vertical Seismic Coefficient, k_v
≤300 kPa	1/10	1/10
>300 kPa	1/20	1/40

Table 2. Factors for defining seismic action according to Art. 152 OGCU-1936.

Class of Ground Foundation	Factor
Sandstone ground	1.0–2.4
Loose sand ground	2.4–4.4
Loose and fill soils	4.4–11.0

In a complementary way, the OGCU also included an analogous criterion that today could be considered a place effect, indicating that "Departments of Construction of City Halls would be able to require that the effect of resonance is studied in special cases". However, according to the reported precedents, this statement was more likely intended to analyse a possible place effect because at this time, engineers lacked the power to consider variable acceleration effects during a seismic event, as OGCU-1936 suggested a constant acceleration magnitude for design [15].

3.2. Construction and Urbanism General Ordinance, OGCU-1940

As a consequence of the devastating earthquake in Chillan on 24 January 1939, which caused more damage and the highest number of fatalities reported in Chile, the government requested the creation of a technical government commission, which submitted an evaluation of the current ordinance with the damage report, highlighting mistakes and ambiguities, proposing new seismic factors and incorporating the first approach to design using a dynamic method; the commission recommended " … *reinforced concrete buildings to be calculated in a more rational way (that is, taking into consideration not only the maximum acceleration of earthquakes as established by the Ordinance but also the amplitude and period)*" [15]. This document is the basis of the main changes related to seismic design and generated a new OGCU that was approved in 1940 [16]. The new OGCU included an adjustment with respect to seismic action, which is now considered a harmonic movement, or followed the use of a static method equivalent to that defined in OGCU-1936, which would subsequently depend on the type and period of a structure. Thus, for structures with the period T_e less than 0.4 s, the seismic action was determined using the static method defined in OGCU-1936. If T_e was greater than or equal to 0.4 s, " … *the action of the earthquake will assimilate to a horizontal vibration in any direction, of a simple harmonic movement, with the following characteristics: Acceleration 0.1 g to 0.2 g (g gravitational acceleration); amplitude 4 to 6 cm, period 1 to 2 s …* ". For this last case, the OGCU-1940 established the possibility of using an approximate method to determine the seismic action, which due to the absence of more precise methods suggests that the government commission utilized such a method [15,17]; based on this reasoning, the provision is interpreted as a seismic classification method, as listed in Table 3.

As shown in Table 3, OGCU-1940 removed the parameter for classifying the admissible resistance of the ground, reducing it to a general characterization of the ground without main specifications. This lack of a main decree with respect to the ground could be attributed to the effects of the 1939 earthquake, which was uniformly devastating throughout the whole area of influence. These effects would make it more difficult to refine particular aspects associated with the ground. Considering the limited geomechanics exploration that was conducted during this period, the normative analysis was centred more on the identification of the characteristic seismic wave, temporarily and partially disregarding the influence of the ground. Nevertheless, participants in the government commission suggested that geological aspects were required to continue investigation [15].

Considering bibliographic antecedents available at the time, the OGCU commission in charge of revision observed that at an international level, prominent investigators concurred, indicating that the evidence of earthquakes, such as those in San Francisco in 1906 or Tokyo in 1923, demonstrated that the destructive aspect of earthquakes was attributed to waves whose periods varied between certain defined limits [17] (p. 87). Therefore, with this assumption, the ordinance adopted a criterion to restrict construction, whose T_e period would range between 1.0 and 2.0 s. The simplified method, previously indicated and suggested by OGCU-1940, did not explicitly include factors for other periods, such as for T_e between 0.75 and 1.0 s and between 2 and 3 s. Therefore, in Table 3, values of seismic action in these period ranges correspond to an interpretation, based on the report of the commission, to complete the analytical range for an accepted T_e in OGCU-1940. The maximum values reported for factors α and β are extrapolated for different site classes.

Table 3. Interpretation of seismic classification criteria according to OGCU-1940.

Ground Foundation	Structure Period, T_e, s	Static Method		Amplitude of the Seismic Action as Simple Harmonic Motion	
		k_h	k_v	a_h, g	δ_h, m
Rock	<0.4	0.10	0.15	-	-
Sandstone or conglomerate		0.12		-	-
Gravel or loose sand		0.15		-	-
Fill		0.20		-	-
Rock	(*1) $0.4 \leq T_e \leq 0.75$ $\alpha = 2.86\left(\frac{T_e}{1\ s}\right) - 0.144$	-	0.15	$0.10 \times \alpha$	-
Sandstone or conglomerate		-		$0.12 \times \alpha$	-
Gravel or loose sand		-		$0.15 \times \alpha$	-
Fill		-		$0.20 \times \alpha$	-
Any type of ground	1.0–2.0	Constructions are not allowed			
Rock	(*2) $2.0 < T_e \leq 3.0 - \beta = 2.0$ $3 > T_e \geq 5.0 - \beta = 3.5 - \left(\frac{T_e}{2\ s}\right)$ $T_e > 5.0 - \beta = 1.0$	-	0.15	-	$0.040 \times \beta$
Sandstone or conglomerate		-		-	$0.052 \times \beta$
Gravel or loose sand		-		-	$0.060 \times \beta$
Fill		-		-	$0.060 \times \beta$

*1. For $0.4\ s \leq T_e < 1.0$ s, according to the modification of the OGCU-1940, the seismic action is considered a uniform acceleration, ah, which is equal to the maximum of the seismic event multiplied by the coefficient. α. *2. For $T_e > 2.0$ s, the design must consider applying the horizontal displacement, δ_h, to the centre of gravity of the structure, which is equal to the maximum amplitude of the seismic wave multiplied by the coefficient β.

3.3. Construction and Urbanism General Ordinance, OGCU-1949

In 1949, a new version of the OGCU was promulgated; however, with respect to seismic design, it contained criteria similar to those in OGCU-1940, which had the government commission's report as technical backup [15].

For seismic action, the recommendation of using the approximate dynamic method proposed by Del Canto et al. [17] was retained, and it was also declared directly that the determination of "… seismic and fatigue solicitation that they produce in materials will be done, in general, by the dynamic equations." This last point represents the ratification of the need to establish a design criterion, the dynamic method, considering dynamic solicitation as the action of waves that spread with amplitude, acceleration and displacement from the ground to the foundation and building.

Based on the previous description, Table 4 shows the classification interpreted according to criteria in OGCU-1949 [18], which still permitted the use of the approximation method and was compatible with the classifying method shown in Table 3. One difference is that the new OGCU-1949, in addition to the period of a structure, incorporated the stiffness of the foundation as a distinguishing parameter of seismic action. In relation to restricted structures with periods between 1.0 s and 2.0 s, OGCU-1949 did not specify; however, to make the approximate method valid, indirect restrictions were retained. Factors α and β shown in Table 3 were deleted, and the values were held constant for cases with periods of $0.75\ s < T_e < 1.0$ s and $2.0\ s < T_e < 3.0$ s as the criteria for the approximate method [17].

Table 4. Interpretation of seismic classification criteria according to OGCU-1949.

Ground Foundation	Structure Period, T_e, s	Foundation Rigidity	k_h *1	Amplitude of Seismic Action, m
Rock	<0.4	-	0.08	-
Conglomerate or very dense ground		-	0.12	-
Loose soils or sand		with raft foundation or similar rigidity	0.10	-
		without raft foundation	0.12	-
Rock	$0.4 \leq T_e \leq 0.75$	-	0.05	-
Conglomerate or very dense ground		-	0.10	-
Loose soils or sand		with raft foundation or similar rigidity	0.12	-
		without raft foundation	0.15	-
Any type of ground	1.0–2.0	Constructions are not allowed		
Rock	$T_e > 3.0$ *2	-	-	0.020
Conglomerate or very dense ground		-	-	0.040
Loose soils or sand		with raft foundation or similar rigidity	-	0.050
		without raft foundation	-	0.060

*1 The vertical component of the seismic wave will only be considered in cases where the nature of the work requires it. Its acceleration and amplitude will be equal to 50% of the horizontal seismic action. *2 Interpretation and limit of the period taken from the approximate method proposed by Del Canto, et al. [17], which the OGCU authorizes can still be applied for the determination of seismic action.

3.4. Earthquake Building Resistance Code NCh 433Of72

Following the modification of OGCU-1949, during the 1950s, there were no changes in seismic regulations, although at least five earthquakes of magnitude Mw higher than 7.5 occurred, culminating with the mega-earthquake in Valdivia in 1960, which reached Mw 9.5. Nevertheless, by the end of the 1950s and prior to the great earthquake in 1960, the National Institute of Investigations, Inditecnor (which in 1973 became the National Institute of Standardization, INN), was requested in 1959 to start developing seismic design criteria independent from OGCU. A committee was created with professionals and academic experts.

During the 1960s, different proposals were formulated; in particular, the work of Arias and Husid [19] had great influence. With respect to the ground effect, they had already indicated that the designer could modify the spectrum shape to consider selective amplifying cases for types of ground. Starting with models and spectrum analyses of real earthquakes, Arias and Petit [20] subsequently presented a theoretical basis to define relationships between the response spectrum and ground properties, directly expressing for the first time the need for seismic site classification analogous to the system employed today.

An antecedent in the form of a 1968 preliminary regulation denoted NCh 433Of68 [21], which was not officially published but accepted the incorporation of the static method and dynamic method. The former is applicable to short structures with simple construction, while the latter is applicable to taller and more morphologically complex buildings. Starting from this basis, a seismic coefficient was defined, and an acceleration spectrum was categorized for different site classes to which the fundamental vibration period, T_0, was assigned. The equations applied for each method are listed as follows:

Seismic coefficient for the static method:

$$C = 0.10 \text{for} T < T_0 \quad (1)$$

$$C = 0.10\frac{2TT_0}{T^2 + T_0^2} < T_0 \tag{2}$$

Acceleration spectrum for the dynamic method:

$$\frac{a}{g} = 0.10K_1K_2 \text{for} T < T_0 \tag{3}$$

$$\frac{a}{g} = 0.10K_1K_2\frac{2TT_0}{T^2 + T_0^2} < T_0 \tag{4}$$

where C is the seismic coefficient; T is the period of the corresponding vibration mode; T_0 is the fundamental vibration soil period; K_1 is a coefficient relative to the intended use of the structure (factor between 0.8 and 1.2); and K_2 is a coefficient relative to the structural form (factor between 0.8 and 1.2).

Taking as references the unpublished proposal from 1968 and recommendations proposal by Saragoni [22] regarding earthquake spectra from Chile, Peru, Japan and the USA, which provided clear knowledge of the foundation ground type where these spectra were recorded, a seismic site classification was proposed. This scheme is shown in Table 5 and was included in the new regulation. All previously indicated criteria were finally consolidated and included in the regulation that was made official as "NCh 433Of72, Earthquake resistance of buildings 1972" [23], which combined with the Argentine code [24], constituted together the first official codes of seismic design at the South American level that incorporated design spectra.

At this time, design codes at the international level, including the Uniform Building Code (USA), Building Law (Japan), and codes in New Zealand, had been continuously revised. However, for 1968, the year in which the seismic site classification criteria in NCh 433 were proposed, every design code proposed defining spectra according to the seismic zones of the country, but codes were not defined according to the type of ground; such criteria were incorporated in the USA in 1976 [25], New Zealand in 1976 [26], and Japan in 1971 [27]. Therefore, NCh 433, with respect to the influence of the ground, was one of the first design codes at the world level that incorporated the influence of ground foundations in the spectra for seismic design in buildings and assigned a seismic classification. Nevertheless, at that time, the Chilean code and international codes still did not have indications with respect to identification parameters to characterize the ground, as ground types for their classification were general and descriptive.

Table 5. First seismic site classification from NCh 433Of72.

Site Class	T_0, s
Rock, dense gravel, dense sandy gravel	0.20
Dense sand, stiff cohesive soils	0.30
Loose granular soils, soft or medium soft cohesive soils	0.90

With code NCh 433, a great step was taken related to regulations for the design of structures considering seismic solicitations; these regulations incorporated not only the points previously indicated but also the first guidelines for design from a probability perspective.

3.5. *Earthquake Resistance of Buildings, Codes NCh 433Of93 and NCh 433Of96*

The earthquake on 3 March 1985, in the central zone of Chile represented a great test for the prevailing seismic laws in Chile, and as usual after a great seismic event, the National Institute of Standardization, INN, in 1986 constituted the Seismic Resistance Coordinator Code Committees. With input from these committees, the first modification to the code was proposed and then approved in 1993, and as a result, code NCh 433Of93 [28] was cancelled and replaced by NCh 433Of72 [23]. The 1993 version included different changes in the design code, highlighting the incorporation of a new system for the seismic site classification of the ground and directly defining parameters to characterize the ground as well as the shear-wave velocity, Vs; SPT N-value; undrained strength, s_u; and the resistance to simple compression, q_u.

Considering the lessons learned after the earthquake of 1985 and great seismic events recorded around this time at an international level (Los Angeles, CA, USA, Mexico, and Peru), these regulations clarified for the first time the application range with respect to the geomorphological conditions of the ground. Parameters for seismic characterization and solicitation were defined considering horizontal surfaces and horizontal stratification and acknowledged that some structures are built on foundations that reveal geomorphological and topographic singularities. This last point was motivated by the effects of topographic amplification observed and categorized for structures built in the area of the Beagle channel, Viña del Mar, during the earthquake and aftershocks in 1985 [29]. Importantly, note that by the publication date of the document that contained the modifications to NCh 433Of72, 1989, no national or international codes had defined specific criteria for the identification of ground types that may suffer topographic effects, leaving their identification to engineers' judgement. This procedure was generally applied in such cases as in regular professional practice, and topographic effects were not considered. Even today, most design codes at the international level still lack criteria for the consideration of topographic effects. Exceptions, such as the French code [30], Eurocode [31], and Italian code [32], share the same criteria.

In relation to special cases, for the first time, associated criteria were introduced for ground types with the potential to liquefy, identifying this phenomenon for sand $(N_1)_{60}$ less than 20 blows/feet, which had been excluded from ground classification systems. In 1989, this point also represented a development, as different design codes, such as the USA [33], Japan [34] and New Zealand codes [35], did not include criteria that conditioned the seismic site classification in cases with possible liquefaction.

As a part of the revision process for the practical use of code NCh 433Of93, the Chilean Association of Seismology and Antiseismic Engineering (ACHISINA) proposed adjustments to the code based on the main conclusions registered by the professional community; in 1996, a second update became official, denoted NCh 433Of96 [36]. With respect to ground and seismic classification, the 1996 version did not involve any changes.

The seismic classification contents for NCh 433Of93 and NCh 433Of96 are presented in Table 6. In this case, four site classes were defined; they were identified starting with defined parameters and helped to reduce subjectivity, which could affect the characterization of ground types. This factor represented an important advance, considering for example, that code UBC 1991 [37] still did not include classification parameters but considered only the shear-wave velocity, Vs, employed to characterize rock. In 1997, when the UBC proposed a seismic classification system with direct classification parameters, the shear wave propagation velocity of the surface 30 m, Vs30, which corresponds to a parameter that many design codes worldwide later incorporated for the seismic classification. Although this system is still in force, in the case of the Chilean standard, it was not incorporated, and work continued until 2009, as shown in Table 6.

NCh 433Of93 also conditioned the classification to the thickness of the superficial strata using the criteria of 10 m and 20 m for site class types III–IV and II, respectively. Although there was no clear antecedent for the origin of criteria recently referenced by members of the code committee, an interpretation was obtained from the criterion defined by UBC 1985 [38] for seismic classification in a minor category corresponding to soft ground; S3 (site class types III and IV in NCh 433Of93) ground types must have a thickness of 30 feet or approximately 10 m. On the other hand, related to measuring the shear-wave velocity in the uppermost 10 m to classify a location in site class type II, there was no clear justification for the criterion, which additionally applied to only this site class, since the shear-wave velocity for types III and IV was not specified. In terms of seismic site classification, NCh 433Of93 was not free of ambiguities, which were later exposed by the 2010 earthquake; consequently, this event led to changes in the seismic code.

Table 6. Seismic site classification NCh 433Of93/NCh 433Of96.

Site Class	Description
I	Rock: Natural material with in-situ Vs $\geq$ 900 m/s, or q_u intact rock $\geq$ 10 MPa and RQD $\geq$ 50%
II	(a) Soil, Vs $\geq$ 400 m/s in the upper 10 m, and increasing with depth; or well, (b) Dense gravel, $\gamma_d \geq 20$ kN/m^3, or relative density $\geq$ 75%, maximum dry density $\geq$ 95%; or well, (c) Dense sand, relative density $\geq$ 75%, $(N_1)_{60} > 40$, maximum dry density $\geq$ 95%; or well, (d) Stiff cohesive soil, $s_u \geq 0.10$ MPa (unfissured samples). Minimum thickness of the horizon layer on the rock, 20 m. If the thickness of the stratum on the rock is less than 20 m, the site class will be classified as type I.
III	(a) Sand permanently unsaturated, 55 < relative density $\leq$ 75%, $N_{60} > 20$; or well, (b) Unsaturated gravel or sand, maximum dry density $\geq$ 95%; or well, (c) Cohesive soil with $0.025 < s_u \leq 0.10$ MPa, regardless of the water table; or well, (d) Saturated sand with $20 < (N_1)_{60} \leq 40$. Minimum horizon layer thickness: 10 m. If the thickness of the stratum on the rock or on the site class corresponding to type II is less than 10 m, the site class will be classified as type II.
IV	Saturated cohesive soil with $s_u \leq 0.025$ MPa. Minimum horizon layer thickness: 10 m. If the thickness of the horizon layer corresponding to certain types I, II or III is less than 10 m, the site class will be classified as type III.

Vs: shear wave velocity; q_u: unconfined compressive strength; RQD: rock quality designation; s_u: undrained shear strength; γ_d: dry unit weight; $(N_1)_{60}$: normalized SPT N-value.

3.6. Resistance of Buildings to Earthquakes, Code NCh 433Of96mod2009

In 2009, a modification of the code denoted NCh 433Of96mod2009 [39], which did not include any changes with respect to seismic site classification, became official, and the criteria in Table 6 remained applicable. Notably, by the time the 2009 version of the code was made official, different countries had incorporated the shear-wave velocity in the upper 30 m, Vs30, as a seismic classification parameter. However, in Latin America, this parameter was not commonly employed in professional practice. The seismic code incorporated the seismic classification parameters from the 1996 version (NCh 433Of96), retaining a certain ambiguity with respect to obligatory measurement of the shear wave propagation velocity.

3.7. Emergency Supreme Decree 117, DS117-2011

Although code NCh 433Of96mod2009 had recently been approved, damage generated by the Mw 8.8 earthquake on 27 February 2010, in Maule, provided evidence for the urgent need to evaluate and adapt the code. The derived changes and immediate adjustments were included in Emergency Supreme Decree number 117-2011 issued by the Urbanism and Buildings Ministry, which modified NCh 433Of96mod2009 [40]. One of the points of special interest at the design code committee level was the seismic site classification, which at the time had undergone no changes for 17 years, as was evidenced from the damage recorded that was considered greater than acceptable; accordingly, the necessity of modifying the classification to incorporate the seismic requirements of the country according to its ground typologies was recognized.

Among the main changes in DS117-2011, new criteria were included for the seismic classification of ground types and associated new design spectra. Therefore, considering the international validation implied by the incorporation of parameter Vs30 proposed by Borcherdt [1,2], the parameter Vs, which included the previous version of the code for certain site classes, was modified to Vs15 and Vs30, considering the lesser value of the two velocities as a parameter for the classification of each type of ground. Here, Vsi represents the velocity of shear waves that travel through the uppermost i metres of the ground. Because of the changes to the regulations, measuring the shear-wave velocity in the field became a professional procedure. At this time, there was not enough equipment to adequately measure the shear-wave velocity. Thus, for a temporary period of three years, the estimation of shear-wave velocity using indirect methods, data analysis or correlations with geotechnical exploration was authorized. Addressing this same point, the parameter Vs15 would represent a conservative criterion of classification in some cases, which could simplify exploration in the short term.

In addition to the four types of site classes in NCh 433Of96, a fifth type, V, which corresponds to ground types denoted "special" for which a special study was needed to define a local spectrum design, was incorporated. Furthermore, the classification shown in Table 7 excludes ground that may potentially liquefy (this class was retained with the criteria from NCh 433Of96 for its definition: sand, saturated sand or silt, $(N_1)_{60} < 20$), which is susceptible to densification because of vibrations.

Table 7. Seismic site classification—DS117-2011.

Site Class	Description	Min (Vs30; Vs15) m/s	RQD	q_u (MPa)	$(N_1)_{60}$ (For Sands)	s_u (MPa) (For Fines Soils)
I	Rock or cemented soil (not soluble in water)	≥900	≥50%	≥10 ($\varepsilon_{qu} \leq 2\%$)		
II	Soft rock or very dense soil or very firm soil	≥500		≥0.4 ($\varepsilon_{qu} \leq 2\%$)	≥50	
III	Medium dense or firm soils	≥180			≥30	≥0.05
IV	Loose or soft soils	<180			≥20	
V	Specials soils	-	-	-	-	-

DS117-2011 had great impacts on practical engineering, confirming the importance of correct and consistent geotechnical exploration. This emergency decree was valid for a period of 9 months, during which analysis continued and whose results were subsequently included in Supreme Decree 61, approved in December 2011 [41].

3.8. Supreme Decree 61, DS61-2011

Following the promulgation of emergency decree 117-2011, the committee on code NCh 433, having clarified the criteria that had been temporarily defined, proceeded to a proposal that included adaptations and adjustments to the seismic site classification system and the design spectra associated with each type of ground; these changes were contained in Supreme Decree 61, approved in December 2011 and known as DS61-2011 [41]. The newly proposed classification defined six types of ground, dividing the previous ground III into two types, defining the shear-wave velocity measured in situ, requiring Vs30 as a parameter for the classification of ground types, and removing the criterion using Vs15. The detailed seismic classification method is shown in Table 8.

DS61-2011 also specified with more detail cases that had previously not been classified and, therefore, that required special studies. These cases are described as follows: potentially liquefiable ground types such as sand, saturated silty sand or silt with $(N_1)_{60} < 20$ blows/foot; ground types susceptible to densification by vibration; collapsible and organic soils; saturated fine soils with liquid limit (LL) >80 and thickness >20 m; fine soils with sensitivity higher than 10; and ground types with irregular topography where local amplification phenomena can occur. Even given these additions, the classification system continued to classify "special cases" that are now categorized as site class type F, which includes the previously indicated cases, except for topographic amplification.

As previously indicated, the regulations adopted a minimum exploration thickness of 100 feet, that is, 30 m, for classification and incorporated it in the International Building Code 1997 [42] and in many international codes; this value was incorporated as a site seismic characterization parameter. Additionally, considering that certain structures were built mainly in the city of Santiago, which had underground levels with depths greater than 15 m, DS61-2011 required the measurement of classification properties to a maximum between 30 m and Df + 15 m, where Df is the embedment depth. Thus, in cases with several basement levels, a minimum exploration depth of 15 m below the level of the foundation system should be guaranteed.

Note that the seismic classification of ground types defined from the response or characterization of the top 30 m has not been exempt from criticism; this depth can be an insufficient criterion for very stiff ground types [43] or for certain stratigraphic configurations, where two places may have the same Vs30 but different stiffness changes with stratum depth, generating a different seismic response for each case. This point will be the topic of a future ground study; thus, the analysis of information obtained from earthquakes and types of ground in the country will eventually incorporate some changes in the most recent proposal from NCh 433.

Table 8. Seismic site classification—DS61-2011.

Site Class	Description	Vs30, m/s	RQD	q_u (MPa)	$(N_1)_{60}$ (For Sands)	s_u (MPa) (For Fines Soils)
A	Rock or cemented soil (not soluble in water)	≥900	≥50%	≥10 ($\varepsilon_{qu} \leq 2\%$)		
B	Soft rock or very dense soil or very firm soil	≥500		≥0.4 ($\varepsilon_{qu} \leq 2\%$)	≥50	
C	Dense or firm soils	≥350		≥0.3 ($\varepsilon_{qu} \leq 2\%$)	≥40	
D	Medium dense or firm soils	≥180			≥30	≥0.05
E	Medium compactness or medium consistence soil	<180			≥20	<0.05
F	Specials soils	-	-	-	-	-

3.9. Earthquake Resistance of Buildings NCh 433—Modified Proposal 2018–2021

After the validation of DS61-2011, members of the code committee continued their meetings with the objective of evaluating the large amount of data obtained and the new data generated by the practical application of DS61-2011. Additionally, two important earthquakes occurred in the northern part of the country with magnitudes of 8.2 and 8.3 in 2014 (Iquique) and 2015 (Illapel), respectively, which provided interesting complementary information for the seismic classification system.

The most recent proposal to code NCh 433 was presented in December 2018, and currently, it is in the final revision stage, which includes a period of public inquiry, where suitable professionals will deliver their observations. This process will enable the consolidation of the final version of the document, the application of which is expected to start during the next few years.

Verdugo [5], based on a rigorous analysis of the seismic site classification procedures identified from available data, including seismic records of earthquakes with a magnitude Mw greater than 8.0, discovered that the H/V spectral ratio (HVR) obtained via ambient vibrations reproduces the predominant periods shown by the response spectra and suggests its use as a complementary parameter for seismic site classification.

Centring the analysis of the proposal only on those revisions related to seismic site classification, one of the main changes proposed by the code committee, taking into account all the antecedents, consists of the incorporation, as a classification parameter of the fundamental period of vibration of the ground, T_g, measured in situ using the HVR method, which is directly linked with the seismic response of the ground and is complementary to Vs30. This criterion allows differentiation of the cases that may present an equal Vs30 but may have stiffness variations with depth that condition the seismic response at the surface; the incorporation of T_g will allow each case to be distinguished. This proposal is innovative in global design codes; although the Japanese code [44] also uses the period as a classification parameter, it is estimated starting with the shear-wave velocity to the named engineering bedrock, and its direct in situ measurement is not required.

Considering the fundamentals of the previous code, the proposed code replaces properties and/or indices associated with the resistance of the ground formerly applied for seismic classification (its density q_u or N-value SPT) for parameters that represent the dynamic stiffness properties of a location, which are correlated with the phenomenon of ground amplification (Vs30 and T_g). The classification system is shown in Table 9, in which Vs30 should be entered, and the T_g value measured in the field should be verified through the HVR method [45].

Table 9. Seismic site classification—NCh 433 proposed for public consultation in 2021.

Site Class	Description	Vs30, m/s	T_g, s *1
A	Rock or cemented soil (not soluble in water)	≥900	<0.15 (or plane HVR) *2
B	Soft or fractured rock; very dense soil or very firm soil	≥500	<0.30 (or plane HVR) *2
C	Dense or firm soils	≥350	<0.40(or plane HVR) *2
D	Medium dense or firm soils	≥180	<1.00
E	Medium compactness or medium consistence soil	<180	-

*1 Soil period, T_g, measured in situ using the HVR method. *2 A flat measurement is considered when the HVR is less than 2.0.

With respect to the validity range or applicability of the classification, the code partially retains the recommendations from DS61-2011; however, special cases are now directly grouped in site class type F, which corresponds to different situations in which a study is required and requested. This provision allows a spectrum of local design to be defined as follows: potentially liquefiable soils corresponding to sandy or silty soils with $(N_1)_{60}$-cs < 30 blows/foot, or $q_{c1\text{-}cs}$ < 17 MPa (updated criterion in this proposal, which should be confirmed or eliminated by means of a study of potential liquefaction); soils sensitive to vibrating densification; collapsible or organic soils, peat, and fine saturated soils with LL > 80 and thickness greater than 20 m; and fine saturated soils with sensitivity higher than 10.

The validity of the code is ratified only for ground types with topography and horizontal stratigraphy and for structures located outside of zones where local amplification phenomena occur or can be generated.

4. Conclusions

The seismic classification of foundation ground types has direct implications for the seismic design of a structure, impacting its construction costs and conditioning its feasibility. Therefore, a clear knowledge of the genesis of regulatory criteria allows engineers to better understand, interpret and apply professional criteria for classifying ground types and following the methodologies in a design code, which defines guidelines that are ultimately intended to safeguard the lives of those who will use the structure.

Great seismic events to which Chilean territory has been subjected have generated important information and experience in the seismic design of structures, which have allowed Chilean regulations to identify and include variations in seismic demand according to the type of ground since the 1930s, prior to most design codes worldwide. The main conclusions have been scattered among the different design codes proposed for Chile, generating seismic classification proposals that could be considered in the vanguard for each period. Along these lines, NCh 433 has recently proposed the replacement of properties associated with ground resistance that are usually applied in different design codes (ground density, q_u, or N-value, SPT) by parameters that represent the dynamic stiffness properties of a specific location, which correlate more consistently with the ground amplification phenomenon. Thus, in addition to the commonly employed Vs30, the code proposes incorporating the fundamental period of ground vibration measured in situ, T_g, as a parameter for seismic classification.

Regarding the predecessor Chilean design codes and homologous regulations at the international level, the most recent proposed version of NCh 433 presents an important advance regarding the selection of seismic site classification parameters in the process of performing a practical characterization that can better correlate with the phenomenon of ground amplification. Likewise, it is recognized that the Vs30 parameter, which is widely employed internationally, should not be substituted but rather should be complemented with dynamic parameters, such as the proposed T_g, which, according to the observations of the last earthquakes, was considered with suitable parameters.

Seismic engineering in its different disciplines has made important advances that have allowed the adjustment of seismic design codes. This review highlighted the clear need to undertake a continuous evaluation of the classification criteria, supported by records of new earthquakes, as well as physical and non-linear numerical models that allow incorporating variables that condition the response of the terrain, such as topography, lateral heterogeneities, and basic effects. Notably, the particularities of each country shape the criteria in their design codes. Nevertheless, acknowledging other criteria, such as those in Chilean codes gained by experience with respect to seismic site classification, allows the consolidation of similar criteria that could be common, at least, at the regional level.

Funding: This research was not external funding.

Institutional Review Board Statement: Not applicable.

Informed Consent Statement: Not applicable.

Acknowledgments: The author would like to thank ACHISINA for allowing the use of the code proposal of 2018–2021 in draft form for academic use.

Conflicts of Interest: The author declares no conflict of interest.

References

1. Borcherdt, R.; Glassmoyer, G. On the characteristics of local Geology and their influence on ground motions generated by the Loma Prieta earthquake in the San Francisco Bay region, California. *Bull. Seismol. Soc. Am.* **1992**, *82*, 603–641. [CrossRef]
2. Borcherdt, R.D. Estimates of site-dependent response spectra for design (methodology and justification). *Earthq. Spectra* **1994**, *10*, 617–653. [CrossRef]
3. Maugeri, M.; Soccodato, C. *Earthquake Geotechnical Engineering Design*; Springer: Berlin/Heidelberg, Germany, 2014.
4. Castellaro, S.; Mulargia, F.; Rossi, P.L. Vs30: Proxy for seismic amplification? *Seismol. Res. Lett.* **2008**, *79*, 540–543. [CrossRef]
5. Verdugo, R. Seismic site classification. *Soil Dyn. Earthq. Eng.* **2019**, *124*, 317–329. [CrossRef]
6. Aleshin, A.S. On the Disadvantages of the NEHRP Soil Classification. *OJER* **2021**, *10*, 1–15. [CrossRef]
7. Carreño, M.L.; Cardona, O.D.; Barbat, A.H. Computational Tool for Post-Earthquake Evaluation of Damage in Buildings. *Earthq. Spectra* **2010**, *26*, 63–86. [CrossRef]
8. Cordero, E. La formación del Derecho urbanístico chileno a partir del siglo XIX. De la legislación urbanística al Derecho urbanístico integrado. *Rev. Derecho (Valdivia)* **2017**, *30*, 127–152. [CrossRef]
9. National Seismological Center of Chile. Available online: https://www.csn.uchile.cl/sismologia/grandes-terremotos-en-chile/ (accessed on 27 July 2021).
10. Ministry of the Interior of Chile. *Construction and Urbanism General Ordinance*; Law 4841; Ministry of the Interior of Chile: Santiago, Chile, 1930. (In Spanish)
11. Architectural Institute of Japan. *Standards for Structural Strength Calculation*; Urban Building Law; The Building Center of Japan: Tokyo, Japan, 1924.
12. International Conference of Building Officials. *Uniform Building Code*; International Conference of Building Officials: Lansing, MI, USA, 1927.
13. New Zealand Standards Institute. *NZSS No. 95, New Zealand Standard Model Building By-Law*; Sections I to X; New Zealand Standards Institute: Wellington, New Zealand, 1935.
14. Ministry of the Interior of Chile. *Construction and Urbanism General Ordinance*; Decree 437; Ministry of the Interior of Chile: Santiago, Chile, 1936. (In Spanish)
15. del Canto, H.; Godoy, P.P.; Aguirre, P.E.; Muñoz, S.J.; Ibáñez, V.J. Report of the Government Commission on the effects produced by the earthquake of January 1939. *An. Inst. Ing. Chile* **1940**, *12*, 434–446. (In Spanish)
16. Ministry of the Interior of Chile. *Construction and Urbanism General Ordinance*; Decree 3388; Ministry of the Interior of Chile: Santiago, Chile, 1940. (In Spanish)

17. del Canto, H.; Aguirre, E.; Ibañez, V. Approximate procedure for determining the action of earthquakes in buildings. *An. Inst. Ing. Chile* **1941**, *2*, 74–91. (In Spanish)
18. Ministry of Public Works and Ways of Communication of Chile. *Construction and Urbanism General Ordinance*; Decree 884; Ministry of Public Works and Ways of Communication of Chile: Santiago, Chile, 1949.
19. Arias, A.; Husid, R. Proposal for a code for the anti-seismic calculation of buildings. *Rev. IDIEM* **1962**, *1*, 121–146. (In Spanish)
20. Arias, A.; Petit, L. A theoretical model for strong earthquake accelerograms. *Rev. IDIEM* **1965**, *4*, 29–60. (In Spanish)
21. Barrientos, M. The Architecture of Earthquakes in Chile (1929–1972). Ph.D. Thesis, Universidad Católica de Chile, Santiago, Chile, 2016.
22. Saragoni, R. Comparison of the design spectrum of the INDITECNOR proposal code "Anti-seismic calculation of buildings". *Rev. IDIEM* **1968**, *7*, 137–150. (In Spanish)
23. INN. *NCh 433Of72, Earthquake Resistance of Buildings*; INN: Santiago, Chile, 1972.
24. INPRES. *Normas Antisísmicas Argentinas—CONCAR 70*; INPRES: Buenos Aires, Argentina, 1972.
25. International Conference of Building Officials. *Uniform Building Code*; International Conference of Building Officials: Whittier, CA, USA, 1976.
26. New Zealand Standards Institute. *Code of Practice for General Structural Design and Design Loadings for Buildings*; NZS 4203; New Zealand Standards Institute: Wellington, New Zealand,, 1976.
27. Building Center of Japan. *Building Standard Law-Revised 1971*; Building Center of Japan: Tokyo, Japan, 1971.
28. INN. *NCh 433Of93, Earthquake Resistance of Buildings*; INN: Santiago, Chile, 1993.
29. Çelebi, M. Topographical and geological amplifications determined from strong-motion and aftershock records of the 3 March 1985 Chile earthquake. *Bull. Seismol. Soc. Am.* **1987**, *77*, 1147–1167. [CrossRef]
30. AFPS. *Guidelines for Seismic Microzonation Studies*; AFPS/DRM, French Association for Earthquake Engineering: Paris, France, 1995.
31. CEN. *EuroCode 8: Design of Structures for Earthquake Resistance—Part 1: General Rules, Seismic Actions and Rules for Buildings*; prEN 1998-5; European Committee for Standardization: Belgium, Brussels, 2004.
32. Ministero delle Infrastrutture e dei Trasporti. *Nuove Norme Tecniche per le Costruzioni*; NTC 2008; Ministero delle Infrastrutture e dei Trasporti: Rome, Italy, 2008.
33. International Conference of Building Officials. *Uniform Building Code*; International Conference of Building Officials: Whittier, CA, USA, 1988.
34. Building Center of Japan. *Building Standard Law-Revised 1981*; Building Center of Japan: Tokyo, Japan, 1981.
35. New Zealand Standards Institute. *Code of Practice for General Structural Design and Design Loadings for Buildings*; NZS 4203; New Zealand Standards Institute: Wellington, New Zealand, 1992.
36. INN. *NCh 433Of96, Earthquake Resistance of Buildings*; INN: Santiago, Chile, 1996.
37. International Conference of Building Officials. *Uniform Building Code*; International Conference of Building Officials: Los Angeles, CA, USA, 1991.
38. International Conference of Building Officials. *Uniform Building Code*; International Conference of Building Officials: Whittier, CA, USA, 1985.
39. INN. *NCh 433Of96mod2009, Earthquake Resistance of Buildings*; INN: Santiago, Chile, 2009.
40. Ministry of Housing and Urban Planning. *Emergency Supreme Decree 117*; DS117-2011; Ministry of Housing and Urban Planning: Santiago, Chile, 2011.
41. Ministry of Housing and Urban Planning. *Supreme Decree 61*; DS61; Ministry of Housing and Urban Planning: Santiago, Chile, 2011.
42. International Code Council. *IBC International Building Code*; International Code Council: Whittier, CA, USA, 1997.
43. Dobry, R.; Borcherdt, R.; Crouse, C.B.; Idriss, L.M.; Joyner, W.B.; Martin, G.R.; Power, M.S.; Rinne, E.E.; Seed, R.B. New Site Coefficients and Site Classification System Used in Recent Building Seismic Code Provisions. *Earthq. Eng.* **2000**, *16*, 41–67. [CrossRef]
44. Building Center of Japan. *Structural Provisions for Building Structures*; Building Center of Japan: Tokyo, Japan, 1997.
45. Nakamura, Y. A method for dynamic characteristics estimation of subsurface using microtremor on the ground surface. *Q. Rep. RTRI* **1989**, *30*, 25–33.

applied sciences

MDPI

Article

Determining the Dynamic Characteristics of a Multi-Story RC Building Located in Chile: A Comparison of the Results between the Nonparametric Spectral Analysis Method and the Parametric Stochastic Subspace Identification Method

Fernando Fuentes, Sebastián Lozano, Miguel Gomez, Juan C. Vielma and Alvaro Lopez *

Civil Engineering School, Pontificia Universidad Católica de Valparaíso, Valparaíso 2340000, Chile; fernando.fuentes.m@mail.pucv.cl (F.F.); sebastian.lozano.a@mail.pucv.cl (S.L.); miguel.gomez.f@pucv.cl (M.G.); juan.vielma@pucv.cl (J.C.V.)

* Correspondence: alvaro.lopez@pucv.cl

Abstract: Estimating the dynamic characteristics of instrumented built structures from seismic vibration data collected from built civil structures is essential input information for structural model updating and assessing the health of structures. This study focuses on the earthquake acceleration time histories obtained from several events recorded during its construction phase by accelerometers placed throughout an office building located in Viña del Mar (Chile) to determine its modal features. To this end, the data obtained were analyzed to compare the building's dynamic properties obtained with two different modal identification techniques. MATLAB programs were developed to implement both methods. The stochastic subspace identification technique for linear systems developed by van Overschee and de Moor was used to study the dynamic properties of the building. In contrast, the nonparametric method employed herein uses correlations and spectral analysis based on the Welch transform in the frequency domain. The investigation demonstrated that both methods identify similar frequencies and that the obtained translational mode shapes exhibit good agreement. Furthermore, the identified frequencies are congruent with the design frequencies.

Keywords: dynamic characteristics; frequency domain; nonparametric method; parametric method; subspace identification

Citation: Fuentes, F.; Lozano, S.; Gomez, M.; Vielma, J.C.; Lopez, A. Determining the Dynamic Characteristics of a Multi-Story RC Building Located in Chile: A Comparison of the Results between the Nonparametric Spectral Analysis Method and the Parametric Stochastic Subspace Identification Method. *Appl. Sci.* **2022**, *12*, 7760. https://doi.org/10.3390/app12157760

Academic Editor: Dario De Domenico

Received: 3 June 2022
Accepted: 28 July 2022
Published: 2 August 2022

Publisher's Note: MDPI stays neutral with regard to jurisdictional claims in published maps and institutional affiliations.

Copyright: © 2022 by the authors. Licensee MDPI, Basel, Switzerland. This article is an open access article distributed under the terms and conditions of the Creative Commons Attribution (CC BY) license (https://creativecommons.org/licenses/by/4.0/).

1. Introduction

Chile is located along the Pacific Ring of Fire, which experiences large earthquakes on a frequently recurring basis. The deep subduction zone earthquake that occurred on 27 February 2010, exhibited a remarkably high magnitude of Mw 8.8 and lasted approximately 3 min. This event affected at least one-third of the nation, the longest country in the world, and released an energy equivalent to 11,780 Hiroshima bombs. Therefore, news of this event spread worldwide. The earthquake generated relatively few cases of structural failure. A study conducted by the Chilean Structural Engineers Association concluded that 11% of the infrastructure in the country's capital city required major repairs to nonstructural elements; 3% presented minor or medium failures in the structure that could be repaired, and a mere 0.4% corresponded to severely damaged or collapsed structures that had to be demolished [1]. The low level of damage suffered by Chilean buildings in the face of such a strong earthquake, in comparison with that of structures in Christchurch, New Zealand, following the 2010 Canterbury earthquake or in Manabí, Ecuador, in response to the 2016 Ecuador earthquake, has intrigued the earthquake engineering community around the world.

Eleven years after the 2010 event, research on its effects on built civil infrastructure is still of great relevance, with studies being continuously performed to improve current structural design codes and thus ensure satisfactory structural performance [2,3]. In particular,

Chile is seen as a natural laboratory that provides the national and international earthquake engineering communities with the opportunity to calibrate structural design standards and serves as a reference in the earthquake-resistant construction industry.

The earthquake-resistant design aims to ensure that structures can adequately resist seismic forces, limiting the seismic risk and guaranteeing peoples' well-being. To this end, standards committees continue to propose design models or methodologies based on experiences such as those of Chile in 2010. Therefore, it is imperative to develop techniques that increase our knowledge and to evaluate the adequacy of proposed design methods through measurements. In this regard, monitoring large structures such as bridges, buildings, and dams is critical for calibrating the existing design methodologies and assessing their safety and health. Typical monitoring approaches for such structures are based on the analysis of either static strain or displacement data or discrete and continuous dynamic acceleration data gathered at specified points in the structures [4–8].

To obtain dynamically measured data, the proper monitoring of a structural system necessitates the collection of vast volumes of data over specified periods, from which mode shapes or other dynamic properties of the structures can be derived. Modal parameter changes in structures provide helpful information about the state of health of structural systems, provided that the dynamic features predicted in the design stage are affected by construction processes as well as seismic phenomena. The dynamic characteristics of structures can be derived either passively from ambient conditions [9] or actively by force excitations [10,11]. Then, modal parameters such as natural frequencies, damping factors, and mode shapes can be obtained based on response data. Thus, the structural properties can be estimated, and the performance and integrity of a structure can be assessed [12,13].

Modal analysis approaches yield essential information on the overall health of buildings. The process of determining the modal parameters of structures typically entails the deployment of accelerometers, gathering vibration data and using one of many analytic techniques. The accuracy of the data is based on not only the number, measurement sites, and resolution of the accelerometers but also the choice of a suitable analytic approach. In particular, the number and positions of sensors significantly impact the outcome. However, as buildings increase in size and become more complicated, the deployment of a slew of accelerometers becomes prohibitively expensive, especially about the number of sensors, the correct sensor placement, and the amount of gathered data (in terms of processing vast volumes of dynamic data), and the analytic technique [14,15].

Fang et al. [16] developed a method to extract the modal parameters of structures based on distributed measurements of dynamic strains using optical fiber sensors in a laboratory beam. The laboratory results indicated that the distributed vibration strain collected was sufficient for convergence. Hence, the first two mode shapes were accurately determined. The dynamic response characteristics of a 51-story building in downtown Los Angeles during the 2019 Mw 7.1 Ridgecrest, California, earthquake were identified from recorded response data, which was possible due to the consideration of other natural hazards, the use of system identification methods, the application of spectral analysis, and computing the coherence-phase angle spectrum [17]. The reinforced concrete (RC) building comprises a dual-core shear wall, perimeter columns, and post-tensioned flat slabs for approximately 80% of the floors. The recorded acceleration levels were trim (0.014 g at the foundation level and 0.069 g at the roof), and the maximum drift ratio computed from the recorded data was ~0.15%, so there were no reports of observed damage. Schanze et al. [18] compared and analyzed several models to study the effects of different underground story modeling methodologies using 11 aftershocks with different hypocenters and magnitudes recorded for a 16-story building office located in the city of Viña del Mar, Chile. The results revealed strong agreement between the model and the actual vibration data for the models considering horizontal springs attached to the retaining walls of the subterranean stories to account for the soil-structure interaction. Wu et al. [19] used a modified discrete-time Fourier transform combined with moving window technology in the time domain to examine the time-varying dynamic features of Shanghai Tower under the stimulation of normal wind and two subsequent typhoons. According to the authors, the findings can help with the wind-resistant design of significantly high-rise structures.

Currently, state-of-the-art modal analyses of already built structures indicate that there is more than one method with which the dynamic properties of buildings can be identified. These methods can be grouped into two main categories: parametric identification, which establishes a simplified mathematical model with which the measured and calculated responses can be compared, and nonparametric identification, which identifies the system by means of transformations, functions, and mathematical processes [20,21]. Some interesting works determining the dynamic response of heritage and modern structures with irregularities using system identification techniques can be consulted in the references [22–25]. In these works, different approaches are used to identify the main dynamic characteristics using accelerometers installed in the studied buildings. The general objective of this research is to compare both system identification philosophies, ascertain how compatible they are, and generate MATLAB code for both approaches that automate the analysis.

2. The Building, Temporary Instrumentation and Methodology

An instrumented structure should provide sufficient information to (a) reconstruct the response of the structure in sufficient detail to compare it with the responses predicted by mathematical models and with those observed in laboratories and, thereby improve the models; (b) make it possible to explain the reasons for any damage to the structure; and (c) facilitate decisions to retrofit/strengthen structural systems when warranted [26].

MATLAB codes were generated for each system identification philosophy based on a set of seismic event records. The building's shaking responses to several aftershocks were recorded during the deployment of the array. Only four of the many response data are utilized in this study. The criterion used for selection was to consider earthquakes with the different magnitudes, hypocenter depths, and epicentral distances. Table 1 lists the specifics seismic events considered herein.

Table 1. List of seismic events recorded in the building (source: USGS).

Location	Date (MM-DD-YYYY)	Date	Depth (km)	Magnitude
Metropolitan Region	04-05-2010	03:32:12 UTC	58.6	4.5 (M_w [(1)])
Offshore Valparaíso	05-01-2010	14:41:08 UTC	35.0	5.0 (M_w)
Coquimbo	10-23-2010	01:38:14 UTC	44.0	5.2 (m_b [(2)])
Araucania	01-02-2011	20:20:18 UTC	25.1	7.2 (M_i/M_{wp} [(3)])
Libertador O'Higgins	01-21-2011	15:36:11 UTC	18.2	5.0 (m_b)
Offshore Maule	02-14-2011	03:40:10 UTC	25.4	6.6 (M_i/M_{wp})

(1) M_w = Moment magnitude; (2) m_b = Body wave magnitude; (3) M_i/M_{wp} = Integrated p-wave.

2.1. Description of the Building and Sensor Layout

The case study corresponds to a sixteen-story building office with three additional basement stories, where the first story is 3.42 in height and stories 2–16 are 3.24 in height. It is situated in Viña del Mar, located in the Chilean central coast area, where several buildings were damaged during the last Mw 8.8 Maule earthquake. A picture of the building and a typical plan view are depicted in Figure 1. The structure was built in 2008, following the 2004 Chilean Building Code, which refers to the ACI–318 concrete design provision [27]. It is designed and built as an RC dual system (i.e., core structural wall and perimeter moment frame-wall), which is a typical construction type for offices in Chile. The monitoring time is between 27 March 2010 and 19 April 2011. During the monitoring period, the building was equipped with 16 force–balance Digitexx uniaxial accelerometers model D–110U (see Table 2 for detailed specifications) distributed with the height and across the floor. Four (4) sensors were installed on the third underground level, three (3) on the ground surface level, and then three (3) sensors on the third, sixth, and top floor. It is worth mentioning that the two vertical sensors 3 and 4 in the 3rd subterranean floor were not activated. The sensors register accelerations along the two horizontal directions of the building, where one sensor measured in the EW direction and two sensors in the NS direction. The criteria followed for the chosen positions of the instruments was in such a way that torsional modes are identified, given that the distribution of walls and columns with non-coincidental mass and rigidity centers naturally is expected to cause torsional behavior. Additionally, this sensor distribution allowed for the monitoring of the structure's spatial motion, the partial capturing of its linear and nonlinear response features since no sensors were available on the intermediate floors (from the 7th to the 16th), The sampling rate for the sensors was 100 Hz. Figure 2 shows a diagram illustrating the array of 16 accelerometers temporarily deployed in the building. General information corresponding to the mounting of the accelerometer array can be found elsewhere [28].

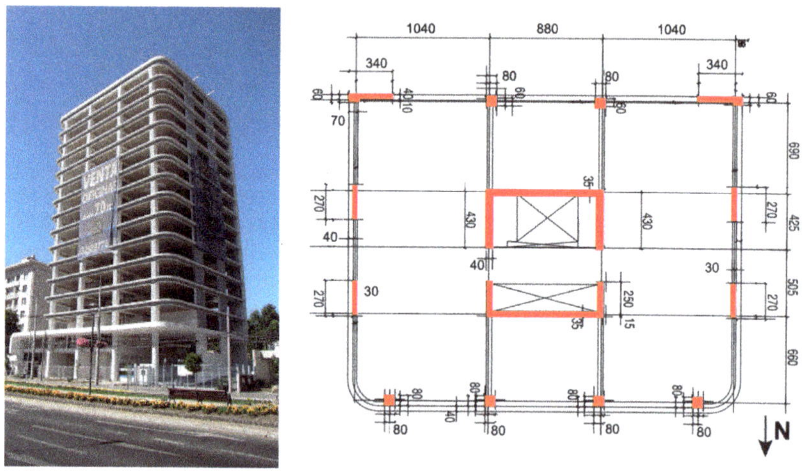

Figure 1. Northwest view (**left**) and typical plan view (**right**) of the building.

Table 2. Specifications of the Digitexx uniaxial accelerometers.

Specification	Parameter
Acceleration range	+/−3 (g)
Output	1.2 (V/g) standard between 2000 (Ω) or greater
Load capacitance	10.000 (pF) maximum
Transverse axis	0.005 (g/g)
Linear error	<±0.2% F.S.
Noise threshold	<1 (μg/$\sqrt{}$Hz) (Band 0.05–1.500 (Hz))
Dynamic interval	>110 (dB)
Bandwidth	DC–2.000 (Hz) to 3 (dB)
Power	+/−12 (Vdc), 9 (mA)
Operating temperature	−40 (°C) to +85 (°C)
Zero axis variation due to change	<200 (μg/°C) outside temperature range
Cover	Pressurized hermetic aluminum housing
Dimensions	108 × 76 × 57 (mm)
Mass	0.5 (kg)

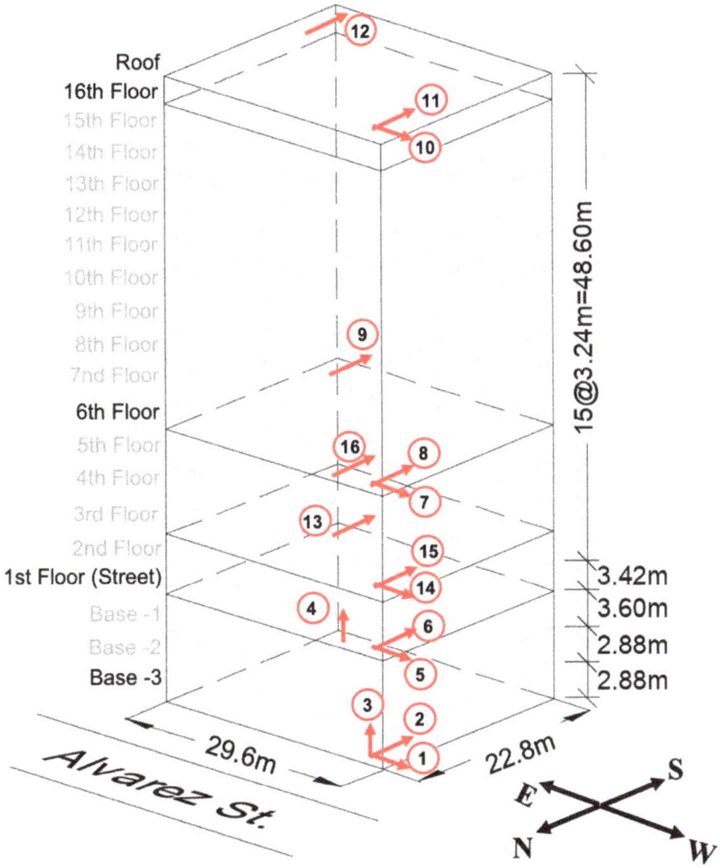

Figure 2. Arrangement of temporary deployment of sensors in the building.

Figure 3 shows the acceleration time-histories of the response recorded by channels 1–2 at the basement and channels 10–12 at the ceiling of the 16th floor (roof level) of the building to the events that occurred on 1 May 2010 and 2 January 2011 (Table 1). The former Mw 5.0 event was located approximately 73 km south-west of Viña del Mar, and the latter Mw 7.2 event was located 42 km north-west of Carahue and approximately 611 km south of Viña del Mar. Response data from these and the other events listed in Table 1 will be used to perform detailed analyses.

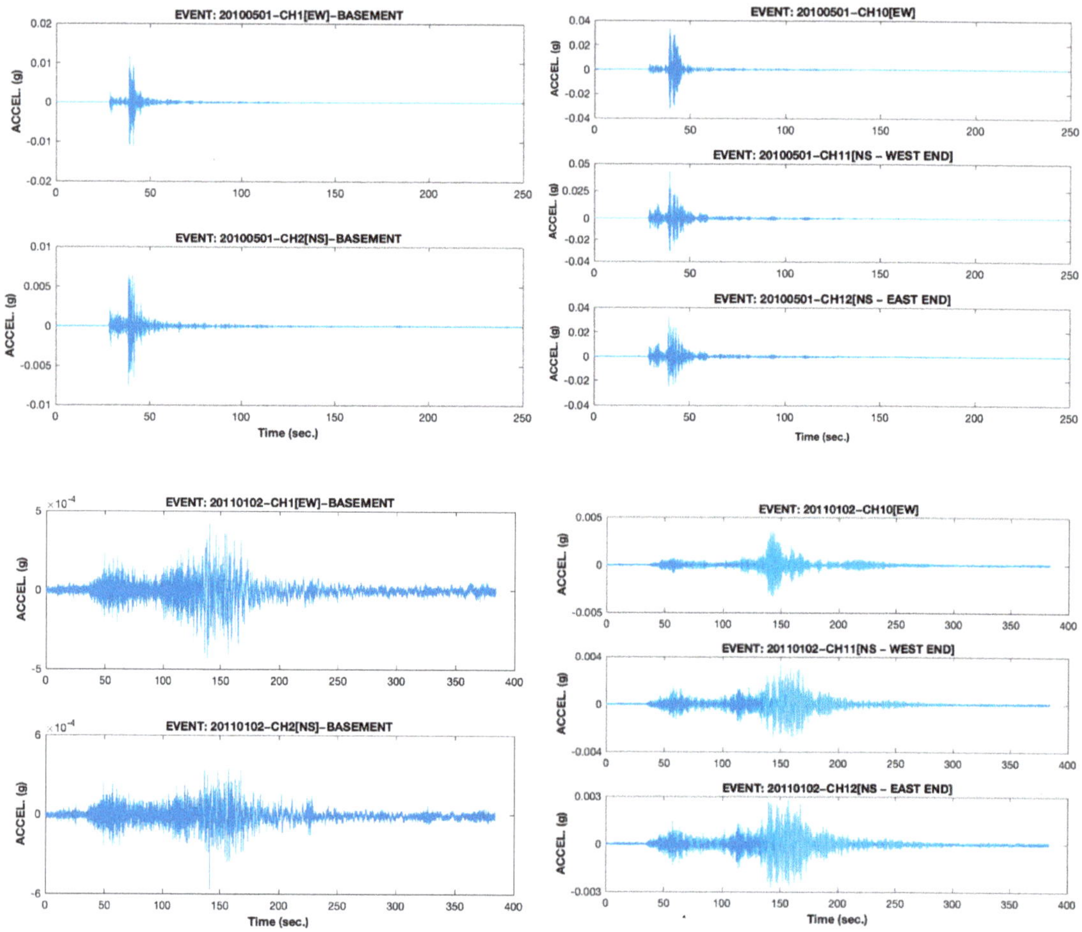

Figure 3. Acceleration time histories recorded at the basement and the roof level of the building on 1 May 2010 (**Top**) and on 2 January 2011 (**Bottom**).

2.2. Signal Processing

To perform the analysis correctly regardless of the method, it is important to verify that the input value is accurate. Many parameters can interfere in the data collection process, such as poorly installed instrumentation, as well as inconsistencies in the reception of data, for example, files that are in bad condition or confusion regarding the units of measurement. It must be made clear that these parameters depend entirely on the instrumentation. With technological advances, the instruments can correct all of these problems automatically, but taking these factors into account, in the case of this article, only one parameter is adjusted to be able to apply the two methods in question, namely the correction of the signal baseline.

Baseline Correction

This occurs when the registers, to perform the analysis, are displaced with respect to the origin. This is a common problem where, due to installation issues, the instruments are not perfectly level. This issue may also arise when the vertical component of an earthquake causes a rebound effect with the surface.

Before the data can be correctly analyzed, they must be prepared via a convenient standardization procedure called a baseline correction, by which the sample values are transformed into a new set of values that have a sample average equal to zero [29].

As previously mentioned, the different methodologies currently used for the identification of dynamic properties in structures are separated into parametric and nonparametric techniques. In the following sections, the procedures of both general methodologies are defined.

2.3. Parametric Analysis Procedure

In this procedure, a simplified mathematical model is used, and the values of the structural parameters necessary to produce an optimal correlation between the measured and calculated responses are estimated. For the latter, the theory of structural analysis and dynamics is used to model the structure and obtain its properties [21,26,30]. Within this category are methods such as the prediction error method (PEM) and stochastic subspace identification (SSI) technique for linear systems. Both are time-domain methods (i.e., they work directly with temporal data without the need to convert them into correlations or spectra) [31]. For this study, we employ the SSI technique, which was published by Peter Van Overschee and Bart De Moor in 1996 [32].

To perform system identification by means of the SSI method, it is necessary to understand how dynamical systems are represented in their state-space form and how it is possible to obtain relevant system information through this representation. First, the dynamic equation of the system must be stated in the traditional form according to the following equation:

$$\boldsymbol{M}_s\ddot{\boldsymbol{x}} + \boldsymbol{C}_s\dot{\boldsymbol{x}} + \boldsymbol{K}_s\boldsymbol{x} = \boldsymbol{\Lambda}u - \boldsymbol{M}_s\boldsymbol{\Gamma}\ddot{\boldsymbol{x}}_g \quad (1)$$

where $\boldsymbol{x} \in \mathbb{R}^{nx1}$ is the vector of displacements relative to the ground of the n floors or degrees of freedom of the structure; $\ddot{x}_g$ is the ground acceleration; $\boldsymbol{\Gamma}$ is the seismic participation vector; and $\boldsymbol{\Lambda}$ indicates the location of the control signal u if any. For the purposes of this study, $\boldsymbol{u} = 0$.

The advantage of the SSI technique is that it performs its analysis only with data that can be directly measured on the structure (only output data), such as the accelerations of the $\boldsymbol{\ell}$ floors where the accelerometers are installed. In addition, the main assumption of this method is that the measurements obtained by the system correspond to natural stochastic vibrations caused by a white noise type of disturbance $\boldsymbol{w}_k$ as well as electrical noise $\boldsymbol{v}_k$ in the measuring devices; however, these types of noise are not related to each other. The development of this technique and its mathematical foundations are realized in the time domain by the state-space representation of the dynamical system under study defined as:

$$\boldsymbol{z}_{k+1} = \boldsymbol{A}_d\boldsymbol{z}_k + \boldsymbol{w}_k \quad (2)$$

$$\boldsymbol{y}_k = \boldsymbol{C}_d\boldsymbol{z}_k + \boldsymbol{v}_k \quad (3)$$

where $\boldsymbol{y}$ is the measured structural response and $\boldsymbol{z}$ is the vector of states. The subscript k of the time variables corresponds to the time instant $\boldsymbol{t}_k = k{\cdot}\Delta$t.

The main objectives of this method are to determine the order $n_s = 2n$ of the system and to obtain the matrices $\boldsymbol{A}_d$ and $\boldsymbol{C}_d$.

2.3.1. Hankel Matrix

The Hankel matrix constitutes the basis of all of the formulations that have been developed with the SSI technique. The matrix has dimensions of $2i\ell$ x j, where $\boldsymbol{\ell}$ is the number of sensors arranged in the structure, $i = 2\frac{n_{smax}}{\ell}$, $j = s - 2i + 1$, s is the number of measurements performed on the structural system, and $n_{s(max)}$ corresponds to the maximum order of the system, which should be neither too small nor too large and is estimated by the engineer. The upper part of the matrix corresponds to the past, and the lower part corresponds to the future.

2.3.2. Projection of Blocks ($\mathcal{P}_i$)

The projection $\mathcal{P}_i$ is defined by the orthogonal projection matrix of the SSI method and is determined as the product between the observability matrix $\boldsymbol{\mathcal{O}}_i$ and the estimated sequence of states of the Kalman filter $\hat{\boldsymbol{X}}_i$. This procedure is depicted in Figure 4. The orthogonal projection of the future outputs $\boldsymbol{Y}_f$ onto past outputs $\boldsymbol{Y}_p$ determines the forward state sequence $\hat{\boldsymbol{X}}_i$. Conversely, the orthogonal projection of past outputs $\boldsymbol{Y}_p$ onto the future outputs $\boldsymbol{Y}_f$ determines the backward state sequence $\hat{\boldsymbol{Z}}_i$ [32]. Figure 4 also presents the variables Δ_i^c and $\boldsymbol{\mathcal{B}}_i$, which correspond to a controllability matrix and the projection of past outputs onto the future outputs, respectively However, these variables are not part of the procedure since they serve only to define and understand the origin of the projection of blocks $\boldsymbol{\mathcal{P}}_i$.

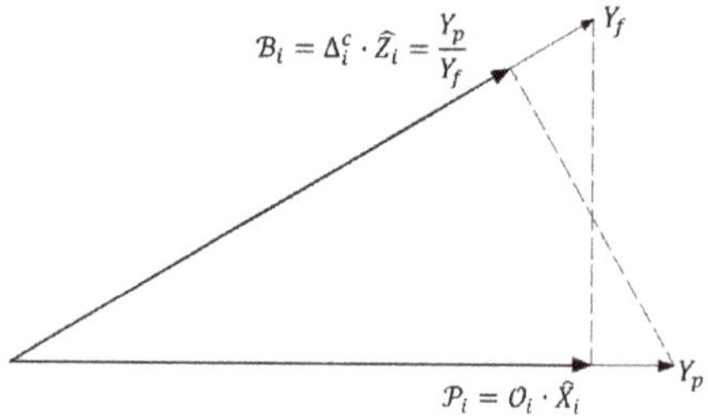

Figure 4. Orthogonal projection matrix.

2.3.3. Singular Value Matrix of the Orthogonal Projection Matrix (S)

The orthogonal projection matrix is decomposed into its singular values considering the pre- and post-multiplication of the so-called weighting matrices. The diagonal matrix S is of order m, which indicates the order n_s of the system with $m = n_s$. The criterion for defining the value of m depends on the assignment of the weighting matrices.

2.3.4. Weighting Matrices of the Orthogonal Projection Matrix ($\boldsymbol{W}_1$ and $\boldsymbol{W}_2$)

There are three algorithms with which the weighting matrices can be defined:

1. The principal component (PC) algorithm.
2. The unweighted principal component (UPC) algorithm.
3. The canonical variant algorithm (CVA).

The first two algorithms calculate the order m of the singular value matrix by detecting the position at which the diagonal of the matrix $\boldsymbol{S}$ starts to become zero. On the other hand, the CVA calculates the order $\boldsymbol{m}$ by detecting the position at which the values of the diagonal of the matrix $\boldsymbol{S}$ stop converging to one, where the tolerance is defined by the engineer.

2.3.5. Eigenvalues (δ) and Eigenvectors ($\boldsymbol{\phi}$)

The eigenvalues and eigenvectors are calculated from the discrete system matrices $\boldsymbol{A}_d$ and $\boldsymbol{C}_d$. The vector δ is used to calculate the identified frequencies of the system, and $\boldsymbol{\phi}$ denotes the vibrational shapes corresponding to the abovementioned frequencies.

2.3.6. Damping Fraction (ξ)

Damping is a property of the system and includes a large number of phenomena that, if the system is left to vibrate freely after an initial excitation has been applied to it, will eventually dissipate [9].

Figure 5 shows a flow chart of the SSI method.

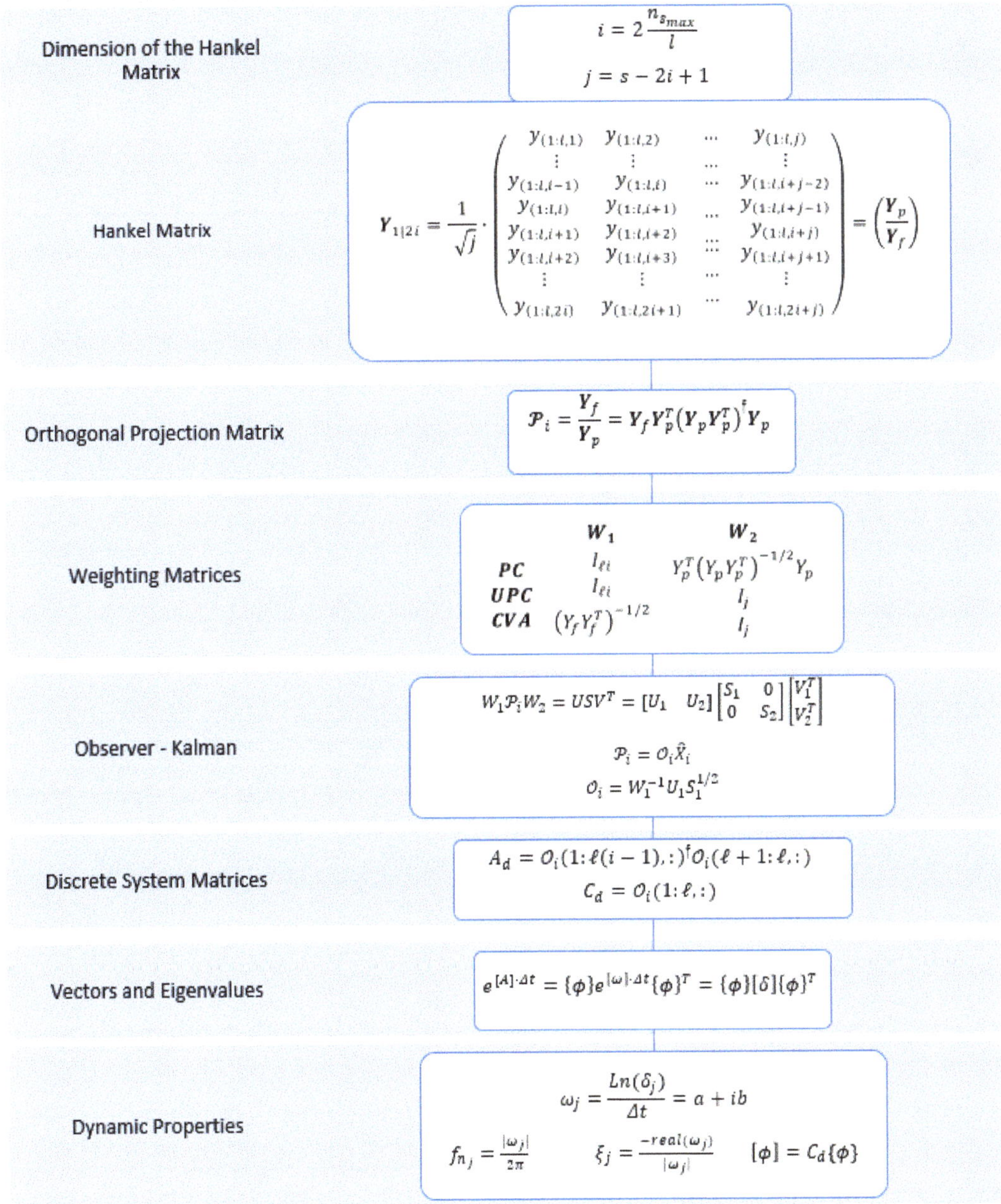

Figure 5. Flow diagram of the SSI method.

2.4. Nonparametric Method

Nonparametric methods are based on applying a set of mathematical procedures, transformations and functions to the signals obtained from the acceleration records. These accelerometer records correspond to variables that are functions of time. One of the main characteristics of nonparametric methods is to perform an analysis of the records as a function of frequency, thereby determining how the energy or power of the signal is distributed. The processes described in Figure 6 allow the specific properties of the system to be identified.

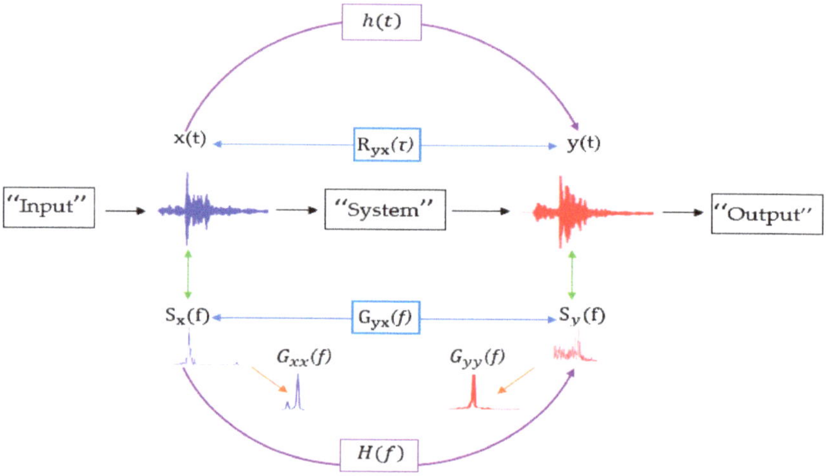

Figure 6. Nonparametric processing functions.

In the above figure, $x(t)$ is the time-domain input signal; $y(t)$ is the time-domain output signal; $h(t)$ is the impulsive system response; $S_x(f)$ is the linear Fourier spectrum of $x(t)$; $S_y(f)$ is the linear Fourier spectrum of $y(t)$; $H(f)$ is the system transfer function; $R_{yx}(\tau)$ is the cross-correlation of $x(t)$ and $y(t)$; $G_{xx}(f)$ is the spectral power of $x(t)$; $G_{yy}(f)$ is the spectral power of $y(t)$; and $G_{yx}(f)$ is the cross-spectral power of $x(t)$ and $y(t)$.

To obtain the dynamic properties of the system, a series of procedures are followed based on the authors' criteria. The process is briefly described in the following subsections.

2.4.1. Obtaining the Spectral Density

As previously mentioned, the Fourier transform identifies important frequencies from an engineering point of view and has become the dominant method [29]. Accordingly, new techniques have been created and improved based on the Fourier transform, as in the case chosen for this analysis, for which the Welch power spectral density estimation (G_{xx}) is used. The Welch power spectrum works with superimposed segments, increasing the frequency resolution and allowing the spectrum to be observed better. Figure 7 shows a spectrum obtained using the Fourier transform superposed onto the Welch spectrum.

2.4.2. Correlations

Correlation functions can acquire parameters to quantify how similar these signals are; these functions include autocorrelation, which analyzes how much a signal varies over time, and cross-correlation, which measures the similarity of two signals. The theoretical processes are as follows:

$$R_{xx}(\tau) = \lim_{T \to \infty} \frac{1}{T} \int_0^T x(t)\, x(t+\tau) dt \tag{4}$$

$$R_{xy}(\tau) = \lim_{T \to \infty} \frac{1}{T} \int_0^T x(t)\, y(t+\tau) dt \tag{5}$$

These can also be estimated by using power spectral density functions (G_{xx} and G_{yx}).

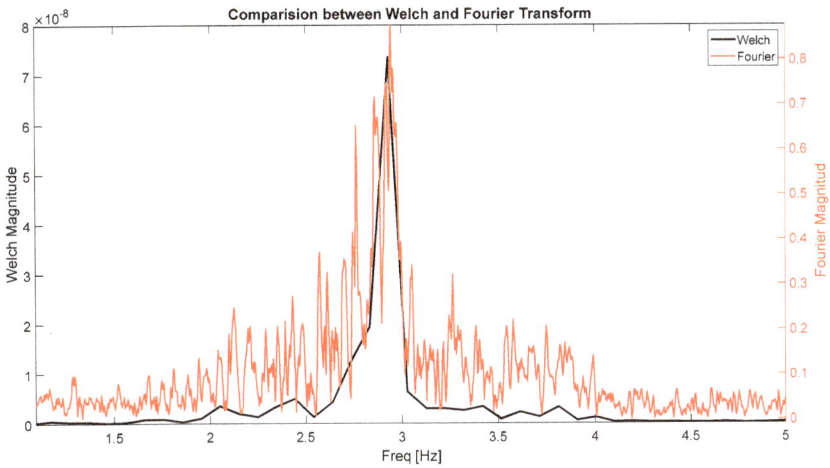

Figure 7. Comparison between the Welch and Fourier spectra.

2.4.3. Phase

It must be taken into consideration that the functions with which the nonparametric method works are sometimes complex variables. Examples include the Fourier transform and cross-power correlations. Complex variables play an important role in calculating vibration shapes since they reveal whether the relative motion between two points occurs in the same direction.

The numerical value given by the phase angle varies only within $\pm\pi$. Thus, a phase angle equal to 0 between 2 different points at an analyzed frequency indicates that the direction of motion is the same, called being in phase. Otherwise, if the phase angle is $\pm\pi$, the relative movement is opposing, known as a phase shift. The formulae used to obtain the phase angle of a complex variable are as follows:

$$G_{xy}(f) = A_{xy}(f) + iB_{xy}(f) \tag{6}$$

$$|Gxy\ (f)| = \sqrt{A_{xy}{}^2(f) + iB_{xy}{}^2(f)} \tag{7}$$

$$\theta_{xy}(f) = \tan^{-1}\frac{B_{xy}(f)}{A_{xy}(f)} \tag{8}$$

2.4.4. Transfer Functions

Transfer function describe the dynamic characteristics of a linear system. These functions can be calculated directly through Fourier spectral densities, but a risk arises for certain frequencies where the mathematical procedure is affected. Instead, the power spectra are used to determine the transfer functions, but this approach is applicable only for random vibrations, and the consistency must be measured by means of the coherence function.

$$H(f) = \frac{S_y(f)}{S_x(f)} \tag{9}$$

$$H1(f) = \frac{G_{xy}(\mathrm{f})}{G_{xx}(\mathrm{f})} \tag{10}$$

$$H2(f) = \frac{G_{yy}(\mathrm{f})}{G_{yx}(\mathrm{f})} \tag{11}$$

2.4.5. Coherence Function

This concept is crucial for differences in frequencies that are particularly important in the analysis. Theoretically, this function provides a good estimate of the energy output due exclusively to the input signal. The coherence function is a dimensionless frequency function that has only the real part and delivers values in the range from 0 to 1. For analysis, this function reflects how consistent the output signal is with the input signal. For example, in a comparison between the records of the second floor and the last floor, the coherence function can be used to determine how consistent the records are and can differentiate a certain frequency of the structure for one of the floors.

$$\gamma^2{}_{xy}(f) = \frac{H_1(f)}{H_2(f)} = \frac{|G_{xy}(f)|^2}{G_{xx}(f)G_{yy}(f)} \tag{12}$$

2.5. Analysis Procedure of the Nonparametric Method

Considering the concepts explained above, this subsection details the use of the nonparametric approach and the criteria applied to obtain the dynamic parameters. To better understand the problem under study, a diagram with sensors ordered by floor is presented in Figure 8.

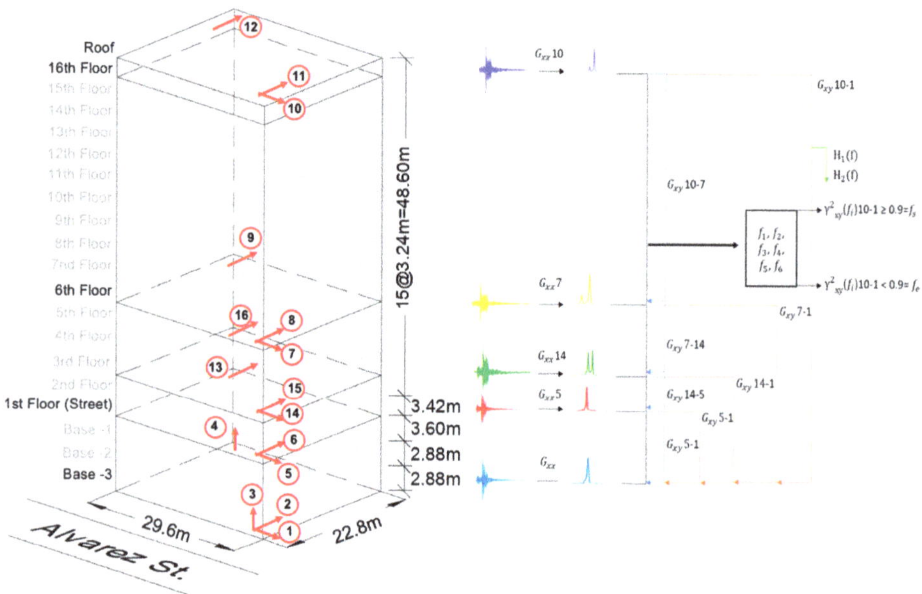

Figure 8. Graphical scheme of the nonparametric process.

With the acceleration records sorted by floor, we proceed to obtain the Welch power spectra (G_{xx}). The first criterion is that only the first six frequencies (i.e., the six frequencies with the highest amplitudes) are identified due to the fact that for structural dynamic analysis, we are interested only in a certain range of frequencies for analysis. Outside that range, for this study case, the frequencies have no physical importance.

The next step is to analyze whether the identified frequencies correspond to the frequencies of the floor or of the structure. To this end, the cross-power correlation between the signal of the first floor and that of the last floor (G_{xy}) is calculated. With the obtained correlation and the Welch power spectrum, it is possible to obtain the transfer functions $H_1(f)$ and $H_2(f)$, and with these values, the coherence function between floor -3 and floor 16 can be acquired. Here, a second criterion is applied to the analysis of the result of the coherence function ($\gamma^2{}_{xy}$). If the value of this variable is greater than 0.9, frequency corresponds to the soil (f_s). This is due to the fact that having a coherence greater than 0.9 means that the movement of the last floor is mostly a product of the movement of floor −3, i.e., the floor in direct contact with the ground. Nevertheless, an analysis of the soil frequencies is beyond the scope of this article.

Once the frequencies of the structure have been identified, only whether they correspond to translational or torsional frequencies remains to be checked. For this purpose, the registers on opposite corners of the same floor are correlated. The phase angle between these signals on the same floor is verified, and the last criterion is used, to verify whether the result of the phase angle is less than 1. If so, the relative movement is in the same direction, which is interpreted as a translational movement.

With the translational frequencies of the structure, we proceed to identify the modal shapes by performing three types of correlations. The first is to correlate all floors with the −3 and obtain the relative displacements with the phase angles. The second option is to take continuous correlations between floors and likewise obtain relative displacements with the phase angles. A third correlation is performed. This is created after appreciating the results between the parametric and non-parametric method in which a way to improve the results is sought, where it is correlated up to floor 3 with -3 and then correlated floor with floor up to 16.

Finally, following these three calculations, the amplitudes of the modes can be obtained through the Welch power spectrum, and the vibration shapes can be normalized. These three ways of correlating the signals are discussed in detail in the results.

3. Results

This section presents the results by date for both methods, including the mode shapes with their normalized amplitudes and their respective relative directions (Figure 9). Table 3 presents the torsional frequencies of the building obtained from each method for the selected seismic events. It is worth noting that the building was in the rough-in stage over this period. The damping value obtained from each method is also presented in Tables 3 and 4. For the parametric case, the three defined algorithms (PC, CVA, and UPC) are presented, while for the nonparametric case, the results are presented using the 3 types of correlations, namely, the Bendat and Piersol floor-floor and Bendat and Piersol base-floor correlations, Bendat and Piersol EIC.

Table 3. Identified torsional frequencies of the building for the selected events.

Date	Torsional Frequency [Hz]	
5 April 2010	3.32	4.00
1 May 2010	3.91	5.08
23 October 2010	5.27	-
2 January 2011	0.88	-
21 January 2011	3.22	-
14 February 2011	1.07	3.52

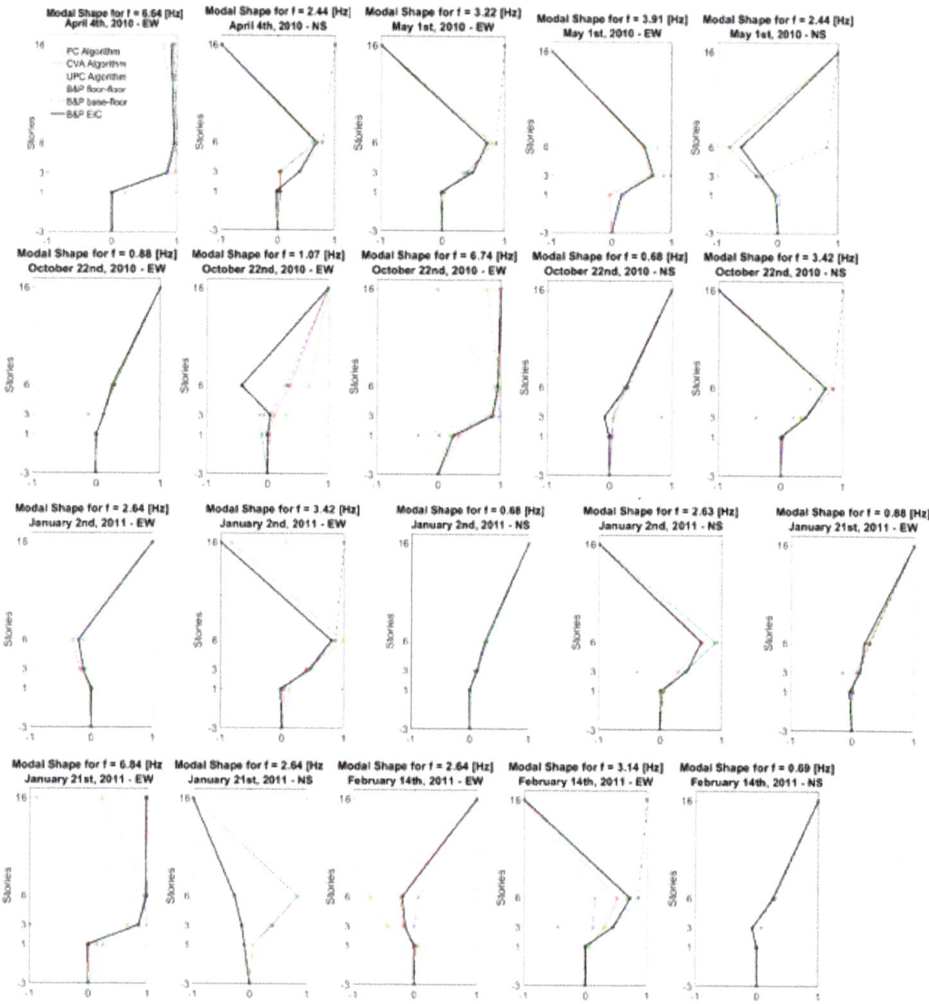

Figure 9. Vibration mode shapes for the translational frequencies of the earthquakes that occurred between 4 April 2010, and 14 February 2011, in the east-west and north-south directions. The blue, red and green curves correspond to the PC, CVA and UPC algorithms, respectively, for the parametric method. The black dotted and segmented curves correspond to the floor-floor and base-floor correlations, respectively, proposed by Nandagopalan et al. [29]. Finally, the thick continuous black lines correspond to the calibrated correlations corresponding to the base-floor correlation up to floor 3 and the floor-floor correlation for floors 6 and 16. This correlation is referred to as Bendat and Piersol EIC.

Table 4. Damping for the parametric method algorithms according to date, direction of analysis, and identified translational frequency.

Date	Direction	Frequency [Hz]	ξ PC [%]	ξ CVA [%]	ξ UPC [%]
5 April 2010	EW	6.64	0.51	5.09	1.4
	NS	2.44	1.77	1.91	1.29
1 May 2010	EW	3.22	0.7	2.24	1.65
	EW	3.91	1.25	−0.55	0.64
	NS	2.44	1.65	10.01	1.25
23 October 2010	EW	6.74	1.7	−0.04	0.35
	NS	0.68	0.78	1.33	0.84
	NS	3.42	1.85	2.36	0.91
2 January 2011	EW	2.64	2.37	2.85	1.45
	EW	3.42	1.8	2.98	4
	NS	0.68	2.61	1.8	2.1
	NS	2.63	2.35	3.24	2.35
21 January 2011	EW	0.88	0.34	0.21	0.42
	EW	6.84	1.11	0.73	0.64
	NS	2.64	2	0.96	1.61
14 February 2011	EW	2.64	2.02	1.12	1.48
	EW	3.14	14.67	1.65	0.23

4. Discussion

4.1. Identified Frequencies

From a total of 11 records of seismic events, 12 frequencies are identified for the east-west direction, and 8 frequencies are identified for the north-south direction. Tables 5 and 6 show the performance of the methods in terms of the number of frequencies identified per direction.

Table 5. Identified frequencies in the east-west direction for the parametric and nonparametric methods.

	Frequencies [Hz]			
Date	**CVA**	**PC**	**UPC**	**Bendat and Piersol**
5 April 2010	6.59	6.55	6.57	6.64
1 May 2010	3.15–3.88	3.21–3.87	3.22–3.94	3.22–3.90
23 October 2010	0.84–1.14–6.60	0.85–1.14–6.75	0.86–1.10–6.72	0.88–1.07–6.74
2 January 2011	2.68–3.33	2.62–3.36	2.61–3.41	2.63–3.42
21 January 2011	0.86–6.86	0.86–6.86	0.85–6.84	0.88–6.84
14 February 2011	2.60–3.13	2.64–3.20	2.61–3.24	2.64–3.14

Table 6. Identified frequencies in the north-south direction for the parametric and nonparametric methods.

	Frequencies [Hz]			
Date	**CVA**	**PC**	**UPC**	**Bendat and Piersol**
5 April 2010	2.42	2.41	2.45	2.44
1 May 2010	2.51	2.51	2.37	2.44
23 October 2010	0.69–3.37	0.70–3.44	0.70–3.41	0.68–3.42
2 January 2011	0.69–2.61	0.69–2.64	0.69–2.64	0.68–2.63
21 January 2011	2.62	2.63	2.62	2.63
14 February 2011	-	-	-	0.69

In the east-west direction for all dates, both the parametric and the nonparametric method algorithms succeed in identifying the frequencies. Moreover, the frequencies identified for each seismic event are similar between the two methods. For the north-south direction on 14 February 2011, the algorithms of the parametric method are not able to identify the frequencies, but the identified frequencies on the previous dates are similar. This allows us to assume that the programs achieve similar results.

Similar frequencies are repeated on different dates. We assume that these represent the frequencies of the fundamental modes of vibration of the structure. A good way to compare these results would be to obtain the design frequencies of the structure. Figure 10 depicts the frequencies with the corresponding dates and their direction.

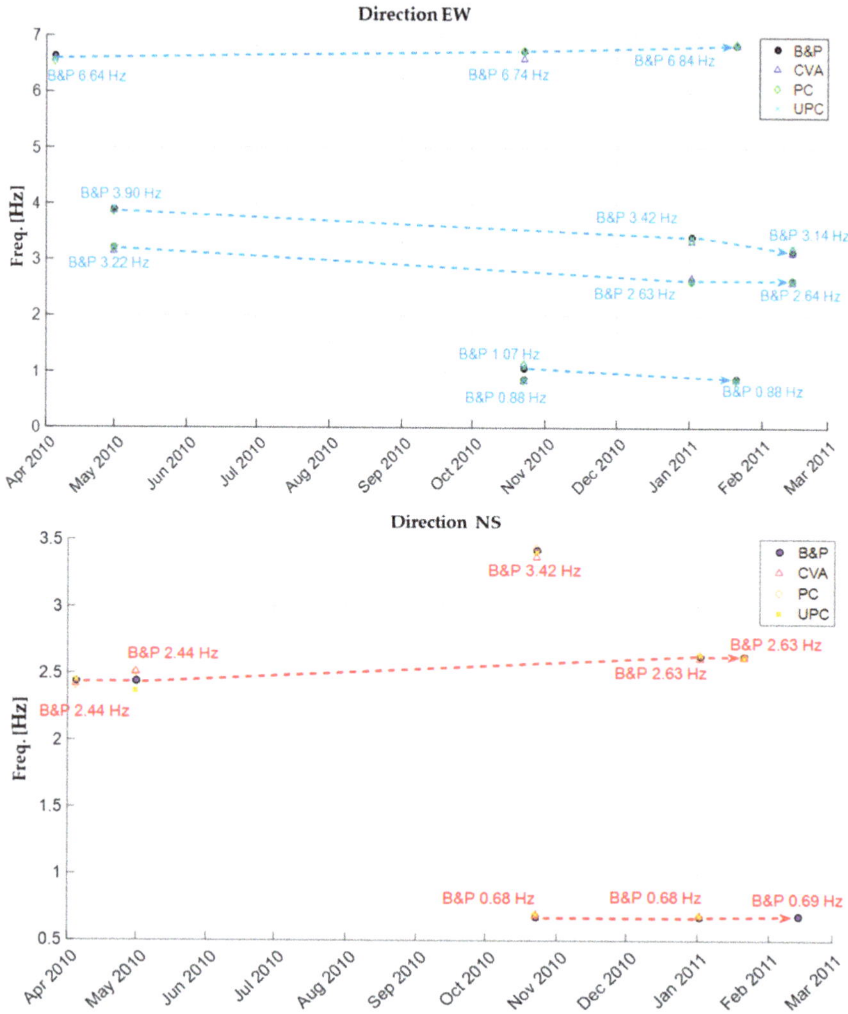

Figure 10. Identified frequencies by date for the east-west and north-south directions showing how the frequency varies during the time the building was instrumented.

We note an interesting trend concerning the identified frequencies—the mass and stiffness of the building change during the construction period. On the one hand, for some lower frequencies, a decrease in the structure's stiffness may be related not only to the change in mass but also as a result of crack development in concrete (including those at the micro-level) due to the previous earthquakes, an increase in the service load and long-term deflections, redistribution of internal forces, and sometimes as a result of moisture shrinkage and thermal strains fluctuations. The structuring is also important since it affects the building's mass distribution and the implementation of non-structural elements, which also contribute to the stiffness differential. Therefore, it can be seen that the higher frequencies in time for both directions increase. On the other hand, the intermediate or low frequencies in the EW direction tend to decrease as time passes. The fact that they increase or decrease may be associated with one of the above factors contributing to the mode of vibrations.

Considering that the design frequencies of the first 6 modes are 0.497, 0.65, 0.88, 2.13, 3.23, and 3.70 Hz [28], it is possible to appreciate that the identified frequencies, in both the east-west and the north-south directions, are similar except for 0.497 Hz.

4.2. Vibration Modes

4.2.1. Parametric Method

For most of the records, the three algorithms have the same relative movement directions. However, the SSI method is not able to distinguish whether the identified frequency or period originates from the ground or from the structure or whether it is torsional or translational. In addition, the number of frequencies it identifies depends entirely on the order of the system assigned to it. Consequently, in this study, the nonparametric method is used to distinguish the translational frequencies of the structure, after which those frequencies are searched for in the SSI algorithm. In most cases, it is sufficient to assign a maximum system order equal to 200. However, for the frequencies on 23 October 2010, and 14 February 2011, system orders equal to 500 and 350, respectively, are used to ensure that the parametric frequencies are as similar as possible to the nonparametric frequencies.

4.2.2. Nonparametric Method

To obtain the relative displacements of the floors, various correlations can be performed. The configurations of the accelerometers from floor −3 to floor 6 are emplaced every 3 floors, while those on floors 16 and 6 are separated by 10 floors. The graphs of the vibration shapes plotted in the previous section illustrate how the structure vibrates according to the correlation of each record of the structure with the record located at the base, represented by floor −3 (the base-floor correlation). On the other hand, we can also observe how the building vibrates according to the correlation of each floor of the structure with the floor that precedes it (the floor-floor correlation). In this study, seeing how these graphs behave for all dates reveals that the best result is obtained by combining the correlations of floors 1 and 3 with the base and the correlations of floors 6 and 16 with the records that precede them. In this way, a third form of vibration is plotted, called the Bendat and Piersol EIC.

4.2.3. Comparison of Methods

The trends of the vibration shapes for both methods on most dates are congruent. That is, the amplitudes present the same relative displacements. In the previous sections, the results in terms of frequency show good agreement, suggesting that both methods manage to identify similar frequencies. Note that for the frequencies of 6.74 and 6.84 Hz on 23 October 2010, and 21 January 2011, respectively, the relative displacements of the 16th floor do not coincide with those from the SSI algorithm. However, in both cases, the CVA corresponds very well with the nonparametric algorithm. This is due to the fact that the frequencies are very high (or the periods are very small).

4.2.4. Damping

The damping values are highly dispersed for all of the dates analyzed. This parameter is, in principle, difficult to obtain. In the design stage, the response spectra of structures are obtained by assuming 5% damping. In this case, the techniques used work directly with the acceleration records of an earthquake, where sudden changes in the magnitudes can be marked in an instant, implying that the methods used to calculate the damping factor are not sufficiently reliable.

5. Conclusions

With the objective of comparing the parametric SSI method with the spectral nonparametric method by means of independent programs, six records of seismic events are analyzed, both methodologies are applied, and compatible results are obtained regarding the identified frequencies and forms of vibration.

Comparing the parametric and nonparametric methods reveals that the SSI method is less effective than the Welch spectrum identification method in the sense that the former is not able to distinguish whether the identified frequencies are rotational or translational and whether they correspond to the structure or the soil. In addition, the identification of frequencies in the SSI algorithm depends on the engineer's criteria to assign an order to the system, which makes this approach unstable since the number of identified frequencies increases as the order of the system increases, whereas most of these frequencies are not recognized by the Welch spectrum.

The models created to determine the dynamic properties of the structure converge to similar results. However, this does not allow us to conclude that they closely approximate the reality. Implicitly, this marks a limitation that, in principle, can be improved. For this case, the seismic records are obtained from institutions, and there are detailed variables (specifically, in the signal processing subsection) for which it is not known whether they are within acceptable ranges. This limitation can be improved for future projects, where the monitoring campaign can be performed by the company itself, although it would require the acquisition of instruments. In this way, it would be possible to have better control of the variables that affect the calculation models, and the methods could be calibrated for future applications.

The application of these two methods allows for a range of results to be obtained, whereas in the case of this project, only the frequencies of structural interest are analyzed. Nevertheless, the fact that seismic-resistant structures are directly coupled to the soil, which transfers seismic excitations, should not be ignored. This article briefly mentions the difference between a soil frequency and a structural frequency and that, by means of the methods used, it is possible to separate and work only with the frequencies of the structure. Future research should analyze these frequencies, perhaps in an approach applied to soil-structure interactions.

Ultimately, the engineering importance of identifying the dynamic properties of structures lies in its potential for structural health monitoring, which is required for the modification of existing structures, structures subject to long-term movement or material degradation, fatigue assessment, and the development of a performance-based design philosophy [33–35].

Author Contributions: Conceptualization, A.L.; methodology, A.L., S.L. and F.F.; validation, A.L., S.L., F.F., M.G. and J.C.V.; formal analysis, S.L. and F.F.; data postprocessing, S.L. and F.F.; writing—original draft preparation, S.L., F.F., M.G. and A.L.; writing—review and editing, A.L., S.L. and J.C.V.; supervision, A.L. and J.C.V.; funding acquisition, A.L. All authors have read and agreed to the published version of the manuscript.

Funding: This research was funded by the Pontificia Universidad Católica de Valparaíso (PUCV), grant number 039.351/2021, and the APC was funded by PUCV.

Institutional Review Board Statement: Not applicable.

Informed Consent Statement: Not applicable.

Data Availability Statement: The data presented in this study are available on request from the corresponding author. The data are not publicly available due to intellectual property.

Acknowledgments: The authors express their gratitude to the PUCV for financing the processing of this article. Further, the authors of this paper gratefully acknowledge Professor Gilberto Leiva for facilitating the building's collected data and structural drawings.

Conflicts of Interest: The authors declare no conflict of interest.

References

1. Boroschek, R.; Bonelli, P.; Restrepo, J.I.; Retamales, R.; Contreras, V. Lessons from the 2010 Chile Earthquake for Performance Based Design and Code Development. *Geotech. Geol. Earthq. Eng.* **2014**, *32*, 143–157. [CrossRef]
2. Martineau, M.O.; Lopez, A.; Vielma, J.C. Effect of earthquake ground motion duration on the seismic response of a low-rise RC building. *Adv. Civ. Eng.* **2020**, *2020*, 8891282. [CrossRef]
3. Ambiado, E.; Lopez, A.; Vielma, J.C. Numerical evaluation of prequalified end-plate connections used in a framed steel industrial structure. *Metals* **2021**, *11*, 243. [CrossRef]
4. Su, H.; Chen, Z.; Wen, Z. Performance improvement method of support vector machine-based model monitoring dam safety. *Struct. Control Heal. Monit.* **2016**, *23*, 252–266. [CrossRef]
5. Li, H.N.; Ren, L.; Jia, Z.G.; Yi, T.H.; Li, D.S. State-of-the-art in structural health monitoring of large and complex civil infrastructures. *J. Civ. Struct. Heal. Monit.* **2016**, *6*, 3–16. [CrossRef]
6. Gregori, A.; Di Giampaolo, E.; Di Carlofelice, A.; Castoro, C. Presenting a New Wireless Strain Method for Structural Monitoring: Experimental Validation. *J. Sens.* **2019**, *2019*, 5370838. [CrossRef]
7. Makoond, N.; Pelà, L.; Molins, C.; Roca, P.; Alarcón, D. Automated data analysis for static structural health monitoring of masonry heritage structures. *Struct. Control Health Monit.* **2020**, *27*, e2581. [CrossRef]
8. Barsocchi, P.; Bartoli, G.; Betti, M.; Girardi, M.; Mammolito, S.; Pellegrini, D.; Zini, G. Wireless Sensor Networks for Continuous Structural Health Monitoring of Historic Masonry Towers. *Int. J. Archit. Herit.* **2021**, *15*, 22–44. [CrossRef]
9. Sony, S.; Laventure, S.; Sadhu, A. A literature review of next-generation smart sensing technology in structural health monitoring. *Struct. Control Health Monit.* **2019**, *26*, e2321. [CrossRef]
10. Iranmanesh, A.; Ansari, F. Energy-Based Damage Assessment Methodology for Structural Health Monitoring of Modern Reinforced Concrete Bridge Columns. *J. Bridg. Eng.* **2014**, *19*, 1061. [CrossRef]
11. Peng, Z.; Li, J.; Hao, H.; Li, C. Nonlinear structural damage detection using output-only Volterra series model. *Struct. Control Health Monit.* **2021**, *28*, e2802. [CrossRef]
12. Astroza, R.; Conte, J.P.; Restrepo, J.I.; Ebrahimian, H.; Hutchinson, T. Seismic response analysis and modal identification of a full-scale five-story base-isolated building tested on the NEES@UCSD shake table. *Eng. Struct.* **2021**, *238*, 112087. [CrossRef]
13. Zai, B.A.; Khan, M.A.; Khan, K.A.; Mansoor, A.; Shah, A.; Shahzad, M. The role of dynamic response parameters in damage prediction. *Proc. Inst. Mech. Eng. Part C J. Mech. Eng. Sci.* **2019**, *233*, 4620–4636.
14. Kalybek, M.; Bocian, M.; Nikitas, N. Performance of optical structural vibration monitoring systems in experimental modal analysis. *Sensors* **2021**, *21*, 1239. [CrossRef] [PubMed]
15. Zahid, F.B.; Ong, Z.C.; Khoo, S.Y. A review of operational modal analysis techniques for in-service modal identification. *J. Brazilian Soc. Mech. Sci. Eng.* **2020**, *42*, 398.
16. Fang, Z.; Su, H.; Ansari, F. Modal analysis of structures based on distributed measurement of dynamic strains with optical fibers. *Mech. Syst. Signal Process.* **2021**, *159*, 107835. [CrossRef]
17. Çelebi, M.; Swensen, D.; Haddadi, H. Response study of a 51-story-tall Los Angeles, California building inferred from motions of the Mw7.1 July 5, 2019 Ridgecrest, California earthquake. *Bull. Earthq. Eng.* **2021**, *19*, 1797–1814. [CrossRef]
18. Schanze, E.; Leiva, G.; Gómez, M.; Lopez, A. Numerical study of the seismic response of an instrumented building with underground stories. *Appl. Sci.* **2021**, *11*, 3190. [CrossRef]
19. Wu, J.; Hu, N.; Dong, Y.; Zhang, Q. Monitoring dynamic characteristics of 600 m+ Shanghai Tower during two consecutive typhoons. *Struct. Control Health Monit.* **2021**, *28*, e2666. [CrossRef]
20. Li, H.J.; Zhu, S.Y.; Celebi, M. Seismic response analysis of an instrumented building structure. *J. Earthq. Eng. Eng. Vib.* **2003**, *23*, 31–36.
21. Murià-Vila, D.; Aldama-Sánchez, B.D.; García-Illescas, M.Á.; Rodríguez Gutiérrez, G. Monitoring of a rehabilitated building in soft soil in Mexico and structural response to the September 2017 earthquakes: Part 1: Structural health monitoring system. *Earthq. Spectra* **2021**, *37*, 2737–2766. [CrossRef]
22. Diaferio, M.; Foti, D.; Sepe, V. Dynamic Identification of the Tower of the Provincial Administration Building, Bari, Italy. In Proceedings of the Eleventh International Conference on Civil, Structural and Environmental Engineering Computing, St. Julians, Malta, 2007; Volume 18, p. 21.
23. Bru, D.; Ivorra, S.; Baeza, F.J.; Reynau, R.; Foti, D. OMA Dynamic Identification of a Masonry Chimney With Severe Cracking Condition. In Proceedings of the 6th International Operational Modal Analysis Conference, Gijón, Spain, 12–14 May 2015; pp. 59–60, ISBN 978-84-617-3880-9.
24. Lerna, M.; Sabbà, M.F.; Diaferio, M.; Carnimeo, L.; Ivorra, S.; Foti, D. *Seismic risk Assessment of a Medieval Tower: The Case Study of Craco, EURODYN 2020, XI International Conference on Structural Dynamics*; Papadrakakis, M., Fragiadakis, M., Papadimitriou, C., Eds.; National Technical University of Athens: Athens, Greece, 2020.
25. Diaferio, M.; Foti, D.; Giannoccaro, N.I.; Sabbà, M.F. Dynamic Identification on an Irregular Structure. *Appl. Sci.* **2022**, *12*, 3445. [CrossRef]
26. Çelebi, M. Seismic Monitoring of Structures and New Developments. In *Earthquakes and Health Monitoring of Civil Structures*; Garevski, M., Ed.; Springer: Heildelberg, Germany, 2013.
27. ACI. *318-08 Building Code Requirements for Structural Concrete*; American Concrete Institute: Farmington Hills, MI, USA, 2008.

28. Çelebi, M.; Sereci, M.; Boroschek, R.; Carreño, R.; Bonelli, P. Preliminary identification of dynamic characteristics of a unique building in chile following 27 february 2010 (Mw = 8.8) earthquake. *RILEM Bookseries* **2012**, *6*, 1071–1077. [CrossRef]
29. Nandagopalan, S.; Bendat, J.S.; Piersol, A.G. Engineering Applications of Correlation and Spectral Analysis. *Technometrics* **1994**, *36*, 220–221. [CrossRef]
30. Henao Ángel, D.; Botero Palacio, J.C.; Muriá Vila, D. Identificación de propiedades dinámicas de un modelo estructural sometido a vibración ambiental y vibración forzada empleando mesa vibradora. *Rev. Ing. Sísmica* **2014**, *91*, 54–73. [CrossRef]
31. Peeters, B.; De Roeck, G.; Andersen, P. Stochastic system identification: Uncertainty of the estimated modal parameters. *Shock Vib. Dig.* **2000**, *32*, 231–237.
32. Van Overschee, P.; De Moor, B. Subspace Identification for Linear System: Theory—Implementation—Applications. *Conf. Proc. Int. Conf. IEEE Eng. Med. Biol. Soc.* **1996**, *2008*, 4427–4430.
33. Gil-oulbé, M.; Al-Shaibani, F.A.N.A.; Lina, A.S. Performance-Based Seismic Design for buildings. *Struct. Mech. Eng. Constr. Build.* **2020**, *16*, 161–166. [CrossRef]
34. Filiatrault, A.; Perrone, D.; Merino, R.J.; Calvi, G.M. Performance-Based Seismic Design of Nonstructural Building Elements. *J. Earthq. Eng.* **2021**, *25*, 237–269. [CrossRef]
35. Sattar, S.; Hulsey, A.; Hagen, G.; Naeim, F.; McCabe, S. Implementing the performance-based seismic design for new reinforced concrete structures: Comparison among ASCE/SEI 41, TBI, and LATBSDC. *Earthq. Spectra* **2021**, *37*, 2150–2173. [CrossRef]

MDPI
St. Alban-Anlage 66
4052 Basel
Switzerland
Tel. +41 61 683 77 34
Fax +41 61 302 89 18
www.mdpi.com

Applied Sciences Editorial Office
E-mail: applsci@mdpi.com
www.mdpi.com/journal/applsci

www.ingramcontent.com/pod-product-compliance
Lightning Source LLC
Chambersburg PA
CBHW061449220926
51850CB00004B/51
9783036564586